GEOPHYSICAL MONOGRAPH SERIES

GEOPHYSICAL MONOGRAPH VOLUMES

1. **Antarctica in the International Geophysical Year** *A. P. Crary, L. M. Gould, E. O. Hulburt, Hugh Odishaw, and Waldo E. Smith (Eds.)*
2. **Geophysics and the IGY** *Hugh Odishaw and Stanley Ruttenberg (Eds.)*
3. **Atmospheric Chemistry of Chlorine and Sulfur Compounds** *James P. Lodge, Jr. (Ed.)*
4. **Contemporary Geodesy** *Charles A. Whitten and Kenneth H. Drummond (Eds.)*
5. **Physics of Precipitation** *Helmut Weickmann (Ed.)*
6. **The Crust of the Pacific Basin** *Gordon A. Macdonald and Hisashi Kuno (Eds.)*
7. **Antarctic Research: The Matthew Fontaine Maury Memorial Symposium** *H. Wexler, M. J. Rubin, and J. E. Caskey, Jr. (Eds.)*
8. **Terrestrial Heat Flow** *William H. K. Lee (Ed.)*
9. **Gravity Anomalies: Unsurveyed Areas** *Hyman Orlin (Ed.)*
10. **The Earth Beneath the Continents: A Volume of Geophysical Studies in Honor of Merle A. Tuve** *John S. Steinhart and T. Jefferson Smith (Eds.)*
11. **Isotope Techniques in the Hydrologic Cycle** *Glenn E. Stout (Ed.)*
12. **The Crust and Upper Mantle of the Pacific Area** *Leon Knopoff, Charles L. Drake, and Pembroke J. Hart (Eds.)*
13. **The Earth's Crust and Upper Mantle** *Pembroke J. Hart (Ed.)*
14. **The Structure and Physical Properties of the Earth's Crust** *John G. Heacock (Ed.)*
15. **The Use of Artificial Satellites for Geodesy** *Soren W. Henriksen, Armando Mancini, and Bernard H. Chovitz (Eds.)*
16. **Flow and Fracture of Rocks** *H. C. Heard, I. Y. Borg, N. L. Carter, and C. B. Raleigh (Eds.)*
17. **Man-Made Lakes: Their Problems and Environmental Effects** *William C. Ackermann, Gilbert F. White, and E. B. Worthington (Eds.)*
18. **The Upper Atmosphere in Motion: A Selection of Papers With Annotation** *C. O. Hines and Colleagues*
19. **The Geophysics of the Pacific Ocean Basin and Its Margin: A Volume in Honor of George P. Woollard** *George H. Sutton, Murli H. Manghnani, and Ralph Moberly (Eds.)*
20. **The Earth's Crust: Its Nature and Physical Properties** *John G. Heacock (Ed.)*
21. **Quantitative Modeling of Magnetospheric Processes** *W. P. Olson (Ed.)*
22. **Derivation, Meaning, and Use of Geomagnetic Indices** *P. N. Mayaud*
23. **The Tectonic and Geologic Evolution of Southeast Asian Seas and Islands** *Dennis E. Hayes (Ed.)*
24. **Mechanical Behavior of Crustal Rocks: The Handin Volume** *N. L. Carter, M. Friedman, J. M. Logan, and D. W. Stearns (Eds.)*
25. **Physics of Auroral Arc Formation** *S.-I. Akasofu and J. R. Kan (Eds.)*
26. **Heterogeneous Atmospheric Chemistry** *David R. Schryer (Ed.)*
27. **The Tectonic and Geologic Evolution of Southeast Asian Seas and Islands: Part 2** *Dennis E. Hayes (Ed.)*
28. **Magnetospheric Currents** *Thomas A. Potemra (Ed.)*
29. **Climate Processes and Climate Sensitivity (Maurice Ewing Volume 5)** *James E. Hansen and Taro Takahashi (Eds.)*
30. **Magnetic Reconnection in Space and Laboratory Plasmas** *Edward W. Hones, Jr. (Ed.)*
31. **Point Defects in Minerals (Mineral Physics Volume 1)** *Robert N. Schock (Ed.)*
32. **The Carbon Cycle and Atmospheric CO_2: Natural Variations Archean to Present** *E. T. Sundquist and W. S. Broecker (Eds.)*
33. **Greenland Ice Core: Geophysics, Geochemistry, and the Environment** *C. C. Langway, Jr., H. Oeschger, and W. Dansgaard (Eds.)*
34. **Collisionless Shocks in the Heliosphere: A Tutorial Review** *Robert G. Stone and Brucc T. Tsurutani (Eds.)*
35. **Collisionless Shocks in the Heliosphere: Reviews of Current Research** *Bruce T. Tsurutani and Robert G. Stone (Eds.)*
36. **Mineral and Rock Deformation: Laboratory Studies—The Paterson Volume** *B. E. Hobbs and H. C. Heard (Eds.)*
37. **Earthquake Source Mechanics (Maurice Ewing Volume 6)** *Shamita Das, John Boatwright, and Christopher H. Scholz (Eds.)*
38. **Ion Acceleration in the Magnetosphere and Ionosphere** *Tom Chang (Ed.)*

MAURICE EWING VOLUMES

1. **Island Arcs, Deep Sea Trenches, and Back-Arc Basins** *Manik Talwani and Walter C. Pitman III (Eds.)*
2. **Deep Drilling Results in the Atlantic Ocean: Ocean Crust** *Manik Talwani, Christopher G. Harrison, and Dennis E. Hayes (Eds.)*
3. **Deep Drilling Results in the Atlantic Ocean: Continental Margins and Paleoenvironment** *Manik Talwani, William Hay, and William B. F. Ryan (Eds.)*
4. **Earthquake Prediction—An International Review** *David W. Simpson and Paul G. Richards (Eds.)*
5. **Climate Processes and Climate Sensitivity** *James E. Hansen and Taro Takahashi (Eds.)*
6. **Earthquake Source Mechanics** *Shamita Das, John Boatwright, and Christopher H. Scholz (Eds.)*

MINERAL PHYSICS VOLUMES

1. **Point Defects in Minerals** *Robert N. Schock (Ed.)*
2. **High-Pressure Research in Mineral Physics** *Murli H. Manghnani and Yasuhiko Syono (Eds.)*

GEOPHYSICAL MONOGRAPH SERIES
Including
Maurice Ewing Volumes
Mineral Physics Volumes

GEOPHYSICAL MONOGRAPH 39
Mineral Physics 2

High-Pressure Research in Mineral Physics

A Volume in Honor of Syun-iti Akimoto

Murli H. Manghnani
University of Hawaii at Manoa
Honolulu, Hawaii

Yasuhiko Syono
Tohoku University
Sendai, Japan

Editors

Terra Scientific Publishing Company
Tokyo, Japan

American Geophysical Union
Washington, D. C.
1987

Library of Congress Cataloging-in-Publication Data

Main entry under title:

High-pressure research in mineral physics.

(Geophysical monograph ; 39. Mineral physics ; 2)
1. Mineralogy and Crystal Chemistry. 2. Phase transformations.
3. High Pressure-High Temperature Research. I. Manghnani,
Murli H., 1936– . II. Syono, Yasuhiko, 1935–
III. Series: Geophysical monograph ; 39. IV. Series:
Geophysical monograph. Mineral physics ; 2.
QE364.H54 1987 549 87-5050
ISBN 0-87590-066-6
ISSN 0065-8448

Published by Terra Scientific Publishing Company (TERRAPUB),
302 Jiyugaoka-Komatsu Building, 24-17 Midorigaoka 2-chome, Meguro-ku,
Tokyo 152, Japan
in co-publication with the American Geophysical Union, 2000 Florida Avenue, N.W., Washington, D.C. 20009, U.S.A.

Sold and distributed in Japan
by Terra Scientific Publishing Company,
302 Jiyugaoka-Komatsu Building, 24-17 Midorigaoka 2-chome, Meguro-ku,
Tokyo 152, Japan
in all other countries, sold and distributed
by the American Geophysical Union,
2000 Florida Avenue, N.W., Washington, D.C. 20009, U.S.A.

Printed in Japan

PREFACE

In recognition of the profound contributions made by Professor Syun-iti Akimoto in high pressure-high temperature research, which have enhanced our understanding of the earth's interior, and for his tireless efforts in furthering the U.S.-Japan cooperative research program, it is most fitting to dedicate this volume to him. This volume in honor of Professor Akimoto is a symbol of cooperation in international science that is exemplified by his contributions.

Mineral physics has rapidly emerged to be recognized as an important interdisciplinary field of the earth sciences. Overlapping with other sister disciplines (mainly physics, chemistry and materials science), it has provided an essential link between the laboratory measurements of the physical and chemical properties of minerals and rocks under extreme conditions of pressure and temperature and the geophysical and geochemical observations of the earth's interior.

Since holding the second U.S.-Japan seminar on High-Pressure Research Applications in Geophysics in Hakone, 12–15 January 1981, the advancements in high pressure-high temperature research have continued impressively. The progress in mineral physics between 1981 and 1985 was prolific and highlighted by a number of breakthroughs in experimental methods: using diamond-anvil cell and large-volume multi-anvil apparatus as exemplified by the synthesis and characterization of a number of high-density mantle phases; use of synchrotron radiation and laser heating techniques to gain a better understanding of properties of minerals at extreme pressures and temperatures on an atomic level; and in applications of results of these experimental data to unravel the nature, composition and dynamic properties of the earth's interior.

In this sense, the U.S.-Japan seminar on High-Pressure Research Applications in Geophysics and Geochemistry held under the auspices of the U.S.-Japan Cooperative Science Program between the U.S. National Science Foundation and the Japan Society for the Promotion of Science (JSPS) at the Turtle Bay Hilton, Kahuku, Hawaii, 13–16 January 1986, was most timely.

There were 25 participants from U.S. and 22 from Japan, and six observers from the other countries, providing an international flavor to the meetings. Fifty-five papers were presented during the regular sessions and a poster session. The areas covered at the symposium were: high-pressure techniques and melting experiments; shock wave experiments; synthesis, phase equilibria and thermodynamic properties of mantle phases; spectroscopy at high pressure; application of synchrotron radiation lattice dynamics studies; geophysical and geochemical constraints; and high-pressure research applications in geophysics and geochemistry (poster session).

This volume is, of course, a direct outgrowth from the meeting; of the fifty-five papers presented, forty-nine appear as contribution in this volume. In writing up their papers, the authors have taken into consideration the discussions at the meeting. This was also greatly assisted by conscientious and painstaking efforts offered by the reviewers. The editors deeply appreciate the cooperation of the following reviewers for the many helpful and constructive suggestions to the authors to improve the quality of the papers and the volume:

T. J. Ahrens	M. Akaogi	D. L. Anderson
C. A. Angell	H. Arashi	W. A. Bassett
P. M. Bell	R. Boehler	L. L. Boyer
J. M. Brown	M. S. T. Bukowinski	F. P. Bundy
D. H. Eggler	S. Endo	J. N. Fritz
Y. Fujii	Y. Fukao	R. G. Gordon
E. K. Graham	N. Hamaya	R. Hazen
D. L. Heinz	R. J. Hemley	W. B. Holzapfel
Y. Ida	R. L. Ingalls	T. Ishidate
E. Ito	H. Iwasaki	I. Jackson

A. Jayaraman
R. Jeanloz
S. Karato
K. Kawamura
S. W. Kieffer
S. H. Kirby
D. Kohlstedt
K. Kondo
Y. Kudoh
I. Kushiro
L. Liu
M. H. Manghnani
H. K. Mao
Y. Matsui
R. G. McQueen
C. Meade
L. C. Ming
H. Mizutani
A. Navrotsky
W. Nellis
M. F. Nicol
E. Ohtani
B. Olinger
G. J. Piermarini
J. P. Poirier
C. T. Prewitt
A. E. Ringwood
H. Sawamoto
A. Sawaoka
D. Schiferl
M. Seal
S. K. Sharma
H. Shimizu
O. Shimomura
E. F. Skelton
K. Suzuki
C. A. Swenson
Y. Syono
E. Takahashi
H. Takei
K. Takemura
S. Uyeda
M. Wakatsuki
G. Will
T. Yagi
H. S. Yoder, Jr.

Grateful acknowledgement is also expressed to the U.S. National Science Foundation and the Japan Society for the Promotion of Science for their financial support. Sincere thanks are due Keiji Oshida, Virginia Pfeifle, Doreen Brom-Bastas, Diane Henderson, Rita Pujalet, Mary Thor, and Henry Bennett for their assistance in editorial work. The financial support for the publication of this volume is gratefully acknowledged from Ministry of Education, Science and Culture, Japan; Hawaii Natural Energy Institute and the University of Hawaii Foundation.

The seminar committee also thanks the East-West Center, in particular Jim McMahan, for the logistics support and the Turtle Bay Hilton management for providing hospitable and enjoyable surroundings for the seminar.

Murli H. Manghnani
Yasuhiko Syono
Editors

A TRIBUTE TO SYUN-ITI AKIMOTO

Syun-iti AKIMOTO

Syun-iti Akimoto is the dean of high pressure research in Japan.

Syun-iti Akimoto and the highly productive laboratory he founded at the Institute for Solid State Physics of the University of Tokyo have provided over the last 30 years an enormous amount of knowledge about the physical properties of minerals. Akimoto and his students' contributions in the last three decades have had profound impact on our understanding of the mineralogy of the earth's mantle as well as on high pressure research in general in both Japan and throughout the world.

His early efforts were directed, among other things, to accurate determination of phase diagrams of major earth constituent minerals, especially olivine-spinel transformation in $(Mg, Fe)_2SiO_4$ solid solution at high temperatures and high pressures of mantle conditions. Akimoto first succeeded in demonstrating quantitatively that seismic discontinuities in the earth's mantle correspond to successive phase transitions in such mineral systems observed in high pressure and high temperature experiments. In 1973 he was awarded the Academy Prize by the Japan Academy for his scientific achievements in the geophysical application of high pressure research. Later Akimoto and his colleagues devoted effort to development of the in situ X-ray diffraction technique at high temperatures and high pressures, which is now extended to a cooperative work using the synchrotron radiation facility at Tsukuba. Thus, he has not only devoted a major portion of his own very active career to high pressure mineral physics research, but also guided the research activities of his students and nurtured the careers of many young scientists in Japan.

He was principally responsible for the concept of the Japan-U.S. Seminars on High-Pressure Research in Geophysics and Geochemistry under the co-sponsorship of the Japan Society for the Promotion of Science and the U.S. National Science Foundation. The success of this series of meetings, now a decade old, bears testimony to his undiminishing enthusiasm for international cooperation in science.

In 1983, Akimoto was awarded the prestigious William Bowie Medal by the American Geophysical Union for "distinguished attainment and outstanding contribution to the advancement of cooperative research in fundamental geophysics". It is befitting to dedicate this volume to Akimoto-sensei.

Murli H. Manghnani
Thomas J. Ahrens
Raymond Jeanloz
Yasuhiko Syono
Takehiko Yagi

CONTENTS

V. CRYSTAL STRUCTURE, ELASTICITY AND LATTICE DYNAMICS

VI. SPECTROSCOPY AT HIGH PRESSURE

VII. SHOCK WAVE EXPERIMENTS

VIII. GEOPHYSICAL AND GEOCHEMICAL CONSTRAINTS

HIGH-PRESSURE RESEARCH IN GEOPHYSICS: PAST, PRESENT AND FUTURE

Syun-iti AKIMOTO[1]

Institute for Solid State Physics, University of Tokyo, Minato-ku, Tokyo 106, Japan

Abstract. The progress of high-pressure geophysics research over the past twenty-five years at the Institute for Solid State Physics, University of Tokyo is summarized by area, as follows: a) development of large-volume, high-pressure apparatuses, b) phase equilibrium investigation using the quenching technique, and c) high-pressure/high-temperature in situ measurements. In this paper, a revised version of the olivine-modified spinel-spinel transformation diagram in the system Mg_2SiO_4–Fe_2SiO_4 is presented. Several phase transformation diagrams of silicate and germanate determined by in situ X-ray measurements, some with synchrotron radiation, are also summarized. These phase boundaries are recommended as a consistent set of high-pressure calibration points at high temperatures. Although the results obtained through these high-pressure/high-temperature laboratory experiments have largely constrained the composition and evolutionary process of the earth's interior, we are still far from having a precise scientific understanding of the structure and dynamics of the earth's deep interior. The outlook of high-pressure geophysics research in the near future is discussed at the end of this paper.

Introduction

High-pressure research in geophysics has evolved over a period of about thirty years from a qualitative attempt to a quantitative science that can directly provide essential insight into the physics and chemistry of the earth's interior. My academic career as an experimental geophysicist has been in this revolutionary era of high-pressure geophysics research. In this paper, I outline the progress of high-pressure geophysics research chiefly based on the experimental results obtained in my laboratory over the past twenty-five years at the Institute for Solid State Physics (ISSP), University of Tokyo.

In the 1950s, BIRCH (1952) and RINGWOOD (1958) built up a sound foundation for the field of high-pressure geophysics and geochemistry. They definitely demonstrated the importance of high-pressure phase transformation of silicate minerals for understanding the structure and properties of the earth's mantle. Thus, when I changed my research subject from rock-magnetism to high-pressure geophysics in 1961, I could unreservedly follow, in a qualitative manner, the track that they had developed. However, at that time, there remained many unsolved technical problems to quantitative investigation, including the generation of very high pressure corresponding to the earth's deep interior, determination of pressure values at high temperature, and temperature measurements at high pressure. BIRCH commented on the uncertainty of high-pressure geophysics research in his classic paper of 1952. He stated: "Unwary readers should take warning that ordinary language undergoes modification to a high pressure form when applied to the interior of the earth." Here are a few examples which he gave: the words "dubious", "perhaps", and "vague suggestion" in ordinary usage transform to the high-pressure forms "certain," "undoubtedly," and "positive proof," respectively. This comparison of language usage symbolically indicates the state-of-the-art of high-pressure geophysics research in the 1950s. To help resolve the real uncertainty in high-pressure research, I decided to make it my aim to improve the capability of the high-pressure apparatus and the accuracy of high-pressure/high-temperature measurements. It has been my constant objective to interpret the properties and processes of the earth's deep interior using modern solid state physics and chemistry.

In the following sections, I outline the progress of high-pressure research in geophysics for the past twenty-five years at the Institute for Solid State Physics, University of Tokyo. I describe some of the significant discoveries and accomplishments related to: a) development of large-volume high-pressure apparatuses, b) phase equilibrium experiments using the quenching technique, and c) high-pressure/high-temperature in situ measurements.

Over the past thirty years, the focus of pioneering high-pressure geophysics research has been shifting to the study of the deeper region of the earth. My interest in high-pressure phase transformation of mantle minerals, which was stimulated by the pioneering studies of BIRCH and RINGWOOD, is now shifting to problems relating to the chemical composition and formation process of the earth's core. The outlook of high-pressure geophysics research in the near future is briefly discussed at the end of this paper.

Development of Large-Volume High-Pressure Apparatus

When I started high-pressure research at ISSP in 1961, my first objective was to construct new high-pressure/high-temperature equipment suitable to achieve my original aim. At that time, the piston-cylinder, belt (or girdle), and

[1]Now at Institute for Study of the Earth's Interior, Okayama University, Misasa, Tottori-ken 682-02, Japan.

High-Pressure Research in Mineral Physics, edited by M. H. Manghnani and Y. Syono, pp. 1–13.

tetrahedral anvil devices had already been popular in the U.S. From among these devices, I adopted the tetrahedral anvil device as the main equipment in my laboratory because of its versatility for use for high-pressure/high-temperature in situ measurements and for phase equilibrium investigations. A possible quasi-hydrostaticity of the pressure environment in the sample space was closely evaluated in developing the capacity to take accurate measurements. I undertook the challenge to build the device using techniques available in Japan. Fortunately, this choice was successful. The tetrahedral anvil press was completed in 1963 making it the first large-volume, high-pressure apparatus in Japan. It has been working for more than twenty-five years as well as I had expected.

The success of the tetrahedral anvil apparatus has stimulated the further construction of similar multianvil type of high-pressure apparatuses in Japan. Many different types of cubic and octahedral anvil apparatuses have come into practical use in several laboratories, including a DIA-type (OSUGI et al., 1964), a split-sphere type (KAWAI and ENDO, 1970), a link type (WAKATSUKI et al., 1971), a MASS type (KUMAZAWA, 1971), a wedge type (ICHINOSE et al., 1975), and a split-cylinder type (KAWADA, 1977). The most significant technological advance in pressure generation was achieved in a cubic-octahedral anvil device using a split-sphere type apparatus originally developed at Osaka University (KAWAI and ENDO, 1970). Modern cubic-octahedral anvil devices attain pressures up to about 30 GPa at high temperatures around 2000 °C (e.g., ITO and WEIDNER, 1986).

I was fortunate to have financial support from the Japanese government in constructing some of these multianvil systems in Japan. All large-volume, high-pressure apparatuses constructed at ISSP for the past twenty-five years are listed in Table 1; a brief description of the specifications of each apparatus is given. Table 1 shows that, over the twenty-five years, the high-pressure technology of multianvil apparatuses has enabled about a fourfold increase in pressures obtained at ISSP. At every stage of development of these apparatuses, remarkable research progress has been made both in the phase equilibrium investigation using the quenching technique and in high-pressure/high-temperature in situ measurements. Some important results are given in later sections.

The large sample volume of the multianvil apparatus was most effective for the large-quantity synthesis of high-pressure phase materials and in growing single crystals of the high-pressure phase. I enjoyed participating in interdisciplinary collaborations with many distinguished scientists within Japan and throughout the world by supplying them with samples of high-pressure phase materials for a wide variety of studies, including crystal structure analysis, electrical and magnetic measurements, static compression using high-pressure X-ray diffraction technique, elasticity measurements using both ultrasonic and Brillouin scattering techniques, infrared and Raman measurements, and calorimetry studies of the stability of high-pressure phases. Usually, several 10s to 100s mg of a powder sample or several grains of a single-crystal of the high-pressure phase materials were requested for performing these studies. Since one run of the multianvil apparatus can produce samples of a few mg to a few 100s mg depending upon the synthesis pressure, these requests were generally fulfilled within the capacity of my laboratory. In one exceptional case, however, I had to synthesize 45 g of a powder sample of the ilmenite-structure phase of $MnGeO_3$ for neutron diffraction measurements by repeating the quenching experiments a few hundred times at 5 GPa and 1000 °C using the tetrahedral anvil apparatus (TSUZUKI et al., 1974).

A number of new techniques were developed at ISSP through these interlaboratory collaborations. For example,

TABLE 1. Development of Large-Volume High-Pressure Apparatus at ISSP

Apparatus	Capability	Date
Tetrahedral anvil (1000 ton × 4)	6 GPa, 2000 °C (with 25-mm-edge WC anvil)	1963
″	10 GPa, 1500 °C (with 9-mm-edge WC anvil)	1969
Bridgman anvil (450 ton)	13 GPa, 1200 °C (with 26-mm-diameter truncated face WC anvil)	1971
DIA-type cubic anvil (250 ton, for X-ray measurements)	7 GPa, room temperature (with 6-mm-edge WC anvil)	1975
″	10 GPa, 1100 °C (with 4-mm-edge WC anvil)	1976
″	20 GPa, room temperature (with 2-mm-edge WC anvil)	1976
″	12 GPa, 1200 °C (with 3-mm-edge WC anvil)	1977
Split-cylinder (oil vessel, cubic-octahedral anvil system)	20 GPa, 1200 °C (with 1.8-mm-edge WC anvil)	1977
Wedge-type cubic anvil (750 ton, single-stage operation)	7 GPa, 2000 °C (with 16-mm-edge WC anvil)	1979
(double-stage operation, cubic-octahedral anvil system)	20 GPa, 1200 °C (with 2.5-mm-edge WC anvil)	1984

the need for high-quality single crystals of high-pressure minerals for Brillouin scattering measurements stimulated the development of a new technique for crystal growth under pressure. Single-crystals of the spinel polymorph of Ni_2SiO_4 (up to 500 μm in size) were successfully grown in association with the olivine-spinel transformation by using mm-sized single-crystals of olivine as starting materials and by taking special care in selecting the path of pressure-temperature treatment (HAMAYA and AKIMOTO, 1982; AKIMOTO et al., 1984). I would like to emphasize that the development of new techniques and devices for research in extreme environments always opens a new research field. This was particularly true in high-pressure geophysics research. We can find many examples that successful synthesis of single-crystals of high-pressure minerals led to a remarkable advancement in crystal chemistry (see MORIMOTO et al., 1970 for modified spinel and HORIUCHI et al., 1982 for spinelloids).

Phase Equilibrium Investigation Using Quenching Technique

It is widely accepted that olivine, pyroxene, and garnet are the major constituent minerals of the upper mantle. Since the suggestion by BERNAL (1936) that common olivine might transform to a new polymorph possessing the spinel structure in the deep mantle, high-pressure phase transformation of the upper mantle minerals have received much attention in the discussion of seismic discontinuities in the mantle's transition zone (at about 400 to 700 km depth).

When the tetrahedral anvil apparatus of ISSP was released for routine use in high-pressure/high-temperature experiments, the synthesis of the spinel polymorph of orthosilicates had already been reported in Fe_2SiO_4, Ni_2SiO_4, and Co_2SiO_4 by RINGWOOD (1958, 1962, 1963). However, no attempt had been made to determine the phase diagram of the olivine-spinel transformation of these silicates over the wide range of pressures and temperatures that might occur in the earth's mantle. Therefore the initial task of my investigation was the accurate determination of the olivine-spinel transformation boundary for Fe_2SiO_4 and Ni_2SiO_4 (AKIMOTO et al., 1965). The excellent capabilities of the tetrahedral anvil apparatus were fully verified by this preliminary investigation. I then gradually expanded the scope of my phase equilibrium investigation of upper mantle minerals from single-component minerals to two-component solid solution systems and then to multicomponent systems including natural mantle rocks.

The magnesium-rich side of the magnesium-iron olivine, $(Mg, Fe)_2SiO_4$, has long been considered to be the most abundant mineral in the upper mantle. In the 1960s all high-pressure mineral physics laboratories in the world aimed to determine the phase diagram of the Mg_2SiO_4–Fe_2SiO_4 system as a function of composition, pressure, and temperature. The first ISSP publication on the olivine-spinel transformation in the system Mg_2SiO_4–Fe_2SiO_4 was issued in 1966 (AKIMOTO and FUJISAWA, 1966). The most significant result of this early work was the first quantitative determination of the location of both the olivine solvus and the spinel solvus of the olivine-spinel transformation diagram over a pressure range up to 8 GPa* at 800 °C. Although the synthesis of the spinel solid solution was limited to an Fe-rich region up to $(Mg_{0.73}Fe_{0.27})_2SiO_4$, this investigation provided detailed information on the width of the two-phase field where spinel solid solutions coexist with olivine solid solutions. This information was valuable to the geophysicists who had been interested in the origin of the high-gradient zone of seismic wave velocities in the mantle. Geophysicists were the first to recognize the importance of obtaining an accurate phase diagram to the correct interpretation of observational data on the earth's deep interior. I believe that this work formed a solid foundation for the later development of high-pressure mineral physics.

In 1966 other important findings were reported by RINGWOOD and MAJOR (1966) on the phase equilibria in the Mg_2SiO_4–Fe_2SiO_4 system. They reported the discovery of an unexpected mode of high-pressure transformation in the Mg-rich side of the system, resulting from the appearance of the β-phase. The crystal structure of the β-phase was soon identified as the modified spinel structure (MORIMOTO et al., 1969, 1970). The successful synthesis of the single-crystal of the β-phase for Co_2SiO_4 and Mn_2GeO_4 enabled us to determine the new structure. It should be mentioned that the large sample volume of our tetrahedral anvil apparatus greatly contributed to the synthesis of single-crystals with a size of several 10s μm, which is suitable for the structural analysis. It is evident that the structure of the β-phase would not have been clarified at such an early stage without the single-crystal.

It took more than ten years of research in my laboratory to complete the olivine-spinel transformation diagram of the Mg_2SiO_4–Fe_2SiO_4 system. In the early 1970s during the course of the investigation, new research equipment, such as the Bridgman anvil with an internal heating system and the split-cylinder type of apparatus, was developed. The final phase diagrams for the Mg_2SiO_4–Fe_2SiO_4 system, which were obtained at ISSP, are shown in Figures 1, 2, and 3; the nature and proportions of phases observed in the high-pressure/high-temperature experiments are presented as isothermal sections at 800, 1000, and 1200 °C. These diagrams were originally prepared by KAWADA (1977) by adding his own data,

*Pressure values in the original publication were revised to reflect the current pressure scale.

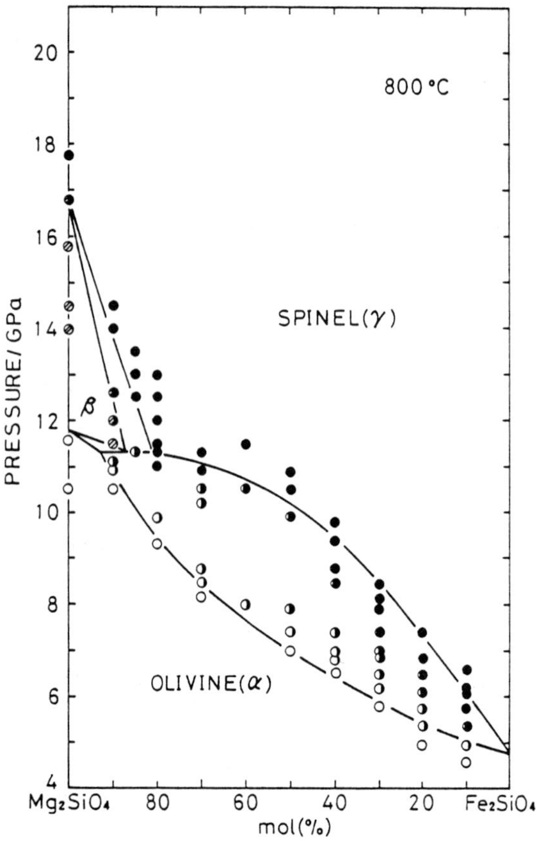

Figure 1. Phase diagram for the olivine-modified spinel-spinel transformation in the system Mg_2SiO_4–Fe_2SiO_4 at 800 °C. Original data are from KAWADA (1977).

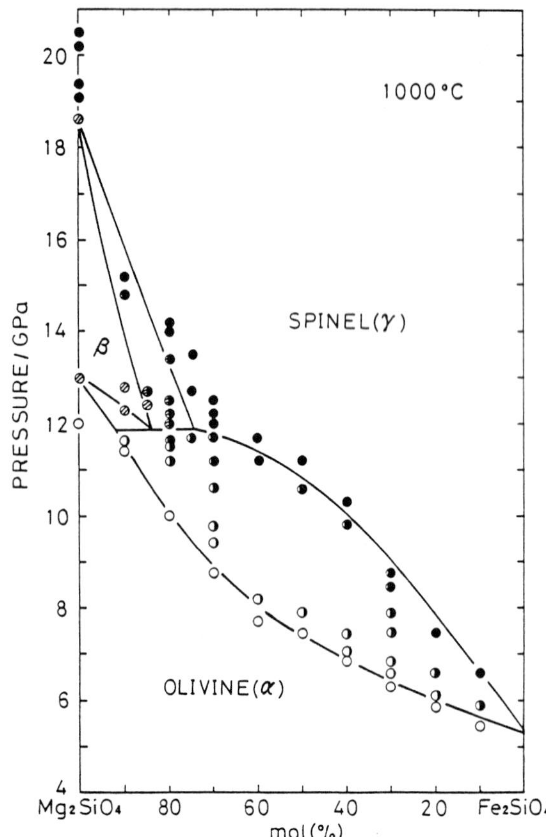

Figure 2. Phase diagram for the olivine-modified spinel-spinel transformation in the system Mg_2SiO_4–Fe_2SiO_4 at 1000 °C. Original data are from KAWADA (1977).

obtained with the aid of a split-cylinder type of apparatus, to the previous data published by AKIMOTO et al. (1976). In this paper, the pressure values in Kawada's diagram are reduced slightly on the basis of the following current values of the pressure fixed points: High Ba (12.6 GPa, AKIMOTO et al., 1975), Pb (14.2 GPa, YAGI and AKIMOTO, 1976a), ZnS (15.4 GPa, BLOCK, 1978), GaAs (18.3 GPa, SUZUKI et al., 1981), GaP (22.0 GPa, PIERMARINI and BLOCK, 1975), and coesite-stishovite (9.1 GPa at 1000 °C, YAGI and AKIMOTO, 1976b). Recent data on the redetermination of the olivine-spinel boundary curve for Fe_2SiO_4, presented in a later section, are also considered.

The success of the split-cylinder type of apparatus enabled us to extend our investigation to pyroxene and garnet, the second and third most abundant groups of minerals in the upper mantle, respectively. When we started the investigation, Ringwood's group had already reported the remarkable increase of the solid solubility of pyroxene into garnet in the system $MgSiO_3$–Al_2O_3 at high pressures and had also pointed out the geophysical importance of the high-pressure reaction between garnet and pyroxene to the formation of the mantle transition zone (e.g., RINGWOOD, 1967). We again attempted to construct the detailed phase diagram for this "pyroxene-garnet transformation" as a function of composition, pressure, and temperature. By the end of 1970s, we were able to complete three sets of phase diagrams for the systems $Mg_4Si_4O_{12}$–$Mg_3Al_2Si_3O_{12}$, $Fe_4Si_4O_{12}$–$Fe_3Al_2Si_3O_{12}$, and $Ca_2Mg_2Si_4O_{12}$–$Ca_{1.5}Mg_{1.5}Al_2Si_3O_{12}$ (AKAOGI and AKIMOTO, 1977, 1979). Recently, we reexamined our previous experimental data based on the current pressure fixed points mentioned above. The final version of the phase relationships in the system $Mg_4Si_4O_{12}$–$Mg_3Al_2Si_3O_{12}$ is presented by AKAOGI et al. (this volume). It is now accepted that no sharp discontinuity in density can be expected from the shape of the phase diagram. This may be the most significant geophysical implication of the series of investigations on the pyroxene-garnet transformation.

Throughout the first fifteen years of research, ISSP provided most fundamental data on the high-pressure

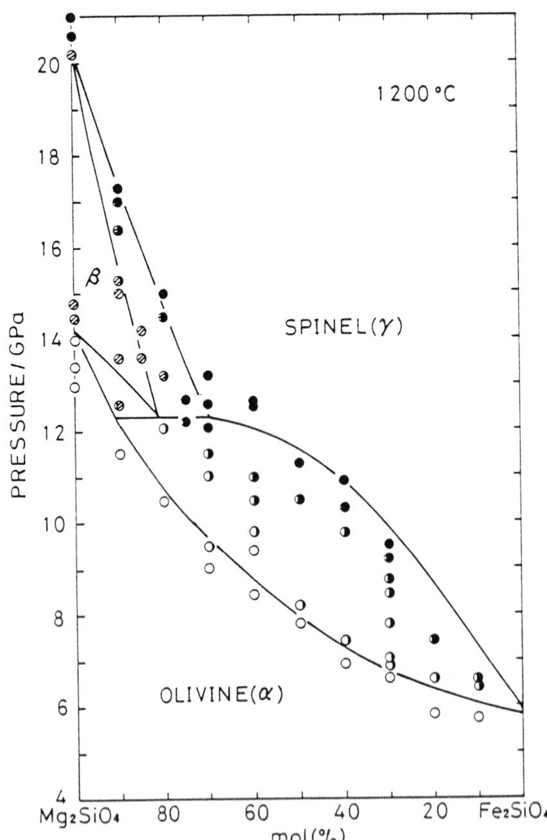

Figure 3. Phase diagram for the olivine-modified spinel-spinel transformation in the system Mg_2SiO_4-Fe_2SiO_4 at 1200 °C. Original data are from KAWADA (1977).

transformation behavior of three important upper mantle minerals, olivine, pyroxene, and garnet, up to 20 GPa. As a next step, ISSP attempted to apply the same experimental procedure to natural mantle rocks. Akaogi and I studied phase equilibria in a natural garnet lherzolite (PHN1611) over a pressure range up to nearly 20 GPa and in the temperature range 1050–1200 °C (AKAOGI and AKIMOTO, 1979). We were able to demonstrate that the data on the transformation behavior of the component minerals of this rock are quite consistent with the information previously collected for the simple binary systems mentioned above. The partition of elements among coexisting high-pressure minerals was directly determined at varying pressures by means of microprobe observation. Based on the experimental data, we could present a model of mineral assemblages in the mantle transition zone as a function of depth to 600 km. I believe that this work was not only a memorable achievement of the ISSP phase equilibrium investigation but also a milestone for high-pressure geophysics.

Although no substantial contribution was made by ISSP to the phase equilibrium investigation relating to the formation of silicate perovskite, the large-volume multianvil apparatus was used at the Institute for Study of the Earth's Interior (ISEI), Okayama University, for the detailed investigation of the mineralogy and chemistry of the lower mantle. ITO et al. (1984) first succeeded in revealing phase relations in the system MgO–FeO–SiO_2 at 26 GPa and 1600 °C, a pressure-temperature condition close to that of 670-km seismic discontinuity. They demonstrated that on the join of Mg_2SiO_4–Fe_2SiO_4, the two-phase assemblage of (Mg, Fe)SiO_3 perovskite and (Mg, Fe)O magnesiowüstite was found to be stable from Mg_2SiO_4 to $(Mg_{0.65}Fe_{0.35})_2SiO_4$. They also found that the maximum solubility of Fe in the (Mg, Fe)SiO_3 was 0.11 in Fe/(Mg+Fe). Very recently, the ISEI group has extended our previous phase equilibrium investigation on the natural garnet lherzolite (PHN1611) to the lower mantle conditions (ITO and TAKAHASHI, this volume). It is a great pleasure to see that the seeds of quantitative high-pressure phase equilibria research planted at ISSP with use of the tetrahedral anvil apparatus and with the model of upper mantle minerals (such as Fe_2SiO_4 fayalite and $FeSiO_3$ ferrosilite) have now yielded advancements in the mineralogical and chemical investigation of the lower mantle with the cubic-octahedral anvil type of apparatus and with natural peridotite.

High-Pressure/High-Temperature In Situ Measurements

In the field of high-pressure/high-temperature in situ measurement, the first contribution of my laboratory was the observation of the electrical conductivity jump associated with the olivine-spinel transformation in Fe_2SiO_4 (AKIMOTO and FUJISAWA, 1965). In the 1960s, in situ measurements of thermal diffusivity were also made for some upper mantle minerals with the aid of the tetrahedral anvil press (FUJISAWA et al., 1968). In the middle of the 1970s, we started a high-pressure/high-temperature in situ X-ray diffraction study. High-pressure/high-temperature in situ X-ray measurements are important to geophysical and geochemical research of the earth's deep interior.

Needless to say, data from laboratory research in high-pressure geophysics and geochemistry must be compared with data from seismological observations in terms of a common pressure scale. In most high-pressure/high-temperature apparatuses with solid pressure media and compressible gaskets, such as the multianvil apparatus, pressure values are conventionally calibrated at room temperature by a number of pressure fixed points that are usually observed as resistance-jump transitions in standard materials such as Bi, Ba, Pb, ZnS, GaAs, and GaP. Even though these transition pressures are fixed with high accuracy, this method is not satisfactory for high-temperature work. Depending upon the type of high-pressure apparatus and the pressure-temperature conditions, the change in shear strength of the solid pressure media, the thermal expansion of the samples and pressure media, and the phase change or chemical alteration of the samples and pressure media are all expected to considerably affect the pressure values. It is very likely that the phase boundaries determined by the quenching method contain uncertainties related to the pressure values at high temperatures. Also, no guarantee is given for the absolute values of pressures in the earlier high-pressure/high-temperature in situ measurements of the electrical conductivity and thermal conductivity of the upper mantle minerals. Thus the establishment of an accurate pressure scale at high temperature has been urgently needed for the quantitative comparison of the experimental data with the observational data.

The high-pressure in situ X-ray diffraction technique is a promising method for measurement of the pressure acting on a sample at elevated temperatures. Since pressure values at high temperature can be determined by measuring the lattice parameter of internal pressure standard materials such as NaCl and Au, the values are accurate to within the uncertainty of the equation-of-state of the standard material used.

The quenching technique is indeed the most useful method for investigating the phase equilibria of the mantle constituent minerals. Fortunately, almost all of the silicates are quenchable. However, the high-pressure phases of some silicates, such as $CaSiO_3$ wollastonite, are unquenchable. Also, even if the high-pressure phase appears to be quenchable, there may still be some possibility that the high-pressure phase varies to a third metastable phase during quenching. We recollect that in the early stage of our investigation, the thermodynamical stability of the modified spinel phase was suspected owing to the peculiarities of its crystal structure (RINGWOOD and MAJOR, 1966; MORIMOTO et al., 1970). For any material, the real thermodynamical stability of its high-pressure phase must be determined by the high-pressure/high-temperature in situ X-ray diffraction technique.

It should be also noted that the phase boundary obtained in the quenching experiments does not always represent the equilibrium boundary. When careful consideration is not given to the reversibility of the phase transformation, the phase diagram determined by the quenching technique represents only a synthesis diagram. Determination of the equilibrium phase boundary is particularly important in geophysics and geochemistry, in which global processes through geological time scale always form the key topics, e.g. subduction of the lithosphere into the deeper region of the mantle for the past 100s Ma. Thermodynamical parameters, such as enthalpy and entropy of mantle minerals, are often used to estimate the pressure of transformation to a high-pressure phase and the slope dP/dT of the phase boundary curve. The comparison of these calculated values with the experimental data is significant only when adequate consideration is given to the determination of the equilibrium phase boundary in high-pressure/high-temperature experiments. Information on the kinetics of the phase transformation is indispensable for obtaining a reliable phase boundary and for determining the mechanism of the transformation. High-pressure/high-temperature X-ray diffraction measurements can directly provide such information through the observation of the change in the X-ray diffraction pattern with time.

The fundamental importance of high-pressure/high-temperature in situ X-ray diffraction measurement to geophysical research is its applicability to the study of equation-of-state (EOS). The most reliable data on the elastic properties of the upper mantle minerals have previously been obtained from ultrasonic measurements on single crystals. Because of the difficulty of growing single crystals, this method cannot be applied to the crystals of high-pressure minerals synthesized in the small vessel of a high-pressure apparatus. However, a very small amount of polycrystalline powder sample is sufficient for an X-ray diffraction study under high pressure and high temperature. Thus the X-ray diffraction technique is more versatile for use in elasticity measurements than the ultrasonic method. We can directly determine the density of the geophysically important materials at high pressure and high temperature through the measurement of their cell parameters. The measurement of cell parameters over a wide range of pressure and temperature also make it possible to estimate compressibility at high temperature and thermal expansivity at high pressure.

I would like to emphasize that high-pressure/high-temperature laboratory experiments are most valuable to geophysics and geochemistry when: a) the pressure values are determined with an absolute accuracy, b) the phase boundary represents the real equilibrium boundary, and c) the density and elastic parameters are determined at high-pressure/high-temperature in situ conditions. The high-pressure/high-temperature X-ray diffraction systems constructed at ISSP fulfilled these criteria. We adopted the DIA-6 type of cubic anvil apparatus originally designed by INOUE and ASADA (1973) for X-ray measurements.

The combination of the DIA-6 type of apparatus and the X-ray diffraction system (angle-dispersive type or energy-dispersive type) with a rotating-anode type of X-ray source was really effective. We could determine the location of the phase boundary between two polymorphs in the pressure-temperature range of interest for several geophysically important materials, including the coesite-stishovite boundary (YAGI and AKIMOTO, 1976b), olivine-spinel boundary in Fe_2SiO_4 (AKIMOTO et al., 1977), pyroxene-ilmenite boundary in $ZnSiO_3$ (AKIMOTO et al., 1977), and αPbO_2-fluorite boundary in PbO_2 (YAGI and AKIMOTO, 1980). The pressure-volume relationship of the mantle minerals and their analog-compounds (including αFe_2SiO_4 fayalite; silicate spinels of γFe_2SiO_4, γCo_2SiO_4, and γNi_2SiO_4; garnets of $Mg_3Al_2Si_3O_{12}$ and $Fe_3Al_2Si_3O_{12}$; ilmenite-type compounds of $ZnSiO_3$ and $MgGeO_3$; stishovite; and corundum-type compounds) was successfully measured under hydrostatic conditions using an angle-dispersive type of X-ray diffraction system (YAGI et al., 1975; SATO, 1977; SATO et al., 1977; SATO et al., 1978; SATO and AKIMOTO, 1979). The thermal expansivity of several alkali halides at high pressure was first estimated directly from the pressure-volume-temperature data obtained for the halides by the in situ X-ray diffraction measurements (YAGI, 1978). Successful observation of the change in diffraction pattern with time made it possible to investigate the kinetics of pressure-induced phase transformation in KCl (HAMAYA and AKIMOTO, 1981). The energy-dispersive type of X-ray diffraction system developed at ISSP greatly contributed to carrying out this kind of detailed investigation. The real structural stability of the modified spinel phase of Mg_2SiO_4 was verified by high-pressure/high-temperature in situ X-ray measurements (FUKIZAWA, 1982) seventeen years after the first reported occurrence of the β-phase by RINGWOOD and MAJOR (1966).

All of the investigations before 1983 were carried out using a classical X-ray source. Recently, the intense coherent X-ray beam of synchrotron radiation has become available for high-pressure research. The X-ray beam of synchrotron radiation is quite suitable for high-pressure experiments because of its excellent characteristics, such as high brightness, small divergence, and continuous energy spectrum. By using synchrotron radiation, we can expect to perform higher quality X-ray diffraction studies at high pressures and temperatures. Synchrotron radiation methods are just now beginning to replace classical X-ray methods.

The synchrotron radiation facility in Tsukuba, Japan, known as Photon Factory, has been used for X-ray diffraction studies since 1982. A working group on high-pressure X-ray diffraction research has constructed an apparatus called MAX80 (multiple anvil type X-ray system) at Photon Factory for high-pressure/high-temperature experiments (SHIMOMURA et al., 1984). The high-pressure vessel MAX80, designed in 1980, is basically the same as the DIA type of cubic anvil apparatus. The expertise accumulated at ISSP in developing the high-pressure X-ray diffraction technology using the DIA-type apparatus was valuable to the practical design of the MAX80 system. The MAX80 apparatus can operate at temperatures much higher than those tolerated by the diamond-anvil cell. With MAX80, we can currently observe energy-dispersive X-ray diffraction patterns up to a pressure of 14 GPa and a temperature of 1700 °C, in the short exposure time of a few 10s to 100s s.

The application of synchrotron radiation to high-pressure geophysics research is a main topic of the present US-Japan Seminar. Many important results that have recently been obtained with the MAX80 system are reported in this volume: e.g., the equation-of-state of majorite (YAGI et al., this volume), in situ measurements of the viscosity of silicate melts (KANZAKI et al., this volume), in situ determination of the $\alpha/\gamma/\varepsilon$ phase boundary of iron (AKIMOTO et al., this volume), and a two-stage operation of MAX80 for generating very high pressure with sintered diamond anvils (ENDO et al., this volume). In this paper I present the equilibrium phase diagram as refined by synchrotron radiation.

We recently attempted to remeasure the equilibrium phase boundary for the olivine-spinel transformation in Fe_2SiO_4 using the MAX80 apparatus at Photon Factory (YAGI et al., 1987). Both constant-load and constant-temperature measurements were carried out in traversing the olivine-spinel phase boundary with starting materials of the mixed phase of olivine and spinel. To identify the stability field during a sudden increase or decrease in load or temperature, data for the change in the diffraction patterns were collected at pressure-temperature conditions extending up to 6.1 GPa and 1200 °C. An equilibrium phase boundary was determined by comparing the relative stability of olivine and spinel at each pressure-temperature condition. The equilibrium olivine-spinel phase boundary in Fe_2SiO_4 so determined is shown in Figure 4. It can be expressed by the linear equation: $P\,(\mathrm{GPa}) = 2.75 + 2.5 \times 10^{-3}\, T(°\mathrm{C})$. The pressure values in this equation are based on the equation-of-state for NaCl (DECKER, 1971), and the uncertainty of the transformation pressure was estimated to be less than ± 0.1 GPa.

A phase equilibrium investigation was also conducted for the garnet-perovskite transformation in $CaGeO_3$ using the MAX80 apparatus combined with synchrotron radiation (SUSAKI et al., 1985). For high-pressure phase transformations in which the perovskite structure phase is associated with a high-density phase, NAVROTSKY (1980) proposed that the phase boundary curve has a negative pressure-temperature slope. With this in mind, we intended to determine the phase boundary for the garnet-perovskite transformation in $CaGeO_3$ as accurately as possible. The experimental procedure for dynamical observation of the phase transformation is the same as that used for Fe_2SiO_4. The equilibrium phase boundary between the garnet and perovskite phases in $CaGeO_3$ is expressed as $P\,(\mathrm{GPa}) = 6.9 - 0.8 \times 10^{-3}\, T(°\mathrm{C})$ and shown in Figure 4. The negative pressure-temperature slope of

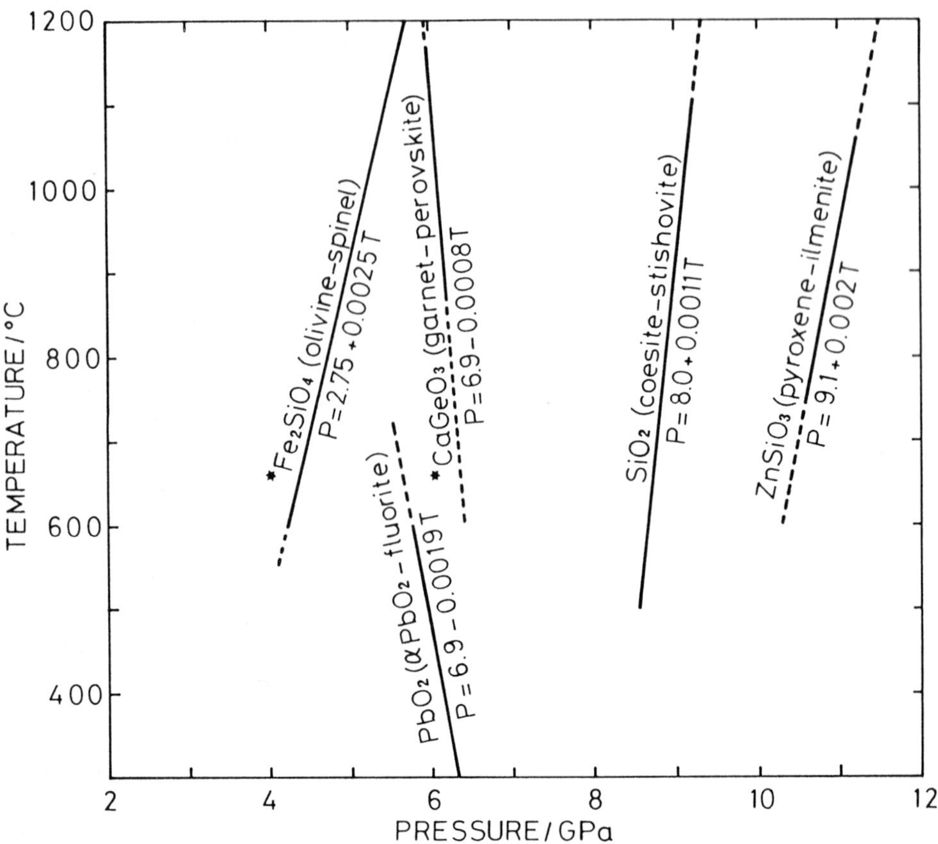

Figure 4. Phase boundaries determined by high-pressure/high-temperature in situ X-ray measurements. Asterisk represents the data with synchrotron radiation X-ray source.

the boundary was definitely established in this investigation.

Figure 4 summarizes all the phase diagrams determined by the ISSP group using the high-pressure/high-temperature in situ X-ray diffraction technique. The pressure scale shown in this figure is accurate within the precision of the NaCl pressure scale. Of the boundary curves shown, only two were recently refined by the synchrotron radiation method. The remaining three curves were obtained with the DIA-6 type of cubic anvil apparatus and with the classical X-ray method. The phase boundaries shown in Figure 4 are recommended as a consistent set of pressure fixed points at high temperature. Since the high-pressure phases of the materials treated in this figure, except that of PbO_2, are quenchable to ambient conditions, their phase boundaries may be useful for calibrating pressure values in a laboratory where no equipment for X-ray diffraction is available. Further, it should be noted that at pressures above 12 GPa there still are no data for development of a phase diagram of which pressure value is well characterized. A reliable phase diagram for the $\alpha/\beta/\gamma$ transformation in Mg_2SiO_4 and its post-spinel transformation into the mixed phase of $MgSiO_3$ perovskite and MgO periclase is urgently needed. Combination of the synchrotron radiation methods and the multi-staged, multianvil system with sintered diamond may make breakthroughs to such challenging problems possible.

Outlook of High-Pressure Geophysics Research in the Near Future

Although the high-pressure geophysics research of the past three decades has provided major constraints to the composition and evolutionary process of the earth's interior, we are still far from obtaining a precise scientific understanding of the structure and dynamics of the earth's deep interior. The main difficulty arises from the absence of basic data on the physical properties of the materials constituting the earth's deep interior under high pressures up to about 400 GPa and under high temperatures up to about 6000 °C. There are several fundamental problems that could be solved in the near future through the combination of experimental investigation of high-pressure/high-temperature behavior of the materials of the earth's mantle and core and the theoretical understanding of properties of these materials on an atomic scale. Suggested research problems are as follows.

How Did the Earth's Mantle Become Layered?

By analogy with the evolution of the moon, it has recently been suggested that the earth may have had an extensive melting stage (magma ocean) at its accretion period. The presence of the magma ocean would have had profound implications for the primary stratification of the mantle into the upper mantle, transition zone, and lower mantle. The origin of the 670-km-depth seismic discontinuity separating the transition zone and lower mantle could be associated with a chemical composition boundary or with a phase boundary through detailed investigation of phase relations of model mantle materials under high pressures around 25 GPa. Precise knowledge of the high-pressure melting behavior of the model mantle materials and the model primordial chondritic materials is particularly important. To this end, a series of investigations have recently been undertaken from the viewpoint of experimental petrology, using cubic-octahedral anvil devices (OHTANI, 1983, this volume; TAKAHASHI and SCARFE, 1985; SCARFE and TAKAHASHI, 1986; TAKAHASHI, 1986; KATO and KUMAZAWA, 1986; OHTANI et al., 1986; ITO and TAKAHASHI, this volume). However, it is still not clear whether the earth underwent an extensive melting stage that produced the 670-km discontinuity by perovskite fractionation. Further research is needed before such conclusion can be reached.

Equation-of-state (EOS) measurements on the pertinent upper-mantle and lower-mantle minerals also play an essential role in our understanding of the layered structure of the mantle. Synchrotron radiation X-ray methods should be used for precise EOS measurements on the high-pressure phases of silicate minerals. High-pressure/high-temperature in situ X-ray measurements of high-precision are particularly important for estimating sound velocity and density versus pressure and temperature and versus depth within the earth. Models of the composition and structure of the mantle are most valuable when reliable values of the pressure and temperature derivatives of the bulk moduli of βMg_2SiO_4, γMg_2SiO_4, majorite, and $MgSiO_3$ perovskite are determined experimentally.

Comparison of laboratory data on high-pressure/high-temperature elastic wave velocities of the pertinent mantle minerals with observed seismic data provides the most direct and strictest constraints to the chemistry and mineralogy of the mantle. A few attempts have previously been made to measure the ultrasonic wave velocities of silicate minerals under high pressure-temperature conditions equivalent to conditions in the upper mantle. Using a large-volume, wedge-type of cubic-anvil apparatus, FUKIZAWA and KINOSHITA (1982) first succeeded in observing the change in shear wave velocity associated with the olivine-spinel transformation in Fe_2SiO_4 at 5.2 GPa and 650 °C. Recently, Fujisawa and Ito (private communication) measured compressional wave velocities of αMg_2SiO_4 and βMg_2SiO_4 up to 14 GPa at room temperature, using the split-sphere type of cubic-octahedral anvil apparatus. Although their data are still preliminary, investigations along this line should be encouraged. High-pressure/high-temperature in situ measurements of the wave velocities of majorite and perovskite would be most valuable.

Information on thermal conductivity of the pertinent mantle minerals is also essential to the understanding of the thermal history of the mantle, which may have had great influence on the mantle's stratification. Although the thermal conductivities of crustal materials have long been used to calculate heat flow from observed temperature gradients, we know very little about the thermal conductivities of mantle minerals under mantle conditions. In the early stages of our high-pressure geophysics research when we used the tetrahedral anvil apparatus, we made preliminary high-pressure/high-temperature in situ measurements of thermal diffusivity for αMg_2SiO_4, αFe_2SiO_4, and γFe_2SiO_4 in the pressure-temperature range up to 5 GPa and 1000 °C (FUJISAWA et al., 1968). Modern large-volume, cubic anvil apparatuses should be suitable for extending this investigation to higher pressures and temperatures. Thermal conductivity data for βMg_2SiO_4, γMg_2SiO_4, majorite, $MgSiO_3$ perovskite are still needed.

How Did the Earth's Core Form?

It is accepted that the density of the earth's outer core is about 10% smaller than that of pure iron under equivalent conditions. Efforts have been made to account for this difference in terms of the dissolution of some lighter elements such as O, Si, S, and H. It is anticipated that the alloying of these light elements with iron not only reduces the density of metal but also lowers the melting temperature. The latter effect has important implications related to the easy separation of core-forming material from the primordial material of the earth.

As a candidate for the light elements, S has been considered to be most plausible. The Fe–S system was thoroughly investigated by USSELMAN (1975) up to 10 GPa, and the Fe–FeS eutectic condition was reported to be 1158 °C and 27.1 wt% S at 10 GPa. More recent experiments on the Fe–O system by OHTANI and RINGWOOD (1984), OHTANI et al. (1984), and M. KATO (1985) have shown that the solubility of O in molten Fe is greatly increased at pressures up to 20 GPa. This strongly suggests that O can also be present in the outer core. Very recently, URAKAWA et al. (this volume) extended the high-pressure melting experiments to the Fe–Ni–O–S system and found that the eutectic temperatures of the

Fe–Ni–O–S system are lower than those of the Fe–FeO–FeS system by about 100 °C in the pressure range of 6–15 GPa.

The possible presence of H in the core due to the reaction $Fe+H_2O \rightarrow FeO+FeH_x$ was suggested by STEVENSON (1977). Recently this idea has acquired strong experimental support. High-pressure experiments carried out at ISSP on the Fe–H, Fe–H_2O, and Fe-hydrous mineral reaction (FUKAI et al., 1982; FUKAI and AKIMOTO, 1983; SUZUKI et al., 1984) have successfully demonstrated that a large amount of hydrogen can be dissolved in iron under high pressure, thereby causing an appreciable reduction in the density and melting temperature. In a detailed investigation of the iron-enstatite-water system with several different kinds of starting materials, we could observe that coagulation of iron particles is greatly facilitated in the partially molten state of silicates at pressures around 5 GPa and at temperatures around 1200 °C. On the basis of these experimental findings, FUKAI and SUZUKI (1986) proposed that the Fe–H_2O reaction played a crucial role in the evolution of the earth, including the core-mantle separation and the dissolution of hydrogen in the earth's core. They also suggested that hydrogen can be a major cause of the density deficit of the outer core. Furthermore, it is expected that the presence of water affects the partition of siderophile elements between core and mantle. The apparent overabundance of siderophile elements in the mantle, as compared to amounts that would be expected based on consideration of equilibrium partitioning between a silicate mantle and a metallic iron core, should be examined experimentally in the presence of water.

Unfortunately, iron hydride produced at high pressures and temperatures cannot be quenched to the ambient conditions. Accordingly, for further development of the model of the iron hydride core, the solubility of H in molten Fe should be quantitatively determined under high-pressure/high-temperature in situ conditions. Information from high-pressure/high-temperature X-ray measurements is indispensable. More comprehensive investigations of the phase relations among Fe-Ni and H, O, S under the higher pressures and temperatures would be especially helpful.

How Much H_2O is Preserved in the Deep Interior of the Earth?

As mentioned above, it is very probable that the presence of water in the primordial material has resulted in the formation of the iron hydride core. It has also been established that the presence of water causes a considerable reduction of the melting temperature of mantle rocks. Further, the presence of water possibly affects the tectonic character of the lithosphere and asthenosphere through the partial melting process and magma genesis process in the mantle. However, the distribution of H_2O within the earth's deep interior is still one of the unsolved problems.

In the early stages of high-pressure geochemical research, RINGWOOD and MAJOR (1967) discovered three new hydrous magnesian silicates, denoted as phases A, B, and C, which were found to be stable at pressure-temperature conditions corresponding to the upper mantle. The chemical compositions of phases A and B were determined to be $Mg_7Si_2O_{14}H_6$ and $Mg_{23}Si_8O_{42}H_6$ (or $Mg_{24}Si_8O_{42}H_4$), respectively, in a series of systematic investigations conducted at ISSP (YAMAMOTO and AKIMOTO, 1974; AKAOGI and AKIMOTO, 1980; AKIMOTO and AKAOGI, 1980). Recently, T. KATO and KUMAZAWA (1985) reported a remarkable stability field of phase B up to 2300 °C at 20 GPa. This result suggests that H_2O can still be preserved in the deep upper mantle, within the possible range of the present geothermal gradient. More extensive and comprehensive investigations of the stability of hydrous silicates, including phase B and phase C, need to be conducted at higher pressures and temperatures corresponding to lower mantle conditions in order to determine the amount and state of H_2O in the earth's deep interior. The crystal structure of the phase B should be characterized first.

How Much Can We Reduce the Uncertainty of Temperature Measurements at High Pressures?

The adoption of ceramic heating elements, such as $LaCrO_3$, in modern cubic-octahedral anvil apparatuses facilitated experiments at very high temperatures above 2000 °C and at very high pressures above 15 GPa. This technique greatly helped refine our knowledge of the melting behavior of the mantle materials. The present discussion of the mechanism of producing the layered mantle completely depends on the results of such melting experiments. In these recent experiments, temperature is usually measured by Pt–Rh or W–Re thermocouples in a conventional manner. In many cases, run temperatures above 2000 °C are estimated by extrapolating the relation between the heating power and thermocouple emfs observed in the initial heating cycle up to about 2000 °C. The effect of pressure on the emfs of the used thermocouples in such high pressure and temperature ranges is not quantitatively known. The uncertainty in temperature measurements above 2000 °C may be as large as ±100 to 200 °C. To establish more definite constraints on the chemistry and mineralogy of the earth's deep mantle and a reliable temperature distribution within the earth, it is necessary to develop a technique to measure temperature at very high pressures with absolute accuracy.

Recently, YAGI et al. (1985) attempted to measure temperature without a thermocouple. The high accuracy of the MAX80 X-ray system made it possible to determine pressure and temperature simultaneously using two pressure markers for which the EOS was characterized with high precision. The volumes of the pressure markers, Au and NaCl, were determined simultaneously at a pressure-temperature condition through the cell parameter measurements. Two isovolumetric lines corresponding to the observed values were plotted on a pressure-temperature diagram, and the values of pressure and temperature were determined as an intersection point. In one particular experiment which was carried out at about 4.5 GPa and about 600 °C, the uncertainty in temperature and pressure was reported to be ±25 °C and ±0.1 GPa. Although this result is still preliminary, we expect that the uncertainty would be reduced considerably if we could find adequate pressure markers with larger thermal expansivities. The pressure effect on the emfs of the Pt–Rh and W–Re thermocouples could be reasonably estimated by applying the present technique.

Acknowledgments. I would like to express my sincere thanks to all my colleagues and former graduate students at ISSP (Y. Syono, Mituko Ozima, H. Fujisawa, Y. Ida, K. Kawada, M. Nishikawa, Y. Sato-Sorensen, T. Yagi, O. Nishizawa, M. Akaogi, N. Hamaya, A. Fukizawa, T. Suzuki, and J. Susaki). All of these people contributed to bringing my laboratory to the frontier of high-pressure geophysics and geochemistry research. Without their assistance, no such work could have been accomplished in my laboratory. I am particularly indebted to Y. Syono, Y. Ida, and T. Yagi, who provided assistance to my laboratory as research associates. I also wish to express my deep appreciation to M.H. Manghnani and Y. Syono for their kind encouragement in writing this paper.

References

Akaogi, M., and S. Akimoto, Pyroxene-garnet solid solution equilibria in the system $Mg_4Si_4O_{12}$–$Mg_3Al_2Si_3O_{12}$ and $Fe_4Si_4O_{12}$–$Fe_3Al_2Si_3O_{12}$ at high pressures and temperatures, *Phys. Earth Planet. Inter.*, *15*, 90–106, 1977.

Akaogi, M., and S. Akimoto, High-pressure phase equilibria in a garnet lherzolite, with special reference to Mg^{2+}–Fe^{2+} partitioning among constituent minerals, *Phys. Earth Planet. Inter.*, *19*, 31–51, 1979.

Akaogi, M., and S. Akimoto, High-pressure stability of a dense hydrous magnesian silicate, $Mg_{23}Si_8O_{42}H_6$ and some geophysical implications, *J. Geophys. Res.*, *85*, 6944–6948, 1980.

Akaogi, M., A. Navrotsky, T. Yagi, and S. Akimoto, Pyroxene-garnet transformation: thermochemistry and elasticity of garnet solid solutions, and application to a pyrolite mantle, this volume.

Akimoto, S., and H. Fujisawa, Demonstration of the electrical conductivity jump produced by the olivine-spinel transition, *J. Geophys. Res.*, *70*, 443–449, 1965.

Akimoto, S., and H. Fujisawa, Olivine-spinel transition in the system Mg_2SiO_4–Fe_2SiO_4 at 800 °C, *Earth Planet. Sci. Lett.*, *1*, 237–240, 1966.

Akimoto, S., and M. Akaogi, The system Mg_2SiO_4–MgO–H_2O at high pressures and temperatures—Possible hydrous magnesian silicates in the mantle transition zone, *Phys. Earth Planet. Inter.*, *23*, 268–275, 1980.

Akimoto, S., H. Fujisawa, and T. Katsura, The olivine-spinel transition in Fe_2SiO_4 and Ni_2SiO_4, *J. Geophys. Res.*, *70*, 1969–1977, 1965.

Akimoto, S., T. Yagi, K. Inoue, and Y. Sato, High-pressure X-ray diffraction study on barium up to 130 kbar, *High Temp.-High Pres.*, *7*, 287–294, 1975.

Akimoto, S., Y. Matsui, and Y. Syono, High-pressure crystal chemistry of orthosilicates and the formation of the mantle transition zone, in *Physics and Chemistry of Minerals and Rocks*, edited by R. G. J. Strens, pp. 327–363, John Wiley, London, 1976.

Akimoto, S., T. Yagi, and K. Inoue, High temperature-pressure phase boundaries in silicate system using in situ X-ray diffraction, in *High-Pressure Research: Applications in Geophysics*, edited by M. H. Manghnani and S. Akimoto, pp. 585–602, Acad. Press, New York, 1977.

Akimoto, S., N. Hamaya, and I. Shirotani, Synthesis of single crystals under high pressure—Crystal growth of Ni_2SiO_4 spinel and black phosphorus, in *Materials Science of the Earth's Interior*, edited by I. Sunagawa, pp. 131–148, Terrapub, Tokyo, 1984.

Akimoto, S., T. Suzuki, T. Yagi, and O. Shimomura, Phase diagram of iron determined by high-pressure/high-temperature X-ray diffraction using synchrotron radiation, this volume.

Bernal, J. D., Discussion, *Observatory*, *59*, 268, 1936.

Birch, F., Elasticity and constitution of the earth's interior, *J. Geophys. Res.*, *57*, 227–286, 1952.

Block, S., Round-robin study of the high pressure phase transition in ZnS, *Acta Cryst.*, A34 suppl., S316, 1978.

Decker, D. L., High pressure equation of state for NaCl, KCl and CsCl, *J. Appl. Phys.*, *42*, 3239–3244, 1971.

Endo, S., N. Toyama, A. Ishibashi, T. Chino, F. E. Fujita, O. Shimomura, K. Sumiyama, and Y. Tomii, Determination of $\alpha \rightarrow \varepsilon$ transition pressure in Fe-V alloy, this volume.

Fujisawa, H., N. Fujii, H. Mizutani, H. Kanamori, and S. Akimoto, Thermal diffusivity of Mg_2SiO_4, Fe_2SiO_4 and NaCl at high pressures and temperatures, *J. Geophys. Res.*, *73*, 4727–4733, 1968.

Fukai, Y., and S. Akimoto, Hydrogen in the earth's core: experimental approach, *Proc. Japan Acad.*, *59*, Ser. B, 158–162, 1983.

Fukai, Y., and T. Suzuki, Iron-water reaction under high pressure and its implication in the evolution of the earth, *J. Geophys. Res.*, *91*, 9222–9230, 1986.

Fukai, Y., A. Fukizawa, K. Watanabe, and M. Amano, Hydrogen in iron: its enhanced dissolution under pressure and stabilization of the γ-phase, *Jpn. J. Appl. Phys.*, *21*, L318–320, 1982.

Fukizawa, A., Direct determination of phase equilibria of geophysically important materials under high pressure and high temperature using in-situ X-ray diffraction method, Ph.D. thesis, 125 pp., University of Tokyo, 1982.

Fukizawa, A., and H. Kinoshita, Shear wave velocity jump at the olivine-spinel transformation in Fe_2SiO_4 by ultrasonic measurements in situ, *J. Phys. Earth*, *30*, 245–253, 1982.

Hamaya, N., and S. Akimoto, Kinetics of pressure-induced phase transformation in KCl at room temperature, *High Temp.-High Pres.*, *13*, 347–358, 1981.

Hamaya, N., and S. Akimoto, Experimental investigation on the mechanism of olivine→spinel transformation: Growth of single crystal spinel from single crystal olivine in Ni_2SiO_4, in *High-Pressure Research in Geophysics*, edited by S. Akimoto and M. H. Manghnani, pp. 373–389, Center Acad. Publ. Jpn., Tokyo, 1982.

Horiuchi, H., M. Akaogi, and H. Sawamoto, Crystal structure studies on spinel-related phases, spinelloids: Implications to olivine-spinel transformation and systematics, in *High-Pressure Research in Geophysics*, edited by S. Akimoto and M. H. Manghnani, pp. 373–389, Center Acad. Publ. Jpn., Tokyo, 1982.

Ichinose, K., M. Wakatsuki, and T. Aoki, A new sliding type cubic anvil high pressure apparatus, *Atsuryoku Gijutsu*, *13*, 244–253, (in Japanese), 1975.

Inoue, K., and T. Asada, Cubic anvil X-ray diffraction press up to 100 kbar and 1000 °C, *Jpn. J. Appl. Phys.*, *12*, 1786–1793, 1973.

Ito, E., and D. J. Weidner, Crystal growth of $MgSiO_3$ perovskite, *Geophys. Res. Lett.*, *13*, 464–466, 1986.

ITO, E., and E. TAKAHASHI, Ultrahigh-pressure phase transformations and the constitution of the deep mantle, this volume.

ITO, E., E. TAKAHASHI, and Y. MATSUI, The mineralogy and chemistry of the lower mantle: An implication of the ultrahigh-pressure phase relations in the system MgO–FeO–SiO_2, *Earth Planet. Sci. Lett.*, *67*, 238–248, 1984.

KANZAKI, M., T. FUJII, K. KURITA, T. KATO, O. SHIMOMURA, and S. AKIMOTO, A new technique to measure the viscosity and density of silicate melts at high pressure, this volume.

KATO, M., Melting experiments on Fe–FeO–FeS up to 20 GPa and its significance on the formation and composition of the earth's core, Ph.D. thesis, 34 pp., Nagoya University, 1985.

KATO, T., and M. KUMAZAWA, Stability of phase B, a hydrous magnesium silicate, to 2300 °C at 20 GPa, *Geophys. Res. Lett.*, *12*, 534–535, 1985.

KATO, T., and M. KUMAZAWA, Melting and phase relations in the system Mg_2SiO_4–$MgSiO_3$ at 20 GPa under hydrous conditions, *J. Geophys. Res.*, *91*, 9351–9355, 1986.

KAWADA, K., The system Mg_2SiO_4–Fe_2SiO_4 at high pressures and temperatures and the earth's interior, Ph.D. thesis, 187 pp., University of Tokyo, 1977.

KAWAI, N., and S. ENDO, The generation of ultrahigh hydrostatic pressures by a split sphere apparatus, *Rev. Sci. Instrum.*, *41*, 1178–1181, 1970.

KUMAZAWA, M., Multiple-anvil sliding system—a new mechanism of producing very high pressure in a large volume, *High Temp.-High Pres.*, *3*, 243–260, 1971.

MORIMOTO, N., S. AKIMOTO, K. KOTO, and M. TOKONAMI, Modified spinel, beta-manganous orthogermanate: Stability and crystal structure, *Science*, *165*, 586–588, 1969.

MORIMOTO, N., S. AKIMOTO, K. KOTO, and M. TOKONAMI, Crystal structures of high pressure modifications of Mn_2GeO_4 and Co_2SiO_4, *Phys. Earth Planet. Inter.*, *3*, 161–165, 1970.

NAVROTSKY, A., Lower mantle phase transitions may generally have negative pressure-temperature slopes, *Geophys. Res. Lett.*, *7*, 709–711, 1980.

OHTANI, E., Melting temperature distribution and fractionation in the lower mantle, *Phys. Earth Planet. Inter.*, *33*, 12–25, 1983.

OHTANI, E., Ultrahigh-pressure melting of a model chondritic mantle and pyrolite compositions, this volume.

OHTANI, E., and A. E. RINGWOOD, Composition of the core, I, Solubility of oxygen in molten iron at high temperatures, *Earth Planet. Sci. Lett.*, *71*, 85–93, 1984.

OHTANI, E., A. E. RINGWOOD, and W. HIBBERSON, Composition of the core, II, Effect of high pressure on solubility of FeO in molten iron, *Earth Planet. Sci. Lett.*, *71*, 94–103, 1984.

OHTANI, E., T. KATO, and H. SAWAMOTO, Melting of a model chondritic mantle to 20 GPa, *Nature*, *322*, 352–353, 1986.

OSUGI, J., K. SHIMIZU, K. INOUE, and K. YASUNAMI, A compact cubic anvil high pressure apparatus, *Rev. Phys. Chem. Jpn.*, *34*, 1–6, 1964.

PIERMARINI, G. J., and S. BLOCK, Ultrahigh pressure diamond anvil cell and several semi-conductor phase transition pressures in relation to the fixed point pressure scale, *Rev. Sci. Instrum.*, *46*, 973–979, 1975.

RINGWOOD, A. E., The constitution of the mantle II. Further data on the olivine-spinel transition, *Geochim. Cosmochim. Acta*, *15*, 18–29, 1958.

RINGWOOD, A. E., Prediction and confirmation of olivine-spinel transition in Ni_2SiO_4, *Geochim. Cosmochim. Acta*, *26*, 457–469, 1962.

RINGWOOD, A. E., olivine-spinel transformation in cobalt orthosilicate, *Nature*, *198*, 79–80, 1963.

RINGWOOD, A. E., Pyroxene-garnet transformation in the earth's mantle, *Earth Planet. Sci. Lett.*, *2*, 255–263, 1967.

RINGWOOD, A. E., and A. MAJOR, Synthesis of Mg_2SiO_4–Fe_2SiO_4 spinel solid solutions, *Earth Planet. Sci. Lett.*, *1*, 241–245, 1966.

RINGWOOD, A. E., and A. MAJOR, High-pressure reconnaissance investigation in the system Mg_2SiO_4–MgO–H_2O, *Earth Planet. Sci. Lett.*, *2*, 130–133, 1967.

SATO, Y., Equation of state of mantle minerals determined through high-pressure X-ray study, in *High-Pressure Research: Applications in Geophysics*, edited by M. H. Manghnani and S. Akimoto, pp. 307–323, Acad. Press, New York, 1977.

SATO, Y., and S. AKIMOTO, Hydrostatic compression of four corundum-type compounds: α-Al_2O_3, V_2O_3, Cr_2O_3 and α-Fe_2O_3, *J. Appl. Phys.*, *50*, 5285–5291, 1979.

SATO, Y., E. ITO, and S. AKIMOTO, Hydrostatic compression of ilmenite phase of $ZnSiO_3$ and $MgGeO_3$, *Phys. Chem. Minerals*, *2*, 171–176, 1977.

SATO, Y., M. AKAOGI, and S. AKIMOTO, Hydrostatic compression of the synthetic garnets pyrope and almandine, *J. Geophys. Res.*, *83*, 335–338, 1978.

SCARFE, C. M., and E. TAKAHASHI, Melting of garnet peridotite to 13 GPa and the early history of the upper mantle, *Nature*, *322*, 354–356, 1986.

SHIMOMURA, O., S. YAMAOKA, T. YAGI, M. WAKATSUKI, K. TSUJI, O. FUKUNAGA, H. KAWAMURA, K. AOKI, and S. AKIMOTO, Multi-anvil type X-ray apparatus for synchrotron radiation, *Mat. Res. Soc. Symp. Proc.*, Vol. *22*, pp. 17–20, Elsevier, Amsterdam, 1984.

STEVENSON, D. J., Hydrogen in the earth's core, *Nature*, *268*, 130–131, 1977.

SUSAKI, J., M. AKAOGI, S. AKIMOTO, and O. SHIMOMURA, Garnet-perovskite transformation in $CaGeO_3$: in situ X-ray measurements using synchrotron radiation, *Geophys. Res. Lett.*, *12*, 729–732, 1985.

SUZUKI, T., T. YAGI, and S. AKIMOTO, Precise determination of transition pressure of GaAs, Abst. 22th High Pressure Conf. Jpn., pp. 8–9, 1981.

SUZUKI, T., S. AKIMOTO, and Y. FUKAI, The system iron-enstatite-water at high pressures and temperatures—Formation of iron hydride and some geophysical implications, *Phys. Earth Planet. Inter.*, *36*, 135–144, 1984.

TAKAHASHI, E., Melting of a dry peridotite KLB-1 up to 14 GPa: Implications on the origin of peridotite upper mantle, *J. Geophys. Res.*, *91*, 9367–9382, 1986.

TAKAHASHI, E., and C. M. SCARFE, Melting of peridotite to 14 GPa and the genesis of komatiite, *Nature*, *315*, 566–568, 1985.

TSUZUKI, K., Y. ISHIKAWA, N. WATANABE, and S. AKIMOTO, Neutron diffraction and paramagnetic scattering from a high pressure phase of $MnGeO_3$ (ilmenite), *J. Phys. Soc. Jpn.*, *37*, 1242–1247, 1974.

URAKAWA, S., M. KATO, and M. KUMAZAWA, Experimental study on the phase relations in the system Fe-Ni-O-S to 15 GPa, this volume.

USSELMAN, T. M., Experimental approach to the state of the core, I, The liquidus relations of the Fe-rich portion of the Fe–Ni–S system from 30 to 100 kb, *Amer. J. Sci.*, *275*, 278–290, 1975.

WAKATSUKI, M., K. ICHINOSE, and T. AOKI, Characteristics of link-type cubic anvil, high pressure-high temperature apparatus, *Jpn. J. Appl. Phys.*, *10*, 357–366, 1971.

YAGI, T., Experimental determination of thermal expansivity of several alkali halides at high pressures, *J. Phys. Chem. Solids*, *39*, 563–571, 1978.

YAGI, T., and S. AKIMOTO, Pressure fixed points between 100 and 200 kbar based on the compression of NaCl, *J. Appl. Phys.*, *47*, 3350–3354, 1976a.

YAGI, T., and S. AKIMOTO, Direct determination of coesite-stishovite transition by in-situ X-ray measurements, *Tectonophys.*, *35*, 259–270, 1976b.

YAGI, T., and S. AKIMOTO, Phase boundary and transition rate of orthorhombic-cubic transformation in PbO_2, *J. Geophys. Res.*, *85*, 6991–6995, 1980.

YAGI, T., Y. IDA, Y. SATO, and S. AKIMOTO, Effect of hydrostatic pressure on the lattice parameters of Fe_2SiO_4 olivine up to 70 kbar, *Phys. Earth Planet. Inter.*, *10*, 348–354, 1975.

Yagi, T., O. Shimomura, S. Yamaoka, K. Takemura, and S. Akimoto, Precise measurement of compressibility of gold at room and high temperatures, in *Solid State Physics under Pressure*, edited by S. Minomura, pp. 363–368, KTK Sci. Publ., Tokyo, 1985.

Yagi, T., M. Akaogi, O. Shimomura, T. Suzuki, and S. Akimoto, In situ observation of the olivine-spinel phase transformation in Fe_2SiO_4 using synchrotron radiation, *J. Geophys. Res.*, *92*, 6207–6213, 1987.

Yagi, T., M. Akaogi, O. Shimomura, H. Tamai, and S. Akimoto, High pressure and high temperature equations of state of majorite, this volume.

Yamamoto, K., and S. Akimoto, High pressure and high temperature investigations in the system MgO–SiO_2–H_2O, *J. Solid St. Chem.*, *9*, 187–195, 1974.

I. HIGH PRESSURE-TEMPERATURE TECHNIQUES

LARGE-VOLUME FLAT BELT APPARATUS

O. FUKUNAGA, S. YAMAOKA, M. AKAISHI, H. KANDA, T. OSAWA, O. SHIMOMURA, and T. NAGASHIMA

National Institute for Research in Inorganic Materials, Namiki, Sakura-mura, Niihari-gun, Ibaraki 305, Japan

M. YOSHIKAWA

Department of Mechanical Engineering, Tokyo Institute of Technology, Meguro-ku, Tokyo 152, Japan

Abstract. Recent development of a large-volume, high-pressure flat belt apparatus provides the opportunity for the synthetic study of large-size minerals and rocks. Apparatuses with 300-ml and 1000-ml chamber volumes were constructed for operation in a pressure range of 5–7 GPa. An apparatus with a 1000 ml chamber volume can obtain polycrystalline samples of 50 mm in diameter and 50 mm in thickness. Polycrystalline diamond of this size will be useful as a new anvil material for generation of ultra high pressure in the near future. This paper describes the design of high-pressure anvils, cylinders, gaskets, and sample chambers for the flat belt type of apparatus with a large volume (FB75 and FB120). A newly constructed, 30,000-ton hydraulic press of the wire-winding type is also briefly reported here.

Introduction

BRIDGMAN (1960) pointed out a quarter century ago that "Whatever the ultimately successful method, it would seem that we must reconcile ourselves to the use of increasingly larger and more complex apparatus, with increasing expense thereby implied, with perhaps, presently, instruments the size of cycrotrons, and all the unwelcome features of government support." However, in the middle of the 1970's, the Institute of High Pressure Physics in Moscow reported a 50,000-ton force press to be operating under the direction of VERESHCHAGIN. In the early 1980's, a liter-volume class, three-stage apparatus was constructed in the Soviet Union (SEMERCHAN et al., 1981).

In Japan, the Science and Technology Agency organized a research project, in which the development of a large-volume, static high-pressure apparatus was included. In 1983, NIRIM constructed a 30,000-ton force press and a modified belt-type apparatus (FB120) for this static high-pressure research project. This paper describes the design and the construction of the apparatus and some preliminary high-pressure experimental results.

Large-Volume Flat Belt (FB) Apparatus

Figure 1 shows the relation between anvil top sizes and the maximum pressures generated in the cubic anvil, the 6-8 type of multianvil, and belt-type apparatus, as reported so far. One can easily recognize that the belt-type apparatus has the best geometry to obtain an extremely large chamber volume in the 5–7 GPa pressure range. In our laboratory, we have continuously worked on a way to increase the chamber volume of the belt-type apparatus. A flat belt (FB) apparatus with a large chamber size (FUKUNAGA et al., 1979) was designed to realize the 5–7 GPa range in pressure needed for materials synthesis studies. The design of the FB apparatus focuses on: 1) minimizing the mass or weight of the total apparatus, exchangeable die, and anvil parts, 2) finding out the optimum condition for the use of steel as a material for the die and anvil to save costs of experiments at 6 GPa pressure, and 3) simplifying the shape and fabrication of the gasket-cell assembly.

The development of a large-volume apparatus will expand high-pressure materials science and earth sciences by providing new subjects for materials or minerals synthesis studies and by allowing the measurement of their physical properties at high pressure. Further, another trial of the apparatus, such as production of a large sintered diamond mass for use as high-pressure generating anvils, is a possible future task for this project.

The FB25 apparatus, with die and anvils both composed of WC and having a bore diameter of 25 mm and chamber volume of about 10 ml, has routinely operated at 5–8 GPa pressures for more than 2000 runs since 1977. The FB75 (YAMAOKA et al., 1985) was designed in 1980, and its operation began in 1983 after many technical problems were solved. This FB75 apparatus utilized high-speed steel for the die materials and could perform more than 200 runs at 6 GPa without failure of the die. In 1983 the FB120 apparatus, which has a 120-mm bore size and a 1000-ml chamber volume, was designed for 6 GPa pressure using a steel die-anvil combination. Figure 2 shows an overall view of the apparatus, and Table 1 lists its main specifications.

High-Pressure Research in Mineral Physics, edited by M. H. Manghnani and Y. Syono, pp. 17–28.

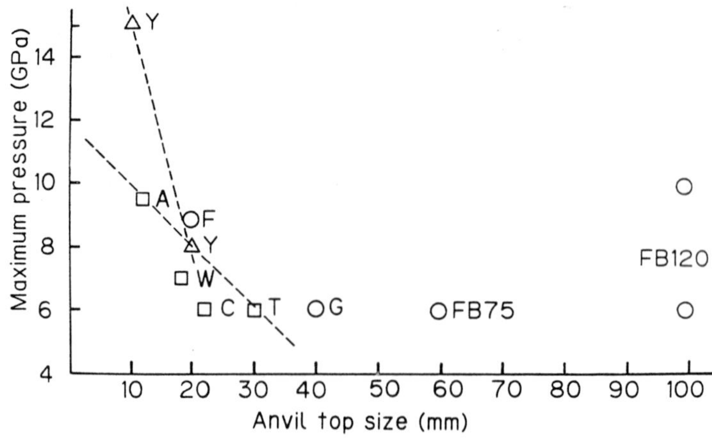

Figure 1. Approximate relation between anvil top sizes and the maximum pressures generated in various high-pressure apparatuses. Y) 6-8 anvil (YOMEDA, 1985), A) cubic anvil (Akimoto, personal communication), T) cubic anvil (TSUJII et al., 1985), W) cubic anvil (WAKATSUKI and ICHINOSE, 1982), G) belt (GE, estimated by author), F) FB25H (FUKUNAGA et al., 1981), and FB75, FB120 (this paper, the pressure for FB120 is based on estimation).

Figure 2. The whole view of the 30,000-ton hydraulic press and the FB120 high-pressure generating apparatus. A) 30,000-ton press frame, B) FB120 high-pressure vessel, C) cylinder moving rail, D) spacer for main hydraulic ram, E) pressure control panel, F) heating power supply, and G) temperature control panel.

TABLE 1. Main Specifications of the 30,000-ton Press and FB120 Apparatus

30,000 Ton Press	
Type	Wire wound frame type single action
Max. press force	30,000 ton
Total net wt.	230 ton
Wt. of the frame	160 ton
Total size	$3800^{w} \times 2200^{d} \times 7600^{h}$ mm
Width between columns	2400 mm
Moving table area	$2200^{w} \times 2200^{d}$ mm
Platen area	$2000^{dia} \times 300^{t}$ mm
Daylight opening	1800 mm (between two platens)
Main ram	diameter 1400 mm, stroke 200 mm
Sub ram	diameter 240 mm, stroke 1000 mm × 8
High pressure pump	max. 2000 kg/cm^2 with 3.2 l/min
Low pressure pump	120 kg/cm^2 with 65 l/min
FB120 Apparatus	
Pressure	6 GPa
Pressurized space	1000 ml
Cylinder size	$120^{id} \times 2{,}000^{od} \times 320^{t}$ mm
Anvil size	$100^{trun.d} \times 1200^{od} \times 350^{t}$ mm
Electric power	10 V AC × 5,000 A
Temperature	2000° C continuous, 4000° C by flash

Design Procedures

Figure 3 shows design procedures for an FB type of high-pressure apparatus. Recently, stress and strain analysis by FEM (finite element method) has provided much information for more accurate design of each high-pressure component. If we can give values for external forces on the component at any given condition, FEM analysis can be carried out throughout the design process. External forces on the die part of the cylinder from the binding rings and from the gasket and the insulating sleeve are acting at the operating conditions. The analysis of the external force from the binding rings can be performed by FEM using the parameter of die shape and known material properties, such as Young's modulus and Poisson's ratio. The external force from the gasket and the insulating sleeve can be estimated only from experimental data for a similar shaped apparatus. The external force cannot be estimated directly from the known

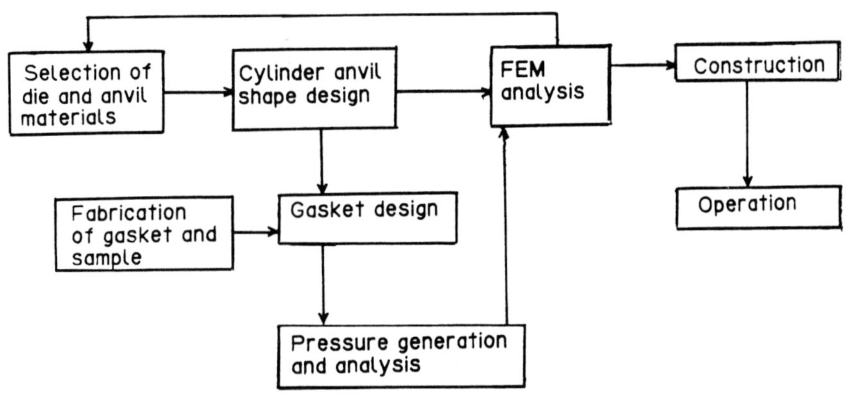

Figure 3. Design procedures for the FB type of high-pressure apparatus.

material properties of the gasket and the insulating sleeve, such as porosity, shear strength, and compressibility. Even in a small scale belt, it is not practical to conduct experiments with many different combinations of die and anvil shapes and gasket and insulating sleeve materials at room and high temperatures. Thus planning the shape of die and the anvil is a very delicate stage of the design process.

The gasket design presents another problem in the development of a large-scale apparatus. If it was possible to conceive an optimum gasket design based on tentative material characteristics, fabrication methods would be limited to large-scale samples and gasket materials. For example, the gasket weight for FB120 would be about 2 kg for each piece. At present, cold die pressing which allows a designed powder mixture composition to be obtained is the only method for daily fabrication of the gasket. Characterization of the gasket and the cell materials is directly linked with gasket design. The amount of shrinkage or expansion during the fabrication process, porosity, machinability, and deformation characteristics (WAKATSUKI et al., 1972) of gasket cell materials are parameters that must be measured for gasket design.

The construction of the apparatus, especially press fitting for each component, must be carried out very carefully, and stress and strain measurements must be made so that the design parameters can be compared during the development process.

Shape of Dies and Anvils

Shapes of the die and the anvil in the belt apparatus can be systematized by giving geometrical parameters, as illustrated in Figure 4. In the design process we first determine the radius R_1 of the bore. Other parameters cannot be easily determined because there are many

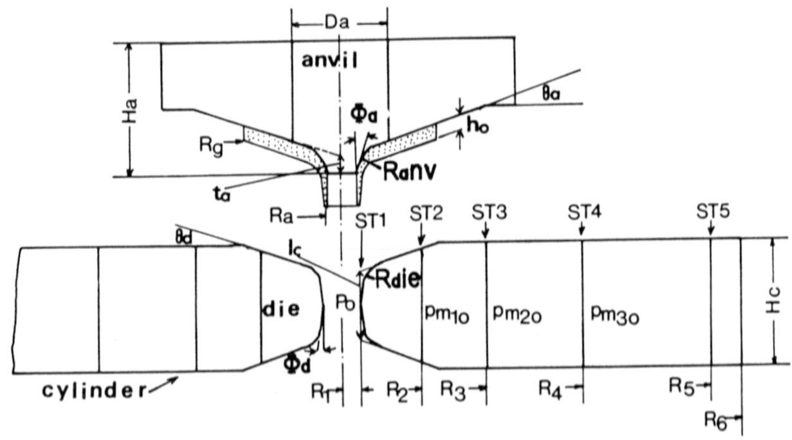

Figure 4. Shape parameters and stresses for the FB apparatus. P_{m10} to P_{m30} designate lateral forces generated by press fitting of the binding rings, and ST1 to ST5 are tangential stresses at the inner wall of the binding rings.

uncertain factors affecting them. The most important yet simple relation to be considered is

$$L_0 = l_c - 2R_1\tan\theta + 2h_0/\sin\theta - 2t_a \ (\theta_d = \theta_a = \theta)$$

where L_0 is the distance between the two anvils just before compression. For designers, it would be better to evaluate relations among l_c, R_1, and t_a in the equation because these parameters cannot be changed after construction of the vessel, whereas L_0 and h_0, an initial thickness of the gasket change with the gasket and sample size.

In designing a large-volume apparatus, careful consideration is given to ways to reduce the total mass of the die and anvils. The volume of the die can be estimated from the simplified shape as

$$V = 3.14(R_2^2 - R_2^2)(l_c + (R_2 - R_1)\tan\theta)$$

Smaller θ and R_2 values are preferable for fixed internal radius R_1 to reduce the mass. The l_c must also be minimized at given internal radius R_1. In a small-scale belt design (HALL, 1960), the value given by l_c/R_1=4.8–3.4 is adopted. This value is reduced to 2.8–2.6 in the FB apparatus to save the mass.

The mass of an anvil is greatly affected by its diameter and the height. When we use a WC anvil, we try to decrease the diameter and the height without reducing the maximum pressure limit and the anvil life. Figure 5 presents summary data on fracture and deformation for anvils of various shapes and composed of different materials. The ratio of the outer diameter to the top diameter, D_a/d_a, and the ratio of the anvil height to the top diameter, H_a/d_a, affect the fracture strength according to the massive support principle. In the fracture experiments (see St(F), F in Figure 5) carried out in the range D_a/d_a=2–3 and H_a/d_a=2–3 for hardened tool steel, no obvious differences in fracture strength were observed. It is possible to reduce the ratios of these parameters down to about 2.

The high-pressure cylinder of the FB apparatus is of the press-fitted, multiring type. The ratio of the outer radius of the cylinder to the inner radius of the die, R_5/R_1, is determined by the number of rings and the outer to inner radius ratios, R_2/R_1, of each binding ring and of the die. Many trial calculations for determining the optimum number and radius ratios of the binding rings were carried out, and three binding rings with the radius ratio of 1.5–1.7 were finally selected. The calculation is based on the assumption that a fracture in a binding ring occurs when the maximum tangential stress at the inner wall of each ring exceeds the limit. Experimental data (FUKUNAGA, 1970; YOSHIKAWA et al., 1979) support this assumption. Maximum tangential stress at the inner wall of each ring is governed by the factor $(k^2+1)/(k^2-1)$ for each radius ratio k, so that a value of more than 1.7 for k does not considerably decrease the tangential stress.

Figure 6 is a schematic section of the FB120 apparatus, and Table 2 presents some of its shape design parameters.

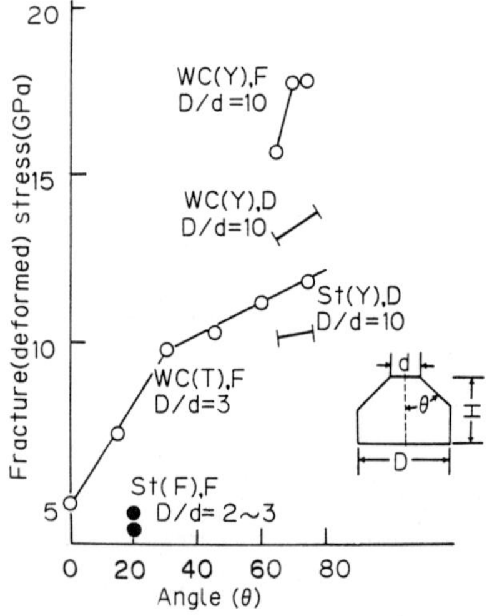

Figure 5. Fracture or deformed stresses of anvils as functions of tapered angle θ and D/d (defined in the figure). WC(Y), F and WC(Y), D are for WC anvil fracture and deformation experiments, respectively, by YOSHIKAWA et al. (1977), and ST(Y), D represents steel anvil deformation experiments by YOSHIKAWA et al. (1977). WC(T), F are for experiments by TSUJII and JINUSHI (1969), and ST(F), F is for experiments by FUKUNAGA et al. (1980).

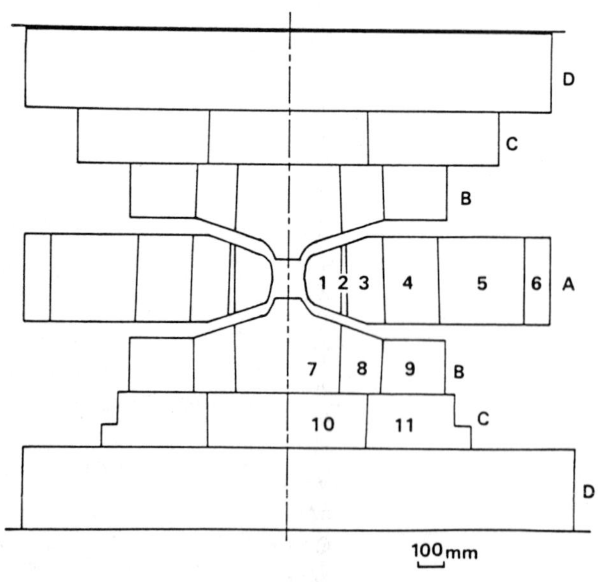

Figure 6. Schematic section of FB120 apparatus: A) cylinder, B) anvil, C) anvil support, and D) anvil base platen. The dimensions and materials for 1–11 are described in Table 2.

TABLE 2. Main Specifications of the FB120 Anvil and Cylinder

	Parts No. in Figure 6	Material	Hardness (H R_c)	Diameter (mm)	Height (mm)
Cylinder	1	YXM-1	61-63	400	250
	2	SGT	58-62	440	261
	3	DAC	45-50	720	320
	4	DAC	44-46	1140	320
	5	DM	39-52	1800	320
	6	DM	—	2000	320
Anvil	7	YXM-1	61-63	400	350
	8	DAC	48-50	700	250
	9	DM	39-42	1200	250
	10	SLD	58-62	600	200
	11	DAC	48-50	1600	200

Interferences

Combinations	dr/r_m ($\times 10^3$)	P_m (kg/mm^2)	Load (tons)
5/4	2.24	10.1	—
6/5 + 4	0.5	—	—
3/4 + 5 + 6	5.3	30.0	2040
1/2	0.5	—	—
1 + 2/3 + 4 + 5 + 6	11.9	108.0	4400
8/9	1.7	9.1	—
7/8 + 9	5.37	50.0	1130
10/11	4.0	34.3	1200

Materials are all specified according to Hitachi Metal Co. Ltd.

FEM Analysis for Die and Anvil

Distribution of the interfacial pressure P_m at the interface of the shrunk-on fitting rings can be accurately calculated by FEM. When the inner ring is press fitted to the outer ring, expansion of the outer ring and contraction of the inner ring result from the interfacial pressure P_{mi}. The amounts of deformation in the outer and inner rings is given by $u_{re}(>0)$ and $u_{ri}(<0)$, respectively. The interference, dr, between the two binded rings is given by

$$dr = u_{re} - u_{ri} = (k_e - k_i)P_{mi}$$

based on known material properties. k, the displacement for the P_{mi} at a given element of the FEM calculation, and dr can be determined from the distribution P_{mi}. However, dr must be obtained first before P_{mi} can be determined. In the design process dr is determined using a tentative calculation of the IFC (infinite cylinder) model.

Figure 7 presents a calculation of tangential stress distribution for given dr values each interface based on FEM and IFC models for the FB cylinder at an internal pressure P_0=6 GPa. The shape of the actual cylinder core is different from that of a simple cylinder. For this reason, the P_{mi} distribution at the outer surface of the core and the stress distribution in the core are greatly different from

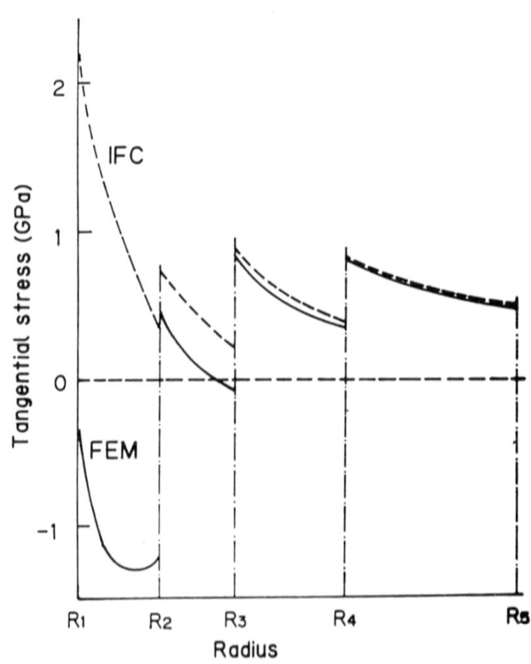

Figure 7. Tangential stress distribution in the cylinder components calculated by FEM and IFC at an internal pressure of 6 GPa.

the distributions determined by the IFC calculation. However, there are only small differences between FEM and IFC calculations of stress distributions in the outer binding rings. The IFC calculation can determine dr using the values R_2 to R_5, which are determined by the selection of a radius ratio for the binding rings in the range 1.5–1.7. An optimum R_2/R_1 cannot be determined by the IFC. We normally adopt values between 3 and 4 for R_2/R_1 in the IFC, but such a value is usually refined by FEM afterward. Tentatively determined radii of the rings R_1-R_5 are used for the calculation of the stress distribution in binding rings based on the IFC model to keep each maximum tangential stress level below about 90 kg/mm^2; at this condition, the dr for each binding ring is determined tentatively. Using values of R_1 to R_5, dr, and core shape factors, the FEM calculation can be accurately carried out for a zero internal pressure state.

Table 2 gives the values of R_1 to R_5 and each dr for the design of the FB120 cylinder which was actually constructed. The FEM calculation is carried out for 0 and 6 GPa internal pressures. The resultant stresses at the central portion of the cylinder are listed in Table 3. Tangential compression of -319 kgf/mm^2 at the inner surface of the core exceeds the elastic limit of the used material, but the high stress level is localized near the central core portion. A stress level as high as -406 kgf/mm^2 has been obtained during the construction and operation of the FB75 apparatus. However, in the actual FB120 construction process, interference of 3+4+5+6/1+2 ring was decreased to 2.62 mm. Therefore the press fitting process was made easier and the equivalent stress at the inner wall of binding ring 3 was decreased by about 23%, compared to the FB75 design.

Stress distribution within the core is greatly affected by the core shape and external forces on its surface. The external forces consist of the P_{m10}, the gasket pressure, and the pressure exerted by the insulating sleeve in the sample chamber. Pressure distribution in the gasket along the die and the anvil wall can be estimated based on the force balance at a given point, $f_c = -(h/2)(dP_c/dl)$. The relation

$$P_c = P_{0,d}\exp - b(l - l_0) = P_a$$

can be assumed, where f_c is internal friction of the gasket, h is the thickness of the gasket, l is the length of the gasket along the anvil wall, l_0 is the length of the gasket from the center of the die to the edge of the anvil, and $P_{0,d}$ is the pressure for the top edge of the anvil wall. The parameter b can be calculated by

$$F = 3.14 r_a^2 P_0/\eta = F_z(P_a, b)$$

where R_a is a radius of the top of the anvil, η is the efficiency of the pressure generation, and P_0 is the internal pressure.

An example of the calculated pressure distribution along the gasket is shown in Figure 8. The tangential stress at the center of the die wall decreases with increasing efficiency of pressure generation, but a larger pressure gradient results along the gasket. Thus the optimum efficiency must be achieved in routine operation through the selection of materials and thickness of the gasket. Although the FEM calculation contains some

TABLE 3. Stresses in the FB120 Cylinder by FEM

		Positions							
	Stress	1^i	1°	3^i	3°	4^i	4°	5^i	5°
	sigma r	0	−103	−103	−68	−68	−27	−27	0
	sigma z	−83	60	−27	6	2	0	0	0
A	sigma c	−319	−127	−6	−24	48	15	64	38
	tau max	160	94	55	37	58	21	46	19
	tau eq	287	176	88	63	101	37	81	38
	sigma r	−500	−142	−142	−83	−83	−29	−29	0
	sigma z	−261	47	−37	7	5	0	0	0
B	sigma c	79	−83	30	−8	66	23	72	43
	tau max	211	95	86	45	75	26	51	22
	tau eq	500	167	150	84	130	45	90	43

All values in the table are in kg/mm^2. The top column of the table denotes the position of the ring which corresponds to the numbers in Figure 6. Superscript of the number, i and o corresponds to the internal wall and the outer wall of the ring, respectively. A and B in the table denote the condition $P_0=0$ and $P_0=6$ GPa, respectively. Calculations were carried out for $P_0=0$ GPa and 6 GPa. When the sample was compressed, the pressure generating efficiency of 35% was used in the calculation.

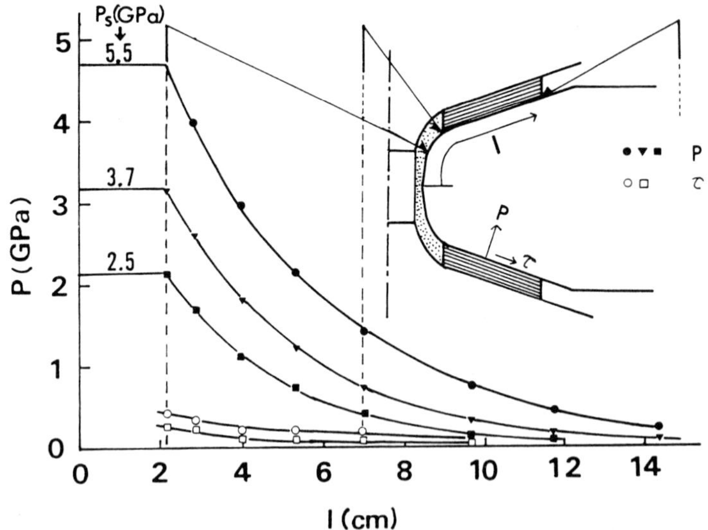

Figure 8. Calculated stress distribution in FB75 gasket. The pressure generating efficiency for each curve was obtained from room temperature calibrations. It must be noted that a pressure drop of about 20% was observed for the center wall of the core, as compared with the pressure at the center of the sample.

uncertain factors, such as the true pressure gradient in the gasket, its results (see Figure 7) indicate that a high-speed steel can be used as a core material at 6 GPa pressure because the maximum stress is considerably below the fracture level, which is estimated from experimental results for a small-scale apparatus.

Figure 9 presents the distribution of shear stress in the WC anvil when the base was constrained. When the WC anvil was not laterally supported by the gasket region, the location of the maximum shear stress value existed on the flank surface, as shown in Figure 9a. Actually, fracture of the anvil often originated near this area. The maximum shear stress at $P_0=6$ GPa was about 2.2 GPa. It has been shown that maximum shear stress can be reduced to about 70% by lateral support from the gasket, but such support causes a higher shear stress at the base of the anvil as shown in Figure 9b. Thus the use of lateral supports by the gasket and the precompression of binding rings, as shown in Figure 9c, would increase the anvil endurance limit. The magnitude of anvil precompression designed for this calculation is discussed later.

Press Fitting

When we consider press fitting the core plus shim member into the binding ring assembly (3+4+5+6 in Figure 6), the initial height difference between these components is determined from the interference (2.62 mm) and the interfacial taper angle (1.5°) to be 100.1 mm. The initial height h decreases to $h-\Delta h$ due to a press-fitting force F, and an interfacial pressure $p=(\Delta h/h)P^{m}$ is generated (where P^{m} is the interfacial pressure at the final press fitting stage). Thus we have the relation

$$F = P\mu = 2\pi R(L - h + \Delta h)\mu\Delta h P^{m}/h$$

or

$$F = A\Delta h + B(\Delta h)^2$$

where F is the protruding press force, P is the total internal force, μ is the friction coefficient, L is the total length of the interface, R is the radius of the core, and A and B are parameters independent of Δh. Based on the FEM calculation, the interference or the initial height difference for the press fitting of the FB120 cylinder and anvil are determined and listed in Table 2.

During the press fitting process, strain at any point of the die can be accurately calculated by FEM, so that the press fitting process can be monitored by measuring the F versus Δh relation and strain at a specific point in the die. If these measurements are omitted, we cannot follow an abrupt change of the friction coefficient during the process. Inspection of the fitness between two faces to be press fitted is always carried out by the paint method. However, it is difficult to certify the fitness only by the paint method. Pressure welding or sticking can easily be detected by the measurement of the F versus h relation during the process. If the coefficient of friction exceeds about 0.14 or a discontinuous change in the value appears, the process must be stopped and the die must be take out for inspection.

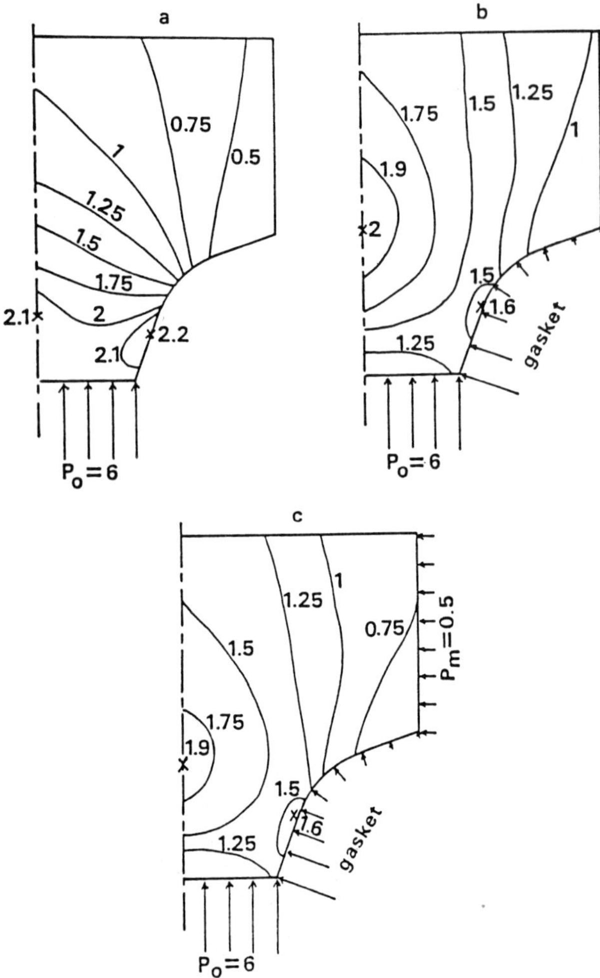

Figure 9. Shear stress distribution in the anvil body based on FEM calculation. Each anvil base is contrained. Numbers are in GPa's.

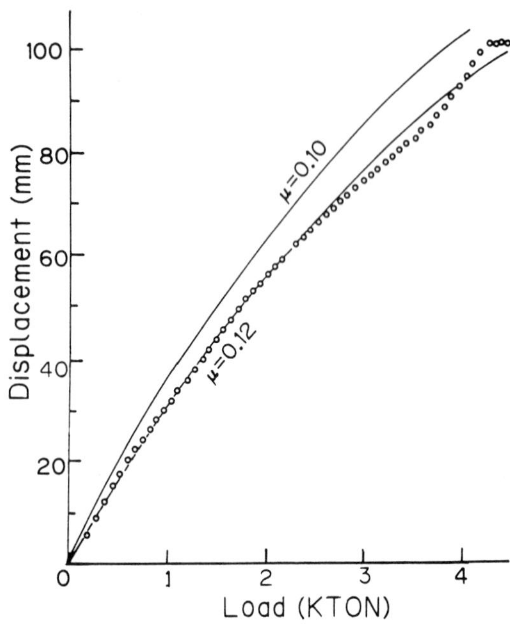

Figure 10. Cylinder core displacement, h, as a function of applied load for the FB120 apparatus. Solid curves denote cases of constant friction coefficients, $\mu=0.10$ and $\mu=0.12$, between the core and the binding rings. Open circles represent experimental data.

The press-fitting record for the FB120 die shown in Figure 10 is a good example of a normal press-fitting process. For almost all of this process, the coefficient of friction is kept at 0.12; only a small deviation to the lower value of the friction coefficient occurred at the final stage. Such a small change can be neglected. In Figure 11, the strain value measured at the center of the die wall is compared with the value calculated by FEM. The two sets of values are in excellent agreement.

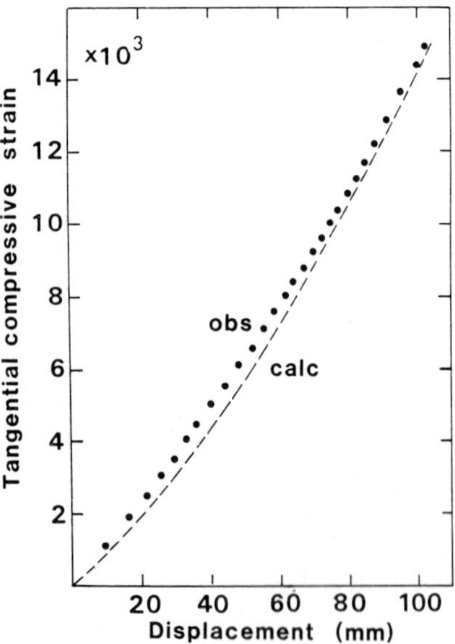

Figure 11. Tangential compressive stress at the center of the internal wall of FB120 core. Dashed line is deduced by an FEM calculation, and the solid circles express observed data.

Sample Chamber Design

The sample and the gasket regions of the FB apparatus are compressed by two anvils, as shown in Figure 12. The outer paper gasket, P, and the inner gasket, G, compressed under the load to provide the stroke for the compression of the sample. The sample container, S, is compressed by the compression of gasket. To increase efficiency of

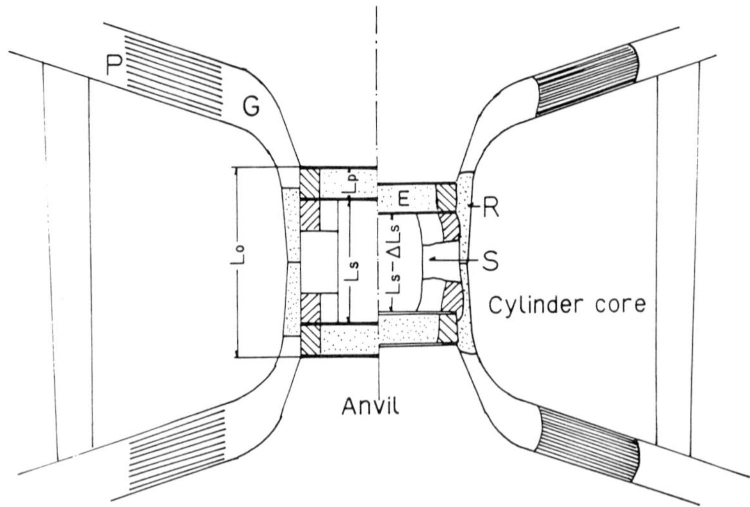

Figure 12. Schematic sample-gasket design in the FB apparatus.

pressure generation, the insulating sleeve, R, and the electrode plate, E, should be rather incompressible. To design and determine the sizes of these parts, we must know the deformation behaviors of their materials at room temperature and at high temperature.

Figure 13 presents pressure calibration curves for different insulating materials. Figure 14 shows experimental data for the relation between relative anvil displacement ($\Delta L/L_s$) and pressure (P) at the center of a

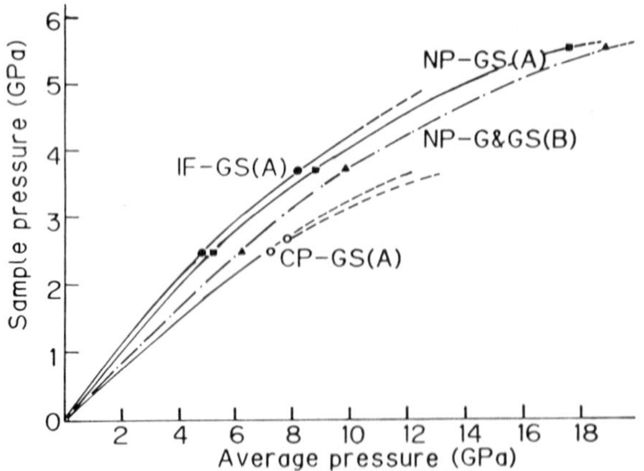

Figure 13. Pressure calibration curves as functions of average pressure to the area of the top surface of the anvil (in GPa) for different insulating sleeves in the FB 25, FB75, and FB120 apparatuses. NP-GS) natural pyrophyllite sleeve plus powder compacted gasket, IF-GS) infiltrated insulating sleeve plus powder compacted gasket, CP-GS) powder compacted insulating sleeve and gasket, NP-G&GS) natural pyrophyllite insulating sleeve and gasket. The curves designated as A are based on measurements in the FB75 and the FB120 apparatuses. Curve B represents FB25 measurements.

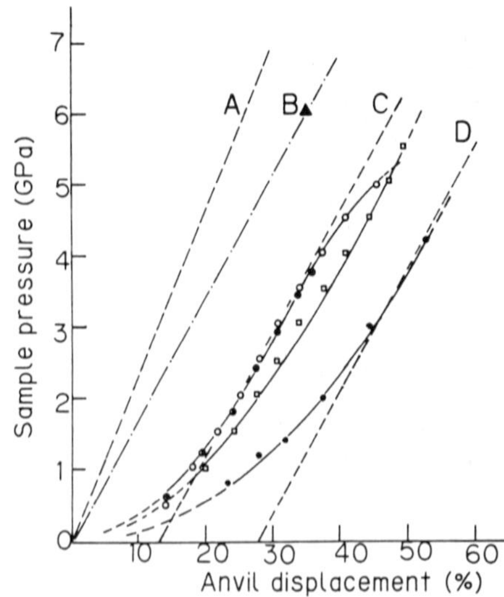

Figure 14. Sample pressure as a function of the anvil displacement for different insulating sleeve materials. The line A denotes a theoretical compression curve for NaCl, which is a main part of the sample chamber. The line B is a calculated NaCl compression based on the measured radial expansion of 6.5% at 6 GPa. Open circles are the data obtained with the sample NP-GS (see Figure 13) semiopen circles which close to open circle are from IF-GS, open squares are from NP-G&GS, and solid circles are from CP-GS. The curves C and D, which are parallel to the curve B, show initial anvil displacement for the different gasket sleeve materials.

sample held at room temperature, for various sample materials in the FB75 and FB25 apparatuses. In these experiments, NaCl disks of about 94% relative density which were fabricated by the powder compaction technique, were used for the material of the sample S. Powder compacted or machined natural pyrophyllite was used for the gasket material and for the insulating sleeve material. When a machined natural pyrophyllite sleeve was used, all data for different gasket fabrication methods (machining or powder compaction) were similar. When a powder compacted insulating sleeve was used, extraordinarily large deformation of the sample chamber was observed. After pressure calibration, the measured average radial expansion of the sample container was about 6.5%. If the sample expands, a much larger displacement is necessary to obtain the same pressure value as that of the unexpanded sample. In Figure 14, curve A is the theoretical NaCl compression and curve B is the calculated NaCl compression for an expansion of 6.5% from the original radius. The observed data include the effect of sample expansion, but the dominant deviation from the calculated compression may be accounted for by the large initial compression in the lower pressure range. It is considered that misfit or clearance between parts of the sample chamber correspond to a porosity of about 5% and will give initial displacement of the anvils at low pressure. In the case of a powder compacted insulating sleeve, initial displacement of the anvil is estimated to correspond to a porosity of about 25% as shown in Figure 14. This large initial displacement is reasonable if we take into account the porosity of the powder compacted sleeve (about 25%), which is larger than the porosity of natural pyrophyllite (about 4%). In this case, the sample must be extraordinarily deformed in order to pressurize the sleeve part from the internal gasket region because of the geometrical difficulty involved.

To get suitable sleeve material for a large-size apparatus, an infiltration technique was developed for a calcined sleeve. A powder mixture of pyrophyllite 52.4%, kaolin 17.4%, ZrO_2 21.1%, and mica 9.1% was ball milled for 48 h, dried, and granulated with 3.5% organic binder. After die pressing, it was calcined at 850 °C in air. The calcined body was dipped into 20 vol% Na_2SiO_3 aqueous solution for 1 h and then dried for 1 h. This infiltration of sodium silicate into the calcined body was repeated 6–7 times until a porosity of about 3% was obtained, as shown in Figure 15. This infiltrated body could easily be machined by a lathe to adjust the size of the sleeve. Figure 14 presents data on anvil displacement versus pressure for the infiltrated sleeve; we can see drastic improvement in the pressure-generating characteristics.

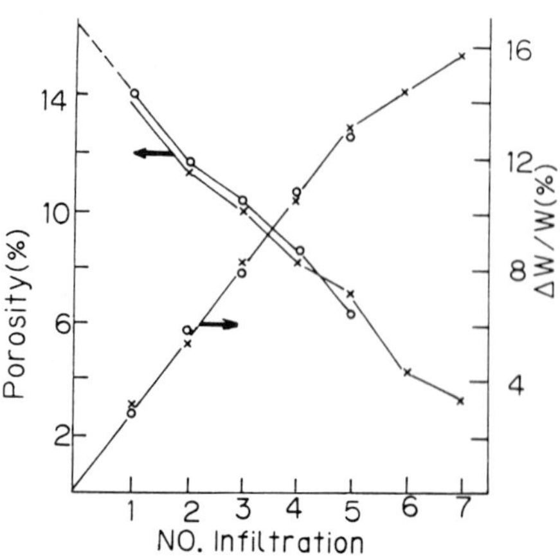

Figure 15. Weight gain and porosity decrease versus infiltration in a calcined insulating sleeve material of the system pyrophyllite-kaolin-zirconia-mica at room temperature.

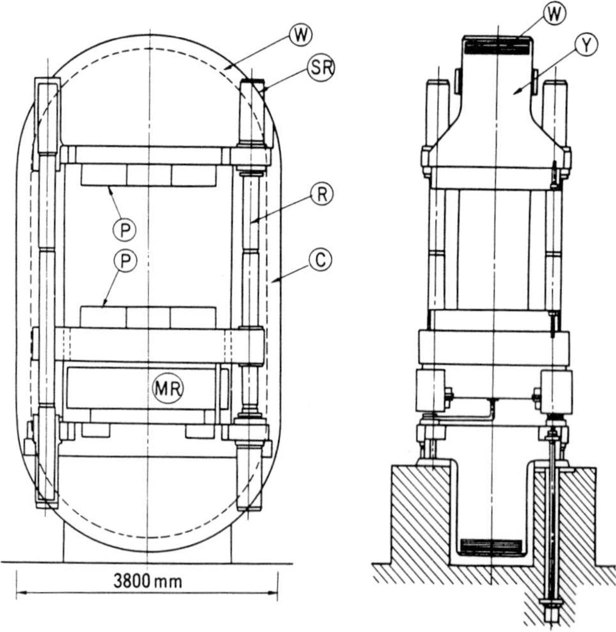

Figure 16. Schematic drawing of 30,000-ton hydraulic press: W) wire winding layer, R) rods for operating the moving plate on the main ram, P) platen, C) column, Y) yoke, MR) main ram, and SR) sub ram.

30,000-Ton Hydraulic Press

Figure 16 shows a general schematic drawing of the 30,000-ton press for operating the FB120 apparatus. The frame of the press is of the wire-winding type to reduce the total press weight. The effective length between the columns and the daylight opening are determined first, according to the FB 120 apparatus design. The cross-section of the column was determined from two types of input data, the lift-off load and the maximum compressive stress at the wire-wound condition (39,000 tons and 30 kg/mm^2, respectively). Generally, the lift-off load is about 1.3–1.5 times larger than the maximum operating load. The maximum compressive stress is determined from data on the strength of the column materials. In our case, a forged steel with yield strength of 45 kg/mm^2 was used. The section of the column A_c is given by

$$A_c = Q_m/2\sigma_m$$

where Q_m is the lift-off load, and σ_m is the maximum compressive stress on the column. If $\sigma_m=30$ kgf/mm^2 and $Q_m=39{,}000$ tons, A_c is calculated to be 0.65 m^2. The column thickness d is determined to be 1000 mm by measuring the difference between the anvil base platen size (2000 mm in diameter) and the column thickness. Because the column width must be larger than 650 mm, a column width w_c of 700 mm was selected. The wire winding layer dimension and the wire tension were then designed. From the design of the wire winding machine, the wire thickness t_p and the width w_p were determined to be 1.5 mm and 6 mm, respectively. Also the maximum wire tension T was determined to be $T=120$ kg/mm^2 from the yield strength of the wire used. (The yield strength of the wire is about 160 kg/mm^2.) The relation between Q_m and a winding layer dimension nm can be expressed as

$$Q_m = 2nmt_pw_pT$$

where n is the turning number of the wire in each layer, and m is the number of the layers. A value of $nm=18{,}000$ is then obtained. From the column thickness (1000 mm), the wire winding width is determined to be 900 mm for roughly $n=150$ turns. Actually, some allowance for wire turns is necessary. If $n=148$, then a value of $m=122$ is obtained. Dimensional design of the frame was carried out according to the above approximate relations, but a more accurate strain calculation was carried out by FEM and the results of strain measurements at many positions were monitored.

Figure 17 shows a schematic drawing of the wire-winding machine designed for 30,000-ton press frame winding. The vertical stroke of the wire was 1000 mm and

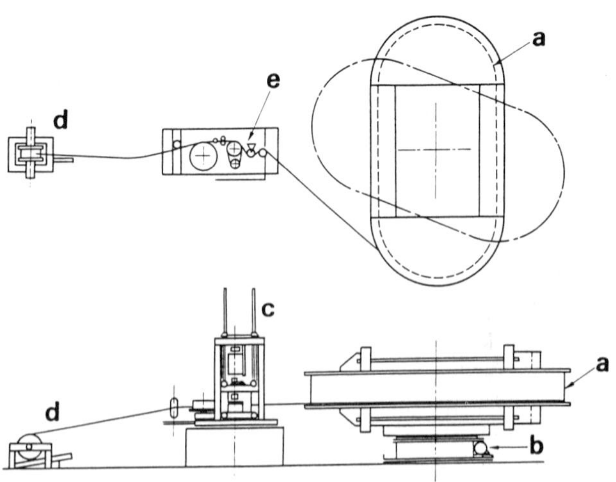

Figure 17. Wire winding for the 30,000-ton press frame: a) frame, b) turning machine, c) wire tension machine, d) wire supply, and e) load cell.

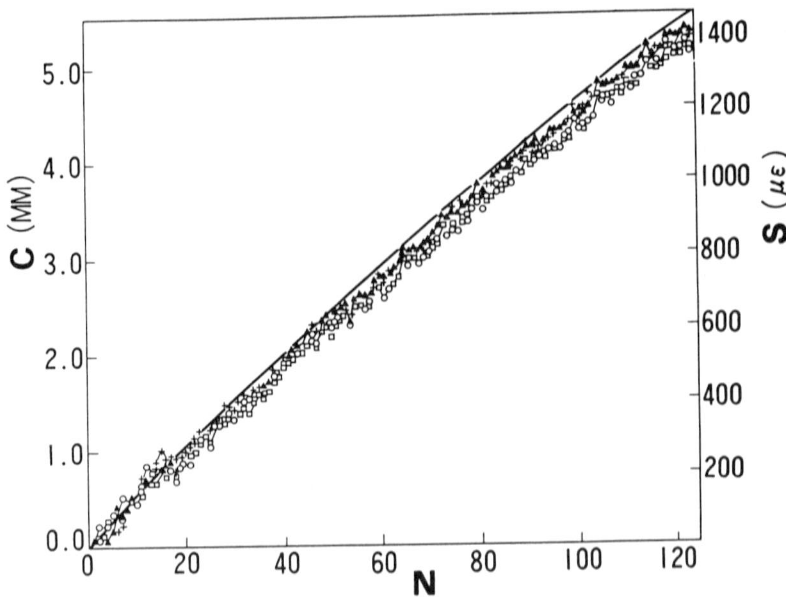

Figure 18. Amount of column shrinkage C and compressive strain S during wire winding. N denotes number of winding layers. The data were obtained at four different points of the column.

the position of the wire was controlled by a stepping motor for each turn. The tension of the wire was supplied by a d.c. motor, and the wire was turned nine times on the supplying dram to compensate for the friction of the wire during supply. Tension to the wire was monitored by a load cell and kept within 3% accuracy of the tensile force. The turning speed of the frame was about 1.4 r.p.m. Figure 18 shows the data on column shrinkage for each layer. Data for 124 layers were obtained by measuring strains at four different positions on the columns. The amount of shrinkage for each layer is approximately the same. Data were corrected for change in temperature of the press frame related to the deviation of room temperature from day to night.

Acknowledgments. This work was supported by a grant of Science and Technology Promotion Fund, 1980–1985, Science and Technology Agency. The authors are grateful to M. Wakatsuki, the chairman of the research promotion committee of the project.

References

Bridgman, P. W., General outlook on the field of high-pressure research, *P. W. Bridgman Collected Papers*, *7*, 4625–4637, Harvard University Press, Cambridge, 1964.

Fukunaga, O., Trial construction of piston-cylinder type high pressure-high temperature apparatus, *J. High Press. Inst.*, *8*, 2–9, 1970.

Fukunaga, O., S. Yamaoka, T. Endoh, M. Akaishi, and H. Kanda, Modification of belt-like high-pressure apparatus, in *High Pressure Science and Technology*, *vol. 1*, edited by K. D. Timmerhaus and M. S. Barber, pp. 846–852, Plenum Pub. Co., New York, 1979.

Fukunaga, O., S. Yamaoka, S. Osawa, and M. Yoshikawa, Stress analysis of pressurized components of belt apparatus, *Proceedings 21st Japan High Pressure Conference*, Tokyo, pp. 181–182, 1980.

Fukunaga, O., S. Yamaoka, and M. Akaishi, FB25-100 type apparatus as a model of large volume 100 kbar apparatus, *Proceedings 22nd Japan High Pressure Conference*, Hiroshima, pp. 108–109, 1981.

Hall, H. T., Ultra-high-pressure, high-temperature apparatus: the "Belt", *Rev. Sci. Instr.*, *31*, 125–131, 1960.

Semerchan, A. A., N. N. Kuzin, and T. N. Davydova, High-pressure, high-temperature apparatus with a large reaction volume for a 50,000-ton press, *Sov. Phys. Dokl.*, *26*, 213–215, 1981.

Tsuji, S., and M. Jinushi, Fundamental experiments on the anvil of high pressure apparatus, *Proceedings 11th Japan High Pressure Conference*, Sendai, pp. 15–17, 1969.

Tsuji, S., M. Wakatsuki, and T. Yamagishi, Development and practical performance of enlarged, multi-anvil link-type cubic high pressure apparatus, *J. High Press. Inst.*, *23*, 57–64, 1985.

Wakatsuki, M., K. Ichinose, and T. Aoki, Note on compressible gasket and Bridgman-anvil type high pressure apparatus, *Japanese J. Appl. Phys.*, *11*, 578–590, 1972.

Wakatsuki, M., and K. Ichinose, A wedge-type cubic anvil apparatus and its application to material synthesis, in *High Pressure Research in Geophysics*, edited by S. Akimoto and M. H. Manghnani, pp. 13–26, Center for Academic Pub. Japan/D. Reidel, Tokyo/Dorderecht, 1982.

Yamaoka, S., H. Kanda, M. Akaishi, T. Osawa, T. Nagashima, and O. Fukunaga, Development of large volume belt type high pressure apparatus (FB75), *J. High Press. Inst.*, *23*, 169–177, 1985.

Yoneda, A., Guiding principles for generation of ultra high pressure in large volume, *J. High Press. Inst.*, *23*, 132–151, 1985.

Yoshikawa, M., S. Takano, and T. Asaeda, Study of piston-cylinder type high pressure generating apparatus,—compressive strength of conical anvil—, *J. Mech. Eng.*, *43*, 3134–3141, 1977.

Yoshikawa, M., T. Suzuki, and I. Suzuki, The breaking strength of the cylinder for high pressure, *J. High Press. Inst.*, *17*, 130–136, 1979.

DETERMINATION OF $\alpha \rightarrow \varepsilon$ TRANSITION PRESSURE IN Fe-V ALLOY

S. ENDO, N. TOYAMA, A. ISHIBASHI, T. CHINO and F. E. FUJITA

Research Center for Extreme Materials, Osaka University, Toyonaka, Osaka 560, Japan

O. SHIMOMURA

National Institute for Research in Inorganic Materials, Sakura-mura, Niihari-gun, Ibaraki 305, Japan

K. SUMIYAMA and Y. TOMII

Department of Metal Science and Technology, Faculty of Engineering, Kyoto University, Kyoto 606, Japan

Abstract. Using a high-pressure X-ray diffraction method developed by combining sintered diamond (SD) anvils and synchrotron radiation, we studied the pressures of the α-ε phase transition in Fe-V alloys in order to confirm pressure-fixed points above 22 GPa. The pressures obtained by simultaneous measurements of the lattice parameters of NaCl and Au and the electric resistance of Fe-V with SD were in agreement with the pressures determined by the ruby fluorescence method and X-ray observation with a diamond-anvil cell. Taking into consideration the pressure distribution within the sample, we determined the pressures at the initiation of the transitions to be 35, 43, and 51 GPa, for Fe-18.8, 22.1, and 25.0 wt%V, respectively. These values are significantly lower than previous values reported for shock-wave and static experiments.

Introduction

Sintered diamond (SD, hereafter) is one of the most promising anvil materials to statically generate ultrahigh pressure. Many of the SD materials obtainable at present, however, contain some kind of binder to cement the diamond particles together. The binders usually intercept the passage of light and X-rays with wide ranges of wavelength, thereby making it difficult to evaluate the generated pressure by an optical ruby fluorescence method or by an X-ray diffraction method using the lattice parameters of standard materials.

SD anvils prepared at Osaka University, were used in the present experiments. These anvils allowed a fair amount of X-rays to pass through them and thus made it possible to perform X-ray diffraction under high pressures (UTSUMI et al., 1986). Using these SD anvils, we studied pressures of the $\alpha \rightarrow \varepsilon$ phase transition in Fe-V alloys, in an attempt to confirm rare pressure-fixed points above 22 GPa (the GaP point). The pressures at which the electrical resistance of the alloys changed due to the transition were determined from the lattice parameters of NaCl and Au which were placed adjacent to the alloys.

X-Ray Diffraction with SD Anvils

X-ray diffraction under ultrahigh pressure generated with SD anvils was developed by combining a cubic-anvil type of "MAX80" press (SHIMOMURA et al., 1985) and the intense synchrotron radiation source at KEK. Details of this method are reported in a previous paper (UTSUMI et al., 1986) and outlined here.

A pair of opposed SD anvils with 14 wt%Co binder were set into a cubic boron-epoxy (B-E) medium as shown in Figure 1. An X-ray beam was collimated to the sample which rested in the center of a disk between the SD anvils. The diffracted beam was detected by a solid-state detector (SSD) set at a glancing angle 2θ and located in an energy-dispersive system. The diffracted beam had to pass

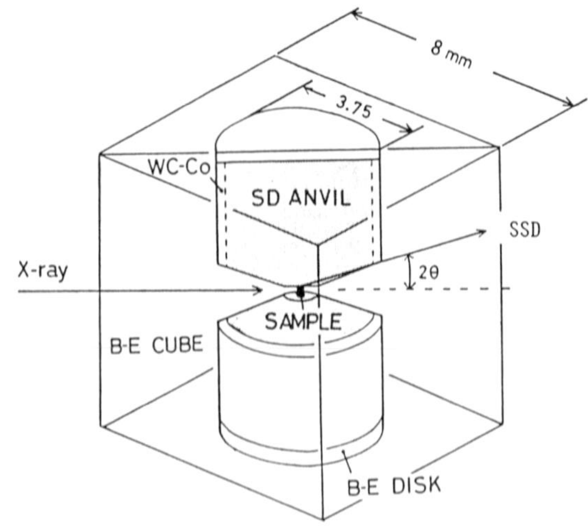

Figure 1. A cross-section of a cell of the "6-2 type" double-stage system for X-ray diffraction with the MAX80 press. A pair of SD anvils with WC-Co sheath are embedded in a boron-epoxy (B-E) cube.

High-Pressure Research in Mineral Physics, edited by M. H. Manghnani and Y. Syono, pp. 29–33.

through the upper SD anvil (e.g., ~2 mm with $2\theta=11°$). The absorption of X-rays by the 2-mm-thick SD anvil was measured to be about 70% for MoKα radiation. The pressure values were determined from the compression of NaCl as recorded in the literature (DECKER, 1971; SATO-SORENSEN, 1983; HEINZ and JEANLOZ, 1984). The generation of 60 GPa pressure was achieved without appreciable deformation of the SD anvils, which had a front face size of 0.7 mm in diameter.

On α-ε Transition in Fe-V Alloys

Pressure calibration based on pressure-fixed points is indispensable to high-pressure experiments since a) the great majority of facilities are not equipped with an X-ray diffraction apparatus and b) the ruby-fluorescent method is only applicable to transparent anvils made of diamond or saphire but not to anvils made of WC-Co alloy or new hard materials expected to be used in the near future.

The highest pressure-fixed point usually adopted at present is 22 GPa, the pressure of semiconductor-metal transition in GaP. There are almost no reliable points in the higher pressure region. As first proposed by BUNDY (1966), α-ε transitions in Fe-Co and Fe-V alloys are useful because they can cover a wide range of pressures according to Co or V content and can be easily analyzed by either X-ray diffraction or electrical resistance methods. Previous studies using Fe-Co alloys have consistently reported transition pressures up to 29 GPa for Fe-40 wt% Co; among these studies are a shock-wave experiment by LOREE et al. (1966), two static experiments using electrical resistance by BUNDY (1975), and an X-ray diffraction study by PAPANTONIS and BASSETT (1977).

Results of previous studies using Fe-V alloys, including a shock-wave experiment by LOREE et al. (1966) and a static experiment by DUNN and BUNDY (1979), were also in agreement. These studies found that the transition pressure steeply increased with increasing V content and reached ~50 GPa for Fe-20 wt%V. However, our calibration, carried out in a Drickamer cell with SD anvils, showed that the efficiency of pressure generation rapidly increased with increasing pressure. Therefore we doubt the correctness of previously reported transition pressures, especially those higher than ~30 GPa.

We have tried to determine the α-ε transition pressures of Fe-V alloys by the following methods: 1) detection of transition by the electrical resistance method and determination of the pressure by X-ray observation of the lattice parameters of NaCl and Au using SD anvils, and 2) detection of the transition by X-ray diffraction and determination of the pressure by the ruby-fluorescent method (MAO et al., 1978) using a diamond-anvil cell (DAC).

Note that, in the present study, the transition pressure means the pressure at the initiation of the α to ε phase transition with increasing pressure.

Sample Preparation

We used an iron sample with 99.9% purity and a vanadium sample with 99.7% purity. Alloys were prepared by argon-arc melting and homogenized by annealing at 1200 °C for one d in evacuated quartz tubes. The alloy compositions were determined by ICP atomic emission spectroscopic analysis. The ingots were filed into fine powders for X-ray diffraction measurements, and thin foils of 50–70 μm in thickness were obtained for electric resistance measurements, by cold rolling. After these mechanical treatments, the samples were annealed at 800 °C for ten h and quenched into water.

Experimental Procedures

Simultaneous Measurements of Electrical Resistance and Lattice Parameter with SD Anvils

The configuration of the central part of the cell (see Figure 1) was modified as shown in Figure 2. An Fe-V foil (0.1-mm in width and 0.03-mm in thickness) and four electrodes were placed on the 1.0-mm-wide top face of the lower SD anvil. The electrodes were pulled through the juncture of the two-part, cubic B-E medium and connect-

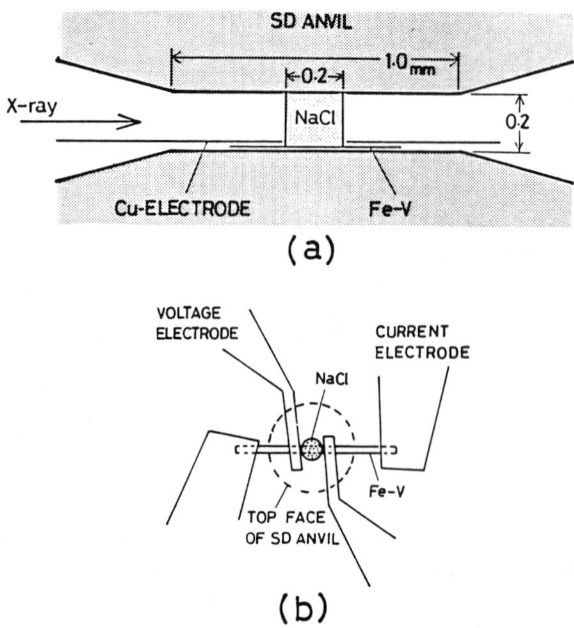

Figure 2. The arrangement of the sample in the center of a cell with 15-mm edge for simultaneous measurement of the lattice parameter of NaCl (and Au) and the electric resistance of Fe-V alloy. a) Cross-sectional view and b) ground view.

ed to the front face of the WC-Co anvils of the MAX80 press.

The mixture of NaCl and Au powder, which was used as a pressure calibrant, was packed into a hole 0.2 mm in diameter in the B-E disk and on the Fe-V foil. As the intensity of the synchrotron radiation beam was increased, exposure times were shortened; the exposure time was only 200 s when an especially strong wiggler beam was used. The electrical resistance of the alloy was continuously measured with increasing load, and the X-ray diffraction patterns of NaCl and Au were recorded at appropriate load intervals.

X-Ray Observation of the Transition and Ruby Fluorescent Determination of Pressure with DAC

A DAC was used with X-ray diffraction to observe the phase transition in Fe-V alloys. The DAC used contained a pair of 1/8 carat single-crystal diamonds, each with a front-face diameter of 0.3 mm. The diamonds were driven toward each other by a screw mechanism. A powdered sample of Fe-V alloy was placed, along with a pressure medium consisting of methanol, ethanol, and water (volume ratio 16:3:1), in a 0.2-mm-wide hole in a metal gasket. The DAC was set into an X-ray camera which had two film cassettes located at different distances from the sample so that the actual distances between the sample and the films could be determined by a simple calculation using the diffraction lines appearing on the both films. MoKα radiation from a rotating anode type generator, with the maximum output of 60 kV and 200 mA, was used. Mean exposure times were 13 and 22 h for the near and far films, respectively. Pressure within a 10 μm spot was measured at many points using the ruby fluorescence technique; a number of fine pieces of ruby was scattered on the sample.

Experimental Results

Results Obtained by SD Anvils

As shown in Figure 3, the diffraction patterns obtained with synchrotron radiation exhibited the 200, 220, and 400 diffraction lines of the B1 phase and the 110 of the B2 phase, for NaCl; and the 111, 200, 220, 311, and 222 lines for Au. Pressure was determined by comparing the molar volumes of NaCl and Au, based on analysis of some of the diffraction lines, with results of previous studies for the B1 phase of NaCl (DECKER, 1971) and for the B2 phase of NaCl (SATO-SORENSEN, 1983; HEINZ and JEANLOZ, 1984), and for Au (JAMIESON et al., 1982). The average value for both materials was adopted as the pressure value. The 220 line of diamond and sometimes, the 110 line of the α phase of Fe-V alloy, appeared in the diffraction patterns.

Figure 4 shows the electrical resistance of the Fe-14.5

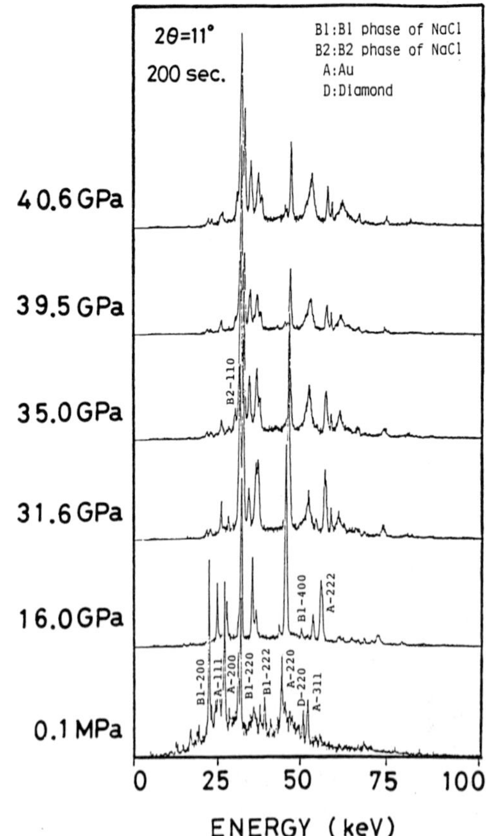

Figure 3. X-ray diffraction patterns of NaCl and Au (using cell shown in Figure 2).

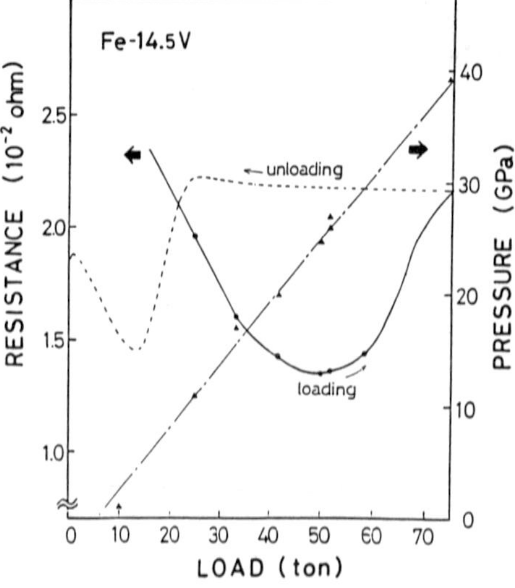

Figure 4. Changes in electric resistance of Fe-14.5 wt%V and pressure as a function of increasing load of MAX80 press. The resistance minimum corresponds to the onset of $\alpha \rightarrow \varepsilon$ transition.

wt% alloy (left ordinate) and the pressure (right ordinate) as a function of the load generated by the MAX80 press. The small circles on the electrical resistance curve correspond to the triangles on the linear pressure plot at the same load at which the X-ray diffraction was performed. As shown in Figure 4, the resistance minimum, considered to be at the onset of the $\alpha \rightarrow \varepsilon$ transition, is at a pressure of ~25 GPa.

Resistance changes in the four alloys Fe-11.4, 14.5, 18.8 and 22.1 wt%V are plotted as a function of pressure in Figure 5. As shown in Figure 5, the resistance minimum shifts to higher pressure with increasing V content of the alloy. Also, the resistance change associated with the phase transition gradually becomes sluggish with increasing V content; this is primarily because of the intrinsic property of the alloy but also, to some extent, because of a larger pressure gradient.

Results Obtained by DAC

Pressures in the gasket hole were measured at more than forty points. Figure 6 shows pressure distribution measurements for Fe-22.1 wt%V alloy; the dotted circle represents the area irradiated by an X-ray beam. The average pressure within the circle was determined to be 37.2 GPa based on the histogram shown in Figure 6b, and the maximum and minimum pressures were observed to be 44.9 and 27.2 GPa, respectively.

Diffraction patterns of the Fe-V alloy exhibited the 110, 200, and 211 lines of the α(bcc) phase and the 100, 002, 101, 102, 110 and 103 lines of the ε(hcp) phase. The average pressures at which one of the diffraction lines of the ε phase was first observed with increasing pressure were 30.4, 37.2 and 49.9 GPa for Fe-18.8, 22.1 and 25.0 wt%V, respectively.

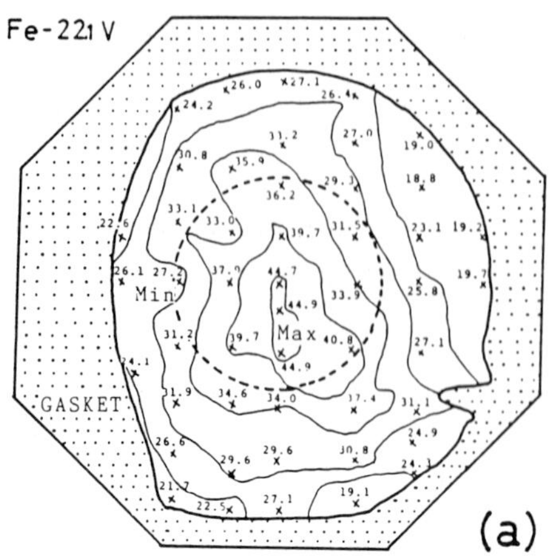

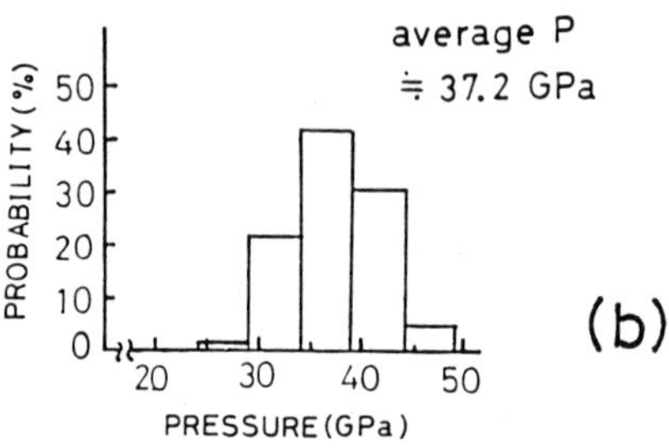

Figure 6. a) An example of pressure distribution in the DAC gasket, and b) the histogram of the X-ray irradiated area (shown as the dotted circle in a)).

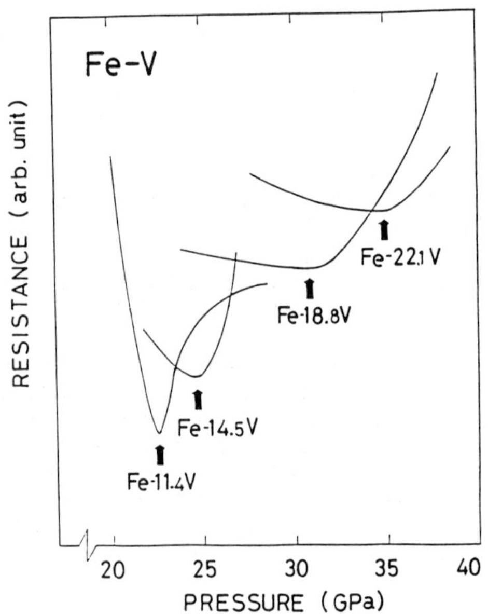

Figure 5. Resistance behaviors near minima in four alloys plotted as a function of pressure.

Discussion

Figure 7 shows the transition pressures obtained by using the lattice parameters of NaCl and Au at the resistance minimum of Fe-V alloys (open circles) and by using the average pressure measured by the ruby fluorescence method at the first appearance of the diffraction line for the ε phase (open squares). These pressures are in close agreement, as expected,since the pressure based on the compression of NaCl and Au is also the average pressure applied to the sample.

However, the definition of "transition pressure" is a serious problem. The average pressure mentioned above cannot simply be adopted as the transition pressure because of the large pressure gradient existing in the sample. The transition initiates at the point of the maximum pressure in the sample: 35.2, 44.7, and 53.0 GPa for Fe-18.8, 22.1, and 25.0 wt%V, respectively.

However, the maximum pressures must actually be a little higher, since the diffraction line of ε phase can barely be observed at the onset of transition. Generally, new diffraction lines are seen when several percent of the alloy has converted to the new phase. Considering this and the pressure distribution histograms, we determined the transition pressures to be 33, 43, and 51 GPa for Fe-18.8, 22.1 and 25.0 wt%V, respectively. The solid curve shown in Figure 7 represents the most reliable determination of the α→ε transition pressures at present.

There are large discrepancies between results of our study and those of two previous studies (see Figure 7). The reasons for these discrepancies have been proposed by the other researchers, LOREE et al. (1966) pointed out that transition pressures determined in their shock-wave experiments with alloys of higher vanadium content lacked sufficient reliability due to an unsuccessful pin technique (see dashed portion of curve in Figure 7). On the other hand, in static experiments using the Drickamer cell with SD anvils by DUNN and BUNDY (1979), the transition pressures were determined on the basis of the relation of linear pressure to applied load extrapolated from the low pressure region; transition pressures so determined might be invalid at pressures higher than ~25 GPa.

Solid solution (bcc) is stable for all compositions in the Fe-V system at high temperatures under ambient pressure. A quenching technique gives all the alloys with desired composition. The present results suggest a simple increase in the α→ε transition pressure with increasing V content. Successive pressure-fixed points up to 100 GPa or more may be determined with the Fe-V system.

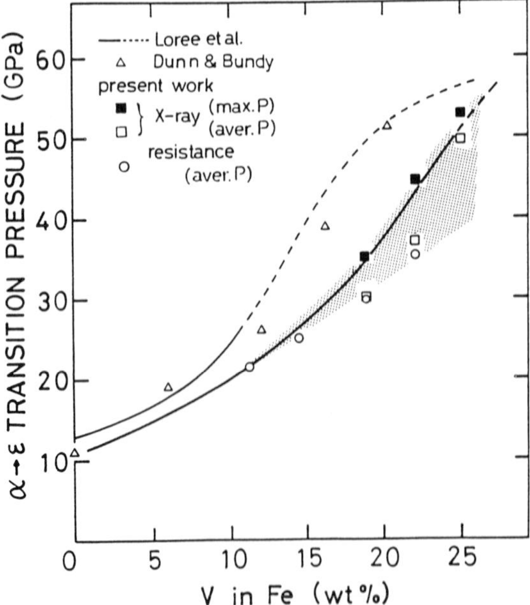

Figure 7. The α→ε transition pressure as a function of V content: results from the present study and two previous studies. Note that the open circle and open square of 18.8 %V are superimposed. A full solid curve represents the most reliable values at present.

Acknowledgments. We are greatly indebted to Y. Akahama, Y. Kozuki, and M. Sugiyama for their help in our high-pressure experiments. We would also like to thank T. Nagashima, NIRIM, for his aid in the computer programming for the MAX80 system. This research was supported in part by funds from Mitsubishi Foundation.

REFERENCES

BUNDY, F. P., Fe-Co and Fe-V alloys for pressure calibration in the 130 to 300 kb region, *Tech. Rep. 66-C-337*, 1–6, Res. Develop. Center, General Electric, Schenectady, New York, 1966.

BUNDY, F. P., Ultrahigh pressure apparatus using cemented tungsten carbide pistons with sintered diamond tips, *Rev. Sci. Instrum., 46*, 1318–1324, 1975.

DECKER, D. L., High-pressure equation of state for NaCl, KCl, and CsCl, *J. Appl. Phys., 42*, 3239–3244, 1971.

DUNN, K. J., and F. P. BUNDY, Materials and techniques for pressure calibration by resistance-jump transitions up to 500 kilobars, in *High-Pressure Science and Technology, vol. 1*, edited by K. D. Timmerhaus and M. S. Barber, pp. 773–778, Plenum, New York, 1979.

HEINZ, D. L., and R. JEANLOZ, Compression of the B2 high pressure phase of NaCl, *Phys. Rev., B30*, 6045–6050, 1984.

JAMIESON, J. C., J. N. FRITZ, and M. H. MANGHNANI, Pressure measurement at high temperature in X-ray diffraction studies: gold as a primary standard, in *High-Pressure Research in Geophysics*, edited by S. Akimoto and M. H. Manghnani, pp. 27–48, Center for Acad. Pub. Jpn., Tokyo/D. Reidel Pub., Dordrecht, 1982.

LOREE, T. R., C. M. FOWLER, E. G. ZUKAS, and F. S. MINSHALL, Dynamic polymorphism of some binary iron alloys, *J. Appl. Phys., 37*, 1918–1927, 1966.

MAO, H. K., P. M. BELL, J. W. SHANER, and D. J. STEINBERG, Specific volume measurements of Cu, Mo, Pd, and Ag and calibration of the ruby R_1 fluorescence pressure gauge from 0.06 to 1 Mbar, *J. Appl. Phys., 49*, 3276–3283, 1978.

PAPANTONIS, D., and W. A. BASSETT, Isothermal compression and bcc→hcp phase transition of iron-cobalt alloys up to 300 kbar at room temperature, *J. Appl. Phys., 48*, 3374–3378, 1977.

SATO-SORENSEN, Y., Phase transitions and equations of state for the sodium halides: NaF, NaCl, NaBr, and NaI, *J. Geophys. Res., 88*, 3543–3548, 1983.

SHIMOMURA, O., S. YAMAOKA, T. YAGI, M. WAKATSUKI, K. TSUJI, H. KAWAMURA, N. HAMAYA, O. FUKUNAGA, K. AOKI, and S. AKIMOTO, Multi-anvil type X-ray system for synchrotron radiation, in *Solid State Physics under Pressure*, edited by S. Minomura, pp. 351–356, KTK Sci. Pub., Tokyo/D. Reidel Pub., Dordrecht, 1985.

UTSUMI, W., N. TOYAMA, S. ENDO, F. E. FUJITA, and O. SHIMOMURA, X-ray diffraction under ultrahigh pressure generated with sintered diamond anvils, *J. Appl. Phys., 60*, 2201–2204, 1986.

DIAMOND ANVIL TECHNOLOGY

Michael SEAL

D. Drukker & Zn. N. V., P.O. Box 15120, Amsterdam 1001 MC, The Netherlands

Abstract. This paper is largely a review of the techniques used in making diamond anvils and the constraints these put on the shapes of anvil. Techniques available for shaping diamonds include cleaving, sawing, polishing, laser cutting, and bruting. At present the shapes most commonly used for anvils are a modification of the brilliant cut derived from the gem industry, and a design based on an octagonal prism with truncated pyramidal top and base, known as the "Drukker standard design". Diamond orientation and material selection are considered as are future possibilities for the attainment of still higher pressures through modifications of the diamond anvil material or design.

Introduction

Diamond anvils owe their usefulness to the extreme hardness, optical transparency, and chemical inertness of this form of crystalline carbon. However, as a crystal diamond has anisotropic mechanical properties, and as carbon it can burn or transform to softer forms at sufficient temperature. These factors set the main limits to the range of conditions attainable in diamond-anvil cells.

Diamond is a fully symmetric cubic material (crystal class m 3 m). As such it should be optically isotropic, though in practice it is never so (all diamonds showing more or less stress birefringence). Mechanically it is anisotropic since the second order tensors stress and strain are related by elastic constants which are fourth order tensors. These have three independent components in this crystal class. Numerical values for these and other physical properties of diamond have been listed by FIELD (1979a). Diamond can be cleaved on {111}, the most widely separated atomic planes in the tetrahedrally co-ordinated structure. In pure crystals such cleavage yields quite smooth surfaces; in less pure crystals the cleavage can be irregular or conchoidal. It is, of course, an extremely strong and hard material, the strongest known with a tensile strength in excess of 1000 kgf mm^{-2} (FIELD, 1979b), and the hardest at Mohs 10.

The hardness of diamond makes it difficult to abrade in that there is nothing harder with which to abrade it. Diamond can, however, be abraded with diamond powder and this makes possible the grinding, sawing, and polishing of diamond against appropriate diamond-loaded surfaces. Such methods are of some antiquity. They have been known in Europe for at least 500 years and in India in all probability for substantially longer. Modern techniques derive from these, but differ from them in the much greater precision of the machinery, in the use of modern materials, and in the use of computer control.

The Abrasion of Diamond

When diamond rubs on diamond there is an anisotropy in the friction (SEAL, 1958, 1981) and there is an anisotropy in the wear rate which may derive from this or may have an independent cause. Despite more than half a century of scientific investigation of the problem, there is no universal agreement on the explanation of how diamond abrades diamond. Some aspects of diamond abrasion have been reviewed by WILKS and WILKS (1979) and the nature of the process may well depend on the exact conditions under which one diamond is presented to the other. It is, however, established fact that the wear of diamond abraded with loose or loosely held diamond powder is very strongly direction dependent. The easy crystal directions abrade perhaps 500 to 1000 times more rapidly than the difficult ones. Further the wear proceeds smoothly yielding flat, smooth surfaces when abrasion is in the easy directions, but with progressive difficulty and increasingly rough, cracked surfaces as the hardest directions are approached. The effect has been studied scientifically (SLAWSON and KOHN, 1950; DENNING, 1953, 1955, 1957) and is part of the basic stock-in-trade of diamond polishers who speak of a "grain" in diamond akin to that in wood. They can detect the grain by inspection of the crystal faces or by sound and feel during polishing.

In brief, the easily polishable directions are [100] and [010] on (001), and [100] on (011). All directions on (111) are very difficult to polish. Crystallographically equivalent directions behave similarly on any crystal form, and in general ⟨100⟩ directions are easy to polish and ⟨110⟩ difficult. However the polishing rates for a particular direction do also depend on which face is being polished. The series of papers by Denning give details.

It seems clear that the statistics of presentation of the loose diamond particles to the diamond being abraded are all-important. Some particles present hard directions, some soft. No particle can present a harder direction than the hardest on the diamond being abraded. Virtually all will appear harder than its softest direction, and in intermediate directions a fraction will present harder

High-Pressure Research in Mineral Physics, edited by M. H. Manghnani and Y. Syono, pp. 35–40.

orientations and a fraction softer. Beyond this, what happens at the actual rubbing interface is unclear. Material may be removed by chipping, graphitization, chemical reaction, vaporization or a combination of these, and differing views have been presented in the scientific literature.

Shaping Diamonds

Cleaving

The popular image of diamond shaping seems to concentrate on cleaving. The idea of the diamond cleaver studying his stone for a week or a month and then—under great psychological stress—splitting it with one blow is a romantic one with a definite foundation in truth as regards very large, very valuable diamonds. Run-of-the-mill cleaving is somewhat more prosaic. It involves orienting the stone from study of its natural crystal faces, abrading a groove which will act as a stress concentrator at an appropriate place, inserting a steel blade in the groove, accurately orienting it along the appropriate {111} plane, and striking the blade a controlled blow to split the diamond. It is a highly skilled job, but skilled cleavers can split substantial numbers of diamonds in a day.

Several features of the process are of interest. It is a true splitting in that a crack is opened up by pushing the sides apart so that stress is concentrated at the crack tip sufficient to break the atomic bonds there. The blade need not be sharp. The crack propagates very rapidly and can travel through a 3 mm diamond in less than a microsecond. Speeds of up to 7 km s^{-1} have been measured and these imply that the atomic bonds break in times comparable to their vibration period (FIELD and HAGAN, 1971). The nature of the material used for supporting the diamond is important as it must provide sufficient acoustic matching and damping to reduce the intensity of the elastic wave reflected at the diamond's lower surface to a level at which it will not cause the diamond to shatter internally. Over the years materials have been developed empirically to satisfy this requirement. Traditionally "diamond cleavers' cement" consists of a mixture of shellac, rosin, wax, and powdered glass (GRODZINSKI, 1953). Such material softens readily when warmed, but is quite hard when cold.

The limitations of cleaving are clear. It can only be used to split diamonds on {111}. Such splitting is generally required only for dividing larger awkwardly shaped stones into pieces of more convenient shape, and for the production of special items of this orientation (for example {111} plates for diamond knife manufacture). It is inappropriate for the vast majority of diamonds and these are shaped by abrasive methods.

Sawing

Diamonds frequently occur with shapes based on the octahedron or rhombic dodecahedron forms (Figure 1a). It is clear that the diamond brilliant shape (Figure 1b), whether for use as gem or anvil, fits most economically into these as inverted pairs (Figure 1c), but that brings with it the need to saw diamonds in two parts. Diamond saws are generally discs of phosphor bronze, 70 mm or so in diameter and 0.06 to 0.15 mm thick. The discs are clamped between flanges, charged with diamond powder applied as a suspension in oil, and rotated at speeds of up to 8,000 r.p.m. Surface speeds are typically 25 to 30 m s^{-1}.

Such saws have the appearance shown in Figure 2. Since cutting rates are low one man can operate many machines. GRODZINSKI (1953) stated that 8 mm^2 h^{-1} would be an average cutting rate, but this was based on rather small diamonds. Larger diamonds take longer, with cutting rates down to 1 mm^2 h^{-1} and less under some conditions.

The attrition of diamond during sawing takes place through abrasion with a loosely bonded powder, this having been pressed into the phosphor bronze saw material as a suspension in a sticky oil. Anisotropy in wear rate is always evident, and in practice it is only possible to saw the cube planes on which the octahedra and dodecahedra are usually divided in cube directions

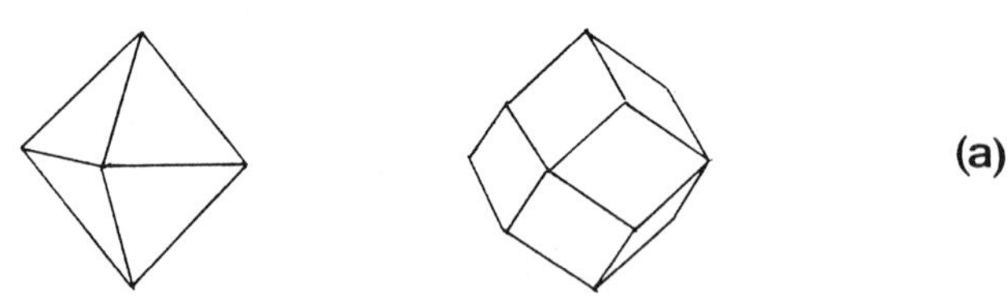

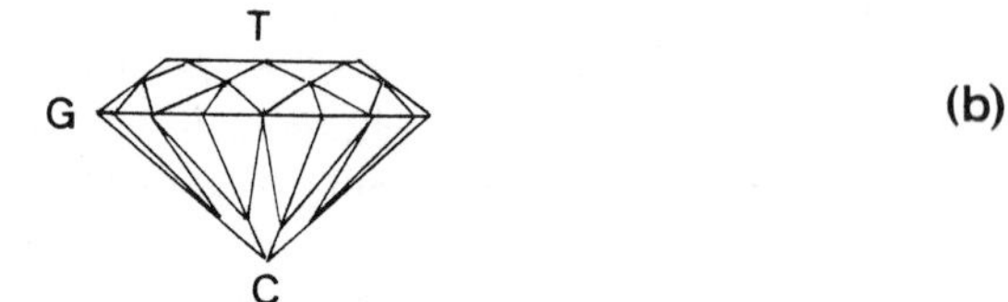

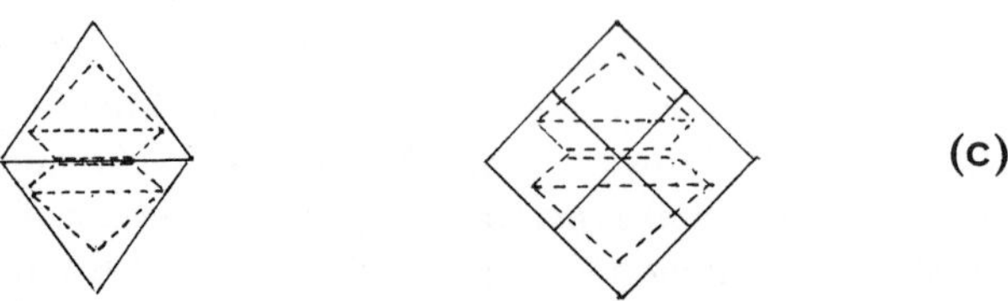

Figure 1. a) Octahedron and rhombic dodecahedron, representing idealised diamond habits. b) The "brilliant" cut: C=culet; G=girdle; T=table. c) Way of making pairs of brilliants from rough octahedra and dodecahedra.

Figure 2. Diamond sawing machine.

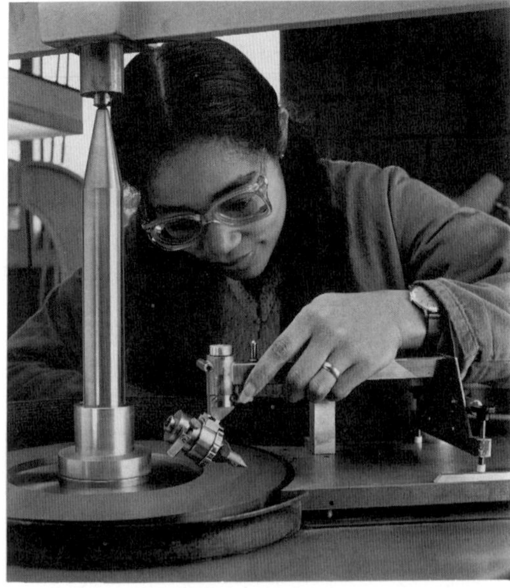

Figure 3. Polishing a small diamond component.

(i.e. {001} in ⟨100⟩ directions). Accordingly the diamond is normally kept stationary except for the slight movement as it sinks on the saw, and the sawing produces circular marks on sawn faces, leaving them comparatively rough with deviations from flat which often amount to many microns. Such faces have to be polished before they are acceptable on most finished products.

Polishing

Diamonds are polished by pressing them against a rotating lap charged with diamond powder. Surface speeds are similar to those used in sawing (25 to 30 m s^{-1}), though the laps are generally of greater diameter (up to 30 cm) and the rotational speeds correspondingly lower. The lap is made traditionally of a porous cast iron, scored with grooves to aid retention of the diamond powder (rubbed into the surface as a suspension in oil). Modern polishing laps are made of a variety of metals and there is a wide range of proprietary surface coating techniques. Proper balancing and smooth running of the wheel are all-important. Figure 3 shows a polishing wheel as used for finishing small diamond components.

The anisotropy of wear rate is again a determining factor and controls the directions which can be polished. Figure 4 is a polar diagram of polishing rate (based on results of TOLKOWSKY, 1920) for the cube sections on which most rough diamonds used for anvils are sawn. Clearly this requires that the diamond be held in a constant azimuth on the lap with a cube axis tangential to the motion. However, this produces a pattern of shallow grooves on the diamond surface, and to correct this the diamond is given a radial reciprocating motion during the latter stages of polishing. The end result is generally a surface flat to one or two fringes of visible light and smooth to better than 50 nm (BAILEY and SEAL, 1956),

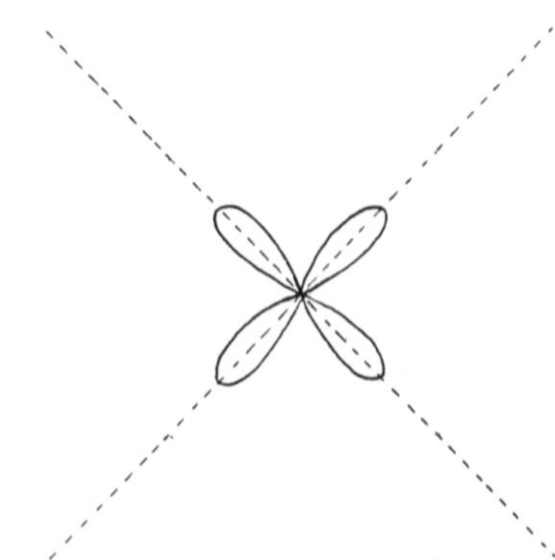

Figure 4. Polar diagram of polishing rate on (001). The dotted lines show [100] and [010] directions.

though it is now possible with special attention to improve these numbers by an order of magnitude.

Laser Cutting

Diamond is a metastable form of carbon at normal pressures and as such will convert to its stable allotrope, graphite, if the temperature is raised sufficiently to overcome the activation energy for this transition. Generally this requires temperatures of about 2000 K (SEAL, 1963), temperatures which though high can easily be generated by a pulse of laser energy. It has been found that this is best produced by a Q-switched neodymium-YAG laser delivering pulses of 1 to 2 mJ. Diamond is transparent to the wavelength of just over 1 μm and so its surface must be blackened with an appropriate paint to couple the laser energy into the diamond. Cutting, once started, proceeds smoothly as the high temperatures generate graphite continuously at the point of cutting and then vaporise it in a plasma.

The advantages of laser cutting are several. The beam can be focussed to a small spot and so kerf losses are lower on small components. The material removal is orientation-independent and complex shapes can be cut. Laser cutting is much faster than sawing, reaching rates of 0.1 $mm^2 s^{-1}$ and higher. The technique lends itself to computer control and a modern installation including such control is shown in Figure 5. The main disadvantage is that the cut surfaces are rather rough and of course blackened by the graphitization. For all critical applications they need polishing smooth.

Bruting

For the sake of completeness bruting should be mentioned. It is a rough turning operation in which the diamond being cut is mounted in the chuck of a lathe and a cylindrical periphery generated by chipping fragments off the edge with a second diamond. Such chipping produces a rough, matt cylindrical surface with a multitude of micro-cracks immediately underneath. It is commonly used in generating the cylindrical edges or girdles of diamond brilliants for gem use, and may have advantages here in providing a rough surface for a claw setting to grip. Because of the micro-cracks it is entirely inappropriate for diamond high pressure anvils.

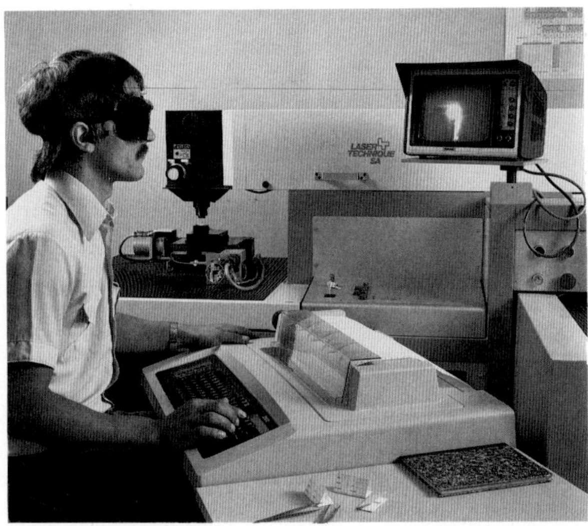

Figure 5. Computer controlled neodymium-YAG laser machine tool used for cutting diamonds.

Anvil Design

The shape of diamond anvils (Figure 6) has derived from the truncated brilliants used in the earliest diamond-anvil cell experiments. Brilliants were used because they were conveniently available, but there is no reason why a shape which was developed to maximise internal reflection and dispersion of light should be optimal as regards strength under load. In fact diamond anvils seem to fail in a variety of ways—under axial tension under the opposed outward couples when the central hole in the support is too large, under Hertzian tension on the shoulders, or under shear beneath the culet. Results of finite element stress analyses of diamond anvil designs have been published by ADAMS and SHAW (1982) and BRUNO and DUNN (1984). It seems that the commonly used angles of 90° to 110° across the culet should be satisfactory and that bevels on the diamond in the gasketed region around

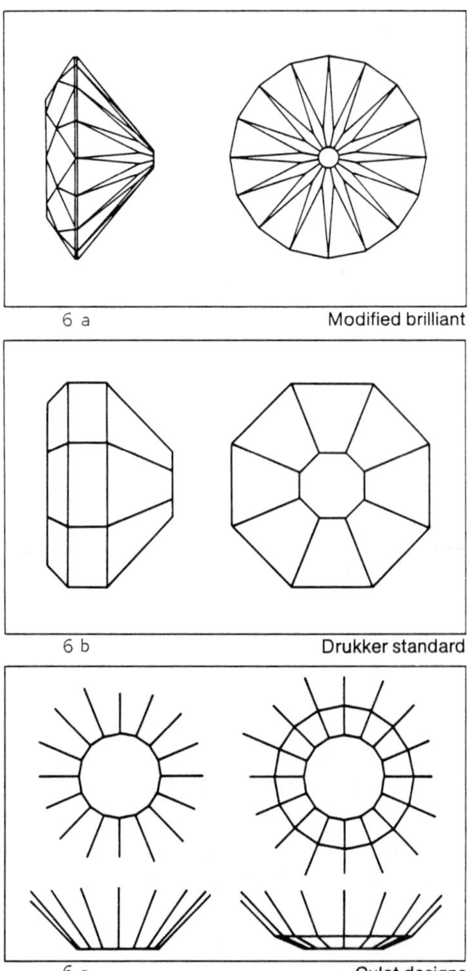

Figure 6. Shapes of diamond anvil. a) Modified brilliant. b) Drukker standard. c) Culet designs.

the culets can reduce stress concentrations. However the material around the edge or girdle of a brilliant-cut diamond anvil is under relatively low stress and serves little purpose.

The parameter determining the overall size and thus cost of a diamond anvil is the table area which must be large enough to provide adequate support. Material outside the table area is not very useful. In the classical brilliant cut the table diameter is 57.5% of the girdle diameter. Larger percentages are possible for anvil use, but steepening the angles of the side facets gives difficulty in polishing as these approach the octahedron angle. Lowering the girdle towards the table is a better approach and this resulted in the "Drukker standard" design (SEAL, 1984). This is derived from a rather flat brilliant cut by reducing the number of facets, thickening the girdle drastically, and lowering the table towards the girdle. Though a very satisfactory solid design, it does probably have a thicker girdle than necessary and a somewhat shallower design would be equally satisfactory and less expensive.

Most diamond anvils are made with the table close to a {100} orientation. This is a convenient one to polish as no face comes close to the difficult {111} orientation. Tilting to other orientations can easily remove this advantage, leading to increased polishing times and costs. A true ⟨111⟩ axis would be impossible to polish.

The {100} orientation is also a strong one for it is very probable that failure occurs on or across {111} planes by shear or tensile splitting and these are symmetrically disposed to {100}. In this orientation the angle between the anvil axis and the {111} planes (ϕ) is 54° 44′. Tilting the axis with respect to [100] decreases ϕ for at least one {111} plane and increases it for at least one other. Failure can be analyzed in term of SCHMIDT's (1924) law of critical shear stress where the maximum resolved component of shear on {111} is proportional to $\cos\theta \sin\theta$ and of tension across it proportional to $\cos^2\theta$, θ being the angle between the principal tension direction and the normal to the {111} plane. Since the stress distribution is axially symmetrical, any tilt from the axially symmetrical arrangement of {111} planes will put at least one set of these planes in a position of greater resolved stress and thus more disposed for failure.

Diamond Selection

The principal parameters affecting choice of the diamond material have been reviewed recently (SEAL, 1984) and so will only be summarized here. The main impurity in natural diamond is nitrogen, which may occur in amounts up to 1/2 wt%. Type 1 and type 2 diamonds differ in ultraviolet and infrared absorption features which derive from nitrogen. Type 1 diamonds contain significantly more nitrogen than type 2 and are subclassified according to the form in which the nitrogen occurs—substitutional single atoms, small groups, larger platelets, etc. Type 2A diamonds are electrical insulators, type 2B semi-conductors (due to traces of boron impurity in material with very low nitrogen content).

There have been reports indicating that the form and amount of nitrogen may affect the strength of diamond, but there do not appear to have been any investigations using numbers of diamonds sufficient to give good statistics. The choice of material has generally been based on other factors, such as a need for transparency in the ultraviolet or infrared, a need for low luminescence for Raman studies, or cost (yellow diamonds are less expensive than white).

Natural diamonds may contain inclusions formed or trapped within the crystal. They may be relatively large crystals of identifiable minerals such as diopside, olivine, pyrrhotite, or garnets (HARRIS and GURNEY, 1979) or they may range in size down to sub-micron dimensions (SEAL, 1966). Being of a different composition to the surrounding diamond they act as centers of stress and as such are obviously undesirable in diamond anvils. Stress may also arise from mis-match of successive diamond growth layers of differing impurity content. Such stress is most easily detected by the resulting birefringence between crossed polarisers and it is common practice to set a stress birefringence specification for diamond anvil material. A specification of stress birefringence <0.0001 is probably a good compromise between material of high internal stress which must be suspect as regards strength, and the rarity of very low birefringence material. It is also desirable to limit gradients of birefringence but these are difficult to measure quantitatively. Perhaps one should avoid diamonds with birefringence patterns showing the rosettes characteristic of individual inclusions, but tolerate those in which reasonable growth layering is visible but the overall birefringence acceptably low.

Future Possibilities

The trends in diamond-anvil cell research are towards ever higher pressures and to combinations of high temperature and high pressure. Attainment of the highest pressures requires careful attention to anvil geometry. Improvements may be possible in diamond tip geometry, double bevelling, and gasket design (MOSS et al., 1986).

Stronger diamond would obviously be desirable and whilst there does not seem to be much likelihood of big improvements, some advances may be made through the use of synthetic diamond. As a manufactured product it is in principle possible to control the impurity content and defect structure of synthetic diamonds. However most synthetic diamonds produced hitherto have contained substantial numbers of metal inclusions arising from the solvent used for crystal growth, together with sufficient nitrogen to give a yellow color and make them spectroscopically type 1B. Modern improvements in diamond synthesis technology are resulting in bigger and better synthetic diamonds and these may well make good high pressure anvils.

The use of polycrystalline diamond (PCD) offers another approach. PCD is a product made by sintering together natural or synthetic diamond powders under high pressure. Because of the randomly oriented intergrown particles it is tougher than natural natural diamond in that it has no cleavage plane. It probably would make an excellent anvil material except that it is optically opaque.

Improvements in strength could also be obtained by going to smaller stressed volumes following AUERBACH's law (1891). FIELD (1979b) has recalculated data from SEAL (1958) to give the values for strength shown in Table 1. These are numbers calculated from the sizes of the ring cracks which result from sufficiently heavy loading of a plane diamond surface with a curved diamond indenter, as a function of indenter radius. It is clear that smaller radii result in smaller stressed volumes and greater strengths, but reduction of the already very small sample size used in diamond-anvil cell work would give obvious practical difficulties.

It is also likely that Griffith cracks (GRIFFITH, 1920) are a source of anvil failure. Such cracks if not vertically disposed to the anvil surface could perhaps be removed by techniques such as ion milling to remove a micron or so of surface, or covered by a layer of diamond-like carbon put down by sputtering, ion implantation, or hydrocarbon-cracking, i.e., by any of the techniques used in this currently very active area of hard coating technology. In both cases the surface would be significantly modified and the load carrying surface would no longer be well ordered crystalline material.

TABLE 1. Strength of Diamond as a Function of Indentor Radius (FIELD, 1979b)

Indentor Radius mm	Strength kgf mm^{-2}
0.2	2750
0.1	3400
0.02	5950
0.01	7400
0.002	12750

REFERENCES

ADAMS, D. M., and A. C. SHAW, A computer-aided design study of the behaviour of diamond anvils under stress, *J. Phys. D, 15*, 1609–1635, 1982.

AUERBACH, F., Absolute Härtemessung, *Ann. Phys. Chem., 43*, 61–100, 1891.

BAILEY, A. I., and M. SEAL, The surface topography of a polished diamond, *Ind. Diam. Rev., 16*, 145–148, 1956.

BRUNO, M. S., and K. J. DUNN, Stress analysis of a beveled diamond anvil, *Rev. Sci. Instrum., 55*, 940–943, 1984.

DENNING, R. M., Directional grinding hardness in diamond, *Amer. Min., 38*, 108–117, 1953.

DENNING, R. M., Directional grinding hardness in diamond: a further study, *Amer. Min., 40*, 186–191, 1955.

DENNING, R. M., The grinding hardness of diamond in a principal cutting direction, *Amer. Min., 42*, 362–366, 1957.

FIELD, J. E., *The Properties of Diamond*, appendix, pp. 641–653, Academic Press, New York, 1979a.

FIELD, J. E., *The Properties of Diamond*, chapter 9, pp. 281–324, Academic Press, New York, 1979b.

FIELD, J. E., and J. T. HAGAN, unpublished work cited by J. E. Field (1979b), pp. 293–294, 1971.

GRIFFITH, A. A., The phenomena of rupture and flow in solids, *Phil. Trans. Roy. Soc. A, 221*, 163–198, 1920.

GRODZINSKI, P., *Diamond Technology*, 2nd edition, N.A.G. Press, London, 1953.

HARRIS, J. W., and J. J. GURNEY, in *The Properties of Diamond*, edited by J. E. Field, chapter 18, pp. 555–591, Academic Press, New York, 1979.

MOSS, W. C., J. O. HALLQUIST, R. REICHLIN, K. A. GOETTEL, and S. MARTIN, Finite element analysis of the diamond anvil cell: Achieving 4.6 Mbar, *Appl. Phys. Lett., 48*, 1258–1260, 1986.

SCHMIDT, E., Neuere Untersuchungen an Metallkristallen, *Proc. Int. Congr. Appl. Mech. Delft*, 342–353, 1924.

SEAL, M., The abrasion of diamond, *Proc. Roy. Soc. A, 248*, 379–393, 1958.

SEAL, M., The effect of surface orientation on the graphitization of diamond, *Phys. State Sol., 3*, 658–664, 1963.

SEAL, M., Inclusions, birefringence and structure in natural diamonds, *Nature, London, 212*, 1528–1531, 1966.

SEAL, M., The friction of diamond, *Phil. Mag., 43*, 587–594, 1981.

SEAL, M., Diamond anvils, *High Temp.-High Press., 16*, 573–579, 1984.

SLAWSON, C. B., and J. A. KOHN, Maximum hardness vectors in the diamond, *Ind. Diam. Rev., 10*, 168–172, 1950.

TOLKOWSKY, M., The abrading, grinding, or polishing of diamonds, doctoral thesis, University of London, 1920.

WILKS, J., and E. M. WILKS, in *The Properties of Diamond*, edited by J. E. Field, chapter 11, pp. 351–382, Academic Press, New York, 1979.

HIGH PRECISION OPTICAL STRAIN MEASUREMENTS AT HIGH PRESSURES

Charles MEADE and Raymond JEANLOZ

Department of Geology and Geophysics, University of California
Berkeley, California 94720, USA

Abstract. We have used a high-resolution video image-shearing system to carry out the first direct measurement of the hydrostatic compression of glass in the diamond cell. Our work is based on a new optical technique for making precise length determinations. From calibration measurements and the diffraction theory of optical imaging we find that our system has the highest precision when measuring the spacing between lines that are thinner than the formal resolution of our objective lens ($\sim 1.5\ \mu m$). By depositing a thin chromium emulsion of this pattern on the glass samples, we can measure sample lengths with an uncertainty of $\pm 0.010\ \mu m$. For typical dimensions of diamond-cell samples ($\sim 100\ \mu m$), this corresponds to measuring linear strains ($\delta l / l_0$) to a precision of 10^{-4}. Our data for fused silica and a Ca-Mg-Na glass, collected at room temperature to about 10 GPa, yield zero pressure bulk moduli (K_0) of 37.0 ± 5.5 and 35.5 ± 3.7 GPa respectively. These values, and our finding that the pressure derivative of the bulk modulus is negative in both cases, are in good agreement with previous ultrasonic measurements.

Introduction

By measuring strain with changes in pressure and temperature one can deduce many of the thermodynamic and elastic properties of a solid. In this way, X-ray diffraction through the high-pressure diamond cell has been an invaluable technique for studying the equations of state of materials to very large compressions. But at the same time, X-ray diffraction can only resolve elastic strains in crystalline materials. Thus to date, it has not been possible to measure either the equation of state of amorphous materials or the nonelastic properties of solids at high pressures.

To perform these experiments, we have developed a new optical technique for measuring strain under pressure in the diamond cell. The advantage of this approach is that the measurement of macroscopic strain does not depend on the periodic nature of the atomic structure, as is the case for X-ray diffraction. Thus, the equation of state of an amorphous material can be determined by optically measuring an averaged bulk strain of the glass structure. Atomic and macroscopic measurements should give the same results for crystalline materials under purely elastic compression. In the presence of shear stresses, however, the two are expected to diverge as the macroscopic measurements include nonelastic strains produced by dislocations and point defects.

Thus X-ray diffraction and optical techniques provide complementary data of strain at high pressures. Under hydrostatic conditions (to ~ 10 GPa), both X-ray and optical methods can be used to measure elastic compression. Above 10 GPa, shear stresses across the sample can produce nonelastic deformation. In this regime, X-ray diffraction is best for measuring the equation of state while optical techniques can resolve the nonelastic properties such as yield strength or effective viscosity.

The most important limitation to making precise optical measurements of strain arises from the finite resolution of microscopes. Typical diamond-cell samples are $\sim 100\ \mu m$ across. To make strain measurements with the same precision as can be achieved by X-ray diffraction ($\delta l / l_0 \sim 10^{-4}$ in linear dimensions) would therefore require an optical resolution of 10^{-8} m. This is two orders of magnitude smaller than the formal resolution of the long working-length objectives used to view samples within the diamond cell.

SCOTT and JEANLOZ (1984) showed that one can optically measure strain by using information from the light intensity across the sample image. They measured the static compression of gold foils by using an image-shearing eyepiece to visually locate the half-intensity point at the edges of the sample. While this technique was a significant improvement over the formal resolution of the microscope objective, the strain measurements had considerable scatter causing large uncertainties in the equation of state.

In this paper we discuss two important improvements that make the precision of the image shearing technique comparable to that of X-ray diffraction methods. First, instead of relying on visual observations we use a high resolution video camera with an oscilloscope to quantitatively measure the light intensity across the image. Second, we prepare one surface of our samples with an emulsion of fine lines which provide reference lengths for our strain measurements. The strain is determined by measuring the change in spacing between the lines. As we show by Fourier diffraction theory, the use of fine lines allows much greater precision than is possible by measuring the distance between the edges of opaque objects (e.g., metal foils). With calibration measurements on a precision graticule and on photomasks, we find that the uncertainty of a particular measurement with this new system is $\pm 0.010\ \mu m$. As an additional confirmation of

High-Pressure Research in Mineral Physics, edited by M. H. Manghnani and Y. Syono, pp. 41–51.

the technique we present the results of the first hydrostatic-compression study of an amorphous material in the diamond cell. We find little scatter in the strain measurements and we determine zero pressure elastic properties in excellent agreement with previous studies.

Experimental Design

The design of the optical measurement system is shown in Figure 1. The sample is viewed in the diamond cell through a Leitz petrographic microscope with long working-length condensers and objective lenses (numerical aperature=0.20, 20X). The samples are illuminated with green-filtered transmitted light (bandpass centered at 530 nm with 100 nm full width at half maximum). Like SCOTT and JEANLOZ (1984), we use a video-based image-shearing technique to enhance the formal resolution of the long working-length objective (~1.5 μm). The magnified sample image is viewed with a high resolution video camera (Cohu 5300) using a scanning slit approximately 50 μm wide to isolate individual raster scans. Projected back to the sample, this corresponds to ~3 μm wide slice across the sample. The advantage of this technique is that instead of analyzing the entire image, discrete sample cross-sections can be repeatedly located, and then precisely focused and measured.

The image-shearing eyepiece contains two prisms that split and partially superpose the image of the sample (DYSON, 1960) (Figure 2). By adjusting a micrometer screw one changes the relative orientations of the prisms and hence the relative positions of the images. Length determinations can then be made by "shearing" the two images, as schematically shown in Figure 3, and reading the difference on the micrometer scale.

Figure 4 illustrates how the image-shearing eyepiece enhances the resolution of the microscope by using information about the intensity distribution of light within the Airy disc. The Airy disc describes the lengthscale over which the image of a point source is blurred due to diffraction of light at the lens apertures

Figure 1. Design of the optical measurement system. The sample is viewed in the diamond cell with a Leitz UM 32/.30 objective on a petrographic microscope. The measurements are made with a Vickers image shearing eyepiece. We use a high resolution video camera (Cohu 5300) and a monitor to view the sample image. A scanning slit isolates individual raster scans which are displayed on the oscilloscope. As described in the text, we use these light intensity profiles to precisely focus the image and set the image shearing eyepiece.

IMAGE SHEARING

J. Dyson (1960) J. Opt. Soc. Am 50, 754

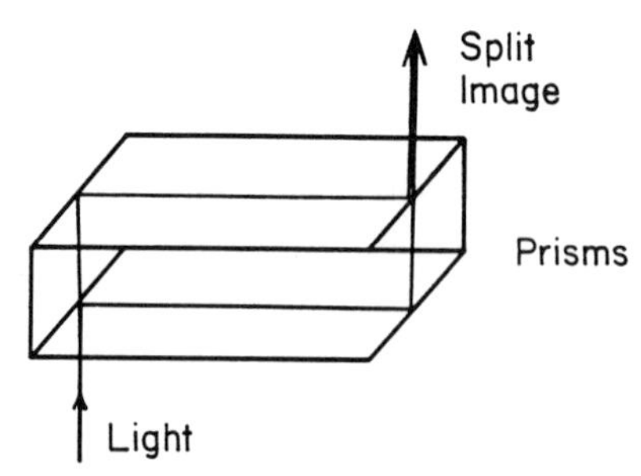

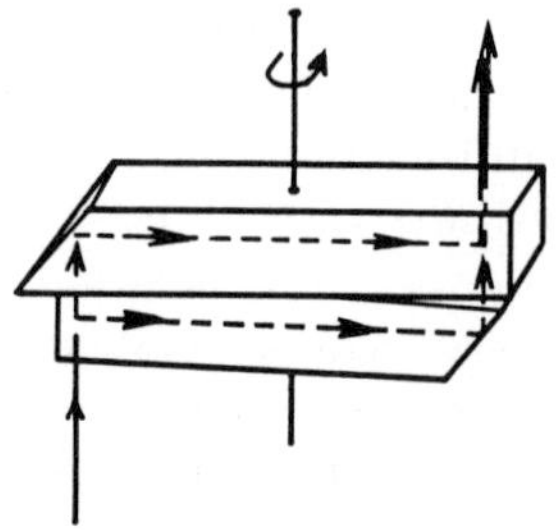

Figure 2. Design of the image shearing eyepiece. When the prisms are exactly aligned, the eyepiece splits and then recombines the light into a single image. By rotating the prisms with a micrometer screw, the split images are offset or sheared (bottom). Sheared images are schematically shown in Figures 1 and 3.

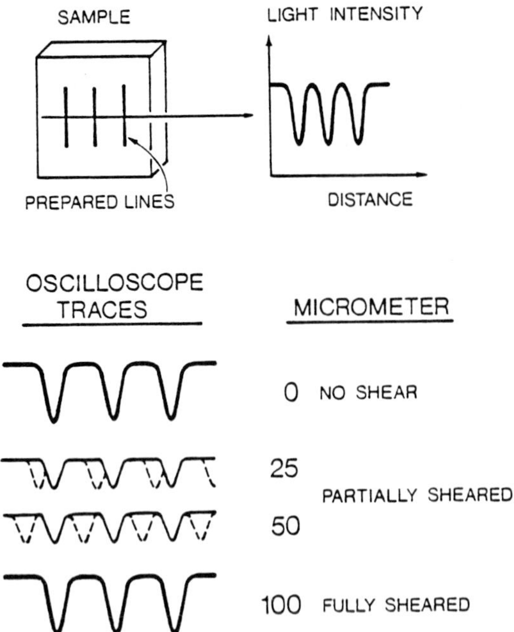

Figure 3. Making measurements with the image shearing eyepiece. The sample is prepared with lines that are narrower than the resolution limit of the objective (top left). The light intensity profile in the image plane consists of sharply resolved peaks (top right). By adjusting the rotation of the prisms, these are offset in the image plane until they are sheared through a distance corresponding to the line spacing (bottom). The precise setting is determined by maximizing the amplitudes of the superimposed peaks. The length is determined by the difference in the micrometer reading (100 units in this example).

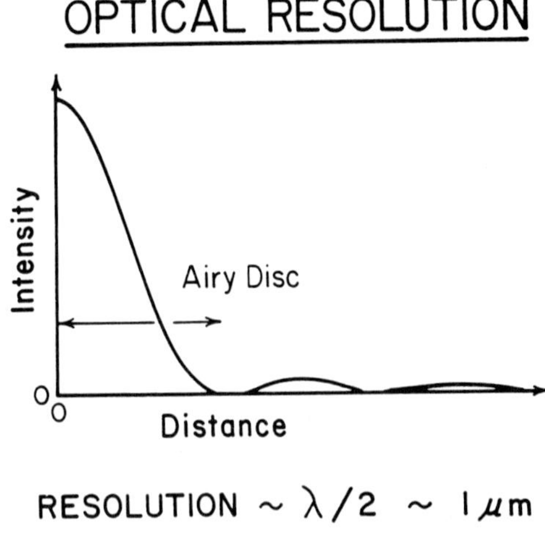

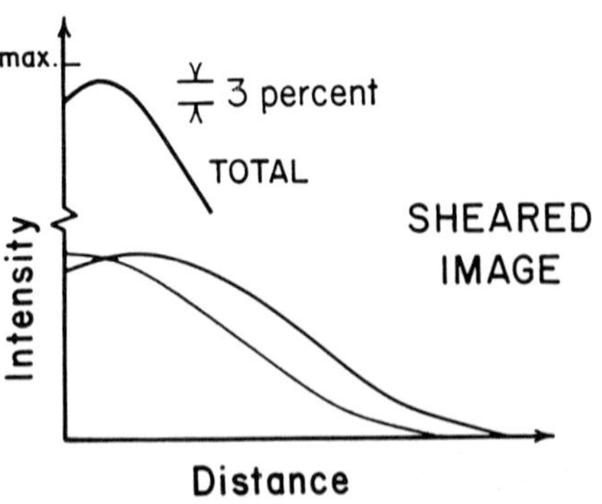

Figure 4. Enhanced resolution with the image shearing eyepiece. The formal resolution of an objective is defined by the scale of the Airy disc (top). By superposing two images, one can adjust their relative positions until the combined intensities are maximized (bottom). Thus intensity changes in the final image can be used to determine the relative positions of the sheared images with a precision much greater than the Airy disc.

within a microscope. This blurring defines the formal resolution of an objective since one cannot discern the boundary between two objects if their Airy discs overlap substantially. However, by superposing two images one can adjust their relative positions so that the combined intensities are maximized (Figures 3 and 4); this corresponds to aligning the peaks of the two Airy discs. If one can discern moderate intensity differences across the image plane (e.g., ~3% with an oscilloscope), the separation of the two objects can be determined with a precision two orders of magnitude finer than the scale of the Airy disc (DYSON, 1960).

Calibration

To quantify the precision of our image-shearing measurements we have carried out a series of calibration experiments at zero pressure. First, we measured a nominal 10 μm line spacing on a precision graticule to calibrate the micrometer scale. Note, however, that absolute length determinations are not necessary for our application, as we only require relative lengths for measuring strain. In addition, we made repeated linewidth measurements on photomasks of chromium emulsion (Figure 5). There are two sets of data from the photomasks: measurements across bars that are wider than the formal resolution of the objective, and measurements between lines that are smaller than the resolution limit. After each measurement, the sample was repositioned and the objective and condenser lenses were refocused.

Individual measurements (corresponding to individual points in Figure 5) were made as follows. The image is sheared in one direction and four edge settings are recorded. The image is subsequently sheared in the other direction and the opposite edge settings recorded. The actual length is then one half the difference between the two averages of four micrometer readings. The uncertainty of a particular measurement is derived from the standard deviations of the average edge settings. The data in Figure 5 are plotted as the deviation about the mean value for a particular run.

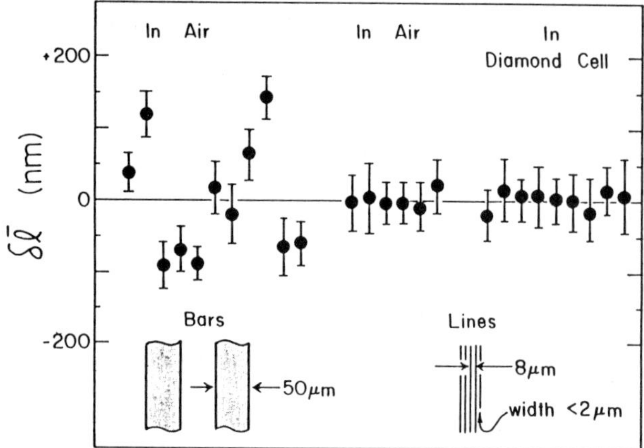

Figure 5. Results of calibration measurements. The data are presented as the deviation about the mean value for a particular run. Measurements across wide bars (standard deviation of ±0.077 μm), have a greater uncertainty than measurements of the spacing of thin lines (±0.010 μm). The resolution of the system is not affected by placing the sample in the diamond cell at 0 pressure.

Figure 6 illustrates the two types of measurements in the calibration experiments. With the bars, we define a threshold intensity (I_t) on the opposing edge profiles and measure the length between these two points. Using the image-shearing eyepiece, this corresponds to matching the 50% level of the two edges so that the sum of the intensities is constant across the superimposed images. Alternatively, the spacing of lines smaller than the resolution limit of the objective can be obtained from the distance between the minima of the intensity profile. These features are resolved as sharp troughs reflecting only the diffraction limit of the microscope lenses and not any details of the lines. With image shearing these can be

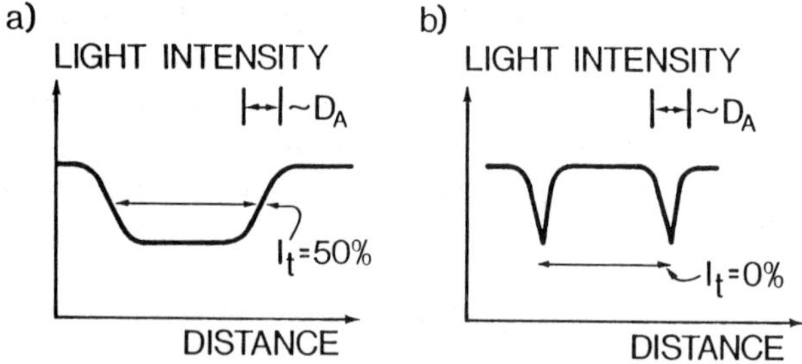

Figure 6. Two types of linewidth measurements: a) For objects larger than the resolution limit of the objective, the width can be determined as the distance between two points of equal intensity along the opposing edge profiles. With the image shearing eyepiece this corresponds to a threshold intensity (I_t) of 50%. b) For objects smaller than the resolution limit, the spacing can be measured as the distance between the intensity minima (I_t=0%). D_A is the scale of the Airy disc.

measured as illustrated in Figure 3 by maximizing the intensity difference across the image.

The calibration experiments show that there are different accuracies associated with the two types of length measurement (Figure 6). The uncertainty of the measurements across wide bars (±0.077 μm) from the standard deviation of the scatter is almost an order of magnitude greater than that between fine lines (±0.010 μm). The latter result is comparable to the best calibration results reported for similar measurements using high quality objectives and focusing mechanisms (SWYT and ROSBERRY, 1977; NYYSSONEN, 1982). This offers a qualitative explanation for the uncertainties in SCOTT and JEANLOZ's (1984) equation of state measurements on gold foils. Although their calibration measurements carried out on a precision graticule showed relatively little scatter, their strain measurements were analogous to our measurements across wide bars and similarly show considerable scatter. That measurements between fine lines are inherently more precise than measurements across wide bars (Figure 6) can be predicted from the theory of diffraction limited optics (see below). The formal uncertainty of a particular measurement (shown by the error bars) is relatively constant for all runs, presumably reflecting the errors inherent in the mechanical components of the image shearing eyepiece and the resolution of the video system. Note that by our use of an oscilloscope and high-quality video system, the uncertainty on each measurement has been reduced ten fold from that obtained in SCOTT and JEANLOZ (1984).

Optical Distortions at High Pressure

The calibration measurements show that the accuracy of a measurement is not affected by placing the sample in the diamond cell at zero pressure (Figure 5). We assess the accuracy of length measurements at pressure through equation of state measurements up to 10 GPa (see below). At high pressures, in the nonhydrostatic regime, it is difficult to accurately calibrate optical strain measurements against other experimental data. Moreover, it has been suggested that stress gradients and deformation of the diamonds may distort the sample image at high pressures (e.g., YAGI et al., 1985). For example, both elastic deformation ("cupping") of the culet faces, and stress-induced gradients in the refractive index can make the diamonds act like a diverging lens. We assess the importance of both of these processes by determining the effective focal length of a "culet lens" and computing the apparent change in magnification of the sample image. In both cases, we find that the effects are small: at the resolution limit of our measurement system.

We can place an upper boundary on the amount of cupping as one half of the sample thickness, ~5 μm (Figure 7). Further cupping makes the diamonds break

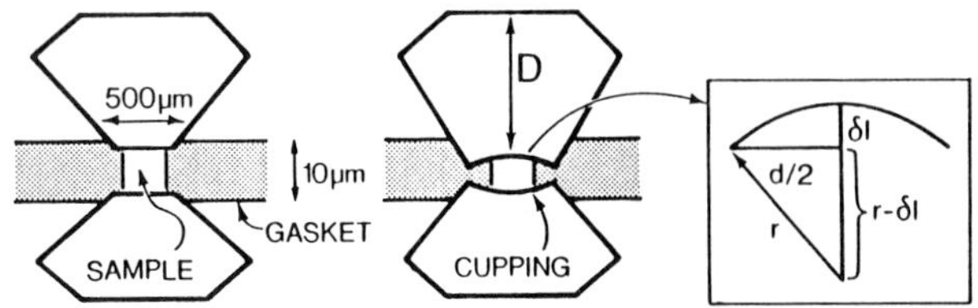

Figure 7. Cupping of the culet faces at high pressures. At low pressures, the sample chamber has a uniform thickness (left). At high pressures, large stress gradients cause the gasket to extrude and thin (center). The inset on the right summarizes the variables used to determine the radius of curvature of the culet face.

through the gasket, which usually terminates the experiment (see YAGI et al., 1985). At present the cupping process is poorly understood, though it seems to be related to the elastic properties of the diamonds and the yield strength of the sample and gasket (MOSS et al., 1986). With increasing pressure, the large stress gradients at the culet edge cause the metal gasket to extrude and progressively thin. As the sample area is observed to be relatively constant at pressures above 10 GPa, however, the sample thickness cannot change appreciably (except for elastic compression) upon further loading.

If M_o is the magnification of the objective (20X), S_o is the distance between the sample and the diamond (~2 μm), and f is the focal length of the culet lens, the magnification of the two lenses together is

$$M = \left| \frac{fM_o}{(S_o - f)} \right| \tag{1}$$

In the limit that S_o goes to 0, (1) defines the magnification of the single microscope objective. Throughout an experiment all of the variables except f are fixed. In the case of cupping (see Figure 7) we can write the radius of curvature, r, of the surface as

$$(r - \delta l)^2 + \left(\frac{d}{2}\right)^2 = r^2 \tag{2}$$

where d is the diameter of a culet face (500 μm) cupped δl relative to the edges. The focal length is then

$$\frac{1}{f_{\text{cupping}}} = (n_d - n_a)\frac{1}{r} \tag{3}$$

where n_d and n_a are the refractive indices of the diamond and the alcohol pressure medium. We use zero pressure values of n_d (2.387) and n_a (1.333) for a conservative estimate of f. In actuality, the difference between the indices of refraction decreases rapidly with pressure (CHEN and VEDAM, 1980).

Substituting values for the most extreme case (δl=5 μm) we find that the magnification ratio M/M_o ~0.9997 (f_{cupping}~−5.93 mm). Thus when measuring a 100 μm object, the uncertainty due to a change in magnification is ~0.030 μm, or as shown above, close to the resolution limit of the measurement system. As we have considered an extreme case, we believe that cupping is unlikely to lead to significant biases in the optical measurement of strain to pressures in the 10–50 GPa range.

If we assume that gradients in refractive index make the diamonds act like a thin Wood lens we can write the effective focal length of the "diamond lens" in terms of the pressure gradient across the sample as (MARCHAND, 1978)

$$f_{\text{RI}} = \frac{r_c}{2ln_0}\left(\frac{dP}{dr}\frac{1}{n}\frac{dn}{dP}\right)^{-1} \tag{4}$$

Here, r_c is the radius of the culet, n_0 is the refractive index at zero pressure, and l is the vertical lengthscale over which the stress gradients decay upward (axially) into the diamonds. Using experimental values for the pressure derivative of the refractive index (FONTANELLA et al., 1977) ($dln(n)/dP$=−0.36×10^{-12}Pa^{-1}) and assuming a large pressure gradient of 0.1 Pa/μm and l~r_c, we find f_{RI}~−5.8 mm. Thus for extreme pressure gradients, the effect of a radially varying refractive index is comparable to cupping at high pressures. Again in the most extreme cases we estimate that refractive index variations only affect the measurements at the limit of resolution. In general, neither "lensing" effect is expected to degrade our relatively low pressure (10^{11} Pa) measurements.

Diffraction Limits of Optical Length Measurements

We use Fourier diffraction theory to model the measurement system and to verify the results of Figure 5. Following BORN and WOLF (1980), we show how an imaging system can be modeled as a linear filter acting on the Fourier representation of the original object (see Figure 8). That is, if $\boldsymbol{I}_o$, $\boldsymbol{I}_i$ are Fourier transformed measures of the light intensity in the object and image planes, then

$$\boldsymbol{I}_i(g, h) = \boldsymbol{K}(g, h)\boldsymbol{I}_o(g, h) \tag{5}$$

$\boldsymbol{K}$ is a Fourier-transformed transfer function which describes the effect of defocus, aberrations, finite apertures, and coherence of the light source (g and h are spatial wave numbers which will be defined below). $\boldsymbol{K}$ is also known as the frequency response function since it defines the spectral range of spatial information passed by the

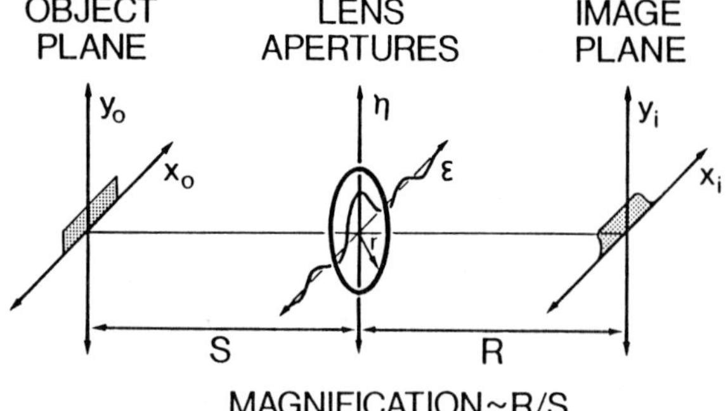

Figure 8. Coordinate systems for Fourier diffraction theory. Diffraction of light at the object in the (x_o, y_o) plane produces the Fourier transform of the original object in the aperture (ξ, η) plane. The aperture with radius r only passes a portion of the diffraction spectra causing the image to be only partially resolved in the (x_i, y_i) plane. The object is magnified approximately R/S.

microscope. In the following, bold notation (e.g., $\boldsymbol{I}_o(g, h)$) refers to the spatial Fourier transform of a two-dimensional scalar valued function (e.g., $I_o(x_o, y_o)$, $I(\xi, \eta)$).

Physically, imaging with the microscope consists of a series of diffraction processes. Illumination of the sample in the object plane produces a diffraction pattern which is transmitted to the lenses. By the Fraunhofer diffraction integral we know that the diffraction pattern is the Fourier transform of the original object. Thus using the coordinates of Figure 8 the amplitude of a wave in the (ξ, η) plane, diffracted by the sample in the (x_o, y_o) plane is

$$I(\xi, \eta) = \frac{-i}{\lambda S} \exp(ikS) \iint f(x_o, y_o)\, e^{-(ik[x_o p + y_o q])} dx_o\, dy_o \tag{6}$$

where p and q are direction cosines of the diffracted rays and $f(x_o, y_o)$ describes the light transmitted by the original object.

We can describe the entire imaging process (and hence derive 5) by sequentially applying (6) as light is diffracted at the object and aperture planes in the microscope. The convolution in real space reduces to a product of terms in Fourier space that are proportional to $\boldsymbol{K}$. Specifically if we consider a point source at (x_o', y_o') $(I_o = \delta(x_o - x_o')\, \delta(y_o - y_o'))$ then the amplitude of the wave in the (ξ, η) plane is

$$I(\xi, \eta) = \frac{i}{\lambda S} \exp(ikS)\, G(\xi, \eta) \tag{7}$$

where

$$\begin{aligned} G(\xi, \eta) &= \text{constant} \qquad \xi^2 + \eta^2 \leq r^2 \\ G(\xi, \eta) &= 0 \qquad \xi^2 + \eta^2 \geq r^2 \end{aligned} \tag{8}$$

The lenses act as apertures that modify a wave (and hence the image of the sample) in two important ways as it is propagated from the object plane through the microscope. First, they limit the spectral range of the diffraction pattern since waves outside the lens aperture cannot contribute to the final image. This process is described by the pupil function, G, in (8). Secondly, defocusing of the lens or aberrations cause the phase of the wave to vary over the aperture. This will be described below.

Using (6) again we can write the amplitude in the image plane as

$$I_i = \frac{i}{\lambda R} \iint I(\xi, \eta)\, e^{ikR}\, e^{-ik(\xi p + \eta q)} d\xi d\eta \tag{9}$$

Substituting (7) into (9) results in an expression like (5) with

$$K(x_o, y_o; x_i, y_i) = \frac{e^{ik(R+S)}}{\lambda^2 RS} \iint G(\xi, \eta) e^{(-ik/S)((x_i - x_o)\xi + (y_i - y_o)\eta)} d\xi d\eta \tag{10}$$

The direction cosines p and q have been replaced by

$$p = \left(\frac{x_i - x_o}{R}\right) \qquad q = \left(\frac{y_i - y_o}{R}\right) \tag{11}$$

Thus, K is defined entirely in terms of the pupil function and the diffraction integral: recalling the definition of the Fourier integral and excluding a constant factor (10) is equivalent to

$$\boldsymbol{K}\left(\frac{\xi}{\lambda R}, \frac{\eta}{\lambda R}\right) = G(\xi, \eta) \tag{12}$$

Hence, the pupil function is just the Fourier transform of K, the frequency response function at the points

$$\xi = \lambda R g \quad \eta = \lambda R h \tag{13}$$

The finite resolution of the objective and the intensity across the Airy disc arise from (12). In the above example, the lens aperture acts like a filter on the Fourier representation of the object, limiting the spatial spectrum of the diffraction pattern passed by the lens. When the image is "reconvolved", it appears blurred because some of the high frequency information has been removed (Figure 8). Specifically, from (12) and (13) the lens limits the diffraction spectrum to wave numbers g and h such that

$$g^2 + h^2 \leq \left(\frac{r}{\lambda R}\right)^2 \tag{14}$$

Intuitively, a small-diameter aperture or lens has low

resolution because it limits the high frequency (large g, h) information in the final image.

One source of scatter in our experiments is a change in the Airy disc profile between measurements. Since the aperture is fixed, this can only occur by defocusing the objective, changing the illumination, or introducing aberrations. As discussed above, these affect the image by modifying the phase of the wave traveling through the apertures. To model a defocused objective we redefine the pupil function, G, as (BORN and WOLF, 1980)

$$G = \exp[i \frac{k}{2} z(\xi^2 + \eta^2)] \qquad \xi^2 + \eta^2 \leq r^2$$
$$G = 0 \qquad \xi^2 + \eta^2 \geq r^2 \qquad (15)$$

where z is the defect of focus in the image plane. Equation (15) shows that defocusing decreases the value of the pupil function across the aperture, and from (12) this leads to a concommitant decrease in $\boldsymbol{K}$ for increasing frequencies. In this way, defocus blurrs the final image by reducing the high frequency information passed by the lens apertures.

In our system, we focus the image by maximizing the intensity difference across the image of the edge of the sample. Using the approximate result of HOPKINS (1955), the decrease in image plane intensity is related to the amount of defocusing by

$$\delta z = \frac{l r \sqrt{2\delta I}}{\pi R} \qquad (16)$$

Here l is the characteristic width of the object and δI is the intensity difference that can be detected in the image plane. Substituting values for our system (R~170 mm, r~2.7 mm, l~0.001 mm, δI~3%), the amount of defocus is δz~4.7 μm. This value is comparable to the resolution of our stage (~2 μm).

To assess the influence of defocusing on the uncertainty of length measurements, we examine the results of NYYSSONEN's (1979) calculations of image-plane intensities. Shown in Figure 9, these illustrate the change in image profiles with defocus of two high-resolution objectives (NA=0.95, 0.90). Evidently small amounts of defocus (~1–2 μm scaled to our system) cause large variations in the image profile, and points of fixed intensity (which are located by the image shearing eyepiece) shift asymmetrically in space. The most important conclusion, however, is that points midway between the dark and light regions (I~50%) shift the most with defocus, whereas the maximum associated with the fine lines is relatively fixed.

When applied to our calibration experiments, the calculations summarized in Figure 9 show that the measurements of object widths (Figure 6a) should always have more scatter than measurements of line spacings (Figure 6b). With small amounts of defocus, the position of the 50% intensity point varies over a lengthscale of up to 10% of the Airy disc. In contrast, the maximum associated with fine lines appears to be fixed in space, even with large amounts of defocusing. These results are consistent with the different scatter observed in the different calibration measurements shown in Figure 5.

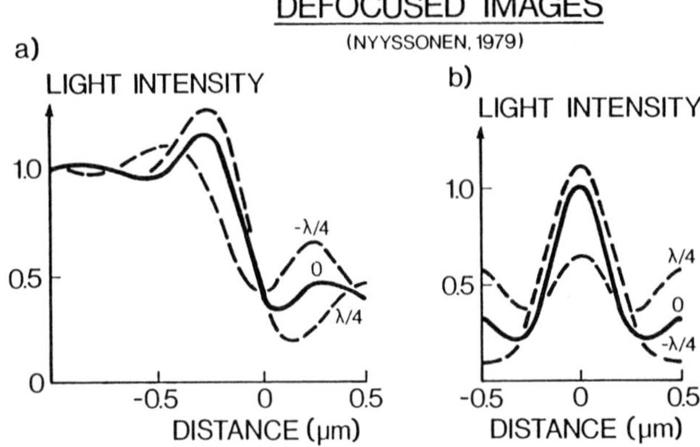

Figure 9. Calculations of image plane intensity with defocus (NYYSSONEN, 1979). The profiles are labeled by the amount of defocus (λ=530 nm). a) Asymmetric change in profile for large object edge (NA=0.95). Points with a fixed intensity shift in space. b) Symmetric variation in profile for periodic grating of 0.5 μm wide lines (NA=0.90). The position of the intensity maximum corresponding to the center of the line spacing is unchanged with defocus.

Hydrostatic Compression of Glasses: Experimental Test of the Image Shearing Technique

As an independent check of our optical technique for measuring strains, we have measured the isothermal, room-temperature equation of state of fused-silica and of a silica rich glass. In both experiments, we determine the linear strains at different pressures by optically measuring the change in spacing between thin lines on glass fragments. Thus the measurements correspond to those shown schematically in Figure 6b. The samples were compressed in a Mao-Bell type diamond cell with a 4:1 mixture of methanol-ethanol as a pressure medium. The pressure remained hydrostatic up to 10 GPa and was measured with the ruby fluorescence technique (MAO et al., 1978). As a check of the hydrostaticity, and as a measure of the shear stresses across the samples, the pressure was measured at four or more points in each run. By monitoring the lack of birefringence in the sample, and by focusing the objective on the upper culet and separately on the sample, we could check that the sample was not pinched between the diamonds. The samples were approximately 10–20 μm thick and 100 μm across and one face was prepared with a thin (0.1 μm) chromium emulsion mask of $\sim 1\ \mu m$ wide lines spaced 33–65 μm apart.

We determined the strain by measuring the line spacing first at zero pressure and then at elevated pressures. The results of our calibration experiments show that when measuring over distances of 50–100 μm our uncertainties in strain ($\delta V/V_0 \sim 10^{-3}$) are comparable to those in X-ray studies through the diamond cell (HEINZ and JEANLOZ, 1984).

Results

Fused Silica

Linear strains were measured on three separate samples at pressures of up to 15 GPa. The samples were portions of fused-silica photomasks manufactured by Hoya Electronics. Because the glass is isotropic, we can convert measurements of linear strains directly into volume strains. The results for all runs are summarized in Figure 10. The lack of birefringence confirms the assumption of isotropy and we note that there is little scatter between strains measured on different samples. Sample lengths (l=a given line spacing) were measured before, during, and after compression and completely reversible elastic strain was measured on increasing and decreasing pressures to 10 GPa.

We use BIRCH's (1978) finite-strain formalism to fit our elastic pressure-volume data to a Birch-Murnaghan equation of state. Defining the Eulerian finite strain parameter

$$f = \frac{1}{2}\left(\left(\frac{l_0}{l}\right)^2 - 1\right) \tag{17}$$

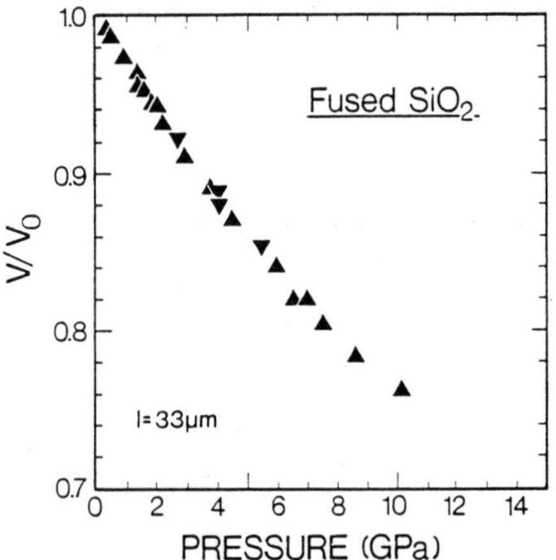

Figure 10. Volume strain for fused silica as a function of hydrostatic pressure. Up and down pointing triangles indicate measurements on increasing and decreasing pressure. The line spacing is 33 μm.

and a normalized pressure

$$F = \frac{P}{3f(1+2f)^{2.5}} \tag{18}$$

we fit the data by a weighted least squares polynomial for F in terms of f:

$$F = af^2 + bf + c \tag{19}$$

In these expressions, P is the experimentally measured pressure and subscript 0 refers to zero-pressure conditions. The coefficients a, b, c in (19) can be expressed in terms of the zero-pressure bulk modulus (K_0) and its pressure derivatives (K_0', K_0'')

$$\begin{aligned} a &= \frac{3}{2}K_0\left(K_0K_0'' + (K_0' - 7) + \frac{143}{9}\right) \\ b &= -\frac{3}{2}K_0(4 - K_0') \\ c &= K_0 \end{aligned} \tag{20}$$

The polynomial expansion of F in terms of f follows directly from the definitions of the strain parameter, normalized pressure, and a Taylor-series expansion for the compressional energy in terms of f. Second-order polynomials in f, which are fourth order in the strain expansion of the energy, successfully describe the equation of state of many crystalline materials up to very large compressions (KNITTLE and JEANLOZ, 1984; HEINZ and JEANLOZ, 1984). The results for fused silica (Figure 11), however, are highly nonlinear, and we find that a weighted quadratic polynomial in f systematically misfits the low strain (low weight) points. Higher order polynomials overfit the data requiring more parameters (higher order derivatives of the bulk modulus) to fit the data than is justified by the number of observations.

Thus, to derive a physically realistic equation of state which reasonably describes the data with a small number of parameters, we fit the F versus f data with quadratic splines. This technique increases the weight of the low pressure points relative to those at high pressure without overfitting the data to higher order polynomials. Also by determining separate high- and low-pressure equations of state we can distinguish between the anomalous high- and low-pressure elastic properties which have been previously documented for fused silica (KONDO et al., 1981). To determine the low pressure equation of state, we performed successive fits with additional points of higher strain (f) in each iteration. In this way, we find that K_0 and K_0' vary systematically as points are added to the fit (Figure 12).

Within the uncertainty of our fits, K_0 is relatively well determined: 37 ($\pm$5.5) in excellent agreement with

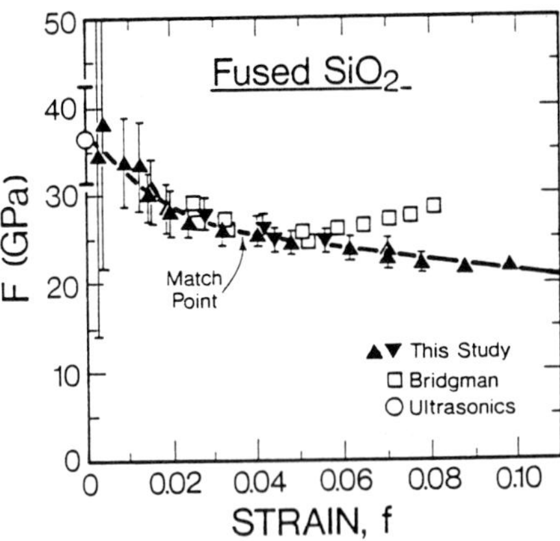

Figure 11. Normalized pressure F versus strain f for fused silica (see text). Symbols are the same as in Figure 10. The line represents the room temperature isotherm (Eulerian fourth-order finite-strain fit to our data). The fit was made with two quadratic splines matched at f=0.039. The zero pressure bulk modulus, K_0, is compared with the ultrasonically determined values (open circle). The open squares show the original static-compression data of BRIDGMAN (1948). On this plot, our fit is indistinguishable from the ultrasonically determined equation of state up to 3 GPa.

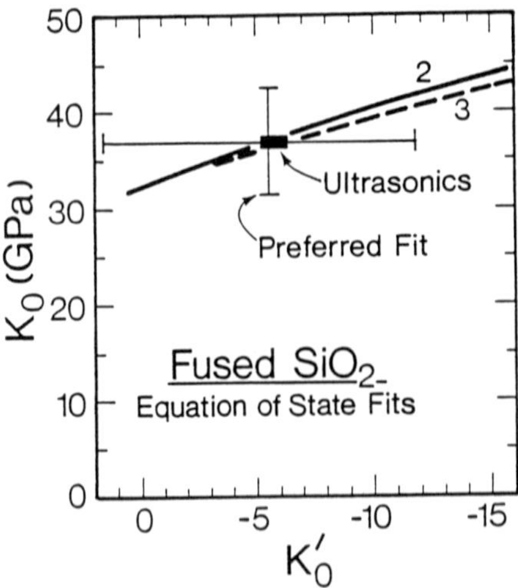

Figure 12. Tradeoff between zero pressure bulk modulus, K_0, and its pressure derivative K_0' for Eulerian fits with increasing range of strain in the data set. K_0 increases and K_0' decreases as the range of the fit is decreased. Also plotted are the range of ultrasonically determined values and the uncertainties for our preferred fit. The two curves compare the results for 2nd and 3rd order fits of f in terms of F.

previous ultrasonic studies (Table 1). The line defining the tradeoff between K_0 and K_0' passes through the ultrasonically determined values of K_0=36.7 (±0.2) GPa and K_0'=−5.8 (±0.5). Although our uncertainty for K_0' is substantial, the prefered value is clearly negative, which is also in good agreement with the ultrasonic data (measured up to 3 GPa: Table 1). Finally, our prefered low pressure fit is indistinguishable in Figure 11 from the ultrasonically determined equation of state.

The uncertainty of comparing zero pressure properties measured under finite and infinitesimal strains reflects important differences in the experimental techniques involved. Static-compression studies determine zero-pressure properties by difference in strain and pressure and extrapolation to P=0. Unless the F versus f relation is strictly linear these values vary significantly with the range of strain in the data. Acoustic velocity techniques, on the other hand, make direct determinations of the elastic constants at a particular pressure and are thus not subject to the uncertainties caused by extrapolation to zero pressure.

TABLE 1. Comparison of K_0 and K_0' Determined from Present Data and Previous Ultrasonic Studies

	K_0	K_0'	Ref.
Fused SiO_2	36.8	−5.9	(MCSKIMIN, 1957)
	36.7	−6.3	(BOGARDUS, 1965)
	36.5	−6.2	(PESELNICK et al., 1967)
	36.9	−5.3	(KONDO et al., 1981)
	37.0 ± 5.5	−5.6 ± 6.2	This Study
Ca-Mg-Na Glass	35.5 ± 3.7	−2.9 ± 4.1	This Study

Glass

We carried out a similar compression study on a Ca-Mg-Na silicate glass. The sample was obtained from a low thermal-expansion photomask manufactured by the IMTEC Corporation. Its composition determined by electron microprobe analysis is shown in Table 2. Its silica content is similar to Pyrex, however, it is mixed with a larger amounts of calcium, sodium, and magnesium relative to aluminum. The deficit in the total is presumably

TABLE 2. Weight Fraction of Ca-Mg-Na Silica Glass Determined by Electron Microprobe Analysis

SiO_2	79.60 (0.67)
CaO	5.71 (0.11)
MgO	4.88 (0.08)
Na_2O	4.97 (0.56)
Al_2O_3	1.05 (0.08)
K_2O	0.32 (0.03)
FeO	0.11 (0.02)
ZnO	0.09 (0.05)
Total	96.71

due to the boron content of the glass (3.29% B_2O_3 by difference). Linear strains were measured on three samples at hydrostatic pressures of up to 11.6 GPa (Figure 13). As before, measurements were made on increasing and decreasing pressures to determine non-elastic effects. The lack of hysteresis confirms that our measurements reflect recoverable elastic strains. We determine the equation of state parameters for the glass in a similar way as above, matching the spline fits at $f=0.05$. The parameters for the fit are shown in Table 1.

These are new data as there have been no high precision studies on a glass with the same composition. We find, however, that our determinations of $K_0=35.5\ (\pm 3.7)$ GPa and $K_0'<0$ are consistent with previous ultrasonic studies on silicate glasses with similar proportions of oxides and alkali cations. For example, alkali silicate glasses with the same mean atomic volume have similar zero pressure bulk moduli ($K_0 \sim 35 \pm 4$ GPa) (SOGA, 1982). K_0' is negative or very small in alkali silicate glasses with similar bulk moduli ($\sim$35 GPa) and mean atomic volume ($\sim 9\ cm^3/gr$-atom) (GAMBERG and UHLMANN, 1973; WEIR and SHARTSIS, 1955). Moreover, ultrasonic studies to 0.9 GPa show K_0' is negative for Pyrex which has a similar silica content (HUGHES and KELLEY, 1953). Therefore, our hydrostatic compression measurements on the Ca-Mg-Na glass are in good agreement with available equation of state data on glasses of similar composition.

Conclusions

We have developed a new technique for making high-precision optical length measurements that involves quantitative determinations of the light intensity across the image plane and the use of special sample preparation techniques. To show that this method can be used for measuring strain at high pressures we have performed the first static-compression study of an amorphous material in the diamond cell. With a series of calibration measurements at zero pressure on a precision graticule and on photomasks, we quantified the uncertainty of a particular measurement and the source of scatter in our experiments. When measuring the spacing between lines that are thinner than the resolution limit of the objective ($\sim$1.5 μm), our uncertainty is $\pm 0.010\ \mu$m. We show from Fourier diffraction theory that this line pattern yields especially precise measurements, and that the resolution is not significantly affected by defocus or optical aberrations in the diamonds at high pressures. Thus, the errors in our strain measurements at high pressures are comparable to high precision X-ray diffraction through the diamond cell. As an independent confirmation of the technique our hydrostatic compression measurements on two glasses show little scatter. Also, we find that our equation of state for fused silica is in excellent agreement with the results of ultrasonic studies.

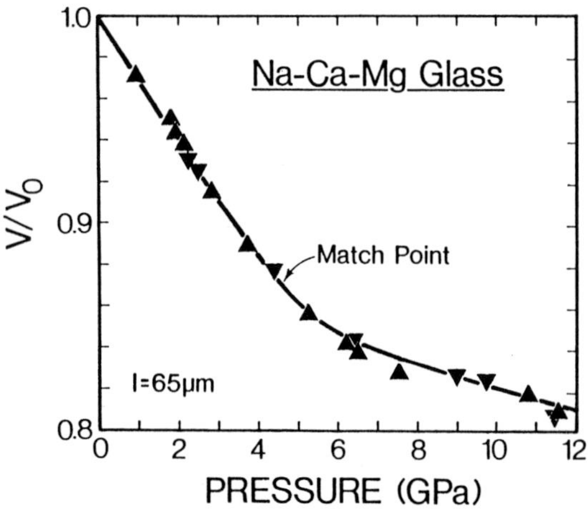

Figure 13. Volume strain and pressure for Ca-Mg-Na silica glass. Symbols are the same as Figure 10.

Acknowledgments. Elise Knittle and Quentin Williams provided helpful comments, and the photomasks were supplied by the University of California at Berkeley Microelectronics Laboratory. This work was supported by the National Science Foundation.

REFERENCES

BIRCH, F., Finite strain isotherm and velocities for single-crystal and polycrystalline NaCl at high pressures and 300 K, *J. Geophys. Res.*, *83*, 1257–1268, 1978.

BRIDGMAN, P. W., The compression of 39 substances to 100,000 Kg/cm², *Proc. Am. Acad. Arts and Sci.*, *76*, 55–70, 1948.

BOGARDUS, E. H., Third order elastic constants of Ge, MgO, and fused SiO_2, *J. Appl. Phys.*, *36*, 2504–2513, 1965.

BORN, M., and E. WOLF, *Principles of Optics*, Pergamon 6th ed., Oxford, 1980.

CHEN, C. C., and K. VEDAM, Piezo- and elastic-optic properties of liquids under high pressure III. Results on twelve more liquids, *J. Chem. Phys.*, *73*, 4577–4594, 1980.

DYSON, J., Precise measurement by image splitting, *J. Opt. Soc. Am.*, *50*, 754–757, 1960.

FONTANELLA, J., R. L. JOHNSTON, J. COLWELL, and C. ANDEEN, Temperature and pressure variation of the refractive index of diamond, *Appl. Optics*, *16*, 2949–2951, 1977.

GAMBERG, E., and D. R. UHLMANN, Pressure dependence of the elastic moduli of glasses in the K_2O-SiO_2 system, *J. Non-Cryst. Solids*, *13*, 399–408, 1973.

GLADKOV, A. V., Influence of the silica modulus on the velocity of ultrasound in the glasses of the system Na_2O-SiO_2, *Zh. Fiz. Khim.*, *31*, 1002–1006, 1957.

HEINZ, D. L., and R. JEANLOZ, The equation of state of the gold calibration standard, *J. Appl. Phys.*, *55*, 885–893, 1984.

HOPKINS, H. H., The frequency response of a defocused optical system, *Proc. Roy. Soc. A*, *231*, 91–103, 1955.

HUGHES, D. S., and J. L. KELLY, Second-order elastic deformation of solids, *Phys. Rev.*, *92*, 1145–1149, 1953.

KNITTLE, E., and R. JEANLOZ, Structure and bonding changes in cesium iodide at high pressures, *Science, 223*, 53–56, 1984.

KONDO, K., S. IIO, and A. SAWAOKA, Nonlinear pressure dependence of the elastic moduli of fused quartz up to 3 GPa, *J. Appl. Phys., 52*, 2826–2831, 1981.

MAO, H. K., P. M. BELL, J. W. SHANER, and D. J. STEINBERG, Specific volume measurements of Cu, Mo, Pd, and Ag and calibration of the ruby R_1 fluorescence pressure gauge from 0.06 to 1 Mbar, *J. Appl. Phys., 49*, 3276–3283, 1978.

MARCHAND, E. W., *Gradient Index Optics*, 166 pp., Academic Press, 1978.

MCSKIMIN, H. J., Ultrasonic pulse technique for measurement of acoustic losses and velocity of propagation in liquids as a function of temperature and hydrostatic pressure, *J. Acoust. Soc. Am., 29*, 1185–1192, 1957.

MOSS, W. C., J. O. HALQUIST, R. REICHLIN, K. A. GOETTEL, and S. MARTIN, Finite element analysis of the diamond-anvil cell; Achieving 4.6 Mbars, *Appl. Phys. Lett., 48*, 1258–1260, 1986.

NYYSSONEN, D., Spatial coherence: The key to optical micrometerology, *SPIE, 194*, 34–44, 1979.

NYYSSONEN, D., Calibration of optical systems for linewidth measurements on wafers, *Opt. Eng., 21*, 882–887, 1982.

PESELNICK, L., R. MEISTER, and W. H. WILSON, Pressure derivatives of elastic moduli of fused quartz to 10 kilobars, *J. Phys. Chem. Solids, 28*, 635–639, 1967.

SCOTT, C., and R. JEANLOZ, Optical length determinations in the diamond-anvil cell, *Rev. Sci. Instrum., 55*, 558–562, 1984.

SOGA, N., Three band theory and elastic moduli of glass, *J. Non-Cryst. Solids, 52*, 365–375, 1982.

SPINNER, Y. M., Elastic moduli of glasses at elevated temperature by dynamic methods, *J. Am. Ceram. Soc., 39*, 113–117, 1956.

SWYT, D. A., and F. W. ROSBERRY, A comparison of some optical measurements of photomask linewidths, *Sol. Stat. Tech., 20*, 70–75, 1977.

WEIR, C. E., and L. SHARTSIS, Compressibility of some binary alkali borate and silicate glasses at high pressures, *J. Am. Ceram. Soc., 38*, 299–318, 1955.

YAGI, T., T. SUZUKI, and S. AKIMOTO, Static compression of Wüstite ($Fe_{0.98}O$) to 120 GPa, *J. Geophys. Res., 10*, 8784–8788, 1985.

MEASUREMENTS OF THE LIFETIME OF THE RUBY R_1 LINE AND ITS APPLICATION TO HIGH-TEMPERATURE AND HIGH-PRESSURE CALIBRATION IN THE DIAMOND-ANVIL CELL

Yosiko SATO-SORENSEN

Geophysics Program, University of Washington
Seattle, Washington 98195, USA

Abstract. The lifetime τ of the R_1 fluorescence line, $^2E(\bar{E}) \rightarrow {}^4A_2$ of ruby (Al_2O_3: Cr^{3+}) has been measured under high-pressure (up to 42.7 GPa) and high-temperature (up to 700 K) conditions using a modulation fluorometry technique. This study shows that τ increases linearly with pressure and that this increase is primarily caused by a rapid decrease in the transition probability of the radiative transition. These results are in agreement with the theoretically predicted decrease in the oscillator strength at high pressure. The rapid decrease in τ previously observed below ~500 K is also shown to continue at higher temperatures. A new method for simultaneously determining the pressure and the temperature inside a diamond-anvil cell (DAC) by using the measured lifetime $\tau(T, p)$ and the measured fluorescence peak position $\nu(T, p)$ is proposed.

Introduction

Many theoretical and experimental studies of the optical properties of ruby have been reported. In particular, the R lines have been extensively studied because of their importance in the ruby laser (for a recent review see: YEN and SELZER, 1981). The development of diamond-anvil-cell (DAC) techniques, on the other hand, have revolutionized high-pressure studies, making routine studies above 1 Mbar possible (JAYARAMAN, 1983, 1986). Recently, ultrahigh-pressures above 5 Mbar have been successfully generated (XU et al., 1986). The rapid development of the DAC methods was greatly facilitated by the pioneering work of BARNETT et al. (1973) in establishing a secondary pressure scale using the pressure-induced wavelength shift of the sharp R_1 line of ruby. Following the initial work by Barnett et al., a number of studies of the R lines under high-pressure conditions and their application to accurate pressure estimation in DAC's have been reported. In addition the hydrostatic limits of different liquid pressure media were examined using the line-broadening of the R lines caused by the nonhydrostatic pressure distribution (PIERMARINI et al., 1973) and the properties of gas pressure media have also been tested (LIEBENBERG, 1979; LESAR et al., 1979; NICOL et al., 1979; BELL and MAO, 1981). The red shift of the R lines was studied in detail (PIERMARINI et al., 1975; MAO et al., 1978) and is now commonly used as the pressure scale in DAC studies. Theoretical work, based on a scaling law (MUNRO, 1977), successfully accounts for the red shift of the R lines.

One of the major issues in DAC research at the present time is the simultaneous determination of both high-pressure and high-temperature conditions inside the cell. To estimate the temperature in DAC's, SHIMOMURA et al. (1982) proposed to use the linewidth of the R_1 line, which becomes broader because of the increase in the two-phonon Raman processes at high temperature (MCCUMBER and STURGE, 1963; POWELL et al., 1966). This technique was examined in detail by MUNRO et al. (1985) and it was shown that high precision simultaneous temperature and pressure measurements can be achieved by the model-line shape analysis of the R lines. However, the application of this technique is limited to hydrostatic experiments because the line shape of the R lines is strongly influenced by the nonhydrostatic pressure distributions which exist in most DAC experiments.

This paper proposes a new method for simultaneous temperature and pressure determination based on the measured fluorescence lifetime and the measured fluorescence peak position.

Lifetime of the R_1 Fluorescence Line of Ruby

Figure 1 is a schematic diagram of the energy levels and the resulting absorption and luminescence of a Cr^{3+} ion which is octahedrally coordinated to oxygen ions in Al_2O_3. In this system there are two strong spin-allowed transitions in the visible range, $^4A_2 \rightarrow {}^4T_2$ (~18000 cm^{-1}, yellow-green) and $^4T_1$ (~25000 cm^{-1}, violet), and three spin-forbidden transitions, $^4A_2 \rightarrow {}^2E$, $^2T_1$, and $^2T_2$. When this system is optically pumped, it decays rapidly via nonradiative transitions until it reaches the lowest excited state ^2E. Because of the large energy gap between ^2E and $^4A_2$, the final transition is radiative and results in the R fluorescence lines. The ^2E level is split by the trigonal crystal field, producing two transitions at room temperature, the R_1 line at 14402 cm^{-1} and the R_2 line at 14432 cm^{-1}. At room temperature this emission has a quantum efficiency near unity and most of the absorbed energy decays through this transition; the transition is a sharp no-phonon transition, and detailed balance holds within the experimental errors of ~5% (NELSON and STURGE,

High-Pressure Research in Mineral Physics, edited by M. H. Manghnani and Y. Syono, pp. 53–59.

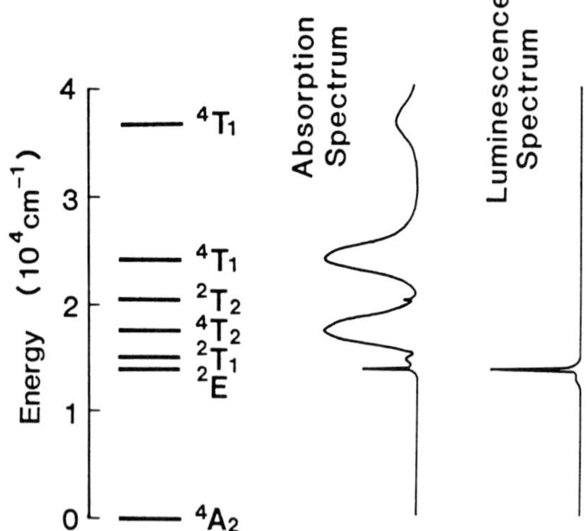

Figure 1. Energy levels and the resulting absorption and luminescence for Cr^{3+} ions in ruby.

1965).

The Einstein spontaneous transition probability w that an excited state i will spontaneously emit a photon and decay to a lower state f is given by

$$w = \frac{8\pi^2 e^2 v^2}{mc^3}\left[\left(\frac{E_{\text{eff}}}{E_0}\right)^2 n\right] f \tag{1}$$

where e and m are the charge and mass of the electron, c is the velocity of light, v is the frequency of the radiation, and n is the refractive index of the sample. E_{eff}/E_0 is the local field correction; for an ionic crystal like ruby $E_{\text{eff}}/E_0=(n^2+2)/3$. The oscillator strength f is given by

$$f = \frac{4\pi m v}{3h}\left|\langle i|\boldsymbol{r}|f\rangle\right|^2 \tag{2}$$

where $\langle i|\boldsymbol{r}|f\rangle$ is the dipole transition matrix element between the initial and final states. The oscillator strength is related to the measured absorption σ by

$$\sigma = \int\sigma(v)dv = \frac{\pi e^2}{mc}\left[\left(\frac{E_{\text{eff}}}{E_0}\right)^2 \frac{1}{n}\right] f \tag{3}$$

For strongly allowed optical transitions the oscillator strength is close to unity, whereas for a spin-forbidden transition such as ${}^2E\rightarrow{}^4A_2$ it is 10^{-6}–10^{-7} (for details see SUGANO et al., 1970 or IMBUSCH, 1978, 1981).

For N_0 Cr^{3+} ions produced at $t=0$, the number of remaining excited ions $N(t)$ decays exponentially because of the fluorescence

$$N(t) = N_0 e^{-t/\tau} \tag{4}$$

where τ is the lifetime of the fluorescence transition. Consequently, by measuring the time dependence of the luminescence intensity, the lifetime τ can then be determined. Because the transition probability w is related to τ by

$$w = \frac{1}{\tau} \tag{5}$$

the transition probability can then be estimated from the measured τ.

Measurement of the Lifetime

The lifetime of the ruby R_1 line was measured using a phase and modulation fluorometry technique (SPENCER and WEBER, 1969). For a time-dependent optical pumping intensity $I(t)$ the resulting time-dependent fluorescence intensity $R(t)$ is given by the convolution of the pumping intensity with the sample fluorescence decay function $F(t)$:

$$R(t) = \int_0^t F(t')I(t - t')dt' \tag{6}$$

If the excitation source is modulated at a frequency ω, and the fluorescence decays exponentially with a single lifetime τ, the response $R(t)$ is the convolution of $I(t)=A+B\cos\omega t$ and $F(t)=F_0e^{-t/\tau}$ so that

$$R(t) = F_0\tau[A + B(1 + \omega^2\tau^2)^{-1/2}\cos(\omega t - \theta)] \tag{7}$$

Here θ is the phase shift between $I(t)$ and $R(t)$ and is given by $\tan\theta=\omega\tau$. Consequently, for a single exponential decay, the lifetime τ can be determined either by measuring the phase shift or by measuring the ratio of the ac to the dc amplitude of the fluorescence light relative to that of the exciting light. This relative ratio is called the modulation ratio m and is given by

$$m = (1 + \omega^2\tau^2)^{-1/2} = \cos\theta \tag{8}$$

The optical system used in this study is shown in Figure 2. This system was a standard pressure measurement system modified to provide lifetime measurements (WHITMORE et al., 1982). A variable speed mechanical light chopper (6 Hz–5.5 kHz, EG & G Model 192) was used to modulate the 442 nm light from a 10 mW HeCd laser (Liconix Model 4210), which excited the ${}^4T_1$ level. The fluorescence from the ruby sample was monochromated (Instruments SA Model H-20) and detected with a photomultiplier (Hamamatsu Model R928). A lock-in amplifier (EG & G Model 124A) was used to measure the amplitude and the phase of the fluorescence signal. The wavelength acceptance for this monochromator was approximately 0.5 nm, and the measurements were made

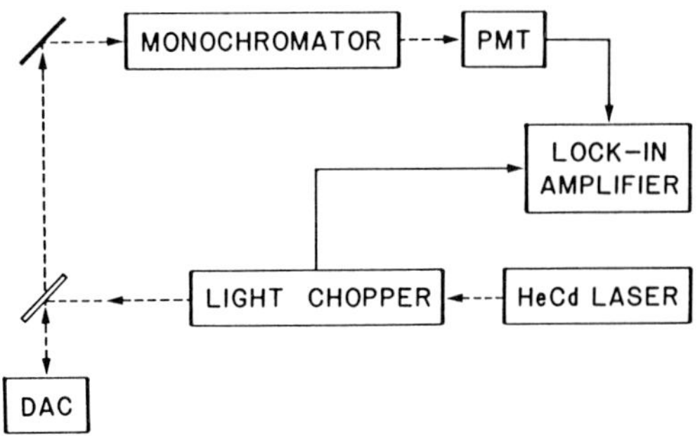

Figure 2. Schematic diagram of the optical system used for the lifetime measurements.

with the monochromator set at the peak of the R_1 line.

A super-pressure DAC was used for the high-pressure measurements, and a Chaixmeca heating stage was used for the high-temperature measurements. The Cr^{3+} concentration of the ruby was determined by electron microprobe analysis to be 0.8–1.0 wt% ($\sim 1.5 \times 10^{20}$ Cr^{3+} ion cm^{-3}). This concentration was higher than the critical concentration (~0.4 wt%); above the critical concentration excited Cr^{3+} ions also transfer energy to nearby Cr^{3+} ions by nonradiative processes (BROWN, 1964). For the lifetime measurements a small piece of ruby $\sim 20 \times 20 \times 10$ μm was pressurized in the super DAC with mineral oil as the pressure medium. Although a broadening of the R lines was observed at high pressure, no splitting typical of applied uniaxial stress was observed (FEHER and STURGE, 1968; MONTEIL et al., 1984). Therefore, this pressure medium produced quasi-hydrostatic conditions. The pressure was estimated from the wavelength shift of the R_1 line of ruby determined by MAO et al. (1978). The radiative lifetime τ_r was also measured for a dilute ruby sample (~0.03 wt% Cr^{3+}) at 1 atm.

Experimental Results

Pressure Dependence of the Lifetime

Figure 3a shows the experimental results for the lifetime measurements at high pressure. The pressure-induced change in the transition probability w calculated from the measured τ's is shown in Figure 3b. The solid line in Figure 3a is the result of a least square fit (see below). The solid line in Figure 3b indicates the resulting $w(p)$ calculated from the least square fit for $\tau(p)$. To determine the effect of a nonhydrostatic pressure distribution on the measured lifetime τ the measurement at 23.5 GPa was conducted on a ruby sample which was mounted directly on the gasket material and pressurized without any pressure medium. The results of this experiment are plotted as solid diamonds in Figures 3a and 3b. As shown, there was no observable difference caused by the non-

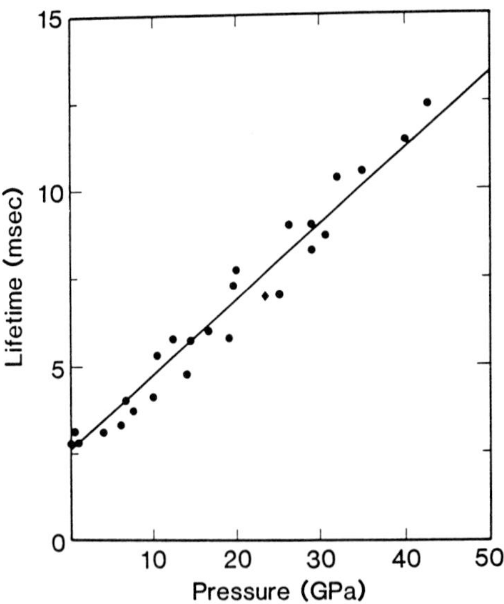

Figure 3a. Experimental results for the high-pressure lifetime measurements. The solid diamond represents the result of the nonhydrostatic experiment. The experimental uncertainty was ~4% in the lifetime, and ~2–5% in the pressure. The solid line is the result of the least square fit: $\tau(p)=2.6+0.22p$.

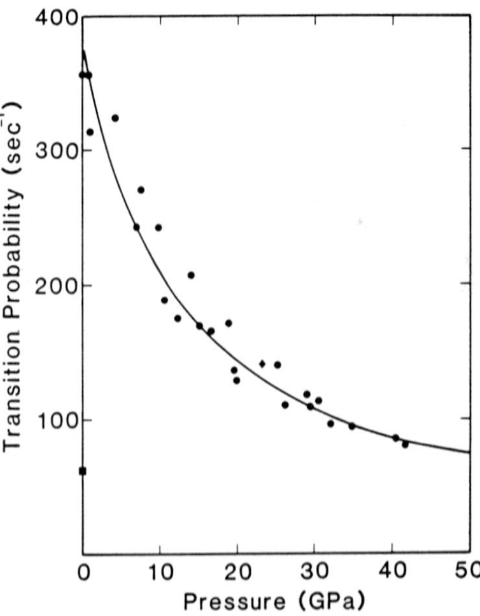

Figure 3b. Pressure-induced change in the transition probability calculated from the measured lifetime. The solid diamond represents the result of the nonhydrostatic experiment. The solid square indicates the nonradiative transition probability at 1 atm. The solid line is the result of the least square fit: $w(p)=(2.6+0.22p)^{-1}$.

hydrostatic conditions produced without the pressure medium. The lifetime was measured both as the pressure was increased and as the pressure was decreased. No hysteresis was observed. Because of the rapid increase in τ with pressure, the present measurements were limited to 42.7 GPa by the lower limit (6 Hz) of the light chopper. The measured pressure dependence of the lifetime was well fit by the form

$$\tau(p) = a + bp \tag{9}$$

with a=2.6±0.1 msec and b=0.22±0.01 msec GPa^{-1}.

Temperature Dependence of the Lifetime

The results of the lifetime measurements at high temperature are shown in Figure 4. In the present experiment measurements were made up to ~700 K. Assuming that there is no change in the transition mechanism over this temperature range, the data were fit by a simple exponential formula,

$$\tau(T) \quad = \quad ae^{-c(T-300)} \tag{10}$$

with c=0.0075±0.0005 K^{-1} and a=2.6±0.1 msec.

Discussion

The lifetime at 1 atm, τ=2.6 msec, obtained in this study for the ~0.8–1.0 wt% Cr^{3+} sample is shorter than the radiative lifetime, τ_r=3.1 msec, measured for dilute ruby

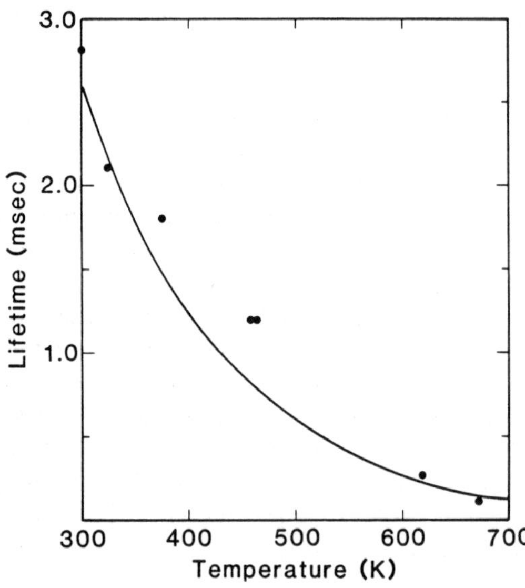

Figure 4. Experimental results for the high-temperature lifetime measurements. The experimental uncertainty was ~4% in the lifetime, and ~1.5% in the temperature. The solid line is the result of the least square fit: $\tau(T)=2.6e^{-0.0075(T-300)}$.

(~0.03 wt% Cr^{3+}). This difference is probably due to additional nonradiative energy transfer between Cr^{3+} ions caused by the high Cr^{3+} concentration of the sample used. In ruby with Cr^{3+} concentrations higher than the critical concentration (~0.4 wt% at room temperature), excited Cr^{3+} ions can efficiently transfer energy to nearby Cr^{3+} ions by nonradiative processes, and the excitation moves around before being emitted as radiation (BROWN, 1964). When both radiative and nonradiative energy processes are considered, the measured decay probability w is the sum of both processes:

$$w = w_r + w_{nr} \tag{11}$$

here $w_r = 1/\tau_r$ and $w_{nr} = 1/\tau_{nr}$ are the probabilities of radiative decay and nonradiative decay, respectively. Consequently, the measured lifetime τ is shorter than the radiative lifetime τ_r. From the measured lifetime τ=2.6 msec and the radiative lifetime τ_r=3.1 msec, the lifetime of the nonradiative decay is estimated as τ_{nr}=~16 msec. This is much longer than the lifetime of the radiative transition and the nonradiative decay probability w_{nr}=~62 sec^{-1} is only ~16% of the total decay probability w=~385 sec^{-1}. The estimated nonradiative probability at 1 atm is shown by the solid square in Figure 3b. At very high pressures the nonradiative processes might become more important and control a larger portion of the energy transfer. However, at 1 atm, the nonradiative transition probability is very small compared with the radiative transition probability and the rapid initial decrease in w above 1 atm is probably caused by a rapid decrease in w_r.

Previously, MERKLE et al. (1981) reported a pressure-induced increase in the lifetime of the ruby R_1 line for pressures up to ~7 GPa. Their measurements were made in the time domain (using a electro-optically chopped Ar ion laser and a boxcar integrator) for a dilute ruby samples (~0.03 wt% Cr^{3+}) in a DAC with a liquid pressure medium (methanol: ethanol=4:1). Although the scatter in their data was too large to allow an accurate determination of $d\tau/dp$, a clear pressure-induced change in the lifetime was observed.

A large pressure-induced change in the lifetime was also observed for the R_1 line of chrysoberyl (Al_2BeO_4: Cr^{3+}) in DAC experiments up to ~7 GPa (JIA et al., 1984). The slope of the pressure-induced change, $d\tau/dp$=0.3 msec GPa^{-1}, estimated for chrysoberyl from the paper by Jia et al. is comparable to the change obtained for ruby in this study, $d\tau/dp$=0.22 msec GPa^{-1}. This agreement is reasonable because these crystals are structurally quite similar: a) the Cr^{3+} ions are octahedrally coordinated to hexagonal close packed oxygen ions, b) the sizes of the octahedral polyhedra are similar, and c) the compressibility of the octahedral polyhedra is nearly the same in ruby and chrysoberyl (WANG et al., 1975; AU and HAZEN, 1985).

Our current understanding of the pressure-induced changes in the ruby energy levels is summarized in Figure 5. The solid lines indicate Munro's predictions for the energy levels (assuming no change in $d\nu/dp$) at high pressure (MUNRO, 1977). The shaded areas indicate the results of the static absorption experiments for the $^4T_2$ and $^4T_1$ levels up to 12 GPa (STEPHENS and DRICKAMER, 1961). The results of shock wave absorption experiments are shown by solid circles for the $^4T_2$ level and by the solid diamond for the $^4T_1$ level (GOTO et al., 1979). As shown, the energy gap between the $^4T_2$ and the ^2E levels increases with pressure. At 1 atm the initial excitation is transferred very quickly from the $^4T_1$ level via the $^4T_2$ level to the ^2E level (EVERETT, 1970; IMBUSCH, 1978). At high pressure the first step in the energy transfer, $^4T_1 \rightarrow {}^4T_2$ level, is not rate limiting, because this transition is a very fast spin-allowed transition. However, the second step in the energy transfer, $^4T_2 \rightarrow {}^2E$ level, is the rate limiting spin-forbidden transition and it will be strongly affected by the pressure-induced change in the energy gap ΔE between the $^4T_2$ and the ^2E level. This effect has been found at 1 atm in the transition probabilities for the ^2E level of Cr^{3+} in various garnet laser crystals, in which the energy gap ΔE is induced by the changes in the chemical composition (STRUVE and HUBER, 1985).

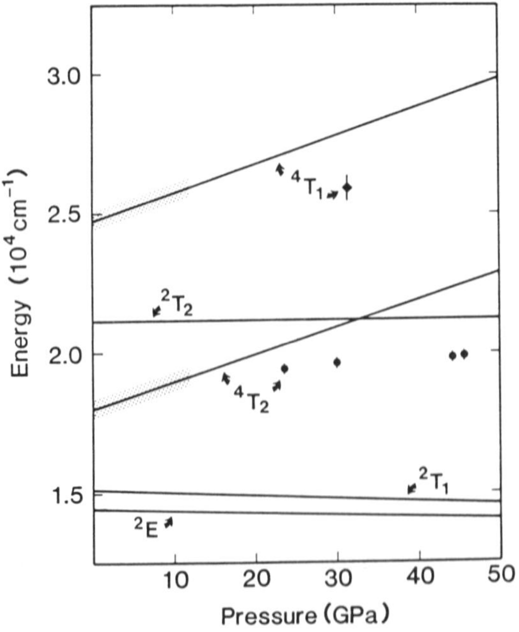

Figure 5. Pressure-induced changes in the ruby energy levels. Solid lines are the theoretical prediction by MUNRO (1977). The shaded areas indicate the results of the static absorption experiments for the $^4T_2$ and $^4T_1$ levels up to 12 GPa (STEPHENS and DRICKAMER, 1961). The results of the shock wave absorption experiments for the $^4T_2$ level are shown as solid circles and for the $^4T_1$ level by the solid diamond (GOTO et al., 1979).

From the measured $d\tau/dp = 0.22$ msec GPa^{-1} and equations (1) and (5), the pressure dependence of the oscillator strength was estimated: $(1/f_0)(df/dp) = -6 \times 10^{-2}$ GPa^{-1} (f_0 is the oscillator strength at 1 atm). This estimate is about 80 times larger than the value, -8×10^{-4} GPa^{-1}, which was theoretically predicted by MUNRO (1977). Although there should be a small correction to the experimental estimate due to the pressure dependence of the refractive index n (SETCHELL, 1979), this correction is much too small to remove the discrepancy. To understand this disagreement the pressure dependence of the oscillator strength must be measured; absorption and lifetime experiments for the direct excitation of the ^2E level from the ground state are required.

The lifetime τ of the R lines shows a rather complicated behavior as the temperature is changed (TOLSTOI and SHUN'-FU, 1962; NELSON and STURGE, 1965). For temperatures below ~300 K, the observed lifetime depends on the thickness of the sample, the Cr^{3+} concentration and whether the sample is a single crystal or polycrystalline. Above ~300 K, the lifetime decreases rapidly as the temperature is raised. Measurements of the absorption intensity of the R lines show that this decrease is due to an increase in the oscillator strength at high temperature (NELSON and STURGE, 1965). This well known decrease causes the severe degradation of the performance of high power ruby lasers at elevated temperatures (WITTKE, 1962).

The temperature-induced lifetime changes obtained in the present study agree qualitatively with previous experimental results; quantitative comparisons are not possible because of the variation in the Cr^{3+} concentrations.

Simultaneous Temperature and Pressure Measurement in DAC's

In this study the measurements were made under either high-pressure or high-temperature conditions. The combined effects of simultaneous high-temperature and high-pressure conditions were not studied, and consequently the full functional form for the lifetime $\tau(T, p)$ has not been determined. We are currently developing the apparatus required for these studies. However, the combined effects of temperature and pressure must produce a functional form $\tau(T, p)$ which reduces to equation (9) when $T = 300$ K, and to equation (10) when $p = 1$ atm. By combining the full form for $\tau(T, p)$ with the known temperature and pressure variation of the fluorescence peak position $\nu(T, p)$ the values of temperature T and pressure p can be determined from the measured values of τ and ν.

In order to evaluate the applicability and accuracy of this method for simultaneous temperature and pressure determinations, we have made preliminary estimates using the known dependence of ν and plausible forms for $\tau(T, p)$. Two possible functions of $\tau(T, p)$ were tested:

$$\tau(T, p) = ae^{-c(T-300)} + bp \quad (12)$$

and

$$\tau(T, p) = (a + bp)e^{-c(T-300)} \quad (13)$$

For the functional form of the fluorescence peak shift, the following equation,

$$\nu(T, p) = \nu_0 \left(\frac{\alpha}{p + \alpha} \right)^{1/5} - \beta(T - 300) \quad (14)$$

with $\nu_0 = 14402$ cm^{-1}, $\alpha = 380.4$ GPa and $\beta = 0.130 \pm 0.006$ cm^{-1} K^{-1} was used. This is a combination of the pressure variation determined by MAO et al (1978) and the temperature variation determined by MUNRO et al. (1985). To estimate the resulting uncertainties in the values of T and p determined from simultaneous measurements of $\tau(T, p)$ and $\nu(T, p)$, equations (12) and (14) and equations (13) and (14) were solved for T and p; for typical experimental uncertainties in τ ($<\sim 4\%$) and ν (<5 cm^{-1}), the resulting estimated temperature and pressure uncertainties were less than 5%.

Conclusion

Our measurements of the lifetime τ of the R_1 line of the ruby show that τ increases rapidly with pressure. The observed pressure-induced change in the lifetime, 0.22 msec GPa^{-1}, corresponds to a pressure-induced change in the oscillator strength, $(1/f_0)(df/dp) = -6 \times 10^{-2}$ GPa^{-1}, which is about 80 times larger than the current theoretical estimate. Further experimental and theoretical studies are needed to understand the origin of this discrepancy.

The proposed $\tau(T, p)$ and $\nu(T, p)$ method should provide rapid and accurate temperature and pressure determinations with fairly simple optical equipment. In addition the proposed technique appears rather insensitive to nonhydrostatic pressure effects which are quite important in the alternative optical line-shape technique. Our new optical method appears very promising and further work is in progress on the combined temperature and pressure dependence of the lifetime.

Acknowledgments. The author wishes to express thanks to R. L. Ingalls, J. E. Whitmore, B. W. Evans, and R. J. Stewart for their help in the experiment, to E. A. Mathez for the microprobe analysis of the sample, and to H. K. Mao for helpful discussions. This work was supported by the National Science Foundation.

References

AU, A., and R. M. HAZEN, Polyhedral modeling of the elastic properties of corundum (α-Al_2O_3) and chrysoberyl (Al_2BeO_4), *Geophys. Res. Lett.*, *12*, 725–728, 1985.

BARNETT, J. D., S. BLOCK, and G. J. PIERMARINI, An optical fluorescence system for quantitative pressure measurement in the diamond-anvil cell, *Rev. Sci. Inst.*, *44*, 1–9, 1973.

BELL, P. M., and H. K. MAO, Degree of hydrostaticity in He, Ne, and Ar pressure-transmitting media, *Year Book, Carnegie Inst. Washington*, *80*, 404–406, 1981.

BROWN, G. C. Jr., Fluorescence lifetime of ruby, *J. Appl. Phys.*, *35*, 3062–3063, 1964.

EVERETT, P. N., Lifetime of $^4T_2$ state in pink ruby, *J. Appl. Phys.*, *41*, 3193–3194, 1970.

FEHER, E., and M. D. STURGE, Effect of stress on the trigonal splittings of d^3 ions in sapphire (α-Al_2O_3), *Phys. Rev.*, *172*, 244–249, 1968.

GOTO, T., T. J. AHRENS, and G. R. ROSSMAN, Absorption spectra of Cr^{3+} in Al_2O_3 under shock compression, *Phys. Chem. Min.*, *4*, 253–263, 1979.

IMBUSCH, G. F., Inorganic luminescence, in *Luminescence Spectroscopy*, edited by M. D. Lumb, pp. 1–92, Academic Press, New York, 1978.

IMBUSCH, G. F., Optical spectroscopy of electronic centers in solids, in *Laser Spectroscopy of Solids*, edited by W. M. Yen and P. M. Selzer, pp. 1–37, Springer-Verlag, New York, 1981.

JAYARAMAN, A., Diamond-anvil cell and high-pressure physical investigations, *Rev. Mod. Phys.*, *55*, 65–108, 1983.

JAYARAMAN, A., Ultrahigh pressures, *Rev. Sci. Instrum.*, *57*, 1013–1031, 1986.

JIA, W., S. SHANG, R. M. TANG, and Z. Y. YAO, Pressure effects of fluorescencer-lines of chrysoberyl $BeAl_2O_4$: Cr^{3+}, *J. Lumin.*, *31 and 32*, 272–274, 1984.

LESAR, R., S. A. EKBERG, L. H. JONES, R. L. MILLS, L. A. SCHWALBE, and D. SCHIFERL, Raman spectroscopy of solid nitrogen up to 374 kbar, *Solid State Com.*, *32*, 131–134, 1979.

LIEBENBERG, D. H., A new hydrostatic medium for diamond-anvil cells to 300 kbar pressure, *Phys. Lett.*, *73A*, 74–76, 1979.

MAO, H. K., P. M. BELL, J. W. SHANER, and D. J. STEINBERG, Specific volume measurements of Cu, Mo, Pd, and Ag and calibration of the ruby R_1 fluorescence pressure gauge from 0.06 to 1 Mbar, *J. Appl. Phys.*, *49*, 3276–3283, 1978.

MCCUMBER, D. E., and M. D. STURGE, Linewidth and temperature shift of the *R* lines in ruby, *J. Appl. Phys.*, *34*, 1682–1684, 1963.

MERKLE, L. D., I. L. SPAIN, and R. C. POWELL, Effects of pressure on the spectra and lifetimes of $Nd_xY_{1-x}P_5O_{14}$ and ruby, *J. Phys. C*, *14*, 2027–2038, 1981.

MONTEIL, A., E. DUVAL, A. ATTAR, G. VILIANI, and R. LACROIX, Splitting of the ruby fluorescence under stress, *J. Physique Lett.*, *45*, 1097–1101, 1984.

MUNRO, R. G., A scaling theory of solids under hydrostatic pressure, *J. Appl. Phys.*, *67*, 3146–3150, 1977.

MUNRO, R. G., G. J. PIERMARINI, S. BLOCK, and W. B. HOLZAPFEL, Model line-shape analysis for the ruby *R* lines used for pressure measurement, *J. Appl. Phys.*, *57*, 165–169, 1985.

NELSON, D. F., and M. D. STURGE, Relation between absorption and emission in the region of the *R* lines of ruby, *Phys. Rev.*, *137*, A1117–A1130, 1965.

NICOL, M., K. R. HIRSCH, and W. B. HOLZAPFEL, Oxygen phase equilibria near 298K, *Chem. Phys. Lett.*, *68*, 49–52, 1979.

PIERMARINI, G. J., S. BLOCK, and J. D. BARNETT, Hydrostatic limits in liquids and solids to 100 kbar, *J. Appl. Phys.*, *44*, 5377–5382, 1973.

PIERMARINI, G. J., S. BLOCK, J. D. BARNETT, and R. A. FORMAN, Calibration of the pressure dependence of the R_1 ruby fluorescence line to 195 kbar, *J. Appl. Phys.*, *46*, 2774–2780, 1975.

POWELL, R. C., B. DIBARTOLO, B. BIRANG, and C. S. NAIMAN, Temperature dependence of the widths and position of the *R* and *N* lines in heavily doped ruby, *J. Appl. Phys., 37*, 4973–4978, 1966.

SETCHELL, R. E., Index of refraction of shock-compressed fused silica and sapphire, *J. Appl. Phys., 50*, 8186–8192, 1979.

SHIMOMURA, O., S. YAMAOKA, H. NAKAZAWA, and O. FUKUNAGA, Application of a diamond-anvil cell to high-temperature and high-pressure experiments, in *High-Pressure Research in Geophysics*, edited by S. Akimoto and M. H. Manghnani, pp. 49–60, Center for Academic Publications, Tokyo, 1982.

SPENCER, R. D., and G. WEBER, Measurements of subnanosecond fluorescence lifetimes with a cross-correlation phase fluorometer, *Annals New York Acad. of Sci., 158, Art 1*, 361–376, 1969.

STEPHENS, D. R., and H. G. DRICKAMER, Effect of pressure on the spectrum of ruby, *J. Chem. Phys., 35*, 427–429, 1961.

SUGANO, S., Y. TANABE, and H. KAMIMURA, *Multiplets of Transition-Metal Ions in Crystals*, pp. 106–125, Academic Press, New York, 1970.

STRUVE, B., and G. HUBER, The effect of the crystal field strength on the optical spectra of Cr^{+3} in gallium garnet laser crystals, *Appl. Phys. B, 36*, 195–201, 1985.

TOLSTOI, N. A., and LIU SHUN'-FU, Luminescence kinetics of chromium luminors (IV), Ruby, Part 2. Interpretation of the relaxation spectrum; Temperature variation of the relaxation time and the intensity of the luminescence, *Opt. Spectr., 13*, 224–228, 1962.

WANG, H., M. C. GUPTA, and G. SIMMONS, Chrysoberyl (Al_2BeO_4): Anomaly in velocity-density systematics, *J. Geophys. Res., 80*, 3761–3764, 1975.

WHITMORE, J. E., C. L. BRUZZONE, and R. INGALLS, Automated spectrometer for pressure measurement using ruby fluorescence, *Rev. Sci. Instrum., 53*, 1602–1603, 1982.

WITTKE, J. P., Effects of elevated temperatures on the fluorescence and optical maser action of ruby, *J. Appl. Phys., 33*, 2333–2335, 1962.

XU, J. A., H. K. MAO, and P. M. BELL, High-pressure ruby and diamond fluorescence: Observations at 0.21 to 0.55 terapascal, *Science, 232*, 1404–1406, 1986.

YEN, W. M., and P. M. SELZER, *Laser Spectroscopy of Solids*, Splinger-Verlag, New York, 1981.

X-RAY DIAMOND ANVIL PRESS FOR STRUCTURAL STUDIES AT HIGH PRESSURES AND HIGH TEMPERATURES

T. KIKEGAWA[1]

The Research Institute for Iron, Steel and Other Metals, Tohoku University Katahira Sendai 980, Japan

Abstract. A high-temperature diamond-anvil press which allows X-ray diffraction investigation of polycrystalline materials at high pressures up to 30 GPa and at high temperatures up to 500 °C is described. Several improvements in the heating efficiency, temperature uniformity, and alignment procedure have been made to the anvil press. Preferred orientation effects from the sample have been eliminated. This device has been used to study the Group V*b* elements phosphorus, arsenic, and antimony at high pressure and temperature to determine the stability of the rhombohedral A7 structure with respect to the closely related primitive simple cubic structure.

Introduction

Development of the diamond-anvil press during last two decades has made a great contribution to the progress of high-pressure materials science (e.g., JAYARAMAN, 1983). Combined with X-ray diffraction equipment, the diamond-anvil press becomes the most powerful tool for structural investigations at nonambient pressures. Since the accessible sample volume is usually very small, an extremely long exposure time (i.e., tens or even hundreds of hours) is required with a conventional X-ray diffraction system. However, this method can generate pressures in excess of 10 GPa without difficulty and gives good diffraction patterns.

Recently, attempts have been made to extend high-pressure structural investigations to high temperature. Various modes of sample heating in the diamond-anvil press have been designed; they can be roughly classified into two alternative types, described below.

With the simplest type of design, the sample is heated in a direct manner, without an actual heating element in the diamond-anvil press. For instance, a laser beam can be used to irradiate the sample through the diamond (MING and BASSETT, 1974; JEANLOZ and HEINZ, 1984), or an electric current can be supplied to the sample for heating purposes (BOEHLER et al., 1986). This direct-heating design is applicable to any type of diamond-anvil press, without special modifications. However, with this design, it is difficult to accurately measure the real temperature of the sample because an extremely large thermal gradient occurs across the sample. In addition, the heating efficiency is very dependent on optical or electrical properties of sample materials. Accordingly, this design does not seem to be suitable for in situ X-ray diffraction experiments.

With the other type of design, the sample is heated indirectly by using a heating device attached to the diamond-anvil press. There are various ways in which a heater can be placed in the press. In one design a metal gasket that encloses the sample materials serves as the heater (MOORE et al., 1970), while in a more elaborate design a resistance wire that is wound around a couple of diamond-anvils serves as the heater (SUNG, 1976; HAZEN and FINGER, 1981; MING et al., 1983; BASSETT et al., 1985). Alternatively, a heating element can be placed outside of the cylinder containing the diamond-anvils, mounting plate, and extended piston (WEIR et al., 1965; BARNETT et al., 1973), or outside of the press body (ASAUMI et al., 1980). In this case, the sample is heated uniformly and the heating efficiency is independent of the sample properties. However, there is a problem with the indirect heating method in that other components of the press, besides the diamonds, heat up. With the loss of strength and hardness of the components, it is difficult to maintain a constant sample pressure during the long exposure time. Furthermore, the diamond rapidly oxidizes at temperatures above 600 °C (SUNG, 1976). There have been many reports describing design, construction, and performance of the high-temperature diamond-anvil press, but only a few reports have described its successful application.

It is necessary to shorten the time that the diamond-anvil press is exposed to high temperatures in order to avoid the difficulties described above, as well as irreversible effects to the sample, such as chemical reactions and grain growth. However, it is difficult with a conventional X-ray source to obtain diffraction intensity data within a short period of time for the very small quantities of sample materials used. Recently, highly brilliant X-ray sources, such as rotating anode-type generators and electron storage rings, have become available for general use. Use of these X-ray sources allow the exposure time to be reduced by three or four orders of magnitude.

[1]Present address: Photon Factory, National Laboratory for High Energy Physics, Oho, Ibaraki 305, Japan.

High-Pressure Research in Mineral Physics, edited by M. H. Manghnani and Y. Syono, pp. 61–68.

Therefore it is now possible to collect diffraction intensity data for materials subjected to high pressure and high temperature without the difficulties encountered previously (MING et al., 1983; BASSETT et al., 1985; SCHIFERL et al., 1986).

This author has studied phase transitions in phosphorus with increasing pressure by using a X-ray diffraction system and a diamond-anvil press at ambient temperature (KIKEGAWA and IWASAKI, 1983). Because there is a wide two-phase mixed region, or hysteresis of high and low pressure phases, it is necessary to conduct structural investigations at high temperatures to determine the true phase boundary.

A new modification of the high-temperature diamond-anvil press has been developed to allow X-ray diffraction experiments on polycrystalline materials to be conducted with more ease and diffraction data of improved quality to be obtained.

The basic design and operation of the high-temperature diamond-anvil press have been previously described (KIKEGAWA and IWASAKI, 1983). This paper describes the construction of a new, modified diamond-anvil press and experimental techniques for this press. X-ray diffraction measurements made on Group V*b* elements show that this high-temperature diamond-anvil press is capable of achieving pressures up to 30 GPa at high temperature.

Design and Operation

In the design of the new high-temperature diamond-anvil press, the principal requirements are as follows:

1) During experimental runs the sample must be heated as uniformly and as efficiently as possible.

2) The press body must be compact so that it can easily be mounted on a standard X-ray goniometer head, and if necessary, it must be rotated around the axis perpendicular to the direction of the incident beam.

3) The diamond-anvil faces must be easily aligned from the outside, without disassembling the components of the press.

The basic design of the high-temperature diamond-anvil press is derived from the original design of BASSETT et al. (1967). Figure 1 shows the assembly of the press in two vertical cutaway cross-sections. A pair of diamond-anvils are mounted on the tapered-disk-shaped plate (A) and the hemispherical plate (B), and fixed by using high-temperature ceramic cement such as SAUEREISEN. The former plate (A) can move along a horizontal plane so that the diamond-anvils can be centered with respect to one another. The latter plate (B) can be rotated with the aid of an annular disk (C) in order to align the faces exactly parallel. Alignment of the diamond-anvils can be easily achieved by turning adjusting screws (D) from the outside. Thus requirement 3, described above, is satisfied.

The needle thrust bearing (E) works to reduce friction between the face of the lower piston (F) and that of the driver screw (G). The set screw (H) with the cone-shaped tip fits into a triangular groove machined into the lower piston, thereby preventing rotation of the piston. There is a conical slot through the center of the lower piston and the driver screw, which is a characteristic of our new design; this slot allows the X-ray beam to strike the sample in the gasket at various incident angles. The tapered slot in the upper stationary piston (I) is for receiving the diffraction beam with the maximum 2θ angle of 45°. The press components are machined from a

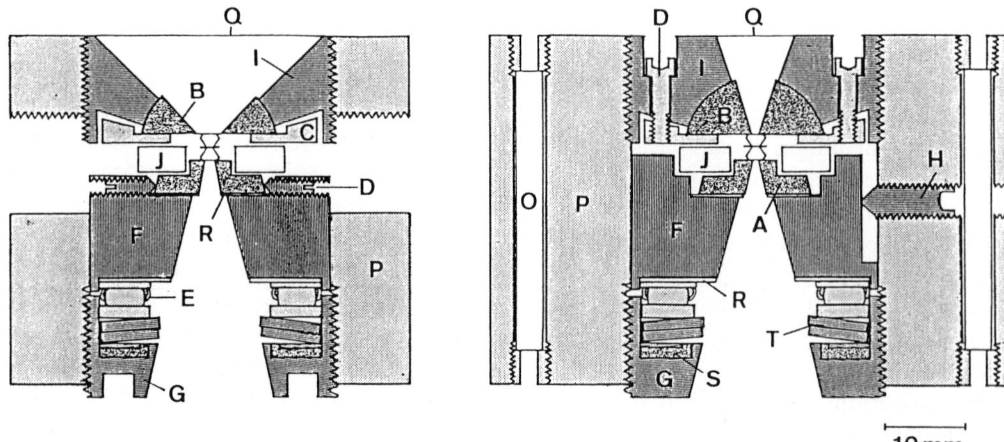

Figure 1. Two different cutaway cross-sections of the high-temperature diamond-anvil press. A) tapered-disk shaped plate, B) hemispherical plate, C) annular disk, D) adjusting screw, E) needle thrust bearing, F) lower sliding piston, G) driver screw, H) set screw, I) upper stationary piston, J) internal heater, O) external heater, P) body, Q) thermal insulating foil, R) mica sheet, S) MACORE plate, T) Belleville washer.

steel, SKH-9, which is hardened to Rockwell C 62 and tempered at 700 °C. The two anvil-supporting plates are made of a sintered tungsten carbide which can maintain hardness up to 700 °C. The severest restriction on the temperature range of the diamond-anvil press, however, is caused by the thermal instability of the diamonds (SUNG, 1976). Diamonds oxidize in air at approximately 600 °C, and graphitize even in an inert gas atmosphere at temperatures as low as 800 °C. The experimental runs are usually carried out at temperatures not exceeding 500 °C without any irreversible effects in the anvil assembly, such as extrusion of the metal gasket.

A heating element (J) is placed between the two anvil-supporting plates. Figure 2 shows the details of this internal heater assembly. Heater (L) is made of Pt-13%Rh allow wire (0.2 mm in diameter and 800 mm in length) in the form of a spiral which is wound in two turns around the hole of the fired pyrophyllite ring (M). The ring (M) fits over the copper rings (K) which hold the diamond-anvils. Heat is conducted efficiently and uniformly from the heater through the copper rings to the diamond-anvils and metal gasket (N).

There is another heater in addition to the internal heater, as shown in Figure 1. This external heater (O) is fixed in the holes drilled vertically at the corners of the press body (P), and it works to compensate the dissipation of heat through the press body. The tapered slot in the upper piston is covered with aluminum and CAPTON foil (Q) in order to suppress thermal radiation and convection from the diamond-anvil face. In addition, 0.2-mm-thick mica sheets (R) and a 1.5-mm-thick MACORE (glass ceramic) insulating plate (S) are put into the circular wells of the lower piston and the driver screw, respectively, to provide thermal insulation for the belleville washers (T). These washers are used to eliminate pressure relaxation due to the thermal expansion of the components.

This press was used to maintain a constant pressure over several weeks at high temperatures and to maintain temperature within ±1 °C over a day at the sample temperature of 500 °C. Simultaneous input of ac power to the internal and the external heaters (40 W and 50 W, respectively, at the sample temperature of 500 °C) improves the heating efficiency and temperature uniformity and stability, because it eliminates the temperature gradient across the anvil and the heat flow through the diamond and press body. Thus requirement 1, described above, is fulfilled by adopting the combined sets of heaters.

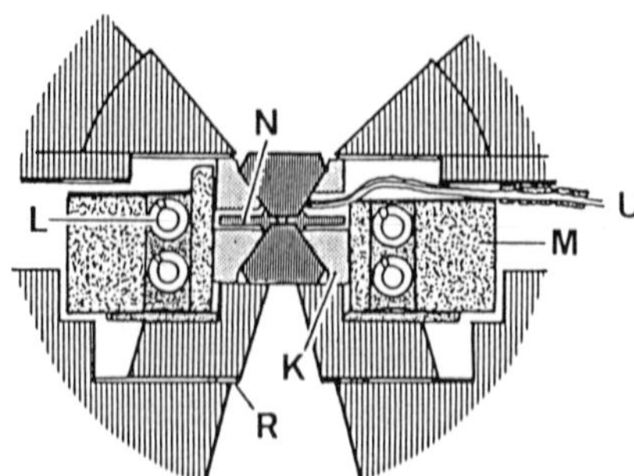

Figure 2. Detailed cutaway cross-view of the internal heater. K) copper ring, L) Pt-13%Rh wire heater, M) phrophyllite ring, N) metal gasket, R) mica sheet, U) alumel-chromel thermocouple.

After the anvil-face alignment (and before sample, gasket, and heating element mounting), the adjustment for precentering the sample in the X-ray beam position is made. Since the press is only 700 g in weight and 50 mm along each edge, one can use the standard adjustment techniques for a single-crystal X-ray diffraction experiment. Furthermore, the instrument required for adjustment can be used interchangeably with various conventional X-ray diffraction equipment. The press is clamped to a standard goniometer head. The center of the sample is positioned to coincide with the vertical axis (i.e., z-axis) perpendicular to the direction of the incident X-ray beam. The axis is at a distance of one half the gasket thickness from the culet face of the upper diamond. This alignment is achieved by moving the press with a translational mechanism of the goniometer head along the beam direction (i.e., y-axis). With increasing pressure, the thickness of the gasket is decreased due to deformation; this results in relative displacement of the sample position with respect to the z-axis and, of course, to the film plate. However, the displacement can be monitored directly by using a depth-micrometer. Thus the effect of the reduction in gasket thickness is corrected from run to run by shifting the goniometer head along the y-axis. After all of the components are set up, the press is moved in the other direction (i.e., the x-axis) so that the incident beam exactly impinges on the center of the sample. The alignment along the z-axis must be made after the sample has been heated to a predetermined temperature, since any appreciable change in temperature causes a shift of the sample position. If necessary, the press is rotated around the z-axis to the maximum angle of ±15 ° (determined by the opening of the slot of the lower piston) during measurement so that more crystallites in the sample have a chance to diffract X-rays. This rotation capability meets requirement 2 described above, and can reduce preferred orientation and grain growth effects. The temperature is measured using an alumel-chromel thermocouple (0.05 mm in diameter). The thermocouple junction is glued to a side face of the lower diamond-anvil. The temperature measured with this arrangement does not necessarily represent the true sample temperature. Thus a calibration must be made at ambient pressure either by inserting another thermocouple head into the gasket hole (the sample space is filled with sodium chloride pressure-transmitting medium) or by replacing the sample with the calibrant materials of known transition temperatures. The temperature difference is within only ±10 °C at the sample temperature of 500 °C.

The pressure generated by turning the driver screw is measured using a mixture of sodium chloride and the sample as the internal standard, and Decker's equation of state (DECKER, 1971); this method of measurement is effective in the pressure range 0–30 GPa and the temperature range 0–800 °C. Since there are few reliable pressure standards available at elevated temperature (MING et al., 1983) the practical limit of pressure is thus determined. With increasing temperature, the sample pressure is often partially released by thermal expansion of the components of the press. The magnitude of the pressure release can be reduced by using the belleville washers (T). However, if the pressure release is too large, the press may require additional compression while the sample is kept at high temperature. It is necessary to adopt such a complicated procedure for isobaric experiments, but not for isothermal experiments.

This high-temperature diamond-anvil press has been used for a variety of X-ray diffraction studies of several polycrystalline materials. The X-ray diffraction patterns are usually recorded on a flat film, but a solid state detector can also be used for a direct counting method. The incident beam, mostly filtered Mo $K\alpha$ radiation from a rotating anode type generator (usually 55 kV, 400 mA), is collimated to form a beam 0.1 or 0.2 mm in cross-sectional diameter. The exposure time of 5–40 h is needed to obtain good diffraction photographs, depending on the scattering power of the sample.

Applications

This high-temperature diamond-anvil press has been used for a variety of X-ray diffraction studies. It has been used in studies of the phase transitions and stabilities of the Group V*b* elements (phosphorus, arsenic, and antimony) to observe the similarities and differences among the high-pressure crystallographic behavior of these elements.

This author has investigated the two pressure-induced phase transitions of phosphorus (black phosphorus) at room temperature (KIKEGAWA and IWASAKI, 1983), which were previously reported by JAMIESON (1963). At low pressure, phosphorus has an orthorhombic structure containing a puckered layer of atoms. At 5.5 GPa, it changes into a rhombohedral structure; this form is of the A7 type found to be common in the other members of the Group V*b* elements at normal pressure. It is stable only in a limited range of pressures. At 10 GPa, its structure becomes the primitive simple cubic form; though this structure is the simplest arrangement of atoms in a crystalline solid, it has been reported to exist only in the low-temperature phase of polonium (DESANDO and LANGE, 1966) and high-pressure phase of antimony (KABALKINA et al., 1964; VERESHCHAGIN and KABALKINA, 1965; KOLOBYANINA et al., 1969). Our study has added accurate structural data and revealed that, in spite of a non-dense packing form, the simple cubic structure remains stable with further increase in pressure; no transition was found to occur up to a pressure of 32 GPa. Another recent investigation by OKAJIMA et al. (1984) has shown that the simple cubic structure is stable up to 70 GPa. Thus it is of interest to see how these transitions and the stability of the simple cubic structure are affected with an increase in temperature.

Pressure-induced structural phase transition in antimony has been investigated by many researchers (KABALKINA et al., 1964; VERESHCHAGIN and KABALKINA, 1965; KOLOBYANINA et al., 1969; KABALKINA et al., 1970; MCWHAN, 1972; SCHIFERL et al., 1981; AOKI et al., 1983). With increasing pressure, the transition sequence is as follows: the rhombohedral form, the simple cubic form, a form with a undetermined structure, and the body-centered cubic form. However, conflicting results have been obtained on the formation of the simple cubic structure at high pressure. KOLOBYANINA et al. (1969) reported that there exists a simple cubic form in the pressure range of 7–8.8 GPa. However, MCWHAN (1972) and SCHIFERL et al. (1981) did not confirm the experimental result of Kolobyanina et al. Since all of these X-ray diffraction studies were carried out at room temperature, the discrepancies in the results may be due to the sluggishness of the transition involved. Also in such compression runs, the formation of the high pressure phase could be missed. If the temperature was raised, atom mobility would be enhanced and one could investigate whether or not antimony, like phosphorus, can take the simple cubic structure as a stable form. The main purpose of applying the high-temperature diamond-anvil press to antimony is to establish a phase relation for the pressure-temperature diagram, with particular emphasis on the existence or nonexistence of the simple cubic phase.

There are only a few papers dealing with the influence of pressure on the crystal structure of arsenic (MOROSIN and SCHIRBER, 1972; MCWHAN, 1972; DUGGIN, 1972). DUGGIN (1972) reported that the high-pressure form of arsenic, quenched from a pressure of 15 GPa, can be classified as a tetragonal lattice. He identified this structure using a low-temperature superconducting phase which had previously been observed at 14 GPa (BERMAN and BRANDT, 1969). However, the identification of this structure is not fully convincing, because there was no in situ crystallographic evidence of it for the high-pressure phase transition of arsenic. Furthermore, a more recent measurement (WITTIG, 1984) suggests that the superconducting phase transition occurs at 20 GPa. The purposes of the present study on arsenic are: 1) to investigate the crystal structure of arsenic under compression, with the emphasis on confirming the existence of the tetragonal phase (DUGGIN, 1972); 2) if the tetragonal phase confirmed, to investigate its stability in a much wider range of pressures than explored in previous studies; 3) to provide reliable crystallographic data and transition pressure; and 4) to investigate the influence of temperature on the crystal structure.

Phosphorus

In a previous work (KIKEGAWA and IWASAKI, 1983), the pressure of the orthorhombic to rhombohedral transition at ambient temperature was determined to be 5.5 GPa. Because of the sluggishness of the transition, there was always a pressure region in which both the high and low pressure phases coexisted. Thus the transition pressure was determined to be the pressure at which the volume fractions of the two phases were nearly the same.

When temperature is raised, the pressure range of the two-phase region becomes narrow. All of the diffraction lines recorded at 3.7 GPa and 340 °C are identified to be those of the orthorhombic structure. A slight increase in pressure results in a complete change in the diffraction pattern. This fact indicates that the transition at 340 °C takes place at a pressure lower than that at room temperature, and that the atoms have sufficient mobility to arrange themselves quickly into the high-pressure form.

At room temperature, the transition from the rhombohedral to simple cubic form is not accompanied by an appreciable hysteresis and is observed to occur at 10 GPa. Figures 3a, b show X-ray diffraction patterns taken at 8.9 GPa and 270 °C and at 9.6 GPa and 270 °C, respectively. A doublet of the diffraction lines 10·4 and 11·0 (see Figure 3a), which is characteristic of the rhombohedral structure, merges into a single line with a pressure increase to 9.6 GPa (see Figure 3b); this indicates an increase in the rhombohedral axial angle and formation of the simple cubic structure. Further increase in pressure to 20 GPa at 270 °C does not result in a change in the diffraction pattern. As shown in Figure 3c, the increase in temperature to 320 °C does not have any effect on the structure of the high-pressure form.

Results of runs conducted at different temperatures and pressures have determined a pressure-temperature phase diagram of phosphorus, as shown in Figure 4.

The most interesting feature of this phase diagram is that the simple cubic form occupies a wide area. There are no other elements which keep the simple cubic structure unchanged with varying pressure and temperature.

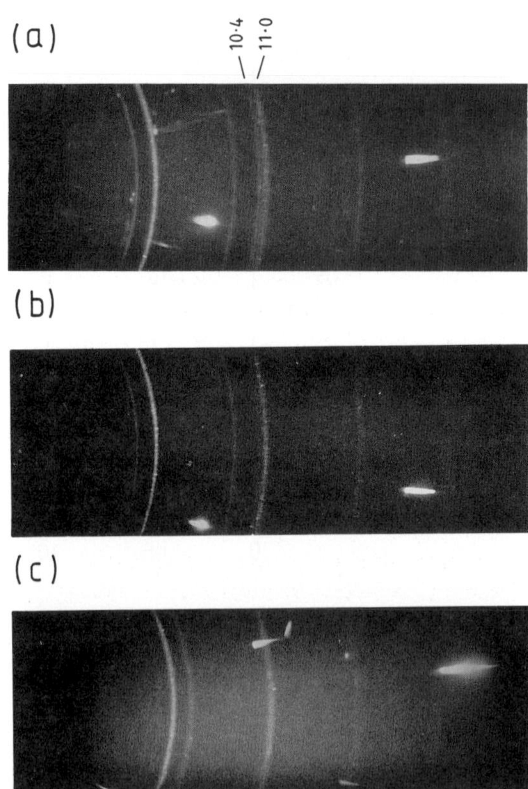

Figure 3. X-ray diffraction patterns of phosphorus. a) the rhombohedral form before transition, 8.9 GPa and 270 °C, b) the simple cubic form after transition, 9.6 GPa and 270 °C, c) the simple form, 29 GPa and 320 °C.

Antimony

Room temperature compression experiments were carried out. No indication of the transition was detected at pressures beyond 7.0 GPa, the pressure at which KOLOBYANINA et al. (1969) observed a transition. Although the rhombohedral axial angle increased continuously with increasing pressure, the doublet (10·4 and 11·0) did not become a single line. The same was true for the other doublet (11·6 and 12·2). The line 01·5, which should have vanished upon transition into the simple cubic structure, remained with appreciable intensity. At 9.5 GPa, a new diffraction pattern which is identical to that of the high-pressure form appeared; this pattern is designated as the second (post simple cubic) high-pressure phase by KABALKINA et al. (1970) and is interpreted to be a tetragonal phase.

The sample temperature was increased to 150–170 °C, and compression runs were made. Figure 5a shows a diffraction pattern taken at 6.4 GPa and 170 °C. A comparison of Figure 5 and Figure 3 indicates a decrease in the rhombohedral distortion of the lattice. However, a

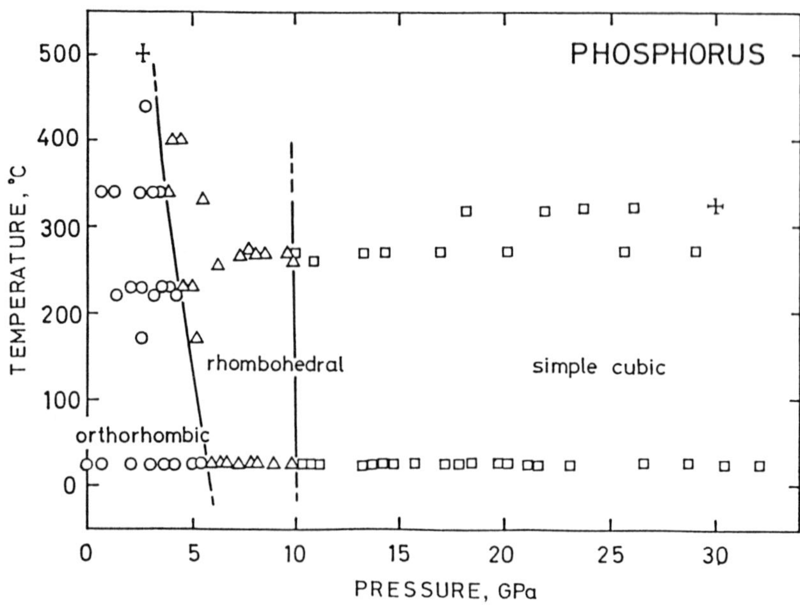

Figure 4. Pressure-temperature phase diagram of phosphorus, two alternative phase boundaries lying in low pressure region are confirmed by using synchrotron radiation and MAX80 at Photon Factory (KIKEGAWA et al., 1986).

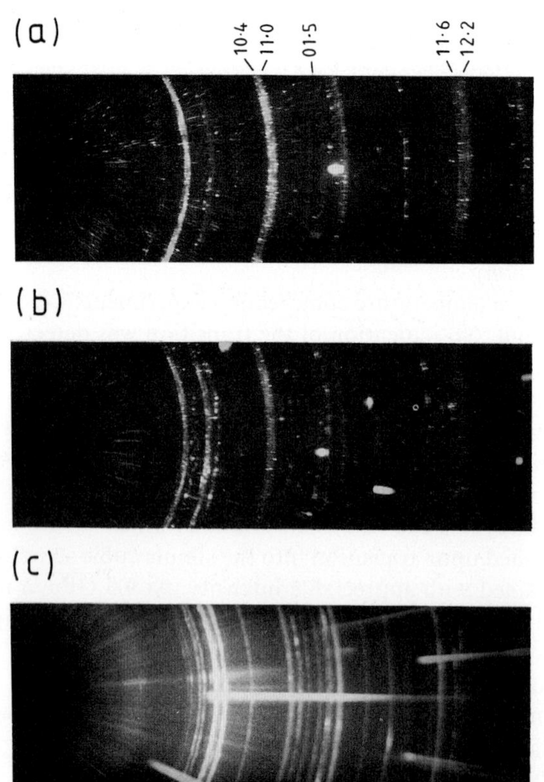

Figure 5. X-ray diffraction pattern of antimony a) rhombohedral form before transition, 6.4 GPa and 170 °C, b) rhombohedral form plus tetragonal form, 6.7 GPa and 170 °C, c) tetragonal form after transition, 11.5 GPa and 150 °C.

close inspection of the pattern shows that there is a small but recognizable splitting of the lines (for example, lines 11·6 and 12·2). The rhombohedral angle is calculated to be 59 °, a little lower than the 60 ° angle corresponding to the cubic lattice. A slight increase in pressure yielded the pattern shown in Figure 5b. One can see that new diffraction lines characteristic of the high-pressure tetragonal phase appear, while the splitting of the lines remains. Additional high-temperature compression experiments have been carried out at 320 °C; the transition into the tetragonal phase was first detected at 6.0 GPa and was almost complete at 8.0 GPa. Similar observations were made at different temperatures. In no case was the formation of the simple cubic structure observed. It can be concluded that, although the distortion of the rhombohedral structure continuously decreases, it does not go to zero and the cubic structure does not form in antimony. Figure 5c shows the X-ray diffraction pattern of antimony at 11.5 GPa and 150 °C. As already has been seen in Figures 5a, b, the reflections do not appear as continuous lines due to grain growth at high temperatures. The pattern in Figure 5c was taken by rotating the sample; the coarse grain effect has considerably been eliminated. Such an improvement enables one to identify a phase and to measure the *d*-values and intensities of diffraction lines. These lines are newly indexed in terms of a tetragonal cell of a=8.09 A, b=3.92 A with ten atoms contained. With measured intensity data, precise atomic positional parameters can be determined (Iwasaki and Kikegawa, to be

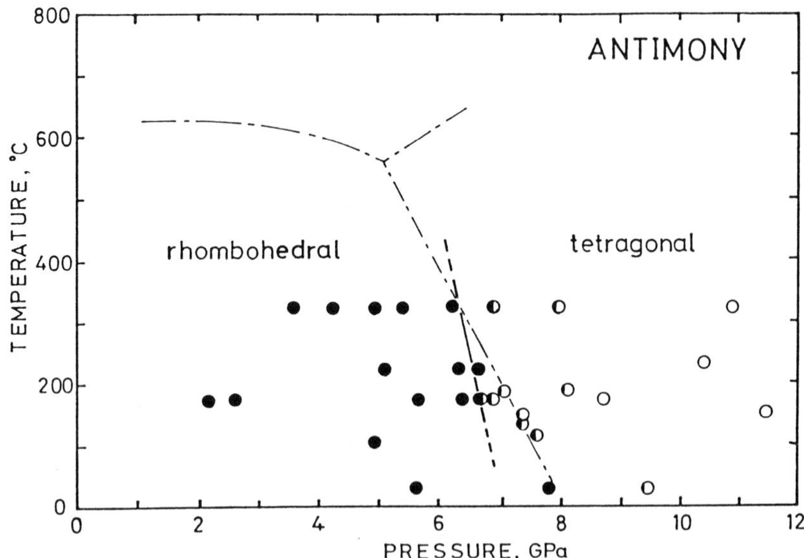

Figure 6. Part of the pressure-temperature phase diagram of antimony. Data points shown by circles are those obtained by the present X-ray diffraction measurements and the dash-dotted lines represent the boundaries determined by KHVOSTANTEV and SINDOROV (1984).

published). Figure 6 shows a part of the pressure-temperature diagram of antimony based on the present X-ray measurements which indicates the nonexistence of the simple cubic phase between the low-pressure rhombohedral phase and the high-pressure tetragonal phase. The conclusion drawn here is in agreement with results of recent thermopower and electrical resistance measurements and differential thermal analysis by KHVOSTANTEV and SIDOROV (1984). The dash-dotted lines in the Figure 6 represent the phase boundaries determined by Khvostantev and Sidorov.

Arsenic

Compression experiments have been conducted up to 26 GPa at room temperature. The X-ray diffraction pattern of arsenic remains unchanged except for the decrease in the distortion of the rhombohedral lattice with increasing pressure; a similar trend has been observed for the other Group V*b* elements. However, does not reach the critical value corresponding to the conversion into the simple cubic structure. There is no indication of a supposed transition at pressures around 15 GPa or 20 GPa. To eliminate problems related to the sluggishness of atomic movements at ambient temperature, high-temperature experiments were carried out around these pressures. However, the same results were obtained. This is in contrast to the behavior of the A7 form of the other V*b* elements; the A7 form is stable only in a limited range of pressures. From the measured lattice parameters, the volume of rhombohedral arsenic, normalized to that at

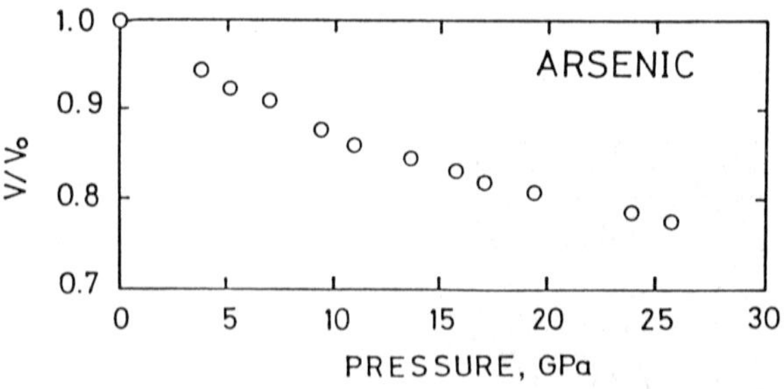

Figure 7. Volume of rhombohedral arsenic, normalized to that at 0.1 MPa, plotted against pressure.

0.1 MPa, has been calculated as a function of pressure (Figure 7).

Acknowledgments. The author wishes to thank H. Iwasaki of the Photon Factory, National Laboratory for High Energy Physics, for his continuous encouragement and critical review of this paper.

REFERENCES

AOKI, K., S. FUJIWARA, and M. KUSAKABE, New phase transition into the b.c.c. structure in antimony at high pressure, *Solid State Commun.*, *45(2)*, 161–163, 1983.

ASAUMI, K., S. KOJIMA, and T. NAKAMURA, Effect of the hydrostatic pressure on the ferroelastic NdP_5O_{14}, *J. Phys. Soc. Jpn.*, *48*, 1298–1306, 1980.

BARNETT, J. D., S. BLOCK, and G. J. PIERMARINI, An optical fluorescence system for quantitative pressure measurement in the diamond anvil cell, *Rev. Sci. Instrum.*, *44(1)*, 1–9, 1973.

BASSETT, W. A., T. TAKAHASHI, and P. W. STOCK, X-ray diffraction and optical observations on crystalline solids up to 300 kbar, *Rev. Sci. Instrum.*, *38(1)*, 37–42, 1967.

BASSETT, W. A., M. D. FURNISH, and E. HUANG, Applications of synchrotron radiation in high pressure-temperature mineralogy, in *Solid State Physics Under Pressure: Recent Advance with Anvil Devices*, edited by S. Minomura, pp. 335–341, KTK Scientific Publishers, Tokyo, 1985.

BERMAN, I. V., and N. B. BRANDT, Superconductivity of arsenic at high pressure, *JETP Lett.*, *10*, 55–57, 1969.

BOEHLER, R., M. NICOL, C. S. ZHA, and M. L. JOHNSON, Resistance heating of Fe and W in diamond-anvil cell, *Physica*, *139 & 140B*, 916–918, 1986.

DECKER, D. L., High-pressure equation of state for NaCl, KCl and CsCl, *J. Appl. Phys.*, *42(8)*, 3239–3244, 1971.

DESANDO, R. J., and R. C. LANGE, The structure of polonium and its compounds-1, *J. Inorg. Nucl. Chem.*, *28*, 1837–1846, 1966.

DUGGIN, M. T., A high-pressure phase in arsenic and its relation to pressure-induced phase changes in group 5B elements, *J. Phys. Chem. Solids*, *33*, 1267–1271, 1972.

HAZEN, R. M., and L. W. FINGER, High-temperature diamond anvil cell for single-crystal studies, *Rev. Sci. Instrum.*, *52(1)*, 75–79, 1981.

JAMIESON, J. C., Crystal structure adopted by black phosphorus at high pressures, *Science*, *139*, 1291–1292, 1963.

JAYARAMAN, A., Diamond anvil cell and high pressure physical investigation, *Rev. Modern Phys.*, *55(1)*, 65–104, 1983.

JEANLOZ, R., and D. L. HEINZ, Experiments at high temperature and pressure: laser heating through the diamond cell, *J. Phys. Colloque C8 Suppl. 11*, *45*, 83–92, 1984.

KABALKINA, S. S., L. F. VERESHAGIN, and V. P. MYLOV, Phase transitions in antimony at high pressures, *Sov. Phys. DOKLADY*, *8(9)*, 917–918, 1964.

KABALKINA, S. S., T. N. KOLOBYANINA, and L. F. VERESHCHAGIN, Investigation of the crystal structure of the antimony and bismuth high pressure phase, *Sov. Phys. JETP*, *31(2)*, 259–263, 1970.

KHVOSTANTSEV, L. G., and V. A. SIDOROV, Phase transitions in antimony at hydrostatic pressure up to 9 GPa, *Phys. Stat. Sol. (a)*, *82*, 389–398, 1984.

KIKEGAWA, T., and H. IWASAKI, An x-ray diffraction study of lattice compression and phase transition of crystalline phosphorus, *Acta Cryst.*, *B39*, 158–164, 1983a.

KIKEGAWA, T., and H. IWASAKI, X-ray diamond-anvil press for use at high temperatures, *Rev. Sci. Instrum.*, *54(8)*, 1023–1025, 1983b.

KIKEGAWA, T., H. IWASAKI, T. FUJIMURA, S. ENDO, Y. AKAHAMA, T. AKAI, O. SHIMOMURA, T. YAGI, S. AKIMOTO, and I. SHIROTANI, Synchrotron radiation study of phase transitions in phosphorus at high pressure and temperature, *J. Appl. Cryst.*, in press, 1987.

KOLOBYANINA, T. N., S. S. KABALKINA, L. F. VERESHCHAGIN, and L. V. FEDINA, Investigation of the crystal structure of antimony at high pressures, *Sov. Phys. JETP*, *28(1)*, 88–90, 1969.

MCWHAN, D. B., The pressure variable in materials research, *Science*, *176*, 751–758, 1972.

MING, L. C., and W. A. BASSETT, Laser heating in the diamond anvil press up to 2000 °C sustained and 3000 °C pulsed at pressures up to 260 kilobars, *Rev. Sci. Instrum.*, *45(9)*, 1115–1118, 1974.

MING, L. C., M. H. MANGHANI, S. B. QADRI, E. F. SKELTON, J. C. JAMIESON, and J. BOLOGH, Gold as a reliable internal pressure calibrant at high temperatures, *J. Appl. Phys.*, *54(18)*, 4390–4397, 1983.

MOORE, M. J., D. B. SORENSEN, and R. C. DEVAIES, A simple heating device for diamond anvil high pressure cell, *Rev. Sci. Instrum.*, *41*, 1665–1666, 1970.

MOROSIN, B., and J. E. SCHIRBER, Linear compressibilities and the pressure dependence of the atomic positional parameter of As, *Solid State Commun.*, *10*, 249–251, 1972.

OKAJIMA, M., S. ENDO, Y. AKAHAMA, and S. NARITA, Electrical investigation of phase transition in black phosphorus under high pressure, *Jpn. J. Appl. Phys.*, *23(1)*, 15–19, 1984.

SCHIFERL, D., D. T. CROMER, and J. C. JAMIESON, Structure determinations on Sb up to 85×10^2 MPa, *Acta Cryst.*, *B37*, 807–810, 1981.

SCHIFERL, D., A. I. KATZ, R. L. MILLS, L. C. SCHMIDT, C. VANDERBORGH, E. F. SKELTON, W. T. ELAM, A. W. WEBB, S. B. QADRI, and M. SCHAEFER, A novel instrument for high-pressure research at ultra-high temperatures, *Physica*, *139 & 140B*, 897–899, 1986.

SUNG, C. M., New modification of the diamond anvil press: a versatile apparatus for research at high pressure and high temperature, *Rev. Sci. Instrum.*, *47(11)*, 1343–1346, 1976.

VERESHCHAGIN, L. F., and S. S. KABALKINA, Phase transitions in antimony at high pressures, *Sov. Phys. JETP*, *20(2)*, 274–277, 1965.

WEIR, C., S. BLOCK, and G. PIERMARINI, Single-crystal X-ray diffraction at high pressure, *J. Res. NBS*, *69C(4)*, 275–281, 1965.

WITTIG, J., Superconductivity in elements and binary systems at high pressure, in *High Pressures Science and Technology*, Proc. 9th AIRAPT International High Pressure Conference, 1983, Part 1, edited by C. Homan et al., pp. 17–28, North-Holland, New York, 1984.

RESISTIVE HEATING IN THE DIAMOND-ANVIL CELL UNDER VACUUM CONDITIONS

L. C. MING, M. H. MANGHNANI, and J. BALOGH

Hawaii Institute of Geophysics, University of Hawaii
Honolulu, Hawaii 96822, USA

Abstract. Modifications of various parts of the high-temperature diamond-anvil cell improved the mechanical stability and the overall performance of the cell. Resistive-wire heating of the cell in 10^{-4}–10^{-5} torr vacuum substantially increased its heating efficiency such that sample temperatures of up to ~1000 °C can be reached at a low input power (about 120 W). With synchrotron radiation, high-quality, energy-dispersive X-ray diffraction spectra were obtained from samples in the improved cell at intervals of 2–5 min and temperatures up to ~1000 °C. High-temperature spectra for two γ-$(Mg,Fe)_2SiO_4$ spinels demonstrate the potential applications of using the improved cell for studying the geophysically important mineral phases under in situ high pressure-temperature conditions.

Introduction

Diamond-anvil cell (DAC) has proven to be one of the most useful and versatile tools for probing various physical, optical, and crystal structure properties under high pressure (JAYARAMAN, 1983; BASSETT, 1979). It is now possible to attain static pressures of 5.5 Mbar (XU et al., 1986) and temperatures of up to 7500 K with a high power YAG laser (HEINZ and JEANLOZ, 1985; BASSETT et al., 1985). For high-pressure X-ray diffraction studies, the extremely small sample in the cell necessitates a long exposure time (20–300 h), making it difficult to maintain the pressure-temperature (P-T) conditions during a typical run. So far no X-ray diffraction study under simultaneously high P-T conditions using the laser heating has been reported. One of the problems is the very small size of laser-heated hot spot (10–30 m in diameter) as compared with that of the collimated X-ray beam (>50 μm in diameter). Another problem is the nonuniform temperature distribution inside the small hot spot itself (HEINZ and JEANLOZ, 1985). Recently a 10-μm beam from a synchrotron radiation source has been successfully employed in high-pressure studies at room temperature (VOHRA et al., 1986). However, the present state of the art is not yet sufficiently developed to conduct in situ high P-T studies with laser-heated DAC and synchrotron radiation.

The resistively heated DAC has been successfully used for several X-ray diffraction studies under simultaneously high pressure and temperature to about 20 GPa and 600 °C using synchrotron radiation (MING et al., 1983; FURNISH and BASSETT, 1983; MING et al., 1984; MANGHNANI et al., 1985). The most difficult problem in heating the sample externally at temperatures higher than 600 °C is the oxidation and/or the creeping of various metal components of the cell. SCHIFERL et al. (1986, this volume) has designed a new cell overcoming some of these problems.

A number of improvements in DAC have been made to the present cell which have enabled us to heat the sample inside the cell to about 1000 °C. In this paper, we describe these improvements and report some preliminary data on the energy-dispersive X-ray diffraction (EDXRD) measurements on two γ-$(Mg,Fe)_2SiO_4$ spinel samples to about 1000 °C using synchrotron radiation.

Experimental Methods

The experiment was carried out in the modified DAC at the Stanford Synchrotron Radiation Laboratory. The experimental method for the high-temperature EDXRD measurements have been described previously (MING et al., 1983).

High-Temperature Diamond-Anvil Cell

The DAC employed in in situ high P-T studies is a lever type (Figure 1) in which two plates are hinged together by a pivot pin at one end and a spring-loading mechanism at the other end. A piston-cylinder assemblage is positioned in the middle. This type of cell was first developed by BASSETT and TAKAHASHI (1965) for both X-ray and optical observations under pressure to about 300 °C. Later, SUNG (1976) modified this cell for his optical observations of the olivine-spinel transition in Fe_2SiO_4 to 30 GPa and 600 °C. Improvements have recently been made on this type of cell in order to carry out in situ high P-T X-ray diffraction studies (MING et al., 1983, 1984). Further modifications of the cell have now been made for augmenting its P-T capabilities and improving overall performance. Major features of the improved cell are shown in Figure 2 and described below:

1. *Inconel pistons and cylinder.* One of the major problems encountered with the initial design of the DAC was the deformation of the stainless steel piston and cylinder during prolonged heating. Difficulties were

High-Pressure Research in Mineral Physics, edited by M. H. Manghnani and Y. Syono, pp. 69–74.

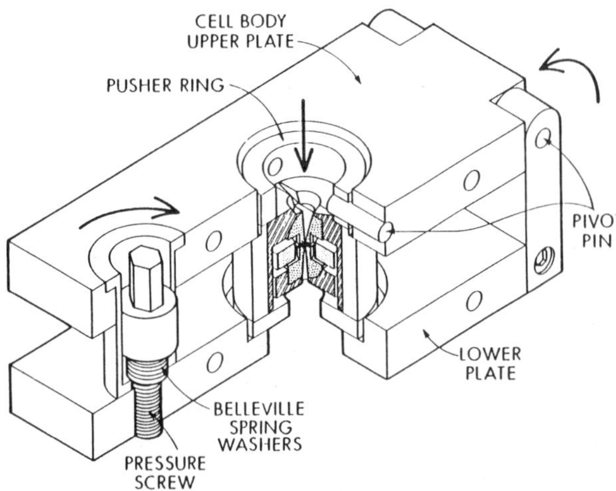

Figure 1. An overview of the improved diamond-anvil cell used for the energy dispersive X-ray studies with synchrotron radiation.

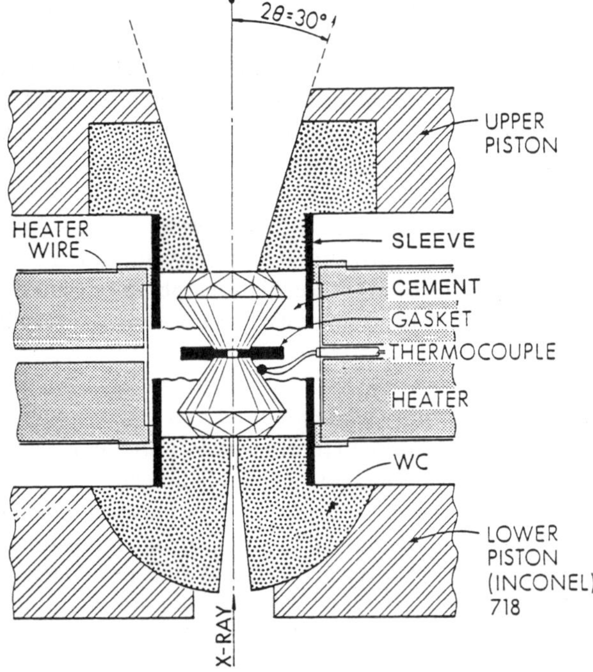

Figure 2. Cross-section in the middle part of the cell showing details of assemblage of pistons, rocker seats, Inconel sleeve, and miniheater.

experienced while both varying the pressure and disassembling the cell after the run. This situation has been greatly improved by replacing the stainless steel pistons and cylinder with those made of a high-temperature alloy (Inconel 718). Also the length of the moving upper piston has been increased by about 60% which has resulted in a far more stable alignment.

2. *Rocker-seat assemblage.* There are two semicylindrical rocker-seat assemblages: one is located on the moving upper piston, and the other is located on the stationary lower piston in a position perpendicular to the first one. The rocker seats are made of tungsten carbide and serve as strong back supports for the diamond anvils. The seat of the lower piston has a 150-μm hole. The seat on the upper moving piston has a conical slot with 30° angle. By virtue of the synchrotron radiation beam being highly collimated, the small hole in the rocker-seat serves advantageously as a collimator and, in conjunction with the slit system outside the radiation hutch, controls the incident beam size. Thus a separate collimator is not needed.

The rocker-seat used previously consisted of two separate WC pieces glued together with a high-temperature cement (Sauereisen P1). One disadvantage of such two-piece seat is the possibility of the failure of the high-temperature cement during unloading. The improved configuration of the rocker-seat which was fabricated as a single piece of WC is shown in Figure 2.

3. *Sleeves for the diamond anvils.* A common problem encountered in using the high P-T cell is that the anvils which are cemented to the rocker-seats tend to pop out during unloading and disassembling of the cell, requiring the realignment of the anvils. To secure the anvils firmly, Inconel sleeves with a height approximately half that of the anvils, are provided around the seats and the anvils. The space between the anvil and the sleeve is filled with high-temperature cement (see Figure 2).

4. *Heating in vacuum.* The miniheater employed in the present study is shown in Figure 3. It is composed of two parts: 1) the base fabricated from talc, heat-treated at 1200 °C for 2 h; and 2) the heating element, which is wired through the talc base with 0.254-mm diameter Pt13%Rh wire. The miniheater has been found to perform very well over prolonged periods of heating (several days) at temperatures of up to 600 °C. However, heating the sample to temperatures higher than 600 °C in air causes serious problems with the WC rocker-seat. WC is readily oxidized as evident from a soft yellowish green layer occurring uniformly around the WC pieces. This soft layering has been analyzed unambiguously by powder X-ray diffraction, using a Deby-Scherrer camera to be $CoWO_4$. It appears that during the oxidation process the binder (i.e., Co) in the carbide combines with W and oxygen to form $CoWO_4$. The formation of $CoWO_4$ not only weakens the WC-rocker seats but also expands the volume to such an extent that the 150-μm hole in the WC seat underneath the lower diamond anvil was completely blocked. One way to prevent such oxidation is to heat the sample either in vacuum or in an inert atmosphere. We have chosen the former. A preliminary test of heating the sample up to ~1000 °C in a vacuum chamber (10^{-4}–10^{-5}

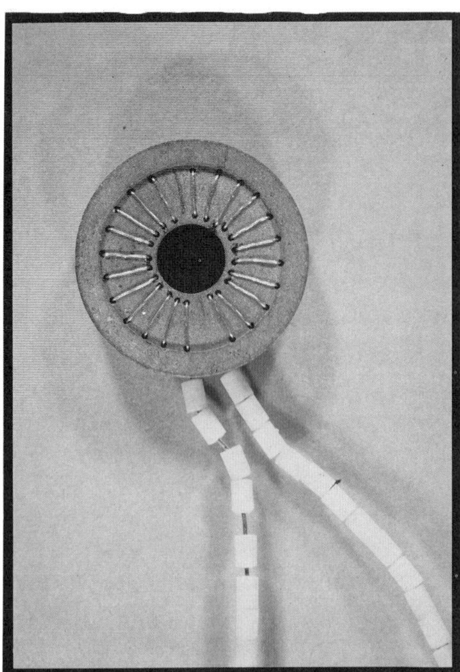

Figure 3. An overview of the miniheater used for heating the sample in the diamond-anvil cell.

torr) was successfully carried out. Heating in a vacuum chamber not only prevented the WC rocker seats from oxidation but also significantly improved the heating efficiency inside the cell. A comparison between the power consumption for heating the sample in air and in a vacuum is made in Figure 4. Two points to be noted are: 1) at a given input power, heating in a vacuum attains much higher temperatures than heating in air (e.g., for an input power of 50 W, the temperature for vacuum heating is higher by a factor of 2 (see Figure 4)); and 2) the high efficiency of the vacuum heating makes it feasible to heat the sample to 1000 °C or higher.

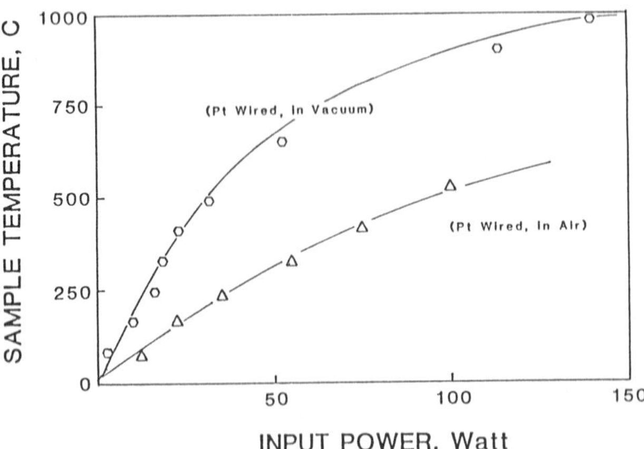

Figure 4. Comparison of the heating efficiency in terms of the input power and the sample temperature reached between the condition in air and that in a vacuum chamber of 10^{-4}–10^{-5} torr using the same Pt wired miniheater.

Temperature Calibration

X-ray diffraction data and the well established thermal expansivity data of NaCl were used in the present study to calibrate the miniheater's thermocouple. NaCl powder was packed loosely into the gasket hole (i.e., 0.4 mm in diameter and 0.25 mm thick) between the two anvils. At each increment of the temperature, the X-ray diffraction pattern of NaCl was acquired when thermal equilibrium was reached. The temperature of the sodium chloride, derived from diffraction data and the published data (AMERICAN PHYSICS HANDBOOK, 1972), was plotted against the thermocouple temperature readout. The calibration thus established up to ~450 °C on the basis of two separate runs is given in Figure 5 and shows clearly the linear relationship between two temperatures. The sample temperatures thus obtained from this calibration are believed to be accurate within ±5 °C. This temperature calibration is also in good agreement with that obtained by the two-thermocouple method described earlier (MING et al., 1983).

Experimental Results

Energy-dispersive X-ray diffraction measurements on two samples of γ-$(Mg,Fe)_2SiO_4$ spinels were successfully

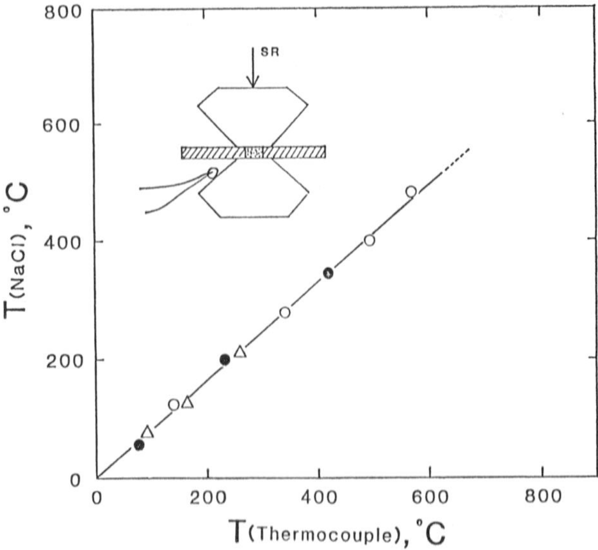

Figure 5. Temperature calibration using NaCl where the circles and triangles represent two different runs. Open and solid circles are obtained by increasing and decreasing temperatures, respectively.

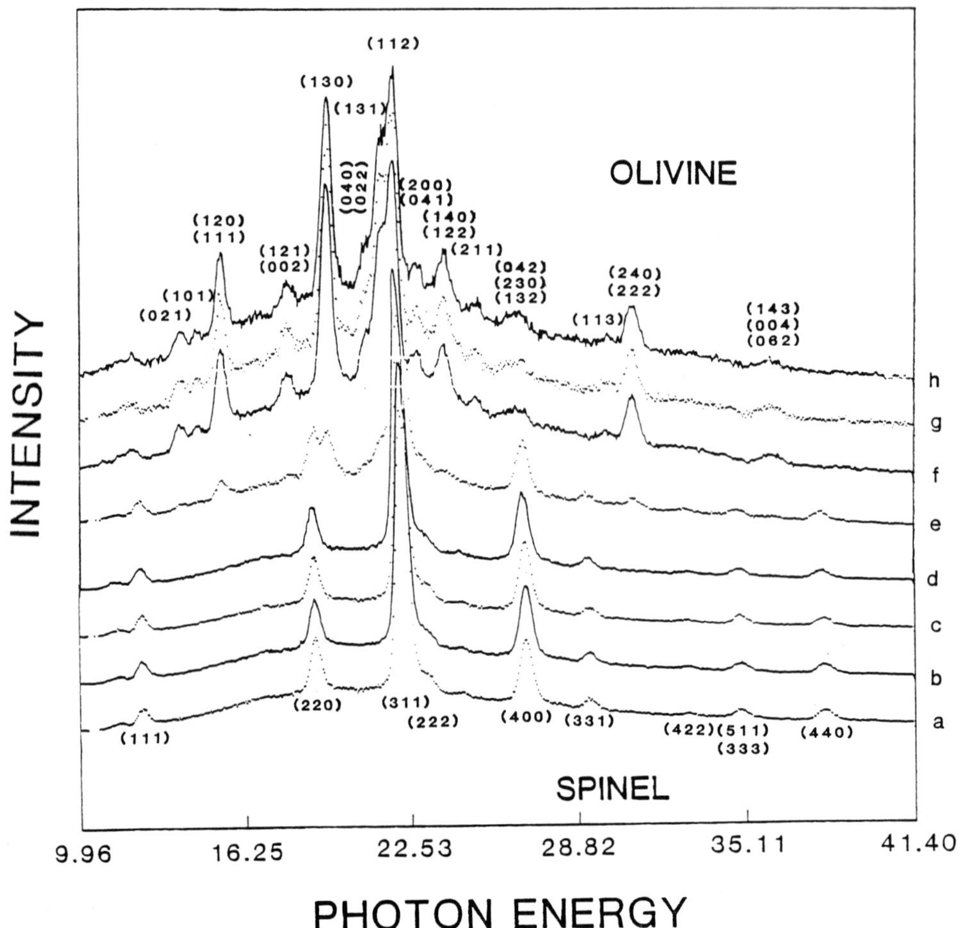

Figure 6. Energy dispersive X-ray diffraction of γ-$(Mg_{0.4}Fe_{0.6})_2SiO_4$ at 1 bar and at temperatures (in °C) of 330 (a), 408 (b), 490 (c), 570 (d), 650 (e), 740 (f), 740 (g, 2 min after f), and 740 (h, 2 min after g). It shows that the back-transformation of $\gamma \rightarrow \alpha$ starts between 575 °C and 650 °C and completes the conversion in ~5 min at 740 °C.

carried out to ~1000 °C at 1 bar using our improved high P-T cell interfaced with the synchrotron radiation at the Stanford Synchrotron Radiation Laboratory (SSRL). With the synchrotron storage ring operating at 3 GeV and 40 mA, a high quality spectrum could be obtained in 2–5 min. Figures 6 and 7 show the spectra at several increasing temperatures of up to 740 °C for the γ-$(Mg_{0.4}Fe_{0.6})_2SiO_4$ and γ-$(Mg_{0.8}Fe_{0.2})_2SiO_4$ spinels, respectively. These diffraction patterns demonstrate a back-transformation from the γ-spinel structure to α-olivine structure at higher temperatures. This is expected because in the solid solution system $(Mg,Fe)_2SiO_4$, the α phase is a stable form at pressures below 6 GPa and at all temperatures up to the melting points, whereas the γ phase is stable at high pressures in the range of 6–15 GPa (as in Fe_2SiO_4) and of 12–25 GPa (as in Mg_2SiO_4) (RINGWOOD and MAJOR, 1970; AKIMOTO and FUJISAWA, 1968; SUITO, 1972; ITO et al., 1974). These preliminary data indicate that at 1 bar, the back-transformation temperature in γ-$(Mg_{0.4}Fe_{0.6})_2SiO_4$ is lower than that of γ-$(Mg_{0.8}Fe_{0.2})_2SiO_4$ by ~100 °C. It is also found that at 735 °C, the γ-$(Mg_{0.8}Fe_{0.2})_2SiO_4$ completes the conversion process in about 1 h. In view of these preliminary results, it would be of interest to carry out systematic study of the back transformation and expansivity for various compositions of γ-$(Mg,Fe)_2SiO_4$ as a function of temperature, pressure, and time. Such data will provide very useful information concerning the kinetics and the mechanism involved in the transition.

As can be seen in Figures 6 and 7, there are 6–9 diffraction peaks for the spinel phase at all temperatures

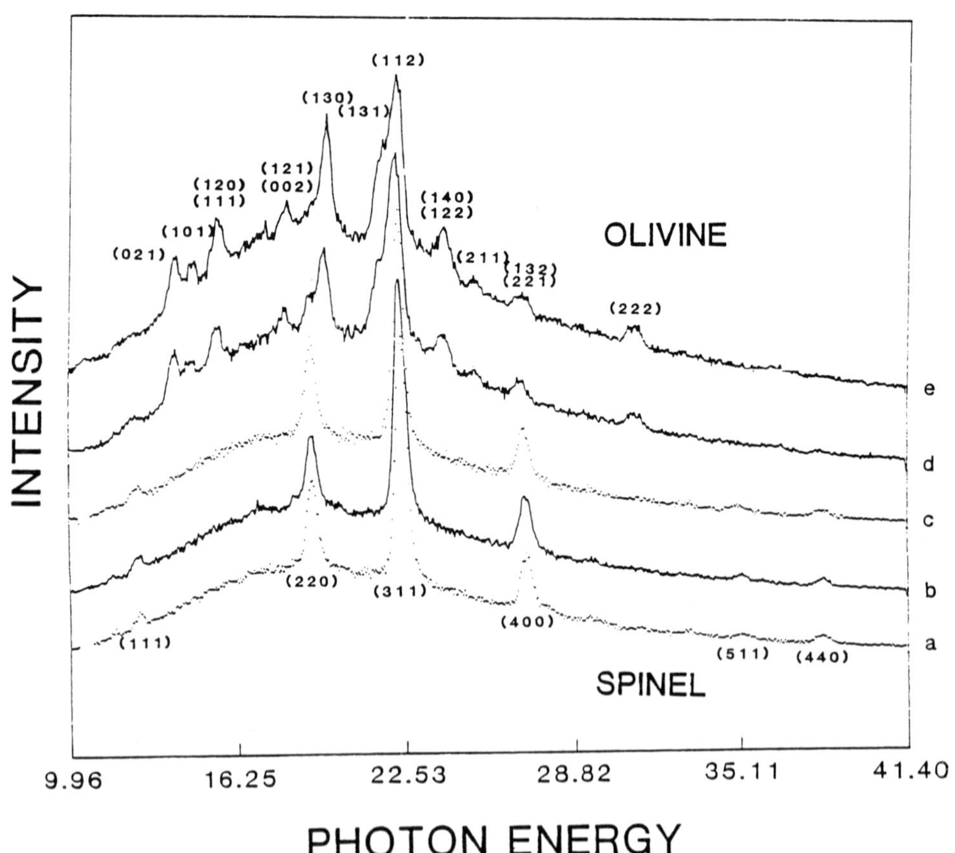

Figure 7. Energy dispersive X-ray diffraction of γ-$(Mg_{0.8}Fe_{0.2})_2SiO_4$ at 1 bar and at temperatures (in °C) of 407 (a), 495 (b), 660 (c), 735 (d), and 735 (e, 1h after d). It shows the back-transformation of $\gamma \rightarrow \alpha$ starts between 660 °C and 735 °C, and completes the conversion at 735 °C in 1h.

before the conversion into the olivine phase. Thus, the lattice parameters and the molar volumes of the γ-spinel phases (cubic structure) are well constrained (within $\pm 0.1\%$ and $\pm 0.3\%$, respectively. A plot of $[(V/V_0)-1]$ versus temperature based on the limited data (Figure 8) shows that the thermal expansivity for γ-$(Mg_{0.8}Fe_{0.2})_2SiO_4$ is slightly higher than that for γ-$(Mg_{0.4}Fe_{0.6})_2SiO_4$. This systematic trend is in contrast with previous results obtained for γ-Fe_2SiO_4 (MAO et al., 1969; TAKEUCHI et al., 1984) and for γ-Mg_2SiO_4 (SUZUKI et al., 1979). However, additional data for γ-spinels of various compositions are needed for evaluating the systematics in the composition dependence of the thermal expansivity of γ-$(Mg,Fe)_2SiO_4$.

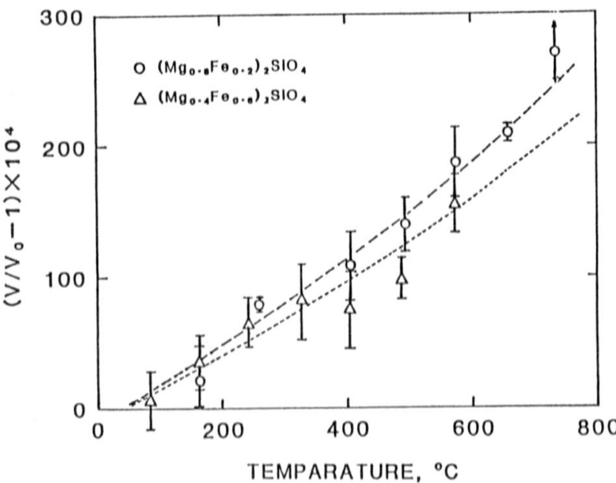

Figure 8. Thermal expansivity of γ-$(Mg_{0.8}Fe_{0.2})_2SiO_4$ and γ-$(Mg_{0.4}Fe_{0.6})_2SiO_4$. The dotted lines are polynomial fit of the experimental data points.

Conclusions

Using an improved DAC design, we have successfully conducted high P-T experiments and obtained high-quality energy-dispersive X-ray diffraction spectra for both the cubic γ-$(Mg,Fe)_2SiO_4$ and orthorhombic α-$(Mg,Fe)_2SiO_4$ to 1000 °C in intervals of 2–5 min. The improved cell has higher pressure-temperature capability and greater mechanical stability, and should serve as a reliable tool for probing the kinetics, the equation of state, and thermodynamic properties of geophysically important mineral phases under in situ high P-T conditions.

Acknowledgments. Thanks are due to E. Ito of Okayama University, Japan, for providing the powder sample of γ-$(MgFe)_2SiO_4$ used in this study and to N. Nakagiri and Y. H. Kim for their assistance in the data collection and discussion. Thanks are also due to the staff of the Stanford Synchrotron Radiation Laboratory, especially A. Bienenstock and K. Cantwell for facilitating our research work. Financial support was provided by the National Science Foundation Grant EAR84-18125, Hawaii Institute of Geophysics Contribution No. 1855.

References

Akimoto, S., and H. Fujisawa, Olivine-spinel solid solution equilibria in the system Mg_2SiO_4, *J. Geophys. Res., 73(4)*, 1467–1479, 1968.

American Institute of Physics, *Handbook*, 3rd ed., 1972.

Bassett, W. A., The diamond cell and the nature of the earth's mantle, *Ann. Rev. Earth Planet. Sci., 7*, 357–384, 1979.

Bassett, W. A., and T. Takahashi, Silver iodide polymorphs, *Am. Mineralogist, 50*, 1576–1594, 1965.

Bassett, W. A., M. S. Weathers, J. M. Bird, M. D. Furnish, and E. Huang, Temperature measurement of laser-heated samples in a diamond anvil cell, *Trans. Am. Geophys. Union, 66(46)*, 1134, 1985.

Furnish, M. D., and W. A. Bassett, Investigation of the mechanism of the olivine-spinel transition in fayalite by synchrotron radiation, *J. Geophys. Res., 88(12)*, 10,333–10,341, 1983.

Heinz, D., and R. Jeanloz, Three-dimensional temperature distribution in the laser-heated diamond cell, *Trans. Am. Geophys. Union, 66(46)*, 1134, 1985.

Ito, E., Y. Matsui, K. Suito, and N. Kawait, Synthesis of γ-Mg_2SiO_4, *Phys. Earth Planet. Inter., 8*, 342–344, 1974.

Jayaraman, A., Diamond anvil cell and high-pressure physical investigations, *Rev. Modern Phys., 55(1)*, 65–104, 1983.

Manghnani, M. H., L. C. Ming, J. Balogh, E. F. Skelton, S. B. Qadri, and D. Schiferl, Use of internal pressure calibrants in situ in X-ray diffraction measurements at high pressure and temperature: review and recent results, *High Temperature-High Pressure, 16*, 563–571, 1984.

Manghnani, M. H., L. C. Ming, J. Balogh, S. B. Qadri, E. F. Skelton, and D. Schiferl, Equation of state and phase transition studies under in situ high P-T conditions using synchrotron radiation, in *Solid State Physics under Pressure: Recent Advance with Anvil Devices*, edited by S. Minomura, pp. 343–350, KTK Scientific Publishers, Tokyo, 1985.

Mao, H. K., T. Takahashi, W. A. Bassett, J. S. Weaver, and S. Akimoto, Effect of pressure and temperature on the molar volumes of wilstite and of three $(Mg,Fe)_2SiO_4$ spinel solid solutions, *J. Geophys. Res., 74*, 1061–1069, 1969.

Ming, L. C., M. H. Manghnani, S. B. Qadri, E. F. Skelton, J. C. Jamieson, and J. Balogh, Gold as a reliable internal pressure calibrant at high temperatures, *J. Appl. Phys., 54(18)*, 4390–4397, 1983.

Ming, L. C., M. H. Manghnani, J. Balogh, S. B. Qadri, E. F. Skelton, A. W. Webb, and J. C. Jamieson, Static P-T-V measurements for MgO: Comparison with shock wave data, in *Shock Waves in Condensed Matter, 1983*, pp. 57–60, North-Holland Physics Publishing, The Netherlands, 1984.

Ringwood, A. E., and A. Major, The system Mg_2SiO_4-Fe_2SiO_4 at high pressures and temperatures, *Phys. Earth Planet. Inter., 3*, 89–108, 1970.

Schiferl, D. S., J. N. Fritz, A. I. Katz, M. Schaefer, E. F. Skelton, S. B. Qadri, L. C. Ming, and M. H. Manghnani, Very high temperature diamond-anvil cell for X-ray diffraction: application to the comparison of the gold and tungsten high-temperature-high-pressure internal standards, this volume.

Schiferl, D., A. I. Katz, R. L. Mills, L. C. Schmidt, E. F. Skelton, W. T. Elam, A. W. Webb, S. B. Qadri, and M. Schaefer, A novel instrument for high-pressure research at ultra-high temperatures, *Physica, 139* and *140-B*, 897–899, 1986.

Suito, K., Phase transformations of pure Mg_2SiO_4 into a spinel structure under high pressures and temperatures, *J. Phys. Earth, 20*, 225–243, 1972.

Sung, C. M., New modification of the diamond anvil press: A versatile apparatus for research at high pressure and high temperature, *Rev. Sci. Instr., 47*, 1343–1346, 1976.

Suzuki, I., E. Ohtani, and N. Kumazawa, Thermal expansion of Mg_2SiO_4, *J. Phys. Earth, 27*, 53–61, 1979.

Takeuchi, Y., T. Yamanaka, N. Haga, and M. Hirano, High-temperature crystallography of olivines and spinels, in *Materials Science of Earth Interior*, edited by I. Sunagawa, pp. 191–231, Terra Scientific Pub. Comp. (TERRAPUB), Tokyo, 1984.

Vohra, Y. K., K. E. Brister, S. T. Weir, S. J. Duclos, and A. L. Ruoff, Crystal structures at megabar pressures determined by use of the Cornell Synchrotron Source, *Science, 231*, 1136–1138, 1986.

Xu, J. A., H. K. Mao, and P. M. Bell, High-pressure ruby and diamond fluorescence: Observations at 0.21 to 0.55 terapascal, *Science, 232*, 1404–1406, 1986.

VERY HIGH TEMPERATURE DIAMOND-ANVIL CELL FOR X-RAY DIFFRACTION: APPLICATION TO THE COMPARISON OF THE GOLD AND TUNGSTEN HIGH-TEMPERATURE–HIGH-PRESSURE INTERNAL STANDARDS

D. SCHIFERL, J. N. FRITZ, and A. I. KATZ

Los Alamos National Laboratory, Los Alamos, NM 87545, USA

M. SCHAEFER

National Research Council/Naval Research Laboratory, Washington, D.C. 20375, USA

E. F. SKELTON

Naval Research Laboratory, Washington, D.C. 20375, USA

S. B. QADRI

Sachs/Freeman Associates, Bowie, MD 20715, USA

L. C. MING and M. H. MANGHNANI

Hawaii Institute of Geophysics, University of Hawaii, Honolulu, HI 96822, USA

Abstract. A high-temperature–high-pressure diamond-anvil cell which has been used previously for X-ray diffraction at over 10 GPa and 1500 K is described. The materials of the cell are listed in a table along with the physical and chemical criteria for choosing them. A major improvement is the use of rhenium gaskets, which are difficult to fabricate but which have worked nearly perfectly during extensive testing. Because the diffraction pattern of Re sometimes interferes with that of the Au in situ high-temperature–high-pressure standard, W is proposed as another such standard. A table of isochores in P, T space for W derived from ultrasonic, shockwave and thermal expansion data is presented. Comparison of X-ray diffraction data from W and Au up to 1300 K shows that there is fairly good agreement between the two standards. However, as the temperature is raised, the pressure according to the Au scale becomes increasingly greater than that from the W scale. Previous workers have reported similar behavior of Au in comparison to other materials at simultaneous high temperatures and pressures. While none of these results are conclusive, it is clear that more careful cross-comparisons of the various high-temperature–high-pressure standards are required to establish them with confidence.

Introduction

In this paper, we describe recent advances with a high-temperature–high-pressure diamond-anvil cell (DAC) which has been successfully used to collect X-ray powder diffraction data at well-characterized, uniform pressures and temperatures over 10 GPa at 1500 K (SCHIFERL et al., 1986). The most important innovation in the work presented here is the use of rhenium gaskets, which allow X-ray studies on properly gasketed samples over the full temperature and pressure ranges that our DAC is capable of achieving.

With this new capability come new problems, the most important of which is the need for a wider range of in situ high-temperature–high-pressure standards. For a number of reasons, the currently most used standards, Au, Pt and NaCl, can be unsuitable at the highest temperatures that we can now achieve. The principal problem is that above about 1100 K, all three standards may be expected to melt or recrystallize for a range of lower pressures. Recrystallization can cause fine-grained X-ray powder samples to become coarse-grained and consequently yield either poor diffraction patterns or none at all.

For these reasons, we decided to investigate tungsten as an additional high-temperature–high-pressure X-ray standard. This element offers many advantages. The compressibility of W is comparable to that of Pt, so that it is an adequately sensitive standard. Its recrystallization temperature (1900–2800 K) is well beyond the range of any resistively heated DAC. While W is not as chemically inert as Au or Pt, its high melting temperatures makes it less likely to form alloys with most metallic samples. It is a strong X-ray scatterer, and its diffraction pattern does not interfere with that obtained from the all-important rhenium gasket.

We have calculated the isotherms of W from ultrasonic and shockwave work, and the tables are given in the Appendix to this paper. Pressures from this W scale were compared with those of the Au scale (JAMIESON et al.,

High-Pressure Research in Mineral Physics, edited by M. H. Manghnani and Y. Syono, pp. 75–83.

1982) by taking X-ray diffraction data of mixtures of the two elements with the high-temperature DAC. As described below, the two pressure scales appear to agree within experimental errors, but the results are not as satisfying as we had hoped they would be. During the analysis of the data it became clear that much more work needs to be done to establish and cross-check high-temperature–high-pressure standards.

Experimental

The high-temperature–high-pressure diamond-anvil cell (DAC) is shown in Figures 1, 2 and 3, and the components are described in Table 1. The DAC is placed in a water-cooled vacuum oven which allows external pressure manipulation. The major design considerations and construction details are described by SCHIFERL et al. (1986).

Heating is accomplished with two sets of resistance heaters. The inner heater is a small coil (5) which heats a cavity around the diamonds from the outside. The inner

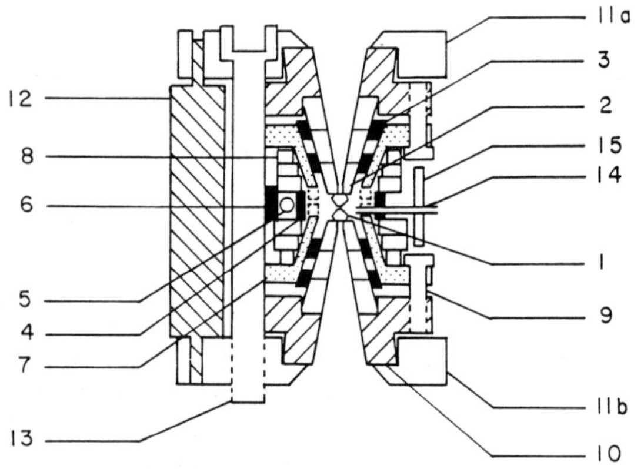

Figure 1. Cross section of high-temperature diamond-anvil cell. Some parts in the vicinity of the diamond anvils are omitted for clarity. The components are described in Table 1.

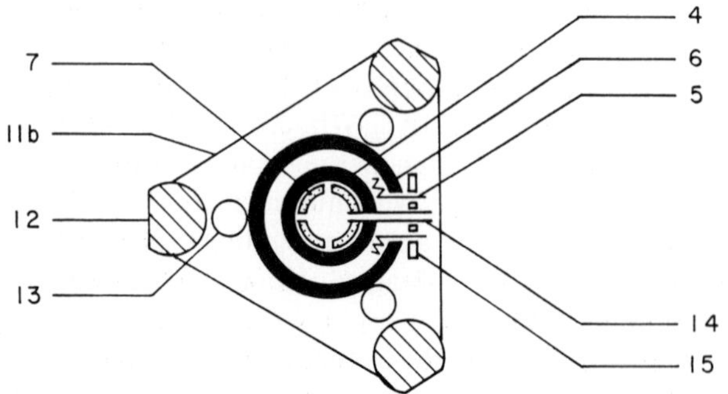

Figure 2. Cross section of high-temperature diamond-anvil cell in the plane of the sample. The sample, gaskets and diamonds are omitted for clarity. The components are described in Table 1.

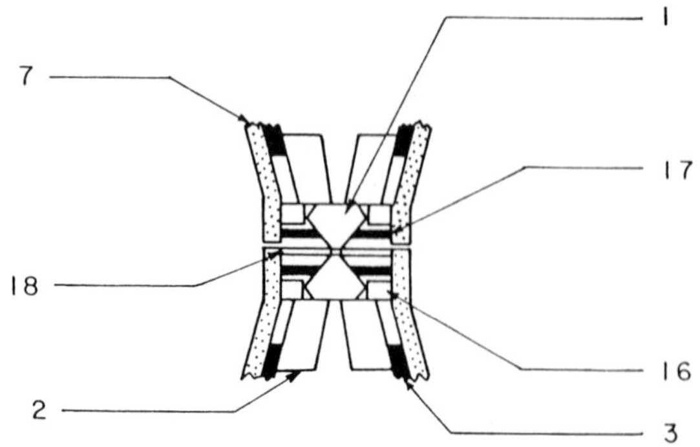

Figure 3. Expanded cross section of the vicinity of the diamond anvils. The components are described in Table 1.

components, including the sample in the gasket (18), diamond anvils (1) and thermocouple, are shielded from direct exposure to the thermal radiation of this heater by the inner insulater (4) and retainer (7), as can be seen in Figure 1 and 2. The temperatures in the vicinity of the sample are thus uniform and well-characterized. This inner heater has been redesigned so that it can now provide enough power to overcome all thermal losses. However, the DAC is still placed in a container which can be heated up to 1000 K by a larger heater coil. For a given sample temperature, this arrangement allows the pressure on the sample to be controlled by changing the temperature gradient in the cell and hence the differential thermal expansion between the various components of the DAC.

The diamond anvils had sixteen-sided culet tips 600 μm in diameter. The gaskets were rhenium disks 250 μm thick, prepressed to 50–80 μm, with a hole diameter of 150 μm. The holes in a number of gaskets were drilled ahead of time with electric-discharge machining techniques. The holes could not be drilled with either high-speed or carbide drills because rhenium work-hardens too rapidly. Even diamond drills are unsatisfactory for such small holes because the large forces required for cutting cause them to break. As described below, however, rhenium gaskets worked spectacularly well, and more than justify the extra effort required to fabricate them.

Samples of roughly equal amounts of W foil and Au in the form of either foil or powder were mixed with a small amount of NaCl, which served as a quasi-hydrostatic pressure medium.

For each run, the sample was taken up to a pressure in the range 14–19 GPa before heating. Unfortunately, the large thermal expansion of the Inconel 718 pressurizing bolts (13) caused the pressure to drop drastically as the cell body was warmed by only a few hundred kelvins. At high temperatures, the highest pressures could only be obtained by keeping these bolts relatively cool. This can be done by heating the sample through the inner heater as

TABLE 1. Description of Components of High-Temperature Diamond-Anvil Cell Shown in Figures 1, 2 and 3

Part #	Fig. #	Description	Material	Comments
1	1, 3	Diamond anvil	Diamond	Environment must be totally free of oxygen.
2	1, 3	Diamond support	Boron Carbide	Must have high compressive stength at high T.
3	1, 3	Spacer	Alumina	Prevents contact and chemical reactions between parts 2 and 7.
4	1, 2, 3	Heater component: Inner insulator	Alumina	Must be electrical insulator. Must not react with parts 5 and 7.
5	1, 2	Heater component: Heater coil	Tantalum	0.010" diameter wire, 40 turn coil. Must not react with parts 4, 6, or 8. Also serves as a getter for oxygen. Must be ductile at 300 K.
6	1, 2	Heater component: Outer insulator	Alumina	Must be electrical insulator. Must not react with part 5.
7	1, 2, 3	Retainer	Ta-10W	Must have moderate strength up to 1600 K. Also serves as a getter for oxygen.
8	1	Heat shield	Alumina	Must not react with part 7. Cut into numerous sections to minimize heat conduction down axis of cell.
9	1	Bolt	Ta-10W	Must have moderate strength up to 1500 K.
10	1	Support piece	Udimet 700	Must have high strength, low creep at 1000K. Must not react with part 2.
11a 11b	1 1, 2	Pressurizing plates	Udimet 700	Must have high strength, low creep at 1000K. Must not gall with parts 12 or 13.
12	1, 2	Guide pin	Inconel 718	Must have high strength at 1000K. Must not gall with parts 11a or 11b. Should be moderately easy to machine.
13	1, 2	Pressurizing bolt	Inconel 718	Must have high strength, low creep, low thermal expansion at 1000K. Must not gall with parts 11a or 11b.
14	1, 2	Thermocouple assembly	W-5% Re--W-26% Re Thermocouple with alumina insulators in Mo-13wt% Re guide tube.	Thermocouple wires must not have high vapor pressure at 1600K. Several layers (not all are shown) of insulators shield thermocouple from blackbody radiation of heater components.
15	1, 2	Insulator	Alumina	Must not react with part 5.
16	3	Centering ring	Machinable Tungsten	Must not react with parts 1, 2, 7 or 17.
17	3	Retainer	Rhenium	Must not react with parts 1, 7, or 16. Must have moderate strength at 1600K. Must be ductile at room temperature and easy to fabricate by punching.
18	3	Gasket	Rhenium	Must not react with parts 1 or 7. Must have the highest possible strength and lowest possible creep at 1600K. Must be ductile at 300K.

quickly as possible and taking data before the temperature of the bolts warmed to more than about 450 K.

Sample temperatures over 1200 K could easily be generated in less than a minute, although the cell was allowed to remain at temperature for two minutes before taking data to allow the volume around the sample to come to thermal equilibrium. On one run temperatures over 1600 K were generated without damage to the load-bearing components of the DAC. However, the pressure dropped to nearly zero due to the expansion of the Inconel bolts. Worse yet, the Ta coil (5) and alumina insulators (4, 6) in the heater melted into a eutectic puddle.

The lattice constants of Au and W were obtained with energy-dispersive X-ray diffraction of "white" X-radiation produced at Beam Line II-4 of the Stanford Synchrotron Radiation Laboratory (SSRL), using techniques developed by SKELTON et al. (1984). To obtain accurate X-ray pressure determinations on such relatively incompressible materials as Au and W, it is necessary for the sample position to be known to high accuracy. The entire high-temperature–high-pressure system is designed so that the sample position is reproducible to better than $\pm 25\ \mu m$ in the directions normal to the X-ray beam. This was checked by placing NaCl on one of the diamond-anvils and exposing it to the narrow, highly collimated X-ray beam for about two minutes. The position of the dark brown spot due to the resulting color centers was then determined with a microscope. Displacement of the sample due to thermal expansion is parallel to the X-ray beam and does not exceed 140 μm even at the highest temperatures. The errors due to sample positioning are discussed below.

The narrow openings in the boron carbide diamond supports require a diffraction angle $2\theta \leq 9.5°$. Thus, the observable diffraction lines, Au(111), Au(200) and W(110) appeared in the energy range 30–45 keV. The Re gasket contributes two lines in this region as well. At low pressure the Re(100) interferes with the Au(111); the Re(101) lies conveniently between the W(110) and Au(200). A third gasket line, the Re(002), which would be superimposed on the W(110), is completely eliminated by preferred orientation due to cold-working of the gasket.

Diffraction patterns were collected with a Princeton Gamma Tech planar intrinsic-Ge detector. Diffraction patterns were collected on different runs for periods varying between ten seconds and ten minutes. Each pattern was stored in a multichannel analyzer with 1024 channel for the spectrum, and also recorded on floppy disks for later analysis.

The counting periods for most of the high temperature runs were for 10–30 seconds. While this reduced the accuracy of the peak positions due to worsened counting statistics, it was necessary because of other experimental problems. The worst of these, the drop in pressure at high temperatures due to the thermal expansion of the Inconel 718 pressurizing bolts, has already been described above.

The energies of the peak positions were determined by a Gaussian profile fitting program and checked visually as well. At a fixed diffraction angle θ, these X-ray photon energies $E(hkl)$ are related to the corresponding interplanar spacings $d(hkl)$ by the Bragg relation

$$d(hkl) = \frac{hc}{2E(hkl)\sin\theta} = \frac{6.1992}{E(hkl)\sin\theta} \tag{1}$$

where h is Planck's constant, c is the speed of light, $d(hkl)$ is in Angstroms and E is in keV. The energy of the W(110) peak under ambient conditions was not hard to determine. However, because the Au(111) overlaps the Re(100) and the Au(200) is very weak, the energies of these peaks were calculated from the known lattice constant a of Au using the measured energies and of the W(110) and Re(101) to fix the value of $\sin\theta$.

Pressures are determined from the measured cubic lattice constant a of Au or W with empirically derived isotherms using the relation

$$\eta \equiv 1 - \frac{V}{V_0} = 1 - \left(\frac{a}{a_0}\right)^3 \tag{2}$$

For gold the isotherms of JAMIESON et al. (1982) were used; and for tungsten, the isotherms are given in the Appendix (Table 3).

Results and Discussion

The new rhenium gaskets have worked extremely well. In all cases, the gasket holes stayed round and did not wander. The rhenium did not become welded to the diamonds at high temperatures and pressures. Finally, the gaskets remained surprisingly thick. The final thickness of the prepressed region was in the range 30–50 μm. In a preliminary run, the pressure was increased to 30 GPa at room temperature, maintained at 25 GPa at 500 K, and the portion of the rhenium gasket between the diamonds was still 30 μm thick! This is the best resistance to extrusion that we have ever encountered in any ductile gasket material, including Inconel 718 and full-hard T301 stainless steel. In retrospect, the hole could have been made 200 μm in diameter to avoid diffraction from the gasket altogether.

The high-temperature DAC worked well although some minor, correctable problems became apparent. The thermal expansion of the Inconel 718 pressurizing bolts caused the pressure to drop entirely too much with increasing temperature. Fortunately, both Mo-13 wt%Re and pure rhenium seem to combine all the desirable mechanical properties of Inconel 718 with about half its thermal expansion.

The small ($2\theta \sim 9.5°$) angle allowed by the conical holes in the boron carbide supports for the diamond-anvils requires the observable Au and W diffraction lines to appear in the energy range 30–45 keV, well past the range of maximum intensity of the synchrotron-produced X-ray beam. The DAC works so well that it is reasonable to open the holes in the boron carbide to obtain a Bragg angle closer to $2\theta \simeq 15°$. The original design was overly conservative to insure that these parts had more than adequate strength. Widening the angle will not compromise the strength significantly but could improve the count rate of the diffraction peaks by several orders of magnitude by bringing the diffracted energy range down to 19–29 keV.

On some runs when the pressures dropped especially low, the Au peaks sometimes disappeared at temperatures over 900 K, although the W(110) peak was unaffected. When the cell cooled, the thermal contraction of the Inconel 718 pressurizing bolts greatly increased the pressure again and the Au peaks reappeared. Apparently, the gold recrystallized at high temperature and was cold-worked as the pressure increased with decreasing temperature. This behavior is not surprising in view of the fact that the annealing temperature for gold is only 575 K. There was no evidence of recrystallization of either the tungsten or rhenium.

Typical energy-dispersive X-ray diffraction patterns for three sets of P–T conditions are shown in Figure 4. In Figure 5, we compare the pressures according to the Au scale (JAMIESON et al., 1982) and those according to the W scale, which is presented in the Appendix.

The pressures for Au are the averages of those determined separately from the Au(111) and Au(200) peaks. The errors bars are dominated by the uncertainties of the peak energies but also include small contributions from the errors in detector calibration and in sample positioning normal to the incident X-ray beam. The systematic error due to the change of the sample position along the beam due to the thermal expansion of the cell is not included. However, the sample displacement is never more than 0.140 mm, which would correspond to pressure corrections of only $\Delta P(\mathrm{W}) = -0.4$ GPa and $\Delta P(\mathrm{Au}) = -0.2$ GPa. Corrections for the effects of non-hydrostatic pressures at lower temperatures and higher pressures are also not

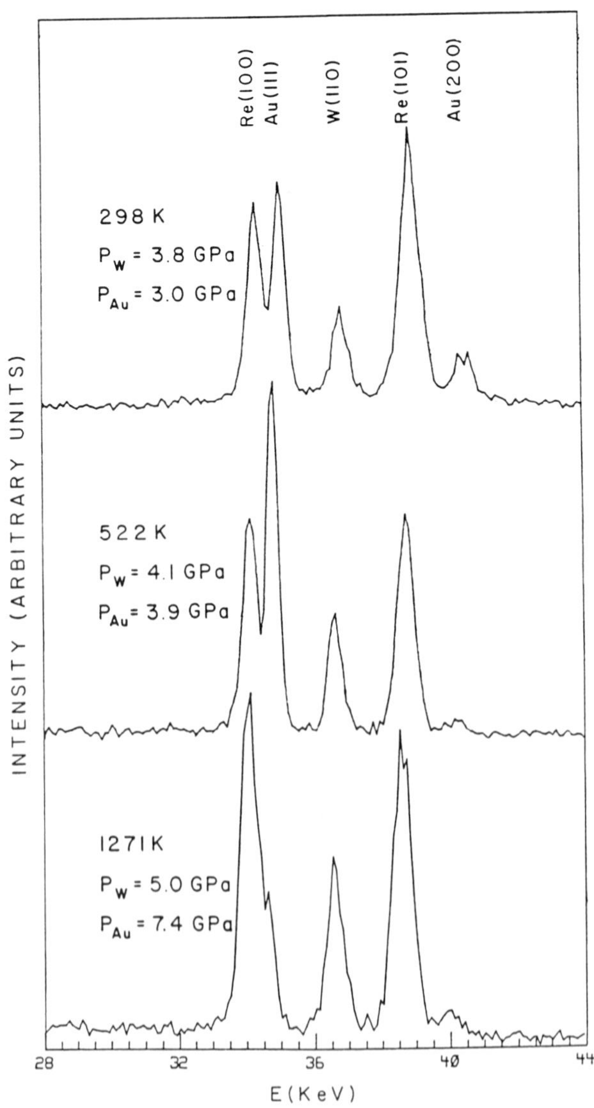

Figure 4. Typical diffraction data at three sets of P, T conditions.

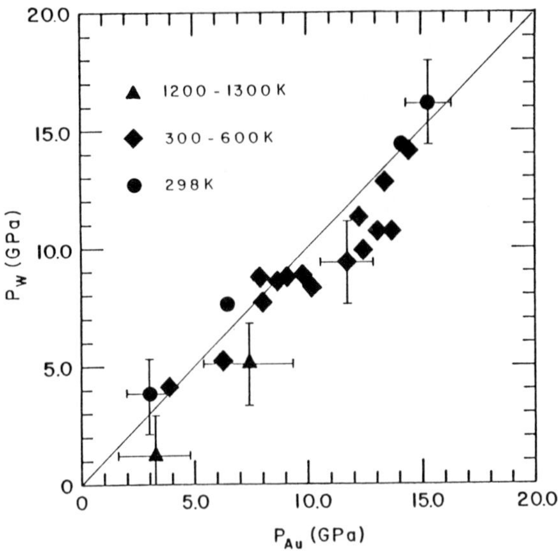

Figure 5. Pressure from Au scale (Jamieson et al., 1982) vs. pressure from W scale given in the Appendix (Table 3).

included.

The tungsten and gold scales appear to be in rough agreement, as can be seen from Figure 5. Despite the scatter in the data, however, there seems to be a discrepancy between the W and Au pressure scales that grows larger with increasing temperature. The cause of this discrepancy is not completely clear, although there are several possibilities that should e explored further. As discussed by Singh and Kennedy (1974) and Singh and Balasingh (1977) and Mao et al. (1978), non-hydrostatic conditions can lead to considerable errors in the pressures determined. For each material, W or Au, the uniaxial stress component t will make the lattice parameters appear to be too large and hence the pressures too low. The magnitude of this effect depends primarily on the value of t each material can support before flowing plastically, the elastic compliance S_{12} and 2θ. At low temperatures both W and Au apparently have similar errors and tend to offset each other. At high temperatures, however, the pressures in the Au approach hydrostatic while those in W remain non-hydrostatic. As a result, P(W) appears lower than P(Au) at high temperatures.

Another possibility is that the pressure scales themselves are not quite correct. Little can be said about errors in the W scale because these are the first comparisons involving it. However, several other workers have found pressures by the gold scale to be somewhat high. Ming et al. (1984) compared MgO to Au, although this work cannot be regarded as conclusive because of the known difficulties with the MgO equation of state. Yagi et al. (1985) compared NaCl to Au in a cubic anvil press and also found pressures from the gold scale to be slightly higher than those from the NaCl scale at elevated temperatures. Heinz and Jeanloz (1984) recalibrated the Au pressure scale. While their room temperature isotherm does not differ significantly from that of Jamieson et al. (1982), their Au scale yields lower pressures at high temperatures. Heinz and Jeanloz (1984) start with a smaller value for the Grüneisen γ_0 and then have it decrease with pressure more rapidly than Jamieson et al. (1982). The consequence of this is to bring their isotherms closer to the Hugoniot. For our data, applying the Heinz and Jeanloz (1984) Au scale would drop P(Au) at high temperatures, which is certainly in the right direction. However, the magnitude of the drop is no more than 1 GPa, and our data are not good

Table 3. P(GPa) vs. η–T for Tungsten

η \ T(K)	200	300	400	500	600	700	800	900	1000	1100	1200	1300	1400	1500
-0.0340	-10.056	-9.645	-9.225	-8.802	-8.377	-7.951	-7.524	-7.097	-6.669	-6.242	-5.814	-5.386	-4.958	-4.529
-0.0330	-9.793	-9.383	-8.963	-8.540	-8.115	-7.689	-7.262	-6.835	-6.407	-5.979	-5.552	-5.123	-4.695	-4.267
-0.0320	-9.530	-9.119	-8.700	-8.276	-7.851	-7.425	-6.998	-6.571	-6.144	-5.716	-5.288	-4.860	-4.432	-4.004
-0.0310	-9.265	-8.854	-8.435	-8.012	-7.586	-7.160	-6.734	-6.306	-5.879	-5.451	-5.023	-4.595	-4.167	-3.739
-0.0300	-8.999	-8.588	-8.169	-7.745	-7.320	-6.894	-6.468	-6.040	-5.613	-5.185	-4.757	-4.329	-3.901	-3.473
-0.0290	-8.731	-8.321	-7.901	-7.478	-7.053	-6.627	-6.200	-5.773	-5.346	-4.918	-4.490	-4.062	-3.634	-3.206
-0.0280	-8.462	-8.052	-7.632	-7.209	-6.784	-6.358	-5.932	-5.504	-5.077	-4.649	-4.221	-3.793	-3.365	-2.937
-0.0270	-8.192	-7.782	-7.362	-6.939	-6.514	-6.088	-5.662	-5.234	-4.807	-4.379	-3.951	-3.523	-3.095	-2.667
-0.0260	-7.921	-7.510	-7.091	-6.668	-6.243	-5.817	-5.390	-4.963	-4.536	-4.108	-3.680	-3.252	-2.824	-2.396
-0.0250	-7.648	-7.238	-6.818	-6.395	-5.970	-5.544	-5.118	-4.690	-4.263	-3.835	-3.407	-2.979	-2.551	-2.123
-0.0240	-7.374	-6.964	-6.544	-6.121	-5.696	-5.270	-4.844	-4.417	-3.989	-3.561	-3.133	-2.705	-2.277	-1.849
-0.0230	-7.098	-6.688	-6.269	-5.846	-5.421	-4.995	-4.568	-4.141	-3.714	-3.286	-2.858	-2.430	-2.002	-1.574
-0.0220	-6.822	-6.412	-5.992	-5.569	-5.144	-4.718	-4.292	-3.865	-3.437	-3.009	-2.581	-2.153	-1.725	-1.297
-0.0210	-6.544	-6.134	-5.714	-5.291	-4.866	-4.440	-4.014	-3.587	-3.159	-2.731	-2.304	-1.876	-1.447	-1.019
-0.0200	-6.264	-5.854	-5.435	-5.012	-4.587	-4.161	-3.734	-3.307	-2.880	-2.452	-2.024	-1.596	-1.168	-0.740
-0.0190	-5.983	-5.573	-5.154	-4.731	-4.306	-3.880	-3.454	-3.027	-2.599	-2.171	-1.744	-1.316	-0.887	-0.459
-0.0180	-5.701	-5.291	-4.872	-4.449	-4.024	-3.598	-3.172	-2.744	-2.317	-1.889	-1.461	-1.033	-0.605	-0.177
-0.0170	-5.417	-5.008	-4.589	-4.166	-3.741	-3.315	-2.888	-2.461	-2.034	-1.606	-1.178	-0.750	-0.322	0.106
-0.0160	-5.132	-4.723	-4.304	-3.881	-3.456	-3.030	-2.603	-2.176	-1.749	-1.321	-0.893	-0.465	-0.037	0.391
-0.0150	-4.846	-4.436	-4.017	-3.594	-3.170	-2.744	-2.317	-1.890	-1.462	-1.035	-0.607	-0.179	0.249	0.677
-0.0140	-4.558	-4.149	-3.730	-3.307	-2.882	-2.456	-2.029	-1.602	-1.175	-0.747	-0.319	0.109	0.537	0.965
-0.0130	-4.269	-3.860	-3.441	-3.018	-2.593	-2.167	-1.740	-1.313	-0.886	-0.458	-0.030	0.398	0.826	1.254
-0.0120	-3.979	-3.569	-3.150	-2.727	-2.303	-1.877	-1.450	-1.023	-0.595	-0.168	0.260	0.688	1.116	1.544
-0.0110	-3.687	-3.277	-2.858	-2.435	-2.011	-1.585	-1.158	-0.731	-0.304	0.124	0.552	0.980	1.408	1.836
-0.0100	-3.393	-2.984	-2.565	-2.142	-1.717	-1.292	-0.865	-0.438	-0.010	0.417	0.845	1.273	1.701	2.129
-0.0090	-3.098	-2.689	-2.270	-1.847	-1.423	-0.997	-0.570	-0.143	0.284	0.712	1.140	1.568	1.996	2.424
-0.0080	-2.802	-2.393	-1.974	-1.551	-1.127	-0.701	-0.274	0.153	0.580	1.008	1.436	1.864	2.292	2.720
-0.0070	-2.504	-2.095	-1.676	-1.254	-0.829	-0.403	0.023	0.451	0.878	1.306	1.733	2.161	2.589	3.018
-0.0060	-2.205	-1.796	-1.377	-0.955	-0.530	-0.104	0.322	0.750	1.177	1.605	2.032	2.460	2.888	3.317
-0.0050	-1.905	-1.496	-1.077	-0.654	-0.230	0.196	0.623	1.050	1.477	1.905	2.333	2.761	3.189	3.617
-0.0040	-1.603	-1.194	-0.775	-0.352	0.072	0.498	0.925	1.352	1.779	2.207	2.635	3.063	3.491	3.919
-0.0030	-1.299	-0.890	-0.472	-0.049	0.376	0.802	1.228	1.655	2.083	2.510	2.938	3.366	3.794	4.222
-0.0020	-0.994	-0.585	-0.167	0.256	0.681	1.106	1.533	1.960	2.387	2.815	3.243	3.671	4.099	4.527
-0.0010	-0.688	-0.279	0.140	0.562	0.987	1.413	1.839	2.266	2.694	3.121	3.549	3.977	4.405	4.833
0.0000	-0.380	0.029	0.448	0.870	1.295	1.721	2.147	2.574	3.002	3.429	3.857	4.285	4.713	5.141
0.0010	-0.070	0.338	0.757	1.180	1.604	2.030	2.456	2.884	3.311	3.738	4.166	4.594	5.022	5.450
0.0020	0.241	0.649	1.068	1.490	1.915	2.341	2.767	3.194	3.622	4.049	4.477	4.905	5.333	5.761
0.0030	0.553	0.962	1.380	1.803	2.227	2.653	3.080	3.507	3.934	4.362	4.789	5.217	5.645	6.074
0.0040	0.867	1.276	1.694	2.117	2.541	2.967	3.393	3.820	4.248	4.675	5.103	5.531	5.959	6.387
0.0050	1.183	1.591	2.009	2.432	2.857	3.282	3.709	4.136	4.563	4.991	5.419	5.846	6.275	6.703
0.0060	1.500	1.908	2.326	2.749	3.173	3.599	4.026	4.453	4.880	5.308	5.735	6.163	6.591	7.020
0.0070	1.818	2.226	2.645	3.067	3.492	3.918	4.344	4.771	5.198	5.626	6.054	6.482	6.910	7.338
0.0080	2.138	2.547	2.965	3.387	3.812	4.238	4.664	5.091	5.518	5.946	6.374	6.802	7.230	7.658
0.0090	2.460	2.868	3.286	3.709	4.133	4.559	4.985	5.412	5.840	6.267	6.695	7.123	7.551	7.979
0.0100	2.783	3.191	3.609	4.032	4.456	4.882	5.309	5.736	6.163	6.590	7.018	7.446	7.874	8.302
0.0110	3.108	3.516	3.934	4.357	4.781	5.207	5.633	6.060	6.487	6.915	7.343	7.771	8.199	8.627
0.0120	3.434	3.842	4.260	4.683	5.107	5.533	5.959	6.386	6.814	7.241	7.669	8.097	8.525	8.953
0.0130	3.762	4.170	4.588	5.011	5.435	5.861	6.287	6.714	7.141	7.569	7.997	8.425	8.853	9.281
0.0140	4.092	4.500	4.918	5.340	5.764	6.190	6.616	7.043	7.471	7.898	8.326	8.754	9.182	9.610
0.0150	4.423	4.831	5.249	5.671	6.095	6.521	6.947	7.374	7.802	8.229	8.657	9.085	9.513	9.941
0.0160	4.756	5.163	5.581	6.004	6.428	6.854	7.280	7.707	8.134	8.562	8.990	9.417	9.846	10.274
0.0170	5.090	5.497	5.915	6.338	6.762	7.188	7.614	8.041	8.468	8.896	9.324	9.752	10.180	10.608
0.0180	5.426	5.833	6.251	6.673	7.098	7.523	7.950	8.377	8.804	9.232	9.659	10.087	10.515	10.943
0.0190	5.763	6.171	6.589	7.011	7.435	7.861	8.287	8.714	9.141	9.569	9.997	10.425	10.853	11.281
0.0200	6.102	6.510	6.928	7.350	7.774	8.200	8.626	9.053	9.481	9.908	10.336	10.764	11.192	11.620
0.0210	6.443	6.851	7.268	7.691	8.115	8.541	8.967	9.394	9.821	10.249	10.676	11.104	11.532	11.960
0.0220	6.786	7.193	7.611	8.033	8.457	8.883	9.309	9.736	10.163	10.591	11.019	11.447	11.875	12.303
0.0230	7.130	7.537	7.955	8.377	8.601	9.227	9.653	10.080	10.507	10.935	11.363	11.791	12.219	12.647
0.0240	7.476	7.883	8.301	8.723	9.147	9.572	9.999	10.426	10.853	11.281	11.708	12.136	12.564	12.992

η \ T(K)	200	300	400	500	600	700	800	900	1000	1100	1200	1300	1400	1500
0.0250	7.823	8.230	8.648	9.070	9.494	9.920	10.346	10.773	11.200	11.628	12.056	12.483	12.911	13.340
0.0260	8.172	8.579	8.997	9.419	9.843	10.269	10.695	11.122	11.549	11.977	12.404	12.832	13.260	13.688
0.0270	8.523	8.930	9.348	9.770	10.194	10.619	11.046	11.473	11.900	12.327	12.755	13.183	13.611	14.039
0.0280	8.876	9.282	9.700	10.122	10.546	10.972	11.398	11.825	12.252	12.680	13.107	13.535	13.963	14.391
0.0290	9.230	9.637	10.054	10.476	10.900	11.326	11.752	12.179	12.606	13.034	13.461	13.889	14.317	14.745
0.0300	9.586	9.992	10.410	10.832	11.256	11.682	12.108	12.535	12.962	13.389	13.817	14.245	14.673	15.101
0.0310	9.943	10.350	10.767	11.189	11.614	12.039	12.465	12.892	13.319	13.747	14.175	14.603	15.031	15.459
0.0320	10.303	10.709	11.127	11.549	11.973	12.398	12.825	13.251	13.679	14.106	14.534	14.962	15.390	15.818
0.0330	10.664	11.070	11.488	11.910	12.334	12.759	13.185	13.612	14.040	14.467	14.895	15.323	15.751	16.179
0.0340	11.026	11.433	11.850	12.272	12.696	13.122	13.548	13.975	14.402	14.830	15.257	15.685	16.113	16.541
0.0350	11.391	11.797	12.215	12.637	13.061	13.486	13.913	14.339	14.767	15.194	15.622	16.050	16.478	16.906
0.0360	11.757	12.164	12.581	13.003	13.427	13.852	14.279	14.706	15.133	15.560	15.988	16.416	16.844	17.272
0.0370	12.126	12.532	12.949	13.371	13.795	14.220	14.647	15.073	15.501	15.928	16.356	16.784	17.212	17.640
0.0380	12.495	12.902	13.319	13.741	14.165	14.590	15.016	15.443	15.870	16.298	16.726	17.153	17.581	18.009
0.0390	12.867	13.273	13.690	14.112	14.536	14.962	15.388	15.815	16.242	16.669	17.097	17.525	17.953	18.381
0.0400	13.240	13.647	14.064	14.485	14.909	15.335	15.761	16.188	16.615	17.043	17.470	17.898	18.326	18.754
0.0410	13.616	14.022	14.439	14.861	15.285	15.710	16.136	16.563	16.990	17.418	17.845	18.273	18.701	19.129
0.0420	13.993	14.399	14.816	15.237	15.662	16.087	16.513	16.940	17.367	17.795	18.222	18.650	19.078	19.506
0.0430	14.372	14.778	15.195	15.616	16.040	16.466	16.892	17.319	17.746	18.173	18.601	19.029	19.457	19.885
0.0440	14.752	15.158	15.575	15.997	16.421	16.846	17.272	17.699	18.126	18.554	18.981	19.409	19.837	20.265
0.0450	15.135	15.541	15.958	16.379	16.803	17.229	17.655	18.081	18.509	18.936	19.364	19.792	20.220	20.648
0.0460	15.519	15.925	16.342	16.763	17.187	17.613	18.039	18.466	18.893	19.320	19.748	20.176	20.604	21.032
0.0470	15.906	16.311	16.728	17.150	17.574	17.999	18.425	18.852	19.279	19.706	20.134	20.562	20.990	21.418
0.0480	16.294	16.699	17.116	17.538	17.961	18.387	18.813	19.240	19.667	20.094	20.522	20.950	21.378	21.806
0.0490	16.684	17.089	17.506	17.927	18.351	18.777	19.203	19.629	20.057	20.484	20.912	21.339	21.767	22.195
0.0500	17.075	17.481	17.898	18.319	18.743	19.168	19.594	20.021	20.448	20.876	21.303	21.731	22.159	22.587
0.0510	17.469	17.874	18.291	18.713	19.137	19.562	19.988	20.415	20.842	21.269	21.697	22.125	22.553	22.981
0.0520	17.865	18.270	18.687	19.108	19.532	19.957	20.383	20.810	21.237	21.665	22.092	22.520	22.948	23.376
0.0530	18.262	18.667	19.084	19.506	19.929	20.355	20.781	21.207	21.635	22.062	22.490	22.917	23.345	23.773
0.0540	18.662	19.067	19.483	19.905	20.329	20.754	21.180	21.607	22.034	22.461	22.889	23.317	23.745	24.173
0.0550	19.063	19.468	19.885	20.306	20.730	21.155	21.581	22.008	22.435	22.862	23.290	23.718	24.146	24.574
0.0560	19.466	19.871	20.288	20.709	21.133	21.558	21.984	22.411	22.838	23.266	23.693	24.121	24.549	24.977
0.0570	19.872	20.277	20.693	21.114	21.538	21.963	22.389	22.816	23.243	23.671	24.098	24.526	24.954	25.382
0.0580	20.279	20.684	21.100	21.521	21.945	22.370	22.797	23.223	23.650	24.078	24.505	24.933	25.361	25.789
0.0590	20.688	21.093	21.509	21.931	22.354	22.779	23.206	23.632	24.059	24.487	24.914	25.342	25.770	26.198
0.0600	21.099	21.504	21.920	22.342	22.765	23.190	23.616	24.043	24.470	24.898	25.325	25.753	26.181	26.609
0.0610	21.512	21.917	22.333	22.754	23.178	23.603	24.029	24.456	24.883	25.311	25.738	26.166	26.594	27.022
0.0620	21.928	22.332	22.748	23.169	23.593	24.018	24.444	24.871	25.298	25.725	26.153	26.581	27.009	27.437
0.0630	22.345	22.749	23.165	23.586	24.010	24.435	24.861	25.288	25.715	26.142	26.570	26.998	27.426	27.854
0.0640	22.764	23.168	23.584	24.005	24.429	24.854	25.280	25.707	26.134	26.561	26.989	27.417	27.845	28.273
0.0650	23.185	23.589	24.005	24.426	24.850	25.275	25.701	26.128	26.555	26.982	27.410	27.838	28.266	28.694
0.0660	23.608	24.012	24.428	24.849	25.273	25.698	26.124	26.551	26.978	27.405	27.833	28.261	28.688	29.117
0.0670	24.033	24.437	24.853	25.275	25.698	26.123	26.549	26.976	27.403	27.830	28.258	28.686	29.114	29.542
0.0680	24.461	24.865	25.281	25.702	26.125	26.550	26.976	27.403	27.830	28.257	28.685	29.113	29.541	29.969
0.0690	24.890	25.294	25.710	26.131	26.554	26.980	27.406	27.832	28.259	28.686	29.114	29.542	29.970	30.398
0.0700	25.321	25.725	26.141	26.562	26.986	27.411	27.837	28.263	28.690	29.118	29.545	29.973	30.401	30.829
0.0710	25.755	26.159	26.574	26.995	27.419	27.844	28.270	28.697	29.124	29.551	29.979	30.406	30.834	31.262
0.0720	26.190	26.594	27.010	27.431	27.854	28.279	28.705	29.132	29.559	29.986	30.414	30.842	31.270	31.698
0.0730	26.628	27.032	27.447	27.868	28.292	28.717	29.143	29.569	29.996	30.424	30.851	31.279	31.707	32.135
0.0740	27.068	27.471	27.887	28.308	28.732	29.157	29.582	30.009	30.436	30.863	31.291	31.719	32.147	32.575
0.0750	27.510	27.913	28.329	28.750	29.173	29.598	30.024	30.451	30.878	31.305	31.733	32.160	32.588	33.016
0.0760	27.954	28.357	28.773	29.194	29.617	30.042	30.468	30.895	31.322	31.749	32.176	32.604	33.032	33.460
0.0770	28.400	28.803	29.219	29.640	30.063	30.488	30.914	31.341	31.768	32.195	32.622	33.050	33.478	33.906
0.0780	28.848	29.251	29.667	30.088	30.511	30.936	31.362	31.789	32.216	32.643	33.071	33.498	33.926	34.354
0.0790	29.299	29.702	30.117	30.538	30.962	31.387	31.813	32.239	32.666	33.093	33.521	33.949	34.376	34.804

enough for preferring one gold scale over the other. It is clear that many more careful and accurate cross-checks between the various high-temperature–high-pressure standards are required.

Conclusions

This work and that previously reported (SCHIFERL et al., 1986) demonstrates that our high-temperature diamond-anvil cell is now a useful research tool capable of generating temperatures as high as 1500 K routinely on gasketed samples. The most important improvement over previous designs is the use of rhenium gaskets. Once fabricated, they have worked perfectly. Their ductility and superior resistance to extrusion suggests that pure rhenium gaskets would also be the material of choice to achieve pressures exceeding 100 GPa at room temperature.

The chief problem with the high-temperature DAC used in this work was that the Inconel 718 pressurizing bolts had too large a thermal expansion, causing the pressure to drop far too much at elevated temperatures. The Udimet 700 bolts used previously showed the same difficulty, but to a lesser degree. The improvisations necessary to minimize this problem, including rapid heating and short counting periods, resulted in greatly reduced accuracy in the diffraction peak positions. Fortunately, this situation appears to be correctable by using different materials for the bolts. Both Mo-13 wt%Re, and pure Re are promising candidates because they have about half the thermal expansion of Inconel 718 or Udimet 700, and even better high-temperature tensile properties. Thus, much longer counting periods can be used. Furthermore, the count rate can be improved by widening the angles of the conical holes in the boron carbide supports to allow the diffraction peaks to be observed at lower energies where the intensity is much greater.

There is considerable scatter in the correlation of *P*(W) versus *P*(Au), and the discrepancy becomes greater with increasing temperature. There are two major possible sources of this difficulty. The effect of non-hydrostatic pressures can make *P*(Au) appear too small only at higher pressures and lower temperatures, but always makes *P*(W) appear too low. It may also be that the gold or tungsten scales may be incorrect. To date, there are no other comparisons for the tungsten scale. However, for several workers the gold scale seems to always yield pressures that are somewhat too high at high temperatures. Unfortunately, neither our data nor that in the literature is conclusive, and more careful comparisons of the high-temperature–high-pressure standards are required to establish them with confidence.

Appendix: Tungsten Pressure Scale Derived from Ultra sonic and Shockwave Work

The calculated isotherms for tungsten are based on room temperature ultrasonic data (S. P. Marsh, unpublished), shockwave data (MCQUEEN et al., 1970) and zero pressure thermal expansion data (KIRBY et al., 1972), following the procedures of JAMIESON et al. (1982). The model parameters defining the equation of state are presented in the Table 2. The isotherms are presented in Table 3.

TABLE 2. Model Parameters Defining the Equation of State of W

Density at T=293 K, P=0,	ρ_0=19.2700 g/cm^3
Bulk sound velocity at T=293 K, P=0	C_0=4.015 km/s S=1.204
Grüneisen parameter at T=293 K, P=0	γ_0=1.640
Debye Temperature	θ_D=0.241 kK 3nk=0.13567 J/mg-kK
Specific Heats from the above parameters T=293 K, P=0	C_V=0.1312 J/g-K C_P=0.1322 J/g-K

Acknowledgments. It is a pleasure to thank the staff of MEC Division at Los Alamos National Laboratory for fabricating the components of the high-temperature diamond-anvil cell. Special credit for making the most difficult parts goes to A. Isensee, P. E. Duran, G. A. Pittel, R. W. Livingston, J. E. Dyson, E. Kenner, P. R. Martinez, R. M. Wheat, Jr., A. D. Montoya, D. J. Hatch, P. Sanchez, and L. J. Salzer. We are also grateful to L. B. Lundberg (Los Alamos) for suggesting the use of pure rhenium as a gasket material. We thank H. K. Mao for very useful referee's comments on the origins of the discrepancies between Au and W pressure scales. One of us (D.S.) thanks J. W. Shaner for encouragement and many useful discussions on this project. We thank the staff of the Stanford Synchrotron Radiation Laboratory for facilitating this research . Financial support for this research was provided by the U.S. Department of Energy, the Condensed Matter Physics and Radiation Sciences Division of the Office of Naval Research, National Research Council, and the National Science Foundation.

REFERENCES

HEINZ, D. L., and R. JEANLOZ, The equation of state of the gold calibration standard, *J. Appl. Phys.*, *55*, 885–893, 1984.

JAMIESON, J. C., J. N. FRITZ, and M. H. MANGHNANI, Pressure measurement at high temperature in X-ray diffraction studies: gold as a primary standard, in *High-Pressure Research in Geophysics*, edited by S. Akimoto and M. H. Manghnani, pp. 27–48, Center for Academic Publishing, Tokyo, 1982.

KIRBY, R. K., T. A. HAHN, and B. D. ROTHROCK, Thermal expansion, in *AIP Handbook*, 3rd Edition, edited by D. E. Gray, pp. 4:119–4:4142, McGraw-Hill, New York, 1972.

McQueen, R. G., S. P. Marsh, J. W. Taylor, J. N. Fritz, and W. J. Carter, The equation of state of solids from shock wave studies, in *High-Velocity Impact Phenomena*, edited by R. Kinslow, pp. 293–417, Academic Press, New York, 1970.

Mao, H. K., P. M. Bell, J. W. Shaner, and D. J. Steinberg, Specific volume measurements of Cu, Mo, Pd, and Ag and calibration of the ruby R_1 fluorescence pressure gauge from 0.06 to 1 Mbar, *J. Appl. Phys., 49*, 3276–3283, 1978.

Ming, L. C., M. H. Manghnani, J. Balogh, S. B. Qadri, E. F. Skelton, A. W. Webb, and J. C. Jamieson, Static P-T-V measurements on MgO: comparison with shock wave data, in *Shock Waves in Condensed Matter 1983*, edited by J. R. Asay, R. A. Graham, and G. K. Straub, pp. 57–60, North Holland, Amsterdam, 1984.

Schiferl, D., A. I. Katz, R. L. Mills, L. C. Schmidt, C. Vanderborgh, E. F. Skelton, W. T. Elam, A. W. Webb, S. B. Qadri, and M. Schaefer, A novel instrument for high pressure research at ultra-high temperatures, *Proc. Xth AIRAPT International High Pressure Conference, 1985*, edited by N. J. Trappeniers et al., pp. 897–900, North Holland, Amsterdam, 1986.

Skelton, E. F., J. D. Ayers, W. T. Elam, T. L. Francavilla, C. L. Vald, A. W. Webb, S. A. Wolf, S. B. Qadri, M. H. Manghnani, L. C. Ming, J. Balogh, C. Y. Huang, D. Schiferl, and R. C. Lacoe, Recent advances in a facility for high pressure structural studies using synchrotron radiation, *High Temp.-High Press., 16*, 527–532, 1984.

Singh, A. K., and G. C. Kennedy, Uniaxial stress component in tungsten carbide high pressure X-ray cameras, *J. Appl. Phys., 45*, 4686–4691, 1974.

Singh, A. K., and C. Balasingh, Uniaxial stress component in diamond anvil high-pressure X-ray cameras, *J. Appl. Phys., 48*, 5338–5340, 1977.

Yagi, T., O. Shimomura, S. Yamaoka, K. Takemura, and S. Akimoto, Precise measurement of compressibility of gold at room and high temperatures, in *Solid State Physics Under Pressure: Recent Advance with Anvil Devices*, edited by S. Minomura, pp. 363–368, KTK Scientific Publishers, 1985.

Note Added in Proof: We have since found that two of the improvements suggested above do in fact work very well. The use of rhenium for the pressurizing bolts eliminated the pressure drop due to thermal expansion. The conical holes in the boron carbide were opened to allow $2\theta<15°$. As expected, the pressure capability remained, and the intensities of the diffraction peaks were greatly increased because they could be observed at lower energies.

II. MELTING EXPERIMENTS AT HIGH PRESSURES

ULTRAHIGH-PRESSURE MELTING OF A MODEL CHONDRITIC MANTLE AND PYROLITE COMPOSITIONS

Eiji OHTANI

Department of Earth Sciences, Faculty of Science, Ehime University
Matsuyama 790, Japan

Abstract. Melting experiments on chondritic mantle and pyrolite compositions for the five-component system CaO-MgO-FeO-Al_2O_3-SiO_2 were conducted to a pressure of 20 GPa. For the chondritic mantle composition, the liquidus phase changes from olivine to majorite between 12 and 15 GPa. For the pyrolite composition, olivine is the liquidus phase at least up to 15 GPa, whereas both modified spinel and majorite coexist around the liquidus temperature and 20 GPa. The majorite crystals coexisting with the partial melt have CaO/Al_2O_3 lower than chondritic for the chondritic mantle and pyrolite compositions. The majorite fractionation by partial melting at the base of the upper mantle could generate an ultrabasic magma with CaO/Al_2O_3 higher than chondritic. Such fractionation in the early stage of terrestrial evolution could have produced the chemical stratification of the mantle composed of a peridotitic upper mantle with high CaO/Al_2O_3 and a majorite-enriched transition zone with a relatively low CaO/Al_2O_3.

Introduction

The possibility of chemical stratification of the mantle has been discussed by some authors (e.g., ANDERSON, 1979, 1981; HERZBERG and O'HARA, 1985; OHTANI, 1985, 1986). Because one of the most effective mechanisms for producing chemical stratification is chemical fractionation by partial melting, the discussion of possible stratification in the mantle is strongly related to the thermal history during the earth stages of terrestrial evolution. Models of accretion and core formation of the earth strongly suggest that the outer layer of the proto-earth was much hotter than today. Recent calculations of the thermal history of the accreting earth (e.g., KAULA, 1979) supported this view of the early stage of terrestrial evolution. These studies suggested that the outer layer of the proto-earth was molten during accretion, and the molten zero might have extended to depths greater than 1000 km (KAULA, 1979; HAYASHI et al., 1979). Such global melting in the early stage may have produced a chemical stratification in the mantle. The nature of the expected stratification has been discussed on the basis of the determined melting relations of silicate systems at a relatively lower pressure range and the estimated melting relations at ultrahigh pressure.

Recent development of techniques for the multiple-anvil high-pressure apparatus has resulted in the generation of temperatures above 2000 °C to a pressure of 20 GPa (OHTANI et al., 1982). Thus, it is now possible to conduct melting experiments on complex silicate systems under these extreme conditions. In order to discuss the nature of the stratification of the mantle formed by this early global melting event, it is essential to clarify the melting relations of the primitive mantle composition to pressures corresponding to the deep upper mantle and lower mantle.

Melting experiments were conducted on the five-component system, CaO-MgO-Al_2O_3-FeO-SiO_2 of the two major candidates of the primitive mantle composition; i.e., the chondritic mantle composition, which contains the elements, Ca, Mg, Al, and Si with the chondritic ratio, and the pyrolite composition (RINGWOOD, 1982) to a pressure of 20 GPa. The purpose of this paper is to present the results of the melting experiments performed on these primitive mantle compositions, and to discuss the nature of the stratification of the mantle expected to result from the proposed melting event in the early stage of terrestrial evolution.

Choice of the Starting Compositions

Cosmochemical studies (e.g., GANAPATHY and ANDERS, 1974) suggest that the composition of the bulk earth is essentially close to chondritic in terms of the major lithophile and refractory elements, such as Mg, Si, Al, and Ca. Because these elements are likely to be primarily concentrated in the mantle, the bulk mantle may contain these elements in a ratio close to chondritic. The mass balance calculation of iron between core and mantle implies that the $Fe/(Mg+Fe)$ of the mantle is likely to be around 0.1 in atomic ratio (e.g., RINGWOOD, 1979; LIU, 1982). Thus, a model chondritic mantle composition of the five-component system CaO-MgO-FeO-Al_2O_3-SiO_2 was adopted as a starting composition, in which the ratios of Ca, Al, Mg, and Si are chondritic (cosmic) with $Fe/(Fe+Mg)=0.1$ in atomic ratio. The cosmic abundance deduced by ANDERS and EBIHARA (1982) was used for calculating the starting composition. The starting composition of the model chondritic mantle used for the experiments is listed in Table 1.

Another possibility for the primitive mantle composition has been derived from petrological consideration:

High-Pressure Research in Mineral Physics, edited by M. H. Manghnani and Y. Syono, pp. 87–93.

TABLE 1. Starting Compositions (Weight Percent) Used for the Melting Experiments

	Chondritic Mantle*	Pyrolite (RINGWOOD, 1982)
SiO_2	50.18	45.20
Al_2O_3	3.62	4.40
FeO	7.16	8.00
MgO	36.18	39.00
CaO	2.86	3.40
Total	100.00	100.00

*The ratios of Si, Al, Mg, and Ca are those of the chondritic values (ANDERS and EBIHARA, 1982). Fe/(Fe+Mg)=0.1 in atomic ratio.

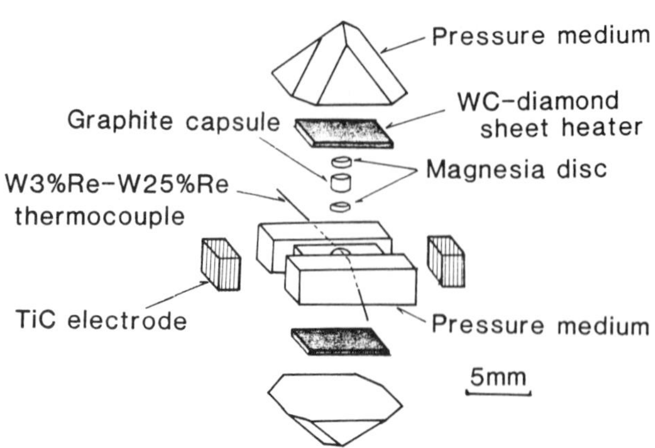

Figure 1. Furnace assembly used for the present experiments.

pyrolite, which can yield basaltic magma (e.g., RINGWOOD, 1966; 1982). Geochemical and petrological estimations of the primitive mantle based on the geochemical nature of komatiites (SUN and NESBIT, 1977) and spinel peridotite nodules (e.g., JAGOUTZ et al., 1979) also give a primitive mantle composition similar to pyrolite. The pyrolite composition is depleted in SiO_2 compared to the model chondritic mantle composition. Ringwood (e.g., RINGWOOD, 1979) explained that the SiO_2 depletion in pyrolite may result from selective evaporation of the earth-forming materials in the solar nebula due to the high volatility of SiO_2 at high temperatures. Pyrolites proposed by Ringwood have various ranges of chemical composition: Especially, the CaO/Al_2O_3 in pyrolite range from chondritic to higher than chondritic. Because fractionation of Ca and Al is not likely to occur in the nebula, the CaO/Al_2O_3 of the bulk mantle will be chondritic. Thus the pyrolite composition proposed by RINGWOOD (1982) is adopted, in which CaO/Al_2O_3 is close to chondritic. The starting composition of pyrolite used for the present experiments is also listed in Table 1.

The starting materials were prepared by mixing the reagents and heating the mixture at about 1100 °C under an atmosphere with controlled oxygen partial pressure ($p_{O_2}=10^{-10}$–10^{-12}, which is the oxygen fugacity between Fe-FeO and FeO-Fe_3O_4 buffers). The starting materials after heating were composed of olivine, pyroxenes, and a small amount of plagioclase for both the chondritic mantle and the pyrolite compositions.

Experimental Procedure

The apparatus used was the MA8-type high-pressure apparatus driven by a 2000-ton uniaxial press, RH3 guide-block system of Nagoya University. The heating assembly was similar to that reported by KATO and KUMAZAWA (1985) and is illustrated in Figure 1. The heating material was a composite material of tungsten carbide and diamond powder, which was packed in the pressure medium to form a sheet heater. The sample capsule made of graphite is put between the twin sheet heaters in the pressure medium, which was made of sintered magnesia containing about 8 mol % cobalt oxide (see Figure 1). A W25%Re-W3%Re thermocouple, 0.2 mm in diameter, was set in contact with the sample capsule without electrical contact with the heater. The thermocouple leads were pulled through pyrophyllite gaskets. No pressure correction was applied for the EMF of the thermocouple.

The furnace components and pressure medium were heated to around 1000 °C for an hour to remove water. The charge was dried again in an oven 170 °C for a few hours immediately before the experiments.

The pressure was calibrated on the basis of the resistance changes associated with the following transitions: Bi I-II, 2.55 GPa, Bi III-V, 7.7 GPa (LLOYD, 1971); PB I-II, 13.0 GPa (TAKAHASHI et al., 1969); ZnS, 15.6 GPa and GaAs, 18.7 GPa (YAGI and AKIMOTO, 1977). The uncertainty in the pressure values of the present experiments was estimated to be ±0.6 GPa at 15 GPa and ±2 GPa at 20 GPa, by comparison of the phase boundaries of silicate systems determined by in situ X-ray diffraction experiments at high pressure and temperature (AKIMOTO et al., 1977; YAGI et al., 1984).

The quenching method was used for the present experiments. The pressure was applied first, and then the temperature was increased to the desired value and held constant for about 0.5–15 minutes, depending on the experimental temperatures. The charge was quenched by shutting off the electric power of the furnace. Examination of the run products adjacent to the thermocouple has been made by X-ray powder diffraction, optical and scanning electron microscopy, and electron probe microanalyser (EPMA) analysis. In typical runs, small pieces of the run products, the compositions of which were determined by EPMA, were picked out and identified by X-ray powder diffraction with the Debye-Scherrer camera with Gandolfi attachment. The melt was identified under the scanning electron microscope by the presence of fibrous or dendritic texture, which has been established as diagnostic of quenching from liquid at high pressures (e.g., OHTANI et al., 1982).

Experimental Results

The melting experiments for the chondritic mantle composition were conducted in the pressure range of 12–20 GPa. Those for the pyrolite composition were carried out at 8–20 GPa. Table 2 summarizes the experimental conditions and the results obtained for the two compositions.

Melting Relations of the Chondritic Mantle Composition

Melting experiments on this composition were carried out at pressures of 12, 15, 17.5, and 20 GPa. Figure 2A shows the back scattered electron image (BEI) of the charge adjacent to the thermocouple junction quenched at 12 GPa and 1900 °C. Large euhedral olivine coexists with quenched liquid as shown in this photograph. This observation clearly indicates that olivine is the liquidus phase. This implies that the melt formed by partial melting at 12 GPa is richer in SiO_2 than the chondritic mantle composition (50.18 wt% SiO_2), which is consistent with the melting experiments on natural peridotite (KLB-1) (TAKAHASHI and SCARFE, 1985; TAKAHASHI, 1986). In the lower temperature portion of the same charge (not

TABLE 2. Experimental Conditions and the Results

Pressure (GPa)	Temperature (°C)	Time (min)	Results
	Chondritic mantle composition		
12	1750±10	3.0	OL+GA+CPX
12	1800±20	2.0	OL+GA+CPX
12	1900±20	2.0	OL+L (Ol+CPX+L)#
12	2010±10	1.5	L
15	1900±10	3.5	OL+GA+tr.CPX
15	2030±30	1.0	GA+L
15	2130±50	0.5	L
17.5	1925±25	1.0	MS+OL+GA+tr.CPX
17.5	2020±20	1.0	GA+L
17.5	2090±30	1.0	L
19	2040±30	2.5	GA+L
20	1925±20	1.0	GA+MS
20	2050±10	1.0	GA+MS+L (GA+L)*
20	2150±30	0.5	L
	Pyrolite composition		
8	1900±50	5.0	OL+L
15	1930±20	2.0	OL+L
15	2030±20	0.5	OL+L
15	2175±25	0.5	L
20	1720± 5	15.0	MS+GA+(tr.CPX?)
20	2100±50	1.0	MS+GA+L

L, liquid; OL, olivine; GA, garnet (or majorite); CPX, Ca-poor clinopyroxene; MS, modified spinel.
#Lower temperature portion of the charge.
*Higher-temperature portion of the charge.

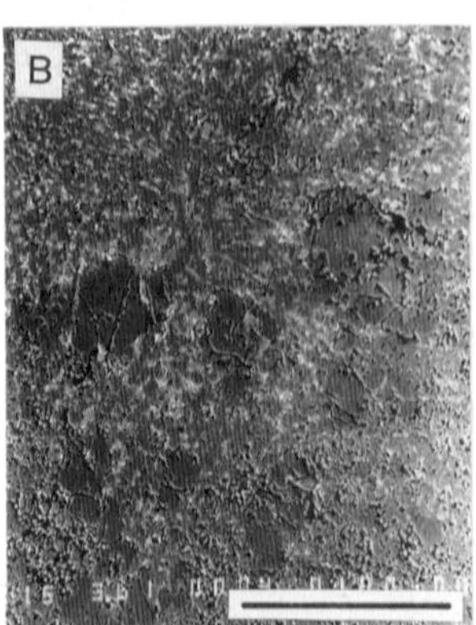

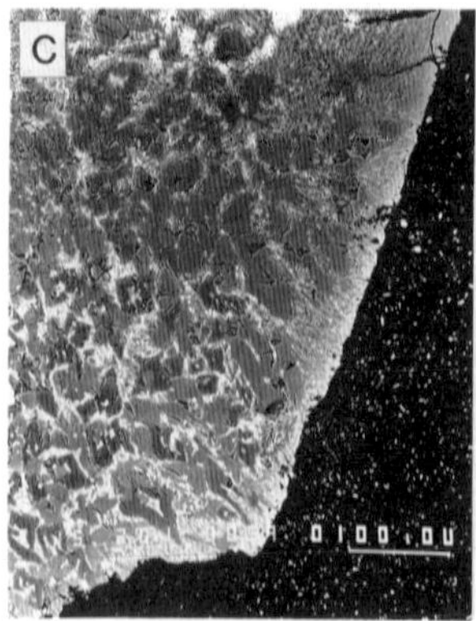

Figure 2. Back scattered electron images of the charges for the chondritic mantle composition. Scale bars represent 100 μm. (A) The charge quenched at 12 GPa and 1900 °C. Large dark crystals are olivine. This texture indicates that the liquidus phase is olivine at 12 GPa. (B) The charge quenched at 17.5 GPa and 2020 °C. Large grey crystals, which were identified as majorite, coexist with the quenched liquid. (C) The charge quenched at 20 GPa and 2050 °C. Modified spinel (dark grey crystals) and majorite (light grey crystals) coexist with the quenched liquid in the lower-temperature portion, whereas majorite coexists with the liquid alone in the higher-temperature portion of the charge.

shown in the figure), Ca-poor clinopyroxene was observed coexisting with olivine and liquid. This observation implies that the second crystallizing phase is Ca-poor clinopyroxene at 12 GPa. The subsolidus mineralogy at this pressure is olivine, Ca-poor clinopyroxene, and garnet.

The temperature gradient in the furnace was measured by IRIFUNE and HIBBERSON (1985). According to their measurement, the temperature difference between the hot and cold spots in the charge is less than 100 °C. In the run product quenched at 20 GPa and 2050 °C, the textural difference due to the temperature gradient of the furnace was observed: in the hot portion of the charge only majorite crystals coexist with quenched liquid, as shown in Figure 2C, whereas both majorite and modified spinel coexist with the quenched liquid in the colder portion of the charge where the thermocouple junction is located. These textural observations clearly indicate that the liquidus phase is majorite and the second phase crystallizing from the liquid is modified spinel. The liquid could not be quenched as glass, but very fine grained crystals with dendritic or lamellae textures were formed upon quenching. They are composed of modified spinel, majorite, and clinopyroxene, which were identified by X-ray powder diffraction with the Debye-Scherrer camera with the Gandolfi attachment. The subsolidus mineralogy at 20 GPa is a mixture of majorite and modified spinel. Majorite was observed as the liquidus phase also at pressures of 15 and 17.5 GPa. The BEI of the charge quenched at 17.5 GPa and 2020 °C is also shown in Figure 2B.

The results of the melting experiments on the chondritic mantle composition are summarized in a phase diagram (Figure 3). The present results clearly indicate that a change in liquidus phase from olivine to majorite occurs at a pressure between 12 and 15 GPa. The compositions of minerals coexisting with the liquid together with those observed under the subsolidus condition at 20 GPa were determined by EPMA and are given in Table 3. It is noteworthy that the CaO/Al_2O_3 ratio of majorite coexisting with the liquid is lower than that observed under the subsolidus conditions. The CaO and FeO contents in majorite decrease with partial melting, whereas the Al_2O_3 content in majorite increases with partial melting. The partition coefficient K_D of Mg and Fe between majorite and modified spinel is close to unity at 20 GPa and temperatures above 1930 °C.

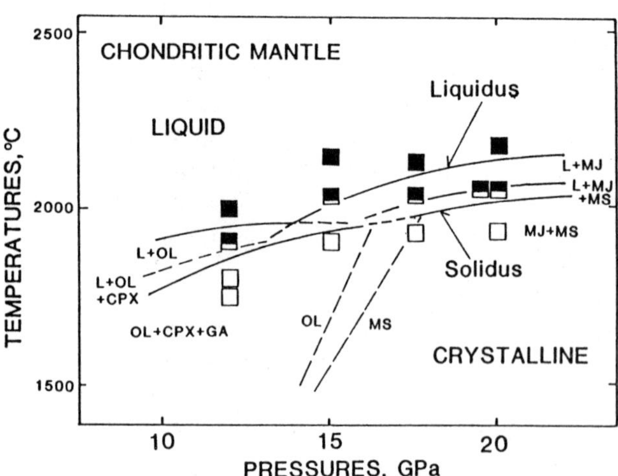

Figure 3. Melting relations of the chondritic mantle composition to 20 GPa. □, subsolidus; ▀, crystal and liquid; ■, superliquidus.

Melting Relations of the Pyrolite Composition

Melting experiments on pyrolite were made at 8, 15, and 20 GPa. Figures 4A, 4B, and 4C show photographs of the back scattered electron images (BEI) of the charges quenched at 8, 15, and 20 GPa, respectively. Figures 4A and 4B clearly indicate that olivine crystals are coexisting with the quenched liquid, i.e., that olivine is the liquidus phase at least up to 15 GPa. Figure 4C indicates that both modified spinel and majorite coexist with the quenched liquid around the liquidus temperature. It is difficult to distinguish whether the first crystallizing phase is modified spinel or majorite. The major phases existing below the solidus for the pyrolite composition are likely to be modified spinel and majorite on the basis of the results of the melting and phase relations of the chondritic mantle composition. The quenching texture at 20 GPa (Figure 4C) implies that the pyrolite composition is close to the eutectic composition between modified spinel and

TABLE 3. Chemical Compositions (weight percent) of Minerals in the Chondritic Mantle and Pyrolite Compositions at 20 GPa

	GA1	MS1	GA2	MS2	GA3	GA4	MS4
SiO_2	56.29	42.89	54.83	42.37	54.24	54.31	42.10
Al_2O_3	4.50	0.29	8.16	1.51	9.70	8.27	1.53
FeO	2.48	4.17	1.88	2.75	1.48	2.35	4.02
MgO	32.60	53.63	34.08	54.22	34.29	30.91	53.30
CaO	3.98	0.09	1.97	0.05	1.14	4.90	0.10
Total	99.85	101.07	100.92	100.90	100.85	100.74	101.05
O	12.0	4.0	12.0	4.0	12.0	12.0	4.0
Si	3.850	1.010	3.668	0.994	3.632	3.657	0.990
Al	0.363	0.008	0.643	0.042	0.765	0.656	0.043
Fe	0.142	0.082	0.105	0.054	0.083	0.132	0.079
Mg	3.323	1.883	3.398	1.895	3.422	3.102	1.869
Ca	0.291	0.002	0.141	0.001	0.082	0.353	0.002

GA1 and MS1, majorite and modified spinel under the subsolidus condition; GA2 and MS2, majorite and modified spinel coexisting with a liquid; GA3, majorite coexisting with a liquid alone for the chondritic mantle composition. GA4 and MS4, majorite and modified spinel coexisting with the liquid for the pyrolite composition.

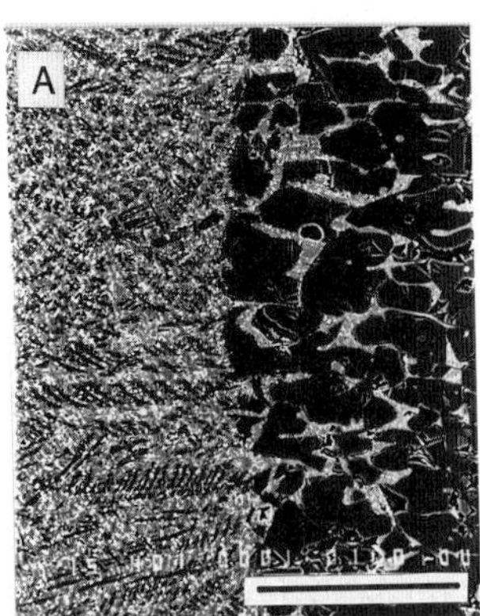

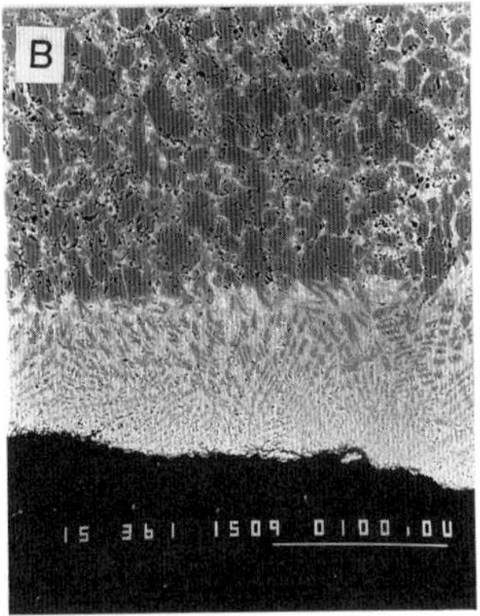

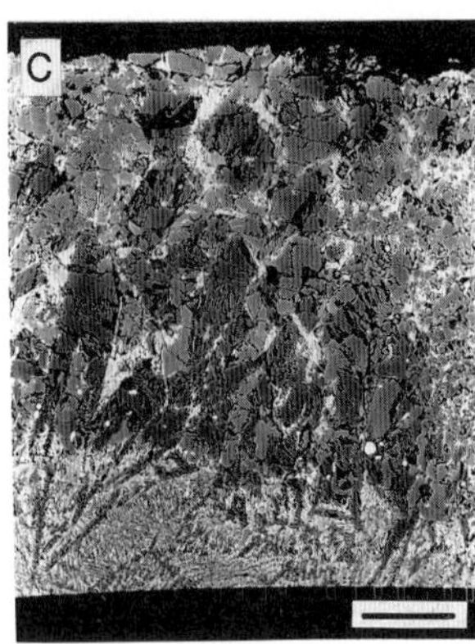

Figure 4. Back scattered electron images of the charges for the pyrolite composition. Scale bars represent 100 μm. (A) The charge quenched at 8 GPa and 1900 °C. Large dark crystals are olivine, which is the liquidus phase at 8 GPa. (B) The charge quenched at 15 GPa and 1930 °C. The large grey crystals coexisting with the quenched liquid are olivine. The texture indicates that olivine is the liquidus phase at 15 GPa. (C) The charge quenched at 20 GPa and 2050 °C. Both modified spinel and majorite coexist with the quenched liquid.

majorite, especially in terms of MgO and SiO_2, although depletion of FeO and CaO was observed in the residual solid upon partial melting. Similar quenching texture in which olivine or modified spinel coexists with majorite at the melt/solid interface was observed under dry conditions at 16–20 GPa (TAKAHASHI, personal communication). The compositions of minerals coexisting with liquid at 20 GPa are shown in Table 3. Majorite coexisting with the liquid and modified spinel has CaO/Al_2O_3 lower than chondritic. This result is similar to that for the chondritic mantle composition. Thus, the peridotitic liquid with CaO/Al_2O_3 higher than chondritic is formed by partial melting of pyrolite at 20 GPa. The partition coefficient K_D of Mg and Fe between modified spinel and majorite is close to unity at above the solidus and 20 GPa, which is consistent with the results observed for the chondritic mantle composition.

The (Mg,Fe)Al_2O_4 Component in Modified Spinel

Compositions of modified spinel coexisting with the liquid listed in Table 3 indicate that these crystals contain about 1.5 wt% of Al_2O_3 above the solidus temperature and 20 GPa. These values are significantly larger than those for olivine observed at the liquidus at 12 GPa. The modified spinel crystals analyzed were larger than 50 μm and no inclusion of aluminous phases was observed in the modified spinel crystals. RINGWOOD and MAJOR (1970) first discussed the possibility of solid solution along the join $(Mg,Fe)_2SiO_4$-(Mg,Fe)Al_2O_4 with modified spinel and spinel structures. Dissolution of a small amount of the (Mg,Fe)Al_2O_4 component in modified spinel structure has been reported by AKAOGI and AKIMOTO (1979). According to their experiments, about 0.3 wt% Al_2O_3, which corresponds to 0.6 mol% (Mg,Fe)Al_2O_4 component, was observed to dissolve at 14.5 GPa and 1200 °C. The present observation of a higher Al_2O_3 content of modified spinel suggests that solubility of Al_2O_3 increases with pressure and temperature. The effect of the solid solubility of the (Mg,Fe)Al_2O_4 component in the modified spinel structure cannot be ignored in discussing the partitioning of the trivalent cations such as Al and Cr between minerals and liquid at the base of the upper mantle.

Discussion

The composition of majorites coexisting with the liquids at 20 GPa (Table 3) together with the bulk composition of the chondritic mantle and pyrolite compositions (Table 1) are projected in the CS-MS-A plane (O'HARA, 1968) and shown in Figure 5. The subsolidus majorite is also projected in this figure for the chondritic mantle composition. The results of KATO and KUMAZAWA (1986) for the spinel lherzolite KLB-1 are shown in the same figure. The majorite crystals coexisting with the liquids have CaO/Al_2O_3 lower than the starting composition. This observation implies that the coexisting liquids have CaO/Al_2O_3 higher than chondritic. These results clearly depict the nature of the majorite fractionation by partial melting at the base of the upper mantle.

Pyrolite and KLB-1 project in the portion close to the diopside-pyrope join in the plane, whereas the chondritic mantle projects to a point close to the enstatite end-component, because of difference in the bulk compositions. Figure 5 clearly indicates that the pyrope component in majorite increases with degree of partial melting for the chondritic mantle composition. The major phases expected under the subsolidus conditions for pyrolite and

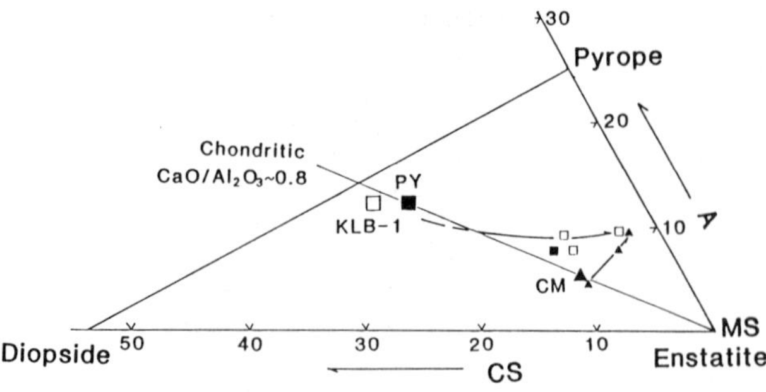

Figure 5. The compositions of majorite coexisting with the liquid together with the bulk compositions of the chondritic mantle and pyrolite. They are projected in the CS-MS-A plane (O'HARA, 1968). The results of KLB-1 (KATO and KUMAZAWA, 1986) are also shown in this figure. The CaO/Al_2O_3 ratio of majorite decreases by partial melting, thus the melt coexisting with majorite is expected to have a CaO/Al_2O_3 higher than chondritic. The chondritic CaO/Al_2O_3 (about 0.8) is also shown in the figure. ▲, bulk composition of the chondritic mantle; ▴, subsolidus and supersolidus majorite for the chondritic mantle (this work); ■, bulk composition of pyrolite; ▪, majorite above the solidus for the pyrolite composition (this work); □, bulk composition of KLB-1; ▫, majorite above the solidus for KLB-1 (KATO and KUMAZAWA, 1986).

KLB-1 will be modified spinel and majorite from the phase relations of the pyroxene-garnet transition (AKAOGI and AKIMOTO, 1977, 1979; AKAOGI and NAVROTSKY, 1986) and the stability field of majorite in the CaO-MgO-Al_2O_3-SiO_2 system (YAMADA et al., 1982). Therefore, the subsolidus majorite projects to the points of the bulk compositions of KLB-1 and pyrolite, although the composition of the subsolidus majorite was not determined in this study. Figure 5 suggests that the pyrope component relative to the diopside component in majorite increases with partial melting, whereas that relative to the enstatite component decreases with partial melting for pyrolite and KLB-1 compositions. It is noteworthy that the liquidus majorites seem to have very similar composition with low CaO/Al_2O_3 for these three different bulk compositions; i.e., the chondritic mantle composition, pyrolite, and KLB-1. The trends of change of the majorite compositions with increasing degree of partial melting are shown as arrows in Figure 5.

The present experiments for the chondritic mantle and pyrolite compositions indicate that majorite fractionation occurs by partial melting at the base of the upper mantle, and the partially molten liquid has a peridotitic composition with high CaO/Al_2O_3. The density of peridotitic liquid has been estimated by OHTANI (1984) theoretically, and the existence of a density cross-over between olivine and the liqud has been suggested at a pressure around 10 GPa. Majorite, however, is likely to be denser than the liquid based on the estimation by OHTANI (1984). Thus, the downward separation of majorite and upward transportation of peridotitic magma are expected to occur in the deep upper mantle.

The melting experiments on the chondritic mantle composition indicate that a change in the liquidus phase occurs at a pressure between 12 and 15 GPa, and majorite continues to be stable as the liquidus phase at least up to 20 GPa. Thus, the majorite fractionation is expected to occur over a wide range of pressures in the chondritic primordial mantle. If the primitive earth was molten to the depth of the base of the upper mantle during accretion and core-formation stages, the majorite fractionation could have produced a layered mantle, with a largely molten peridotitic upper mantle and a majorite-enriched transition zone. The CaO/Al_2O_3 ratio of the upper mantle is expected to be higher than the chondritic value, whereas that of the transition zone is likely to be lower than the chondritic value. The CaO/Al_2O_3 of the upper mantle was studied by PULME and NICKEL (1985) on the basis of the geochemical features of spinel lherzolite nodules. They suggested a high CaO/Al_2O_3 ratio for the upper mantle. Such a feature of the upper mantle is consistent with that resulting from majorite fractionation in the deep upper mantle.

It is well known that HREE are partitioned into pyropic garnet relative to the liquid (e.g., SHIMIZU and KUSHIRO, 1975). Because majorite has the garnet structure, the HREE is likely to be enriched in majorite compared to the coexisting liquid. If majorite is enriched in the transition zone by the majorite fractionation, HREE may be enriched in this region and depleted in the upper mantle, which was formed by accumulation of partially molten ultrabasic magmas in the early stages of terrestrial evolution.

Acknowledgments. The author is grateful to M. Kumazawa of Tokyo University and M. Kato of Ehime University for constant encouragement during this work. He thanks H. Sawamoto of Nagoya University for providing the opportunity to use the high-pressure system of Nagoya University. He also thanks H. Sawamoto and T. Kato of Nagoya University, E. Takahashi and E. Ito of Okayama University, and C. T. Herzberg of Rutgers University for useful discussions during various stages of this work. He is indebted to K. Fujino of Ehime University for providing the opportunity to use the Debye-Scherrer camera with the Gandolfi attachment. The author thanks two anonymous reviewers for critical reading and improving the manuscript. This work was supported by a grant-in-aid from the Ministry of Education, Science and Culture, Japan.

REFERENCES

AKAOGI, M., and S. AKIMOTO, Pyroxene-garnet solid solution equilibria in the systems $Mg_4Si_4O_{12}$-$Mg_3Al_2Si_3O_{12}$ and $Fe_4Si_4O_{12}$-$Fe_3Al_2Si_3O_{12}$ at high pressures and temperatures, *Phys. Earth Planet. Inter., 15*, 90–106, 1977.

AKAOGI, M., and S. AKIMOTO, High-pressure phase equilibria in a garnet lherzolite, with special reference to Mg^{2+}-Fe^{2+} partitioning among constituent minerals, *Phys. Earth Planet. Inter., 19*, 31–51, 1979.

AKAOGI, M., and A. NAVROTSKY, Phase diagram of the pyroxene-garnet transition in the $Mg_4Si_4O_{12}$-$Mg_3Al_2Si_3O_{12}$ system based on the thermochemical data (abstract), *Programme and Abstracts of the Seismological Society of Japan, No. 1*, p. 244, 1986.

AKIMOTO, S., T. YAGI, and K. INOUE, High temperature-pressure phase boundaries in silicate systems using in situ X-ray diffraction, in *High-Pressure Research Applications in Geophysics*, edited by M. H. Manghnani and S. Akimoto, pp. 595–602, Academic Press, New York, 1977.

ANDERS, E., and M. EBIHARA, Solar-system abundances of the elements, *Geochim. Cosmochim. Acta, 46*, 2363–2380, 1982.

ANDERSON, D. L., Chemical stratification of the mantle, *J. Geophys. Res., 84*, 6297–6298, 1979.

ANDERSON, D. L., A global geochemical model for the evolution of the mantle, in *Evolution of the Earth, Geodyn. Ser.*, Vol. 5, pp. 6–18, Am. Geophys. Union, Washington, D.C., 1982.

GANAPATHY, R., and E. ANDERS, Bulk compositions of the moon and earth estimated from meteorites, *Proc. Lunar Sci. Conf. 5th*, 1181–1206, 1974.

HAYASHI, C., K. NAKAZAWA, and H. MIZUNO, Earth's melting due to the blanketing effect of the primodial dense atmosphere , *Earth Planet. Sci. Lett., 43*, 22–28, 1979.

HERZBERG, C. T., and M. J. O'HARA, Origin of mantle peridotite and komatiite by partial melting, *Geophys. Res. Lett., 12*, 541–544, 1985.

IRIFUNE, T., and W. O. HIBBERSON, Improved furnace design for multiple anvil apparatus for pressures to 18 GPa and temperature to 2000 °C, *High Temp. High Pressures, 17*, 575–579, 1985.

JAGOUTZ, E., H. PALME, H. BADDENHAUSEN, K. BLUM, M. CENDALES, G. DREIBUS, B. SPETTEL, V. LORENZ, and H. WANKE, The abundances of major, minor and trace elements in the earth's mantle as derived from primitive ultramafic nodules, *Proc. Lunar Planet. Sci. Conf. 10th*, 2031–2050, 1979.

KATO, T., and M. KUMAZAWA, Incongruent melting of Mg_2SiO_4 at 20 GPa, *Phys. Earth Planet. Inter., 41*, 1–5, 1985.

KATO, T., and M. KUMAZAWA, Melting experiment of natural lherzolite at 20 GPa: Formation of phase B coexisting with garnet, *Geophys. Res. Lett., 13*, 181–184, 1986.

KAULA, W. M., Thermal evolution of the earth and moon growing by planetesimal impacts, *J. Geophys. Res., 84*, 999–1008, 1979.

LIU, L. G., Speculations on the composition and origin of the earth, *Geochem. J., 16*, 287–310, 1982.

LLOYD, E. C., Accurate characterization of the high pressure environment, *NBS Spec. Publ. No. 326*, pp.1–3, Washington D.C., 1971.

OHTANI, E., Generation of komatiite magma and gravitational differentiation in the deep upper mantle, *Earth Planet. Sci. Lett., 67*, 261–272, 1984.

OHTANI, E., The primodial terrestrial magma ocean and its implication for stratification of the mantle, *Phys. Earth Planet. Inter., 38*, 70–80, 1985.

OHTANI, E., Chemical stratification of the mantle formed by melting in the early stage of the terrestrial evolution, *Tectonophysics*, in press, 1987.

OHTANI, E., M. KUMAZAWA, T. KATO, and T. IRIFUNE, Melting of various silicates at elevated pressures, in *High-Pressure Research in Geophysics: Advances in Geophysics, Adv. in Earth Planet. Sci.*, Vol. 12, edited by S. Akimoto and M. H. Manghnani, pp. 259–270, Center for Academic Publications, Tokyo, 1982.

O'HARA, M., The bearing of phase equilibria studies in synthetic and natural systems on the origin and evolution of basic and ultrabasic rocks, *Earth Sci. Rev., 4*, 69–133, 1968.

PULME, H., and K. G. NICKEL, Ca/Al ratio and composition of the earth's upper mantle, *Geochim. Cosmochim. Acta, 49*, 2123–2132, 1985.

RINGWOOD, A. E., Mineralogy of the mantle, in *Advance in Earth Sciences*, edited by P. M. Hurley, pp. 357–399, M.I.T. Press, Cambridge, Mass., 1966.

RINGWOOD, A. E., *Origin of the Earth and Moon*, 295 pp., Springer-Verlag, New York, 1979.

RINGWOOD, A. E., Phase transformations and differentiation in subducted lithosphere: Implication for mantle dynamics, basalt petrogenesis and crustal evolution, *J. Geol., 90*, 611–643, 1982.

RINGWOOD, A. E., and A. MAJOR, The system Mg_2SiO_4-Fe_2SiO_4 at high pressures and temperatures, *Phys. Earth Planet. Inter., 3*, 89–108, 1970.

SHIMIZU, N., and I. KUSHIRO, The partitioning of rare earth elements between garnet and liquid at high pressures: preliminary experiments, *Geophys. Res. Lett., 2*, 413–416, 1975.

SUN, S. S., and R. W. NESBITT, Chemical heterogenity of the Archaean mantle, composition of the earth and mantle evolution, *Earth Planet. Sci. Lett., 35*, 429–448, 1977.

TAKAHASHI, E., Melting of a dry peridotite KLB-1 up to 14 GPa: Implications on the origin of peridotite upper mantle, *J. Geophys. Res., 91*, 9367–9382, 1986.

TAKAHASHI, E., and C. M. SCARFE, Melting of peridotite to 14 GPa and the genesis of komatiite, *Nature, 315*, 566–568, 1985.

TAKAHASHI, T., H. K. MAO, and W. A. BASSET, Lead: X-ray diffraction study of a high pressure polymorph, *Science, 165*, 1352–1353, 1969.

YAGI, T., and S. AKIMOTO, Pressure calibration above 100 kbar based on the NaCl internal standard, in *High Pressure Research: Applications in Geophysics*, edited by M. H. Manghnani and S. Akimoto, pp. 573–583, Academic Press, New York, 1977.

YAGI, T., M. ARASHI, T. OKAI, K. KAWAMURA, K. SHINO, M. SHIMOMURA, T. SUZUKI, K. TABATA, and S. AKIMOTO, Precise determination of olivine-spinel phase transformation in Fe_2SiO_4 (abstract), *High Pressure Conf. Jpn. 25th*, 30–31, 1984.

YAMADA, H., E. TAKAHASHI, and E. ITO, High-pressure and high-temperature experimental study in the system CaO-MgO-Al_2O_3-SiO_2 (abstract), *High Pressure Conf. Jpn. 23rd*, 124–125, 1982.

EXPERIMENTAL STUDY ON THE PHASE RELATIONS IN THE SYSTEM Fe-Ni-O-S UP TO 15 GPa

S. URAKAWA and M. KATO

Department of Earth Sciences, Nagoya University, Chikusa-ku, Nagoya 464, Japan

M. KUMAZAWA

Department of Geophysics, University of Tokyo, Bunkyo-ku, Tokyo 113, Japan

Abstract. The effects of nickel, oxygen, and sulfur alloying on the melting relations of iron have been experimentally investigated up to a pressure of 15 GPa to provide information on problems involved in investigation of the earth's core. The Fe-rich portion of the system Fe-Ni-O-S is a eutectic system with a wide region of liquid immiscibility which is reduced with increasing pressure and which disappears above 20–25 GPa. The oxygen solubility in the eutectic liquid increases with pressure, from 1 atomic percent at 6 GPa to about 2–3 atomic percent at 15 GPa. The sulfur content is about 34 atomic percent at 6 GPa and decreases with pressure. The change of the eutectic composition is related to reduction of liquid immiscibility. The eutectic temperature in the system Fe-Ni-O-S is about 825 °C at 6 GPa and increases with a small pressure gradient of 6 K/GPa.

A sufficient amount of oxygen, sulfur, and other light elements (such as carbon and hydrogen) are available in the primordial earth, and the melting temperature of the actual core-forming material is less than 800 °C. The oxygen and sulfur bearing metallic liquid can easily penetrate the polycrystalline texture of silicates by chemical corrosion because of the low interfacial energy. Separation of the core material should have started in the accretionally growing earth. Thus core formation is considered to be a process of chemical equilibration between the mantle material and the molten iron alloy that is travelling down to the core.

Introduction

The earth's core is supposed to consist of metallic iron and some nickel. BIRCH (1952, 1964) showed that the density of the molten outer core is about 10 percent less than that of iron-nickel alloy under the relevant conditions of pressure and temperature. He also suggested that the outer core must contain some light elements to compensate for this density discrepancy, but that the solid inner core is Fe-Ni alloy and contains no lightening elements. Birch's conclusion seems quite reasonable and has been accepted by later investigators. However, there have been controversies regarding 1) the alloying elements that lighten the core density and 2) the formation processes of the core. Therefore we need to study a) the solubility of all the candidate lightening elements in the molten iron and b) all possible processes in which these elements are incorporated into the core, particularly with regard to the sinking of the core material and the chemical interaction of the core material with the mantle following core formation.

BIRCH (1952, 1964) proposed oxygen as a possible lightening elements. This possibility was also considered on geochemical grounds by RINGWOOD (1977). The system Fe-FeO was experimentally studied by OHTANI et al. (1984) and KATO (1987), and its liquidus relations were clarified up to 20 GPa. The nature of liquid immiscibility in the system Fe-FeO was also investigated thermodynamically (OHTANI et al., 1984; KATO, 1987). MASON (1966) suggested the possible presence of sulfur in the earth's core based on a H-chondrite model of the earth. MURTHY and HALL (1970) proposed that sulfur is the most adequate light element in the core because: 1) it is more depleted than the other volatiles (C, N, etc.) in the mantle and crust, and 2) it can easily alloy with iron and thus cause a lowering of the melting temperature by about 500 °C at 0 GPa. The first experimental study on the system Fe-FeS was made by BRETT and BELL (1969) at 3 GPa. The eutectic temperature and composition in the system Fe-FeS and Fe-Ni-S were experimentally determined up to 10 GPa by USSELMAN (1975a). USSELMAN (1975b) also extrapolated the temperature and composition at the eutectic point in the system Fe-FeS under pressure and temperature conditions of the core based on his data. Recently FUKAI and SUZUKI (1986) have examined the possibility that hydrogen is an important lightening element, on the basis of high pressure experiments on metal hydrides up to 6 GPa.

However, no study has yet been made the effects of the plural elements that are most likely to actually occur. The effect of nickel, in particular, on the melting of iron has seldom been systematically studied, except by USSELMAN (1975a), probably because nickel is supposed to behave the same as iron. Therefore the physicochemical properties of actual core materials at high pressure have not been sufficiently clarified in previous experiments to allow elucidation of the core problem.

In this study, we extend the composition and pressure range of the experimental study of core materials and

High-Pressure Research in Mineral Physics, edited by M. H. Manghnani and Y. Syono, pp. 95–111.

investigate the interaction of core and mantle materials. We assume, on the basis of meteoritic analogue and solar abundances of elements, that the candidate light elements are oxygen and sulfur. In preceding study on the system Fe-FeO-FeS pressure was restricted to only 3 GPa (HILTY and CRAFTS, 1952; NALDRETT, 1969; WENDLANDT and HUEBNER, 1979). Recently KATO (1987), in a experimental study of melting relations in the system Fe-FeO-FeS, extended the pressure range to 20 GPa by using an MA8 type of high pressure device. Based on Kato's work, we experimentally investigate the system Fe-FeO-FeS in more detail and also study the effect of the addition of nickel on the melting temperature and liquidus relations up to 15 GPa. The composition and formation process of the earth's core is discussed below, on the basis of the chemical and physical properties of metallic melt under high pressure.

Experiments

High Pressure Experiments

Pressure-generating apparatus and calibration. High pressure was generated by means of the MA8 type of apparatus (e.g., SAWAMOTO, 1986). This apparatus was also used for a series of melting experiments in our laboratory (e.g., OHTANI, 1979; KATO and KUMAZAWA, 1986; KATO, 1987). The MA8 apparatus consists of eight tungsten carbide cubic anvils with one truncated corner. Semisintered magnesia was used as a pressure transmitting medium, and filled the octahedral space among the eight cubic anvils. Preformed gaskets of pyrophyllite were placed between the anvils to provide lateral support. In runs up to 10 GPa, anvils with a truncated edge length of 8 mm were used, and the pressure was calibrated by measuring the electric resistivity change at room temperature due to the phase transitions of BiI-II (2.55 GPa), BiIII-V (7.7 GPa), and SnI-II (9.2–9.6 GPa) and due to the metallic transition of ZnTe (12.9 GPa, ONODERA and OHTANI, 1980). For runs at 15 GPa, anvils with a truncated edge length of 3.5 mm were used, and the pressure was calibrated at room temperature with metallic transition of ZnS (15 GPa).

Sample assembly and heating system. The cell assemblies used in these experiments are shown in Figure 1. The heating element used in experiments up to 10 GPa was composed of a graphite sleeve. The sample was enclosed in a polycrystalline MgO capsule in order to avoid any reaction between the sample and the heating material. The two sample capsules were packed into one graphite heater. Temperatures were measured at the center and the end of the heater with a W97%Re3%-W75%Re25% thermocouple which contacted the truncated top of the anvils as a cold junction. The temperature of the anvil-top was also simultaneously monitored by a Chromel-Alumel thermocouple to correct the e.m.f. of two W97%Re3%-W75%Re25% thermocouples. The temperature of the sample was estimated from the e.m.f. of three thermocouples by a simple thermal conduction model. The temperature difference between the center and end of the heater is less than 100 °C in runs at

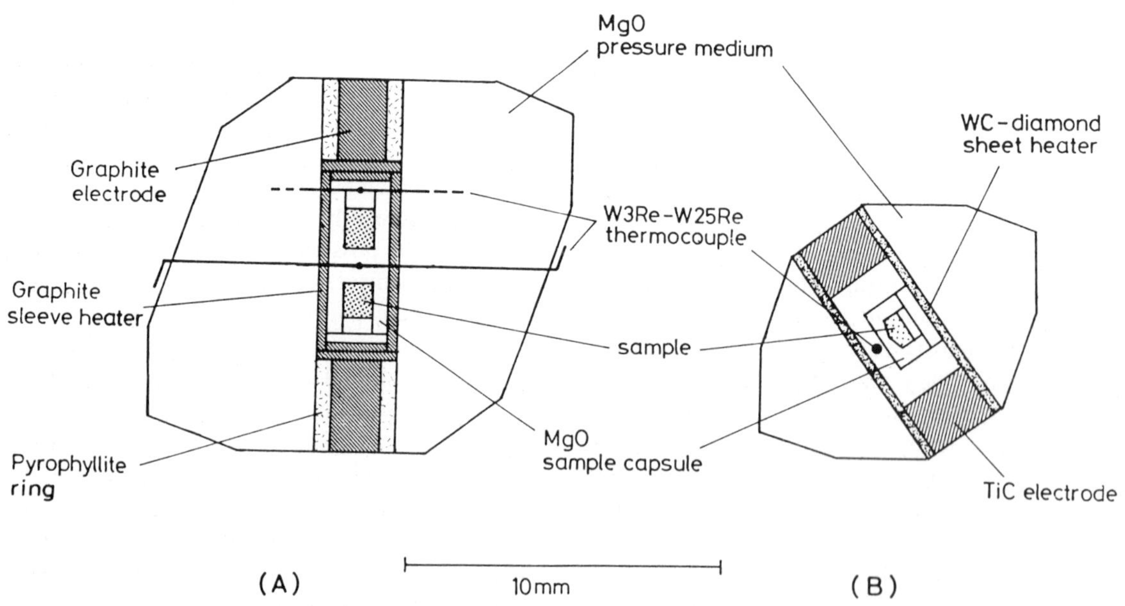

Figure 1. Cross-section of the pressure medium and furnace assembly for experiments up to 10 GPa (A) and at 15 GPa (B).

temperatures lower than 1500 °C (except for runs 602, 606, 612, N605, N606, and N607), and is less than 200 °C in runs above 2000 °C.

A mixture of tungsten carbide and diamond powder with a volume ratio of 1:1 was used for the heater at 15 GPa instead of graphite because graphite undergoes conversion to diamond at this pressure (KATO and KUMAZAWA, 1986). The pressure medium is shown in Figure 1B; the middle part contained the MgO sample capsule and was bounded by the two sheet heaters and was sandwiched by two MgO caps. A W97%Re3%-W75%Re25% thermocouple was inserted into the middle part of the pressure medium without contacting the sheet heaters, and was extended out of the cell through the gaskets. Thus temperature could be measured directly without requiring correction for the temperature on the truncated-top of the anvils. The correction for the pressure effect on the e.m.f. of the thermocouple was not made for all experiments.

Experimental procedures. The starting material was enclosed in an MgO sample container, and then dried in an oven at about 110 °C for a few hours to prevent the effect of water on the melting temperature. After pressure was applied, the sample was heated to the desired temperature. Heating duration was limited to 1–5 min, except in several low temperature runs, in order to reduce reaction of sample with the MgO capsule. The sample was quenched by cutting electrical power.

Sample Preparation

For experiments on the system Fe-FeO-FeS, the starting materials were prepared by mixing the powders of iron, pyrrhotite, and wüstite in the desired ratios listed in Table 1. The powders of iron (99.99% pure) and pyrrhotite (99.9% pure) were obtained from Rare Metallic Co., Ltd. Wüstite was synthesized by thermal decomposition of 99.99% pure iron oxalate ($FeC_2O_4{\cdot}2H_2O$) in Ar carrier gas for 2 h at 1000 °C. An X-ray diffraction of wüstite shows the composition to be $Fe_{0.94}O$ (JETTE and FOOTE, 1933), and optic microscopic observation of wüstite reveals that it contains a trace amount of metallic iron.

In the Ni-bearing system, the ratio of Ni/(Fe+Ni) in the starting material was fixed at 0.1. First, the powder samples of ($Fe_{0.9}Ni_{0.1}$), ($Fe_{0.9}Ni_{0.1}$)O, and ($Fe_{0.9}Ni_{0.1}$)S were prepared by mixing Fe, Ni, FeO, NiO, FeS, and NiS. Next, the starting samples were made by mixing metal, oxide, and sulfide in the desired ratios listed in Table 2. The powder samples were 99.99% pure (in metal and NiO) and 99.9% pure (in sulfide). Iron oxide was prepared by the same method as in the case of the system Fe-FeO-FeS.

Examination of Run Products

Recovered run products were polished to observe their texture and to analyze their chemical composition. Textural observation was carried out by using a reflecting microscope and a scanning electron microscope (SEM). Chemical analyses were subsequently made using a JEOL JCXA-733 electron probe microanalyzer (EPMA). Four elements (i.e., Fe, Ni, O, and S) were analyzed in all samples, and Mg was analyzed in some samples. For EPMA measurements the following conditions were used: an accelerating voltage of 15 kV, a beam current of 15–40 nA, and a beam diameter of 3–30 μ. The standard samples for Fe, Ni, O, and S are metallic iron (Fe), metallic nickel (Ni), hematite (Fe_2O_3), and chalcopyrite ($CuFeS_2$). The measured values were corrected by the ZAF method in which corrections for atomic number, absorption, and fluorescence were made by Duncumb-Reed, Philibert, and Reed methods, respectively.

Results

Textural Observations by SEM

The polished sections of run products were first examined by SEM. Figure 2 shows several examples of the back-scattered electron images of polished sections. Melting is indicated by the presence of a specific texture that is characteristic of quenching from melt: that is, in this case, fine intergrowth or dendrites of the constituent phases, metal, oxide, and sulfide. It is observed that the effect of the Ni addition does not alter the texture and the phases present.

Run products are classified into 5 groups: 1) solid-state products composed of irregularly shaped metal, oxide, and sulfide grains; 2) a cotectic liquid coexisting with solid metal and oxide as liquidus phases; 3) a liquid coexisting with solid oxide as liquidus phase; 4) a two-liquid state which is composed of metallic liquid rich in iron and nickel and of an ionic liquid rich in oxygen; and 5) the single liquid phase with the starting composition. The cotectic liquid coexisting with solid metal and sulfide, or oxide and sulfide does not appear in this study, because the starting composition is rich in metal and oxide relative to the eutectic composition.

The eutectic liquid which is the first melt of this ternary system is quenched into the eutectic intergrowth. The eutectic intergrowth consists of fine acicular crystals of metal, oxide, and sulfide (Figure 2A-2).

The liquid with the composition on the cotectic line which connects the eutectic point with the (Fe, Ni)-(Fe, Ni)O eutectic point, coexists with solid metal and oxide as liquidus phases. This cotectic liquid contains excess iron and nickel relative to the eutectic liquid. First the excess iron and nickel precipitate as metallic dendritic crystals during quenching, and then the residuum is quenched in a matrix which has the same texture as the eutectic intergrowth. The cotectic liquid positioned near the eutectic point shows the quench texture which is compos-

TABLE 1. Experimental Conditions, Starting Compositions and the Resulting Phase Assembly in the System Fe-FeO-FeS

Run No.	Pressure (GPa)	Temperature (°C)	Duration (min)	Composition (mol%) (Fe:FeO:FeS)[a]	Results[b]
601A	6	965±20	30	53: 7:40	Fe+FeO+liq(m)
B				43:14:43	FeO+liq(m)
602A	6	1110±75	10	53: 7:40	FeO+liq(m)
B				43:14:43	FeO+liq(m)
603A	6	1085±20	1	53: 7:40	FeO+liq(m)
B				61: 8:31	Fe+FeO+liq(m)
604A	6	1505±10	1	90: 5: 5	FeO+liq(m)+liq(i)
B				90: 5: 5	FeO+liq(m)+liq(i)
605A	6	940±10	5	53: 7:40	Fe+FeO+liq(m)
B				90: 5: 5	Fe+FeO+liq(m)
606A	6	1195±85	2	70: 5:25	Fe+FeO+liq(m)
B				80: 5:15	Fe+FeO+liq(m)
607A	6	1400±10	1	90: 5: 5	Fe+FeO+liq(m)
B				80: 5:15	liq(m)+liq(i)
608A	6	920±10	5	53: 7:40	Fe+FeO+FeS
B				43:14:43	Fe+FeO+FeS
609A	6	1410±15	1.5	53: 7:40	liq(m)+liq(i)
B				70: 5:25	liq(m)+liq(i)
610A	6	1295±20	4	80: 5:15	Fe+FeO+liq(m)
B				70: 5:25	FeO+liq(m)
611A	6	1450±20	2	61: 8:31	liq(m)+liq(i)
B				0:50:50	liq
612	6	1395±35	3	57:14:29	liq(m)+liq(i)
613	6	2100±55	1	61:36: 3	liq(m)+liq(i)
1001A	10	1060±10	3	53: 7:40	FeO+liq(m)
B				90: 5: 5	Fe+FeO+liq(m)
1002A	10	995±10	5	90: 5: 5	Fe+FeO+liq(m)
B				70: 5:25	Fe+FeO+liq(m)
1003A	10	965±10	2	90: 5: 5	Fe+FeO+liq(m)
B				70: 5:25	Fe+FeO+liq(m)
1501	15	1100[c]	5	53: 7:40	FeO+liq(m)
1502	15	1055	5	70: 5:25	Fe+FeO+liq(m)
1503	15	1010	5	70: 5:25	Fe+FeO+liq(m)
1504	15	960	2.6	70: 5:25	Fe+FeO+FeS
1505	15	1230	2	70: 5:25	Fe+FeO+liq(m)
1506	15	1405	3	70: 5:25	FeO+liq(m)
1507	15	1410	3	80: 5:15	Fe+FeO+liq(m)
1508	15	1600	2	90: 5: 5	Fe+FeO+liq(m)
1509	15	2000	1.5	50:25:25	liq(m)+liq(i)

[a]Starting composition of experiment: Fe=99.99% pure iron, FeO=$Fe_{0.94}O$, FeS=$Fe_{0.9}S$.

[b]Fe is iron, FeO is iron oxide ($Fe_{1-x}O$, liq(m) is metallic liquid phase (eutectic liquid, cotectic liquid, and metallic liquid of two-liquid), liq(i) is ionic liquid of two-liquid, and liq is one liquid with the starting composition.

[c]The relative temperature is well determined at 15 GPa, but uncertainty within ±50 degrees remains in the absolute temperature.

ed of a small amount of metallic dendrites distributed in a large matrix (Figure 2B-1), because the liquid's excess metallic component is low. The cotectic liquid near the (Fe, Ni)-(Fe, Ni)O join dissolves a considerable amount of iron and nickel at high temperatures. This liquid displays a quench texture which is composed of thick cross-stripped metallic dendrites and a matrix of network structure (Figure 2B-2).

The textural characteristics of the liquid coexisting with the solid oxide appears almost the same as for the cotectic

TABLE 2. Experimental Conditions, Starting Compositions and the Resulting Phase Assembly in the System Fe-Ni-O-S

Run No.	Pressure (GPa)	Temperature (°C)	Duration (min)	Composition[a] (M:O:S)	Results[b]
N601A	6	810± 15	3	6:1:3	M+O+S
B				7:1:2	M+O+S
N602A	6	840± 10	5	6:1:3	M+O+liq(m)
B				7:1:2	M+O+liq(m)
N603A	6	850± 25	3	6:1:3	M+O+liq(m)
B				7:1:2	M+O+liq(m)
N604A	6	890± 10	4	6:1:3	M+O+liq(m)
B				7:1:2	M+O+liq(m)
N605A	6	1160± 30	3	6:1:3	O+liq(m)
B				7:1:2	M+O+liq(m)
N606A	6	1360± 35	4	8:1:1	M+O+liq(m)
B				6:1:3	O+liq(m)
N607A	6	1470± 65	3.5	8:1:1	O+liq(m)
B				7:1:2	O+liq(m)
N608	6	2000±100	3	7:1:2	liq(m)+liq(i)
N609	6	2100± 55	2	6:3:1	liq(m)+liq(i)
N1001	10	900± 10	4	7:1:2	M+O+liq(m)
N1002	10	910± 25	5	7:1:2	M+O+liq(m)
N1501	15	865[c]	3	8:1:1	M+O+S
N1502	15	900	4	7:1:2	M+O+liq(m)
N1503	15	950	4	7:1:2	M+O+liq(m)
N1504	15	1225	2	7:1:2	O+liq(m)
N1505	15	1500	5	8:1:1	M+O+liq(m)
N1506	15	1830	2	5:3:2	O+liq(m)
N1507	15	2000	2	7:1:2	liq
N1508	15	2150	1	6:3:1	liq(m)+liq(i)

[a]M is metal, O is oxide, and S is sulfide, all with Ni/(Fe+Ni)=0.1.

[b]M is metallic phase, O is oxide phase, liq(m) is metallic liquid (eutectic liquid, cotectic liquid, and metallic liquid of two liquid), liq(i) is ionic liquid, and liq is one liquid with the starting composition.

[c]The relative temperature is well determined at 15 GPa, but uncertainty within ±50 degrees remains in the absolute temperature.

liquid. When the oxygen solubility in the melt is high, the occurrence of oxide dendrites is recognized in the liquid phase.

In the broad immiscible liquid field on the (Fe, Ni)-(Fe, Ni)O side of the system Fe-Ni-O-S, liquid separates into two phases: 1) metallic liquid rich in iron and nickel, and 2) ionic liquid rich in oxygen. The quenched phases of the two liquids have different textures, as shown in Figure 2C. The ionic liquids are shown in Figure 2C as round dark spherules. These spherules are mainly composed of iron-nickel oxide, but also contain certain metal and metallic sulfide components. These metal and metallic sulfide components of the ionic liquid are also shown in Figures 2C-2 and 2C-3 as light inclusions within dark spherules. The metallic liquid shows the same texture as the cotectic liquid and the liquid which coexists with solid oxide.

The single liquid phase appears above the liquidus temperature and has the same composition as the starting material (Figure 2D). The sample N1507 totally melted at 15 GPa and 2000 °C. The quenched liquid phase contains many oxide dendrites and metallic dendrites.

Chemical Analyses by EPMA

Liquid phase in the system Fe-FeO-FeS. The chemical compositions of liquid phases in run products, which were analyzed by EPMA, are listed in Table 3. Run 605A, which had a starting composition close to the eutectic point, is quenched from just above the eutectic temperature. The liquid phase of run 605A, which has the texture of the eutectic intergrowth, gives the eutectic composition $Fe_{66.7}O_{1.6}S_{32.8}$ at 6 GPa. In the same way, the eutectic composition at high pressures is determined from products of runs 1002 ($Fe_{68.1}O_{1.8}S_{29.9}$ at 10 GPa) and 1503 ($Fe_{69.5}O_{3.4}S_{26.6}$ at 15 GPa).

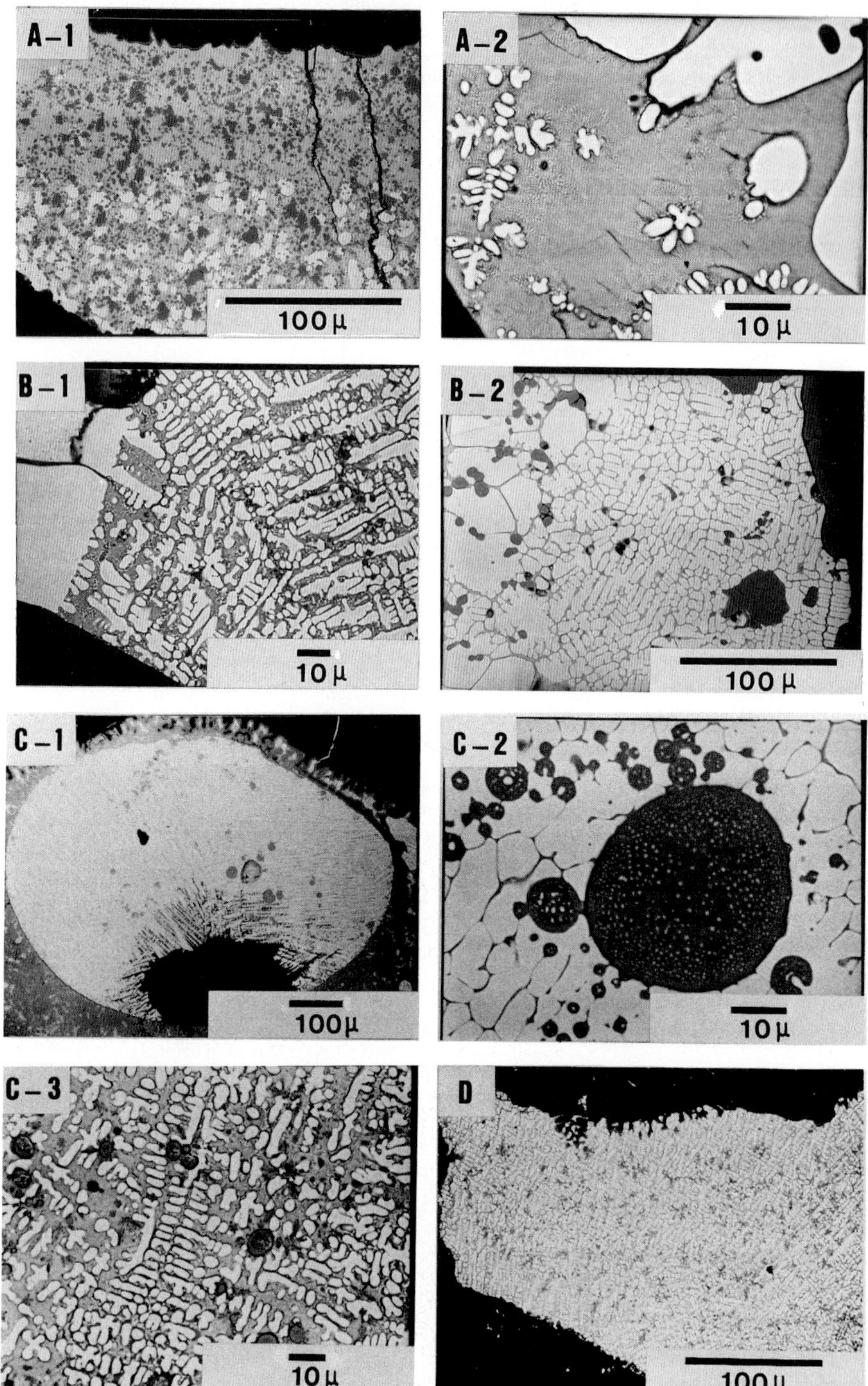

Figure 2. Photomicrographs (back-scattered electron images) of the recovered samples. A) Eutectic liquid coexisting with solid metal and oxide. In A-1 for run N1503, white round grain is solid metal, dark gray grain is solid oxide, and light gray part is quenched eutectic melt. Liquid phase of A-2 for run 605B is composed of metallic dendrites (white) and the eutectic intergrowth (gray), which has a fine texture in submicron size. B) Cotectic liquid coexisting with solid metal and oxide. In B-1 and B-2 for runs N1505 and 607A, the cotectic liquid, which consists of the metallic dendrites (white bars) and matrix (light gray part), coexists with solid metal (white part) and oxide (dark gray part). C) Immiscible two-liquid. In C-1 for run N609, the round dark spherules are the ionic liquid and the remaining light part is the metallic liquid. In C-2 for run 614 and C-3 for run 609B, the ionic liquid droplets contain metallic and sulfide inclusions, which are also liquid roplets. D) The single liquid for run N1507. Oxide dendrites (dark gray) and metallic dendrites (white) are recognized.

TABLE 3. Chemical Composition of Liquid in the System Fe-FeO-FeS

Run No.	P (GPa)	T (°C)	Composition of liquid (atomic %) Fe	(S.D.)[a]	O	(S.D.)[a]	S	(S.D.)[a]	Total	Analyzed[b] phase	Coexisting solid phase
605A	6	940	66.7	(1.1)	1.6	(0.8)	32.8	(0.5)	101.1	liq(m)	Fe+FeO
606A	6	1195	70.3	(0.9)	2.0	(0.3)	27.0	(0.9)	99.3	liq(m)	Fe+FeO
606B	6	1195	74.4	(1.3)	1.7	(0.5)	23.8	(0.9)	99.9	liq(m)	Fe+FeO
610A	6	1295	82.2	(1.1)	2.1	(0.7)	16.0	(0.9)	100.3	liq(m)	Fe+FeO
610B	6	1295	74.1	(1.6)	3.0	(1.0)	23.5	(1.3)	100.6	liq(m)	FeO
607A	6	1400	93.0	(3.1)	1.0	(0.7)	7.7	(2.8)	101.7	liq(m)	Fe+FeO
607B	6	1400	85.4	(4.6)	1.0	(0.6)	15.2	(4.8)	101.6	liq(m)	
609A	6	1410	77.1	(2.2)	2.3	(1.1)	22.4	(2.8)	101.8	liq(m)	
609B	6	1410	68.2	(0.6)	3.5	(1.5)	28.1	(8.1)	99.8	liq(m)	
612	6	1395	64.6	(1.1)	2.3	(0.8)	32.3	(1.1)	99.2	liq(m)	
611A	6	1450	66.8	(1.2)	2.4	(0.6)	28.4	(1.3)	97.6	liq(m)	
611A	6	1450	48.2	(1.1)	32.9	(1.1)	14.6	(1.8)	95.7[c]	liq(i)	
611B	6	1450	45.7	(0.4)	25.5	(0.6)	25.1	(0.5)	96.3[c]	liq	
604B	6	1505	94.8	(1.1)	0.8	(0.8)	3.7	(0.9)	99.3	liq(m)	FeO
604B	6	1505	53.0	(3.0)	43.3	(3.9)	5.2	(1.9)	101.5	liq(i)	FeO
613	6	2100	95.7	(0.7)	1.5	(0.5)	2.3	(0.4)	99.5	liq(m)	
613	6	2100	50.8	(0.2)	46.7	(0.1)	1.1	(0.1)	98.6	liq(i)	
1002B	10	965	68.1	(0.7)	1.8	(0.2)	29.9	(0.4)	99.8	liq(m)	Fe+FeO
1503	15	1010	69.5	(0.5)	3.4	(0.5)	26.6	(0.4)	99.5	liq(m)	Fe+FeO
1505	15	1230	76.9	(2.0)	2.4	(0.7)	22.5	(1.5)	101.8	liq(m)	Fe+FeO
1506	15	1405	75.4	(0.8)	2.7	(0.8)	23.5	(0.7)	101.6	liq(m)	FeO
1507	15	1410	84.2	(1.3)	1.8	(0.5)	15.8	(1.1)	101.8	liq(m)	Fe+FeO
1508	15	1600	87.9	(1.3)	1.2	(0.8)	9.8	(1.0)	98.9	liq(m)	Fe+FeO
1509	15	2000	65.7	(1.4)	7.8	(2.5)	25.0	(1.3)	98.5	liq(m)	
1509	15	2000	48.2	(1.3)	45.8	(1.4)	3.6	(1.4)	97.6	liq(i)	

[a]S.D. is the standard deviation of analytical values.

[b]liq(m) is metallic liquid (eutectic liquid, cotectic liquid and metallic liquid of two-liquid), liq(i) is ionic liquid of two liquid, liq is total melt of the starting composition.

[c]It is noted that the chemical analysis is not reliable beyond the deficiency from 100%. This is probably caused by the mechanical cracking and tipping of the sample during the polishing.

The composition of cotectic liquids is determined by using a defocused electron beam with a diameter of 20–30 μ. Six cotectic liquids produced at 6 GPa and 15 GPa (runs 606, 607A, 610A, 1505, 1507, and 1508) are analyzed. The oxygen content of the six cotectic liquids ranges from 1.0 to 2.4 atomic percent.

In the case of the two-liquid state, the texture of a quenched metallic liquid is similar to that of the cotectic liquid. Therefore we also use the defocused electron beam to determine composition by EPMA. The chemical analyses are carried out for seven metallic liquids produced at 6 GPa (runs 604B, 607A, 609A, 609B, 611A, 612, and 613) and for one produced at 15 GPa (run 1509).

EPMA measurement of the composition of ionic liquids is not reliable for several cases, because ionic liquids are generally too small in size to be determined by an electron probe. The ionic liquid droplets (larger than 20 μ in diameter) in metallic liquid are analyzed by the defocused beam method (runs 611A, 613, and 1509). The holes in the ionic liquid droplets, which were made by plucking during polishing, reduce the reliability of EPMA analysis due to the absorption of X-ray at the edge of holes. This affects large the analytical value of run 611A. The ionic liquid of run 604B is analyzed by the narrow beam method, because the size of droplets is less than 10 μ in diameter. The result shows an iron content value about 5–6% higher than that of other ionic liquids, probably because of the fluorescence effect of an environmental iron-rich metallic liquid. Only one datum (run 1509) is available at 15 GPa; its composition is near the FeO end.

Liquid phase in the system Fe-Ni-O-S. The liquid composition of the Ni-bearing system, as shown in Table 4, is determined by the same method as for the system Fe-FeO-FeS. The liquid phase in run N604A gives $Fe_{55.9}Ni_{8.5}O_{0.8}S_{34.7}$ as the eutectic composition of the system Fe-Ni-O-S at 6 GPa. The eutectic composition at 15 GPa is determined by using data from runs N1502 and N1503; the oxygen content is found to be the very small value of 0.5 atomic percent. The liquid phase in both runs N1502 and N1503 is not well separated from the oxide phase (see Figure 2A-1). Thus the oxide component of the liquid may precipitate on the surface of the coexisting

TABLE 4. Chemical Composition of Liquid in the System Fe-Ni-O-S

Run No.	P (GPa)	T (°C)	Composition of liquid (atomic %) Fe	(S.D.)[a]	Ni	(S.D.)[a]	O	(S.D.)[a]	S	(S.D.)[a]	Total	Analyzed[b] phase	Coexisting solid phase
N604A	6	890	55.9	(1.5)	8.5	(1.2)	0.8	(0.4)	34.7	(1.1)	99.9	liq(m)	metal+oxide
N605A	6	1160	61.5	(1.0)	7.7	(0.4)	1.4	(0.4)	28.0	(1.0)	98.6	liq(m)	oxide
N605B	6	1160	64.7	(1.5)	8.6	(0.2)	1.2	(0.4)	24.2	(0.9)	98.8	liq(m)	metal+oxide
N606A	6	1360	72.7	(0.7)	9.6	(0.1)	1.2	(0.3)	14.6	(0.9)	98.1	liq(m)	metal+oxide
N606B	6	1360	61.4	(0.8)	7.6	(0.4)	3.2	(0.6)	22.7	(1.0)	95.0[c]	liq(m)	oxide
N607A	6	1470	73.3	(0.7)	9.2	(0.2)	1.8	(0.8)	10.7	(0.6)	94.9[c]	liq(m)	oxide
N607B	6	1470	63.8	(1.2)	8.2	(0.3)	3.1	(1.3)	19.9	(0.6)	94.9[c]	liq(m)	oxide
N608	6	2000	68.6	(1.6)	8.8	(0.3)	2.4	(1.1)	21.6	(1.3)	101.4	liq(m)	
			47.9	(1.1)	1.1	(0.2)	38.4	(2.4)	13.4	(1.9)	100.7	liq(i)	
N609	6	2100	74.3	(1.3)	13.1	(0.4)	1.6	(0.7)	10.0	(1.6)	99.0	liq(m)	
			50.8	(1.1)	0.8	(0.4)	41.3	(0.9)	4.5	(0.5)	97.4	liq(i)	
N1502	15	900	61.8	(0.9)	9.2	(0.3)	0.5	(0.1)	26.7	(0.3)	98.2	liq(m)	metal+oxide
N1503	15	950	61.7	(1.2)	9.4	(0.5)	0.4	(0.2)	26.6	(0.5)	98.1	liq(m)	metal+oxide
N1504	15	1225	65.2	(2.5)	8.8	(0.7)	1.0	(0.7)	25.1	(2.4)	100.1	liq(m)	oxide
N1505	15	1500	71.9	(0.6)	9.3	(0.2)	0.9	(0.5)	18.3	(0.8)	100.4	liq(m)	metal+oxide
N1506	15	1830	65.9	(1.6)	9.6	(0.4)	3.9	(2.5)	20.4	(1.2)	99.8	liq(m)	oxide
N1507	15	2000	67.4	(0.9)	8.2	(0.2)	5.3	(1.1)	16.6	(1.1)	97.5	liq	
N1508	15	2150	64.9	(1.0)	9.9	(0.5)	8.6	(2.0)	12.3	(1.4)	95.7[c]	liq(m)	
			49.9	(0.6)	0.9	(0.3)	40.7	(0.8)	1.5	(0.4)	93.0[c]	liq(i)	

[a]S.D. is the standard deviation of analytical values.

[b]liq(m) is metallic liquid (eutectic liquid, cotectic liquid, and metallic liquid of two-liquid), liq(i) is ionic liquid, and liq is total melt with the starting composition.

[c]It is noted that the chemical analysis is not reliable beyond the deficiency from 100%. This is probably caused by the mechanical cracking and tipping of the sample during the polishing.

solid oxide during quenching. The extremely low oxygen content of the eutectic liquid at 15 GPa may be a result of this precipitation process. Therefore the actual oxygen solubility must be higher than the measured value.

The cotectic liquid in the system Fe-Ni-O-S was analyzed in products of three runs (N605B, N606A, N1505). The oxygen content of the cotectic liquids was found to be as low as about 1 atomic percent.

The composition of the liquid coexisting with solid oxide above cotectic temperature shows a large oxygen solubility. Runs N606B and N607B, which were conducted at 6 GPa, show that about 3 atomic percent of oxygen dissolves into the metallic liquid at 1360–1470 °C. At 15 GPa, the oxygen content of the metallic liquid increases with temperature; the content ranges from 1.0 atomic percent at 1225 °C (run N1504) to 3.9 atomic percent at 1830 °C (run N1506), as shown in Figure 4B.

In the immiscible liquid region, the oxygen content of the metallic liquid is about 9 atomic percent at 15 GPa and 2150 °C (run N1508), but is about 2 atomic percent at 6 GPa above 2000 °C (runs N608 and N609).

The Ni/(Fe+Ni) ratio is almost constant and lies within a range of 0.11–0.13 in 15 metallic liquids, but has a slightly higher value of 0.15 in the metallic liquid of run N609. The Ni/(Fe+Ni) ratio is a little higher in the metallic liquids than in the starting materials, indicating that nickel concentrates in the metallic liquid phase.

Data for ionic liquids are available from three runs (N608, N609, and N1508). The nickel content of the ionic liquids is about 1 atomic percent, and is lower than that of the metallic liquids by one order of magnitude.

Solid phase in the system Fe-Ni-O-S. The composition of solid phases coexisting with liquid is shown in Table 5. The metallic phase which coexists with the eutectic or cotectic liquid is mainly composed of iron and nickel. The oxygen content of the solid metal is so low that the data are not reliable. The metallic phase contains about 0.1–0.2 atomic percent sulfur. The sulfur increases with pressure to 0.45 atomic percent at 15 GPa.

Oxide almost has the composition of wüstite and contains trace amounts of bunsenite (NiO) and periclase (MgO) used as the sample capsule. The nickel content of oxides shows a tendency to increase with pressure; the content ranges from less than 0.1 atomic percent at 6 GPa to 0.1–0.2 atomic percent at 15 GPa. Nickel content is also shown to increase slightly with temperature at 15 GPa, but no definite trend is indicated at 6 GPa because of technical limitations of the present study.

TABLE 5. Chemical Composition of Solid in the System Fe-Ni-O-S

Run No.	P (GPa)	T (°C)	Composition of solid (atomic %) Fe	(S.D.)[a]	Ni	(S.D.)[a]	O	(S.D.)[a]	S	(S.D.)[a]	Total	Analyzed[b] phase
N604A	6	890	88.8	(0.4)	10.8	(0.1)	0.2	(0.4)	0.10	(0.01)	99.9	metal
			48.9	(0.5)	0.08	(0.01)	51.0	(0.5)	0.10	(0.02)	100.1	oxide
N605A	6	1160	46.5	(0.5)	0.06	(0.01)	50.0	(0.4)	0.03	(0.01)	96.5	oxide
N605B	6	1160	84.0	(0.4)	9.8	(0.1)	0.3	(0.1)	0.20	(0.01)	94.3[b]	metal
			45.9	(0.7)	0.07	(0.01)	49.7	(0.3)	0.04	(0.01)	95.7[b]	oxide
N606A	6	1360	85.6	(0.9)	10.3	(0.1)	0.0	(0.01)	0.17	(0.01)	96.0	metal
			46.6	(0.9)	0.09	(0.01)	49.4	(0.3)	0.03	(0.01)	96.1	oxide
N606B	6	1360	44.2	(1.5)	0.07	(0.01)	49.7	(0.4)	0.04	(0.2)	94.0[b]	oxide
N1503	15	950	48.6	(0.5)	0.12	(0.01)	49.6	(0.3)	0.07	(0.01)	98.4	oxide
N1504	15	1225	48.6	(0.5)	0.13	(0.01)	46.6	(0.8)	0.05	(0.01)	95.3[b]	oxide
N1505	15	1500	87.5	(0.3)	10.1	(0.1)	0.01	(0.02)	0.45	(0.03)	98.1	metal
			48.5	(0.2)	0.17	(0.03)	47.7	(0.2)	0.02	(0.01)	96.3	oxide
N1506	15	1830	41.4	(0.2)	0.21	(0.02)	50.1	(0.3)	0.03	(0.01)	91.8[c]	oxide

[a]S.D. is the standard deviation of analytical values.

[b]It is noted that chemical analysis is not reliable beyond the deficiency from 100%. This is probably caused by the mechanical cracking and tipping of the sample during the polishing.

[c]This sample apparently contains Mg, for which the analysis was not made.

Phase Relations at 6 GPa and 15 GPa

The system Fe-FeO-FeS. The phase diagrams of the system Fe-FeO-FeS at 6 GPa and 15 GPa are constructed by interpreting the coexisting phases and their chemical compositions (Figure 3). The Fe-FeO-FeS join at these pressures is a ternary eutectic system with a wide immiscible liquid region on the Fe-FeO side.

The ternary eutectic point is very close to the Fe-FeS binary eutectic point.

The cotectic line from the ternary eutectic point to the Fe-FeO binary eutectic point directs to Fe-rich side of immiscible liquid region. The oxygen content of cotectic liquids at 15 GPa increases with sulfur content toward the eutectic point, while the oxygen content of cotectic liquid at 6 GPa is rather constant.

The immiscible liquid region is poorly defined in this study. The immiscible liquid region occurs above 1400 °C at 6 GPa. The oxygen content of metallic liquid at 6 GPa increases only slightly with temperature near the Fe end; it ranges from 0.8 atomic percent at 1505 °C to 1.5 atomic percent at 2100 °C. The oxygen content increases with sulfur content near the bottom of immiscible liquid region, ranging from 0.8 atomic percent near the Fe end to about 3.0 atomic percent near the eutectic point. In contrast to temperature effect, the pressure effect is very significant. The oxygen content of metallic liquid increases from 2–3 atomic percent at 6 GPa to about 8 atomic percent at 15 GPa. On the other hand, the composition of ionic liquid at 15 GPa is not much different from that at 6 GPa. The composition range of the immiscible liquid region, however, certainly narrows with pressure.

The system Fe-Ni-O-S. The phase relations in the Fe-rich portion of the system Fe-Ni-O-S are mostly the same as those in the system Fe-FeO-FeS. The phase diagrams of the system Fe-Ni-O-S at 6 GPa and 15 GPa are also constructed by interpreting the coexisting phases and their chemical compositions (Figure 4). In these diagrams, the sum total of iron and nickel is used to represent the metallic end-member.

The oxygen content of the metallic side of the immiscible liquid region is almost the same as that of the cotectic liquid at 6 GPa, except it is slightly higher in the sulfur-rich region. The location of ionic side of the immiscible liquid region is near the (Fe, Ni)O-(Fe, Ni)S join. However, the ionic side is only poorly determined in this study. It is apparent that the metallic side of immiscible liquid region at 15 GPa is farther from the cotectic line than that at 6 GPa. Therefore the increase of pressure depresses and narrows the extent of the immiscible liquid region.

Eutectic Temperatures

The eutectic temperature of this system is determined by the lowest temperature of runs with diagnostic melting (i.e. the quenched texture in run products) and also by the highest temperature of runs without melting. Figure 5 shows eutectic melting curves for several systems involving Fe.

The eutectic temperatures of the system Fe-FeO-FeS are 930±10 °C at 6 GPa, 960 °C at 10 GPa, and 985±25 °C at 15 GPa. The solidus is fitted to these data points for

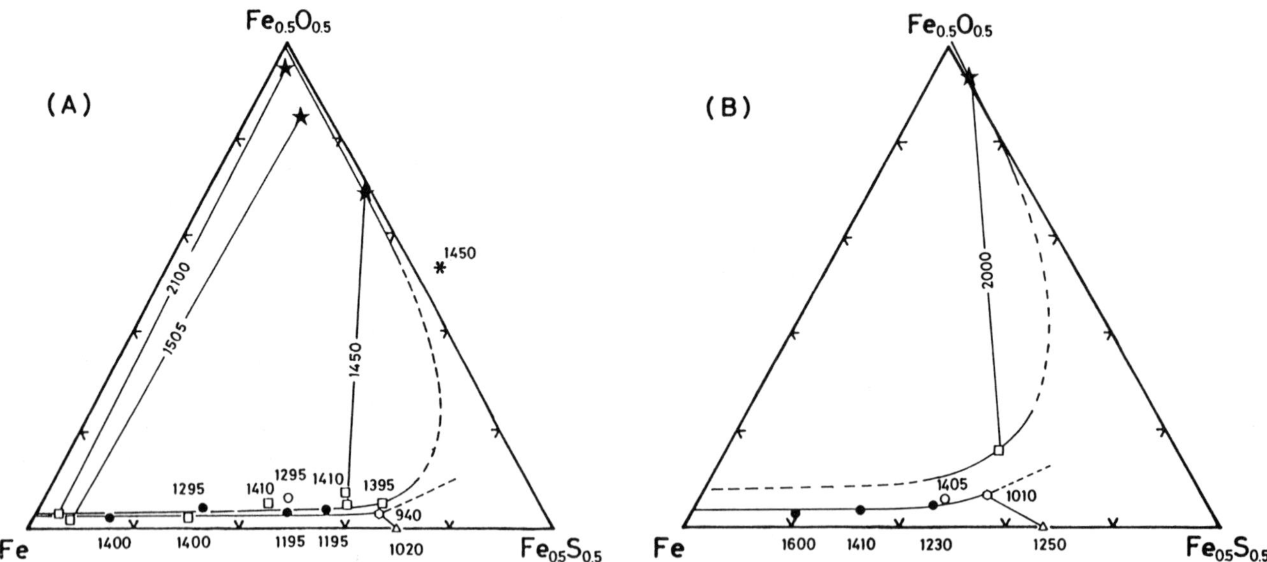

Figure 3. Liquidus surface in the system Fe-FeO-FeS shown in atomic percent of Fe, O, and S at 6 GPa (A) and at 15 GPa (B). Open hexagon is the ternary eutectic point, open triangle is the Fe-FeS binary eutectic point (USSELMAN, 1975a, 1975b), closed circles indicate the liquid on the cotectic line coexisting with solid iron and solid iron oxide, open circles indicate the liquid coexisting with solid iron oxide, open squares tied with closed stars indicate metallic liquid in contact with ionic liquid in the two-liquid region, and asterisk is the single liquid. Numerals indicate temperature in degrees centigrade.

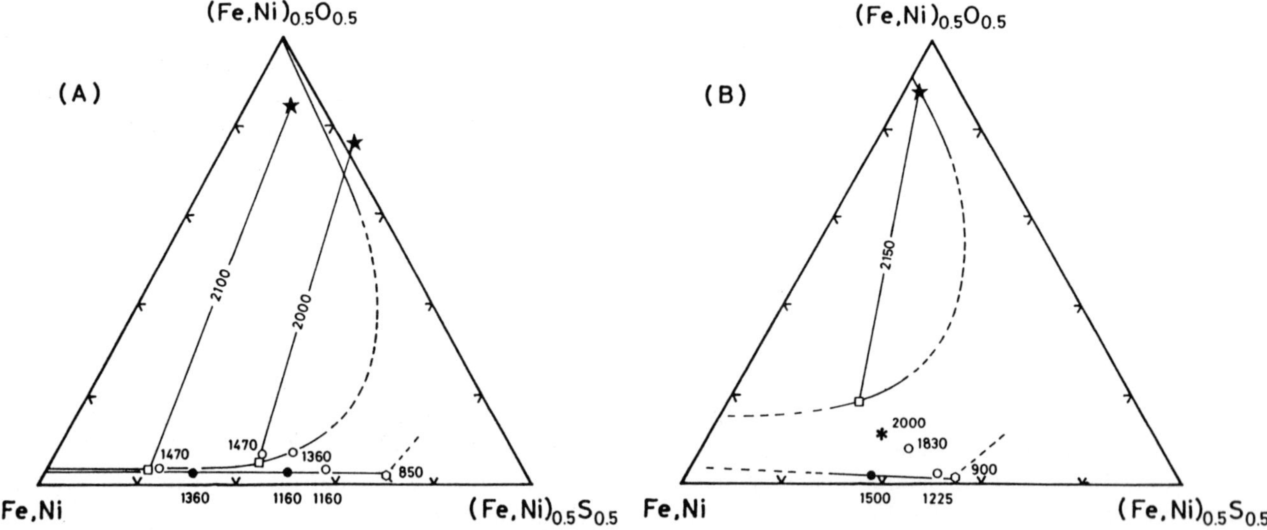

Figure 4. Liquidus surfaces in the system Fe-Ni-O-S illustrated in atomic percent of Fe+Ni, O, and S at 6 GPa (A) and at 15 GPa (B). The symbol notations are the same as Figure 3.

high pressures and to the eutectic temperature at 0 GPa reported by NALDRETT (1969); a possible inflection point occurs at 5.5 GPa due to the phase transformation of FeS, an end-member crystalline phase. Actually, the Fe-FeS eutectic melting curve has an inflection point at 5.5 GPa and 980 °C which is related to the FeS phase transformation (USSELMAN, 1975a). This phase transformation is thought to affect the Fe-FeO-FeS eutectic temperature, but evidence of the inflection of the eutectic melting curve has not yet been revealed.

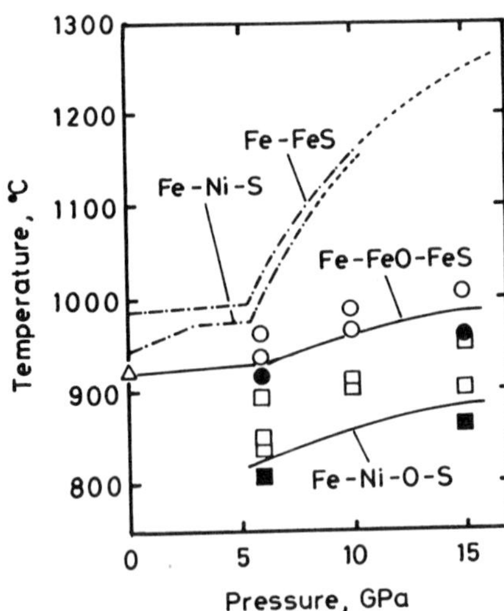

Figure 5. Eutectic temperatures in the system Fe-Ni-O-S at pressures up to 15 GPa. The results of melting experiments in the system Fe-FeO-FeS are shown by open circles (liquid) and closed circles (solid), and those of experiments in the system Fe-Ni-O-S by the open squares (liquid) and closed squares (solid). Open triangle is the eutectic temperature in the ternary system Fe-FeO-FeS at 0 GPa (NALDRETT, 1969). The eutectic melting curves in the systems Fe-FeS and Fe-Ni-S (dash-dotted lines) that were reported by USSELMAN (1975a, 1975b) are also shown for comparison.

Addition of another component is expected to lower the eutectic temperature. The addition of 10% nickel to the system Fe-FeO-FeS reduces the solidus temperature by about 100 °C. The eutectic temperatures in the system Fe-Ni-O-S are 825±10 °C at 6 GPa and 880±20 °C at 15 GPa (Figure 5).

Discussion

Phase Relations in the System Fe-FeO-FeS

Change of phase relations at low pressures. Previous low pressure studies showed that the ternary system Fe-FeO-FeS has the following characteristics up to 3 GPa (HILTY and CRAFTS, 1952; NALDRETT, 1969; WENDLANDT and HUEBNER, 1979): 1) the ternary eutectic point at 0 GPa is located close to the FeO-FeS join (Figure 6); 2) the cotectic line which connects eutectic point with the Fe-FeO join extends directly from this point almost to the FeO end; 3) oxygen solubility in the ternary eutectic point decreases with increasing pressure (Figures 6 and 7); and 4) the eutectic point at 3 GPa is near the Fe-FeS binary eutectic point (Figure 6), but the cotectic line still extends toward the FeO-FeS side of the immiscible liquid region.

We find drastic changes in phase relations at pressures between 3 and 6 GPa, based on a comparison of the low pressure data from previous studies with present results. The position of the cotectic line switches from the FeO-FeS side to the Fe-FeS side of the immiscible liquid region at 3.3 GPa. This change is related to a change in the melting mode of the system Fe-FeO. LINDSLEY (1966) determined the melting curve of wüstite coexisting with pure iron up to 4 GPa. At low pressures, the melting temperature of wüstite is lower than the eutectic temperature in the Fe-rich portion of the system Fe-FeO, whereas

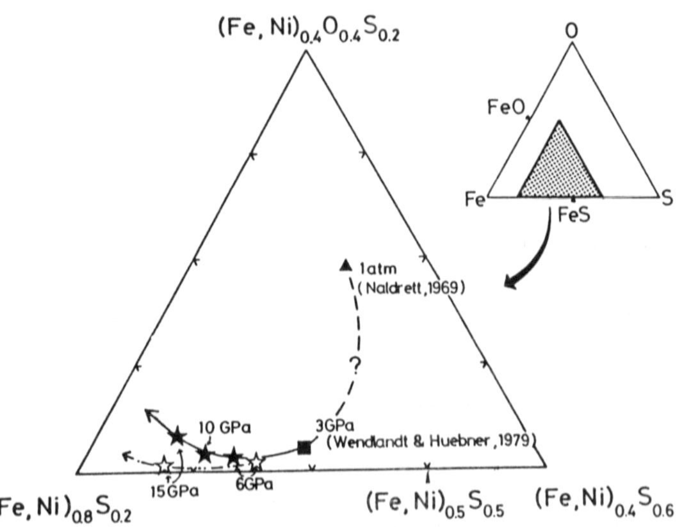

Figure 6. The pressure shift of the eutectic point in the system Fe-Ni-O-S is plotted in a stippled area of the system Fe-Ni-O-S. Triangle and square show the eutectic point at 0 GPa (NALDRETT, 1969) and at 3 GPa (WENDLANDT and HUEBNER, 1979), respectively. Stars represent the results of this study. Closed symbols represent the case of no Ni, and open symbols show the quarternary eutectic point in the system Fe-Ni-O-S.

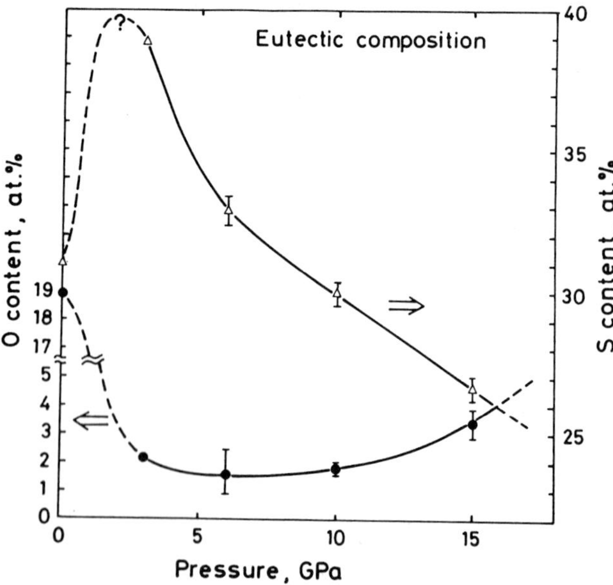

Figure 7. The pressure variation of oxygen and sulfur content at the ternary eutectic point in the system Fe-FeO-FeS.

the pressure derivative of the wüstite melting curve is greater than that of the Fe-rich eutectic melting curve. The two curves intersect at 3.3 GPa and 1590 °C, as reported by OHTANI et al. (1984). Therefore the minimum melting point of the iron-wüstite system changes from the FeO side to the Fe-rich side of the immiscible liquid region at 3.3 GPa. KATO (1987) demonstrated that this high-pressure phase relation of the system Fe-FeO is maintained up to 20 GPa. The discrete shift of the minimum melting point in the Fe-FeO binary system directly causes the drastic movement of the cotectic line, because the cotectic line connects the ternary and binary eutectic points.

Eutectic point. Figures 6 and 7 show the variation in the ternary eutectic point with pressure. The ternary eutectic point is located at $Fe_{49.9}O_{19.2}S_{31.0}$ at 0 GPa (NALDRETT, 1969). The oxygen solubility at the eutectic point decreases with increasing pressure to 3 GPa, and is less than 2 atomic percent at 3 GPa (WENDLANDT and HUEBNER, 1979). On the other hand, the sulfur solubility at the ternary eutectic point increases with pressure to 3 GPa. This change in oxygen and sulfur solubility is related to the rapid increase of the wüstite melting temperature and to the change of phase relations in the system Fe-FeO-FeS. The iron and oxygen solubilities at the Fe-FeO-FeS eutectic point increase as pressure increases from 3 to 15 GPa, as shown in Figures 6 and 7. This is consistent with experimental and thermodynamical studies of the system Fe-FeO which found that the oxygen solubility at the Fe-FeO eutectic point increases with pressure (OHTANI et al., 1984; KATO, 1987), and also with the results of a Fe-FeS melting experiment to 10 GPa which found that the sulfur solubility at the binary eutectic point decreases with pressure (USSELMAN, 1975a).

Cotectic line. The chemical compositions of cotectic liquids are shown in Figures 8A and 8B. The oxygen solubility at 15 GPa is larger than that at 6 GPa, particularly for the sulfur-rich composition close to the eutectic point (Figure 8B). This is related to an increase in oxygen solubility with pressure at the Fe-FeO eutectic point. However, the measured oxygen solubility in liquid may actually show a lower limit, because the oxygen in the melt rapidly diffuses into the capsule oxide during quenching (OHTANI and RINGWOOD, 1984). This quenching effect on oxygen solubility becomes more significant

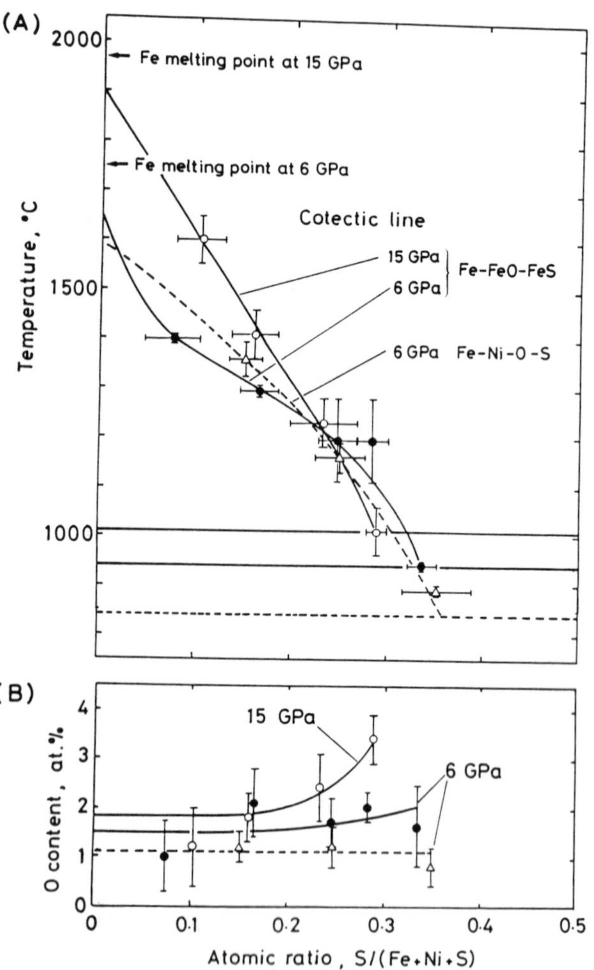

Figure 8. A) The cotectic line projected from the FeO end to the (Fe, Ni)-(Fe, Ni)S join. B) The oxygen content of cotectic liquid. Closed circles and open circles are the cotectic liquid compositions in the system Fe-FeO-FeS at 6 GPa and 15 GPa, respectively. Open triangles and dashed line show the cotectic line in the system Fe-Ni-O-S.

as the run temperature increases. Therefore the oxygen solubility in the cotectic liquid near the Fe end is thought to be higher than measured value.

The variation of cotectic lines at 6 GPa and 15 GPa with temperature and composition is shown in Figure 8A. The cotectic line at 6 GPa suggests a remarkable nonideality. However, the degree of nonideality decreases for the cotectic line at 15 GPa.

The low pressure structure (MnP-type) of $Fe_{1-x}S$ is found to transform into an unquenchable unknown structure at 6.7 GPa and room temperature (KING and PREWITT, 1982). At high temperatures, $Fe_{1-x}S$ has a NiAs-type structure and is supposed to transform into an unknown phase. This phase transformation should take place at 5.5 GPa and 980 °C based on the inflection of the Fe-FeS eutectic melting curve by USSELMAN (1975a). There has been no direct observation of this phase transformation in $Fe_{1-x}S$ at high temperature. Therefore the phase relations of FeS polymorphs around 6 GPa remain to be clarified. We suppose that the change in nonideality of cotectic line in the system Fe-FeO-FeS may be related to the $Fe_{1-x}S$ phase transformation.

Immiscible liquid region. The liquid immiscibility of the system Fe-FeO-FeS occurs at temperatures above 1350 °C and atmospheric pressure (HILTY and CRAFTS, 1952). HILTY and CRAFTS (1952) reported that the immiscible liquid region at 0 GPa extends from near the Fe-FeS join to beyond the FeO-FeS join. This characteristic of the immiscible liquid region is insensitive to pressures up to 6 GPa (Figure 3), although the phase relations change significantly with pressure. Our work has revealed that this liquid immiscibility occurs above 1400 °C at 6 GPa, which is essentially the same temperature range as at 0 GPa. Our work has also shown that the oxygen solubility of the metallic liquid slightly increases with temperature at 6 GPa. However, the same quenching effect as in the case of cotectic liquid is thought to lower the oxygen solubility at high temperature. Thus the increase of oxygen in the metallic melt with pressure is expected to occur more rapidly.

This immiscible gap is related to the difference in metallic bonding and ionic bonding between Fe and O in the liquid. The immiscible liquid gap in the Fe-FeO binary system provides information on the gap in this ternary system.

In the system Fe-FeO at 0 GPa, the immiscible liquid region occurs above 1523 °C (HANSEN and ANDERKO, 1958), up to 2800–3500 °C (RINGWOOD, 1977; OHTANI and RINGWOOD, 1984; RINGWOOD, 1984). The onset temperature increases with pressure; it is 2000 °C at 10 GPa and 2200 °C at 15 GPa (KATO, 1987). KATO (1987) showed experimentally that oxygen solubility in the metallic liquid increases with pressure and the immiscible liquid region narrows with pressure. KATO (1987) and OHTANI et al. (1984) conducted thermodynamic calculations based on a simple, nonideal solution model. Results of the calculations indicated that the immiscible liquid region is suppressed by the decrease in maximum temperature with increasing pressure. It completely vanishes around 20 GPa (OHTANI et al., 1984) or 25 GPa (KATO, 1987); the difference between these two results is due to the difference in adopted values of the partial molar volume of FeO in metallic liquid.

Therefore one can reasonably predict that the immiscible region of the system Fe-FeO-FeS narrows with pressure and disappears above 25 GPa. This tendency is certainly supported by our measurements of the oxygen solubility of metallic liquid in the immiscible region up to 15 GPa.

Phase Relations in the System Fe-Ni-O-S

Eutectic point. The quarternary eutectic points are clarified by only two points at 6 GPa and 15 GPa. The movement of this eutectic point with increasing pressure is quite different from that in the system Fe-FeO-FeS. The oxygen solubility at the quaternary eutectic point decreases with pressure, so that this point moves to the metal-rich and oxygen-poor direction as shown in Figure 6. This is contrary to the predicted effect of nickel addition. In the Ni-NiO binary system at 0 GPa, the oxygen solubility at the eutectic point is 0.87 atomic percent which is higher than that of the system Fe-FeO (HANSEN and ANDERKO, 1958). Therefore the addition of nickel to iron is expected to enhance the oxygen solubility at the eutectic point. However, the difference of the oxygen solubility at the eutectic points at 15 GPa and at 0 GPa is not clear yet, because of remaining experimental problems with sample quenching.

Cotectic line. The composition of the cotectic liquids in the system Fe-Ni-O-S are shown in Figure 8A. The addition of 10% nickel increases the solubility of sulfur in the cotectic liquid by about 5 atomic percent at 1350 °C. Although detailed characteristics of the cotectic line in the Ni-bearing system have not yet been clarified due to a lack of data, we stress that the nonideality of the cotectic line at 6 GPa is reduced by the addition of nickel.

Immiscible liquid region. The system Ni-NiO-NiS would be ternary eutectic without an immiscible liquid region, even at 0 GPa, because the system Ni-NiO is completely miscible above the solidus (HANSEN and ANDERKO, 1958). RINGWOOD and MAJOR (1982) showed that the complete miscibility of the system Ni-NiO persists up to 3 GPa. These findings suggest that the addition of nickel to the system Fe-FeO-FeS reduces the immiscibility of the metallic liquid and the ionic liquid.

Although the immiscible liquid region appears above 1400 °C in the system Fe-FeO-FeS at 6 GPa (Figure 3A), the liquid in the 10% Ni-bearing system does not separate into two phases, at least up to 1470 °C (Figure 4A). The addition of nickel raises the onset temperature of immiscible melting. The composition range of the immiscible liquid region is not affected much by addition of 10% nickel, contrary to predictions (Figures 3B and 4B), although as in the case of the system Fe-FeO-FeS, a quenching problem related to the oxygen content of the metallic liquid may occur. However, reduction of the composition range with increasing pressure is clearly seen and is as large as in the system Fe-FeO-FeS. With both pressure and nickel addition, there is a remarkable reduction in the immiscible liquid region, both in composition and temperature. The immiscible liquid region in the Ni-bearing system is expected to disappear at pressures lower than 25 GPa.

Core Formation Process

Eutectic temperature in the system Fe-Ni-O-S. A significant problem in core formation is the effect of adding other components to iron on the melting temperature (MURTHY and HALL, 1970; RINGWOOD, 1984). The melting temperature of iron to 20 GPa has been determined by STERRETT et al. (1965) and LIU and BASSETT (1975); the melting temperatures are about 1750 °C at 6 GPa and about 2050 °C at 20 GPa. In the pressure range of 6–20 GPa, the addition of oxygen lowers the melting temperature of iron by about 100 °C (KATO, 1987), and that of sulfur by about 700 °C (USSELMAN, 1975a, 1975b). The combined effect of adding plural elements is to significantly lower the melting temperature of iron. The solidus temperature in the system Fe-Ni-O-S is less than 900 °C in the pressure range of 0–15 GPa (Figure 5). Further, FUKAI and SUZUKI (1986) have shown that the dissolution of hydrogen into iron alloys, as a light element, greatly lowers the melting temperature of the iron alloys. Therefore the actual melting temperature of iron alloy in the primitive earth material is expected to be lower than 800 °C at pressures up to 20 GPa.

USSELMAN (1975b) and KATO (1987) extrapolated the melting curves of iron alloys to the core pressure, using the Kraut-Kennedy equation. They showed that the eutectic temperature in the system Fe-FeS is lower than 2000 °C even at the pressure of the present core-mantle boundary; the addition of oxygen lowers the eutectic temperature by an additional 500 °C. Present experiments to 15 GPa have indicated that the pressure derivative of the eutectic melting temperature in the system Fe-Ni-O-S is very small: that is 6 K/GPa (a value even smaller than the adiabatic temperature gradient). All the data presented above suggest that the actual core-forming material possesses a melting temperature much lower than previously thought, and that melting starts at less than 1500 °C even at the pressure of the present core-mantle boundary.

Effect of oxygen and sulfur dissolution on the texture of silicate-liquid metal composite. KATO (1987) proposed two mechanisms for promoting the separation of core material from mantle silicate: 1) chemical and stress corrosion of silicates and oxides by molten iron alloy, and 2) hydrofracturing action of liquid wedges driven by gravitational force. Here we reexamine the former process with regard to the additional effect of oxygen and sulfur.

The alloying of oxygen and sulfur with iron-nickel melt has been found to reduce the surface energy (ELLIOTT et al., 1963). The surface energy of pure iron (1.8 J/m^2) is very much different from that of oxides and silicates (0.2–0.6 J/m^2), and the silicates do not wet by iron melt with low oxygen solubility at low pressures. However, high pressure lowers the difference in surface energy between the silicates and molten iron alloy because of increasing chemical affinity between these components, as indicated by the reduction of the immiscible liquid region. At high pressures in the mantle, silicates and oxides can be wet by the coexisting molten iron alloy. Therefore the iron-silicate aggregate is expected to show the grain-boundary networks of the liquid phase. This situation has been suggested to occur for partially molten ice and peridotite (FRANK, 1968; WAFF, 1980; TORAMARU and FUJII, 1986).

Figure 9 shows the cross-section of the sample recovered from the run in the system Fe-Ni-O-S at about 2500 °C and 6 GPa. The texture of this cross-section shows that metallic liquid is connected along the grain boundaries of capsule oxides. The metallic liquid loosens the adhesion of oxide grains by chemical corrosion and permeates the grain boundaries of oxides due to the low interfacial energy. This texture has been observed under hydrostatic conditions maintained for less than 1 min at about 2500 °C. According to a simple calculation assuming an activation energy of 40 J/mol, the same texture as shown in Figure 9 would be achieved for a duration of about 1 yr and a temperature of only 1000 °C (i.e., the temperature of the Fe-Ni-O-S solidus).

In the primordial mantle during the growth of the earth, there should have been various types of stress due to impacts of planetesimals and to tectonic processes such as convective movements. Stress corrosion must have also loosened the boundaries between the silicate and oxide grains by intrusion of molten iron alloy. In addition, hydrofracturing action of a liquid wedge of dense metallic liquid could work efficiently in such a texture as shown in Figure 9. In particular, a gravity field can enhance the loosening of the grain boundaries between oxides and silicates. Such textural considerations suggest that pressure effects that enhance the solubility of oxygen in molten iron alloy can cause the core materials to be equilibrated with mantle materials during the growth of the core. The formation and fast sinking of extremely

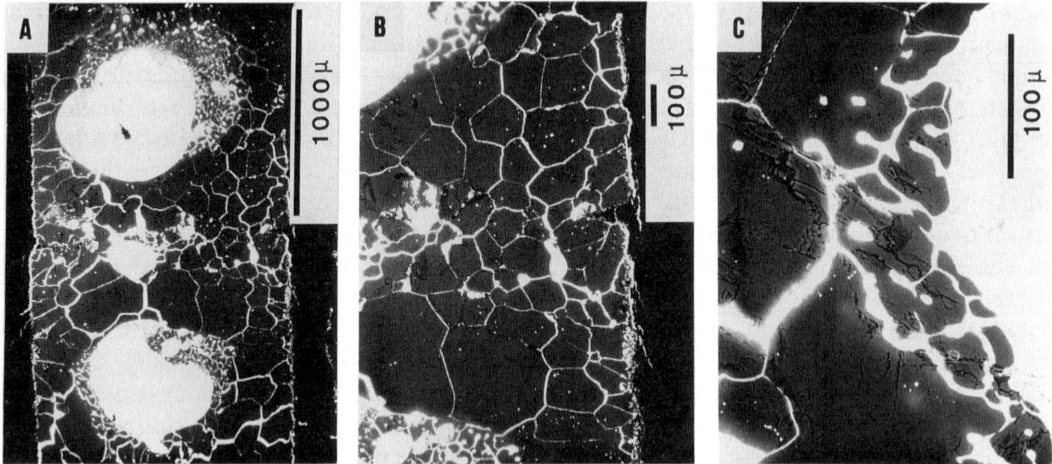

Figure 9. The photomicrograph (back-scattered electron images) of the recovered sample at 6 GPa and 2500 °C. A) Whole view of sample cell. White part is metal, gray part is MgO, and black periphery is part of the graphite heater. Two large round parts of metal are original sample space in a dual cell, and small round area at center is the place where the thermocouple is inserted. All metallic components, including the thermocouple, are melted and connected along the grain boundaries of the grown MgO crystals of the capsule. B) Enlarged picture of the central region. MgO grains contact each other through the melted iron alloy. Note the small metallic spots dispersed or chained in the MgO grains. These spots are interpreted as having been trapped during grain growth of MgO in iron melt. C) Enlarged picture of the left side-wall of an iron blob (at bottom). Metallic melt is shown to react with MgO at the metal-oxide boundary and to intrude into the polycrystalline texture of MgO at the grain boundaries. MgO component is supposed to be dissolved into and transported through the molten iron alloy for recrystallization of MgO.

large metallic blobs may not be important, and the core formation process can be considered to be a process of chemical equilibration which is significantly dependent on pressure.

Formation of the core. The earth is believed to have been formed by the accretion of planetesimals, which were composed of silicates and metals, about 4.6 billion years ago. The liberation of gravitational energy from the infalling planetesimals effectively raised the temperature in the growing earth (e.g., KAULA, 1979). The calculations show that the temperature of the surface and shallow region of the earth exceeded 1000 °C, which is higher than the melting temperature in the system Fe-Ni-O-S-H, when the earth had grown to 2000–3000 km in radius (e.g., KAULA, 1979; CORADINI et al., 1983; ABE and MATSUI, 1986). As a result, the dense metallic liquid started to break down the silicate to grain size by chemical and stress corrosions, and to sink down to the deeper interiors where sulfur and other volatiles are expected to coexist with iron-nickel alloy in primordial earth.

The continuous accretional heating and subsequent heating in the deeper horizon by gravitational energy liberation (i.e., due to the sinking dense metallic liquid) must have raised the temperature of the earth's interior higher than the solidus of the silicate system and formed a deep magma ocean (100–1000 km in depth) on the surface of the growing earth (e.g., RINGWOOD, 1984; ABE and MATSUI, 1986; SASAKI and NAKAZAWA, 1986). According to ABE and MATSUI (1986) and SASAKI and NAKAZAWA (1986), the magma ocean should have been in a partially molten state composed of metallic and silicate melts and silicate solid residue. Once the magma ocean was formed on the growing earth, the metallic component of accreting material would have easily sunk, even without a significant degree of partial melting of silicate matrix.

Suppose that the molten iron alloy had sunk to the bottom of the magma ocean and coagulated as a metallic layer (SASAKI and NAKAZAWA, 1986). Before the earth could grow larger, such a metallic layer would have to collapse and replace the underlying light, undifferentiated material because of the Rayleigh-Taylor instability (STEVENSON, 1981). STEVENSON (1981) suggested that a catastrophic overturn of the metallic layer occurred, but this seems quite unlikely because the underlying layer would have been too soft to sustain the dense and thick metallic layer until the catastrophe occurred. Because of the chemical and stress corrosion of oxides and silicates by molten iron alloy in the underlying, undifferentiated layer, even a thin metallic layer would collapse and sink very quickly and continuously in the form of small metallic blobs or through the veinlike channels among the grain boundaries in association with growth of the earth. Even if a catastrophic overturn had occurred, it would have been only once or a few times during the very early stage of growth. After the certain degree of core growth, the silicate layer below the magma ocean would have heated up and been softened by the liberation of gravitational energy from the sinking core material.

Therefore the late-accreting metallic component which contributes the largest fraction of the core material, is expected to have sunk down to the core continuously without a catastrophic overturn, so that the chemical equilibrium between the sinking core-forming material and the matrix mantle material was maintained at each depth as the core material descended downward. The missing ingredient essential to our understanding of core formation must be the pressure-dependent and equilibrium partitioning data of elements between the core and mantle materials (JONES and DRAKE, 1985).

Composition of the Earth's Core

The first separation of metallic liquid from silicate and oxide in equilibrium occurred at the near surface region of the growing earth before formation of the deep magma ocean. The solubility of oxygen in this molten iron alloy is only a few atomic percent due to the existence of a large immiscible liquid region. Sulfur is found to be the dominant alloying component in iron alloy. Thus the early core material must have consisted of three main components: iron, nickel, and sulfur; this material would not have contributed much to the density deficit of the present earth's core because of its small amount.

Once the deep magma ocean was formed on the growing earth, the metallic liquid was easily separated in equilibrium with the silicate liquid in the shallow region of the molten zone. This metallic liquid has almost the same composition as the iron alloy which separated first. However, this sulfur-rich iron alloy cannot directly sink down to the core without chemical reaction with mantle material. Most of the iron alloy separated in the magma ocean would have suffered the reequilibration with the silicate and oxide that constituted the protomantle at each depth down to the core. This reequilibriated iron alloy would have contained a significant amount of oxygen and sulfur as principal light elements because of the complete miscibility between iron-nickel alloy, oxide, and sulfide at the pressures above 20–25 GPa.

USSELMAN (1975b) extrapolated the sulfur solubility in the eutectic trough of the system Fe-FeS to high pressure on the basis of experimental data. He estimated that the eutectic melt dissolves about 27 atomic percent of sulfur at 140 GPa. A simple thermodynamic calculation for the ideal solution provides an estimation of the oxygen solubility in iron melt at very high pressure. The iron melting curve was extrapolated to core pressure by ANDERSON (1982) on the basis of the Lindemann law. The extrapolation of the FeO melting point is conducted up to 140 GPa using the Kraut-Kennedy equation and disregarding the phase transformation which occurs around 70 GPa, as reported by JEANLOZ and AHRENS (1980). The calculation which uses these melting temperatures and the entropy change of melting shows that the eutectic liquid in the system Fe-FeO contains 13–22 atomic percent of oxygen at the present core-mantle boundary.

Therefore the amount of oxygen and sulfur dissolving into iron during the growth stage of the earth is large enough to reduce the iron-alloy density to the present core density. Although we have discussed only oxygen and sulfur as possible light elements in the earth's core, other light elements such as hydrogen and carbon are available to alloy with iron and thereby reduce its density. This suggests that light elements can be incorporated into the core in greater amounts than indicated by the seismic and shock data.

The present data do not constrain the detailed core composition. Additional information, such as data on the partitioning of siderophile elements, is required to determine the core composition and formation processes. The key core formation process is the expected chemical equilibration in the iron-silicate composite during the sinking of iron alloy. The next problem to investigate is the clarification of the chemical interaction of core and mantle materials during core formation. This problem will be discussed in a separate paper with regard to the pressure dependence of nickel partitioning data acquired in this study.

Acknowledgments. The authors are very grateful to H. Sawamoto and T. Kato for their discussions and technical support in the high pressure experiments, and to H. Mizutani and A. Fujimura for their helpful discussions. The authors also thank S. Yamamoto for his assistance in sample preparation, and K. Suzuki and H. Noro for their help in the EPMA analysis. This research was partially supported by the Japanese DELP project.

REFERENCES

ABE, Y., and T. MATSUI, Early evolution of the earth: accretion, atmosphere formation, and thermal history, Lunar Planet. Sci. Conf. 17th Proceedings, *J. Geophys. Res.*, **91**, E291–E302, 1986.

ANDERSON, O. L., The earth's core and the phase diagram of iron, *Phil. Trans. R. Soc. Lond.*, *A306*, 21–35, 1982.

BIRCH, F., Elasticity and constitution of the earth's interior, *J. Geophys. Res.*, *57*, 227–286, 1952.

BIRCH, F., Density and composition of mantle and core, *J. Geophys. Res.*, *69*, 4377–4388, 1964.

BRETT, R., and P. M. BELL, Melting relations in the Fe-rich portion of the system Fe-FeS at 30 kb pressure, *Earth Planet. Sci. Lett.*, *6*, 479–482, 1969.

CORADINI, A., C. FEDERICO, and P. LANCIANO, Earth and Mars: early thermal profiles, *Phys. Earth Planet. Inter.*, *31*, 145–160, 1983.

ELLIOTT, J. F., M. GLEISER, and V. RAMAKRISHNA, *Thermochemistry*

for Steelmaking, II, 846 pp., Addison-Wesley, Massachusetts, 1963.
FRANK, F. C., Two-component flow model for convection in the earth's upper mantle, *Nature, 220,* 350–352, 1968.
FUKAI, Y., and T. SUZUKI, Iron-water reaction under high pressure and its implication in the evolution of the Earth, *J. Geophys. Res., 91,* 9222–9230, 1986.
HANSEN, M., and K. ANDERKO, *Constitution of Binary Alloys,* 2nd ed., 1305 pp., McGraw-Hill, New York, 1958.
HILTY, D. C., and W. CRAFTS, Liquidus surface of the Fe-S-O system, *J. Metals, 4,* 1307–1312, 1952.
JEANLOZ, R., and T. J. AHRENS, Equation of state of FeO and CaO, *Geophys. J. R. Astron. Soc., 62,* 505–528, 1980.
JETTE, E. R., and F. FOOTE, An X-ray study of the wüstite (FeO) solid solutions, *J. Chem. Phys., 1,* 29–36, 1933.
JONES, J. H., and M. J. DRAKE, Experiments bearing on the formation and primordial differentiation of the earth, *Lunar Planet. Sci., XVI,* 412–413, 1985.
KATO, M., Melting experiment on Fe-FeO-FeS system up to 20 GPa and its significance on the formation and composition of the earth's core, *J. Geophys. Res.,* in press, 1987.
KATO, T., and M. KUMAZAWA, Melting and phase relations in the system Mg_2SiO_4-$MgSiO_3$ at 20 GPa under hydrous conditions, *J. Geophys. Res., 91,* 9351–9355, 1986.
KAULA, W. M., Thermal evolution of earth and moon growing by planetesimal impact, *J. Geophys. Res., 84,* 999–1008, 1979.
KING, H. E., Jr., and C. T. PREWITT, High-pressure and high-temperature polymorphism of iron sulfide (FeS), *Acta Cryst., B38,* 1877–1887, 1982.
LINDSLEY, D. H., Pressure-temperature relations in the system FeO-SiO_2, *Carnegie Year Book, 65,* 226–230, 1966.
LIU, L. G., and W. A. BASSETT, The melting of iron up to 200 kbar, *J. Geophys. Res., 80,* 3777–3782, 1975.
MASON, B., Composition of the earth, *Nature, 211,* 616–617, 1966.
MURTHY, V. R., and H. T. HALL, The chemical composition of the earth's core: possibility of sulfur in the core, *Phys. Earth Planet. Inter., 2,* 276–282, 1970.
NALDRETT, A. J., A portion of the system Fe-S-O between 900 and 1080 °C and its application to sulfide ore magmas, *J. Petrology, 10,* 171–201, 1969.
OHTANI, E., Melting relation of Fe_2SiO_4 up to about 200 kbar, *J. Phys. Earth, 27,* 189–208, 1979.
OHTANI, E., and A. E. RINGWOOD, Composition of the core, I. Solubility of oxygen in molten iron at high temperature, *Earth Planet. Sci. Lett., 71,* 85–93, 1984.
OHTANI, E., A. E. RINGWOOD, and W. HIBBERSON, Composition of the core, II. Effect of high pressure on solubility of FeO in molten iron, *Earth Planet. Sci. Lett., 71,* 94–103, 1984.
ONODERA, A., and A. OHTANI, Fixed points for pressure calibration above 100 kbar related to semiconductor-metal transitions, *J. Appl. Phys., 51,* 2581–2585, 1980.
RINGWOOD, A. E., Composition of the core and implications for origin of the earth, *Geochem. J., 11,* 111–135, 1977.
RINGWOOD, A. E., The earth's core: its composition, formation and bearing upon the origin of the earth, *Proc. R. Soc. Lond., A395,* 1–46, 1984.
RINGWOOD, A. E., and A. MAJOR, Mutual solubilities of molten transition metals and oxides, *Lunar Planet. Sci., XIII,* 651–652, 1982.
SASAKI, S., and K. NAKAZAWA, Metal-silicate fractionation in the growing earth: energy source for the terrestrial magma ocean, *J. Geophys. Res., 91,* 9231–9238, 1986.
SAWAMOTO, H., Single crystal growth of the modified spinel (β) and spinel (γ) phases of $(Mg, Fe)_2SiO_4$ and some geophysical implications, *Phys. Chem. Minerals, 13,* 1–10, 1986.
STERRETT, K. F., W. KLEMENT, Jr., and G. C. KENNEDY, Effect of pressure on melting of iron, *J. Geophys. Res., 70,* 1979–1984, 1965.
STEVENSON, D. J., Models of the earth's core, *Science, 214,* 611–619, 1981.
TORAMARU, A., and N. FUJII, Connectivity of melt phase in a partially molten peridotite, *J. Geophys. Res., 91,* 9239–9252, 1986.
USSELMAN, T. M., Experimental approach to the state of the core: part I. The liquidus relations of the Fe-rich portion of the Fe-Ni-S system from 30 to 100 kb, *Am. J. Sci., 275,* 278–290, 1975a.
USSELMAN, T. M., Experimental approach to the state of the core: part II. Composition and thermal regime, *Am. J. Sci., 275,* 291–303, 1975b.
WAFF, H. S., Effects of the gravitational field on liquid distribution in partial melts within the upper mantle, *J. Geophys. Res., 85,* 1815–1825, 1980.
WENDLANDT, R. F., and J. S. HUEBNER, Melting relations of portion of the system Fe-S-O at high pressure and applications to composition of the earth's core, *Lunar Planet. Sci., X,* 1329–1331, 1979.

TEMPERATURE MEASUREMENTS IN THE LASER-HEATED DIAMOND CELL

Dion L. HEINZ* and Raymond JEANLOZ

Department of Geology, University of California
Berkeley, California 94720, USA

Abstract. A spectroradiometer has been developed to measure peak temperatures (1500–6000 K) and temperature gradients of laser-heated samples contained at high pressures in the diamond cell. It is necessary to measure temperature gradients because the high thermal conductivity of diamond results in large thermal gradients in the sample. A narrow sampling slit is scanned across the image of the radially symmetric heated spot. This gives a set of line integrals across the distribution of emitted thermal radiation, which can be inverted by an Abel transformation to obtain the radial intensity distribution. Collection and inversion of line integrals at two or more wavelengths allow the radial temperature distribution to be determined. The effect of using a finite-width sampling slit is to slightly bias the high spatial-frequency components of the temperature distribution. If the temperature distribution is spherically symmetric a second Abel transformation directly yields the three-dimensional temperature distribution. Deviations from spherical symmetry (horizontal:vertical dimensions up to 2:1) can be corrected for, and larger deviations reduce the three-dimensional distribution to a two-dimensional problem.

Introduction

Diamond possesses a unique set of physical properties: hardness, strength, transparency to electromagnetic radiation, and high thermal conductivity (e.g., ORLOV, 1977). The strength has allowed the development of the high-pressure diamond cell and has made pressures in excess of one Megabar (100 GPa) readily accessible in the laboratory (BELL et al., 1984; GOETTEL et al., 1985). One aspect of the transparency of diamond was exploited by MING and BASSETT (1974), who pioneered the use of laser heating of samples in the diamond cell. Current lasers are capable of generating temperatures in excess of 6000 K while the sample is contained at high pressure in the diamond cell (JEANLOZ and HEINZ, 1984, 1986). Figure 1 summarizes the pressure-temperature range that is experimentally accessible and compares this with a family of estimated geotherms and melting curves. We note that, to date, no material properties have been measured outside the temperature range accessible to cryogenic or external-heating experiments. In particular, the quantitative study of physical-chemical properties of matter at the conditions of the earth's interior has been hampered by the inability to accurately measure the temperature in ultra-high-pressure experiments.

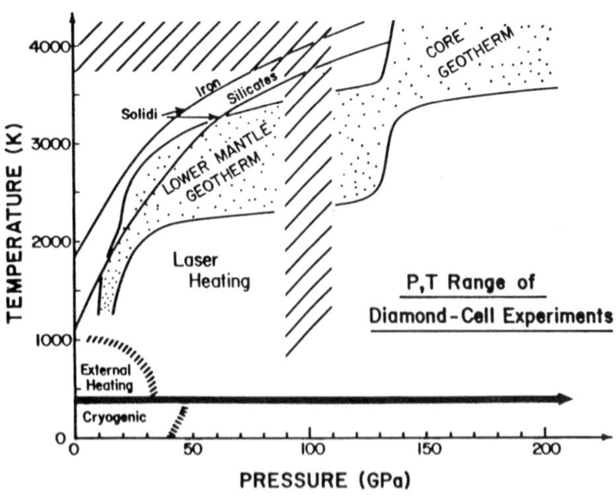

Figure 1. Range in pressures and temperatures that is accessible in diamond-cell experiments: with external heating or cooling (cryogenic experiments) pressures of 30–50 GPa have been achieved, and at room temperature pressures exceeding 200 GPa have been documented. With laser heating simultaneous pressures and temperatures exceeding 100 GPa and 4000 K, respectively, have been achieved. For comparison, the range of estimated temperature profiles through the earth's mantle and core (geotherms: stippled), and current estimates for the melting curves of iron and mantle silicates are shown as functions of pressure (after JEANLOZ and HEINZ, 1984).

This paper describes techniques that we have developed to make the first quantitative measurements of the temperature distribution in a laser-heated diamond cell. Our approach includes the measurement of peak temperatures obtained in the sample as well as the spatial variation of temperature. The thermal gradients across the sample are large because the high thermal conductivity of diamond, along with the small size of the sample relative to the diamond anvils, causes the anvils to become large heat sinks. Figure 2 shows schematically the expected temperature variations in a laser-heated sample. Note especially the existence of temperature gradients in the vertical (on-axis) as well as the horizontal (radial) directions. These thermal gradients represent one of the main obstacles to quantifying experimental conditions at ultra-high pressures and temperatures. Nevertheless, the temperature gradients present an important advantage

*Current address: Department of Geophysical Sciences University of Chicago, Chicago, IL 60637.

High-Pressure Research in Mineral Physics, edited by M. H. Manghnani and Y. Syono, pp. 113–127.

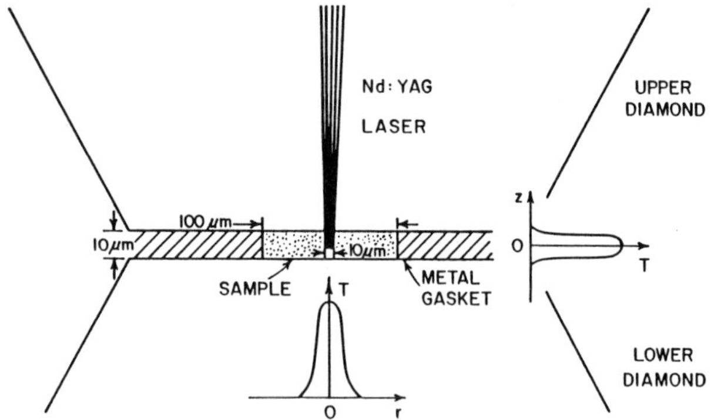

Figure 2. The sample is contained on the top and bottom by the diamonds, and on the sides by a metal gasket. The Nd:YAG laser beam is focused down into the sample through the upper diamond. Schematic representations of the horizontal (r) and vertical (z) temperature (T) distributions are given, along with typical dimensions of the sample and focal spot (JEANLOZ and HEINZ, 1984).

because cooler sample material surrounds, and hence chemically buffers, the hot sample material from the diamond anvils.

In the following sections we discuss the physical basis of thermal radiation, the measurement of temperature from thermal radiation, and the importance of thermal gradients in biasing the measurements. Then we cover the experimental techniques used to measure both temporal and spatial variations of temperature in laser-heated samples. Finally the analysis of experimental data from the diamond cell is summarized. The application of the techniques described here to investigating the conditions required for melting in the lower mantle are described in a companion paper (HEINZ and JEANLOZ, 1987).

Blackbody Radiation and the Optical Properties of Solids

Our approach for measuring temperature in the laser-heated diamond cell is based on observing the thermal radiation from the sample (JEANLOZ and HEINZ, 1984). According to Planck the intensity of light radiated from a material at temperature T is given as a function of wavelength λ by

$$I(\lambda, T) = (2\pi c^2 h)\, \varepsilon\lambda^{-5}\, [\exp((hc/k)/\lambda T) - 1]^{-1} \qquad (1)$$

Here ε, h, k, and c are emissivity, Planck's constant, Boltzmann's constant and the velocity of light, respectively (for this discussion see, e.g., BORN and WOLF, 1975; LOUDON, 1973; WOOTEN, 1972). Kirchhoff's law, or its microscopic equivalent, detailed balancing, states that a body at thermal equilibrium absorbs and emits radiation at equal rates at all frequencies. Thus the emissivity takes on its maximum value, $\varepsilon = 1$, in the case of a blackbody: a material that absorbs all incoming radiation and hence appears black. What about real materials, such as the minerals that are the subject of our experiments (HEINZ and JEANLOZ, 1987)?

Because the silicates and oxides comprising the earth's mantle are relatively transparent dielectrics (absorption coefficient $\alpha \sim 0$), we ignore the effects of reflection or of scattering of light. Thus the intensity of light transmitted through a transparent sample (I_T) is related to the incident light intensity (I_0), and to the absorption coefficient and thickness of the sample (X):

$$I_T(\lambda, X) = I_0 \exp(-\alpha(\lambda)X) \qquad (2)$$

Regardless of whether or not I_0 depends on wavelength, the effect of the absorption coefficient is to make the transmitted intensity depend on both thickness and wavelength. The total light absorbed is therefore given by the nondimensional optical thickness, αX, which is related to the emissivity by Kirchhoff's law and (2):

$$\varepsilon(\lambda) = 1 - \exp(-\alpha X) \qquad (3)$$

The wavelength dependence of ε comes from α (cf. (2)), and it should be emphasized that (3) is only valid if reflection and scattering can be ignored. According to (3), the emissivity is between 0 and 1. The case of a blackbody corresponds to $\alpha X \rightarrow \infty$, whereas a material of unit optical thickness has an emissivity $\varepsilon = 0.632$. These two cases are specific examples of greybodies, samples with emissivity independent of wavelength but not necessarily equal to one.

As our measurements of thermal radiation are carried out at visible and near-infrared wavelengths, we need to consider the physical mechanisms of absorption and emission over this spectral range. For example, free electrons in a simple metal can absorb and give up energy at any frequency, thus providing a mechanism to emit and absorb thermal radiation at all wavelengths. In a dielectric, however, the main absorption mechanisms for photons are valence-to-conduction interband transitions at ultraviolet wavelengths and lattice vibrations at mid- to far-infrared wavelengths. Weaker absorption mechanisms, including charge-transfer and crystal-field absorptions in transition-metal bearing samples, overtone or combination transitions, and defect-related absorptions, are generally present as well. These lead to a finite but small value of α over the range of our spectral measurements. In combination with the small dimensions of our samples (Figure 2), this transparency yields a small, highly non-blackbody emissivity according to (3). Because many of the weak absorption processes are distributed over a range of wavelengths, however, our samples can be accurately treated as greybodies despite the fact that ε

may be near zero.

We discuss below how the greybody assumption can be tested as part of our temperature measurement. For mantle silicates heated in the diamond cell, we find $\varepsilon \sim 10^{-3}$ to 10^{-2} at visible to near-infrared wavelengths. Although significant amounts of absorption can be present at other wavelengths (e.g., the 1.06 μm wavelength of the heating laser), this does not affect our use of the greybody model for deriving temperatures from radiation measured over a limited spectral range.

Optical Properties in Relation to Temperature Measurement

Previous workers (MING and BASSETT, 1974; YAGI et al., 1979) have used optical pyrometry to measure the temperature in laser-heated experiments at high pressure. An optical pyrometer compares the absolute intensity of thermal radiation from the sample with that of a known and variable source. The comparison is made at a single wavelength so the emissivity of the sample must be known in order to derive a temperature from (1). The variable source that is used as a standard is usually a blackbody and the sample emissivity is also assumed to be 1 unless known to be otherwise.

The error involved in assuming $\varepsilon = 1$ for the sample can be evaluated from

$$T_{\text{actual}} = \frac{T_{\text{apparent}}}{1 + \dfrac{\lambda T_{\text{apparent}}}{(hc/k)} \ln\varepsilon} \qquad (4)$$

where T_{actual} is the actual temperature of the sample and T_{apparent} is the temperature of the sample as observed with the optical pyrometer. Figure 3 was calculated from the above relation, with contours of actual temperature being plotted as a function of apparent temperature and emissivity. For a given emissivity the upper axis gives the required equivalent absorption coefficient of the material, assuming a sample thickness of 10 μm (see Figure 2).

For a transparent sample the emissivity is much less than one, as noted above. As an example, the absorption coefficient of olivine measured at high temperatures and visible wavelengths is less than 10 cm^{-1} (FUKAO et al., 1968). According to Figure 3, an emissivity $\varepsilon \sim 10^{-2}$ would be observed in the diamond cell, and this could lead to an actual temperature differing by 50 to 100 percent from the apparent temperature. In addition, reflection and scattering can be significant for absorbing materials such as metals. These processes also result in low emissivities. The conclusions is that $\varepsilon \lesssim 0.1$ can often arise in high-temperature diamond-cell experiments, thus resulting in significant biases in the temperatures estimated by optical pyrometry. Low emissivity always leads to an under-

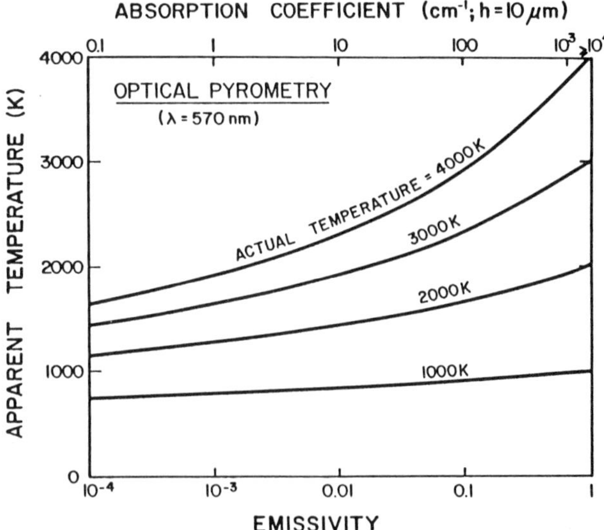

Figure 3. Contours of actual temperature as a function of emissivity and apparent temperature, as observed by an optical pyrometer centered at a wavelength of 570 nm. The upper axis is the equivalent absorption coefficient for a transparent sample assuming a thickness of 10 μm. As discussed in the text, reflection and scattering are not taken into account in this figure.

estimate of the temperature in this instance, and Figure 3 suggests that apparent temperatures derived by optical pyrometry from the diamond cell can be off by a factor of two.

A relatively simple way to surmount the problem of emissivity is to ratio the intensities of thermal radiation observed at two wavelengths, so that the emissivity cancels if it is independent of wavelength. This two-wavelength pyrometry involves comparing the wavelength dependency of the thermal radiation with that of the blackbody curve. For a greybody the temperature is determined independently of emissivity, but there is a trade off in the case of wavelength-dependent emissivity: it is important to choose two wavelengths close enough together that the emissivity does not vary much, but not so closely together that the intensities at the two wavelengths appear to be the same (for which no temperature information can be obtained). Although this method is far superior to an optical pyrometer, it cannot be used to verify that the emissivity is independent of wavelength. Also the measurement is prone to error because of the lack of any redundancy.

Spectroradiometry, or the measurement of emitted intensity at several wavelengths, is the next step to be taken for greater accuracy and precision. The additional information obtained in this case can be used to determine if there is variable emissivity. By comparing the data with the shape of the blackbody relation both the absolute magnitude and the wavelength dependence of ε can be measured. As more information is collected in spectroradiometry it is less susceptible to noise than either optical pyrometry or two-color pyrometry. The adaptation of spectroradiometry to measuring temperature distributions in laser-heated samples is the subject of the rest of this paper.

Temperature Gradients

So far only uniform temperature fields have been considered, although the temperature distribution in the laser-heated diamond cell is expected to be strongly peaked as illustrated in Figure 2. This spatial variation of temperature is induced both by the variation of laser intensity with radial distance from the focal spot and by the high thermal conductivity of diamond. Because of the temperature variation, the intensity of thermal radiation must also vary across the sample (see (1)) and any optically-based measurement across this intensity distribution must yield an average value for the temperature distribution.

The nature of the averaging induced by observing an entire temperature distribution through a pyrometer or spectroradiometer can to some degree be understood by way of the Stefan-Boltzmann relation between the total intensity of thermal radiation and temperature. Integrating (1) yields this relation:

$$I_{\text{total}} = \int_0^{\infty} I(\lambda, T)d\lambda = \varepsilon_t \sigma T^4 \tag{5}$$

with the total emissivity (ε_t) being a weighted average over the wavelength-dependent emissivity and σ being the Stefan-Boltzmann constant. According to (5), the optically determined temperature would be expected to be strongly biased towards the peak value across any temperature distribution because of the strong increase of light intensity with temperature. We now show that the optically derived average temperature across a peaked distribution is indeed close to the maximum value, but not always as close as might be inferred directly from (5).

The effect of temperature gradients upon the average temperature can be illustrated by the following calculation. For both optical pyrometry and spectroradiometry, the average temperature ($\bar{T}$) refers to the value obtained by way of the spatially averaged light intensity observed across the field of view of the particular system. In this approach we assume that the temperature variations that are of interest occur on a spatial scale much smaller than the field of view, as is the case in practice.

The basis of measuring temperature with either an optical pyrometer or a spectroradiometer is the measurement of the intensity of the thermal radiation at one or more wavelengths. Assuming that both systems have the same detector and field of view, we calculate the intensity of thermal radiation that the detector observes at each wavelength from a nonuniform temperature distribution and use these intensities to solve for the average temperature, $\bar{T}$ as defined above. For instance, consider a Gaussian temperature distribution in two dimensions

$$T(r) = T_0 \exp\left[-\left(\frac{r}{\rho}\right)^2\right] \tag{6}$$

where ρ is the characteristic (e-folding) distance of the Gaussian curve. Integration of the intensity distribution obtained by combining (1) and (6) yields the value of the intensity $L(\lambda)$ that the detector sees over the field of view of the detector

$$L(\lambda) = \int_0^R 2\pi I(\lambda, T(r))r dr \tag{7}$$

The upper limit R in (7) defines the field of view, and for our purposes we set it equal to any value beyond which $I(\lambda, T(r))$ is negligible (e.g., $R \gtrsim 1.5\rho$). For comparative purposes, we assume $\varepsilon = 1$ (blackbody emission) in deriving L from T, and thence I in (7).

The justification for using the Gaussian distribution (6) for simulating the two-dimensional intensity distribution seen by the detector is twofold. In the near field (near the focal point of the laser beam, $r \to 0$), the steady-state temperature distribution achieved in the diamond cell can reasonably be assumed to be proportional to the laser-beam intensity. This intensity distribution is Gaussian for the TEM_{00} mode that we use. In the far field, (6) represents the two-dimensional Green's function for the heat-flow problem, assuming that radiative transfer can be treated as a perturbation on lattice conduction. Although the actual temperature distribution is three dimensional in the diamond cell (see Figure 2), we show below that the divergence from a two-dimensional distribution is not severe. Physically this results from the small thickness of the sample. Finally, the use of (6) is justified *a posteriori* from our actual measurements of temperature distributions in the diamond cell (JEANLOZ and HEINZ, 1986; HEINZ and JEANLOZ, 1987).

From (7), the average temperature obtained with an optical pyrometer is

$$\bar{T}_{OP} = \frac{hc/k}{\lambda \ln\left(\dfrac{2\pi c^2 h}{L(\lambda)\lambda^5} + 1\right)} \tag{8}$$

For comparison, the average temperature measured with a spectroradiometer is obtained numerically from (1), (6), and (7). To do this we use Wien's approximation and linear least squares fitting, as discussed below in the section on experimental techniques. The two systems, optical pyrometry and spectroradiometry, measure a weighted-average temperature when temperature gradients are present. Figure 4a illustrates this point for the case of a Gaussian temperature distribution. The shape of the assumed temperature distribution is shown by the heavy curve, and the apparent (measured) average temperatures are indicated for different peak temperatures (T_0) by the two types of systems. The cooler surrounding material contributes to the total intensity, thus ensuring that the average temperature (derived from the field-of-view average intensity) is less than the peak value. Also, both types of systems measure average temperatures that depend upon the magnitude of the peak temperature of the distribution.

There are two points to consider here: a) the spectroradiometer system measures an average temperature closer to the peak temperature than an optical pyrometer does and b) the average temperatures are closer in both cases to the peak temperature at lower temperatures than they are at higher temperatures. These conclusions are summarized in Figure 4b, a plot of average temperature versus peak temperature for Gaussian distributions as they would be measured by both types of systems. The first point derives from the redundancy of information obtained in spectroradiometry as compared with pyrometry. That is, although the emitted light intensity is assumed to be that of a blackbody ($\varepsilon=1$), the effective emissivity observed by averaging the light intensity across the field of view is generally less than one because of the low-intensity contribution from the cooler regions. As the spectroradiometer independently measures the emissivity of the field of view, it approximates the peak temperature better than the optical pyrometer (for which an emissivity of one is assumed).

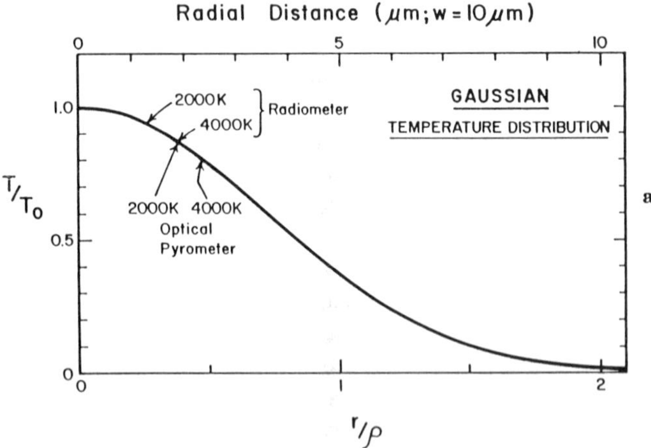

Figure 4a. Temperature (T) normalized to the peak temperature (T_0) is shown as a function of radial distance (r) for a Gaussian distribution. The e-folding or characteristic distance (ρ) of the temperature distribution is used to nondimensionalize r and a typical value of 10 μm has been assumed for the beam waist ($\omega=2\rho$) on the upper scale. The arrows along the Gaussian temperature distribution indicate the field-of-view average temperatures ($\bar{T}$; normalized to T_0) that would be measured with an optical pyrometer (characteristics given in Figure 3) and a spectroradiometer for peak temperatures of T_0=2000 K and 4000 K.

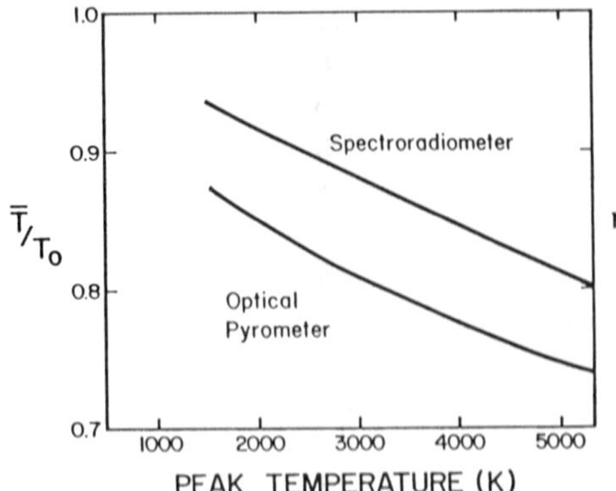

Figure 4b. The ratio of average to peak temperature ($\bar{T}/T_0$), as given in Figure 4a, is shown as a function of peak temperature for both optical pyrometer and spectroradiometer systems.

To understand why the peak temperature is better approximated by the average temperature at low peak temperatures as opposed to high peak temperatures one must look at how the intensity changes at each wavelength. A measure of the change of intensity at each wavelength is shown in Figure 5 as $\partial \ln I/\partial \ln T$ as a function of wavelength and temperature as derived from (1). The intersection of the thin horizontal line (labeled "peak") and the temperature contours indicates the wavelength of the intensity maximum at each temperature. Thus it can be seen from Figure 5a that the intensity increases more rapidly with temperature on the short- than on the long-wavelength side of the intensity peak of the blackbody spectrum.

A more general form of Figure 5a is shown in Figure 5b, with $\partial \ln I/\partial \ln T$ being plotted as a function of $hc/(k\lambda T)$ for both the Planck and Wien functions. The Wien approximation is used for most of our data reduction, as discussed in the next section. Note that the derivative of emitted intensity with wavelength increases for both functions with increasing values of $hc/(k\lambda T)$, and the two functions are indistinguishable for $hc/(k\lambda T)\gtrsim 4$. For a given temperature the peak in the emitted radiation occurs at λ_{max} such that $hc/(kT\lambda_{max})=4.97$. The magnitude of the derivative in Figure 5 expresses how quickly the intensity varies with temperature. Thus, at a given wavelength, the intensity at 2000 K is varying faster with temperature than it is at 4000 K, implying that a peak temperature of 2000 K can outshine the surrounding cooler material more easily than can the 4000 K peak temperature. This explains the better approximation of

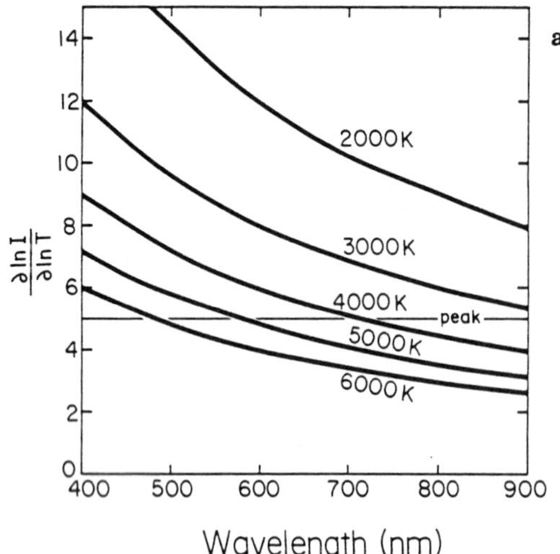

Figure 5a. The sensitivity of the blackbody thermal emission to temperature changes is shown as a function of temperature and wavelength. The axis $\partial \ln I/\partial \ln T = m$ gives the effective power-law dependency of emitted intensity (I) with temperature (T) at a given wavelength (λ): $I(\lambda, T) \propto T^m$. At short wavelengths and low temperatures the intensity varies more rapidly with temperature than would be expected from the Stefan-Boltzmann relation (see text): $I_{total} \propto T^4$. The wavelength with peak intensity (given by $m=4.97$) is indicated as a function of temperature.

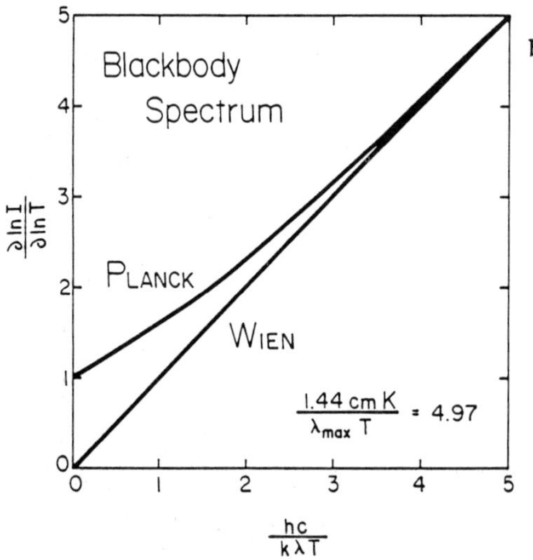

Figure 5b. A nondimensional version fo Figure 5a, with h, c, and k being Planck's constant, the velocity of light, and Boltzmann's constant; λ_{max} is the wavelength with peak intensity at a given temperature. The Planck spectrum for blackbody radiation (equation (1)), from which Figure 5a is derived, is accurately reproduced at short wavelengths and low temperatures by the Wien approximation (equation (9); see also Figure 7). Beyond the nondimensional frequency $hc/(k\lambda T)=4$ the Planck value for $\partial \ln I/\partial \ln T = m$ is indistinguishable from the Wien result $mk\lambda T = hc$.

peak temperature by average temperature at lower peak temperatures.

Experimental Technique

The laser-heating and spectroradiometric system has been described previously (Jeanloz and Heinz, 1984, 1986) and is shown schematically in Figure 6. Samples are heated using a Quantronix 117 Nd-YAG laser with a maximum output of 25 watts at $\lambda = 1.064$ μm when operating in the cw TEM_{00} mode. The TEM_{00} mode has a radially symmetric, Gaussian intensity profile at the focus, and the geometry of the diamond cell is such that the symmetry along the optic path is radial (see Figure 2). Thus, neglecting small scale heterogeneities, we expect that the temperature distribution in the diamond cell would also be radially symmetric. We first take the laser output to be truly cw (constant), and return later to the problem of power fluctuations and the feed-back circuit that we have developed.

As the output of the laser is polarized, the power reaching the sample can be controlled by passing the beam through a $\lambda/4$ plate to rotate the plane of polarization and then through a brewster polarizer. We do this in order to control the laser power without changing any of the characteristics of the laser cavity (there is a slight reflection from the polarizer-attenuator back into the cavity, but this can be ignored for the present discussion). Instabilities due to heating and cooling of the laser crystal are thereby minimized. Note also that a low-pass filter (F_1) is included in the polarizer-attenuator so that none of the visible and near-infrared light from the pump lamp can reach the sample and spectroradiometer.

The laser beam is reflected off of a dichroic "hot" mirror (M), which is designed to reflect infrared and transmit visible radiation (Schott 116). The beam is then focused onto the sample with a Lietz UM-20 lens (L_1), with a numerical aperture of 0.33, (focal length of 12 mm and working distance of 14 mm). The focal spot is between about 10 and 30 μm in radius, depending on details of the focusing conditions and optics used. For a perfectly absorbing sample the laser power corresponds to a heating rate of $\sim 10^4$ K/s in the diamond cell. Thus a thermal steady state is achieved on time scales of seconds or less, at which time the light emitted from the sample can be used to determine temperature.

The thermal radiation collected from the sample by the objective lens (L_1) is focused with a biconvex lens (L_2: focal length of 125 mm) onto the entrance slit of the monochromator. The dichroic filter (F_2), which is used to filter out any stray laser radiation, consists of three layers each of which is identical to the mirror M. Unlike M, which is oriented at 45° to the optic axis, F_2 is at an angle of 10° to prevent internal reflections. The image of the

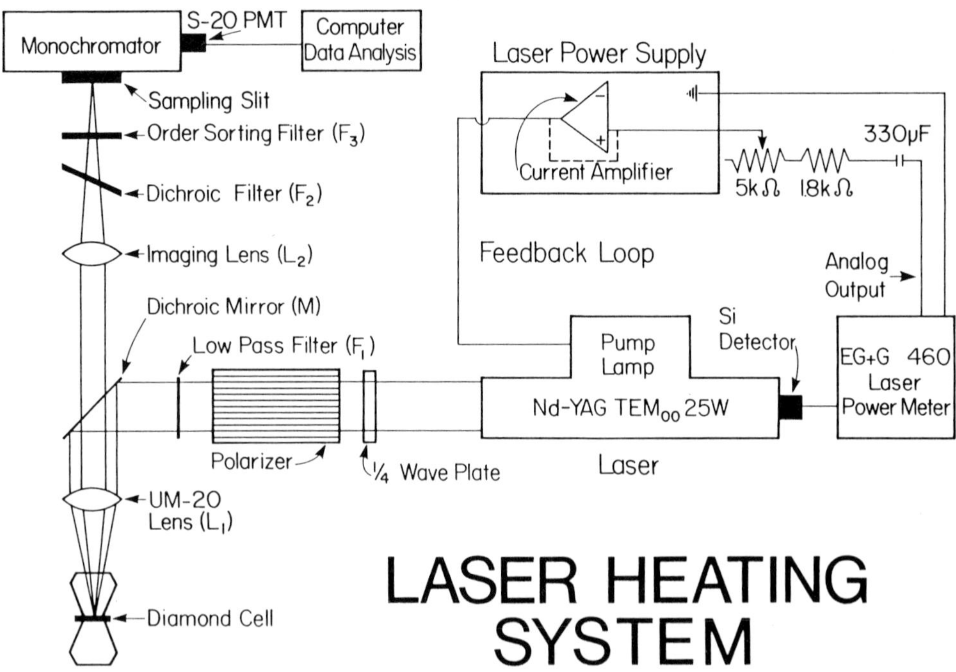

Figure 6. Schematic diagram of the laser-heating and temperature-measurement system. The details of this system are described in the text.

entrance slit is focused onto the exit slit by a concave holographic grating (f 3.5 and 12 nm/mm dispersion) and an order-sorting filter (F_3) prevents the overlap of different orders of the dispersed spectrum. A slit width of 0.83 mm results in a spectral resolution of 10 nm and a diffuser is further placed between the exit slit and detector in order to reduce the effects of inhomogeneous sensitivity across the detector element. An S-20 type photomultiplier tube (PMT) was used for most experiments although a silicon detector is generally adequate for measuring average temperatures (JEANLOZ and HEINZ, 1984).

Our spectroradiometer has been calibrated by way of a tungsten-ribbon filament lamp with spectral-radiance values that are traceable to the National Bureau of Standards. The known irradiance of the lamp was fitted using a fourth-order polynominal to allow for interpolation between the known irradiances at given wavelengths. The calibration was performed by placing the lamp below a pinhole located at the focus of the UM-20 lens and by using the calculated spectral radiances at the pinhole. The pinhole was used to simulate the spot size and location of a laser-heated sample. The calculated temperature of the fit to the measured lamp spectrum differed from the calculated temperature of the known irradiances by $<1\%$.

The response of the spectroradiometer system is plotted as a function of wavelength in Figure 7. For comparison, the wavelength of the Nd-YAG laser and the Planck and Wien functions for 2000, 3000, and 4000 K are also

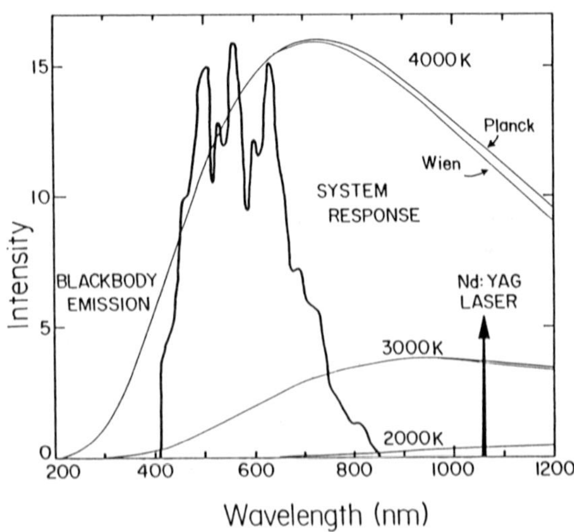

Figure 7. Spectral response of the present spectroradiometer system compared with the laser wavelength (1064 nm) and the thermal emission from blackbodies (both Planck and Wien laws) at 2000 K, 3000 K, and 4000 K. The system response (intensity in arbitrary units) includes the effects of filters, dichroic mirrors, lenses, and the detector. There is additionally an electronic scale factor ranging over 10^4 for the system response. (JEANLOZ and HEINZ, 1986).

shown. The system response can be electronically changed by a scale factor between 1 and 10^4, which is not shown. The measured response function is the result of the convolution of the spectral sensitivity of the S-20 photocathode in the PMT and the transmission spectra of the optical components (i.e., dichroic mirror, filters, diffuser, and lenses). The S-20 photocathode sensitivity decreases around 800 nm and the transmission of the dichroic mirror and filter drops rapidly near 900 nm. The figure demonstrates two points: a) the spectroradiometer system can accurately measure temperatures in the range of 2000–6000 K, and b) the Wien approximation is good over this temperature range.

To test the calibration of our laser-heating and spectroradiometer system, we have examined the melting of several metals at zero pressure. Figure 8 is a plot of the experimentally determined melting points of several high-purity metal wires at zero pressure, using the above system and calibration (emission spectra are typically collected at 20 nm intervals between 600 and 780 nm). The melting temperature of the metals is bracketed by the observation (or lack thereof) of flowage and the formation of a bead of melt at the end of the wire sample. A controlled flow of argon past the wire is used to prevent oxidation of the metal in these experiments. Thus the main difference between the melting of metals at zero pressure and the conditions of our high-pressure experiments is that the region that is uniformly heated is an order of magnitude larger in the former than in the latter case. Again, the presence of diamonds around a thin sample at elevated pressures produces large temperature gradients.

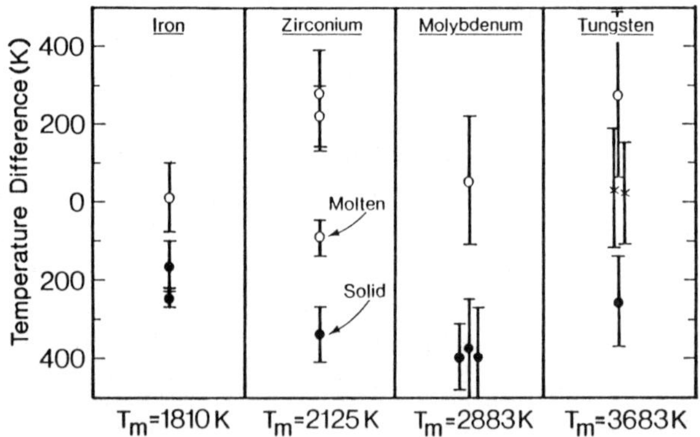

Figure 8. Melting of four metals under the laser beam at zero pressure. The difference between the spectroradiometrically determined temperature and the known melting temperature (T_m, given at the bottom of each box) is plotted for Fe, Zr, Mo, and W. Open symbols refer to the direct observation of flow, whereas closed symbols indicate that the samples appear solid. The cross symbols for W are for observations in which the interface was seen to be fluctuating back and forth over the bead of metal. In all cases the metals were contained in a stream of Argon. (JEANLOZ and HEINZ, 1986).

The experimentally measured melting points shown in Figure 8 are consistent with the known melting points of the metals except in the case of zirconium. The melting temperature of zirconium is underestimated by 5 to 15%, but this is within the uncertainties expected from the laser fluctuations (see below). The main difficulty with these experiments is that because of the non-linearity of the laser-sample coupling it is not always possible to achieve temperatures close to the melting point in both the liquid and the solid. One exception to this is the tungsten experiment, in which the liquid-solid interface was directly observed because the absorption of the laser beam and the heat loss from the sample were fortuitously balanced. In addition, the fluctuations in the intensity of the heating laser are absorbed in the enthalpy of melting of the tungsten. This resulted in estimates of the melting temperature that are within thirty degrees of the known value for tungsten. Based on Figure 8, we estimate the absolute calibration of the spectroradiometer system to be within ± 200 K, which is in accordance with a previous calibration using a different detector (JEANLOZ and HEINZ, 1984).

Wien's law

$$I(\lambda) = (2\pi c^2 h)\varepsilon\lambda^{-5}\exp(-(hc/k)/(\lambda T)) \qquad (9)$$

is a convenient approximation to Planck's law because it allows the data to be analyzed by linear statistics. Defining a normalized intensity

$$J = \ln[I\lambda^5/(2\pi c^2 h)] \qquad (10a)$$

and a normalized frequency

$$\omega = (hc/k)/\lambda \qquad (10b)$$

which are both observable, Wien's law can be expressed as a linear equation in T^{-1}:

$$J = \ln\varepsilon - \omega T^{-1} \qquad (11)$$

The difference between the Planck function and the Wien function for the system discussed in this paper is plotted in Figure 9. For temperatures of interest in the earth the deviation is less than 4%, thus justifying the use of (11) in our data analysis. With this form we solve for the greybody temperature and emissivity using linear least squares; uncertainties in the observed intensities are quantitatively propagated to uncertainty estimates for ε and T. Furthermore, curvature of the data in the normalized intensity versus normalized frequency plane of (11) can be directly interpreted in terms of a wavelength-depen-

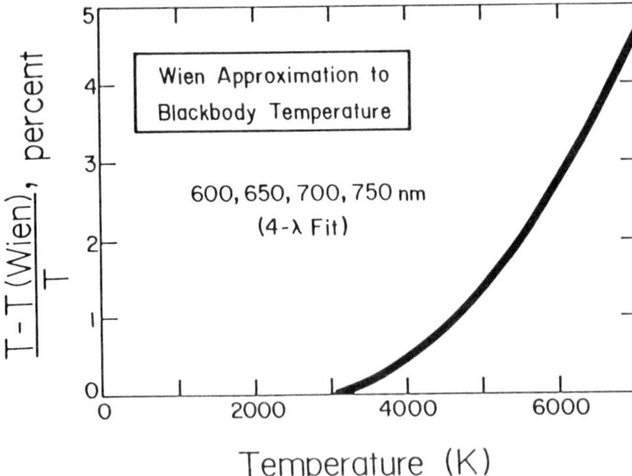

Figure 9. The deviation of temperature derived from a Wien's-law fit to a blackbody (Planck) spectrum is shown in percent as a function of the blackbody temperature. A least-squares fit of the spectrum at four wavelengths (600, 650, 700, and 750 nm) yields a temperature estimate that is accurate to within 5 percent for temperatures less than 7000 K.

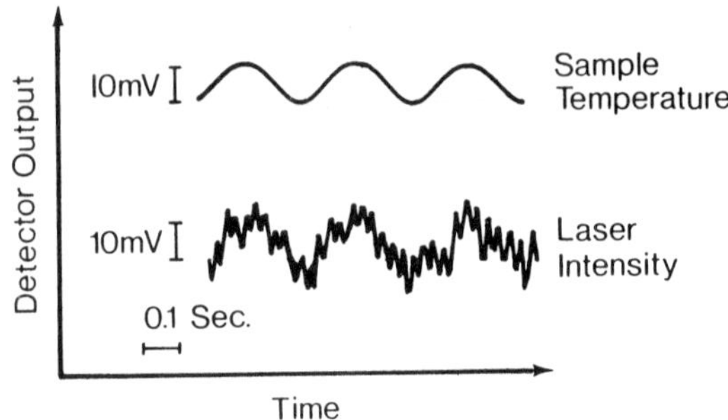

Figure 10. Oscilloscope traces illustrating the temporal fluctuations of the laser intensity (lower trace) and the sample intensity (upper trace). The sample is FeO contained at 14 GPa in an NaCl matrix. The duration of the trace is one second and time increases to the right. Note that there is a slight phase lag of the sample intensity behind the laser. The sample acts as a low-pass filter on the high-frequency laser fluctuations. (JEANLOZ and HEINZ, 1986).

dent emissivity or in terms of noise in the observations. To the extent that the curvature is not due to the Wien approximation, the quality of the fit of (11) to the data directly reflects the appropriateness of the greybody model (see HEINZ and JEANLOZ, 1987).

Temperature Fluctuations and Their Determination

When a laser-heated sample is observed visually it can be seen to flicker due to instabilities in the laser output. This phenomenon has been documented by looking at the correlation between the laser intensity fluctuations and the sample intensity fluctuations (Figure 10). The main fluctuations appear to be on the order of 1 to 10 Hz and are probably due to building vibrations. Note that the sample acts as a low pass filter to the higher frequency components of the noise in the laser signal (JEANLOZ and HEINZ, 1984). We have followed the recommendations of KOECHNER (1972) to stabilize these fluctuations, in particular by mechanically isolating the laser-spectrometer system on a granite table. We have also installed soft plastic hoses in the closed cooling system to reduce the transmission of vibrations from the water pump in the closed system.

In addition to the vibrational damping we have inserted a simple electronic feed-back loop to help reduce the laser-output fluctuations. The laser power is continuously monitored through the back mirror of the cavity by means of a Si detector (Figure 6). We strip off the dc component of the detector output with a 330 μF capacitor and scale the remaining ac component of the output with a 1.8 kΩ fixed resistor plus a 5 kΩ potentiometer. This scaled output is fed back into the over-current operational amplifier of the laser power supply. As a result the output is stabilized to about 1–3% (rms), which is well below the manufacturer's specifications (5% rms).

Aside from stabilizing the laser, we have modified the detector system to average the signal over time in order to obtain reproducible spectra. This involves placing high-quality polystyrene capacitors across the calibration resistors in the detector electronics. We adjusted the time constant (τ) to be on the order of 1 to 3 seconds. That is, τ (in seconds) is approximately equal to the capacitance (in farads) times the resistance (in ohms): capacitors of 0.00156, 0.0156, 0.147, 1.5, and 15 μF are used for resistors of 10^9, 10^8, 10^7, 10^6, and 10^5 ohms, respectively. Thus our measurements are temporal averages over the temperature fluctuations and this must be taken into account when we want to determine the peak temperature obtained in a sample.

Based upon the correlation of the oscilloscope traces in Figure 10 it is evident that most of the low frequency fluctuations in sample temperature are due to noise in the laser signal. We have used the output of the PMT and a digital voltmeter sampling at 20 samples per second to measure the flickering of the sample. Collecting intensities at four wavelengths (600, 650, 700, and 750 nm) and using the relation

$$\Delta T = \Delta I(\partial T/\partial I) \tag{12}$$

with

$$\frac{\partial T}{\partial I} = \frac{(hc/k)}{\lambda I(\ln[I\lambda^5/(2\pi c^2 h)])^2} \tag{13}$$

(where we have used Wien's approximation) we have confirmed that 3% rms noise from the laser produces fluctuations on the order of 200–1000 K in the sample.

Measurement of Temperature Gradients and the Radon Transform

For quantitative experiments at high temperatures it is necessary to know the spatial variation of temperature or to have information about the temperature gradients across the sample. We have developed a technique to measure the temperature gradients in a laser-heated diamond cell (JEANLOZ and HEINZ, 1984, 1986). We focus the image of the laser-heated spot on the fixed entrance slit of the monochromator and then measure the intensity of light as we scan a small sampling slit across the image of the spot, as shown in Figure 11. This yields line integrals of intensity across the spot and, given a sufficient number of line integrals collected along the x axis, it is possible to invert for the radial distribution of the intensity. If the radial distribution of intensity is known at two or more wavelengths it is then possible to invert for the radial temperature profile using Wien's law and linear least squares. In our experiments we collect the line integrals at 600, 650, 700, and 750 nm to invert for the radial temperature distribution. The light intensity coming from the small sampling slit is monitored by the PMT and is digitally recorded and processed by computer. No capacitors are used in these experiments because this would require us to run the slit at very low speeds causing the experiments to be prohibitively long. Points are collected at the rate of 1.2 per second, with each scan across the laser-heated spot requiring 2–3 minutes. Once the profiles are collected at the four wavelengths they are smoothed using a running average of 5 points. As the temperature and intensity distributions are radially symmetric, the profiles contain redundant information and the two halves (data collected over positive and negative x) can be averaged to further increase the signal-to-noise ratio. To do this it is necessary to determine the center ($x=0$) of each profile. Two measures of the center are the maximum value of the profile and the center of gravity of the half-maximum value on each side of the profile. An average of these two criteria is used but the difference is not significant.

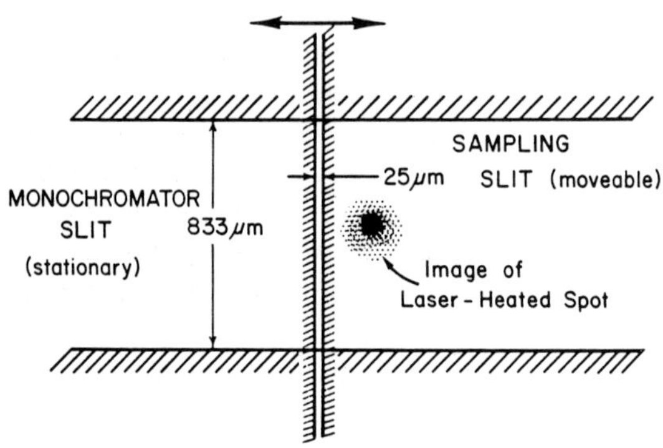

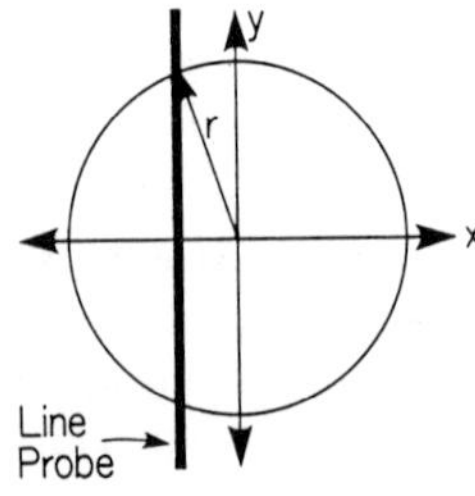

Figure 11. A vertical sampling slit (mathematically represented as a line probe) is scanned across the image of the radially symmetric laser-heated spot with the image being focused on the fixed entrance slit of the monochromator. The spatial variation of temperature in the sample is determined by monitoring the intensity of light at two or more wavelengths as the moveable slit is scanned across the image of the hot spot. The coordinate axes shown in the lower half of the figure illustrate the coordinates used in the equations in the text. The x coordinate corresponds to the position of the slit and r is the radial coordinate. (after JEANLOZ and HEINZ, 1984).

A set of line integrals can be inverted for the radial distribution using a degenerate form of the Radon transform, the Abel transform (e.g., DEANS, 1983). The Radon transform of a two-dimensional function f is equivalent to first taking the two-dimensional Fourier transform of f and then taking the inverse one-dimensional Fourier transform of the radial component of the two-dimensional Fourier transform. Our slit-sampling technique provides a Radon transform of the temperature field, which must be inverted to yield the actual temperature distribution in the sample.

We take advantage of the radial symmetry to reduce the general Radon transform to its one-dimensional version, the Abel transform. Thus, a set of line integrals of the intensity collected by the sampling slit, $I(\lambda, x)$ (see Figure 11), is transformed to the radial intensity $I(\lambda, r)$ by

$$I(\lambda, r) = \frac{-1}{\pi} \int_r^R (x^2 - r^2)^{-1/2} \frac{dI(\lambda, x)}{dx} dx \tag{14}$$

where it is assumed that the magnitude of $I(\lambda, r)$ is negligible beyond some distance R. Following NESTOR and OLSEN (1960) and GOODERUM and WOOD (1950), (14) is integrated numerically. First, divide the x axis into zones of equal width a such that $x_n = na$. Using the transformation $v = r^2$ and $u = x^2$, (14) becomes

$$I(\lambda, v) = \frac{-1}{\pi} \int_r^{R^2} \frac{dI(\lambda, u)}{du} (u - v)^{-1/2} du \tag{15}$$

If $I(\lambda, v)$ is assumed to be linear in v across each zone (labelled by n), then (15) becomes

$$I(\lambda, v_k) = \frac{-1}{\pi} \sum_{n=k}^{N-1} \frac{dI(\lambda, u)}{du} \int_{an^2}^{[a(n+1)]^2} [u - (ak)^2]^{-1/2} du \quad (16)$$

where

$$\frac{dI(\lambda, u)}{du} = \frac{I_{n+1}(u) - I_n(u)}{a^2[(n+1)^2 - n^2]} \quad (17)$$

Combining (16) and (17), and retransforming to the original coordinates, (16) becomes

$$I(\lambda, v_k) = \frac{-2}{\pi a} \sum_{n=k}^{N-1} A_{k,n}[I_{n+1}(x) - I_n] \quad (18a)$$

with

$$A_{k,n} = [(n+1)^2 - k^2]^{1/2} - \frac{[n^2 - k^2]^{1/2}}{2n+1} \quad (18b)$$

NESTOR and OLSEN (1960) avoided taking the differences of the experimental data by modifying (18) to

$$I(\lambda, v_k) = \frac{-2}{\pi a} \sum_{n=k}^{N} B_{k,n} I_n \quad (19a)$$

where

$$B_{k,n} = -A_{k,k} \quad \text{for} \quad n = k \quad (19b)$$

$$B_{k,n} = A_{k,n-1} - A_{k,n} \quad \text{for} \quad n \geq k+1 \quad (19c)$$

The $B_{k,n}$ are tabulated in NESTOR and OLSEN (1960) or may be calculated as needed. As can be seen by (16), the radial distribution of intensity at a distance k from the center ($k=0$) depends only upon values of the line integrals further from the center (i.e., $n>k$). Due to the differentiation step (17), the numerical solution of (14) can be unstable in the presence of noise. This is why a running average is used to smooth the experimental profiles.

Experimentally it is impossible to collect line integrals. Instead a finite slit width must be used, which amounts to strip integration. The finite size of the sampling slit blurs out the high spatial frequency content of any measured intensity gradients because the signal is averaged over the width of the slit. This effect can be documented by calculating strip integrals of different widths:

$$L(\lambda, \Delta x) = \int_x^{x+\Delta x} \int_0^Y I(x, y, \lambda) dy\, dx \quad (20)$$

where Δx is the width of the strip, Y is a value beyond which I is negligible, and (14) is used to invert for the filtered version of the original function. Figure 12 illustrates the filtering properties of various sized slits upon different functional forms for the original temperature distribution. Temperature is shown as a function of radial distance for the original distribution and for the filtered versions, with W/ρ being the slit width relative to the characteristic length for each curve. In the limiting cases of W/ρ going to either 0 or ∞, either the original temperature distribution or the average temperature $\bar{T}$ are recovered. This figure demonstrates that as the slit width W/ρ gets larger the higher spatial frequencies of the temperature distribution are increasingly less well resolved. The corresponding emissivities are plotted in the lower portion of each figure and it can be noted that the r-dependence of the observed (apparent) emissivity contains information about the degree to which the original distribution has been smoothed.

With an opaque solid only the surface is visible and thus the Abel transform yields the surface temperature directly. If the absorption depth of the laser is the same as for visible light then the peak temperature should occur at the surface, because that is where the heat source is. Thus in order to achieve high temperatures it can be important to imbed opaque solids in a transparent matrix to isolate them from the diamond heat sinks.

For transparent solids the temperature distribution is inherently three-dimensional. A spherically symmetric temperature distribution can be recovered by the slit-scanning method described above provided that the emissivity is constant. That is, the length-scale characterizing the temperature distribution in the radial (r) dimension is equivalent to that in the vertical dimension (z) in this case (cf. Figure 2). The intensity for a spherically symmetric distribution is

$$L(\lambda, r) = 4\pi \int_0^R I(T(r), \lambda) r^2 dr \quad (21)$$

The integral at one wavelength only depends upon one variable (r) and it is sufficient to scan a slit across only one direction in order to obtain the required information. The process of recovering the spherical distribution is as follows (VEST and STEEL, 1978). First the distribution is scanned with a sampling slit and these data are inverted for the vertically averaged radial distribution. Then the vertically averaged radial distribution is transformed once again to obtain the spherically symmetric distribution. To summarize the above process, a spherically symmetric distribution can be recovered from a scanned slit by the application of two successive Abel transforms, but this is only true for transparent solids with constant emissivity.

To test the effect of the third dimension in determining temperature gradients across transparent samples we

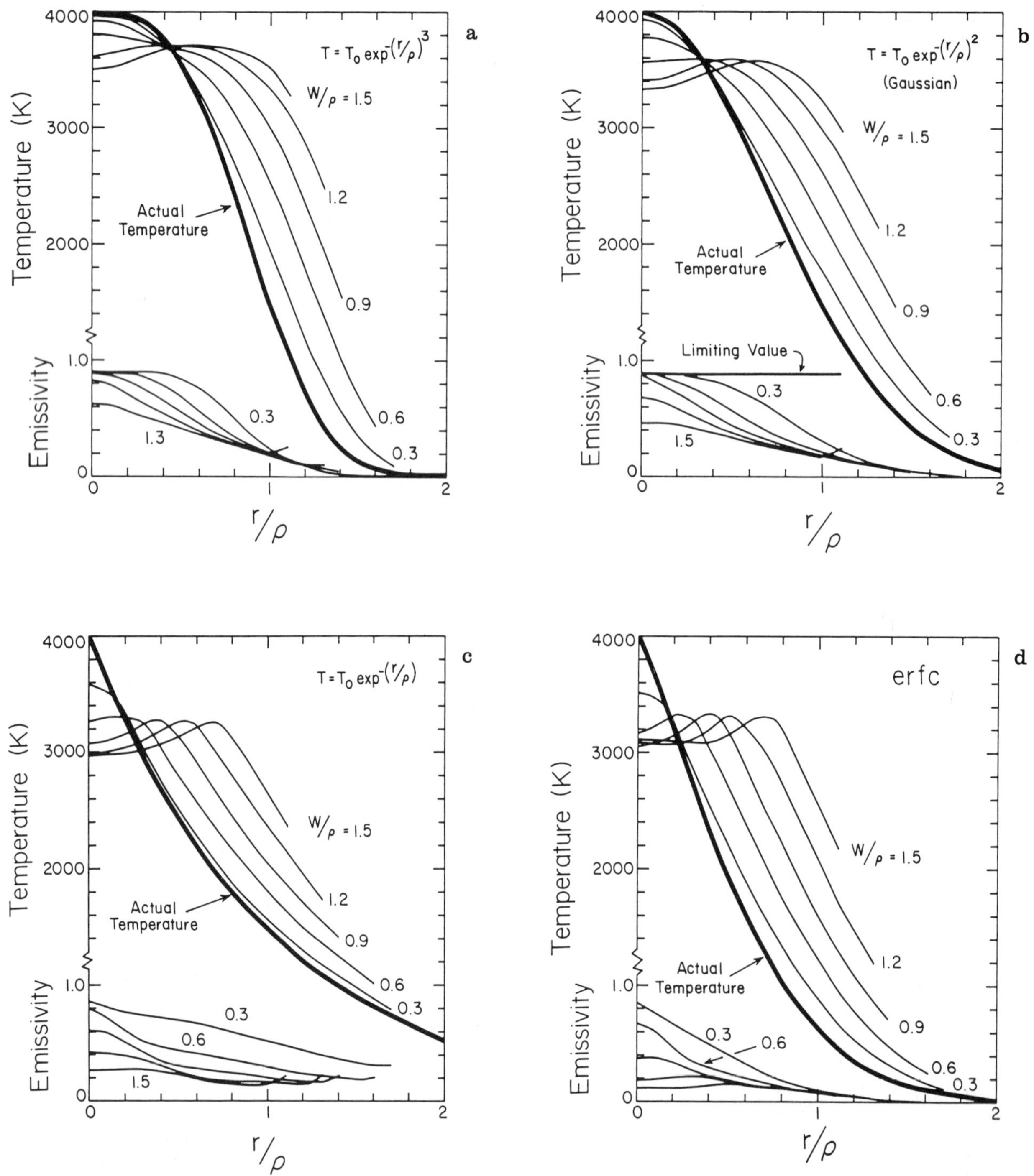

Figure 12.

took a variety of functional forms for different temperature distributions and calculated the average temperatures for one-, two-, and three-dimensional distributions. We also took the same functional forms as two-dimensional distributions and used an Abel transform on the radial component to mimic the process of inverting for a spherically symmetric distribution as described above. The two calculations give us the difference between peak

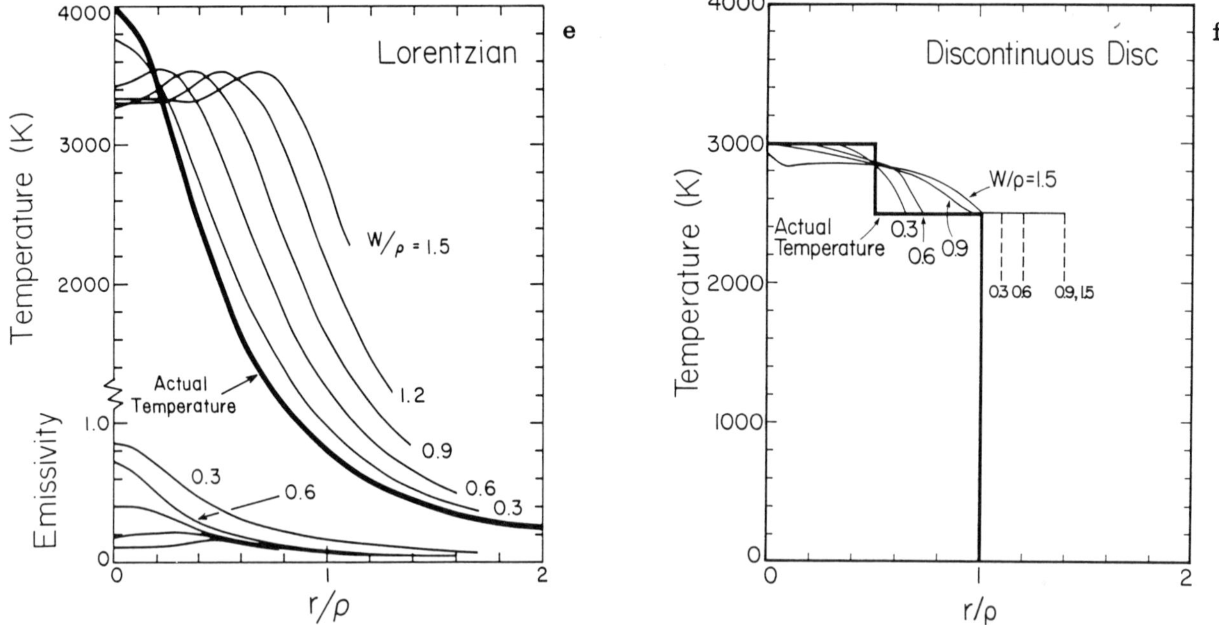

Figure 12. Temperature and emissivity as functions of nondimensionalized radial distance (see Figure 4a). The heavy curves represent the original two-dimensional temperature distributions, and in every case but the last the peak temperature is 4000 K. The light curves show the distributions that are recovered after the original distribution is sampled using strip integrals of various widths. The values of W/ρ represent the width of the sampling slit (see Figure 11). Typically $W/\rho \sim 0.3$ in our experiments. The following six temperature distributions are considered: a, $n=3$: $T=T_0\exp[-(r/\rho)^3]$. b, $n=2$: $T=T_0\exp[-(r/\rho)^2]$. c, $n=1$: $T=T_0\exp[-(r/\rho)]$. d, erfc: $T=T_0\mathrm{erfc}(r/\rho)$. e, Lorentzian: $T=T_0/[4(r/\rho)^2+1]$. f, Discontinuous Disc: $T=3000$ K, $0<r/\rho<0.5$; $T=2500$ K, $0.5<r/\rho<1.0$; $T=0$ K, $r/\rho>1.0$.

and average temperatures and the difference between a vertically averaged two-dimensional peak temperature and the actual peak temperature, respectively.

Figure 13 is a plot of the average temperature that would be observed for one-, two-, and three-dimensional radially symmetric temperature distributions with a peak temperature of 4000 K. These are the same functional forms as those plotted in Figure 12a–e. Note that the average temperature is closest to the peak temperature for gentle thermal gradients and farthest for steep thermal gradients at the center of the distribution. The three-dimensional temperature distribution and the original two-dimensional Gaussian distribution from which it was derived by an Abel transform are compared in Figure 14. The two results to note here are: a) the shapes of the two-and three-dimensional distributions are similar, and b) the difference in peak temperature between the two-dimensional distribution and the resulting three-dimensional distribution (ΔT) is consistent with the difference seen between plane and volume distributions in Figure 13. Figures 13 and 14 are the result of forward and inverse calculations, respectively. Because the forward and the inverse calculations produce results that are consistent with each other we consider the inversion to be stable.

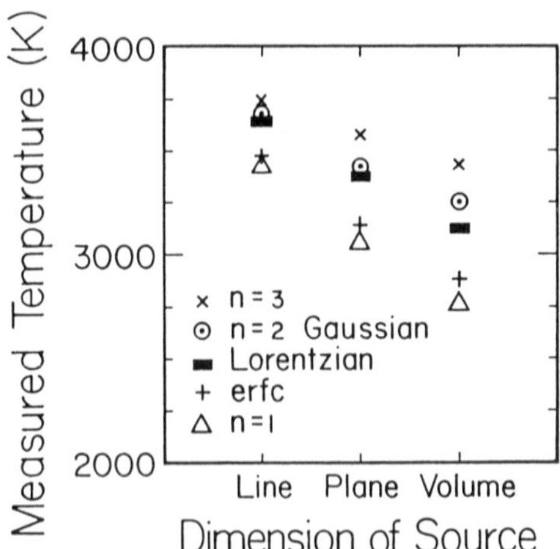

Figure 13. The spatially averaged temperature ($\bar{T}$) that is observed as a function of the number of dimensions of the source: line, plane, and volumetric are one-, two-, and three-dimensional temperature distributions, respectively. Opaque and transparent samples correspond to plane and volume sources (see text). Five types of distributions are considered, all with a peak temperature of 4000 K; these are defined and plotted in Figure 12. The average temperatures are calculated in a manner analogous to those in Figure 4. Details are discussed in the text.

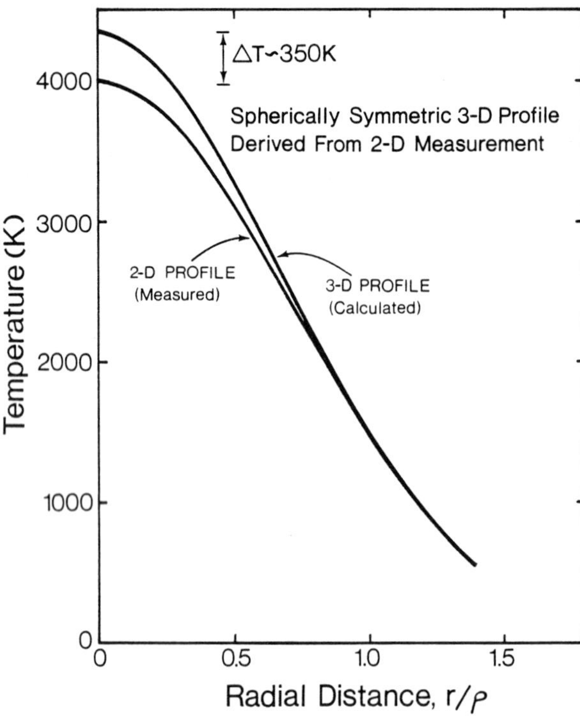

Figure 14. Temperature as a function of nondimensionalized radial distance (see Figures 4a and 12). The lower line is a two-dimensional Gaussian distribution. The upper line is the spherically symmetric three-dimensional temperature distribution that results from acting on the two-dimensional distribution with an Abel transform. This calculation makes the assumption that the emissivity is constant. The difference between the peak temperatures of the two curves, labeled ΔT, is ΔT=230 K, 350 K, 380 K, 370 K, and 380 K for n=3, n=2, n=1, erfc, and Lorentzian distributions respectively (see Figure 12). (JEANLOZ and HEINZ, 1986).

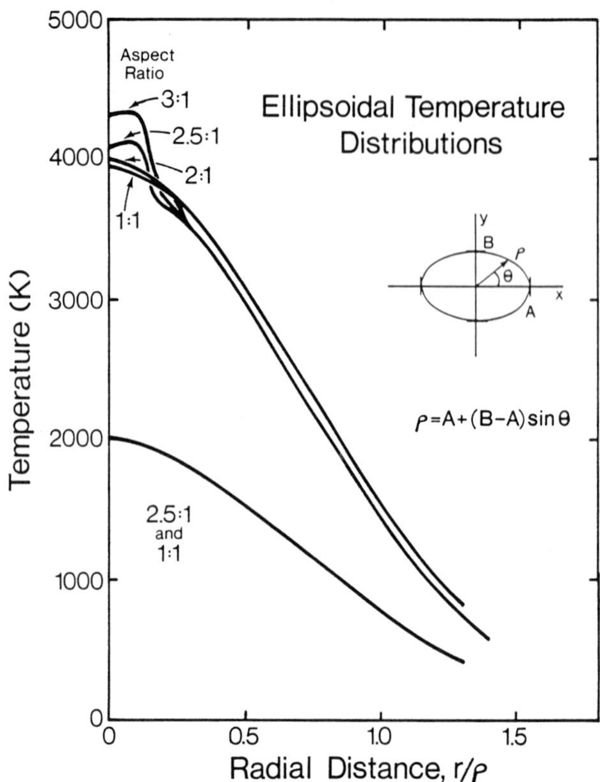

Figure 15. Abel transforms of two-dimensional Gaussian temperature distributions that deviate from radial symmetry. The characteristic distance ρ now becomes a function of direction, as indicated in the right side of the figure. A and B are the major and minor axes of the ellipsoidal distribution, with y and x being vertical and horizontal (radial) directions, respectively (cf. Figure 2). The curves are labelled by the ratio $A{:}B$. The upper set of curves is for Gaussian distributions with 4000 K peak temperature and the lower curves are for a peak temperature of 2000 K. The oscillations in the upper curves near the origin are related to the width of the sampling slit, which in this case is $W/\rho=1$. Similar oscillations are evident in Figure 12.

Spherical symmetry is not usually obtained in a diamond cell, however. The horizontal length scale is two-to-four times the vertical length scale, based upon the observed thicknesses of samples and the observed horizontal extents of heated material (Figure 2). In order to determine how far from a radially symmetric distribution one can go before the Abel transform breaks down when inverting for the vertical gradients we have calculated the Abel transform of ellipsoidal Gaussian temperature distributions for aspect ratios up to 3:1. This aspect ratio includes the ratio of the horizontal to vertical length scale obtained in the diamond cell. As can be seen from Figure 15, there is no significant deviation from spherical symmetry for profiles of aspect ratios up to 2:1. At an aspect ratio of 3:1 the deviation is on the order of 300–400 K for peak temperatures of 4000 degrees. This effect is much less at lower temperatures: for example, at 2000 K a 2.5:1 ratio overestimates the peak temperature by only 10 K. This overestimate must be subtracted from the three-dimensional distribution derived after the second transformation.

To conclude, the correction from two- to three-dimensional distributions is largest for the spherically symmetric case, but this can be calculated exactly. In the limit of a very flat ellipsoidal distribution, the two-dimensional distribution is correct. Intermediate cases can be corrected by interpolation, and this amounts to no more than a few hundred Kelvin in peak temperature for samples in the laser-heated diamond cell.

Summary and Conclusions

We have shown that both the emissivity and thermal gradients are important factors that must be determined in quantitative high-temperature/high-pressure experiments with the laser-heated diamond cell. The combined problems of emissivity and temperature gradients have forced us to use a modified spectroradiometric technique capable of measuring peak temperatures (1500–6000 K) and temperature gradients, as well as average temperatures, of samples contained in the diamond cell. This method is applicable to either opaque or transparent solids. It involves scanning a narrow sampling-slit (effectively a line probe) across the image of a radially symmetric heated spot. This gives a set of line integrals across the distribution of emitted thermal radiation. These line integrals are inverted by an Abel transform to give the radial intensity distribution. Collection and inversion of these line integrals at several wavelengths allows the radial temperature distribution to be determined. Laser-heated transparent solids are found to have significant temperature gradients in the vertical (on-axis) direction: up to about 10^3 K/μm, depending on the magnitude of the temperature peak. If the temperature distribution is spherically symmetric the application of a second Abel transform directly yields the spherical distribution. Moderate deviations (up to $\approx$2:1) from spherical symmetry can be compensated for if the aspect ratio is known in the third dimension. Thus we have shown that temperature variations across laser-heated diamond-cell samples must and can be measured in quantitative experiments.

Acknowledgments. We thank T. J. Ahrens for loaning the W calibration lamp used in this study. This study was supported by the National Science Foundation.

References

Bell, P. M., H. K. Mao, and K. Goettel, Ultrahigh pressure: beyond 2 Megabars and the ruby fluorescence scale, *Science, 226*, 542–544, 1984.

Born, M., and E. Wolf, *Principles of Optics*, 5th edition, 808 pp., Pergamon, New York, 1975.

Deans, S. R., *The Radon Transform and Some of its Applications*, 289 pp., Wiley-Interscience, New York, 1983.

Fukao, Y., H. Mizutani, and S. Uyeda, Optical absorption spectra at high temperatures and radiative thermal conductivity of olivines, *Phys. Earth Planet. Inter., 1*, 57–62, 1968.

Goettel, K. A., H. K. Mao, and P. M. Bell, Generation of static pressures above 2.5 Megabars in a diamond anvil pressure cell, *Rev. Sci. Instrum., 56*, 1420–1427, 1985.

Gooderum, P. B., and G. P. Wood, *NASA Technical Note 2173*, 1950.

Heinz, D. L., and R. Jeanloz, Melting of (Mg, Fe) SiO_3 perovskite: measurement at lower mantle conditions and geophysical implications, *J. Geophys. Res.*, in press, 1987.

Jeanloz, R., and D. L. Heinz, Experiments at high temperature and pressure: laser heating through the diamond cell, *J. de Physique, C8*, 83–92, 1984.

Jeanloz, R., and D. L. Heinz, Measurement of the temperature distribution in CW-laser heated materials, in *Laser Welding, Machining and Materials Processing, Proc. Int. Congress Applics. Lasers Electro-Optics*, edited by C. Albright, pp. 239–244, Springer-Verlag, New York, 1986.

Koechner, W., Output fluctuations of c.w.-pumped Nd-YAG lasers. *IEEE J. Quantum Electron. QE-8*, 656–661, 1972.

Loudon, R., *The Quantum Theory of Light*, 350 pp., Oxford, New York, 1973.

Ming, L. C., and W. A. Bassett, Laser heating in the diamond anvil press up to 2000 °C sustained and 3000 °C pulsed at pressures up to 260 Kilobars, *Rev. Sci. Instrum., 45*, 1115–1118, 1974.

Nestor, O. H., and H. N. Olsen, Numerical methods for reducing line and surface probe data, *SIAM Rev., 2*, 200–207, 1960.

Orlov, Y. L., *The Mineralogy of the Diamond*, 235 pp., J. Wiley, New York, 1977.

Vest, C. M., and D. G. Steel, Reconstruction of spherically symmetric objects from slit-imaged emission: application to spatially resolved spectroscopy, *Opt. Lett., 3*, 54–56, 1978.

Wooten, F., *Optical Properties of Solids*, 260 pp., Academic, New York, 1972.

Yagi, T., P. M. Bell, and H. K. Mao, Phase relations in the system MgO-FeO-SiO_2 between 150 and 700 Kbar at 1000 °C, *Carnegie Inst. Washington Year Book, 78*, 614–618, 1979.

TEMPERATURE MEASUREMENT IN A LASER-HEATED DIAMOND CELL

William A. BASSETT and Maura S. WEATHERS

Department of Geological Science, Cornell University
Ithaca, New York 14853, USA

Abstract. Infrared light from a YAG laser is used to heat samples under pressure in a diamond-anvil cell. The incandescent light from the hot sample is analyzed by a diffraction grating spectrometer and the spectra are collected by a photodiode array and stored in a multichannel analyzer. Each spectrum is corrected for response of the detector and then fit to a blackbody curve to determine the temperature. Laser pulses from 10 to 50 msec duration are used in the temperature range from 1500 K to 3500 K and pulses of 200 nsec duration produced by a Q-switched YAG laser are used for temperatures in the range from 3500 K to 10,000 K. Pressure is measured by means of the ruby fluorescence method using green light from frequency-doubled radiation from the YAG laser to excite the fluorescence.

Methods for detecting phase changes in the samples include optical microscopy, transmission and scanning electron microscopy, and X-ray diffraction.

Introduction

Ever since the first studies made with laser heating in the diamond-anvil cell (BASSETT and MING, 1972) there has never been an accurate method for measuring temperature in this process. Attempts to measure the temperature by brightness-optical-pyrometry suffered from the extremely small size of the sample ($\sim 10\ \mu m$) as well as the lack of information about the emissivity of the sample and loss of intensity due to the diamonds. Although it was recognized many years ago that more accurate temperature measurements would be possible using color-optical-pyrometry, no real progress toward this goal was made until recently (BASSETT and WEATHERS, 1986; JEANLOZ and HEINZ, 1984).

Most of our knowledge of temperature within earth's interior depends on experimental work on the temperatures of phase transitions at high pressures and especially on the melting temperatures of mantle and core materials. Although many such measurements have been made, long extrapolations of data to conditions of the core and lower mantle are usually required. More accurate models of temperature, pressure, and composition of the deep earth will depend on direct experimental measurements at those pressure-temperature conditions.

Instrumentation

Infrared radiation ($\lambda = 1.06\ \mu m$) from a YAG laser is reflected through a microscope objective and focused onto the sample through the upper diamond of a diamond-anvil cell (Figure 1). The incandescent light from the hot sample, which passes back through the objective lens, is split; part is directed into a TV camera so that the process can be monitored visually while most of the incandescent light enters the spectrometer for spectral analysis. The spectrum produced by the spectrometer is received by a photodiode array with 1024 pixels and is recorded in a multichannel analyzer. It is then stored on floppy disk for later analysis by curve fitting to calculated blackbody curves for temperature determination.

For heating samples in the range 1500 K to 3500 K, an intracavity rotating shutter is used to produce a pulse of infrared light having a duration of 10 to 50 msec. As soon as the shutter has been released, the photodiode array makes multiple rapid scans. The computer can be programmed to control the number of scans and their duration. With this procedure it is possible to accurately measure the initiation and duration of the heating as well as the peak temperature (Figure 2). There is an 80 msec delay between the triggering of the detector and the opening of the shutter. The intensity of the incandescent light then rises to its maximum value in much less than one msec, remains at the maximum intensity for the duration of the pulse (usually 10–20 msec) and then decreases in less than a msec (Figure 3a). There is always a cooling tail but the intensity of the tail is usually a few hundredths of the maximum. The decrease of intensity sometimes takes place in steps, i.e., the intensity may fall to about a third of the maximum before falling to the tail (Figure 3b). The cause of this is not understood but it appears not to have a significant effect on temperatures calculated on the basis of the maximum intensity.

For heating samples in the range 3500 K to 10,000 K an intracavity Q-switch is used. The Q-switch produces pulses of approximately 200 nsec duration having a power three-to-four orders of magnitude greater than the pulses produced by the rotating shutter. As a result these pulses generally produce much higher temperatures. Because the intensity of light increases as the fourth power of the temperature, a duration of 200 nsec is sufficient to produce as good spectra as are produced by the longer pulses at lower temperatures. On the basis of time profiles as well as some indirect observations, the light from these

High-Pressure Research in Mineral Physics, edited by M. H. Manghnani and Y. Syono, pp. 129–133.

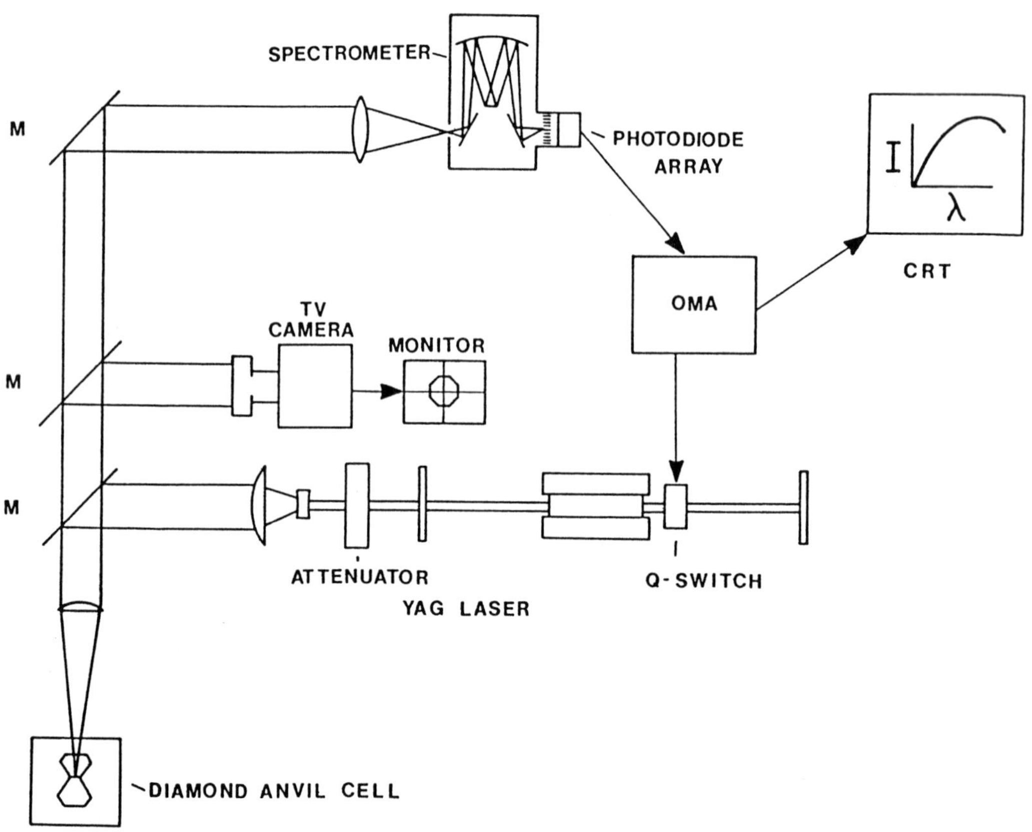

Figure 1. Apparatus for measurement of temperature in a laser-heated diamond-anvil cell.

pulses appears to increase within nanoseconds and decrease in less time than one microsecond.

Pressure is measured by means of the ruby fluorescence method (PIERMARINI et al., 1975). This method consists of placing chips of ruby in the sample and then exciting them to fluoresce by means of green light ($\lambda=0.53\ \mu m$) produced by frequency doubling the light from the YAG laser. The pressure measured this way is the pressure at ambient temperature. The effect of laser heating on the pressure is unknown. However, because the heating is very local and the pulse durations are long relative to the time required for sound to travel across the sample, the effect of laser heating on the pressure is believed to be minor.

Data

Figure 4a shows a raw spectrum of a "red" hot sample of fayalite. The spectrum was produced by laser pulse of approximately 20 msec. Figure 4b is the same spectrum

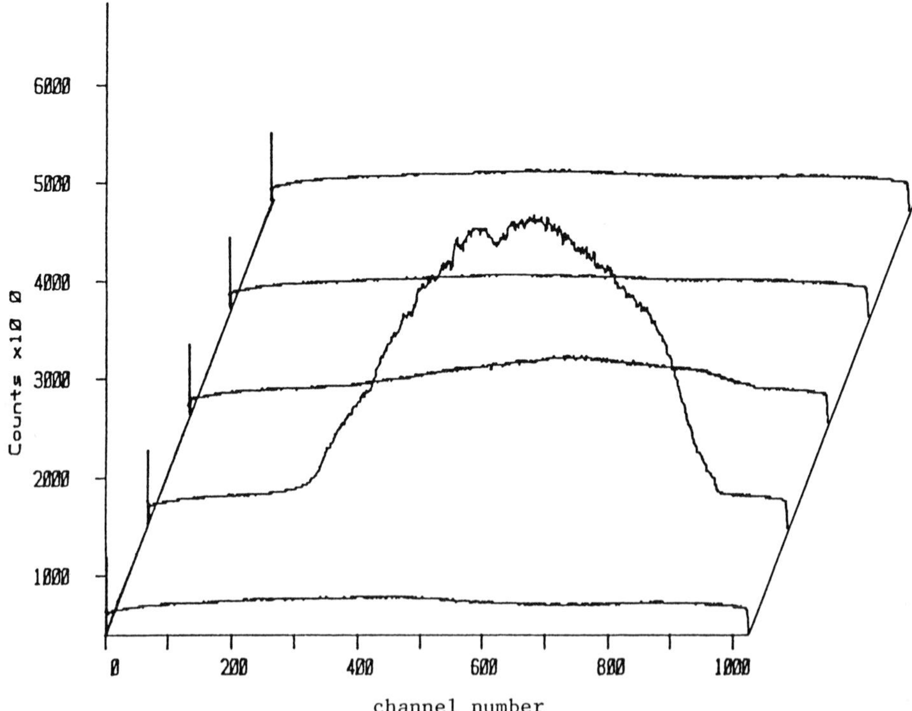

Figure 2. Series of five 80-msec scans of a sample of graphite in an NaCl medium at approximately 100 kbar in a diamond cell. The rotating shutter was released by the computer at the start of the first scan, but the flash did not occur until the second scan. In this series the flash started and stopped entirely within the second scan. This procedure allows us to minimize dark current background by collecting the spectrum only while the sample is hot and emitting light.

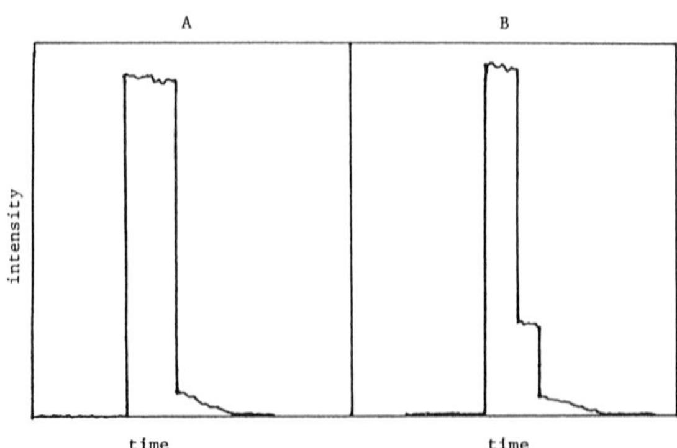

Figure 3. Time profiles of flashes of incandescent light produced by pulses from the rotating intracavity shutter: a) Most flashes are of this form. The temperature rises in less than 1 msec, remains at a fairly constant intensity for approximately 20 msec (this depends on the size of the slit in the rotating shutter), and drops in less than 1 msec. The last few percent of intensity appear to decay exponentially during the next 20 to 30 msec indicating cooling of the sample. b) Some of the flashes showed a more complex structure. These are not yet understood. For temperature determination the photodiode array is turned on just before the pulse and off just after the pulse. The shortest exposure time possible without using gating electronics is 16 msec.

after corrections have been made. The corrections consist of first subtracting the background from the raw spectrum and then correcting those results using a spectrum of a calibrated lamp. The light from the calibrated lamp passes through the same optical system and is received by the same photodiode array; therefore, corrections are made for absorption in the optical components and wavelength sensitivity of the photodiodes. The solid line in Figure 4b is the calculated blackbody curve for 1850 K fit to the spectrum by a least squares method.

Sample Analysis

We use several analytical methods for examining our samples after they have been removed from the diamond-anvil cell. These include optical microscopy, transmission and scanning electron microscopy, and X-ray diffraction. Criteria for melting include change in morphology from angular to rounded grains (Figure 5), severing of a filament (Figure 6), melting of a hole in foil (Figure 7), production of a groove in a surface (Figure 8) (GOLD et al., 1984), and loss of diffraction spots or lines.

A

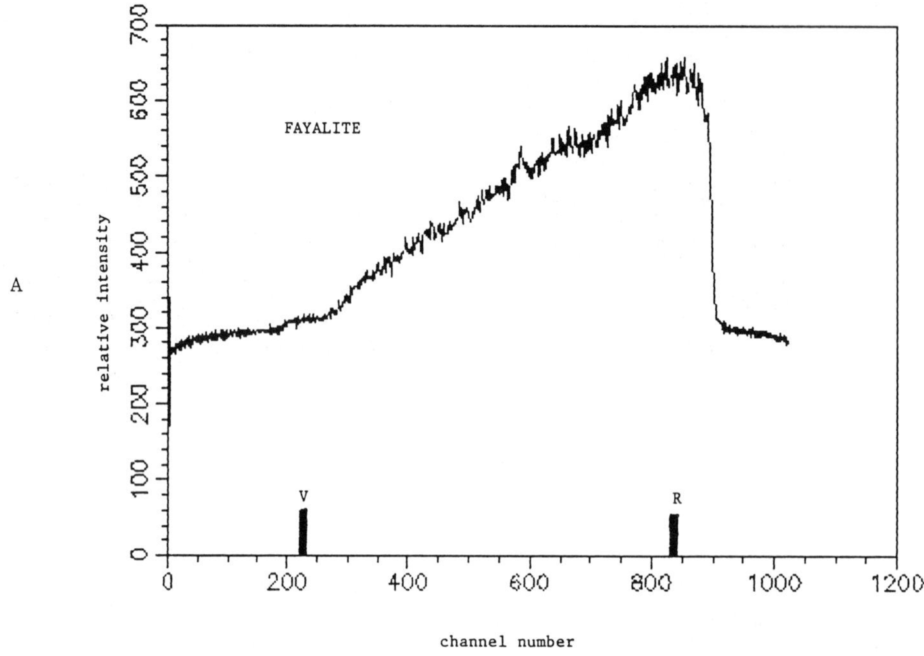

B

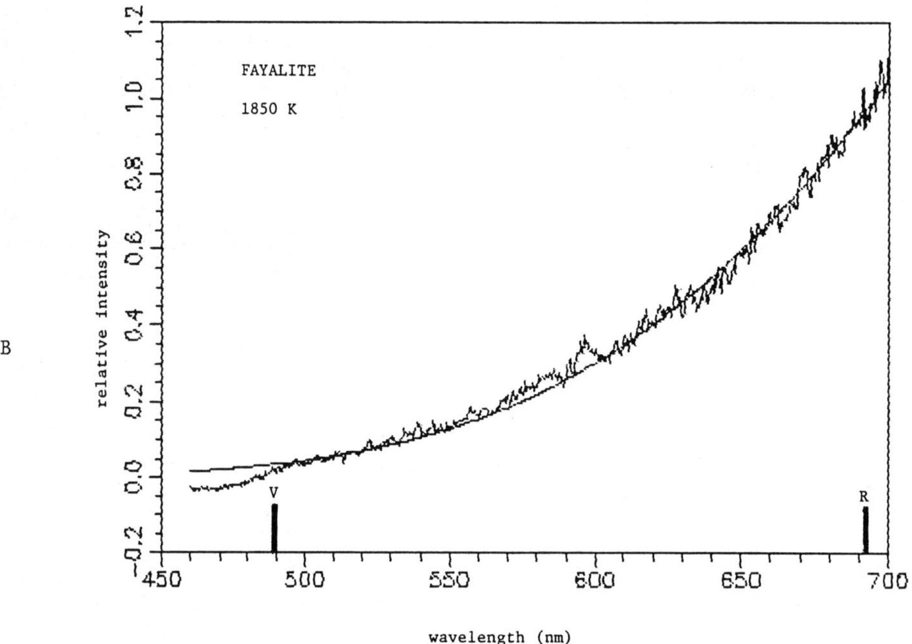

Figure 4. a) Raw spectrum of a "red" hot sample of fayalite heated for a few tens of msec, b) the same spectrum after correction based on a calibrated lamp. The solid line is the calculated blackbody curve for 1850 K fit to the spectrum by a least squares procedure. The temperature determination program was run at intervals of 50 K. The temperature indicated by the fit of the spectrum to the blackbody curve is better than 50 K. "V" represents violet light and "R" represents red light. Only channels 220–860 are exposed to light. Channels lying outside of this range record only dark current background. Therefore, the blackbody curve fit in Figure 4b applies only from 488 nm to 700 nm.

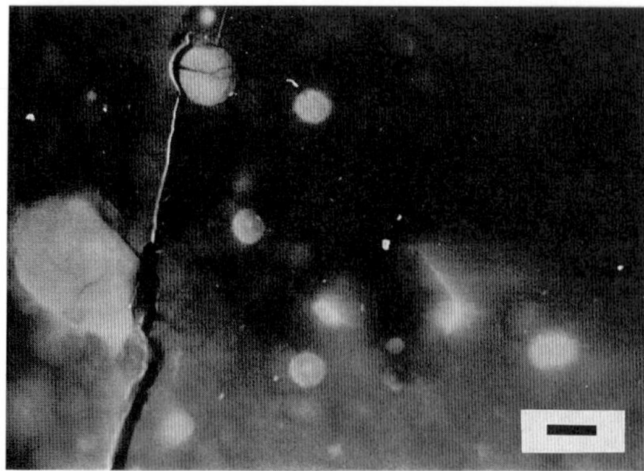

Figure 5. Scanning electron micrograph of rounded droplets of silicon produced by laser heating in the diamond cell. The material surrounding each droplet is NaCl. (bar scale=2 μm).

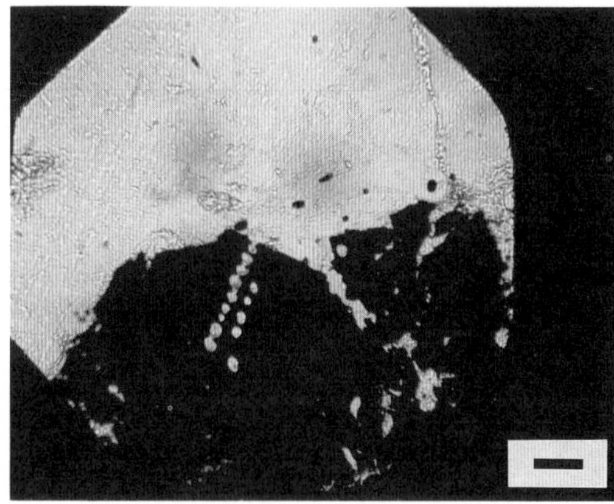

Figure 7. Optical micrograph of W foil (black) mounted between two layers of LiF in a diamond cell. The holes in the W foil were produced by single pulses of the Q-switched laser. (bar scale=50 μm).

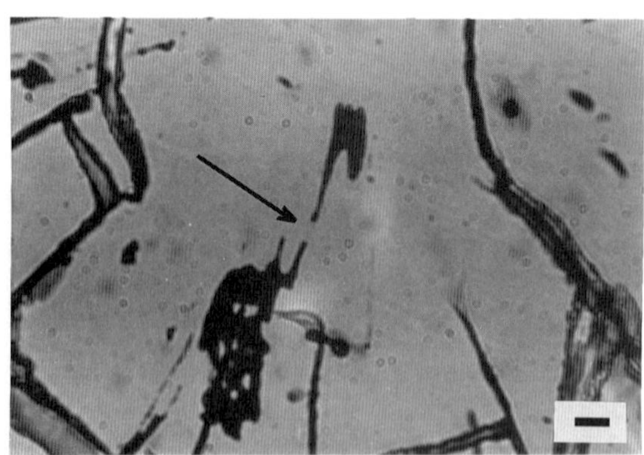

Figure 6. Optical micrograph of filaments of platinum metal severed by a laser pulse (arrow). Surrounding material is NaCl. (bar scale=10 μm).

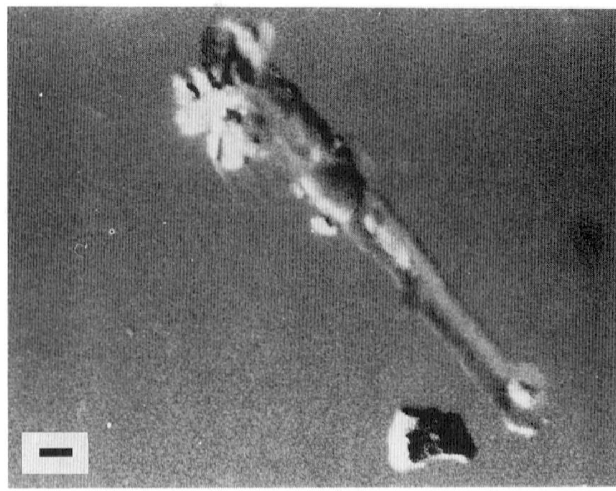

Figure 8. Scanning electron micrograph of a "groove" on the surface of one of the diamond anvils. The raised rims and the arcuate markings indicate that it formed by melting of the diamond at high pressure. The upper part of the groove was at lower pressure and, therefore, was graphitized (Gold et al., 1984). (bar scale=10 μm).

Acknowledgments. The authors wish to thank John M. Bird, Michael D. Furnish, and Eugene Huang for their many contributions to the development of the techniques described in this paper. This research was supported by grant DMR82-17227-AO1 from the Cornell Materials Science Center and grant EAR83-13569 from the National Science Foundation.

References

Bassett, W. A., and L. C. Ming, Disproportionation of Fe_2SiO_4 to $2FeO+SiO_2$ at pressures up to 250 kbar and temperatures up to 3000 °C, *Phys. Earth Planet. Inter.*, *6*, 154–160, 1972.

Bassett, W. A., and M. S. Weathers, Temperature measurement in laser heated diamond anvil cells, *Physica, 139 & 140B*, 900–902, 1986.

Gold, J., W. A. Bassett, M. S. Weathers, and J. M. Bird, Melting of diamond, *Science, 225*, 921–922, 1984.

Jeanloz, R. and D. L. Heinz, Experiments at high temperature and pressure: laser heating through the diamond cell, *J. de Phys., 45, Colloque C8*, 83–92, 1984.

Piermarini, G. J., S. Block, J. F. Barnett, and R. A. Forman, Calibration of the pressure dependence of the R1 ruby fluorescence line to 195 kbar, *J. Appl. Phys., 46*, 2774–2780, 1975.

EXPERIMENTAL PHASE RELATIONS OF IRON TO 360 KBAR, 1400 °C, DETERMINED IN AN INTERNALLY HEATED DIAMOND-ANVIL APPARATUS

H. K. MAO, P. M. BELL, and C. HADIDIACOS

Geophysical Laboratory, Washington, DC 20008, USA

Abstract. Internal resistance heating experiments were conducted in a diamond-anvil apparatus to determine the γ-ε-iron transition. Accurate values of the transition temperature at high pressure were obtained with a diode-array spectrometer, which was used to measure the blackbody emission of the heated iron sample at high pressure. The transition curve was observed in the range 190–360 kbar, 800–1375 °C. The slope of the transition was 3.5 °C/kbar. The γ-ε transition curve, if extrapolated, would project an iron field stability under conditions of the earth's Core.

Introduction

The phase diagram of iron has long had important implications for geophysical models (MAO and BELL, 1979). BIRCH (1952) discussed the importance of liquid and solid iron and iron alloy phases in the earth's core. ANDERSON (1986) reviewed theoretical analyses of the phase relations of iron at core pressures and temperatures. The core-mantle boundary is assumed to be at a temperature above the liquidus temperature of iron or its alloy, and the inner core boundary is thought to be at the iron or alloy liquidus temperature at core pressure.

At high temperatures and high pressures, iron exists as either the ε-phase (hcp) or the γ-phase (fcc). Problems arise in theoretically estimating the phase boundaries because of the apparent closeness of the pressure and temperature of the γ-ε-liquid triple point to the pressure and temperature of the inner core boundary. Experimental data are needed to determine the crystal structure of the iron phase that is stable at the pressure of the core-mantle boundary (estimated at 1.5 Mbar). Also, experimental determination of the crystal structure of solid iron at the inner core boundary and the determination the liquidus temperatures of iron at core pressures are important goals of geophysics (SPILIOPOULOS and STACEY, 1984).

The present study presents data on the ε-γ iron transition as a function of pressure and temperature to maximum conditions of 360 kbar and 1400 °C. A significant part of the study was to develop techniques to generate the required conditions and to obtain reliable measurements of pressure and temperature. The resulting techniques may potentially be extended in range (XU et al., 1986) so that core conditions can be duplicated in the laboratory.

Experimental Technique

Internal resistance techniques are preferred for heating metallic samples to high temperatures in the diamond-anvil apparatus. External furnace techniques, which involve inert atmospheres, are limited to maximum temperatures of approximately 900 °C at 300 kbar and 1200 °C at 60 kbar (SCHIFERL, this volume) to avoid damage of the diamond anvils. Internal laser heating techniques are subject to difficulties caused by reflections and opacity of metallic surfaces. The internal resistance method is based on the ease of passing a current through a small filament in the sample chamber. LIU and BASSETT (1975) stated that their internal resistance heating techniques had been developed to routinely achieve "a few thousand degrees Celsius while the sample is under pressure." The basic heating techniques and the combined methods of measurement and calibration of pressure and temperature introduced by BOEHLER et al. (this volume) and by this laboratory (XU et al., 1986) were employed.

A diagram of the experimental design is shown in Figure 1; a cross-section of the diamond-anvils and the gasket is shown, and the iron foil electrodes and iron sample/resistance element can be seen in plane view. Iron of purity 99.999% (Goodfellow Metal Inc.), in the form of a foil of 10 μm in thickness and a wire of 10 μm in thickness, was employed in these experiments. The gasket and the diamond anvils were insulated from the iron by layers of compacted alumina powder. Current flow through the iron sample was from a D.C. power supply. The power supply was regulated over a power range of 0–25 W. Measurements of applied voltage and current in the iron sample were used to calculate resistance and power in the circuit. Thus, for a given pressure and as the temperature was cycled, the resistance-power curve served as a monitor of temperature and of the phase of iron at pressure and temperature. Figure 2 shows plots of resistance versus power of an iron sample held at 192 kbar. The change in slope marks the phase change from epsilon to gamma. The first few points beyond the break in slope show a slight curvature which indicates a temperature gradient and thus a two-phase region along the wire. The wire was uniformly heated, except for the 5–10 μm cooled sections touching the iron foil leads.

High-Pressure Research in Mineral Physics, edited by M. H. Manghnani and Y. Syono, pp. 135–138.

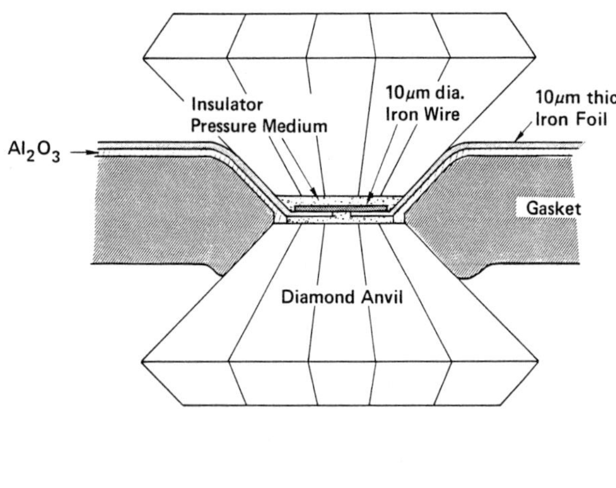

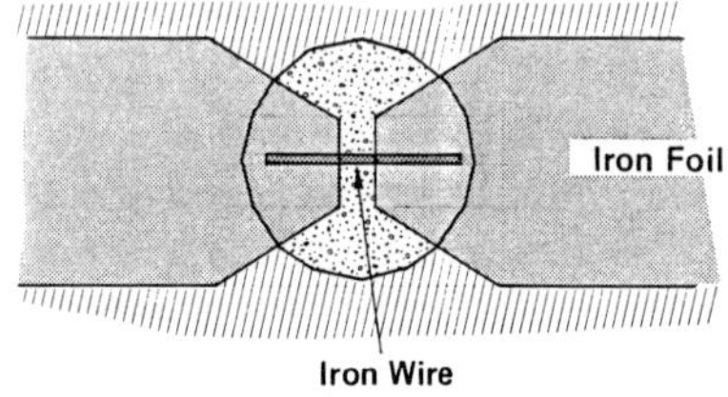

Figure 1. Diagramatic sketches of the cross-section of an internally heated diamond-anvil, high-pressure apparatus and of the iron sample which also served as the resistance element.

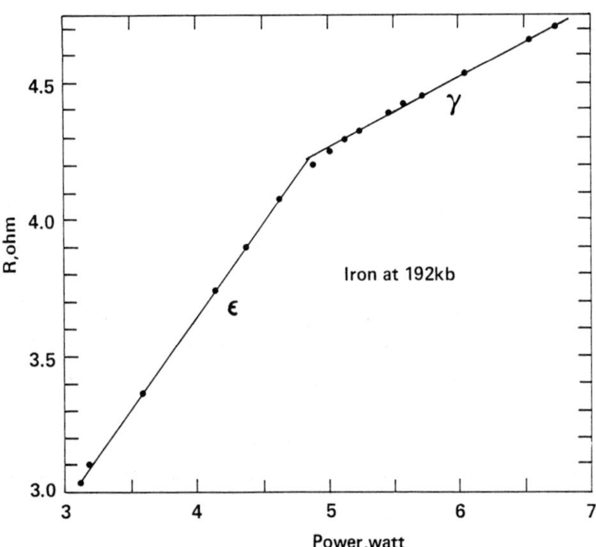

Figure 2. Plot of the resistance versus power of the sample/heater, showing the ε-γ iron transition point at 192 kbar.

Measurement of the blackbody emission of the heated sample at high pressure was made in this study with an optical microscope coupled with a spectrometer. The microscope was equipped with a special imaging pentaprism. It was possible to select a 3×3 μm area of the sample for the measurements. The blackbody emission for the temperature calculation was measured at the central area of the heated iron wire to minimize the edge effect. The emission was constant within the 5 μm central region of the 10 μm wire.

The spectrometer consisted of a Spex Industries Triple-Mate spectrograph attached to a Princeton Instruments diode array detector (OSMA, Optical Spectral Multichannel Analyzer). The system was calibrated against a standard tungsten emission source supplied by Optronic Laboratories, Inc. traceable to the National Bureau of Standards (NBS). The uncertainty in the transfer from the NBS standard is estimated to be less that $\pm 1\%$ in the 350–2500 nm region. The standard lamp was calibrated to NBS values at two temperatures, 2000 and 2600 K.

Temperature was calculated by fitting the blackbody Planck radiation equation for spectral radiance (L) to the observed spectra as follows

$$L_\lambda = C_1/n^2\lambda^5 \, [\exp(C_2/n\lambda T) - 1)] \qquad (1)$$

where C_1 is 1.19066×10^{-12} W cm^2 steradian^{-1}, n is the index of refraction of air, λ is the wavelength (in air) in centimeters, C_2 is 1.4388 cm Kelvins, and T is the blackbody temperature in Kelvins. The ratio of spectral radiance (L_1/L_2) at a series of wavelengths taken progressively along the spectrum was used to eliminate effects due to departures of emissivity of the iron sample from that of an ideal blackbody. Above 1000 K the errors of measurement diminish. Typical errors in temperature measurement of ± 20 K at 1000 K and ± 8 K at 2300 K were reported by BOEHLER et al. (this volume) and are essentially the same in the present experiments. Below 1000 K the errors increase markedly because of the rapid fall-off of the radiative component of heat which is the basis of temperature measurements in the present study. A series of blackbody spectra of the iron sample at a fixed pressure of 313 kbar is shown in Figure 3. Each curve yields a specific temperature by calibration and calculation of the ratio with the NBS-derived standard. The experimental procedure involved making the emission measurements at several fixed points of power, at fixed pressures. The correspondence of temperature to power was found to be linear. Figure 4 shows a plot of this relationship in the temperature range 1100–1850 K for one of the experimental runs.

The pressure of the iron sample was measured by the ruby fluorescence technique at room temperature (MAO et al., 1978). LIU and BASSETT (1975) noted that the

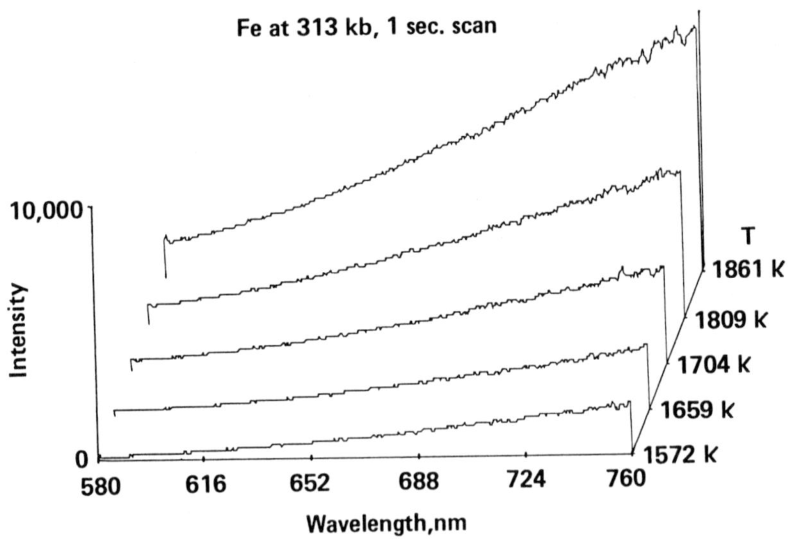

Figure 3. Diode array spectral scans of the thermal emission of iron at several temperatures and 313 kbar.

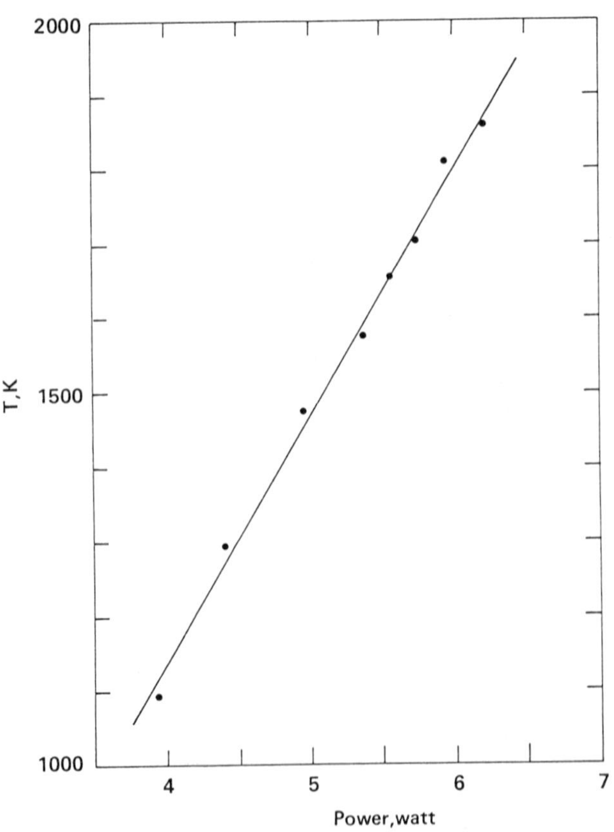

Figure 4. Plot of the power versus temperature in the internal heater/sample of the high-pressure apparatus.

pressure could be determined accurately by comparing values before and after a heating cycle (although they used an X-ray diffraction technique to calculate the pressure, rather than the fluorescence method used in this study). Our resulting ruby fluorescence scale was calibrated in part against the equation of state of sodium chloride, which was indirectly related to the calibration used by LIU and BASSETT (1975).

Results and Discussion

The data are plotted as experimental points on the γ-ε pressure-temperature transition curve in Figure 5. The error bars indicate the uncertainty in determining the point of phase transitions (Figure 2) due to temperature gradients. The present curve has a smaller displacement to lower pressure than the curve of BUNDY (1965). This displacement may be the result of revisions to calibration of the working pressure scale that occurred subsequent to Bundy's study.

The region of the γ-ε boundary in the range of the present experiments can be described by the linear relationship

$$T\,(^\circ\mathrm{C}) = 3.5P(\mathrm{kbar}) + 150 \tag{2}$$

The slope of the transition curve (3.5 °/kbar) is steeper than previous determinations (2–3 °/kbar). This slope was measured at higher pressures (360 kbar) and over a greater range in pressure (165 kbar) than in the previous experiments of BUNDY (1965), which LIU and BASSETT (1975) used to calibrate their experiments. The curve also

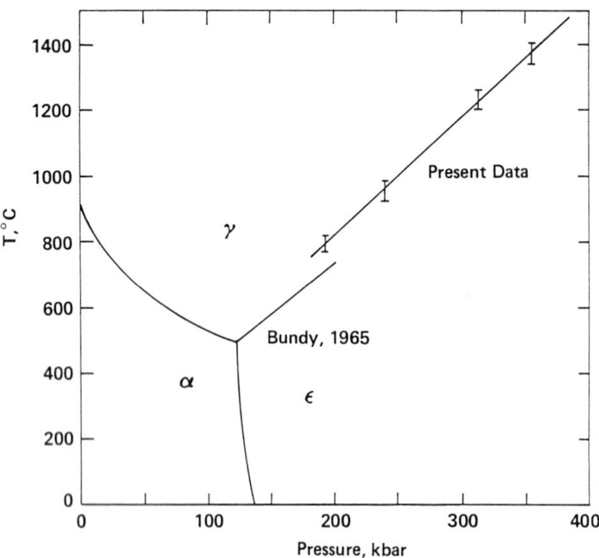

Figure 5. Phase diagram of iron showing present results on the ε-γ transition curve.

may be considered somewhat more reliable for extrapolation to core temperatures because of the precision and accuracy of temperature calibration of the experimental data.

Geophysical models of the earth's core are highly dependent on the accuracy of experimental data on ε and γ iron. Shockwave data (BROWN and MCQUEEN, 1986) on iron at megabar pressures are the best available data at this time, but the temperature uncertainty is high and it has not been feasible to identify the phases with certainty. Location of the ε-γ-liquid triple point and the phase stability fields are hypothetical in current geophysical models. SPILIOPOULOS and STACEY (1984) calculated both liquid-solid and solid-solid phase boundaries with the Gilvarry-Lindemann melt relationship, and concluded that the inner core has the γ structure. ANDERSON (1986) used approximations of thermodynamic functions to extrapolate low pressure data. He estimated the triple point and concluded that the inner core is composed of ε iron. BROWN and MCQUEEN's (1982) calculations suggested iron stability relations similar to ANDERSON's (1986) model. The present results also tend to support the concept that the inner core has the ε structure, depending on the method chosen for extrapolation. The advantage of the present experimental diamond-anvil technique is that the sample temperature and pressure are exceedingly stable. It may be possible in the near future to identify the iron phases directly by synchrotron-X-ray diffraction techniques.

REFERENCES

ANDERSON, O. L., Properties of iron at the earth's core conditions, *Geophys. J. R. Astron. Soc.*, *84*, 561–579, 1986.

BIRCH, F., Elasticity and constitution of the earth's interior, *J. Geophys. Res.*, *57*, 227–286, 1952.

BOEHLER, R., M. NICOL, and M. L. JOHNSON, Internally-heated diamond-anvil cell: Phase diagram and *P-V-T* of iron, this volume.

BROWN, J. M., and R. G. MCQUEEN, The equation of state for iron and the earth's core, in *High-Pressure Research in Geophysics*, vol. 12, edited by S. Akimoto and M. H. Manghnani, pp. 611–623, Center for Academic Publications, Tokyo, 1982.

BROWN, J. M., and R. G. MCQUEEN, Phase transitions, Gruneisen parameter, and elasticity for standard iron between 77 GPa and 400 GPa, *J. Geophys. Res.*, *91*, 7485–7494, 1986.

BUNDY, F. P., Pressure-temperature phase diagram of iron to 200 kbar, 900°C, *J. Appl. Phys.*, *36*, 616–620, 1965.

LIU, L., and W. A. BASSETT, The melting of iron to 200 kbar, *J. Geophys. Res.*, *80*, 3777–3782, 1975.

MAO, H. K., and P. M. BELL, Equations of state of MgO and Fe under static pressure conditions, *J. Geophys. Res.*, *84*, 4533–4536, 1979.

MAO, H. K., P. M. BELL, J. SHANER, and D. STEINBERG, Specific Volume measurements of Cu, Mo, Pd, and Ag and calibration of the ruby R1 fluorescence pressure gauge form 0.06 to 1 Mbar, *J. Appl. Phys.*, *49*, 3276–3283, 1978.

SCHIFERL, D., this volume.

SPILIOPOULOS, S., and F. D. STACEY, The earth's thermal profile: is there a mid-mantle thermal boundary layer?, *J. Geodynam.*, *1*, 61–77, 1984.

XU, J. A., H. K. MAO, and P. M. BELL, High-pressure ruby and diamond fluorescence: observations at 0.21 to 0.55 terpapscale, *Science*, *232*, 1404–1406, 1986.

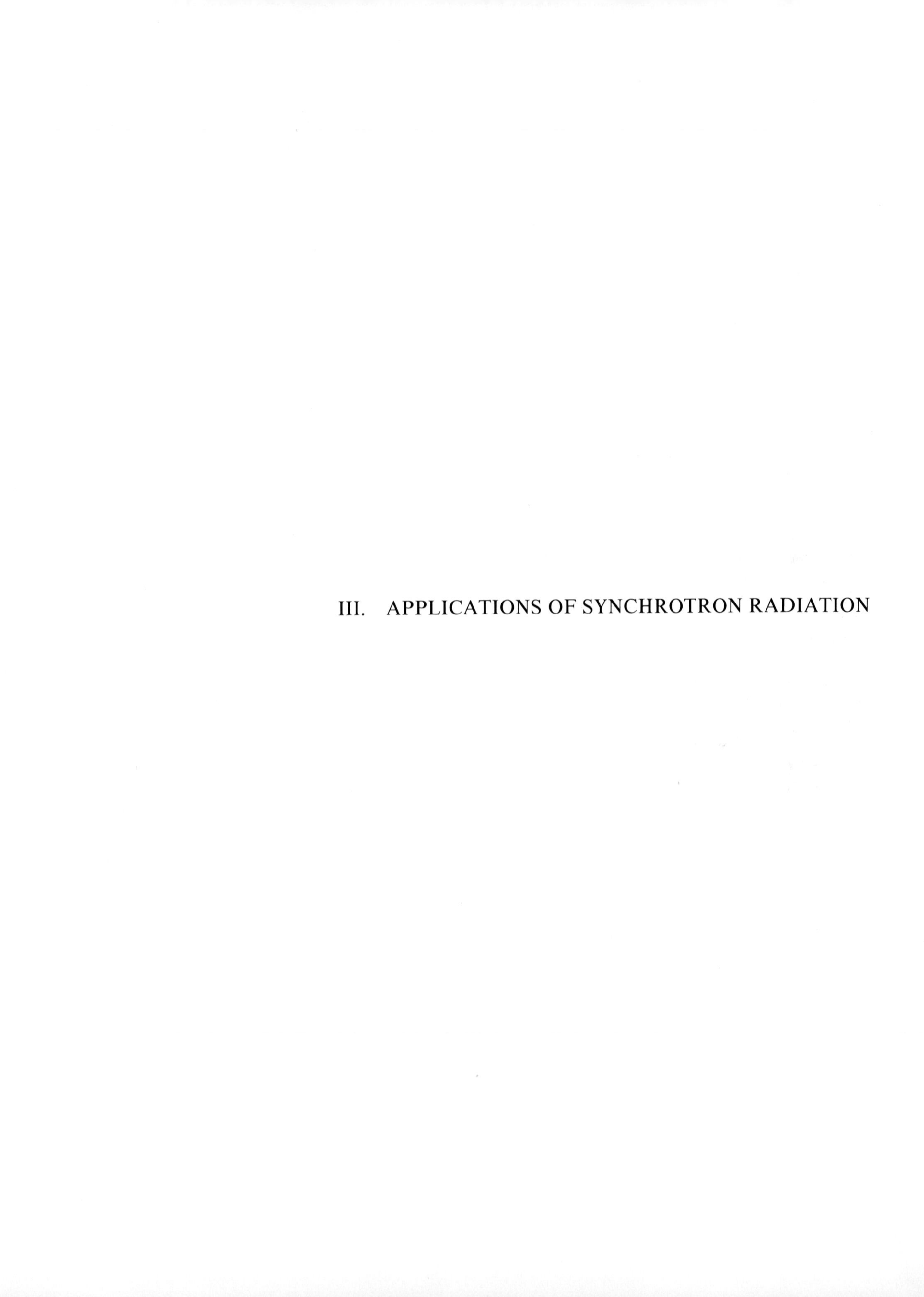

III. APPLICATIONS OF SYNCHROTRON RADIATION

HIGH PRESSURE AND HIGH TEMPERATURE EQUATIONS OF STATE OF MAJORITE

Takehiko YAGI[1]

The Research Institute for Iron, Steel and Other Metals, Tohoku University, Katahira, Sendai 980, Japan

Masaki AKAOGI

Department of Earth Sciences, Kanazawa University, Kanazawa 920, Japan

Osamu SHIMOMURA

National Institute for Research in Inorganic Materials, Sakura-mura, Niihari-gun, Ibaraki 305, Japan

Hiroshi TAMAI

Department of Earth Sciences, Chiba University, Chiba 260, Japan

Syun-iti AKIMOTO

Institute for Solid State Physics, University of Tokyo, Minato-ku, Tokyo 106, Japan

Abstract. Bulk modulus and thermal expansion of pyrope, almandine, and two majorites, with the compositions 58% enstatite-42% pyrope solid solution and 18% ferrosilite-82% almandine solid solution, were measured using a cubic-anvil type of high-pressure and high-temperature X-ray diffraction apparatus combined with synchrotron radiation. Based on these measurements, we established the systematics of the elastic property in pyroxene-garnet join.

It was found that dissolution of the pyroxene component in garnet decreases the bulk modulus. This result is contrary to previously reported results, and thus gives new constraints to the model of the earth's mantle. It was also found that thermal expansion of garnet changes very little with dissolution of the pyroxene component. The volume dependence of thermal expansion for garnet was also measured.

Introduction

In many petrological models of the mantle, garnet-structured pyroxene-garnet solid solution, also known as majorite, is an important component in the region between the lower portion of the upper mantle and the transition zone (e.g., RINGWOOD, 1975; BASS and ANDERSON, 1984). This is because both pyroxene and garnet are abundant minerals in the upper mantle, and the solid solubility of pyroxene in garnet increases drastically with pressure (AKAOGI and AKIMOTO, 1977). Accurate knowledge of the elastic property of majorite, therefore, is indispensable to the comparison of various seismic models of the earth. Although numerous laboratory data on the elastic property of common garnets have accumulated, little is known about majorite.

So far majorite has been formed only in a polycrystalline form, either as a quench product of a high-pressure experiment above 15 GPa (AKAOGI and AKIMOTO, 1977) or in veins within a naturally shocked meteorite (MASON et al., 1968). JEANLOZ (1981) studied the bulk modulus of majorite obtained from a meteorite by means of high-pressure X-ray diffraction using a diamond-anvil cell. He reported a bulk modulus which is much higher than that expected from the systematics in garnet solid solutions.

Application of synchrotron radiation to a high-pressure and high-temperatue X-ray diffraction study made it possible to measure equations of state for tiny polycrystalline samples with a much higher accuracy than before (YAGI et al., 1985). The purpose of the present study is to measure the P-V-T relations of four garnet specimens in the systems pyrope (Py:$Mg_3Al_2Si_3O_{12}$)-enstatite (En:$MgSiO_3$) and almandine (Alm:$Fe_3Al_2Si_3O_{12}$)-ferrosilite (Fs:$FeSiO_3$), and to establish the systematics of the elastic property in pyroxene-garnet solid solutions. A cubic-anvil type of high-pressure apparatus, combined with synchrotron radiation, was used.

[1]Now at Institute for Solid State Physics, University of Tokyo, Minato-ku, Tokyo 106, Japan.

High-Pressure Research in Mineral Physics, edited by M. H. Manghnani and Y. Syono, pp. 141–147.

TABLE 1. Lattice Parameters and Compositions of Garnets and Majorites

Material	Lattice Parameter (Å)	Composition
pyrope	11.455(3)	$Mg_3Al_2Si_3O_{12}$
En-Py solid solution	11.474(3)	$(Mg_4Si_4O_{12})_{0.58}(Mg_3Al_2Si_3O_{12})_{0.42}$
almandine	11.533(4)	$Fe_3Al_2Si_3O_{12}$
Fs-Alm solid solution	11.547(4)	$(Fe_4Si_4O_{12})_{0.18}(Fe_3Al_2Si_3O_{12})_{0.82}$

Experimental Methods

Sample Preparation

All the specimens used in the present study were prepared using methods described before (see AKAOGI and AKIMOTO, 1977). Pure pyrope was synthesized from a mixture of brucite, silicic acid, and corundum at about 4 GPa and 1000 °C. For preparing solid solution in the system enstatite-pyrope, a uniform glass of nominal composition 58 mol% En-42 mol% Py was prepared at atmospheric pressure. This glass was crystallized into garnet at 18 GPa and 900 °C using a double-stage, cubic-octahedral press. In the system ferrosilite-almandine, mixtures of fayalite, silicic acid, and aluminum hydroxide were used in desired compositions as starting materials. Pure almandine and 18 mol% Fs-82 mol% Alm solid solution were prepared at 900 °C and at 3 GPa and 10 GPa, respectively. All the materials were examined by X-ray diffraction and confirmed to be a single phase of garnet. Parameters of these starting materials are summarized in Table 1. Compositions were determined from the lattice parameters, using relationships between composition and lattice parameter of these systems (AKAOGI and AKIMOTO, 1977).

Experiments were performed using "MAX80", a cubic-anvil type of high-pressure and high-temperature X-ray diffraction apparatus, combined with the "Photon Factory", a synchrotron radiation facility at the National Laboratory for High Energy Physics (KEK), Tsukuba. The apparatus consists of a cubic-anvil type of high-pressure vessel, a 500-ton hydraulic ram, and an X-ray diffractometer system. Details of this apparatus have been described elsewhere (SHIMOMURA et al., 1984, 1985). Tungsten carbide anvils with 6 mm×6 mm cube faces were used throughout this study.

Sample Assembly

In high-pressure X-ray diffraction experiments, hydrostatic pressure is essential for reliable equations of state (SATO et al., 1975). In order to maintain a purely hydrostatic environment, a liquid pressure-transmitting medium was employed for a room temperature experiment, using a sample assembly similar to that described before (YAGI et al., 1985). A mixture of methanol and ethanol (4:1 by volume) was sealed in a teflon container, together with a powder specimen and sodium chloride pressure marker. This container was then embedded in a cubic boron/epoxy resin pressure-transmitting medium.

The sample assembly for the high-temperature experiment is shown in Figure 1. A mixture of the specimen and pressure marker was placed in a cylindrical graphite furnace and then embedded in a boron/epoxy resin cube. The specimen mixture was directly compressed with a solid pressure-transmitting medium, and reliable compression curves were obtained only at high temperatures where sodium chloride becomes soft enough to give a quasihydrostatic environment. Temperature was measured by a chromel/alumel thermocouple inserted into the furnace.

Experimental Procedure

During the present experiments, the Photon Factory operated at an energy of 2.5 GeV and a ring current between 150 and 80 mA. The incident X-ray was

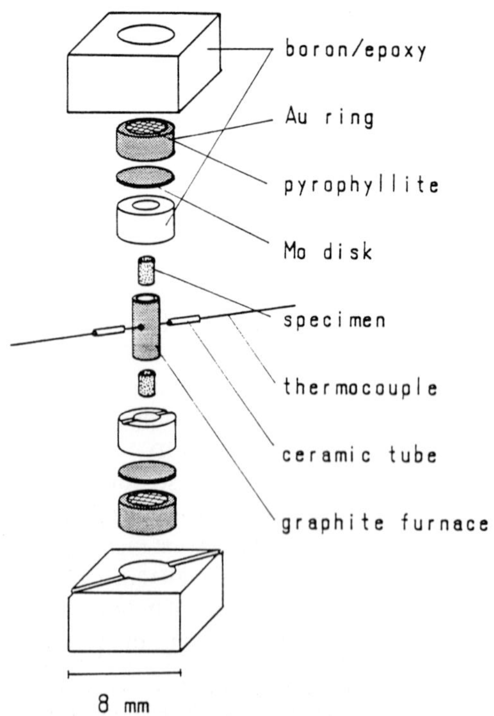

Figure 1. Sample assembly used for high-temperature experiments.

collimated to 0.3 mm in height and approximately 0.5 mm in width (determined by the clearance between anvils), and the diffracted X-ray was measured by a pure germanium solid-state detector (SSD) which was located about 260 mm from the specimen. An energy dispersive X-ray diffraction technique was employed. The SSD was fixed at $2\theta = 15°$. Signals from the SSD were analyzed by a 2048-channel pulse height analyzer. The energy range used for the analysis was approximately 15–30 keV. Exposure time for each observation was 200 or 300 s. An example of the diffraction profile is shown in Figure 2.

Both garnet and sodium chloride have cubic symmetry and most of the diffraction lines did not overlap in a pressure range studied. Two diffraction lines of NaCl, (200) and (220), were used for the pressure measurement. Five to 9 diffraction lines of garnet were used for measuring the unit cell volume of the specimen. Pressure was calculated using Decker's equation of state for sodium chloride (DECKER, 1971). An example of the data analysis is shown in Table 2. As is evident from this example, unit-cell volumes of both the NaCl pressure marker and garnet specimen were determined with an accuracy of $+/-0.04\%$. Uncertainty in the pressure calculated from this volume measurement is $+/-0.3\%$ in this pressure range.

Hydrostaticity of the pressure is a very important factor in obtaining reliable results. It was checked by the method summarized below and has been discussed in detail in a previous study (SATO et al., 1975). X-ray measurements were made with both increasing and decreasing pressure cycles, and the hysteresis of the P-V relation in these cycles were carefully observed. When a liquid pressure-transmitting medium was used, no hysteresis was observed at all; this confirmed that a hydrostatic environment existed in the sample chamber. In case of the high-temperature sample assembly, considerable hysteresis was observed at room temperature, but disappeared above 500 °C.

In order to measure thermal expansions at high pressure, observations were made at constant ram load. Real pressure in a sample chamber, however, changed when temperatue varied. Isobaric data were obtained by correcting the effect of pressure change on the analysis.

Results and Analysis

Room-Temperature Compressions

Room-temperature compression curves for pure pyrope and enstatite-pyrope solid solution are shown in Figure 3. The Birch-Murnaghan equation of state was fitted to these compression curves to determine the zero-pressure bulk modulus K_0 and its pressure derivative K_0' for each material. When these two parameters are determined by the least-squares calculation, the K_0' value was found to

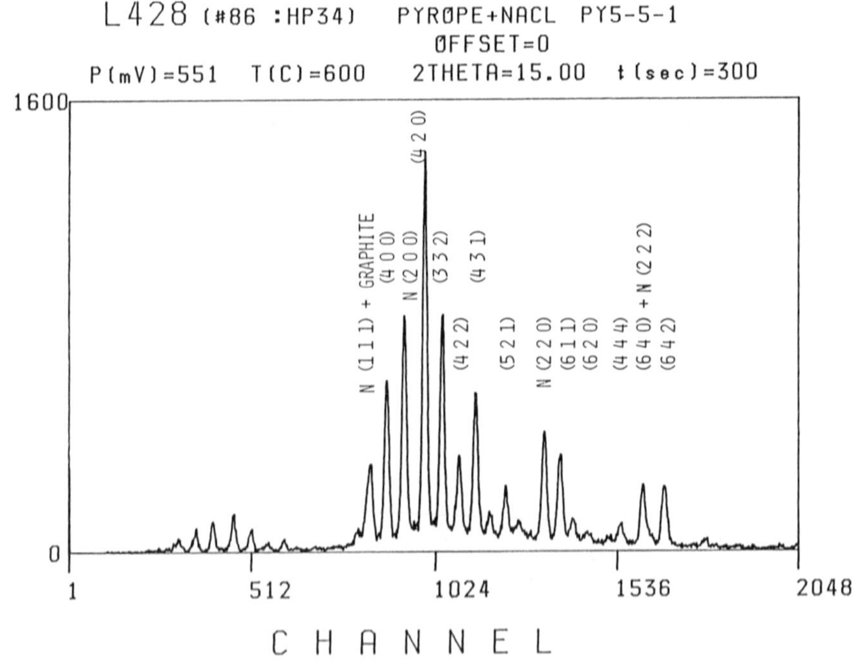

Figure 2. An example of diffraction spectra of pyrope and sodium chloride obtained at 5.8 GPa and 600 °C. Exposure time was 300 s. Diffractions from sodium chloride and pyrope are indexed as "N(XXX)" and "(XXX)", respectively. Other peaks are either from materials around the sample or escape peaks. Analysis of this data is shown in Table 2.

TABLE 2. An Example of Data Analysis at 5.8 GPa and 600 °C

RUN NO. L428

NACL

a(Å) = 5.4044 + − 0.87E-03
V(Å^3) = 157.8563 + − 0.76E-01
V/V_0 = 0.8797 + − 0.43E-03

h	k	l	d_{obs}(Å)	d_{cal}(Å)	$(d_o/d_c)-1$
2	0	0	2.7018	2.7022	−0.00016
2	2	0	1.9111	1.9108	0.00016

Decker Pressure Scale at T=600 °C: P(GPa)=5.82+−0.02

GARNET

a(Å) = 11.3900 + − 0.12E-02
V(Å^3) = 1477.6826 + − 0.46E-00
V/V_0 = 0.9831 + − 0.31E-03

h	k	l	d_{obs}(Å)	d_{cal}(Å)	$(d_o/d_c)-1$
4	0	0	2.8471	2.8475	−0.00016
4	2	0	2.5467	2.5469	−0.00009
3	3	2	2.4283	2.4284	−0.00003
4	2	2	2.3246	2.3250	−0.00015
4	3	1	2.2328	2.2338	−0.00044
5	2	1	2.0800	2.0795	0.00024
6	1	1	1.8486	1.8477	0.00047
6	2	0	1.8014	1.8009	0.00027
4	4	4	1.6436	1.6440	−0.00022
6	4	0	1.5800	1.5795	0.00032
6	4	2	1.5224	1.5221	0.00019

* d_{cal} were calculated from the lattice constant which is obtained by the least squares fitting of the observed d-values.

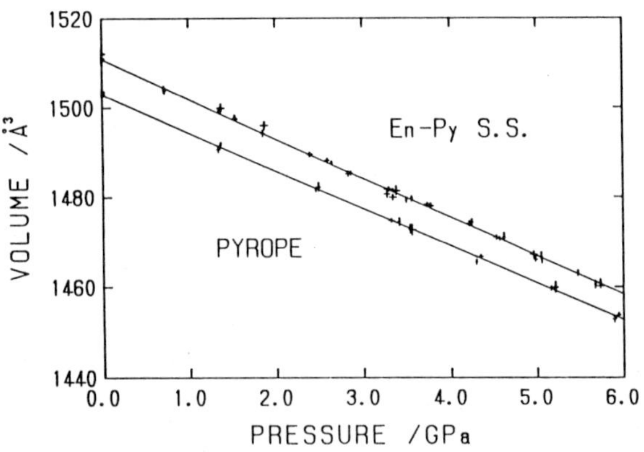

Figure 3. Isothermal compression curves of pyrope and En-Py solid solution at room temperature. Length of cross represents error bar in each measurement. Solid lines are best fit of Birch-Murnaghan equations for pyrope (K_0=172 GPa and K_0'=1.6) and En-Py s.s. (K_0=165 GPa and K_0'=1.6).

be 1.6 for pyrope. This value is small compared to the values reported for other garnets: 4.25 for grossular (HALLECK, 1973), 4.5 for pyrope (LEVIEN et al., 1979), and 5.45 for almandine-pyrope solid solution (SOGA, 1967). In order to see the correlation between K_0 and K_0', values from 1 to 4 were given to K_0' and only K_0 was calculated by least-squares fitting. Results of this analysis are summarized in Table 3. Results for pyrope and the enstatite-pyrope solid solution are also plotted in Figure 4. It became clear that, within present accuracy, any combination of K_0 and K_0' can explain the experimental results.

Many measurements, using various techniques, have been made on pyrope to determine K_0. Figure 4 compares

TABLE 3. Bulk Modulus K_0 vs dK_0/dP of Garnets and Majorites

dK_0/dP	1	2	3	4
Material				
pyrope	173.5(8)	171.1(7)	168.7(8)	166.4(8)
58% En-48% Py	166.5(5)	164.2(6)	162.0(6)	159.8(6)
almandine	176.0(22)	173.4(22)	170.8(22)	168.3(23)
18% Fs-82% Alm	172.7(11)	169.9(11)	167.2(11)	164.6(11)

* Unit of bulk modulus is GPa.

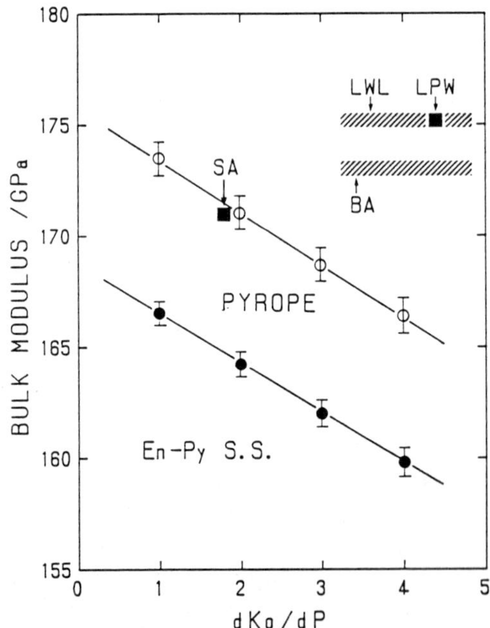

Figure 4. Bulk modulus versus dK_0/dP for pyrope and En-Py s.s. with comparisons in pyrope and other measurements. Solid squares represent isothermal compressions on pyrope using the powder diffraction technique (SA: SATO et al., 1978) and single-crystal analysis (LPW: LEVIEN et al., 1979). Hatched areas are the results of Brillouin scattering measurement (LWL: LEITNER et al., 1980) and estimation based on the systmatics of ultrasonic measurements on various garnets (BA: BABUŠKA et al., 1978).

the present result with those of previous studies. SATO et al. (1978) performed powder X-ray diffraction study up to 10 GPa using a conventional X-ray source and cubic-anvil type of high-pressure apparatus. He reported K_0 and K_0' values which were almost identical to the present values. LEVIEN et al. (1979) determined that $K_0=175$ GPa and $K_0'=4.5$, using single-crystal X-ray diffraction analysis up to 5 GPa. LEITNER et al. (1980) measured acoustic wave velocity using the Brillouin scattering technique and calculated the isothermal bulk modulus to be 175 GPa. BABUŠKA et al. (1978) estimated the bulk modulus of pure pyrope to be 173 GPa, based on systematic ultrasonic measurements in various natural garnets. Results of two other reports on pyrope (TAKAHASHI and LIU, 1970; HAZEN and FINGER, 1978) differed too much to be compared with the present figure. Possible reasons for these discrepancies have already been discussed by LEVIEN et al. (1979).

Based on these comparisons, we conclude that, if K_0' is assumed to be about 2, the present bulk modulus for pyrope is in agreement with previous observations. However, it is difficult to determine at present, whether or not the small K_0' observed in the present study is meaningful. Although the precision of the present measurements is much higher than for the previous powder X-ray diffraction studies under pressure, further studies are still required to obtain a reliable K_0' value.

In the present study garnets with four different compositions were studied using an identical technique. Therefore relative comparisons of these four garnets are most meaningful and are discussed later.

High-Temperature Experiments

Measurements under high temperature were made between room temperature and 1100 °C. However, below 500 °C, X-ray measurements were not reliable because of the nonhydrostatic nature of the pressure. Also, above 1000 °C, temperature measurements became less reliable because of the chemical reaction of the thermocouple with surrounding materials. From the X-ray measurements, we found that a temperature gradient as large as 20% exists from the center to end of the cylindrical graphite furnace, which was approximately 3.0 mm long under pressure. Since the area observed by X-ray was only 0.3 mm in the present system, the position of the measurement was adjusted to be close to the hot junction of the thermocouple so that possible errors in temperature measurements were less than 1%.

Even when the ram load was kept constant, pressure in the sample chamber changed as much as 5% when temperature was adjusted between 600 °C and 800 °C. In order to get thermal expansion at a constant pressure, the small difference in the unit cell volumes at observed and calculated pressures were corrected using the bulk modulus of the specimens at room temperature. Since corrections were minor, the difference in the bulk modulus at room temperature and at high temperature was negligible. Examples of the present results for Fs-Alm solid solution is shown in Figure 5. For pyrope and En-Py solid solution, thermal expansions were measured at 6.0 GPa. For Fs-Alm solid solution, thermal expansion measurements were made at two different pressures, 2.9 GPa and 5.5 GPa. By combining these high-pressure and high-temperature data with hydrostatic-compression curves for room temperature, we calculated thermal expansions at high pressures using the least-squares fitting of a second-order polynomial to the volume expansion data collected between room temperature and 800 °C. Reported thermal expansions at atmospheric pressure for pyrope and almandine (SKINNER, 1956; Ohsako, private communication) were also fitted to the same equations. All results are summarized in Table 4.

Discussion

Systematics of K_0 in Pyroxene-Garnet Join

In the present study isothermal compression curves of four different garnets were measured. From these measurements, the systematics of the bulk modulus in the pyroxene-garnet join can be constructed. For comparisons of bulk moduli, K_0 is assumed to be 2 for all the garnets; this gives a reasonable K_0 value for pyrope. Bulk moduli of enstatite-pyrope and ferrosilite-almandine joins are plotted as a function of composition in Figure 6. Results of previous studies are also shown for comparison.

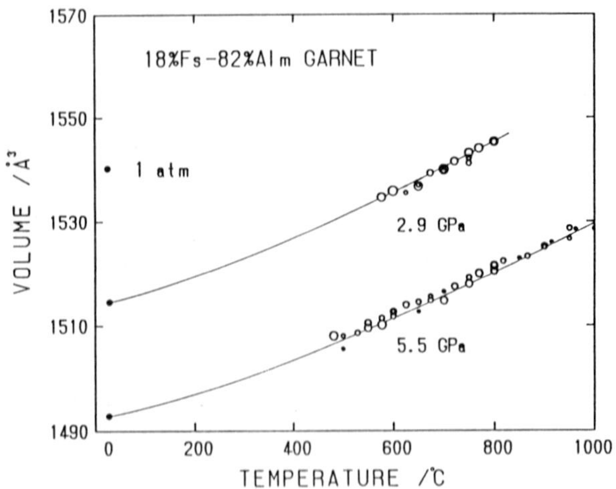

Figure 5. Thermal expansions of 18% Fs-82% Alm solid solution at 2.9 GPa and at 5.5 GPa. Radius of open circle represents error bar of each volume measurement. Solid circles are obtained from room temperature compression curve.

TABLE 4. Thermal Expansion* of Garnets and Majorites at 1 atm and at High Pressures

Material	P(GPa)	A_0	$A_1(\times10^{-5})$	$A_2(\times10^{-8})$	$\alpha_0(\times10^{-5})$	Reference
Pyrope	0	1.000	2.6	0.6	2.6	OSAKO (unpublished data)
Pyrope	0	1.000	2.2	0.5	2.2	SKINNER (1956)
Pyrope	6.0	0.968	1.8	0.8	1.9	present study
58% En-42% Py	6.0	0.967	1.6	0.8	1.7	present study
Almandine	0	1.000	1.8	0.9	1.8	SKINNER (1956)
18% Fs-82% Alm	2.9	0.983	1.5	1.4	1.5	present study
18% Fs-82% Alm	5.5	0.970	1.4	1.0	1.4	present study

*Volume expansion data were fitted to the following equation:

$$V/V_0=A_0+A_1T+A_2T^2 \ (T: °C)$$

where V_0 is a unit cell volume at 1 atm and 25 °C. Expansion at 0 °C is calculated as $\alpha_0=A_1/A_0$.

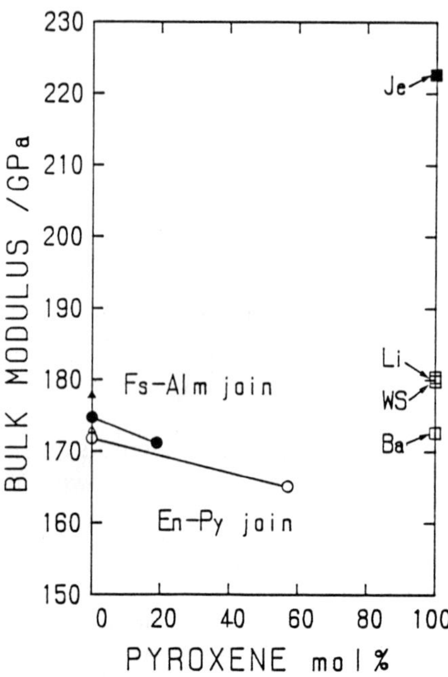

Figure 6. Variation of bulk modulus with composition in pyroxene-garnet join. Open and solid circles represent present experiments on En-Py join and on Fs-Alm join, respectively. Open and solid triangles represent adiabatic bulk moduli on pyrope and almandine by BABUŠKA et al. (1978), respectively. Solid square is the reported bulk modulus on majorite by JEANLOZ (1981) and open squares are estimated adiabatic bulk moduli of garnet structured $MgSiO_3$ (Li: LIEBERMANN, 1974; WS: WANG and SIMMONS, 1974; Ba: BABUŠKA et al., 1978).

Two conclusions can be derived from analysis of Figure 6. One is that substitution of magnesium with iron slightly increases the bulk modulus. For the garnet end-member, K_0 increases approximately 1.4% from pyrope to almandine; this result is in agreement with the systematics in this join established by previous ultrasonic measurements (BABUŠKA et al., 1978).

The second conclusion is that dissolution of the pyroxene component in garnet decreases the bulk modulus. This tendency is contrary to that proposed by JEANLOZ (1981). It has been reported that in garnet solid solutions, the bulk modulus varies linearly with composition (BABUŠKA et al., 1978). When the present result is extrapolated linearly, the bulk modulus of garnet-structured $MgSiO_3$ becomes 160 GPa. In the present systems, there is only one observation in an intermediate composition, and much ambiguity still exists in this extrapolation. Nevertheless, Figure 6 clearly shows that the bulk modulus of majorite of pyroxene composition is unlikely to be as large as that reported by JEANLOZ (1981). Based on systematic ultrasonic measurements on various garnets, the adiabatic bulk modulus of majorite has been estimated by several authors (LIEBERMANN, 1974; WANG and SIMMONS, 1974; BABUŠKA et al., 1978). If we take into account the ambiguities of the extrapolations, the present result is in harmony with these estimated values.

Thermal Expansion of Garnets Under Pressure

When thermal expansions of pure pyrope and En-Py solid solution are compared at 6.0 GPa (Table 4), it is clear that dissolution of pyroxene in garnet has very little effect on thermal expansion. On the other hand, substitution of iron and magnesium in the pyrope-almandine join seems to have a large effect. However, the thermal expansion values for garnets at atmospheric pressure reported by SKINNER (1956) and recent remeasurement by Osako (private communication) differ considerably. Because of this ambiguity, it is difficult to make detailed comparisons of the pressure effects of thermal expansion in garnet and in this paper only a brief discussion will be made.

BIRCH (1952, 1968) and ANDERSON (1967) discussed volume dependence of thermal expansion of materials and derived several equations. In the case of alkali halides, it was observed that volume dependence of thermal expansion could be expressed by the following equation (YAGI, 1978)

$$\alpha / \alpha_0 = (V / V_0)^{\delta_0} \quad (1)$$

where $\delta_0 = -(1/\alpha K)(dK/dT)_P$. In this case, δ_0 was directly calculated from static P-V-T measurement and a systematic difference of δ_0 was observed between alkali halides with NaCl structure (δ_0=2~3) and those with CsCl structure (δ_0=6~7). In the present study, the volume dependence of thermal expansion is reasonably well expressed by Equation 1 when δ_0 is assumed to be 5 for pyrope. For iron-rich majorite, no data are available for α at atmospheric pressure, but when α is assumed to be similar for both almandine end-member and 18% Fs-82% Alm solid solution, experimental results can also be expressed well by Equation 1 with δ_0=7. If δ_0 is calculated using the $(\partial K/\partial T)_P$ reported by SOGA (1967), δ_0 becomes 5.3 for pyrope and 6.4 for almandine. These results suggest that the pressure effect on thermal expansion in majorite is reasonably well expressed by Equation 1.

In conclusion, the present study clarified the elastic property of the pyroxene-garnet solid solution, as follows:

1) In the present system the bulk modulus decreased with increasing pyroxene component and the garnet-structured $MgSiO_3$ was found to probably be much more compressible than previously expected. This result presents new constraints to the model of the earth's mantle; a detailed discussion is provided elsewhere (AKAOGI et al., this volume).

2) Thermal expansion of majorite is expected to be similar to that of garnet with same iron magnesium ratio.

3) Thermal expansion under pressure is reasonably well expressed by Equation 1.

Acknowledgments. The authors are grateful to E. Ito (Okayama University), M. Kato (Nagoya University), and K. Omura (Tokyo University) for their help in the synchrotron radiation experiments, and to T. Nagashima (NIRIM) for preparing computer programs used in the MAX80 system. Thanks are also due to M. Osako (National Science Museum) who kindly provided thermal expansion data on pyrope prior to publication, and to R. C. Liebermann and M. Kumazawa for their helpful comments and discussions.

REFERENCES

AKAOGI, M., and S. AKIMOTO, Pyroxene-garnet solid-solution equilibria in the system $Mg_4Si_4O_{12}$-$Mg_3Al_2Si_3O_{12}$ and $Fe_4Si_4O_{12}$-$Fe_3Al_2Si_3O_{12}$ at high pressures and temperatures, *Phys. Earth Planet. Inter.*, *15*, 90–106, 1977.

AKAOGI, M., A. NAVROTSKY, T. YAGI, and S. AKIMOTO, Pyroxene-garnet transformation: Thermochemistry and elasticity of garnet solid solutions, and application to a pyrolite mantle, this volume.

ANDERSON, O. L., Equation for thermal expansivity in planetary interiors, *J. Geophys. Res.*, *72*, 3661–3668, 1967.

BABUŠKA, V., J. FIALA, M. KUMAZAWA, I. OHNO, and Y. SUMINO, Elastic properties of garnet solid-solution series, *Phys. Earth Planet. Inter.*, *16*, 157–176, 1978.

BASS, J. D., and D. L. ANDERSON, Composition of the upper mantle: geophysical tests of two petrological models, *Geophys. Res. Lett.*, *11*, 229–232, 1984.

BIRCH, F., Elasticity and constitution of the earth's interior, *J. Geophys. Res.*, *57*, 227–286, 1952.

BIRCH, F., Thermal expansion at high pressures, *J. Geophys. Res.*, *73*, 817–819, 1968.

DECKER, D. L., High-pressure equation of state for NaCl, KCl, and CsCl, *J. Appl. Phys.*, *42*, 3239–3244, 1971.

HALLECK, P. M., The compression and compressibility of grossular garnet: a comparison of X-ray and ultrasonic methods, Ph. D. thesis, University of Chicago, 1973.

HAZEN, R. M., and L. W. FINGER, Crystal structures and compressibilities of pyrope and grossular to 60 kbar, *Am. Mineral.*, *63*, 297–303, 1978.

JEANLOZ, R., Majorite: vibrational and compressional properties of a high-pressure phase, *J. Geophys. Res.*, *86*, 6171–6179, 1981.

LEITNER, B. J., D. J. WEIDNER, and R. C. LIEBERMANN, Elasticity of single crystal pyrope and implications for garnet solid solution series, *Phys. Earth Planet. Inter.*, *22*, 111–121, 1980.

LEVIEN, L., C. T. PREWITT, and D. J. WEIDNER, Compression of pyrope, *Am. Mineral.*, *64*, 805–808, 1979.

LIEBERMANN, R. C., Elasticity of pyroxene-garnet and pyroxene-ilmenite phase transformations in germanates, *Phys. Earth Planet. Inter.*, *8*, 361–374, 1974.

MASON, B., N. NELEN, and J. S. WHITE, Jr., Olivine-garnet transformation in a meteorite, *Science*, *160*, 66–67, 1968.

RINGWOOD, A. E., *Composition and Petrology of the Earth's Mantle*, 618 pp., McGraw-Hill, New York, 1975.

SATO, Y., T. YAGI, Y. IDA, and S. AKIMOTO, Hysteresis in the pressure-volume relation and stress inhomogeneity in composite material, *High Temp. High Pressures*, *7*, 315–323, 1975.

SATO, Y., M. AKAOGI, and S. AKIMOTO, Hydrostatic compression of the synthetic garnets pyrope and almandine, *J. Geophys. Res.*, *83*, 335–338, 1978.

SHIMOMURA, O., S. YAMAOKA, T. YAGI, M. WAKATSUKI, K. TSUJI, O. FUKUNAGA, H. KAWAMURA, K. AOKI, and S. AKIMOTO, Multi-anvil type X-ray apparatus for synchrotron radiation, *Mat. Res. Soc. Symp. Proc.*, *22*, 17–20, 1984.

SHIMOMURA, O., S. YAMAOKA, T. YAGI, M. WAKATSUKI, K. TSUJI, H. KAWAMURA, N. HAMAYA, O. FUKUNAGA, K. AOKI, and S. AKIMOTO, Multi-anvil type X-ray system for synchrotron radiation, in *Solid State Physics Under Pressure: Recent Advance with Anvil Devices*, edited by S. Minomura, pp. 351–356, KTK/Reidel, Tokyo/London, 1985.

SKINNER, B. J., Physical properties of end-members of the garnet group., *Am. Mineral.*, *41*, 428–436, 1956.

SOGA, N., Elastic constants of garnet under pressure and temperatue, *J. Geophys. Res.*, *72*, 4227–4234, 1967.

TAKAHASHI, T., and L. G. LIU, Compression of ferromagnesian garnets and the effect of solid solutions on the bulk modulus, *J. Geophys. Res.*, *75*, 5757–5766, 1970.

WANG, H., and G. SIMMONS, Elasticity of some mantle crystal structures 3. Spessartite-almandine garnet, *J. Geophys. Res.*, *79*, 2607–2613, 1974.

YAGI, T., Experimental determination of thermal expansivity of several alkali halides at high pressures, *J. Phys. Chem. Solids*, *39*, 563–571, 1978.

YAGI, T., O. SHIMOMURA, S. YAMAOKA, K. TAKEMURA, and S. AKIMOTO, Precise measurement of compressibility of gold at room and high temperature, in *Solid State Physics Under Pressure: Recent Advance with Anvil Devices*, edited by S. Minomura, pp. 363–368, KTK/Reidel, Tokyo/London, 1985.

PHASE DIAGRAM OF IRON DETERMINED BY HIGH-PRESSURE/TEMPERATURE X-RAY DIFFRACTION USING SYNCHROTRON RADIATION

Syun-iti AKIMOTO[1] and Toshihiro SUZUKI[2]

Institute for Solid State Physics, University of Tokyo, Minato-ku, Tokyo 106, Japan

Takehiko YAGI[3]

Research Institute for Iron, Steel and Other Metals, Tohoku University, Katahira, Sendai 980, Japan

Osamu SHIMOMURA

National Institute for Research in Inorganic Materials, Sakura-mura, Niihari-gun, Ibaraki 305, Japan

Abstract. The phase diagram of iron was investigated in the pressure-temperature region up to 12.5 GPa and 800 °C using a cubic-anvil type of high-pressure apparatus combined with synchrotron radiation. In situ measurements with an energy-dispersive X-ray diffraction system enabled us to carry out dynamical observations of the phase transformation. It was found that iron transformed rather rapidly when the temperature was raised to the stability field of the γ-phase. Preliminary α-γ and ε-γ phase boundaries were determined. The $\alpha/\gamma/\varepsilon$ triple point, which was tentatively defined as a minimum temperature of the phase boundary, was located at 8.3 GPa and 440 °C. It was difficult to precisely determine the equilibrium phase boundary between the α- and ε-phases, but the present experiments suggest a positive dP/dT for this α-ε phase boundary.

Introduction

A precise knowledge of the pressure-temperature phase diagram of iron is indispensable to investigations of the origin and evolution of the earth. A few attempts have been made to clarify the stability field of three iron polymorphs, α (bcc), γ (fcc), and ε (hcp) phases (BUNDY, 1965; FUKIZAWA, 1982; MANGHNANI et al., 1985). BUNDY (1965) studied the pressure-temperature phase diagram of iron to 20 GPa and 900 °C by means of in situ electrical resistivity measurements using a belt-type, high-pressure apparatus; he determined the $\alpha/\gamma/\varepsilon$ triple point to be at 11.0 GPa and 490 °C. In the recent investigations by FUKIZAWA (1982) and MANGHNANI et al. (1985), in situ high pressure-temperature X-ray diffraction measurements were used to observe the phase transformation. These investigations tentatively assigned the triple point to be at 9 GPa and 470 °C; however, no careful consideration was given to the kinetics of phase transformation. Consequently no reliable equilibrium phase boundaries have yet been determined. The purpose of this study is to establish the phase boundaries as accurately as possible by applying the strong X-ray source of synchrotron radiation to the dynamical observation of the phase transformation.

Experimental Methods

The cubic-anvil, high-pressure and high-temperature X-ray diffraction apparatus (MAX80), which is installed at the Photon Factory (a synchrotron radiation source at the National Institute for High Energy Physics (KEK) in Tsukuba), was used. Details of the apparatus have been described elsewhere (SHIMOMURA et al., 1984, 1985). Tungsten carbide anvils with truncated 3-mm square faces were used for the experiments above 6 GPa, and 6-mm edge anvils were used for those below 6 GPa. A cube of amorphous boron/epoxy resin mixture (4:1), with a size 2 mm larger than the anvils adopted, was used as a pressure-transmitting medium.

A sample assembly used for the experiments in the pressure range above 6 GPa is shown in Figure 1. A powder sample of high-purity, reagent-grade iron was compressed to a small flake, about 0.05 mm in thickness and 0.8 mm in width. The sample was then inserted between a pair of semi-cylinders, which were composed of a mixture of powdered NaCl and NaF, in order to avoid contamination of boron or carbon by the pressure-transmitting medium. NaCl and NaF were simultaneously used as pressure markers. This assembly was placed in the upper half of the furnace. The lower half of the furnace was not used as a sample chamber in the present measurement, instead it was filled with a boron nitride disk. The furnace was constructed from a pair of alumel

[1]Present address: Institute for Study of the Earth's Interior, Okayama University, Misasa, Tottori-ken 682-02, Japan.

[2]Present address: R and D Laboratory I, Nippon Steel Corporation, Nakahara-ku, Kawasaki 211, Japan.

[3]Present address: Institute for Solid State Physics, University of Tokyo, Minato-ku, Tokyo 106, Japan.

High-Pressure Research in Mineral Physics, edited by M. H. Manghnani and Y. Syono, pp. 149–154.

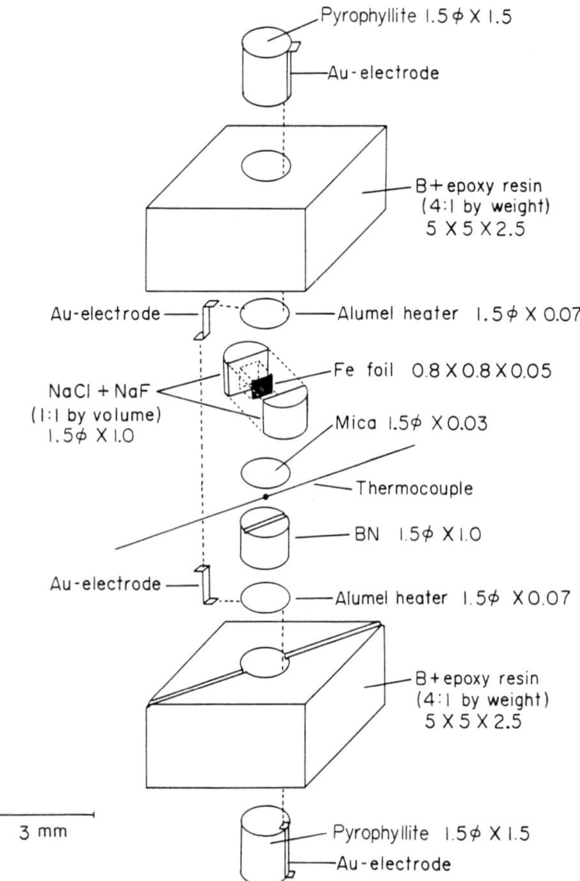

Figure 1. Sample assembly for the high-pressure and high-temperature in situ X-ray measurements.

disk heaters (1.5 mm in diameter and 0.07 mm in thickness) and embedded in the center of the amorphous boron/epoxy resin cube. Temperature was measured with a chromel/alumel thermocouple which was placed at the center of the furnace.

Energy-dispersive technique was used throughout the present in situ X-ray diffraction measurements. An extremely intense white X-ray beam from the synchrotron radiation source, operating at 2.5 GeV and ≦150 mA, was directed into the sample chamber through the amorphous boron/epoxy resin pressure-transmitting medium. Diffracted X-rays were detected by an intrinsic germanium solid-state detector at 2θ angle of 11–12 °. The incident X-ray beam was 0.3 mm in height and approximately 0.5 mm in width. Beam width was determined by the clearance between the anvils. The position of the incident beam was monitored by observing the shadow profile of the direct beam. The beam was adjusted close to the hot junction of the thermocouple so as to minimize error in the temperature measurement. A double slit (0.2 mm in height) was used as a receiving slit. In this geometry, diffraction patterns of iron with satisfactory resolution were obtained in the exposure times of 200 or 300 s.

It has been repeatedly reported that the phase transformation of iron is very sluggish, especially in a pressure cycle corresponding to the α-ε transformation under constant temperature (e.g., MAO et al., 1967). Accordingly, constant-load paths were adopted to alter the pressure-temperature conditions of the iron sample. An attempt to construct the equilibrium phase diagram was made by comparing the relative stability of three iron polymorphs at various pressure-temperature conditions up to 12.5 GPa and 800 °C. To identify the stability field, data on the change in the diffraction patterns were collected at each step of a sudden increase or decrease in temperature (and occasionally in pressure), (a total of more than fifty points in the pressure-temperature space). The pressure value at each point was calculated using Decker's equation of state for NaCl (DECKER, 1971). Two or three diffraction lines of NaCl, (200), (220), and (222), were used for determining the pressure values. The accuracy of the pressure determination was estimated to be ±0.1–0.2%. NaF was used as an auxiliary pressure standard with reference to the equation of state reported by YAGI (1978).

Results and Discussion

Pressure and temperature data used to determine the phase boundary are summarized in Figure 2. Typical examples of the experimental paths traversing the phase boundary are shown in Figures 3a–c. Even when the ram load was kept constant, pressure in the sample chamber varied in a complicated manner with temperature, depending on the sample assembly and previous history of the pressure-temperature conditions. Changes in the X-ray diffraction profile along these paths are illustrated in Figures 4–6. Diffraction profiles designated by A, B, and C in Figure 4 were observed at the pressure-temperature conditions A, B, and C in Figure 3a, respectively. It was found that, with a sudden increase in temperature at about 7 GPa (from 400 °C at A to 500 °C at B), the α-phase became unstable and the new diffraction peaks designated as the γ-phase appeared. With a subsequent decrease in temperature to 450 °C (at C), the growth of α-phase, at the expense of the γ-phase, was clearly observed. These measurements indicate that an equilibrium phase boundary, between the α- and γ-phases, is located at an intermediate point between B and C.

Figure 5 is concerned with the γ-ε transformation in a different run. The series of diffraction profiles D–H in Figure 5 represent the measurements along the path D–H in Figure 3b. It must be noted that before the present measurements shown in Figure 5 were performed, the

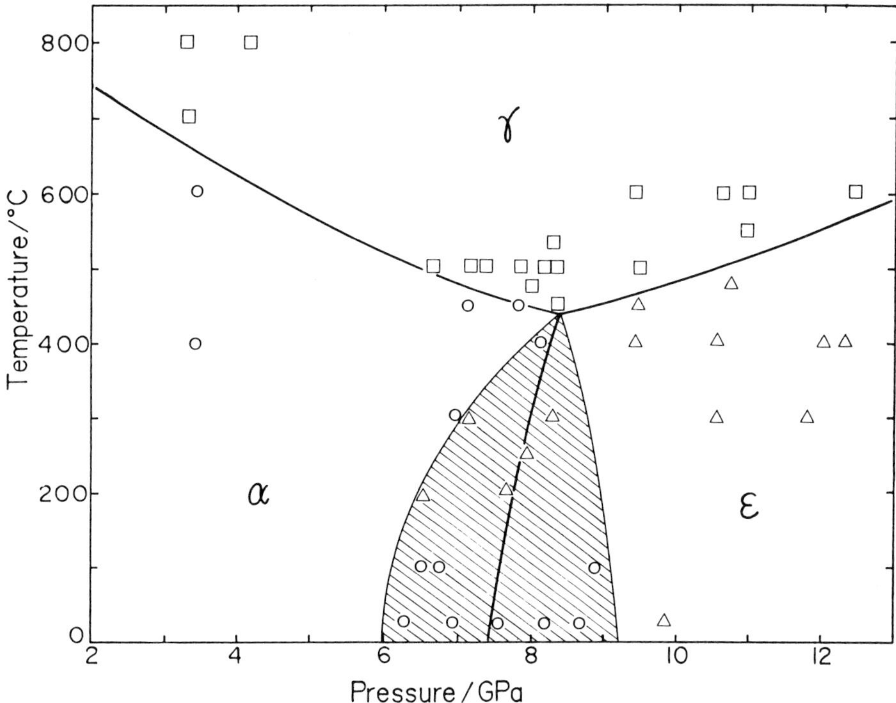

Figure 2. Phase diagram of iron determined by in situ X-ray diffraction measurements. Circles, squares, and triangles, represent the pressure-temperature condition where the growth of the α-, γ-, and ε-phases of iron were observed, respectively. Hatched area indicates the transition zone for the α-ε transformation. In this area, stability of α- and ε-phases are affected not only by pressure and temperature, but also by some other factors (see text). Present data suggest a positive slope ($dP/dT>0$) for the α-ε transformation.

iron sample occasionally experienced temperature cycles traversing the γ-ε phase boundary at pressures ranging from about 10–12.5 GPa. Just prior to the measurement at point D, the iron sample was at about 9 GPa and room temperature. This history of sample behavior results in a mixed phase, in which the γ-phase occurs as a metastable phase in addition to the stable ε-phase, as observed in diffraction profile D. With a sudden increase in temperature from 450 °C (at D) to 500 °C (at E) along the constant load path at about 9.4 GPa, a significant change in the relative intensity of the ε-phase to the γ-phase was observed. It is shown in Figure 5 that the γ-phase is more stable than the ε-phase at the pressure-temperature conditions designated by E (9.4 GPa and 500 °C). It is also shown in Figure 5 that further increase in temperature to 600 °C (at F) results in the complete disappearance of the ε-phase. Reverse transformation from γ-phase to ε-phase was found to be rather sluggish. An ε-phase could not be identified until the temperature decreased to 300 °C (point H in Figure 3b). Results shown in Figure 5 suggest that an equilibrium γ-ε boundary is passing through an intermediate point (shown between D and E in Figure 3b).

Figure 3c is the continuation of the experimental path shown in Figure 3b. The corresponding diffraction profiles are illustrated in Figure 6. As seen in diffraction profile I, a small amount of the metastable γ-phase, in addition to the main ε-phase, still existed at room temperature. When the pressure was reduced from 8.2 GPa (at I) to 7.4 GPa (at J) at room temperature, growth of α-phase at the expense of ε-phase was observed. With a subsequent increase in temperature to 200 °C (at K), a slight increase in the relative intensity of the ε-phase to the α-phase was observed. The pressure-temperature conditions designated by point K in Figure 3c (7.8 GPa and 200 °C), may be located within the stability field of the ε-phase.

All experimental results used to determine the phase boundary are summarized in Figure 2. As previously stated, it was established that iron transformed rather rapidly when the temperature was raised to about the stability field of the γ-phase. Preliminary determinations of the α-γ and ε-γ phase equilibrium boundaries are shown by thick, solid lines in Figure 2. The α/γ/ε triple point, which was tentatively determined from these boundaries as a minimum temperature point, is located at 440 °C and 8.3 GPa. These temperature and pressure values are significantly lower than those previously determinated by Bundy (1965), 490 °C and 11 GPa. The lower pressure value in the present study is reasonable because Bundy

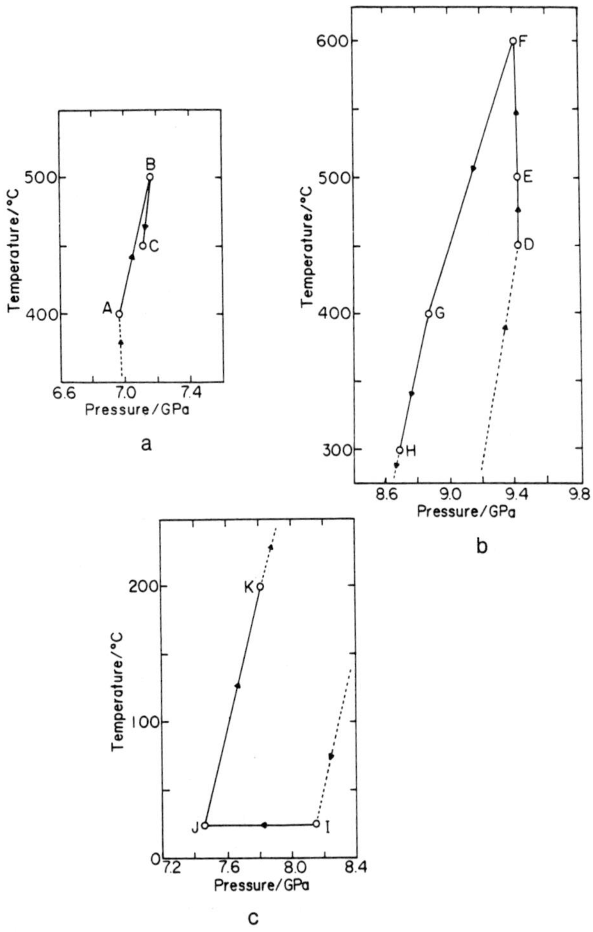

Figure 3. Typical examples of the experimental path in the pressure-temperature space, where a series of in situ X-ray diffraction measurements were performed.

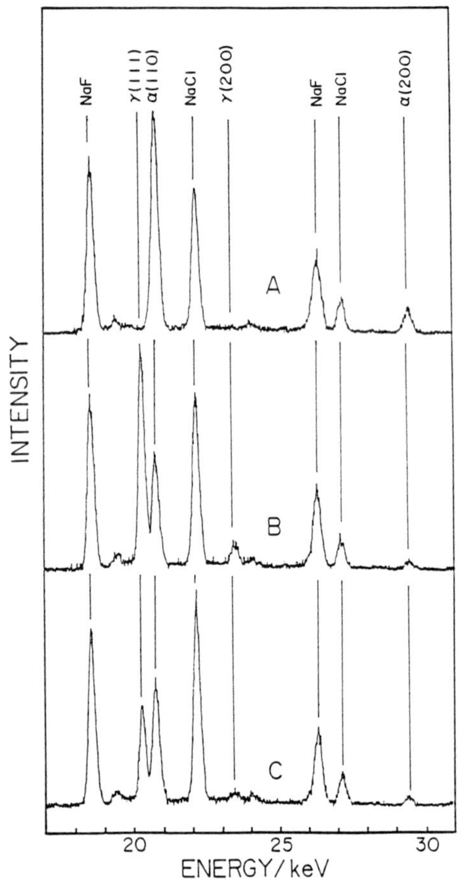

Figure 4. A series of X-ray diffraction profiles observed across the α-γ phase boundary of iron. Diffraction profiles A, B, and C correspond to the pressure-temperature conditions designated by A, B, and C in Figure 3a.

calibrated his pressure values at room temperature based on the older pressure fixed points. A slight difference was also found between the present values and those reported by FUKIZAWA (1982), 470 °C and 9 GPa. Although these two investigations used the similar technique of in situ X-ray diffraction measurement, no careful consideration was given to the reverse transformation in Fukizawa's work. The present determination may be a closer estimation of the final equilibrium triple point.

The present study also confirmed that the transformation rate of iron becomes very slow when temperature is below 400 °C. Coexistence of the three phases was often observed at about 8 GPa and below 400 °C. Hence, it was still difficult to determine the α-ε phase boundary precisely. In previous investigations a negative slope ($dP/dT<0$) was suggested for this phase boundary (BUNDY, 1965; FUKIZAWA, 1982), and the α-ε transformation pressure at room temperature was reported to be 11.0–11.5 GPa (DRICKAMER, 1970). Since the transformation rate of iron is very sensitive to temperature, as indicated in the present study, it is natural to expect that the transformation from the α-phase to the ε-phase needs larger over-pressure at low temperatures. In fact, in previous experiments (FUKIZAWA, 1982) performed under a pressure-increasing cycle, the α-ε transformation occurred at higher pressures under lower temperatures, and the slope of the phase boundary was determined to be negative.

Figure 2 shows the pressure-temperature conditions at which the growth of α-, γ-, and ε-phases of iron were observed. These points should represent the stable phase of each condition. However, as shown in Figure 2, the observed stability fields of α- and ε-phases were complicated. The growth of the α-phase was often observed when the pressure was changed, whereas the growth of ε-phase was observed when the temperature

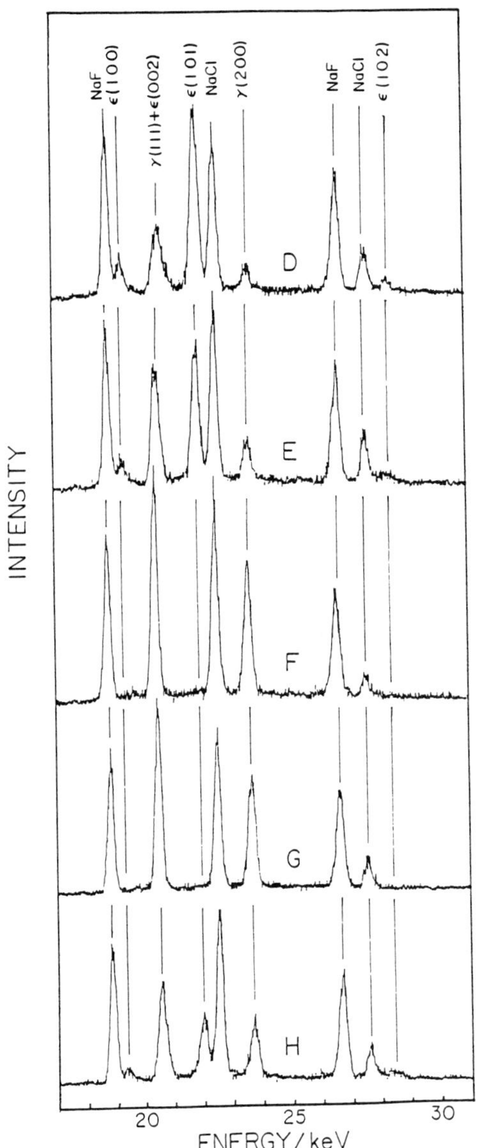

Figure 5. A series of X-ray diffraction profiles observed across the ε-γ phase boundary of iron. Diffraction profiles D–H correspond to the pressure-temperature conditions designated by D–H in Figure 3b.

was changed (see hatched area in Figure 2). Such observations suggest that the relative stability of the α- and ε-phases may be largely affected not only by the P-T conditions, but also by the stress condition of the sample. This fact made it difficult to determine the α-ε phase boundary precisely. However, present observation suggests that the stability field of the ε-phase extends to a lower pressure region than reported in the previous study (Bundy, 1965; Fukizawa, 1982; Manghnani et al., 1985). Although it is still difficult to draw a definite α-ε phase boundary based only on these observations, a

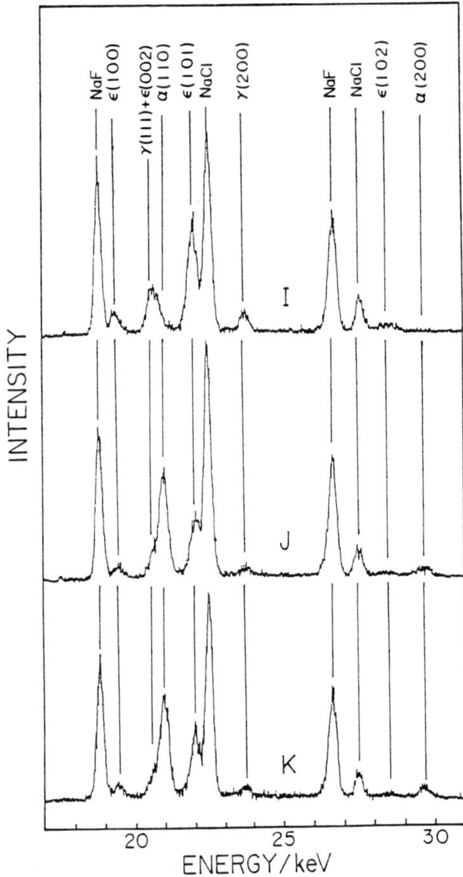

Figure 6. A series of X-ray diffraction profiles used to identify the stability field of the α-phase and ε-phase of iron. Diffraction profiles I, J, and K were observed at the pressure-temperature conditions designated by I, J, and K in Figure 3c.

tentative boundary with a positive slope is considered to be an intermediate line of the transition zone (see the hatched area in Figure 2). More detailed investigations are required to definitively establish the phase boundary of α-ε transformation in iron.

Acknowledgments. The authors are grateful to the staff of the Photon Factory, KEK, for facilitating our work and to T. Nagashima of NIRIM for preparing computer programs used in the MAX80 system. We are also indebted to J. Susaki, K. Kusaba, K. Tsuji, T. Kikegawa, and T. Tamai for their help in carrying out the present experiments at the Photon Factory.

References

Bundy, F. P., Pressure-temperature phase diagram of iron to 200 kbar, 900 °C, *J. Appl. Phys.*, *36*, 616–620, 1965.

Decker, D. L., High-pressure equation of state for NaCl, KCl, and CsCl, *J. Appl. Phys.*, *42*, 3239–3244, 1971.

Drickamer, H. G., Revised calibration for high pressure electrical resistance cell, *Rev. Sci. Instrum. 41*, 1667–1668, 1970.

FUKIZAWA, A., Direct determinations of phase equilibria of geophysically important materials under high pressure and high temperature using in situ X-ray diffraction method, Ph. D. thesis, 125 pp., University of Tokyo, 1982.

MANGHNANI, M. H., L. C. MING, J. BALOGH, S. B. QADRI, E. F. SKELTON, and D. SCHIFERL, Equation of state and phase transition studies under in situ high P-T conditions using synchrotron radiation, in *Solid State Physics under Pressure: Recent Advance with Anvil Devices*, edited by S. Minomura, pp. 343–350, Terra Sci. Pub. Co., Tokyo, 1985.

MAO, H. K., W. B. BASSETT, and T. TAKAHASHI, Effect of pressure on crystal structure and lattice parameters of iron up to 300 kbar, *J. Appl. Phys.*, *38*, 272–276, 1967.

SHIMOMURA, O., S. YAMAOKA, T. YAGI, M. WAKATSUKI, K. TSUJI, O. FUKUNAGA, H. KAWAMURA, K. AOKI, and S. AKIMOTO, Multi-anvil type X-ray apparatus for synchrotron radiation, *Mat. Res. Soc. Symp. Proc., Vol. 22*, III, pp. 17–20, Elsevier Sci. Pub. Co., New York, 1984.

SHIMOMURA, O., S. YAMAOKA, T. YAGI, M. WAKATSUKI, K. TSUJI, H. KAWAMURA, N. HAMAYA, O. FUKUNAGA, K. AOKI, and S. AKIMOTO, Multi-anvil type X-ray system for synchrotron radiation, in *Solid State Physics under Pressure: Recent Advance with Anvil Devices*, edited by S. Minomura, pp. 351–356, Terra Sci. Pub. Co., Tokyo, 1985.

YAGI, T., Experimental determination of thermal expansivity of several alkali halides at high pressures, *J. Phys. Chem. Solids*, *39*, 563–571, 1978.

INVESTIGATION OF THE α-Fe$\rightleftharpoons\varepsilon$-Fe PHASE TRANSITION BY SYNCHROTRON RADIATION

Murli H. MANGHNANI, Li Chung MING, and Nobuyuki NAKAGIRI[†]

Hawaii Institute of Geophysics, University of Hawaii, Honolulu, Hawaii 96822, USA

Abstract. The α(bcc)$\rightleftharpoons\varepsilon$(hcp) transition in iron was investigated in situ under simultaneously high pressure and temperature (to ~27.5 GPa and 540 °C) in a diamond-anvil cell using synchrotron radiation and the energy-dispersive X-ray diffraction technique. The data for five isothermal runs (at 190, 270, 360, 450, and 540 °C) were interpreted on the basis of the intensity ratio of the ε-Fe(101) peak and the overlapped {ε-Fe(002)+α-Fe(110)} peak to constrain the pressures for the starts of the $\alpha\rightarrow\varepsilon$ and $\varepsilon\rightarrow\alpha$ transitions (i.e., $P^s_{\alpha\rightarrow\varepsilon}$ and $P^s_{\varepsilon\rightarrow\alpha}$, respectively). The resultant "equilibrium" dT/dP slope is approximately −435 deg/GPa ($dP/dT=-0.0023$ GPa/deg).

The isothermal P-V relationships for the α-Fe and ε-Fe phases at temperatures up to 442 °C were deduced to yield the coefficients of volumetric thermal expansion as a function of pressure as

$\alpha_v(\text{deg}^{-1}\times 10^5)=(5.53\pm0.04)-(0.070\pm0.006)P$ (GPa) for the α-phase, and

$\alpha_v(\text{deg}^{-1}\times 10^5)=(5.29\pm0.13)-(0.090\pm0.009)P$ (GPa) for the ε-phase, resulting in a lower value of α_v for the ε-phase at any given pressure.

Introduction

Iron has been the subject of experimental and theoretical high pressure-temperature studies for three decades. It is essential, from a geophysical standpoint, to understand the phase relations in iron and the properties of those phases. To this end, several studies concerned with the phase transitions and the equations of state of the polymorphs of iron under dynamic and static loading have been conducted. BANCROFT et al. (1956) were the first to observe the shock-induced phase transition in iron at 13 GPa; this transition was subsequently confirmed by the electrical resistivity and X-ray methods (e.g., BALCHAN and DRICKAMER, 1961; JAMIESON and LAWSON, 1962; TAKAHASHI and BASSETT, 1964) and recognized as the low-pressure α(bcc)$\rightarrow$high-pressure ε(hcp) transition.

Using the electrical resistivity method, BUNDY (1965) investigated the phase relationships among the three polymorphs (α, ε and γ (fcc)) of iron to 20 GPa and 900 °C. He found the triple point to be at 11±0.3 GPa and 490±10 °C and the dT/dP slope of the α-ε boundary to be from −200 to −500 deg/GPa.

Employing the cubic-anvil, high P-T apparatus and the energy-dispersive X-ray diffraction (EDXRD) technique, FUKIZAWA (1982) studied the α-ε-γ phase boundaries and estimated the dT/dP slope of the α-ε transition to be −169 deg/GPa. The result of a study by MANGHNANI et al. (1985), using the first appearance of the ε-phase in the transition, were in good agreement with Fukizawa's results.

The α-ε transition is known to be very sluggish and to exhibit appreciable hysteresis; this is probably responsible for the somewhat inconsistent results for the slope dT/dP of the phase boundary. Table 1 lists slopes reported by various investigations. It is evident that the dT/dP slopes are steep and mostly negative. In a recent in situ high pressure-temperature EDXRD study using the large-volume multianvil X-ray apparatus (MAX80) and synchrotron radiation, AKIMOTO et al. (this volume) have found that the stability field of the ε-phase extends to lower pressures than previously reported; they have tentatively determined that the slope of the α-ε phase boundary is positive.

To investigate the nature of the $\alpha\rightleftharpoons\varepsilon$ transition, the present study performed both forward ($\alpha\rightarrow\varepsilon$) and backward ($\varepsilon\rightarrow\alpha$) transitions in situ under simultaneously high pressure and temperature (to ~27.5 GPa and 540 °C) using a resistively heated diamond-anvil cell interfaced with synchrotron radiation. The results of this study are compared with the other recent α-ε transition studies in order to better understand the nature of this transition and to tightly constrain the phase boundary.

The equation of state studies on Fe have been carried out by shock wave measurements (e.g., JEANLOZ, 1979; BROWN and MCQUEEN, 1982) and by static-compression measurements using X-ray methods (e.g., MAO et al., 1967; MAO and BELL, 1979; ZOU et al., 1981; JEPHCOAT et al., 1986; HUANG et al., 1987). Except for the shock wave and Huang et al. studies, all other investigations have been limited to room temperature. Another purpose of this study is therefore to evaluate the effect of pressure on the thermal expansivity for both α and ε phases at high temperatures.

Experimental Methods

In the present study, an improved high-temperature diamond-anvil cell (MING et al., this volume) was interfaced with the synchrotron radiation and energy-

[†]Present address: Research Development Corporation of Japan, Tsukuba, Ibaraki 300-26, Japan.

High-Pressure Research in Mineral Physics, edited by M. H. Manghnani and Y. Syono, pp. 155–163.

TABLE 1. Various Investigations of the α-Fe⇌ε-Fe Transition

P_0, GPa	dT/dP, GPa/deg	dT/dP, deg/GPa	Method & Instrumentation	Reference
11.1	− 0.0020 to 0.0050	− 500 to − 200	RES, BAP	BUNDY (1965)
11.8	− 0.0059	− 169	XRD, RA, CAP	FUKIZAWA (1982)
11.2	− 0.0056	− 179	XRD, SR, DAC	MANGHNANI et al. (1985)
13.5	− 0.0035	− 286	XRD, SR, DAC	HUANG et al. (this volume)
9.0	− 0.0022	− 455	XRD, SR, DAC	This study
7.4	+ 0.0022	455	XRD, SR, CAP	AKIMOTO et al. (this volume)

Notation: P_0: Equilibrium transition pressure at room temperature, RES: Resistivity measurement, XRD: X-ray diffraction, RA: Rotating anode X-ray generator, SR: Synchrotron radiation, BAP: Bridgman anvil press, DAC: Diamond-anvil cell, CAP: Cubic-anvil press.

dispersive X-ray diffraction (EDXRD) system at the Stanford Synchrotron Radiation Laboratory (SSRL). The experimental setup and the data acquisition system used have been described in detail previously (MING et al., 1983, this volume). High-pressure runs up to 28 GPa were conducted on polycrystalline foil of iron of 99.9975% purity held isothermally at five temperatures (190, 270, 360, 450, and 540 °C). The sample temperature was maintained within ±3 °C and monitored by a precalibrated Pt-Pt10%Rh thermocouple placed adjacent to the sample.

MgO was chosen as the internal pressure calibrant for three reasons: 1) there is no overlapping of the X-ray diffraction peaks of MgO, α-Fe, and ε-Fe, thus making it easier to unambiguously detect the ε-Fe phase; 2) MgO is stable at high temperatures, and hence reaction between MgO and Fe under high pressure and temperature is unlikely; and 3) *P-V-T* relations for MgO based on shock wave data are available for a wide *P-T* range (JAMIESON et al., 1982).

In our previous equation-of-state study on MgO and Au under simultaneously high pressure and temperature conditions to 20 GPa and 407 °C (MING et al., 1984), we found that the pressures determined from the molar volume ratio $(V/V_0)_{MgO}$ and *P-V-T* relations for MgO were lower than those determined from $(V/V_0)_{Au}$ and *P-V-T* relations for Au. The relationship between P_{Au} and P_{MgO} is expressed by: $P_{Au}=0.212+1.104\ P_{MgO}$ (GPa). Accordingly, the pressure values reported here have been adjusted to the P_{Au} scale (JAMIESON et al., 1982).

For high-pressure runs at each of the five temperatures, pressure was gradually applied to the sample until the α-Fe was partially or completely converted into ε-Fe and then slowly unloaded until ε-Fe was completely transformed back to α-Fe. Energy-dispersive X-ray diffraction data were collected for the five increasing and decreasing pressure cycles at 85 *P-T* points, using a Si(Li) detector set at $2\theta \approx 13°$. With the storage ring operating at 3 GeV and 80–30 mA, the data for each *P-T* point were acquired in 10 min in live-time mode.

Experimental Results

Typical energy-dispersive X-ray diffraction patterns obtained at various pressures and temperatures are shown in Figure 1. Below the α⇌ε transition pressure only two diffraction peaks, α-Fe(110) (peak No. 3) and MgO(200) (peak No. 2), are observed. Above the transition pressure two discrete peaks, ε-Fe(100) (peak No. 1) and ε-Fe(101) (peak No. 4), are observed. However, another peak, ε-Fe(002), overlaps with the α-Fe(110) peak (see peak No. 3) and is not discretely distinguishable. The overall precision of the molar volumes of MgO, α-Fe, and ε-Fe calculated from the above diffraction peaks is ±0.3%.

The most intense peak in ε-Fe phase is known to be (101), although this is not indicated in our 190 °C run (Figure 1). In general the quality of ε-Fe(101) peak is better than that of ε-Fe(100). Therefore we have used the intensity data for peaks No. 3 and No. 4 to evaluate the relative abundances of α and ε phases for each *P-T* data point. In the energy-dispersive method the integrated intensity of a diffraction peak at photon energy E (BURAS et al., 1977) is given by

$$I(E) = c \cdot \eta(E) \cdot I_0(E) \cdot m \cdot d^2 \cdot |F|^2 \cdot A(E, \theta) \cdot P(E) \sin\theta \quad (1)$$

where c is a constant; $\eta(E)$ is the detector efficiency; $I_0(E)$ is the intensity per unit energy range of the incident beam; m and $|F|^2$ are the multiplicity and the structure factors,

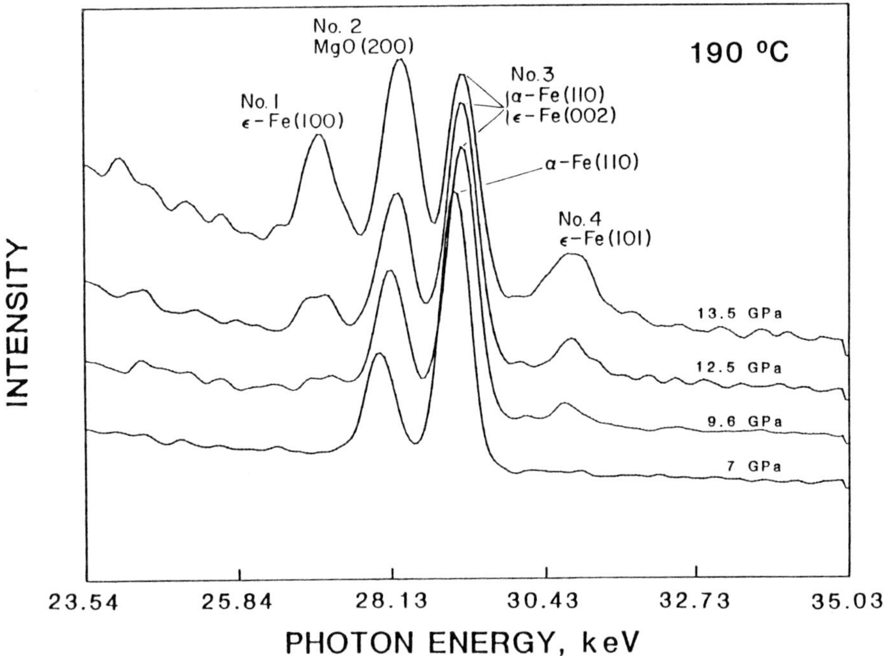

Figure 1. Typical energy-dispersive X-ray diffraction spectra for Fe as a function of pressure at 190 °C showing pure α-Fe phase at 7 GPa and a mixture of α-Fe and ε-Fe phases at higher pressures. Peak Nos. 1, 2, 3, and 4 represent, respectively, the ε-Fe(100), MgO(200), compounded {α-Fe(110)+ε-Fe(002)}, and ε-Fe(101) peaks.

respectively; A and P are the absorption and polarization factors, respectively; θ is the diffraction angle; and d is the interplanar spacing ($=6.199/E\sin\theta$). Since the diffraction peaks α-Fe(110), ε-Fe(002), and ε-Fe(101) are observed in a very small energy range, $\eta(E)$, $I_0(E)$, $P(E)$, and $A(E, \theta)$ are assumed to be the same for all of these reflections. Thus Equation (1) can be written as

$$I(E) = Cm\,|F|^2/E^2 = C{\cdot}K \quad (2)$$

where C is a new constant and $K=m|F|^2/E$. For calculating the intensity of any discrete i^{th} peak of a specific reflection in the spectrum, Equation (2) can be rewritten as

$$I_i = C{\cdot}K_j \quad (3)$$

Here the subscript $i=(1, 2, 3, 4)$ denotes the ith assigned peak and $j=$a, b, c represents the specific diffraction peak ε-Fe(101), ε-Fe(002), and α-Fe(110), respectively. If the fraction of ε-Fe is x, then the intensity of the ε-Fe(101) peak (peak No. 4) is

$$I_4 = CxK_a \quad (4)$$

and the intensity of the overlapping {ε-Fe(002)+α-Fe(110)} peak (peak No. 3) is

$$I_3 = C[xK_b + (1-x)K_c] \quad (5)$$

Taking the ratio of the Equations (4) and (5), we get

$$I_4/I_3 = xK_a/[xK_b + (1-x)K_c] \quad (6)$$

In our experiments the intensity ratio I_4/I_3 is directly measurable. Further, K_a, K_b and K_c can be calculated from Equation (2) since the multiplicity factor m, the structure factor $|F|^2$, and the photon energy E for each of the diffraction peaks are known. The relationship between I_4/I_3 and x can thus be evaluated (see Figure 2) and used for estimating the relative amounts of ε-Fe and α-Fe in the α-Fe⇌ε-Fe transition and for locating the "equilibrium" phase boundary.

Discussion

Results of the five high-pressure isothermal runs at 190, 270, 360, 450 and 540 °C are presented in Figure 3. The phases observed at each pressure during the increasing and decreasing pressure cycles are indicated. Because of the sluggish nature of the α⇌ε transition, significant hysteresis is observed; the pressure for the forward

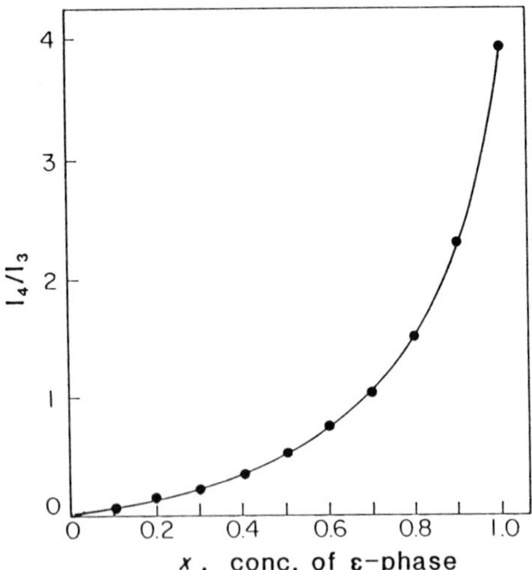

Figure 2. A calculated relationship between the intensity ratio (I_4/I_3) and the concentration of the ε-Fe phase (see Equation (5) in the text).

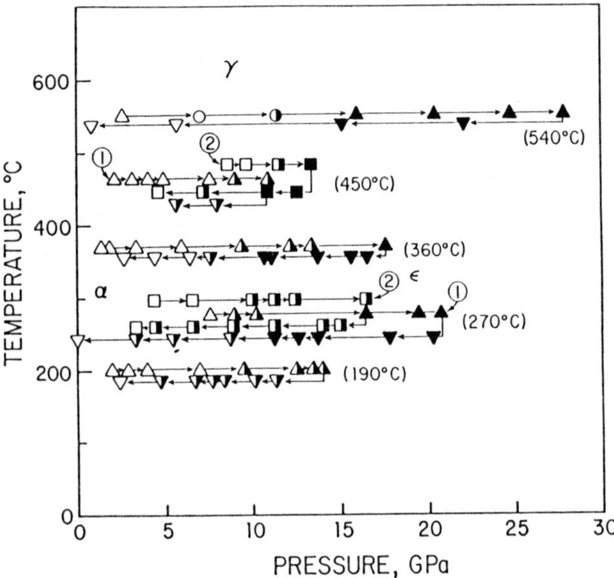

Figure 3. Isothermal runs for the α⇌ε transition at four temperatures (190, 270, 360 and 450 °C) and for the α⇌γ⇌ε transition at 540 °C. The triangles Δ and ▲ represent the single phases of α-Fe and ε-Fe, respectively, for the increasing pressure cycle; half-filled symbols indicate a mixture of α-Fe+ε-Fe; and the circles represent the γ-Fe. The inverted triangles represent the data for the decreasing pressure cycles and the arrows indicate the experimental sequence. In order to show two runs at both 270 °C and 450 °C, an offset in the vertical axis is thus introduced.

transition (α→ε) is higher than for the backward (ε→α) transition. Along each isotherm, four pressures are assigned to each of the four transition stages in a complete pressure cycle: 1) the starting of α→ε at pressure $P^{s}_{\alpha\to\varepsilon}$, 2) the completion of α→ε at pressure $P^{c}_{\alpha\to\varepsilon}$, 3) the starting of ε→α at pressure $P^{s}_{\varepsilon\to\alpha}$, and 4) the completion of ε→α at pressure $P^{c}_{\varepsilon\to\alpha}$, The four stages are characterized by the appearance of ε phase, disappearance of α phase, reappearance of α phase, and disappearance of ε phase, respectively. The equilibrium transition pressure (P_e) is assumed to be between stages (1) and (3), at a pressure which is an average of $P^{s}_{\alpha\to\varepsilon}$ and $P^{s}_{\varepsilon\to\alpha}$ (GILES et al., 1971).

If experimental data related to the appearance of the ε and α phases can be closely monitored and also if the data points along an isotherm are closely spaced, then $P^{s}_{\alpha\to\varepsilon}$ and $P^{s}_{\varepsilon\to\alpha}$ at that temperature will be tightly constrained. Furthermore if such data are available for several isotherms, the equilibrium boundary constrained by $P^{s}_{\alpha\to\varepsilon}$ and $P^{s}_{\varepsilon\to\alpha}$ can be reliably established. However, most of the previous in situ P-T studies on the α→ε transition in Fe, including one of our own (MANGHNANI et al., 1985), have not evaluated the equilibrium boundary in this manner, as no such complete sets of data were obtained for the forward α→ε and the backward ε→α transitions. The upper and lower bounds for $P^{s}_{\alpha\to\varepsilon}$ and $P^{s}_{\varepsilon\to\alpha}$ based on the present study are shown in Figure 4. Also shown are the bounds for $P^{s}_{\alpha\to\varepsilon}$ from our previous study (MANGHNANI et al., 1985). Another uncertainty arises from the overlapping of the α-Fe(110) and ε-Fe(002) peaks. If the conversion from α-Fe to ε-Fe is not completed, then $P^{s}_{\varepsilon\to\alpha}$ determined by the first appearance of the α-Fe phase during the decreasing pressure cycle is subject to uncertainty. For this reason, the bounds of $P^{s}_{\varepsilon\to\alpha}$ for 190, 270 (2nd run), and 450 °C (1st run) are not included in Figure 4. Note that the right-hand end of the ε-in bounds indicates the pressure (the first data point) at which the

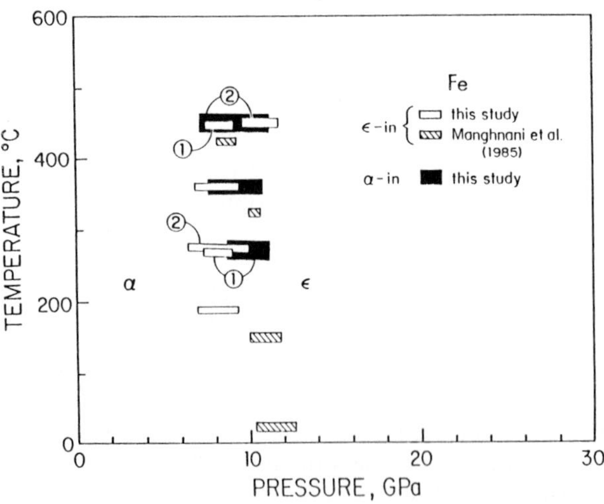

Figure 4. The pressure ranges for α⇌ε transition at various temperatures found in this and previous studies (MANGHNANI et al., 1985). Note that, because of the wide pressure ranges, the transition boundaries for ε-in and α-in cannot be well constrained (see text for discussion).

ε-phase was first observed during the increasing pressure cycle, and the left-hand end indicates the highest pressure at which *only* the α-phase was present. Thus $P^s_{\alpha\to\varepsilon}$ should lie between the two ends of the bounds. Parallel reasoning holds for the α-in bounds. As will be seen in our discussion, the 1st run at 270 °C and the 2nd run at 450 °C are abnormal runs and are therefore not considered in evaluating the $P^s_{\alpha\to\varepsilon}$ and $P^s_{\varepsilon\to\alpha}$. It is evident from Figure 4 that the bounds are somewhat too wide for an unambiguous determination of the equilibrium phase boundary.

In order to determine $P^s_{\alpha\to\varepsilon}$ and $P^s_{\varepsilon\to\alpha}$ more reliably, the relative abundance of the ε phase for each *P-T* data point along the four isotherms (190, 270, 360 and 450 °C) is determined from the curve in Figure 2 and plotted versus pressure as shown in Figures 5, 6, and 7. We note that in general, the relative abundance of the ε-phase is negligible in the initial stage (i.e., below the α→ε transition) but increases with increasing pressure above $P^s_{\alpha\to\varepsilon}$ revealing the progressive transformation of the α-phase into the ε-phase. Further, during all the decreasing pressure cycles, the initial amount of the ε-phase remains more or less the same, depending on how much of the α-phase has been converted into the ε-phase, and then decreases indicating the ε-phase is progressively converted back into the α-phase. The ε-phase amount becomes zero again when the conversion of ε→α is completed. Assuming that the amount of the ε-phase increases (or decreases) linearly with pressure during the phase transitions, we obtained the values of $P^s_{\alpha\to\varepsilon}$ and $P^s_{\varepsilon\to\alpha}$ (i.e., pressures for ε-in and α-in) by extrapolation of the growth and nongrowth lines as shown in Figures 5, 6, and 7; results for the four pressure stages along the four isotherms are listed in Table 2.

Some important features of phase transitions during each experimental run are shown in Figures 5, 6, and 7. Compared to the data at 190 and 360 °C (Figure 5), the data at 450 °C (Figure 7) are limited and display some scatter and therefore are less reliable for extrapolating transition pressures for ε-in and α-in. A comparison of Figures 5, 6, and 7 also shows that there is no uniformity in the growth patterns of the ε-phase with increasing pressure and of the α-phase with decreasing pressure. Also there is no clear evidence of any systematic relationship between the growth patterns and temperature and overpressure. The results of the 1st run at 270 °C (Figure 7) are abnormal in that the transition pressure for α-in is higher than that for ε-in, contrary to the normal hysteresis effect. Similar results obtained by HUANG et al. (this volume) have been attributed to a "stress mechanism" due to a unique crystallographic orientation. In the present study, the five isothermal runs were conducted in order of increasing temperatures at 190, 360 and 540 °C, and then decreasing to temperature to 270 °C (two runs) and finally at 450 °C (two runs). Whether decreasing the temperature from 540 to 270 °C could have introduced any such crystallographic orientation effect in the 1st run at 270 °C (Figure 7) is unclear and needs further investigation.

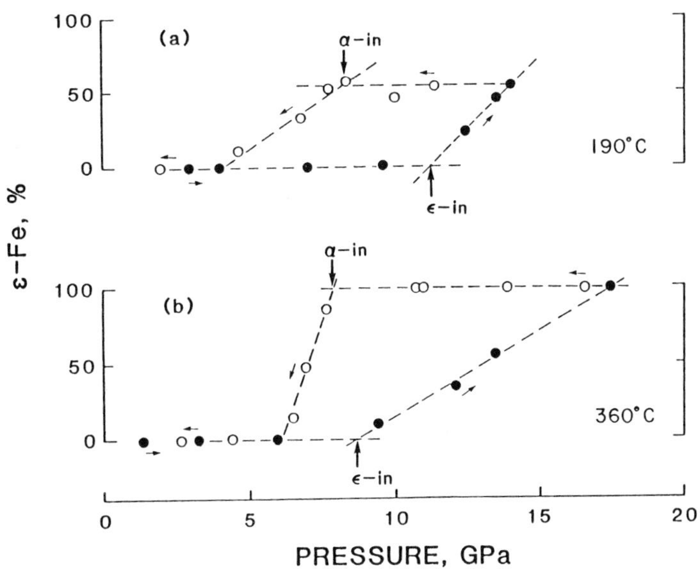

Figure 5. Percent ε-Fe as a function of pressure at (a) 190 °C and (b) at 360 °C. The transition pressure for ε-in and α-in are obtained at the intersection of the growth and the nongrowth lines, as denoted by the heavy arrows. The light arrows indicate the experimental sequence. Closed and open circles are the data points for increasing and decreasing pressures, respectively.

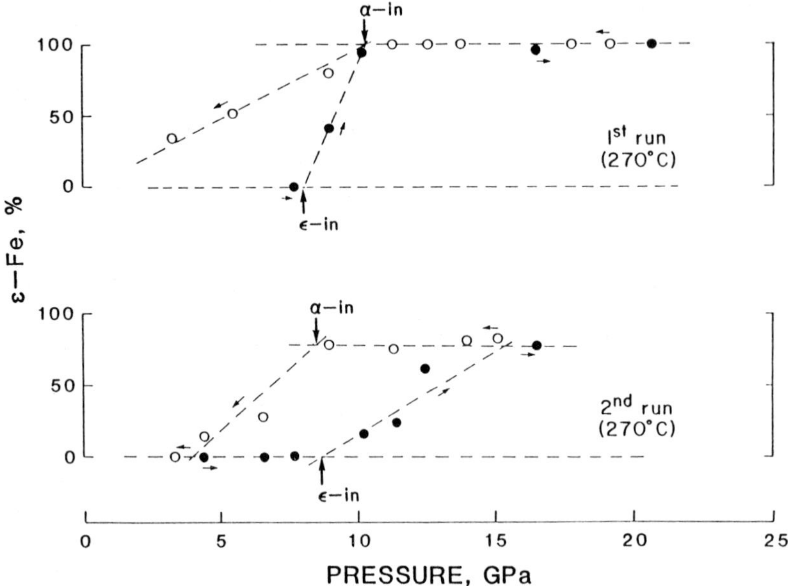

Figure 6. Percent ε-Fe as a function of pressure for two separate runs at 270 °C. The transition pressures for ε-in and α-in are obtained at the intersection of the growth and the nongrowth lines, as denoted by the heavy arrows. The light arrows indicate the experimental sequence. Closed and open circles are the data points for increasing and decreasing pressures, respectively.

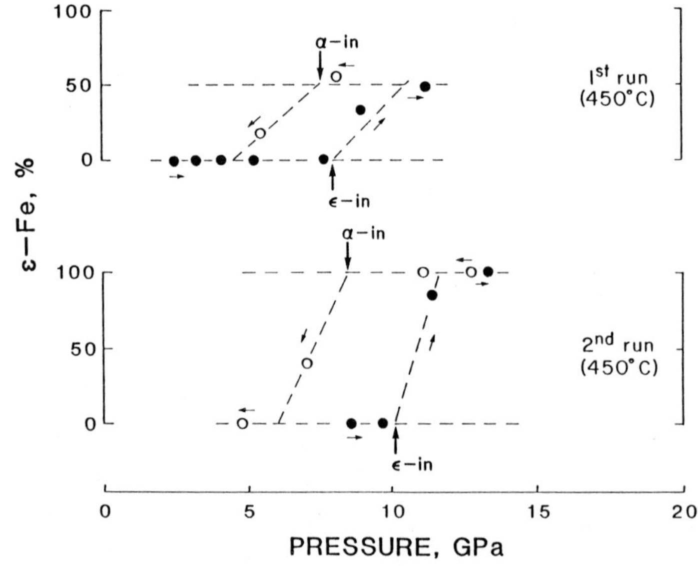

Figure 7. Percent ε-Fe as a function of pressure for two separate runs at 450 °C. The transition pressures for ε-in and α-in are obtained at the intersection of the growth and the nongrowth lines, as denoted by the heavy arrows. The light arrows indicate the experimental sequence. Closed and open circles are the data points for increasing and decreasing pressures, respectively.

Furthermore, for some unknown reason, the 2nd run at 450 °C shows abnormally high transition pressure for the ε-in as compared with the other runs in this study. The inconsistent results for the 1st run at 270 °C and the 2nd run at 450 °C have not been taken into account in evaluating the transition boundaries for ε-in, α-in, and P_e.

The transition boundaries thus obtained for ε-in, α-in, and P_e have negative slopes as shown in Figure 8, and are

TABLE 2. Pressure for Four Stages in the α-Fe⇌ε-Fe Transition at Four Temperatures

	Transition Pressure, GPa				
T, °C	$P^s_{\alpha\to\varepsilon}$ (ε-in)	$P^c_{\alpha\to\varepsilon}$ (α-out)	$P^s_{\varepsilon\to\alpha}$ (α-in)	$P^c_{\varepsilon\to\alpha}$ (ε-out)	P^*_e
190	9.0	—	8.2	4.0	8.60
360	8.5	17.5	8.0	6.0	8.25
270(a)	8.2	10.4	10.2	3.3	9.2
(b)	8.7	—	8.3	3.7	8.5
450(a)	8.2	—	7.5	4.7	7.85
(b)	10.2	11.7	8.5	6.0	9.35

(a) and (b) are the 1st and 2nd runs, respectively.
*P_e is the equilibrium pressure, assumed to be an average of the ε-in and α-in pressures.

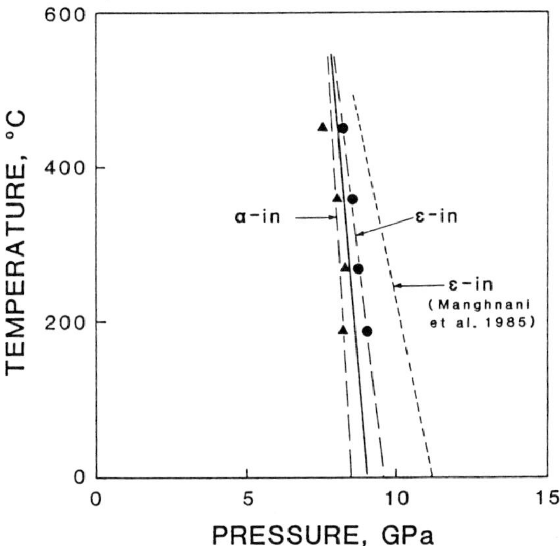

Figure 8. Phase boundaries for the α⇌ε transition in Fe, where the ε-in and α-in dashed lines represent the start of the α→ε and ε→α transitions. The solid line, the average of ε-in and α-in boundaries, is the tentative "equilibrium" phase boundary for α⇌ε transition. The solid circle and triangle points are the $P^s_{\alpha\to\varepsilon}$ and $P^s_{\varepsilon\to\alpha}$ values, respectively, as estimated from Figures 5, 6, and 7.

represented, respectively, by the following equations:

$$P(\text{GPa}) = 9.57 - 0.0032T(°\text{C}) \quad (7)$$

$$P(\text{GPa}) = 8.50 - 0.0016T(°\text{C}) \quad (8)$$

and

$$P(\text{GPa}) = 9.03 - 0.0023T(°\text{C}) \quad (9)$$

Also shown in Figure 8 is the ε-in line from our previous study (MANGHNANI et al., 1985). Our present results give consistently lower ε-in transition pressures than those of our previous study. This discrepancy is probably a result of two independent conditions. First, the pressure standard used in our previous study was platinum, and the diffraction peak Pt(200) overlapped somewhat with the ε-Fe(101) peak. The initial appearance of the ε-Fe(101) peak was thus masked and first appeared at somewhat higher pressures. Secondly, because of the overlapping of the Pt(200) and ε-Fe(101) peaks, it was difficult to evaluate the relative amounts of the α- and the ε-phases present in each run, and hence the pressure at which the ε-phase first appeared could only be estimated. The large window in the previous experimental data (Figure 4) inherently introduces a large uncertainty in such an estimation. In view of these facts, the transition pressures for ε-in estimated in the present study are considered more reliable than our previous values.

Comparison of α-Fe⇌ε-Fe Transition Studies

Results of various investigations of the α⇌ε transition in Fe are listed in Table 1. An overview of all the data shows that: 1) the values of dP/dT range from +0.002 to −0.006 GPa/deg; 2) the equilibrium phase boundary for the α⇌ε transition is not sensitive to temperature; and 3) except for the results of AKIMOTO et al. (this volume), all the experimental data have yielded a negative slope.

Figure 9 also shows a comparison among the three most recent determinations of this transition using synchrotron radiation (HUANG et al., this volume; AKIMOTO et al., this volume; this study). The main features that can be noted are: 1) In all three studies the hysteresis decreases with

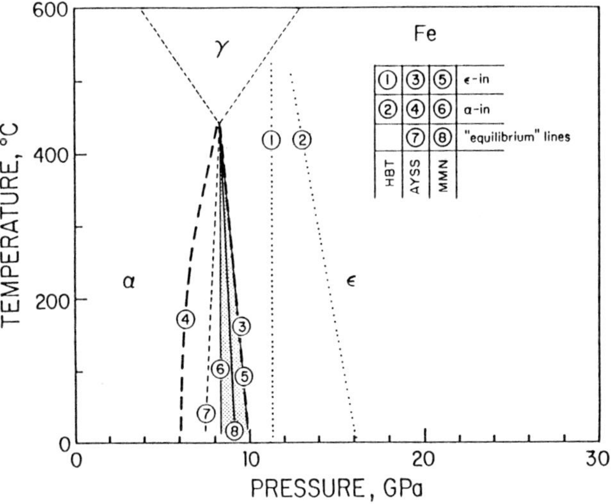

Figure 9. A comparison of the results from three most recent investigations of the α⇌ε transition in Fe: HBT (HUANG et al., 1987), AYSS (AKIMOTO et al., this volume), and MMN (stippled area, present study).

temperature; 2) As compared to the other studies, the hysteresis (i.e., $|P^s_{\alpha\to\varepsilon} - P^s_{\varepsilon\to\alpha}|$) observed in our study is smaller; 3) Our ε-in line is in excellent agreement with that of Akimoto et al.; however, their α-in lies in much lower pressure range; 4) Huang et al. data are in substantial discord with the two other studies in two ways: (a) the α-in and ε-in lines are reversed in sequences and (b) both fall on the higher pressure side.

In view of the sluggish nature of this transition, as indicated by the hysteresis in the data reported so far, including data from the three synchrotron studies, the value of the slope for this transition is still considered to be tentative. The large discrepancy among the transition pressure values, P_0, reported in the three synchrotron studies is probably due to the differences in samples used (i.e., variations in impurity levels, degree of preferred orientation, dislocations, and grain size), sample environments (e.g., pressure gradient), the pressure vessels employed (i.e., diamond-anvil cell or cubic-anvil press), and also the size of the incident radiation beam, which gives an average pressure over the nonhydrostatic area covered. It is obvious that more detailed experiments and interlaboratory study on the same sample are needed to establish the phase boundary of the $\alpha \rightleftharpoons \varepsilon$ transition in Fe.

Equation of State of α-Fe and ε-Fe

Figure 10 presents the *P-V* data for the two phases at several temperatures obtained from previous (MANGHNANI et al., 1985) and present studies. Our room temperature data are in good agreement with the results of MAO et al. (1967). Despite some scatter in the data, it is evident that the slopes of the isotherms for the ε-phase are gentler than those of the α-phase indicating a higher value of K_T for the ε-phase. However, quantitative evaluation of K_T at high pressure and temperature was not pursued because reliable values of $V(P, T)$ could not be extrapolated from our present *P-V-T* data.

Pressure dependence of thermal expansivity of α- and ε-phases. Using the smoothed data at 270 and 442 °C for the ε-phase and at 150, 330, and 435 °C for the α-phase as shown in Figure 10, we determined the pressure dependencies of the α_v (in deg$^{-1}\times 10^5$) for the two phases to be $\alpha_v=(5.53\pm0.04)-(0.070\pm0.006)P$ (GPa) for α-phase and $\alpha_v=(5.29\pm0.13)-(0.090\pm0.009)P$ (GPa) for the ε-phase (Figure 11). Thus $(\partial\alpha/\partial P)$ for the ε-phase is significantly lower, resulting in a smaller volumetric thermal expansion for the ε-phase than that of the α-phase at high pressure. (For example, α_v for α and ε phases at 10 GPa is 4.8×10^{-5} and 4.4×10^{-5} deg^{-1}, respectively).

Conclusions

1) The hysteresis in the α-ε transition (i.e., $|P^s_{\alpha\to\varepsilon} - P^s_{\varepsilon\to\alpha}|$ observed in this study is small compared to other studies, and it decreases with temperature.

2) The tentative slope of the α-ε "equilibrium" line, based on the complete observations of the forward and backward transitions, is negative and is expressed as: P (GPa)$=9.03-0.0023T$ (°C).

3) $(\partial\alpha/\partial P)_T$ is negative, as expected, for both phases; however, it is significantly lower for the ε-phase, resulting in a smaller α_v for the ε-phase at any given high pressure.

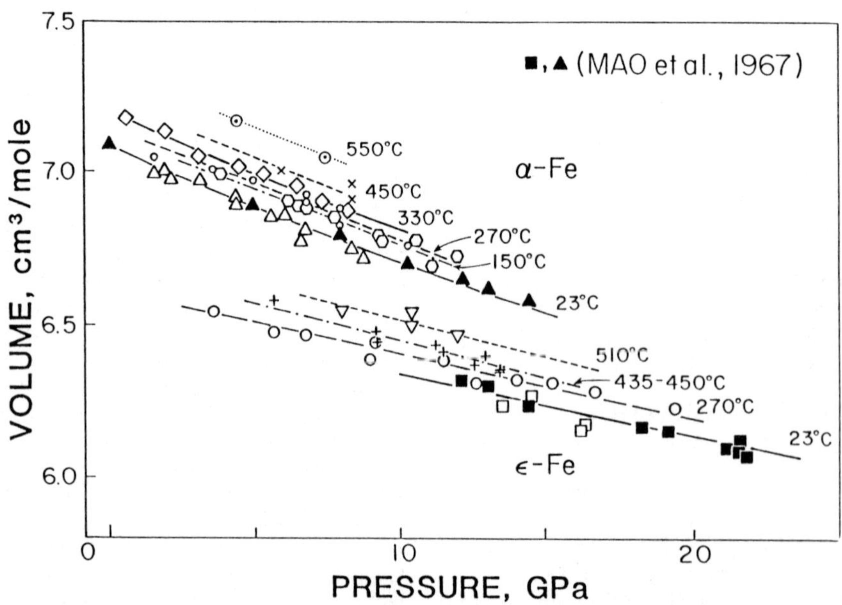

Figure 10. Pressure-volume isotherms for α-Fe and ε-Fe at various high temperatures.

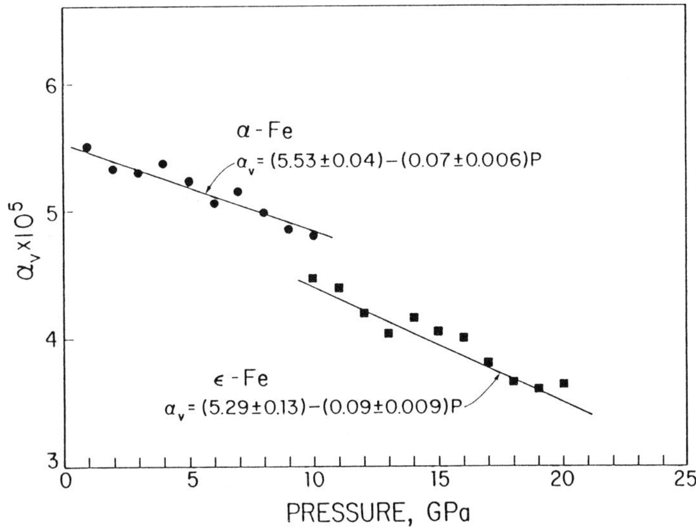

Figure 11. Volumetric thermal expansion of α-Fe and ε-Fe as a function of pressure.

Acknowledgments. This research was supported by the U.S. National Science Foundation Grant EAR84-18125. The authors are indebted to S. Usha Devi for helpful discussion and to R. Jeanloz, W. A. Bassett, and T. Yagi for their critical reviews and constructive suggestions. Hawaii Institute of Geophysics contribution No. 1889.

REFERENCES

AKIMOTO, S., T. SUZUKI, T. YAGI, and O. SHIMOMURA, Phase diagram of iron determined by high-pressure/temperature X-ray diffraction using synchrotron radiation, this volume.

BALCHAN, A. S., and H. G. DRICKAMER, High-pressure electrical resistance cell, and calibration points above 100 kilobars, *Rev. Sci. Instrum.*, *32*, 308–313, 1961.

BANCROFT, D., E. L. PETERSON, and S. MINSHALL, Polymorphism of iron at high pressures, *J. Appl. Phys.*, *27*, 291–298, 1956.

BROWN, J. M., and R. G. MCQUEEN, The equation of state for iron and the earth's core, in *High Pressure Research in Geophysics*, edited by S. Akimoto and M. H. Manghnani, pp. 611–623, Center for Academic Publications Japan, Tokyo, 1982.

BUNDY, F. P., Pressure-temperature phase diagram of iron to 200 kbar, 900 °C, *J. Appl. Phys.*, *36(2)*, 616–620, 1965.

BURAS, B., J. S. OLSEN, L. GERWARD, G. WILL, and E. HINZE, X-ray energy-dispersive diffractometry using synchrotron radiation, *J. Appl. Cryst.*, *10*, 431–438, 1977.

FUKIZAWA, T., Direct determination of phase equilibria of geophysically important materials under high pressure and high temperature using in-situ X-ray diffraction method, Ph.D. thesis, 125 pp., University of Tokyo, 1982.

GILES, P. M., M. H. LONGENBACH, and R. A. MANDER, High pressure α⇌ε Martensic transformation in iron, *J. Appl. Phys.*, *42*, 4290–4295, 1971.

HUANG, E., W. A. BASSETT, and P. TAO, Study of bcc-hcp iron phase transition by synchrotron radiation, this volume.

HUANG, E., W. A. BASSETT, and P. TAO, Pressure-temperature-volume relationship for hexagonal close packed iron determined by synchrotron radiation, *J. Geophys. Res.*, *92*, 8129–8135, 1987.

JAMIESON, J. C., J. N. FRITZ, and M. H. MANGHNANI, Pressure measurements at high temperature in X-ray diffraction studies: gold as a primary standard, in *High Pressure Research in Geophysics*, edited by S. Akimoto and M. H. Manghnani, pp. 27–48, Center for Academic Publications Japan, Tokyo, 1982.

JAMIESON, J. C., and A. W. LAWSON, X-ray diffraction studies in the 100 kbar range, *J. Appl. Phys.*, *33*, 776–780, 1962.

JEANLOZ, R., Properties of iron at high pressures and the state of the core, *J. Geophys. Res.*, *84*, 6059–6069, 1979.

JEPHCOAT, A. P., H. K. MAO, and P. M. BELL, Static compression of iron to 78 GPa with rare gas solids as pressure-transmitting media, *J. Geophys. Res.*, *91*, 4677–4684, 1986.

MANGHNANI, M. H., L. C. MING, J. BALOGH, S. B. QADRI, E. F. SKELTON, and D. SCHIFERL, Equation of state and phase transition studies under in situ high *P-T* conditions using synchrotron radiation, in *Solid State Physics under Pressure-Recent Advances with Anvil Devices,* edited by S. Minomura, pp. 343–350, KTK Scientific Publisher, Tokyo, 1985.

MAO, H. K., W. A. BASSETT, and T. TAKAHASHI, Effect of pressure on crystal structure and lattice parameters of iron up to 300 kbar, *J. Appl. Phys.*, *38*, 272–276, 1967.

MAO, H. K., and P. M. BELL, Equations of state of MgO and ε Fe under static pressure conditions, *J. Geophys. Res.*, *84*, 4533–4536, 1979.

MING, L. C., M. H. MANGHNANI, J. BALOGH, S. B. QADRI, E. F. SKELTON, and J. C. JAMIESON, Gold as a reliable internal pressure calibrant at high temperature, *J. Appl. Phys.*, *54*, 4390–4397, 1983.

MING, L. C., M. H. MANGHNANI, J. BALOGH, S. B. QADRI, E. F. SKELTON, A. W. WEBB, and J. C. JAMIESON, Static *P-T-V* measurements on MgO: comparison with shock wave data, in *Shock Waves in Condensed Matter-1983*, edited by J. R. Asay, R. A. Graham, and G. K. Straub, pp. 57–60, Elsevier Science Publishers, B.V., 1984.

MING, L. C., M. H. MANGHNANI, and J. BALOGH, Resistive heating in the diamond-anvil cell under vacuum conditions, this volume.

TAKAHASHI, T., and W. A. BASSETT, High pressure polymorph of iron, *Science*, *145*, 483–486, 1964.

ZOU, G., P. M. BELL, and H. K. MAO, Application of the solid-helium pressure medium in a study of the α-ε Fe transition under hydrostatic pressure, *Year Book Carnegie Inst. Washington*, *80*, 272–274, 1981.

STUDY OF BCC-HCP IRON PHASE TRANSITION BY SYNCHROTRON RADIATION

E. HUANG, W. A. BASSETT, and P. TAO

Department of Geological Sciences, Cornell University
Ithaca, New York, 14853, USA

Abstract. Experiments on the bcc-hcp phase transition in iron were carried out in a high-temperature, high-pressure diamond-anvil cell using synchrotron radiation. Four isothermal runs at room temperature, 150 °C, 300 °C, and 450 °C were accomplished using gold as a pressure calibrant. The onset of the phase transition with increasing pressure was observed to fall in the pressure range of 11.0–11.7 (±0.5) GPa and over the temperature range of 25–450 °C. Hence an essentially vertical phase boundary for the onset of transition (hcp-in) was obtained. A two-phase zone was found to exist over a wide pressure range in each isothermal run. Beyond the transition pressure, the extent of the phase transition (measured by the ratio of the two phases) increased with further application of the load. The extent to which bcc iron converted to its high pressure phase was not found to change significantly with time at 18.0 GPa and room temperature. This observation, therefore, is consistent with a stress-dependent martensitic transformation. At 25 °C, 150 °C, and 300 °C, the first appearance of the hcp phase occurred at lower pressure than the reappearance of the bcc phase (bcc-in) with decreasing pressure. The Clapeyron boundary of the bcc-hcp phase transition is determined by averaging the hcp-in and bcc-in pressures at various temperatures. Accordingly, the initial appearance of the hcp phase occurs at pressures below the Clapeyron boundary, and the initial appearance of the bcc phase with decreasing pressure occurs at pressures above the Clapeyron boundary, suggesting that the phase transition is strongly controlled by the deviatoric stress. The Clapeyron boundary has a slope (dT/dP) of −283 °C/GPa with a transition pressure of 13.5 (±1.0) GPa at room temperature.

Introduction

Elemental iron has been studied as part of a wide variety of disciplines. It is of particular interest in high-pressure, high-temperature experimental geophysics because of its importance as a major element in the interior of the earth (LIU, 1975; JEANLOZ, 1979) and its simple crystal structure. At ambient temperature and pressure, iron possesses a body-centered cubic (bcc) structure, sometimes termed the alpha phase. Bancroft et al. first reported the transformation of iron at high pressure on the basis of shock-wave evidence. This high-pressure phase was interpreted by BANCROFT et al. (1956) to be the gamma phase of iron, sometimes termed the face-centered cubic (fcc) phase which is known to be stable at high temperatures. However, the phase was soon reinterpreted to be a high-pressure polymorph of iron by BRIDGMAN (1956). This high-pressure polymorph of iron was later confirmed and determined to be a hexagonal close-packed (hcp) structure by JAMIESON and LAWSON (1962).

The bcc-hcp iron phase transition has been investigated with the use of various methods in addition to shock experiments (LOREE et al., 1966), including electrical resistance (BALCHAN and DRICKAMER, 1961), X-ray diffraction (TAKAHASHI and BASSETT, 1964), neutron diffraction (SHIL'SHTEIN et al., 1983), and direct optical observations (BASSETT et al., 1967). In the X-ray diffraction experiments, the bcc-hcp phase transition of iron was studied using a diamond cell held mostly at room temperature and a conventional X-ray source.

Previous studies have reported controversial results on the transition pressure (for discussion, see GILES et al., 1971; and DECKER et al., 1972), although there is general agreement on the sluggishness of the phase transition at room temperature. Another consideration in studying phase transition is the nature of the applied stress (ZOU et al., 1981).

The pressure-temperature phase boundary of the bcc-hcp phase transition was determined by BUNDY (1965) using the electrical resistance method. This phase boundary was later questioned by GILES et al. (1971) on the basis of their X-ray results. However, the phase boundary has never been determined with the X-ray diffraction method using conventional X-ray sources, due to the difficulty in maintaining high temperature in a high pressure cell for the prolonged exposure times that are needed. With the introduction of synchrotron radiation, an energy dispersive X-ray diffraction pattern could be obtained within a very short time (BURAS et al., 1976). This has made it feasible to undertake high temperature experiments on a very small amount of sample in a diamond cell (BASSETT, 1980). Recently, synchrotron radiation has proven to be a powerful tool in studying the kinetics of phase transitions at simultaneously high temperature and high pressure (SKELTON et al., 1982; HUANG and BASSETT, 1986).

In this paper, we describe the application of synchrotron radiation to the study of the phase transition of iron up to 450 °C and 25.0 GPa. We first report our experimental results on the bcc-hcp phase transition, and then discuss the kinetics and nature of this phase transition. Finally, we propose the phase boundary based on the kinetics and mechanism of the transition.

High-Pressure Research in Mineral Physics, edited by M. H. Manghnani and Y. Syono, pp. 165–172.

Experimental Method

The experiment was carried out at the Wilson Laboratory using the facility of the Cornell High Energy Synchrotron Source (CHESS) (BATTERMAN and ASHCROFT, 1979). We used a long, uniaxial type of diamond cell with resistance heaters surrounding the diamond supports. The instrumentation used in the experiment is similar to that used by FURNISH and BASSETT (1983) and HUANG and BASSETT (1986) in studying the mechanism of the olivine-spinel phase transition and in mapping the Fe_3O_4 phase diagram, respectively.

The samples used in this experiment were mixtures of iron and gold powders. The iron powder was of reagent grade (Alfa Products, 400020) and had a purity of 99.999% according to manufacturer's analysis. Each grain of the iron powder was polycrystalline, and ~50 μm across. Gold served as pressure calibrant throughout the runs (JAMIESON et al., 1982). The gold powder used was an Alfa Products (00767) reagent grade chemical which had a purity of 99.95% and an average grain size of 2 μm. The lattice parameters of the iron and gold were determined to be 2.866±0.002 and 4.079±0.002 Å, respectively, using the Debye-Scherrer powder camera method. The iron and gold powders (3:1 in volume) were gently ground by mortar and pestle to ensure a homogeneous mixture.

The experiment was carried out in the B cave at CHESS using white radiation. An X-ray beam of approximately 100 μm diameter was centered on the diamond-anvil faces. The size of the beam was limited by means of a pinhole in a 0.43-cm-thick piece of lead which was inserted in the piston.

A selsyn motor which could be cranked from outside the X-ray cave was used to adjust pressure. The selsyn motor, through a series of gears, turned the driver of the diamond cell and thus generated pressure on the sample. Several washer-type springs were placed between the driver and piston in order to lower the torque and to smooth the applied forces on the sample.

Temperature was controlled by a variable transformer outside the cave. A pair of heaters were made by winding chromel wires (0.27 mm in diameter) around the diamond seats. A thermocouple made of very fine alumel and chromel wires (0.03 mm in diameter) was attached to one of the diamonds for temperature measurement. The heaters and thermocouple are described in more detail in FURNISH and BASSETT (1983) and HUANG and BASSETT (1986).

Four isothermal runs at room temperature, 150 °C, 300 °C, and 450 °C, were carried out. The sample was replaced after each run. In all but the 450 °C run, pressure was applied after the temperature was brought to the desired value. In the 450 °C run, some pressure was applied before the temperature was raised in order to avoid serious oxidation. Once the desired temperature was reached, pressure was gradually increased through the range of interest. After each adjustment of pressure and temperature, the diffraction signals were collected by an intrinsic Ge detector at a fixed Bragg angle, 2θ. The signals were then processed and displayed on the screen of the multichannel analyzer (Tracor Northern, TN-1710) as an energy dispersive spectrum. The spectra were stored on a floppy disk for later analysis. The acquisition time for each spectrum was approximately 1 min or occasionally longer for better statistics.

The application of load was well controlled, and the onset of the phase transition was easily recognized and recorded. The shift of the gold diffraction peaks allowed rapid and rough estimation of pressure. In each isothermal run load was increased far beyond the transition pressure in order to determine the extent of the phase transition with pressure. During unloading, which proceeded more rapidly than loading pressure was reduced to 0.1 MPa and spectra were collected. The sample was quenched to room temperature once the pressure was removed. The Bragg angle for each run was determined from the diffraction patterns of the sample taken at ambient conditions. The lattice parameters of both iron and gold were measured at ambient conditions before and after each run to determine if any reaction between them had taken place.

In order to test the effect of time on the bcc-hcp iron phase transition, an additional run was carried out at room temperature. This was accomplished by first bringing the sample into the pressure range where both the bcc and hcp iron were present. The sample was then left sitting in the diamond cell for over 78 h. Diffraction patterns were taken both before and after this period of time for comparison.

During all of the experiments, temperature was controlled within ±2 °C of the desired value. Pressure was calculated by the molar volume of gold at various temperatures using the tabulated equation of state of JAMIESON et al. (1982). The error in temperature was less than ±5 °C and the uncertainty in pressure reading was estimated to be ±5%.

Results

During each run, diffraction patterns were collected at a carefully determined Bragg angle, 2θ, in the range of 10.5–10.7°. The lattice parameters of iron and gold were found, within experimental error, to be the same before and after the 25 °C, 150 °C, and 300 °C runs. Thus the geometry of the experiment, as well as the purities of iron and gold, remained essentially constant during each of these three runs. A small increase in the lattice parameter of gold was found after the 450 °C run indicating that some alloying between iron and gold may have taken place.

In all runs the diffraction peaks of iron and gold were seen clearly in the energy dispersive diffraction patterns (Figure 1a). Iron showed reflections of the (110), (200), and (211) planes, while gold showed the (111) and (311) peaks. The (220) and (222) peaks of gold were never observed in the spectra because they overlap the (200) and (211) peaks of bcc iron, respectively. Although the (200) peak of gold and the (110) peak of bcc iron have d-spacings that are close, they can be resolved when a Bragg angle, 2θ, of 11° is used. However, the (200) peak of gold was never observed because of the small quantity of gold used in the sample and possibly because of preferred orientation (BAUBLITZ et al., 1981). Overlapping of gold and iron peaks does not influence the results because the amount of gold used is small and because the affected peaks are not used as an index for the phase transformation. Regardless of the effect of overlapping, the relative intensity of the bcc-iron diffraction peaks (110):(200):(211) was approximately 100:10:5 at room temperature (Figure 1a). The ratio was slightly different in each isothermal run but it remained constant as the pressure was increased (Figure 1b, c). Because the (311) reflection of gold is weak at high pressure and the other diffraction peaks are obscured, only the (111) reflection of gold was used to determine the pressure.

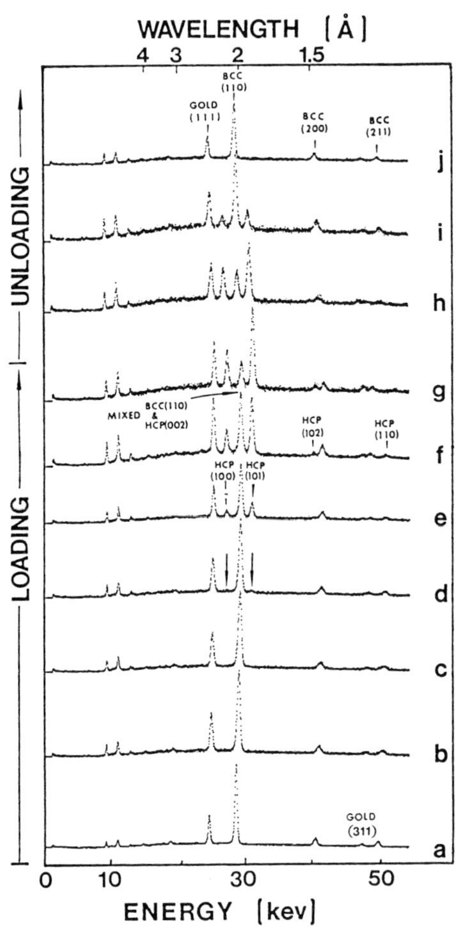

Figure 1. Series of selected spectra showing the bcc-hcp phase transition of iron.

According to the index scheme of TAKAHASHI and BASSETT (1964), the hcp phase of iron had five recognizable diffraction peaks (the (100), (002), (101), (102) and (110) reflections) above the transition pressure (Figure 1e, f). However, the (002) of the hcp iron overlaps the (110) of the bcc iron. Hence the first appearance of the hcp iron with increasing pressure was recognized by the simultaneous appearance of the hcp(100) and hcp(101) on the flanks of the overlapping bcc(110) and hcp(002) reflections (Figure 1d).

As the pressure was further increased, the intensity of the hcp iron peaks increased while that of the bcc iron decreased (Figure 1e, f). The complete disappearance of the bcc phase was not easy to recognize because all the diffraction peaks of the bcc phase suffered interference from peaks of the other phases. The peak height ratio of the mixed phase reflections (bcc(110) and hcp(002)) and the most intense reflection of the hcp phase, (hcp(101)) decreased with further application of the load (Figure 1f). The ratio showed no further change above a certain pressure (Figure 1g); this is taken as an indication that the bcc phase was completely eliminated. At high temperatures complete elimination of the bcc phase was not achieved during loading. However, the bcc phase was eliminated as soon as the load was slightly decreased. The bcc-out pressures (Table 1) were determined to be the pressures beyond which relative intensity of the mixed peak and hcp(101) reached a minimum or remained a constant. During unloading, the hcp phase persisted until the reappearance of the bcc phase was recognized by a sudden increase in the peak height ratio of the mixed phase and hcp(101) (Figure 1h, i). The hcp phase was completely eliminated at much lower pressure (Figure 1j).

The transition behavior at high temperatures was similar to that at room temperature. However, there are differences in the pressures for the first appearance of the

TABLE 1. Transition Pressures of the bcc-hcp Phase Transition of Iron at Various Temperatures in GPa

Temperature	hcp-in	bcc-out	bcc-in	hcp-out
25 °C	10.4–11.5	21.2–22.0	16.6–15.7	4.8–2.5
150 °C	10.5–11.8	22.0–22.3	15.9–14.1	4.1–1.9
300 °C	11.4–11.9	16.4–17.1	14.3–13.8	3.7–2.7
450 °C	11.0–11.4	19.0–20.5	9.1– 8.4	4.3–2.2

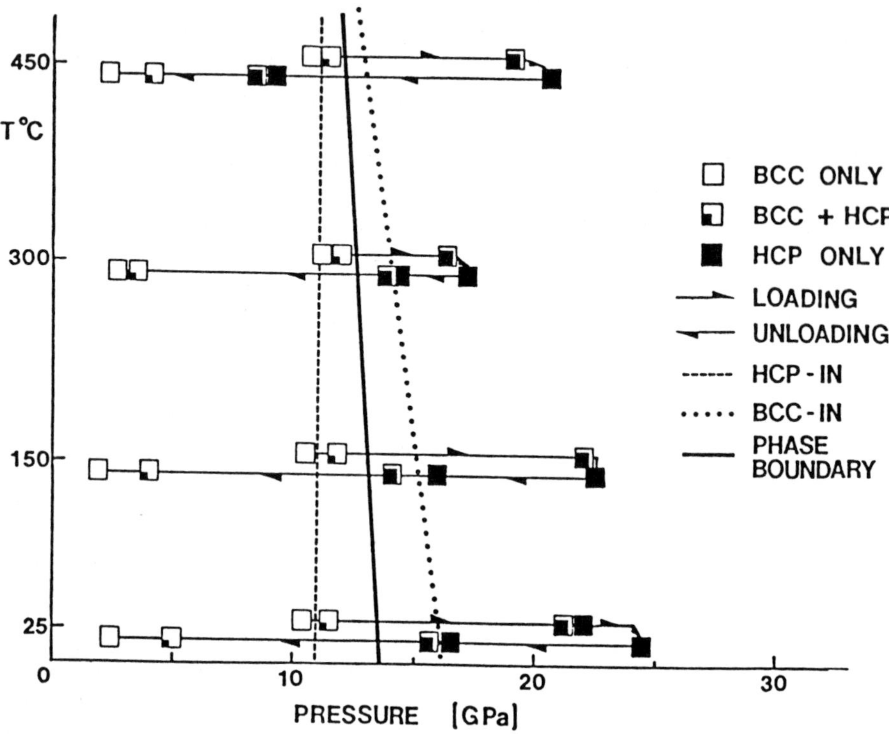

Figure 2. Pressure-temperature phase diagram showing sequences of the bcc-hcp phase transition of iron and the boundaries for hcp-in (dashed line), bcc-in (dotted line), and the proposed Clapeyron boundary (solid line).

high-pressure phase (hcp-in) and elimination of the low-pressure phase (bcc-out) with increasing load, and for the reappearance of the low-pressure phase (bcc-in) and elimination of the high-pressure phase (hcp-out) with decreasing load (Table 1). The pressures for the first appearance of hcp iron at various temperatures were observed to cluster in a narrow range of 11.0–11.7 GPa (Table 1). The slope of the hcp-in boundary, therefore, was determined to be nearly vertical (Figure 2). The hcp-out boundary is also nearly vertical based on the present results. The transition pressures for bcc-out and bcc-in are somewhat irregular. The bcc-in slope is determined to be −131 °C/GPa (Figure 2) based on the results of the 25 °C, 150 °C, and 300 °C runs. The results at 450 °C might be inaccurate due to the reaction between gold and iron. Hence they are not used. It is important to note that the bcc-in boundary is at lower pressures than the hcp-in boundary. The observation is vital in inferring the mechanism of the transition (see discussion). The determination of the bcc-out boundary is not attempted because the elimination of the bcc phase might be affected by the nature of the applied stress and therefore is not easily estimated.

Figure 3 summarizes the time-dependent experimental results on the bcc-hcp phase transition in iron. A sample

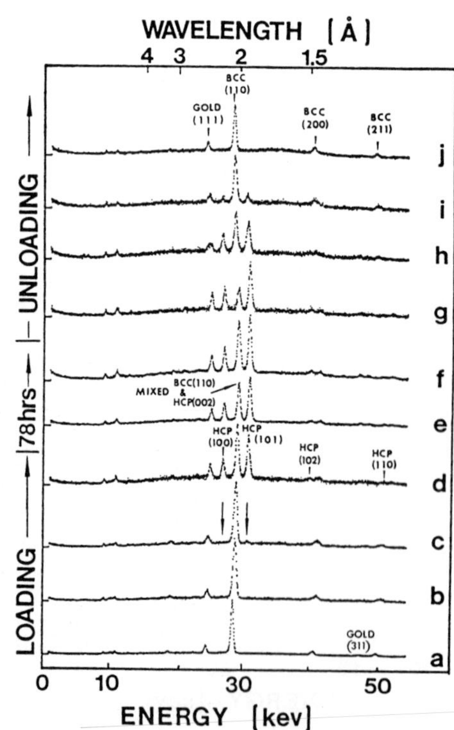

Figure 3. Series of selected spectra showing results of the time-dependent experiment.

was first loaded in the diamond cell at room temperature (Figure 3a) and the pressure was gradually increased (Figure 3b). The hcp phase appeared between 9.0 and 12.7 GPa (Figure 3b, c). As pressure was further increased, the minor peaks of hcp phase became clearer (e.g., at 16.8 GPa, Figure 3d). The pressure was raised to 18.0 GPa where the hcp/bcc ratio was approximately 1:0.9 (Figure 3e). After 78 h at 18.0 GPa, the relative intensities of the two phases were found to be unchanged (Figure 3f). However, as soon as the load was slightly decreased (Figure 3g), the hcp/bcc ratio was significantly increased and the pressure (18.5 GPa) was increased slightly. Further unloading caused an increase in the bcc phase and the elimination of the hcp phase as described above (Figure 3h, i, j).

Discussion

Transition Pressures for the bcc-hcp Transition in Iron

Synchrotron radiation, due to its high intensity, has made it possible to rapidly acquire diffraction patterns. Because of this feature, it has the ability to detect the onset of the phase transition more precisely than the conventional X-ray diffraction methods (TAKAHASHI and BASSETT, 1964; GILES et al., 1971) and the neutron diffraction method (SHIL'SHTEIN et al., 1983).

Using synchrotron radiation, we determined the first appearance of the hcp phase to occur between 10.4 and 11.5 GPa (most likely 11.0 GPa) at room temperature.

The phase transition has also been studied by various other techniques and researchers. The results of other studies are listed in Table 2. Our result is in good agreement with most of those obtained by the electrical resistance method. It is also consistent with that of MANGHNANI et al. (1985) using synchrotron radiation diffraction method. However, some discrepancies do occur.

The transition pressure proposed by VERESHCHAGIN et al. (1969), based on a sharp increase in the electrical resistance, is much higher than other determinations based on similar techniques. We do not have an explanation for this. The higher transition pressures determined by shock loading (SINGH, 1985) might be attributed to the dynamic nature of the technique (i.e., overshooting is required to reveal the phase transition). The discrepancy between our result and the X-ray diffraction result of ZOU et al. (1981) is attributed to difference in the pressure medium.

Using a method similar to our method, MANGHNANI et al. (this volume) determined the transition pressure to be 8.4 GPa at 270 °C; this is also inconsistent with our present result. AKIMOTO et al. (this volume), on the basis of synchrotron radiation analysis, reported a transition pressure that is lower than any other reported value. The reason for these discrepancies is not yet clear. Some discrepancies may be due to the effect of sample orientation and deviatoric stress in the pressure cell.

Hysteresis of the Phase Transition

The coexistence of both bcc and hcp phases over a wide pressure range was observed at room temperature (MAO et al., 1967; GILES et al., 1971). In our present experiment, this two-phase zone was found not only at room temperature but also at high temperatures (Figure 2). The two-phase zone occurred during both loading and unloading cycles, and hence revealed the hysteresis of the transition. MANGHNANI et al. (this volume) and AKIMOTO et al. (this volume) also found similar hysteresis in their synchrotron high pressure, high temperature experiments.

Due to the presence of the hysteresis, the bcc-hcp phase transition of iron can be divided into four stages: the appearance of the high-pressure phase (hcp-in) and the elimination of the low-pressure phase (bcc-out), both during loading; and the appearance of the low-pressure

TABLE 2. List of the Transition Pressures for bcc-hcp Phase Transition of Iron at Room Temperature

Researcher		Trans. Pressure (GPa)	Method
BANCROFT et al.	(1956)	13.0	Shock
LOREE et al.	(1966)	12.7±.1	Shock
STARK and JURA	(1964)	11.8±.6	Elect. resist.
VERESHCHAGIN et al.	(1969)	15.2±.2	Elect. resist.
DRICKAMER	(1970)	11.0–11.5	Elect. resist.
SHIL'SHTEIN et al.	(1983)	15.0	Neutron diff.
TAKAHASHI and BASSETT	(1964)	13.0	X-ray diff.
GILES et al.	(1971)	10.7±.8	X-ray diff.
ZOU et al.	(1981)	15.3	X-ray diff.
MANGHNANI et al.	(1985)	11.0	Synchrotron
AKIMOTO et al.	(this volume)	8.6–9.8	Synchrotron
HUANG et al.	(this paper)	10.4–11.5	Synchrotron

phase (bcc-in) and disappearance of the high-pressure phase (hcp-out), both during unloading (Table 1). These four stages are shown in the pressure-temperature diagram (Figure 2).

GILES et al. (1971) claimed that the bcc-hcp phase transition in iron is martensitic and "abaric"; this indicates that the extent of the reaction is determined by the amount of pressure in excess of the transition pressure and is relatively independent of time. In this experiment, we arrive at a similar conclusion.

The fraction of transition (x) at various pressures in both loading and unloading processes was determined in each isothermal run (Figure 4). In all runs a hysteresis loop is obviously present. The pressures for hcp-in, bcc-out, bcc-in, and hcp-out are clearly seen from each loop. Complete elimination of the bcc phase with increasing load was accomplished only in the room temperature run. Hence the room temperature results are used for quantitative determination of the fraction of transition with pressure.

The variation of the fraction of transition with pressure during the loading and unloading processes is shown in Figure 5. An empirical formula may thus be fit to these data for estimating the proportion of the transition as

$$x = A \exp[-b(P - P_0)^n]$$

where P is the applied pressure, P_0 is the hcp-in pressure, and A, b, and n, are constants. In the loading process, P_0 is 11.0 GPa, A, b, and n are 0.014, 1.3, and 0.5, respectively. A similar formula is used for the unloading process using 4.8 GPa for P_0. The curves for these formulae are plotted in Figure 5. It is noted that in the above equation P is not purely hydrostatic. In addition, the deviatoric stress plays an important role and therefore must be considered (for discussion, see next section). TAKAHASHI and BASSETT (1964) reported the elimination of bcc phase at 19.2 GPa, a value which is slightly lower than our present value of 21.7 GPa (Figure 5). The ratio of hcp(002) to hcp(101) at 19.2 GPa is 80:100 judging from X-ray data (TAKAHASHI and BASSETT, 1964). It is likely that the relatively high intensity of hcp(002) could be partly attributed to its overlapping with bcc(110) which still remained at that pressure. GILES et al. (1971) reported the proportion of the bcc iron to be approximately 40% at 16.3 GPa; this is in agreement with our present observation (Figure 5).

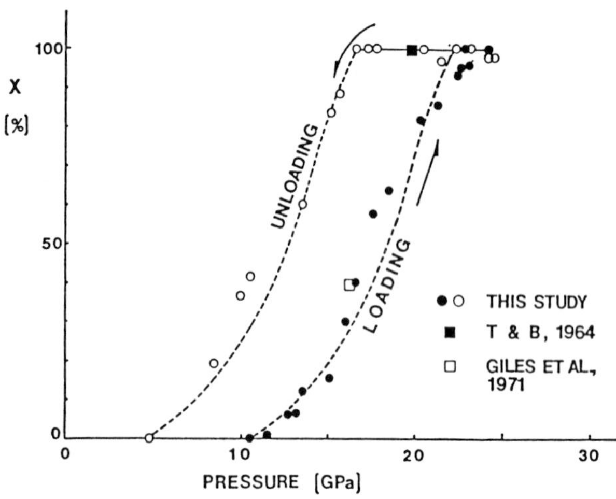

Figure 5. Plot of the fraction of transformation of iron (X) vs. pressure at room temperature. The dashed curves are based on the equation given in the text. The results of TAKAHASHI and BASSETT (1964) (solid square) and GILES et al. (1971) (open square) are also plotted for comparison.

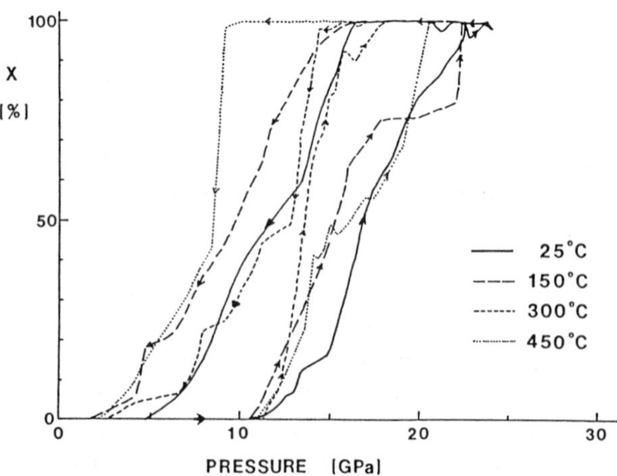

Figure 4. Plot of the fraction of transformation of iron (X) vs. pressure at various temperatures. Arrows show the loading (→, and ↑) and unloading (←, and ↓) processes of the run.

Mechanism of the bcc-hcp Phase Transition

The coexistence of two phases over a wide pressure range is generally an indication that the transition is sluggish. In other words, there exists a kinetics problem that prohibits the reaction from going to completion at the transition point. However, the existence of the two-phase zone might be due to effects of the pressure gradient and the sample orientation in the diamond cell.

For our runs the size of the synchrotron beam was only 100 μm in diameter where it hit the sample. The difference in pressure throughout the incident area was only about 10% according to the empirical formula proposed by LIPPINCOTT and DUECKER (1964). Hence pressure gradient is probably not an important factor in the existence of the two-phase zone. This argument is supported by the experiment of GILES et al. (1971) which found the two-phase zone to be present even when the incident area was greatly reduced.

Since hcp phase bears a crystallographic relationship with the bcc phase (MAO et al., 1967), the orientation of the sample might be an important factor in controlling the phase transition in iron. MAO et al. (1967) proposed that bcc iron may transform to hcp phase by a simple displacive mechanism during which the two structures maintain an epitaxial relationship. In a uniaxial compression, some crystallites of iron may be more advantageously oriented for the deviatoric stress to induce the phase transition. These grains undergo the transition at a lower pressure than those having less advantageous orientations. Thus the range of pressure over which crystallites transform is caused by the randomness of orientation in the polycrystalline iron.

In a purely hydrostatic medium the transition goes to completion over a small pressure range (ZOU et al., 1981). Although the transition pressure may be higher than that observed in the nonhydrostatic case, the two-phase zone is essentially eliminated, at least during the loading process. In a hydrostatic environment orientation does not affect the phase transition because the sample is compressed isotropically. The phase transition may still have a kinetic problem since the reaction takes place at higher pressure.

The growth of the high pressure phase is independent of time judging from the experiment at room temperature (Figure 3). The effect of time also appears to be insignificant at high temperatures. Hence the mechanism of the bcc-hcp phase transition in iron may be different from other reactions in which the extent of phase transition varies with time.

Introduction of shear stress when the load is slightly decreased has been observed in this experiment since the pressure, as well as the amount of the phase transition, is increased. We have observed optically that the area of the hcp phase increases as the load on the diamond anvils is slightly decreased. The pressure enhancement in the cell is attributed to extrusion at the edge of the sample as the load is slightly decreased. This additional stress was large enough to convert most, if not all, of the remaining bcc iron in the central part of the cell to the hcp phase. This elimination of the bcc phase due to enhanced stress is also an indication that the phase transition is controlled primarily by stress instead of time. Using the neutron diffraction method to study the bcc-hcp phase transition in iron, SHIL'SHTEIN et al. (1983) determined that the amount of phase transition increased with time over a period of months. This observation is not necessarily in contradiction with our present result since the effect of stress is still the dominant factor in controlling the extent of the phase transition. Moreover, the phase transition observed during the prolonged experimental period might have been affected by variation of stress configuration in the pressure cell. Variation in stress configuration could result the observed change in the relative proportion of the two phases.

The fact that bcc-hcp phase transition is not found to be a time-dependent but rather a stress-dependent reaction is indicative of a martensitic type transformation which is in agreement with the conclusion of GILES et al. (1971).

The hysteresis loops appear at all temperatures indicating that the mechanism for the bcc-hcp iron phase transition is similar up to 450 °C. In Figure 2, the hysteresis appears somewhat narrower at higher temperatures. This indicates that the phase transition may be affected, to some extent, by temperature. This effect, however, is not necessarily due to kinetics but may be related to the mechanism of the phase transition.

Phase Boundary of the bcc-hcp Transition in Iron

At room temperature, the first appearance of the hcp phase was determined to be between 10.4 and 11.5 GPa in the present study. The points of first appearance of the hcp phase at other temperatures (Table 1) can be connected by a line that has a nearly vertical slope (Figure 2). This line, however, does not represent the Clapeyron boundary.

On the basis of our present results, the phase transition of bcc-hcp in iron is determined to be a stress-dependent, martensitic transformation. The Clapeyron boundary of the phase transition is tentatively determined to be the average of the pressures for hcp-in and bcc-in at various temperatures as proposed by GILES et al. (1971). The transition pressures for the bcc-hcp phase transition in iron, as determined by this method, are:

13.5 (±1.0) GPa at 25 °C,
13.1 (±1.0) GPa at 150 °C,
12.6 (±1.0) GPa at 300 °C.

The transition pressure at 450 °C is not considered to be reliable due to the possible alloying of approximately 7% of iron in gold. Therefore it is not listed. The phase boundary has a slope (dT/dP) of −283 °C/GPa (Figure 2).

In most phase transitions the appearance of the high pressure phase occurs at a higher pressure with loading than the reappearance of the low pressure phase with unloading. Our results at 25 °C, 150 °C, and 300 °C, however, show the opposite. In each case the hcp-in occurs at a lower pressure than the bcc-in. Accordingly, the Clapeyron boundary, taken as the average of the hcp-in and bcc-in, lies at pressures above the hcp-in and below the bcc-in boundary (Figure 2). That is, the hcp phase must appear metastably in the stability field of bcc phase and/or the first appearance of the bcc phase with decreasing pressure must occur metastably in the stability field of hcp phase.

The explanation for this behavior must lie in the

mechanism of transformation described above. The crystallites of bcc iron which are advantageously oriented with respect to the deviatoric stress can apparently undergo the transformation to hcp phase even when the pressure is lower than the Clapeyron boundary. Likewise, crystallites of hcp iron that are advantageously oriented may undergo transformation to bcc phase at pressure higher than Clapeyron boundary during decrease of pressure.

Conclusions

We conclude that the bcc-hcp phase transition of iron is more stress-dependent than time-dependent. The extent of the phase transition is controlled by the amount of pressure. The phase transition is also orientation-dependent, and the transition pressure may be controlled primarily by the deviatoric stress.

The Clapeyron boundary is determined to have a slope of −283 °C/GPa with a transition pressure of 13.5 (±1.0)GPa at room temperature.

Acknowledgments. The authors would like to express their appreciation to the CHESS staff and Jens Otto for their help in the B cave experiment. This work was supported by NSF grants EAR-7804388 and EAR-8116083.

References

Akimoto, S., T. Suzuki, T. Yagi, and O. Shimomura, Phase diagram of iron determined by high-pressure/temperature X-ray diffraction using synchrotron radiation, this volume.

Balchan, A. S., and H. G. Drickamer, High pressure electrical resistance cell, and calibration points above 100 kilobars, *Rev. Sci. Instrum., 32*, 308–313, 1961.

Bancroft, D., E. L. Peterson, and S. Minshall, Polymorphism of iron at high pressure, *J. Appl. Phys., 27*, 291–298, 1956.

Bassett, W. A., Synchrotron radiation, an intense X-ray source for high pressure diffraction studies, *Phys. Earth Planet. Inter., 23*, 337–340, 1980.

Bassett, W. A., T. Takahashi, and P. W. Stook, X-ray diffraction and optical observations on crystalline solids up to 300 kbar, *Rev. Sci. Instrum., 38*, 37–42, 1967.

Batterman, B. W., and N. W. Ashcroft, CHESS: the new synchrotron radiation facility at Cornell, *Science, 206*, 157–161, 1979.

Baublitz, M. A., Jr., V. Arnold, and A. Ruoff, Energy dispersive X-ray diffraction from high pressure polycrystalline specimens using synchrotron radiation, *Rev. Sci. Instrum., 52*, 1616–1624, 1981.

Bridgman, P. W., High pressure polymorphism of iron, *J. Appl. Phys., 27*, 659, 1956.

Bundy, F. P., Pressure-temperature phase diagram of iron to 200 kbar, 900 °C, *J. Appl. Phys., 36*, 616–620, 1965.

Buras, B., J. Staun Olsen, and L. Gerward, X-ray energy-dispersive powder diffractometry using synchrotron radiation, *Nucl. Instrum. Methods, 135*, 193–195, 1976.

Decker, D. L., W. A. Bassett, L. Merrill, H. T. Hall, and J. D. Barnett, High pressure calibration, a critical review, *J. Phys. Chem. Ref. Data, 1*, 773–836, 1972.

Drickamer, H. G., Revised calibration for high pressure electrical resistance cell, *Rev. Sci. Instrum., 41*, 1667–1668, 1970.

Furnish, M. D., and W. A. Bassett, Investigation of mechanism of the olivine-spinel transition in fayalite by synchrotron radiation, *J. Geophys. Res., 88*, 10333–10341, 1983.

Giles, P. M., M. H. Longenbach, and A. R. Marder, High pressure alpha-epsilon martensitic transformation in iron, *J. Appl. Phys., 42*, 4290–4295, 1971.

Huang, E., and W. A. Bassett, Rapid determination of Fe_3O_4 phase diagram by synchrotron radiation, *J. Geophys. Res., 91*, 4697–4103, 1986.

Jamieson, J. C., and A. W. Lawson, X-ray diffraction studies in the 100 kilobar pressure range, *J. Appl. Phys., 33*, 776–780, 1962.

Jamieson, J. C., J. N. Fritz, and M. H. Manghnani, Pressure measurement at high pressure in X-ray diffraction studies: gold as a primary standard, in *High-Pressure Research in Geophysics*, edited by S. Akimoto and M. H. Manghnani, pp. 27–48, Center for Academic Publications Japan, Tokyo, 1982.

Jeanloz, R., Properties of iron at high pressures and the state of the core, *J. Geophys. Res., 84*, 6059–6069, 1979.

Lippincott, E. R., and H. C. Duecker, Pressure distribution measurements in fixed-anvil high-pressure cells, *Science, 146*, 1119–1121, 1964.

Liu, L., On the (γ, ε, l) triple point of iron and the Earth's core, *Geophys. J. R. Astron. Soc., 43*, 697–705, 1975.

Loree, T. R., C. M. Fowler, E. G. Zukas, and F. S. Minshall, Dynamic polymorphism of some binary iron alloys, *J. Appl. Phys., 37*, 1918–1927, 1966.

Mao, H. K., W. A. Bassett, and T. Takahashi, Effect of pressure on crystal structure and lattice parameters of iron up to 300 kbar, *J. Appl. Phys., 38*, 272–276, 1967.

Manghnani, M. H., L. C. Ming, J. Balogh, S. B. Qadri, E. F. Skelton, and D. Schiferl, Equation of state and phase transition studies under in situ high P-T conditions using synchrotron radiation, in *Solid State Physics Under Pressure: Recent Advance with Anvil Devices*, edited by S. Minomura, pp. 335–341, KTK Scientific Publishers, Tokyo, 1985.

Manghnani, M. H., L. C. Ming, and N. Nakagiri, Investigation of the α-Fe⇌ε-Fe phase transition by synchrotron radiation, this volume.

Shil'shtein, S. S., V. P. Glazkov, I. N. Makarenko, V. A. Somenkov, and S. M. Stishov, Neutron diffraction experiments carried out using diamond anvils at ultrahigh pressures, *Sov. Phys. Solid State, 25*, 1907–1909, 1983.

Singh, A. K., The kinetics of some pressure-induced transformations, in *Materials Science Forum*, Vol. 3, edited by G. E. Murch, pp. 291–306, Trans Tech Publications Ltd., Switzerland, 1985.

Skelton, E. F., J. Kirkland, and S. B. Qadri, Energy-dispersive measurements of diffracted synchrotron radiation as a function of pressure: applications to phase transitions in KCl and KI, *J. Appl. Cryst., 15*, 82–88, 1982.

Stark, W., and G. Jura, Some high-pressure calibration points above 80 kbars, *ASME Publication No. 64-WA/Pt-28*, 1964.

Takahashi, T., and W. A. Bassett, High pressure polymorph of iron, *Science, 145*, 483–486, 1964.

Vereshchagin, A. L. F., A. A. Semerchan, N. N. Kuzin, and Y. S. Sadkov, On the real value of the pressure of the polymorphic transformation in iron, *Soviet Phys.*, Eng. transl., *14*, 340–342, 1969.

Zou, G., P. M. Bell, and H. K. Mao, Application of the solid-helium pressure medium in a study of the α-ε Fe transition under hydrostatic pressure, *Carnegie Inst. Washington Yearbook, 80*, 272–274, 1981.

INTERNALLY-HEATED DIAMOND-ANVIL CELL: PHASE DIAGRAM AND *P-V-T* OF IRON

R. BOEHLER, M. NICOL, and M. L. JOHNSON

Department of Chemistry and Biochemistry, University of California Los Angeles, California 90024, USA

Abstract. In situ pressure, temperature, and resistance measurements of resistively heated iron wires in gasketed diamond-anvil cells are used to locate solid-solid and solid-liquid phase transitions with high accuracy. Pressures are determined by using the ruby pressure scale and temperatures are measured by fitting a blackbody function to the spectrum of the hot wire taken with an optical multichannel analyzer. In this manner the α-γ transition line has been measured to 29 GPa. With a synchrotron X-ray source and resistively heated wires the powder diffraction patterns of α-, γ-, and ε-Fe have been obtained at pressures between 7 and 16 GPa and temperatures as high as 1000 K. The average volume thermal expansivity of α-Fe between 300 and 815 K at 9.5 GPa has been found to be $14\pm4\times10^{-6}$ K^{-1} or about one-third of the value at atmospheric pressure. Preliminary data for γ-Fe indicate that the molar volume decreases by 6% along the α-γ transition line between atmospheric pressure and 9.5 GPa.

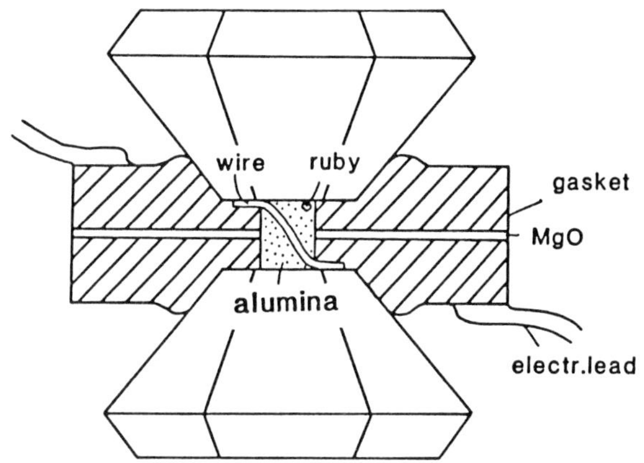

Figure 1. A cross-sectional view of the sample assembly in the diamond-anvil cell.

Introduction

This report describes a new technique of internal heating in diamond cells that allows in situ measurements of phase transitions, pressures, temperatures, and densities. The method adopts two previously published techniques form independent measurements of electrical resistivity using a sandwiched-gasket arrangement (BLOCK et al., 1977) and measurements of melting of a resistively-heated iron wire between two ungasketed diamonds (LIU and BASSETT, 1975). Pressures in the latter experiment were obtained by measuring the lattice constants of iron at room temperature by X-ray diffraction and then calculating the pressure from a previously measured equation of state for iron. This method yields large uncertainties in the pressure because of the small compressibility of iron. Temperatures were measured with an optical pyrometer. The emissivities of the iron phases had to be calculated by fitting low-pressure melting data to results obtained from piston-cylinder work.

Experiment

The high pressure cell is shown in Figure 1. Two 301 stainless steel disks are electrically insulated by an alumina-based cement of about 0.05 mm thickness. A fine iron wire of 0.025 mm diameter is mechanically connected to the two halves of the gasket, which act as electrical leads. In most experiments alumina is used as a pressure transmitter. The hot part of the electrically-heated wire is located in the center of the cell, thermally insulated from the diamonds. Thus because of the low power requirements stable temperature conditions can be maintained.

Ruby chips were located near the cooled diamond anvil and pressures were determined from the shift of the ruby luminescence spectrum using the conversion factor 0.36 nm/GPa. Pressures were measured both with the wire at room temperature and during heating. No significant change in pressure as indicated by the ruby spectra as observed.

Optical emission spectra from the hot Fe wires were collected with a 1024-channel Princeton Instruments IRY/1024 diode-array detector attached to a 1/4 meter SPEX monochromator with a 300 line/mm diffraction grating. Various spectral regions 240-nm-wide were sampled with the 680-to-930-nm region giving the best results (Diode response diminishes at longer wavelengths; intensities are much lower at shorter wavelengths.). The detector was calibrated with a 2800 K blackbody source (Optronic Laboratories Model 16). Corrected spectra were then fed into a locally-produced least-squares blackbody program in order to determine the corresponding temperatures. Typical errors in temperatures were ±20 K at 1000 K and ±8 K at 2300 K. This method of temperature measurement neither provides nor requires

High-Pressure Research in Mineral Physics, edited by M. H. Manghnani and Y. Syono, pp. 173–176.

quantitative information on the emissivity of the sample as the diamond cell is not calibrated and whole spectral regions, rather than single point intensities, are acquired. A schematic of the optical arrangement is shown in Figure 2.

Diffraction patterns were obtained with the triple-axis diffractometer at HASYLAB, Hamburg, West Germany (BESSON et al., 1985). The incident beam of 0.745 Å photons was collimated with Ta slits which were narrowed until no scattering was detected from an empty 100 μm diameter hole in a stainless steel gasket. Scattered photons were detected with a position sensitive counter at a distance of 1 m from the sample giving an angular resolution of 5×10^{-5} rad. Typical counting times were 30-to-60 minutes.

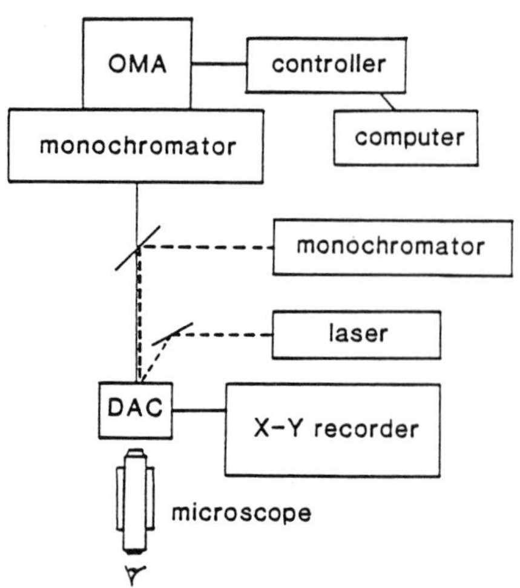

Figure 2. Schematic diagram of the optical arrangement.

Results

Phase Diagram of Fe

Though experiments on the phase diagram of Fe are still in progress, preliminary results on the α-γ and γ-ε transitions are reported here. The phase diagram of Fe (based on work by STRONG et al., 1973, BUNDY, 1965; and LIU and BASSETT, 1975) is shown in Figure 3. A wire of pure iron was resistively heated at constant load to either the α-γ or γ-ε transition, which is accompanied by a discontinuous change in resistance. Temperature and pressure are measured at the onset of these transitions. The hystereses of the transitions, typically less than 10 K, are measured by reversing the temperature cycles and repeating the measurements at the onset of the reverse transitions. The average of the two measurements was taken as the transition pressure.

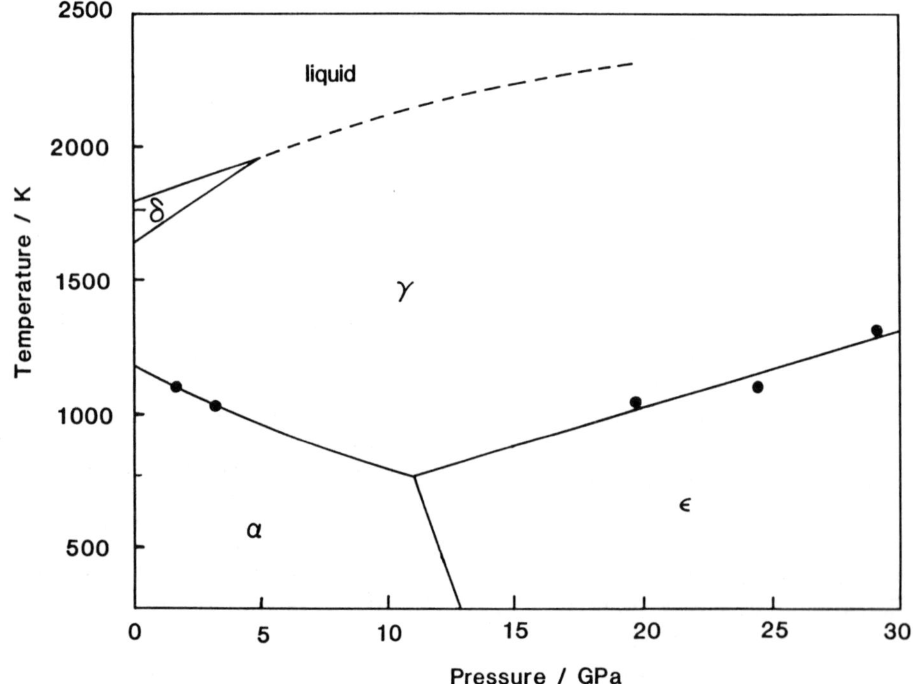

Figure 3. Phase diagram of iron. The solid melting line and the γ-δ transitions are from STRONG et al. (1973) The solid α-γ, γ-ε, and α-ε lines are from BUNDY (1965). The dashed curves are from LIU and BASSETT (1975). The circles represent present data.

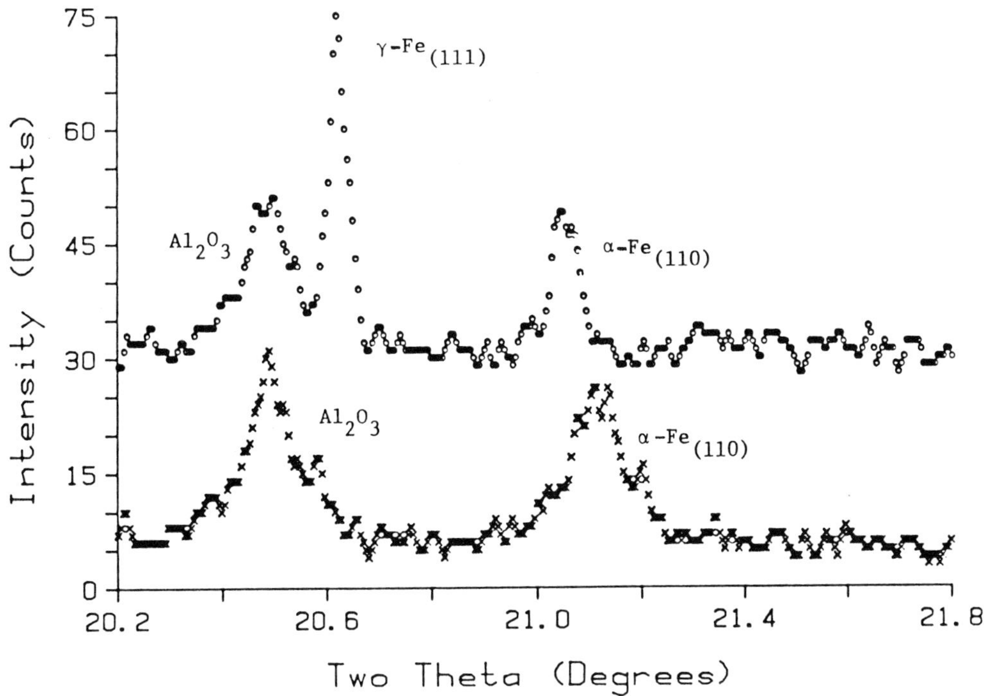

Figure 4. Diffraction patterns for: (upper trace), α-Fe near 300 K and 9.5 GPa; (lower trace), coexisting α- and γ-Fe at 815 K and 9.5 GPa. The peak near 20.49° is due to the alumina that surrounds the iron wire. The upper data set is shifted by 25 counts.

The measured temperatures of the α-γ transitions were compared with those from previous studies and were found to be in excellent agreement with results obtained in piston-cylinder apparati. The γ-ε transition was measured to 29 GPa and, using a α-γ-ε triple point at 11.0 GPa and 773 K, the slope of the transition line was found to be 24 K/GPa. This is in good agreement with LIU and BASSETT (1975).

P-V-T of Fe

Examples of the diffraction patterns obtained at HASYLAB are shown in Figure 4. The patterns are for α-Fe at room temperature and 9.5 GPa and the mixture of α-Fe and γ-Fe when the sample was at the α-γ transition at 815 K. This γ-Fe feature provides the first measurement, 6.90 cm^3, of the molar volume of γ-Fe at high pressure, which is 94% of the molar volume of the α-γ transition line near 1 atmosphere and 1200 K (LYMAN, 1961).

From the variation of the position of the α-Fe feature at 9.5 GPa with heating power between room temperature and the α-γ transition, the average volume thermal expansivity of α-Fe was determined to be $14 \pm 4 \times 10^{-6}$ K^{-1}. This value is between one-third and one-half of the value at atmospheric pressure (WEAST, 1984). YAGI (1978) has shown that the pressure dependences of the thermal expansivities of several alkali halides can be represented by:

$$\alpha/\alpha_0 = (V/V_0)^{\delta_0}$$

The values of δ_0, approximately 2 for halides with NaCl-like structures and 7 for those with CsCl-like structures, clearly appear to be sensitive to structure. The preliminary result for Fe reported here indicates that, if the same functional form is appropriate, the value of δ_0 would be much larger than for the halides; that is between 11 and 17.

Acknowledgments. Work at HASYLAB would not have been possible without valuable contributions by J. M. Besson, J.-P. Itie, C. Kunz, G. Materlik, and M. Nielsen and assistance provided by J. Gonzales, F. Grey, S. Johnson, P. Stanitzeck, Y. Talmi, C. S. Zha, and the staff of HASYLAB. We are pleased to acknowledge that financial support and equipment for these experiments was derived from NSF Grants DMR 83-18812 (MFN) and EAR 85-07755 (RB), NASA Grant NAGW-104 (MLJ), and grants from the LLNL-IGPP and the UCLA Research Committee.

REFERENCES

BESSON, J. M., R. BOEHLER, J. GONZALEZ, J. P. ITIE, S. JOHNSON, M. NICOL, F. GREY, and M. NIELSON, Equation of state of iron at high pressures and temperatures, *HASYLAB Jahresbericht 1985*, 217–218, 1985.

BLOCK, S., R. A. FORMAN, and G. J. PIERMARINI, Pressure and electrical resistance measurements in the diamond cell, in *High Pressure Research, Applications in Geophysics*, edited by M. H. Manghnani, S. Akimoto, p. 503, Academic Press, New York, 1977.

BUNDY, F. P., Pressure-temperature phase diagram of iron to 200 kbar, 900 °C, *J. Appl. Phys. 36*, 616, 1965.

LIU, L. G., and W. A. BASSETT, The melting of iron up to 200 kbar, *J. Geophys. Res., 80*, 3777, 1975.

LYMAN, T., ed., *Metals Handbook*, 18th Ed., v. 1, p. 1211, American Society for Metals, Metals Park, Ohio, 1961.

STRONG, H. M., R. E. TUFT, and R. E. HANNEMAN, The iron fusion curve and α-γ-ε triple point, *Met. Trans., 4*, 2657, 1973.

WEAST, R. C., ed., *CRC Handbook of Chemistry and Physics, 65th Ed.*, p. D-188, CRC Press, Boca Raton, Florida, 1984.

YAGI, T., Experimental determination of thermal expansivity of several alkali halides at high pressures, *J. Phys. Chem. Solids, 39*, 563–571, 1978.

THE KINETICS OF THE PRESSURE INDUCED OLIVINE-SPINEL PHASE TRANSITION Mg_2GeO_4

G. WILL and J. LAUTERJUNG

Mineralogical Institute, University Bonn, Bonn, West Germany

Abstract. The kinetics of phase transitions in solids were studied with energy dispersive diffraction techniques. Complete diffraction spectra were measured in rapid sequence every 45 s. Also, by using the window method, the intensity variation of a narrow pre-selected energy window, containing one reflection, was passed along the channels in a multichannel analyzer over time. The multichannel analyzer was thus operated as a single-channel analyzer. The minimal recording time was 0.1–0.5 s.

Using the full X-ray diffraction pattern method, the kinetics of the phase transformation in Mg_2GeO_4, which is isostructural to the mantle-relevant mineral forsterite Mg_2SiO_4, was investigated in the experimental range 0–20 kbar and 800–1200 °C. Phase transition was found to be accompanied by a density increase of 8.76%. Kinetic behavior of the olivine spinel phase transformation was investigated in seven transformation cycles with different P-T conditions. A total of 109 spectra were recorded and evaluated. The gradual growth of the spinel phase with time could be analyzed by fitting the data to the Avrami equation. Several full spectra (i.e., those recorded at every time step) were analyzed by profile fitting, thereby yielding integrated intensities. These intensity values were consequently used in structure factor and least-squares calculations to determine the crystal structure, particularly the cation occupancy. This procedure allowed a detailed analysis of the behavior of the cations Ge and Mg as they migrate into the oxygen frame. The transformation (i.e., the crystallization of the spinel phase) began with a rearrangement of the anion lattice from a distorted hcp arrangement to a cubic, close-packed arrangement. After a considerable delay of about 15 min, the octahedral and tetrahedral voids were filled; this could be observed experimentally through the measurement of reflections with contributions from Mg and Ge in the A- and B-sites.

Introduction

It has been standard practice for phase diagrams and phase transitions to be determined in static high-pressure experiments, with the samples held for many hours at elevated conditions. In these experiments the reaction products are quenched from the high-pressure/high-temperature conditions and analyzed. The more direct approach of in situ X-ray diffraction allows one to directly study the high-pressure phases and phase transformations. Despite this intrinsic advantage, in situ X-ray diffraction has not been applied as widely as one would expect because the experiments take many hours or days with conventional sealed X-ray tubes, and even alternatively with rotating anode generators. In general, since only peak positions can be evaluated, only changes of the unit cell with pressure can be safely assessed. A very few crystal structures have been refined from high-pressure data collected from single-crystal measurements at room temperature and at pressures below 100 kbar. No crystal structure of a high-pressure phase has been reported in situ after a phase transformation has taken place, probably because such phase transformation would destroy the crystal. This type of information can only be determined from powder diffraction data.

The availability of synchrotron radiation sources now makes it possible for scientists to perform such studies on a broader basis. However it is not yet routine practice to study crystal structure under in situ high-pressure conditions. With synchrotron radiation, such studies are feasible (WILL, 1981; WILL et al., 1983) and will certainly become more common in the future.

By investigating the kinetics of phase transformations, it is possible to add another dimension to the investigation of phase transformation time. Since the transformation zones in the earth are not sharp and may extend over many kilometers and since the transformations often proceed slowly (e.g., subduction mechanisms with velocities of about <8 cm/year), it is of great interest to study how transitions take place and what intermediate crystal structures or phases are formed.

Olivines are considered to be the major constituents of the upper mantle and therefore have been much studied. A phase transition from the olivine- to spinel-type structure in $(Mg, Fe)_2SiO_4$ is widely accepted today as an explanation for the observed seismic discontinuity at 420 km (RINGWOOD, 1979). The phase diagram of the $(Mg, Fe)_2SiO_4$ olivine system has been studied by RINGWOOD and MAJOR (1970), RINGWOOD (1979), and AKIMOTO et al. (1976), among others. The mechanism of the transition from the olivine to the spinel structure, however, is not fully understood. It is difficult to follow the phase transition experimentally since the pressures and temperatures needed for the phase transition in the Mg/Fe olivines are still not readily accessible in our laboratory for in situ X-ray diffraction. We have therefore selected the structural analog Mg_2GeO_4, instead of $(Mg, Fe)_2SiO_4$.

Experimental

To reliably follow the phase transformations over time, several requirements must be met: 1) experimental methods must be developed to measure and record

High-Pressure Research in Mineral Physics, edited by M. H. Manghnani and Y. Syono, pp. 177–186.

diffraction diagrams within short time intervals; 2) the data evaluation must be modernized to use profile analysis, and programs to calculate and refine crystal structures from powder data (texture and low counting statistics now pose serious problems); and 3) high-pressure cells with larger sample volumes than are commonly available in the widely used diamond-anvil cells must be developed. This last requirement, unfortunately, limits the pressure-temperature range accessible today.

Our experiments were conducted in the HASYLAB (Hamburg Synchrotron Laboratory) at the DESY (Deutsches Elektronen Synchrotron) in Hamburg, West Germany, using storage ring X-rays from DORIS. The operating conditions of 3.7 GeV and 50 mA gave X-rays up to about 80 keV. Data were collected in the energy-dispersive mode with a Ge-intrinsic solid state detector (WILL, 1981). Data were recorded on a multichannel analyzer (MCA) and after each run, transferred to a PDP 11/33 computer and permanently stored on floppy discs for later evaluation on a host computer.

Methods and Techniques

Storage ring X-rays, generally called synchrotron radiation, are characterized by high energies and by intensities several orders of magnitude higher than X-rays from a conventional X-ray tube. Therefore these X-rays make it possible to directly study time-dependent phenomena by X-ray diffraction. Reasonably good diffraction spectra can be obtained in one second or less, or within several minutes if necessary, for limited structural analysis. Depending on the transformation mechanism and speed, we have developed two methods to study phase transitions with time: a) the full pattern method, and b) the window method.

In present studies conducted as a function of time there are limitations in data handling, especially in the transfer of data from multichannel analyzer to a computer. Data transfer in our equipment takes about 25 s. If we allow a minimum time of 20 s for collecting data for a full diffraction diagram, then the total time is about 45 s. This longer time, however, has the advantage that we have recorded and stored a complete diffraction diagram at each time step. We call this procedure the "full pattern method"; that is, a complete diffraction diagram is recorded, read into the computer, and the next measurement is started.

There are some extremely fast time-dependent phenomena (such as nucleation growth, recrystallization, and some phase transformations, as in KCl) which require a time resolution of one second or less. It is not feasible to follow such fast changes because of the limitations in the data storage capacity and in the data analysis. In such cases it is sufficient to look at a small segment of the diagram (i.e., a "window") preferably containing just one diffraction peak, and then to follow the development of this "window" with time. We call this method the "window method" (see Figure 1).

Output of the solid-state detector is transmitted to a pulse-height analyzer system (e.g., a single-channel analyzer, SCA) or a window discriminator (not a multichannel analyzer, MCA). The upper and lower levels of the discriminator are set in such a way that a narrow, integrating energy window ΔE is stored in one channel, while the pulses are counted for a fixed time Δt. The channels in the MCA are switched from an energy mode to a time mode; that is, the channels now represent consecutive times which can be increased stepwise by increasing Δt from the beginning of the experiment at $t=t_0$ to $t=t_{end}$. This mode of operation allows one to follow a very fast phase transition with a lower time limit, Δt, of about 0.3 s.

High-Pressure Apparatus

The highest pressures are currently obtained in diamond-anvil cells. (for example, see JAYARAMAN, 1983). A well-known and serious limitation in using such cells for crystallographic studies is the limited sample volume

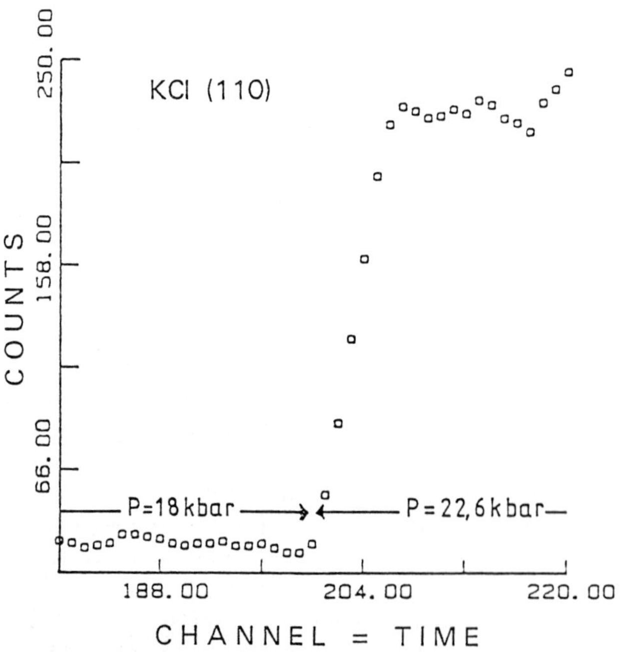

Figure 1. Demonstration of the window method. As an example we show the time dependence of the phase transition in KCl by monitoring the (110) reflection, as KCl goes from the NaCl-type phase into the CsCl-type phase by raising the pressure from 18 to 22.6 kbar. The multichannel analyzer is operated in the SCA-mode (singlechannel analyzer), in which the depicted transition 1 channel is equivalent to 0.1 s.

illuminated by the incoming X-ray beam. Using an average beam diameter of 100 μm, a typical sample thickness of 10 μm, and grain sizes of around 1 μm, we cannot get the necessary statistical average required to give an undisturbed intensity distribution in the powder diffraction diagram. Because of some degree of preferred orientation of the crystallites in the specimen (WILL et al., 1983), the observed intensities cannot be used for evaluating crystal structures; even peak splittings due to symmetry changes can hardly be extracted. Preferred orientation is generated to a high degree when pressure is applied (this is not the same as the lack of statistical particle distribution); it poses a second serious problem in using diamond-anvil cells and makes it practically impossible to evaluate diffraction intensities for determining or refining crystal structures from powder diffraction data.

Pressure cells with much larger volumes are therefore needed. The Japanese high-pressure research community has built a large press for X-ray diffraction with synchrotron radiation, called MAX80, at Tsukuba, Japan (AKIMOTO, this volume). We have also built, at reasonable cost, high-pressure cells with fairly large sample volumes for X-ray diffraction. Because of limited space in HASYLAB at the DESY in Hamburg, our pressure cells had to be small. This size limitation would not allow the setup of a press similar to the MAX80.

Two cells were used in this investigation. One is a miniature piston-cylinder press (NUDING et al., 1980), which has a limited pressure/temperature capability. For experiments on Mg_2GeO_4, we used a newly developed piston-cylinder apparatus which has been described by HINZE et al. (1983). The size of the inner insert is 6 mm in diameter and 7 mm in height. The insert holds a graphite cylinder which acts as a furnace. This cylinder, in turn, contains the sample, a thermocouple, and some amorphous carbon. Figure 2 shows the inner part of this assembly. The actual sample volume is 6 mm^3 (2 mm in

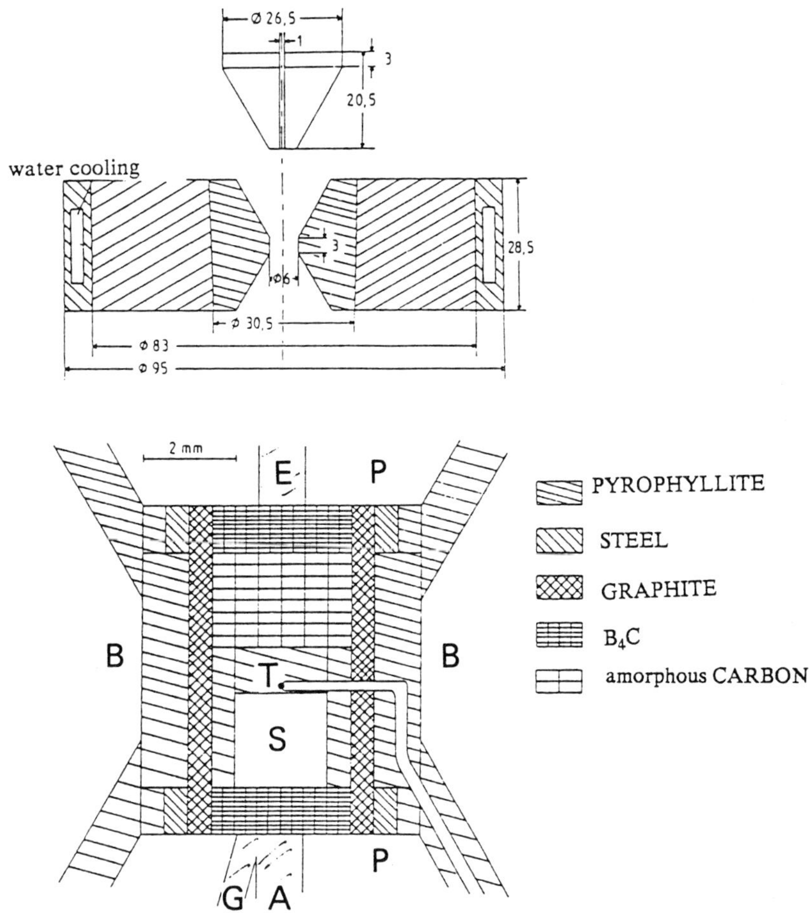

Figure 2. Schematic drawing of the piston cylinder high-pressure/high-temperature insert in the belt apparatus. B is belt rings; P is piston; S is sample; T is thermocouple, E is entrance hole for the primary X-ray beam; A is exit hole; and G is hole for the beam diffracted under the constant angle $2\theta_0$.

height and 2 mm in diameter). The insert is sealed at both ends by two small discs made of diamond or B_4C. Force is produced on the pistons by a 60-ton hydraulic ram. The X-ray path through the pistons is parallel to the pressure/force direction. The X-rays strike and leave the sample through holes of about 1 mm in diameter, which have been drilled into the pistons. These holes are sealed against the sample by the diamond or B_4C discs. In addition, the second piston has a second drilled hole to allow the diffracted beam to leave the sample at fixed diffraction angle 2θ.

Synthesis and Sample Preparation

The α-phase (olivine-phase) was prepared by firing the constituent oxides MgO and GeO_2 at 900 °C for 10 h. The reaction product was ground by mortar and pestle into a fine powder with grain sizes $<10\ \mu m$, and then pressed into pellets and tempered in air for 3 days at 1200 °C. X-ray patterns revealed a small amount (about 4%) of MgO, which was left in the sample. It is our experience that the cation distribution and the composition in olivine systems is strongly dependent on the oxygen fugacity (WILL et al., 1979), especially at elevated temperatures. For this reason a small amount of MgO will act as a buffer in this system by defining the MgO activity and will therefore stabilize the compound at higher temperatures. The γ-phase (spinel-phase) was prepared by high-pressure/high-temperature treatment of olivine-phase samples.

Phase Diagram of Mg_2GeO_4

The P-T phase diagram of Mg_2GeO_4 has been previously studied by Shiota et al. (private communication). Shiota's group was primarily interested in the melting behavior and conducted experiments mostly in high-temperature ranges up to 2000 °C. We have supplemented these data by conducting experiments at lower P-T conditions between 0 and 20 kbar and up to 1200 °C. Our experiments were done in a conventional high-pressure cell (WILL et al., 1979). α- and γ-phase materials were mixed in a 1:1 ratio, pressed into pellets 2 mm in height and 2 mm in diameter, and then enclosed in Pt-foil for high pressure runs. The P-T conditions were held for 60 h, for temperatures below 1000 °C, and for 6 h, for temperatures greater than 1000 °C. Phases were determined by powder X-ray diffraction. Results of our experiments are shown in Figure 3, together with results of Shiota's group. It has been found that the γ-phase of Mg_2GeO_4 with the spinel-type structure is the stable phase

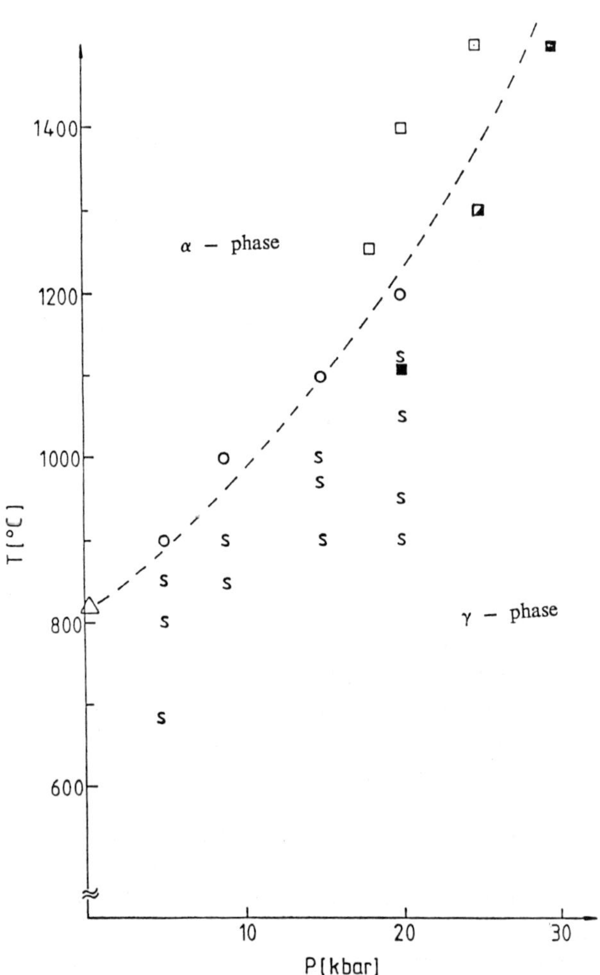

Figure 3a. P-T phase diagram of Mg_2GeO_4. △=transformation at 1 bar (DACHILLE and ROY, 1960); □=olivine-type, (Shiota, personal communication, 1984); ■=spinel-type, (Shiota, personal communication, 1984); O=olivine-type, this paper; S=spinel-type, this paper.

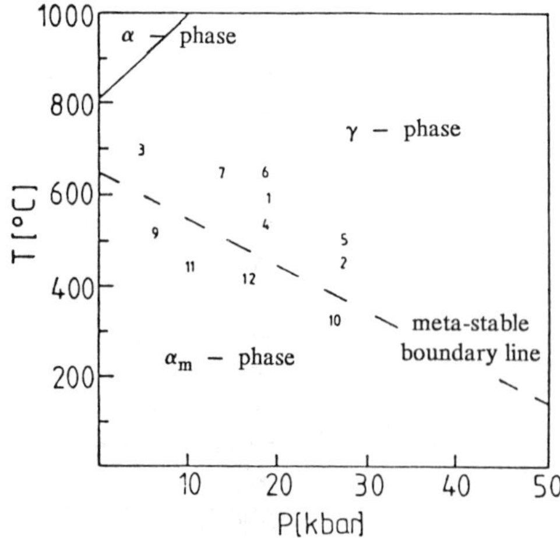

Figure 3b. P-T phase diagram of Mg_2GeO_4 at lower conditions showing the phase line for metastable conditions (dashed line). The numbers indicate the experimental conditions used for the kinetics measurements (see also Table 4).

at ambient conditions, and transforms at 810 °C and atmospheric pressure into α-Mg_2GeO_4 with the olivine structure (DACHILLE and ROY, 1960). The analogue to the β-phase in forsterite with a distorted spinel structure has not yet been found.

The α-phase remains in a metastable condition, α_m-Mg_2GeO_4, over a wide P-T range to about 500 °C at 10 kbar and 200 °C at 40 kbar. We have established a boundary line for this metastable α_m-region, which we used for studying the kinetics of the phase transformation (see Figure 3b, dashed line). The phase boundary was established by maintaining a given P-T condition for 2–3 h and then analyzing the reaction product by X-ray diffraction. We considered the phase to be metastable α_m when no transformation had taken place.

Crystal Structure

The mineral α-Mg_2GeO_4 crystallizes with the olivine structure in space group D_{2h}^{16}-Pbnm. Lattice constants and crystal structure parameters were calculated and refined from the powder diffraction data by least-squares analysis using the two-step integration method developed earlier by WILL (1979, 1981) and WILL et al. (1983). Data were collected by the energy dispersive method using synchrotron radiation, as well as by conventional angle-dispersive diffractometry on a Philips diffractometer with Cu-K_α radiation. Data were evaluated by profile fitting analysis, thus given the d_{hkl} and I_{hkl} values (WILL et al., 1983). Figure 4 shows an example of the energy dispersive diffraction diagram of γ-Mg_2GeO_4, including the profile fitted curves. The *d*-values were used for determining the lattice parameters, while the integrated intensities were the input data for the crystal structure refinement using

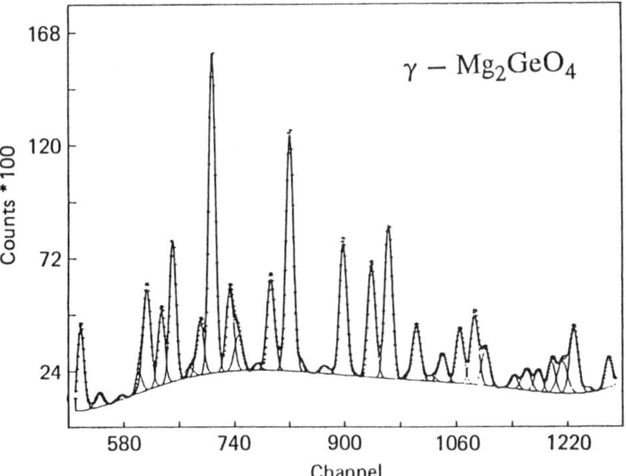

Figure 4. Diffraction diagram of γ-Mg_2GeO_4 taken in the belt high pressure cell at 1 bar with $2\theta_0=15°$. The diagram has been fitted by profile analysis, as shown by the solid lines.

the powder least-squares program POWLS (WILL, 1979). The *R*-values obtained in the crystal structure refinement were $R=5.3\%$ and $R=4.8\%$ for the energy- and angular-dispersive methods, respectively. Octahedral bonds Mg-O were the same in germanate as in forsterite, while the tetrahedral bonds were larger in germanate than in the silicate due to the larger Ge-ionic radius.

γ-Mg_2GeO_4 crystallizes in the spinel structure. The structure was previously refined by VAN DREELE and NAVROTZKY (1977) from neutron diffraction data. Our own analysis was in agreement with those results.

Compressibilities

Compressibilities were measured at room temperature for α- and γ-phases up to 40 and 60 kbar, respectively, using NaCl and Decker's equation of state (DECKER, 1971) for the pressure calibration. Because of the high quality of the diffraction diagrams and good separation of the diffraction peaks from the profile analysis, it was possible to observe up to 15 reflections for NaCl and up to 20 reflections for the germanate. The standard deviations are estimated to be 0.003 A for NaCl and 0.01 A for Mg_2GeO_4. The pressure uncertainty is estimated to be $P=\pm 0.5$ kbar. Results are shown in Figure 5. Data were fitted with the Murnaghan equation

$$P = \frac{K_0}{K_0'}\left[\left(\frac{V_0}{V_P}\right)^{K_0'} - 1\right] \qquad (1)$$

where P is pressure in kbar; V_0 is unit cell volume at $P=1$ bar; V_P is unit cell volume at pressure P; K_0 is isothermal compression module; and $K_0'=(\partial K_0/\partial P)_T$.

K_0' was set at a constant value of 4. This value is justified, since it is based on measurements from a previous compressibility investigation of forsterite (a compound isostructural and homologous to Mg_2GeO_4) to up 300 kbar (WILL et al., 1986). Our data are summarized in Table 1, along with data for Mg_2SiO_4 from the literature.

Kinetics of the Phase Transformation

The kinetics of the olivine- to spinel-phase transformation of Mg_2GeO_4 was studied using the full pattern method. We investigated this transition in seven cycles, and recorded a total of 109 diffraction spectra. Experimental conditions are listed in Table 2. Time intervals varied between $\Delta t=80$ s to $\Delta t=430$ s in the different runs. The number of runs (N) at each cycle is also listed in Table 2.

Analysis of these data was performed by: a) the integration method, and b) the full data POWLS (Powder Least Squares) method.

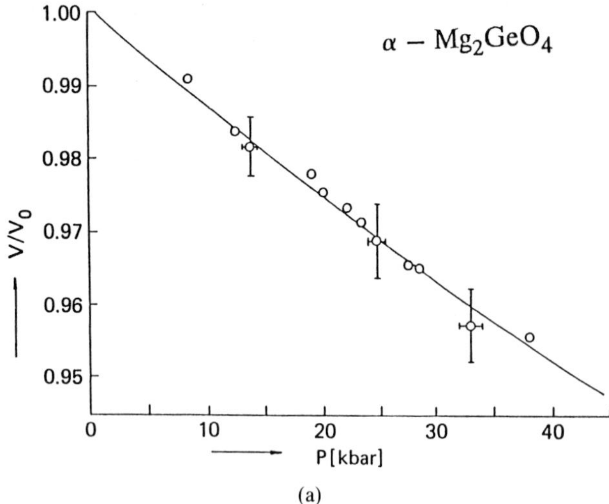

(a)

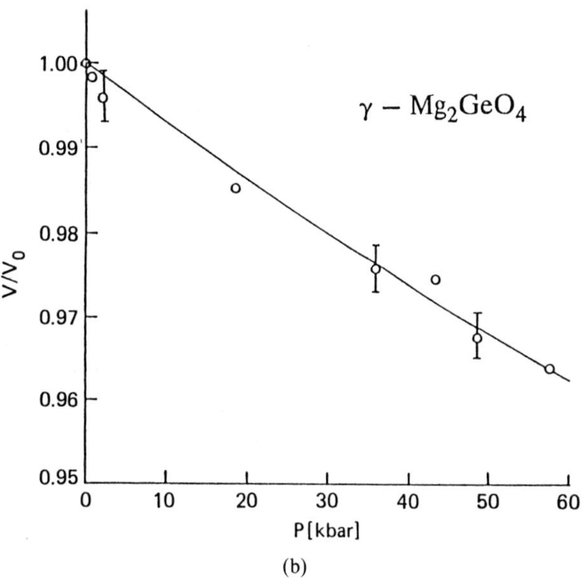

(b)

Figure 5. Compressibilities of Mg_2GeO_4 plotted as the relative volumes V/V_0 versus pressure. a) for α-Mg_2GeO_4; b) for γ-Mg_2GeO_4.

TABLE 1. Bulk Modulus K_0 of α and γ Phases in Mg_2GeO_4, Derived by Least-Squares Fitting of the Present P-V Data with the Murnaghan Equation (The respective values for α and γ phases in Mg_2SiO_4 are taken from the literature.)

	Mg_2GeO_4		Mg_2SiO_4	
K_0(α phase)	746 ± 7	kbar[1]	1360	kbar[2]
K_0(γ phase)	1445 ± 15	kbar[1]	1840	kbar[3]
Q_A(activation energy)	49.7	kcal/mol[1]	83	kcal/mol[4]

[1)] this work.
[2)] WILL *et al.* (1986).
[3)] SUNG (1979).
[4)] WEIDNER et al. (1984).

TABLE 2. Experimental Conditions, Pressure and Temperature, Used for the Kinetics Runs with the Derived Growth Rate Parameter k Derived by Least Squares Fitting with the Avrami Equation (Δt gives the time interval between consecutive measurements, t_g is the total time (in minutes) spent, and N=number of data points.)

Run. No.	P (kbar)	T (K)	Δt(s)	t_g(min)	N	k $10^4(s^{-1})$
1	19.1	870	130	90	18	5.61
2	27.2	720	230	45	8	0.63
3	4.7	975	430	105	11	1.03
4	18.1	820	230	150	31	1.07
5	30.2	774	230	105	21	2.06
6	17.8	920	80	30	14	11.78
7	13.8	927	230–430	20	6	5.61

The Integration Method

We monitored total integrated intensities, I_{obs}, of 5 reflections—(422), (440), (620), (642), and (800)—of the newly forming spinel phase as a function of time. The intensities were derived by profile fitting.

$$I_{obs} - I_{\gamma\text{-phase}} = I_{422} + I_{440} + I_{620} + I_{642} + I_{800} \qquad (2)$$

Because of the gradual decay of the storage ring current i_0 with time (the lifetime is about 60 min at 5 GeV, our operating voltage), the observed intensities have to be normalized. We did this through the Ge-fluorescence line (Ge K_α=9.87 keV), which also monitors the sample volume irradiated by the beam

$$I_{fluoresc} \propto i_0(t) \text{ and } \propto [\mathrm{Ge}] \qquad (3)$$

where $i_0(t)$ is the electron current in the storage ring and equals the primary X-ray beam intensity, and [Ge] represents the concentration of Germanium. The result is

$$I_{norm} = I_{obs}/I_{fluoresc} = I(t) \qquad (4)$$

The analysis is based on the Avrami equation (CAHN, 1956)

$$X(t) = 1 - e^{-k \cdot t^n} \qquad (5)$$

where $X(t)$ is degree of transformation; k is the rate function; t is time, and n is an exponent which is determined by the mechanism of nucleation. The degree of transformation $X(t)$ can be determined from the observed intensities of the spinel phase during the phase transition. Since the diffracted intensity is proportional to the amount of spinel phase, we can write, using equation 2 and 4

$$X(t) = I_{norm}(t)/I_0 \qquad (6)$$

where I_0 is the total and normalized intensity according to equation 2 and 4 after the transition is completed.

We plot I_{norm} versus time in two representative plots, as shown in Figure 6, and then fit the Avrami equation to the experimental data points by a least-squares analysis. In the first evaluation both parameters k and n were varied. This gave values for n between 0.8 and 1.1. Consequently n was set to $n=1$. The velocity constant k was then left as the only changing variable to be determined. Therefore, from the seven transitions measured, we calculated seven values of k for the olivine-spinel transition at different temperatures and pressures. Results are listed in Table 3.

For further interpretation of k, we rewrote equation 5 for the case of heterogeneous nucleation, according to CAHN (1956) as follows

$$X(t) = 1 - e^{-\pi/3 \cdot I \cdot Y^3 \cdot t^4} \tag{7}$$

where Y is growth rate of (idealized) spherical nuclei and t

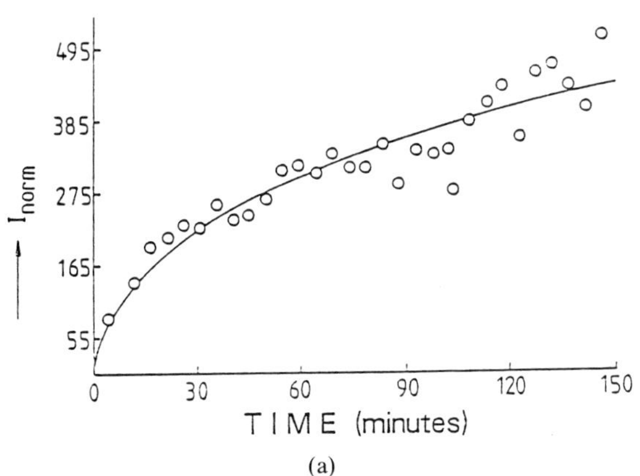

(a)

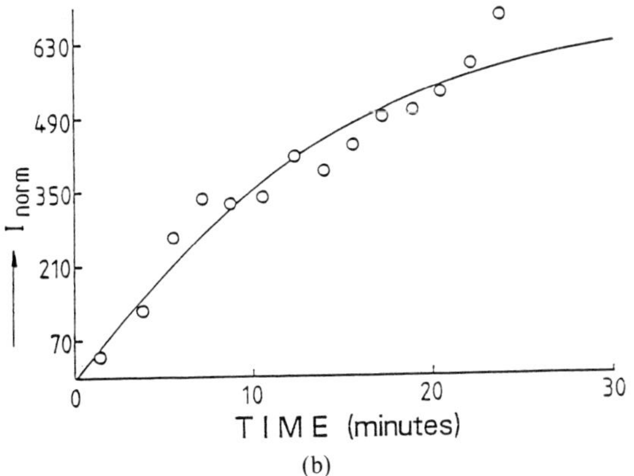

(b)

Figure 6. Phase transition from the α-phase to the γ-phase. Plotted are I_{norm} (see Equations 2 and 4) as the sum of 5 reflections coming from the spinel lattice. The solid line is calculated by fitting the Avrami equation, Equation 5, to the data. Experimental conditions were: a) $P=18.1$ kbar, $T=820$ K, $\Delta t=230$ s; b) $P=17.8$ kbar, $T=920$ K, $\Delta t=80$ s.

TABLE 3. The Growth Parameter k Determined from the Fitting of the Experimental Data to the Avrami Equation (k-obc), in Comparison to k-calc as Calculated with the Thermodynamic Quantities

Run No.	P(kbar)	T(K)	$k_{obs}(s^{-1})10^4$	$k_{obs}(s^{-1})10^4$
1	19.1	870	5.61	4.30
2	27.2	720	0.63	0.18
3	4.7	975	1.03	1.10
4	18.1	820	1.07	0.70
5	30.2	774	2.06	2.53
6	17.8	920	11.78	12.00
7	13.8	927	5.61	5.94

$Q_a=47.9$ (9) kcal/mol
$\Delta V_A=13.0$ (1.5) cm
$\Delta H=2.3$ (1.0) kcal/mol

is steady state nucleation rate. This equation (with $n=4$ in t^n) describes the kinetics for nucleation before the available sites are filled. In powdered samples these sites are on grain boundaries. Therefore equation 7 describes the very early and fast stage of the transformation that could not be observed in the experiments reported here.

After the nucleation is exhausted, further growth is then described by CAHN (1956) as

$$X(t) = 1 - e^{-2S \cdot Y \cdot t} \tag{8}$$

where $S=3.4/D$ and D is grain diameter. When $n=1$ in equation 5, the Avrami equation is identical to equation 8. Therefore we deal here only with equation 8. The growth rate can then be expressed according to TURNBULL (1953) by

$$Y = \frac{\lambda k_B T}{h} \times e^{-(Q_A + \Delta V_A)/RT} \times [1 - e^{-\Delta v \Delta p + \Delta H \Delta T)}/RT] \tag{9}$$

where λ is thickness of the olivine-spinel interface; k_B is the Boltzman constant; R is the universal gas constant; h is Planck's constant; T is temperature; Q_A is activation energy; ΔV_A is activation volume; P is pressure; Δv is volume change during the transition; ΔP is overpressure; ΔH is heat of transition at $P=0$; and finally $\Delta T=(T-T_0)/T_0$, where T_0 is transition temperature at $P=0$.

The thickness of the interface can be calculated from the data available using the relationship given by CAHN (1956): $A=2S\cdot\lambda\cdot k_B T/h$. With $S=3.4/D$ (CAHN, 1956), $D\sim500\,\mu$m, diameter of the grains, and $k_B T/h=1.8\times10^{13}$ (for $T=500$ °C), we calculate the thickness of the interface, λ, as $1.1\cdot10^{-8}$ cm (i.e., about 1 Å).

From the experimentally determined k values, we can derive the thermodynamic quantities Q_a, ΔV_A and ΔH of equation 9. (ΔV can be calculated from the known PV data of the two phases. We calculated these values by a least-squares fitting of equation 9 to the k-values. Table 4 lists the results of this calculation for different temperatures and pressures.

In addition to this determination from the kinetics experiments, we also can calculated ΔH from the data of the P-T phase diagram using the Clausius-Claperyon relation (LEVINE, 1978) given by

$$\frac{dP}{dT} = \Delta H/(T \cdot \Delta V) \tag{10}$$

Based on the known lattice constants and the phase diagram, we get ΔV=3.6 cm^3/mol and dP/dT=0.05 kbar/°C, respectively. These values give ΔH=3.5 kcal/mol. Shiota (personal communication, 1984) calculated from his data a similar value, ΔH=4.0 kcal/mol. Therefore there is reasonable agreement between our method and Shiota's method.

Full Data POWLS Method

Figure 7 shows an example of a sequence of diffraction diagrams recorded during the phase transition. (The time interval between the plotted diagrams is 2–3 min.) Profile analysis of these diagrams yielded intensities, that in turn, could be used for crystal structure calculations, or refinements of model structures with the program POWLS (WILL, 1979; WILL et al., 1983).

The intensity diffracted from a polycrystalline sample is described by BURAS et al. (1977) and WILL (1981) as

$$I_{hkl} = C \cdot i_0(E) \cdot \eta(E) \cdot j \cdot |F_{hkl}|^2 \cdot d_{hkl}^2 \cdot A(E, \theta) \cdot C_p \cdot (1/\sin \theta_0) \tag{11}$$

where C is scale factor, $i_0(E)$ is the primary intensity, $\eta(E)$ is the energy dependent detector response, F is the crystallographic structure factor including the temperature factor; d_{hkl} is the spacing, C_p is the polarization factor (in general, C_p=1 with synchrotron radiation); and 1/sin θ_0 is the Lorentz-factor. The primary beam distribution and detector response were measured at the beginning by running a standard sample. $A(E, \theta)$, the absorption,

TABLE 4. Kinetics of the Phase Transition as Determined from Crystal Structure Evaluation Using Least Squares Refinement with POWLS

Experimental Conditions		Results from POWLS analysis		Results from integration analysis
P(kbar)	T(K)	d	n	k (from Table 3) determined with the Avrami equation
13.8	927	0.055	0.7	5.61×10^{-4}
17.8	920	0.111	1.0	11.78×10^{-4}

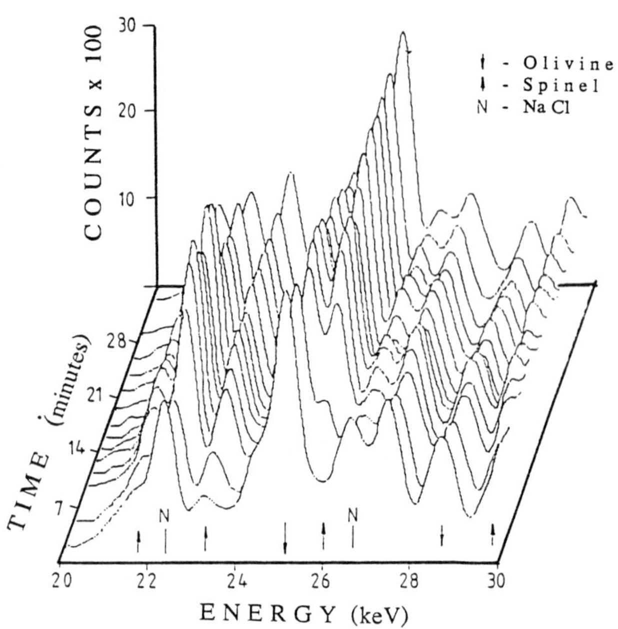

Figure 7. A sequence of diffraction diagrams versus time during the phase transition showing the growth of the spinel phase.

was calculated using Bond's formula (BOND, 1959) as follows

$$A(E) = e^{-\mu/\rho \cdot \rho \cdot d} \tag{12}$$

and

$$\mu/\rho = B' \cdot \lambda^3 - D \cdot \lambda^4 + \sigma_{NA}(\lambda) \tag{13}$$

where d is sample thickness, B' and D are atom specific constants, and σ_{NA} is the Klein-Nishina cross-section for Compton scattering.

In the structure factor F_{hkl} we have, as the important variables for this investigation and our ensuing least-squares analysis, the numbers C_i attached to each atomic site i, which describe the site occupancy. C_i can be calculated quite accurately by least-squares methods with the program POWLS, using the observed integrated intensities and a model structure of a spinel lattice that is highly depleted in the cations in the octahedral and tetrahedral sites. Since such calculation is done at each time step, the transition process, (i.e., the recrystallization) can be studied as it develops with time. However, there is one limitation in that we could not distinguish between Ge and Mg because of the limited number of peaks and also the limited accuracy of the measured intensities. Therefore we assumed that the octahedral and tetrahedral sites are being filled at the same rate, and we determined only an averaged value C (Mg, Ge) for the cation occupacy. This is shown in Figure 8 for the two runs made in this experimental manner. (Data are plotted on a logarithmic scale.) We can describe the migration of the cations into the sites by

$$C(t) = d\cdot(t - t_0)^n \tag{14}$$

where $C(t)$ is the fraction of the occupied, coherently scattering cations; d is a parameter; and t_0 is the beginning of the transition process. By taking the logarithms we get

$$\ln C(t) = \ln d + n\cdot(t - t_0). \tag{15}$$

The slope yields n, and d indicates how many sites are occupied immediately after the transformation has been started at $\Delta t = t - t_0 = 0$. Therefore d is a measure of the cation occupancy directly after the phase transformation has taken place. Since we find d not equal zero, this indicates that a small fraction of the cation sites are occupied immediately during the phase transition. The results indicate that this fraction becomes larger as the overpressure increases, as one would expect. Results are given on Table 4. The two experiments gave $d=5.5\%$ for the first run at $P=13.8$ kbar and $T=927$ K, and $d=11\%$ for the second run at $P=17.8$ kbar and $T=920$ K. The occupation of the cation sites proceeds faster with higher overpressure (i. e., the difference to the transition pressure). As Table 4 shows, there is a proportionality between the growth rate k and the parameter d.

Results and Conclusions

γ-Mg_2GeO_4 crystallizes in the cubic spinel structure with a closed-packed oxygen arrangement, and a stacking sequence ABC along ⟨111⟩. The Mg^{2+} ions are in six-fold coordinated octahedral sites and Germanium is in four-fold coordinated tetrahedral sites. γ-Mg_2GeO_4 with the olivine-type structure has a distorted hcp oxygen arrangement (stacking sequence AB) with Ge in tetrahedral sites and Mg^{2+} in octahedral sites. The transformation therefore produces no change in the cation coordination. Only the oxygens are rearranged, with a gain in density by 8.8%.

In the distorted hcp arrangement of oxygen in the olivine-type phase, the layers are stacked along the a-axis. Therefore we can go from the α-phase to the γ-phase by rotating every third oxygen plane (100). This view is supported by our compressibility measurements on forsterite up to 300 kbar (WILL et al., 1986), in which we observed a higher compressibility for the b- and c-axis than for the a-axis. Also the present in situ experiments, what evaluate the crystal structure with least-squares intensity refinements, support this view. At the beginning of the phase transformation, we observed only the coherent scattering from the cubic closed-packed oxygen ions. As much as 15 min after the transformation commenced, additional reflections coming from the coherently scattering cations in octahedral and tetrahedral sites could be observed.

Further evidence is given by experiments of FURNISH and BASSETT (1983) on fayalite, Fe_2SiO_4. Their study suggests that the rearrangement of oxygen atoms appreciably precedes cation reordering during the $\alpha \rightarrow \gamma$ transition.

Therefore we come to a very important conclusion that both studies on silicate and the germanate are consistent with a model of the transition in which the anions are transformed from the nearly hcp arrangement of the α-structure to the ccp arrangement of the γ-structure by a coherent restacking, while the cations diffuse to the appropriate sites over a significant span of time. The findings of these investigations are not consistent with a transition mechanism based on incoherent nucleation of the γ-phase, nor with a mechanism involving anion restacking with a simultaneous reordering (synchroshear) as proposed by POIRIER (1981).

Finally, the small energy difference of about 2.3–4 kcal/mol between the α- and γ-phases explains quite well the large metastable field of the olivine phase up to high pressures and temperatures, as indicated in Figure 3.

In summary, our study concerns a phase transformation from the metastable olivine phase to the spinel structure. This limitation is imposed by the experimental conditions presently available to us. Our results are applied to an interpretation of a mechanism of the olivine to spinel transformation in the earth's mantle. It would be interesting to repeat such a study in the higher temperature range of the transition from the stable olivine phase to the spinel

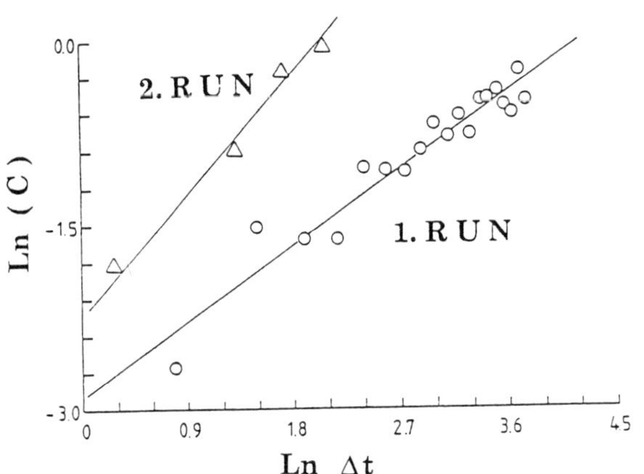

Figure 8. Plot in logarithmic scale of the occupation of the coherently scattering cation sites (Mg and Ge) in the spinel phase versus time after the phase transition. First run: $P=13.8$ kbar, $T=927$ K. Second run: $P=17.8$ kbar, $T=920$ K.

phase; some changes in the kinetic parameters might be observed. At present, we are planning such a study.

Acknowledgments. This work was supported by the Bundesminister fuer Forschung und Technologie, Bonn, which is gratefully acknowledged.

References

AKIMOTO, S., High-pressure research in geophysics: Past, present and future, this volume.

AKIMOTO, S., Y. MATSUI, and Y. SYONO, High-pressure crystal chemistry of orthosilicates and the formation of the mantle transition zone, in *The Physics and Chemistry of Minerals and Rocks*, edited by R. G. J. Strens, pp. 328–363, John Wiley, New York, 1976.

BOND, W. L., *International Tables for X-ray Crystallography*, III, pp. 157–174, Kynoch Press, 1959.

BURAS, B., J. St. OLSEN, L. GERWARD, G. WILL, and E. HINZE, X-ray energy-dispersive diffractometry using synchrotron radiation, *J. Appl. Cryst.*, *10*, 431–438, 1977.

CAHN, J. W., The kinetics of grain boundary nucleated reactions, *Acta Metallurgica*, *4*, 449–459, 1956.

DACHILLE, F., and R. ROY, High pressure studies of the system Mg_2GeO_4-Mg_2SiO_4 with special reference to the olivine-spinel transition, *Americ. J. Sci.*, *32*, 225–246, 1960.

DECKER, D. L., High-pressure equation of state for NaCl, KCl and CsCl, *J. Appl. Phys.*, *42*, 3239–3244, 1971.

FURNISH, M. D., and W. A. BASSETT, Investigation of the mechanism of the olivine-spinel transition in fayalite with synchrotron radiation, *J. Geophys. Res.*, *88*, 10.333–10.341, 1983.

HINZE, E., J. LAUTERJUNG, and G. WILL, A new belt-type apparatus for energy-dispersive X-ray diffraction under high pressure and temperature, *Nucl. Instr. and Methods*, *208*, 569–572, 1983.

JAYARAMAN, A., Diamond anvil cell and high-pressure physical investigation, *Rev. Mod. Physics*, *55*, 65–108, 1983.

JEANLOZ, R., and A. B. Thomson, Phase transitions and mantle discontinuities, *Rev. Geophys. Space Phys.*, *21*, 51–74, 1983.

LEVINE, I. N., *Physical Chemistry*, pp. 173–176, Mc Graw Hill, New York, 1978.

NUDING, W., E. HINZE, and G. WILL, A miniature piston cylinder apparatus for high pressure X-ray diffraction in conjunction with energy dispersion, *J. Appl. Cryst.*, *13*, 46–49, 1980.

POIRIER, J. P., Martensitic olivine-spinel transformation and plasticity of the mantle transition zone, in *Anelasticity in the Earth, Vol. 4*, edited by F. D. Stacey et al., pp. 113–117, AGU, Washington, D. C., 1981.

RINGWOOD, A. E., *Composition and Petrology of the Earth's Mantle*, pp. 387–426, McGraw Hill, New York, 1979.

RINGWOOD, A. E., and A. MAJOR, *The System Mg_2SiO_4-Fe_2SiO_4 at High Pressure and Temperatures in Phase Transformations and the Earth Interior*, edited by A. E. Ringwood and D. J. Green, pp. 89–108, North-Holland Publishing Company, Amsterdam, 1970.

SUNG, C. M., Kinetics of the olivine-spinel transition under high pressure and temperature: experimental results and geophysical implications, in *Proceedings of the 6th AIRAPT Conference*, pp. 31–42, New York, 1979.

TURNBULL, D., Phase changes, in *Solid State Physics*, *Vol. 3*, edited by F. Seitz and D. Turnbull, pp. 225–306, Academic Press, New York, 1953.

VAN DREELE, R. B., and A. NAVROTZKY, Refinement of the crystal structure of Mg_2GeO_4 spinel, *Acta Cryst.*, *B33*, 2287–2288, 1977.

WEIDNER, D. J., H. SAWAMOTO, S. SASAKI, and M. KUMAZAWA, Single-crystal elastic properties of the spinel phase of Mg_2SiO_4, *J. Geophys. Res.*, *89*, 7852–7860, 1984.

WILL, G., POWLS: a powder least squares program, *J. Appl. Cryst.*, *12*, 483–485, 1979.

WILL, G., Energiedispersion und Synchrotronstrahlung: eine neue Methode und eine neue Strahlenquelle fuer Roentgenbeugung, *Fortschr. Miner.*, *59*, 31–94, 1981.

WILL, G., L. CEMIC, E. HINZE, K. F. SEIFERT, and R. VOIGT, Electrical conductivity measurements on olivines and pyroxens under defined thermodynamic activities as a function of temperature and pressure, *Phys. Chem. Minerals*, *4*, 189–197, 1979.

WILL, G., E. HINZE, and W. NUDING, Energy-dispersive X-ray diffraction applied to the study of minerals under pressure up to 200 kbar, in *High Pressure Research in Geoscience*, edited by W. Schreger, pp. 177–201, E. Schweizerbart, Stuttgart, 1982a.

WILL, G., E. HINZE, K. F. SEIFERT, L. CEMIC, E. JANSEN, and A. KIRFEL, A device for electrical conductivity measurements on mantle relevant minerals at high pressures and temperatures under defined thermodynamic conditions, in *High Pressure Research in Geoscience*, edited by W. Schreger, pp. 201–219, E. Schweizerbart, Stuttgart, 1982b.

WILL, G., W. PARRISH, and T. C. HUANG, Crystal structure refinement by profile fitting and least squares analysis of powder diffraction data, *J. Appl. Cryst.*, *16*, 611–622, 1983.

WILL, G., W. HOFFBAUER, E. HINZE, and J. LAUTERJUNG, The compressibility of forsterite up to 300 kbar measured with synchrotron radiation, *Physica*, *139* and *140B*, 193–197, 1986.

EXAFS AND XANES STUDY UNDER PRESSURE

Osamu SHIMOMURA

National Institute for Research in Inorganic Materials Namiki 1-1, Sakura-mura, Niihari-gun Ibaraki 305, Japan

Takaaki KAWAMURA

Department of Physics, Yamanashi University, Takeda 4-4-37, Kofu, Yamanashi 400, Japan

Abstract. EXAFS and XANES are applied to the study of semiconductor-metal transitions of tetrahedrally bonded crystals under high pressure. When the variation of single-bond length is to be determined, EXAFS data can be easily and conveniently analyzed by a phase difference method. For more general cases, EXAFS data can be analyzed by a curve fitting method. With both methods, the accuracy of the estimated bond length is 0.01 A and that of the coordination number is 0.2. Calculations based on full multiple scattering theory show that the XANES spectrum is quite sensitive to the arrangements and electronic states of the atoms surrounding the absorbing atom. It is shown that EXAFS is a powerful tool for analyzing structural parameters under pressure, and that XANES is potentially useful for stodying structural and electronic states of a material under pressure.

Introduction

Since the development of synchrotron radiation (SR), many studies of X-ray absorption spectra have been performed. Just above the absorption edge of a specified atom in a material, there appears a fine structure overlapping a smooth absorption spectrum for about 1000 eV. The fine structure is a final-state electron effect, arising from the interference of the outgoing photoelectron from the specified atom, which was excited by the X-ray, with its back-scattered component from the neighboring atoms.

The fine structure is divided into two parts, based on the energy range. In the low energy range ($E<50$ eV), the fine structure consists of a complicated feature, termed XANES (X-ray Absorption Near Edge Structure). In this region the cross-scchon of electron-atom scattering is relatively large, and multiple scattering of the electron by the reighboring atoms plays an important role (see Figure 1a). Thus the XANES spectrum just above the edge provides information on the three-dimensional atomic configuration around the absorbing atom.

On the other hand, the structure in the energy region above 50 eV shows a sum of sinusoidal oscillations, termed EXAFS (Extended X-ray Absorption Fine Structure). The cross-section in this energy region is so small that the electron is scattered primarily in the forward direction; that is, the interference occurs between the outgoing and the singly back-scattered electrons (see Figure 1b). Therefore EXAFS spectra provide information on the atomic distances and coordination numbers around the absorbing atom.

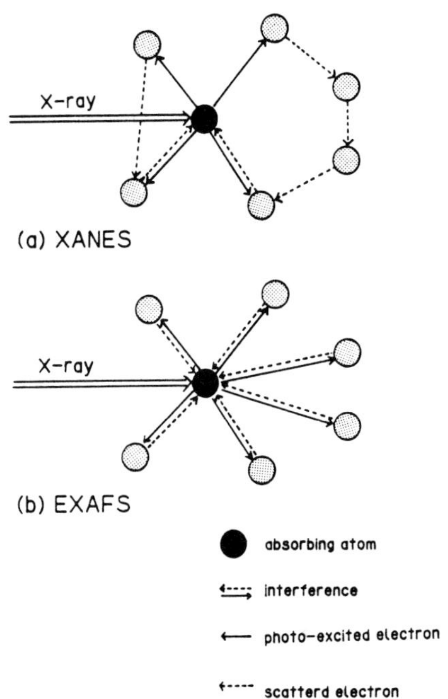

Figure 1. Scattering of a photo-excited electron by at energy levels of a) XANES and b) EXAFS.

EXAFS and XANES are useful for the following types of high-pressure studies:

1) Structural investigation of amorphous and liquid materials under high pressure and high temperature.

2) Determination of bond length and coordination number in unknown high-pressure phases.

3) Determination of the local compressibility around a specified atom in complex materials.

4) Study of the electronic structure of high-pressure phases or its variation with pressure.

In the present paper, we describe the application of EXAFS and XANES to high-pressure research. First, we

High-Pressure Research in Mineral Physics, edited by M. H. Manghnani and Y. Syono, pp. 187–193.

investigate the accuracy of EXAFS analysis. We analyze the structural parameters of three polymorphs of Ge, based on the assumption that the composition of the material stays the same under pressure. We can make this assurnphon in high pressure studies, because even when the structure changes with pressure, the constituent atoms remain the same. Second, we report the results of a high-pressure EXAFS study using a diamond-anvil cell and a cubic-anvil apparatus. Finally, we describe the XANES spectrum under pressure and the calculation based on full multiple scattering theory.

Samples and Experimental Apparatus

We measured the absorption spectra of: three polymorphs of Ge at normal pressure (diamond and ST-12 types, KASPER and RICHARDS, 1964; and amorphous type, SCHEVCHICK and PAUL, 1972); a high-pressure form of Ge (β-Sn type, JAMIESON, 1963); GaAs at normal pressure (zinc blend type); and a high-pressure form of GaAs (orthorhombic type, BAUBLITZ and RUOFF, 1982). The structural characteristics for each material are as follows:

1) for the diamond type Ge (semiconductor): O_h^7-$Fd3m$, $Z=8$, $a=5.658$ A; stable at normal pressure; exact tetrahedral coordination with a bond angle of 109.3°; and bond length of 2.45 A.

2) For the ST-12 type Ge (semiconductor): D_4^8-$P4_32_12$, $Z=12$, $a=5.93$ A, $c=6.98$ A; obtained by releasing the pressure above the metallic transition pressure (11 GPa); tetrahedrally bonded, but bond angles are scattered discretely between 88° to 135°; and bond lengths of 2.48 A(2), 2.486 A(1), and 2.488 A(1).

3) For amorphous Ge (semiconductor): obtained by evaporation onto the glass substrate; tetrahedrally bonded, but the bond angle is distributed uniformly around 109° with half width of about 30°; and mean bond length of 2.46 A, and mean coordination number of 3.9.

4) For β-Sn type Ge (metal): D_{4h}^{19}-$I4_1/amd$, $Z=4$, $a=4.884$ A, $c=2.694$ A; body-centered tetragnal; and two bond lengths, 2.56 A and 2.69 A.

5) For the zinc-blend type GaAs (semiconductor): T_d^2-$F43m$, $Z=4$, $a=5.653$ A; stable at normal pressure; the same structure as with the diamond type structure, except for the alternative configuration of Ga and As; and bond length of 2.45 A.

6) For orthorhombic type GaAs (metal): $Fmmm$, $Z=1$, $a=4.946$ A, $b=4.628$ A, $c=5.493$ A; stable above 18 GPa; and three bond lengths, 2.473 A(2), 2.314 A(2), and 2.747 A(2).

The procedures are the same for EXAFS and XANES experiments except for the energy range scanned. Intensities of the monochromatized incident X-ray and of the transmitted X-ray are measured by ionization chambers. When the transmitted intensity is weak, a scintillation counter is used instead of an ionization chamber.

The EXAFS experiments with three polymorphs of Ge at normal pressure were done at beam line IV (which is equipped with a six-pole wiggler), SSRL, Stanford University; the operating conditions of 3.0 GeV and 100 mA were used (KAWAMURA et al., 1981). The X-rays were monochromatized by a two-crystal Si 220 monochromator at a parallel setting, with a energy resolution of ± 1 eV. The powder samples were finely ground and mixed uniformly with epoxy-resin to make a thin plate.

The high-pressure experiment on GaAs was done in the EXAFS hutch at SSRL; operating conditions of 2.5 GeV and 15 mA were used (SHIMOMURA et al., 1978). X-rays from a bending magnet were monochromatized by a channel cut Si 220 monochromator, with a energy resolution of ± 1 eV. GaAs powder was compressed by a diamond-anvil cell. The pressure-transmitting medium was an alcohol mixture. The pressure was calibrated by the ruby fluorescent method.

A cubic-anvil apparatus named MAX80 (SHIMOMURA et al., 1985) was used for the EXAFS study of Ge under pressure (SHIMOMURA et al., in preparation). The pressure-transmitting medium was a 4:1 mixture of boron and epoxy-resin. A fine Ge powder was uniformly sandwiched between the plastic adhesives and put into the pressure-transmitting medium. The pressure was determined using the NaCl scale. This experiment was performed at BL-4C, Photon Factory, National Institute for High Energy Physics. The operating conditions were 2.5 GeV and 100 mA. The X-ray beam from a bending magnet was monochromatized by a sagittal focusing monochromator (Si 111) with a energy resolution of ± 1 eV.

Accuracy of the EXAFS Analysis

In the single scattering approximation, the EXAFS oscillation $\chi(k)$ is expressed by (SAYERS et al., 1971) as

$$\chi(k) = \sum_j \frac{N_j}{kR_j^2} |f_j(\pi)| \sin(2kR_j + \delta_j) \exp(-2R_j/\lambda) \times \exp(-\sigma_j^2 k^2) \quad (1)$$

where N_j is the number of atoms at a distance R_j from the absorbing atom; $f_j(\pi)$ is the back-scattering amplitude for the j-th atom; δ_j is the sum of the phase shifts of the photo-excited electron due to the absorbing and j-th back-scattering atoms; λ is the mean free path of the electron; and σ_j the root-mean-square fluctuation of the j-th atom relative to the absorbing atom. The wave number **k** of the excited photoelectron is given by

$$k = \sqrt{2m(\hbar\omega - E_0)}/\hbar \quad (2)$$

where $\hbar\omega$ is the X-ray energy ($\hbar$ is the Planck's constant divided by 2π) and E_0 the energy of the K core electron.

Prior to high-pressure experiments, we checked the accuracy of the EXAFS analysis using the three polymorphs of Ge mentioned above (KAWAMURA et al., 1981). Bond lengths differed slightly for the three structures, whereas bond angles differed substantially. These three polymorphs are semiconductors and are supposed to have similar electronic states since all are tetrahedrally coordinated. These materials are appropriate for checking the accuracy of EXAFS, because EXAFS depends on bond lengths and coordination numbers but not on bond angles.

Figure 2 shows three absorption spectra near the Ge K edge of the polymorphs. We focused our attention on determining the distance to the first nearest neighbor and the coordination number. We analyzed the data by the phase difference method and curve fitting method.

With both methods, the EXAFS part of the absorption spectrum was identified by subtracting the smooth background. The EXAFS oscillations were Fourier transformed to real space. This transformation corresponds to the radial distribution function (RDF) shown in Figure 3. In order to separate the contribution of the first nearest neighbor from those of others, a window function was multiplied by the first peak of the RDF. We used a window function with a minimum r of 1.68 A and a maximum r of 2.76 A, for the three cases (see the broken curve in Figure 3). This filtered RDF was Fourier transformed back to the k-space as shown in Figure 4 (solid curves). The Fourier-filtered EXAFS (the nearest-neighbor EXAFS) obtained corresponds to the component $j=1$ in Equation 1, and is expressed as a single sine function multiplied by some envelope functions.

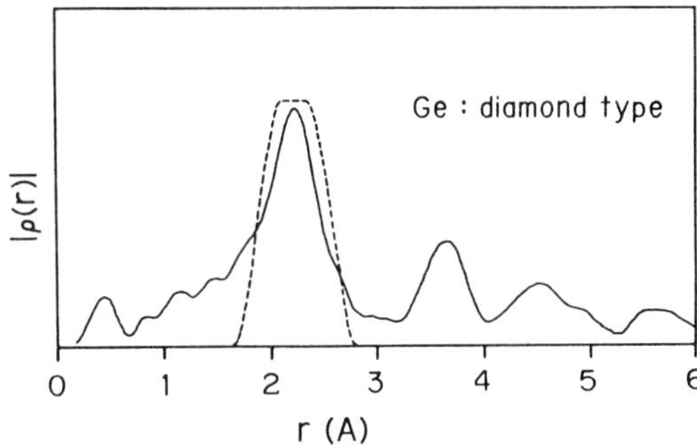

Figure 3. The radial distribution function of the diamond type Ge, which is obtained by Fourier transformation of the curve a) in Figure 1. The broken curve shows the window function which is used to separate the contribution of the nearest neighbor from those of other atoms and from the background noise. (Note that the peak positions are different from the actual values because phase shifts were ignored in the transformation.)

Beyond this experimental stage, procedures varied for the phase difference and curve fitting methods.

In the phase difference method, the nearest-neighbor EXAFS spectrum was normalized to show a single sine function

$$\Phi(k) = \sin(2kR + \delta)$$

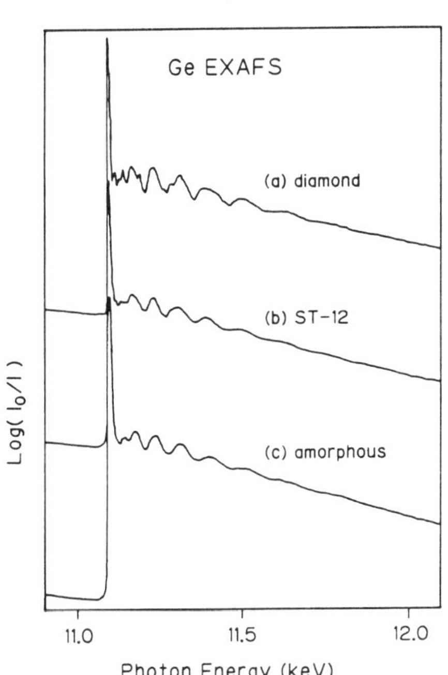

Figure 2. X-ray absorption spectra near the Ge K edge, for three Ge polymorphs: a) diamond type, b) ST-12 type, and c) amorphous type.

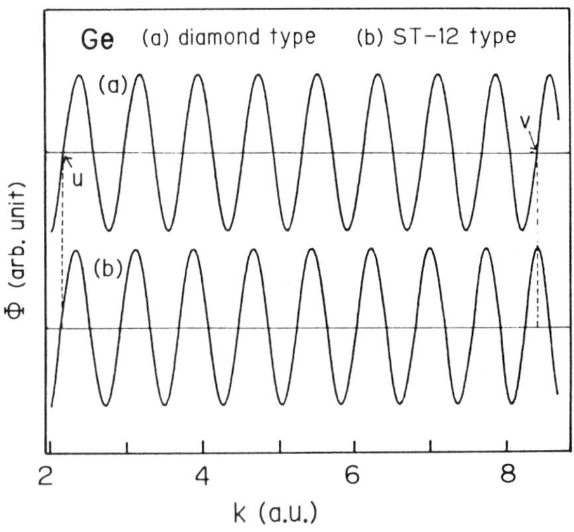

Figure 4. The normalized EXAFS profiles of diamond and ST-12 types of structures. (See text for explanation of 'u' and 'v'.)

The electronic structures were similar and the differences in bond lengths among the polymorphs were at most, 0.03 A. Thus we consider the phase shift δ and the energy origin E_0 in Equation 2 to be accurate. The normalized nearest-neighbor EXAFS spectra are given as $\Phi_{dia}(k)=\sin(2kR_{dia}+\delta)$ for the diamond type structure, and $\Phi_{ST\text{-}12}(k)=\sin(2kR_{ST\text{-}12}+\delta)$ for the ST-12 type structure where R_{dia} and $R_{ST\text{-}12}$ are the nearest-neighbor distances of the diamond and ST-12 type structures, respectively. These spectra are shown in Figure 4. At the point marked by 'u', where the phase of $\Phi_{dia}(k)=0$, the phase of $\Phi_{ST\text{-}12}$ is 0.01. At 'v', where again the phase of $\Phi_{dia}=0$, the phase of $\Phi_{ST\text{-}12}$ is 0.26. Here, the phases are expressed in unit of a period. Eight periods for the diamond type structure corresponds to 8.25 periods for the ST-12 type structure. The difference in phase ($\Delta\phi$) is related to the difference in bond length (ΔR) as $\Delta\phi/n=2\Delta R/R$, where n is the number of period and R is the bond length of the standard state. If we take the diamond type structure as the standard state ($R=R_{dia}=2.45$ A), then,

$$\Delta R = R_{ST\text{-}12} - R_{dia} = (\Delta\phi/n)(R_{dia}/2) = 0.04 \text{ A}$$

This finding is in excellent agreement with the X-ray diffraction data reported by KASPER and RICHARDS (1964). Using the same procedure, the mean bond length of the amorphous structure is determined to be 2.46 A; this is in excellent agreement with the X-ray diffraction data reported by SCHEVCHICK PAUL (1972).

In the curve fitting method, we compare the first-neighbor EXAFS with results of a calculation using the calculated phase shifts and the back-scattering amplitude. We changed the parameters N, r, and E_0 so as to obtain the smallest reliability factor, defined as

$$R = \frac{\Sigma\,(I_{exp} - I_{cal})^2}{\Sigma\,(I_{exp} - I_{cal}^2)}$$

where the summation is taken all over k-points of the nearest EXAFS. We changed the energy origin E_0 of the EXAFS until the first-neighbor distance of the diamond-type structure was in agreement with the known value of 2.45 A. Using this same E_0, the nearest-neighbor distance of the ST-12 structure was determined to be 2.49 A; this value is in excellent agreement with the value obtained by phase difference method. Note that, as shown in Figure 5, the nearest-neighbor EXAFS of the diamond and ST-12 types of structures are in excellent agreement with results of calculations which use the same phase shifts for both structures. This confirms that parameters such as phase shifts and the energy origin are transferable between these polymorphs.

In both methods, the standard deviation for the nearest-neighbor distance is estimated to be 0.01 A.

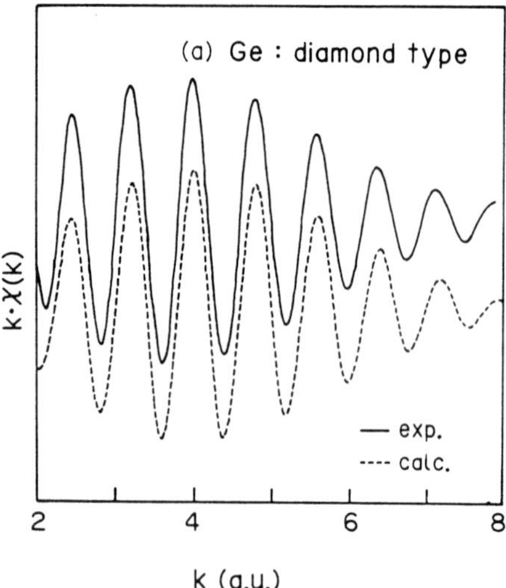

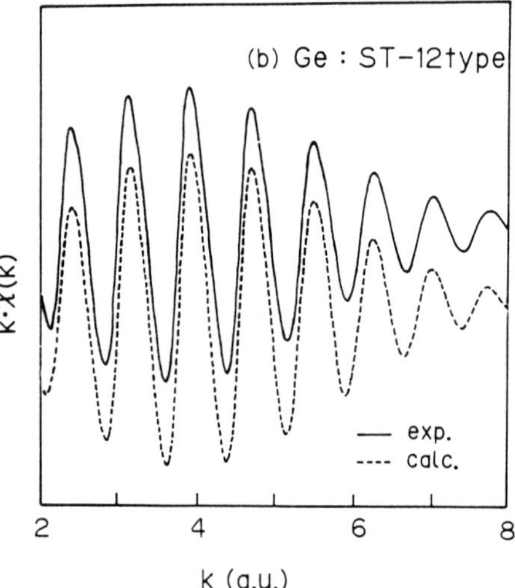

Figure 5. The nearest EXAFS profiles of a) diamond type and b) ST-12 type of structures. Dotted curves show results of the calculations which give the minimum value of R-factor for each case.

It is usually difficult to obtain the coordination number from a EXAFS spectrum without a standard sample, because there are some factors which affect the amplitude of EXAFS oscillation. When the coordination number of a certain structure is known, that of a similar type of structure can be estimated by comparing the amplitude of EXAFS oscillation after normalizing the absorption coefficient by its jump at the absorption edge. The coordination numbers thus obtained are 4.0 ± 0.2 for the ST-12 structure and 3.8 ± 0.2 for the amorphous structure, assuming the standard coordination number of the diamond-type structure to be 4.0. These values are in good agreement with the diffraction data of KASPER and RICHARDS (1964) and of SCHEVCHICK and PAUL (1972).

EXAFS Under Pressure

We found a serious problem with use of a diamond-anvil cell for the high-pressure EXAFS study (see also SHIMOMURA, 1978). Figure 6 shows the X-ray absorption spectra near the Ga K edge of metallic GaAs at 22 GPa, with a diamond-anvil cell. In these spectra, many sharp peaks, other than the EXAFS signal appear. These peaks are due to the diamond anvils. Since the anvils are single crystals, Bragg reflection occurs at a certain energy, causing the diffracted X-rays to not be received at the receiving detector. The detector behaves as if some absorption occurred at that energy. Although the energy corresponding to the Bragg reflection can be changed by adjusting the position of diamond, it is difficult to completely eliminate the apurious peaks. If two diamonds are precisely cut parallel to the (100) plane and aligned in the same direction, these peaks would degenerate. However, the accuracy of diamond cutting is only about 27°, and it is not possible to align the two diamonds precisely in the same direction.

Diffraction caused by the diamonds is a serious problem in EXAFS measurements using the diamond-anvil cell. INGALLS et al. (1980) successfully used a B_4C anvil cell up to 10 GPa to overcome this problem. In addition, WERNER and HOCHHEIMER (1982) have constructed a high-pressure cell made of Be cylinder for EXAFS measurements up to 0.8 GPa. Recently, SUENO et al. (1986) measured EXAFS near the Ni K edge using a diamond-anvil cell without obtaining diffraction peaks.

We used of a cubic-anvil system, named MAX80, to avoid the spurious peak problem. In MAX80, all the materials through which the X-ray beam passes consist of powdered materials. Bragg reflections occur at every energy level. The signal obtained is the sum of absorption due to the sample and the escape due to Bragg reflections. The Bragg reflection component is a smooth function of energy and does not affect the EXAFS oscillation. The EXAFS signal obtained in this manner is comparable in quality to the signal measured for the usual powder sample.

Figure 7 shows the X-ray absorption spectrum of Ge at 8 GPa (semiconducting phase) and at 13 GPa (metallic phase). We analyzed the data by the curve-fitting method, using the reliability factor mentioned above. For data of 8 GPa, the smallest value of the reliability factor (R=4.5%) was obtained when r=2.38 A; this is in agreement with the value of 2.39 A obtained by a piston displacement experiment (BRIDGMAN, 1948).

Using the data at 8 GPa as a standard, we analyzed the data at 13 GPa; that is, we assumed that the phase shift and the back-scattering amplitude at 13 GPa are the same as at 8 GPa (phase shift transferability assumption). The value with the best fit (R=3.4%) was obtained for quasi six-fold coordination with 3.8 atoms at r=2.56 A and 1.9 atoms at r=2.69 A. The coordination number at 13 GPa was determined by taking the coordination number at 8 GPa (N=4) as the standard. In the X-ray diffraction experiments (JAMIESON, 1963), the coordination numbers of the β-Sn structure of Ge were determined to be 4 at r=2.56 A and 2 at r=2.69 A. The values determined in the present EXAFS study are in excellent agreement with these values. Assuming exact six-fold coordination, we obtained the value of R=6.8% when the nearest-neighbor

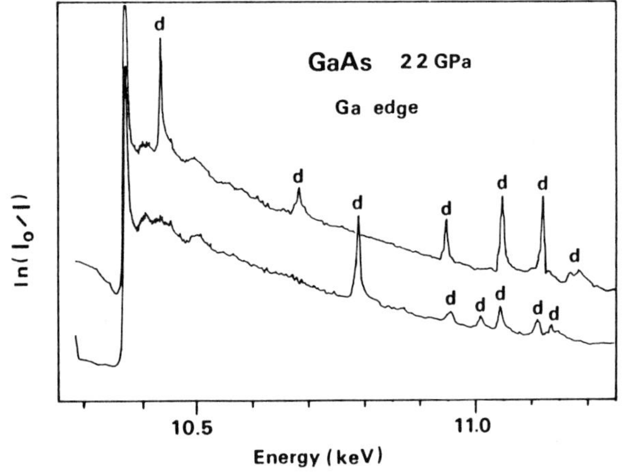

Figure 6. The X-ray absorption spectrum near the Ga K edge of GaAs at 22 GPa, obtained with a diamond-anvil cell. The sharp peaks denoted by 'd' are due to Bragg reflections of the single-crystal diamond. Upper and lower profiles were taken with different alignments of the anvil to the X-ray beam.

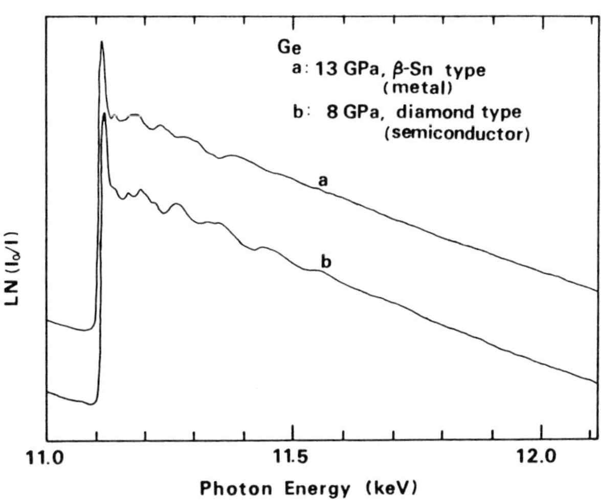

Figure 7. X-ray absorption spectra near the Ge K edge obtained using the MAX80 system at a) 13 GPa and b) 8 GPa.

distance was 2.6 A. It is worthwhile to note that the weighted average value of the nearest-neighbor distance is 2.60 A; we obtain the same value when we assume exact six-fold configuration.

XANES

Figure 8 shows the XANES spectra of GaAs near the Ga and As K edges at normal pressure and at 22 GPa (SHIMOMURA et al., 1978). At normal pressure, the profiles at the Ga and As edges consist of two peaks. The separation between the two peaks is larger at the Ga edge. In the zinc-blend type structure, bond charge is distributed towards the As atoms. This distribution may cause the difference in the edge profiles at normal pressure. The absorption profiles of Ga and As at 22 GPa each consist of a single peak and are very similar to one another. This observation indicates that the electronic surroundings of both Ga and As atoms, in the metallic state may be similar.

In order to verify the qualitative explanation of edge profiles given above, we calculated XANES based on multiple scattering theory, using a program called "XANES" (DURHAM et al., 1982). We assumed that there is a cluster of atoms around the absorbing atom, and divided this cluster into shells of atoms according to the distance from the central absorbing atom. First, we determined the multiple scattering of the excited photoelectron by atoms in a single shell, and then the multiple scatterings between shells. This gave the final wave function of the scattered electron at the absorbing atom. Since we considered all of the multiple scatterings with atoms around the absorbing atom, the final wave function can be considered to provide information on the three-dimensional atomic cnfiguration around the absorbing atom. In other words, the final electron state can be determined by considering the interaction of the electron with atoms surrounding the absorbing atom.

As a first step in our analysis, we checked the atomic shell size. We assumed that the muffin-tin radii of Ga and As atoms are also the ionic radii of those atoms and that both atoms are neutral. As is shown in Figure 9a, the calculation up to the 4th shell is apparently consistent with results of our experiments. However, the calculation up to the 6th shell differs from the present results. This indicates that the shell size up to the 4th shell is not enough for the model calculation. The calculation up to the 8th shell is almost the same as that for the six-shell. In this way, we have determined the number of shells in the calculation. We, then, change the ionicity of Ga and As atoms. Figure 9c shows the results of this analysis when

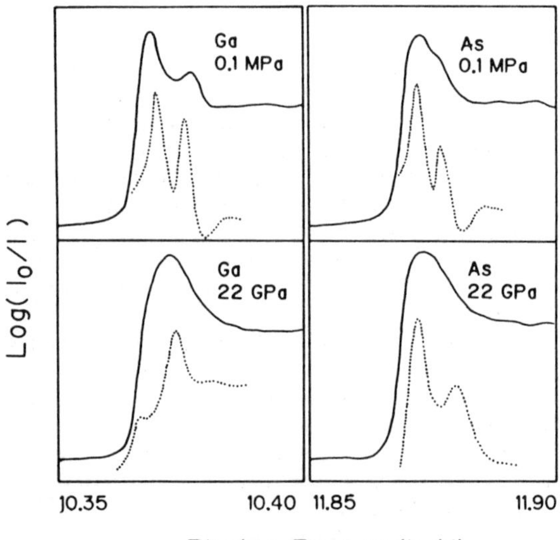

Figure 8. The XANES spectra of GaAs at the Ga and As edges, for 0.1 MPa and 22 GPa. The dotted curves are results of the calculation.

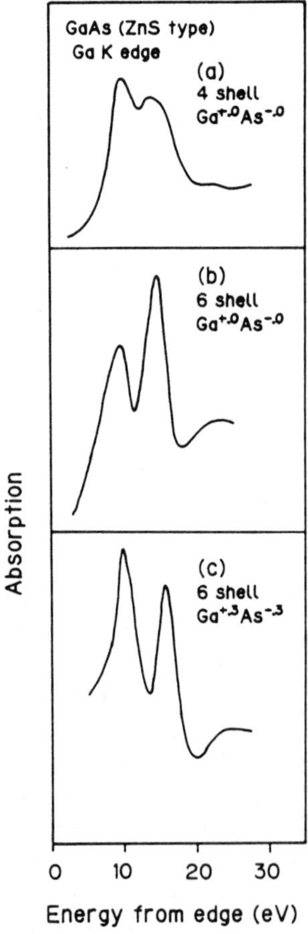

Figure 9. Calculated XANES spectra at the Ga edge for a Zns type structure: a) $Ga^{0.0}As^{0.0}$ configuration, up to the 4th shell, b) $Ga^{0.0}As^{0.0}$ configuration, up to the 6th shell, and c) $Ga^{0.3}As^{-0.3}$ configuration, up to the 6th shell.

Ga is positively charged by 0.3 and As is negatively charged by −0.3. The intensity ratio of two peaks and the separation between peaks are in better agreement with the experiment. Using the same shell size and the same ionicity, the profile near the As K edge was calculated (see Figure 8).

For the calculation at high pressure, we assumed that the muffin-tin radii for Ga and As atoms are the same and that both Ga and As atoms are neutral. Results are shown in Figure 8 (dotted curves).

The calculation is generally consistent with the experimental data. Additional systematic experiments using similar materials and a more detailed analysis are now in progress.

Summary

We have shown that EXAFS and XANES are applicable to the structural study of materials under pressure. In the EXAFS study, we have proposed use of the simple a phase difference method. When we use this method, we do not need to know parameters such as phase shift, energy origin, and mean free path, thus making the calculation quite simple. However we assume that there is only one nearest-neighbor distance and that the parameters of a known phase are transferable to an unknown phase. The curve fitting method is slightly more complicated than the phase difference method, but it can be used for a more general system. In fact curve fitting method must be used for a material which has two or more nearest-neighbor distances and/or atomic species, such as the β-Sn type of structure. With both methods, the accuracy in the bond length is 0.01 A and the accuracy in the coordination number is 0.2. If we refer to the normal pressure state as standard, the accuracy improves, especially for the determination of coordination number.

For samples used in the present study, the atoms occupy the special positions in a unit cell. The bond lengths can only be determined with ease by using the lattice constant data. But when the atoms are not at their special positions, we have to determine their atomic positions in order to calculate the bond length. For this reason, a precise intensity measurement is required as part of the X-ray diffraction analysis; such prescision requires careful experimental setup and an X-ray system with high efficiency. However, EXAFS analysis directly determines the bond length, regardless of the crystal structure. This means that the atomic positions can be determined by combining the lattice constant data and EXAFS data. Thus we conclude that the EXAFS is a promising and powerful tool for high-pressure studies.

The calculations for XANES spectra are preliminary. As discussed above, the spectra are quite sensitive to atomic structure and electronic states. Since the XANES energy range is less than 50 eV, use of XANES makes it easier to avoid spurious peaks due to Bragg diffractions, which appear when a diamond-anvil cell is used. Therefore we expect that the XANES method will be a useful tool for high-pressure structural research, when accompanied by a precise analytical method.

Acknowledgments. Part of this work was carried out as an activity of the Japan-US Cooperative Scientific Program. Thanks are due to Japan Society for the Promotion of Science (JSPS) and the National Science Foundation (NSF). The present work was partially supported by NSF grant no. DMR 73-07692, in cooperation with Stanford Linear Accelerator Center and the U.S. Department of Energy.

References

Baublitz, M., and A. L. Ruoff, Diffraction studies of the high pressure phases of GaAs and GaP, *J. Appl. Phys.*, *53*, 6179–6185, 1982.

Bridgman, P. W., The compression of 39 substances to 100,000 kg/cm^2, *Proc. Am. Acad. Arts Sci.*, *76*, 71–87, 1948.

Durham, P. J., J. B. Pendry, and C. H. Hodges, Calculation of X-ray absorption near-edge structure and XANES, *Computer Phys. Commun.*, *25*, 193–205, 1982.

Ingalls, R., E. D. Crozier, J. E. Whitmore, A. J. Seary, and J. M. Tranquada, Extended X-ray absorption fine structure of NaBr and Ge at high pressure, *J. Appl. Phys.*, *51*, 3158–3163, 1980.

Jamieson, J. C., Crystal structure at high pressure of metallic modification of silicon and germanium, *Science*, *139*, 762–764, 1963.

Kasper, J. S., and S. H. Richards, The crystal structure of new forms of Silicon and Germanium, *Acta Cryst.*, *17*, 752–755, 1964.

Kawamura, T., O. Shimomura, T. Fukamachi, and P. H. Fuoss, A phase difference approach to EXAFS analysis of germanium polymorphs, *Acta Cryst.*, *A37*, 653–658, 1981.

Sayers, G. E., E. A. Stern, and F. W. Lytle, New technique for investigating noncrystalline structures: Fourier analysis of the extended X-ray—absorption fine structure, *Phys. Rev. Letters*, *27*, 1204–1207, 1971.

Schevchick, N. J., and W. Paul, The structure of amorphous Ge, *J. Non-Cryst. Solids*, *8-10*, 381–387, 1972.

Shimomura, O., T. Fukamachi, T. Kawamura, S. Hosoya, S. Hunter, and A. Bienenstock, EXAFS measurement of high pressure metallic phase of GaAs by use of a diamond anvil cell, *Jpn. J. Appl. Phys.*, *17 Suppl. 17-2*, 221–223, 1978.

Shimomura, O., S. Yamaoka, T. Yagi, M. Wakatsuki, K. Tsuji, H. Kawamura, N. Hamaya, O. Fukunaga, K. Aoki, and S. Akimoto, Multi-anvil type x-ray system for synchrotron radiation, in *Solid State Physics Under Pressure: Recent Advance with Anvil Devices*, edited by S. Minomura, pp. 351–356, KTK, Tokyo and Reidel, Dordrecht, 1985.

Sueno, S., I. Nakai, M. Imafuku, H. Morikawa, M. Kimata, K. Ohsumi, M. Nomura, and O. Shimomura, EXAFS measurements under high pressure conditions using a combination of a diamond anvil cell and synchrotron radiation, *Chem. Letters*, 1663–1666, 1986.

Werner, A., and H. D. Hochheimer, High-pressure cell for pressure up to 8 kbar designed for X-ray absorption measurements, *Rev. Sci. Instrum*, *53*, 1467–1469, 1982.

A NEW TECHNIQUE TO MEASURE THE VISCOSITY AND DENSITY OF SILICATE MELTS AT HIGH PRESSURE

M. KANZAKI and K. KURITA

Geophysical Institute, University of Tokyo, Tokyo, Japan

T. FUJII

Earthquake Research Institute, University of Tokyo, Tokyo, Japan

T. KATO

Department of Earth Science, Nagoya University, Nagoya, Japan

O. SHIMOMURA

National Institute for Research in Inorganic Materials, Sakura, Ibaraki, Japan

S. AKIMOTO

Institute for Solid State Physics, University of Tokyo, Tokyo, Japan

Abstract. A new method to measure viscosity and density of silicate melts at high pressure has been developed. With this method, the movement of the spheres falling within silicate melts that are contained in a high-pressure apparatus can be monitored in real time through an X-ray shadowgraph, and the settling velocity can be measured without quenching the charge. Because of the difference between the X-ray absorption coefficients of metal spheres and silicate melts, the image of the falling spheres can be clearly traced through an X-ray TV camera. The experiment was conducted with high-pressure and high-temperature apparatus (MAX80) using synchrotron radiation. The high intensity and the parallelism of the X-rays of synchrotron radiation and of the high-resolution Saticon X-ray TV camera allow accurate measurement of the falling velocity of the spheres, even for melts with low viscosity. With the present method, the measurable range of viscosity can be extended down to 10^{-3} Pa s. Because of the multianvil type of high-pressure apparatus used, the experimental range of pressure can be extended up to 10 GPa.

Introduction

It is important to study the viscosity and density of magmas at high pressure to understand magmatic processes such as segregation of magmas from the source mantle and gravitational separation of crystals within the magma chamber. Viscosity and density would also have been critical parameters in controlling the gravitational differentiation processes that played significant roles in large-scale melting during the early stages of the earth's history (OHTANI, 1985).

The viscosity of silicate magmas has been measured up to 3 GPa by Kushiro and his coworkers (e.g., KUSHIRO, 1986) using a piston-cylinder apparatus. Their results indicate that the viscosity of most silicate magmas decreases with increasing pressure. This unexpected behavior at high pressure indicates that magmatic transport may be greatly enhanced in the earth's deep interior (KUSHIRO, 1986). Based on the ion dynamics calculation, ANGELL et al. (1982) determined that most ions in jadeite ($NaAlSi_2O_6$) melt have maximum mobilities at about 25 GPa. This suggests that the viscosity of jadeite melt decreases with increasing pressure but has a minimum value at some pressure. However, this finding has not yet been confirmed experimentally because of the limitation of the high-pressure apparatus used in previous experiments.

The effect of pressure on the density of magmas has been determined by FUJII and KUSHIRO (1977) using the falling sphere methods with a piston-cylinder apparatus. Their results indicate that silicate magmas are more compressible than crystals. This was confirmed by RIGDEN et al. (1984) using the shock-wave technique. The results of these previous studies suggest that the density of basaltic magma could exceed the density of mantle peridotite at a depth between 200 and 300 km (STOLPER et al., 1981; OHTANI, 1985). This possible density crossover within the upper mantle may have played a significant role in the early stages of formation of the present stratified earth. However, silicate melts coexisting with peridotite at those depths are more enriched than basaltic melt in the olivine component (TAKAHASHI and SCARFE, 1985). It is necessary to extend experimental measure-

High-Pressure Research in Mineral Physics, edited by M. H. Manghnani and Y. Syono, pp. 195–200.

ments to higher pressure and over a wider range of melt compositions to evaluate the above possibilities.

The falling sphere method is considered to be the most suitable for the measurement of viscosity and density of magmas in a high-pressure apparatus. In the previous studies the settling velocity of a sphere was estimated from the distance settled over a certain experimental time, for the quenched sample (e.g., KUSHIRO, 1976). Therefore the accuracy of the results depended on the measurement of the run duration. In the case of low-viscosity melts, spheres with only small density differences from melts have been used to extend the run duration and to increase the accuracy of the estimated settling velocity. These spheres, which are mostly silicate minerals, sometimes react with the magmas, especially at high temperatures. Consequently, this type of experiment becomes difficult to conduct at higher pressure because the liquidus temperature of silicate materials increases with increasing pressure. In the present study, we use noble metals for the falling spheres, and thereby overcome problems related to the reactivity of the spheres with silicate melts. Because of the large density difference between silicate melts and metals, the settling velocity of the spheres becomes quite large. Thus Kushiro's method of estimating the settling velocity is inadequate for low-viscosity melts. In order to accurately measure the velocity of such quickly settling spheres, a method which measures the velocity of falling spheres in real time has been developed. By using the present method, as compared to the method developed by KUSHIRO (1976), the accuracy of the measurements can be increased, especially for the melts with low viscosity. Also the experimental range of pressures in which we can measure viscosity and density can be expanded because of the multianvil type of high-pressure apparatus used.

Experimental Methods

The experiments were conducted with the MAX80 high-pressure system of the Photon Factory, National Research Institute for High Energy Physics in Tsukuba (SHIMOMURA et al., 1985). The present method is based on observation of the X-ray shadowgraph of falling spheres. Two important advances in experimental techniques have made this method possible. One is the excellent X-ray source of synchrotron radiation. Due to the high intensity and parallelism of the incident X-ray beam, we can directly observe a clear shadowgraph of metal spheres in silicate melts that are embedded in a pressure medium. Another important advance is the Saticon TV camera which is sensitive to X-rays (ANDO et al., 1984). This Saticon camera (amorphous Se-As alloy 2-D sensor) has a high spatial resolution (6 μm) and low lag characteristics. We also used an optical TV camera with a fluorescent screen for indirect observation of the spheres. The spatial resolution in this indirect system which depends on grain size of the fluorescent material, is about 10 μm. Compared to the direct Saticon X-ray TV camera, the results of this fluorescent camera are quite unsatisfactory with regard to both the spatial resolution and image contrast.

General features of the experimental system are shown in Figure 1. White X-rays generated by the synchrotron (2.5 GeV and 60–120 mA) penetrate the pressure medium and sample assembly and are detected by the Saticon TV camera. Due to the large difference in X-ray absorption coefficients between the silicate melt and metal spheres, high-contrast images of the spheres can be observed on a video monitor. The area of the high-pressure cell visible by X-ray is limited by the opposing anvils, and depends on both anvil size and pressure. The typical visible area is about 0.5 mm in width and 6 mm in height, in the case of a 6-mm anvil (see Figure 3).

Figure 2 shows the sample assembly used in our experiments. The pressure medium is a mixture of boron and epoxy resin. In the early stages of our experiments a straight cylindrical carbon heater (Figure 2) was used, and a large temperature gradient within the sample cell was detected. In later stages of our experiments we used a specially designed carbon heater, which had a central portion of greater thickness than other portions, to obtain a more uniform temperature distribution within the sample cell. With this heater, the temperature variation within a sample was limited to 30 °C at a run temperature of 1500 °C.

Temperature was monitored by a Pt/PtRh 13% thermocouple. In some high-temperature experiments, however, temperature measurement was not successful because of the reaction (described below) between the thermocouple (Pt/PtRh 13%) and the pressure medium.

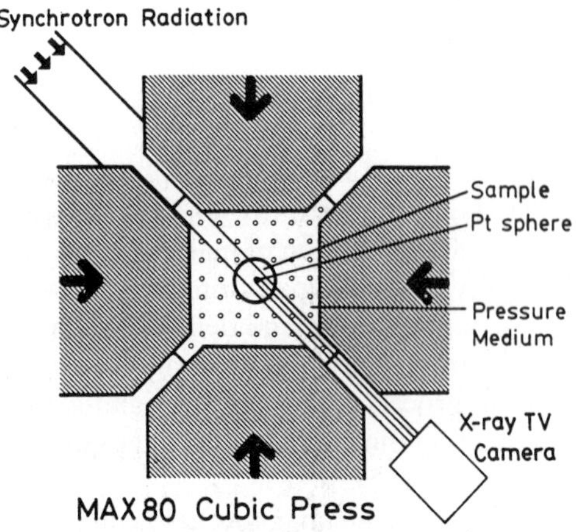

Figure 1. Schematic illustration of the present in situ high-pressure X-ray system for measurement of viscosity and density.

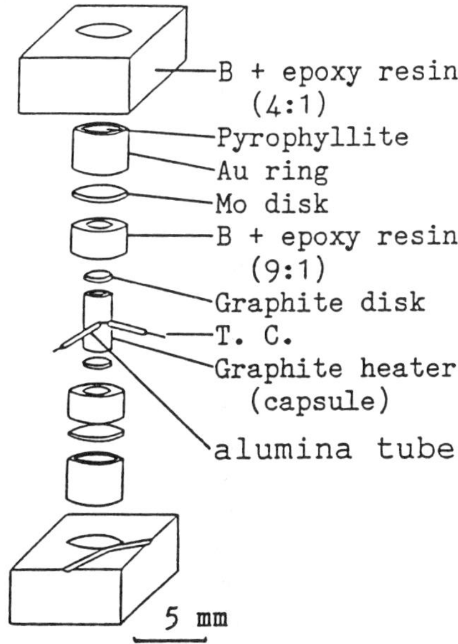

Figure 2. High-pressure cell assembly for in situ measurement.

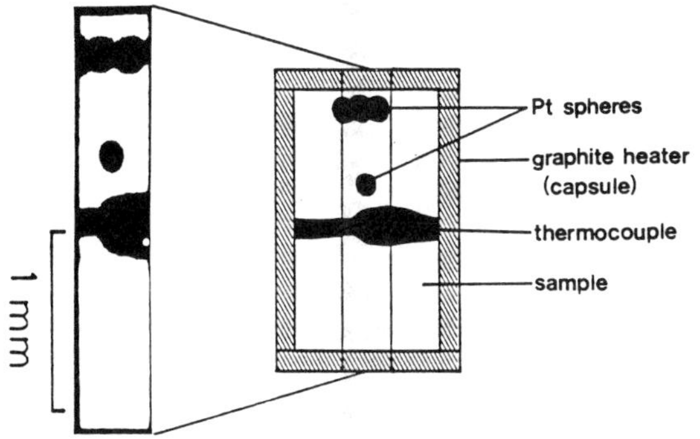

Figure 3. An example of an X-ray shadow image of platinum spheres in silicate melt at 3 GPa. Figure of right-hand side indicates cross-section of the graphite capsule (see Figure 2). This figure shows relation between visible area and capsule (i.e., graphite heater, see Figure 2). The visible area is limited by opposed anvils.

For each experiment, a powdered sample was placed within the central part of the carbon heater. Platinum and/or molybdenum spheres ranging from 100–300 μm in radius were embedded in the upper portion of the sample.

The sample assembly was then set in the MAX80 cubic press. After pressure was increased to the desired level, the sample was quickly heated beyond its melting point. Movement of the sphere within the silicate melt was observed on the TV screen and recorded by a VTR. The falling velocity, measured from the TV image, gives the viscosity, and the difference between the velocities of platinum and molybdenum spheres gives the density.

Results

An example of the X-ray TV image of falling spheres within a silicate melt is shown in Figure 3. Circular shadows indicate the platinum spheres. The horizontally elongated shadow at the center of the image indicates the central thermocouple. One of the platinum spheres, (i.e., the sphere with a diameter of 180 μm) is settling very slowly in the albite ($NaAlSi_3O_8$) melt at a pressure of 3 GPa and a temperature of 1100 °C±200 °C (estimated from heating power). The settling of this platinum sphere is clearly shown in the sequential photos of Figure 4. The time interval between each photo is 20 s. For this run, we used a straight carbon furnace (Figure 2) instead of the specially designed furnace. Due to the temperature gradient within the capsule during this run, the upper three spheres (see Figure 3) did not settle because they are in the temperature range below solidus.

From the shadow image we can determine the settling velocity of the spheres. Figure 5 shows the falling distance as a function of time for spheres in the albite melt (see Figure 4). Excepting the acceleration stage at the beginning and the deceleration stage at the end, there is a constant terminal velocity of 7 μm/s. From the observed terminal velocity we can calculate the viscosity, if the density of the melt is known. The density of the albite glass quenched at high pressure was measured by KUSHIRO (1978); it is about 2.5 g/cm^3 at 2–3 GPa. With this density of 2.5 g/cm^3, viscosity of the melt is calculated to be 31 Pa s. Because the difference between the densities of the silicate melt and noble metal is large, the error in the viscosity estimation which is attributable to the density change caused by temperature and pressure, is negligibly small.

Figure 6 shows results of another run. In this run, platinum spheres 250 μm in diameter settle very quickly

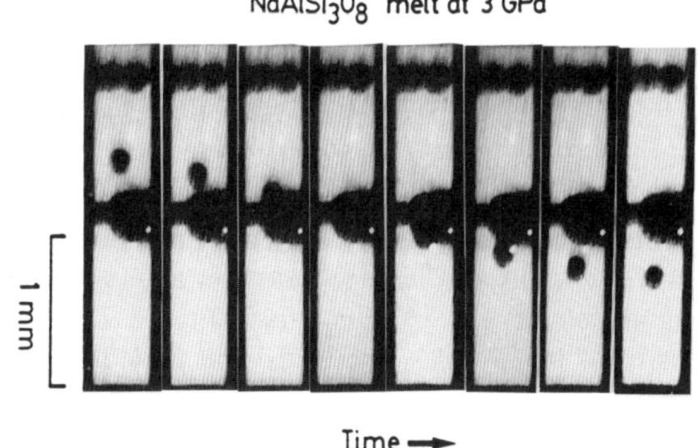

Figure 4. Sequential photographs of a settling platinum sphere (rounded shadow) in $NaAlSi_3O_8$ melt at 3 GPa and 1100 °C±200 °C. The time interval between each photo is 20 s.

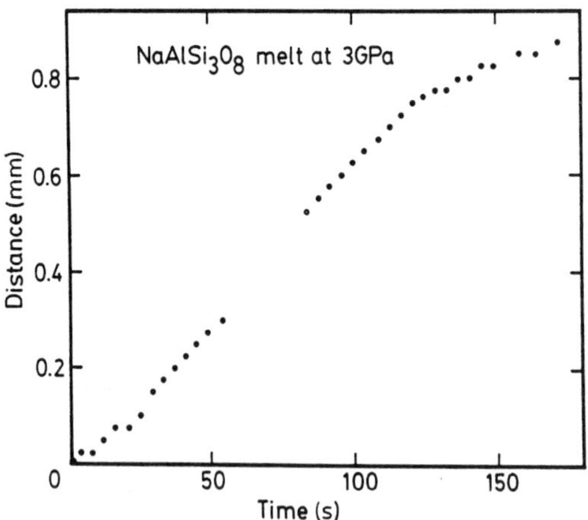

Figure 5. The settled distance of a platinum sphere as a function of time in $NaAlSi_3O_8$ melt (see also Figure 3).

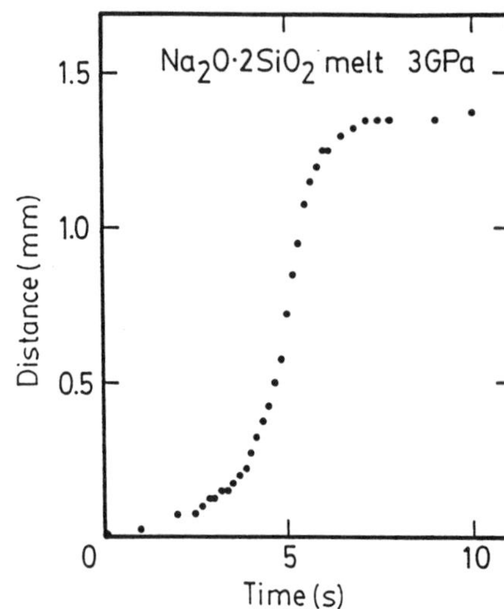

Figure 7. The settled distance of a platinum sphere as a function of time in $Na_2O{\cdot}2SiO_2$ melt (see also Figure 6).

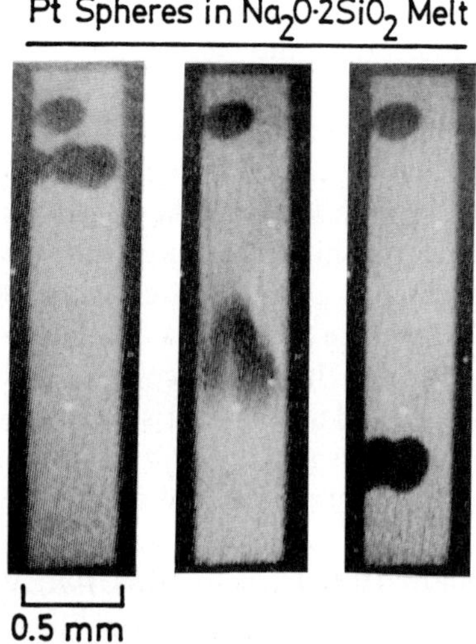

Figure 6. Sequential photographs of falling platinum spheres in $Na_2O{\cdot}2SiO_2$ melt at 3 GPa and 1000 °C±200 °C. Time interval between the sequential photo is about 2 s.

within a sodium disilicate ($Na_2O{\cdot}2SiO_2$) melt at 3 GPa and 1000 °C±200 °C. The time difference between the sequential images is only 2 s. Based on a density of 2.8 g/cm^3 for sodium disilicate (SCARFE et al., 1979) and a terminal velocity of 730 μm/s (see Figure 7), a melt viscosity of 0.6 Pa s is obtained. The previous falling sphere method cannot measure such a high velocity. Figure 7 demonstrates the exceptional ability of the present method for measuring high settling velocities.

The lowest viscosity measured by the present system was 0.04 Pa s. This viscosity was measured in a $CaGeO_3$ melt at 1.6 GPa and 1225 °C using a falling sphere of platinum. It is one to two orders of magnitude lower than the lowest viscosity previously measured by researchers, even those using silicate minerals as the falling sphere.

As demonstrated by FUJII and KUSHIRO (1977), the density of a melt can be measured with falling spheres of different densities. Movement of two kinds of spheres in an albite melt was monitored with the present method. Results for an albite melt at 2 GPa and 1100 °C±200 °C is shown in Figure 8. Because the density of molybdenum is about one half that of platinum, the settling velocity of the molybdenum sphere is much lower than that of the platinum sphere. Based on these velocities, the melt

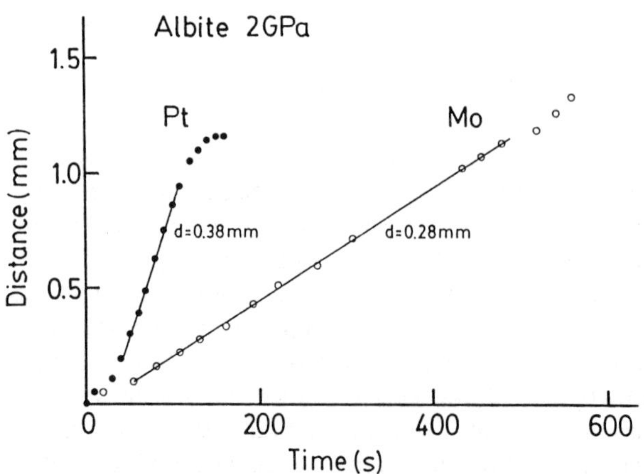

Figure 8. Time versus settled distance of platinum and molybdenum spheres in $NaAlSi_3O_8$ melt at 2 GPa and 1100 °C±200 °C.

density and viscosity are calculated to be 2.6 g/cm^3 and 77 Pa s, respectively. The melt density obtained is similar to that of quenched glass at 2 GPa (2.5 g/cm^3). However, the viscosity obtained by present method seems to be much smaller than that expected from the results of KUSHIRO (1978) at 1400 °C (1600 Pa s). This discrepancy could be due to the contamination of the melt by a volatile component released from the pressure medium (discussed below). Thus the measured values may not be applicable to an albite melt without the volatile component.

Discussion

Our results are still quite preliminary. We have mainly discussed the technical aspects of our experiments.

In a low-viscosity melt, remarkable acceleration and deceleration stages are observed (see Figure 7). These nonlinear parts of runs are explained by either the end boundary effect at the top and bottom of the sample cell or by a temperature drop at the both ends of the cell. This end effect probably occurred because the metal spheres used in the experiment were not small enough, as compared to the size of the container. This may indicate that a correction only for the wall effect is not sufficient, and that a correction for the effect of container height is also necessary for the falling sphere in a container of finite length. This finding of a nonlinear distance-time relation suggests that the application of the previous falling sphere method to low-viscosity melts might cause an overestimation of viscosity. The maximum attainable pressure for this method depends on the anvil size of the MAX80 system. The upper and lower limits of the measurable viscosity depends on the range of the measurable falling velocity. Therefore the upper and lower limits are constrained by the spatial and temporal resolution of the X-ray imaging system, respectively. The present Saticon X-ray TV camera has a spatial resolution of 6 μm and temporal resolution of several tens of milliseconds. When we use an anvil of standard 6 mm size (the maximum pressure would be about 7 GPa), we can measure viscosity down to 10^{-3} Pa s (Figure 9) using a platinum sphere. This viscosity is almost the same as that of water at ambient conditions. The hatched area shows the experimental range studied by the conventional piston-cylinder apparatus. Our present system greatly extends the experimental ranges of both pressure and viscosity.

Although the method shown here proved to work well, there still remain several problems to overcome. The most significant problem is the selection of the pressure medium. In the present experiments we used a pressure medium of boron and epoxy resin to reduce X-ray absorption. However, platinum thermocouples reacted with the pressure medium when the temperature exceeded 1000 °C. Sometimes platinum spheres within samples

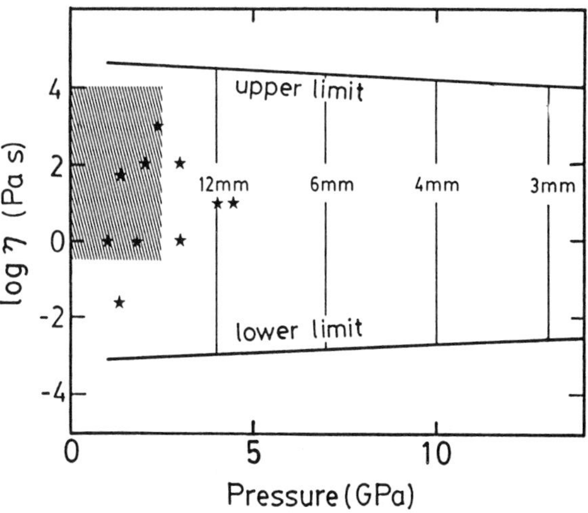

Figure 9. Measurable range of viscosity with pressure in the present system. Vertical lines indicate attainable pressures using various anvil sizes. Hatched area and stars show experimental conditions in previous works (SCARFE et al., 1979) and present work, respectively.

fused at relatively low temperatures of about 1400 °C. In the case of basalt and dacite melts, the platinum fused at such low temperatures that two separate phases were formed. Analysis of the sphere which was in dacite liquid, indicates that one phase contained 4 weight% of silicon and that the other phase contained 3 weight% of iron. Boron was detected in neither platinum nor silicate glass. We believe that the samples were reduced too much to form silicon and iron from a dacite melt, but rather those reduced metals formed alloys with the platinum. This reducing reaction could have been caused by a volatile component created by the breakdown of epoxy resin. This volatile component may have also reduced the liquidus temperature and viscosity of the melt. This could be the reason that our results on the viscosity and density of albite melt are lower than those expected from the results of KUSHIRO (1978). Use of a baked pyrophyllite sleeve between the heater and the pressure medium improved the situation somewhat, but it was still not perfect. A good material for the pressure medium still needs to be found before we can perform reliable experiments at high temperatures.

Acknowledgments. The authors are grateful to H. Kawata of Photon Factory and F. Sato of NHK Broadcasting Science Research Laboratories for their assistance and permission to use the Saticon camera. The authors are also indebted to M. Kato and T. Yagi for their numerous suggestions and considerable advice, and to T. Watanabe and A. Yasuda of the University of Tokyo for their assistance.

References

Ando, M., J. Chikawa, H. Hashizume, K. Hayakawa, T. Imura, T. Ishikawa, S. Kishino, S. Nagakura, O. Nittono, and S. Suzuki, High speed X-ray topography, *Photon Factory Activity Report*, *82/83*, VI-100, 1984.

Angell, C. A., P. A. Cheeseman, and S. Tamaddon, Pressure enhancement of ion mobilities in liquid silicates from computer simulation studies to 800 kilobars, *Science*, *218*, 885–887, 1982.

Fujii, T., and I. Kushiro, Density, viscosity, and compressibility of basaltic liquid at high pressures, *Carnegie Inst. Wash. Yearb.*, *76*, 419–424, 1977.

Kushiro, I., Changes in viscosity and structure of melt of $NaAlSi_2O_6$ composition at high pressure, *J. Geophys. Res.*, *81*, 6347–6350, 1976.

Kushiro, I., Viscosity and structural changes of albite ($NaAlSi_3O_8$) melt at high pressures, *Earth Planet. Sci. Lett.*, *41*, 87–90, 1978.

Kushiro, I., Viscosity of partial melts in the upper mantle, *J. Geophys. Res.*, *91*, 9343–9350, 1986.

Ohtani, E., The primordial terrestrial magma ocean and its implication for stratification of the mantle, *Phys. Earth Planet. Inter.*, *38*, 70–80, 1985.

Rigden, S. M., T. J. Ahrens, and E. M. Stolper, Densities of liquid silicates at high pressures, *Science*, *226*, 1071–1074, 1984.

Scarfe, C., B. O. Mysen, and D. Virgo, Changes in viscosity and density of melts of sodium disilicate, sodium metasilicate, and diopside composition with pressure, *Carnegie Inst. Wash. Yearb.*, *78*, 547–551, 1979.

Shimomura, O., S. Yamaoka, T. Yagi, M. Wakatsuki, K. Tsuji, H. Kawamura, N. Hamaya, O. Fukunaga, K. Aoki, and S. Akimoto, Multi-anvil type X-ray system for synchrotron radiation, in *Solid State Physics under Pressure*, edited by S. Minomura, p.351–356, Terra Sci. Pub. Co., Tokyo, 1985.

Stolper, E. M., D. Walker, B. H. Hager, and J. F. Hays, Melt segregation from partially molten source region: the importance of melt density and source region size, *J. Geophys. Res.*, *86*, 6261–6271, 1981.

Takahashi, E., and C. M. Scarfe, Melting of peridotite to 14 GPa and the genesis of komatiite, *Nature*, *315*, 566-568, 1985.

IV. SYNTHESIS, PHASE EQUILIBRIA AND THERMODYNAMIC PROPERTIES OF MANTLE PHASES

SUPPRESSION OF SPONTANEOUS NUCLEATION AND SEEDED GROWTH OF DIAMOND

Masao WAKATSUKI and Kaoru J. TAKANO

Institute of Materials Science, The University of Tsukuba, Sakura-mura, Ibaraki-ken 305, Japan

Abstract. In the flux method of growing diamonds from graphite, the spontaneous nucleation of the diamond is suppressed, if graphite particles are regrown and cover the surface of the starting graphite prior to the diamond nucleation. The minimum pressure necessary for diamond formation from the regrowth-treated graphite (P^*) is higher than that from fresh graphite (P_0), by 0.19 GPa at 1350 °C with an alloy of Fe–Ni–Co employed as the flux. As an application of this suppressive effect, a seed crystal of diamond, located between the graphite and the flux, can be grown epitaxially by a two-stage pressure cycle. The first stage of the cycle is at a pressure lower than P_0 allowing the regrowth of graphite. The second stage is for the growth of the seed to occur at a sustained elevated pressure level. Crystals up to 2 mm in diameter have been grown.

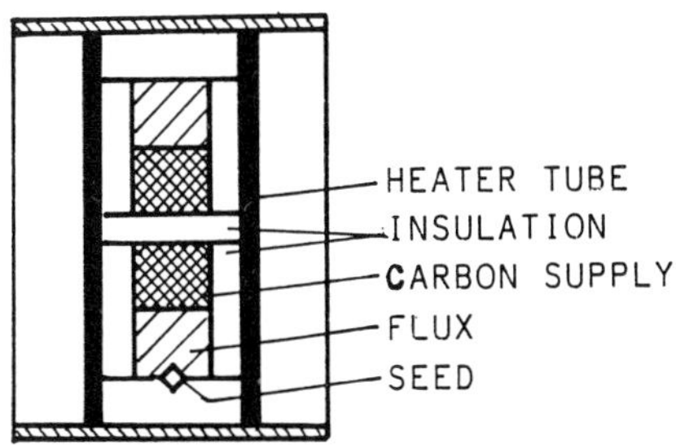

Figure 1. A high-pressure reaction cell for the growth of diamond crystals (after STRONG and WENTORF, 1972).

Introduction

Synthesis or growth of materials at high pressure has direct implications for the experimental characterization of the materials of the earth's interior. The most essential points for successful synthesis are, first, to understand the conditions of the synthesis clearly and, second, to control them precisely. An apparatus for high-pressure synthesis has been reported previously (WAKATSUKI and ICHINOSE, 1982); as an example of its application, a precisely controlled process of nucleation of diamond was discussed. In this extended study we discuss the roles of pressure and temperature to drive change in materials and a new method of growing single-crystal diamonds.

Driving Forces for Growing Crystals of Diamond

In the conventional method of growing single-crystal diamonds (STRONG and WENTORF, 1972; KANDA and FUKUNAGA, 1982) materials are arranged as shown in Figure 1. This is called "the temperature-gradient method." The carbon source consists of small diamonds and is located at the higher-temperature side of a flux layer; the seed is located at the lower-temperature side. Thus, the driving force for growth is the temperature difference between the source and the seed. The transportation of dissolved carbon from the source to the seed usually governs the rate of growth; this transport occurs by diffusion. As a result, the growth rate is proportional to the temperature difference δT ($=T_s - T_c$). The subscripts s and c correspond to the source and the seed, respectively. The solubility difference δX ($=X_s - X_c$) is almost proportional to δT according to the theory of regular solutions, and is written

$$\delta X / X_c = (\Delta S_f / R)(\delta T / T_c) \tag{1}$$

where ΔS_f is the molar entropy of fusion of diamond. The advantage of this method is the ease of keeping a small driving force δX or δT for a long time, as long as the reaction pressure P and the temperature T_s guarantee thermodynamical stability of diamond.

On the other hand, solubility difference between graphite and diamond depends not only on temperature but also on "excess pressure," δP, which is defined as

$$\delta P = P - P_e \tag{2}$$

where P is the reaction pressure, which is uniform throughout the growing cell. P_e is the equilibrium pressure between graphite and diamond. Assuming that temperature is uniform in the cell, the solubility difference is proportional to δP and written as

$$\delta X / X_c = (\Delta V / RT) \cdot \delta P \tag{3}$$

where ΔV is the difference in molar volumes of graphite and diamond. Thus a seed may be grown even in uniform temperature if graphite is used as the source material with a very small excess pressure, as shown by earlier workers (BUNDY, HALL, STRONG, and WENTORF, 1955; BUNDY, BOVENKERK, STRONG, and WENTORF, 1961). In this case

High-Pressure Research in Mineral Physics, edited by M. H. Manghnani and Y. Syono, pp. 203–207.

the driving force is the excess pressure. However, δX can play a role of an excess concentration for the spontaneous nucleation of diamond, especially in the neighborhood of the source graphite. According to STRONG and HANNEMAN (1967) and BUNDY, STRONG and WENTORF (1973), spontaneous nucleation of diamond with graphite as the source material begins at a pressure very near to the equilibrium and is slightly suppressed in the neighborhood of a seed crystal if the excess pressure is small ($\lesssim 0.5$ GPa). Metastable growth of graphite together with diamond is also described. It seems, however, that many of the nucleation behaviors are left for further research. The growth of a single crystal can be realized easily if there is a sufficiently large range of excess pressure in which the nucleation is inhibited.

Related to this point, the previous study (WAKATSUKI and ICHINOSE, 1982) suggested a possible method of intentional suppression of nucleation. In the previous study the sample was fabricated from graphite and a flux material (such as Ni, Co, Fe, and their alloys) in contact with each other. The sample was heated to a reaction temperature and kept for an appropriate period (ca. one minute) at a pressure slightly lower than the minimum pressure for formation of diamond (regrowth stage for graphite). Then pressure was raised and kept for a sufficient time (nucleation and growth stage for diamond). Introduction of the earlier stage decreased the number of diamond nuclei and gave diamonds of a large and uniform size. The controlling mechanism for number of diamond nuclei was explained by assuming that the presence of regrown graphite particles covering the surfaces of the graphite plate (starting material) suppresses the diamond nucleation.

Experiments

Suppression of Diamond Nucleation by Regrown Graphite

We tried to confirm the suppressive effect by comparing the minimum pressure needed to form diamond on a fresh graphite surface with that on a surface of graphite covered with the regrown particles. The sample is shown in Figure 2. The graphite used was a synthetic material with a density of 1.6 g/cm^3. The flux was an alloy of Fe–Ni–Co (55:29:16 by weight). A cubic anvil apparatus, described previous by (WAKATSUKI and ICHINOSE, 1982), was used. Pressure and temperature reproducibility in the apparatus was 0.05 GPa and about 50 °C, respectively, as reported. The temperature difference across the flux layer was estimated to be less than 20 °C. Two kinds of experiments (OP. 1 and OP.2) were done to determine the two characteristic pressures mentioned above (see Figure 3).

First the sample was compressed to a reaction pressure

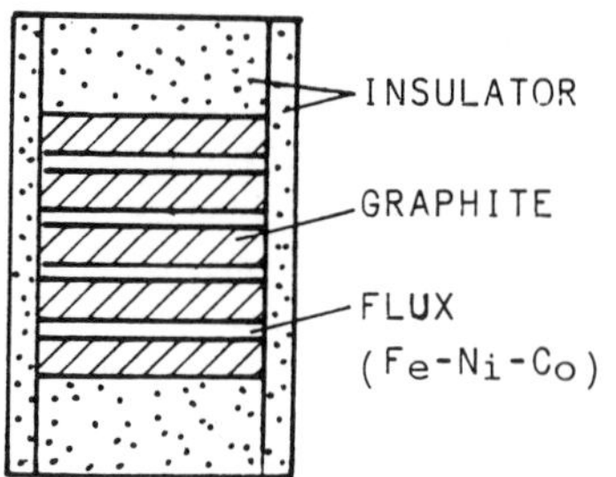

Figure 2. Sample for evaluation of the suppression of diamond nucleation.

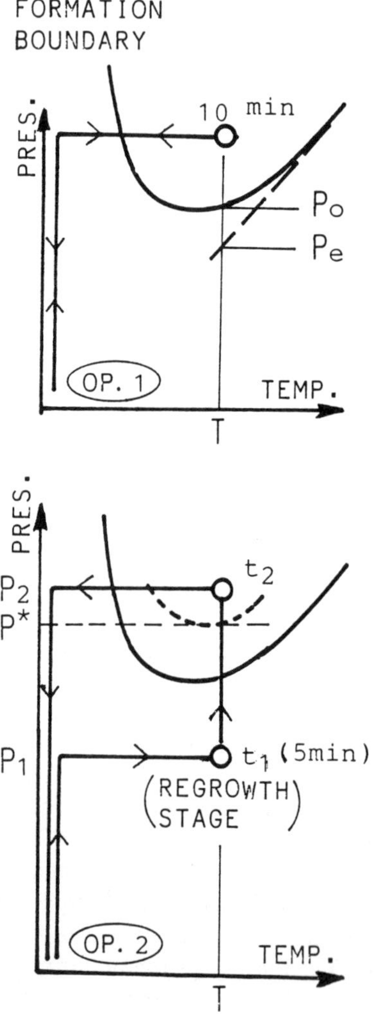

Figure 3. Pressure and temperature paths in experiments.

P and then the temperature was raised to a reaction temperature T, in a sufficiently short time so that fresh surfaces of graphite reacted with flux at the condition P and T. This condition was held for 10 min. This is shown as OP. 1 in Figure 3. For all these experiments the reaction temperature T, measured at the center of the cell, was kept at 1350 °C, while P was varied from sample to sample. The recovered sample was acid-treated and examined with a microscope. The minimum pressure P_0 required to form diamond on a fresh surface of graphite was determined as 5.33 GPa. This must be a little higher than the equilibrium pressure P_e at the same temperature. The difference, $P_0 - P_e$, will be evaluated later.

The experimental procedure was then changed to that shown as OP. 2 in Figure 3. In an experimental run of OP. 2, pressure was applied in two stages, as shown in Figure 3. For the first stage at P_1, it was anticipated that graphite particles would be regrown and cover the surface of the graphite plate. Thus P_1 had to be lower than P_0. Actually $P_0 - 0.26$ GPa was selected. Reaction temperature was again 1350 °C. A heating time of 5 min was obviously sufficient for this purpose according to the previous study. After heating for 5 min at P_1, pressure was raised to a higher level, P_2, in less than half a min and sustained there for 10 min, while the reaction temperature was not changed. The second stage pressure, P_2, was changed from sample to sample. Formation of diamond was examined in the recovered material. The pressure P^*, the minimum level of P_2 required to form diamond from the "regrowth-treated" graphite, was found to exceed P_0 by 0.19 GPa, or

$$P_0 = 5.33 \text{ GPa} \tag{4}$$

$$P^* = P_0 + 0.19 \text{ GPa} \tag{5}$$

Thus we confirm the suppressive effect on the diamond nucleation brought about by the presence of regrown graphite particles on the surface of the graphite plate. The magnitude of the effect may be evaluated with a pressure difference of 0.19 GPa.

Application of the Suppressive Effect to the Seeded Growth of Diamond

Based on the above experiments, growing a single crystal of diamond was expected to be possible with a seed-containing sample by a procedure like OP. 2 with P_2 in a region between P_e and P^*. The sample is shown in Figure 4. The starting materials, graphite and flux, were the same as described previously (WAKATSUKI and ICHINOSE, 1982). Seeds (0.3–0.4 mm) were selected from a commercial product synthesized for abrasive use.

Trials were made with various P_2 and t_2, while T, P_1, and t_1 were same as those in OP. 2 of the preceding experiments, and the selective growth of the seed was confirmed. Figure 5 shows examples of single crystals epitaxially grown on the seeds.

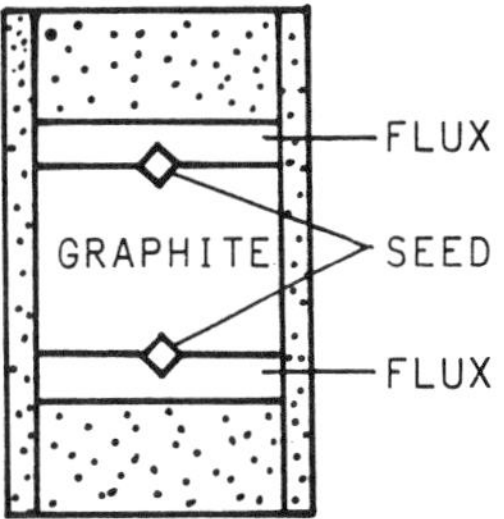

Figure 4. Sample for seeded growth of diamond.

Growth experiments made at a pressure P_2 of 5.37 GPa showed a linear relationship between diametral increase of the grown crystal and the growth time, as shown in Figure 6. The designations, "BOTTOM" and "TOP", in Figure 6 correspond to the positions of the seeds, as illustrated in Figures 4 and 8. The diametral growth rate of the bottom crystal changes linearly with pressure, as seen in Figure 7, but the top crystal behaves curiously. Different behaviors between the seed positions are also

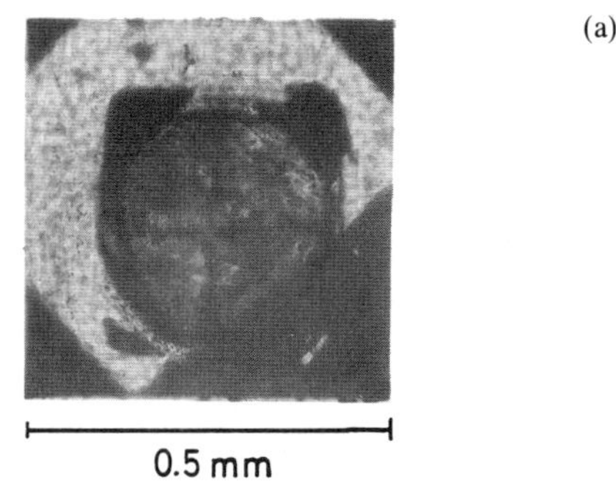

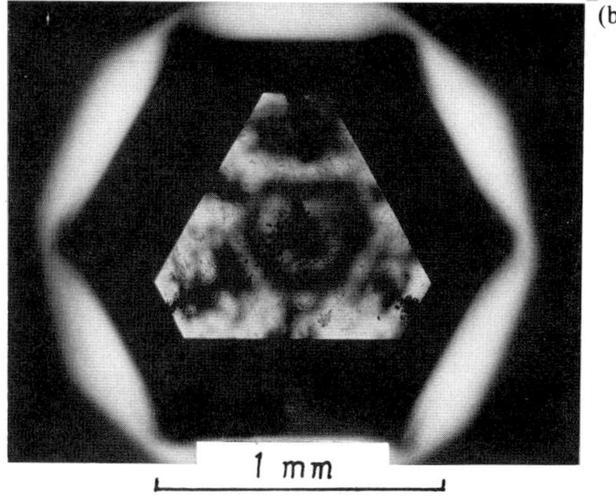

Figure 5. Grown crystals of diamond. a) $P_2 - P_0 = 0.04$ GPa and $t_2 = 0.5$ hr at the top site. b) $P_2 - P_0 = 0.05$ GPa and $t_2 = 3.0$ hr at the bottom site.

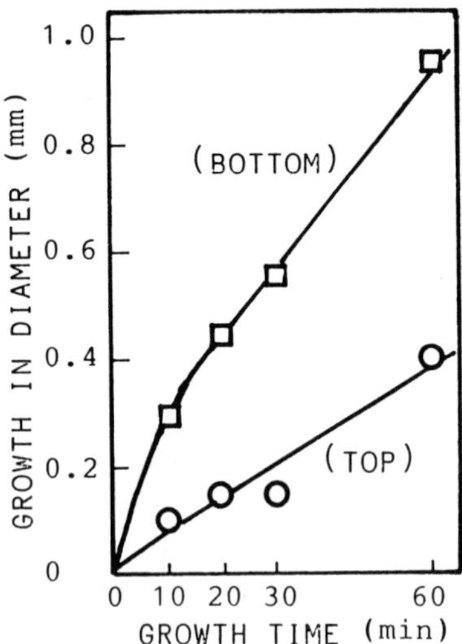

Figure 6. Dependence of diametral increase on growth time.

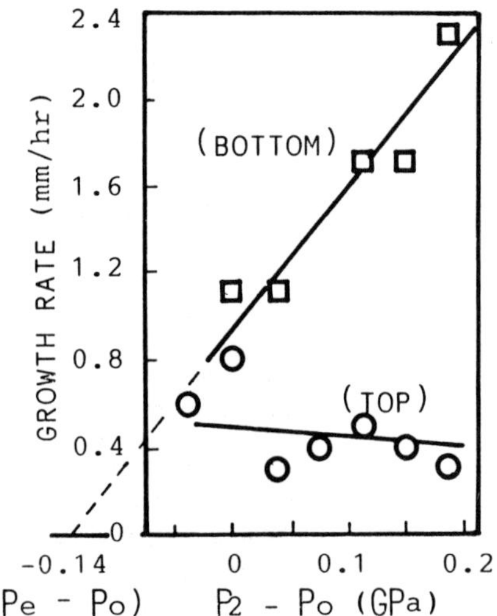

Figure 7. Dependence of growth rate on pressure.

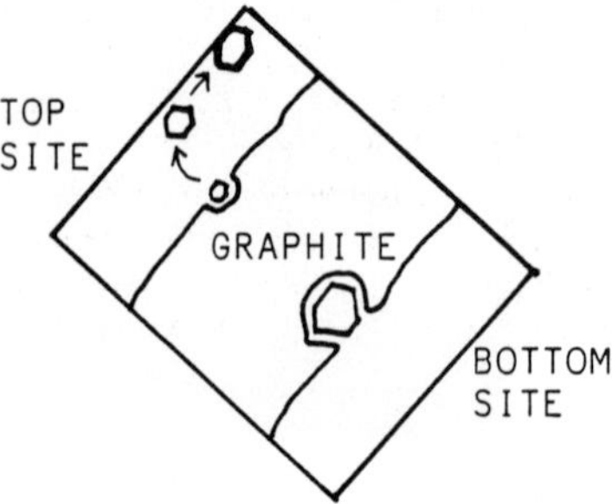

Figure 8. Effect of seed position.

shown in Figure 6. Detailed observation of recovered samples revealed that the top seed was not held on the graphite surface but moved to a different place by floating into molten flux. Thus the geometrical condition around the top seed changes remarkably during its growth. On the other hand, the bottom seed did not change its place and kept contact with graphite through a thin layer of flux. Crystals up to about 2 mm in diameter have been grown. At present it is difficult to understand and control the growth condition around the top seed, but this position seems more suitable than the other for growing crystals of good quality.

Discussion

Equilibrium Pressure and Growth Behavior

As described in the preceding section, quantitative treatment of the growth rate should be based on data for the bottom seed, which can be fitted to a straight line, as shown in Figure 7. The line intersects the pressure axis at -0.14 GPa, and the equilibrium pressure at 1350 °C is derived, that is, $P_e = P_0 - 0.14$ GPa. Using a value of 5.33 GPa evaluated for P_0, we get

$$P_e = 5.19 \text{ GPa} \quad (\text{for } T = 1350\,°\text{C}) \tag{6}$$

which agrees reasonably with Berman's calculation (BERMAN, 1979). This suggests that diametric growth rate is proportional to excess pressure, $\delta P (= P_2 - P_e)$. According to observation of the recovered samples, the bottom crystal apparently grows into graphite, and it is covered with a film of flux on the interface between itself and graphite, as illustrated in Figure 8. If transportation of dissolved carbon (which must occur almost entirely by diffusion) governs the growth rate, and if thickness of the flux film changes little during the growth, then constant diametral growth rate and its proportionality to the excess pressure are expected. Growth behavior characterized in Figures 6 and 7 leads us to such an image of the growth process.

Solubility Difference between Source Graphite and Diamond

The solubility-difference ratio, given in Eq. (3), can be written as

$$\delta X / X_c = 0.1075 \cdot \delta P \quad (\delta P \text{ in GPa}) \tag{7}$$

as $\Delta V = 1.45$ cm^3 (BERMAN, 1979), $R = 8.31$ J/K·mol, and $T = 1350\,°\text{C} = 1623$ K. At this temperature the solubility of diamond, X_c, in the present flux has been evaluated as 4.5 wt%, or 0.183 in atomic fraction (unpublished data). Using these parameters, growth conditions corresponding to the two critical pressures, P_0 and P^* at $T = 1350\,°\text{C}$, are summarized as follows:

$$\delta P_0 = 0.14, \quad \delta X_0 / X_c = 0.015, \quad \delta X_0 = 2.75 \times 10^{-3}$$

$$\delta P^* = 0.33, \quad \delta X^* / X_c = 0.035, \quad \delta X^* = 6.50 \times 10^{-3}$$

where δP's are in GPa and δX's in atomic fraction.

There exists an excess concentration δX for diamond on the surface of graphite and it corresponds to a supersaturation $\sigma = \delta X / X_c$ for diamond. Therefore, the values of supersaturation required for nucleation of diamond on a fresh surface and on a regrowth-treated surface of graphite are evaluated as

$$\sigma_0 = 0.015 \quad \text{and} \quad \sigma^* = 0.035 \tag{8}$$

respectively. Thus the suppressive effect on diamond nucleation can be expressed in terms of supersaturation as "an increase in supersaturation by 0.020."

There is a width of $(P^* - P_e)$ allowed for P_2 in which a seed can be grown selectively. It is not difficult for modern high-pressure techniques to reproduce a reaction pressure in a width of about 0.3 GPa, though the most suitable region of P_2 is narrower and reproduction thereof may be difficult.

Advantages and Disadvantages of the Present Method

Although the growth conditions have not yet been optimized, we can compare, in principle, a few features of the present method to those of the temperature-gradient method:

a) Temperature may be uniform throughout the reaction cell in the present method, and growth units containing a source graphite and a seed can be stacked to realize an efficient use of the volume of the reaction cell. In the temperature-gradient method each of the sources and seeds has only one suitable position along the cell axis.

b) Growth rate in the present method can easily be changed, even during an experiment; it is governed by the excess pressure or by the reaction pressure. On the other hand, geometry of the reaction cell must be changed, and additional calibrating runs must be made, to generate a different growth rate in the conventional method.

c) Regarding pressure drift during an experiment, the ratio of change in δP_2 (the driving force) is larger than that in P_2 itself by a factor of 30 or more. The driving force, δT, in the conventional method is independent of pressure as long as thermodynamical stability of diamond is maintained. Further, δT drifts with a ratio about the same as that of T. Thus stable maintenance of a small driving force is much more difficult in the present method than in the conventional method.

Summary

a) Driving forces for growing seeds were discussed for two cases: diamond → solution → diamond and graphite → solution → diamond.

b) In reaction of graphite with a metal flux at high pressure, nucleation of diamond is effectively suppressed by the regrowth treatment of the graphite surface, the strength of suppression corresponding to an increase by 0.19 GPa in the excess pressure required for the nucleation of diamond.

c) A new method of seeded growth of diamond, with a reaction of graphite → solution → diamond, has been realized by using this suppressive effect. The growth is driven by the excess pressure instead of the temperature difference

Acknowledgments. Sincere thanks are due Hassel M. Ledbetter, Institute for Materials Science and Engineering, National Bureau of Standards, USA, for kindly giving valuable advice for preparation and correction of the paper, and F. P. Bundy and M. Seal for reviewing the paper and offering many helpful suggestions for improving it.

References

Berman, R., Thermal properties, in *The Properties of Diamond*, edited by J. E. Field, pp. 4–22, Academic Press, London, 1979.

Bundy, F. P., H. T. Hall, H. M. Strong, and R. H. Wentorf, Jr., Man-made diamonds, *Nature, 176*, 51–55, 1955.

Bundy, F. P., H. P. Bovenkerk, H. M. Strong, and R. H. Wentorf, Jr., Diamond-graphite equilibrium line from growth and graphitization of diamond, *J. Chem. Phys., 35*, 383–391, 1961.

Bundy, F. P., H. M. Strong, and R. H. Wentorf, Jr., Methods and mechanisms of synthetic diamond growth, *Chem. and Phys. of Carbon, 10*, 213–263, 1973.

Kanda, H., and O. Fukunaga, Growth of large diamond crystals, in *High-Pressure Research in Geophysics*, edited by S. Akimoto and M. H. Manghnani, pp. 525–535, CAPJ/D. Reidel, Tokyo/Dordrecht, 1982.

Strong, H. M., and R. E. Hanneman, Crystallization of diamond and graphite, *J. Chem. Phys., 46*, 3668–3676, 1967.

Strong, H. M., and R. H. Wentorf, Jr., The growth of large diamond crystals, *Naturwissenschaften, 59*, 1–7, 1972.

Wakatsuki, M., and K. Ichinose, A wedge-type cubic anvil high-pressure apparatus and its application to material synthesis research, in *High-Pressure Research in Geophysics*, edited by S. Akimoto and M. H. Manghnani, pp. 13–26, CAPJ/D. Reidel, Tokyo/Dordrecht, 1982.

PHASE DIAGRAM OF $MgSiO_3$ AT PRESSURES UP TO 24 GPa AND TEMPERATURES UP TO 2200 °C: PHASE STABILITY AND PROPERTIES OF TETRAGONAL GARNET

Hiroshi SAWAMOTO

Department of Earth Sciences, Nagoya University, Chikusa, Nagoya 464, Japan

Abstract. The phase diagram for pure $MgSiO_3$ has been obtained at pressures up to 24 GPa and at temperatures up to 2200 °C using a MA8 type of high-pressure apparatus. It is found that $MgSiO_3$ has many transformation modes at different temperatures. At temperatures above 1800 °C, $MgSiO_3$ has a polymorphic sequence with five phases in order of increasing pressure: orthopyroxene, clinopyroxene, tetragonal garnet, ilmenite, and perovskite. At lower temperature, it exhibites seven phases or phase-assemblages: orthopyroxene, clinopyroxene, tetragonal garnet, modified spinel plus stishovite, spinel plus stishovite, ilmenite, and perovskite, in order of increasing density. All phase boundaries are determined by the present experiments, with consideration to thermodynamic constraints. The pressure-temperature slope of the phase boundary between tetragonal garnet and perovskite is determined to be positive by using parameters representing well-constrained phase boundaries and the volume differences between the related phases.

The powder X-ray diffraction analysis of $MgSiO_3$ with tetragonal symmetry shows that it is isostructural with $CaGeO_3$, $CdGeO_3$, and $MnSiO_3$; these compounds possess an ordered cationic distribution on the noequivalent sites. The unit cell dimensions of the tetragonal garnet are $a=11.470(1)$ Å, $c=11.398(2)$ Å and $c/a=0.994$, and the molar volume is about 1% smaller than that estimated for a hypothetical cubic garnet. Therefore tetragonal garnet is a high-pressure and low-entropy phase. The elastic properties of $MgSiO_3$ with tetragonal symmetry are predicted from elasticity systematics: bulk modulus $K=177$ GPa and rigidity $\mu=101$ GPa, their pressure derivatives $dK_s/dP=4.5$ and $d\mu/dP=1.44$, and seismic wave velocities $V_p=9.37$ km/s and $V_s=5.32$ km/s.

The tectonic significance of the garnet phase with a higher entropy than perovskite is also discussed.

Introduction

It is widely accepted that the seismic discontinuity at 400 km depth is due to the transformation from the olivine (α) to modified spinel (β) phase of (Mg, Fe)$_2$SiO$_4$. If the seismic discontinuity at 400 km depth is a result of the phase transformation in (Mg, Fe)$_2$SiO$_4$, the magnitude of the discontinuity should be related to the difference in acoustic velocities of these two phases and to the volume fraction of the (Mg, Fe)$_2$SiO$_4$ component at this depth. According to the recent measurement of elastic moduli of modified spinel by Brillouin spectroscopy (SAWAMOTO et al., 1984), the velocity jump across the α-β transformation is 1.06 km/s for compressional wave velocity and 0.71 km/s for shear wave velocity at room temperature and pressure. The observed seismic discontinuity of a compressional wave of 0.43 km/s velocity (BURDICK, 1981) indicates a 40% volume fraction of the (Mg, Fe)$_2$SiO$_4$ component, and that of a shear wave of 0.23 km/s velocity (GRAND and HELMBERGER, 1984) indicates a 35% volume fraction (SAWAMOTO et al., 1984). Assuming appropriate pressure and temperature derivatives of bulk and shear moduli for α and β phases along the geotherm, WEIDNER (1985) concluded that the (Mg, Fe)$_2$SiO$_4$ content of the mantle at 400 km depth is 40–60% by volume. Therefore there must be dilution of (Mg, Fe)$_2$SiO$_4$ in the mantle by other minerals that do not undergo phase transitions at around this depth. From the cosmochemical constraints, the minerals that could such dilution must be those in the system (Mg, Fe)SiO$_3$-(Mg, Fe)$_3$Al$_2$Si$_3$O$_{12}$.

AKAOGI and AKIMOTO (1977) studied the phase relations in the systems $Mg_4Si_4O_{12}$-$Mg_3Al_2Si_3O_{12}$ and $Fe_4Si_4O_{12}$-$Fe_3Al_2Si_3O_{12}$, and LIU (1977, 1980) studied the system of enstatite-pyrope at high pressures and high temperatures. (Mg, Fe)SiO$_3$ has a complicated transformation mode with six phase-assemblages involving seven phases: orthopyroxene, clinopyroxene, modified spinel plus stishovite, spinel plus stishovite, ilmenite, and perovskite. However, the presence of stishovite, unless it is a minor phase in the mantle , is not consistent with any seismic velocity model because the acoustic wave velocities of stishovite are too high (WEIDNER et al., 1982).

Recently SAWAMOTO and KOHZAKI (1984) and KATO and KUMAZAWA (1985) reported the presence of $MgSiO_3$ with a garnet-like structure at temperatures higher than 2000 °C. This finding has raised important problems regarding the complex transformation modes in the system (Mg, Fe)SiO$_3$-(Mg, Fe)$_3$Al$_2$Si$_3$O$_{12}$, the interpretation of seismic velocity profiles, and mantle tectonics.

In this paper we focus on the phase relations of $MgSiO_3$ at high-pressures and high-temperatures and on the physical properties of tetragonal garnet in order to provide basic data on mantle problems, including the contribution of $MgSiO_3$ to the seismic wave velocities in the earth's mantle.

Experimental Procedure

High Pressure and High Temperature Apparatus

A MA8 type of high-pressure and high-temperature

High-Pressure Research in Mineral Physics, edited by M. H. Manghnani and Y. Syono, pp. 209–219.

apparatus used in this study was driven by a pair of guide blocks in a 2000 ton uniaxial press (SAWAMOTO, 1986). The components that are used for generating high pressure are eight anvils (cemented tungsten carbide) and twelve sheets of preformed gaskets (raw pyrophyllite). The truncated edge length of each anvil is 2.5 mm (for pressures up to 20 GPa) and 2.0 mm (for pressures up to 25 GPa). Three flank surfaces are tapered by 1° in the direction of (111) of each anvil, for pressures above 22 GPa. These tapered surfaces provide lateral support according to an elastic or plastic design (KUMAZAWA, 1973). After generating pressure above 15 GPa, plastic deformation of the anvil takes place in the region adjacent to the pressure medium; thus a new set of anvils is required for each run at these pressures. The initial volume of the pressure medium is about 290 mm^3, which is equivalent to the volume of an octahedron of 9.5-mm-edge length for anvils with 2.5-mm-edge length and to a volume of 240 mm^3 (an octahedron with 8.0-mm-edge length) for an anvil with a 2.0-mm-edge length. The initial thickness of the preformed gaskets depends on the oversizing of the initial pressure medium, and the width of preformed gaskets is determined on a trial and error basis.

The pressure generated in the MgO pressure medium is calibrated as a function of the press load at room temperature based on changes in electric resistance associated with phase changes in Bi (III–V; 7.6 GPa), ZnS (semiconductor metal; 15 GPa), GaAs (semiconductor metal; 19 GPa), and GaP (semiconductor metal; 22 GPa). Pressure can be reproduced within ±0.3 GPa up to 22 GPa. Pressures above 22 GPa are estimated from by extrapolated calibration curve, giving rise to an error larger than 0.5 GPa. The calibration curves for the apparatus are shown in Figure 1.

Generation of High Temperature

The furnace assembly is made of a thick lanthanum-chromite sleeve, a thin graphite tube with two end caps,

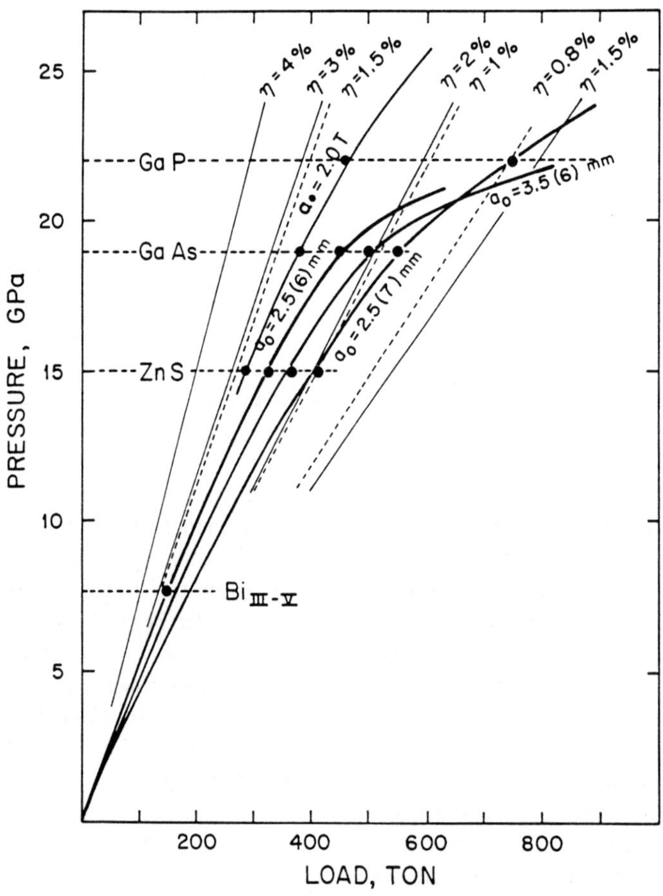

Figure 1. Pressure calibration curves for the MA8-type high-pressure apparatus. The number in parentheses denotes the initial width (mm) of the preformed garnets, and a_0 is the edge-length of the triangle for the truncated anvil. The straight lines represent the efficiency of the pressure generation; solid lines are for a truncated anvil edge-length of 3.5 mm, and dashed lines are for edge lengths of 2.5 mm.

two tantalum rods (electrode), and a thermocouple. The tube heater consists of an inner graphite tube and an outer lanthanum chromite sleeve. The graphite tube and end caps have two roles: 1) preventing contamination of the specimen by the lanthanum-chromite, and 2) preheating the lanthanum-chromite up to temperatures above 1000 °C. At low temperatures an electric current flows through the graphite tube. When the temperature exceeds about 1000 °C, however, the AC current through the graphite suddenly decreases as a result of direct conversion of graphitic carbon to diamond. The current begins flowing through the lanthanum-chromite sleeve in this temperature range, and the semiconducting lanthanum-chromite operates as an ohmic heater above these temperatures.

The temperature of the sample is measured with a 97%tungsten-3%rhenium and 75%tungsten-25%rhenium thermocouple. The thermocouple is in direct contact with the graphite and lanthanum-chromite at the center of the heater. The ends of the thermocouple are connected to the opposing surfaces of two anvils. Therefore the temperature of the anvil surface will influence recorded temperatures, but no correction was made for this effect in the present study.

A pressure calibration at high temperature was done by determining the pressure of the coesite-stishovite transition at 1500 °C. The transition pressure was determined to be 9.6 GPa, which is consistent with the extrapolated value of the boundary determined by in situ X-ray diffraction measurements (AKIMOTO et al., 1977). Therefore the calibration curve at room temperature can be used at high temperature.

A stable temperature of 2200 °C±70 °C was generated at 24.0±0.5 GPa for 1 h.

Preparation of Starting Materials

The starting material for phase transformation experiments was synthesized from a mixture of MgO (purity 99.9%) and SiO_2 (purity 99.9%) in a pyroxene ratio. The mixed sample was then heated at 1800 °C and 200 bar in a hot press. The synthesized pyroxene was examined by X-ray diffraction and found to be a singe phase with orthopyroxene structure. An EPMA (Electron Probe Micro Analyzer) analysis shows that MgO=50.5±1.0 mol%, SiO_2=49.5±1.0 mol%, and Al_2O_3 less than 0.1 mol%.

Determination of Phase Equilibria

The experimental method used in the present study for determining the equilibrium phase diagram was a conventional quenching technique. The sample, embedded directly in a graphite capsule, was maintained at a desired pressure and temperature for a time interval sufficient for equilibrium, and quenched isobarically by turning off the heating power on the furnace. The temperature was reduced to 100 °C within 0.5 s. Thus the effective quenching rate is faster than 2000 °C/s.

After each run, the quenched sample was examined by optical microscopy and X-ray powder diffraction. The unit cell parameters of a garnet-like structure were determined by means of X-ray powder diffractometry with 2 θ step scanning of 1/36° using $FeK\alpha_1$ radiation.

Experimental Results and Discussions

Phase Equilibrium of $MgSiO_3$ at High Pressure and High Temperature

Phase relations of $MgSiO_3$. Experimental data on the phase relations of $MgSiO_3$ were obtained at pressures of 15–24 GPa and at temperatures of 1100–2200 °C. The results are summarized in Table 1 and shown in Figure 2.

$MgSiO_3$ shows many different transformation modes at different temperatures, as explained below:

1) At temperatures below 1600 °C, $MgSiO_3$ has six phase assemblages: orthopyroxene, clinopyroxene, modified spinel plus stishovite, spinel plus stishovite, ilmenite, and perovskite in order to increasing pressure.

2) At temperatures between 1600 °C and 1650 °C, the stability field of spinel plus stishovite disappears: and $MgSiO_3$ has five phase assemblages: orthopyroxene, clinopyroxene, modified spinel plus stishovite, ilmenite, and perovskite.

3) At temperatures between 1650 °C and 1800 °C, a garnet-like phase appears between clinopyroxene and modified spinel plus stishovite: and $MgSiO_3$ has six phase assemblages: orthopyroxene, clinopyroxene, garnet, modified spinel plus stishovite, ilmenite, and perovskite in order to increasing pressure.

4) At temperatures between 1800 °C and 2060 °C, as the stability field of modified spinel plus stishovite disappears, $MgSiO_3$ exhibits five polymorphic phases: orthopyroxene, clinopyroxene, garnet, ilmenite, and perovskite.

5) At temperatures above 2060 °C, the ilmenite is unstable. $MgSiO_3$ exhibits four polymorphic phases: orthopyroxene, clinopyroxene, garnet, and perovskite.

Phase boundary curves. There are several phase boundaries, some of which are well-constrained whereas others are not constrained at all by the present experimental results. The phase boundaries constrained by the experimental data have been represented by the following linear equation,

$$P_{ij} = (dP/dT)_{ij}T + Po_{ij}$$

for the equilibrium between two coexisting phase fields i and j.

For the phase boundary between clinoensatite and modified spinel plus stishovite the equation is tentatively

TABLE 1. Experimental Conditions and Results

Run No.	Pressure (GPa)	Temperature (°C)	Time (min)	Phases present
MA2.5-20	15.5	1400	30	Cpx
MA2.5-19	16.0	1200	30	Cpx
MA2.5-18	16.0	1400	30	Beta+St
MA2.5-17	16.0	1600	20	Cpx
MA2.5-30	16.5	1100	40	Gamma+St
MA2.5-20	16.9	1700	20	Gar, Cpx
MA2.5-03	17.0	1340	40	Gamma
MA2.5-02	17.0	1500	30	Gamma
MA2.5-01	17.0	1800	20	Cpx
MA2.5-14	17.0	2000	15	Cpx
MA2.5-16	17.3	1900	20	Gar
MA2.5-43	17.5	1500	30	Gamma+St
MA2.5-07	18.0	1300	40	Gamma+St
MA2.5-06	18.0	1400	30	Beta+St
MA2.5-05	18.0	1700	20	Gar, Beta+St
MA2.5-04	18.0	1800	20	Gar
MA2.5-21	18.0	2100	10	Gar
MA2.5-09	18.5	1650	25	Beta+St
MA2.5-10	18.5	1700	20	Beta+St
MA2.5-11	18.5	1750	20	Gar, Beta+St
MA2.5-08	18.5	1800	20	Gar
MA2.5-13	19.0	1500	30	Beta+St
MA2.5-12	19.0	1700	20	Beta+St
MA2.5-29	19.6	1550	30	Gamma, Beta, St, Ilm
MA2.5-27	19.6	1700	20	Beta+St
MA2.5-28	19.6	1820	20	Gar
MA2.0-25	20.0	1250	40	Gamma+St
MA2.0-26	20.0	1400	30	Gamma, St, Ilm
MA2.0-24	20.0	1500	30	Ilm
MA2.0-23	20.0	1760	20	Ilm
MA2.0-15	20.0	2000	10	Gar
MA2.0-22	20.0	2100	10	Gar
MA2.0-31	20.5	1200	40	Ilm
MA2.0-33	21.0	1800	20	Ilm
MA2.0-32	21.0	1900	20	Gar, Ilm(trace)
MA2.0-37	21.7	2000	10	Gar, Ilm(trace)
MA2.0-36	21.7	2100	10	Gar
MA2.0-34	22.0	1860	20	Ilm
MA2T1-35	23.0	1900	10	Ilm
MA2T1-40	23.0	2050	10	Pev
MA2T1-42	23.0	2100	10	Gar
MA2T1-41	23.7	1600	25	Ilm
MA2T1-45	23.7	1810	15	Pev
MA2T1-38	24.0	1450	25	Ilm
MA2T1-39	24.0	2000	10	Pev

Cpx, clinopyroxene; Beta, modified spinel; Gamma, spinel; St, stishovite; Gar, tetragonal garnet; Ilm, ilmenite; Pev, perovskite

given by

$$P = 0.005(2)T + 7.4 \ (T < 1920)$$

where P is in GPa and T is in Kelvins. However, this is not accurate enough to constrain the thermodynamic properties. For example, this phase boundary is about 3.0 GPa lower than that reported by AKAOGI and AKIMOTO (1977), and the present slope (0.005(2) GPa/K) is considerably larger than the slopes of 0.003 GPa/K reported by AKAOGI and AKIMOTO (1977) and fundamentally inconsistent with 0.0015 GPa/K reported by ITO and NAVROTSKY (1985). The volume change in the clinoensatite to modified spinel and stishovite transformation in $MgSiO_3$ is 4.02 cm^3/mol. Therefore the calculated entropy change based on the Clausius-Clapeyron relation is 20.10 J/mol K for 0.005 GPa/K, 12.1 J/mol K for 0.003 GPa/K, and 6.0 J/mol K for 0.0015 GPa/K. Thus no reliable entropy change is obtained. The problem arising from the difference in the slopes given by different sources is discussed later. The volume changes quoted in this discussion are for ambient conditions; the effects of pressure and temperature on the volume changes and the calculated entropy changes are numerically small in the present case.

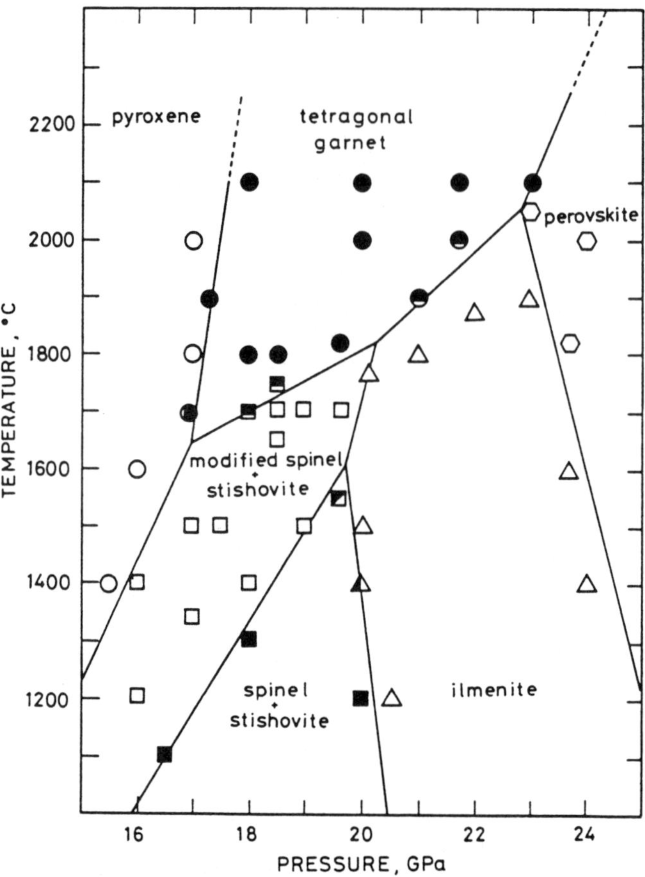

Figure 2. Phase diagram of $MgSiO_3$ at 15–24 GPa and 1100–2200 °C. ○: pyroxene, ●: tetragonal garnet, □: modified spinel plus stishovite, ■: spinel plus stishovite, Δ: ilmentite, ⬡: perovskite.

The boundary of the clinoensatite to tetragonal garnet transformation is given by

$$P = 0.0015(1)T + 14.0$$

where $T>1920$. This is the first determination of this phase boundary. The volume change in this transformation is 3.05 cm^3/mol, and the entropy change is calculated to be 4.58 J/mol K.

The boundary between the modified spinel plus stishovite and the spinel plus stishovite phases is given by

$$P = 0.0063T + 7.8$$

where $T<1880$. The volume change is 0.49 cm^3/mol; thus the entropy change is 3.09 J/mol K.

The disproportionation of tetragonal garnet to modified spinel plus stishovite takes place with a phase boundary given by

$$P = 0.019(2)T - 20.2$$

where $1920<T<2090$. In this disproportionation, the volume change is 0.97 cm^3/mol, and the entropy change is calculated to be 18.43 J/mol K.

The phase boundary between spinel plus stishovite and ilmenite has a negative slope,

$$P = -0.0015T + 22.5$$

where $T<1880$. The volume change in the transformation is 0.46 cm^3/mol, and the entropy change is -0.69 J/mol K. The run product at 1823 K and 19.5 GPa (Run MA2.5-29) includes modified spinel, spinel, stishovite, and ilmenite. Therefore this P-T condition is suggested to be very close to the triple point involving all phases listed above.

The phase boundary of the tetragonal garnet-ilmenite transition is given by

$$P = 0.011T - 2.8$$

where $2090<T<2330$. This is also the first determination of this phase boundary. The differences in molar volume and entropy between tetragonal garnet and ilmenite are 1.92 cm^3/mol and 21.12 J/mol K, respectively.

In the present work a positive slope is suggested for the boundary between the garnet and perovskite phases, based on results of runs MA2T1-40 and MA2T1-42 (see Table 1). The two runs, which had a temperature difference of only 50 K but the same pressure of 23 GPa, have yielded very different results; the high temperature run yielded tetragonal garnet, whereas perovskite was found to crystallize in the lower temperature run. Although the purpose of such experiments is not to enable us to trace the phase boundary up to higher temperatures and pressures, the slope of the phase boundary can be thermodynamically calculated by using the differences in molar volumes and slopes of the other boundaries between related phase assemblages. This problem is discussed in the next section.

The transformation of ilmenite to perovskite has been known to occur. The phase boundary determined for this transformation has a negative slope given by

$$P = -0.0026(5)T + 28.9$$

where $T<2330$. This phase boundary is almost the same as that reported by ITO and YAMADA (1982). However, the pressure determined by extrapolating the calibration curve includes some errors as a result of a plastic deformation of the anvils. Thus we have no way of calibrating pressures higher than 22 GPa. The slope of the P-T boundary of the ilmenite and perovskite transition in $MgSiO_3$ is determined to be 0.0026(5) GPa/K (present work) and 0.002 GPa/K (ITO and YAMADA, 1982). The volume change in the transition is 1.85 cm^3/mol. Therefore the entropy change is calculated to be either 4.81 J/mol K (present work) or 3.70 J/mol (ITO and YAMADA, 1982).

Calculation of phase boundaries and entropy changes. The phase boundary between tetragonal garnet and perovskite in $MgSiO_3$ is very important to the understanding of the basic characteristics of the 670-km discontinuity, but this boundary has never been clarified. As shown in Figure 2, threre are nine phase boundaries, six phases, and four triple points in this pressure and temperature range. Some phase boundaries are well constrained, whereas others are not constrained by the present experiments. However, the unconstrained boundaries can be calculated by means of thermodynamic considerations related to the slopes of the well-constrained phase boundaries and the difference in molar volumes between related phases.

The slope of a boundary between i and j phases or phase assemblages is expressed by the Clausius-Clapeyron relation as follows

$$(\mathrm{d}P/\mathrm{d}T)_{ij} = \Delta S_{ij}/\Delta V_{ij} \tag{1}$$

where $\Delta V_{ij}=V_i-V_j$ and $\Delta S_{ij}=S_i-S_j$.

When three phases (i, j, and k) coexist with equilibrium at a triple point, the summation of the volume differences and of the entropy differences are zero as follows

$$\Delta V_{ij} + \Delta V_{jk} + \Delta V_{ki} = 0 \tag{2}$$

$$\Delta S_{ij} + \Delta S_{jk} + \Delta S_{ki} = 0 \tag{3}$$

Therefore we have from (1) and (3)

$$\Sigma(\mathrm{d}P/\mathrm{d}T)_{ij}\Delta V_{ij} = 0 \tag{4}$$

for each triple point. There are four triple points and six phase-assemblages in the present case, and we have the following four constraints

$$(dP/dT)_{12}\Delta V_{12} + (dP/dT)_{23}\Delta V_{23} + (dP/dT)_{31}\Delta V_{31} = 0 \quad (5)$$

$$(dP/dT)_{32}\Delta V_{32} + (dP/dT)_{53}\Delta V_{53} + (dP/dT)_{25}\Delta V_{25} = 0 \quad (6)$$

$$(dP/dT)_{43}\Delta V_{43} + (dP/dT)_{54}\Delta V_{54} + (dP/dT)_{35}\Delta V_{35} = 0 \quad (7)$$

$$(dP/dT)_{65}\Delta V_{65} + (dP/dT)_{26}\Delta V_{26} + (dP/dT)_{52}\Delta V_{52} = 0 \quad (8)$$

where 1=clinopyroxene, 2=garnet, 3=β+stishovite, 4=γ+stishovite, 5=ilmenite, and 6=perovskite. The four equations, (5) to (8), relate the nine slopes through the parameters ΔV_{ij} which are easily determined.

If we employ five well-constrained slopes of phase boundaries, $(dP/dT)_{ij}$ with ij=12, 45, 34, 25, and 56, we can obtain the other slopes of phase boundaries and also the entropy changes, as summarized in Table 2.

The value of ΔV_{ij} used for the calculation is determined from crystallographic data at ambient conditions, and it includes some deviations due to thermal expansion by temperature and compression by pressure. The value of ΔS_{ij} is also dependent on temperature due to the difference in specific heats and on pressure due to thermal expansion. However, these effects are numerically small in the present case. Each of the experimental $(dP/dT)_{ij}$ includes some errors originating mostly from ambiguities in the pressure scale above 22 GPa and the arbitrariness of locating these phase boundaries by based on a small number of high-pressure runs. Therefore it is important to know how the errors propagates to the other parameters when a specific set of input data are used. After several tests, we concluded that a set of $(dP/dT)_{ij}$ (marked with (d) in Table 2) is appropriate with regard to internal consistency.

A simple check is provided by the comparison of observed and calculated values of $(dP/dT)_{ij}$ in two cases (see Table 2): 1) The calculated value, 0.0193 GPa/K, is compared with an observed value of 0.019±0.002 GPa/K for the phase boundary (ij=23) between tetragonal garnet and stishovite plus modified spinel, and 2) the calculated value, 0.0058 GPa/K, is compared with 0.005±0.002 GPa/K given by experiments for the phase boundary (ij=13) between clinopyroxene and stishovite plus modified spinel. A value as small as 0.0015 GPa/K (ITO and NAVROTSKY, 1985) for $(dP/dT)_{13}$ is not acceptable because it gives rise to a large inconsistency among the existing data.

TABLE 2. Thermodynamic Data to Constrain Entropy Changes, Volume Changes and Slopes of Phase Boundaries at the Four Triple Points. Suffices i and j denote phases; 1=clinopyroxene, 2=garnet, 3=β+stishovite, 4=γ+stishovite, 5=ilmenite and 6=perovskite, Note that $\Delta V_{ij}=-\Delta V_{ji}$ and $\Delta S_{ij}=-\Delta S_{ji}$

Triple Point	Phase Boundary ij	Observed $(dP/dT)_{ij}$ (GPa/K)	Observed ΔV_{ij} ($10^{-6}m^3$/mol)	Calculated or Adjusted ΔS_{ij} (J/mol)	Calculated or Adjusted $(dP/dT)_{ij}$ (MPa/K)	Calculated or Adjusted Po_{ij} (GPa)
1 2 3	1 2	0.0015	3.05	4.58	(1.5)[d]	14.0
	2 3	0.019	0.97	18.72[h], 18.43[j]	19.3	−20.2
	3 1	0.005	−4.02	−23.28[i], − 20.10[j]	5.8	5.8
		0.003[a]		−12.06[a]		
		0.0015[b]		−6.03[b]		
2 5 3	2 5	0.011	1.92	21.12	(11.0)[d]	−2.8
	5 3	—	−0.95	−2.40[f]	2.5	15.0
	3 2	0.019	−0.97	−18.72[g]	19.3	−20.2
3 4 5	3 5	—	0.95	2.40[e]	2.5	15.0
	5 4	− 0.0015	−0.46	0.69	(−1.5)[d]	2.3
	4 3	0.0063	−0.49	−3.09	(6.3)[d]	7.8
5 2 6	5 2	0.011	−1.92	−21.12	(11.0)[d]	−2.8
	2 6	—	3.77	16.31[e], 17.42[k]	4.3, 4.6[k]	12.8
	6 5	−0.0026	−1.85	4.81	(−2.6)[d]	28.9
		−0.002[c]		3.70[c]		

[a] AKAOGI and AKIMOTO (1977), [b] ITO and NAVROTSKY (1985), [c] ITO and YAMADA (1982), [d] given from the experimental data, [e] directly determined from Equation (3), [f] transferred from revalue of (e), [g] secondary determination by Equation (3) from the value of (f) and experimental data, [h] transferred from the value of (g), [i] secondary determination by Equation (3), [j] determined by Equation (1) from the experimental data, [k] secondarily calculated from (c) instead of the present data on $(dP/dT)_{56}$.

The thermodynamic considerations discussed above lead us to relocate the phase boundaries and triple points in this system. The revised parameters for the phase boundaries are shown in the last two columns of Table 2, and four triple points are given by

$$P = 16.9 \text{ GPa and } T = 1921 \text{ K for } (1,2,3)$$

$$P = 19.7 \text{ GPa and } T = 1884 \text{ K for } (3,5,4)$$

$$P = 20.2 \text{ GPa and } T = 2094 \text{ K for } (2,5,3)$$

$$P = 22.8 \text{ GPa and } T = 2330 \text{ K for } (2,6,5)$$

As a result of the above analysis, the phase boundary between modified spinel plus stishovite and ilmenite is given by

$$P = 0.0025T + 15.0$$

where $1880<T<2090$. The associated entropy change is 2.4 J/mol K and the volume change is 0.95 cm^3/mol. The value of Po_{ij}, the nominal T=0 pressure along the Clapeyron curve, is given by referring to the triple points given above.

The phase boundary between tetragonal garnet and perovskite is now calculated as follows

$$P = 0.0043T + 12.8$$

where $T>2330$. The associated entropy change is calculated to be 16.3 J/mol K.

When we use −0.002 GPa/K (ITO and YAMADA, 1982) for the ilmenite-perovskite transition (see (c) in Table 2), we have 0.0046 GPa/K and 17.4 J/mol K for the slope and entropy change, respectively, for the garnet-perovskite transition. This fact indicates that the slope of the garnet-perovskite transition has a positive slope, contrary to the previous supposition presented by NAVROTSKY (1980) on the basis of the germanate analogue. This is one of the most important results of the present work, although the values of slope and entropy change given by this work are not accurate enough to be final and are the subject of further investigation.

Phase Stability and Property of Garnet-Like Structure

Crystal structure of tetragonal garnet. In 1980, we had grown single crystals (~100 μm) of tetragonal garnet $MnSiO_3$ from $MnSiO_3$ rhodonite at 1500 °C and 12 GPa by means of a MA8 type of high-pressure apparatus. We suggested the possibility of an isostructural garnet for $MgSiO_3$ (FUJINO et al., 1981). The results of detailed structure analysis and crystal chemistry of the tetragonal garnet have been reported by FUJINO et al. (1986).

This tetragonal $MnSiO_3$ has seven different cation sites: Mn(1)[×16] and Mn(2)[×8] at dodecahedral sites; Mn(3)[×8] and Si(1)[×8] at octahedral sites; and Si(2)[×4], Si(3)[×4], and Si(4)[×16] at tetrahedral sites. A structural characteristic of this tetragonal garnet is that the atoms in the octahedral sites (16a in cubic (Ia3d) garnet) are ordered into two noequivalent sites; 8e and 8d in $I4_1/a$. On polymorphic transition from clinopyroxene to the tetragonal garnet, a partial increase in the coordination number occurs both for Mn^{2+} (6–8) and Si^{4+} (4–6).

The analogous garnet of $MgSiO_3$ with tetragonal symmetry has been shown to exist at high temperatures above 1600 °C and at a pressure of 17 GPa by SAWAMOTO and KOHZAKI (1984). Some petrologic significance has been discussed by KATO and KUMAZAWA (1985), and the compositional stability has been discussed by KATO (1986).

In the present work the stability field of this tetragonal $MgSiO_3$ garnet in pressure and temperature space is clarified, and crystallographic and physical properties are discussed. The X-ray diffraction pattern of this phase is similar to those of garnet structures, except that each peak is split into a multiplet. Such a garnet-like structure with tetragonal symmetry has been found in high-pressure phases of $CaGeO_3$, $CdGeO_3$ (PREWITT and SLEIGHT, 1969), and $MnSiO_3$ (FUJINO et al., 1986). The present X-ray diffraction patterns were indexed by referring to those of other tetragonal garnets. In Table 3, both observed and calculated d-spacings for $MgSiO_3$ are given on the basis of a tetragonal unit cell by means of the least squares method. For almost all indices, the differences between observed and calculated d-spacings are less than 0.001 Å. Such good agreement indicates that the new

TABLE 3. Powder X-ray Diffraction Data for Tetragonal Garnet

h k l	*d*(obs)	*d*(cal)	*I*/*Io*	*h k l*	*d*(obs)	*d*(cal)	*I*/*Io*
2 1 1	4.684	4.678	15	1 0 5	2.2423	2.2358	10
1 1 2	4.665	4.663	15	5 2 1	2.0981	2.0940	35
3 2 1	3.064	3.064	15	5 1 2	2.0918	2.0926	25
3 1 2	3.059	3.060	15	2 1 5	2.0824	2.0832	25
2 1 3	3.051	3.053	15	6 1 1	1.8614	1.8607	15
4 0 0	2.8674	2.8680	75	1 1 6	1.8499	1.8496	10
0 0 4	2.8500	2.8495	35	6 2 0	1.8140	1.8139	5
4 2 0	2.5646	2.5652	100	6 0 2	—	1.8127	—
4 0 2	—	2.5619	—	2 0 6	1.8028	1.8033	5
2 0 4	2.5526	2.5520	50	4 4 4	1.6531	1.6522	15
3 3 2	2.4424	2.4429	50	6 4 0	1.5901	1.5909	15
3 2 3	2.4398	2.4393	50	6 0 4	—	1.5877	—
4 2 2	2.3400	2.3392	20	4 0 6	1.5846	1.5838	15
2 2 4	2.3306	2.3316	20	6 4 2	1.5322	1.5322	50
5 1 0	—	2.2498	—	6 2 4	1.5294	1.5301	35
5 0 1	2.2484	2.2492	25	4 2 6	1.5258	1.5266	20

a=11.472(1) Å, c=11.398(2) Å, V=1499.6(4) $Å^3$, ρ=3.557 g/cm^3.

$MgSiO_3$ phase is the expected tetragonal garnet which is isostructural with the tetragonal $MnSiO_3$.

The unit cell dimensions of the tetragonal lattice are determined to be a=11.470(1) Å, c=11.398(2) Å, V=a^2c=1499.6(3.6) Å^3, and ρ=3.557 g/cm^3. On the other hand, the unit cell dimensions of tetragonal $MnSiO_3$ garnet are a=11.774(1) Å, c=11.636(2) Å, V=a^2c=1613.1(3) Å^3, and ρ=4.32 g/cm^3. The ratio of lattice parameters, c/a for $MgSiO_3$ is 0.994; this compares well with the value 0.988 for $MnSiO_3$. A consistent variation is expected with increasing difference in cation radii between Oc1(8c) and Oc2(8d).

Comparison of cubic and tetragonal garnets. The lattice parameters of garnet solid solutions in the $Mg_4Si_4O_{12}$-$Mg_3Al_2Si_3O_{12}$ system are plotted against composition in Figure 3 in order to compare them with that of tetragonal garnet. The lattice parameter of cubic garnet increases linearly with increasing content of $Mg_4Si_4O_{12}$, from the $Mg_3Al_2Si_3O_{12}$ end to 90%[$Mg_4Si_4O_{12}$]-10%[$Mg_3Al_2Si_3O_{12}$]. The end member $Mg_4Si_4O_{12}$ with tetragonal symmetry, has lattice parameters smaller than that estimated for cubic symmetry by extrapolation. The extrapolated value of molar volume, for a virtual garnet with cubic symmetry (28.48 Å^3/mol) is larger than that of tetragonal garnet (28.23 Å^3/mol). Additional support for tetragonal garnet with a smaller volume is provided by an empirical equation of NOVAK and GIBBS (1971) for calculating cell edge, a, of silicate garnet from the mean radius of cations in the dodecahedral site, $\langle r|D|\rangle$, and the mean radius of cations in the octahedral site, $\langle r|O|\rangle$, as fellows

$$a = 9.04(2) + 1.61(4) \langle r|D| \rangle + 1.98(8) \langle r|O| \rangle$$

The lattice parameter of virtual cubic garnet is calculated to be a=11.78 Å and V=a^3=1634 Å^3 for $MnSiO_3$, and a=11.60 Å and V=a^3=1559 Å^3 for $MgSiO_3$, whereas the equivalent volume of tetragonal garnet is given by V=a^3=1613(3) Å^3 for $MnSiO_3$ and V=a^3=1499.6(3.6) Å^3 for $MgSiO_3$. In both cases, the volume of tetragonal garnet is smaller by 1–4% than the volume that of the virtual cubic garnet. Although the density contrast is not yet accurately known, we conclude that the density of tetragonal $MgSiO_3$ garnet is larger than that of the cubic equivalent by more than 1%, indicating that the tetragonal structure is high-pressure phase.

The cubic garnet (majorite) possesses a disordered structure with a mixing of Mg and Si ions in the octahedral site, whereas the tetragonal garnet possesses an ordered structure with Mg and Si ions partitioned into two crystallographically independent sites. Therefore cubic garnet is expected to be a high-entropy phase with additional configurational entropy due to mixing of cations, and the tetragonal garnet is a low-entropy phase. These considerations indicate that the phase boundary between tetragonal and cubic garnet has a positive slope, if both coexist. Here we note that even the tetragonal garnet with an ordered cation distribution possesses a larger entropy than perovskite as discussed in the previous section. This may be a structural consequence of the garnet structure which has a larger unit cell size and a complex structure with more independent sites, and hence a larger configurational entropy.

Elastic properties of $MgSiO_3$ with tetragonal and cubic structures. BABUSKA et al. (1978) measured the elastic constants of many garnet single crystals with a wide variety of chemical compositions, using the rectangular

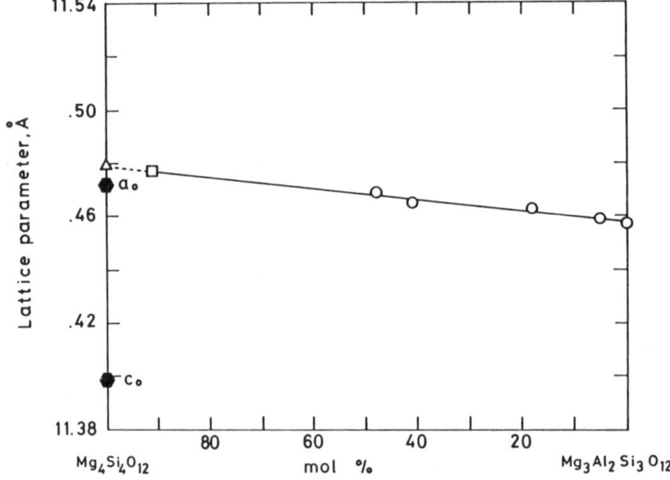

Figure 3. Lattice parameter versus composition of homogeneous $Mg_4Si_4O_{12}$-$Mg_3Al_2Si_3O_{12}$ garnet solid solutions. ○: after AKAOGI and AKIMOTO (1977), □: after RINGWOOD (1967), Δ: extrapolated lattice parameter of virtual garnet with cubic symmetry, ●: lattice parameters of tetragonal garnet (present work).

parallelepiped resonance method (OHNO, 1976). The elastic properties of the garnet single crystals are summarized in Table 4. Empirical relations derived from these data are used to estimate the elastic properties of $MgSiO_3$ with tetragonal garnet structure and also those of hypothetical cubic $MgSiO_3$ garnet for comparison. The bulk modulus $K=177$ GPa is obtained from KV_m and $V_\phi M^{1/2}$=constant, by using the density of tetragonal garnet (3.556 g/cm^3) determined in the present study. The evaluated properties of tetragonal garnet structure V_p=9.37 km/s, V_s=5.32 km/s, K=177 GPa, and μ=101 GPa, are consistent with those of hypothetical cubic garnet, V_p=9.17 km/s, V_s=5.11 km/s, K=173 GPa, and μ=91.8 GPa, given by BABUSKA et al. (1978).

Pressure derivative of shear moduli. The elastic and thermal properties of silicates have been previously summarized (ANDERSON et al., 1968; JEANLOZ and THOMPSON, 1983; SUMINO and ANDERSON, 1984). The pressure derivatives of shear moduli for high-pressure phases are essential parameters for studying the constitution of the earth's mantle through the comparison of laboratory and seismological data. However, these pressure derivatives have not been measured, and thus no conclusive information is available yet.

ANDERSON (1968) suggested that the dimensionless parameter $d\mu/dp$ for a number of oxides and silicates can be characterized by two parameters, that is, the coordination number for pairs in the lattice and the ratio of the shear modulus to the bulk modulus. We can estimate the pressure derivative of the shear modulus of high pressure forms by the following method (ANDERSON, 1968).

Poisson's ratio, σ, is related to μ/K_s by

$$\sigma = (3K_s - 2\mu)/(6K_s + 2\mu) \tag{9}$$

where K_s is the adiabatic bulk modulus and μ is the shear modulus (rigidity). Differentiation of Equation (9) with respect to pressure yields

$$\partial\sigma/\partial P = [9K_s\mu/2(3K_s + \mu)^2\{(1/K_s)\partial K_s/\partial P - (1/\mu)\partial\mu/\partial P\}]$$

$(\partial\sigma/\partial P)_T=0$ implies that

$$(\partial\mu/\partial P)_T = (\mu/K_s)\partial K_s/\partial P$$

But $(\partial\sigma/\partial P)_T\neq0$ implies that

$$(\partial\mu/\partial P)_T = (\mu/K_s)(\partial K_s/\partial P) + b \tag{10}$$

Figure 4 shows how the value of $d\mu/dP$ varies with μ/K_s by using calculated aggregate properties of single-crystal data for four AO type compounds with rock salt structure (coordination numbers (6-6)), four AO_2 type compounds with rutile structure (6-3), three A_2O_3 type compounds with corundum structure (6-4), orthopyroxene (6-4-4), olivine (6-4), four garnets (8-6-4-4), and six ABF_3 with perovskite structure (12-6-6) (SUMINO and ANDERSON, 1984).

We can observe the following relationships from Figure 4:

a) A large value of μ/K_s tends to result in a larger value of $d\mu/dP$.

b) The pressure derivative of the bulk modulus for silicate minerals is 5±1, so that the slope of Equation (10) is expected to be close to 5.

c) Simple oxides, both types of AO (rock salt), except BaO and AO_2 (rutile), are represented by Equation (10) with the expected slope

TABLE 4. Elasticity of Garnet

Sample	ρ (g/cc)	M (g)	V_m (cc/mol)	K_s (GPa)	μ (GPa)	KV_m (GPa cc/mol)	V_ϕ (Km/s)	V_p (Km/s)	$V_\phi M^{1/2}$ (Km g/s)	$V_p M^{1/2}$ (Km g/s)
PY-2	3.699	21.47	28.93	171.6	92.1	4964	6.81	8.92	31.6	41.3
PY-1	3.705	21.46	28.90	171.3	92.6	4951	6.80	8.92	31.5	41.3
PY-0	3.704	21.42	28.90	170.0	92.6	4913	6.77	8.90	31.3	41.1
PY-A	3.723	21.54	28.91	170.8	92.0	4938	6.77	8.88	31.4	41.2
AL-6	3.916	22.48	28.68	173.7	95.4	4982	6.66	8.77	31.6	41.5
AL-4	3.930	22.78	28.89	174.9	95.5	5052	6.67	8.77	31.8	41.8
AL-X	3.945	22.85	28.91	173.4	95.9	5013	6.63	8.74	31.7	41.8
AL-Y	3.976	22.93	28.81	173.6	95.6	5001	6.61	8.70	31.7	41.7
AL-5	4.043	23.46	29.02	175.4	96.2	5090	6.59	8.69	31.9	42.0
SP-1	4.172	24.60	29.46	176.4	96.5	5197	6.50	8.55	32.2	42.4
SP-2	4.185	24.75	29.44	171.8	93.3	5058	6.41	8.41	31.9	41.8
GR-1	3.659	22.96	31.48	161.6	102.6	5087	6.65	9.02	31.9	43.2
AN-2	3.775	24.50	32.60	147.3	92.7	4802	6.25	8.47	30.9	41.9
Average						5004			31.7	42.0
$MgSiO_3$	3.556	20.08	28.23	177.2	100.7		7.07	9.37		

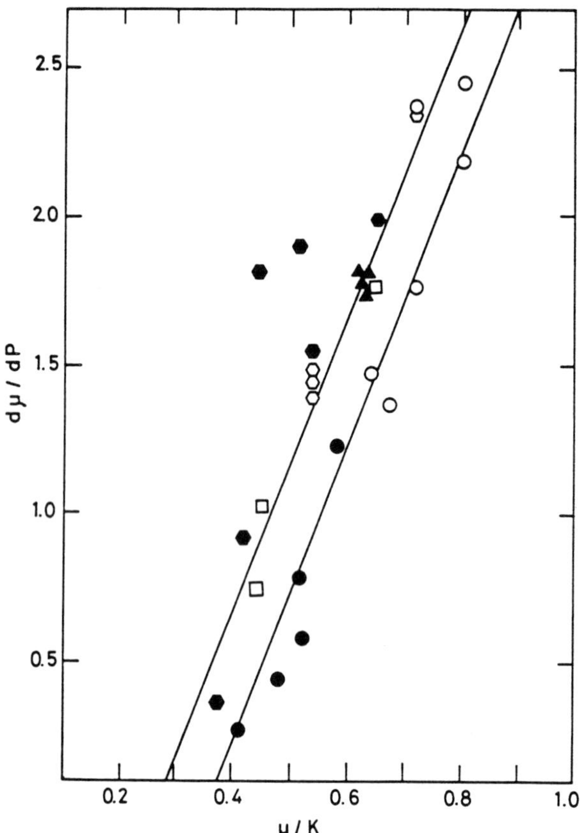

Figure 4. Relation between $d\mu/dP$ and μ/K_s. ○: rock salt, ●: rutile, □: corundum, ○: garnet, ▲: olivine and pyroxene, ●: perovskite.

$$d\mu/dP = 5(\mu/K_s) - 1.77 \tag{11}$$

d) Silicate, corundum, and perovskite (ABF_3 type, except for $CsCdF_3$ and $RbMnF_3$) are expressed as well by Equation (10) with slope

$$d\mu/dP = 5(\mu/K_s) - 1.40 \tag{12}$$

even though the coordinate numbers of these structures are quite different.

Equation (12) is an empirical relation that fits well with the data for most mantle minerals, and the values of $d\mu/dP$ for high-pressure phases of silicate are estimated by using μ and K_s.

The physical properties of tetragonal garnet $MgSiO_3$ are summarized in Table 5.

TABLE 5. Physical Properties of Tetragonal Garnet $MgSiO_3$

Properties	Estimated Values	Properties	Estimated Values
V_p (km/s)	9.37	dK_s/dP	4.5
V_s (km/s)	5.32	$d\mu/dP$	1.45
K_s (GPa)	177	$d\mu/dT$ (GPa/K)	− 0.0101
μ (GPa)	101	α (K^{-1})	24.0×10^{-6}
V_m ($10^{-6}m^3$/mol)	28.23	$d\alpha/dT$	0.39×10^{-8}
ρ (g/cm^3)	3.556	δ_s	3.5
V_Φ (km/s)	7.07		

Conclusions

The phase diagram of $MgSiO_3$ is determined in the pressure range of 15–25 GPa and temperature range of 1000–2200 °C on the basis of high-pressure experiments combined with thermodynamic constraints. The ordered garnet of $MgSiO_3$ with tetragonal symmetry occurs stably above 1600 °C and 17 GPa up to its temperature of congruent melting and to the pressure of transformation to perovskite. The tetragonal garnet is a denser polymorph (ρ=3.556 g/cm^3) of cubic garnet (majorite) without alumina, and it possesses a higher entropy than perovskite by 16 J/K mol indicating that the slope of the boundary with the perovskite phase is not negative but positive. The seismic wave velocities of the tetragonal garnet are estimated as V_p=9.37 km/s and V_s=5.32 km/s, and other elastic and thermal properties are also determined by empirical rules in order to provide the basic data for interpreting the velocity profile of seismic waves in the upper mantle.

The finding of new garnet phase of $MgSiO_3$ leads us to reexamine many problems involving the constitution of mantle minerals, chemical evolution by partial melting, and deep tectonics in the earth and planetary interiors. Among others, we discuss only the tectonic implication in this paper as a conclusion.

The key process in mantle evolution is mantle convection, which may or may not penetrate through the 670 km discontinuity. The seismic discontinuity has been through to correspond to the depth of the perovskite transition from ilmenite or garnet of pyroxene-stoichiometric mineral, and also to the depth of spinel decomposition to perovskite and magnesiowüstite in the case of olivine-stoichiometric mineral. All phase boundaries of these transitions have been considered to have a negative slope: dP/dT=−2.0 MPa/K for spinel-perovskite plus magnesiowüstite (ITO and YAMADA, 1982), −2.6 MPa/K (present work) and −2.0 MPa (ITO and YAMADA, 1982) for ilmenite-perovskite. Therefore convective flow of the mantle penetrating through this boundary has been supposed to be difficult, suggesting that the mantle convection is separated at 670 km in depth. If this is the case, the overall rate of material migration between the upper and lower mantles is very slow, and the nature of mantle fractionation and tectonics are very much constrained by the layering of the convection.

However, the presence of the garnet phase, which has a phase boundary of positive slope (4.3 MPa/K), with the perovskite phase requires a reexamination of 670 km discontinuity described above. If the temperature around the 670 km discontinuity is higher than 2000 °C, the minerals could easily migrate between the upper and lower mantle, giving rise to whole mantle convection. A high-temperature state appears to be inevitable in the mantle, particularly at the early stage (OHTANI, 1985).

In other words, we have a new view of mantle tectonics and fractionation; whole mantle convection for material mixing may have occured at the early stage, followed by layered convection for material fractionation between the upper and lower mantle at later stages of the earth and planetary mantles. One may question whether the situation would change if cubic garnet instead of tetragonal garnet is stabilized as a result of iron and alumina addition (KATO, 1986) in the actual mantle. However, the situation does not change, because the cubic garnet is a high-entropy phase relative to the tetragonal garnet, indicating that the negative slope of the garnet-perovskite boundary should persist.

The interpretation of velocity profiles of seismic waves will be reported in a separate paper, based on the elastic properties of the garnet described in this paper.

Acknowledgments. The author is grateful to M. Kumazawa for his critical discussion and invaluable comments on this work, and to Y. Shono, M. H. Manghnani and R. Jeanloz for their critical reviewing on the manuscript. The author wish to express to M. Kohzaki for his cooperation in the experiment at the early stage of this work. The work was financially supported by a Grant-in-Aid from the Ministry of Education, Science and Culture and by the Japanese DELP Project.

REFERENCES

AKAOGI, M., and S. AKIMOTO, Pyroxene-garnet solid-solution equilibria in the systems $Mg_4Si_4O_{12}$-$Mg_3Al_2Si_3O_{12}$ and $Fe_4Si_4O_{12}$-$Fe_3Al_2Si_3O_{12}$ at high pressures and temperatures, *Phys. Earth Planet. Inter.*, *15*, 90–106, 1977.

AKIMOTO, S., T. YAGI, and K. INOUE, High temperature-pressure phase boundaries in silicate systems using in situ X-ray diffraction, in *High-Pressure Research*, edited by M. H. Manghnani and S. Akimoto, pp. 585–692, Academic Press, New York, 1977.

ANDERSON, O. L., Comments on the negative pressure dependence of the shear modulus found in some oxides, *J. Geophys. Res.*, *73*, 7707–7712, 1968.

ANDERSON, O. L., E. SCHREIBER, R. C. LIEBERMANN, and N. SOGA, Some elastic constant data on minerals relevant to geophysics, *Rev. Geophys.*, *6*, 491–524, 1968.

BABUSKA, V., J. FIALA, M. KUMAZAWA, I. OHNO, and Y. SUMINO, Elastic properties of garnet solid-solution series, *Phys. Earth Planet. Inter.*, *16*, 154–176, 1978.

BURDICK, L. J., A comparison of upper mantle structure beneath North America and Europe, *J. Geophys. Res.*, *86*, 5926–5936, 1981.

FUJINO, K., H. MOMOI, H. SAWAMOTO, and M. KUMAZAWA, Crystal structure and chemistry of $MnSiO_3$ tetragonal garnet, *Am. Mineral.*, *71*, 781–785, 1986.

FUJINO, K., H. SAWAMOTO, and H. Momoi, High pressure type tetragonal garnet (Crystal structure of $MnSiO_3$) (abstract), *Mineralogical Society of Japan*, 13, 1981.

GRAND, S. P., and D. V. HELMBERGER, Upper mantle shear structure of North America, *Geophys. J. R. Atron. Soc.*, *76*, 399–438, 1984.

ITO, E., and A. NAVROTSKY, $MgSiO_3$ ilmenite: calorimetry, phase equilibria, and decomposition at atmospheric pressure, *Am. Mineral.*, *70*, 1020–1026, 1985.

ITO, E., and H. YAMADA, Stability relations of silicate spinels, ilmenites and perovskites, in *High-Pressure Research in Geophysics*, edited by S. Akimoto and M. H. Manghnani, pp. 405–419, Center for Academic Publishing, Tokyo, 1982.

JEANLOZ, R., and A. B. THOMPSON, Phase transition and mantle discontinuities, *Rev. Geophy. Space Phy.*, *21*, 51–74, 1983.

KATO, T., Stability relation of (Mg, Fe)SiO_3 garnet, major constituents in the earth's interior, *Earth Planet. Sci. Lett.*, *77*, 399–408, 1986.

KATO, T., and M. KUMAZAWA, Garnet phase of $MgSiO_3$ filling the pyroxene-ilmenite gap at very high temperature, *Nature*, *316*, 803–807, 1985.

KUMAZAWA, M., Theory of generation of very high static pressures by an external force, *High Temp.-High Press.*, *5*, 599–619, 1973.

LIU, L., The system enstatite-pyrope at high pressures and temperatures and the mineralogy of the earth's mantle, *Earth Planet. Sci. Lett.*, *36*, 237–245, 1977.

LIU, L., The pyroxene-garnet transformation and its implication for the 200-km seismic discontinuity, *Phys. Earth Planet. Inter.*, *23*, 286–291, 1980.

NAVROTSKY, A., Lower mantle phase transitions may generally have negative pressure-temperature slopes, *Geophys. Res. Lett.*, *7*, 709–711, 1980.

NOVAK, G. A., and G. V. GIBBS, The crystal chemistry of the silicate garnets, *Am. Mineral.*, *56*, 791–825, 1971.

OHNO, I., Free vibration of rectangular parallelepiped crystal and its application to determination of elastic constants of orthorhombic crystal, *J. Phy. Earth.*, *24*, 355–379, 1976.

OHTANI, E., The primodial terrestrial magma ocean and its implications for stratification of the mantle, *Earth Planet. Sci. Lett.*, *38*, 70–80, 1985.

PREWITT, C. T., and A. W. SLEIGHT, Garnet-like structure of high-pressure cadmium germanate and calcium garnet, *Science*, *163*, 386–387, 1969.

RINGWOOD, A. E., The pyroxene-garnet transformation in the earth's mantle, *Earth Planet. Sci. Lett.*, *2*, 255–263, 1967.

SAWAMOTO, H., Single crystal growth of modified spinel(β) and spinel(γ) phases of (Mg, Fe)$_2SiO_4$ and some geophysical implications, *Phys. Chem. Minerals*, *13*, 1–10, 1986.

SAWAMOTO, H., and M. KOHZAKI, Stability field of garnet-like $MgSiO_3$ and its geophysical significance, *Seismological Society of Japan, Programs and Abstracts*, *2*, 135, 1984.

SAWAMOTO, H., D. J. WEIDNER, S. SASAKI, and M. KUMAZAWA, Single-crystal elastic properties of the modified spinel (beta) phase of magnesium orthosilicate, *Science*, *224*, 749–751, 1984.

SUMINO, Y., and O. L. ANDERSON, Elastic constants of minerals, in *CRC Handbook of physical properties of rocks, Volume III*, edited by R. S. Carmichael, pp. 39–138, CRC Press Inc., Florida, 1984.

WEIDNER, D. J., A mineral physics test of a pyrolite mantle, *Geophys. Res. Lett.*, *12*, 417–420, 1985.

WEIDNER, D. J., J. D. BASS, A. E. RINGWOOD, and W. SINCLAIR, The single-crystal elastic moduli of stishovite, *J. Geophys. Res.*, *87*, 4740–4746, 1982.

ULTRAHIGH-PRESSURE PHASE TRANSFORMATIONS AND THE CONSTITUTION OF THE DEEP MANTLE

Eiji ITO and Eiichi TAKAHASHI

Institute for Study of the Earth's Interior, Okayama University, Misasa, Tottori-ken 682-02, Japan

Abstract. In order to clarify the nature of the 670-km discontinuity and the state of the lower mantle, phase equilibria in the systems MgO-FeO-SiO_2 and $CaSiO_3$-$MgSiO_3$-Al_2O_3 were investigated in the pressure range of 10–27 GPa and at temperatures up to 1600 °C, using a uniaxial split-sphere apparatus (USSA-5000). We found that the dissociation of the spinel phase into the assemblage of perovskite and magnesiowüstite could be responsible for the 670-km discontinuity. The dissociation was completed in a narrow pressure interval (less than 1 GPa), thereby making the discontinuity very sharp. By considering the detailed phase relations and the estimated physical properties of the lower mantle, we determined possible compositions of the lower mantle in the system MgO-FeO-SiO_2. In the system $CaSiO_3$-$MgSiO_3$-Al_2O_3, the stability field of majorite was observed to expand rapidly towards the $CaSiO_3$-$MgSiO_3$ join in the pressure range of 10–18 GPa, and to retrograde towards the grossular-pyrope join at pressures higher than 23 GPa, dissociating $MgSiO_3$-rich perovskite and an unquenchable "CaO-rich phase" with diopsidic composition. Therefore majorite is expected to be an important constituent in the transition zone, and the dissociation of majorite could contribute to the 670-km discontinuity. The complete dissociation of majorite, however, requires a fairly large pressure interval (~3 GPa), which would produce $MgSiO_3$-rich perovskite, a "CaO-rich phase" with a trace amount of stishovite, and an "Al_2O_3-rich phase" as the lower mantle constituents. Based on the above results and assuming a peridotitic mantle composition, we describe the mineralogical constitution of the deep mantle down to the lower mantle.

Introduction

Recent seismic earth models have found large, sharp increases in velocity and density at the depth of around 670 km. This discontinuity has been accepted as the boundary between the upper and lower mantle (e.g., DZIEWONSKI and ANDERSON, 1981). Therefore the interpretation of the 670-km discontinuity is of vital importance for understanding the mineralogical and chemical constitution of the lower mantle and probably, for understanding the evolution and dynamics of the whole mantle as well. For this reason, recent ultrahigh-pressure phase equilibrium studies have focused on the nature of the discontinuity and the state of the lower mantle in terms of mineralogy and chemistry.

Along this line, we have studied ultrahigh-pressure phase equilibria in various systems and a natural rock, including MgO-SiO_2 (ITO and MATSUI, 1977; ITO and YAMADA, 1982), $MgSiO_3$-$FeSiO_3$ (ITO and YAMADA, 1982), $MgSiO_3$-$Mg_3Al_2Si_4O_{12}$ (KANZAKI, 1986), Mg_2SiO_4-Fe_2SiO_4 (Ito and Takahashi, unpublished), MgO-FeO-SiO_2 (ITO, 1984; ITO et al., 1984), CaO-MgO-Al_2O_3-SiO_2 (YAMADA and TAKAHASHI, 1984), and mantle-derived garnet lherzolite PHN-1611 (TAKAHASHI and ITO, this volume).

In this article we summarize our results on the ultrahigh-pressure phase equilibria in the system MgO-FeO-SiO_2 and some preliminary results in the system $CaSiO_3$-$MgSiO_3$-Al_2O_3 up to 27 GPa, with emphasis on special implications to the mantle structure. Based on these results, we describe the mineralogical constitution of the mantle with a peridotitic composition along a model geotherm, down to the lower mantle.

Experimental Procedures

High-pressure and high-temperature experiments were performed using a uniaxial split-sphere apparatus (USSA-5000) at the Institute for Study of the Earth's Interior. The cubic anvil assembly, composed of tungsten carbide anvils, was compressed with the aid of a 5000-ton hydraulic press (Figure 1). The experimental procedures, including pressure calibration, have been described elsewhere (ITO and YAMADA, 1982; ITO et al., 1984) and are summarized here.

Powdered starting materials were put directly into a cylindrical tantalum heater embedded in the center of a semi-sintered magnesia or pyrophyllite octahedron. Temperatures were monitored with a Pt/Pt-13%Rh thermocouple with cold junctions at ambient conditions. No correction was made for the pressure effect on the emf of the thermocouple. During each experimental run, the temperature and pressure were kept constant, within ±10 °C and ±0.5 GPa of the nominal values, respectively. The sample was kept at a desired pressure and temperature for a specified duration, and then quenched by shutting off the electric power supply. Subsequently, the pressure was released at the rate of 1 GPa/h, and the run product was recovered.

The central portion of each run product was examined by the powder X-ray diffraction method to identify phases present. Small chips adjacent to the thermocouple junction were extracted from some run products and made into polished sections for examination by optical

High-Pressure Research in Mineral Physics, edited by M. H. Manghnani and Y. Syono, pp. 221–229.

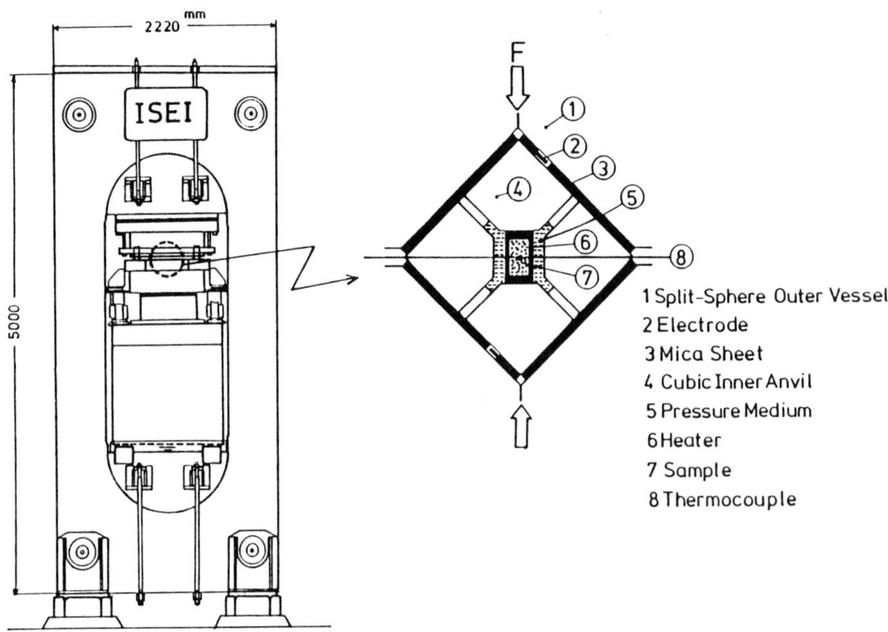

Figure 1. A whole plane of the uniaxial split-sphere apparatus (USSA-5000) at the Institute for Study of the Earth's Interior (left) and a schematic cross-section of the split-sphere vessel with the furnace assembly (right). See ITO et al. (1984).

microscope, SEM, and EPMA.

Ultrahigh-Pressure Phase Equilibria and Their Geophysical Implications

The System MgO-FeO-SiO$_2$

Since MgO, FeO, and SiO$_2$ have been considered to be the dominant components of the earth's mantle (e.g., RINGWOOD, 1975; GANAPATHY and ANDERS, 1974), data on the high-pressure phase equilibria in the system MgO-FeO-SiO$_2$ provide a basis for understanding the constitution of the deep mantle. Based on detailed studies in the pseudobinary systems of MgSiO$_3$-FeSiO$_3$ (ITO and YAMADA, 1982) and Mg$_2$SiO$_4$-Fe$_2$SiO$_4$ (Ito and Takahashi, unpublished), the phase diagrams for the system MgO-FeO-SiO$_2$ at 1100 °C were constructed. Figure 2 shows the phase diagrams in order of increasing pressure (ITO, 1984). Dotted lines indicate data of a reconnaissance nature.

As shown in Figure 2a for a pressure of 19 GPa, the spinel phase is unstable for compositions with iron content higher than (Mg$_{0.35}$Fe$_{0.65}$)$_2$SiO$_4$ that dissociate to magnesiowüstite (Mg$_{1-x}$Fe$_x$)O plus stishovite, and any compound with the (Mg$_{1-x}$Fe$_x$)SiO$_3$ stoichiometry is nonexistent. At 23 GPa (Figure 2b), the stability region of spinel phase diminishes to (Mg$_{0.6}$Fe$_{0.4}$)$_2$SiO$_4$ composition and the ilmenite phase is stabilized for compositions close to MgSiO$_3$. The ilmenite region expands towards the FeSiO$_3$ end-member with increasing pressure, but does not significantly exceed the (Mg$_{0.9}$Fe$_{0.1}$)SiO$_3$ composition. At 25 GPa (Figure 2c), the spinel region becomes even narrower and approaches the composition near (Mg$_{0.8}$Fe$_{0.2}$)$_2$SiO$_4$; the ilmenite phase which has a composition close to MgSiO$_3$, transforms into perovskite. With further increase in pressure, the equilibrium state shown in 2f, is achieved in a very small pressure interval (less than 1 GPa) after passing through the transitional states (see 2d and 2e). At 26 GPa (2f), the ilmenite is completely replaced by perovskite and the surviving spinel dissociates into the assemblage of perovskite and magnesiowüstite. The solubility of FeSiO$_3$ component in perovskite is limited to approximately 8 mol%. The pressure needed to dissociate spinel decreases slightly with increasing iron content, as indicated in Figure 2c.

It is widely accepted that, if the mantle chemistry is represented by MgO, FeO, and SiO$_2$ components, the molar fractions of (MgO+FeO)/SiO$_2$ and FeO/(MgO+FeO) are in the ranges of 1–2 and 0.1–0.2, respectively (e.g., RINGWOOD, 1975). Based on the stability relations shown in Figure 2, any material with such composition exists as the assemblage of spinel plus stishovite at 23 GPa. This assemblage successively transforms with increasing pressure and eventually is completely replaced by perovskite plus magnesiowüstite (with or without a small

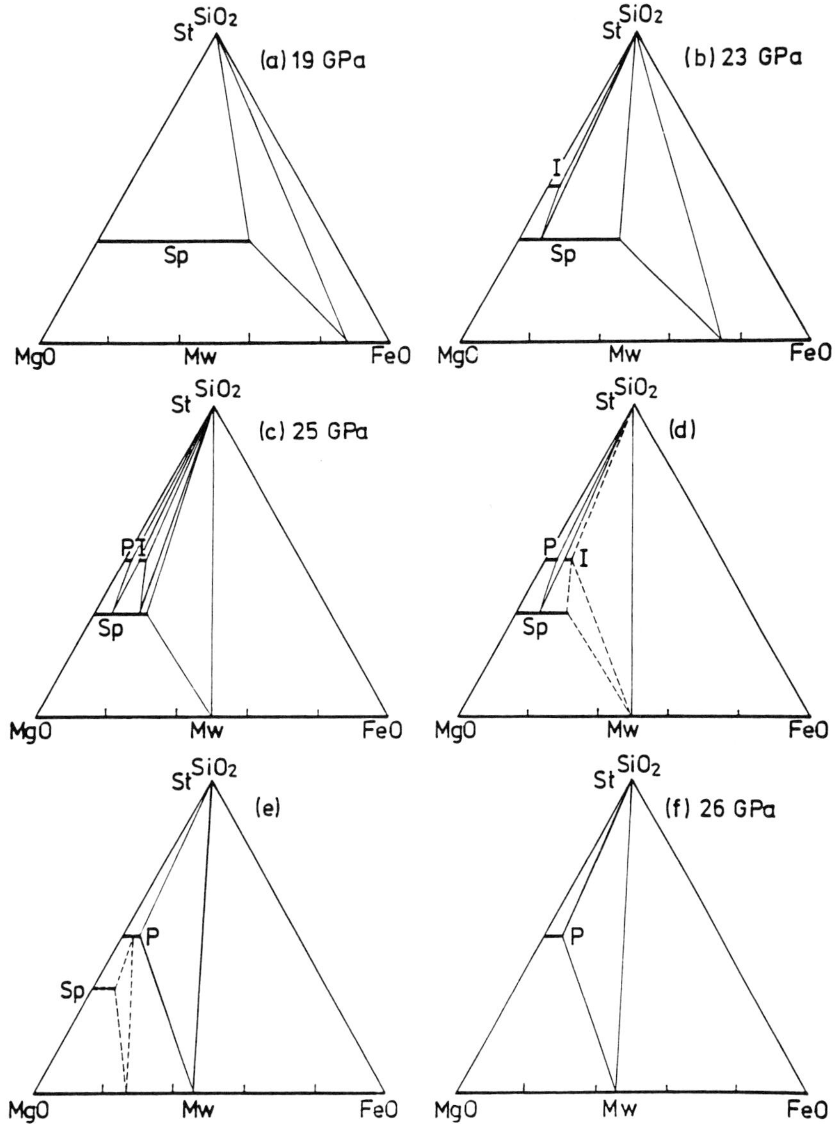

Figure 2. Phase diagrams for the system MgO-FeO-SiO_2 at 1100 °C and various pressures. Sp=$(Mg_{1-x}Fe_x)_2SiO_4$ spinel; Mw=$(Mg_{1-x}Fe_x)O$ magnesiowüstite; I=$(Mg_{1-x}Fe_x)SiO_3$ ilmenite; P=$(Mg_{1-x}Fe_x)SiO_3$ perovskite. See ITO (1984).

amount of stishovite). This transformation scheme may not be directly applicable to the mantle structure, because the pyroxene components mostly dissolve into the garnet phase at pressures higher than 10 GPa when a certain amount of Al_2O_3 is present (RINGWOOD, 1967; AKAOGI and AKIMOTO, 1977, 1979; KANZAKI, 1986). However, it is likely that the dissociation of spinel would be the dominant cause of the 670-km discontinuity. Taking into account that the temperature around a depth of 670 km would be in the range of 1600–2000 °C (BROWN and SHANKLAND, 1981; STACEY, 1977) and that the dissociation has a negative Clapeyron slope (ITO and YAMADA, 1982), the dissociation pressure can be reconciled with the pressure at the discontinuity (approximately 24 GPa) (DZIEWONSKI and ANDERSON, 1981). It should be noted that, since the dissociation is completed within a narrow pressure interval, it will result in a large increase in density (11%) and probably also in seismic velocities over a small interval of depth. Thus there would be a sharp discontinuity at 670 km depth.

The phase relations shown in Figure 2 suggest that the perovskite and magnesiowüstite would be the major constituents of the lower mantle. In order to better understand the state of the lower mantle, ITO et al. (1984) performed a detailed study of the system at 26 GPa and 1600 °C; results are reproduced in Figure 3. The overall topology is similar to that at 1100 °C (Figure 2f). However, the maximum solubility of $FeSiO_3$ in perovskite increases to 11 mol% and the iron content of magnesiowüstite coexisting with perovskite and stishovite shifts from 46 mol% at 1100 °C to 58 mol% at 1600 °C. The apparent Fe-Mg partition coefficients between coexisting perovskite and magnesiowüstite $K'=(Fe/Mg)^{Pv}/(Fe/Mg)^{Mw}$ were determined from the unit-cell dimensions and EPMA analysis (see Figure 3).

ITO et al. (1984) examined what compositions in each stability field would satisfy the estimated physical properties of the lower mantle (i.e., density and bulk modulus extrapolated to the ambient conditions). Such possible lower mantle compositions, shown in Figure 4 (screen toned area), are compared with other various estimated mantle compositions and chondrite compositions. The mineral assemblages shown are perovskite plus stishovite above the En-Fs join and perovskite plus magnesiowüstite below the join. The central line passing through the screen toned area indicates the pair of $(Mg_{0.96}Fe_{0.04})SiO_3$ perovskite and $(Mg_{0.76}Fe_{0.24})O$ magnesiowüstite; this assemblage had the best fit to the adopted lower-mantle density. It should be noted that pyrolite and other upper-mantle compositions based on a peridotitic model are well within the acceptable region. In these cases the magnesian perovskite and relatively iron-rich magnesiowüstite are the major phases. On the other hand, a lower mantle of almost pure perovskite or the assemblage of perovskite plus stishovite cannot be ruled out either. However, these silica-rich lower mantles are required to be fairly magnesian especially in the latter case; i.e., Mg/(Mg+Fe) >0.93 (see BUTLER and ANDERSON, 1978; WATT and AHRENS, 1982).

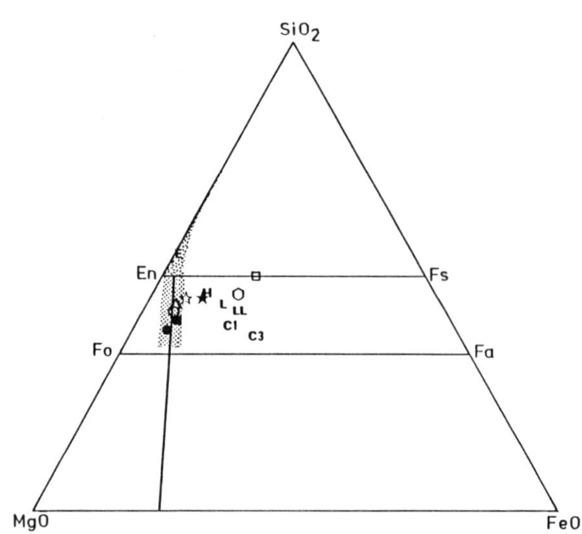

Figure 4. Possible compositions for the lower mantle (screened area). Estimated mantle compositions plotted are: open circle=pyrolite (RINGWOOD, 1975); hexagon=lower mantle (ANDERSON, 1970); open square=lower mantle (ANDERSON et al., 1972); triangle=lower mantle (ANDERSON, 1982); open star=bulk mantle (GANAPATHY and ANDERS, 1974); solid star=bulk mantle (MASON, 1966); solid circle=average spinel lherzolite xenoliths in alkali basalt (MAALØE and AOKI, 1977); and solid square=PHN-1611, a fertile garnet lherzolite xenolith from kimberite (NIXON and BOYD, 1973). E, H, L, LL, C1, and C3 denote bulk silicate compositions of various chondrite groups (MASON, 1962). Also see ITO et al.(1984).

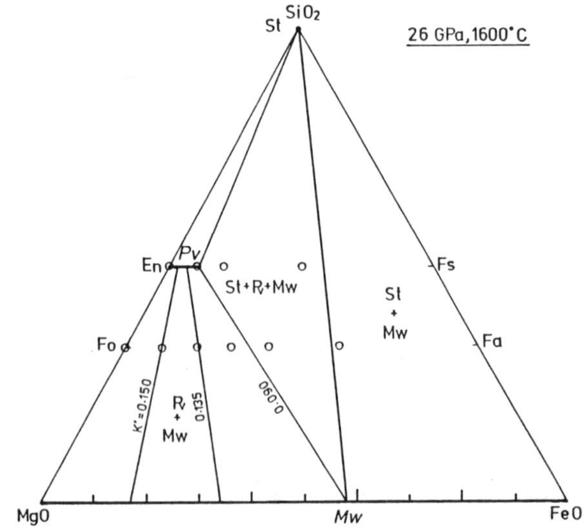

Figure 3. Phase relations in the system $MgO\text{-}FeO\text{-}SiO_2$ at 26 GPa and 1600 °C. Open circles show the compositions of the starting materials. Coexisting perovskite and magnesiowüstite are connected with tie lines. $K'=(Fe/Mg)^{Pv}/(Fe/Mg)^{Mw}$. See ITO et al. (1984).

The System $CaSiO_3$-$MgSiO_3$-Al_2O_3

The subsolidus phase relations of the system $CaSiO_3$-$MgSiO_3$-Al_2O_3 up to 10 GPa were recently investigated in detail by YAMADA and TAKAHASHI (1984) for their study of the petrogenesis of mantle-derived lherzolites. In the present study the experimental pressure was extended up to 27 GPa in order to examine how the CaO and Al_2O_3 components affect the phase transformations in the deep mantle and what phases would accommodate these components in the deep mantle. In our series of experiments, the glasses EWC ($0.725MgSiO_3+0.2CaSiO_3+0.075Al_2O_3$ in mol ratio) and EWC10 ($0.64MgSiO_3+0.18CaSiO_3+0.18Al_2O_3$) were used as starting materials. EWC is very close to the olivine-subtraced peridotitic mantle composition, if iron is considered to be equivalent to magnesium (YAMADA and TAKAHASHI, 1984).

Figure 5 shows preliminary phase diagrams for 1600 °C, in order of increasing pressure. At 10 GPa, the assemblage of garnet solid solution, clinopyroxene and orthopyroxene is stable (YAMADA and TAKAHASHI, 1984). The stability of garnet solid solution expands towards the $CaSiO_3$-$MgSiO_3$ join with increasing pressure. This observation is in agreement with the stability relations observed in the pseudobinary systems $Mg_4Si_4O_{12}$-$Mg_3Al_2Si_3O_{12}$ (AKAOGI and AKIMOTO, 1977; KANZAKI, 1987) and $Ca_2Mg_2Si_4O_{12}$-$Ca_{1.5}Mg_{1.5}Al_2Si_3O_{12}$ (AKAOGI and AKIMOTO, 1979). At 20 GPa, EWC composition was found to be in the single-phase of garnet, suggesting the conversion of almost all of the upper mantle components (except olivine) to the garnet solid solution, or majorite.

More detailed stability relations of EWC composition up to 20 GPa are shown in Figure 6. At pressures of 15–17 GPa, two phases of majorite and diopsidic clinopyroxene are stable. At lower pressures, the assemblage of majorite and two pyroxenes is defined. The composition of majorite coexisting with pyroxene(s) changes from $(Mg_{0.88}Ca_{0.12})_3Al_2Si_3O_{12}$ to nearly $(MgSiO_3)_{0.78}(CaSiO_3)_{0.11}(Al_2O_3)_{0.11}$, almost parallel to the Al_2O_3-$MgSiO_3$ join up to 15 GPa and then rapily moves towards the EWC composition with increasing pressure.

The stability relations would not change significantly with further increase in pressure because the product of EWC quenched at 22 GPa was also the single phase of majorite. At pressures under which $MgSiO_3$ perovskite is stable, however, the stability field of majorite rapidly retrogrades towards the grossular-pyrope join, dissociating $MgSiO_3$-rich perovskite and an unquenchable "CaO-rich" phase shown as Ca-P in Figure 5. Powder X-ray diffraction and EPMA analyses of the products of EWC and EWC10 quenched at 24 GPa showed that majorite, perovskite, and the "CaO-rich phase" were the dominant stable phases. However, a trace amount of stishovite was also found in the products. The results of the EPMA analyses of all phases, except stishovite and perovskite, were distributed over a rather wide range mainly because of grain sizes (3–5 μm) which were too small and the intergrowth texture of the products. However, the phase diagram for 24 GPa (upper right portion of Figure 5) can be proposed as a preliminary diagram, if the small amount of stishovite is ignored. The composition of perovskite was found to be very close to $MgSiO_3$ with only 1.3 mol% Al_2O_3 and 0.14 mol% CaO components.

With further increase in pressure, the stability field of majorite continue to diminish and the perovskite and stishovite become more dominant. Powder X-ray diffraction, however, revealed that a new crystalline phase appears and becomes pronounced as majorite diminishes. In the product of EWC10 quenched at 27 GPa, the majorite was completely absent and the observed phases

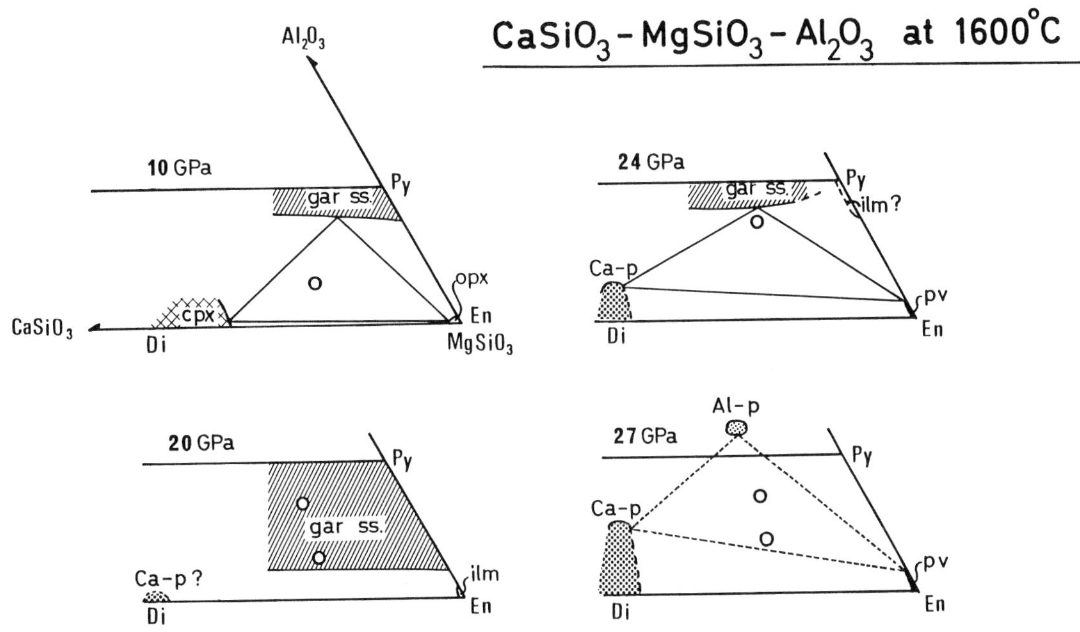

Figure 5. Preliminary phase diagrams in the system $CaSiO_3$-$MgSiO_3$-Al_2O_3 at 1600 °C and various pressures. Open circles denote the composition of starting materials. En, Di, Py, Gr represent enstitite, diopside, pyrope, and grossular compositions, respectively. gar ss=garnet solid solution; Ca-P=unquenchable "Ca-rich phase" (see text); Al-P="Al_2O_3-rich phase" (see text); Ilm=Ilmenite; and Pv=perovskite.

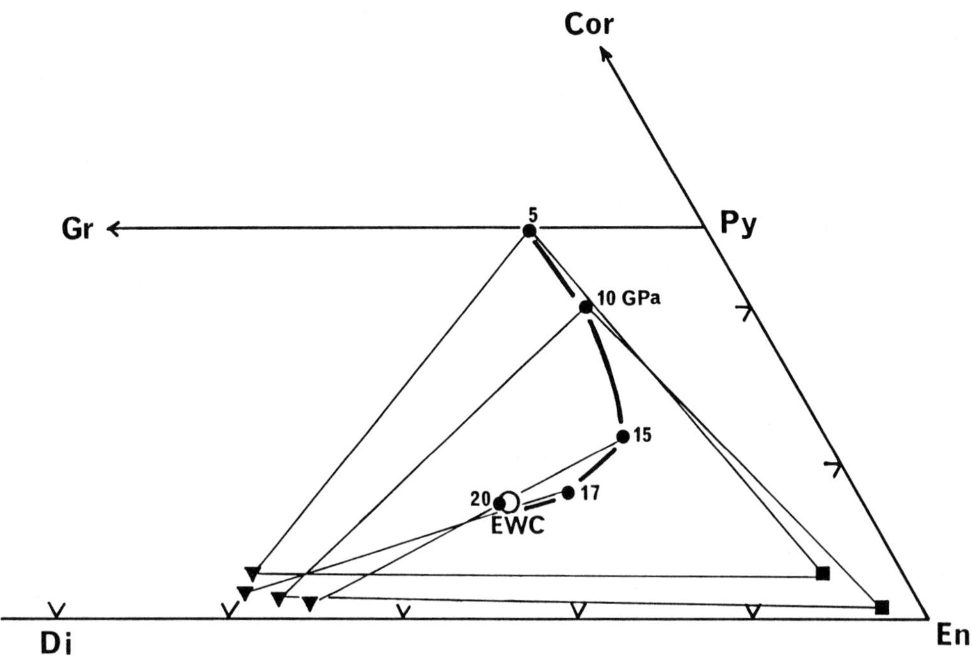

Figure 6. Stability relations of the EWC composition at 1600 °C. Open circle denotes the EWC composition. Solid circles, squares, and triangles show the compositions of garnet, orthopyroxene, and clinopyroxene, respectively, at various pressures. The bold line indicates a change in the garnet composition with pressure.

were perovskite, "CaO-rich phase", stishovite, and the new crystalline phase. The crystal structure of the new phase is still unknown. EPMA analyses of this phase showed scattering within the compositional region shown as Al-p. It was suggested that the phase have a composition richer in Al_2O_3 than grossular-pyrope join, and thus the new phase was tentatively named "Al_2O_3-rich phase" and designated as Al-p in Figure 5. The lower-right diagram in Figure 5 shows the assumed compositional relations among perovskite, "CaO-rich", and "Al_2O_3-rich phases" at 27 GPa. The existence of stishovite in the product suggests that the $SiO_2/(MgO+CaO)$ molar ratio in the total mass of other phases is less than unity. The ratios of "Al_2O_3-" and "CaO-rich phases" (0.97 and 0.94, respectively) show this tendency. However, some of the stishovite might be formed metastably from the "CaO-rich phase" through retrogressive transformation (LIU, 1979). Further investigation, including the TEM analysis of both phases, is required for any quantitative discussion.

Evidence for the existence of the "CaO-rich phase" was not obtained by powder X-ray diffractometry. However, many bright grains of several microns in size were observed in the back-scattered electron image of the products. The compositions of these grains were found to be diopsidic. Therefore it was judged that the unquenchable phase existed at high-pressure conditions but reverted to an amorphous phase or glass upon pressure release. Similar results have been obtained in high-pressure experiments on natural peridotite (TAKAHASHI and ITO, this volume). For the high-pressure transformations of $MgCaSi_2O_6$, results of experiments using a diamond-anvil cell coupled with a laser heating system have been inconsistent. LIU (1979) suggested that $MgCaSi_2O_6$ might transform to an unquenchable single phase, possibly with the perovskite structure. On the other hand, MAO et al. (1977) reported the mixture of $MgSiO_3$ perovskite and $CaSiO_3$ perovskite (LIU and RINGWOOD, 1975) as the high-pressure assemblage of $MgCaSi_2O_6$. The present study did not confirm whether the "CaO-rich phase" recovered at ambient conditions is strictly homogeneous or not. However, the fact that the "CaO-rich phase" and $MgSiO_3$-rich perovskite were found to form individual grains, in spite of using homogeneous glasses as starting materials, suggests that the "CaO-rich phase" might be a single phase at high-pressure conditions.

Although the stability field of the ilmenite phase, which was inferred from the phase relations of $MgSiO_3$ (ITO and NAVROTSKY, 1985) and the system $MgSiO_3$-Al_2O_3 (KANZAKI, 1986), are shown in the phase diagrams for 20 GPa and 24 GPa (Figure 5), no ilmenite phase was observed in the series of experiments conducted at 1600 °C. However, in experiments at 1200 °C and in the pressure range of 22–23 GPa, the ilmenite phase of $MgSiO_3$-rich composition was found as to be a dissociation product of majorite; this product was replaced by the perovskite phase at higher pressures (Ito and Takahashi, 1982, unpublished). Therefore it is not expected that the ilmenite phase would be a major constituent of the mantle, so far as modern thermal models of the mantle (e.g., STACEY, 1977; BROWN and SHANKLAND, 1981) are accepted. However, the ilmenite phase could be important in the low-temperature regions, such as in subducting slabs.

The phase equilibria described in this section indicate that almost all pyroxene components dissolve into the garnet solid solution (or majorite) which would be an important constituent in the lower part of the transition zone (compare with AKAOGI and AKIMOTO, 1977, 1979; KANZAKI, 1986). It is also demonstrated that the majorite dissociates into the perovskite-bearing assemblage at pressures higher than 24 GPa. This dissociation should result in fairly large increases in density and seismic velocities, and therefore could be a major cause of the 670-km discontinuity. For the EWC composition, the amount of majorite is suggested to diminish rapidly after the onset of dissociation. It should be noted, however, that the complete dissociation requires a large pressure interval of approximately 3 GPa, which corresponds to a depth interval of about 80 km. The dissociation products, $MgSiO_3$-rich perovskite, diopsidic "CaO-rich phase", stishovite, and "Al_2O_3-rich phase", are considered to be the lower mantle constituents. The last two phases, especially the "Al_2O_3-rich phase", might be present in quantities less than the other two by an order of magnitude.

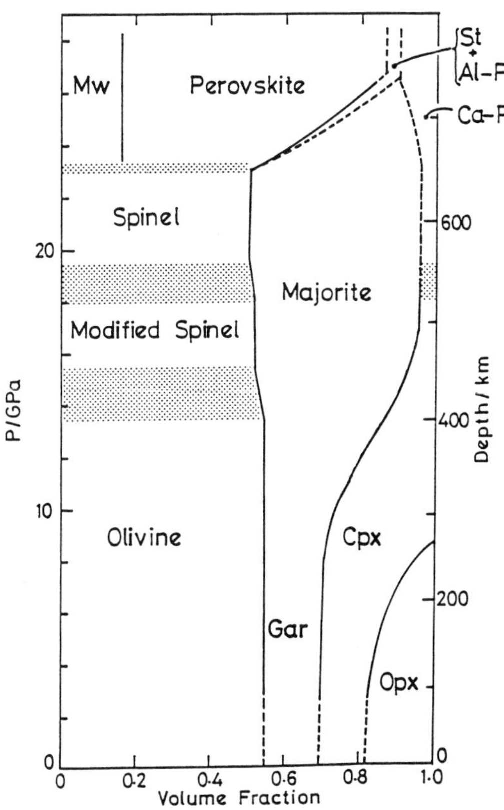

Figure 7. Volumetric mineral constitution of a peridotitic mantle down to the lower mantle. Gar=garnet; Opx=orthopyroxene; Cpx=clinopyroxene; Mw=magnesiowüstite; St=stishovite; Ca-P="CaO-rich phase" (see text); and Al-P="Al_2O_3-rich phase" (see text).

Constitution of the Deep Mantle

The phase relations described above provide principal features necessary for understanding the constitution of the region between the transition zone and the lower mantle. Another series of experiments on the high-pressure transformations of natural garnet lherzolite (PHN-1611) have been carried out up to 26 GPa (TAKAHASHI and ITO, this volume). Based on all of these results, the volumetric mineralogical constitution of the peridotitic mantle (e.g., pyrolite, RINGWOOD, 1975) or a mantle-derived lherzolite (i.e., PHN-1611, BOYD, 1973), is determined down to the lower mantle region along a model geotherm (see Figure 7). In the model geotherm, the equilibrium pressure and temperature of PHN-1611 (BOYD, 1973) was adopted as the starting point, and an adiabatic temperature gradient of 0.3 deg/km (RICHTER and MCKENZIE, 1981) was assumed in the deeper region (see TAKAHASHI and ITO, this volume).

At low pressures, the dominant constituents are olivine, clinopyroxene, garnet, and orthopyroxene. The olivine transforms to the modified spinel and spinel phases successively. The olivine-modified spinel transformation are considered to be responsible for the 400-km discontinuity (e.g., RINGWOOD, 1975; AKIMOTO et al., 1976). The spinel phase dissociates into magnesian perovskite plus iron-rich magnesiowüstite at higher pressure.

The pyroxenes dissolve extensively into the garnet phase at pressures higher than 10 GPa. The majorite is considered to be an important constituent of the transition zone. A small amount of CaO-rich clinopyroxene, however, survives up to fairly high pressures and transforms into the unquenchable "CaO-rich phase" (see TAKAHASHI and ITO, this volume). At pressures higher than 24 GPa, the majorite dissociates into the assemblage of magnesian perovskite, the "CaO-rich phase", stishovite, and the "Al_2O_3-rich phase".

The 670-km discontinuity is considered to be caused by the dissociations of spinel and majorite in the model mantle, because these dissociations start at a pressure of about 23.5 GPa. The mineralogical constitution of the peridotitic lower mantle is inferred to be 70% perovskite, 18% magnesiowüstite, and 8% "CaO-rich phase" with a small amount of stishovite and "Al_2O_3-rich phase" in volumetric ratio.

It is difficult to make a direct comparison of the present model of the 670-km discontinuity to seismological models, because the velocity data for all the phases concerned are not yet available. However, the increase in bulk density across both dissociations is estimated to be ~0.4 g/cm^3 (or ~10%) and the increase with a seismic parameter at zero pressure ($\phi_0 = K_0/\rho_0$) is ~8 (km/s)^2 (or ~15%), based on the bulk modulus data for γ-Mg_2SiO_4 (WEIDNER et al., 1984), $MgSiO_3$ perovskite (YAGI et al., 1978), and $(MgSiO_3)_{0.58}(Mg_3Al_2Si_3O_{12})_{0.42}$ majorite (AKAOGI et al., this volume). It should be noted that the dissociation of spinel is completed within a very small depth interval; this would make the discontinuity very sharp, especially in its upper portion. The dissociation of majorite, on the other hand, proceeds over a fairly large depth range, approximately 80 km. All of these features are in agreement with the recent seismic models in which the 670-km discontinuity is large and sharp, but there is still a high gradient of velocities to depths of 750 km or 800 km (DZIEWONSKI and ANDERSON, 1981; GRAND and HELMBERGER, 1984; WALCK, 1984).

Acknowledgments. The authors are grateful to Y. Matsui and S. Akimoto for their support and encouragement. This work was supported by Grant-in-Aid for Scientific Research from the Ministry of Education, Science, and Culture of the Japanese Government (No. 60221014 and No. 59460048).

REFERENCES

AKAOGI, M., and S. AKIMOTO, Pyroxene-garnet solid solution equilibria in the systems $Mg_4Si_4O_{12}$-$Mg_3Al_2Si_3O_{12}$ and $Fe_4Si_4O_{12}$-$Fe_3Al_2Si_3O_{12}$ at high pressures and temperatures, *Phys. Earth Planet. Inter., 15*, 90–106, 1977.

AKAOGI, M., and S. AKIMOTO, High-pressure phase equilibria in a garnet lherzolite, with special reference to Mg^{2+}-Fe^{2+} partitioning among constituent minerals, *Phys. Earth Planet. Inter., 19*, 31–51, 1979.

AKAOGI, M., A. NAVROTSKY, T. YAGI, and S. AKIMOTO, Pyroxene-garnet transformation: thermochemistry and elasticity of garnet solid solutions and application to a pyrolite mantle, this volume.

AKIMOTO, S., Y. MATSUI, and Y. SYONO, High-pressure crystal-chemistry of orthosilicates and the formation of the mantle transition zone, in *Physics and Chemistry of Minerals and Rocks*, edited by R. G. J. Strens, pp. 327–363, Wiley Interscience, London, 1976.

ANDERSON, D. L., Petrology of the mantle, *Mineral. Soc. Am. Spec. Pap., 3*, 85–93, 1970.

ANDERSON, D. L., Chemical composition and evolution of the mantle, in *High-Pressure Research in Geophysics*, edited by S. Akimoto and M. H. Manghnani, pp. 301–318, D. Reidel, Dordrecht, 1982.

ANDERSON, D. L., C. SAMMIS, and T. JORDAN, Composition of the mantle and core, in *The Nature of the Solid Earth*, edited by E. C. Robertson, pp. 41–66, McGraw-Hill, New York, N.Y., 1972.

BOYD, F. R., A pyroxene geotherm, *Geochem. Cosmochim. Acta, 37*, 2533–2546, 1973.

BROWN, J. M., and T. J. SHANKLAND, Thermodynamic parameters in the earth as determined from seismic profiles, *Geophys. J. R. Astron. Soc., 66*, 579–596, 1981.

BUTLER, R., and D. L. ANDERSON, Equation of state fits to the lower mantle and outer core, *Phys. Earth Planet. Inter., 17*, 147–162, 1978.

DZIEWONSKI, A. M., and D. L. ANDERSON, Preliminary reference earth model, *Phys. Earth Planet. Inter., 25*, 297–356, 1981.

GANAPATHY, R., and E. ANDERS, Bulk compositions of the moon and earth, estimated from meteorites, *Proc. Fifth Lunar Sci. Conf., 2*, 1181–1206, 1974.

GRAND, S., and D. HELMBERGER, Upper-mantle shear structure of North America, *Geophys. J. R. Astron. Soc., 76*, 399–438, 1984.

ITO, E., Ultra-high pressure phase relations of the system MgO-FeO-SiO_2 and their geophysical implications, in *Material Science of the Earth's Interior*, edited by I. Sunagawa, pp. 387–394, TERRAPUB, Tokyo, 1984.

ITO, E., and Y. MATSUI, Silicate ilmenites and the post-spinel transformations, in *High Pressure Research Applications in Geophysics*, edited by M. H. Manghnani and S. Akimoto, pp. 193–208, Academic Press, New York, 1977.

ITO, E., and A. NAVROTSKY, $MgSiO_3$ ilmenite: Calorimetry, phase equilibrium, and decomposition at atmospheric pressure, *Am. Mineral., 70*, 1020–1026, 1985.

ITO, E., and H. YAMADA, Stability relations of silicate spinels, ilmenites, and perovskites, in *High-Pressure Research in Geophysics*, edited by S. Akimoto and M. H. Manghnani, pp. 405–419, D. Reidel, Dordrecht, 1982.

ITO, E., E. TAKAHASHI, and Y. MATSUI, The mineralogy and chemistry of the lower mantle: an implication of the ultrahigh-pressure phase relations in the system MgO-FeO-SiO_2, *Earth Planet. Sci. Lett., 67*, 238–248, 1984.

KANZAKI, M., Ultrahigh-pressure phase relations in the system $MgSiO_3$-$Mg_3Al_2Si_3O_{12}$, *Phys. Earth Planet. Inter.*, in press, 1987.

LIU, L., The system enstatite-wollastonite at high pressures and temperatures, with emphasis on diopside, *Phys. Earth Planet. Inter., 19*, 15–18, 1979.

LIU, L., and A. E. RINGWOOD, Synthesis of a perovskite-type polymorph of $CaSiO_3$, *Earth Planet. Sci. Lett., 28*, 209–211, 1975.

MAALØE, E., and K. AOKI, The major element composition of the upper mantle estimated from the composition of lherzolites, *Contr. Min. Petrol., 63*, 161–173, 1977.

MAO, H. K., T. YAGI, and P. M. BELL, Mineralogy of the earth's deep mantle: quenching experiments on mineral compositions of high pressure and temperature, *Carnegie Inst. Wash. Year B., 76*, 502–504, 1977.

MASON, B., *Meteorites*, 274 pp., John Wiley and Sons, New York, 1962.

MASON, B., Composition of the earth, *Nature, 211*, 616–618, 1966.

NIXON, P. H., and F. R. BOYD, Petrogenesis of the granular and sheared ultrabasic nodule suites in kimberite, in *Lesotho Kimberites*, edited by P. H. Nixon, pp. 48–56, Lesotho Natl. Dev. Corp., 1973.

RICHTER, F. M., and D. P. MCKENZIE, On some consequences and possible causes of layered mantle convection, *J. Geophys. Res., 86*, 6133–6142, 1981.

RINGWOOD, A. E., The pyroxene-garnet transformation in the earth's mantle, *Earth Planet. Sci. Lett., 2*, 255–263, 1967.

RINGWOOD, A. E., *Composition and Petrology of the Earth's Mantle*, 618 pp., McGraw-Hill, New York, 1975.

STACEY, F. D., A thermal model of the earth, *Phys. Earth Planet. Inter., 15*, 341–348, 1977.

TAKAHASHI, E., and E. ITO, Mineralogy of mantle peridotite along a model geotherm up to 700 km depth, this volume.

WALCK, M. C., The p-wave upper mantle structure beneath an active spreading center: the Gulf of California, *Geophys. J. R. Astron. Soc., 76*, 697–723, 1984.

WATT, J. P., and T. J. AHRENS, The role of iron partitioning in mantle composition, evolution, and scale of convection, *J. Geophys. Res., 87*, 5631–5644, 1982.

WEIDNER, D. J., H. SAWAMOTO, S. SASAKI, and M. KUMAZAWA, Single-crystal elastic properties of spinel phase of Mg_2SiO_4, *J. Geophys. Res., 89*, 7852–7860, 1984.

YAGI, T., H. K. MAO, and P. M. BELL, Hydrostatic compression of $MgSiO_3$ perovskite structure, *Carnegie Inst. Washington Year B., 78*, 613–614, 1978.

YAMADA, H., and E. TAKAHASHI, Subsolidus phase relations between coexisting garnet and two pyroxenes at 50 to 100 kbar in the system CaO-MgO-$Al_2O_3SiO_2$, in *Kimberlites, II: The Mantle and Crust-Mantle Relationship*, edited by J. Kornprobst, pp. 247–255, Elsevier Science Publishers B. V., Amsterdam, 1984.

PHASE TRANSFORMATIONS IN PRIMITIVE MORB AND PYROLITE COMPOSITIONS TO 25 GPa AND SOME GEOPHYSICAL IMPLICATIONS

T. IRIFUNE and A. E. RINGWOOD

Research School of Earth Sciences, Australian National University, Canberra, ACT 2601, Australia

Abstract. The mineralogies adopted by primitive MORB and "pyrolite minus olivine" compositions have been studied over the pressure range of 4–25 GPa at 1200–1400 °C. Both compositions crystallize to eclogite assemblages (garnet+clinopyroxene) with and without orthopyroxene between 4 and 10 GPa, and there is little change in the relative proportions of garnet and pyroxene over this range. The proportion of garnet, however, increases rather rapidly above 10 GPa as pyroxene dissolves in the garnet structure, so that pyroxene-free garnetites are formed at 14–16 GPa. A Ca-rich glassy phase+garnet are recovered from 20 GPa in both compositions. The glassy phase has a composition close to $CaSiO_3$ with 2–8 mol% $CaTiO_3$ and 5–10 mol% $MgSiO_3$ in solid solution, and is believed to be the retrogressive transformation product of a Ca-rich perovskite formed on the release of pressure. The zero-pressure density of this Ca-rich perovskite is estimated as 4.31 g/cm^3, which is substantially higher than that of $MgSiO_3$ perovskite (4.11 g/cm^3). The amount of $CaSiO_3$-rich perovskite increases and residual garnet again becomes aluminum-rich with increasing pressure above 20 GPa. Significant amounts of ilmenite (at 22.5 GPa) and perovskite (at 24.5 GPa) forms of $MgSiO_3$ are also found to coexist with garnet and $CaSiO_3$-rich perovskite in the "pyrolite minus olivine" composition. Density calculations based on these experimental results confirm that the subducted oceanic crust is substantially denser than surrounding pyrolite throughout the mantle, except for a limited region between 670 and 710 km. In this depth interval, a large amount of garnet remains stable in the MORB composition, whereas pyrolite transforms to a denser assemblage $(Mg,Fe)SiO_3$-perovskite plus magnesiowüstite. Seismic velocity and density profiles calculated for the pyrolite composition are consistent with those obtained from seismic observations for the upper mantle and transition zone.

Introduction

A wide range of petrological, geochemical, and geophysical evidence has been employed to support the hypothesis that the earth's mantle is mainly composed of peridotite or pyrolite compositions, containing about 60 wt% normative olivine and 40 wt% normative pyroxene and garnet components (e.g., RINGWOOD, 1975). An alternative model has been proposed by ANDERSON (1979, 1982) and ANDERSON and BASS (1984, 1986). The latter authors maintain that the mantle between depths of 200 (or 400) km and 670 km is composed predominantly of basaltic (or "piclogite") material in the eclogite or garnetite facies.

In order to test these models IRIFUNE (1987) and IRIFUNE et al. (1986, 1987) determined the phase equilibria of basalt and pyrolite compositions up to pressures of 18 GPa so as to study the detailed nature of the pyroxene-garnet transformation. Pressures up to 25 GPa can now be obtained in our laboratory; we report herein the preliminary results on primitive MORB and pyrolite compositions in this range. Implications for the constitution and dynamics of the upper mantle are also discussed in light of our new experimental data.

Experimental Methods

Starting Materials

Two starting compositions, representative of basalt and pyrolite, were prepared as glasses for the high-pressure experimental runs. The first composition, "DSDP", is identical to the "primitive MORB" composition studied by GREEN et al. (1979) and is also the same as "olivine tholeiite A" used by IRIFUNE et al. (1986). The second composition "pyrolite minus olivine" is the residual pyroxene+garnet component of pyrolite (SUN, 1982) obtained by subtracting 62 wt% olivine (Fo_{88}) from the pyrolite composition. The experimental results on this latter composition up to 18 GPa have been already reported in IRIFUNE (1987). The chemical compositions of these glasses, as analyzed by electron microprobe, are listed in Table 1. Amphibolite starting materials of the same compositions with about 1 wt% H_2O were also prepared to provide hydrous conditions in some runs. This procedure was found to facilitate garnet nucleation

TABLE 1. Analyzed Compositions of Starting Materials

	DSDP	Pyrolite Minus Olivine
SiO_2	50.39	51.78
TiO_2	0.57	0.46
Al_2O_3	16.08	11.44
Cr_2O_3	—	0.92
FeO	7.68	3.12
MgO	10.49	23.14
CaO	13.05	9.50
Na_2O	1.87	0.94
Total	99.26	101.30
Mg*	70.9	91.9

Mg*, 100 Mg/(Mg+Fe).

High-Pressure Research in Mineral Physics, edited by M. H. Manghnani and Y. Syono, pp. 231–242.

and to suppress associated metastability problems (IRIFUNE et al., 1986). Starting materials were sealed in platinum capsules.

Experimental Techniques

High-pressure experiments were carried out with an MA8-type apparatus (KAWAI and ENDO, 1970), which was driven by truncated split-sphere guide blocks and a 1200-ton uniaxial press (OHTANI et al., 1987). Sets of eight cubic anvils, truncated on all corners by either 3.5 mm ($P<18$ GPa), 2.5 mm (18 GPa$<P<$22 GPa) or 2.0 mm (22 GPa$<P<$25 GPa), were used according to the pressure range required. Pressure was calibrated against the press load on the basis of room-temperature transformations of appropriate reference materials (IRIFUNE et al., 1986). Pressure calibration at high temperature was also carried out by comparison with the olivine-modified spinel transformation in Mg_2SiO_4 (AKAOGI et al., 1984) and the pyroxene-ilmenite-perovskite transformations in $MgSiO_3$ (ITO and YAMADA, 1982; SAWAMOTO, 1986a). Internal consistency in pressures generated by three sets of anvils with different truncations was monitored by the rapidly increasing lattice parameters of garnet in the "pyrolite minus olivine" composition between 14 and 16 GPa (Figure 1).

Two thin sheets of TiC+MgO were employed as heaters to provide a very uniform temperature distribution throughout the sample (IRIFUNE and HIBBERSON, 1985). Temperature was measured by a Pt-Pt10%Rh thermocouple whose hot junction was placed at the end of the platinum capsule. The thermocouple sometimes did not work satisfactorily because of breakage of lead wires or contact with the anvil surface, particularly in runs at pressures above 22 GPa. In these cases, the run temperature was estimated from the supplied electric power. This procedure introduces some uncertainty in temperature estimates, but these probably lie within ±10% of the nominal value according to established power-temperature correlations.

Runs were carried out at 1200–1400 °C for 20–30 min and then quenched by turning off the power supply. The products were examined by optical microscopy, X-ray powder diffraction, and electron microprobe techniques. The lattice parameter of garnet was determined from the (14,4,0) and (14,4,2) back reflections using Cu $K_{\alpha 1}$ radiation. In some runs, particularly those above 20 GPa, these reflections were rather weak and diffuse, and the lattice parameter was determined from other lower d-spacing lines. This latter procedure introduces somewhat larger uncertainties in the measurement of the lattice parameter.

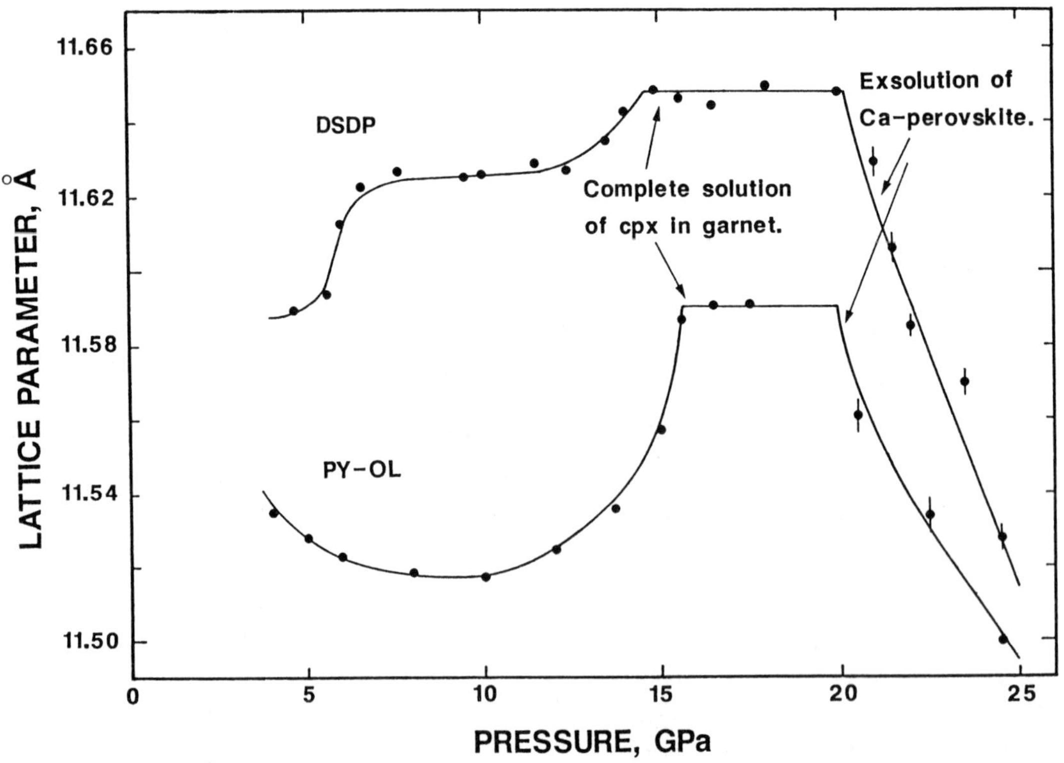

Figure 1. Lattice parameters of garnets in the DSDP and pyrolite minus olivine compositions as a function of pressure.

Experimental Results

Table 2 lists the run conditions, phase assemblages and garnet lattice parameters for the present series of experiments. Experimental conditions and results at pressures lower than 20 GPa are taken from IRIFUNE et al. (1986) and IRIFUNE (1987). Figure 1 shows lattice parameters of garnets in the primitive MORB (DSDP) and pyrolite minus olivine compositions as a function of pressure.

In the primitive MORB (DSDP) composition, a normal eclogite assemblage was formed between 4.6 and 10 GPa, and there was little change in the relative proportion of garnet and pyroxene over this pressure range. The proportion of garnet increases rather rapidly thereafter so that pyroxene-free garnetite (plus stishovite) is produced by 14–15 GPa. The lattice parameter of garnet leveled off around this pressure, indicating that there is no further change in the chemical composition of garnet. At pres-

TABLE 2. Experimental Conditions and Results

Run No.	Press. (GPa)	Temp. (°C)	Time (min.)	Phases Present[a]	Lattice Parameter of Garnet (Å)
(a) DSDP					
S376	4.6	1200	80	Ga+Cpx	11.590(2)
S331	5.7	1200	60	Ga+Cpx	11.594(2)
S378	6.0	1200	70	Ga+Cpx	11.613(2)
S375	6.6	1100	90	Ga+Cpx	11.623(2)
S377	7.6	1200	6	Ga+Cpx	11.627(2)
S335	7.6	1200	70	Ga+Cpx	11.624(2)
S340	9.5	1200	70	Ga+Cpx	11.625(2)
MA71	10.0	1200	30	Ga+Cpx	11.626(1)
S341	10.7	1200	73	Ga+Cpx	11.626(2)
MA51	11.5	1200	30	Ga+Cpx	11.629(1)
S354	12.4	1200	80	Ga+Cpx+St(tr)	11.627(2)
MA49	13.5	1200	30	Ga+Cpx+St	11.635(1)
MA70	14.0	1200	30	Ga+Cpx+St	11.643(1)
MA76	14.9	1200	10	Ga+St+Cpx(tr)	11.649(2)
MA85	15.6	1200	30	Ga+St	11.646(1)
MA84	16.5	1200	30	Ga+St	11.644(1)
MA79	18.0	1200	10	Ga+St	11.650(1)
MA135	20.0	1200	15	Ga+St	11.649(1)
MA136	21.0	1200	20	Ga+St+Ca-per	11.629(4)
MA138	21.5	1200	25	Ga+St+Ca-per	11.606(4)
MA137	22.0	1200	10	Ga+St+Ca-per	11.585(3)
MA161	23.5	1300	5	Ga+St+Ca-per	11.570(4)
MA165	24.5	1200	20	Ga+St+Ca-per	11.528(3)
(b) Pyrolite Minus Olivine					
9965	4.0	1200	30	Ga+Cpx+Opx	11.535(1)
MA98	5.0	1200	30	Ga+Cpx+Opx	11.528(1)
MA97	6.0	1200	30	Ga+Cpx+Opx(tr)	11.523(1)
MA96	8.0	1200	30	Ga+Cpx+Opx(tr)	11.519(1)
MA92	10.0	1200	30	Ga+Cpx	11.517(1)
MA112	10.0	1200	30	Ga+Cpx+Opx(tr)	11.517(1)
MA93	12.0	1200	30	Ga+Cpx	11.525(1)
MA113	12.0	1400	35	Ga+Cpx	11.522(1)
MA94	13.7	1200	30	Ga+Cpx	11.536(1)
MA119	15.0	1200	30	Ga+Cpx	11.557(1)
MA122	15.0	1400	30	Ga+Cpx	11.554(1)
MA95	15.6	1200	30	Ga+Cpx(tr)	11.587(1)
MA124	16.0	1400	30	Ga+Cpx(tr)	11.584(1)
MA120	16.5	1200	30	Ga	11.591(1)
MA121	17.5	1200	30	Ga	11.591(1)
MA174	20.5	1200	20	Ga+Ca-per	11.560(4)
MA169	22.5	1200	10	Ga+Ca-per+Ilm?	11.534(5)
MA183	24.5	1300	20	Ga+Ca-per+Mg-per	11.500(1)

[a]Ga, garnet; Cpx, clinopyroxene; Opx, orthpyroxene; St, stishovite; Ca-per, $CaSiO_3$ perovskite; Ilm, Ilmenite; Mg-per, $MgSiO_3$ perovskite.

sures immediately above 20 GPa, however, the lattice parameters of garnets begin to decrease rapidly with increasing pressure. In this pressure range, a glassy, low-refractive-index (<1.7) phase was found to coexist with garnet. Electron microprobe analysis revealed that this phase has a composition close to $CaSiO_3$, as shown in Table 3; It is interpreted as a retrogressive transformation product formed on pressure release from a Ca-rich cubic perovskite (RINGWOOD and MAJOR, 1971; LIU and RINGWOOD, 1975) as discussed below. It should be noted that this Ca-perovskite has less than 10 mol% $MgSiO_3$ and contains about 5 mol% of $CaTiO_3$ in solid solution. Corresponding to the exsolution of $CaSiO_3$ perovskite, the residual garnet becomes Al-rich and Ca-poor with increasing pressure, approaching normal stoichimetric pyrope-almandine in composition. The phase assemblage and composition changes of individual phases in the DSDP composition are also shown in Figure 2a as a function of pressure.

Analogous changes in the mineral assemblage are observed in the pyrolite minus olivine composition as a function of pressure, as shown in Table 2 and Figure 2b, except for the presence of an $MgSiO_3$-rich phase in this bulk composition. A single-phase garnetite was produced by 16 GPa, which is slightly higher than that observed for the DSDP composition. The lattice parameter of garnet becomes smaller with increasing pressure above 20 GPa, and the $CaSiO_3$-rich glassy phase is again found coexisting with garnet. Thereafter garnet becomes Al-rich and approaches pyrope-almandine stoichiometry as shown in Figures 2b and 3. At 22.5 GPa, the presence of the ilmenite phase of $MgSiO_3$ (e.g., LIU, 1976; ITO and YAMADA, 1982) is suggested by a mass-balance calculation for coexisting phases, although it has not yet been confirmed by X-ray diffraction or electron microprobe analyses. The ilmenite phase is expected to have a composition close to pure $MgSiO_3$ and contain a substantial amount of Cr as indicated by the low Cr content of coexisting garnet at this pressure (Figure 3). At a still higher pressure of 24.5 GPa, the coexistence of $MgSiO_3$-rich perovskite with garnet and Ca-rich perovskite was confirmed by X-ray diffraction and electron microprobe analyses of the run products. Analyses of these phases are given in Table 3. It should be noted that the Ca-perovskite in the pyrolite minus olivine composition, like that in the DSDP composition, contains only a limited amount of $MgSiO_3$ in solid solution.

TABLE 3. Representative Analyses of Coexisting Phases Above 20 GPa

	(a) DSDP (21.5 GPa)		(b) Pyrolite Minus Olivine (24.5 GPa)		
	Ga	Ca-per	Ga	Ca-per	Mg-per
SiO_2	43.35	47.92	46.58	51.21	52.29
TiO_2	0.41	5.86	0.09	1.48	1.81
Al_2O_3	20.47	2.50	18.71	1.90	5.28
Cr_2O_3	0.11	0.04	1.84	0.16	0.82
FeO	10.71	1.15	3.01	0.60	3.70
MgO	11.27	1.93	25.26	3.06	34.78
CaO	12.03	39.52	4.06	41.29	0.97
Na_2O	1.52	1.04	0.93	0.29	0.07
Total	99.87	100.00**	100.48	100.00**	99.72
No. of Oxygens	12	3	12	3	3
Si	3.175	0.922	3.213	0.974	0.903
Ti	0.022	0.085	0.005	0.021	0.024
Al	1.767	0.057	1.521	0.043	0.108
Cr	0.006	0.001	0.100	0.003	0.011
Fe	0.656	0.019	0.174	0.009	0.054
Mg	1.230	0.055	2.598	0.087	0.896
Ca	0.994	0.813	0.300	0.841	0.018
Na	0.216	0.039	0.124	0.011	0.002
Sum	8.066	1.991	8.035	1.989	2.016
Mg*	65.2	74.3	93.7	90.6	94.3

Mg*, 100 Mg/(Mg+Fe).
**, recalculated to 100.00%.

Discussion

$CaSiO_3$ Perovskite in the Upper Mantle

The existence of an unquenchable perovskite form of $CaSiO_3$ at high pressure was suggested by RINGWOOD and MAJOR (1967, 1971) and confirmed by in situ X-ray diffraction measurements in a diamond-anvil cell by LIU and RINGWOOD (1975). The optical properties of the present $CaSiO_3$-rich phase are very similar to those described by RINGWOOD and MAJOR (1967, 1971). Moreover this phase contains a substantial amount of the $CaTiO_3$ perovskite molecule in solid solution. RINGWOOD and MAJOR (1971) demonstrated complete solid solubility between $CaTiO_3$ and $CaSiO_3$ perovskite phases. Thus it is very probable that the present $CaSiO_3$-rich glassy phase likewise represents the retrogressive transformation product of $CaSiO_3$ perovskite on the release of pressure.

Recently ITO and TAKAHASHI (this volume) and TAKAHASHI and ITO (this volume) have found that Ca-rich perovskite is a stable phase in the $CaSiO_3$-$MgSiO_3$-Al_2O_3 system and in a natural peridotite composition (PHN-1611) at pressures of about 20 GPa. Their data indicate, however, that these "Ca-rich phases" contain subequal quantities of $MgSiO_3$ in solid solution (i.e., close to diopside composition), contrary to our own results that show a much more $CaSiO_3$-rich phase. As ITO and TAKAHASHI (this volume) have mentioned and we have also noticed in our experiments, the Ca-rich phase was generally very small (<5 μm) and thus it is quite difficult

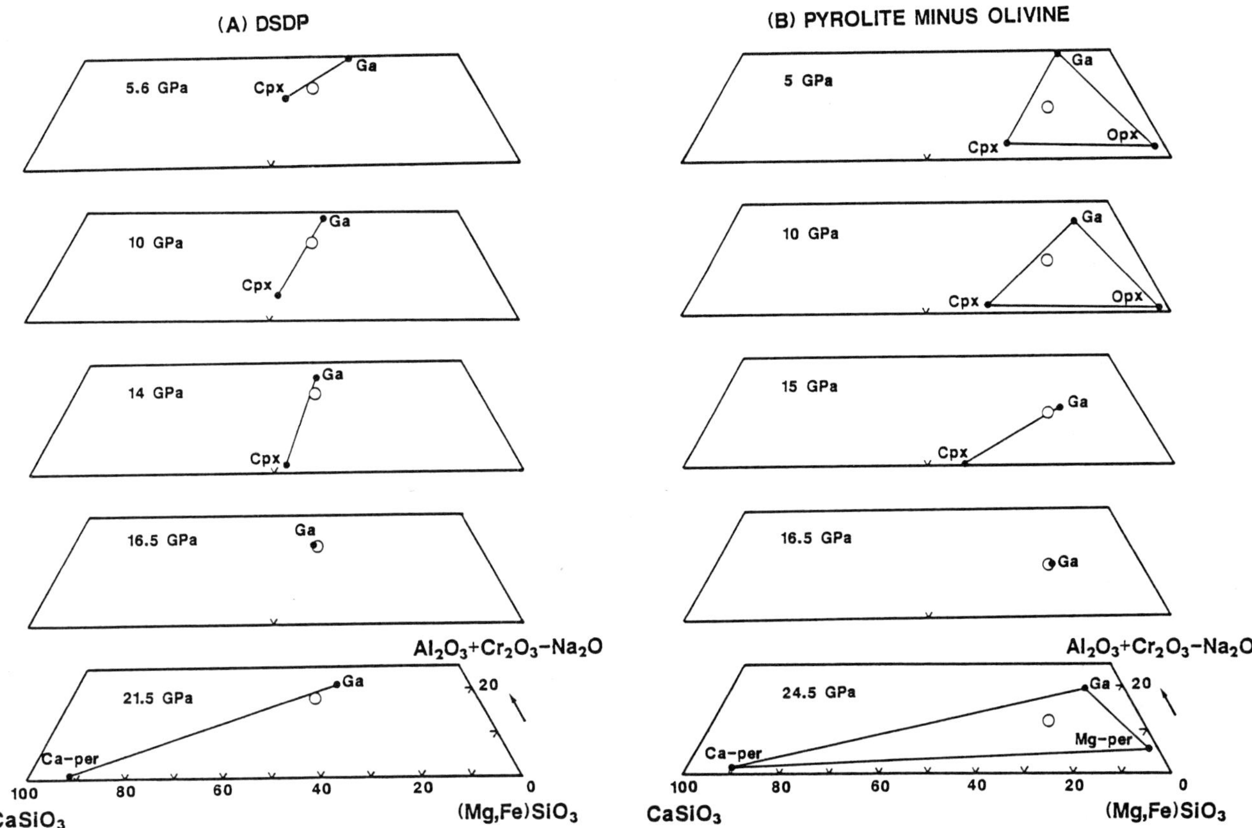

Figure 2. Variations of chemical compositions of coexisting phases in a) DSDP and b) pyrolite minus olivine compositions plotted onto the $CaSiO_3$-(Mg, Fe)SiO_3-Al_2O_3 (+Cr_2O_3-Na_2O) diagram. Open circles indicate compositions of starting materials.

to obtain satisfactory electron microprobe analyses. This is believed to be the reason why analytical data for the "Ca-rich phase" by ITO and TAKAHASHI (this volume) are rather scattered and include some magnesian compositions. We obtained some analyses with slightly higher MgO and Al_2O_3 than those shown in Table 3, but attributed them to beam overlap onto $MgSiO_3$ or pyrope-rich garnet and therefore excluded them. Thus we believe that unquenchable phase has a composition close to pure $CaSiO_3$. Further detailed study, however, is clearly necessary to determine the composition of the Ca-rich phase and how it varies with pressure and temperature changes.

LIU and RINGWOOD (1975) reported that $CaSiO_3$ perovskite has a density of 4.56 ± 0.03 g/cm^3 at 16 GPa and room temperature, on the basis of an in situ X-ray diffraction study with a diamond-anvil cell. The zero-pressure and ambient-temperature density of this phase is accordingly calculated to be 4.34 g/cm^3, using the Birch-Murnaghan equation of state and assuming the bulk modulus and its pressure derivative are 274 GPa and 3.85, respectively (WOLF and JEANLOZ, 1985). This value is substantially greater than that of $MgSiO_3$ perovskite (4.108 ± 0.01; ITO and MATSUI, 1978). Assuming ideal mixing of $CaSiO_3$ (density=4.34 g/cm^3), $MgSiO_3$ (4.108 g/cm^3), $CaTiO_3$ (4.036 g/cm^3), and other hypothetical perovskite end members of minor quantities (such as $FeSiO_3$ and $NaAlSi_2O_6$), we obtained a zero-pressure density of 4.31 g/cm^3 for Ca-perovskite in both the DSDP and pyrolite compositions.

The appearance of Ca-rich perovskite in both pyrolite and basaltic compositions at pressures of about 20 GPa strongly suggests that it is a stable phase in the deep mantle at depths below about 570 km depth. As shown above, this phase has a very high zero-pressure density compared with other high-pressure phases (e.g., for Mg_2SiO_4 spinel, density=3.55 g/cm^3; for $Mg_3Al_2Si_3O_{12}$ garnet, 3.56 g/cm^3; for $MgSiO_3$ ilmenite, 3.80 g/cm^3; for $MgSiO_3$ perovskite, 4.11 g/cm^3; for SiO_2 stishovite, 4.29 g/cm^3), and is thus expected to play an important role in the constitution and dynamics of the earth's mantle. These topics will be discussed below in more detail.

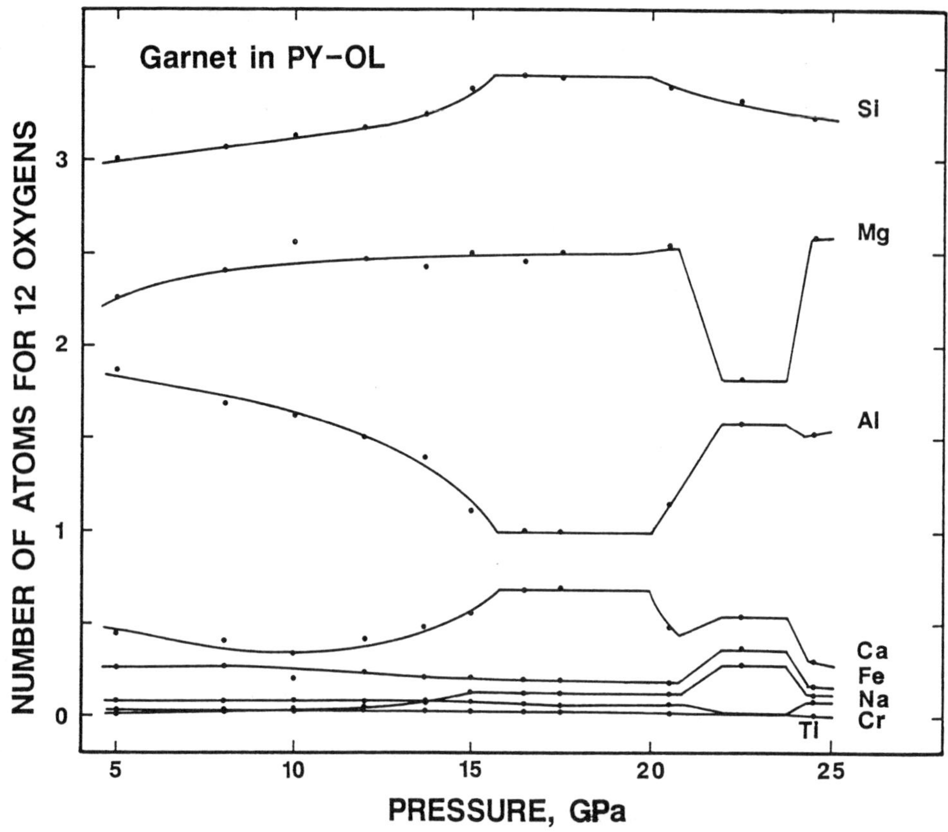

Figure 3. Variations in chemical compositions of garnets in the pyrolite minus olivine composition as a function of pressure. The anomalous Mg- and Cr-poor composition at 22.5 GPa is probably due to coexistence of the highly refractory ilmenite forms of $MgSiO_3$ (see text).

Mineral Assemblages and Density Changes in DSDP and Pyrolite Minus Olivine Compositions at High Pressure

Mineral proportions and zero-pressure bulk densities in the DSDP and pyrolite minus olivine compositions are calculated on the basis of a mass-balance method described previously by IRIFUNE et al. (1986). The results are shown in Figures 4a, b.

In the DSDP composition, a gradual but major density increase occurs between 10 and 15 GPa owing to the increase in the model abundance of garnet due to solid solution of pyroxene in the garnet structure. The bulk density levels off at 3.75 g/cm^3 at 14.5 GPa, where the assemblage of garnetite and ~10% stishovite becomes stable. The bulk density begins to increase gradually above 20 GPa; this increase is associated with the appearance of Ca-perovskite. The proportion of Ca-perovskite increases with increasing pressure, amounting to about 20 vol% at 24.5 GPa, where the bulk density reaches 3.95 g/cm^3. The chemical compositions of coexisting garnet and Ca-perovskite were not obtained because of poor crystallization of these phases. The composition of garnet at 24.5 GPa was therefore estimated from its lattice parameter and by an extrapolation of its chemical variation at lower pressures, while that of Ca-perovskite was assumed to be the same as that obtained at 21.5 GPa. If Ca-perovskite were to contain substantial amounts of Mg and Al, as suggested by ITO and TAKAHASHI (this volume), both its modal abundance and the bulk density at 24.5 GPa would be higher than those estimated above. Nevertheless, it was confirmed by optical and X-ray observations that there is abundant garnet present even at these extremely high pressures, which are equivalent to the depth of the major seismic discontinuity near 670 km.

In the pyrolite minus olivine composition, the pyroxene garnet transformation proceeds rapidly in the pressure interval between 14 and 16 GPa, and a single phase garnetite with a density of 3.61 g/cm^3 is stable thereafter as shown in Figure 4b. The bulk density begins to increase further at about 20 GPa where the garnetite begins to disproportionate to a more Mg-, Fe-, and Al-rich garnet + Ca-perovskite assemblage. A small amount of stishovite was found to occur in this assemblage, but it was less than 5% based on mass-balance calculations, thus it has been omitted from the phase assemblage shown in Figure 4b.

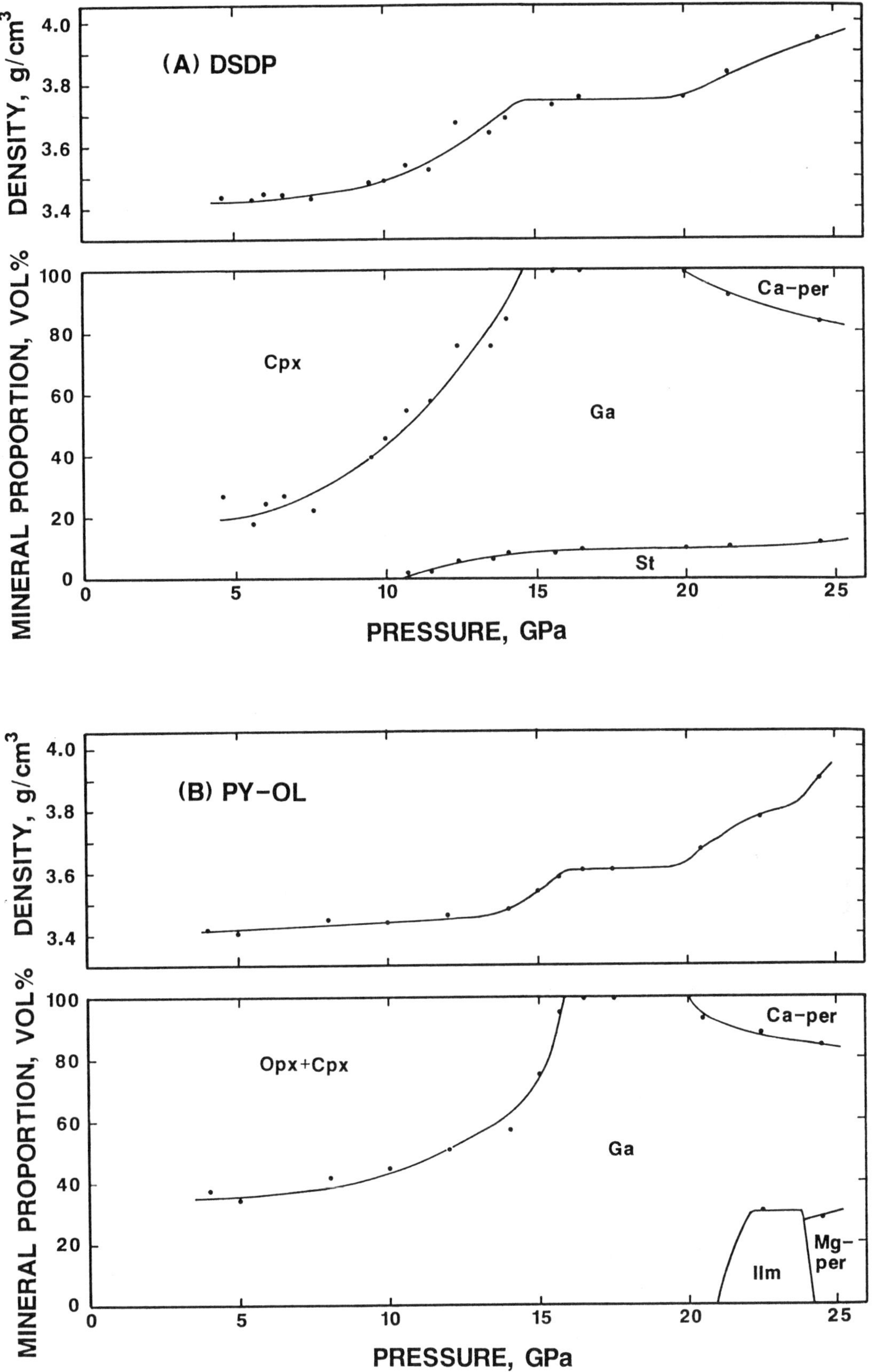

Figure 4. Calculated zero-pressure densities and mineral proportions in a) DSDP and b) pyrolite minus olivine compositions.

(It should be noted that the actual composition of the "pyrolite minus olivine" starting material used in this study has about 1 wt% more SiO_2 as compared to the reference composition given by IRIFUNE (1987).

Chemical analyses of coexisting phases at 22.5 GPa revealed that the garnet phase was relatively Al-rich and Ca-poor. Mass-balance calculations based on this analyzed composition imply the presence of a significant amount of an undetected third phase that is rich in Mg and Si. This phase is believed to be $MgSiO_3$ ilmenite according to the recent phase equilibria studies (e.g., LIU, 1976; ITO and YAMADA, 1982). Because of small crystal size and overlap of X-ray d-spacings with garnet, it was difficult to identify this phase unequivocally. This interpretation, however, is strongly supported by the results of our run at 24.5 GPa, in which the presence of $MgSiO_3$ perovskite was established unequivocally by electron-probe microanalysis and X-ray diffraction. Complete analyses of coexisting Mg-perovskite, Ca-perovskite, and residual garnet were obtained in this run. The mass balance shows that over 50 vol% of garnet still persists at this very high pressure in the Mg-perovskite stability field (Figure 4b). The density of the mineral assemblage formed in the 24.5 GPa run is calculated to be 3.90 g/cm^3. Satisfactory compositional data for Ca-perovskite were obtained only in the 24.5 GPa run. This composition has been used in calculating densities for the runs at 20 and 22.5 GPa.

Implications for the Constitution and Dynamics of the Mantle

Phase transformations along the Mg_2SiO_4-Fe_2SiO_4 compositional join have been well documented by the extensive studies carried out during the last two decades (e.g., RINGWOOD and MAJOR, 1970; ITO et al., 1984; AKAOGI et al., 1984; SAWAMOTO, 1986; ITO and TAKAHASHI, this volume). Combining these data and the present results on the pyrolite minus olivine composition, we can calculate the mineral proportions and bulk density changes in pyrolite composition as a function of pressure. We have used the results of NAVROTSKY and AKAOGI (1984) for the olivine-modified spinel-spinel transformations and those of ITO and TAKAHASHI (this volume) for spinel-perovskite+magnesiowüstite transformation in Mg_2SiO_4-Fe_2SiO_4 solid solutions. In the following discussion, we assume that the whole mantle has the pyrolite composition and that oceanic crust is modelled by the primitive DSDP composition. The composition of oceanic crust may be somewhat altered during subduction, but this results in only minor effects on its mineralogy and density according to IRIFUNE et al. (1986, 1987), which have shown the eclogite-garnetite transformation pressures and densities of the resultant garnetites are similar in several basaltic compositions.

Density Relationships. Figure 5 depicts the zero-pressure density changes in DSDP and pyrolite compositions calculated as a function of pressure on the basis of the present experiments at 1200 °C. Although the actual upper mantle temperature may be somewhat higher than this temperature, the temperature dependences of the major phase transformations involved in these compositions are relatively small (e.g., about ±0.002 GPa/°C for pyroxene-garnet and spinel-perovskite+magnesiowüstite transformations; ITO and YAMADA, 1982; IRIFUNE,

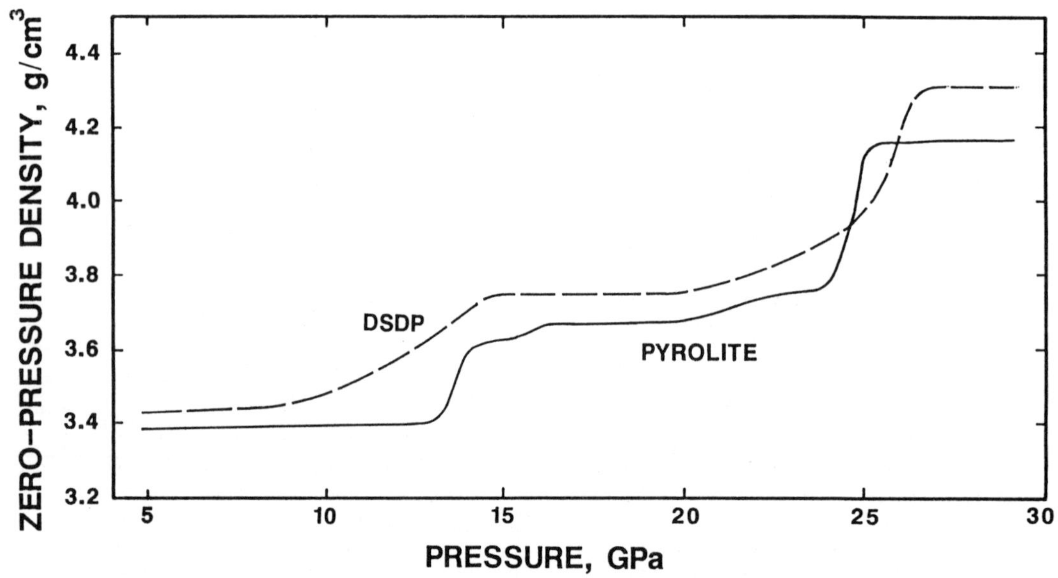

Figure 5. Comparison of zero-pressure densities in the DSDP and pyrolite compositions.

1987), and are unlikely to significantly affect the present conclusions. In fact, the density calculation along an appropriate geotherm (see Figure 6) provides density profiles similar to those described in Figure 5. These results show clearly that the oceanic crust modelled by the DSDP composition is substantially denser (by 0.1–0.2 g/cm^3) than surrounding mantle through the depth interval between 100 km and the major seismic discontinuity occurring near 670 km.

There is a small reversal of these relationships between 670 and 710 km, where the DSDP composition is about 0.1 g/cm^3 less dense than pyrolite. This relationship results from the presence of a large amount of garnet in the DSDP composition at this depth interval. The pyrolite composition, however, transforms completely to a denser assemblage of Mg-perovskite, Ca-perovskite, and magnesiowüstite below 670 km. Our experiment at 24.5 GPa on the "pyrolite minus olivine" (11.4 wt% Al_2O_3) shows that immediately below the 670 km discontinuity, this composition would consist of Ca-perovskite (16 vol%), Mg-perovskite (27 vol%), and pyrope-rich garnet (57 vol%). In this composition, the latter phase would remain stable to depth in the vicinity of 740 km where the uncharacterized Al-rich phase (ITO and TAKAHASHI, this

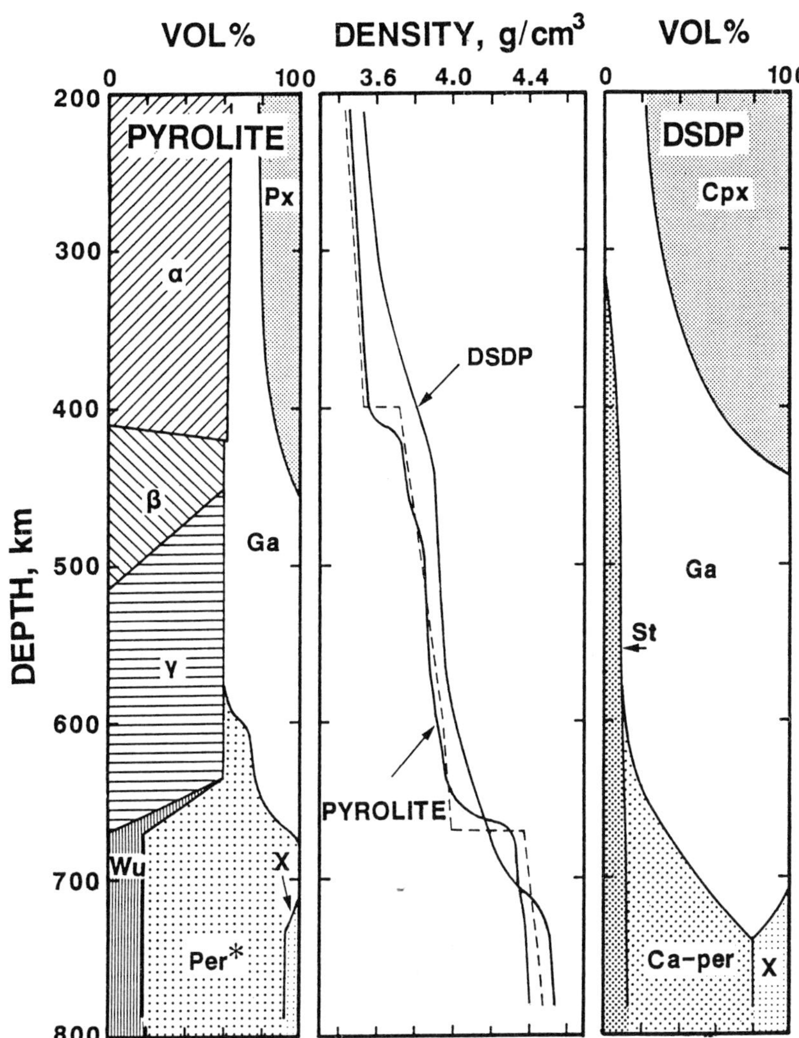

Figure 6. Calculated mineral proportions and densities in DSDP and pyrolite as a function of depth along a representative geotherm. Dashed line indicates density profile obtained from seismic observations (PREM; DZIEWONSKY and ANDERSON, 1981). Minerals are: α, olivine, β, modified spinel, and γ, spinel forms of $(Mg, Fe)_2SiO_4$; Px, orthopyroxene+clinopyroxene; Cpx, clinopyroxene; Ga, garnet; Wu, magnesiowüstite; St, Stishovite; Ca-per, $CaSiO_3$ perovskite (+minor $MgSiO_3$ perovskite below about 700 km); Per*, $CaSiO_3$ plus ilmenite (600–650 km) or perovskite (>650 km) forms of $MgSiO_3$. X denotes an unidentified Al-rich phase found by ITO and TAKAHASHI (this volume) or an ultra-dense form of $MgAl_2O_4$ (LIU, 1978).

volume) is stabilized. However, the ideal pyrolite composition (4.3 wt% Al_2O_3) contains a much larger proportion of perovskite (~80 vol%). Because of the substantial amount of alumina which enters Mg-perovskite under these pressure and temperature conditions, pyrope garnet may not remain stable substantially below 670 km in the pyrolite composition. The transformation of spinel to perovskite plus magnesiowüstite is accompanied by solid solution of residual pyrope garnet into the predominant Mg-perovskite phase. The partial molar volume of the Al_2O_3 component dissolved in perovskite is almost identical to that of corundum (WENG et al., 1982). At higher pressures equivalent to depths exceeding 740 km, the results of ITO and TAKAHASHI (this volume) indicate that Al_2O_3 is exsolved to form an uncharacterized highly aluminous phase "X" (Figure 6).

In the DSDP composition, the results of LIU (1980) and ITO and TAKAHASHI (this volume) show that pyrope-rich garnet transforms mainly to denser perovskite solid solutions over a substantial depth interval below 670 km. Consequently, the density of the DSDP composition exceeds that of pyrolite below a depth of about 710 km. At still greater depths, in the vicinity of 740 km, the ultradense aluminous phase "X" is expected to form from residual garnet and by exsolution of Al_2O_3 from perovskite (ITO and TAKAHASHI, this volume). By this stage, the density of the DSDP composition exceeds that of pyrolite by about 0.1–0.2 g/cm^3 (Figures 5 and 6); this relation is expected to prevail throughout the lower mantle (RINGWOOD, 1982, 1987).

These density relationships have important implications for mantle dynamics. Because of its higher density between 100–670 km, subducted oceanic crust provides a significant contribution to the body forces which cause subducted lithosphere to sink through this depth interval. Between 670 and 710 km, however, the density relationship reverses and the oceanic crust becomes slightly less dense than pyrolite. ANDERSON (1979, 1982) has suggested that this effect will result in the trapping of the subducted oceanic crust above 670 km and ultimate filling of the region between 200 (or 400) km and 670 km. This process is considered unlikely. Because of the high positive density contrast above 670 km and the small negative density contrast below 670 km, isostatic considerations indicate that a large body of oceanic crust collected in this region would penetrate below 670 km. After the "root" of this body had reached 710 km, it would again become denser than surrounding mantle and would then experience body forces tending to pull it deeper into the mantle. On the other hand, small bodies of former oceanic crust could indeed be trapped and would spread to form thin, localized lenses (e.g., ~10 km) immediately overlying the 670 km discontinuity. This might explain why the 670 km discontinuity in this region is sharp enough to cause seismic reflections (RICHARDS, 1972), whereas in some other regions, it is not sharp and seems to possess a complex structure (MUIRHEAD, 1985; SOURIAU, 1986).

The subducted slab is believed to be composed principally of former oceanic crust overlying a thicker layer of former residual harzburgite. Studies by RINGWOOD (1982, 1987) and new experimental investigations by Irifune and Ringwood (unpublished) show that the former harzburgite is intrinsically less dense than pyrolite below 670 km, but is actually as dense or denser than pyrolite between depths of 600–670 km. Because of the lower densities of both components of the slab immediately beneath 670 km, the top of the slab experiences strong resistance to its entry into the lower mantle. This has some complex and interesting dynamical implications which are discussed further by RINGWOOD (1986) and Irifune and Ringwood (unpublished).

Seismic Structure. The mineral assemblages and density change in pyrolite and DSDP compositions are shown in Figure 6 as a function of depth to 800 km along the geotherm given by BROWN and SHANKLAND (1981). The mineral proportions and bulk densities were calculated by the method described by IRIFUNE (1987). Figure 6 shows that the density profile for the pyrolite composition throughout the upper mantle and transition zone is in reasonable agreement with the representative model PREM (DZIEWONSKI and ANDERSON, 1981). Moreover IRIFUNE (1987) has shown that the seismic velocities calculated for the pyrolite composition as a function of depth are consist with the velocity profiles obtained from seismic observations, within uncertainties in the current measurements of physical properties of high-pressure phases and seismic velocities.

The present work, together with the results of TAKAHASHI and ITO (this volume) and ITO and TAKAHASHI (this volume), shows that the transformation of the garnet component of pyrolite into perovskites and other dense phases occurs over a broad depth interval between about 570 and 680 km. Because the garnet component represents about 40 wt% of the pyrolite composition, this transformation will produce significant positive seismic velocity gradients over this depth interval; these positive gradients are superimposed on the major seismic discontinuity near 670 km which results from the disproportion of spinel to perovskite plus magnesiowüstite.

ANDERSON (1979, 1982) proposed that the composition of the mantle between depths of 200 and 670 km is predominantly eclogitic. We have previously shown that the properties of this composition do not match the observed seismic velocity distribution in the mantle between 200 and approximately 600 km (IRIFUNE, 1987; IRIFUNE et al., 1986). For example, the transformation of eclogite to garnetite is rather gradual and does not provide a seismic discontinuity at 400 km. Furthermore, the eclogite-garnetite transformation is completed by 450 km, and the resultant garnetite assemblage does not display any further transformation until a depth of 570 km.

Our present results show that an eclogitic or basaltic composition would display rapid increases in density and seismic velocities between 570 and 670 km because of the increasing degree of exsolution of $CaSiO_3$ perovskite from garnet that occurs in this depth interval. Consequently, the density of a basaltic or eclogitic composition immediately above 670 km differs by about 0.2 g/cm^3 from that of a pyrolite mineral assemblage immediately below the major seismic discontinuity near 670 km (Figure 6). This is in disagreement with the much larger change in density (0.4 g/cm^3) that occurs at 670 km according to the PREM model. Moreover, because seismic velocities are approximately proportional to density (e.g., BIRCH, 1964), it seems that the model for the structure of the mantle proposed by ANDERSON (1979, 1982) may not provide a satisfactory explanation of the magnitude of the major seismic discontinuity observed near 670 km.

More recently, ANDERSON and BASS (1984, 1986) have proposed an alternative model with a "piclogite" composition (essentially about 80 wt% eclogite plus 20 wt% olivine) for the transition region between 400 and 670 km. This composition is believed to provide velocity-density profiles which are in better accord with the seismic observations than pyrolite. Their calculation, however, relies on the phase relations studied by AKAOGI and AKIMOTO (1977, 1979), which are applicable to peridotitic rather than to eclogitic compositions and are now known to require some revisions in pressure scale and in compositions of coexisting phases (KANZAKI, 1987). The present results on the DSDP composition, which are closer to the piclogite composition and have been studied at pressures much higher than those by AKAOGI and AKIMOTO (1977, 1979), show that piclogite would display low-velocity gradients between 450 and 570 km and high gradients between 570 and 670 km. The PREM model (on which ANDERSON and BASS (1984, 1986) base their case) provides an essentially opposite distribution of velocity gradients between 400 and 670 km. In this model, velocity gradients are high between 400 and 600 km and low between 600 and 670 km. On the other hand, in a pyrolite-model mantle composition, the transformations of pyroxene to garnet and of modified spinel to spinel provide high gradients in the upper part of the transition zone, whereas subsequent transformations provide relatively small gradients in the lower part of the transition zone. We conclude that pyrolite provides a closer match to the inferred distribution of seismic velocities and densities between 300 and 700 km than eclogite or piclogite.

Acknowledgments. We are indebted to S. Kesson, E. Takahashi, and T. Kato for their helpful discussions and useful comments. We thank E. Takahashi and E. Ito for preprints. Thanks are also due to W. O. Hibberson and N. Ware for their technical assistance in high-pressure experiments and electron microprobe analyses. Critical comments by the reviewers were helpful in improving the manuscript.

REFERENCES

AKAOGI, M., and S. AKIMOTO, Pyroxene-garnet solid solution equilibria in the systems $Mg_4Si_4O_{12}$-$Mg_3Al_2Si_3O_{12}$ and $Fe_4Si_4O_{12}$-$Fe_3Al_2Si_3O_{12}$ at high pressures and temperatures, *Phys. Earth Planet. Inter., 15*, 90–106, 1977.

AKAOGI, M., and S. AKIMOTO, High-pressure phase equilibria in a garnet lherzolite, with special reference to Mg^{2+}-Fe^{2+} partitioning among constituent minerals, *Phys. Earth Planet. Inter., 19*, 31–51, 1979.

AKAOGI, M., M. L. ROSS, P. MCMILLAN, and A. NAVROTSKY, The Mg_2SiO_4 polymorphs (olivine, modified spinel and spinel): thermodynamic properties from oxide melt solution calorimetry, phase relations, and models of lattice vibrations, *Am. Mineral., 69*, 499–512, 1984.

ANDERSON, D. L., The upper mantle transition region: eclogite?, *Geophys. Res. Lett., 6*, 433–436, 1979.

ANDERSON, D. L., Chemical composition and evolution of the mantle, in *High-Pressure Research in Geophysics*, edited by S. Akimoto and M. H. Manghnani, pp. 301–308, D. Reidel, Dordrecht, 1982.

ANDERSON, D. L., and J. D. BASS, Mineralogy and composition of the upper mantle, *Geophys. Res. Lett., 11*, 637–640, 1984.

ANDERSON, D. L., and J. D. BASS, Transition region of the earth's upper mantle, *Nature, 320*, 321–328, 1986.

BIRCH, F., Density and composition of mantle and core, *J. Geophys. Res., 69*, 4377–4388, 1964.

BROWN, J. M., and T. J. SHANKLAND, Thermodynamic properties in the earth and determined from seismic profiles, *Geophys. J. R. Astron. Soc., 66*, 579–596, 1981.

DZIEWONSKI, A. M., and D. L. ANDERSON, Preliminary reference earth model, *Phys. Earth Planet. Inter., 25*, 297–356, 1981.

GREEN, D. H., W. O. HIBBERSON, and A. L. JAQUES, Petrogenesis of mid-ocean ridge basalts, in *The Earth, Its Origin, Structure and Evolution*, edited by M. W. Elhinney, pp. 265–299, Academic Press, London, 1979.

IRIFUNE, T., An experimental investigation of the pyroxene-garnet transformation in a pyrolite composition and its bearing on the constitution of the mantle, *Phys. Earth Planet. Inter.*, in press, 1987.

IRIFUNE, T., and W. O. HIBBERSON, Improved furnace design for multi-anvil apparatus for pressures to 18 GPa and temperatures to 2000 °C, *High Temp.-High Press.*, 575–579, 1985.

IRIFUNE, T., W. O. HIBBERSON, and A. E. RINGWOOD, Eclogite-garnetite transformations in basaltic and pyrolitic compositions at high pressure and high temperature, *Proceedings of the Fourth International Kimberlite Conference*, in press, 1987.

IRIFUNE, T., T. SEKINE, A. E. RINGWOOD, and W. O. HIBBERSON, The eclogite-garnetite transformation at high pressure and some geophysical implications, *Earth Planet. Sci. Lett., 77*, 1986.

ITO, E., and Y. MATSUI, Synthesis and crystal-chemical characterization of $MgSiO_3$ perovskite, *Earth Planet. Sci. Lett., 38*, 443–450, 1978.

ITO, E., and H. YAMADA, Stability relations of silicate spinels, ilmenites, and perovskites, in *High-Pressure Research in Geophysics*, edited by S. Akimoto and M. H. Manghnani, pp. 405–419, D. Reidel, Dordrecht, 1982.

ITO, E., and E. TAKAHASHI, Ultrahigh-pressure phase transformations and the constitution of the deep mantle, this volume.

ITO, E., E. TAKAHASHI, and Y. MATSUI, The mineralogy and chemistry of the lower mantle: an implication of the ultrahigh-pressure phase relations in the system $MgO-FeO-SiO_2$, *Earth Planet. Sci. Lett., 67*, 238–248, 1984.

KANZAKI, M., Ultrahigh-pressure phase relations in the system $MgSiO_3$-$Mg_3Al_2Si_3O_{12}$, *Phys. Earth Planet. Inter.*, in press, 1987.

KAWAI, N., and S. ENDO, The generation of ultrahigh pressure by a split sphere apparatus, *Rev. Sci. Instrum., 41*, 1178–1181, 1970.

LIU, L.-G., The high-pressure phase of $MgSiO_3$, *Earth Planet. Sci. Lett., 31*, 200–208, 1976.

LIU, L.-G., A new high-pressure phase of spinel, *Earth Planet. Sci. Lett., 41*, 398–404, 1978.

LIU, L.-G., The mineralogy of an eclogitic earth mantle, *Earth Planet. Sci. Lett., 23*, 262–267, 1980.

LIU, L.-G., and A. E. RINGWOOD, Synthesis of a perovskite-type polymorph of $CaSiO_3$, *Earth Planet. Sci. Lett., 28*, 209–211, 1975.

MUIRHEAD, K., Comments on "Reflection properties of phase transitions and compositional change models of the 670 km discontinuity" by A. Lees, M. Bukowinski, and R. Jeanloz, in *J. Geophys. Res., 89*, 10,135–10,140, 1985.

NAVROTSKY, A., and M. AKAOGI, The α, β, γ phase relations in Fe_2SiO_4-Mg_2SiO_4 and Co_2SiO_4-Mg_2SiO_4: calculation from thermochemical data and geophysical application, *J. Geophys. Res., 89*, 10,135–10,140, 1984.

OHTANI, E., T. IRIFUNE, W. O. HIBBERSON, and A. E. RINGWOOD, Modified split-sphere guide-block for practical operation of a multiple-anvil apparatus, *High Temp.-High Press.*, in press, 1987.

RICHARDS, S. M., Seismic waves reflected from velocity gradient anomalies within the earth's upper mantle, *Zeitschrift fur Geophysik, 38*, 517–527, 1972.

RINGWOOD, A. E., *Composition and Petrology of the Earth's Mantle*, 618 pp., McGraw Hill, New York, 1975.

RINGWOOD, A. E., Phase transformations and differentiation in subducted lithosphere: implications for mantle dynamics, basalt petrogenesis, and crustal evolution, *J. Geology, 90*, 611–643, 1982.

RINGWOOD, A. E., Constitution and evolution of the mantle, *Proceedings of the Fourth International Kimberlite Conference*, in press, 1987.

RINGWOOD, A. E., and A. MAJOR, Some high-pressure transformations of geophysical significance, *Earth Planet. Sci. Lett., 2*, 106–110, 1967.

RINGWOOD, A. E., and A. MAJOR, The system Mg_2SiO_4-Fe_2SiO_4 at high pressures and temperatures, *Phys. Earth Planet. Inter., 3*, 89–108, 1970.

RINGWOOD, A. E., and A. MAJOR, Synthesis of majorite and other high pressure garnets and perovskites, *Earth Planet. Sci. Lett., 12*, 411–418, 1971.

SAWAMOTO, H., Single crystal growth of the modified spinel (β) and spinel (γ) phases of $(Mg, Fe)_2SiO_4$ and some geophysical implications, *Phys. Chem. Mineral., 13*, 1–10, 1986.

SAWAMOTO, H., Phase diagram of $MgSiO_3$ at pressures up to 24 GPa and temperatures up to 2200°C: Phase stability and properties of tetragonal garnet, this volume.

SOURIAU, A., First analyses of broadband records on the geoscope network: potential for detailed studies of mantle discontinuities, *Geophys. Res. Lett., 13*, 1011–1014, 1986.

SUN, S.-S., Chemical composition and origin of the earth's primitive mantle, *Geochim. Cosmochim. Acta, 16*, 179–192, 1982.

TAKAHASHI, E., and E. ITO, Mineralogy of mantle peridotite along a model geotherm up to 700 km depth, this volume.

WENG, K., H. K. MAO, and P. M. BELL, Lattice parameters of the perovskite phase in the system $MgSiO_3$-$CaSiO_3$-Al_2O_3, *Carnegie Inst. Washington Year Book, 81*, 273–277, 1981.

WOLF, G. H., and R. JEANLOZ, Lattice dynamics and structural distortions of $CaSiO_3$ and $MgSiO_3$ perovskites, *Geophys. Res. Lett., 12*, 413–416, 1985.

THE ACTIVATION ENERGY OF THE BACK TRANSFORMATION OF SILICATE PEROVSKITE TO ENSTATITE

E. KNITTLE and R. JEANLOZ

Department of Geology and Geophysics, University of California, Berkeley, CA 94720, USA

Abstract. The activation energy for the back transformation of $(Mg_{0.9}Fe_{0.1})SiO_3$ in the perovskite structure to the enstatite structure has been measured at zero pressure. The activation energy is 70 (±20) kJ/mole, small in comparison with values for creep, diffusion and phase transitions in many other silicates. We suggest that the transition may be activated by movement of the central Mg ion (or the corresponding vacancy) through the four oxygen ions of the dodecahedral face of the perovskite structure. From a simple harmonic model of the amplitudes of vibration of Mg–O bonds at elevated temperature, this mobility of the Mg ions is expected to occur with an activation energy of 60 kJ/mole (in accord with our measurement) and an activation volume near 1 cm^{-1}/mole. Implications of the small value of the activation energy for silicate perovskite are as follows: 1) there should be no kinetic hindrance to the perovskite phase transition in the earth; 2) any silicate perovskite emplaced on the earth's surface in a xenolith would revert to enstatite in a geologically brief time (3–100 years), therefore minimizing the chances of finding this high-pressure phase as a natural sample on the earth's surface.

Introduction

Information on the kinetics and mechanisms of high-pressure phase transformations is available for only a small number of mineral systems. The olivine to spinel transition has been studied for $(Mg,Fe)_2SiO_4$ (SUNG and BURNS, 1976; SUNG, 1979; POIRIER, 1981) and for nickel olivine (KASAHARA and TSUKHARA, 1971; HAMAYA and AKIMOTO, 1982). Among the high-pressure phase transformations in the SiO_2 system, the kinetics of both the forward and back reactions of quartz and amorphous silica to coesite and stishovite, have been extensively studied (STISHOV and POPOVA, 1961; SKINNER and FAHEY, 1963; DACHILLE et al., 1963; GIGL and DACHILLE, 1968; NAKA et al., 1976). However, no information is known about the kinetics or mechanisms of the high-pressure mineral reactions involving the silicate-perovskite phase of $(Mg,Fe,Ca)(Si,Al)O_3$. As it is now generally established that all major upper mantle minerals, olivine, pyroxene and garnet, transform to perovskite-structured minerals (or perovskite-dominated assemblages) at pressures exceeding 22 GPa and temperatures exceeding 1700 K (LIU, 1979; JEANLOZ and THOMPSON, 1983), silicate perovskite is probably the most abundant mineral in the earth's mantle. In order to obtain the first kinetic information for this important mineral phase, we have measured the activation energy of the back transformation from $(Mg_{0.9},Fe_{0.1})SiO_3$ perovskite to enstatite at zero pressure.

The relationship between our measurements on metastable perovskite and the activation energy for the transformation of stable perovskite at high pressure is illustrated in Figure 1. For the purpose of our discussion we ignore the activation entropy, including it in the preexponential frequency factors described below (see CHRISTIAN, 1975). Therefore, we need not distinguish between the activation enthalpy and Gibbs free energy. Furthermore, the activation enthalpy and (internal) energy are identical at zero pressure.

The effect of pressure is to change the relative Gibbs free energies, and hence the stability, between enstatite and

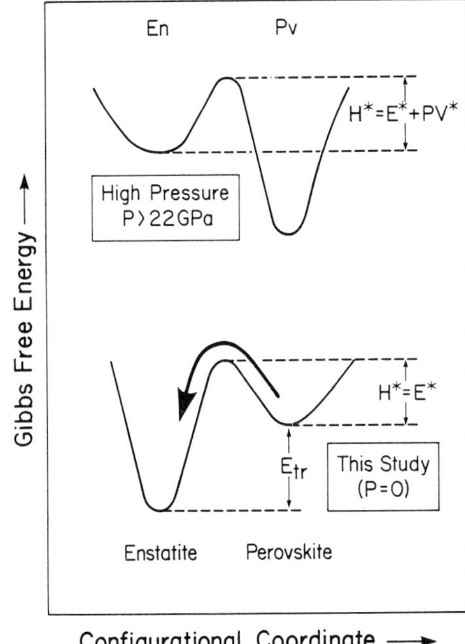

Figure 1. The relationship between the Gibbs free energies for the perovskite and enstatite phase of $(Mg,Fe)SiO_3$ at low and high pressure are schematically illustrated as a function of the configurational coordinate describing the phase transformation. For this study, we assume that the value of E^*, the activation energy, is equivalent at low and high pressure. The energy of transformation at zero pressure, E_{tr} is shown for comparison.

High-Pressure Research in Mineral Physics, edited by M. H. Manghnani and Y. Syono, pp. 243–250.

perovskite phases. In accord with the activated-state model, the effect of pressure on the activation energy is separately accounted for through the activation volume, V^* (CHRISTIAN, 1975). Thus, the activation enthalpy for transformations involving perovskite at pressure P is given by the value that we measure at zero pressure (E^*) plus a correction of PV^* (Figure 1).

In this model, the activation energy is identical for either forward or backward reaction: it does not matter which phase is stable. In support of this conclusion, we note that the activation energies for the reactions from quartz to coesite and from coesite to cristobalite have been separately measured (NAKA et al., 1976; DACHILLE et al., 1963). Nearly identical values are found, ~150 kJ/mole and ~140 kJ/mole, with the difference being partly accounted for by the enthalpy difference between quartz and cristobalite (3 kJ/mole according to ROBIE et al., 1978). Thus, we interpret our measurements as describing the fundamental kinetics of ion mobilities in the silicate-perovskite phase, despite the fact that the diffusive motions act to disassemble the perovskite structure in our experiments.

Experimental Technique

The silicate perovskite used in these experiments was synthesized in a Mao-Bell type diamond-anvil cell using a laser-heating system (MAO et al., 1979; HEINZ and JEANLOZ, 1987, this volume). A natural enstatite of composition $(Mg_{0.88}Fe_{0.12})SiO_3$ (Bamble, Norway) was used as the starting material, and the synthesis conditions were typically above 30 GPa and 2000 K. The synthesis pressures were determined from the spring length of the diamond cell (independently calibrated using the ruby fluorescence method: BARNETT et al., 1973; MAO et al., 1978), and the temperature was determined by spectroradiometric measurements (HEINZ and JEANLOZ, 1987, this volume). The resultant perovskite samples were polycrystalline aggregates (~1 μm-scale grain size) between 100 and 200 μm in diameter and ~10 μm thick. As much unconverted starting material (enstatite) as possible was removed from the run products, and the perovskite samples synthesized in this manner did not show the lines of any other phase when analyzed by X-ray diffraction.

Three types of experiments on the back transformation of perovskite to enstatite were carried out. In one type, data were obtained as a consequence of thermal expansion measurements of perovskite (KNITTLE et al., 1986), and the time-temperature interval over which the perovskite reverted to enstatite could be directly observed by X-ray diffraction. In the second type of experiments, samples of perovskite were heated in a furnace for specified lengths of time, then the samples were quenched in temperature and the run products analyzed by X-ray diffraction. In these latter experiments, the back transformation is not directly observed but a limiting time required for transformation is obtained at each temperature. In the third type of experiment, optical absorption spectra were collected at high temperature at various time intervals during the back transformation from perovskite to enstatite. In this way, the amount of transformation with time could be directly determined.

For the first type of experiment, the perovskite samples were loaded in a 200 μm quartz capillary tube and surrounded by a yoke-shaped, corundum-insulated heater. The sample temperature was measured using an external thermocouple placed next to the capillary tube and with an internal thermal-expansion standard (NaCl). Typical temperature uncertainties were ± 3 °C. The sample temperature was raised every six hours and a Guinier X-ray diffraction film was obtained at each successive temperature. Although the samples were heated in air, we observed no evidence of oxidation either by microscopy or by X-ray diffraction. The X-ray diffraction lines observed for perovskite were (002), (110), (111), (020), (112), (200), (210) and (211). The back reaction does not appear to be sluggish since, for the two X-ray patterns between which the inversion takes place, one X-ray diffraction pattern is of perovskite and the next consists entirely of enstatite lines with only a very faint trace of the strongest (112) perovskite line. The enstatite diffraction pattern observed after reversion was the same as that for the starting material.

Most of the data points for this study were obtained by the second type of experiment: heating perovskite in a furnace. The samples were loaded into small aluminum crucibles or 200 μm quartz capillary tubes. The experiments were conducted in vacuum or in an argon atmosphere to prevent oxidation of the perovskite (potentially a problem because higher temperatures and longer times were attained than in the first experiments described above). The accuracy of the furnace temperatures were determined from separate calibration experiments, using the melting temperature of AgCl (455 °C) as a temperature fixed point. The temperature accuracy of the furnaces was ± 5 °C. The run products were analyzed by X-ray diffraction using a Debye-Scherrer configuration, so that the run duration for each temperature only places a limit on the transition time for the perovskite-to-enstatite reaction. For samples that revert to enstatite, the reaction could have taken place at an earlier time and we obtain an upper limit for the transformation time; for samples showing no evidence of reversion, the run duration provides a lower limit (the sample requires more time to transform). The error bars quoted on the run durations are estimated from the uncertainty in the time of removal of the sample from the furnace. Typically,

24-hour runs are accurate to ± 20 minutes, with the start time being counted from the time the furnace reaches its specified temperature. In the shortest run, the quartz capillaries containing the samples were lowered into the furnace and removed 30 seconds after the furnace temperature maximized. The total time from introduction to removal of the sample ranged from 3 to 5 minutes. Except for two cases, the run product was found to be either entirely perovskite or entirely enstatite in all of the X-ray diffraction patterns. In the two cases with coexisting phases, only the strongest perovskite (112) line was seen faintly superimposed on the enstatite pattern. The run products were examined with a polarizing microscope in order to check whether any glass is produced upon back transformation. As no portions of any sample are observed to be isotropic, we believe that less than 1 percent glass is present in the run products.

For the optical absorption measurements, the third type of high-temperature experiments, spectra were collected through a microscope using a 100 W tungsten-halogen lamp as the light source. The light was passed through a holographic-grating monochromator (12 nm/mm dispersion) and the intensity as a function of wavelength measured using a photomultiplier tube. Data points were collected from 400 to 800 nm with spectral resolution of 10 nm. Zero-pressure, room-temperature spectra were taken for both perovskite and for the enstatite starting material for comparison with the high-temperature results. Both samples were approximately 200 μm in diameter and the spectra were referenced to a 50 μm pinhole, to a compressed pellet of MgO and to a compressed pellet of SiO_2. Since the spectra were quantitatively similar for all of the reference materials, the 50 μm pinhole was chosen as the reference for the high-temperature measurements.

For the high-temperature optical-absorption experiments, the samples were kept in the stainless steel gaskets which were cemented to a heater made of platinum wire wound in a spiral and embedded in cement. The gasket and heater were placed in a brass holder designed to allow light to pass through the sample and to blow argon over the sample surface to prevent oxidation. The approximate temperature at the sample was measured with Pt/Pt 10%Rh thermocouples. One thermocouple was attached to the gasket as close to the sample as possible and one thermocouple was cemented directly to the heater; however, separate calibration experiments showed that accurate measurement of the temperature in the actual sample chamber was not possible in these experiments. Problems arose from the fact that the thermocouple could not be securely attached to the sample without obstructing the spectral measurement, and the argon flow reduced the temperature measured by the thermocouples in an irreproducible way. We could determine however, that the temperature was greater than 850 K and that the temperature remained stable to within 10 K during the measurement of the absorption spectra.

The perovskite and enstatite absorption were found to differ significantly between 450 and 800 nm, meaning that in the high-temperature experiments where both phases are present, the fractions of perovskite and enstatite can be estimated from the measured absorption spectra. The absorbance of the sample is given by:

$$\log(I/I_0) = (-\alpha_{pv}C_{pv} - \alpha_{en}C_{en})\,D \qquad (1)$$

where I/I_0 is the intensity of light transmitted through the sample at a given wavelength normalized to a reference spectrum, α_{pv} and α_{en} are the absorption coefficients of the perovskite and enstatite phases respectively, C_{pv} and C_{en} are the volume fractions of the two phases and D is the total sample thickness ($10\pm 2\,\mu$m). Data were obtained for one sample in various stages of reversion to enstatite using this method. Because of the uncertainty in the absolute temperature, the data cannot be used to estimate the activation energy. However, these measurements can be directly used to determine the time dependence of the transformation of perovskite to enstatite.

Results

The time-temperature data for the back transformation of $(Mg,Fe)SiO_3$ perovskite to enstatite at zero pressure is given in Table 1 and Figure 2. The activation energy of the back-transformation can be described by the equation: $t = t_0 \exp(-E^*/RT)$ where t is the time of the experiment for a given amount of transformation, t_0 is a preexponential factor, E^* is the activation energy, R is the gas constant and T is the temperature in degrees Kelvin. The rate of transformation can be calculated from the slope of the relationship of the logarithm of the time of the experiment to the reciprocal of the temperature. Thus, for the back transformation of perovskite to enstatite, we find an activation energy of 70 (± 20) kJ/mole or 17 kcal/mole.

From the optical-absorption experiments, the degree of transformation of the perovskite to enstatite at a constant high temperature is determined as a function of time. In Figure 3, data points with different degrees of transformation at a constant high temperature are plotted in accord with Avrami's equation:

$$X = 1 - \exp(-kt^n) \qquad (2)$$

where X is the volume fraction of the transformed phase, k is the rate of transformation, t is the transition time and n is a constant which is determined by the mechanism of nucleation and growth (Christian, 1975; Cahn, 1956). Data were collected at five times with increasing degrees

TABLE 1. Back-Transformation Data for $(Mg_{0.9}Fe_{0.1})SiO_3$ Perovskite to Enstatite

Time	Temperature, °C	Phase[a]
30 sec.	770(±5)	pv
5 min.	960(±5)	en
12.5(±5) min.	>580(±10)	90%pv[b]
17.5(±5) min.	>580(±10)	50%pv[b]
22.5(±5) min.	>580(±10)	40%pv[b]
27.5(±5) min.	>580(±10)	25%pv[b]
20 min.	770(±5)	en
20 min.	960(±5)	en
32.5(±5) min.	>580(±10)	10%pv[b]
6 hrs.	555(±3)	pv
6 hrs.	565(±3)	pv
6 hrs.	606(±3)	en
8 hrs.	598(±5)	en
20 hrs.	570(±5)	en
24 hrs.	444(±5)	pv
24 hrs.	520(±5)	en
24 hrs.	575(±5)	en
24 hrs.	505(±5)	pv+en
24 hrs.	476(±5)	pv+en

[a]pv=perovskite structure; en=enstatite structure. [b]Samples were not used to constrain the activation energy but only to determine the value of n.

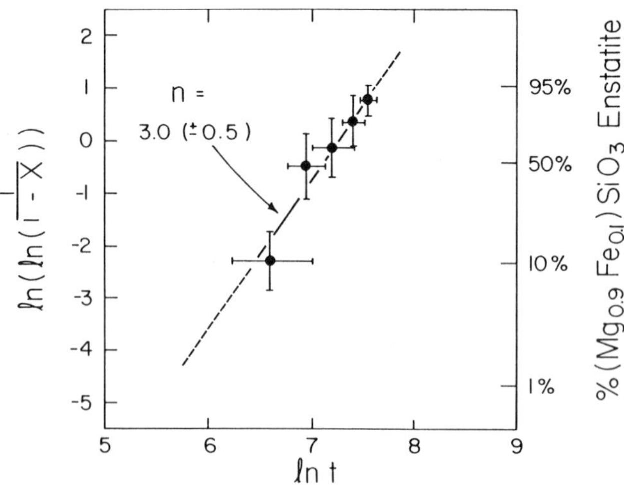

Figure 3. The degree of transformation as a function of time for the perovskite to enstatite transition is plotted in accord with Avrami's equation. The data set was collected spectroscopically at a temperature greater than 850 K (constant to ± 10 K). A linear least squares fit through the data yields a value n=3.0 (±0.5). The value of the slope gives an indication of the mechanism of nucleation and growth as discussed in the text.

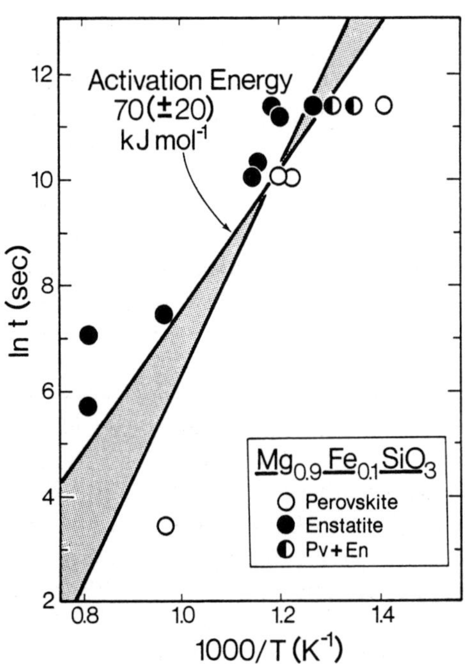

Figure 2. The back-transformation data for the $(Mg,Fe)SiO_3$ perovskite to enstatite transition at zero pressure are plotted as the logarithm of the time of transformation (t) versus the reciprocal of temperature (T). The activation energy is calculated from the transformation rate-temperature data, and is found to be 70 (±20) kJ/mole.

of transformation to enstatite. As discussed in the experimental section, the geometry of the absorption-spectrum experiments made an accurate determination of the temperature impossible, but we expect it to be greater then 850 K. Since the temperature remained constant during the experiment, the data are well suited to determining the value of n. A weighted least squares fit yields a value of n=3.0 (±0.5), which, following the model of CAHN (1956), suggests that our measurements involve the kinetics of growth after the saturation of point nucleation sites.

The activation energy for the transformation of perovskite is small in comparison with values for diffusion, creep or transformation of other silicates (Table 2). Given the result for Mg_2SiO_4, the similarity between the activation energy for the olivine-spinel transition in nickel silicate with that which we find for perovskite may be fortuitous. One possible explanation for our low value is that the Mg-silicate perovskite becomes a solid electrolyte, with the anion (O^{2-}) sublattice "melting" at the high temperatures of the present experiments. In favor of this idea are the experimental results of O'KEEFFE and BOVIN (1979) and POIRIER et al. (1983) suggesting that at elevated temperatures, analog fluoride perovskites exhibit fast ion conduction (solid electrolyte behavior) with an activation energy near 70 kJ/mole.

We do not believe, however, that our results can plausibly be explained in terms of solid-electrolyte behavior for the following reasons. First, if we extrapolate the melting data of HEINZ and JEANLOZ (1987) to zero

TABLE 2. Silicate Activation Energies

	Activation Energy, kJ/mole	Reference
$(Mg,Fe)SiO_3$ pv → en	70(±20)	This study
Mg_2SiO_4 ol → sp	325(±17)	SUNG and BURNS, 1976
Ni_2SiO_4 ol → sp	113(±20)	HAMAYA and AKIMOTO, 1982
SiO_2 qtz → coes	~ 150	NAKA et al., 1976
SiO_2 amorphous → coes	~ 200	NAKA et al., 1976
Olivine (creep)	520(±40)	see summaries by GOETZE, 1978; KIRBY, 1983
Forsterite (creep)	669(±29)	JAOUL, et al., 1981a
Olivine (oxygen self-diffusion)	372(±13)	REDDY et al., 1980
	320(±40)	JAOUL et al., 1981b
Olivine (silicon self-diffusion)	376(±42)	JAOUL et al., 1981a
Olivine (static annealing)	323(±18)	KOHLSTEDT et al., 1980 KARATO, 1981

pressure, our experiments are found to be uniformly at low temperatures relative to the melting point: $T/T_m \lesssim 0.3$ to 0.4. In contrast, fast ion conduction characteristically occurs at temperatures above about 0.8 T_m (cf. MARCH and TOSI, 1984). Second, we would expect the activation energy for mobility in an oxide perovskite to be larger, due to the larger bond charge, than that in a fluoride perovskite. Hence, the similarity with our value may be coincidental. Finally, there have been recent experiments that call into question the evidence that fluoride perovskites exhibit fast ion conduction at all (ANDERSEN et al., 1985). We conclude that further documentation is required before a solid-electrolyte mechanism can be considered in silicate perovskite.

An alternative possibility is that a high cation mobility is achieved at elevated temperatures. For example, the energy barrier associated with the magnesium ion jumping into or out of the central, dodecahedral site of the perovskite unit cell may be relatively small. In this case, the activated state corresponds to the Mg ion passing between the four oxygen ions of a dodecahedron face (or of a cation vacancy making this jump). If there is no relaxation around the cation, the Mg–O distance would be about 0.17 nm in the activated state (YAGI et al., 1978). For comparison, the equilibrium Mg–O distance in crystals is 0.19 nm and 0.21 nm for four-fold and six-fold coordination, respectively. Thus, a relaxation in oxygen packing of only 10 to 20 percent is required around the magnesium as it moves out of the central site.

That such relaxation around the diffusing cation can occur is suggested by an approximate calculation of the vibrational amplitudes in the harmonic approximation (e.g., REISSLAND, 1973). At 1000 K, we obtain a root-mean-square variation in bond distance of about 0.02 nm, assuming an average lattice vibrational frequency of 285 cm^{-1} derived from the Debye temperature ($\Theta_D=525$ K) for perovskite obtained by KNITTLE et al. (1986). In fact, the octahedral-tilting vibrations of the lattice are expected to occur at 200 to 300 cm^{-1} for $MgSiO_3$ perovskite (WILLIAMS et al., 1987); the corresponding bond-length variations are between 0.02 and 0.03 nm, thus providing ample relaxation for passage of the cation through the dodecahedron face. We note that zero-point motions already amount to 0.012 nm, according to this model. In addition to the excitation of vibrational states, further relaxation due to anharmonicity might be expected at high temperatures.

Following MUKHERJEE (1965), the work associated with moving the cation or vacancy through the activated state is expected to be proportional to $\kappa \cdot \Delta r^2$ in the harmonic model. The spring constant κ is related to the vibrational frequency of the atomic masses, and the amount of relaxation Δr is taken to be proportional to the cube root of the unit cell volume. Assuming that Mukherjee's empirically derived constant of proportionality can be used here, we obtain a predicted activation energy of 60 kJ/mole for the mobility of Mg ions in silicate perovskite.

We emphasize that this model is highly speculative and, although it yields an activation energy close to our measured value, it involves serious approximations. For example, MUKHERJEE's (1965) analysis is for close-packed metals. Nevertheless, a similar argument can be outlined for Si (jumping through octahedron faces in the perovskite structure), and this model provides an alternative to the solid-electrolyte mechanism outlined above. If valid, the present model suggests that high-pressure phases with high-coordination (large polyhedron) sites may commonly exhibit relatively easy cation mobility.

Discussion

Figure 4 illustrates the importance of determining the activation volume (V^*) for perovskite in order to assess

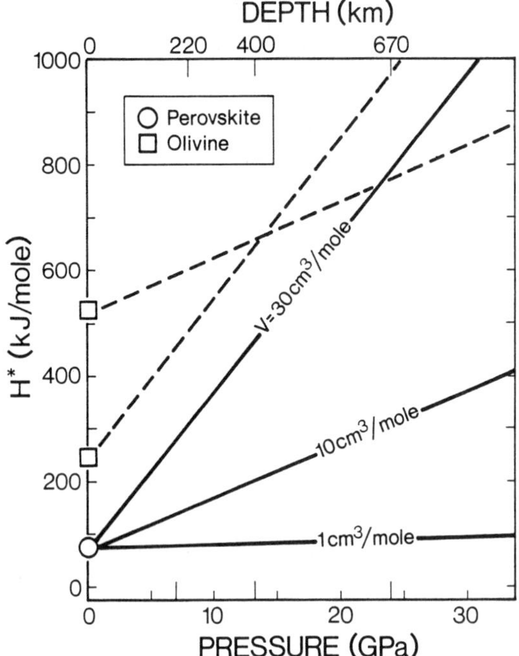

Figure 4. The activation enthalpy ($H^*=E^*+PV^*$) is plotted as a function of pressure for silicate perovskite assuming that the activation volume varies between 1 and 30 cm^3/mole. The top scale indicates the depth in the earth that corresponds to the pressure given on the bottom scale. The extrapolated values of H^* for perovskite are compared with a range of values found for olivine (see text). If V^* for perovskite is small (1–10 cm^3/mole), then at 670 km depth, H^* is still small ($\lesssim$300 kJ/mole) in comparison with a representative silicate such as olivine at comparable pressures. The figure illustrates the importance of determining V^* in order to determine H^* for minerals deep in the earth.

the value of the activation enthalpy of the perovskite phase transformation in the mantle. A relatively small value of V^* (1–10 cm^3/mole), means that the activation enthalpy remains low ($\leq$400 kJ/mole) even at mantle pressure (24 GPa at 670 km depth). We note that our application of Mukherjee's model described above yields an activation volume of about 1 cm^3/mole for $MgSiO_3$ perovskite, given WILLIAMS et al. (1987) measurement of the shift in vibrational frequencies with pressure (2.6 cm^{-1}/GPa). However, if the activation volume is large (~30 cm^3/mole), as has recently been suggested for olivine (GREEN and HOBBS, 1984), then the activation enthalpy may be as high as 1000 kJ/mole at 670 km depth. As shown in Figure 4 from the range of estimates for both V^* and E^* of olivine, even in this well-studied mineral there is a considerable uncertainty in the activation energy at depth.

The small value of the activation energy for perovskite indicates that there should be no kinetic hindrance to the perovskite phase transformation in the earth's mantle, using the assumption that the transition to perovskite follows the same kinetics as the transition from perovskite to enstatite. For instance, even at the coldest temperatures in the interior of a downgoing slab (~1000 °C at 670 km depth: SHUBERT et al., 1975), once the slab reaches the pressure of the perovskite stability field, the transition would occur in less than 1 sec.

In contrast, upper-mantle phase transitions, including the olivine-spinel transformation, may be kinetically hindered in the coldest slabs (SUNG and BURNS, 1976; SUNG, 1979). If deep-focus earthquakes are related to the occurrence of metastable transformations in the subducted slab (e.g., RUBIE, 1984), then the rapid transformation to perovskite inferred from our data might explain the cessation of earthquakes near 650 km depth. We also note that the small (slightly negative) Clapeyron slope for the transformation to perovskite (ITO and YAMADA, 1982; KATO and KUMAZAWA, 1985), along with the rapid kinetics involved, implies that the transformation occurs at constant depth in the mantle. Unlike the olivine-spinel transition, there is no deflection expected for the transformation across cold regions of the mantle (e.g., subducted slabs) and the perovskite transformation does not contribute a significant buoyancy force to mantle convection.

A final consequence and a possible future corroboration of our data are suggested by Figure 5. Here we summarize

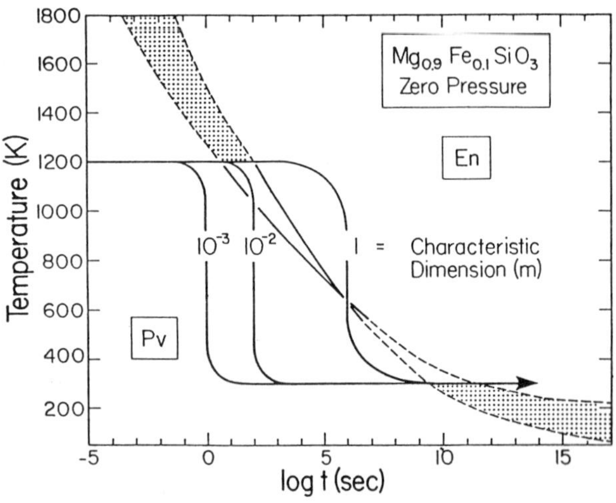

Figure 5. The time-temperature data for the back transformation of silicate perovskite to enstatite are replotted over a geologically relevant range of time and temperature. The broken lines surrounding the shading are extrapolations of the data in Figure 2. At high temperature (>1400 K), the back transformation to enstatite occurs in less than 1 second. At room temperature (~300 K), the back transformation is predicted to occur at times as short as three years and certainly less than 10^3 years. Also plotted are thermal-conduction cooling curves for xenoliths of sizes ranging from 1 mm to 1 m, and that have been emplaced on the earth's surface at 1200 K. For the small sizes, the cooling rate is rapid enough that the samples would quench in the perovskite structure on the earth's surface. The metastable perovskite would be expected to revert on a geologically short time scale, however.

the reaction rate expected for perovskite transforming to enstatite at zero pressure and as a function of temperature. First, we consider whether there is any possibility that silicate perovskite might be successfully transported to the earth's surface in a xenolith carried from lower-mantle depths. According to the figure, perovskite could survive in a centimeter-sized xenolith erupted at 1200 K or less; for comparison, only a xenolith smaller than 100 μm could retain the perovskite phase if erupted at 1600 K. Extrapolating our kinetic measurements, however, we would expect (Mg,Fe)SiO_3 perovskite to revert on a geologically short time scale of 3 to 100 years at zero pressure and 300 K. Thus, we hold little hope of finding any natural samples of silicate perovskite at the earth's surface. In fact, the time scale for reversion of perovskite at 300 K is expected to be short enough that it may be useful to periodically check the identity of a synthetic perovskite sample over a period of years or decades. If our data are properly extrapolated in Figure 5, the chances are good that back transformation can be observed within the lifetime of one researcher.

Acknowledgments. This work was supported by NASA and the University of California Institute of Geophysics and Planetary Physics. We thank G. H. Brimhall, R. Lange, G. L. Smith, D. Snyder and X. Li for assistance with the experiments, and Q. Williams, C. Meade, A. Navrotsky and especially S. Karato for comments on the manuscript.

References

Andersen, N. H., J. K. Kjems, and W. Hayes, Ionic conductivity of the perovskite $NaMgF_3$, $KMgF_3$ and $KZnF_3$ at high temperatures, *Solid State Ionics, 17*, 143–145, 1985.

Barnett, J. D., S. Block, and G. J. Piermarini, An optical fluorescence system for quantitative pressure measurement in the diamond-anvil cell, *Rev. Sci. Instrum., 44*, 1–9, 1973.

Cahn, J. W., The kinetics of grain boundary nucleated reaction, *Acta Met., 4*, 449–459, 1956.

Christian, J. W., *The Theory of Transformations in Metals and Alloys*, 586 pp., Pergamon Press, New York, 1975.

Dachille, F., R. J. Zeto, and R. Roy, Coesite and stishovite: stepwise reversal transformations, *Science, 140*, 991–993, 1963.

Durham, W. B., C. Froidevaux, and O. Jaoul, Transient and steady-state creep of pure forsterite at low stress, *Phys. Earth Planet. Inter., 19*, 263–274, 1979.

Gigl, P. D., and F. Dachille, Effect of pressure and temperature on the reversal transitions of stishovite, *Meteoritics, 4*, 123–136, 1968.

Goetz, C., The mechanisms of creep in olivine, *Phil. Trans. Roy. Soc. London, Ser. A, 288*, 99–119, 1978.

Green, H. W., and B. E. Hobbs, Pressure dependence of creep in dry polycrystalline olivine, *EOS, 65*, 1107, 1984.

Hamaya, N., and S. Akimoto, Experimental investigation on the mechanism of olivine→spinel transformation: growth of single crystal spinel from single crystal olivine in Ni_2SiO_4, in *High-Pressure Research in Geophysics*, edited by S. Akimoto and M. H. Manghnani, pp. 373–390, Center for Academic Publications, Tokyo, 1982.

Heinz, D. L., and R. Jeanloz, Melting of (Mg, Fe)SiO_3 perovskite: measurements at lower mantle conditions and geophysical implications, *J. Geophys. Res.*, (in press), 1987.

Heinz, D. L., and R. Jeanloz, Temperature measurements in the laser-heated diamond cell, this volume.

Ito, E., and H. Yamada, Stability relations of silicate spinels, ilmenites and perovskites, in *High-Pressure Research in Geophysics*, edited by S. Akimoto and M. H. Manghnani, pp. 405–419, Center for Academic Publications, Tokyo, 1982.

Jaoul, O., M. Poumellec, C. Froidevaux, and A. Havette, Silicon diffusion in forsterite: a new constraint for understanding mantle deformation, in *Anelasticity in the Earth, Geodynamics Series*, vol. 4, edited by F. D. Stacey, M. S. Paterson and A. Nicholas, pp. 95–100, AGU Publications, Washington, D. C., 1981a.

Jaoul, O., C. Froidevaux, W. B. Durham, and M. Michaut, Oxygen self-diffusion: implications for the high-temperature creep mechanism, *Earth Planet. Sci. Lett., 47*, 391–397, 1981b.

Jeanloz, R., and A. B. Thompson, Phase transitions and mantle discontinuities, *Rev. Geophys. Space Phys., 21*, 51–74, 1983.

Karato, S., Comment on "The effect of pressure on the rate of dislocation recovery in olivine" by D. L. Kohlstedt, H. P. K. Nichols and P. Hornack, *J. Geophys. Res., 86*, 9319, 1981.

Kasahara, J., and H. Tsukahara, Experimental measurement of reaction rate at the phase change of nickel olivine to nickel spinel, *J. Phys. Earth, 19*, 79–88, 1971.

Kato, T., and M. Kumazawa, Garnet phase of $MgSiO_3$ filling the pyroxene gap at very high temperature, *Nature, 316*, 803–805, 1985.

Kirby, S. H., Rheology of the lithosphere, *Rev. Geophys. Space Phys., 21*, 1458–1486, 1983.

Knittle, E., R. Jeanloz, and G. L. Smith, The thermal expansion of silicate perovskite and stratification of the Earth's mantle, *Nature, 319*, 214–216, 1986.

Kohlstedt, D. L., H. P. K. Nichols, and P. Hornack, The effect of pressure on the rate of dislocation recovery in olivine, *J. Geophys. Res., 85*, 3122–3130, 1980.

Liu, L.-G., Phase transformations and the constitution of the deep mantle, in *The Earth: Its Origin, Structure and Evolution*, edited by M. W. McElhinny, pp. 117–202, Academic, New York, 1979.

Mao, H.-K., P. M. Bell, K. J. Dunn, R. M. Chrenko, and R. C. De Vries, Absolute pressure measurements and analysis of diamonds subjected to maximum static pressures of 1.3–1.7 Mbar, *Rev. Sci. Instrum., 50*, 1002–1009, 1979.

Mao, H.-K., P. M. Bell, J. W. Shaner, and D. J. Steinberg, Specific volume measurements of Cu, Mo, Pd and Ag and calibration of the ruby R_1 fluorescence pressure gauge from 0.06 to 1 Mbar, *J. Appl. Phys., 49*, 3276–3283, 1978.

March, N. H., and M. P. Tosi, *Coulomb Liquids*, 351 pp., Academic Press, New York, 1984.

Mukherjee, R., Monovacancy formation energy and Debye temperature of closepacked metals, *Phil. Mag., 12*, 915, 1965.

Naka, S., S. Ito, T. Kameyama, and M. Inagaki, Crystallization of coesite, *Memoirs Faculty Engineering, Nagoya University, 28*, 265–316, 1976.

O'Keeffe, M., and J. -O. Bovin, Solid electrolyte behavior of $NaMgF_3$: geophysical implications, *Science, 206*, 599–600, 1979.

Poirier, J. P., On the kinetics of olivine spinel transition, *Phys. Earth Planet. Int., 26*, 179–187, 1981.

Poirier, J. P., J. Peyronneau, J. Y. Gesland, and G. Brebec, Viscosity and conductivity of the lower mantle, an experimental study on a $MgSiO_3$ perovskite analog, $KZnFe_3$, *Phys. Earth Planet. Inter., 32*, 273–287, 1983.

Reddy, K. P. R., S. M. Oh, L. D. Major, and A. R. Cooper, Oxygen diffusion in forsterite, *J. Geophys. Res., 85*, 322–326, 1980.

Reissland, J. A., *The Physics of Phonons*, 319 pp., Wiley & Sons, New York, 1973.

Robie, R. A., B. S. Hemmingway, and J. R. Fisher, Thermodynamic properties of minerals and related substances to 298.15 K (25.0 °C) and one (1.013 bars) atmosphere pressure and at higher temperatures, *U. S. Geol. Survey Bull., 1452*, 1978.

RUBIE, D. C., The olivine→spinel transformation and the rheology of the subducting lithosphere, *Nature, 308*, 305, 1984.

SCHUBERT, G., D. A. YUEN, and D. L. TURCOTTE, Role of phase transitions in a dynamic mantle, *Geophys. J. R. Astr. Soc., 42*, 705–735, 1975.

SKINNER, B. J., and J. J. FAHEY, Observations on the inversion of stishovite to silica glass, *J. Geophys. Res., 68*, 5595–5604, 1963.

STISHOV, S. M., and S. V. POPOVA, New dense modifications of silica, *Geokhimiya, 10*, 923–926, 1961.

SUNG, C.-M., Kinetics of the olivine→spinel transition under high pressure and temperature: experimental results and geophysical implications, in *High Pressure Physics and Technology*, vol. 2, edited by K.D. Timmerhaus and M. S. Barber, pp. 31–42, Plenum Press, New York, 1979.

SUNG, C.-M., and R. G. BURNS, Kinetics of high-pressure phase transformations: implications to the evolution of the olivine→spinel transition in the down-going lithosphere and its consequences on the dynamics of the mantle, *Tectonophysics, 31*, 1–32, 1976.

WILLIAMS, Q., R. JEANLOZ, and P. MCMILLAN, The vibrational spectrum of $MgSiO_3$-perovskite: zero pressure Raman and mid-infrared spectra to 27 GPa, *J. Geophys. Res.*, in press, 1987.

YAGI, T., H-K. MAO, and P. M. BELL, Structure and crystal chemistry of perovskite-type $MgSiO_3$, *Phys. Chem. Minerals, 3*, 97–110, 1978.

PYROXENE-GARNET TRANSFORMATION: THERMOCHEMISTRY AND ELASTICITY OF GARNET SOLID SOLUTIONS, AND APPLICATION TO A PYROLITE MANTLE

Masaki AKAOGI

Department of Earth Sciences, Kanazawa University, Kanazawa 920, Japan

Alexandra NAVROTSKY

Department of Geological and Geophysical Sciences, Princeton University, Princeton, NJ 08544, USA

Takehiko YAGI[1]

Research Institute for Iron, Steel, and Other Metals, Tohoku University, Sendai 980, Japan

Syun-iti AKIMOTO[2]

Institute for Solid State Physics, University of Tokyo, Tokyo 106, Japan

Abstract. Thermochemical properties and equations of state of garnet solid solutions in the system $Mg_4Si_4O_{12}$-$Mg_3Al_2Si_3O_{12}$ were investigated. Based on high-temperature solution calorimetry experiments, the enthalpy of the pyroxene-garnet transition of $Mg_4Si_4O_{12}$ was estimated to be about 146 kJ/mol. The phase relations, calculated from the thermochemical data, among pyroxene, garnet, and spinel (or modified spinel) plus stishovite in the system $Mg_4Si_4O_{12}$-$Mg_3Al_2Si_3O_{12}$ were generally consistent with the phase diagrams determined experimentally. The bulk modulus of the garnet solid solution in this system was measured by hydrostatic compression experiments using synchrotron radiation and X-ray diffraction techniques. With increasing $Mg_4Si_4O_{12}$ component, the bulk modulus of the garnet solid solution was found to decrease slightly relative to pyrope. Using the new phase equilibria and elasticity data, density and bulk sound velocity profiles were calculated for the pyroxene-garnet system, as well as for pyrolite. The bulk sound velocity distribution of the pyroxene-garnet system showed no sharp increase in velocity in the mantle down to 670 km. The density and bulk sound velocity of pyrolite at depths of 220–670 km were in general agreement with the seismological velocity and density profiles when the new elasticity data for the garnet solid solution were used.

Introduction

Experimental studies of the pyroxene-garnet system at high pressures and high temperatures have important implications for the understanding of mineralogy and chemistry of the mantle. RINGWOOD (1967) first reported extensive solid solubility of pyroxene in garnet in the system $MgSiO_3$-Al_2O_3 at high pressures, and discussed the geophysical importance of this "pyroxene-garnet transition". AKAOGI and AKIMOTO (1977) carried out detailed studies on the systems $Mg_4Si_4O_{12}$-$Mg_3Al_2Si_3O_{12}$ and $Fe_4Si_4O_{12}$-$Fe_3Al_2Si_3O_{12}$, and concluded that pyroxenes in the upper mantle dissolve gradually in garnet with increasing depth, resulting in formation of a homogeneous garnet solid solution in the transition zone. AKAOGI and AKIMOTO (1979) extended investigations to the $Ca_2Mg_2Si_4O_{12}$-$Ca_{1.5}Mg_{1.5}Al_2Si_3O_{12}$ system and to a natural garnet lherzolite composition which is very close to pyrolite. LIU (1977) studied phase equilibria in the $Mg_4Si_4O_{12}$-$Mg_3Al_2Si_3O_{12}$ system at high pressures, confirming the formation of the garnet solid solution. He also reported that this garnet solid solution transforms to ilmenite, and subsequently to perovskite at higher pressures.

Precise determination of phase equilibria using high-pressure experiments still involves several problems. There is considerable uncertainty in the pressure calibration at room and high temperatures at pressures above 10–15 GPa. Also, it is difficult to demonstrate equilibrium for reactions by reversed runs when the reactions are very sluggish. These problems suggest that an independent approach is highly desirable for determining phase relations more accurately. In this paper, we show experimental results of high-temperature solution calorimetry of the $Mg_4Si_4O_{12}$-$Mg_3Al_2Si_3O_{12}$ garnet solid solutions. The phase relations of the pyroxene-garnet system are calculated from the thermochemical data and compared with those determined experimentally, for further refinement of the phase diagrams.

In comparison to the phase equilibria data, our knowledge of the elasticity of the garnet solid solution with pyroxene component has been quite limited. Only a

[1]Now at Institute for Solid State Physics, University of Tokyo, Tokyo 106, Japan.

[2]Now at Institute for Study of the Earth's Interior, Okayama University, Misasa, Tottori-ken 682-02, Japan.

High-Pressure Research in Mineral Physics, edited by M. H. Manghnani and Y. Syono, pp. 251–260.

single study on the elasticity of majorite, a garnet with approximately (Mg, Fe)SiO_3 pyroxene composition, was reported by JEANLOZ (1981). He obtained a bulk modulus of 221 GPa for the majorite, which is about 50 GPa higher than the value for pyrope.

In this paper we also briefly summarize results of our current hydrostatic compression experiments in the $Mg_4Si_4O_{12}$-$Mg_3Al_2Si_3O_{12}$ garnet solid solution using high-pressure X-ray diffraction techniques and synchrotron radiation.

New phase diagrams based on our current thermochemical data are used to clarify details of the pyroxene-garnet transition in the mantle. Density and bulk sound velocity distributions under mantle P, T conditions of the pyroxene-garnet system are calculated from the new elasticity data of the garnet solid solution. These distributions are also calculated for the olivine system and pyrolite from the available elasticity data. By comparing these profiles with seismic velocity distributions, we can discuss mantle mineralogy in the pyrolitic composition.

Experimental Methods

Garnets and pyroxenes in the system $Mg_4Si_4O_{12}$-$Mg_3Al_2Si_3O_{12}$ were synthesized using a tetrahedral-anvil and a double-stage cubic-octahedral anvil type of apparatus at pressures of 4–17 GPa and temperatures of 900–1000 °C. Starting materials used were glasses or an oxide mixture (for pyrope) with compositions of 0.0, 17.7, 40.6, and 57.6 mol% $Mg_4Si_4O_{12}$. Chemical compositions of the glasses were analyzed with an electron microprobe.

Calorimetric experiments were performed using a Calvet type of high-temperature microcalorimeter (NAVROTSKY, 1977). Enthalpies of solution of the garnets and pyroxenes in $2PbO \cdot B_2O_3$ solvent were measured at 714 °C by solution calorimetric techniques, because the metastable persistence of all samples at the calorimeter temperature for about 15 h was confirmed.

Isothermal compression experiments of pyrope and $[Mg_4Si_4O_{12}]_{0.58}$ $[Mg_3Al_2Si_3O_{12}]_{0.42}$ garnet (mol%) were carried out using a high-pressure, high-temperature X-ray diffraction apparatus named MAX80 interfaced with a synchrotron radiation source at Photon Factory, Tsukuba, Japan. Powder samples were compressed at room temperature under hydrostatic conditions using a methanol/ethanol mixture as the pressure transmitting medium. Sodium chloride powder mixed with the garnet sample was used as a pressure standard. More details of the high-pressure X-ray diffraction experiments are given in a separate paper (YAGI et al., this volume).

Thermochemical Data and Calculation of Phase Relations in the System $Mg_4Si_4O_{12}$-$Mg_3Al_2Si_3O_{12}$

Enthalpies of solution of garnet and pyroxene samples at 714 °C are shown in Table 1 and Figure 1. The heat of solution of garnet decreases from pyrope almost linearly with increasing $Mg_4Si_4O_{12}$ component. The enthalpy of solution of $Mg_4Si_4O_{12}$ garnet was estimated to be about 0 kJ/mol, assuming ideal mixing of both end-members. Using the heat of solution of $Mg_4Si_4O_{12}$ pyroxene determined by ITO and NAVROTSKY (1985), the enthalpy of the pyroxene-garnet transition is calculated as 146 ± 8 kJ/mol per 12-oxygen mol. Table 2 shows the enthalpies of transition in $Mg_4Si_4O_{12}$ from our present study and our previous works (AKAOGI et al., 1984; AKAOGI and NAVROTSKY, 1984; ITO and NAVROTSKY, 1985).

Phase relations in $Mg_4Si_4O_{12}$ at high pressures and high temperatures are shown in Figure 2, based on three sets of experimental data by AKAOGI and AKIMOTO (1977), ITO and NAVROTSKY (1985), and SAWAMOTO (1986). All of these experimental runs are plotted on the basis of the same pressure-fixed points: ZnS (15.5 GPa, BLOCK, 1978), GaAs (18.3, SUZUKI et al., 1981) and GaP (22.0, PIERMARINI and BLOCK, 1975). At 1000 °C, $Mg_4Si_4O_{12}$ pyroxene decomposes into β-Mg_2SiO_4 plus stishovite at 16 GPa and to γ-Mg_2SiO_4 plus stishovite at 17 GPa.

TABLE 1. Enthalpies of Solution of Garnets and Pyroxenes in the System $Mg_4Si_4O_{12}$-$Mg_3Al_2Si_3O_{12}$ at 714 °C

Composition		ΔH°_{sol} (kJ/mol)	
$Mg_4Si_4O_{12}$ (mol%)	$Mg_3Al_2Si_3O_{12}$ (mol%)		
	Garnet		
0.0	100.0	108.80 (4)[a]	s.d. = 3.49[b] s.m. = 1.74
17.7	82.3	93.16 (6)	s.d. = 3.83 s.m. = 1.56
40.6	59.4	65.94 (4)	s.d. = 4.31 s.m. = 2.15
57.6	42.4	43.04 (2)	s.m. = 6.32
	Pyroxene		
57.6	42.4	126.12 (4)	s.d. = 4.40 s.m. = 2.21
100.0	0.0	145.35[c] (9)	s.d. = 3.82 s.m. = 1.27

a) Number in parentheses is number of runs, b) s.d. is standard deviation of the data, s.m. is standard deviation of the mean, c) From ITO and NAVROTSKY (1985).

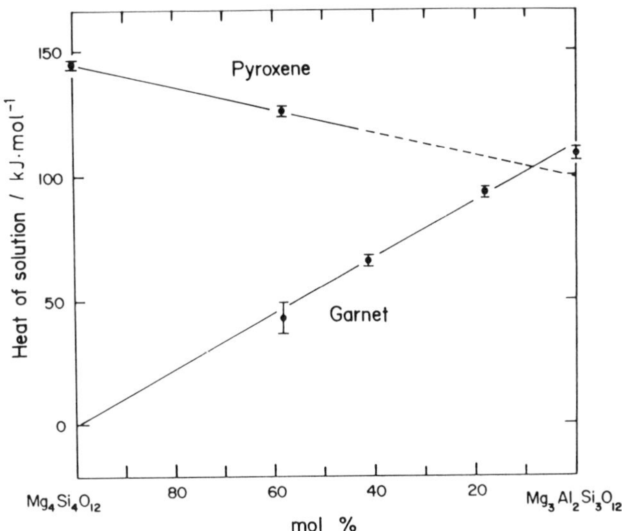

Figure 1. Heats of solution of garnets and pyroxenes in the system $Mg_4Si_4O_{12}$-$Mg_3Al_2Si_3O_{12}$ at 714 °C.

TABLE 2. Thermochemical Data for Transitions in $Mg_4Si_4O_{12}$

Transition	ΔV°_{298} (cm^3/mol)	ΔH° (kJ/mol)	ΔS° (J/mol·K)	Ref.
Px→Gar	− 11.15	146.4 ± 7.8	− 14.6 ± 2.9	a
Gar→β + St	− 4.95	34.4 ± 10.9	− 13.8 ± 6.7	a, b, c
Gar→γ + St	− 6.99	48.0 ± 13.3	− 33.1 ± 8.8	a, b, c
Px→β + St	− 16.10	180.8 ± 7.7	− 28.5 ± 5.9	b, c, d
β + St→γ + St	− 2.04	13.6 ± 7.5	− 19.2 ± 5.9	b, c

References: a) this work, b) AKAOGI et al. (1984), c) AKAOGI and NAVROTSKY (1984), d) ITO and NAVROTSKY (1985).

However, KATO and KUMAZAWA (1985) and SAWAMOTO (1986) have found that, at temperatures above about 1700 °C and pressures of about 17 GPa, $Mg_4Si_4O_{12}$ pyroxene directly transforms to garnet, the structure of which is slightly distorted from cubic to tetragonal symmetry. Using the enthalpies of these transitions and observed transition pressures, the entropies of transition and the phase boundaries are calculated (see Table 2 and Figure 2). Effects of compression and thermal expansion are corrected, using the parameters in Table 3, by the same method as in the study by NAVROTSKY et al. (1979). The calculated boundaries are almost consistent with the experimental data. Figure 2 clearly shows that the stability field of garnet intervenes between those of the β-phase plus stishovite (or ilmenite) and pyroxene at temperatures above about 1550 °C and pressures above 17 GPa. In the following discussion, we assume the enthalpies and entropies of transition shown in Table 2 to be independent of temperature in the temperature range of interest.

Using the thermochemical data in Table 2, phase boundary curves for the pyroxene-garnet equilibria in the system $Mg_4Si_4O_{12}$-$Mg_3Al_2Si_3O_{12}$ are calculated as follows. Equilibrium of pyroxene and garnet is expressed as

$$Mg_4Si_4O_{12}(\text{pyroxene}) = Mg_4Si_4O_{12}(\text{garnet}) \qquad (1)$$

for which we have at equilibrium

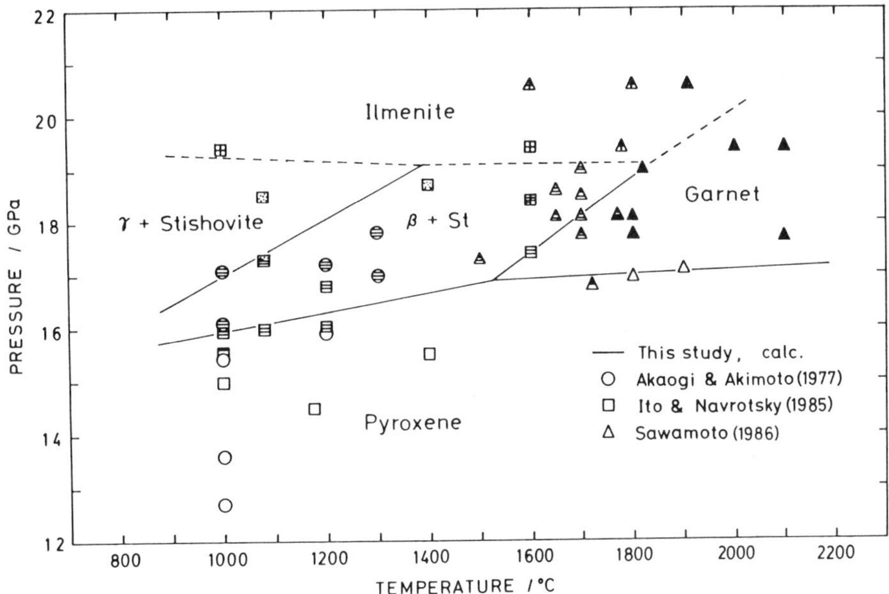

Figure 2. Phase relations of $Mg_4Si_4O_{12}$. Solid boundaries are calculated from thermochemical data. Dashed boundaries are determined by experimental data points.

TABLE 3. Parameters Used for Calculation of Phase Equilibria and of Density and Bulk Sound Velocity Profiles

Phase	V_0 (cm^3/mol)	α (10^{-5}/K)	K_0 (GPa)	K_0'	γ	δ
α-Mg_2SiO_4	43.79[a]	3.72[f]	129	5.2	1.25	4
β-Mg_2SiO_4	40.51[b]	3.07[g]	174[l]	4	1.3	3[s]
γ-Mg_2SiO_4	39.49[c]	2.55[h]	184[m]	4[s]	1.35	3[s]
$Mg_4Si_4O_{12}$(Px)	125.15	2.92[f]	108	7	1.1	6
$Mg_4Si_4O_{12}$(Gar)	114.00[d]	2.57[i]	154[n]	4.5[i]	1.1	6
$Mg_3Al_2Si_3O_{12}$(Gar)	113.21[d]	2.57[f]	166[n]	4.5[t,u]	1.1	6
$Ca_2Mg_2Si_4O_{12}$(Px)	132.18[a]	2.92[f]	107[o]	8[u]	1.1[v]	6[v]
$Ca_2Mg_2Si_4O_{12}$(Gar)	123.18[e]	2.46[j]	156[p]	4.5[j]	1.1[i]	6[i]
$Ca_{1.5}Mg_{1.5}Al_2Si_3O_{12}$(Gar)	119.45[e]	2.46[f]	167[q]	4.5[i]	1.1[i]	6[i]
SiO_2(St)	14.01[a]	2.00[k]	306[r]	4	1.7	3

a) ROBIE et al. (1978), b) HORIUCHI and SAWAMOTO (1981), c) SASAKI et al. (1982), d) AKAOGI and AKIMOTO (1977), e) AKAOGI, unpublished, f) SKINNER (1966), g) SUZUKI et al. (1980), h) SUZUKI et al. (1979), i) same as pyrope, j) same as $Ca_{1.5}Mg_{1.5}Al_2Si_3O_{12}$ garnet, k) ITO et al. (1974), l) SAWAMOTO et al. (1984), m) WEIDNER et al. (1984), n) this work, o) LEVIEN et al. (1979b), p) estimated from $Ca_{1.5}Mg_{1.5}Al_2Si_3O_{12}$ garnet using the relationship in Fig. 5, q) average of pyrope (this work) and grossular (BABUŠKA et al., 1978), r) WEIDNER et al. (1982), s) assumed, t) LEVIEN et al. (1979a), BONCZAR et al. (1977), u) BASS et al. (1981), v) same as enstatite, others: JEANLOZ and THOMPSON (1983).

$$\Delta G_{P,T} = -RT\ln K \quad (2)$$

where $\Delta G_{P,T}$ is free energy change at P and T, and K is the equilibrium constant. The equation (2) is also expressed as

$$\Delta G_{P,T} = \Delta H^\circ_T - T\Delta S^\circ_T + \int_{1\,\mathrm{atm}}^{P} \Delta V_{P,T}\,dP = -RT\ln\frac{a_i^{gar}}{a_i^{px}} \quad (3)$$

where ΔH°_T, ΔS°_T are enthalpy and entropy of transition at 1 atm and T, $\Delta V_{P,T}$ is volume change at P and T, and a_i^j is the activity of i(=$Mg_4Si_4O_{12}$) component in the j phase. Experimental studies show that the $Mg_3Al_2Si_3O_{12}$ component of pyroxene in equilibrium with garnet decreases with increasing pressure, and that above about 4 GPa and at 1000 °C, the pyroxene composition is very close to $Mg_4Si_4O_{12}$ (PERKINS et al., 1981). Therefore, we assume that a_i^{px} is equal to unity. We also approximate the garnet phase as a statistically ideal solid solution and assume completely disordered distribution of Mg^{2+}, Si^{4+} and Al^{3+} ions in the octahedral sites in the garnet structure. This gives, relative to garnet standard state for both end-members in the system $x(Mg_3)[MgSi]Si_3O_{12}-(1-x)(Mg_3)[Al_2]Si_3O_{12}$, where ( ) denotes dodecahedral and [ ] octahedral sites

$$a_i^{gar} = x^2 \quad (4)$$

Using equations (3) and (4) with the thermochemical data in Table 2, the boundary for reaction of pyroxene plus garnet to single-phase garnet solid solution is calculated and shown in Figure 3. The boundaries for the decomposition of garnet solid solution to β- or γ-Mg_2SiO_4 plus stishovite plus garnet with a smaller amount of pyroxene component are also calculated in a similar manner from the thermochemical data in Table 2. In the above calculation, effects of compression and thermal expansion are corrected using the parameters in Table 3. A boundary curve for single-phase field of pyroxene (see Figure 3) is based on the results by GASPARIK and NEWTON (1984). The error in the transition pressures of the boundary of pyroxene plus garnet to garnet solid solution is estimated to be about ±0.5 GPa based on the errors of the thermochemical data in Table 2, while that for decomposition of garnet solid solution to β (or γ)-phase plus stishovite plus garnet is about ±1 GPa.

Experimental data in Figure 3 are taken from AKAOGI and AKIMOTO (1977). However, pressures of runs above about 13 GPa are reduced on the basis of the revised pressure-fixed points at room temperature, as described

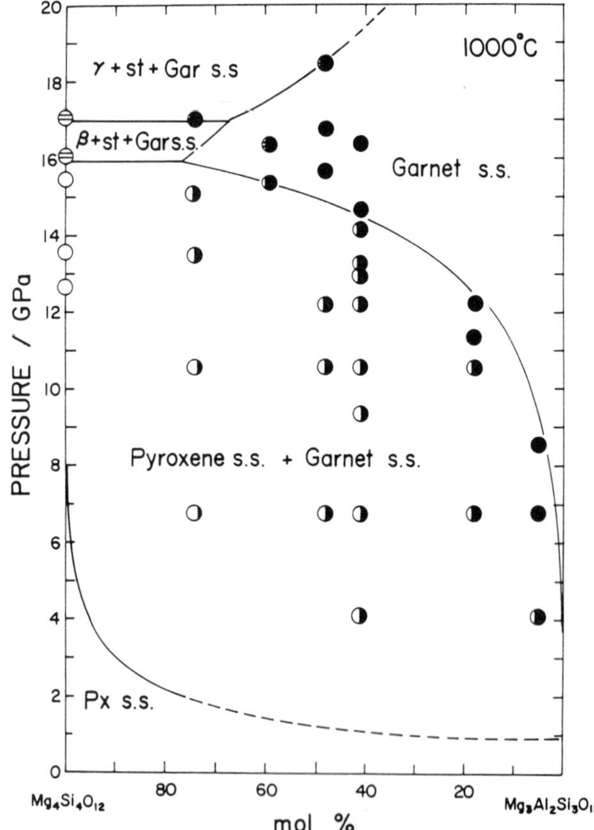

Figure 3. Phase relations in the system $Mg_4Si_4O_{12}$-$Mg_3Al_2Si_3O_{12}$ at 1000 °C. The boundaries are calculated from thermochemical data. The experimental data points are from AKAOGI and AKIMOTO (1977).

above. A high-temperature correction to pressure was made using the coesite-stishovite transition (YAGI and AKIMOTO, 1976; AKAOGI and NAVROTSKY, 1984), the pyroxene-ilmenite transition in $ZnSiO_3$ (AKIMOTO et al., 1977), and the modified spinel-spinel transition in Mg_2SiO_4 (SUITO, 1977; AKAOGI et al., 1984). The calculated boundaries are generally consistent with the experimental runs. The maximum solubility of the $Mg_4Si_4O_{12}$ component in garnet at 1000 °C is about 77±10 mol% (see Figure 3), considering the uncertainties in the boundaries described above. This value is consistent with the value of 80% determined by LIU (1977) and KANZAKI (1987), but is significantly larger than 60% determined by AKAOGI and AKIMOTO (1977). The boundary for decomposition of single-phase garnet into β-phase plus stishovite plus garnet, which restricts the maximum solubility of the $Mg_4Si_4O_{12}$ component in garnet, is highly temperature dependent, as described below. Therefore, the temperature gradient observed in the sample by AKAOGI and AKIMOTO (1977) might explain the discrepancy in the maximum solubility of $Mg_4Si_4O_{12}$ in garnet between this study and that of Akaogi and Akimoto.

KANZAKI (1987) has reported the phase relations in the system $Mg_4Si_4O_{12}$-$Mg_3Al_2Si_3O_{12}$ at pressures 9–24 GPa at 1000 °C. His phase diagram, below about 16 GPa, is generally consistent with our phase diagram shown in Figure 3, if the difference of pressure calibration points used are taken into account. In his diagram, however, the boundaries for the decomposition of garnet solid solution to β (or γ)-phase plus stishovite plus garnet are almost pressure-independent; the boundaries are located at the composition of 75–80 mol% $Mg_4Si_4O_{12}$. In Kanzaki's experiments, single-phase garnet solid solutions were synthesized from glass materials in several runs made at 18–22 GPa and 1000 °C with the composition of 60 mol% $Mg_4Si_4O_{12}$. These run products are inconsistent with the boundaries shown in Figure 3. In Kanzaki's experiments, both pressure and temperature were increased simultaneously to a desired P, T condition at which the sample was held for 1–2 h and then quenched (Kanzaki, personal communication). The P-T paths of the above runs of the 60 mol% $Mg_4Si_4O_{12}$ might pass through the single-phase field of garnet shown in Figure 3. Our study frequently observed that single-phase garnet solid solutions rapidly crystallized from glasses within 2–3 min in the garnet field at temperatures as low as 800 °C. Since decomposition of the garnet crystals once synthesized is quite sluggish, it would be likely that the garnet phase synthesized at 18–22 GPa and 1000 °C in Kanzaki's study is a metastable phase.

Phase boundaries at 1300 and 1600 °C are also calculated, and shown in Figure 4 together with those at 1000 °C. The pressure for the decomposition of garnet solid solution to β- or γ-Mg_2SiO_4 plus stishovite plus garnet increases rapidly with increasing temperature: the slope is 0.007–0.009 GPa/K at the compositions of 40–90 mol% $Mg_4Si_4O_{12}$. In contrast, the slope of the boundary of pyroxene plus garnet to single phase garnet solid solution has a relatively small value, 0.0003–−0.001 GPa/K in the same compositional range. Therefore, the stability field of single-phase garnet solid solution expands rapidly with temperature, resulting in formation of continuous garnet solid solutions in the whole compositional range above about 1550 °C.

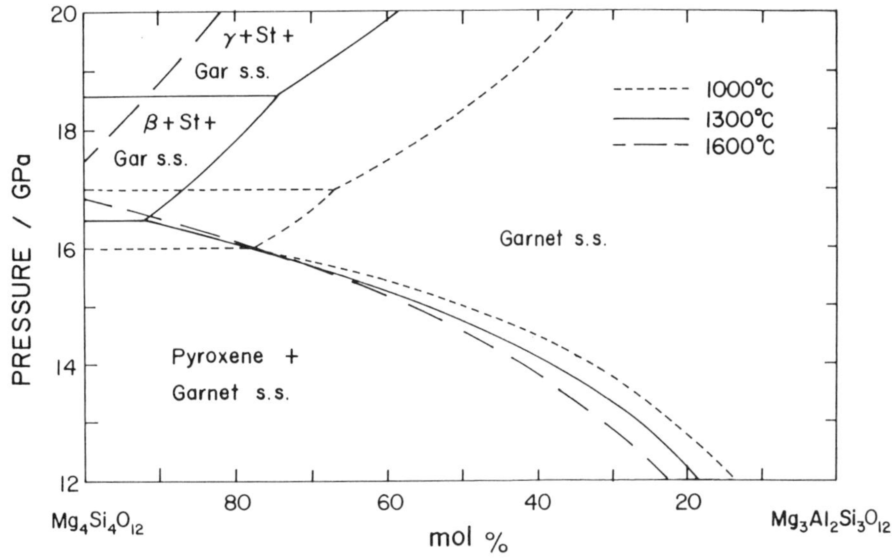

Figure 4. Calculated phase boundaries in the system $Mg_4Si_4O_{12}$-$Mg_3Al_2Si_3O_{12}$ at 1000, 1300, and 1600 °C.

Elasticity of Garnets in the System $Mg_4Si_4O_{12}$-$Mg_3Al_2Si_3O_{12}$

The detailed results of hydrostatic compression experiments of $Mg_3Al_2Si_3O_{12}$ garnet (pyrope) and 58 mol% $Mg_4Si_4O_{12}$·42 mol% $Mg_3Al_2Si_3O_{12}$ garnet are described in a separate paper (YAGI et al., this volume). The bulk moduli of the garnets are calculated from a least squares fit of the hydrostatic compression data with the Birch-Murnaghan equation of state, assuming pressure derivative values of 0–4. The results are illustrated in Figure 5. The bulk modulus of pyrope in this study is a little smaller than, but close to, the data (K_0=171–175 GPa) measured by static compression or acoustic experiments (BABUŠKA et al., 1978; SATO et al., 1978; LEVIEN et al., 1979a; LEITNER et al., 1980). Figure 5 clearly shows that the garnet with 58 mol% $Mg_4Si_4O_{12}$ has a K_0 which is smaller by about 7 GPa than pyrope. This suggests that the substitution of two Al^{3+} ions by Mg^{2+} and Si^{4+} at octahedral sites in the structure makes the garnet phase a little more compressible. Bulk modulus versus its pressure derivative of $Mg_4Si_4O_{12}$ garnet (majorite) can be estimated, assuming K_0 changes linearly with the octahedral cation substitution (as observed in $Ca_3Al_2Si_3O_{12}$-$Ca_3Fe_2Si_3O_{12}$ garnet solid solutions) (BABUŠKA et al., 1978; LEITNER et al., 1980; HANIFORD and WEIDNER, 1985). From Figure 5, a K_0 value of 154 GPa for $Mg_4Si_4O_{12}$ garnet was estimated assuming K_0'=4.5, which is the same as measured for pyrope in ultrasonic and static compression experiments (BONCZAR et al., 1977; LEVIEN et al., 1979a; BASS et al., 1981). The present results disagree with those of JEANLOZ (1981), who determined a K_0 of 221 GPa and K_0' of about 4.4 for natural magnesium-rich majorite, based on hydrostatic compression experiments using diamond-anvil cells. In general, hydrostatic compression data using the diamond-anvil cells may have systematic errors due to change in the sample position during compression. The structural formula of Jeanloz's majorite still has some ambiguities in the relative distribution of Fe and Mg between the dodecahedral and octahedral sites. In our experiments, however, the synthetic samples with the analyzed compositions in the $Mg_4Si_4O_{12}$-$Mg_3Al_2Si_3O_{12}$ system were used. The compression experiments of pyrope and garnet with 58 mol% $Mg_4Si_4O_{12}$ were made using identical techniques, i.e., the cubic-anvil apparatus interfaced with synchrotron radiation. In our experiments on garnet solid solutions in the system $Fe_4Si_4O_{12}$-$Fe_3Al_2Si_3O_{12}$, a smaller bulk modulus of garnet solid solution with 18 mol% $Fe_4Si_4O_{12}$ than that of almandine was observed by the same experimental techniques (YAGI et al., this volume). These findings suggest that the results in our study should be more reliable than those of JEANLOZ (1981) and that $M_4Si_4O_{12}$ garnets (M=Mg and Fe) have smaller bulk moduli than $M_3Al_2Si_3O_{12}$ garnets.

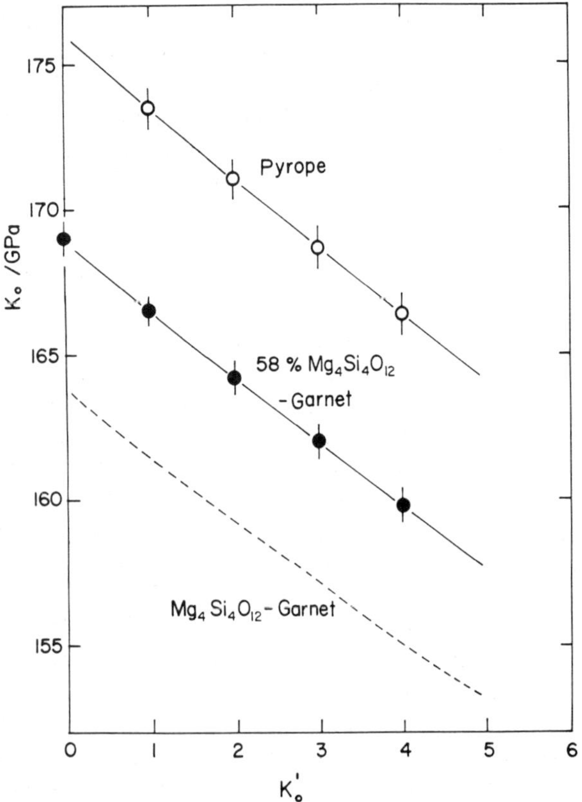

Figure 5. Bulk modulus versus its pressure derivative for garnets of $Mg_3Al_2Si_3O_{12}$ (pyrope) and 58 mol% $Mg_4Si_4O_{12}$·42 mol% $Mg_3Al_2Si_3O_{12}$. The dashed curve is estimated for $Mg_4Si_4O_{12}$ garnet (majorite).

Pyroxene-Garnet Transition in the Mantle

Figure 6 compares the calculated phase boundaries for the $Mg_4Si_4O_{12}$-$Mg_3Al_2Si_3O_{12}$ system at 1200 °C with those for the systems $M_4Si_4O_{12}$-$M_3Al_2Si_3O_{12}$ (M=Fe at 1000 °C; M=$Ca_{0.5}Mg_{0.5}$ at 1200 °C). The boundaries for the latter two systems are taken from AKAOGI and AKIMOTO (1977, 1979), after revision of pressure values based on the pressure calibration points adopted in this paper. In Figure 6, the pyroxene-garnet transition pressure increases in the order $Fe_4Si_4O_{12}$-$Fe_3Al_2Si_3O_{12}$, $Mg_4Si_4O_{12}$-$Mg_3Al_2Si_3O_{12}$, $Ca_2Mg_2Si_4O_{12}$-$Ca_{1.5}Mg_{1.5}Al_2Si_3O_{12}$. Therefore substitution of Mg by Fe reduces the transition pressure, while substitution by Ca increases it.

AKAOGI and AKIMOTO (1979) carried out phase equilibrium experiments on a garnet lherzolite (PHN1611) up to 20.5 GPa and 1200 °C, and examined partition of elements among coexisting phases at various pressures. The chemical composition of the lherzolite is very close to pyrolite. Akaogi and Akimoto also demonstrated a continuous increase in the solid solubility of pyroxene

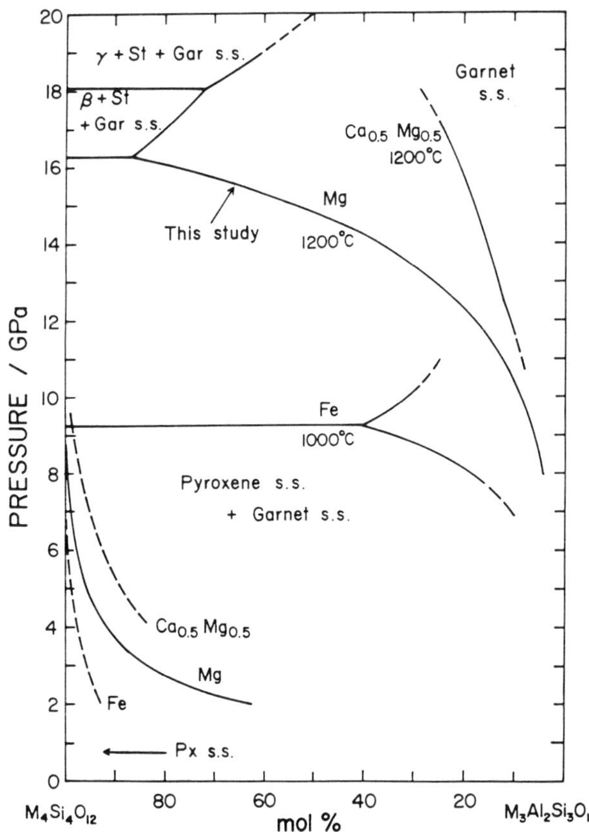

Figure 6. Phase relations in the systems $Mg_4Si_4O_{12}$-$Mg_3Al_2Si_3O_{12}$ and $Ca_2Mg_2Si_4O_{12}$-$Ca_{1.5}Mg_{1.5}Al_2Si_3O_{12}$ at 1200 °C and $Fe_4Si_4O_{12}$-$Fe_3Al_2Si_3O_{12}$ at 1000 °C.

components in garnet in the lherzolite system. As shown by WEIDNER (1985), the solubility of pyroxene in garnet in the lherzolite composition is very close to that interpolated from the three simple systems in Figure 6. These results suggest that the phase relations shown in Figure 6, as well as in Figures 3 and 4, are useful for estimating the pyroxene-garnet equilibria in the pyrolitic mantle.

In the following discussion, we assume a homogeneous pyrolitic composition for the mantle in the region 220–670 km. The mineral assemblage of pyrolite at upper mantle conditions is 57 wt% olivine, 14% garnet, 17% Ca-poor pyroxene, and 12% Ca-rich pyroxene (RINGWOOD, 1975). We also use a temperature profile calculated by BROWN and SHANKLAND (1981), in which the temperature at 670 km is 1600 °C. For chemical compositions of coexisting phases, particularly Mg/(Mg+Fe) and Ca/(Ca+Mg+Fe) ratios, we adopt the mineral compositions in the lherzolite system in AKAOGI and AKIMOTO (1979) at three different pressures where olivine, modified spinel, and spinel are stable. Considering the pyroxene/garnet ratio of pyrolite, together with the compositions described above, we determine from Figures 3, 4, and 6 that Ca-poor pyroxene disolves in garnet gradually with depth, resulting in complete dissolution at about 450 km (15 GPa). Figure 6 shows that Ca-rich pyroxene in equilibrium with garnet is stable at higher pressures than Ca-poor pyroxene. We tentatively assume complete dissolution of Ca-rich pyroxene in garnet at about 620 km (22 GPa). The single-phase garnet solid solution is not expected to decompose into a mixture of β (or γ)-phase, stishovite, and garnet at the P, T conditions of the transition zone, because of the rapid expansion of the single-phase field of the garnet solid solution with increasing temperature (as shown in Figure 4).

Phase transitions of olivine in the mantle are estimated using phase diagrams of AKIMOTO (this volume), who revised KAWADA's (1977) phase diagrams on the basis of the same pressure scale used in this work. According to the compositions mentioned above, $(Mg_{0.88}, Fe_{0.12})_2SiO_4$ olivine in the upper mantle transforms to $(Mg_{0.87}, Fe_{0.13})_2SiO_4$ modified spinel through the mixed region of 390–420 km. The β-phase further changes to $(Mg_{0.87}, Fe_{0.13})_2SiO_4$ spinel through the region of $\beta+\gamma$ at 480–550 km.

Density and Bulk Sound Velocity Profiles in the Mantle

Using the new phase diagrams and elasticity data, density (ρ) and bulk sound velocity (V_ϕ) profiles for the pyroxene-garnet system in the pyrolitic mantle are calculated. The ρ and V_ϕ profiles of the olivine system are also calculated, in order to combine both results to form the profiles of pyrolite. The bulk modulus (K) and the density (ρ) are first calculated at an appropriate pressure and room temperature by the Murnaghan equation of state. Temperature correction on the bulk modulus and density are made by a similar method to LEES et al. (1983). Table 3 summarizes the bulk moduli (K_0), their pressure derivatives (K_0'), thermal expansion coefficients (α), molar volumes (V_0), Gruneisen parameters (γ), and Anderson-Gruneisen constants (δ) used for the calculation of ρ and V_ϕ profiles. The bulk moduli of pyrope and $Mg_4Si_4O_{12}$ garnet in this work are used, together with bulk moduli of β- and γ-phases recently measured by SAWAMOTO et al. (1984) and WEIDNER et al. (1984), respectively. The effects of iron on the molar volumes and densities were corrected using the data from JEANLOZ and THOMPSON (1983); however, effects on the bulk moduli were very small for the magnesium-rich mantle minerals (LEITNER et al., 1980; WEIDNER et al., 1984; BASS and WEIDNER, 1984). At each P, T condition, the coexisting phases were mixed to form pyrolite. The Hill average of the Voigt and Reuss bounds was used for bulk moduli.

Figures 7 and 8 compare the calculated density and bulk sound velocity profiles with those from the earth model

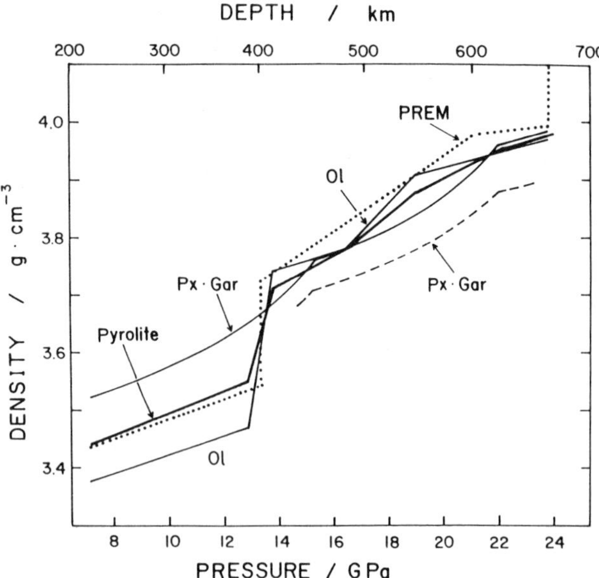

Figure 7. Density profiles of the olivine and pyroxene-garnet systems and of pyrolite. The dashed curve is for the pyroxene-garnet system using JEANLOZ's (1981) bulk modulus of majorite. The dotted curve is from PREM model.

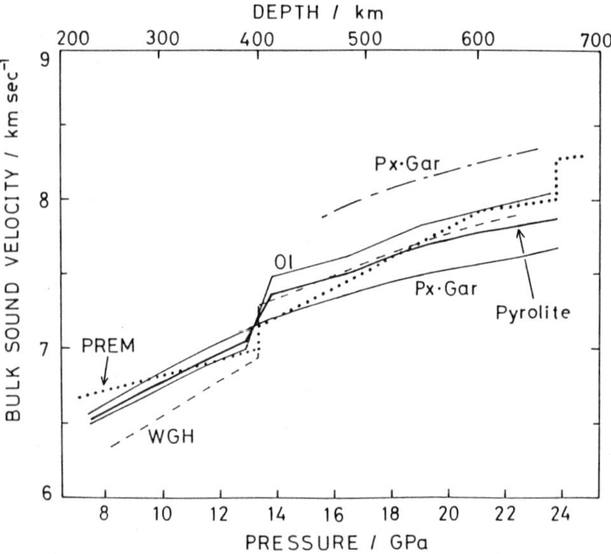

Figure 8. Bulk sound velocity profiles of the olivine and pyroxene-garnet systems and of pyrolite. The dashed curve is for the pyroxene-garnet system using JEANLOZ's (1981) bulk modulus of majorite. WGH is V_ϕ profile calculated from V_p of WALCK (1984) and V_s of GRAND and HELMBERGER (1984). The dotted curve is from PREM model.

PREM by DZIEWONSKI and ANDERSON (1981). Although the PREM model has a discontinuity at 220 km in the mantle, this sharp discontinuity is not clearly observed at 220 km for most of seismic velocity models. The velocity distribution above 220 km varies in different models. Therefore, we compare the calculated bulk sound velocity and density profiles with PREM in the depth range of only 220–670 km. As shown in Figure 7, the density of the pyroxene-garnet system in pyrolite gradually increases in the region of 220–670 km. Small changes of slope are observed at about 450 and 620 km; the former corresponds to completion of Ca-poor pyroxene dissolution in garnet, and the latter to Ca-rich pyroxene dissolution in garnet. Although such small changes of gradient are present, no sharp density increase is shown. The pyrolite density calculated from both the olivine and pyroxene-garnet systems is generally consistent with the density profile estimated by the PREM model. In the transition zone, the ρ of pyrolite is about 1% smaller than the density of PREM, but it is still within a perturbation of the PREM density profile, as shown by BASS and ANDERSON (1984). This agreement in densities, however, does not suggest that pyrolite is the best compositional model, because density is rather insensitive to the mantle composition. The density profile of the pyroxene-garnet system calculated using a K_0 of 221 GPa for majorite (JEANLOZ, 1981) is also shown with the dashed curve in Figure 7. This curve shows considerably smaller density than the profile calculated based on a K_0 of 154 GPa for $Mg_4Si_4O_{12}$ garnet.

Figure 8 shows bulk sound velocity profiles for the olivine and pyroxene-garnet systems and that of pyrolite. The V_ϕ of the pyroxene-garnet system increases very gradually with depth, without any sharp velocity change, because of gradual dissolution of pyroxenes in garnet. The slope V_ϕ decreases slightly with depth. These features of the bulk sound velocity distribution are very similar to the V_ϕ profiles calculated by BINA and WOOD (1984) for two eclogitic compositions. The V_ϕ profile of the olivine system has a large velocity jump at around 400 km due to the olivine-modified spinel transition, and a small velocity increase at 480–550 km caused by the modified spinel-spinel transition. The bulk sound velocity profile of pyrolite is generally compatible with the V_ϕ of PREM, but the gradient in the 400–670 km region is smaller than PREM. The bulk sound velocity at the depths 220–670 km is also tentatively calculated from the compressional-wave velocity model (GCA) by WALCK (1984) and the shear-wave velocity model (TNA) by GRAND and HELMBERGER (1984). Both velocity models are for the similar tectonic regions in North and Central America and are similar in general character, particularly in the region deeper than 300 km. It is interesting that the calculated V_ϕ profile (WGH in Figure 8) from the P- and S-velocity models has a smaller gradient in the transition zone than PREM and is very close to the calculated profile of pyrolite. The velocity jump of pyrolite at about 400 km is larger than that of PREM. However, the velocity increment strongly depends not only on the amount of olivine in the mantle but also on the pressure and temperature derivatives of bulk moduli of $(Mg, Fe)_2SiO_4$ polymorphs.

When the bulk modulus (221 GPa) determined by JEANLOZ (1981) for majorite is used, the calculated V_ϕ for the pyroxene-garnet system in the transition zone is about 0.6–0.7 km/s higher than the profile based on our bulk modulus (154 GPa) of $Mg_4Si_4O_{12}$ garnet, as shown in Figure 8. Therefore, the bulk sound velocity of pyrolite calculated from Jeanloz's K_0 is systematically higher by about 0.2–0.3 km/s than the V_ϕ of PREM in the transition zone. This result is compatible with calculations by BASS and ANDERSON (1984). ANDERSON and BASS (1984) and BASS and ANDERSON (1984) compared the PREM model with the compressional (V_p) and shear (V_s) wave velocities and densities calculated for pyrolite and picritic eclogite (piclogite) compositions. Piclogite consists of 44 wt% Ca-rich pyroxene, 37% garnet, 16% olivine, and 3% Ca-poor pyroxene at upper mantle P, T conditions (BASS and ANDERSON, 1984). Bass and Anderson showed that the V_p and V_s for pyrolite are about 0.2–0.4 km/s higher than estimated by PREM and do not match the high velocity gradient in the PREM model in the depth interval 400–670 km. However, they indicated that the V_p and V_s for piclogite closely match the observed seismic velocities. However, it should be noted that Bass and Anderson's calculations were based on the high bulk modulus for majorite determined by Jeanloz. When the K_0 for majorite from this study is used, the bulk sound velocity of pyrolite in the transition zone is reduced and is found to be much more consistent with PREM; it is very close to WGH as shown in Figure 8. We therefore suggest that pyrolite is still an acceptable model for depths of 220–670 km, in contrast with Bass and Anderson's conclusions. It is also noteworthy that the V_ϕ estimated for the pyroxene-garnet system using our K_0 for $Mg_4Si_4O_{12}$ garnet is lower in absolute value and less steep in slope than PREM in the transition zone, although solid solubility of pyroxenes in garnet increases with depth. For different ratios of Ca-poor pyroxene, Ca-rich pyroxene, and garnet of model mineral assemblages (for example, piclogite), the V_ϕ in the transition zone is expected to be smaller and have a shallower slope than PREM, when the small bulk modulus of majorite is adopted. Therefore, if a very high velocity gradient in the transition zone, like in PREM, is a strong constraint in the mantle velocity profiles, another broad phase transition or continuous compositional change ranging over the entire region of 400–670 km would be required.

Finally, it should be noted that the above calculations of densities and bulk sound velocities are based in part on the estimated, not measured, elastic properties of several phases, including pressure and temperature derivatives of bulk moduli of modified spinel, spinel, and majorite. Since no experimental data on shear modulus of majorite have been reported, calculation of V_p and V_s profiles of the pyroxene-garnet system is still less accurate than that of V_ϕ. Further experimental measurements of these elastic properties are indispensable to put tighter constraints to the chemistry and mineralogy of the mantle.

Acknowledgments. We thank H. Arashi for providing the glasses used and H. Tamai for his assistance in the synthesis of the garnet samples. We are grateful to O. Shimomura for his support in high-pressure experiments using synchrotron radiation, to M. Kanzaki for the preprint of his paper, and to D. L. Anderson, E. Ito, and D. J. Weidner for their helpful comments and discussions. We also thank E. Ito and two anonymous reviewers for their constructive review. M. Akaogi is grateful to Yoshida Science Foundation for paying his travel expenses. This work was supported by a National Science Foundation grant (DMR 8521562) and by Grants-in-Aid for Scientific Research from the Ministry of Education, Science and Culture, Japan (Nos. 60840023 and 60540519).

REFERENCES

AKAOGI, M., and S. AKIMOTO, Pyroxene-garnet solid solution equilibria in the systems $Mg_4Si_4O_{12}$-$Mg_3Al_2Si_3O_{12}$ and $Fe_4Si_4O_{12}$-$Fe_3Al_2Si_3O_{12}$ at high pressures and temperatures, *Phys. Earth Planet. Inter.*, *15*, 90–106, 1977.

AKAOGI, M., and S. AKIMOTO, High-pressure phase equilibria in a garnet lherzolite, with special reference to $Mg^{2+}-Fe^{2+}$ partitioning among constituent minerals, *Phys. Earth Planet. Inter.*, *19*, 31–51, 1979.

AKAOGI, M., and A. NAVROTSKY, The quartz-coesite-stishovite transformations: new calorimetric measurements and calculation of phase diagrams, *Phys. Earth Planet. Inter.*, *36*, 124–134, 1984.

AKAOGI, M., N. L. ROSS, P. MCMILLAN, and A. NAVROTSKY, The Mg_2SiO_4 polymorphs (olivine, modified spinel, and spinel): thermodynamic properties from oxide melt solution calorimetry, phase relations, and models of lattice vibrations, *Am. Mineral.*, *69*, 499–512, 1984.

AKIMOTO, S., High-pressure research in geophysics: past, present and future, this volume.

AKIMOTO, S., T. YAGI, and K. INOUE, High temperature-pressure phase boundaries in silicate systems using in situ X-ray diffraction, in *High-Pressure Research: Applications in Geophysics*, edited by M. H. Manghnani and S. Akimoto, pp. 585–602, Acad. Press, New York, 1977.

ANDERSON, D. L., and J. D. BASS, Mineralogy and composition of the upper mantle, *Geophys. Research Lett.*, *11*, 637–640, 1984.

BABUŠKA, V., J. FIALA, M. KUMAZAWA, I. OHNO, and Y. SUMINO, Elastic property of garnet solid solution series, *Phys. Earth Planet. Inter.*, *16*, 157–176, 1978.

BASS, J. D., and D. L. ANDERSON, Composition of the upper mantle: geophysical tests of two petrological models, *Geophys. Res. Lett.*, *11*, 229–231, 1984.

BASS, J. D., and D. J. WEIDNER, Elasticity of single-crystal orthoferrosilite, *J. Geophys. Res.*, *89*, 4359–4371, 1984.

BASS, J. D., R. C. LIEBERMANN, D. J. WEIDNER, and S. J. FINCH, Elastic properties from acoustic and volume compression experiments, *Phys. Earth Planet. Inter.*, *25*, 140–158, 1981.

BINA, C. R., and B. J. WOOD, The eclogite to garnetite transition: experimental and thermodynamic constraints, *Geophys. Res. Lett.*, *11*, 955–958, 1984.

BLOCK, S., Round-robin study of the high pressure phase transition in ZnS, *Acta Cryst.*, *A 34*, *Suppl.*, 316, 1978.

BONCZAR, L. J., E. K. GRAHAM, and H. WANG, The pressure and temperature dependence of the elastic constants of pyrope garnet, *J. Geophys. Res.*, *82*, 2529–2534, 1977.

BROWN, J. M., and T. J. SHANKLAND, Thermodynamic parameters in the earth as determined from seismic profiles, *Geophys. J. R. Astron. Soc.*, *66*, 579–596, 1981.

DZIEWONSKI, A. M., and D. L. ANDERSON, Preliminary reference earth model, *Phys. Earth Planet. Inter.*, *25*, 297–356, 1981.

GASPARIK, T., and R. C. NEWTON, The reversed alumina contents of orthopyroxene in equilibrium with spinel and forsterite in the system MgO-Al_2O_3-SiO_2, *Contr. Min. Petrol.*, *85*, 186–196, 1984.

GRAND, S. P., and D. V. HELMBERGER, Upper mantle shear structure of North America, *Geophys. J. R. Astron. Soc.*, *76*, 399–438, 1984.

HANIFORD, V. M., and D. J. WEIDNER, Elastic properties of grossular-andradite garnets $Ca_3(Al, Fe)_2Si_3O_{12}$, *EOS, Trans. AGU*, *66*, 1063, 1985.

HORIUCHI, H., and H. SAWAMOTO, β-Mg_2SiO_4: single crystal X-ray diffraction study, *Am. Mineral.*, *66*, 568–575, 1981.

ITO, E., and A. NAVROTSKY, $MgSiO_3$ ilmenite: calorimetry, phase equilibria, and decomposition at atmospheric pressure, *Am. Mineral.*, *70*, 1020–1026, 1985.

ITO, H., K. KAWADA, and S. AKIMOTO, Thermal expansion of stishovite, *Phys. Earth Planet. Inter.*, *8*, 277–281, 1974, Erratum, ibid., *9*, 371, 1974.

JEANLOZ, R., Majorite: vibrational and compressional properties of a high-pressure phase, *J. Geophys. Res.*, *86*, 6171–6179, 1981.

JEANLOZ, R., and A. B. THOMPSON, Phase transitions and mantle discontinuities, *Rev. Geophys. Space Phys.*, *21*, 51–74, 1983.

KANZAKI, M., Ultrahigh-pressure phase relations in the system $Mg_4Si_4O_{12}$-$Mg_3Al_2Si_3O_{12}$, *Phys. Earth Planet. Inter*, in press, 1987.

KATO, T., and M. KUMAZAWA, Garnet phase of $MgSiO_3$ filling the pyroxene-ilmenite gap at very high temperature, *Nature*, *316*, 803–805, 1985.

KAWADA, K., The system Mg_2SiO_4-Fe_2SiO_4 at high pressures and temperatures and the earth's interior, Ph.D. thesis, Univ. of Tokyo, Tokyo, 187 pp., 1977.

LEES, A. C., M. S. T. BUKOWINSKI, and R. JEANLOZ, Reflection properties of phase transition and compositional change models of 670-km discontinuity, *J. Geophys. Res.*, *88*, 8145–8159, 1983.

LEITNER, B. J., D. J. WEIDNER, and R. C. LIEBERMANN, Elasticity of single-crystal pyrope and implications for garnet solid solution series, *Phys. Earth Planet. Inter.*, *22*, 111–121, 1980.

LEVIEN, L., C. T. PREWITT, and D. J. WEIDNER, Compression of pyrope, *Am. Mineral.*, *64*, 805–808, 1979a.

LEVIEN, L., D. J. WEDNER, and C. T. PREWITT, Elasticity of diopside, *Phys. Chem. Min.*, *4*, 105–113, 1979b.

LIU, L. G., The system enstatite-pyrope at high pressures and temperatures and the earth's mantle, *Earth Planet. Sci. Lett.*, *36*, 237–245, 1977.

NAVROTSKY, A., Progress and new directions in high temperature calorimetry, *Phys. Chem. Min.*, *2*, 89–104, 1977.

NAVROTSKY, A., F. S. PINTCHOVSKI, and S. AKIMOTO, Calorimetric study of the stability of high-pressure phases in the systems CoO-SiO_2 and "FeO"-SiO_2 and calculation of phase diagrams in MO-SiO_2 systems, *Phys. Earth Planet. Inter.*, *19*, 275–292, 1979.

PERKINS, D., T. J. B. HOLLAND, and R. C. NEWTON, The Al_2O_3 contents of enstatite in equilibrium with garnet in the system MgO-Al_2O_3-SiO_2 at 15–40 kbar and 900–1600 °C, *Contr. Min. Petrol.*, *78*, 99–109, 1981.

PIERMARINI, G. J., and S. BLOCK, Ultrahigh pressure diamond-anvil cell and several semi-conductor phase transition pressures in relation to the fixed point pressure scale, *Rev. Sci. Instrum.*, *46*, 973–979, 1975.

RINGWOOD, A. E., The pyroxene-garnet transformation in the earth's mantle, *Earth Planet. Sci. Lett.*, *2*, 255–263, 1967.

RINGWOOD, A. E., *Composition and Petrology of the Earth's Mantle*, pp. 488, McGraw-Hill, New York, 1975.

ROBIE, R. A., B. S. HEMINGWAY, and J. R. FISHER, Thermodynamic properties of minerals and related substances at 298.15 K and 1 bar (10^5 Pascals) pressure and at higher temperatures, *U.S. Geol. Surv. Bull. 1452*, 456 pp., 1978.

SASAKI, S., C. T. PREWITT, Y. SATO, and E. ITO, Single crystal X-ray study of γ-Mg_2SiO_4, *J. Geophys. Res.*, *87*, 7829–7832, 1982.

SATO, Y., M. AKAOGI, and S. AKIMOTO, Hydrostatic compression of synthetic garnets pyrope and almandine, *J. Geophys. Res.*, *83*, 335–338, 1978.

SAWAMOTO, H., Phase equilibrium of $MgSiO_3$ under high pressures and high temperatures: garnet-perovskite transformation (abstract), paper presented at U.S.-Japan Seminar, High-Pressure Research: Application in Geophysics and Geochemistry, Honolulu, Hawaii, pp. 50–51, 1986.

SAWAMOTO, H., D. J. WEIDNER, S. SASAKI, and M. KUMAZAWA, Single-crystal elastic properties of the modified spinel (beta) phase of Mg_2SiO_4, *Science*, *224*, 749–751, 1984.

SKINNER, B. J., Thermal expansion, in *Handbook of Physical Constants*, edited by S. P. Clark, Geol. Soc. Am., *97*, pp. 75–96, 1966

SUITO, K., Phase relations of pure Mg_2SiO_4 up to 200 kilobars, in *High-Pressure Research: Applications in Geophysics*, edited by M. H. Manghnani and S. Akimoto, pp. 255–266, Acad. Press, New York, 1977.

SUZUKI, I., E. OHTANI, and M. KUMAZAWA, Thermal expansion of γ-Mg_2SiO_4, *J. Phys. Earth*, *27*, 53–61, 1979.

SUZUKI, I., E. OHTANI, and M. Kumazawa, Thermal expansion of modified spinel, β-Mg_2SiO_4, *J. Phys. Earth*, *28*, 273–280, 1980.

SUZUKI, T., T. YAGI, and S. AKIMOTO, Precise determination of transition pressure of GaAs (abstract), paper presented at 22th High Pressure Conf. Japan, pp. 8–9, 1981.

WALCK, M. C., The P-wave upper mantle structure beneath an active spreading centre: the Gulf of California, *Geophys. J. R. Astron. Soc.*, *76*, 697–723, 1984.

WEIDNER, D. J., A mineral physics test of a pyrolite mantle, *Geophys. Res. Lett.*, *12*, 417–420, 1985.

WEIDNER, D. J., J. D. BASS, A. E. RINGWOOD, and W. SINCLAIR, The single crystal elastic moduli of stishovite, *J. Geophys. Res.*, *87*, 4740–4746, 1982.

WEIDNER, D. J., H. SAWAMOTO, S. SASAKI, and M. KUMAZAWA, Single-crystal elastic properties of spinel phase of Mg_2SiO_4, *J. Geophys. Res.*, *89*, 7852–7860, 1984.

YAGI, T., and S. AKIMOTO, Direct determination of coesite-stishovite transition by in situ X-ray diffraction, *Tectonophys.*, *35*, 259–270, 1976.

YAGI, T., M. AKAOGI, O. SHIMOMURA, H. TAMAI, and S. AKIMOTO, High pressure and high temperature equations of state of majorite, this volume.

SILICATE AND GERMANATE GARNETS, ILMENITES AND PEROVSKITES: THERMOCHEMISTRY, LATTICE VIBRATIONS, AND SPECTROSCOPY

Alexandra NAVROTSKY

Department of Geological and Geophysical Sciences, Princeton University
Princeton, New Jersey 08544, USA

Abstract. Thermochemical studies by high-temperature calorimetry have been performed on the high-pressure polymorphs of $CaGeO_3$, $CdGeO_3$, $MgGeO_3$, $MnSiO_3$, and on $MgSiO_3$ ilmenite. The enthalpies of transition show systematic trends. The enthalpies of transition among pyroxene and pyroxenoid polymorphs are uniformly small. The pyroxenoid (or pyroxene) to garnet transition has a $\Delta H°$ of less than 10 kJ mol^{-1} in magnitude for $CaGeO_3$ and $CdGeO_3$, 36 kJ mol^{-1} in $MnSiO_3$ and (estimated) 50–60 kJ mol^{-1} in $MgSiO_3$. The entropies of these transitions, obtained both by combining calorimetry and phase studies and from lattice-vibrational modelling using Kieffer's approach, are small and negative, confirming shallow positive *P-T* slopes. The transition from pyroxene to ilmenite occurs stably in $MgGeO_3$, with $\Delta H°=6$ kJ and metastably (with β+stishovite or γ+stishovite as intervening phases) in $MgSiO_3$ with $\Delta H°=70$ kJ. It has small negative $\Delta S°$ values but $MgSiO_3$ ilmenite may have a slightly higher entropy than a mixture of spinel and stishovite. The transition pyroxenoid (or pyroxene) to perovskite has $\Delta H°$ near 40 kJ mol^{-1} for $CaGeO_3$, $CdGeO_3$, and $MgGeO_3$. The garnet to perovskite transition has a positive $\Delta S°$ for $CaGeO_3$. The ilmenite to perovskite transition in $CdGeO_3$ has a positive $\Delta S°$, as is probably also the case for the ilmenite to perovskite transition in $MgSiO_3$. The vibrational modelling supports this conclusion and the idea that perovskite-forming transitions have negative *P-T* slopes.

Introduction

The transformation of silicates and their analogues from olivine and pyroxene structures to high-pressure phases form the basis of a family of phase transitions probably responsible for the seismic variations separating upper mantle, transition zone, and lower mantle. In addition to direct synthesis and phase relations determined in situ or on quenched samples, thermodynamic characterization by high-temperature calorimetry of quenched samples has proved very useful in delineating the systematics of these transitions (NAVROTSKY, 1979). At the last U.S.–Japan seminar, we presented data relevant to the olivine-spinel transition (NAVROTSKY et al., 1982). This paper is a review of our work on "post-spinel" phases of stoichiometry $MSiO_3$ or $MGeO_3$, where M is a divalent ion, with emphasis on transitions from pyroxenoid or pyroxene structures to garnet, ilmenite, and perovskite structures or to isochemical binary oxide mixtures. In addition, this contribution discusses the application of vibrational modelling to estimate the entropies of these transitions.

Calorimetric Results and Phase Relations

$MgSiO_3$

The enthalpy, $\Delta H°$, of the pyroxene→ilmenite (px→il) reaction in $MgSiO_3$ was measured (ITO and NAVROTSKY, 1985) using a combination of drop and solution calorimetry on less than 15 mg of $MgSiO_3$ (il). The results are shown in Table 1. Three sets of internally consistent data for enthalpy, $\Delta H°$; entropy, $\Delta S°$; and volume, $\Delta V°$; for reactions involving the formation of $MgSiO_3$(il) from Mg_2SiO_4 (modified spinel, β, or spinel, γ)+SiO_2 (stishovite, st) were generated using slightly different assumptions concerning thermochemical data and the location of the β-γ boundary. These three calculations illustrate the effect of uncertainties in the data on the calculated phase relations (Figures 1 a–c).

The computed diagrams show several common features. a) The direct transition of $MgSiO_3$ (px) to $MgSiO_3$ (il) is metastable at temperatures of 1000–1700 K. The single phase field (px) is separated from the single phase field (il) by two-phase fields (γ+st or β+st). The thermochemical data suggest a triple point (px, il, β+st) to occur near 16–17 GPa and a temperature between about 1700 and 1900 K. At higher temperatures pyroxene could transform to ilmenite directly (unless melting or transformation to other phases occurred first). b) A triple point (β+st, γ+st, px) is suggested to occur near 1000 K and 14 GPa. Its exact location is sensitive to the precise position of the β-γ boundary. c) Another triple point (β+st, γ+st, il) is suggested near 1400 K and 18 GPa. The two triple points are related and cannot be moved independently of each other in *P-T* space by varying the thermodynamic parameters. d) The boundary γ+st→il appears to have a strongly negative *P-T* slope. The boundary β+st→il also appears to have a somewhat negative slope. It is these negative slopes which lead to the predicted (px, il, β+st) triple point. Other phase boundaries have positive *P-T* slopes, with the β-γ boundary being the steepest (and probably most uncertain). The topology of the phase relations is constrained by the thermochemical data although the exact location of the triple points is subject to relatively large error limits.

Recently, E. ITO (unpublished) performed synthesis

High-Pressure Research in Mineral Physics, edited by M. H. Manghnani and Y. Syono, pp. 261–268.

TABLE 1. Thermochemical Data for Reactions Relevant to $MgSiO_3$ Transitions

	Data Set	ΔH° (kJ mol^{-1})	ΔS° (J mol^{-1} K^{-1})	ΔV° (cm^3 mol^{-1})
$Mg_2SiO_4(\beta) \rightarrow Mg_2SiO_4(\gamma)$	A[a]	6.8	− 9.3	− 1.10
	B	6.8	− 6.3	− 0.89
	C	6.8	− 9.3	− 1.10
$2MgSiO_3(px) = Mg_2SiO_4(\beta) + SiO_2(st)$	A	93.1	− 23.6	− 7.99
	B	93.1	− 21.8	− 8.03
	C	93.1	− 23.6	− 7.99
$2MgSiO_3(px) \rightarrow Mg_2SiO_4(\gamma) + SiO_2(st)$	A	99.9	− 32.9	− 9.09
	B	99.9	− 28.0	− 8.92
	C	99.9	− 32.9	− 9.09
$Mg_2SiO_4(\gamma) + SiO_2(st) = 2MgSiO_3(il)$	A	43.7	+ 20.5	− 0.79
	B	43.7	+ 18.6	− 0.96
	C	31.2	+ 11.3	− 0.79
$Mg_2SiO_4(\beta) + SiO_2(st) = 2MgSiO_3(il)$	A	50.6	+ 11.2	− 1.89
	B	50.6	+ 12.0	− 1.85
	C	38.0	+ 2.0	− 1.89[c]
$MgSiO_3(px) \rightarrow MgSiO_3(il)$	A	71.8	− 6.2	− 4.94
	B	71.8	− 4.9	− 4.94
	C	65.5	− 10.8	− 4.94

[c] data sets A and B refer to slightly different values of parameters for the β-γ transition, data set C uses a value of ΔH°(px→il) two standard deviations lower than the average value. See ITO and NAVROTSKY (1985).

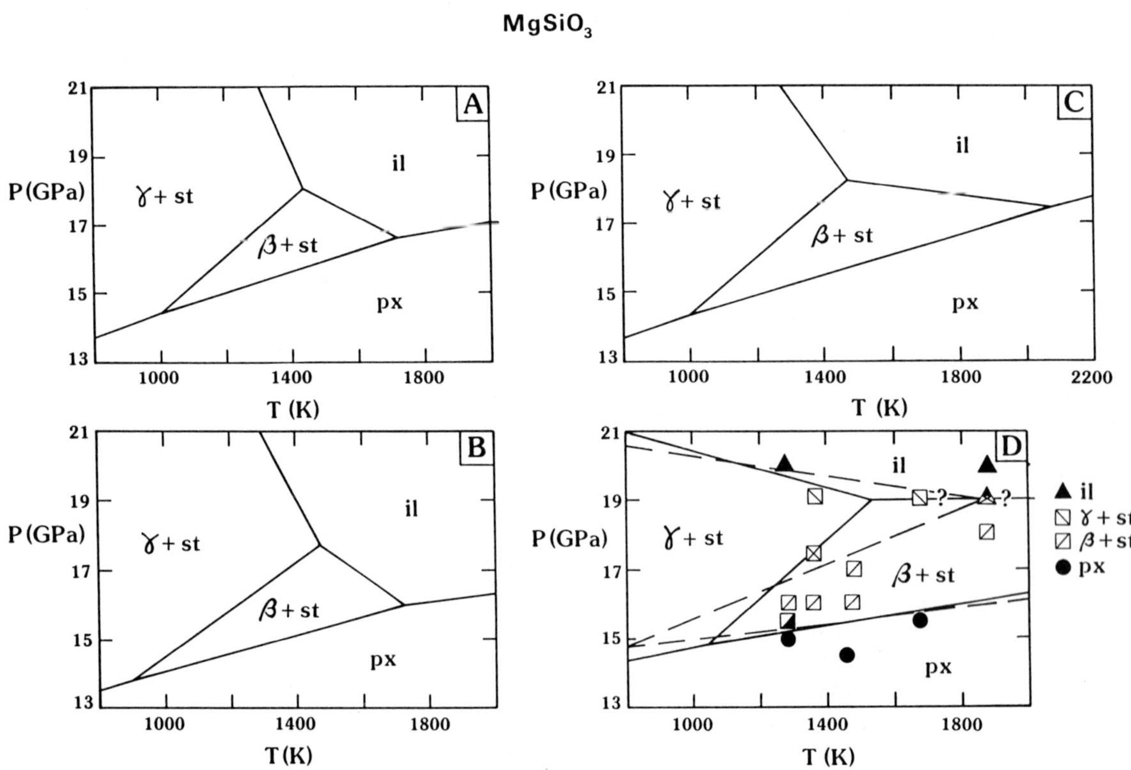

Figure 1. Phase relations at $MgSiO_3$ composition. Figures 1a, b, and c correspond to calculations based on data sets A, B, and C in Table 1. Figure 1d shows experimental data. See ITO and NAVROTSKY (1985).

experiments, using a uniaxial split-sphere apparatus, on this system. The results are shown in Figure 1d. Two possible sets of phase boundaries are sketched. Solid lines are consistent with SUITO's (1977) β-γ boundary. Dashed lines show a β-γ boundary consistent with Ito's synthesis data but having a slope much shallower than found in previous studies. The two points which suggest that shallower slope are annotated with question marks.

A comparison of Ito's experimental data and the thermochemical calculations show them to be generally consistent with respect to overall topology. The only significant discrepancy is the following. Although Ito suggests a slightly negative dP/dT for the transition γ+st→il, the thermochemistry suggests a much more strongly negative slope. The experimental data suggest that the triple point (px, il, β+st), if it exists at all, would lie at much higher temperatures than the calculations imply. Thus, although both high-pressure synthesis and calorimetry suggest that ilmenite is a phase of relatively high entropy, the exact value of the positive $\Delta S°$ of its formation from low-pressure phase assemblages is in doubt. Given the uncertainties in both the high-pressure work and the thermochemical calculations, one cannot say whether the experimental or the calculated phase relations is likely to be quantitatively closer to the "real" phase diagram.

Recently calorimetric studies of $MgSiO_3$-rich garnets on the enstatite-pyrope join (AKAOGI et al., this volume) permit an estimate of the enthalpy of the enstatite to garnet transition in $MgSiO_3$ as ~37 kJ mol^{-1}. Although $MgSiO_3$ garnet, if it exists, is limited to very high temperatures (1700–2800 K) and pressures (17–24 GPa) (KATO and KUMAZAWA, 1985), the calorimetric data suggest it to be energetically comparable to the spinel plus stishovite or beta plus stishovite assemblage.

$MnSiO_3$

Calorimetric studies were performed in $MnSiO_3$ rhodonite, pyroxmangite, clinopyroxene, and garnet (AKAOGI and NAVROTSKY, 1985). Thermodynamic data for the transition to garnet are summarized in Table 2. $\Delta H°$, $\Delta S°$, $\Delta V°$, and $\Delta G°$ for the transitions among the three chain-silicate polymorphs are very small. Since $\Delta H°$ and $T\Delta S°$ are comparable in magnitude at temperatures of 900–1500 K, stability fields of rhodonite, pyroxmangite, and clinopyroxene are determined by very finely balanced energetic and entropic factors. The data suggest that, in the order rhodonite, pyroxmangite, and clinopyroxene (increasing number of tetrahedra in the repeat unit), enthalpy increases while entropy and volume decrease, although the uncertainties in $\Delta H°$ and $\Delta S°$ are relatively large. The enthalpy of the clinopyroxene-garnet transition is 34.6±2.5 kJ mol^{-1}. Using a point on the experimental phase boundary and estimates of thermal expansion and compressibility, we calculate $\Delta S°$ to be −6.3±2.1 J K^{-1} mol^{-1} and dP/dT to be 1.5 (±0.7)×10^{-3} GPa K^{-1}. The calculated phase boundary is compared to experiment in Figure 2. The calculated boundary satisfies all the data points in AKIMOTO and SYONO's (1972) unreversed synthesis experiments.

Enthalpy and entropy data for $MnSiO_3$ garnet allow calculation of the boundary of the decomposition reaction of garnet into MnO plus stishovite and that of Mn_2SiO_4 olivine into garnet plus MnO. The results are shown in Figure 2. ITO et al. (1974) and ITO and MATSUI (1977) observed that Mn_2SiO_4 olivine disproportionates

TABLE 2. Thermodynamics of ABO_3 Phase Transitions

	$\Delta H°$(kJ mol^{-1})	$\Delta S°$(kJ mol^{-1} K^{-1})	$\Delta V°$(cm^{-1} mol^{-3})
Pyroxene or pyroxenoid→garnet			
$CaGeO_3$	− 4.9 ± 4.2	− 5.9 ± 1.5	− 5.97
$CdGeO_3$	+ 0.5 ± 2.7	− 8.4 ± 2.0	− 5.30
$MnSiO_3$	+ 34.6 ± 2.5	− 6.7 ± 2.1	− 4.00
$MgSiO_3$	+ 37.0 ± 3.0		
garnet→perovskite			
$CaGeO_3$	+ 43.3 ± 5.0	+ 10.9 ± 3.8	− 5.35
$CdGeO_3$	+ 43.1 ± 5.0	− 1.7 ± 3.0	− 4.88
$MgSiO_3$	(83)		
pyroxene→ilmenite			
$MgGeO_3$	+ 7.5 ± 0.6	− 6.3	− 5.11
$MgSiO_3$	+ 71.8 ± 6.3	− 10.9 ± 4.0	− 4.94
ilmenite→perovskite			
$CdTiO_3$	+ 15.0 ± 0.8	+ 14.2	− 2.94
$CdGeO_3$	+ 34.3 ± 4.0	+ 2.6 ± 2.0	− 3.00
$MgSiO_3$	(48.2)	positive	− 1.90

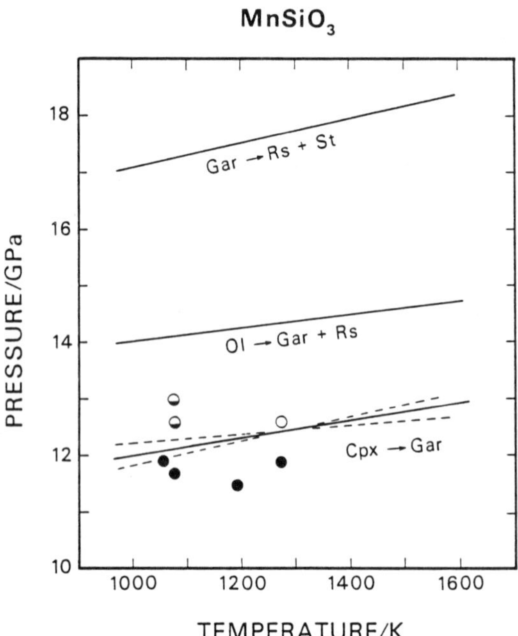

Figure 2. Phase relations in manganese silicates. Lines are calculated from thermochemical data. See AKAOGI and NAVROTSKY (1985).

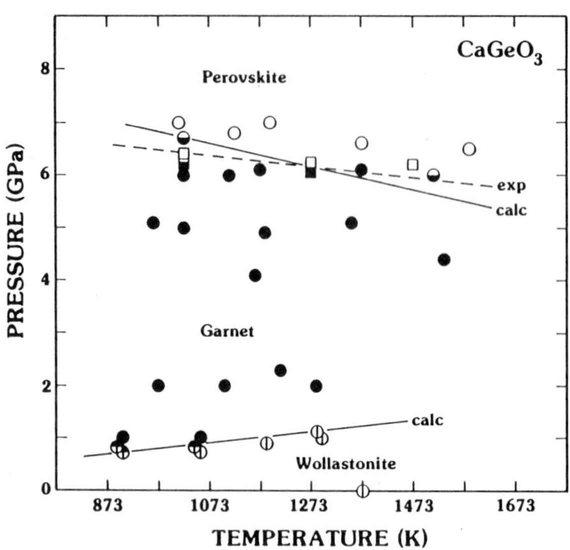

Figure 3. Phase relations in $CaGeO_3$. See ROSS et al. (1986).

completely into $MnSiO_3$ garnet plus MnO at approximately 14 GPa at about 1273 K, but did not report a value for the *P-T* slope. The calculated boundary, with an uncertainty of about 2 GPa, agrees well with the experimentally determined transition pressure. Figure 2 indicates that olivine decomposes into garnet plus MnO at pressures higher by about 2 GPa than the pressure at which the garnet phase is first stabilized at $MnSiO_3$ composition. A mixture of 2MnO plus SiO_2 (stishovite) is the assemblage for a Mn_2SiO_4 composition which is stable at pressures higher than the garnet decomposition boundary. It is interesting that no high-pressure phases of stoichiometry Mn_2SiO_4 (e.g. β or γ) appear to exist.

The uncertainty in absolute pressures for the transitions shown in Figure 2 is about ± 2 GPa. However, the position of the phase boundaries relative to each other is considerably better constrained. The thermochemical data require the sequence of reactions shown and strongly suggest shallow positive slopes. The boundaries do not intersect in the accessible temperature range.

$CaGeO_3$ and $CdGeO_3$

These systems are noteworthy for their wealth of phases (wollastonite, garnet, ilmenite, and perovskite). Thermodynamic data (ROSS et al., 1986; AKAOGI and NAVROTSKY, 1987) are summarized in Table 2 and the calculated and experimental phase relations are shown in Figures 3 and 4. Several similarities are evident in the behavior of

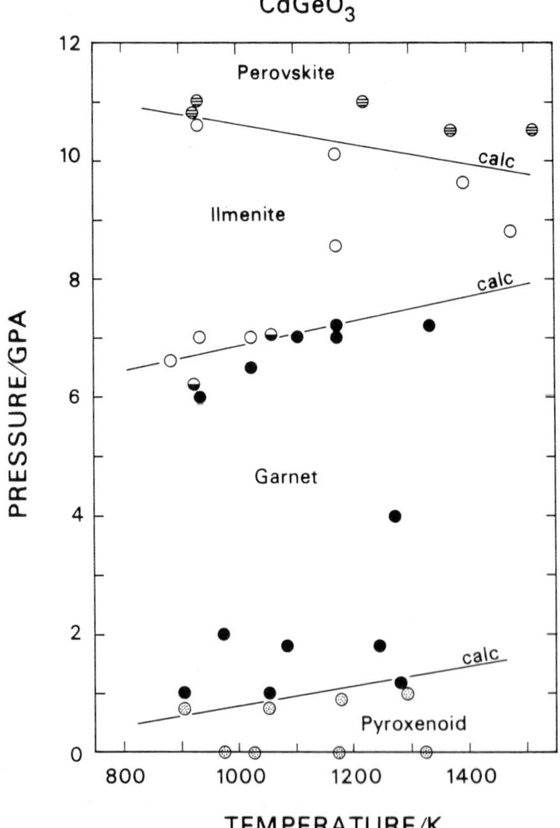

Figure 4. Phase relations in $CdGeO_3$ from AKAOGI and NAVROTSKY (in preparation). Lines are calculated from thermochemical data.

$CaGeO_3$ and $CdGeO_3$. In both the garnet phase is energetically virtually equivalent to the pyroxenoid but it has a slightly lower entropy, resulting in a shallow positive dP/dT for the transition. The perovskite, on the other hand, is a phase of considerably higher enthalpy but also of higher entropy. The reaction boundary for the formation of perovskite (garnet→perovskite in $CaGeO_3$, ilmenite→perovskite in $CdGeO_3$) has a small negative dP/dT.

$MgGeO_3$

Thermochemical data are summarized in Table 2. The thermodynamics of the various pyroxene (ortho, clino, etc.) polymorphs is still unclear (KIRFEL and NEUHAUS, 1974; OZIMA and AKIMOTO, 1983; ROSS, 1985) and will not be discussed further here. The transition from pyroxene to ilmenite has a ΔH_{973} of 7.52 kJ mol^{-1}. Calculation of the phase boundary using that value and an equilibrium pressure near 3–5 GPa at 973 K (KIRFEL and NEUHAUS, 1974) gives a much shallower slope than that reported by those authors, see Figure 5. Such a shallow slope may be supported by the data of RINGWOOD and SEABROOK (1963) and RINGWOOD and MAJOR (1968) (triangles in Figure 5). This system contains major unresolved discrepancies between experimental high pressure results and thermochemical predictions. This is the only $MO–SiO_2$ or $MO–GeO_2$ system studied thus far where the difference between high-pressure and thermochemical results is clearly beyond the estimated uncertainties in each. Problems in the kinetics of transformation and of phase retention during quenching are one possible explanation. This problem needs further study.

Energy Systematics

Figure 6 shows the enthalpies of various high-pressure $MSiO_3$ and $MGeO_3$ phases relative to their atmospheric-pressure polymorphs. In the germanates, the wollastonite and garnet phases are very similar in energy; indeed the garnet is a high-pressure phase not so much because of unfavorable energetics but because of its lower entropy and volume. $MnSiO_3$ and $MgSiO_3$ garnets, in contrast, are energetically much higher than the pyroxenes, with $MgSiO_3$ garnet estimated to be comparable in energy to the assemblage spinel plus stishovite. $MgSiO_3$ ilmenite is much higher in energy relative to pyroxene than is $MgGeO_3$ ilmenite. The perovskites $CaGeO_3$, $CdGeO_3$, and $MgSiO_3$ are all energetically rather unfavorable phases.

Vibrational Entropies: Systematics and Models

Entropies of transition obtained from a combination of calorimetry and phase equilibria are summarized in Table 2. Transitions among α-, β-, and γ-M_2SiO_4 polymorphs are characterized by small negative $\Delta S°$ values (NAVROTSKY et al., 1979; AKAOGI et al., 1984). Those

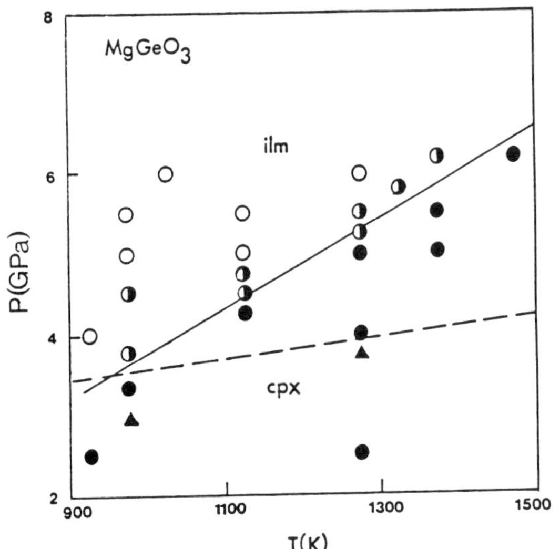

Figure 5. Phase relations in $MgGeO_3$, from ROSS (1985). Circles represent KIRFEL and NEUHAUS (1974) data and solid line is their orthopyroxene-ilmenite boundary. Triangles represent data of RINGWOOD and SEABROOK (1962) and RINGWOOD and MAJOR (1968). Dashed line is boundary calculated from thermochemical data.

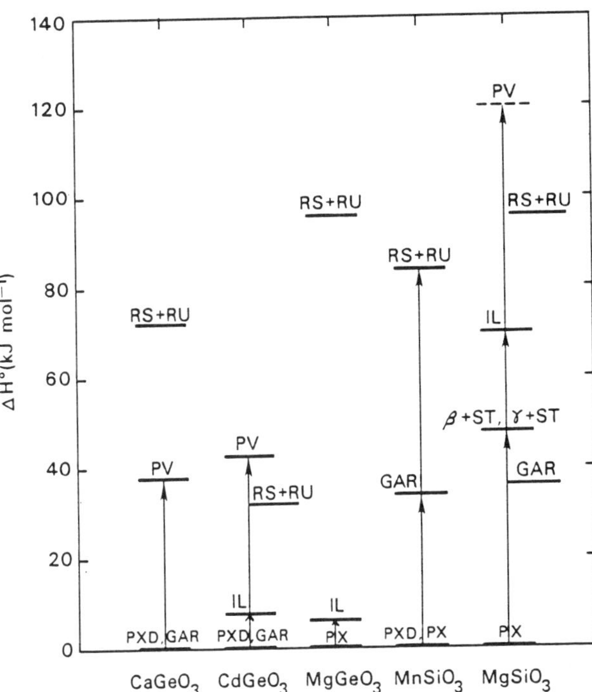

Figure 6. Enthalpies of high-pressure polymorphs relative to phases stable at 1 atm. Solid lines indicate relatively well constrained data, dashed lines indicate estimates.

involving change in coordination number show more complex entropy behavior. Coesite-stishovite transition has a negative $\Delta S°$ of -4.2 J K^{-1} (AKAOGI and NAVROTSKY, 1984). This $\Delta S°$ value is perhaps surprisingly close to zero when one considers the large volume contraction (32%). Transitions to form the post-spinel perovskite phases (γ+st→pv, gar→pv, il→pv) have positive $\Delta S°$ values and negative P-T slopes. The higher vibrational entropy of the denser phase results from a shift in the vibrational density of states toward generally lower frequencies, so that more states are populated at lower temperatures. Qualitatively, such behavior can be rationalized as follows (NAVROTSKY, 1980). In a phase such as $CaGeO_3$ or $MgSiO_3$ perovskite the large central site is occupied by a relatively small cation having rather large M–O distances, loose bonding, and, one surmises, rather easy vibration involving it and neighboring oxygens. In addition, the octahedral Si–O or Ge–O bond is longer and has a lower vibrational frequency than the strong covalent tetrahedral Si–O or Ge–O bond. Indeed, while the tetrahedral Si–O or Ge–O stretching vibrations can be identified relatively unambiguously in the vibrational spectrum, those of octahedrally coordinated Si or Ge occur amidst many other lattice vibrations and probably are not simple isolated modes (ROSS and MCMILLAN, 1984). Nor can M–O vibrations be unambiguously identified. Both these factors lead to greater excitation of vibrations at lower temperatures and a higher value of $S°$ (at 298 K and higher T) for the perovskite.

To quantify and test such reasoning one needs either complete knowledge of the vibrational density of states (generally unavailable for high-pressure phases) or models which approximate this density of states in detail sufficient for accurate thermodynamic calculations. Because the phases in question contain several atoms of very different masses and bonding tendencies, they do not act as simple Debye solids (KIEFFER, 1979a) and the Debye approximation is insufficient to calculate their thermodynamic properties. The model approach pioneered by KIEFFER (1979a, b, c, 1980) offers a useful middle ground between theoretical rigor and practical use of available data. In it the vibrational density of states is considered to consist of three contributions (see Figure 7). Three acoustic modes assumed to be Debye-like with cutoff frequencies calculated from elastic data contribute at low frequency (usually below 150 cm^{-1}). Remaining $3n-3$ vibrational modes (for a primitive unit cell containing n atoms) are optic modes. Some number of them can be relatively unambiguously assigned to specific vibrations (tetrahedral Si–O stretching modes or O–H stretching modes in hydrous phases) generally seen at high frequency ($>\sim 900$ cm^{-1}) in infrared and Raman spectra. These are treated as "Einstein oscillators," spikes in the vibrational

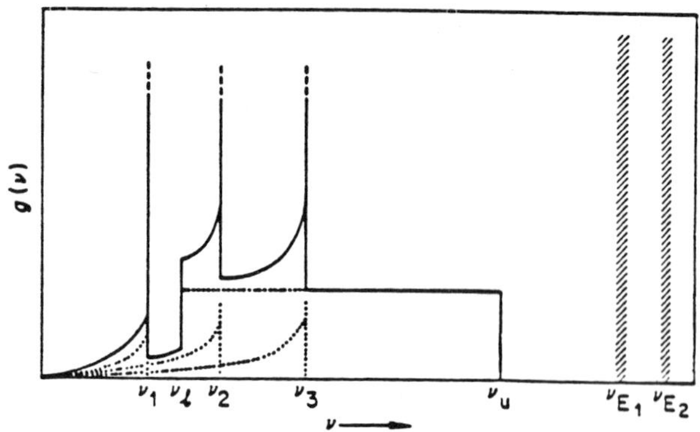

Figure 7. Schematic representation of vibrational density of states (VDOS) in the Kieffer model. The ν_1, ν_2, and ν_3 are all acoustic modes. Optic continuum spans ν_l to ν_u. Einstein oscillators are at ν_{E_1}, ν_{E_2}. Total number of modes, $\int g(\nu)d\nu$, is $3n$. In this sketch acoustic modes overlap optic continuum; this is not always the case.

density of states, each at a certain single frequency. The number of modes assigned to each Einstein oscillator is deduced from the crystal structure. The remaining modes (in the mid-infrared) generally cannot be assigned readily; they correspond to various stretches, bends, and lattice deformations. The model treats them as an "optic continuum," a constant density of states between a low- and a high-frequency cutoff. These cutoffs are chosen to be consistent with the infrared and Raman spectra of each phase. Because such spectra give information on the vibrational density of states only at the center of the Brillouin zone, some assumptions about dispersion must also be made. The approximate vibrational density of states can then be integrated to give the heat capacity at constant volume, C_v. Thus, at constant pressure, C_p is obtained from the relation $C_p = C_v + TV\alpha^2K$ (with T absolute temperature, V molar volume, α thermal expansion coefficient, and K bulk modulus). The integral of $C_p dT$ gives the enthalpy, that of $(C_p/T)dT$ the entropy (at atmospheric pressure). A comparison of $S°$ between different structures gives the entropy of transition. The assumptions inherent in these calculations have been discussed previously (KIEFFER, 1979a, b, c, 1980; AKAOGI et al., 1984).

This approach is particularly applicable to high-pressure phases. First, the input data needed to constrain the model are space group, molar volume, elastic constants, bulk modulus, thermal expansion, and vibrational (IR and Raman) spectra. These data are all feasible to obtain for ultrahigh pressure phases and, indeed, are of interest to geophysics for other reasons. Second, the models work best when comparing phases with quite different crystal structures and vibrational spectra. Then, although one does not need to assign all vibrational bands, one can see how changes in the spectra (reflecting structural changes) influence the entropy. Third, for hypothetical high-pressure phases, or those not readily quenched to ambient conditions for detailed measurements, one can judiciously invent or "synthesize" a likely vibrational density of states based on crystal chemical systematics. One can then use this to calculate an entropy and then to decide whether this phase is likely to appear, for example, at high temperature and high pressure. The vibrational calculations are illustrated by the $CaGeO_3$ polymorphs (wollastonite, garnet, and perovskite) (NAVROTSKY, 1985; ROSS et al., 1986). On the basis of vibrational spectra, a set of models can be constructed for each phase. Though the models differ in detail, each set constrains the entropy of that polymorph rather closely (ROSS et al., 1986). A representative model and its calculated thermodynamic parameters is shown in Figure 8 for each phase, wollastonite, garnet, and perovskite. From the models one concludes the following. The wollastonite is a phase of higher entropy (at 400–1500 K) than the garnet primarily because the optic continuum of the wollastonite is generally shifted to lower frequencies than that of garnet; in particular there are more vibrations below 200 cm^{-1} which lead to higher S°_{298} since they are readily excited at low temperature. This effect is partially offset because the garnet does not have as large a fraction of its vibrations at >800 cm^{-1} and in part because half its Ge is in octahedral coordination with vibrations hidden in the optic continuum. The perovskite, on the other hand, is a higher entropy phase than the garnet because, with no tetrahedral Ge, it has a higher density of states in the optic continuum and a fairly low cutoff at the low frequency end. The calculations confirm a negative ΔS° and positive dP/dT for the wollastonite-garnet transition and a positive ΔS° and negative dP/dT for the garnet-perovskite transition in $CaGeO_3$.

The fact that $MgSiO_3$ (il) has a higher entropy than the corresponding low pressure assemblage (γ+st) probably reflects the change in silicon coordination (4-fold to 6-fold). The structural changes are reflected in the vibrational

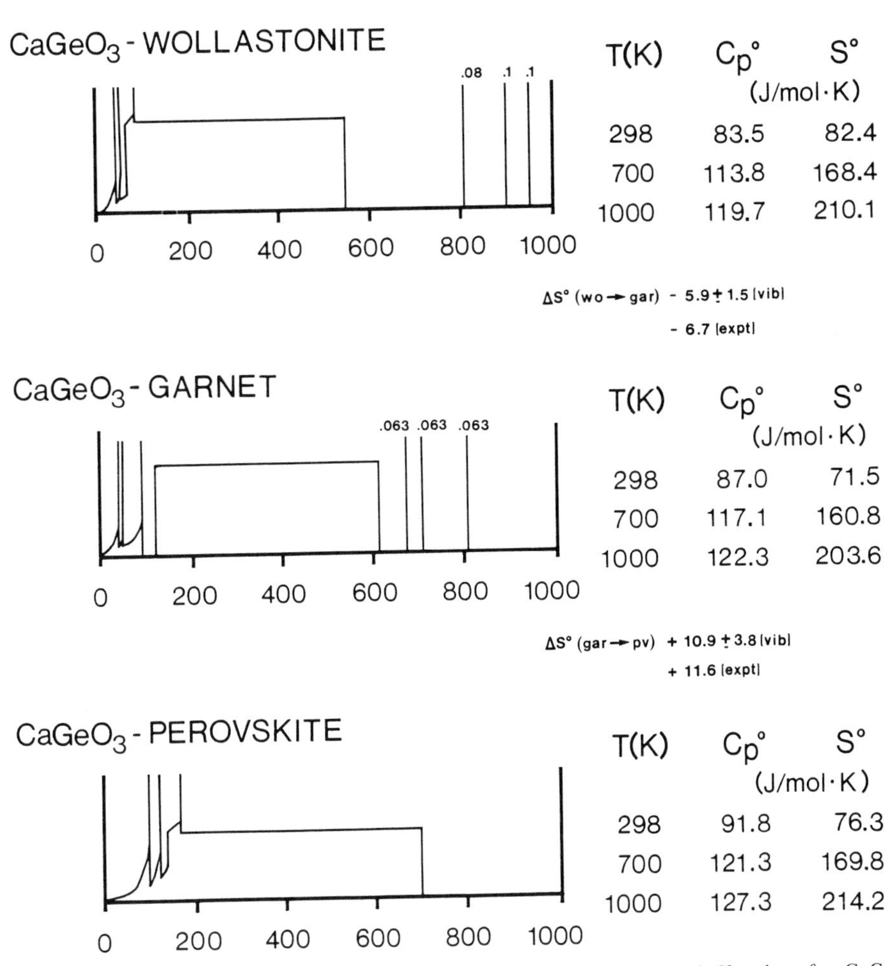

Figure 8. Representative schematic vibrational models and calculated C_p and S° values for $CaGeO_3$ polymorphs, after ROSS (1985) and ROSS et al. (1986).

spectrum of $MgSiO_3$ ilmenite (ROSS and MCMILLAN, 1984) which shows the absence of high-frequency tetrahedral Si–O stretching modes and the concentration of modes at lower frequencies compared to vibrations of tetrahedral silicates Lattice vibrational calculations (MCMILLAN and ROSS, 1987) support the conclusion that the transition γ+st→2 il has a positive $\Delta S°$, but exact numerical values are sensitive to parameters (dispersion of low frequency modes and the compressibility and thermal expansion needed to convert C_v to C_p) that are poorly known. Calculations for $MgSiO_3$ perovskite also suggest a negative dP/dT for the il→pv transition.

Acknowledgments. Masaki Akaogi, Eiji Ito, and Nancy Ross were actively involved in the work reviewed here. S. Akimoto, Y. Matsui, S. Kume, and P. McMillan provided encouragement and valuable discussion. This work was supported by the Solid State Chemistry Program of the National Science Foundation (Grants DMR 8106027 and 8521562).

REFERENCES

AKIMOTO, S., and Y. SYONO, High-pressure transformation in $MnSiO_3$, *Am. Min., 57*, 76–84, 1972.

AKAOGI, M., N. L. ROSS, P. MCMILLAN, and A. NAVROTSKY, The Mg_2SiO_4 polymorphs (olivine, modified spinel, and spinel)—thermodynamic properties from oxide melt calorimetry, phase relations, and models of lattice vibrations, *Am. Min., 69*, 499–512, 1984.

AKAOGI, M., and A. NAVROTSKY, The quartz-coesite-stishovite transformations: New calorimetric measurements and calculation of phase diagrams, *Phys. Earth Planet. Inter., 56*, 124–134, 1984.

AKAOGI, M., and A. NAVROTSKY, Calorimetric study of high pressure polymorphs of $MnSiO_3$, *Phys. Chem. Min., 12*, 317–323, 1985.

AKAOGI, M., and A. NAVROTSKY, Calorimetric study of high pressure phase transitions among the $CdGeO_3$ polymorphs (pyroxenoid, garnet, ilmenite, and perovskite structures). *Phys. Chem. Min.*, in press, 1987.

AKAOGI, M., A. NAVROTSKY, T. YAGI, and S. AKIMOTO, Pyroxene-garnet transformation: thermochemistry and elasticity of garnet solid solutions, and application to a pyrolite mantle, this volume.

ITO, E. and Y. MATSUI, Silicate ilmenites and the post-spinel transformations, in *High Pressure Research—Application to Geophysics*, edited by M. H. Manghnani and S. Akimoto, pp. 193–208, Academic Press, New York, 1977.

ITO, E., T. MATSUMOTO, and N. KAWAI, High pressure decompositions in manganese silicates and their geophysical significance, *Phys. Earth Planet. Inter., 8*, 241–245, 1974.

ITO, E., and A. NAVROTSKY, $MgSiO_3$ ilmenite: Calorimetry, phase equilibria, and decomposition at atmospheric pressure, *Am. Min., 70*, 1020–1026, 1985.

KATO, T., and M. KUMAZAWA, Garnet phase of $MgSiO_3$ filling the pyroxene-ilmenite gap at very high temperature, *Nature, 316*, 803–805, 1985.

KIEFFER, S. W., Thermodynamics and lattice vibrations of minerals: 1. Mineral heat capacities and their relationships to simple lattice vibrational models, *Rev. Geophysics and Space Phys., 17*, 1–19, 1979a.

KIEFFER, S. W., Thermodynamics and lattice vibrations of minerals: 2. Vibrational characteristics of silicates, *Rev. Geophysics and Space Phys., 17*, 20–34, 1979b.

KIEFFER, S. W., Thermodynamics and lattice vibrations of minerals: 3. Lattice dynamics and an approximation for minerals with application to simple substances and framework silicates, *Rev. Geophysics and Space Phys., 17*, 35–58, 1979c.

KIEFFER, S. W., Thermodynamics and lattice vibrations of minerals: 4. Application to chain and sheet silicates and orthosilicates, *Rev. Geophysics and Space Phys., 18*, 862–886, 1980.

KIRFEL, A., and A. NEUHAUS, Zustandsverhalten une elektrische Leitfahigkeit von $MgGeO_3$ bei Drucken bis 65 kbar und Temperaturen bis 1300 °C (mit Folgerungen fur das Druckverhalten von $MgSiO_3$), *Z. Physikal Chemie Neue Folge, 91*, 121–152, 1974.

MCMILLAN, P., and N. L. ROSS, Heat capacity calculations for Al_2O_3 corundum and $MgSiO_3$ ilmenite, *Phys. Chem. Min., 14*, 225–234, 1987.

NAVROTSKY, A., Calorimetry: its application to petrology, *Ann. Rev. Earth Planet. Sci., 7*, 93–115, 1979.

NAVROTSKY, A., F. S. PINTCHOVSKI, and S. AKIMOTO, Calorimetric study of the stability of high pressure phases in the systems CoO–SiO_2 and "FeO"–SiO_2 and calculation of phase diagrams in MO–SiO_2 systems, *Phys. Earth Planet. Inter., 19*, 275–292, 1979.

NAVROTSKY, A., Lower mantle phase transitions may generally have negative pressure-temperature slopes, *Geophys. Res. Lett., 7*, 709–711, 1980.

NAVROTSKY, A., H. FUKUYAMA, and P. K. DAVIES, Calorimetric studies of crystalline and glassy high pressure phases, in *High Pressure Research in Geophysics*, edited by S. Akimoto and M. Manghnani, pp. 465–478, Center for Academic Publication, Tokyo, 1982.

NAVROTSKY, A., Crystal chemical constraints on thermochemistry of minerals, in *Reviews in Mineralogy*, vol. 13, edited by S. W. Kieffer and A. Navrotsky, P. H. Ribbe, Series Ed., *Min. Soc. Amer.*, 225–275, 1985.

OZIMA, M., and S. AKIMOTO, Flux growth of single crystals of $MgGeO_3$ (orthopyroxene, clinopyroxene, and ilmenite) and their phase relations and crystal structures, *Am. Min., 68*, 1199–1205, 1983.

RINGWOOD, A. E., and M. SEABROOK, High pressure phase transformations in germanate pyroxenes and related compounds, *J. Geophys. Res., 68*, 4601–4609, 1963.

RINGWOOD, A. E., and A. MAJOR, High-pressure transformations in pyroxenes II, *Earth Planet. Sci. Lett., 5*, 76–78, 1968.

ROSS, N. L., and P. MCMILLAN, The Raman spectrum of $MgSiO_3$ ilmenite, *Am. Min., 69*, 719–721, 1984.

ROSS, N., A thermochemical and lattice vibrational study of high pressure phase transitions in silicates and garmanates, Ph.D. thesis, Arizona State University, 1985.

ROSS, N. L., A. AKAOGI, J. SUSAKI, and P. MCMILLAN, Phase transitions among the $CaGeO_3$ polymorphs (wollastonite, garnet, and perovskite structures): Studies by high pressure synthesis, high temperature calorimetry, and vibrational spectroscopy and calculation. *J. Geophys. Res., 91*, 4685–4696, 1986.

SUITO, K., Phase relations of pure Mg_2SiO_4 up to 200 kilobars, in *High Pressure Research: Applications to Geophysics*, edited by M. Manghnani and S. Akimoto, pp. 255–266, Academic Press, New York, 1977.

THERMODYNAMIC ASPECTS OF PHASE BOUNDARY AMONG α-, β-, AND γ-Mg_2SiO_4

Toshifumi ASHIDA and Shoichi KUME

College of General Education, Osaka University, Toyonaka, Osaka, Japan

Eiji ITO

Institute for Study of the Earth's Interior, Okayama University, Misasa, Tottori, Japan

Abstract. Molar heat capacities of three polymorphs (α, β, γ) of Mg_2SiO_4 were measured in the temperature range of 180–700 K by means of differential scanning calorimetry (DSC). The results were extrapolated towards low and high temperatures, and the entropy of each phase was evaluated in the range of 0–2000 K. The phase boundaries for α-β and for β-γ were determined by thermodynamic calculations using these entropies and additional reported data on enthalpies of formations, compressibilities, and thermal expansions. The boundaries of α-β and β-γ above 1000 K were roughly approximated by the following linear relations

$$P=11.68+0.0025\ T,\ \text{for}\ \alpha\text{-}\beta$$
$$P=11.08+0.0048\ T,\ \text{for}\ \beta\text{-}\gamma$$

where P is in GPa and T is in degrees K.

The triple point, where the three phases coexist, was found to be located at 13±0.5 GPa and 400–600 K. The results for the boundaries were compared with those obtained by two synthesis experiments and found to be in agreement with the results reported by SUITO (1977) for above 1000 K.

Introduction

Thermodynamics is a useful approach for determining phase equilibrium states of minerals at various pressure and temperature conditions. Recently, there have been improvements in experimental techniques for measuring thermodynamic quantities, for example, differential scanning calorimetry (DSC) measurements of the heat capacities of samples weighing several milligrams.

WATANABE (1982) measured the temperature dependencies of heat capacities of minerals above 350 K, including three polymorphs (α, β, γ) of Mg_2SiO_4, and evaluated the entropies by applying the Debye model to his results.

AKAOGI et al. (1984) measured the formation enthalpies and the IR and Raman spectra of the three polymorphs mentioned above. The optical data were used to calculate the differences in entropy and enthalpy among the phases, and the calculated values were compared to those obtained by calorimetry to assess the accuracy of the calculations. The phase boundaries were determined by combining these results with other data on thermodynamic quantities and with results of synthesis experiments on equilibrium conditions.

In the present study, the molar heat capacities at constant pressure (C_p) of α-, β-, and γ-Mg_2SiO_4 were measured by the DSC method in the temperature range of 180–700 K and converted to the heat capacities at constant volume (C_v). The temperature dependencies of C_v were extrapolated to 0 K and to 2000 K using the lattice vibration model presented by KIEFFER (1979, 1980). These temperature dependencies were then used to calculate the entropy and the enthalpy of each phase. These values were thermodynamically related to the reported data on compressibility and thermal expansion, and the phase equilibria among three phases were determined without referring to any result of synthesis experiments.

Experimental Methods

The α-Mg_2SiO_4 was synthesized by calcining a mixture of MgO and SiO_2 (mixed molar ratio 2:1) at 1673 K for 4 days. The β- and γ-Mg_2SiO_4 was obtained through conversions of the α-phase at 15 GPa and 1173 K and at 19 GPa 1273 K, respectively, using a uniaxial, split-sphere type apparatus. Although the pressure-temperature conditions were high enough to convert the whole sample to the desired phases, only the central part of sample was used for C_p measurements in order to avoid any contamination by undesirable phases. The purity and homogeneity of all samples were confirmed by X-ray powder diffraction.

The C_p was measured by DSC (a Rigaku DSC-8131L) in the temperature range of 180–700 K. Measurement procedures and data analysis techniques were the same as those previously described by ASHIDA et al. (1985). Powdered samples of the α-phase (19.20±0.01 mg) and the β-phase (16.52±0.01 mg) were sealed in preweighed aluminum pans. For the γ-phase, a sintered sample (8.80±0.01 mg) was used to avoid loss of the small amount of sample during preparation. Each sample was heated at rate of 10 K / min, above 350 K, and at the rate of 5 K / min, below 350 K. Powdered Al_2O_3 (16.00±0.01 mg)

High-Pressure Research in Mineral Physics, edited by M. H. Manghnani and Y. Syono, pp. 269–274.

was used as the calibrant of specific heat.

Results and Considerations

The C_p values obtained are listed in Table 1. The measurement for the α-phase is plotted in Figure 1, along with results reported by ROBIE et al. (1982) and ORR (1953). The mean deviations from the reported values were ±1.4% at temperatures between 170 and 350 K and ±0.3% between 350 and 700 K. No systematic error was observed. From these results and also the results of the previous work (ASHIDA et al., 1985), the error for this method is estimated to be less than ±1% above room temperature and ±2% at lower temperatures.

The results for β- and γ-phases are shown in Figure 2. The values above 350 K are consistent, within ±1%, with those of WATANABE (1982). The C_p values in the temperature range of 300–700 K are fitted by the equation

$$C_p = a + bT + cT^{-2}$$

The three parameters, a, b, and c, were determined by the least squares method and are listed in Table 2.

The extrapolation of heat capacity at low temperatures is not so easy as that at high temperatures. No simple polynomial has been found for this purpose, but the Debye model is often used. The Debye model is useful for substances with simple structure or compositions such as atomic molecules or metals; however, the fitting of actual

TABLE 1. Observed Heat Capacities of Three Polymorphs of Mg_2SiO_4

T(K)	C_p(J mol^{-1} K^{-1})			T(K)	C_p(J mol^{-1} K^{-1})		
	α	β	γ		α	β	γ
170	72.14	—	—	440	141.98	139.25	137.39
180	74.73	72.42	70.40	450	143.26	140.32	137.83
190	82.86	76.23	76.14	460	144.43	141.26	139.62
200	87.33	84.81	78.64	470	145.53	142.65	140.76
210	89.70	87.43	79.35	480	146.62	143.75	140.80
220	98.42	87.43	84.84	490	147.60	144.74	140.77
230	100.78	94.92	90.21	500	148.58	145.73	142.59
240	102.57	98.58	95.01	510	149.42	146.74	143.38
250	105.93	101.36	94.89	520	150.32	147.81	144.23
260	106.62	101.81	98.85	530	151.16	148.82	145.02
270	111.26	106.72	102.07	540	151.88	149.71	146.40
280	113.76	110.42	106.99	550	152.52	150.37	147.18
290	119.47	112.67	110.40	560	153.04	150.92	147.15
300	120.04	114.14	110.89	570	153.62	152.04	147.18
310	123.88	114.10	113.14	580	154.08	151.51	147.40
320	124.65	120.32	119.76	590	155.10	152.07	148.49
330	125.87	122.36	117.63	600	156.01	152.52	149.46
340	127.99	126.18	118.28	610	156.98	153.02	150.48
350	129.48	129.73	122.43	620	157.89	153.28	149.31
360	130.97	129.92	123.75	630	158.33	153.56	150.61
370	132.30	130.99	124.95	640	158.77	153.85	151.92
380	133.52	131.96	127.79	650	159.21	154.42	153.40
390	135.12	133.37	128.81	660	159.59	156.06	153.93
400	136.66	134.72	129.76	670	160.20	154.06	154.06
410	138.01	135.88	130.52	680	160.81	156.03	154.17
420	139.29	136.98	133.44	690	161.37	157.07	154.21
430	140.69	138.17	136.92	700	161.94	157.13	154.24

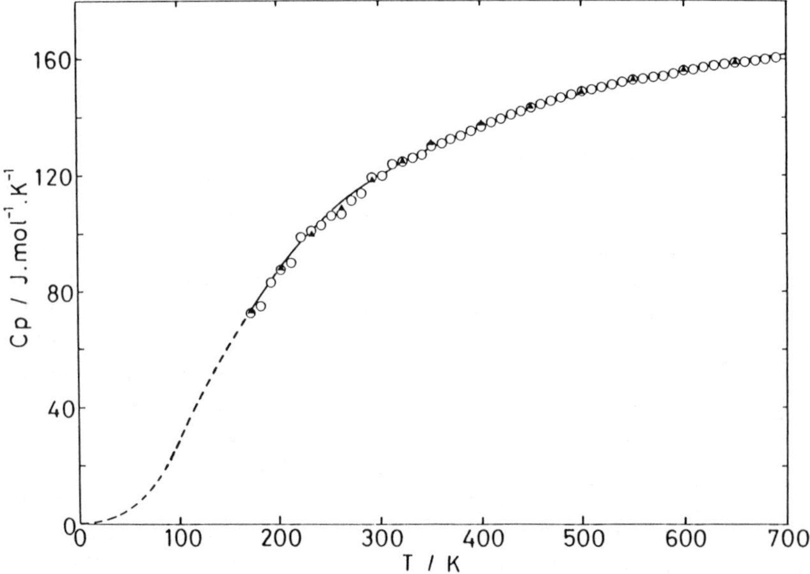

Figure 1. Temperature dependencies of C_p of α-Mg_2SiO_4. Open circles indicate present results. Closed triangles indicate results of previous works (ORR, 1953; ROBIE et al., 1982).

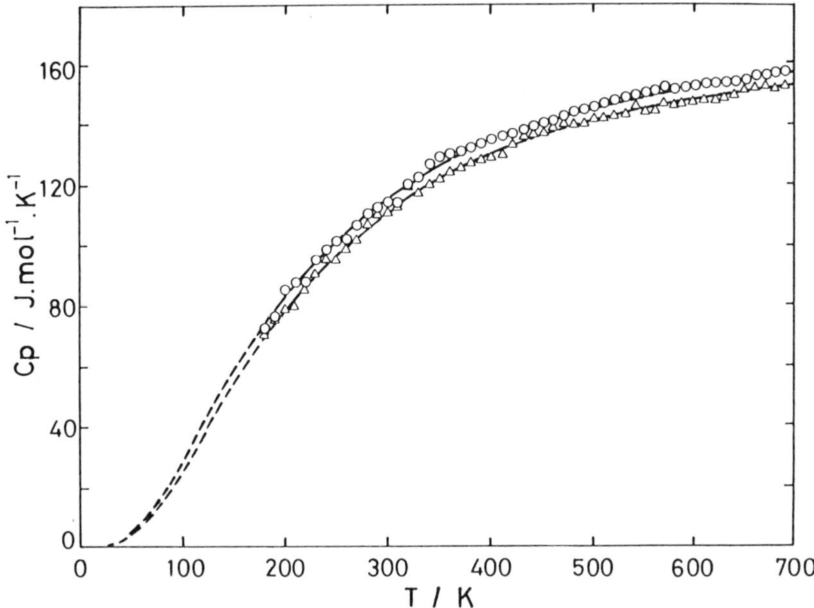

Figure 2. Temperature dependencies of C_p of β- and γ-Mg_2SiO_4. Open circles indicate the β-phase. Open triangles indicate the γ-phase.

TABLE 2. Parameters of Heat Capacity Function (300–700 K)

	C_p(J mol^{-1} K^{-1}) = $a + bT + cT^{-2}$			deviations (%)	
	a	$b \times 10^2$	$c \times 10^{-6}$	maximum	average
α-phase	137.7	4.351	− 2.852	0.5	0.26
β-phase	158.9	1.084	− 4.570	1.3	0.61
γ-phase	142.4	2.941	− 3.748	2.3	0.79

measurements to the model is not good in silicates, as shown in Figure 3.

KIEFFER (1979, 1980) attempted to relate the lattice vibrational spectra to the thermodynamic properties of minerals. She proposed a model to calculate the temperature dependencies of heat capacities using elastic, crystallographic, and spectroscopic data. AKAOGI et al. (1984) collected the IR and Raman spectra of α-, β-, and γ-Mg_2SiO_4 and evaluated the entropies and C_p of these phases using Kieffer's model. This model was used for the fitting and extrapolations of C_p measurements, because it was considered to be better than the Debye model used in this study.

In the present calculation of specific heat, the frequency distributions of lattice vibrations were assumed to consist of the following modes: 1) three sinusoidal acoustic branches (ω_1, ω_2, ω_3), which were determined from elastic wave velocities (KIEFFER, 1980; WEIDNER et al., 1984); 2) one optical continuum, with upper and lower limits (ω_u, ω_l); and 3) one Einstein mode (ω_e). The fraction of the ω_e mode, (q), was estimated crystallographically since this mode was considered to be attributed to the Si-O stretching. The three parameters (ω_u, ω_l, ω_e) were determined by least squares fitting of the temperature dependencies of C_v. C_v was converted to C_p as follows

$$C_v = C_p - a^2 KT/V$$

where a is the coefficient related to the thermal expansion, K the bulk modulus, T temperature in K, and V the molar volume at T. Thermal expansion and bulk modulus values are listed in Table 3.

Specific heat calculations were made by iteration, using the initial values of the parameters (ω_u, ω_l, ω_e) reported by AKAOGI et al. (1984). The frequencies finally obtained are listed in Table 4, and the relations between C_v and T are shown in Figure 3. Figure 3 also shows results obtained by the Debye model, for comparison.

The entropies calculated from the data above are listed in Table 5. The uncertainties are estimated to be within ±1 J/mol K. The result for the α-phase agrees with that measured by ROBIE et al. (1982). The entropies obtained for the α- and β-phases are also consistent with those calculated by AKAOGI et al. (1984). The result for the γ-phase is compatible with that of Model 1, as reported by AKAOGI et al. (1984), although slight differences are seen when compared to results of other models.

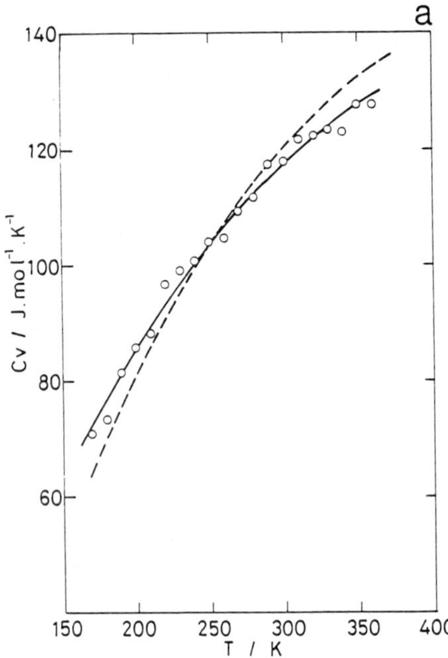

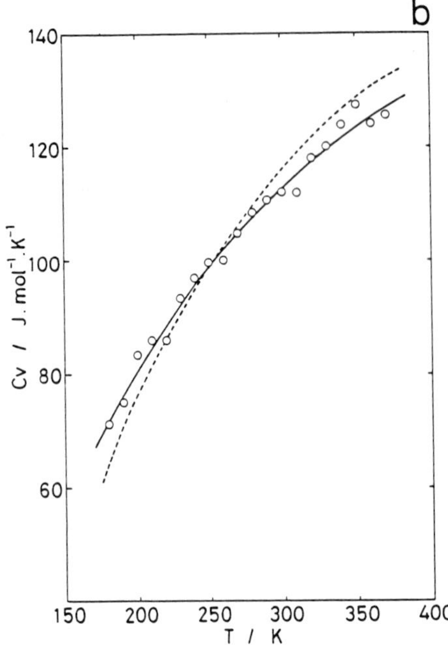

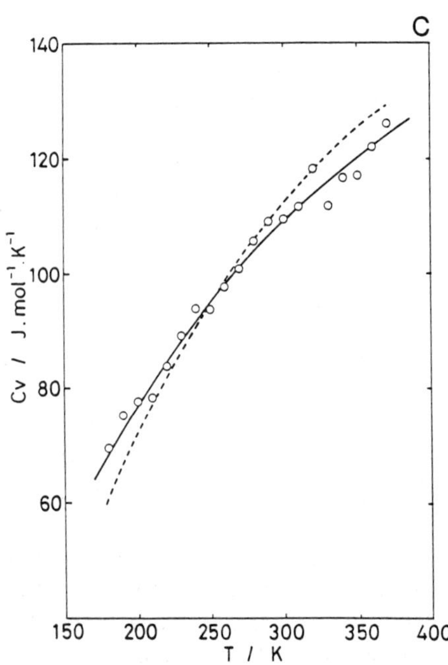

Figure 3. C_v at low temperatures, fitted by two kinds of approximations. Solid and broken lines correspond to results of Kieffer and Debye models, respectivey. a) α-phase, Deby temperature (θ_D) is assumed to be 850 K; b) β-phase, θ_D=899 K; and c) γ-phase, θ_D=935 K.

TABLE 3. Thermodynamic Properties of Three Polymorphs of Mg_2SiO_4

	Bulk modulus		Thermal expansion	
	K_0 (GPa)	K'	V_0 ($cm^3\ mol^{-1}$)	a ($cm^3\ mol^{-1}\ K^{-1}$)
α-phase	129[a]	5.2[a]	43.67[d]	0.00162[g, h]
β-phase	174[b]	4[c]	40.54[d,e]	0.00128[g, i]
γ-phase	184[b]	4[c]	39.65[d, f]	0.00104[g, j]

[a]GRAHAM and BARSCH (1969), KUMAZAWA and ANDERSON (1969); [b]WEIDNER et al. (1984); [c]assumed; [d]JEANLOZ and THOMPSON (1983); [e]AKIMOTO et al. (1976); [f]ITO et al. (1974); [g]calculated by AKAOGI et al. (1984) by fitting reported value to a linear equation; [h]HAZEN (1976), SMYTH and HAZEN (1973), SUZUKI (1975); [i]SUZUKI et al. (1979); and [j]SUZUKI et al. (1980).

K_0, V_0 are values at room temperature, 1 atm.

Discussion

The extrapolations of C_p above 1000 K were reexamined based on the same assumptions of the frequency distribution model as used for low temperatures with the ω_e determined by C_p measurements above 400 K. A good fit was found for expression, at high temperatures

$$C_p = a + bT + cT^{-2} + dT^{1/2}$$

The parameters a, b, c, and d are listed in Table 7. This C_p function was used only for temperatures above 700 K in the calculations of entropies and enthalpies discussed below.

The change in Gibbs free energy between two phase, (ΔG), is given as

TABLE 4. Parameters in Frequency Distribution Model

	ω_1 (cm^{-1})	ω_2 (cm^{-1})	ω_3 (cm^{-1})	ω_l (cm^{-1})	ω_u (cm^{-1})	ω_e (cm^{-1})	q	Z
α-phase	97[a], *	99[a], *	170[a], *	180	560	953	0.19**	4
β-phase	116[b], *	117[b], *	197[b], *	224	521	973	0.23**	4
γ-phase	141[b], *	157[b], *	289	260	475	928	0.3**	2

* calculated from sound wave velocity; [a] from KIEFFER (1980); [b] from WEIDNER et al. (1984); ** calculated from structural consideration. Z is the number of molecules in one vibrational unit.

TABLE 5. Calculated Entropies (J mol^{-1} K^{-1}) of Three Polymorphs of Mg_2SiO_4 (at 300 K)

	S_{dby}	S_{fd}	S_{lit}
α-phase	83.7	94.0	94.85
β-phase	76.9	88.0	
γ-phase	71.9	83.8	

S_{dby}: calculated by Debye model; S_{fd}: calculated by frequency distribution model; S_{lit}: reported value by ROBIE et al. (1982)

TABLE 6. Entropy and Enthalpy Changes for the Phase Transformations at 1000 K

	ΔS[a] (J mol^{-1} K^{-1})	ΔH[b] (J mol^{-1})
α → β	− 10.09	29957
β → γ	− 7.78	6820
α → γ	− 17.87	36777

[a] this work; [b] reported by AKAOGI et al. (1984)

$$\Delta G(T, P) = \Delta H(T, O) - T\Delta S(T, O) + \int_0^P \Delta V(T, P)\, dP$$

ΔG=0 at equilibrium. ΔH and ΔS are the enthalpy and entropy changes between the reactant and product at T, respectively, and are given by

$$\Delta H(T, O) = \Delta H(T_0, O) + \int_{T_0}^T \Delta C_p\, dT$$

$$\Delta S(T, O) = \int_0^T \Delta C_p / T\, dT$$

where $\Delta H(T_0, 0)$ is enthalpy change at T_0. $\Delta H(T_0, 0)$ is determined from the formation enthalpy of each phase, as measured by calorimetry (AKAOGI et al., 1984). The values for each phase transformation at 1000 K are listed in Table 6.

The molar volume at (T, P) is approximated by

$$V(T, P) = [V_0 + a(T - 297)]\,[PK'/K + 1]^{-1/K'}$$

where V_0 is the volume at room termperature and 1 atm, K is the bulk modulus at room temperature and 1 atm, and K' is the pressure derivative. These values are listed in Table 3.

Using all of the values obtained, the phase diagram of Mg_2SiO_4 was drawn, as shown in Figure 4 (see solid lines). Note that below 1000 K, the phase boundaries are not straight, but rather display an increasing curvature with increasing temperature. Above 1000 K, the boundaries are approximated by the following linear expressions

$$P = 11.7 + 0.0025\ T, \quad \text{for } \alpha\text{-}\beta \qquad (1)$$
$$P = 11.1 + 0.0048\ T, \quad \text{for } \beta\text{-}\gamma \qquad (2)$$

where P is in GPa and T is in degrees K.

SUITO (1977) and KAWADA (1977) proposed the phase boundaries from the results of their synthesis experiments:

$$P = 9.8 + 0.0035\ T \text{ (SUITO, 1977)} \qquad (3)$$
$$P = 5.7 + 0.0058\ T \text{ (KAWADA, 1977) both for } \alpha\text{-}\beta \qquad (4)$$

and

$$P = 10.0 + 0.0055\ T \text{ (SUITO, 1977)} \qquad (3')$$
$$P = 7.0 + 0.0103\ T \text{ (KAWADA, 1977) both for } \beta\text{-}\gamma \qquad (4')$$

As shown in Figure 4, the present results for the equilibrium conditions are in better agreement with results of SUITO (1977) than with those of KAWADA (1977).

If the error arises only from the thermodynamic determinations of enthalpies and entropies, the uncertainty in the equilibrium pressure would be ±1 GPa at the α-β phase transformation. The discrepancy between values calculated using Equations (1) and (3) is less than this

TABLE 7. Parameters of the Heat Capacity Function (700–2000 K)

	C_p(J mol^{-1} K^{-1}) = $a + bT + cT^{-2} + dT^{-1/2}$				deviations (%)	
	a	$b\times10^3$	$c\times10^{-6}$	$d\times10^{-3}$	maximum	average
α-phase	220.8	− 0.862	− 1.86	− 1.443	0.2	0.10
β-phase	237.1	− 3.562	0.33	− 2.036	0.8	0.25
γ-phase	268.9	− 12.33	2.08	− 2.876	1.2	0.33

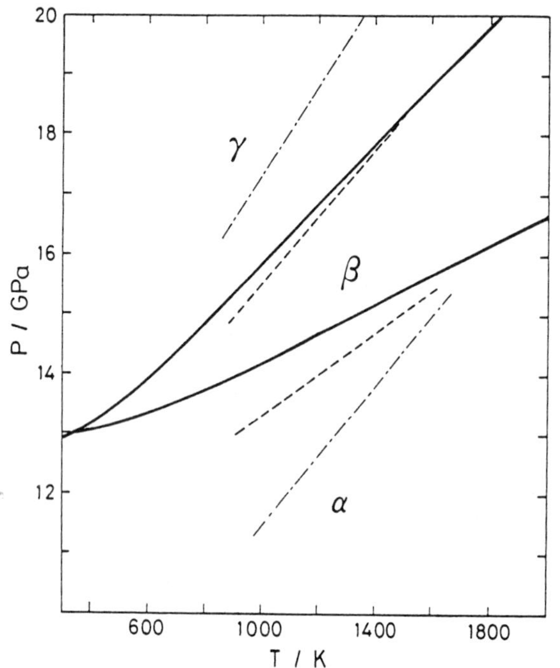

Figure 4. Phase diagram of Mg_2SiO_4. Solid line, our results; broken line, results of SUITO (1977); and dot-and-dash line, results of KAWADA (1977).

amount.

For the β-γ phase transformation, the discrepancy in equilibrium pressure is possibly attributed to the errors of all the thermodynamic data used, since the accuracy of these data seems to be less than that of data for the α-β transformation. If the reported errors in the enthalpies and the bulk moduli are considered, the uncertainty would be ± 2 GPa at pressures above 1000 K; however, the influence of such an uncertainty on the slope would be small. The error in the entropy also gives an uncertainty of ± 2 GPa for the equilibrium pressure; this uncertainty could change the slope from 0.0048 to 0.006 GPa/K. Errors in the thermal expansion coefficients also affect the slope.

The discrepancy between equilibrium pressures calculated by Equations (2) and (3′), above 1000 K, is less than the uncertainties stated above. However, the difference between pressures from Equations (2) and (4′) is greater than the uncertainties stated above; the thermal expansion coefficients listed in Table 4 would have to be reduced by one-half to make these results consistent. The differences in equilibrium conditions determined by SUITO (1977) and KAWADA (1977) have already been discussed by AKAOGI et al. (1984).

The triple point, where the three phases α-, β-, and γ-Mg_2SiO_4 coexist, is located at 13 ± 0.5 GPa and 400–600 K. Since the α-β phase boundary is more reliable and has a smaller slope than the β-γ phase boundary below 1000 K, the pressure is considered to be more accurate than the temperature for the triple point.

REFERENCES

AKAOGI, M., N. L. ROSS, P. MCMILLAN, and A. NAVROTSKY, The Mg_2SiO_4 polymorphs (olivine modified spinel and spinel): thermodynamic properties from oxides melt solution calorimetry, phase relations, and models of lattice vibrations, *Am. Mineral.*, *69*, 499–512, 1984.

AKIMOTO, S., Y. MATSUI, and Y. SYONO, High pressure crystal chemistry in orthosilicates and formation of the mantle transition zone, in *The Physics and Chemistry of Minerals and Rocks*, edited by R. G. J. Strens, pp. 327–364, Wiley, London, 1976.

ASHIDA, T., Y. MIYAMOTO, and S. KUME, Heat capacity, compressibility and thermal expansion coefficient of ilmenite-type $MgGeO_3$ *Phys. Chem. Minerals*, *12*, 129–131, 1985.

GRAHAM, E. K., and G. R. BARSCH, Elastic constants of single crystal forsterite as a function of temperature and pressure, *J. Geophys. Res.*, *75*, 5949–5960, 1969.

HAZEN, R. M., Effect of temperature and pressure on the crystal structure of forsterite. *Am. Mineral.*, *61*, 1280–1293, 1976.

ITO, E., Y. MATSUI, K. SUITO, and N. KAWAI, Synthesis of γ-Mg_2SiO_4, *Phys. Earth Planet. Inter.*, *8*, 342–344, 1974.

JEANLOZ, R., and A. B. THOMPSON, Phase transitions and mantle discontinuities, *Rev. Geophys. Space. Phys.*, *21*, 51–74, 1983.

KAWADA, K., The system Mg_2SiO_4-Fe_2SiO_4 at high pressures and temperatures and the earth's interior, Ph.D. thesis, University of Tokyo, Institute for Solid State Physics, 1977.

KIEFFER, S. W., Thermodynamics and lattice vibrations of minerals: 3. Lattice dynamics and an approximation for minerals with application to simple substance and framework silicates, *Rev. Geophys. Space Phys.*, *17*, 35–58, 1979.

KIEFFER, S. W., Thermodynamics and lattice vibrations of minerals: 4. Application to chain and sheet silicates and orthosilicates, *Rev. Geophys. Space Phys.*, *18*, 862– 886, 1980.

KUMAZAWA, M., and O. L. ANDERSON, Elastic moduli, pressure derivatives and temperature derivatives of single crystal olivine and single crystal forsterite, *J. Geophys. Res.*, *74*, 5961–5972, 1969.

ORR, R. L., High temperature heat contents of magnesium orthosilicate and ferrous orthosilicate, *J. Am. Chem. Soc.*, *75*, 528–529, 1953.

ROBIE, R. A., B. S. HEMIGWAY, and H. TAKEI, Heat capacities and entropies of Mg_2SiO_4, Mn_2SiO_4, and Co_2SiO_4 between 5 and 380 K, *Am. Mineral.*, *67*, 470–482, 1982.

SMYTH, J. R., and R. M. HAZEN, The crystal chemistry of forsterite and holtonolite at several temperatures up to 900 °C, *Am. Mineral.*, *58*, 588–593, 1973.

SUITO, K., Phase relationships of pure Mg_2SiO_4 up to 200 kilobars, in *High-pressure Research: Applications to Geophysics*, edited by M. H. Manghnani and S. Akimoto, pp. 255–266, Academic Press, New York, 1977.

SUZUKI, I., Thermal expansion of periclase and olivine and their unharmonic properties, *J. Phys. Earth*, *23*, 145– 149, 1975.

SUZUKI, I., E. OHTANI, and M. KUMAZAWA, Thermal expansion of γ-Mg_2SiO_4, *J. Phys. Earth*, *27*, 53–61, 1979.

SUZUKI, I., E. OHTANI, and M. KUMAZAWA, Thermal expansion of modified spinel, β-Mg_2SiO_4, *J. Phys. Earth*, *28*, 273–280, 1980.

WATANABE, H., Thermochemical properties of synthetic high pressure compounds relevant to the earth's mantle, in *High-pressure Research in Geophysics*, edited by M. H. Manghnani and S. Akimoto, p. 441, Center for Academic Publications, Japan, 1982.

WEIDNER, D. J., H. SAWAMOTO, S. SASAKI, and M. KUMAZAWA, Single crystal elastic properties of the spinel phase of Mg_2SiO_4, *J. Geophys. Res.*, *89*, 7852–7860, 1984.

PHYSICO-CHEMICAL PROPERTIES OF OLIVINE AND SPINEL SOLID SOLUTIONS IN THE SYSTEM Mg_2SiO_4-Fe_2SiO_4

Hiroshi WATANABE

Osaka Industrial University, Daito, Osaka 574, Japan

Abstract. Molar heat capacities of synthetic spinel (γ-phase) solid solutions (γ-$(Mg_{0.7},Fe_{0.3})_2SiO_4$, γ-$(Mg_{0.5},Fe_{0.5})_2SiO_4$, and γ-$(Mg_{0.3},Fe_{0.7})_2SiO_4$) and synthetic olivine (α-phase) solid solutions (α-$(Mg_{0.8},Fe_{0.2})_2SiO_4$, α-$(Mg_{0.5},Fe_{0.5})_2SiO_4$, and α-$(Mg_{0.2},Fe_{0.8})_2SiO_4$) were measured between 350 and 700 K using differential scanning calorimetry. It was found that the heat capacities of both types of solid solutions increase with increasing mole percent of the Fe_2SiO_4 component. It was also found for both types of solid solutions that, between 350 and 700 K, the mean thermal Grüneisen constants have a small composition dependence and the mean thermal Debye temperatures decrease with increasing mole percent of the Fe_2SiO_4 component.

Introduction

Phase relationships in the system Mg_2SiO_4-Fe_2SiO_4 at high temperature and high pressure have been extensively studied (AKIMOTO and FUJISAWA, 1968; KAWAI et al., 1970; RINGWOOD and MAJOR, 1970; AKIMOTO, 1972; SUITO, 1977; KAWADA, 1977). Recently, heat capacities for α-, β-, and γ-Mg_2SiO_4 and for α- and γ-Fe_2SiO_4 were measured by use of differential scanning calorimetry (DSC) (WATANABE, 1982; ASHIDA et al., this volume).

Using the DSC technique, we determined high-temperature heat capacities between 350 and 700 K for γ-$(Mg_{0.7},Fe_{0.3})_2SiO_4$, γ-$(Mg_{0.5},Fe_{0.5})_2SiO_4$, γ-$(Mg_{0.3},Fe_{0.7})_2SiO_4$, α-$(Mg_{0.8},Fe_{0.2})_2SiO_4$, α-$(Mg_{0.5},Fe_{0.5})_2SiO_4$, and α-$(Mg_{0.2}, Fe_{0.8})_2SiO_4$. Our methods and results are discussed below.

Experimental Methods

The high-pressure phases γ-$(Mg_{0.7},Fe_{0.3})_2SiO_4$, γ-$(Mg_{0.5},Fe_{0.5})_2SiO_4$, and γ-$(Mg_{0.3},Fe_{0.7})_2SiO_4$ were synthesized from individual olivine (α-phase) solid solution under high-pressure and high-temperature conditions by means of a double-stage, cubic-octahedral anvil type of high-pressure apparatus (KAWADA, 1977). The products were determined to be of a single phase, using X-ray diffraction and microscopic observation. The heat capacity was measured using a Perkin-Elmer Model DSC-2 differential scanning calorimeter. The weights of samples used were in the range of 10–15 mg. The heating rate was 20 K/min. We used dried nitrogen as purge gas. Correction for the difference in the aluminium pan's weight was made with heat capacity data reported by EASTMAN et al. (1924). Pure synthetic sapphire, α-Al_2O_3, was used in powder form as a standard sample for the heat capacity measurement. The values for α-Al_2O_3 reported by GINNINGS and FURUKAWA (1953) were used for our measurement. We calibrated the temperature of the calorimeter by measuring melting temperatures of indium, tin, lead, and zinc and comparing our measurements to standards provided by Perkin-Elmer. To check the accuracy of the calorimeter, we measured heat capacity of MgO and compared our result with the value reported by PANKRATZ and KELLEY (1963). The error in the present measurement was estimated to be less than 1.0%.

Results and Discussion

Molar heat capacity versus temperature data for olivine (α-phase) and spinel (γ-phase) solid solutions with end-members in the system Mg_2SiO_4-Fe_2SiO_4 are shown in Figures 1 and 2, respectively. The symbols in Figures 1 and 2 represent mean values, as determined from several experimental runs. These mean values were fitted by the least-squares method to the conventional equation,

$$C_p(T) = a + bT + cT^{-2} \quad (1)$$

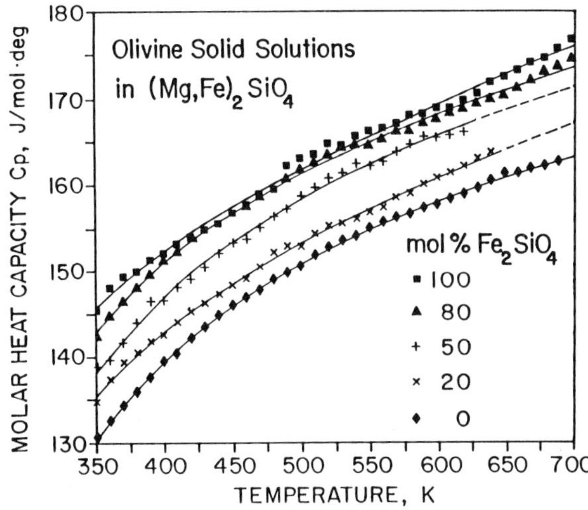

Figure 1. Molar heat capacity versus temperature for olivine solid solutions in the system Mg_2SiO_4-Fe_2SiO_4. Data for α-Mg_2SiO_4 and α-Fe_2SiO_4 are reproduced from WATANABE (1982).

High-Pressure Research in Mineral Physics, edited by M. H. Manghnani and Y. Syono, pp. 275–278.

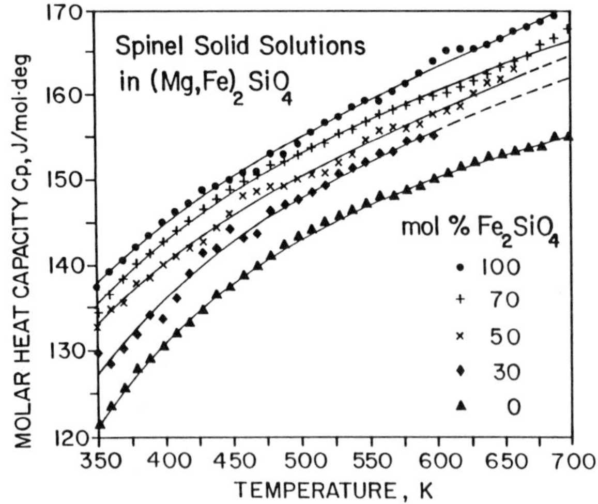

Figure 2. Molar heat capacity versus temperature for spinel solid solutions in the system Mg_2SiO_4-Fe_2SiO_4. Data for γ-Mg_2SiO_4 and γ-Fe_2SiO_4 are reproduced from WATANABE (1982).

where $C_p(T)$ is the heat capacity at constant pressure, T is absolute temperature (K), and a, b, and c are constants. The solid lines in Figures 1 and 2 represent the least-squares-fitted curves. The values of a, b, and c for each compound are listed in Table 1.

In general, we found that C_p increases monotonically with increasing temperature and shows no abrupt changes, that would suggest the occurrence of phase transitions, in the temperature range studied. The molar heat capacities of both olivine and spinel solid solutions were observed to increase with increasing mole percent of the Fe_2SiO_4 component. It has been established that α-Fe_2SiO_4 has a heat capacity maximum which is accompanied by a magnetic transition at 65.0 K (KELLEY, 1941). However, we did not find any anomaly due to the magnetic transition in the $C_p(T)$ curves for three olivine solid solutions and α-Fe_2SiO_4. It has also been reported that the Néel temperature of γ-Fe_2SiO_4 is 12 K (SUITO et al., 1984), but we did not find any anomaly in the $C_p(T)$ curves for three spinel solid solutions and γ-Fe_2SiO_4. Thus the magnetic transition may not influence the $C_p(T)$ in the higher temperature range studied by the present method.

Thermal Grüneisen constants were calculated between 350 and 700 K using the present heat capacity data, based on the following thermodynamic relationship,

$$\gamma_{th}(T) = \frac{\alpha_v(T)V(T)K_s(T)}{C_p(T)} \tag{2}$$

where $\alpha_v(T)$ is the volume coefficient of thermal expansion, $V(T)$ is the molar volume, $K_s(T)$ is the adiabatic bulk modulus. The molar volume values used were those reported by MATSUI and SYONO (1968) for α-Mg_2SiO_4 and olivine solid solutions, by AKIMOTO et al. (1976) for α-Fe_2SiO_4, by SASAKI et al. (1982) for γ-Mg_2SiO_4, by AKIMOTO (1972) and KAWADA (1977) for spinel solid solutions, and by MARUMO et al. (1977) for γ-Fe_2SiO_4. The values of thermal expansion and bulk moduli used were limited to those available for the end-members, that is, α-Mg_2SiO_4, α-Fe_2SiO_4, γ-Mg_2SiO_4, and γ-Fe_2SiO_4. Values used were reported by MATSUI and MANGHNANI (1985) for α_v of α-Mg_2SiO_4, by SUZUKI et al. (1981) for α_v of α-Fe_2SiO_4, by SUZUKI et al. (1979) for α_v of γ-Mg_2SiO_4 and γ-Fe_2SiO_4, by SUZUKI et al. (1983) for K_s of α-Mg_2SiO_4, by SUMINO (1979) for K_s of α-Fe_2SiO_4, by WEIDNER et al. (1984) for K_s of γ-Mg_2SiO_4, and by SATO (1977) for K_T of γ-Fe_2SiO_4. The conversion from the isothermal bulk modulus K_T to the adiabatic bulk modulus K_s was made from the equation,

$$K_s = K_T(1 + \alpha_v\gamma_{th}T) \tag{3}$$

TABLE 1. Molar Heat Capacity, Mean Thermal Grüneisen Constant, and Mean Thermal Debye Temperature between 350 and 700 K

	Constants in C_p (J/mol·K) = a + bT + cT^{-2}				
	a	b × 10^2	c × 10^{-6}	$\bar{\gamma}_{th}$	$\bar{\theta}_{th}$
α-Mg_2SiO_4*	155.9	2.22	− 4.09	1.20 ± 0.02	914 ± 9
α-$(Mg_{0.8}, Fe_{0.2})_2SiO_4$	131.9	5.81	− 2.01	1.18 ± 0.02	885 ± 10
α-$(Mg_{0.5}, Fe_{0.5})_2SiO_4$	156.7	3.70	− 3.88	1.16 ± 0.03	800 ± 11
α-$(Mg_{0.2}, Fe_{0.8})_2SiO_4$	160.3	2.84	− 3.38	1.15 ± 0.03	720 ± 11
α-Fe_2SiO_4*	144.2	5.10	− 2.02	1.15 ± 0.04	691 ± 12
γ-Mg_2SiO_4*	156.5	1.45	− 4.79	1.20 ± 0.02	1019 ± 8
γ-$(Mg_{0.7}, Fe_{0.3})_2SiO_4$	147.8	3.47	− 3.75	1.21 ± 0.03	926 ± 9
γ-$(Mg_{0.5}, Fe_{0.5})_2SiO_4$	137.9	4.79	− 2.46	1.23 ± 0.02	864 ± 10
γ-$(Mg_{0.3}, Fe_{0.7})_2SiO_4$	146.9	3.88	− 2.88	1.25 ± 0.03	821 ± 10
γ-Fe_2SiO_4*	134.8	5.92	− 1.94	1.29 ± 0.04	781 ± 10

*Reproduced from WATANABE (1982).

This correction is less than 1%. In the present study, it was assumed that the volume coefficient of thermal expansion and the adiabatic bulk modulus, for both olivine and spinel solid solutions in the system Mg_2SiO_4-Fe_2SiO_4, change monotonically with the mole percent of the Fe_2SiO_4 component. The insensitivity of bulk moduli of olivines and spinels to the Mg/Fe ratio was reported by WANG (1978), and the similarity of thermal expansion values for γ-Mg_2SiO_4, γ-Fe_2SiO_4, and $MgAl_2O_4$ was noted by SUZUKI et al. (1979). However, small errors in assumed values, as suggested by these reported findings, should not significantly affect our results. Since the temperature variations of γ_{th} were quite similar for both solid solutions and very small in the temperature range studied, the mean value, $\bar{\gamma}_{th}$, was calculated between 350 and 700 K. The values of $\bar{\gamma}_{th}$ are listed in Table 1 and shown in Figure 3. We found that: 1) the thermal Grüneisen constants for both solid solutions have a small composition dependence; 2) the $\bar{\gamma}_{th}$ of spinel solid solutions increases with increasing mole percent of the Fe_2SiO_4 component, while the $\bar{\gamma}_{th}$ of olivine solid solutions decreases; and 3) the calculated value of $\bar{\gamma}_{th}$ for γ-Mg_2SiO_4 agrees with that for α-Mg_2SiO_4, that is, $\bar{\gamma}_{th}=1.20\pm0.02$. The above findings indicate that the difference between olivine and spinel phases in the values of the thermal Grüneisen constant may be quite small, less than 12%.

Thermal Debye temperature was calculated between 350 and 700 K by fitting the present heat capacity data to the Debye's theoretical expression for heat capacity,

$$C_v(T) = 9Rn\left(\frac{T}{\theta}\right)^3 \int_0^{\theta/T} \frac{x^4 e^x}{(e^x - 1)^2}\, dx \tag{4}$$

where C_v is the heat capacity at constant volume, R is the gas constant, n is the number of atoms/formular unit, and θ is the thermal Debye temperature. The conversion of C_p to C_v was made using the present γ_{th} values, based on the following thermodynamic relationship,

$$C_p = V_v(1 + \alpha_v \gamma_{th} T) \tag{5}$$

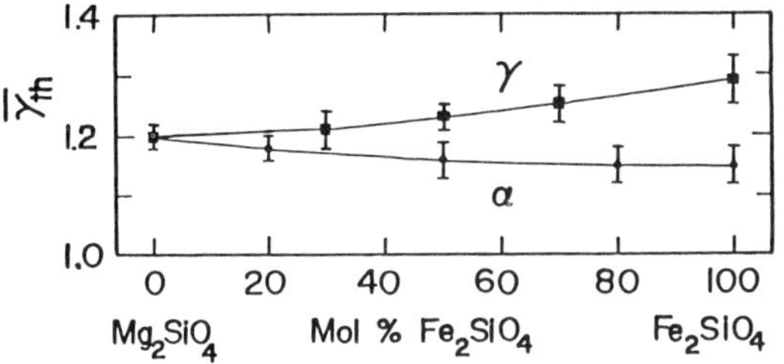

Figure 3. Mean thermal Grüneisen constants between 350 and 700 K, $\bar{\gamma}_{th}$, versus mole percent of the Fe_2SiO_4 component for both olivine(α) and spinel(γ) solid solutions in the system Mg_2SiO_4-Fe_2SiO_4.

In our determination of thermal Debye temperatures, errors in the measured $C_p(T)$ were found to significantly affect the results. The weighted mean value, $\bar{\theta}_{th}$, was calculated between 350 and 700 K based on the changes in the $\theta_{th}(T)$ value which was deduced from the 1% change in $C_p(T)$, the experimental error. The values of $\bar{\theta}_{th}$ for both solid solutions are listed in Table 1 and shown in Figure 4. We found that $\bar{\theta}_{th}$, for both olivine and spinel solid solutions, decreases with increasing mole percent of the Fe_2SiO_4 component, and that the difference between olivine and spinel phases in the values of $\bar{\theta}_{th}$ is about 100 K. The above findings indicate that entropy for olivine and spinel solid solutions increases with increasing mole percent of the Fe_2SiO_4 component and that the difference in the values of entropy between olivine and spinel phases is about 20 J/mol·K. As far as we know, no experimental data on the entropy of olivine and spinel solid solutions in the system Mg_2SiO_4-Fe_2SiO_4 have been reported.

A further understanding of the physico-chemical properties of olivine and spinel solid solutions requires accurate determination of heat capacities in the lower temperature range below 350 K, and of thermal expansion coefficients and bulk moduli.

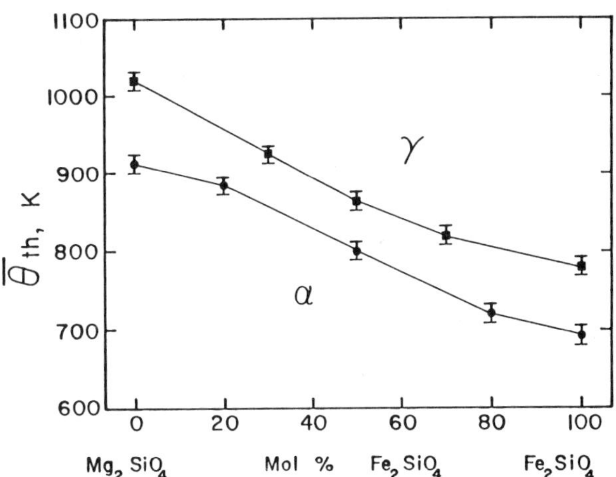

Figure 4. Mean thermal Debye temperature between 350 and 700 K, $\bar{\theta}_{th}$, versus mole percent of the Fe_2SiO_4 component for both olivine(α) and spinel(γ) solid solutions in the system Mg_2SiO_4-Fe_2SiO_4.

Acknowledgments. I express my sincere gratitude to S. Akimoto and K. Kawada for their help in sample preparation. I also thank Y. Kumanotani and H. Kanetsuna for their help in the calorimetric measurements.

REFERENCES

AKIMOTO, S., The system MgO-FeO-SiO_2 at high pressures and temperatures: phase equilibria and elastic properties, *Tectonophysics*,

13, 161–187, 1972.

AKIMOTO, S., and H. FUJISAWA, Olivine-spinel solid solution equilibria in the system Mg_2SiO_4-Fe_2SiO_4, *J. Geophys. Res.*, *73*, 1467–1479, 1968.

AKIMOTO, S., Y. MATSUI, and Y. SYONO, High pressure crystal chemistry in orthosilicates and formation of the mantle transition zone, in *The Physics and Chemistry of Minerals and Rocks*, edited by R. G. J. Strens, pp. 327–363, John Wiley, London, 1976.

ASHIDA, T., S. KUME, and E. ITO, Thermodynamic aspects of phase boundaries among α-, β-, and γ-Mg_2SiO_4, this volume.

EASTMAN, E. D., A. D. WILLIAMS, and T. F. YOUNG, The specific heats of magnesium, calcium, zinc, aluminium and silver at high temperatures, *J. Am. Chem. Soc.*, *46*, 1178–1183, 1924.

GINNINGS, D. C., and G. T. FURUKAWA, Heat capacity standards for the range 14 to 1,200 K, *J. Am. Chem. Soc.*, *85*, 522–527, 1953.

KAWADA, K., The system Mg_2SiO_4-Fe_2SiO_4 at high pressures and temperatures and the earth's interior, Ph.D. thesis, Univ. of Tokyo, Japan, 1977.

KAWAI, N., S. ENDO, and K. ITO, Split sphere high pressure vessel and phase equilibrium relations in the system Mg_2SiO_4-Fe_2SiO_4, *Phys. Earth Planet. Inter.*, *3*, 182–185, 1970.

KELLEY, K. K., Specific heats at low temperatures of ferrous silicate, manganous silicate and zirconium silicate, *J. Am. Chem. Soc.*, *63*, 2750–2752, 1941.

MARUMO, F., M. ISOBE, and S. AKIMOTO, Electron density distributions in crystals for γ-$Fe_2SiO_4$and γ-Co_2SiO_4, *Acta Crystallogr., Sect B*, *33*, 713–716, 1977.

MATSUI, T., and M. H. MANGHANI, Thermal expansion of single-crystal forsterite to 1,023 K by Fizeau interferometry, *Phys. Chem. Minerals*, *12*, 201–210, 1985.

MATSUI, Y., and Y. SYONO, Unit cell dimensions of some synthetic olivine group solid solutions, *Geochemical J.*, *2*, 51–59, 1968.

PANKRATZ L. B., and K. K. KELLEY, Thermodynamic data for magnesium oxide (Periclase), *U.S. Bur. Mines, Rept. Invest.*, *6295*, 1963.

RINGWOOD, A. E., and A. MAJOR, The system Mg_2SiO_4-Fe_2SiO_4 at high pressures and temperatures, *Phys. Earth Planet. Inter.*, *3*, 89–108, 1970.

SASAKI, S., C. T. PREWITT, Y. SATO, and E. ITO, Single-crystal x ray study of γ-Mg_2SiO_4, *J. Geophys. Res.*, *87*, 7829–7832, 1982.

SATO, Y., Equation of state of mantle minerals determined through high-pressure x-ray study, in *High-Pressure Research: Applications to Geophysics*, edited by M. H. Manghnani and S. Akimoto, pp. 307–323, Academic Press, New York, 1977.

SUITO, K., Phase relations of pure Mg_2SiO_4 up to 200 kilobars, in *High-Pressure Research: Applications to Geophysics*, edited by M. H. Manghnani and S. Akimoto, pp. 255–266, Academic Press, New York, 1977.

SUITO, K., Y. TSUTSUI, S. NASU, A. ONODERA, and F. E. FUJITA, Mössbauer effect study of the γ-form of Fe_2SiO_4, *Mat. Res. Soc. SYmp. Proc.*, *22*, 295–298, 1984.

SUMINO, Y., The elastic constants of Mn_2SiO_4, Fe_2SiO_4 and Co_2SiO_4, and the elastic properties of olivine group minerals at high temperature, *J. Phys. Earth*, *27*, 209–238, 1979.

SUZUKI, I., E. OHTANI, and M. KUMAZAWA, Thermal expansion of γ-Mg_2SiO_4, *J. Phys. Earth*, *27*, 53–61, 1979.

SUZUKI, I., K. SEYA, H. TAKEI, and Y. SUMINO, Thermal expansion of fayalite, Fe_2SiO_4, *Phys. Chem. Minerals*, *7*, 60–63, 1981.

SUZUKI, I., O. L. ANDERSON, and Y. SUMINO, Elastic properties of a single-crystal forsterite Mg_2SiO_4, up to 1,200 K, *Phys. Chem. Minerals, 10*, 38–46, 1983.

WANG, H. F., Elastic constant systematics, *Phys. Chem. Minerals*, *3*, 251–261, 1978.

WATANABE, H., Thermochemical properties of synthetic high-pressure compounds relevant to the earth's mantle, in *High Pressure Research in Geophysics*, edited by S. Akimoto and M. H. Manghnani, pp. 441–464, Center for Academic Publications, Tokyo, Japan, 1982.

WEIDNER, D., H. SAWAMOTO, and S. SASAKI, Single-crystal elastic properties of the spinel phase of Mg_2SiO_4, *J. Geophys. Res.*, *89*, 7852–7860, 1984.

V. CRYSTAL STRUCTURE, ELASTICITY AND LATTICE DYNAMICS

CRYSTAL STRUCTURE OF MgF_2 AND FeF_2 UNDER HIGH PRESSURE

N. Nakagiri[1], M. H. Manghnani, Y. H. Kim, and L. C. Ming

Hawaii Institute of Geophysics, University of Hawaii
Honolulu, Hawaii, 96822, USA

Abstract. The crystal structure and the unit cell parameters of MgF_2 and FeF_2 have been studied at room temperature up to a pressure of 4.8 GPa using a diamond-anvil cell and a four-circle X-ray diffractometer. From the pressure-volume data, the isothermal bulk moduli are determined to be 97.1 GPa and 100.0 GPa for MgF_2 and FeF_2, respectively. The ratio of the cell parameters, c/a, increases with pressure in both MgF_2 and FeF_2. The positional parameter x decreases with pressure in MgF_2, and increases with pressure in FeF_2. The distortion of the octahedron has been discussed on the basis of the present experimental results. The pressure dependence of x has been theoretically calculated using the rigid-ion model for the six difluorides MgF_2, NiF_2, ZnF_2, CoF_2, FeF_2 and MnF_2.

Introduction

Many of the dioxides and difluorides crystallize in the rutile structure at room temperature and pressure. The tetragonal unit cell (space group $P4_2/mnm$) contains two cations at (0, 0, 0) and (1/2, 1/2, 1/2), and four anions at $(x, x, 0)$, $(-x, -x, 0)$, $(1/2+x, 1/2-x, 1/2)$ and $(1/2-x, 1/2+x, 1/2)$. The structure, therefore, is defined by only these three parameters: two cell parameters (a and c) and one positional parameter (x). Each cation is surrounded by six anions, forming an octahedron whose distortion depends on x and c/a.

The elastic properties (Manghnani et al., 1980), high-pressure polymorphs (Ming et al., 1980; Ming and Manghnani, 1982) and thermal properties (Rao, 1969) of the rutile-structure dioxides and difluorides have been studied extensively. Among the dioxides, SiO_2 (stishovite) is geophysically important as it is considered to be one of the major phases in the earth's mantle (Birch, 1964).

Hazen and Finger (1981) investigated the crystal structure of TiO_2, SnO_2, GeO_2, RuO_2 and MnF_2 to 6 GPa at ambient temperature. They stated that the positional parameter x does not vary systematically with pressure in TiO_2, RuO_2 or MnF_2, but that the parameter may slightly increase in SnO_2 and decrease in GeO_2 with pressure. However, in view of the large scatter and appreciable standard deviations in x values in their data, the pressure dependence of x for these compounds is still uncertain. Jorgensen et al. (1978) studied the crystal structure of NiF_2 using a powder sample and neutron diffraction. Because of the low transition pressures of MnF_2 and NiF_2 (1.5 and 1.83 GPa, respectively), the experimental data do not provide a clear understanding of the pressure dependence of the parameter x. For this reason, we have chosen MgF_2 and FeF_2 for studying the pressure dependence of x.

Among the rutile-structure difluorides, MgF_2 is the most ionic (Baur, 1971) and does not have any phase transition up to about 25 GPa (Ming and Manghnani, 1979). The transformation pressure, estimated to be about 4.5 GPa in FeF_2 (Ming and Manghnani, 1978), would provide a pressure range wide enough for this kind of study. Another motivation for studying FeF_2 was that the distortion of the octahedron in FeF_2 is believed to be influenced by the Jahn-Teller effects (Baur, 1976). Furthermore, it shows a negative thermal expansion along the c axis (Rao, 1969).

The purpose of this paper is three-fold: 1) to report the effect of pressure on the crystal structure of MgF_2 and FeF_2 up to 5 GPa at room temperature, 2) to discuss the pressure-induced distortion of the octahedron in the rutile structure, and 3) to present theoretical calculations of the pressure derivative of x (that is, $\partial x/\partial P$).

Experiments

The single-crystal specimen of MgF_2 was obtained from Harshaw Chemical Co., and that of FeF_2 was supplied by R. Feigelson, Stanford University. These single crystals were also used in ultrasonic measurements in a previous study (Manghnani et al., 1980). For the present measurements the single-crystal specimens were prepared in the form of rectangular fragments with the following dimensions: $120\times120\times70$ μm (for 1 bar measurements) and $90\times90\times60$ μm (for high-pressure measurements), in the case of MgF_2; and $150\times140\times70$ μm (for 1 bar) and $110\times90\times60$ μm (for high-pressure), in the case of FeF_2.

The sample and a few chips of ruby (30 μm) were loaded in a Merrill-Bassett type diamond-anvil pressure cell (Merrill and Bassett, 1974) with a stainless steel gasket (0.25 mm in thickness and with a hole of 0.3 mm in diameter). A mixture of 4:1 methanol-ethanol was used as the pressure medium. The ruby fluorescence technique (Barnett et al., 1973) was used for pressure determina-

[1]Now at Research Development Corporation of Japan, Tsukuba, Ibaraki 300-26, Japan.

High-Pressure Research in Mineral Physics, edited by M. H. Manghnani and Y. Syono, pp. 281–287.

tion.

A Picker automated four-circle diffractometer system with MoK_α radiation was used for the X-ray measurements. The bisecting and the fixed ϕ mode (FINGER and KING, 1978) were used for these measurements at 1 bar and at high pressures, respectively. Intensity data were collected up to 90° in 2θ (with $h \geq 0$ and $k \geq 0$) for the 1-bar measurements without the pressure cell, and up to 60° in 2θ for all measurable reflections with the cell.

The intensities were corrected for the absorption by the pressure cell and by the crystal itself. For each data set the average structure factor was calculated for reflections related by symmetry. The least-squares refinement of the structure parameters was performed by the program RFINE (FINGER and PRINCE, 1975) using the neutral scattering factors from CROMER and MANN (1968) and the anomalous scattering coefficients from CROMER and LIBERMANN (1970). The anisotropic temperature factor and an isotropic correction for the secondary extinction were used in each refinement. The pressure on the specimen was measured before and after the X-ray measurements at each pressure. All of the experiments were conducted at room temperature. The other experimental details have been previously published (NAKAGIRI et al., 1986).

Results

The unit-cell parameters a and c, the axial ratio c/a, and the unit-cell volume V of MgF_2 and FeF_2 (determined at various pressures) are tabulated in Table 1. The ratios a/a_0, c/c_0, and V/V_0 are plotted in Figures 1 and 2. The values of the lattice parameters obtained at 1 bar without pressure cell are: a=4.6233(1) Å and c=3.0522(1) Å for MgF_2, and a=4.6960(1) Å and c=3.3085(1) Å for FeF_2. These values are in good agreement with the following values determined by NAIDU (1966): a=4.6213(1) Å and c=3.0519(1) Å for MgF_2, and a=4.6945(4) Å and c=3.3097(1) Å for FeF_2.

The values of a and c for FeF_2 were also obtained in this study with the sample inside the pressure cell at 1 bar. These values were used to determine the regression

TABLE 1. The Cell Parameters a and c, Ratio c/a and Unit-Cell Volume V at Various pressures

	P, GPa	a, Å	c, Å	c/a	V, Å^3
MgF_2	0.0001*	4.6233(1)	3.0522(1)	0.66018(1)	65.242(2)
	0.96(5)	4.6069(2)	3.0439(7)	0.6607(1)	64.60(1)
	4.77(5)	4.5516(2)	3.0166(8)	0.6628(2)	62.49(2)
	3.80(5)	4.5638(2)	3.0209(7)	0.6619(1)	62.92(2)
	2.85(5)	4.5768(1)	3.0293(6)	0.6619(1)	63.45(1)
	2.08(5)	4.5901(2)	3.0352(7)	0.6612(2)	63.95(2)
	1.03(5)	4.6063(2)	3.0441(9)	0.6608(2)	64.59(2)
	0.29(5)	4.6172(1)	3.0499(3)	0.6606(1)	65.02(1)
FeF_2	0.0001*	4.6960(1)	3.3085(1)	0.70454(2)	72.961(5)
	0.44(5)	4.6884(2)	3.3064(2)	0.70523(2)	72.676(6)
	3.96(5)	4.6269(1)	3.2868(1)	0.71036(2)	70.364(5)
	3.37(5)	4.6364(2)	3.2902(2)	0.70966(4)	70.727(9)
	2.73(5)	4.6486(2)	3.2939(2)	0.70858(3)	71.181(8)
	1.61(5)	4.6669(2)	3.2997(2)	0.70706(4)	71.866(9)
	0.80(5)	4.6809(2)	3.3047(2)	0.70599(2)	72.409(6)
	0.0001	4.6965(2)	3.3093(2)	0.70463(4)	72.991(9)

The data are presented in the same sequence as that of measurements. Parenthesized figures represent standard deviations of the least unit cited.

*Measured without the pressure cell.

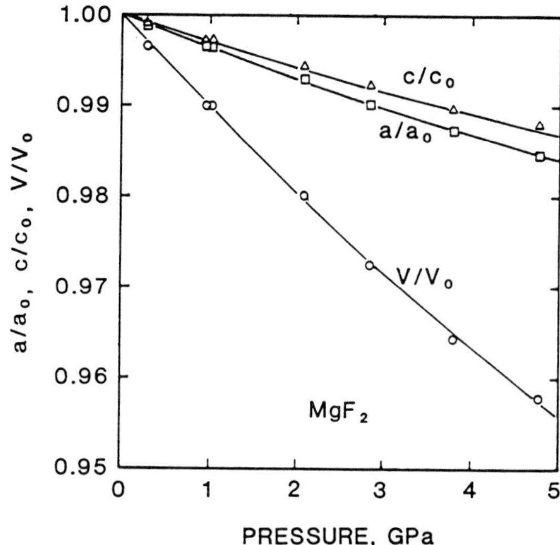

Figure 1. Pressure dependence of the ratios a/a_0, c/c_0, and V/V_0 for MgF_2.

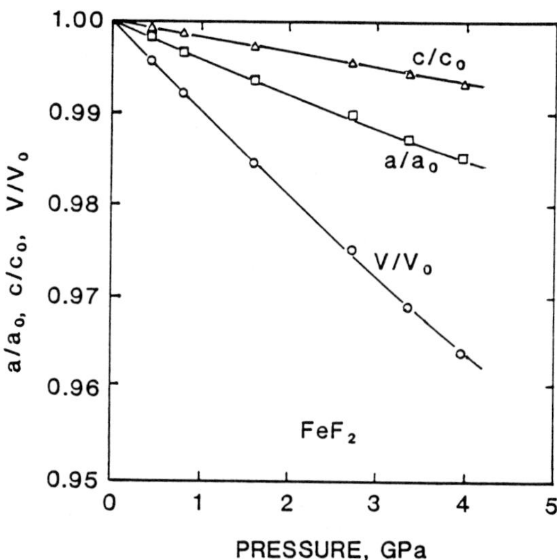

Figure 2. Pressure dependence of the ratios a/a_0, c/c_0, and V/V_0 for FeF_2.

curves. The regression curves are given by the following equations, for MgF_2

$$a=4.6230(2)-1.739(6)\times10^{-2}\ P+4.5(2)\times10^{-4}\ P^2$$
$$c=3.0525(2)-0.888(5)\times10^{-2}\ P+2.2(2)\times10^{-4}\ P^2$$
$$c/a=0.66029(5)+5.7(1)\times10^{-4}\ P-2.4(5)\times10^{-5}\ P^2$$

and for FeF_2

$$a=4.6962(3)-1.848(10)\times10^{-2}\ P+2.6(4)\times10^{-4}\ P^2$$
$$c=3.3090(1)-0.570(4)\times10^{-2}\ P+0.3(2)\times10^{-4}\ P^2$$
$$c/a=0.70461(5)+1.56(1)\times10^{-3}\ P-2.7(6)\times10^{-5}\ P^2$$

where P is in GPa.

The linear compressibilities along the a and c axes, β_a and β_c, are

$$\beta_a=3.37(11)\times10^{-3}\ \mathrm{GPa}^{-1} \text{ and } \beta_c=2.60(7)\times10^{-3}\ \mathrm{GPa}^{-1}$$

for MgF_2, and

$$\beta_a=3.77(6)\times10^{-3}\ \mathrm{GPa}^{-1} \text{ and } \beta_c=1.71(4)\times10^{-3}\ \mathrm{GPa}^{-1}$$

for FeF_2.

The bulk moduli were determined to be 97.1 and 100.0 GPa for MgF_2 and FeF_2, respectively, by fitting the P-V data to the Birch-Murnaghan equation assuming ultrasonic values of K_0' (i.e. 5.1 for MgF_2 and 4.65 for FeF_2; see Table 2). The regression curves thus obtained are shown in Figures 1 and 2 by the solid lines. The present results are in good agreement with the results of ultrasonic measurements (Table 2).

The refinement conditions and the refined parameters are given in Table 3. The values of x obtained at 1 bar are 0.3032(2) for MgF_2 and 0.3011(4) for FeF_2; these values are in good agreement with the value 0.30293(16) of BAUR (1976) and 0.3032(2) of VIDAL-VALAT et al. (1979) for MgF_2, and with the value 0.3010(8) of BAUR and KHAN (1971) for FeF_2. The positional parameter x is plotted against pressure in Figure 3 for MgF_2 and FeF_2. The solid lines are regression curves given by

TABLE 2. Isothermal Bulk Modulus and the Pressure Derivatives for the Rutile-Structured MgF_2 and FeF_2

	K_T, GPa	K'	References
MgF_2	85.8		CUTLER et al. (1968)
	100.2		HAUSSUHL (1968)
	99.6		ALEKSANDROV et al. (1969)
	99.6	5.06	RAI and MANGHNANI (1976)
	99.5	5.10	
	100.6		JONES (1977)
	101.7	5.1	DAVIES (1977)
	101.6	5.17	MANGHNANI et al. (1980)
	97.1	5.1(assumed)	Present study
FeF_2	99.6	4.65	MANGHNANI et al. (1980)
	100.0	4.65(assumed)	Present study

Note: The adiabatic bulk moduli obtained by ultrasonic methods were converted to the isothermal values by using the relation $K_S=K_T(1+\alpha\gamma T)$.

TABLE 3. Refinement Conditions and Refined Parameters

	Pressure GPa	Number of Reflections	R(%)	wR(%)	Extinct 10^{-5}	x	B_M[b]	B_F[b]	σ_θ^2	$\langle\lambda\rangle$
MgF_2	0.0001[a]	125	3.2	2.7	0.73(13)	0.3032(2)	0.64(2)	0.82(2)	34.285	1.00948
	0.96(5)	36	2.8	2.7	0.49(42)	0.3028(4)	0.76(8)	1.00(27)	33.933	1.00967
	4.77(5)	34	2.8	3.1	0.03(44)	0.3015(5)	0.62(9)	0.87(11)	32.611	1.00961
	3.80(5)	34	2.5	2.4	0.52(35)	0.3023(4)	0.83(7)	0.99(8)	33.723	1.00937
	2.85(5)	34	2.2	2.1	0.29(31)	0.3021(4)	0.67(6)	0.88(7)	33.165	1.00965
	2.08(5)	34	2.6	2.3	0.20(36)	0.3026(4)	0.81(8)	0.86(8)	33.793	1.00938
	1.03(5)	34	2.2	2.2	0.48(35)	0.3030(4)	0.78(7)	0.89(7)	33.427	1.00973
	0.29(5)	34	2.5	2.6	0.21(39)	0.3030(4)	0.79(9)	0.97(9)	34.285	1.00984
FeF_2	0.0001[a]	165	2.8	3.3	1.24(9)	0.3011(4)	0.66(1)	1.06(3)	59.392	1.02298
	0.44(5)	43	2.5	2.7	0.82(18)	0.3011(7)	0.94(5)	1.26(10)	59.951	1.01882
	3.96(5)	39	2.1	1.8	0.31(8)	0.3030(6)	0.85(5)	1.20(7)	69.151	1.02114
	3.37(5)	39	2.4	2.2	0.28(11)	0.3026(7)	0.79(5)	1.24(9)	67.357	1.02021
	2.73(5)	42	2.7	2.6	0.33(11)	0.3020(7)	0.78(6)	1.20(10)	64.808	1.01952
	1.61(5)	43	1.9	1.9	0.42(10)	0.3012(5)	0.78(4)	1.25(7)	61.549	1.01908
	0.80(5)	45	2.5	2.2	0.40(11)	0.3020(6)	0.79(4)	1.29(7)	62.690	1.01897

a) Measured without the pressure cell.
b) Isotropic temperature factor equivalent to the anisotropic values.
c) σ_θ^2 and $\langle\lambda\rangle$ indicate the octahedral angle variance and mean octahedral elongation, respectively. (ROBINSON et al., 1987)

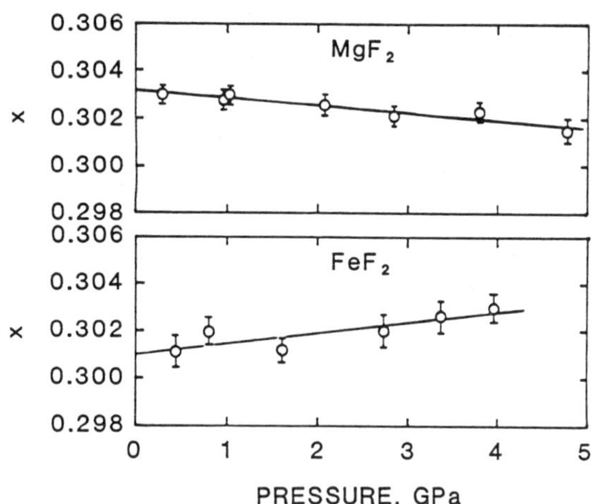

Figure 3. Variation of x with pressure. The solid lines are regression curves.

TABLE 4. Interatomic Distances and Angle

	P, GPa	R_1, Å	R_2, Å	R_3, Å	R_4, Å	α, deg
MgF_2	0.0001*	1.982(1)	1.996(1)	2.8133(3)	2.574(3)	80.29(6)
	0.29	1.979(3)	1.995(2)	2.8098(7)	2.572(5)	80.29(13)
	0.96	1.973(3)	1.992(2)	2.8034(7)	2.570(5)	80.34(11)
	1.03	1.974(2)	1.991(2)	2.8035(7)	2.567(5)	80.27(11)
	2.08	1.964(3)	1.986(2)	2.7934(7)	2.563(5)	80.36(12)
	2.85	1.955(2)	1.984(2)	2.7854(6)	2.562(5)	80.45(11)
	3.80	1.951(3)	1.977(2)	2.7779(7)	2.552(5)	80.37(12)
	4.77	1.941(3)	1.977(2)	2.7702(9)	2.555(7)	80.53(15)
FeF_2	0.0001*	2.000(2)	2.117(1)	2.9120(5)	2.624(5)	77.22(10)
	0.44	1.996(5)	2.115(3)	2.9082(11)	2.637(9)	77.16(19)
	0.80	1.999(4)	2.109(2)	2.9060(9)	2.622(8)	76.87(17)
	1.61	1.988(3)	2.108(2)	2.8975(8)	2.624(7)	76.99(15)
	2.73	1.985(5)	2.099(3)	2.8894(12)	2.604(10)	76.65(21)
	3.37	1.984(4)	2.093(3)	2.8841(11)	2.589(9)	76.39(19)
	3.96	1.983(4)	2.089(2)	2.8799(9)	2.578(7)	76.21(16)

*Measured without the pressure cell.

$$x=0.3032(2)-3.1(5)\times10^{-4}\,P$$

for MgF_2 and

$$x=0.3010(4)+4.5(14)\times10^{-4}\,P$$

for FeF_2.

The values of x obtained at 1 bar without the pressure cell have not been included in the least-squares fit, because, as discussed in an earlier paper (NAKAGIRI et al., 1986), the data obtained under high pressure might contain some systematic errors. However, the values of x at 1 bar, determined from the least-squares fit, are in good agreement with the values measured at 1 bar without the pressure cell. This agreement indicates that the systematic errors are not significant in the present experiments. The interatomic distances were calculated from the a, c, and x determined at various pressures (Table 4).

A list of observed and calculated structure factors may be obtained upon request from authors.

Discussion

Distortion of the Octahedron

In the rutile structure each cation is surrounded by six anions, forming a co-ordination octahedron (Figure 4). Among the six cation-anion bonds, two apical bonds have the same length R_1, and the remaining four equatorial bonds have the same length R_2. The length R_1 is slightly different from R_2, resulting in a distortion of the octahedron. This distortion has been mainly discussed based on the relative lengths of R_1 and R_2. With the data available at that time, BAUR (1961) pointed out that all the dioxides appeared to have $R_1>R_2$ and that all the difluorides had $R_2>R_1$. However, this trend was violated as more data were added (BAUR and KHAN, 1971; BAUR, 1976). This is shown in Figure 5 with some updated values. The solid line in Figure 5 corresponds to the values of (x, c/a) for which $R_1=R_2$. The compounds located above the line have $R_1>R_2$, while the compounds located below the line have $R_1<R_2$.

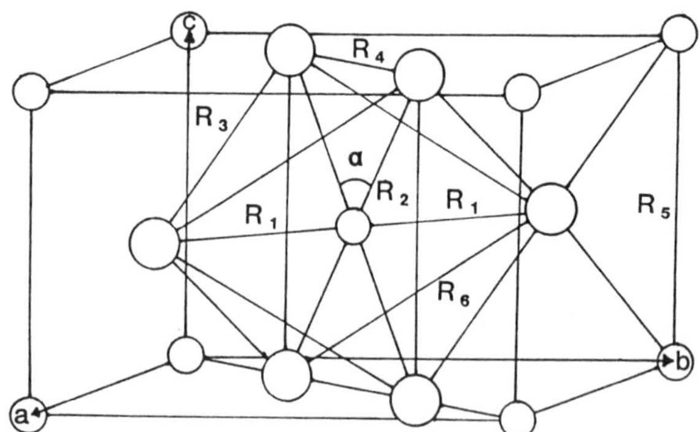

Figure 4. The crystal structure. The large circles and the small circles show anions and cations, respectively.

The pressure dependences of R_1 and R_2 are shown in Figures 6 and 7 for MgF_2 and FeF_2, respectively. The bond length R_2 is longer than R_1 in both MgF_2 and FeF_2. The difference between R_1 and R_2 increases with pressure in MgF_2, and decreases with pressure in FeF_2. Based on the pressure dependence of the difference, R_2-R_1, the distortion of the octahedron seems to increase in MgF_2 and decrease in FeF_2 with increasing pressure.

The effect of pressure on the distortion for MgF_2 and FeF_2, can also be shown on the x versus c/a plot (Figure

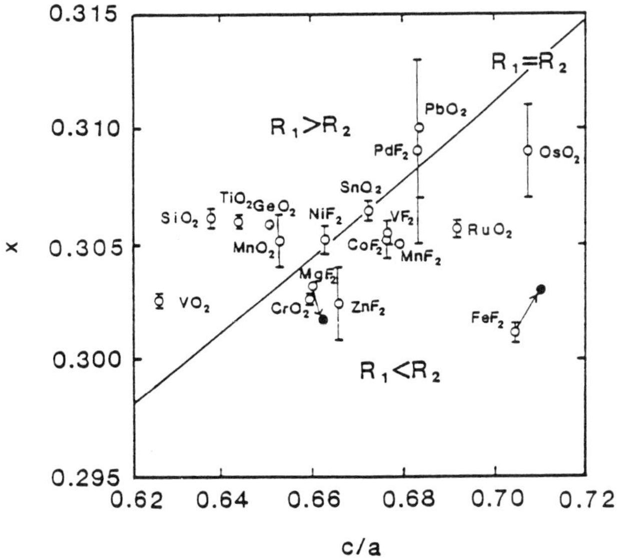

Figure 5. x versus c/a for the rutile-structure dioxides and difluorides. MgF_2 and FeF_2, the present data; MnF_2, TiO_2, SnO_2 and RuO_2, HAZEN and FINGER (1981); NiF_2, TAYLOR and WILSON (1974); SiO_2, HILL et al. (1983); the others, BAUR (1976).

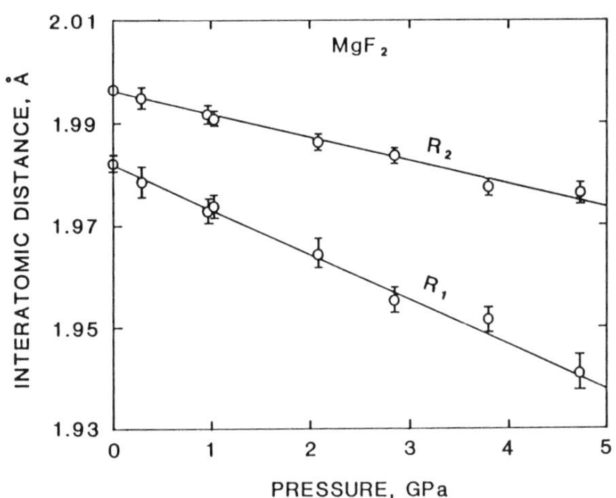

Figure 6. Pressure dependence of the interatomic distances for MgF_2.

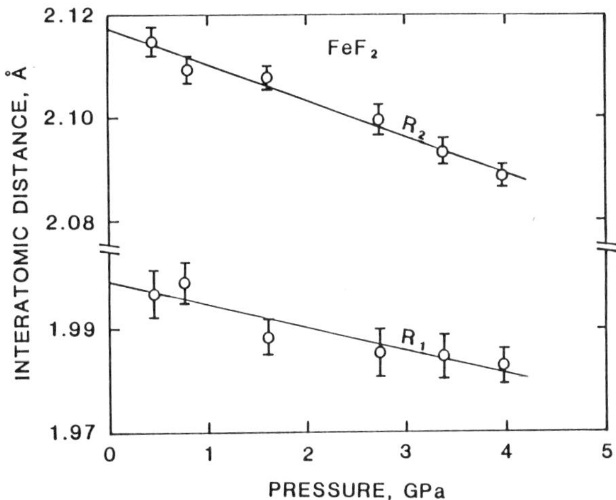

Figure 7. Pressure dependence of the interatomic distances for FeF_2.

5). The origins of the arrows indicate the experimental values at 1 bar and the tips of the arrows the measured values at the highest pressure reached in the present experiments. The point (x, c/a) moves away from the $R_1 = R_2$ line in the case of MgF_2, but moves towards the line in the case of FeF_2. This indicates that the distortion of the octahedron increases in MgF_2 and decreases in FeF_2 with increasing pressure. This conclusion is consistent with the conclusions mentioned in the preceding paragraph. However, the octahedral angle variance and mean octahedral elongation (ROBINSON et al., 1971) indicate that the distortion increases with pressure in both MgF_2 and FeF_2 (see Table 3). A detailed discussion on the distortion in the RX_2 compounds with the rutile structure will be given elsewhere.

Calculation of $(\partial x/\partial P)$

The rutile structure is characterized by three parameters: the lattice parameters a and c, and the positional parameter x. The lattice energy is, therefore, a function of these three parameters. We can derive a relation between the pressure derivative of x and the pressure derivatives of a and c, as follows

$$\frac{\partial x}{\partial P} = \frac{1}{\dfrac{\partial^2\Phi}{\partial x^2}}\left\{\left(\frac{\partial^2\Phi}{\partial x\partial a}\right)\left(-\frac{\partial a}{\partial P}\right)+\left(\frac{\partial^2\Phi}{\partial x\partial c}\right)\left(-\frac{\partial c}{\partial P}\right)\right\} \quad (1)$$

where Φ represents the lattice energy, and can be written as

$$\Phi = -\frac{M\,(Ze)^2}{a} + \sum_{i=1}^{4} n_i\,\Phi_i\,(R_i) \quad (2)$$

where M is the Madelung constant, e the electronic charge, Z an effective charge indicative of the degree of ionicity, and n_i the number of nearest neighbor atoms. The first term in the equation represents the Coulomb energy, and the second term the short-range repulsive energy. The interatomic distance R_i is a function of a, c and x and given by the following equations

$$R_1 = \sqrt{2}\,x\,a \quad (3\text{-}1)$$

$$R_2 = \left\{2(x - \frac{1}{2})^2 a^2 + \frac{1}{4} c^2\right\}^{1/2} \tag{3-2}$$

$$R_3 = \left\{\frac{1}{4} a^2 + (\frac{1}{2} - 2x)^2 a^2 + \frac{1}{4} c^2\right\}^{1/2} \tag{3-3}$$

$$R_4 = \sqrt{2}\,(1 - 2x)\,a \tag{3-4}$$

The second derivatives of the lattice energy can be derived as follows

$$\left(\frac{\partial^2\Phi}{\partial x^2}\right) = \sum_{i=1}^{4} n_i \left\{\left(\frac{\partial^2 R_2}{\partial x^2}\right)\left(\frac{\partial\phi_i}{\partial R_i}\right) + \left(\frac{\partial R_i}{\partial x}\right)^2\left(\frac{\partial^2\phi_i}{\partial R_i^2}\right)\right\} \tag{4-1}$$

$$\left(\frac{\partial^2\Phi}{\partial x\partial a}\right) = \frac{(Ze)^2}{a^2}\left(\frac{\partial M}{\partial x}\right) + \sum_{i=1}^{4} n_i \left\{\left(\frac{\partial^2 R_i}{\partial x\partial a}\right)\left(\frac{\partial\phi_i}{\partial R_i}\right) + \left(\frac{\partial R_i}{\partial x}\right)\left(\frac{\partial R_i}{\partial a}\right)\left(\frac{\partial^2\phi_i}{\partial R_i^2}\right)\right\} \tag{4-2}$$

$$\left(\frac{\partial^2\Phi}{\partial x\partial c}\right) = \sum_{i=1}^{4} n_i \left\{\left(\frac{\partial^2 R_i}{\partial x\partial c}\right)\left(\frac{\partial\phi_i}{\partial R_i}\right) + \left(\frac{\partial R_i}{\partial x}\right)\left(\frac{\partial R_i}{\partial c}\right)\left(\frac{\partial^2\phi_i}{\partial R_i^2}\right)\right\} \tag{4-3}$$

where the second derivatives of the Madelung constant like $(\partial M/\partial x\partial a)$ are omitted to make the calculation simpler. This is justified because the contribution of these are supposed to be small enough to be neglected. The derivatives of the interatomic distances can be calculated from the Equations (3-1) to (3-4). To calculate $(\partial x/\partial P)$, STRIEFLER and BARSCH's values (1973) were used for Z, $(\partial M/\partial x)$, $(\partial\phi_i/R_i)$ and $(\partial^2\phi_i/R_i^2)$. The values for $\partial a/\partial P$ and $\partial c/\partial P$ are from the present data (MgF_2 and FeF_2), HAZEN and FINGER (1981, MnF_2), WU (1977, NiF_2, sonic measurements), and HART (1978, CoF_2 and ZnF_2, sonic measurements). The sonic data were used to get $\partial a/\partial P$ and $\partial c/\partial P$ as follows

$$\frac{\partial a}{\partial P} = -a_0 \frac{c_{33} - c_{13}}{(c_{11} + c_{12})\,c_{33} - 2c_{13}^2} \tag{5-1}$$

$$\frac{\partial c}{\partial P} = -c_0 \frac{c_{11} + c_{12} - 2c_{13}}{(c_{11} + c_{12})\,c_{33} - 2c_{13}^2} \tag{5-2}$$

The calculated values of $(\partial x/\partial P)$ (Figure 8) seem to show a systematic variation with the unit-cell volume at 1 bar; that is, $(\partial x/\partial P)$ increases linearly from -1.3×10^{-4} GPa^{-1} in MgF_2 to 0.30×10^{-4} GPa^{-1} in MnF_2, with a change of sign from negative to positive.

The experimentally observed results on $(\partial x/\partial P)$ indicate a similar pattern having a negative sign in MgF_2 and a positive sign in FeF_2. However, the magnitudes are different from the calculated values. Further experimental and theoretical studies are needed to compare $\partial x/\partial P$ in the various difluorides. Studies on CoF_2 and ZnF_2 are in progress and the results will be published shortly.

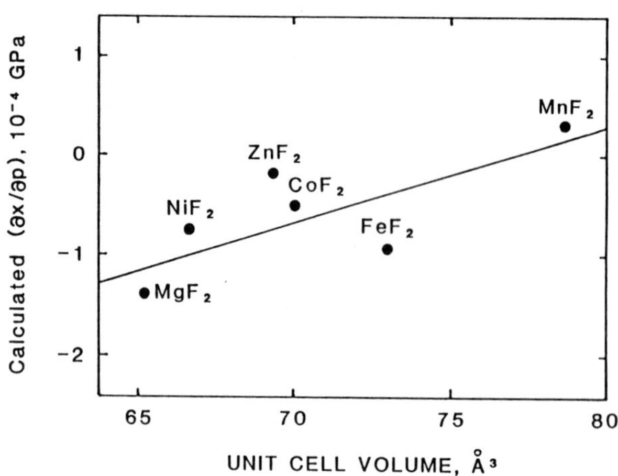

Figure 8. The calculated $(\partial x/\partial P)$ against the unit-cell volume at 1 bar for six difluorides.

Acknowledgments. John Balogh and Barry Lienert maintained the equipment and computer system used in this research. Support was provided by NSF grants EAR82-19201 and EAR84-18125. Hawaii Institute of Geophysics Contribution No. 1856.

REFERENCES

ALEKSANDROV, K. S., L. A. SHABANOVA, and V. I. ZINEMKO, Elastic constants of MgF_2 single crystals, *Phys. Stat. Solidi, 33*, K1–K3, 1969.

BARNETT, J. D., S. BLOCK, and G. J. PIERMARINI, An optical fluorescence system for quantitative pressure measurement in diamond anvil cell, *Rev. Sci. Instrum., 44*, 1–9, 1973.

BAUR, W. H., Uber die Verfeinerung der kristallstrukturbestimmung einiger vertreter des rutiltypes. III. Zur gittertheorie des rutiltypes, *Acta Cryst., 14*, 209–213, 1961.

BAUR, W. H., Rutile-type compounds. V. Refinement of MnO_2 and MgF_2, *Acta Cryst., B32*, 2200–2204, 1976.

BAUR, W. H., and A. A. KHAN, Rutile-type compounds. IV. SiO_2, GeO_2 and a comparison with other rutile-type structures, *Acta Cryst., B27*, 2133–2139, 1971.

BIRCH, F., Density and composition of mantle and core, *J. Geophys. Res., 69*, 4377–4388, 1964.

CROMER, D. T., and D. LIBERMAN, Relativistic calculation of anomalous scattering factors for X-rays, *J. Chem. Phys., 53*, 1892–1893, 1970.

CROMER, D. T., and J. B. MANN, X-ray scattering factors computed from numerical Hartree-Fock wave functions, *Acta. Cryst., A24*, 321–324, 1968.

CUTLER, H. R., J. J. GIBSON, and K. A. MCCARTHY, The elastic constants of magnesium fluoride, *Solid State Commun., 6*, 431–433, 1968.

Davies, G. F., Elasticity of single-crystal MgF_2 (rutile structure) under pressure, *Earth Planet. Sci. Lett.*, *34*, 300–306, 1977.

Finger, L. W., and H. E. King, A revised method of operation of the single-crystal diamond cell and refinement of the structure of NaCl at 32 Kbar, *Am. Mineral.*, *63*, 334–342, 1978.

Finger, L. W., and E. Prince, A system of Fortran IV computer programs for crystal structure computations, *Natl. Bur. Stand. (U.S.) Tech. Note 854*, 1–128, 1975.

Hart, S., The elastic constants of rutile structure fluorides, *S.-Afr. Tydskr. Fis.*, *1*, 65–68, 1978.

Haussuhl, S., Elastic and thermoelastic behavior of MgF_2 and MnF_2, *Phys. Stat. Solidi.*, *28*, 127–130, 1968.

Hazen, R. M., and L. W. Finger, Bulk moduli and high-pressure crystal structures of rutile-type compounds, *J. Phys. Chem. Solids*, *42*, 143–151, 1981.

Hill, R. J., M. D. Newton, and G. V. Gibbs, A crystal chemical study of stishovite, *J. Solid State Chem.*, *47*, 185–200, 1983.

Jones, L. E. A., High temperature elasticity of rutile-structure-MgF_2, *Phys. Chem. Minerals*, *1*, 179–197, 1977.

Jorgensen, J. D., T. G. Worlton, and J. C. Jamieson, Pressure-induced strain transition in NiF_2, *Phys. Rev. B*, *17*, 2212–2214, 1978.

Manghnani, M. H., L. C. Ming, and T. Matsui, Elasticity and phase transformations in rutile-structured compounds, in *High Pressure Science and Technology*, Vol. 2, edited by B. Vodar and Ph. Marteau, pp. 1092–1100, 1980.

Merrill, L., and W. A. Bassett, Miniature diamond anvil pressure cell for single crystal diffraction studies, *Rev. Sci. Instrum.*, *45*, 290–294, 1974.

Ming, L. C., and M. H. Manghnani, High pressure phase transformations in FeF_2 (rutile), *Geophys. Res. Lett.*, *5*, 491–494, 1978.

Ming, L. C., and M. H. Manghnani, High pressure phase transformations in MgF_2 (rutile), *Geophys. Res. Lett.*, *6*, 13–16, 1979.

Ming, L. C., M. H. Manghnani, T. Matsui, and J. C. Jamieson, Phase transformations and elasticity in rutile-structured difluorides and dioxides, *Phys. Earth Planet. Inter.*, *23*, 276–285, 1980.

Ming, L. C., and M. H. Manghnani, High-pressure phase transformations in rutile-structured dioxides, in *Advances in Earth and Planetary Science, Vol. 12, High-pressure Research in Geophysics,* edited by S. Akimoto, M. H. Manghnani, pp. 329–347, Center for Academic Publications, Japan, 1982.

Naidu, S. Y. N., X-ray studies on rutile-type compounds, Ph.D. thesis, Osmania University, India, 1966.

Nakagiri, N., M. H. Manghnani, L. C. Ming, and S. Kimura, Crystal structure of magnetite under pressure, *Phys. Chem. Minerals*, *13*, 238–244, 1986.

Rai, C. S., and M. H. Manghnani, Pressure and temperature dependence of the elastic moduli of polycrystalline MgF_2, *J. Amer. Ceram. Soc.*, *59*, 499–502, 1976.

Rao, K. V. Krishna, Thermal expansion of crystals, in *Physics of the Solid State*, edited by S. Balakrishna, M. Krishnamurthy, and B. Ramachandra Rao, pp. 415–426, Academic Press, London, 1969.

Robinson, K., G. V. Gibbs, and P. H. Ribbe, Quadratic elongation: A quantitative measure of distortion in coordination polyhedra, *Science*, *172*, 567–570, 1971.

Striefler, M. E., and G. R. Barsch, Elastic and optical properties of rutile-structure fluorides in the rigid-ion approximation, *Phys. Stat. Solidi*, *59*, 205–217, 1973.

Taylor, J. C., and P. W. Wilson, The structures of fluorides. V. The x-parameter in NiF_2, *Acta Cryst.*, *B30*, 554–555, 1974.

Vidal-Valat, G., J. Vidal, C. M. E. Zeyen, and K. K. Suonio, Neutron diffraction study of magnesium fluoride single crystals, *Acta Cryst.*, *B35*, 1584–1590, 1979.

Wu, A. Y., The pressure dependence of elastic moduli of NiF_2 to 10 kBar, *Phys. Lett.*, *60A*, 260–262, 1977.

THE INTERRELATIONSHIP OF THERMODYNAMIC PROPERTIES OBTAINED BY THE PISTON-CYLINDER HIGH PRESSURE EXPERIMENTS AND RPR HIGH TEMPERATURE EXPERIMENTS FOR NaCl

Orson L. ANDERSON and Shigeru YAMAMOTO*

Department of Earth and Space Sciences and Institute of Geophysics and Planetary Physics
University of California, Los Angeles, CA 90024, USA

Abstract. Thermodynamic properties (e.g., K_T, $(\partial K_T/\partial P)_T$, and $(\partial T/\partial P)_S$) for compressible solids can be measured by the experimental equation of state determined by the piston-cylinder experiment. However, this experiment is more limited for relatively incompressible solids such as MgO. For incompressible solids, the rectangular parallelepiped resonance (RPR) method has been useful for finding the thermodynamic properties such as K_S, K_T, γ, δ_S, δ_T, $(\partial K_T/\partial T)_V$, and $(\partial\gamma/\partial T)_P$ of materials at high temperatures. The measured properties derived from both types of experiments are connected by the standard thermodynamic formulas.

In this paper, we show that results from high temperature elasticity measurements, such as the RPR measurement, although having no compression information, nevertheless yield many of the piston-cylinder compression results when combined with the data of $K_{T0}'=(\partial K_T/\partial P)_T$ versus T at $P=0$.

By using the high temperature elasticity data for NaCl along with K_{T0}', we show that $(\partial K_T/\partial T)_V$ is very close to zero at high temperatures. As a consequence many thermodynamic identities are simplified and we find: 1) the thermal pressure is independent of volume at high temperature, confirming previous conclusions (ANDERSON et al., 1982) about the BOEHLER and KENNEDY NaCl data (1980); 2) the change of γ with T at constant V is negative at high T; 3) the variation of the parameter δ_T with temperature is quite small, and close in value to $(\partial^2 K_T/\partial T\partial P)$ as measured by SPETZLER et al. (1972).

Further simplifications are possible for NaCl because we find that $(\partial\alpha K_T/\partial T)_P=0$ at high T. This yields: 1) $q=-(\partial\ln\gamma/\partial\ln\rho)_T$ is close to unity (and thus $\gamma\rho=$constant) confirming several previous studies of this parameter; 2) the product $\alpha K_T=(\partial P/\partial T)_V$ is independent of both P and T (or V and T) as long as T is above the Debye temperature, θ. We further conclude that even through there is no anharmonicity in $(\partial P/\partial T)_V$, there is measurable anharmonicity in the temperature variation of C_V and $\gamma(V=V_0)$; 3) the thermodynamic isothermal equation of state, derived by BRENNAN and STACEY (1979), is valid for NaCl because $\gamma\rho=$constant, and is shown to follow the pattern of shock wave experiments up to 20 GPa.

In this paper, our RPR method results were obtained to nearly 800 K. This happens to be the upper limit of the acoustic experiment measurements. Also, our elasticity data has confirmed the previous work on NaCl by SPETZLER et al. (1972). Spetzler's results could have been used to find the same conclusions described above. Our emphasis on NaCl arises from the fact that there is abundant thermodynamic data on this solid considerably above its Debye temperature, where $\theta=300$ K. For minerals where the Debye temperature is much higher (geophysically interesting minerals, $600<\theta<1000$ K), the acoustic method fails to give elasticity measurements in the high temperature region. Here the RPR method succeeds, for it has been used at temperatures as high as 1300 K (SUMINO et al., 1983) for MgO. The ultrasonic experiment is limited to a maximum of about 800 K because of glues and the acoustic transducer problems at high T.

Using the RPR experiment, the methods and equations described herein will be particularly suitable for geophysically interesting minerals with high values of θ. But the proposed method can best be illustrated by applying it to NaCl because pertinent high pressure data now exist for this solid so that exact calculations can be made in the high pressure-high temperature field (see detailed calculations of BIRCH, 1986).

Introduction

Thermodynamic data for minerals of interest to geophysics are very much in demand. In principle, the piston-cylinder method is the ideal experiment obtaining data where values of the isothermal bulk modulus, K_T, and its pressure derivative, $(\partial K_T/\partial P)_T$, the thermal expansivity, $\alpha=[1/V](\partial V/\partial T)_P$, and the adiabatic relaxation, $(\partial T/\partial P)_S=(\gamma T)/K_S$, can be precisely defined as a function of P and T (BOEHLER and KENNEDY, 1980; BOEHLER, 1981).

The piston-cylinder method is especially valuable for compressible solids like alkali metals and most alkali halides. But the data found by this experiment for solids with a K_T modulus equal to or larger than that of LiF, are not very accurate because ΔV is quite small in the pressure range available. So, with a few exceptions (e.g., BOEHLER, 1981), this valuable experimental tool has not been applied to important minerals lying in the earth's interior.

One major difficulty is that the Debye temperature, θ, of the minerals appropriate to the earth's interior is high (from 600 K to 900 K), so that high-temperature properties appropriate to mantle and core conditions cannot be measured easily by the piston-cylinder method. On the other hand, θ values of alkali metals and most alkali halides are near room temperature or below, thus allowing access to high temperature along with high compression properties.

New experimental techniques are beginning to allow high temperature measurement of some of these properties on incompressible minerals. For example, the measurement of thermal expansivity versus temperature of perovskite of a very small sample made in the diamond cell has been achieved (KNITTLE et al., 1986).

*Present address: National Laboratory for High Energy Physics (KEK), Tsukuba, Japan.

High-Pressure Research in Mineral Physics, edited by M. H. Manghnani and Y. Syono, pp. 289–298.

This paper is concerned with the application of the measurements of high temperature properties using the rectangular parallelepiped resonance (RPR) method. High temperature measurements (at $P=0$) of the elastic constants of a number of incompressible solids of interest to geophysics have been achieved: MgO to 1300 K (SUMINO et al., 1983); Mg_2SiO_4 to 1200 K (SUZUKI et al., 1983); pyrope garnet to 1000 K (SUZUKI and ANDERSON, 1983); NaCl to 766 K (YAMAMOTO et al., 1986).

Our object is to demonstrate that by combining the high temperature (at 1 bar) elastic moduli measurements with a minimum amount of high pressure data (e.g., $(\partial K_T/\partial T)_P$ at $P=0$), we can obtain information on mantle and core materials at simultaneous high P and high T conditions. This result is possible because of a number of identities, arising from calculus and from thermodynamics, which allow the shift of variables. Thus, in certain cases, it is possible to calculate what we cannot measure from what we can measure.

We will limit the discussion in this paper to NaCl. For this solid, quality measurements exist from both high pressure experiments (see BOEHLER and KENNEDY, 1980) and high temperature experiments (YAMAMOTO et al., 1986); further, high temperature-low pressure quality acoustic measurements exist (SPETZLER et al., 1972), along with well tested equations of state (DECKER, 1971; BIRCH, 1978, 1986). The method described below can be used on solids with higher values of θ, providing sufficient quality elasticity information exists at temperatures above θ.

The Thermodynamic Identities

We begin by listing the thermodynamic identities used in this paper, many which are well known (see STACEY, 1978; BRENNAN and STACEY, 1979; ZEMANSKY, 1943; BASSETT et al., 1968).

The thermodynamic identities are:

Identities Arising from Definitions of Parameters K_T, α and C_V

$$[\partial(\alpha K_T)/\partial V]_T = (-1/V)(\partial K_T/\partial T)_V \quad (1)$$

$$[\partial(\alpha K_T)/\partial P]_T = (1/K_T)(\partial K_T/\partial T)_V \quad (2)$$

$$[\partial(\alpha K_T)/\partial T]_V = (1/T)(\partial C_V/\partial V)_T = (-\rho K_T/T)(\partial C_V/\partial P)_T \quad (3)$$

$$[\partial(\alpha K_T)/\partial T]_P = K_T(\partial\alpha/\partial T)_V \quad (4)$$

$$(\partial K_T/\partial T)_P = K_T^2(\partial\alpha/\partial P)_T = -K_T V(\partial\alpha/\partial V)_T \quad (5)$$

Identities Arising from Calculus

$$(\partial W/\partial T)_V = (\partial W/\partial T)_P + \alpha K_T(\partial W/\partial P)_T \quad (6)$$

where W can be the bulk modulus, K_T, the Grüneisen parameter, γ, specific heat at constant volume, C_V, and αK_T.

$$K_T = K_S/(1+\alpha\gamma T) \quad (7)$$

where (GRÜNEISEN, 1926)

$$\gamma = \alpha V K_S/C_P = \alpha V K_T/C_V \quad (8)$$

$$\gamma = (K_S/T)(\partial T/\partial P)_S \quad (9)$$

$$\alpha K_T = (\partial P/\partial T)_V \quad (10)$$

Similarly,

$$(\partial W/\partial P)_S = (\partial W/\partial P)_T + (\gamma T/K_S)(\partial W/\partial T)_P \quad (11)$$

where S stands for entropy.

Identities Involving the Temperature Derivative of K_T

An important anharmonic parameter at high temperature is

$$\delta_X = (-1/\alpha K_X)(\partial K_X/\partial T)_P \quad (12)$$

where $X=S$ represents the adiabatic case and $X=T$, the isothermal case. From equations (6) and (12), we have

$$\delta_T - K_{T0}' = (-1/\alpha K_T)(\partial K_T/\partial T)_V \quad (13)$$

and similarly, when replacing T by S and where

$$K_{X0}' = (\partial K_X/\partial P)_T \quad \text{as} \quad P\to 0 \quad (14)$$

Taking the derivative of equation (7) with respect to Y at constant Z, we find

$$(\partial K_S/\partial Y)_Z = (1+\alpha\gamma T)(\partial K_T/\partial Y)_Z + \alpha\gamma K_T[(T/\gamma)(\partial\gamma/\partial Y)_Z + (T/\alpha)(\partial\alpha/\partial Y)_Z + (\partial T/\partial Y)_Z] \quad (15)$$

When $Y=T$ and $Z=P$, we use equation (12) to find

$$\delta_T - \delta_S = \gamma[(\partial\ln\alpha/\partial\ln T)_P + (\partial\ln\gamma/\partial\ln T)_P + 1] - \alpha\gamma T\delta_T \quad (16)$$

Now $\delta_T > \delta_S$ because $(\partial\ln\alpha/\partial\ln T)_P$ is a positive number, $(\partial\ln\gamma/\partial\ln T)_P$ is close to zero, α increases strongly with T, and $\alpha\gamma T\delta_T$ is small compared to 1.

When $Y=P$ and $Z=T$, we use equation (5) in equation (15) to find

$$K_{T0}' = K_{S0}' + \alpha\gamma T(\delta_T + q - K_T') \quad (17)$$

where

$$q = (\partial \ln \gamma/\partial \ln V)_T \quad (18)$$

Since $\alpha\gamma T > 0$, and if, as is usually found, $\delta_T \simeq K_{T0}'$ and $q > 0$; then $K_{T0}' > K_{S0}'$: The difference is small—close to 0.1.

Identities Involving the Pressure Derivative of γ

A modification of a general equation derived by BASSETT et al. (1968), is

$$\begin{aligned} q &= (\partial \ln \gamma/\partial \ln V)_T \\ &= [1 + \gamma + (1 + \alpha\gamma T)\delta_S - K_S' \\ &\quad + T(\partial\gamma/\partial T)_P]/(1 + \alpha\gamma T) \end{aligned} \quad (19)$$

An equivalent identity for q using equation (1) with the derivative of equation (8) is

$$\begin{aligned} q = (\partial \ln \gamma/\partial \ln V)_T = 1 - (\partial \ln C_V/\partial \ln V)_T \\ - [1/(\gamma\rho C_V)](\partial K_T/\partial T)_V \end{aligned} \quad (20)$$

From equation (5), we have

$$\delta_T = -K_T(\partial \ln \alpha/\partial P)_T \quad (21)$$

Evaluating equation (21) by using equations (8), (18) and (3), we find that

$$q = \delta_T + 1 - K_{T0}' - (T/\gamma C_V)[\partial(\alpha K_T)/\partial T]_V \quad (22)$$

From equation (6), we see that whenever αK_T is both independent of pressure and temperature, the last term in equation (22) is zero. For this special case,

$$q = \delta_T + 1 - K_T' \quad (23)$$

Further, if δ_T is close to K_{T0}', q is close to unity.

Identities Involving the Thermal Pressure Using the General Equation of State

Using the general equation of state,

$$P = P_0(V) + P_{TH}(V, T) \quad (24)$$

where P_{TH} is the thermal pressure, we see that $(\partial P/\partial T)_V = (\partial P_{TH}/\partial T)_V$, and from equation (10), we find

$$\alpha K_T = (\partial P_{TH}/\partial T)_V \quad (25)$$

Anharmonicity Identities

Taking the derivative of equation (8) with respect to temperature at constant volume and using equation (4), we find

$$\begin{aligned} (\partial \ln \gamma/\partial T)_V = [1/\alpha K_T][\partial(\alpha K_T)/\partial T]_P \\ - [(\partial \ln C_V/\partial T)_V + (\partial \ln K_T/\partial T)_V] \end{aligned} \quad (26)$$

Assuming that $(\partial K_T/\partial T)_V = 0$, equation (26) becomes

$$(\partial\gamma/\partial T)_V = [\gamma/\alpha K_T][\partial(\alpha K_T)/\partial T]_P - (\gamma/C_V)(\partial C_V/\partial T)_V \quad (27)$$

The term on the left is the anharmonicity in the Grüneisen parameter, the first term on the right is the anharmonicity in the thermal pressure, and the last term is the anharmonicity in the specific heat. Note that the disappearance of any of the three anharmonicities does not by itself guarantee the disappearance of the remaining two (anharmonicity in specific heat is defined as the departure from the Dulong-Petit limit).

The Strategy

Important simplifications of the equations in the previous sections result if, as in the case for NaCl,

$$(\partial K_T/\partial T)_V = 0 \quad (28)$$

This parameter can be measured in a pressure experiment (BOEHLER and KENNEDY, 1980), but equation (13) shows that it can also be found, if K_{T0}' is known, in a temperature experiment where elastic constants and the thermal expansivity are measured versus temperature at 1 bar (e.g., the RPR method, or the acoustic method).

Supposing that equation (28) is valid, we see from equation (1) that

$$[\partial(\alpha K_T)/\partial V]_T = 0 \quad (29)$$

further we find from equation (25) that

$$(\partial P_{TH}/\partial T)_V \text{ is not equal to } f(V) \quad (30)$$

Ordinarily, we would expect that a proof would require an elaborate high pressure experiment. But the only pressure data input we use is K_{T0}'.

If equation (28) is valid, it can be shown from equations (2) and (6) that

$$[\partial(\alpha K_T)/\partial T]_V = [\partial(\alpha K_T)/\partial T]_P \quad (31)$$

Thus, measurements of either quantities in equation (31) give information on $(\partial\alpha/\partial T)_V$ and $(\partial C_V/\partial V)_T'$, but note that the quantity on the right of equation (31) is measured at constant pressure.

We see from previous equations that if $(\partial K_T/\partial T)_V=0$ and if αK_T is independent of temperature, then q is equal to unity. Thus, as a close approximation, (γ/V) is almost constant. Conversely, if it is found that (γ/V)=constant is a good approximation and $(\partial K_T/\partial T)_V=0$, we may deduce that both C_V and $(\gamma C_V/V)$ are independent of volume, an assumption sometimes used in the reduction of shock wave Hugoniot data. In any case, the value of q can be determined by high temperature elastic constant data when used with $(\partial K_T/\partial P)_T$ versus T.

If equation (28) is verified, then equation (13) becomes $\delta_T = K_{T0}'$. This means that

$$(\partial \delta_T/\partial T)_P = \partial^2 K_T/\partial T \partial P = \partial K_{T0}'/\partial T \tag{32}$$

and

$$(\partial \delta_T/\partial P)_T = \partial^2 K_T/\partial P^2 = K_{T0}'' \tag{33}$$

The Experimental Elastic Constant Data on NaCl and Basic Parameters

The experimental data for elastic constants and associated parameter, up to 766 K for NaCl as measured by the RPR experiment, are reported elsewhere in a companion paper; for experimental details, see YAMAMOTO et al. (1986).

Table 1 shows the data on thermal expansivity, specific volume and specific heat (taken from other sources), used in the calculation of elastic constants at high temperatures. Table 2 shows the data (primary and calculated) for the elastic constants and the parameters γ, $\gamma(V=V_0)$, δ_S,

TABLE 1. Temperature Variation of α, V and C_P in NaCl

T	α	$V=1/\rho$	C_P
294	117.0	0.4701	8.635
338	121.1	0.4726	8.792
381	125.1	0.4751	8.913
419	128.9	0.4774	9.015
449	131.9	0.4793	9.095
479	135.0	0.4812	9.173
511	138.3	0.4833	9.260
544	141.9	0.4856	9.350
563	143.9	0.4869	9.400
571	144.8	0.4874	9.420
596	147.6	0.4892	9.485
637	152.3	0.4922	9.605
675	156.7	0.4951	9.715
702	159.9	0.4972	9.800
745	165.1	0.5008	9.950
766	167.6	0.5025	10.030

T: Temperature, K. α: Volume thermal expansion coefficient reported by ENCK and DOMMEL (1965), 10^{-6}/K. V: Specific volume calculated on the basis of α, cm^3/g. C_P: Specific heat at constant pressure summarized in JANAF thermochemical tables by STULL and PROPHET (1971), 10^6 erg/(gK).

TABLE 2. Temperature Variation of Thermodynamic Parameters in NaCl. K_S Observed from RPR Experiment. All Other Variables Calculated from K_S Using Values in Table 1

T	K_S	K_T (7)	$K_T(V=V_0)$ (6)	γ (8)	$\gamma(V=V_0)$ (6)	δ_S (12)	δ_T (12)	αK_T	ΔP_{TH}^{294} (34)	q (19)
294	25.35	24.02	24.02	1.614	1.614	3.48	5.24	2.809	0.000	1.02
338	24.89	23.34	24.00	1.620	1.612	3.56	5.52	2.826	0.124	1.11
381	24.41	22.65	23.97	1.628	1.613	3.61	5.64	2.834	0.245	1.18
419	24.03	22.07	23.98	1.640	1.617	3.68	5.75	2.845	0.353	1.21
449	23.69	21.59	23.97	1.646	1.618	3.70	5.79	2.847	0.438	1.30
479	23.41	21.14	24.00	1.658	1.623	3.60	5.73	2.854	0.523	1.17
511	23.09	20.66	24.03	1.667	1.626	3.69	5.82	2.857	0.614	1.03
544	22.54	19.98	23.89	1.661	1.612	3.80	5.94	2.835	0.707	1.20
563	22.27	19.63	23.85	1.660	1.608	3.65	5.82	2.825	0.761	1.04
571	22.16	19.48	23.83	1.661	1.606	3.71	5.88	2.822	0.783	1.00
596	21.74	18.98	23.73	1.655	1.595	3.88	6.02	2.801	0.854	1.30
637	21.24	18.30	23.72	1.657	1.588	3.80	5.94	2.786	0.969	1.17
675	20.76	17.66	23.71	1.658	1.580	3.74	5.89	2.767	1.075	1.13
702	20.45	17.24	23.73	1.659	1.575	3.74	5.84	2.756	1.150	1.02
754	19.91	16.55	23.74	1.654	1.559	3.74	5.81	2.731	1.268	1.00
766	19.67	16.23	23.77	1.652	1.552	3.74	5.79	2.720	1.326	1.05

T: Temperature, K. K_S: Adiabatic bulk modulus, GPa. K_T: Isothermal bulk modulus, GPa. $K_T(V=V_0)$: Isothermal bulk modulus at constant volume, where V_0 is volume at ambient conditions. γ: Grüneisen parameter at 1 atmosphere. $\gamma(V=V_0)$: Grüneisen parameter at constant volume, V_0, when q=1.0 is assumed. δ_T: Isothermal Anderson-Grüneisen parameter. αK_T: Temperature coefficient of thermal pressure at constant volume, $(\partial P_{TH}/\partial T)_V$, 10^{-3} GPa/deg. ΔP_{TH}^{294}: Thermal pressure above 294 K, $\Delta P_{TH}^{294}=\int_{294}^{T}\alpha K_T dT$, GPa. q: calculated by equation (19).

δ_T, αK_T, found from the high temperature ΔP_{TH}^{294} experiment, where $\gamma(V=V_0)$ is the Grüneisen parameter at constant volume, V_0 (volume at ambient conditions). The calculated variables use differences and are not smoothed.

It is apparent from Table 1 that γ, δ_T and δ_S are reasonably constant and approximately independent of temperature at high temperatures (e.g., above the Debye temperature, θ, which for NaCl is 308 K). BIRCH (1986) also found that γ is nearly independent of temperature. Our attention is restricted to this high temperature range, sometimes called the classical range. Below θ, we observe rapid variation of the parameters with temperature because they are affected by quantum details (see solid lines in Figures 2 and 3 determined by using low temperature elasticity data).

Figure 1 shows the variation of K_S and K_T versus T as well as the variation of shear modulus, μ, versus T.

Computation of Thermodynamic Parameters which Ordinarily Involve Compression

The Vanishing of $(\partial K_T/\partial T)_V$

We need $K_{T0}'=(\partial K_T/\partial P)_T$ as $P\rightarrow 0$. This has been measured for NaCl by several experiments, but we shall use two reports: K_{T0}' as reported by SPETZLER et al. (1972) which requires a correction from the acoustically measured K_S'; and K_T' as reported by CHHABILDAS and RUOFF (1976), which is directly determined by measurement of length (metric method). The values at room temperature are respectively 5.35 and 5.80±0.15. In addition, SPETZLER et al. (1972) report that $K_{T0}'=6.13$ at 800 K; lower values of K_{T0}' have been reported by the piston-cylinder method: 4.92 by VAIDYA and KENNEDY (1971).

From Table 2, we see that at high temperatures, $\delta_T\rightarrow 5.8$; so that if we use the SPETZLER et al. (1972) high temperature value for K_{T0}' (6.13), and considering the uncertainty in that parameter, we have approximately

$$\delta_T - K_{T0}' = 0 \tag{34}$$

Consider equation (32). Using the value of $\partial K_T^2/\partial T\partial P$ which is 1.57×10^{-3}/deg (found by SPETZLER et al. (1972)), we find that δ_T should increase by about 0.11 between 294 and 756 K. The same trend is found in the

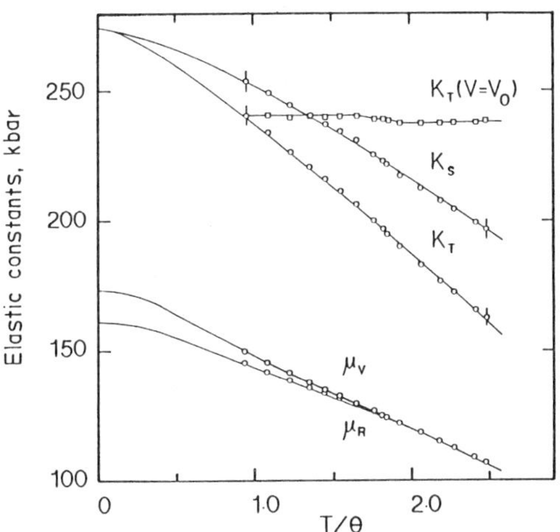

Figure 1. Temperature dependency of the bulk and shear moduli of sodium chloride. The measured K_S and the calculated isothermal bulk modulus, K_T, is at 1 atmosphere. Isothermal bulk modulus at constant volume, $K_T(V=V_0)$, where V_0 is volume at ambient conditions, is shown by squares, Note that $K_T(V=V_0)$ is very nearly independent of temperature. Two rigidities, μ_V and μ_R, are calculated by the Voigt and Reuss formulas, respectively.

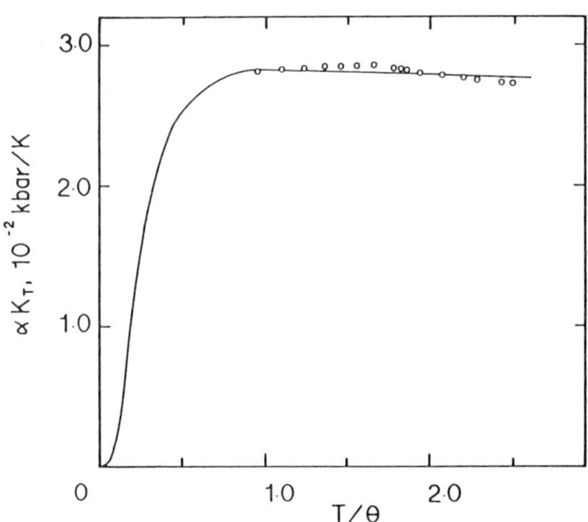

Figure 2. Temperature dependency of product αK_T of sodium chloride. αK_T is identical to the temperature coefficient of thermal pressure at constant volume, $(\partial P/\partial T)_V$. A slight decrease of αK_T above the Debye temperature, θ, is not sufficient to affect the apparent linearity of P_{TH} at high T.

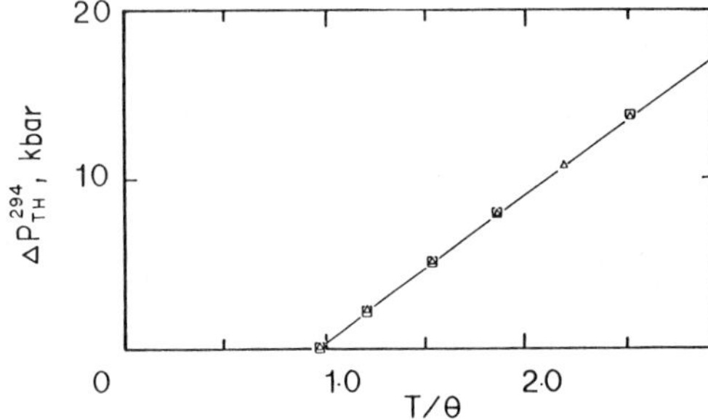

Figure 3. Thermal pressure above 294 K, ΔP_{TH}^{294}, versus temperature in sodium chloride. Note the agreement of the present results (shown as a straight line) with the data from compression experiment (BOEHLER and KENNEDY, 1980) denoted by triangles and from theoretical calculation (DECKER, 1971) denoted by squares.

RPR determined δ_T (see Table 2) but the amount of the increase is smaller. We take this to mean that our uncertainty in δ_T at high T is ± 0.4. In all, we find that $(\partial K_T/\partial T)_V$ is close to zero (see also SMITH and CAIN, 1980).

Another method is to numerically integrate equation (6) for K_T at constant volume versus temperature, where $(\partial K_T/\partial T)_P$ and αK_T vary slightly with temperature as listed in Table 2. In this integration, we use the two values of K_{T0}' estimated from SPETZLER et al. (1972) at 294 K and 800 K to make K_{T0}' a linear function of temperature. The integration begins with $T=294$ K (which defines V_0), and the result is shown by the curve of $K_T(V=V_0)$ in Figure 1. We see that K_T at constant volume is approximately independent of T over a wide range in T/θ. The lack of exact equality between δ_T and K_T' at high T, shown in Table 1, may be due to the inaccuracy of K_{T0}' at high T or to the fact that $(\partial K_T/\partial T)_V$ is not exactly zero, but only nearly so. BIRCH (1986) also found $(\partial K_T/\partial T)_V \simeq 0$ and $K_T' \simeq \delta_T$ for NaCl.

The Thermal Pressure

Since equation (28) is valid for NaCl, we predict that P_{TH} at high T is independent of volume, according to equation (29), and that P_{TH} is linear in temperature.

The value of P_{TH} at a given T data is found from integrating αK_T versus T. The plot of αK_T versus T is shown in Figure 2. This parameter tends to be independent of temperature above θ. We find $P_{TH}=-0.243+0.00283T$ GPa for temperatures above $T=294$ K. Since the evaluation of equations of state is often made with room temperature conditions as the standard reference state, it is convenient to define ΔP_{TH} between 294 K and T:

$$\Delta P_{TH}^{294} = P_{TH}(T) - P_{TH}(294) = 0.00283(T-294) \text{ GPa} \tag{35}$$

The data represented by (35) are graphed in Figure 3. The slope, $(\partial P_{TH}/\partial T)_V=0.00283$ GPa/deg, is in agreement with BOEHLER and KENNEDY's (1980) piston-cylinder experiment results at $P=0$ (see below). Equation (35) agrees well with the theoretical calculation of P_{TH} by DECKER (1971).

Now according to equation (29), ΔP_{TH}^{294} should be independent of volume. This result has been established by ANDERSON et al. (1982), who analyzed the data of the piston-cylinder experiment taken by BOEHLER and KENNEDY (1980) (see Table 3). From the Table 3 data, we find $(\partial P_{TH}/\partial T)_P=0.00284$ GPa/deg. The results show that there is complete consistency between the present elasticity and the piston-cylinder data. BIRCH (1986) found $(\partial P_{TH}/\partial T)_V=0.00286$ GPa/deg: In his approach, Birch found an equation of state compatible with the shock compression data (see FRITZ et al., 1971), the ultrasonic data (see SPETZLER et al., 1972), and the piston-cylinder data (see BOEHLER and KENNEDY, 1980).

TABLE 3. Change of Thermal Pressure, P_{TH}, from Room Temperature to Reference Temperature T, $\Delta P_{TH}^{298}=P_{TH}(T)-P_{TH}(298\text{ K})$, at Various Compressions for NaCl

V/V_0	($\Delta P_{TH}^{298} = P_{TH}(T) - P_{TH}(298$ K), kbar)					
	298 K	373 K	473 K	573 K	673 K	773 K
1.00	0.0	0.216	0.501	0.785	1.067	1.349
0.99	0.0	0.22	0.500	0.784	1.067	1.348
0.98	0.0	0.223	0.510	0.783	1.068	1.348
0.97	0.0	0.215	0.499	0.783	1.069	1.348
0.96	0.0	0.214	0.499	0.784	1.071	1.348
0.95	0.0	0.214	0.499	0.786	1.071	1.349
0.94	0.0	0.214	0.502	0.789	1.078	1.351
0.93	0.0	0.213	0.502	0.792	1.084	—
0.92	0.0	0.213	0.50	—	—	—
0.91	0.0	0.213	—	—	—	—

Note that ΔP_{TH}^{298} is substantially independent of volume. Data measured by BOEHLER and KENNEDY (1980) as reported by ANDERSON et al. (1982).

The Variation of γ versus Temperature

We see from Table 2 that $(\partial\gamma/\partial T)_P$, although small, is not quite zero. Assuming it to be zero, we find from equation (5) that

$$(\partial\gamma/\partial T)_V = (\partial\gamma/\partial P)_T\alpha K_T = -\alpha\gamma q \tag{36}$$

Thus $(\partial\gamma/\partial T)_V$ should be negative at high temperature for positive q. It is definitely not zero as often assumed in some papers on shock wave analysis. If q remains constant with temperature, the slope shown by equation (35) should increase with increasing T because α increases with temperature.

We see this effect when the data on $\gamma(V=V_0)$ versus T are found from integrating $(\partial\gamma/\partial T)_V$ according to equation (6). The result is plotted in Figure 4. The value of q is sometimes found by high pressure experiments in which γ is found as a function of V according to equation (18). Our own calculations for q are shown in the subsection on q. BIRCH (1986) also found that γ is virtually independent of T at constant P.

The Temperature Variation of Bulk Modulus

It is convenient to use δ_T and δ_S from equation (12), to find the temperature variation of the bulk modulus. Equation (16) is simplified when $(\partial\gamma/\partial T)_P$ is small. From Table 1, $(\partial \ln\alpha/\partial \ln T)_P$ is about 0.4, so that according to equation (16), $\delta_T-\delta_S=1.4\gamma$ for NaCl, as found in determinations of δ_T and δ_S separately. This invalidates for NaCl the often used approximation $\delta_T-\delta_S=\gamma$.

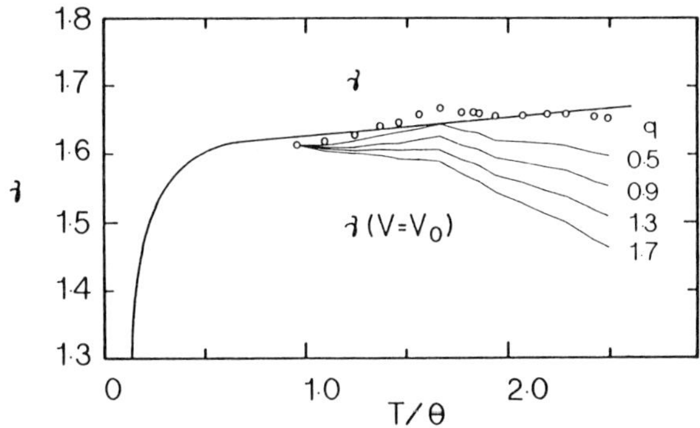

Figure 4. Temperature dependency of the Grüneisen parameter, γ, of sodium chloride at constant pressure, and $\gamma(V=V_0)$ at constant volume. The value of $\gamma(V=V_0)$ at high T depends on the value of $q=(\partial \ln \gamma/\partial \ln V)_T$.

The Value of q

The published value of q for NaCl varies in the high pressure experiments of Boehler and his colleagues: ($q=0.9$, BOEHLER and KENNEDY, 1980; $q=1.3$, BOEHLER and RAMAKRISHNAN, 1980; $q=1.9$, BOEHLER, 1981). This shows that q is sensitive to the fine details of the pressure experiment.

However, we can estimate q from our RPR data, combined with the acoustically determined value of K_{T0}'. The value of q is computed from equation (19) using SPETZLER's (1972) value of K_{T0}' versus T (from 5.35 to 6.1). The computed results are listed in the last column in Table 2 (here we see that q is about 1.1 and independent of temperature).

DECKER (1971) adopted the value $q=0.93$, which is close to our values. Our upper bound for q at 300 K, is slightly lower than found by SPETZLER et al. (1972), who reported $q=1.4$ for 300 K. But we agree with their values at 700 K ($q=1.1$).

We conclude that $q=1$ is reasonable and for many applications, it is safe to assume $\gamma/V=$constant for NaCl. If Spetzler's data for K_{T0}' is accurate, $q=0.9\pm0.2$ and virtually independent of T, which is similar to BIRCH's (1986) conclusion.

Anharmonicity

Since $[\partial(\alpha K_T)/\partial T]_P=0$, there is no anharmonicity in P_{TH}: it increases steadily and linearly with temperature above θ, with no evidence of a T^2 term at least up to $T=2.5\theta$. But this does not necessarily mean that anharmonicity is completely absent in NaCl. From equation (27), we see that the absence of anharmonicity in thermal pressure yields

$$(\partial\gamma/\partial T)_V = -(\gamma/C_V)(\partial C_V/\partial T)_V \qquad (37)$$

From Table 1, above 500 K, $(\partial\gamma/\partial T)_V$ approximately equals -3×10^{-3} at $q=1$.

From equations (28) and (2), $(\partial C_V/\partial V)_T=0$. Using this along with equation (6) yields $(\partial C_V/\partial T)_P=(\partial C_V/\partial T)_V$. Thus, $(\partial \ln C_V/\partial T)_P=(\partial \ln C_V/\partial T)_V=2\times10^{-3}$/deg which shows a positive T^2 term in C_V above about 500 K.

This number can be compared to the equivalent number computed directly from C_P data versus T listed in Table 1 (see STULL and PROPHET, 1971). Making the correction from C_P to C_V, we find $C_V(766)-C_V(294)=8.276-8.182$, or $(\partial \ln C_V/\partial T)_P=2.53\times10^{-3}$/deg which agrees with the above.

Thermal Expansivity Versus Temperature and Pressure

Using equations (5) and (12) we see that

$$(\partial \ln \alpha/\partial \ln V)_T = \delta_T \qquad (38)$$

This result is quite general depending only on an identity and a definition. If δ_T is independent of V along an isobar, by integrating equation (38)

$$\alpha(T, P)/\alpha(T, 0) = [V(T, P)/V(T, 0)]^{\delta_T} \qquad (39)$$

(see also ANDERSON, 1967; BIRCH, 1968; GUILLERMET, 1986; D. L. ANDERSON, 1987). If $(\partial^2 K_T/\partial P^2)=0$ is a reasonable assumption (which is rigorously true for the Murnaghan equation of state, and holds for any equation of state at pressures small compared to K_{T0}), then along an isotherm

$$K_T(T, P)/K_T(T, 0) = [V(T, P)/V(T, 0)]^{K_{T0}} \qquad (40)$$

In a solid, such as NaCl, where equation (34) is a close approximation, the exponential in equation (40), K_{T0}', can be replaced by δ_T. Comparing equations (39) and (40), we see that along an isotherm, αK_T is independent of P if equation (34) holds. The experiments show that for NaCl along an isobar, αK_T is independent of T at high T, or nearly so (see Figure 2). Considering this experiment and (40), it should therefore hold that in a portion of P, T space

$$\alpha K_T = \text{constant} \qquad (41)$$

which is independent of P and T (or V and T). This approximation has been discussed by BIRCH (1968), where he proposed that it might be useful for the estimation of α in the earth's interior (see also ANDERSON, 1984).

For equation (41) to be rigorously valid, δ_T has to be independent of T and P, and equal to K_{T0}'. How good is this approximation? SPETZLER et al. (1972) found from acoustic measurements that for NaCl, K_T' varied between

5.25 and 6.13 from 300 to 800 K, so that $\partial^2 K_T/\partial P \partial T = 1.6 \times 10^{-3}$deg. This should be the value of $\Delta\delta_T/\delta T$, according to equation (32); but we find a slightly smaller value; 1.2×10^{-3}, using Table 1. Thus, by assuming that δ_T is independent of T we could have an underestimated error in α of about two percent for $\delta T = 1000$ K. This error is of the order of the difference in measurements of α from one experimenter to another at higher temperatures. SPETZLER et al. (1972) found $(\partial^2 K_T/\partial P^2) = 0.94/$GPa at room temperature up to 1 GPa, close to the recommended value of −1 from CHHABILDAS and RUOFF (1976). BIRCH (1986) found a slightly smaller value (up to 3 GPa) using his equation of state: $\Delta\delta_T/\Delta P = 0.87/$GPa. We might expect δ_T to decrease by 1 for an increase of $\Delta P = 1$ GPa, along an isotherm.

The error $\Delta\alpha$ induced by a change of 1000 K is about the value of $\Delta\alpha$ induced by equation (40) at 1.5 GPa, but of opposite sign. Note, however, that if ΔP and ΔT are both positive increasing values, as on an adiabat, the temperature and pressure errors induced by using equation (41) will tend to cancel. Thus, along an adiabat or a Hugoniot, equation (41) may be quite good (see BIRCH's (1968) predictions of αK_T for the earth's mantle).

The Isothermal Equation of State (EOS)

In the case, such as for NaCl, where equation (28) is verified, and also $[\partial(\alpha K_T)/\partial T]_P = 0$, from which $q = 1$, it is possible to construct a differential equation from the definition of γ. As pointed out by BRENNAN and STACEY (1979) many theoretical formulations of γ are of the type,

$$\gamma = f(\partial^2 P/\partial\rho^2, \partial P/\partial\rho) \qquad (42)$$

Now, if $q = 1$, (or $\rho\gamma$=constant) then by equation (18)

$$\gamma = \gamma_0(\rho_0/\rho) \qquad (43)$$

By equating equations (42) and (43), a differntial equation results, which can be solved for $P(\rho/\rho_0)$, where one of the parameter is γ_0. BRENNAN nd STACEY (1979) used the free volume formation of γ for equation (42) and found

$$P(x) = (K_{T0}/2\gamma_0)x^{4/3}[\exp 2\gamma_0(1 - x^{-1}) - 1] = P_{B\text{-}S} \qquad (44)$$

where $x = \rho/\rho_0$. In this formation, γ_0 is constrained by K_{T0}, and

$$\gamma_0 = (K_{T0}' - 5/3)/2 \qquad (45)$$

Thus equation (44) is a two parameter equation of state, and while K_T'' is not zero (see BRENNAN and STACEY, 1979), it is at the same time not an arbitrary variable subject to adjustment. BRENNAN and STACEY (1979) called equation (44) the thermodynamic formulation of the equation of state since it is derivable from strictly thermodynamic conditions which when valid lead to $\gamma\rho$=constant.

Since $\gamma\rho$=constant is consistent with the thermodynamic properties of NaCl, $P_{B\text{-}S}$ can be tested against the 300 K isotherm data (for example, the isotherm deduced from shock wave measurements reported by FRITZ et al. (1971)).

We use $K_T(300) = 23.84$ GPa and $K_T' = 5.35$ from the SPETZLER et al. (1972) acoustic measurements to evaluate equation (44), $P_{B\text{-}S}$. The assignment of a value of γ_0 requires a choice: using equation (45), γ_0 is 1.814, but our measurements yield $\gamma = 1.615$. We use both values of γ_0 for two determinations of $P_{B\text{-}S}$, to be compared to the FRITZ et al. (1971) experimental isotherm, P_F, as well as to the third order Birch-Murnaghan EOS, $P_{B\text{-}M1}$, and the DECKER (1971) EOS for NaCl, P_D.

BIRCH (1986) found that the fourth order Birch-Murnaghan EOS, $P_{B\text{-}M2}$ (where $K_0'' = -0.059/$kbar and $K_0' = 5.35$), was required to best satisfy the FRITZ et al. (1971) shock wave data. The comparisons are made in Table 4. We see that $P_{B\text{-}S}$ follows the shock wave deduced isotherm, P_F, rather well. $P_{B\text{-}S}$ is closer to the shock data at high compressions than is the third order $P_{B\text{-}M1}$. $P_{B\text{-}S}$ tracks the experimental data much better than the Decker equation of state.

We conclude that when there is no access to K_0'', the Brennan-Stacey isothermal EOS is quite satisfactory, provided that the solid under question approximately satisfies the condition $\rho\gamma$=constant, (or satisfies the two equivalent thermodynamic identities).

In our research on high temperature elasticity, we find that the thermodynamic conditions given by equation (28) are satisfied for NaCl and Mg_2SiO_4. But equation (28) is not a good approximation for MgO and KCl (see YAMAMOTO and ANDERSON (1987)). We conclude therefore that $P_{B\text{-}S}$ may not be applicable to MgO or KCl.

The Thermal Equation of State

The thermal EOS of NaCl, using $P_{B\text{-}S}$ for the isothermal part is

$$P(\rho/\rho_0, T) = P_{B\text{-}S} + 0.00284(T - 294) \qquad (46)$$

found by combining equations (43) and (35), and where the standard condition state is taken to be 294 K. A comparison of the equation (45) results with the BOEHLER and KENNEDY (1980) data (Table 5) shows good agreement, even up to 773 K.

Acknowledgments. We are grateful for the helpful remarks of three reviewers. We acknowledge financial support provided by the Department of Energy (DE-FG03-84ER 13203) and the National Science Foundation (EAR83-12938). IGPP contribution no. 2906.

TABLE 4. Comparison of the Brennan-Stacey Equation of State Pressure, $P_{B\text{-}S}$, for NaCl with Measurements by FRITZ et al. (1971), P_F, and with Other Theoretical Equations of State. ($P_{B\text{-}M1}$, 3rd Birch-Murnaghan; $P_{B\text{-}M2}$, 4th Order Birch-Murnaghan, BIRCH, 1986; P_D, DECKER, 1971). $P_{B\text{-}S}$ Calculated from Equation (44), with a Small Correction (294 to 300 K) from (35)

Compression	Pressure, Exp. (GPa)	Pressure, Analytical Equation of State (GPa)				
V/V_0	P_F at 300 K	$P_{B\text{-}S}$		$P_{B\text{-}M1}$	$P_{B\text{-}M2}$	P_D
		$\gamma_0=1.84$	$\gamma_0=1.614$	3rd order	4th order	
1.0	0	0	0	0	0	0
0.95	1.403	1.42	1.41	1.402	1.398	
0.90	3.468	3.34	3.30	3.326	3.289	3.028
0.85	5.996	5.95	5.73	6.006	5.838	5.185
0.8	9.395	9.51	9.04	9.671	9.266	7.850
0.75	14.080	14.37	13.47	14.83	13.871	11.277
0.7	20.350	21.05	19.43	22.12	20.05	15.510

Note: $K_0=23.84$ GPa and $K_0'=5.35$ used for $P_{B\text{-}M1}$ and $P_{B\text{-}M2}$. The latter also uses $K_0''=-0.0059/\text{GPa}$.

TABLE 5. V/V_0 for NaCl as a Function of P and T. Comparison of Experimental Data with Theoretical Calculations. Experimental Values Taken from Table 3, BOEHLER and KENNEDY (1980). Theory Represents the Isothermal P, Given by $P_{B\text{-}S}$, Equation (44), Plus the Thermal Pressure, Equation (35), All Represented by Equation (46)

P GPa	25 °C		100 °C		200 °C		300 °C		400 °C		500 °C	
	Exper.	Theor.	Exper.	Theor.	Exper.	Theor.	Exper.	Theor.	Exper.	Theor.	Exper.	Theor.
0	1.00000	—	1.00928	1.0097	1.02253	1.0227	1.03682	1.0368	1.05226	1.0523	1.06908	1.0694
0.5	0.98020		0.98833		0.99989		1.01247		1.02575		1.04014	
1.0	0.96268	0.9629	0.96999	0.9702	0.98024	0.9805	0.99124	0.9914	1.02290	1.0335	1.01538	1.0156
1.5	0.94690		0.95356		0.96277		0.97248		0.98287		0.99382	
2.0	0.93248	0.9321	0.93857	0.9382	0.94699	0.9467	0.95571	0.9556	0.96509	0.9650	0.97478	0.9749
2.5	0.91913		0.92471		0.93251		0.94056		0.94913		0.95776	
3.0	0.90667	0.9060	0.91172	0.9112	0.91907	0.9184	0.92678	0.9260	0.93468	0.9338	0.94241	0.9421
3.5	0.89492	0.8942	0.89943	0.8991	0.90647	0.9056	0.91416	0.9128	0.92150	0.9201	0.92845	0.9289

REFERENCES

ANDERSON, D. L., The seismic equation of state: II Shear properties and thermodynamics of the lower mantle, *Phys. Earth Planet. Inter.*, submitted, 1987.

ANDERSON, O. L., Equation for thermal expansivity in interiors, *J. Geophys. Res.*, *72*, 3661–3668, 1967.

ANDERSON, O. L., The determination of the volume dependence of the Grüneisen parameter, γ, *J. Geophys. Res.*, *79*, 1153–1155, 1974.

ANDERSON, O. L., A universal thermal equation of state, *J. Geodynamics*, *1*, 185–214, 1984.

ANDERSON, O. L., R. BOEHLER, and Y. SUMINO, Anharmonicity in the equation of state at high temperature for some geophysically important solids, in *Advances in Earth and Planetary Sciences, 12, High Pressure Research in Geophysics*, edited by S. Akimoto and M. H. Manghnani, pp.273–283, Center for Academic Publications, Tokyo, 1982.

BASSETT, W. A., T. TAKAHASHI, H. K. MAO, and J. S. WEAVER, Pressure induced phase transition in NaCl, *J. Appl. Phys.*, *39*, 319–325, 1968.

BIRCH, F., Thermal expansion at high pressures, *J. Geophys. Res.*, *73*, 817–819, 1968.

BIRCH, F., Equation of state and thermodynamic parameters of NaCl to 300 kbar in the high temperature domain, *J. Geophys. Res.*, *91*, 4949–4954, 1986.

BOEHLER, R., Adiabats $(\partial T/\partial P)_S$ and Grüneisen parameter of NaCl up to 50 kilobars at 800 °C, *J. Geophys. Res.*, *86*, 7159–7162, 1981.

BOEHLER, R., and G. C. KENNEDY, Equation of state of sodium chloride up to 32 kbar and 400 °C, *J. Phys. Chem. Solids*, *41*, 517–523, 1980.

BOEHLER, R., and J. RAMAKRISHNAN, Experimental results on the pressure dependence of the Grüneisen parameter: a review, *J. Geophys. Res.*, *85*, 6991–7002, 1980.

BRENNAN, B. A., and F. D. STACEY, A thermodynamically based equation of state for the lower mantle, *J. Geophys. Res.*, *84*,

5535–5539, 1979.

CHHABILDAS, L. C., and A. L. RUOFF, Isothermal equations of state for sodium chloride by the length change measurement technique, *J. Appl. Phys., 47*, 4182–4187, 1976.

DECKER, D. L., High pressure equation of state for NaCl, KCl, and CsCl, *J. Appl. Phys., 42*, 3239–3244, 1971.

DEMAREST, H. H., Jr., Cube resonance method to determine the elastic constants of solids, *J. Acoust. Soc. Am., 49*, 768–775, 1969.

ENCK, F. D., and J. G. DOMMEL, Behavior of the thermal expansion of NaCl at elevated temperatures, *J. Appl. Phys., 36*, 839–844, 1965.

FRITZ, J. N., S. P. MARSH, W. J. CARTER, and R. G. MCQUEEN, The Hugoniot equation of state of sodium chloride in the sodium chloride structure: accurate characterization of the high pressure environment, edited by E. C. Lloyd, *NBS Spec. Publ., 326*, 201–208, 1971.

GRÜNEISEN, E., State of a solid body, NASA Publ. No. RE2-18-59W, Translation of *Handbuch der Physik*, vol. 10, pp. 1–52, 1926.

GUILLEMET, A. F., The pressure dependence of the expansivity and of the Anderson-Grüneisen parameter in the Murnaghan approximation, *J. Phys. Chem. Solids, 47*(6), 605–607, 1986.

KNITTLE, E., R. JEANLOZ, and G. L. SMITH, The thermal expansion of silicate perovskite and stratification of the earth's mantle, *Nature*, in press, 1986.

SMITH, C. S., and L. S. CAIN, Temperature derivatives at constant volume of the elastic constants of the alkali halides, *J. Phys. Chem. Solids, 41*, 199–203, 1980.

SPETZLER, H., C. G. SAMMIS, and R. J. O'CONNELL, Equation of state of NaCl: ultrasonic measurements to 8 kbar and 800 °C and static lattice theory, *J. Phys. Chem. Solids, 33*, 1727–1750, 1972.

STACEY, F. D., Applications of thermodynamics to fundamental earth physics, *Geophys. Survey, 3*, 175–204, 1977.

STULL, D. R., and H. PROPHET (eds.), *JANAF Thermochemical Tables*, 2nd ed., U.S. Department of Commerce, National Bureau of Standards, Washington, D.C., 1971.

SUMINO, Y., O. L. ANDERSON, and I. SUZUKI, Temperature coefficients of elastic constants of single-crystal MgO between 80 and 1,300 K, *Phys. Chem. Minerals, 9*, 38–47, 1983.

SUZUKI, I., and O. L. ANDERSON, Elasticity and thermal expansion of a natural garnet up to 1,000 K, *J. Phys. Earth, 31*, 125–138, 1983.

SUZUKI, I., O. L. ANDERSON, and Y. SUMINO, Elastic properties of a single-crystal forsterite Mg_2SiO_4, up to 1,200 K, *Phys. Chem. Miner., 10*, 38–46, 1983.

VAIDYA, S. N., and G. C. KENNEDY, Compressibility of 27 halides to 45 kbar, *J. Phys. Chem. Solids, 32*, 951–964, 1971.

YAMAMOTO, S., I. OHNO, and O. L. ANDERSON, High temperature elasticity of sodium chloride, *J. Phys. Chem. Solids, 48*, 143–151, 1987.

YAMAMOTO, S., and O. L. ANDERSON, Elasticity and anharmonicity of potassium chloride, *J. Phys. Chem. Miner.*, in press, 1987.

ZEMANSKY, M. W., *Heat and Thermodynamics*, p. 73, McGraw Hill, New York, 1943.

HIGH PRESSURE RAMAN AND X-RAY STUDIES OF SULFUR AND ITS NEW PHASE TRANSITION

Lizhong WANG, Yongnian ZHAO, Ren LU, Yue MENG, Yuguo FAN*, Huan LUO, Qiliang CUI, and Guangtian ZOU

*Institute of Atomic and Molecular Physics, *Institute of Theoretical Chemistry, Jilin University, Changchun, China*

Abstract. High pressure Raman spectra reveal that orthorhombic sulfur probably undergoes a structural transition at ambient temperature and about 5 GPa. In situ single-crystal X-ray diffraction data confirm that the orthorhombic structure transforms into a triclinic structure at 5.3 GPa. The dependence of mode-Grüneisen parameter (γ_i) on vibrational frequency (ν_i) was determined at several pressures from the Raman spectra and compressibilities measured with a piston-cylinder apparatus. The behavior of γ_i with ν_i is roughly described by the expression $\gamma_i \sim \nu_i^{-2}$.

Introduction

The equilibrium phase diagram of sulfur at moderate pressures, as synthesized by PISTORIUS (1976) from a number of studies, is very complex and ill-determined, with considerable discrepancies between results of different workers.

Pressure induced polymorphic transitions on sulfur at 25 °C below 10 GPa have been studied by volumetric, electrical conductivity, and optical and X-ray diffraction methods. PAUKOV et al. (1965) found a single transition at 2.2 GPa; VEZZOLI et al. (1969) found five separate transitions; and SUCHAN et al. (1959), BRIDGMAN (1945), and VAIDYA and KENNEDY (1972) found no evidence of any transitions. Similarly, the results of studies of the insulator-to-metal transition in sulfur, as summarized by RUOFF and GUPTA (1979), show disagreement about the transition pressure, which has been reported from as low as 20 GPa up to 50 GPa. It must be concluded that much of the work reported thus far may be questionable and that considerable work is still required on all aspects of the sulfur phase diagram.

We set ourselves the task of trying to resolve the ambiguities in these previous studies of polymorphic transitions at ambient temperature below 10 GPa by utilizing Raman scattering, single-crystal X-ray diffraction, and volume compression.

Experimental Methods

Raman Scattering

High-pressure Raman spectra were obtained with a gasketed diamond-anvil cell with 16:3:1 methanol-ethanol-water mixture for the hydrostatic pressure medium. The pressure was calibrated with the well-known ruby fluorescence scale. A thin plate of orthorhombic sulfur (Ø150 μm×75 μm thick), ground from a perfect single crystal, was used in order to have completely randomly oriented crystallites and to avoid spurious intensity dependence from a partially-oriented sample. The backscattering Raman spectra were recorded using a Spex-1403 double monochromator with a conventional photon counting system and a Datamate. The 5145 Å radiation from a Spectra Physics Model 165 Argon Ion Laser was used for excitation at an output power of 350 mW.

Single-Crystal X-ray Diffraction

Single-crystal X-ray diffraction work at high pressures was carried in a gasketed Mao-Bell type miniature diamond-anvil cell (MAO and BELL, 1979). The gasket was prethinned before sample mounting to decrease changes in size of the gasket hole (Ø200 μm×150 μm thick) under high pressure. Pressures measured before and after the diffraction data collection were calibrated using the ruby scale. Measurements of unit cell dimensions and integrated intensities were performed on a computer-controlled Nicolet XRD R3 four-circle diffractometer using a modified program of data collection. A fixed-ϕ mode rather than the normal mode (ω=0) was adopted, and angles between the X-ray axis and the symmetry axis of the diamond cell were limited to less than 32° in order to reduce the attenuation of X-ray. Molybdenum radiation (λ=0.71069 Å), and a θ-2θ scan technique were used.

Volume Compression

Volume compression experiments were carried out in an end-loaded piston-cylinder apparatus by monitoring piston displacements. Details of the experimental technique have been described by VAIDYA and KENNEDY (1970). Machined cylindrical specimens slightly less than 11 mm diameter and 16 mm length were sheathed in indium. Samples of rhombic sulfur were prepared in a special way in order to insure the uniformity throughout the material. A specimen of appropriate length was

High-Pressure Research in Mineral Physics, edited by M. H. Manghnani and Y. Syono, pp. 299–304.

obtained by melting crushed sulfur (industrial purity) enclosed in a cylindrical indium jig and then recrystallizing it in the air. The cylinder obtained was used for the compressibility measurements.

Results and Discussion

Raman Scattering

Group theory predicts 48 allowed Raman-active modes for vibrations within the unit cell consisting of 4 S_8 puckered octagonal molecules. Usually many of these 48 Raman bands will be either too weak to be observed or too close to other bands to be resolved. In fact, a total of 25 peaks were observed in spectra at zero pressure, which agrees well with spectra reported from earlier studies (ANDERSON and LOH, 1969). The frequencies of scattering intensity maxima are collected in Table 1. For comparison, data from previous work are also included in this table. Further, 21 peaks have been measured under high pressure. The intramolecular and rigid-molecule vibrational modes (i.e., internal and external modes) are distinguished according to the work of ANDERSON and LOH (1969). The internal modes span the range from 473 to 74.6 cm^{-1} and the external modes crowd into the frequency region below about 69 cm^{-1} without any appreciable frequency gap separating the internal and external modes.

Raman spectra of sulfur at several pressures are reproduced in Figure 1 and pressure dependences of the Raman frequencies are plotted in Figure 2. It can be seen that although some modes (69, 183, 245.8, and 249 cm^{-1}) are pressure insensitive, the normal behavior is for

TABLE 1. Raman Spectra of Rhombic Sulfur at Zero Pressure

Mode	Raman frequencies (cm^{-1})					
	This work (300 K)	ZALLEN (1974) (300 K)	ANDERSON and LOH (1969) (300 K)	(100 K)	Assignments	
External	27.2	26.5	28	29	R_z	$B_{2g}B_{3g}$
	29.8					
	43.2	43	44	39	T_xT_y	$B_{1g}B_{2g}B_{3g}$
				44		A_g
	50.8	51		53	R_xR_y	A_g
	(53)	54.5	51	59		B_{1g}
	62.8	62	63	63		B_{2g}
	[69]	[65]		68		B_{3g}
Internal	74.6	83.5	84	77	ν_9	B_{2g}
	83.2	89		79		B_{3g}
	88			85		A_g
				91		B_{1g}
	148.4	152	157	149	ν_8	B_{2g}
	151.6			152		B_{3g}
	154.2			156		A_g
	156			159		B_{1g}
	(158.4)					
	183		186	182	ν_6	B_{1g}
	186.4			187		A_g
	(194.4)			197		$B_{2g}B_{3g}$
	214.6	215	217	214	ν_2	B_{1g}
	219.2	220	221	218		A_g
	(236)		241	236	ν_4	$B_{2g}B_{3g}$
	245.8	249	249	246	ν_{11}	A_gB_{1g}
	249			250		$B_{2g}B_{3g}$
				417	ν_3	A_gB_{1g}
	439.2	438	442	441	ν_{10}	A_gB_{1g}
						$B_{2g}B_{3g}$
	468			464	ν_7	(or ν_5)?
				466		
	[470]	474	471	470	ν_1	B_{1g}
	473			477		A_g

Note: The numerals in parentheses are the frequencies for the weak lines or shoulders seen only at zero pressure; the numerals in square brackets are the extrapolated $p=0$ frequencies for the lines seen only at high pressure.

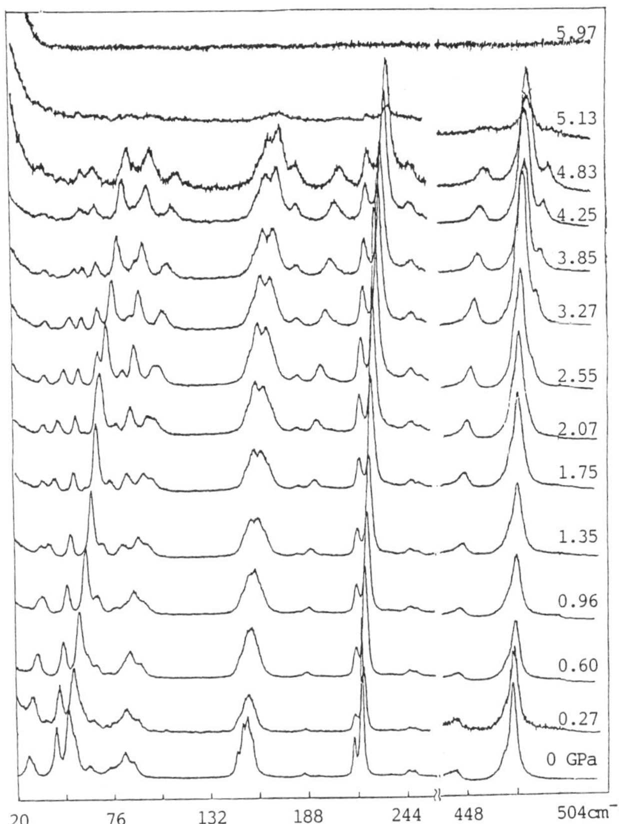

Figure 1. Raman spectra of sulfur at several pressures and room temperature. No Raman peaks were observed at 5.97 GPa.

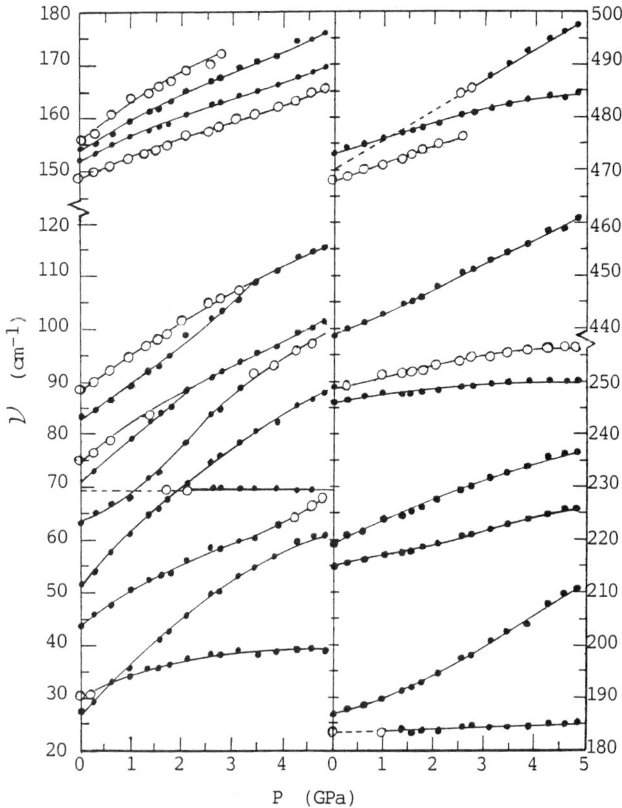

Figure 2. Pressure dependence of the Raman frequencies in rhombic sulfur up to 5 GPa. Solid circles correspond to peak positions of strong or well-resolved Raman lines; open circles correspond to weak lines or shoulders. The dashed lines represent the linear extrapolation to zero pressure from high pressure.

Raman frequencies to increase appreciably with pressure. These two figures show a variety of features, but perhaps most important is that the separation of the pairs of low-frequency modes, 43.2–69 cm^{-1}, 62.8–74.6 cm^{-1}, and 83.2–88 cm^{-1}, decreases appreciably with increasing pressure. At 4.83 GPa they cross or mix each other, leading to an abrupt decrease in the number of Raman modes. In addition, all of the Raman peaks become very weak in intensity at 4.83 GPa. At 5.13 GPa most of the peaks vanish but the highest mode, at 473 cm^{-1}, is still seen. In the spectrum taken at 5.97 GPa no Raman peaks could be seen, while the sample appears orange in color. These changes presumably are due to a phase transition and the shift in the optical absorption edge below the 5145 Å laser used as the exciter source in our Raman measurements. The former speculation has been confirmed by our X-ray diffraction experiments and the latter is in good agreement with data obtained by SUCHAN et al. (1959).

In order to reveal that pressure induced shifts of Raman frequencies in rhombic sulfur are strikingly inconsistent with a one-parameter Grüneisen frequency-volume scaling law, the frequency dependence of the mode-Grüneisen parameter γ_i at 0, 2, and 4 GPa is plotted in Figures 3a, b, and c, respectively. The mode-Grüneisen parameters were calculated using the formula

$$\gamma_i = 1/\beta(dln\ \nu_i/dP)$$

where the compressibility β was determined from our volume compression experiments using the piston-cylinder apparatus. This formula, strictly speaking, is true only for isotropic crystals. As in our case, however, rhombic sulfur shows a quasi-isotropy in compression as seen in Figure 4 (see following section). Thus the use of γ_i in terms of compressibilities for rhombic sulfur is not unreasonable.

Figures 3a, b, and c show clearly that, far from being frequency independent, γ_i is a monotonically decreasing function of the frequency, falling steeply from values of order 1 at low frequencies to values of order 0.1-to-0.01 at high frequencies. The gross behavior of γ_i with ν_i is roughly $\gamma_i \sim \nu_i^{-2}$, first obtained and preliminarily interpreted with a simple model by ZALLEN (1974). In comparing

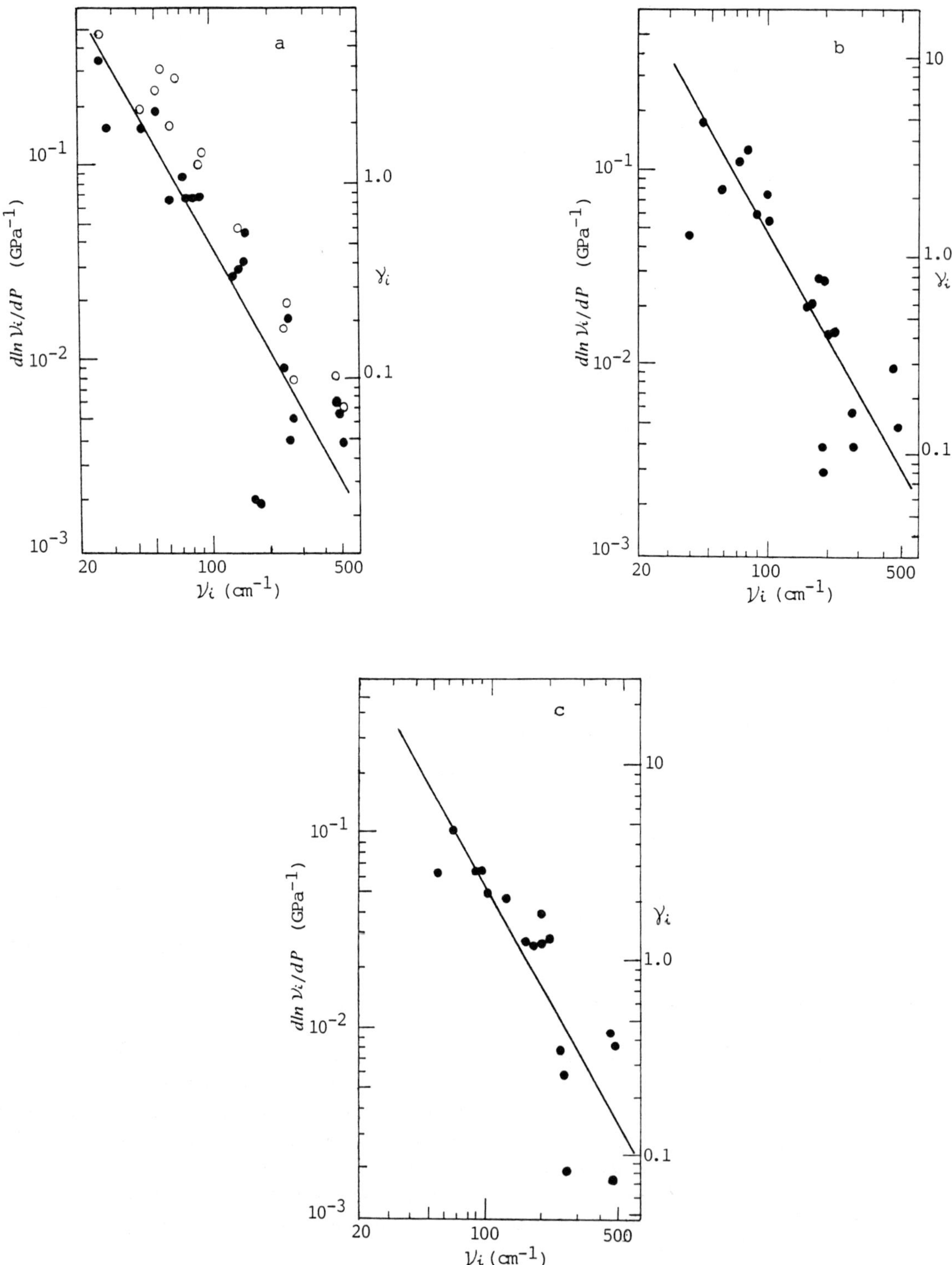

Figure 3. Correlation between mode Grüneisen parameter and Raman frequency for rhombic sulfur. a. $P=0$ GPa, $\beta=0.083$ GPa^{-1}; b. $P=2$ GPa, $\beta=0.037$ GPa^{-1}; c. $P=4$ GPa, $\beta=0.023$ GPa^{-1}. Solid circles are obtained in this work; open circles represent Zallen's data; solid curves are based on Zallen's theory.

Figure 3a with Figures 3b and 3c, we can find that, although it is not sensitive to pressure, the range of magnitudes spanned by the observed γ_i tends to decrease with increasing pressure, which implies the discrepancy between intra- and inter-molecular bonds in rhombic sulfur tends to shrink slightly. In view of the range of γ_i at a given pressure, reflecting a hierarchy of forces in the crystal, the above broad features imply that the intra- and inter-molecular bonds in rhombic sulfur do not change essentially before phase transition.

In addition, for comparison, Zallen's data observed to 1 GPa are included in Figure 3a. It can be seen that although the number of ν_i measured by Zallen is less than ours (see Table 1) and Zallen's slopes ($d\ln\nu_i/dP$) are systematically higher than ours, the overall conclusions are identical on the behavior of γ_i with ν_i under pressures up to 1 GPa. We believe that the reason for the above differences is probably the different experimental equipment used for the two works, including hydrostatic-pressure systems, exciter sources, monochromators, and recording systems.

Single-Crystal X-ray Diffraction

The unit cell parameters of sulfur at several pressures are listed in Table 2. Numerals in parentheses were obtained by an index transformation program. Table 2 illustrates that the crystal structure of sulfur at 3.45 GPa is still rhombic and that at 6.71 GPa it has been transformed into triclinic. It should be mentioned here that the X-ray diffraction pattern in a photograph taken at 5.30 GPa is similar to that at 6.71 GPa. Unfortunately the unit cell parameters at 5.30 GPa could not be calculated by the computer of the four-circle diffractometer because of the absence of diffraction data. As mentioned above, it is clear that the phase transition pressure from rhombic to triclinic is to be expected between 3.45 and 5.30 GPa.

The variations of the unit cell parameters, *a*, *b*, and *c*, with pressure for rhombic sulfur are plotted together in Figure 4. It shows that the three curves are nearly parallel, which indicates a quasi-isotropy in compressibility for sulfur before the phase transition. This result provides a powerful backing for using the mode-Grüneisen parameter's formula as mentioned under "Raman Scattering."

TABLE 2. Unit Cell Parameters of Sulfur at Several Pressures

Pressure (GPa)	*a* (Å)	*b* (Å)	*c* (Å)	α	β	γ	*V* (Å^3)
0.0001	11.21	13.76	26.18	89.87	89.98	89.98	4038
	(11.21	13.80	26.25	90.00	90.00	90.00	4061)
1.06	11.12	13.31	25.32	88.15	89.31	89.16	3745
	(10.87	13.40	25.78	90.00	90.00	90.00	3755)
1.99	10.60	13.09	25.64	90.30	90.50	90.10	3551
	(10.57	13.16	25.51	90.00	90.00	90.00	3548)
3.45	10.64	12.86	24.69	88.60	89.58	88.62	3377
	(10.34	12.84	25.13	90.00	90.00	90.00	3336)
6.71	13.95	15.98	16.06	76.80	69.30	68.50	3098

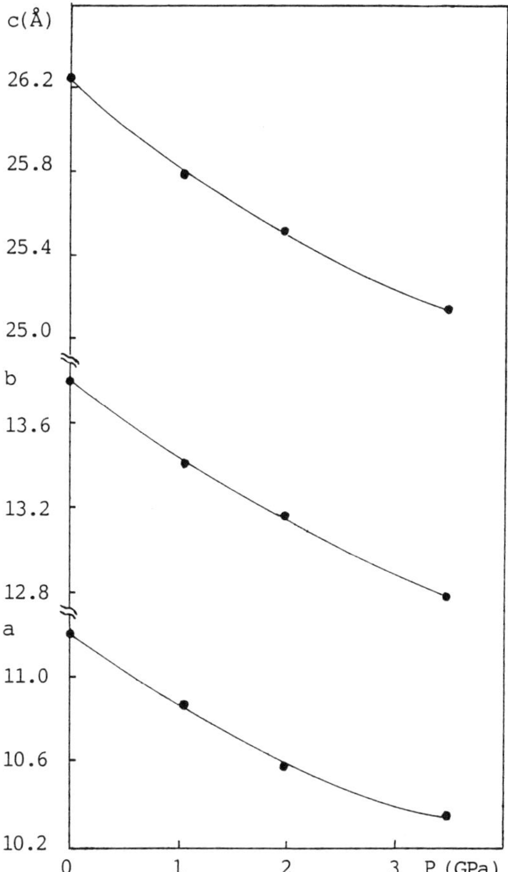

Figure 4. The variation of the unit cell parameters, *a*, *b* and *c*, with pressure for rhombic sulfur.

Volume Compression

The measured values of V/V_0 by using the piston-cylinder apparatus are given in Table 3 and shown in Figure 5 along with our X-ray diffraction results and those of VAIDYA and KENNEDY (1972). Agreement between our results and results from Vaidya and Kennedy is, in general, excellent.

In the rhombic phase the pressure dependence of $\Delta V/V_0$ $(=1-V/V_0)$ is represented by the expression

$$\Delta V/V_0 = aP + bP^2 + cP^3$$

where $a=9.527\times10^{-2}$, $b=-2.061\times10^{-2}$, $c=1.903\times10^{-3}$, and P is in GPa. Using this equation, we can calculate the Grüneisen constant:

$$\gamma_0 = b/a^2 - 2/3 = 1.60,$$

TABLE 3. Comparison of V/V_0 for Rhombic Sulfur at Several Pressures

Pressure	This work		VAIDYA and KENNEDY (1972)
(GPa)	Piston-cylinder	X-ray	Piston-cylinder
0.5	0.9585		0.9545
1.0	0.9229	0.9260	0.9196
1.5	0.8971		0.8923
2.0	0.8759	0.8765	0.8703
2.5	0.8598		0.8520
3.0	0.8498		0.8361
3.5	0.8377	0.8288	0.8221
4.0	0.8266		0.8100

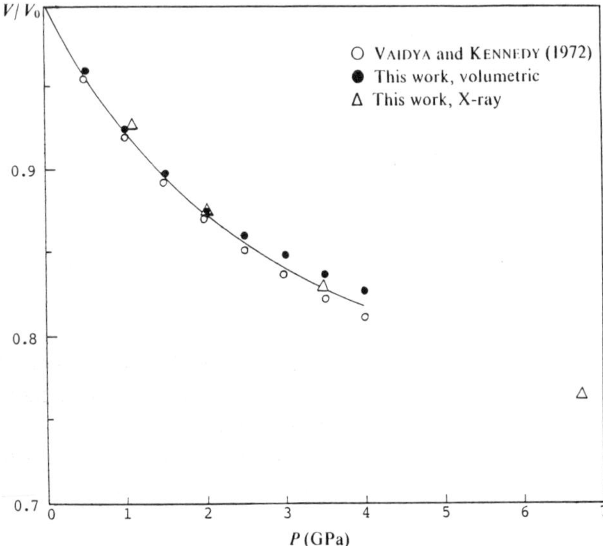

Figure 5. Comparison of V/V_0 vs. pressure from the present work with those of VAIDYA and KENNEDY (1972).

which is the same order of magnitude with and somewhat larger than the overall Grüneisen γ figures of 0.91, 1.35, and 1.32, determined from the weighted average of mode-Grüneisen parameters γ_i at 0, 2, and 4 GPa respectively. This is reasonable because the unmeasured Raman-active modes and all $k \neq 0$ modes were missed in the weighted average.

Summary

We now summarize the important observations of our experiments:

a) Rhombic sulfur appears to undergo a structural transition between 4.83 and 5.13 GPa based on our Raman data. Some Raman modes cross or mix and all of the Raman peaks become very weak at 4.83 GPa; also, most of Raman peaks vanish suddenly at 5.13 GPa. No Raman signals could be measured at 5.97 GPa, where the samples color changed from yellow to orange.

b) There is no phase transition in the pressure range from 0 to 4 GPa, as illustrated by values of V/V_0 measured using the piston-cylinder apparatus. In this pressure range the volume compression curve is continuously smooth without any clear discontinuity.

c) X-ray data suggest that the rhombic-to-triclinic phase transition pressure is expected to be between 3.45 and 5.30 GPa because single-crystal X-ray diffraction data indicate that the structure of sulfur at 3.45 GPa is still rhombic and X-ray diffraction photograph shows that it has changed into triclinic at 5.30 GPa.

d) The above results show the rhombic-to-triclinic phase transition pressure is about 5 GPa. This is a new phase transition, and the triclinic phase has never before been seen.

e) In the rhombic phase, the mode-Grüneisen parameter is a monotonically decreasing function of Raman frequency described by the expression first proposed by Zallen of $\gamma_i \sim \nu_i^{-2}$. The spanned range of γ_i tends to shrink with increasing pressure, reflecting the decreasing discrepancy between intra- and inter-molecular bonds in the crystal.

REFERENCES

ANDERSON, A., and Y. T. LOH, Low temperature Raman spectrum of rhombic sulfur, *Can. J. Chem.*, *47*, 879–884, 1969.

BRIDGMAN, P. M., The compression of twenty-one halogen compounds and eleven other simple substances to 100,000 kg/cm^2, *Proc. Am. Acad. Arts Sci.*, *76*, 1–7, 1945.

MAO, H. K., and P. M. BELL, Design and operation of a diamond-window, high pressure cell for the study of single-crystal samples loaded cryogenically, *Carnegie Inst. Washington Yearb.*, *79*, 409–411, 1980.

PAUKOV, I. E., E. Y. TONKOV, and D. S. MIRINSKII, Phase diagram of sulfur under high pressure, *Dokl. Akad. Nauk SSSR*, *164*, 588–589, 1965.

PISTORIUS, C. W. F. T., Phase relations and structures of solids at high pressures, in *Progress in Solid State Chemistry*, Vol. 11, edited by J. McCaldin and G. Somorjai, pp. 1–151, Pergamon Press, London, 1976.

RUOFF, A. L., and M. C. GUPTA, Studies of the high pressure transition in sulfur, in *High-Pressure Science and Technology*, Vol. 1, Physical Properties and Material Synthesis, edited by K. D. Timmerhaus and M. S. Barber, pp. 161–171, Plenum Press, New York, 1979.

SUCHAN, H. L., S. WIEDERHORN, and H. G. DRICKAMER, Effect of pressure on the absorption edges of certain elements, *J. Chem. Phys.*, *31*, 355–357, 1959.

VAIDYA, S. N., and G. C. KENNEDY, The compressibility of 18 metals to 45 kbar, *J. Phys. Chem. Solids*, *31*, 2329–2345, 1970.

VAIDYA, S. N., and G. C. KENNEDY, Compressibility of 22 elemental solids to 45 kb, *J. Phys. Chem. Solids*, *33*, 1377–1389, 1972.

VEZZOLI, G. C., F. DACHILLE, and R. ROY, Sulfur melting and polymorphism under pressure: Outlines of fields for 12 crystalline phases, *Science*, *166*, 218–221, 1969.

ZALLEN, R., Pressure-Raman effects and vibrational scaling laws in molecular crystals: S_8 and As_2S_3, *Phys. Rev.*, *B9*, 4485–4496, 1974.

COMPUTER-EXPERIMENTAL SYNTHESIS OF SILICA WITH THE α-PbO_2 STRUCTURE

Yoshito MATSUI

Institute for Study of the Earth's Interior, Okayama University, Misasa, Tottori-ken 682-02, Japan

Katsuyuki KAWAMURA

Department of Chemistry, Faculty of Science, Hokkaido University, Sapporo 060, Japan

Abstract. A series of molecular dynamics (MD) calculations on model SiO_2 systems with the fluorite configuration have been carried out over the pressure range of 0–300 GPa. A pressure-controlling, noncubic cell MD program (XDORTHO) was used, assuming two-body central potential functions. We observed a very rapid transformation from cubic to orthorhombic lattice structure below pressures of dynamic instability of SiO_2-fluorite, and found the "product" to have the α-PbO_2 structure.

The density (4.34 Mg/m^3) and cohesive (lattice) energy (−13109 kJ/mol) of computer-synthesized SiO_2 with the α-PbO_2 structure are very close to the values for SiO_2 with the rutile structure (4.30 Mg/m^3 and −13135 kJ/mol, respectively), with the IF-2 potential at 0 GPa and 300 K. In view of the absence of experimental evidence for existence of SiO_2 with the α-PbO_2 structure, we suppose that the polymorph, if it does exist, is metastable. However, if SiO_2-fluorite exists at ultrahigh-pressure, we anticipate having only a reverted material in the α-PbO_2 structure.

Introduction

It has been recognized that moleculer dynamics (MD) calculations are useful in simulating dynamic behaviors of atoms in molten, vitreous, and crystalline materials. During the course of our MD study of "hypothetical" compounds of potential geophysical importance, we encountered a rather unexpected phase transformation from the fluorite to α-PbO_2 structure in SiO_2.

Silica with fluorite structure has long been expected to be the densest polymorph. However, no experimental confirmation for the existence of SiO_2-fluorite has been obtained. Very recently, HEMLEY et al. (1985) have determined, on the basis of quantum mechanical and quasiharmonic lattice dynamics calculations, that SiO_2 with the fluorite structure is dynamically unstable under low pressures. BUKOWINSKI and WOLF (1986) have carried out similar calculations and concluded that SiO_2-fluorite is unstable below 160 GPa. These studies, however, do not indicate the fate of the SiO_2-fluorite below the critical pressure.

MD calculations provide us with an especially powerful tool to survey behaviors of atoms that are not suitable for study by the quantum mechanical and lattice dynamics approach. MD calculations have been successfully applied to the study of structural transformations in model superionic conductors by PARRINELLO et al. (1983). Our study, though less sophisticated in technique, is believed to further demonstrate the importance of MD calculations.

Outline of Computational Method

The XDORTHO Program

In the MD calculations, we assumed a pairwise additive, two-body central force. Because the MD is virtually the only way to obtain a dynamic picture of structures and structural changes in liquid, vitreous, and crystalline materials (for description of principles and techniques, see MATSUI et al., 1982; MATSUI and KAWAMURA, 1984), the MD program has been extended to incorporate the basic cells of rectangular parallelepiped and to operate at constant pressure (P) and temperature (T) conditions (i.e., to become the XDORTHO program, KAWAMURA, 1984). The XDORTHO program is able to treat crystals of cubic, hexagonal/trigonal, and orthorhombic systems.

Pressure control is achieved simply by changing the edge length of the basic cell while the fractional coordinates of the atoms are kept constant, as follows

$$\frac{L_x + \delta L_x}{L_x} = 1 + k_1 (\bar{P}_{calc} - P_{presc}) + k_2 (\bar{P}_x - [\bar{P}_x + \bar{P}_y + \bar{P}_z]/3) \qquad (1)$$

where δL is change in basic cell edge length, k_1 and k_2 are arbitrary constants of around 0.001, P_{presc} is the prescribed pressure, and $\bar{P}_{calc}$ and $\bar{P}_x$ are averaged instantaneous pressure and normal stress, respectively. The averages are taken over 20 to 50 steps as described below. The instantaneous pressure is given as

$$P_{calc} = \frac{NkT}{V} - \frac{1}{3V}\sum_{i<j} r_{ij}F_{ij} \qquad (2)$$

where N is number of particles in a basic cell with a volume V, k is the Boltzmann's constant, and r_{ij} and F_{ij} are

High-Pressure Research in Mineral Physics, edited by M. H. Manghnani and Y. Syono, pp. 305–311.

interatomic distance and force, respectively.

The instantaneous normal stress is simply defined by

$$P_x = \frac{NkT_x}{V} - \frac{1}{3V}\sum_{j<j} r_{ij}F_{ij,x} \qquad (3)$$

where T_x means the x-component of instantaneous temperature (temperatue obtained from x-component of particle velocities), and $r_{ij,x}$ and $F_{ij,x}$ are the x-components of r_{ij} and F_{ij}, respectively. Basic cells containing 27 (3×3×3) or 125 (5×5×5) unit cells of the cubic fluorite (number of ions totalling 324 or 1500) were used. The time increment (Δt) in the MD calculation was 1.25 fs (1 fs$=10^{-15}$ s) in most runs.

The adjustment of basic-cell edge lengths by Equation 1 was made intermittently so as not to interfere with spontaneous quasi-periodical fluctuations in instantaneous pressure of the system. The adjustment was carried out every 25 steps (~30 fs) for the systems with $N=324$ and every 50 steps (~60 fs) for systems with $N=1500$.

A physically more sophisticated algorithm for constant P and T calculations has been proposed (ANDERSEN, 1980; PARRINELLO and RAHMAN, 1980). However, for the silica and silicate systems, there is no significant discrepancy between the rigorous and simple algorithms (I. Okada, personal communication, 1986). Presumably this may be due to the fact that the adjustments of positions and velocities of particles are quite similar in effect for both methods, so far as k_1 and k_2 values in Equation 1 are small and the adjustment are carefully timed.

The Coulomb energy and force were obtained using Ewald's method (EWALD, 1921) with a cut-off distance of $L_{min}/2$, where L_{min} is the shortest basic-cell edge length. The Ewald's parameter for convergence, α, was chose so as to minimize the squared sum of differences between interatomic forces (not potentials) and the reference values obtained by the calculation using a cut-off distance of $L_{min}\times 2$ and 871 reciprocal cell vectors. In the present study errors are estimated to be less than 0.08% for force and less than 0.5% for energy.

The pair correlation functions (i.e., radial distribution functions of interatomic distances) were optionally recorded up to a distance of twice the cut-off distance.

Choice of Potential Function

Interatomic potentials used have been of Gilbert-Ida type (GILBERT, 1968; IDA, 1976), with or without the dispersion term

$$U(r_{ij}) = \frac{q_iq_j}{r_{ij}} + f_0(b_i + b_j)\exp\left(\frac{a_i + a_j - r_{ij}}{b_i + b_j}\right) + c_ic_jr_{ij}^{-6} \quad (4)$$

The relevant parameters are given in Table 1. IF-2 has been successfully used to simulate structures of aluminosilicate melts (DEMPSEY and KAWAMURA, 1984), and MS-1 has recently been developed by M. Matsui (unpublished) to reproduce structural and elastic properties in SiO_2 (rutile).

TABLE 1. Potential Parameters* for Equation 4

Set-ID		O	Si	Mg
IF-2	$q/\lvert e\rvert$	−2	+4	+2
	$a/$Å	1.900	1.047	1.174
	$b/$Å	0.180	0.035	0.053
MS-1	$q/\lvert e\rvert$	−1.3	+2.6	—
	$a/$Å	1.727	0.962	—
	$b/$Å	0.150	0.050	—
	$c/$Å^{6}kJ mol^{-1}	45.6	0	—

* $f_0=1$ kcal Å^{-1} mol$^{-1}=6.948\times10^{-6}$ dyn

In order to calibrate the pressure scale, bulk moduli were calculated for SiO_2 (rutile) from 0 to 100 GPa. Use of the MS-1 gives a quite reasonable bulk modulus value, whereas use of the IF-2 gives a value that makes SiO_2 (rutile) too stiff by a factor of two. Accordingly, the nominal pressure in the IF-2 runs is considered to be too high by factors of similar magnitude. Both potentials are able to maintain the SiO_2 in the fluorite structure above 230 GPa (300 K). This is consistent with results of a quasiharmonic lattice dynamics calculation for the fluorite structure using IF-2 (R. J. Hemley, personal communication, 1986), in which the dynamic instability at 0 K was shown to occur at 210 GPa.

Table 2 presents the calculated zero-pressure bulk moduli (K_0), along with crystallographic data, for SiO_2 (rutile) and SiO_2 (α-PbO_2) at 0 GPa, 300 K. Because the fluorite-α PbO_2 transformation was first observed in a run using the IF-2 potential, most calculations have been carried out using IF-2.

Results and Discussion

The Fluorite to α-PbO_2 Transition

At 0 GPa, the transition (or decomposition) of SiO_2 (fluorite) occurs violently, presumably due to great differences in pressures under which SiO_2 (fluorite) is dynamically stable and in cohesive energies of the fluorite-type phase and of possible phases to which SiO_2 (fluorite) transforms. It is surprising, therefore, to see that the successful transformation into the "single-crystal" SiO_2 with the α-PbO_2 structure is not a rare event, provided that: first, the initial velocity distribution has been carefully controlled by a provisional equilibration at very high-pressure runs in which SiO_2 (fluorite) behaves stably; and second, the control of P and T has allowed

TABLE 2. Crystallographic and Related Data of MD-synthesized Polymorphs of SiO_2 at 0 GPa, 300 K

α-PbO_2 structure

Potential	IF-2	MS-1	Obs.
a/Å	4.123(2)	4.152(1)	—
b/Å	4.927(2)	4.988(1)	—
c/Å	4.526(2)	4.516(1)	—
ρ/Mg m^{-3}	4.336(1)	4.268(1)	—
O(x)	0.280	0.276	
O(y)	0.380	0.379	
O(z)	0.419	0.419	
Si(x)	0	0	
Si(y)	0.167	0.165	
Si(z)	0.25	0.25	
K_0/GPa	663	318	—
U/kJ mol^{-1}	−13109	−5599	—

Rutile structure

Potential	IF-2	MS-1	Obs.*
a/Å	4.110(1)	4.128(1)	4.1772
c/Å	2.747(1)	2.758(1)	2.6651
ρ/Mg m^{-3}	4.301(1)	4.246(1)	4.2915
O(u)	0.308(1)	0.305(1)	0.3062
K_0/GPa	656	325	298
U/kJ mol^{-1}	−13135	−5609	—

* Data sources: Crystallographic parameters from SINCLAIR and RINGWOOD (1978); Bulk modulus from WEIDNER et al. (1982).

sufficient release of thermal and strain energies during the course of the reaction. The SiO_2 with the α-PbO_2 structure, once formed, shows no tendency to convert further into other structures, based on prolonged MD calculations for more than 6000 steps (7.5 ps). In a majority of runs, however, the product was found to be a mosaic of small portions of the α-PbO_2 structure accompanying stacking and orientational faults, as will be shown later.

Geometrical Relation Among the Fluorite, the α-PbO_2, and Related Structures

The space group *Pbcn*, to which the α-PbO_2 structure belongs, is one of the subgroups of the space group *Fm3m* of the fluorite structure. In the α-PbO_2 structure all anions are geometrically equivalent at the 8d position, forming a hexagonal close-packed array. This 8d position corresponds to the special position 8c of *Fm3m*, the position for anions in simple cubic arrangement in the fluorite structure. Cations in the α-PbO_2 structure are also mutually equivalent at the 4b position, which corresponds the special position 4a for cations in fluorite. The crystallographic orbit theory (ENGEL et al., 1984) indicates that this sort of perfect correspondence (i.e., of equivalent anions and cations in *Fm3m* to equivalent anions and cations in other space groups) can be found among the subgroups of *Fm3m*, *Pa3* and *Pbcn*.

The fluorite structure can be regarded as a special case of the structure with cubic-*Pa3* symmetry. In hypothetical SiO_2 with *Pa3* symmetry, all anions are at the general position 8c and have an irregular packing mode, and all cations are at the special position 4a (which corresponds to the 4a position in *Fm3m*) forming an fcc array. When the coordinate of the 8c position (u, u, u) assumes the special value (0.25, 0.25, 0.25), the packing mode of anions becomes simple cubic and the whole structure is identical to that of fluorite. Because of this geometric relation, the structure with the *Pa3* symmetry will be called "modified fluorite" hereafter.*

The relation between the fluorite and α-PbO_2 structures is not as simple as in the case described above. The similarity of the structures is evident when the conventional *Fm3m* setting is displaced to the *Pbcn* setting, as shown in Table 3. Table 3 shows that the fluorite structure is an extreme case for *Pbcn* with particular atomic coordinates and an equal cell-edge lengths. It is worthwhile noting that the orthorhombic cell edge lengths in α-PbO_2 are also similar to one another, and the atomic coordinates are not very different from those in the "extreme" (fluorite) structure. The α-PbO_2 structure, therefore, may be regarded as a "deformed fluorite". Upon deformation of the cubic structure, "threefold degenerate" cell edges should split into three different ones with six degrees of freedom in choosing the orthorhombic orientation; it is thought that this causes frequent orientation disorders in the resultant structures.

It is important to note that the "modified fluorite" provides the nearest potential minimum to that of the "true" fluorite. The SiO_2-modified fluorite can maintain its structural identity even at low pressures and 300 K, for both IF-2 and MS-1 potentials (Table 3). The existence of this potential minimum is considered to be one of the main reasons why the fluorite to α-PbO_2 transition does not proceed in a straightforward manner in most runs.

Successful and Unsuccessful Runs

In "successful" runs, the transition was completed within 0.5 ps (1 ps$=10^{-12}$s) over a pressure range of 0–200 GPa. The unusually quick transition may be explained by

* Pyrite, $Fe^{2+}[S_2]^{2-}$, belongs to the same space group. However, because the hypothetical SiO_2 does not contain peroxide group (O_2^{2-}), the name of "pyrite structure" is not adequate for this compound.

TABLE 3. Comparison of Silica with Fluorite and Related Structures (IF-2 Potential)

α-PbO_2 structure (*Pbcn*)

	0 GPa	240 GPa
a/ A	4.123(2)	3.897(1)
b/ A	4.927(2)	4.491(1)
c/ A	4.526(2)	4.188(1)
O(x)	0.280	0.270
O(y)	0.380	0.398
O(z)	0.414	0.424
Si(y)*	0.167	0.180
ρ/ Mg m^{-3}	4.336(1)	5.446(1)
U/ kJ mol^{-1}	−13109	−12830

"Modified fluorite" structure (*Pa3*)

	0 GPa	240 GPa
a/ Å	4.418(1)	4.123(1)
O(u)	0.331	0.334
ρ/ Mg m^{-3}	4.627(1)	5.694(1)
U/ kJ mol^{-1}	−12951	−12714

Fluorite structure (*Fm3m*; *Pbcn* setting)

	0 GPa**	240 GPa
a/ A	4.372(4)	4.087(1)
O(x)	0.25	0.25
O(y)	0.5	0.5
O(z)	0.5	0.5
Si(y)*	0.25	0.25
ρ/ Mg m^{-3}	4.780(10)	5.845(1)
U/ kJ mol^{-1}	−12885	−12646

* Si(x)=0, Si(z)=0.25.
** Uncertain due to the instability (see text).

the geometric relationships between the two structures, as described above. An example of atomic trajectories upon transformation is shown in Figure 1. In this case, the transition seems to have occurred straightforwardly, but even in this exceptionally simple case, an awkward inflexion on trajectories suggests that the transition occurs in two stages: first, displacements of oxide ions with direction, as predicted by BUKOWINSKI and WOLF (1986, Figure 3); and second, cooperative movements towards the better arrangement (in this case, α-PbO_2), involving displacements of Si ions. In another case, the α-PbO_2 structure was eventually established after a complicated "trial-and-error" wandering of atoms, as shown in Figure 2. This can be explained only by assuming the existence of shallow, local potential minima, which may be related to the "modified fluorite" arrangement. It follows that the proximity of the initial and the final structures might not be a sole reason of the unique transition. In this context it is important to note that the fluorite to α-PbO_2 conversion is an irreversible process.

Conversion from the fluorite to α-PbO_2 structure in "single crystal" requires the cooperative motion of all atoms. When random, noncooperative motion prevails and offsets cooperative movements, the product tends to become aggregates of domains in which the local structure is the α-PbO_2 type. Stacking faults and orientational disorders accompany lower density and higher cohesive energy; the latter are diagnostic features of unsuccessful, imperfect structures. An example of a product with orientational mismatch is shown in Figure 3.

At higher pressures, "successful" (nonrandom) conversion becomes more common. This may be explained by enhanced similarity in both geometry and energy between the two structures, as shown in Table 3. It should be noted, however, that the α-PbO_2 structure does not eventually merge into the fluorite structure, even under severe compression up to 320 GPa. It follows that, above the pressure of dynamic instability of the SiO_2-fluorite, we cannot observe any transition from the α-PbO_2 to fluorite structure and vice versa, based on the MD calculations. Thus the MD-observed transition is irreversible and monotropic (fluorite to α-PbO_2), indicating that the transition is not an equilibrium process. Thermodynamically, the SiO_2-fluorite may be enantiotropic with SiO_2-rutile (or, with SiO_2-α-PbO_2 or modified fluorite if these structures are stable) and conversion may proceed as a reversible process, at pressures far beyond the dynamic instability of the SiO_2-fluorite.

Disordered α-PbO_2-type Structures

It is worthwhile to consider the "unsuccessful" products that have a lower density and higher cohesive energy (hereafter called the disordered phases), because they show definite similarity to SiO_2 with "modified NiAs structure" or Fe_2N structure (LIU et al., 1978; LIU, 1982). The silica with Fe_2N structure has been poorly characterized; only three diffraction lines (or series of spots) have been documented, and identification of the structure has only been based upon and justified by the absence of lower-angle diffraction spots. Needless to say, the absence of diffraction spots is more difficult to confirm than the presence of spots. Considering the variety of disordered phases of α-PbO_2-related structures that contain stacking faults and orientational disorders, it seems likely that the phase described by LIU et al. (1978) might be a member of a disordered α-PbO_2 related group, where presence of face-shared SiO_6 octahedra (which should be abundant in

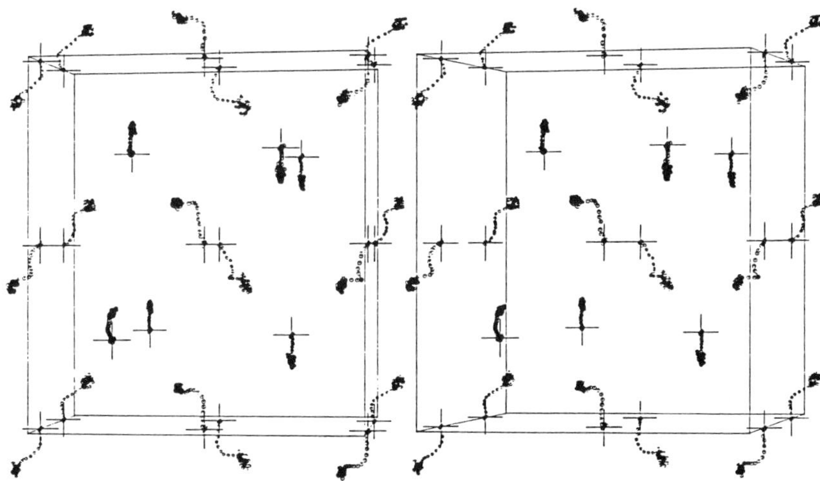

Figure 1. Stereoscopic-pair view showing simple trajectories of atoms at every 5 steps (Δt=1.5 fs) in the fluorite-α-PbO_2 transformation. Only one unit cell of the basic cell, which comprises 27 unit cells (N=324), is shown. The initial fluorite configuration is indicated by crosses. Ions with original heights 1/4 and 3/4 are Si; the rest are O.

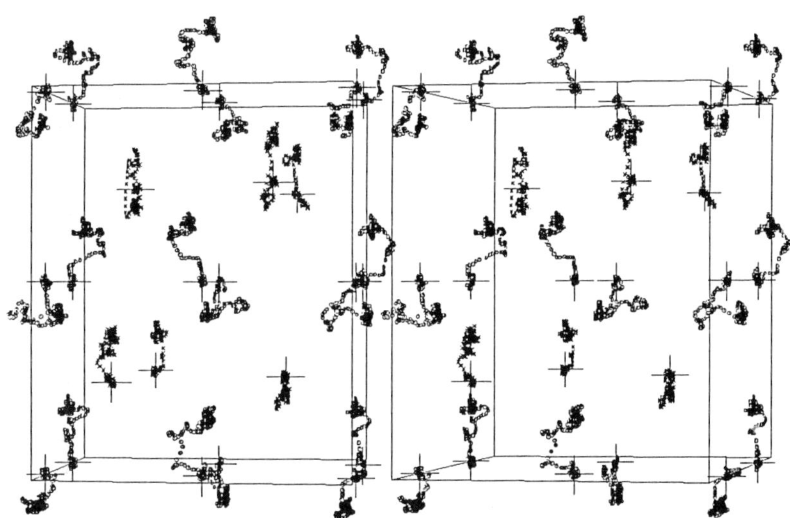

Figure 2. Stereoscopic-pair view showing complicated trajectories of atoms at every 5 steps (Δt=1.25 fs) in the fluorite-α-PbO_2 transformation. Of the 27-unit cells in the basic cell, only one unit cell in shown. Orientation is the same as shown in Figure 1.

NiAs-related structures) is expected to be much rarer. In Table 4 our calculated X-ray diffraction profiles are compared to the data of LIU et al. (1978) and to data on silica with the α-PbO_2 structure reported by GERMAN et al. (1974). The calculation of X-ray intensity was carried out in the same manner as in the case of noncrystalline materials (i.e., through Fourier transforms of pair correlation functions for Si-O, Si-Si, and O-O over a distance up to 20A). It is noted that the three peaks reported by LIU et al. (1978) are also common in perfect and disordered α-PbO_2 structures. There should be some reason for the reduction in the apparent intensities of several peaks.

The intensity of the peak(s) at $s \doteqdot 2.4$ is almost exclusively determined by the pair-correlation function for Si-Si. Therefore it is expected that the intensity is sensitive to the degree of long-range order in the arrangement of Si. Orientational faults, especially on the (010)

TABLE 4a. Calculated X-ray Diffraction Profiles of MD-synthesized Products (IF-2 potential, 0 GPa, 300 K)*

1			2			3			4			5		
SiO_2(α-PbO_2) ρ=4.336 perfect			SiO_2(α-PbO_2) ρ=4.308 nearly perfect			SiO_2(α-PbO_2) ρ=4.260 twinned on (010)			SiO_2(α-PbO_2) ρ=4.215 twinned on (100)			SiO_2(rutile) ρ=4.301 perfect		
$s/Å^{-1}$	$d/Å$	I	$s/Å^{-1}$	$d/Å$	I	$s/Å^{-1}$	$d/Å$	I	$s/Å^{-1}$	$d/Å$	I	$s/Å^{-1}$	$d/Å$	I
1.98	3.17	70	1.98	3.17	60	1.96	3.21	55	1.93	3.26	55	2.17	2.90	100
2.42	2.60	100	2.41	2.61	80	2.33	2.70	70**	2.38	2.64	100	2.69	2.34	35
3.25	1.93	80	3.22	1.95	75	3.23	1.95	80	3.29	1.91	50	3.16	1.99	40
4.16	1.51	100	4.15	1.51	100	4.08	1.54	100	4.28	1.47	60**	4.11	1.53	80
4.93	1.27	60	4.94	1.27	60	4.94	1.27	50**	4.82	1.30	55	5.11	1.23	55

* Density (ρ) in Mg m^{-3}; $s=4\pi\sin\theta/\lambda$, where θ=diffraction angle, λ=wavelength; $d=2\pi/s$.
** Broad peak(s).

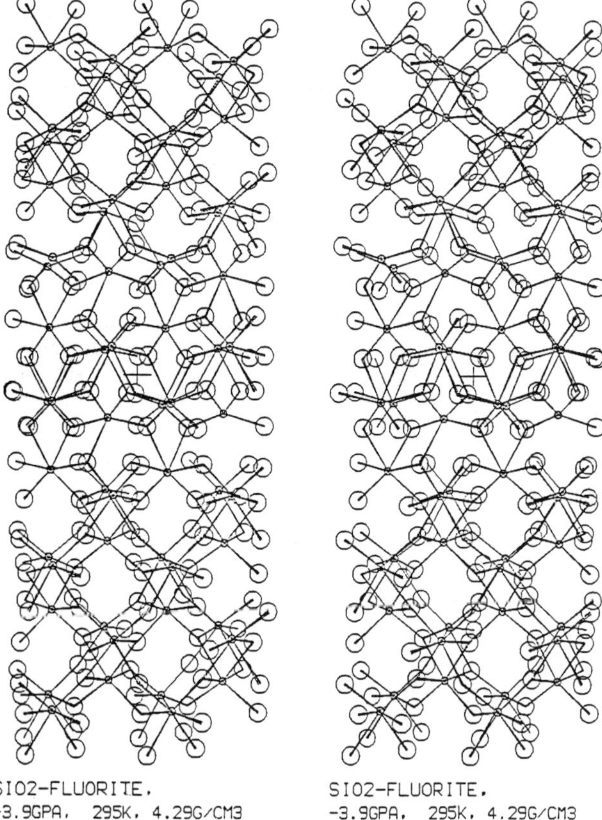

Figure 3. An example of "unsuccessful" product (N=1500) with orientational disorder (stereo). The pressure, temperature, and density shown on the bottom are instantaneous values, within the fluctuation ranges in equilibration at 0 GPa, 300 K. The central part is projected on the a-c plane while the rest is on the b-c plane. The corresponding X-ray diffraction profile is given in Table 4a (No. 3).

TABLE 4b. Reported X-ray diffraction Profiles of SiO_2 with Non-rutile Structures

1		2	
"α-PbO_2 structure" GERMAN et al. (1973)		"Fe_2N structure" LIU et al. (1978)	
$d/Å$	I	$d/Å$	I
2.59	strong	1.952	70
2.35	moderate	1.508	100
2.25	moderate	1.208	30
2.15	moderate		
2.08	strong		
1.99	weak		
1.88	very weak		
1.58	weak		
1.55	weak		
1.49	moderate		
1.35	very weak		
1.31	very weak		

plane, make the peak broader and less intense, as shown in Table 4a, Nos. 1 to 3. A similar situation can be seen for the peak at $s\doteqdot4.95$. It appears very likely that the singular nature of the diffraction profile observed by LIU et al. (1978) can be explained by this observation, without invoking the "Fe_2N structure" or "the NiAs structure with half of the cations removed randomly".

On the other hand, the "SiO_2 with α-PbO_2 structure" of GERMAN et al. (1973, 1974) shows no similarity with our calculated profile. That material is considered to be an artifact, perhaps arising from misidentification of diffraction peaks of material(s) with composition presumably different from pure SiO_2 (see also SYONO et al., this volume).

Acknowledgments. We are grateful to S. Akimoto, A. E. Ringwood, and Y. Syono for their encouragement and stimulating discussions at various stages of this study. We also thank M. Matsui (Kanazawa

Medical University) for his generous permission to use the MS-1 potential which he developed, prior to publication; R. J. Hemley for carrying out the quasiharmonic lattice dynamics calculation for the IF-2 potential; and T. Matsumoto (Kanazawa University) and M. Tokonami (University of Tokyo) for their advice on crystallographic aspects. Comments from three reviewers, including M. S. T. Bukowinski and R. G. Gordon, were of great help in improving the text of this paper.

Our work was supported by Grant-in-Aid for Scientific Research, Ministry of Education, Science and Culture of Japan (Project Nos. 60121013 and 61113012).

References

Andersen, H. C., Molecular dynamics simulations at constant pressure and/or temperature, *J. Chem. Phys., 72*, 2384–2393, 1980.

Bukowinski, M. S. T., and G. H. Wolf, Equation of state and stability of fluorite-structured SiO_2, *J. Geophys. Res., 91*, 4704–4710, 1986.

Dempsey, M. J., and K. Kawamura, Molecular dynamics simulation of the structure of aluminosilicate melts, in *Progress in Experimental Petrology*, edited by C. M. B. Henderson, N.E.R.C. Publications, Ser. D, No. 25, 49–56, 1984.

Engel, P., T. Matsumoto, G. Steinmann, and H. Wondratschek, The non-characteristic orbits of the space groups, *Z. Kristallogr., Suppl. No. 1*, 217 pp., 1984.

Ewald, P. P., Die Berechung optischer und elektrostatischer Gitterpotentiale, *Ann. Phys., 64*, 253–287, 1921.

German, V. N., M. A. Podurets, and R. F. Trumin, Synthesis of a high-density phase of silicon dioxide in shock waves, *Sov. Phys.-JETP, 37*, 107, 1973.

German, V. N., N. N. Orlova, L. A. Tarasova, and R. F. Trunin, Synthesis of the orthorhombic phase of silica in dynamic compression, *Izv. Akad. Sci., Earth Phys.*, No. 7, 50–56, 1974.

Gilbert, T. L., Soft-sphere model for closed-shell atoms and ions, *J. Chem. Phys., 49*, 2640–2642, 1968.

Hemley, R. J., M. D. Jackson, and R. G. Gordon, Lattice dynamics and equations of state of high-pressure mineral phases studied with electron-gas theory (Abstract), *EOS Trans. Am. Geophys. Union, 66*, 357, 1985.

Ida, Y., Interionic repulsive force and compressibility of ions, *Phys. Earth Planet. Inter., 13*, 97–104, 1976.

Kawamura, K., Simulation of crystal structures of silicates by molecular dynamics calculations, Ph.D. thesis, Univ. Tokyo (in Japanese) 1984.

Liu, Lin-gun, High-pressure phase transformations of the dioxides: implications for structures of SiO_2 at high pressure, in *High-Pressure Research in Geophysics*, edited by S. Akimoto and M. H. Manghnani, 349–360, Center for Academic Publications Japan, Tokyo, 1982.

Liu, Lin-gun, W. A. Bassett, and J. Sharry, New high-pressure modification of GeO_2 and SiO_2, *J. Geophys. Res., 83*, 2301–2305, 1978.

Matsui, Y., and K. Kawamura, Computer simulation of structures of silicate melts and glasses, in *Materials Science of the Earth's Interior*, edited by I. Sunagawa, 3–23, Terra Scientific Publications, Tokyo, 1984.

Matsui, Y., K. Kawamura, and Y. Syono, Molecular dynamics calculations applied to silicate systems: molten and vitreous $MgSiO_3$ and Mg_2SiO_4, in *High-Pressure Research in Geophysics*, edited by S. Akimoto and M. H. Manghanani, 511–524, Center for Academic Publications Japan, Tokyo, 1982.

Parrinello, M., and A. Rahman, Crystal structure and pair potentials: a molecular dynamics study, *Phys. Rev. Lett., 45*, 1196–1199, 1980.

Parrinello, M., A. Rahman, and P. Vashista, Structural transitions in superionic conductors, *Phys. Rev. Lett., 50*, 1073–1076, 1983.

Sinclair, W., and A. E. Ringwood, Single crystal analysis of the structure of stishovite, *Nature, 272*, 714–715, 1978.

Syono, Y., K. Kusaba, M. Kikuchi, K. Fukuoka, and T. Goto, Shock-induced phase transitions in rutile single crystal, this volume.

Weidner, D. J., J. D. Bass, A. E. Ringwood, and W. Sinclair, The single-crystal elastic moduli of stishovite, *J. Geophys. Res., 87*, 4740–4746, 1982.

THEORETICAL STUDY OF THE STRUCTURAL PROPERTIES AND EQUATIONS OF STATE OF $MgSiO_3$ AND $CaSiO_3$ PEROVSKITES: IMPLICATIONS FOR LOWER MANTLE COMPOSITION

George H. WOLF* and Mark S. T. BUKOWINSKI

Department of Geology and Geophysics, University of California, Berkeley, CA 94720, USA

Abstract. We present a detailed theoretical study on the stability and equations of state of $MgSiO_3$ and $CaSiO_3$ perovskites. Results are obtained as a function of temperature and pressure through a minimization of the free energy with respect to the structural parameters, as given by self-consistent quasiharmonic lattice dynamics. Bonding forces are derived from the parameter-free modified electron-gas theory. A cubic structure for $CaSiO_3$ and an orthorhombic structure for $MgSiO_3$ are predicted at zero pressure in accord with observation. The calculated compressibility and thermal expansivity of $MgSiO_3$ are in very good agreement with available data. The model predicts that the degree of distortion in $MgSiO_3$ increases weakly with pressure but this increase in distortion has only a minor effect on its compressibility. At high temperatures, model $MgSiO_3$ perovskite exhibits critical soft-mode behavior and undergoes successive second-order transitions to tetragonal and cubic phases. The transition temperatures increase with pressure such that the orthorhombic phase is stable throughout most of lower mantle, although an adiabatic extrapolation of the lower mantle to zero pressure may approach or even cross these phase boundaries. We also find that an adiabatic extrapolation of the lower mantle seismic properties will overestimate its inferred zero-pressure density and bulk modulus. Hence, compositional constraints on the lower mantle that are based on comparisons of decompressed seismological properties with zero-pressure mineralogic data will overestimate the relative proportion of perovskite to magnesiowüstite. Supplementing available data with theoretical estimates for the high-pressure and high-temperature properties of relevant mantle phases, we find that both a pyrolitic composition, at relatively low geotherm temperatures appropriate to whole-mantle convection, and a pyroxene composition, at the higher geotherm temperatures appropriate to layered-mantle convection, are compatible with the seismic data.

Introduction

From experimental evidence it is known that the major chemical components of the upper mantle crystallize primarily as a mixture of (Mg, Fe)SiO_3 perovskite and (Mg, Fe)O magnesiowüstite at the pressure and temperature conditions of the lower mantle (see review by JEANLOZ and THOMPSON, 1983). However, it is not known whether phase transitions alone, in an otherwise isochemical mantle, account for the observed variations in the inferred density and seismic velocities or if, in addition, there exist chemical differences between the upper and lower mantle. Chemical stratification in the mantle implies separately convecting layers and a relatively large thermal inertia for the earth, in contrast to an isochemical mantle which implies whole mantle convection and a substantially lower thermal inertia. Comparisons of the seismic properties of the lower mantle with those derived from model mineral assemblages have not been able to discriminate between uniform and stratified chemical models of the mantle. This inability is primarily due to the paucity of experimental data on the physical properties of these phases at the combined temperature and pressure conditions of the lower mantle, and the significant tradeoffs between compositions, temperature and iron concentration.

Theoretical approaches may be able to provide useful constraints on the high-temperature and high-pressure properties of minerals that are necessary for mantle models. Considerable advancements have been made in the ability to predict properties of solids from first-principles. Perhaps the most fruitful of the *ab initio* approaches to mineralogic applications has been the electron-gas theory of GORDON and KIM (1972), and variants of this original theory (COHEN and GORDON, 1976; MUHLHAUSEN and GORDON, 1981; HEMLEY and GORDON, 1985; MEHL et al., 1986; WOLF and BUKOWINSKI, 1987). In this paper we present results of a detailed lattice dynamical study on the stability and equations of state of Mg and Ca silicate perovskites over a substantial range of temperatures and pressures employing the modified electron-gas approach (MEG) in the quasiharmonic approximation. This work expands on our earlier theoretical studies of $MgSiO_3$ and $CaSiO_3$ perovskites and, for purposes of modelling the chemical composition of the lower mantle, complements recent electron-gas calculations on MgO and CaO (HEMLEY et al., 1985; WOLF and BUKOWINSKI, 1987). Although other formulations of the electron-gas theory of crystals give better estimates of the absolute densities of crystals, the MEG method has proven to give the most reliable estimates of equations of state and thermal expansivity of oxides when expressed in terms of normalized densities (WOLF and BUKOWINSKI, 1987). This approach has also given excellent results on the structural properties of some fluoridic perovskites (BOYER and HARDY, 1981; BOYER, 1984) and we find very good agreement of the observed

*Now at the Department of Chemistry, Arizona State University, Tempe, AZ 85287, USA.

High-Pressure Research in Mineral Physics, edited by M. H. Manghnani and Y. Syono, pp. 313–331.

room-temperature equation of state and thermal expansivity of $MgSiO_3$ perovskite with those computed using the MEG theory.

The model predicts that at high temperatures $MgSiO_3$ perovskite exhibits critical soft-mode behavior and undergoes successive second-order transitions to tetragonal and cubic phases. The transition temperatures increase with pressure such that the orthorhombic phase is stable throughout most of the lower mantle. However, it is possible that an adiabatic extrapolation of the lower mantle to zero pressure and the extrapolation of perovskite's zero-pressure properties to high temperature will approach or even cross these phase boundaries and we discuss the validity of extrapolating material properties into or through the critical regions associated with these phase transitions.

We also summarize here our attempts to place constraints on the lower mantle composition by supplementing available mineralogic data with theoretical estimates for the high-temperature and high-pressure properties of perovskite and magnesiowüstite. Direct comparisons are made between observed seismic properties and those of model mineral assemblages at mantle temperature and pressure conditions. We find that, because of tradeoffs between physical properties, composition and temperature, at least two compositional-thermal models are compatible with the observed density and seismic parameter of the lower mantle: 1) a pyrolitic composition at the relatively low geotherm temperatures appropriate to whole-mantle convection; and 2) a pyroxene composition at higher geotherm temperatures appropriate to layered-mantle convection. This conclusion differs from that of KNITTLE et al. (1986) who found that the adiabatically decompressed density of the lower mantle is incompatible with the high-temperature zero-pressure density of a pyrolitic composition lower-mantle assemblage but is compatible with that for a perovskite mineralogy. However, in addition to requirements of adiabaticity and chemical homogeneity, zero-pressure constraints on the composition of the lower mantle rely on the validity of the adiabatic extrapolation of the seismic properties and the approximately twofold extrapolation of perovskite's thermal expansivity to the inferred high temperature at the foot of the mantle adiabat. We point out that a finite strain decompression of high-temperature adiabats systematically errs in the estimation of zero-pressure properties. In addition, we show that an adiabatic extrapolation of the lower mantle to zero pressure and the extrapolation of perovskite's zero-pressure properties to high temperature may approach or even cross polymorphic phase boundaries in perovskite. Thus, as a result of the systematic errors in the adiabatic extrapolation and possible critical behavior in perovskite, zero-pressure constraints on the lower mantle composition are necessarily broadened and cannot yet discriminate between pyrolite and pyroxene composition models of the lower mantle.

Perovskite Structure

$CaSiO_3$ crystallizes in a perovskite structure at pressures in excess of 15 GPa and temperatures near 1500 K (LIU and RINGWOOD, 1975). Although $CaSiO_3$ perovskite has not been successfully quenched to ambient conditions, a cubic structure is inferred at high pressures from the in situ powder diffraction data. In the cubic structure (shown in Figure 1), fully linked SiO_6 octahedra are located at the corners of a cube surrounding the central Ca^{+2} ion which is dodecahedrally coordinated with oxygen. There are five atoms in the cubic perovskite unit cell ($Z=1$) and only one parameter (i.e., a single lattice parameter) is required to fully characterize its structure. Many perovskite compounds, including $MgSiO_3$, are distorted from the ideal cubic structure. This distortion occurs primarily through the cooperative rotation of the octahedra, although the octahedra need not remain regular and a further displacement of the cations can also occur (GLAZER, 1972).

At ambient conditions, powder diffraction data for $MgSiO_3$ perovskite, quenched from synthesis pressures between 28–38 GPa and temperatures up to approximately 2000 K, indicate an orthorhombic structure of $Z=4$ having either a Pbnm (centrosymmetric) or $Pbn2_1$ (non-centrosymmetric) space group (ITO and MATSUI, 1978; YAGI et al., 1978). Electron microscopy on a similarly prepared sample indicates a superstructure with $Z=32$

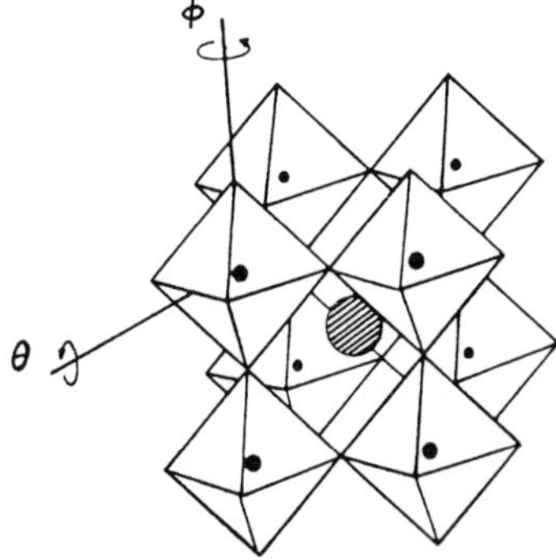

Figure 1. In the ideal cubic structure for ABX_3 perovskite, BX_6 octahedra are located at the corners of a cube surrounding the central A cation. The orthorhombic (Pbnm) symmetry results from the cooperative rotation of octahedra about the indicated Θ and Φ axial directions: the rotational sense about the Θ axis is opposite for adjacent octahedra and is in the same sense for rotations along a particular Φ axes. In orthorhombic $MgSiO_3$ perovskite, the SiO_6 polyhedra are slightly distorted from regular octahedra and the Mg^{+2} cation is displaced from its central position.

and lattice parameters twice as large as those obtained from powder diffraction (MADON et al., 1980). As we discuss later, little difference is expected in the computed thermoelastic properties among these distorted perovskite polytypes; hence, we have limited our theoretical study of distorted perovskites to that of the Pbnm space group (Z=4).

The relation of the Pbnm structure type to the cubic perovskite structure is shown in Figure 1. There are 20 atoms in the orthorhombic unit cell and 10 parameters, collectively referred to as ξ, are required to fully characterize the structure: 3 lattice parameters and 7 internal coordinates. If the SiO_6 polyhedra are constrained to regular octahedra, the total number of structural parameters can be reduced to 5: the Si–O distance, two parameters describing displacement of the central Mg^{+2} or Ca^{+2} cation within the (001) plane, and two Eulerian angles, Θ and Φ, needed to describe the coupled octahedral rotations about the respective (110) and (001) pseudocubic axes.

Calculations and Numerical Methods

The structural properties of $MgSiO_3$ and $CaSiO_3$ perovskite were computed as function of pressure, P, and temperature, T, by a minimization of the total Gibbs free energy in the quasiharmonic approximation, with respect to the structural parameters, ξ. The Gibbs free energy is given by

$$G(P, T) = U(\xi) + F_{Th}(\xi, T) + PV(\xi) \qquad (1)$$

where U is the static lattice energy, F_{Th} is the thermal contribution to the free energy, and V is the volume. Although we find that there is significant irregularity in the SiO_6 polyhedron, especially at high pressures, the computed equations of state and thermal expansivity are relatively insensitive to whether or not the SiO_6 polyhedra are allowed to distort from perfect octahedra. Therefore, in most of the minimizations we have assumed regular SiO_6 octahedra, thus reducing the number of variational parameters from ξ=10 to ξ=5.

The static lattice energy was calculated by assuming pairwise additive central potentials derived from the modified electron gas (MEG) formulation (GORDON and KIM, 1972). In this approach, pair interactions result from the overlap of rigid-ion charge densities that are derived from Hartree-Fock wavefunctions of isolated closed shell ions (CLEMENTI and ROETTI, 1974). The interaction potential includes contributions from Coulombic, kinetic, exchange and correlation energies. The Coulombic contribution is treated exactly, while the kinetic, exchange, and correlation contributions are approximated with local electron-gas density functionals which are scaled using the correction factors derived by WALDMAN and GORDON (1979).

Since O^{-2} is not a stable free ion, it must be stabilized by artificial means, thus introducing a certain amount of ambiguity into the model. Following COHEN and GORDON (1976), we chose to stabilize its wavefunction using a Watson-sphere having a charge of +1 and a radius of 2.66 atomic units (WATSON, 1958). These parameters have been found to give good estimates of the equation of state and averaged thermal properties of many simple oxide minerals, although equilibrium bond lengths tend to be overestimated (COHEN and GORDON, 1976; WOLF and BUKOWINSKI, 1987).

We include nearest-neighbor cation–oxygen and up to the third shell in the oxygen–oxygen pair interactions (coordination shells refer to the cubic perovskite structure). Cation–cation interactions were found to be negligible and hence, are not included. The long-ranged point Coulomb energy was computed exactly using the Ewald method (EWALD, 1921).

In the quasiharmonic approximation, the thermal contribution to the free energy is given by

$$F_{Th}(\xi, T) = \sum_{k\lambda} \left[\frac{1}{2} \hbar\omega\binom{k}{\lambda} + k_B T \ln\left(1 - e^{-\hbar\omega\binom{k}{\lambda}/k_B T}\right) \right] \qquad (2)$$

where $\hbar$ is Planck's constant divided by 2π, k_B is Boltzmann's constant and the summation extends over all $\boldsymbol{k}$ points and polarization branches, λ, in the first Brillouin zone. The phonon angular frequencies $\omega\binom{k}{\lambda}$ were computed by diagonalizing the dynamical matrix at each required $\boldsymbol{k}$ point. Contributions from the short-ranged interactions, obtained from derivatives of the MEG pair potentials, and long ranged point-Coulomb interactions were included in the dynamical matrix. Phonon frequencies at 529 points in the cubic Brillouin zone and at 128 points in the orthorhombic Brillouin zone were required for sufficient convergence of the thermal properties. Furthermore, it is important to point out that the total Gibbs free energy was minimized at each P and T with F_{Th} computed self-consistently with respect to all structural parameters, ξ.

Results and Discussion

Static Lattice Energy

The temperature and pressure dependence of the structural properties are determined by the minimum in the total Gibbs free energy (Equation (1)). However, in order to understand the factors which control the distortional and dynamical properties of perovskites it is important to explore the static lattice potential hypersurface with respect to the structural ξ parameters. We find that the volume, V, and the octahedral rotation angles, Θ and Φ, are the principal parameters which describe the general distortion properties in these perovskites and the static lattice energy dependence on these parameters is discussed below.

In Figure 2, we have plotted contributions to the total static potential from the short-ranged Si–O, O–O and Mg–O (Ca–O) overlap energies for the cubic perovskite structure as a function of the Si–O bond length. In the modified electron gas model, the equilibrium lattice constant for the cubic structure is predominantly determined by a balance between the Madelung attraction and Si–O repulsion. Mg–O and Ca–O repulsions have a smaller effect on the equilibrium structure in cubic perovskite, although the relatively larger Ca–O repulsion accounts for the slightly larger lattice constant in $CaSiO_3$.

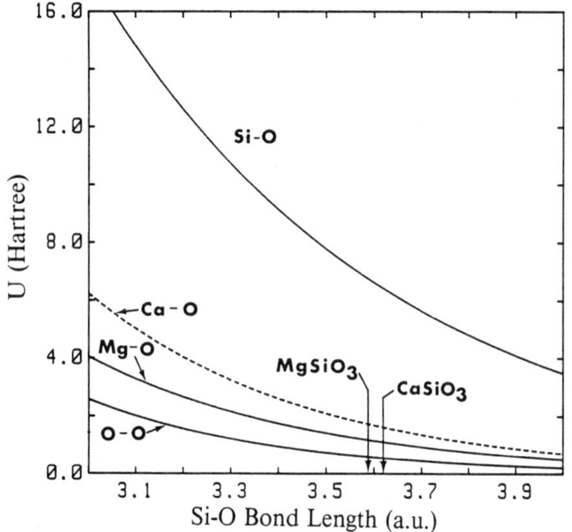

Figure 2. Contributions to U from Si–O, Mg–O (Ca–O), and O–O overlap interactions for the cubic perovskite structure as a function of the Si–O bond length. Equilibrium Si–O bond lengths at zero pressure are indicated for the cubic structures. Note that the Si–O overlap repulsion is considerably larger than for the other pair interactions. The equilibrium lattice constant for cubic perovskite is predominantly determined by a balance between the Madelung attraction and Si–O overlap repulsion.

In Figures 3 and 4, we show contours of the total potential energy with respect to the octahedral rotation angles, Θ and Φ, at several *fixed* volumes for $MgSiO_3$ and $CaSiO_3$, respectively. At V_0, the volume which minimizes the static lattice energy in the cubic structure, $MgSiO_3$ exhibits four structurally equivalent minima at finite values of Θ and Φ, whereas $CaSiO_3$ exhibits a single potential minimum corresponding to the undistorted cubic structure. With a decrease in volume, the potential minima for $MgSiO_3$ deepen and occur at larger values of Θ and Φ. For $CaSiO_3$, minima in the potential develop for finite values of Θ at high compressions followed by the appearance of Φ minima at slightly higher compression.

In Figures 5 and 6, individual contributions to the total potential due to the long-ranged Madelung energy and short-ranged Si–O, O–O and Mg–O (Ca–O) overlap energies are plotted as a function of Θ and Φ at constant volume. Because of an increase in its bond length, the Si–O overlap repulsion decreases with distortion. This decrease in energy is opposed by an increase in the Madelung energy and Mg–O (Ca–O) repulsion. Interestingly, the distortional energy is relatively insensitive to the O–O interactions at either low or high compressions and, in

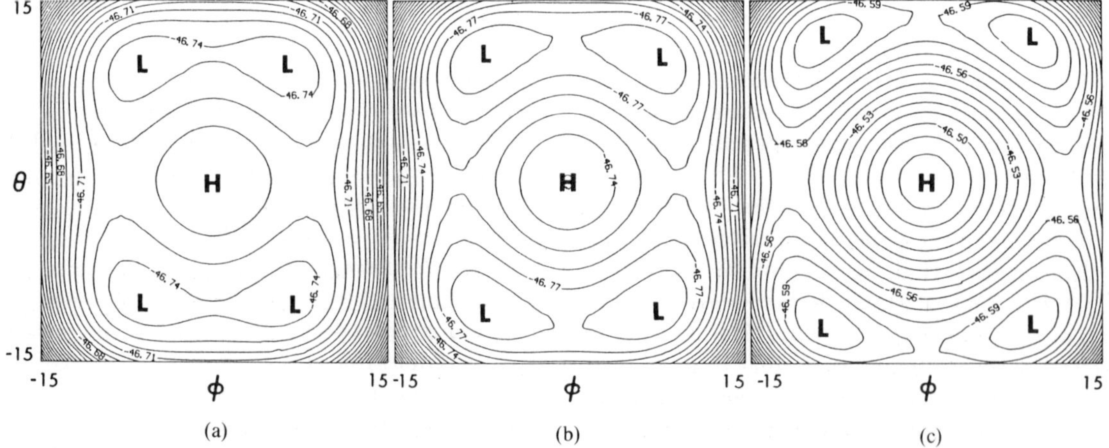

Figure 3. Contour plots of the total static lattice potential, U, for $MgSiO_3$ perovskite as a function of the octahedral rotation angles, Θ and Φ, at constant volume. (a) $V=1.05\ V_0$; (b) $V=V_0$; and (c) $V=.885\ V_0$, where V_0 is the zero-pressure static lattice volume for the cubic structure. With decreasing volume, the potential minima deepen and occur at larger values of Θ and Φ.

CaSiO$_3$ Perovskite

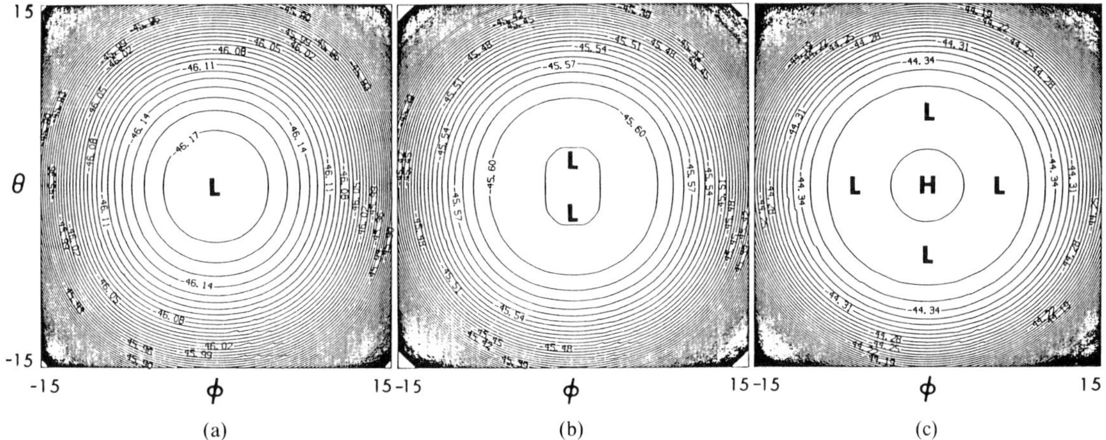

Figure 4. Contour plots of the total static lattice potential, U, for CaSiO$_3$ perovskite as a function of the octahedral rotation angles, Θ and Φ, at constant volume. (a) $V = V_0$; (b) $V = .828\ V_0$; and (c) $V = .723\ V_0$. Note that at high compressions minima in the potential surface develop: First for Θ, followed by minima along Φ at higher compressions.

fact, decreases with small distortions. We note that these results are qualitatively similar to those computed for the fluoridic perovskites (BOYER and HARDY, 1981; BOYER, 1984).

The existence of multiple minima on the potential surface indicates that the cubic perovskite lattice is *mechanically* unstable relative to octahedral rotations. This instability is also related to quasiharmonic *dynamic* instabilities which occur in the cubic lattice for particular octahedral rotation vibrations as discussed later. However, these instabilities do not necessarily imply that the cubic perovskite phase is *thermodynamically* unstable. The potential barriers between minima can be overcome by thermal excitation at high temperatures. We show later that due to large excursions of the octahedral rotation vibrations at high temperatures the cubic phase is *thermodynamically* stable relative to the distorted phases even though the cubic perovskite lattice is *dynamically* unstable in the quasiharmonic approximation.

Lattice Dynamics of Cubic Perovskites

In Figure 7, we show the computed quasiharmonic phonon frequency spectrum of CaSiO$_3$ in the cubic perovskite structure, along selected high-symmetry directions in the Brillouin zone. At the volume which minimizes U, all mode-frequencies are positive and CaSiO$_3$ is dynamically stable in the cubic perovskite structure. The high-frequency branches exhibit a large dispersion across the Brillouin zone and the density of states in this region can be reasonably modeled as an optic continuum (see KIEFFER, 1979). At low frequencies, however, branches related to octahedral rotation vibrations show little dispersion and will produce large peaks in the density of states (BOYER and HARDY, 1981).

For CaSiO$_3$, the lowest frequency R_{25}–T–M_2 branch, along the Brillouin zone boundary, softens with pressure. Ultimately, over a pressure interval between 70 and 90 GPa, mode-frequencies associated with this branch go to zero in the quasiharmonic approximation. This indicates that at high pressure there is no restoring force to infinitesimal atomic displacements associated with these phonons in the cubic perovskite lattice.

In Figure 8, we show the corresponding phonon spectrum for MgSiO$_3$ in the cubic perovskite structure at zero pressure. Mode-frequencies along the R_{25}–T–M_2 branch are imaginary at zero pressure, in contrast to CaSiO$_3$ where this branch is unstable only at high pressures. In both compounds, the associated atomic displacements for these phonons are related to the cooperative rotation of SiO$_6$ octahedra about the pseudo-cubic axes. Each mode along this R_{25}–T–M_2 branch has a different repeat period for the rotational sense in successive octahedral layers perpendicular to the rotation axis. The repeat distance is shortest for the R_{25} mode where adjacent octahedra along the axial rotation direction have an opposite sense of rotation and is longest for the M_2 mode where the rotations of adjacent axial octahedra are in phase.

Lattice Dynamics of Distorted Perovskite

A variety of space group symmetries and unit cell sizes can result from the mechanical instabilities that occur in the cubic perovskite structure. Resultant lower-symmetry structures can be related to the cubic structure through

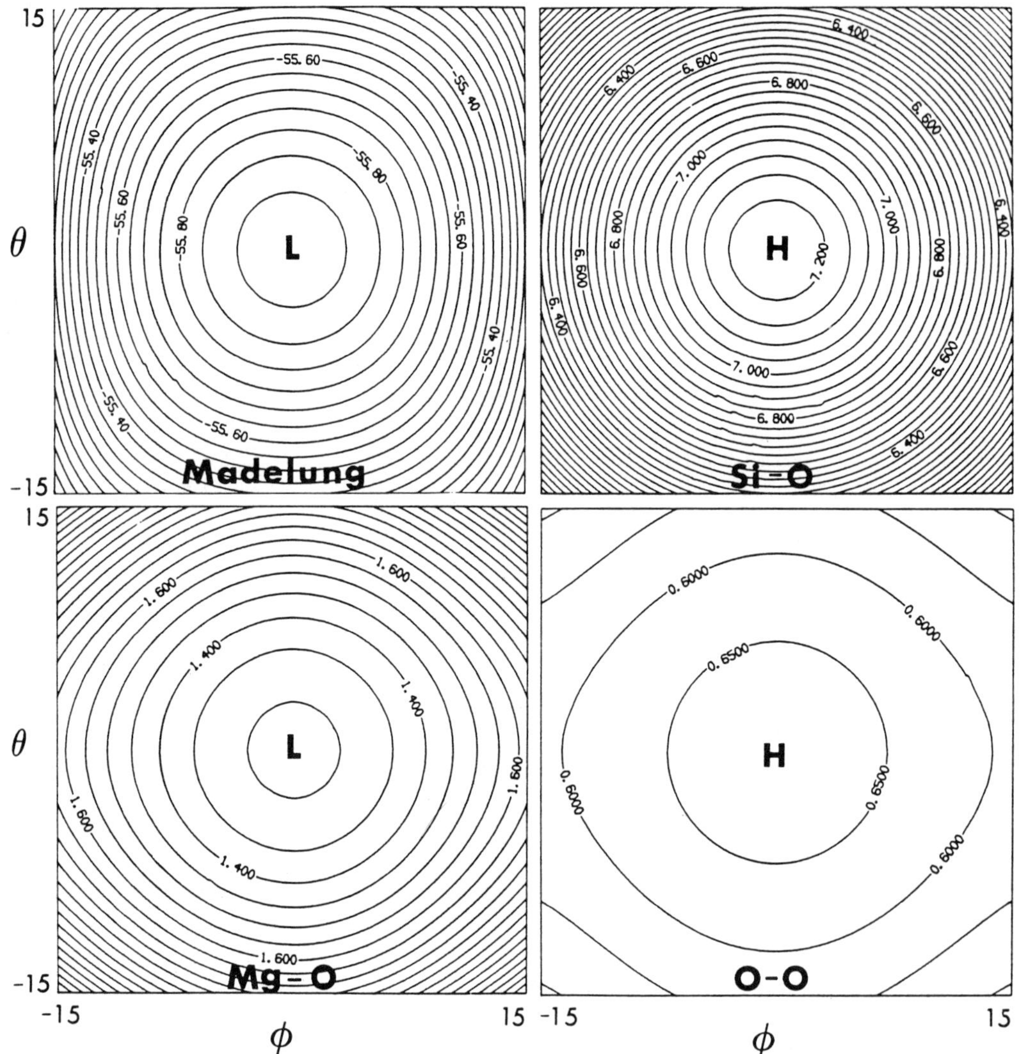

Figure 5. Contributions to U from the Madelung energy and short-ranged Si–O, Mg–O and O–O overlap energies for $MgSiO_3$ perovskite as a function of the distortional parameters, Θ and Φ, at constant volume, V_0. The Si–O overlap repulsion decreases with distortion. The degree of distortion is limited by the increase in Madelung energy and Mg–O overlap repulsion. Note that the O–O overlap energy is relatively insensitive to the degree of distortion.

the "condensation" of particular unstable phonons and a "freezing-in" of their associated octahedral rotations (GLAZER, 1972). Free energy differences between these various polymorphs are likely to be small, as suggested by the fact that all modes along the unstable R_{25}–T–M_2 branch have about the same frequency. In this paper we investigate only a small subset of the possible distorted perovskite structures, in particular, the Pbnm orthorhombic structure, which has been proposed for $MgSiO_3$ perovskite at zero pressure on the basis of powder diffraction data (ITO and MATSUI, 1978; YAGI et al., 1978). This structure can be related to the cubic structure through a condensation of the unstable R^x_{25}, R^y_{25}, and M^z_2 phonons (the pseudocubic axial rotation directions are indicated by superscripts) as previously discussed by WOLF and JEANLOZ (1985a).

In Figure 9, we show the computed phonon spectrum of orthorhombic (Pbnm) $MgSiO_3$ perovskite for the equilibrium structure at zero pressure and 300 K. Because there are 20 atoms in this unit cell, the phonon spectrum is

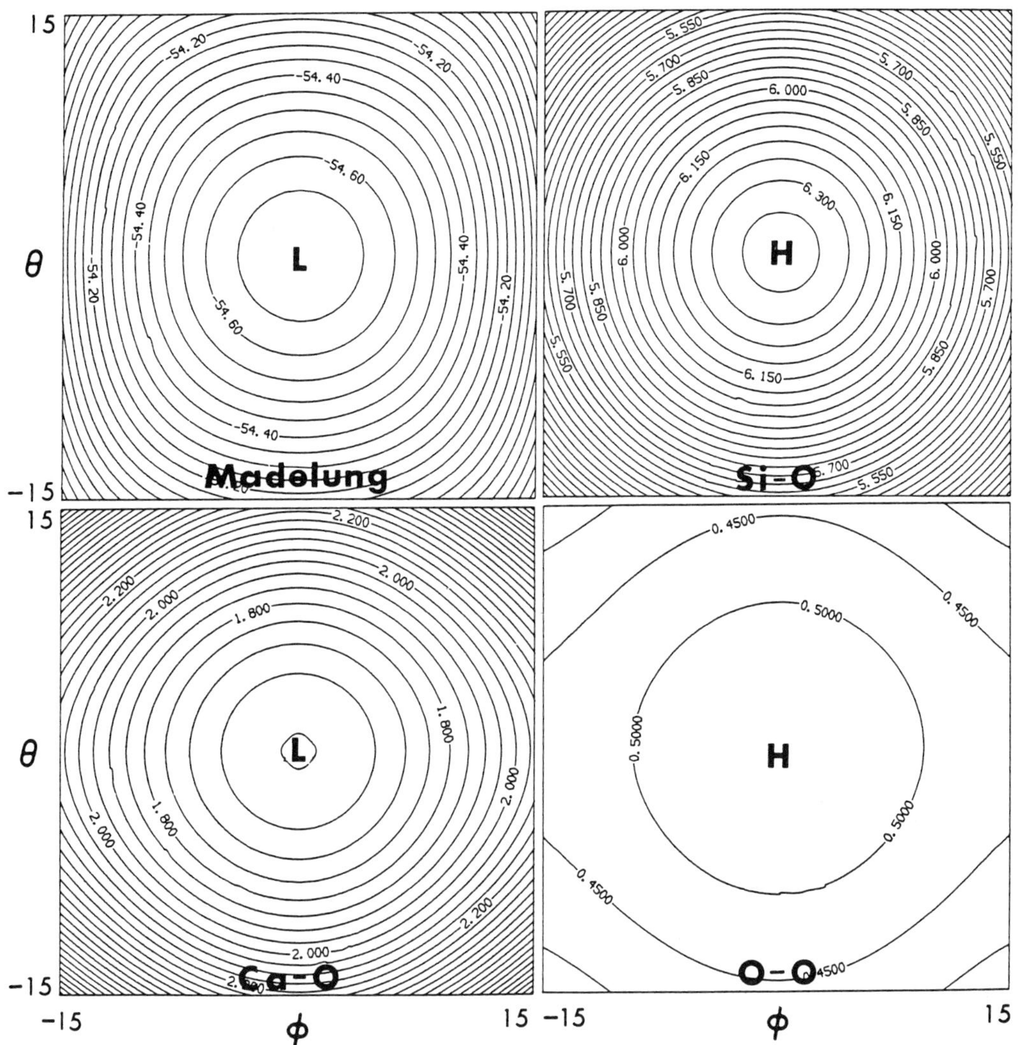

Figure 6. Contributions to U from the Madelung energy and short-ranged Si–O, Ca–O and O–O interaction energies for $CaSiO_3$ perovskite as a function of the distortional parameters Θ and Φ, at constant volume, V_0. As in $MgSiO_3$ perovskite, Si–O repulsion favors distortion. This distortion is opposed by the increase in Madelung energy and Ca–O repulsion. Note that the Ca–O repulsion is considerably larger than that for Mg–O; hence, $CaSiO_3$ perovskite is cubic at zero pressure while $MgSiO_3$ is distorted.

considerably more complex than for the cubic perovskites. At ambient conditions, all mode frequencies are positive and the distorted perovskite structure is dynamically stable in the quasiharmonic approximation. For comparison, we have also plotted absorption frequencies that are observed in the mid-infrared region for $MgSiO_3$ perovskite (WENG et al., 1983). We note however, that since we have employed a rigid-ion potential model, the computed phonon spectrum should be interpreted with caution since charge deformation and polarization effects could be significant in some of the modes. In particular, we expect that the longitudinal optic mode frequencies are overestimated. However, based on our experience with the alkali halides and simple oxides, averages over the density of states, and hence thermal properties, are reliably predicted.

With an increase in pressure, at 300 K, all the phonon frequencies increase and the orthorhombic $MgSiO_3$ lattice remains dynamically stable to at least 150 GPa. However, with increasing temperature, phonons related

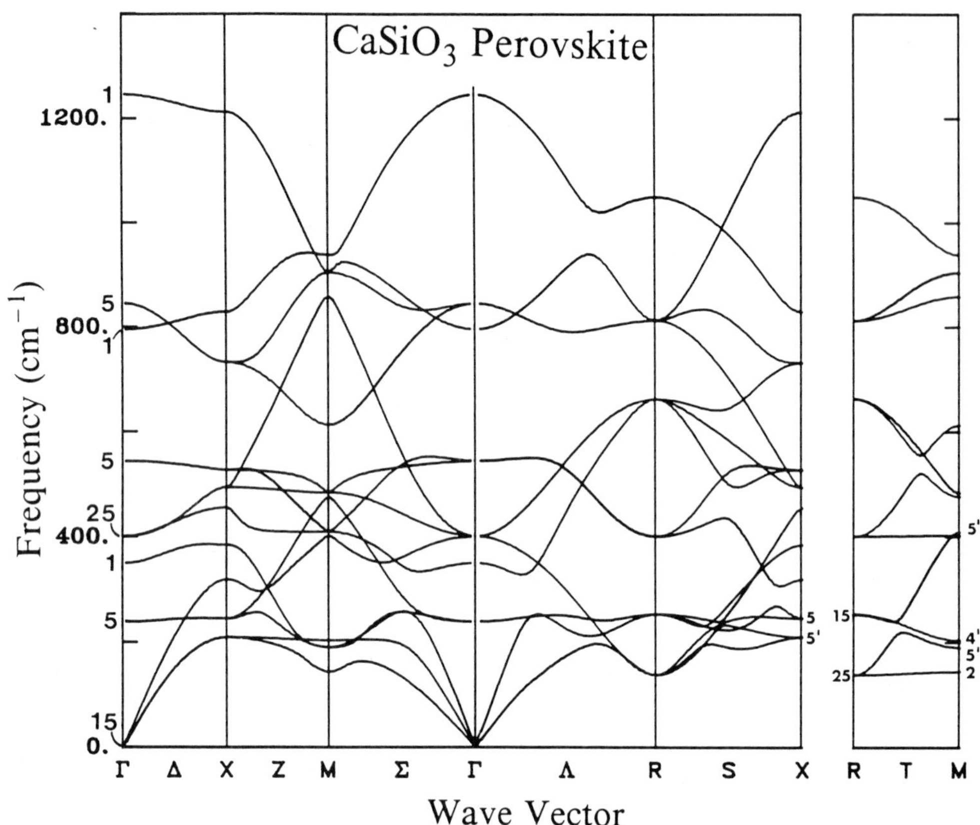

Figure 7. Quasiharmonic frequency spectrum of $CaSiO_3$ in the cubic perovskite structure for selected high-symmetry directions in the Brillouin zone. Volume is equal to V_0, the zero pressure static lattice volume.

to the "condensed" R_{25}^x, R_{25}^y, and M_2^z phonons exhibit critical behavior and the quasiharmonic frequencies approach zero. This behavior is typical of antiferroelectric perovskites. For example, the low temperature tetragonal phase of $SrTiO_3$ perovskite exhibits critical behavior with increasing temperature and undergoes a second-order transition to the cubic phase (COWLEY et al., 1969). This critical behavior is a result of a large octahedral rotation fluctuations that occur in the vibrational modes associated with the "condensed" M_2^z phonons. At the critical temperature these modes "unfreeze" and the structure becomes cubic. For $MgSiO_3$ perovskite at zero pressure, we find that at a temperature near 1000 K, at zero pressure, the M_2^z phonons "unfreeze" first, followed by an "unfreezing" of the R_{25}^x and R_{25}^y phonons at a slightly higher temperature. This behavior is indicative of successive critical transformations to tetragonal and cubic polymorphs and is discussed below.

Temperature and Pressure Dependence of Structural Properties

In Table 1, we list the experimentally derived structural parameters and bond distances of Pbnm orthorhombic $MgSiO_3$ perovskite obtained at ambient conditions by ITO and MATSUI (1978) and YAGI et al. (1978). We also tabulate our theoretical results derived from the MEG theory at zero pressure and 150 GPa. Results for both the regular octahedra ($\xi=5$) model and the full distortion ($\xi=10$) model are listed. As discussed earlier, bond lengths derived from the MEG model are overestimated: At zero pressure, the MEG models predict an average value of 1.904 Å for the Si–O bond length at zero pressure and 300 K compared to the experimental value of 1.79 Å. Likewise a value of 2.661 Å is computed for the average Mg–O bond length compared to the experimental average of 2.48 Å.

Although the MEG model correctly predicts that the perovskite structure of $MgSiO_3$ is distorted from cubic symmetry at zero pressure, the degree of this distortion is underestimated. One measure of this distortion is obtained from the ratio of the orthorhombic unit cell parameters. In an undistorted structure, the unit cell parameters satisfy the relations $b/a=1.0$ and $c/a=1.414$. The observed values of these ratios at ambient conditions are 1.032 and 1.444, respectively. The computed values are 1.017 and 1.425 if regular octahedra are assumed ($\xi=5$) or 1.012 and

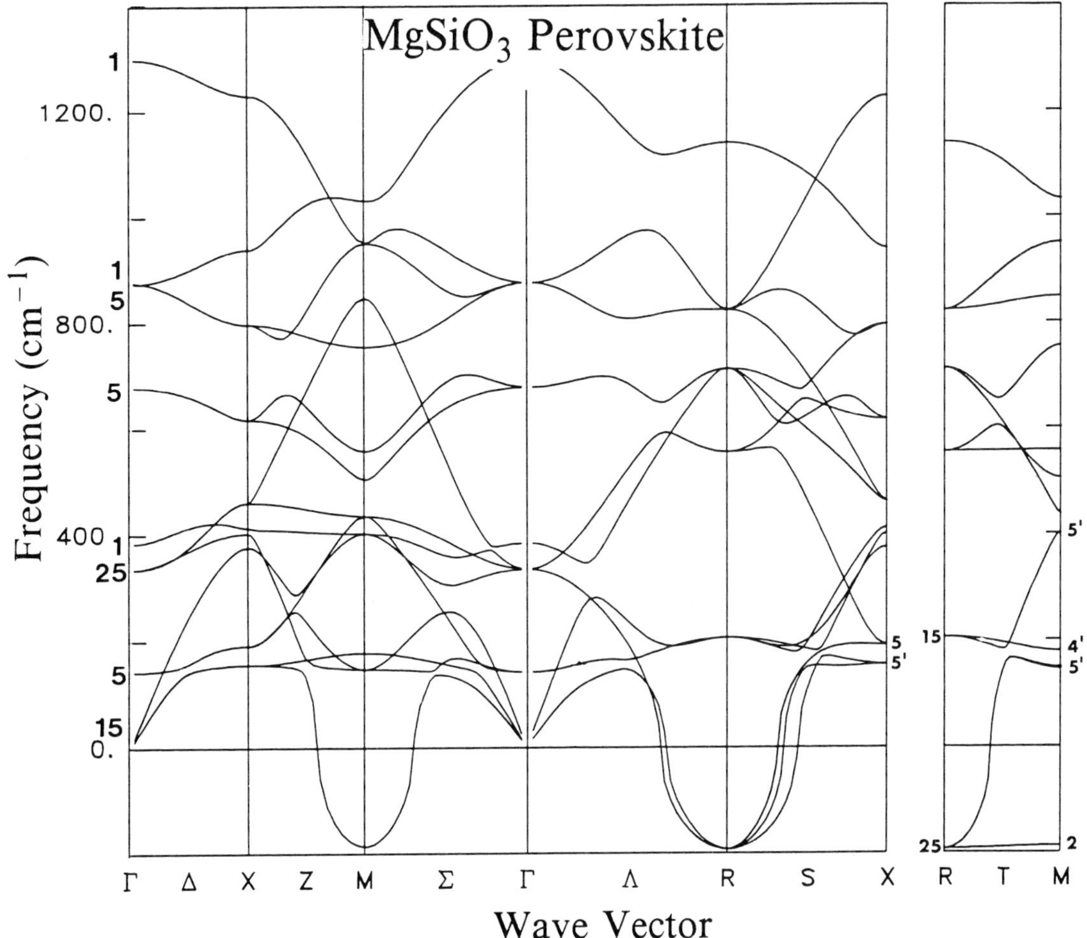

Figure 8. Quasiharmonic frequency spectrum of $MgSiO_3$ in the cubic perovskite structure for selected high-symmetry directions in the Brillouin zone. Volume is equal to V_0, the zero pressure static lattice volume. Note that mode frequencies along the R_{25}–T–M_2 branch are imaginary but are plotted here as negative.

1.427 if full octahedral distortions are allowed ($\xi=10$). The degree of distortion can also be indicated in terms of the octahedral rotation angles. We note, however, that this measure is somewhat ambiguous if the SiO_6 polyhedra are distorted from perfect octahedra. Nevertheless, if we assume regular octahedra the rotation angles can be obtained from the ratios of the unit cell parameters as follows:

$$a/b = \cos\Theta$$

$$a/c = \cos\Phi/\sqrt{2}. \tag{3}$$

Therefore, under the assumption of regular octahedra, the observed rotation angles in $MgSiO_3$ perovskite at ambient conditions are $\Theta=14.3°$ and $\Phi=11.7°$ compared with computed rotation angles of 10.49° and 7.05° for $\xi=5$ or 8.83° and 7.68° for $\xi=10$. In order to fully characterize the distortion in perovskites, the deviation of all the structural parameters from their respective values in the cubic structure must be considered. These values are listed in Table 1 for both the experimental and theoretical results.

There is some controversy as to whether pressure will increase or decrease the degree of distortion in silicate perovskites (YAGI et al., 1982; O'KEEFFE et al., 1979). In Figure 10, we have plotted the computed pressure dependence of the octahedral tilt angles and Si–O bond length for $MgSiO_3$ perovskite at 300 K. If the SiO_6 octahedra are constrained to be regular, there is a marked increase in the rotation angles with pressure: Over a 150 GPa interval Θ increases by 34% and Φ increases by 66%. However, if the SiO_6 polyhedra are allowed to distort, Θ and Φ (calculated from Equation (3)) remain nearly constant with pressure. This is also apparent in Figure 11 where the computed c/a and b/a for $MgSiO_3$ perovskite

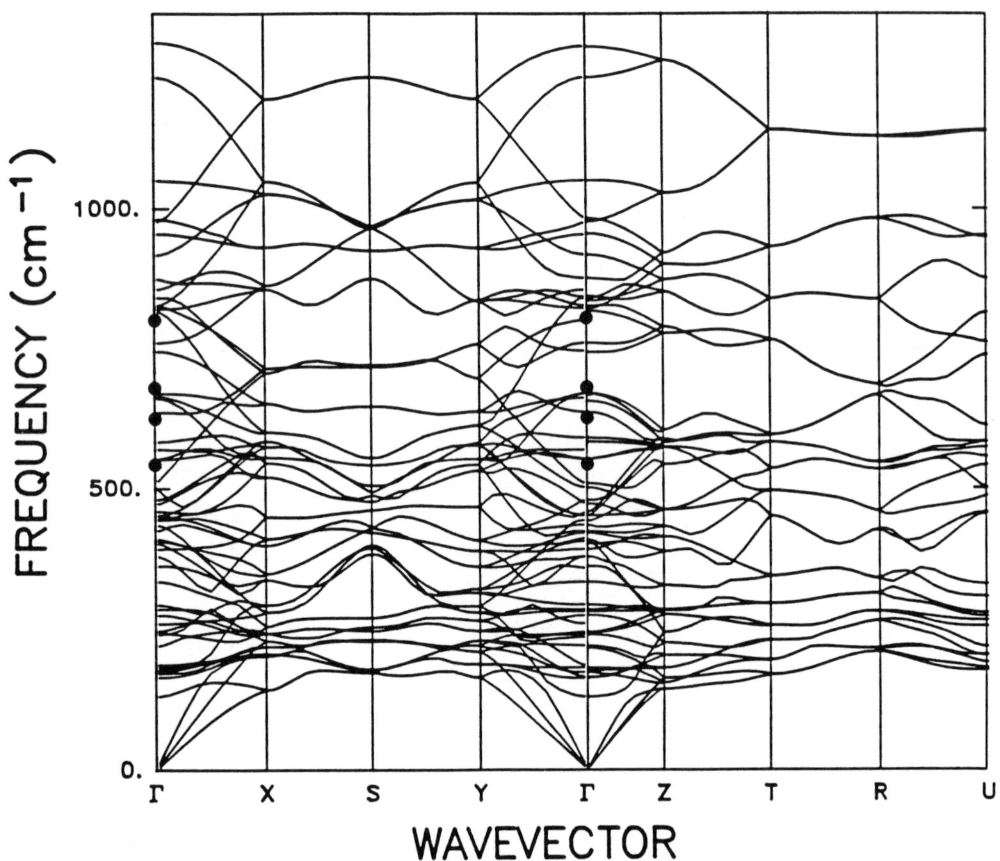

Figure 9. Quasiharmonic frequency spectrum of orthorhombic (Pbnm) $MgSiO_3$ perovskite computed for the equilibrium structure at zero pressure and 300 K. All mode frequencies are positive and the lattice is dynamically stable. Observed infra-red vibrational frequencies are indicated at Γ (WENG et al., 1983).

are plotted as a function of pressure. For the $\xi=5$ model, c/a and b/a increase with pressure. However, for the $\xi=10$ model, the ratios of the orthorhombic cell parameters remain nearly constant with pressure. It is important to point out, however, that even though c/a and b/a are nearly independent of pressure there is still a substantial increase in the degree of distortion of $MgSiO_3$ perovskite from the cubic structure with increasing pressure. This is evident from Table 1 where the parameters describing the atomic positions in $MgSiO_3$ perovskite show a substantial increase with pressure in their deviation from values corresponding to the cubic structure. Thus, although $MgSiO_3$ perovskite becomes increasingly distorted from the cubic structure with pressure, the concomitant increase in the irregularity of the SiO_6 octahedra with pressure has the effect of keeping c/a and b/a nearly constant. This irregularity is primarily exhibited by a rectangular distortion of the octahedral oxygens in the a–b plane while maintaining equivalent Si–O bond lengths.

The temperature dependence of the octahedral rotation angles and Si–O bond length for $MgSiO_3$ perovskite, computed assuming regular octahedra, are plotted in Figure 12 for both zero pressure and 30 GPa. In Figure 13, we have plotted the temperature dependence of b/a and c/a computed for the two distortion models. As is typical for other perovskites, $MgSiO_3$ becomes less distorted with increasing temperature. At low temperatures the dependence of these parameters on temperature is relatively small but increases more rapidly at higher temperatures. With a further increase in temperature, and concomitant decrease in distortion, the quasiharmonic mode-frequencies of the rotation vibrations decrease and eventually become imaginary for *both* the $\xi=5$ and $\xi=10$ models. It is important to emphasize that imaginary phonon frequencies do not actually exist in a real crystal and their occurrence here is only a consequence of using the

TABLE 1. Experimental and Theoretical Unit Cell Parameters, Atomic Positions and Bond Distances for Pbnm Orthorhombic $MgSiO_3$ Perovskite at 300 K. All Distances are Reported in Å

	Data $P=0$		MEG Theroy $P=0$		$P=150$ GPa	
	IM[a]	YMB[b]	$\xi=5$[c]	$\xi=10$[d]	$\xi=5$[c]	$\xi=10$[d]
Unit Cell Parameters:						
a	4.7754(3)	4.780(1)	5.257	5.262	4.692	4.737
b	4.9292(4)	4.933(1)	5.345	5.324	4.837	4.800
c	6.8969(5)	6.902(1)	7.491	7.508	6.776	6.760
$b{:}a$	1.0322	1.032	1.017	1.012	1.031	1.013
$c{:}a$	1.4443	1.444	1.425	1.427	1.444	1.427
Θ[e]	14.4°	14.3°	10.49°	8.7°	14.1°	9.3°
Φ[e]	11.7°	11.6°	7.05°	7.6°	11.7°	7.7°
Mg_x	.521(6)	.527(7)	.5012	.5019	.5060	.5049
Mg_y	.561(5)	.563(5)	.5053	.5083	.5332	.5314
$O1_x$	.096(8)	.096(10)	.0650	.0666	.0888	.0909
$O1_y$	.468(12)	.477(11)	.4921	.4878	.4821	.4698
$O2_x$	.201(8)	.196(7)	.2187	.2195	.1965	.1986
$O2_y$	.205(8)	.209(7)	.2197	.2204	.1997	.2017
$O2_z$	.558(4)	.556(4)	.5325	.5335	.5444	.5469
Bond Distances:						
Si-O1 (2x)	1.79	1.79(1)	1.904	1.911	1.747	1.750
Si-O2 (2x)	1.79	1.75(3)	1.904	1.901	1.747	1.743
Si-O2 (2x)	1.79	1.82(3)	1.904	1.902	1.747	1.754
Mg-O1	2.08	2.10(6)	2.294	2.294	1.973	1.983
Mg-O1	2.08	2.12(5)	2.625	2.578	2.217	2.153
Mg-O2 (2x)	2.05	2.06(5)	2.326	2.331	2.002	1.995
Mg-O2 (2x)	2.20	2.20(4)	2.643	2.623	2.357	2.345
Mg-O2 (2x)	2.48	2.47(4)	2.670	2.677	2.352	2.372
Mg-O1	2.98	2.95(5)	2.765	2.795	2.703	2.734
Mg-O1	2.78	2.75(5)	2.965	2.973	2.745	2.792
Mg-O2 (2x)	3.15	3.16(4)	3.002	3.015	2.948	2.939
Mg-O average	2.47	2.48	2.661	2.661	2.413	2.414

[a] ITO and MATSUI, 1978.
[b] YAGI et al., 1978.
[c] Computed assuming regular octahedra.
[d] Computed allowing full octahedral distortions.
[e] Computed from Equation (3) assuming regular octahedra.

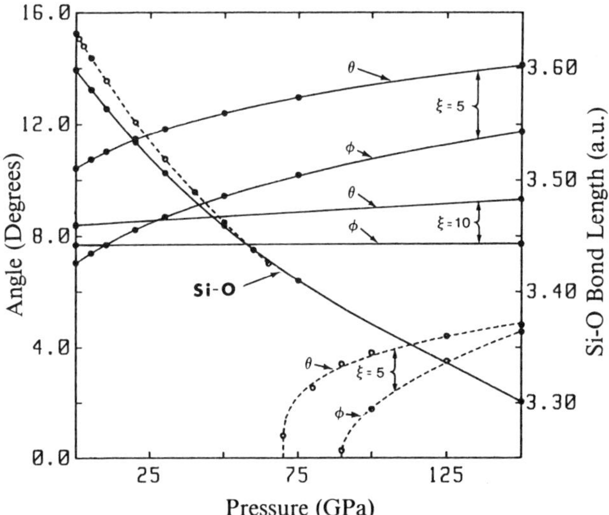

Figure 10. Pressure dependence of the computed Si–O bond length and octahedral tilt angles. All $MgSiO_3$ properties, plotted with solid lines, are for 300 K. $CaSiO_3$ properties are plotted using dashed lines: Si–O bond length is for 300 K and distortion parameters correspond to static lattice potential minima. If the SiO_6 octahedra are constrained to remain regular ($\xi=5$), Θ and Φ in $MgSiO_3$ increases with pressure. However, if the octahedra are allowed to shear ($\xi=10$), the Θ and Φ distortional parameters (computed from Equation (3)) remain nearly constant with pressure.

quasiharmonic approximation. As discussed earlier, the static lattice potential surface of $MgSiO_3$ perovskite exhibits minima that correspond to the same octahedral rotations that are associated with the dynamically unstable rotation vibrations; hence, the structure is distorted at low temperatures. These minima, however, are separated by low potential barriers. As the temperature is increased the amplitude of the rotation-vibrations increases and eventually the low potential barriers can be overcome. Thermal "well-hopping" between the potential minima can occur resulting in a fluctuation of the structural properties over time scales much longer than those that correspond to the vibrational frequencies. At sufficiently high temperatures the rotation-vibrations are no longer impeded by the potential barrier and the average structure, over a time scale on the order of the thermal vibrations, is cubic. This critical behavior is analogous to that which occurs in any second-order phase transition, although usually this behavior is interrupted by the onset of a first-order phase transition to the higher-symmetry phase.

In the context of critical phase transition theory, the tilt angles, Φ and Θ, can be viewed as order parameters between the perovskite structural polymorphs. For $MgSiO_3$, the temperature dependence of each of these order parameters, obtained from our quasiharmonic calculations using the $\xi=5$ model, was fit with a Landau expression of the form $\Theta(T)=\Theta_0(T_c-T)^\beta$, where T_c is the critical temperature and β is the critical exponent. The fits were made over the entire temperature range where the quasiharmonic frequencies were found to be real, as indicated in Figure 12. Extrapolating these equations to zero, we obtain critical temperatures of 930 K and 1115 K for the zero pressure transitions to the tetragonal and cubic phases, respectively. For both order parameters, the computed critical exponents are approximately equal to 0.1, considerably smaller than values between 0.2–0.5 which are commonly observed in other perovskites (FLEURY et al., 1968; MULLER and BERLINGER, 1971).

It is clear that an exact description of the structural behavior in the critical region cannot be derived from

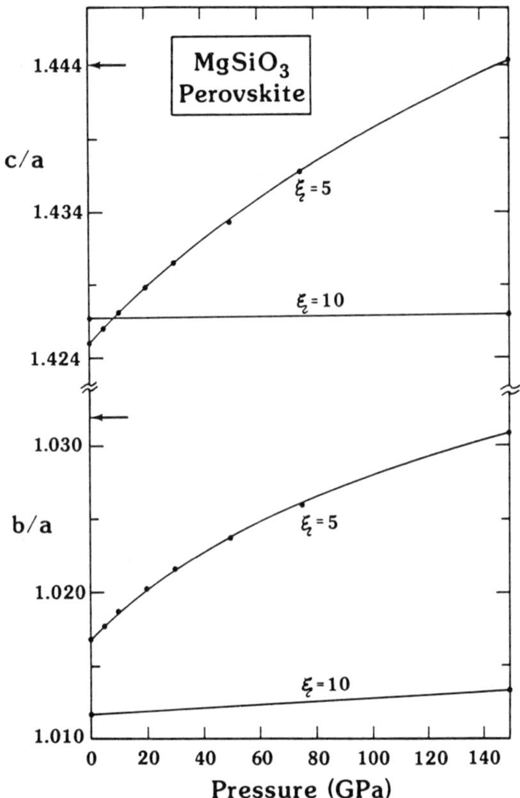

Figure 11. Pressure dependence of c/a and b/a computed for orthorhombic $MgSiO_3$ perovskite at 300 K. Experimental values at zero pressure are indicated by the arrows (YAGI et al., 1978; ITO and MATSUI, 1978). If the SiO_6 constrained to be perfect octahedra ($\xi=5$), c/a and b/a increase with pressure. However, if the octahedra are allowed to shear ($\xi=10$), then c/a and b/a are nearly independent of pressure.

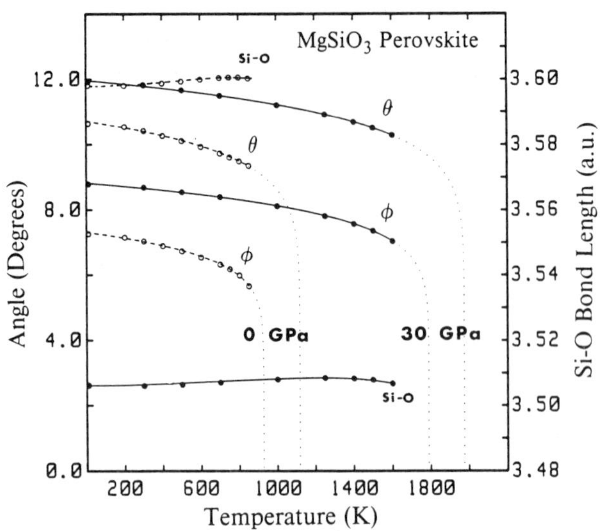

Figure 12. Temperature dependence of computed Θ, Φ and Si–O bond length for $MgSiO_3$ at $P=0$ (dashed lines) and $P=30$ GPa (solid lines) assuming regular octahedra. With increasing temperature the degree of distortion decreases. At high temperatures, perovskite exhibits critical behavior with successive transformations to tetragonal and cubic polymorphs. Dotted lines represent Landau extrapolations of the computed parameters fit over the indicated temperature ranges.

quasiharmonic lattice dynamics and the coarse sampling of the Brillouin zone that we have made here. Changes in the structure are extremely sensitive to details in the behavior of the soft phonon branches and their coupling with other modes. The quasiharmonic approximation assumes independent lattice-modes whose frequencies correspond to restoring forces that are derived from infinitesimal displacements. In the critical region, however, specific modes exhibit large-amplitude anharmonic motions with restoring forces derived over large excursions on the potential energy surface. Moreover, the phonons are not independent but exhibit a large degree of inter-mode coupling. Thus, a complete account of these anharmonic effects along with a finer sampling of the Brillouin zone is required to obtain a precise description of the structural properties in the critical region.

As an additional uncertainty, the electron-gas potential model that we have employed here underestimates the observed degree of distortion in $MgSiO_3$ perovskite by about 30% at ambient conditions. On the basis of this

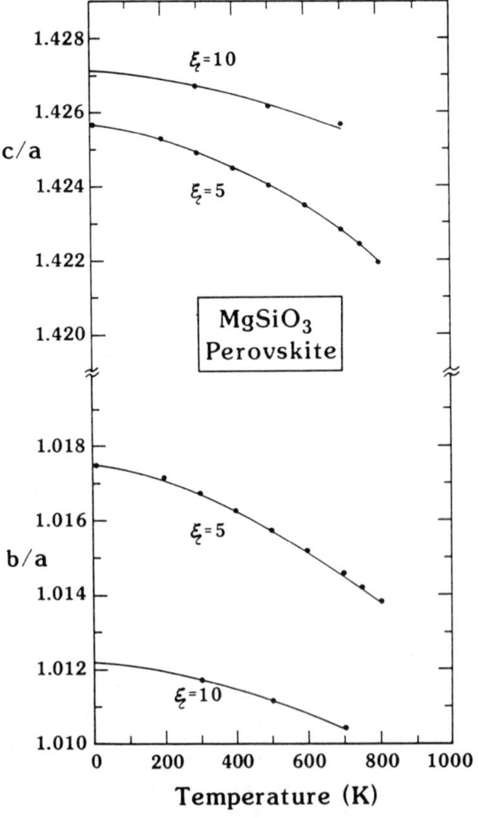

Figure 13. Temperature dependence of c/a and b/a computed for orthorhombic $MgSiO_3$ perovskite at zero pressure allowing full octahedral distortions ($\xi=10$) and assuming regular octahedra ($\xi=5$).

error and the low values obtained for the critical exponents, we expect that the computed transition temperatures are lower bounds for the transitions in real perovskite. It is interesting to note, however, that the lower computed critical temperature is close to the temperature (840 K) at which ($Mg_{.9}$, $Fe_{.1}$)SiO_3 perovskite is observed to undergo a retrograde reversion to enstatite at zero pressure (KNITTLE et al., 1986). Large thermal fluctuations which occur in the critical region may help nucleate this reversion. In addition, the computed critical behavior of $MgSiO_3$ perovskite is quite similar to that observed for its isoelectronic fluoride analogue $NaMgF_3$ (CHAO et al., 1961). At ambient conditions, $NaMgF_3$ exhibits an orthorhombic (Pbnm) structure with approximately the same degree of distortion from cubic symmetry as that observed for $MgSiO_3$ perovskite and undergoes successive critical transformations to the tetragonal and cubic polymorphs at 1023 K and 1173 K, respectively. Although the computed transition temperatures are certainly dependent on details of the potential model and anharmonic effects (see COWLEY, 1977), these results suggest that the silicate perovskite phase could undergo critical phase transformations at high temperatures, the implications of which for the earth are discussed later.

With increasing pressure, the computed critical transition temperatures in $MgSiO_3$ perovskite increase by approximately 28.6 K/GPa. Interestingly, the pressure derivatives in $MgSiO_3$ are nearly identical for both transitions and have values in the range of those observed for other perovskites (SAMARA et al., 1975). These trends are in accord with the universal prediction that pressure raises critical temperatures associated with ferroelastic phase transitions (BOYER and HARDY, 1981).

Thermoelastic Properties

The computed thermal expansivity of orthorhombic $MgSiO_3$ perovskite is plotted in Figure 14 along several isobars. Although we have constrained the SiO_6 to regular octahedra in these calculations, we find that the change in thermal expansivity is negligible if the octahedra are allowed to shear; for example, at 700 K, V/V_{293} computed for the $\xi=10$ model is only .0002 larger than that for the $\xi=5$ model at zero pressure. The agreement of our results with the zero pressure thermal expansion data for ($Mg_{.9}$, $Fe_{.1}$)SiO_3 perovskite is quite good (KNITTLE et al., 1986). The effect of iron substitution on the thermal expansivity is not known but is expected to be small at low concentrations. From a fit of the computed isobars for $MgSiO_3$, using the Mie-Grüneisen formulation of SUZUKI et al. (1979), we obtain values of 1.8×10^{-5} K^{-1} and 2.9×10^{-5} K^{-1}, respectively, for the coefficients of thermal expansivity at 300 K and 800 K, respectively. The computed thermal expansivity of cubic $CaSiO_3$ is also plotted in Figure 14. We find that the thermal expansion

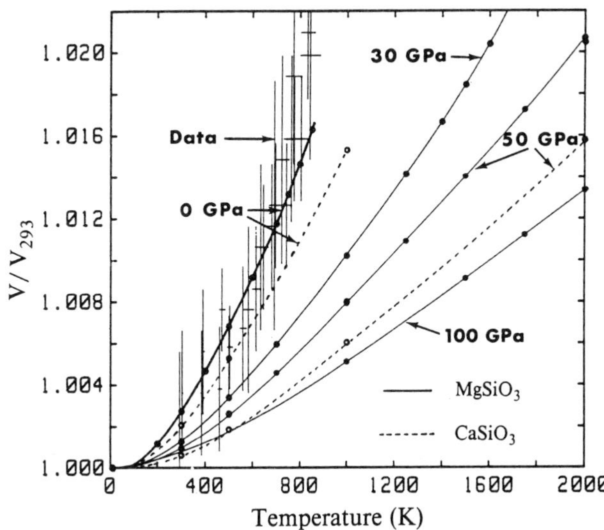

Figure 14. Thermal expansivity of orthorhombic $MgSiO_3$ perovskite computed for several isobars are indicated by the solid lines. Dashed lines are computed results for cubic $CaSiO_3$ perovskite. Experimental data at zero-pressure for ($Mg_{.9}$, $Fe_{.1}$)SiO_3 perovskite to 840 K (where retrograde reversion to enstatite occurred) is also plotted (KNITTLE et al., 1986).

is significantly less than that of orthorhombic $MgSiO_3$: at zero pressure we obtain values of 1.4×10^{-5} K^{-1} at 300 K and 2.0×10^{-5} K^{-1} at 800 K for the thermal expansion coefficients. As also shown in Figure 14, the thermal expansion in these silicate perovskites is strongly dependent on pressure. For example, in $MgSiO_3$ perovskite the calculated expansivity is reduced by about 50% over 30 GPa.

The equation for state of $MgSiO_3$ perovskite has been experimentally measured at room temperature under hydrostatic conditions to 8.5 GPa (YAGI et al., 1982). A Birch-Murnaghan equation of state fitted to this data yield best-fit values for the bulk modulus, K_{OT}, between 255 and 262 GPa, assuming that the pressure derivative of the bulk modulus, K'_{OT}, is between 3 and 5. In Figure 15, we plot the experimental compression data and our computed equation of state for orthorhombic $MgSiO_3$ perovskite for the static lattice and several isotherms. The computed 300 K isotherm is in excellent agreement with the data: A fourth-order Birch-Murnaghan equation of state fitted to the theoretical points gives $K_{OT}=260.0$ GPa and $K'_{OT}=4.01$ assuming regular octahedra, or $K_{OT}=258.2$ GPa and $K'_{OT}=4.05$ for irregular octahedra.

Experimental data on the compressional properties of $CaSiO_3$ perovskite is currently unavailable. Our computed equation of state of $CaSiO_3$ for the static lattice and along several isotherms is shown in Figure 16. At 300 K, we obtain values of $K_{OT}=263.0$ GPa and $K'_{OT}=4.13$ from a

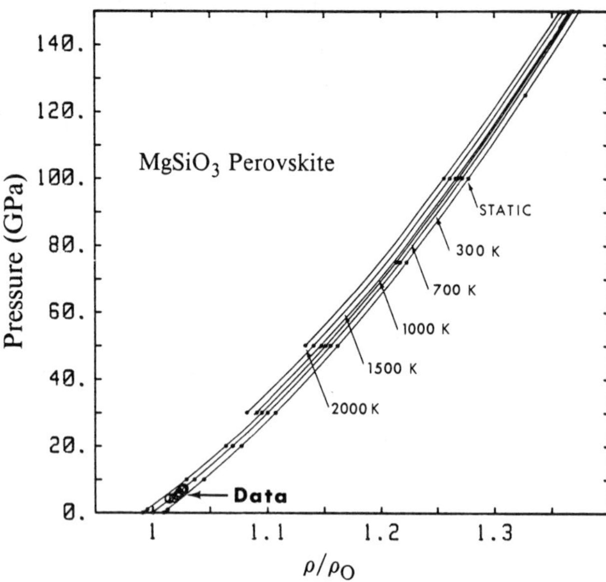

Figure 15. Computed equation of state for orthorhombic $MgSiO_3$ perovskite for the static lattice and several isotherms assuming regular octahedra. Compression data of YAGI et al. (1982) obtained at 300 K is also plotted (plotted symbols are larger than reported errors).

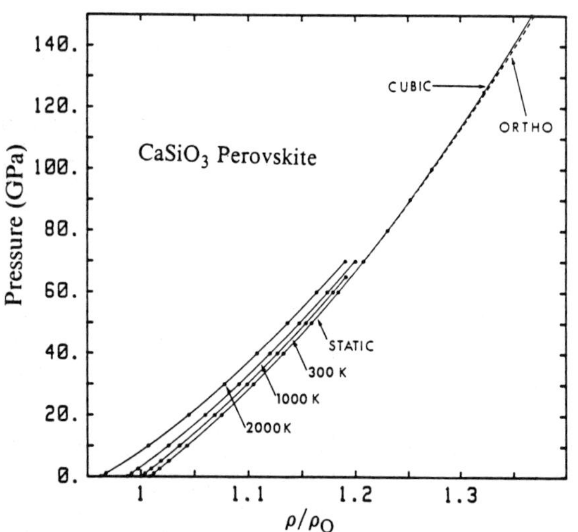

Figure 16. Computed equation of state for cubic $CaSiO_3$ perovskite for the static lattice and several isotherms. Above 70 GPa the static lattice is distorted, resulting in only a slight increase in the compressibility.

fourth-order Birch-Murnaghan fit to the equation of state.

The response of the SiO_6 octahedra to pressure in $MgSiO_3$ is qualitatively different to that often observed in low-pressure phases with SiO_4 tetrahedra. SiO_4 tetrahedra are found to be relatively incompressible and the equations of state of low-pressure phases are strongly controlled by angle bending forces and compression of lower-valent cation–oxygen polyhedra. However, for $MgSiO_3$ (as illustrated earlier in Figure 10), we find a substantial compression of the Si–O bond with pressure. Conversely, the effect of distortions on the equation of state of perovskites is relatively small and can be demonstrated by a comparison of the computed bulk moduli. In the cubic phases, the static lattice values of K_{OT} are 274 GPa for $CaSiO_3$ and 281 GPa for $MgSiO_3$ (WOLF and JEANLOZ, 1985a). In distorted $MgSiO_3$ perovskite, we obtain a value of K_0=271.9 GPa for the static lattice compared to a value of 288 GPa if the octahedra are not allowed to rotate but are fixed at their position that minimizes U at zero pressure. Only a slight further reduction in the bulk modulus occurs (K_0=270.6 GPa) if the octahedra are also allowed to shear. The insensitivity of the compressional properties of perovskite to distortion is also exhibited by $CaSiO_3$, as shown in Figure 16. At pressures above 70 GPa, $CaSiO_3$ perovskite distorts with only a negligible increase in its compressibility over the cubic phase.

The temperature dependence of K_{OT} and K'_{OT} for $MgSiO_3$ and $CaSiO_3$ perovskite are compared in Figures 17 and 18. For $CaSiO_3$, K_{OT} and K'_{OT} exhibit a nearly linear dependence on temperature. At low temperatures, the temperature dependence of K_{OT} and K'_{OT} for $MgSiO_3$ is approximately the same as that for $CaSiO_3$. However, at high temperatures there is a large departure from linearity in both K_{OT} and K'_{OT}. This results from the critical behavior which occurs in the orthorhombic perovskite

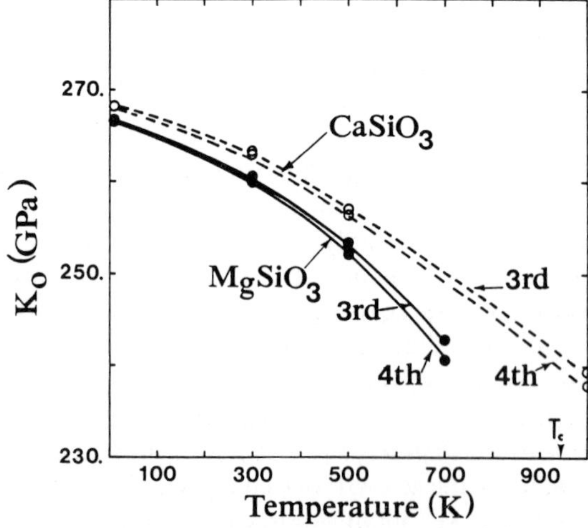

Figure 17. Calculated temperature dependence of K_{OT} for orthorhombic $MgSiO_3$ perovskite (solid lines) and cubic $CaSiO_3$ perovskite (dashed lines). The upper and lower lines for each compound indicate values obtained for respective third and fourth-order Birch-Murnaghan fits to the equation of state.

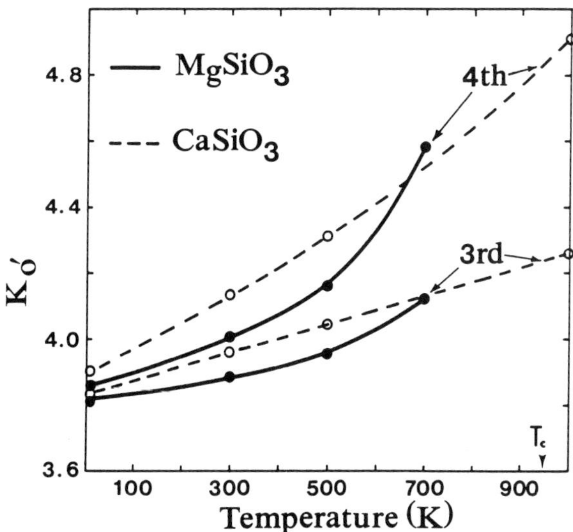

Figure 18. Calculated temperature dependence of K'_{0T} for orthorhombic $MgSiO_3$ perovskite (solid lines) and cubic $CaSiO_3$ perovskite (dashed lines). The upper and lower lines for each compound indicate values obtained for respective third and fourth-order Birch-Murnaghan fits to the equation of state.

structure at high temperatures. Near the critical transition temperatures, soft modes at the Brillouin zone center in the orthorhombic lattice approach the acoustic branches. As a result of intermode coupling, the acoustic modes are pushed down, anomalously decreasing the elastic moduli. This rapid change of the elastic constants near a critical transition has been well documented in other perovskite compounds and is most pronounced for shear moduli (BELL and RUPPERCHT, 1963; ROUSSEAU et al., 1975). Again we must stress however, that anharmonic effects will be important near the critical transition and must be included for an accurate description of the temperature dependence of the elastic moduli.

Composition Models of the Lower Mantle

One approach to constrain the composition of earth's deep interior, which removes the necessity of knowing high-pressure mineralogic properties at high-temperature, is to compare adiabatically decompressed seismic properties with those obtained from zero-pressure model mineralogies (BIRCH, 1952; JACKSON, 1983; JEANLOZ and THOMPSON, 1983). For the lower-mantle, the temperature at the zero-pressure foot of an adiabatic extrapolation would likely be in excess of 1600 K. Thus, to make these comparison, the high-temperature properties of silicate perovskites and magnesiowüstite (believed to be the relevant lower mantle phases) are required. Magnesiowüstite is thermodynamically stable at zero pressure and many of its physical properties have been measured to relatively high temperatures. The silicate perovskite phase, however, is not thermodynamically stable at pressures less than about 22 GPa so that any comparisons made at zero-pressure are actually for a thermodynamically metastable phase. Recently, KNITTLE et al. (1986) measured the 1 bar expansivity of $(Mg_{.9}, Fe_{.1})SiO_3$ perovskite that was quenched from a high-pressure synthesis. The expansivity could only be measured to about 840 K because of a retrograde reversion to the low-pressure enstatite phase. By an extrapolation of their thermal expansivity data to higher temperatures, they found that a pure perovskite mineralogy was compatible with the adiabatically decompressed density and temperature of the lower mantle. Further, a pyrolitic composition was incompatible with the decompressed lower-mantle properties, thus inferring chemical stratification and layered connection within the mantle.

In order to assess the validity of these conclusions, it is important to investigate the possible errors that may occur in an extrapolation of the lower mantle properties to zero pressure and in the extrapolation of the thermal expansion data of metastable perovskite to high temperatures. In Figure 19, we have plotted a range of possible geotherm temperatures for the lower mantle along with their adiabatic extrapolations to zero pressure. The lower-temperature geotherms would likely be associated with a chemically homogeneous mantle and whole-mantle convection, while the relatively higher-temperature geotherms would be associated with chemical stratification and layered convection (BROWN and SHANKLAND, 1981; RICHTER and MCKENZIE, 1981). Superimposed on these geotherms we have plotted our computed stability fields of the $MgSiO_3$ perovskite polymorphs. The perovskite phase is predicted to exist in an orthorhombic structure throughout most of the lower mantle, and its degree of distortion from cubic symmetry to increase slightly with depth along any reasonable geotherm. For the high-temperature geotherms, it is possible that the perovskite phase at the top of the lower mantle exists in one of the higher symmetry polymorphs, although, as noted earlier, we expect that our computed phase boundaries are low-temperature bounds. Also, the presence of Fe, Ca and Al components will complicate these stability fields. Nevertheless, an adiabatic decompression of the lower mantle density may approach or even cross perovskite polymorph phase boundaries. An extrapolation of the zero-pressure properties of $MgSiO_3$ perovskite to high temperatures will also cross these phase boundaries. For properties that are derived from higher-order derivatives of the free energy, (i.e., elastic moduli) their variation in the critical region can be extreme (see, for example, the experimental data of BELL and RUPPERCHT, 1963, and ROUSSEAU et al., 1975). Hence,

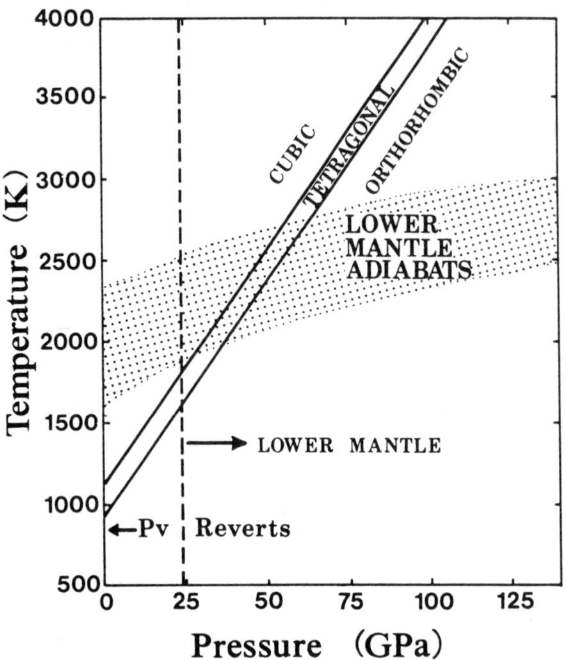

Figure 19. Representative range of geotherm temperatures for the lower mantle, and their adiabatic extrapolations to zero pressure, is indicated by the stippled region. Superimposed on these geotherms are the computed stability fields for the orthorhombic, tetragonal, and cubic $MgSiO_3$ perovskite polymorphs. Both the adiabatic decompression of the lower mantle geotherm and the extrapolation of $MgSiO_3$ perovskite's thermal expansivity to high temperatures cross the computed phase boundaries. The observed retrograde reversion temperature of perovskite to enstatite is approximately equal to the computed lower critical temperature.

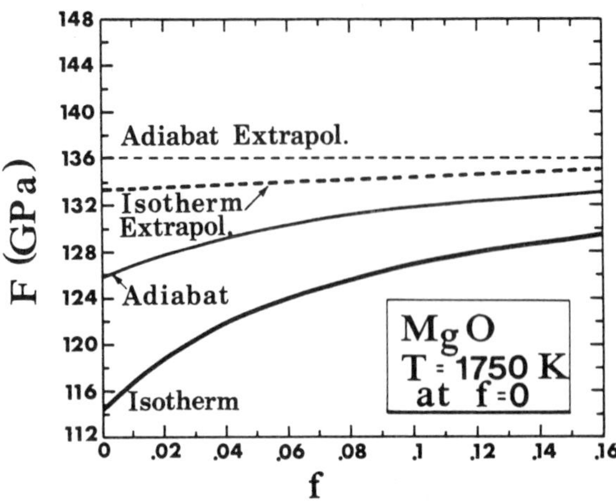

Figure 20. F vs f high-temperature equation of state for MgO computed from Monte-Carlo simulations (WOLF et al., 1987). The 1750 K isotherm and adiabat are indicated by the solid lines. At $f=0$, the F intercept is equal to K_0 and the slope is proportional to $3/4(K_0'-4)$. Dashed lines show the inferred equations of state that result from a third-order Birch-Murnaghan fit to the computed isotherm and adiabat densities between 30–125 GPa: Note that their extrapolation to zero pressure does not reproduce the curvature in F at low strains. As a result, K_0 is overestimated and K_0' is underestimated by these high-pressure fits.

the extrapolation of material properties through or into these critical regions can result in substantial errors.

In addition to the possible critical behavior in perovskite, the extrapolation of the lower mantle presents another problem that has until now escaped notice: a finite strain decompression of high temperature isotherms or adiabats, regardless of the occurrence of critical behavior, systematically overestimates the zero pressure density and K_0 and underestimates K_0'. A good example of this effect is provided by the high-temperature equation of state of MgO computed by WOLF et al. (1987) using a Monte-Carlo simulation to insure that anharmonic effects were properly accounted for. The computed equation of state, thermal expansivity and temperature dependence of K_{OT} are all in excellent agreement with available data; hence, the conclusions that follow are realistic. Figure 20 shows the 1750 K isotherm and corresponding adiabat obtained from the Monte Carlo simulations of MgO. We have plotted the equations of state in terms of BIRCH's (1978) normalized pressure, F, given by $F=P[3f(1+2f)^{5/2}]^{-1}$ where f is the Eulerian strain measure defined by, $f=1/2[(V_0/V)^{2/3}-1]$. A third-order Birch-Murnaghan fit to the high-temperature isotherm densities between 30–125 GPa overestimates the zero-pressure density by .05 gm/cc, overestimates K_{OT} by 20 GPa, and underestimates K_{OT}' by about 1.0. For the adiabat, errors in the extrapolated properties are reduced by about 50%, although they clearly remain significant. These errors are primarily a consequence of the large volume dependence of the thermal pressure that occurs in materials at high temperatures and low strains (e.g., WOLF and JEANLOZ, 1985b) and, since high-pressure data contain no information about the strain dependence of the thermal properties at low strain, a finite-strain fit to high-pressure data will underestimate the temperature effect on the zero-pressure properties.

Consequently, because the zero-pressure density and bulk modulus of $MgSiO_3$ perovskite is greater than that of MgO, compositional constraints for the lower mantle that are based on an adiabatic decompression of its seismic properties will systematically overestimate the proportional abundance of the pyroxene component relative to olivine. This error in the extrapolated zero-pressure high-temperature properties of the lower mantle is indeed suggested by the differences in the third and fourth-order finite strain fits to the lower mantle density. A third-order fit to the PREM density model of the lower mantle (DZIEWONSKI and ANDERSON, 1981) yields values of $\rho_0=4.00$ gm/cc, $K_{OS}=222$ GPa and $K_{OS}'=3.77$, whereas a fourth-order fit, which would be more sensitive to the curvature in F at low strains, gives values of $\rho_0=3.99$ gm/cc, $K_{OS}=215$ GPa and $K'_{OS}=4.03$. That even the fourth-order extrapolation is inaccurate is indicated by the unrealistically low value of K'_{OS} that is inferred for the lower mantle mineral assemblage which would be at a temperature near 1750 K at the foot of the adiabat. This high temperature value is less than the experimentally determined value of K'_{OS} for MgO at room temperature and is also less than our computed 300 K value for $MgSiO_3$ perovskite. Since K'_{OS} is expected to increase with temperature, an assemblage of perovskite and magnesiowüstite requires a substantially larger value of K'_{OS} at high temperatures. For example, from the high temperature Monte-Carlo results we obtain a value of $K'_{OS}=4.52$ for MgO at 1750 K. As discussed earlier, this effect will be even more substantial for perovskite bearing assemblages if the silicate perovskite phase in the earth exhibits critical behavior along the decompression path.

The theoretical results that we have presented for the thermal expansivity and equation of state of $MgSiO_3$ perovskite, and those recently computed for MgO (WOLF and BUKOWINSKI, 1987), are in very good agreement with the available data. Assuming that our results are also valid at the combined temperature and pressure conditions appropriate for the lower mantle, we are able to make direct comparisons of the seismic properties with those obtained from model mineral assemblages. In the remainder of this section, we summarize the results of comparisons we have made between the inferred density and seismic parameter of the lower mantle at a depth corresponding to 30 GPa and theoretically derived estimates for model compositions ranging from pure pyroxene to pure olivine, with an assumed Fe/(Mg+Fe) ratio, X_{Fe}, between 0.10 and 0.12.

In Figures 21 and 22, we plot the computed density and seismic parameter for composition models of the lower mantle as a function of temperature at 30 GPa and $X_{Fe}=0.10$. The predicted properties were obtained from a combination of our theoretical results on $MgSiO_3$ perovskite and MgO with experimental data on zero-pressure densities, Fe/Mg partitioning coefficients, and the effect of Fe substitution on thermodynamic properties. The error bars reflect the allowance of a generous variance in the computed properties: $\pm30\%$ in the mean thermal expansion, ±10 GPa in K_{OT}, ±0.5 in K'_{OS}, and ±0.5 in $(\partial \ln K_s/\partial \ln \rho)_P$. Details of these calculations are presented in another publication (Bukowinski and Wolf, in preparation). The arrows in Figures 21 and 22 indicate the change in properties of the model assemblages that would result from an increase in the value of X_{Fe} to 0.12. Superimposed on these plots is the seismologically inferred density and seismic parameter of the earth at a depth corresponding to 30 GPa (approximately 800 km).

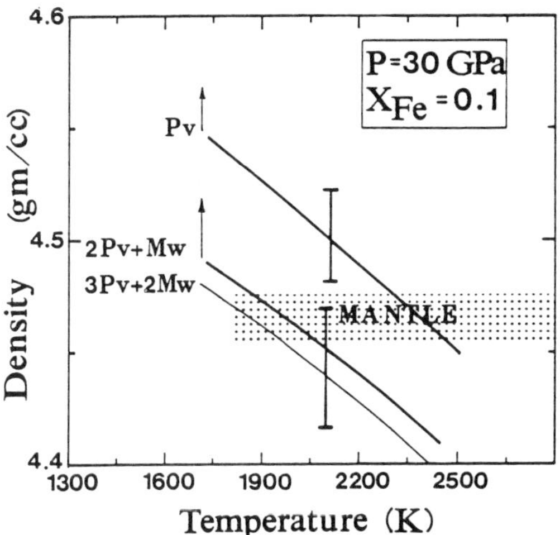

Figure 21. Computed density of model lower mantle perovskite (Pv) and magnesiowüstite (Mw) mineral assemblages for $X_{Fe}=0.10$ at 30 GPa as a function of temperature. Error bars reflect the allowed variances in physical properties, as specified in the text. The arrows indicate the change in density of the mineral assemblages that results from an increase in X_{Fe} to 0.12. The superimposed shaded region is the density of the lower mantle at a depth corresponding to 30 GPa.

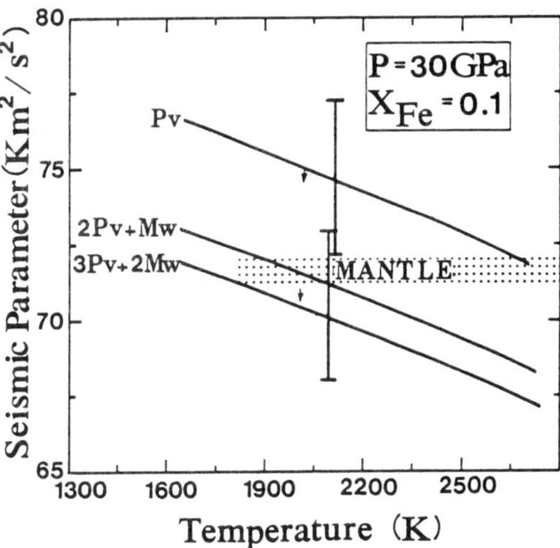

Figure 22. Computed seismic parameter of model lower mantle perovskite (Pv) and magnesiowüstite (Mw) mineral assemblages for $X_{Fe}=0.10$ at 30 GPa as a function of temperature. Error bars reflect the allowed variances in physical properties, as specified in the text. The arrows indicate the change in seismic parameter of the mineral assemblages that results from an increase in X_{Fe} to 0.12. The superimposed shaded region is the seismic parameter of the lower mantle at a depth corresponding to 30 GPa.

From these results, we conclude that for the lower mantle to have a pure perovskite mineralogy with $X_{Fe}=0.10$, a geotherm temperature near 2400 K is required at a depth of 800 km, consistent with temperature estimates for layered mantle convection (RICHTER and MCKENZIE, 1981). However, for a 3:2 or 2:1 assemblage of perovskite and magnesiowüstite, approximately corresponding to a pyrolitic composition, a geotherm temperature near 2000 K is required which is consistent with temperature estimates for whole mantle convection (BROWN and SHANKLAND, 1981). Therefore, the density and seismic parameter of the lower mantle are compatible with *either* a pyrolitic *or* pyroxene composition given the different convection modes and geotherm temperatures that are appropriate for these mineralogies.

These conclusions are different than those reached by KNITTLE et al. (1986) who found that the adiabatically decompressed density and temperature of the lower mantle was incompatible with a pyrolitic composition, and whole mantle convection. As we have pointed out, the inconsistent conclusions that are reached by these different approaches are largely due to the systematic errors which occur in the adiabatic decompression of the lower mantle and the extrapolation of the zero-pressure thermal expansion of perovskite to high temperatures.

Conclusions

We have presented a detailed study of the stability and thermodynamic properties of $CaSiO_3$ and $MgSiO_3$ perovskites as a function of temperature and pressure. The theoretical properties were derived from quasi-harmonic lattice dynamics employing parameter-free pair potentials derived from the MEG formulation. In accord with observations, the orthorhombic structure is predicted for the equilibrium phase of $MgSiO_3$ perovskite at ambient conditions. For $CaSiO_3$, although a perovskite phase has not yet been successfully quenched from its high-pressure stability field, our theoretical results indicate that the cubic perovskite structure is dynamically stable for this compound at zero pressure and to pressures of at least 70 GPa. At higher pressures, the static potential energy surface of $CaSiO_3$ exhibits minima that correspond to a distorted perovskite structure. However, these minima are separated by small potential barriers so that large amplitudes in the octahedra rotation vibrations are likely to occur even at low temperatures such that the average structure remains cubic.

At high temperatures, our model predicts that the orthorhombic $MgSiO_3$ perovskite phase will exhibit critical soft-mode behavior resulting in successive transformations to tetragonal and cubic phases. Whether or not this critical behavior is interrupted by first-order phase transitions cannot be determined from our present treatment. Moreover, a detailed description of the critical behavior cannot be made without an inclusion for anharmonic effects and our computed critical temperatures (~1000 K at zero pressure) represent lower bounds for real $MgSiO_3$ perovskite.

The computed stability fields for the $MgSiO_3$ perovskite polymorphs indicate that the orthorhombic phase is stable throughout the conditions of the lower mantle and that the degree of distortion in this phase from cubic symmetry will slightly increase with depth along any reasonable geotherm. However, it is probable that an extrapolation along the adiabatic decompression path of the lower mantle will approach, or even cross, the phase boundaries to the tetragonal and cubic polymorphs. As observed in other perovskites, large variations in the physical properties can occur as the second-order phase boundaries are approached. Therefore, an adiabatic extrapolation of the lower mantle properties, or the extrapolation of perovskite's zero pressure thermal expansivity to higher temperatures, is clearly dangerous.

In addition to the possible critical behavior in perovskite, finite strain decompressions of high-temperature adiabats universally overestimate the zero-pressure density and bulk modulus and underestimate the pressure derivative of the bulk modulus. Therefore, in order to tighten compositional contraints for the lower mantle it will be necessary to make direct comparisons of the seismological properties with mineralogic properties at the appropriate temperature and pressure conditions of the mantle, or, to make an explicit account for the strain dependence of the thermal pressure along the mantle's decompression path.

Our theoretical results on the equation of state and thermal expansivity of $MgSiO_3$ perovskite are in very good agreement with available data. Assuming these results are valid at the temperature and pressure conditions of the lower mantle we have shown that both the density and seismic parameter of the lower mantle are compatible with either a pyrolite or pyroxene mineralogy given the different geotherm temperatures that would be appropriate for these mineralogies.

Acknowledgments. We would like to thank Raymond Jeanloz, Amotz Agnon and Lars Stixrude for a critical reading of the manuscript. We also appreciate helpful comments received by several referees. This research was funded in part by the National Science Foundation under grant EAR8416778, the Institute of Geophysics and Planetary Physics at Lawrence Livermore Laboratories, and the Center for Computational Seismology at Lawrence Berkeley Laboratories under contract DE-AC03-76SF00098.

References

BELL, R. O., and G. RUPPERCHT, Elastic constants of strontium titanate, *Phys. Rev., 129*, 90–94, 1963.

BIRCH, F., Elasticity and constitution of the earth's interior, *J. Geophys.*

Res., *51*, 227–286, 1952.

BIRCH, F., Finite strain isotherm and velocities for single-crystal and polycrystalline NaCl at high pressures and 300 K, *J. Geophys. Res.*, *83*, 1257–1268, 1978.

BOYER, L. L., Parameter-free equation of state calculation for $CsCaF_3$, *J. Phys. C.: Solid State Phys.*, *17*, 1825–1832, 1984.

BOYER, L. L., and J. R. HARDY, Theoretical study of the structural phase transition in $RbCaF_3$, *Phys. Rev. B*, *24*, 2577–2591, 1981.

BROWN, J. M., and T. J. SHANKLAND, Thermodynamic parameters in the earth as determined with seismic profiles, *Geophys. J. R. astr. Soc.*, *66*, 579–596, 1981.

CHAO, E. C. T., H. EVANS, B. SKINNER, and C. MILTON, Neighborite, $NaMgF_3$, a new mineral from the Green River Formation, South Ouray, Utah, *Am. Mineral.*, *46*, 379–393, 1961.

CLEMENTI, E., and C. ROETTI, *Atomic Data and Nuclear Tables*, 478 pp., Academic Press, New York, 1974.

COHEN, A. J., and R. G. GORDON, Modified electron-gas study of the stability, elastic properties, and high-pressure behavior of MgO and CaO crystals, *Phys. Rev. B*, *14*, 4593–4605, 1976.

COWLEY, R. A., Structural phase transitions, in *Proceedings of the International Conference on Lattice Dynamics*, edited by M. Balkanski, pp. 625–631, Flammarion Sciences, Paris, 1977.

COWLEY, R. A., W. J. L. BOYERS, and G. DOLLING, Relationship of normal modes of vibration of strontium titanate and its antiferroelectric phase transition at 110 K, *Solid State Commun.*, *7*, 181–184, 1969.

DZIEWONSKI, A. M., and D. L. ANDERSON, Preliminary reference earth model, *Phys. Earth Planet. Int.*, *25*, 297–386, 1981.

EWALD, P. P., The calculations of optical and electrostatic lattice potentials, *Ann. Phys.* (Leipzig), *64*, 253–287, 1921.

FLEURY, P. A., J. F. SCOTT, and J. M. WARLOCK, Soft phonon modes and the 110 K phase transition in $SrTiO_3$, *Phys. Rev. Lett.*, *21*, 16–19, 1968.

GLAZER, A. M., The classification of tilted octahedra in perovskites, *Acta Cryst.*, *B28*, 3384–3391, 1972.

GORDON, R. G., and Y. S. KIM, A theory for the forces between closed shell atoms and molecules, *J. Chem. Phys.*, *56*, 3122–3133, 1972.

HEMLEY, R. J., and R. G. GORDON, Theoretical study of solid NaF and NaCl at high pressures and temperatures, *J. Geophys. Res.*, *90*, 7803–7813, 1985.

HEMLEY, R. J., M. D. JACKSON, and R. G. GORDON, First-principles theory for the equation of state of minerals at high-pressures and temperatures: applications to MgO, *Geophys. Res. Lett.*, *12*, 247–250, 1985.

ITO, E., and Y. MATSUI, Synthesis and crystal-chemical characterization of $MgSiO_3$ perovskite, *Earth Planet. Sci. Lett.*, *31*, 443–449, 1978.

JACKSON, I., Some geophysical constraints on the chemical composition of the earth's lower mantle, *Earth Planet. Sci. Lett.*, *62*, 91–1033, 1983.

JEANLOZ, R., and A. B. THOMPSON, Phase transitions and mantle discontinuities, *Rev. Geophys. Space Phys.*, *21*, 51–74, 1983.

KIEFFER, S. W., Thermodynamics and lattice vibrations of minerals: 1. Mineral heat capacities and their relationship to simple lattice vibrational models, *Rev. Geophys. Space Phys.*, *17*, 1–19, 1979.

KNITTLE, E., R. JEANLOZ, and G. L. SMITH, Thermal expansion of silicate perovskite and stratification of the earth's mantle, *Nature*, *319*, 214–216, 1986.

LIU, L., and A. E. RINGWOOD, Synthesis of a perovskite-type polymorph of $CaSiO_3$, *Earth Planet. Sci. Lett.*, *28*, 209–211, 1975.

MADON, M., P. M. BELL, H. K. MAO, and J. P. POIRIER, Transmission electron diffraction and microscopy of a synthetic high pressure $MgSiO_3$ phase with perovskite structure, *Geophys. Res. Lett.*, *7*, 629–632, 1980.

MEHL, M. J., R. J. HEMLEY, and L. L. BOYER, Potential induced breathing model for the elastic moduli and high-pressure behavior of the cubic alkaline earth oxides, *Phys. Rev.*, *B23*, 900–923, 1981.

MUHLHAUSEN, C., and R. G. GORDON, Electron-gas theory of ionic crystals, including many-body effects, *Phys. Rev.*, *B23*, 900–923, 1981.

MULLER, K. A., and W. BERLINGER, Static critical exponents at structural phase transitions, *Phys. Rev. Lett.*, *26*, 13–16, 1971.

O'KEEFFE, M., B. G. HYDE, and J. BOVIN, Contributions to the crystal chemistry of orthorhombic perovskite: $MgSiO_3$ and $NaMgF_3$, *Phys. Chem. Minerals*, *4*, 299–305, 1979.

RICHTER, F. M., and D. P. MCKENZIE, On some consequences and possible causes of layered mantle connection, *J. Geophys. Res.*, *86*, 6133–6142, 1981.

ROUSSEAU, M., J. V. JULLIARD, and J. NONET, Crystallographic, elastic, and Raman scattering investigation of structural phase transition for $RbCdF_3$ and $TlCdF_3$, *Phys. Rev. B.*, *12*, 1579–1590, 1975.

SAMARA, G. A., T. SAKUDO, and K. YOSHIMITSU, Important generalization concerning the role of competing forces in displacive phase transitions, *Phys. Rev. Lett.*, *35*, 1767–1769, 1975.

SUZUKI, I., S. OKAJIMA, and K. SEYA, Thermal expansion of single-crystal manganosite, *J. Phys. Earth.*, *27*, 63–69, 1979.

WALDMAN, M., and R. G. GORDON, Scaled electron-gas approximation for intermolecular forces, *J. Chem. Phys.*, *71*, 1325–1339, 1979.

WATSON, R. E., Analytic Hartree-Frock solution for O^{-2}, *Phys. Rev.*, *111*, 1108–1110, 1958.

WENG, K., J. XU, H. K. MAO, and P. M. BELL, Preliminary Fourier-transform infrared spectral data on the SiO_6 octahedral group in silicate perovskite, *Carnegie Inst. Washington, Year Book 82*, 355–356, 1983.

WOLF, G. H., and R. JEANLOZ, Lattice dynamics and structural distortions of $CaSiO_3$ and $MgSiO_3$ perovskites, *Geophys. Res. Lett.*, *12*, 413–416, 1985a.

WOLF, G. H., and R. JEANLOZ, Vibrational properties of model monatomic crystals under pressure, *Phys. Rev. B*, *32*, 7798–7810, 1985b.

WOLF, G. H., and M. S. T. BUKOWINSKI, Ab initio structural and thermoelastic properties of orthorhombic $MgSiO_3$ perovskite, *Geophys. Res. Lett.*, *12*, 809–812, 1985.

WOLF, G. H., and M. S. T. BUKOWINSKI, Variational stabilization of the ionic charge densities in electron-gas theory: applications to MgO and CaO, *Phys. Chem. Minerals*, in press, 1987.

WOLF, G. H., M. S. T. BUKOWINSKI, and M. ROSS, Anharmonic equation of state of MgO, in preparation, 1987.

YAGI, T., H. K. MAO, and P. M. BELL, Structure and crystal chemistry of perovskite-type $MgSiO_3$, *Phys. Chem. Minerals*, *3*, 97–110, 1978.

YAGI, T., H. K. MAO, and P. M. BELL, Hydrostatic compression of perovskite-type $MgSiO_3$, in *Advances in Physical Geochemistry*, edited by S. K. Saxena, Chap. 10, pp. 317–325, Springer-Verlag, New York, 1982.

VI. SPECTROSCOPY AT HIGH PRESSURE

RAMAN SPECTROSCOPIC STUDIES AT HIGH TEMPERATURES AND HIGH PRESSURES: APPLICATION TO DETERMINATION OF *P-T* DIAGRAM OF ZrO_2

Haruo ARASHI

Research Institute for Scientific Measurements, Tohoku University
1-1, 2-chome, Katahira, Sendai, Japan

Abstract. A diamond-anvil cell (DAC) of the lever-arm type has been modified for the measurement of Raman spectra at high temperatures and high pressures. Using the modified DAC temperatures up to 700 °C have been obtained at pressures up to 7 GPa. Raman spectra were measured at various temperatures and pressures up to 350 °C and 7 GPa. However, Raman spectra could not be measured above 410 °C due to the strong background. From the temperature dependence of the signal from the diamond anvils it was found that the origin of the strong background in Raman spectra is the fluorescence from the diamond anvils, which increases in intensity with increasing temperature. Raman spectra of ZrO_2 measured below 350 °C and 7 GPa were used to determine the *P-T* diagram in this compound.

Introduction

Raman scattering measurements provide a useful means for investigating phase transitions in crystals as they reveal the dynamical properties associated with the structural changes. Therefore, Raman scattering measurements using a diamond-anvil cell (DAC) have been undertaken by many workers at room temperature to investigate the phase transition induced at high pressures. (WEINSTEIN and PIERMARINI, 1975; ADAMS et al., 1977; MAMMONE et al., 1980)

Investigations of phase transition induced at high temperatures and high pressures are very important in the earth sciences research for the elucidation of the seismic velocity variations separating upper mantle, transition zone, and lower mantle. For this reason Raman scattering measurements at high temperatures and high pressures were carried out (using a modified DAC) and these measurements were applied to investigate the phase transitions in ZrO_2 at high temperatures and high pressures.

Recently zirconia ceramics have been investigated with much interest from a point of view of their high toughness (GUPTA, 1978). It is reported that this toughness may be explained by a stress-induced phase transition in zirconia ceramics. In order to clarify the toughening mechanism of zirconia ceramics it is important to investigate the phase transition of ZrO_2 at high pressure. ARASHI and ISHIGAME (1982a) have indicated from the pressure dependence of Raman spectra measured at room temperature that the high-pressure phase of ZrO_2 at room temperature belongs to the orthorhombic system. However, there are some discrepancies in the proposed phase relationships for ZrO_2 at high temperatures and high pressures (KULCINSKI, 1968; BOCQUILLON and SUSSE, 1969; LITYAGINA et al., 1978; LIU, 1980). These discrepancies arise from the difficulty associated with quenching the high-temperature and high-pressure phase of ZrO_2. Therefore, it is desirable to do in situ experiments to determine the phase relationships for ZrO_2 at high temperatures and high pressures.

Experimental

A diamond-anvil cell of the lever-arm type, designed and developed by MAO and BELL (1978), is suitable for investigation of pressure-induced phase transitions in crystal because it allows fine control of pressure at high pressure. We constructed a diamond-anvil cell of this type (ARASHI et al., 1984). To use the cell in high-temperature and high-pressure experiments it has been modified as shown Figure 1. Modified components are as follows: minimize the heat flow through the diamond anvils, the anvils are mounted directly on a disk made of stabilized zirconia ceramics (Yttria-stabilized zirconia YTZ) and affixed with ceramic cement. The ceramic; fabricated by Nippon Kagaku Togyo Co., Ltd.; has thermal conductivity as low as 0.008 cal/cm/sec/°C. Its fracture toughness of 15 MN/m is much higher than that of alumina ceramic. One-fifth carat diamond anvils of type I were used. Anvil surfaces were 450 μm in diameter. A small platinum-wire heater was wound around two diamond anvils to heat a sample contained in the hole of a metal gasket. Temperature was measured by an R-thermocouple of 0.1 mm in diameter in contact with the pavilion of the diamond anvil. Pressure measurements at room temperature are usually performed using the pressure-dependence of the ruby fluorescence line. This method, however, is limited to the temperature range below 400 °C because of the decrease in intensity of the fluorescence with increasing temperature (SHIMOMURA et al., 1982). Therefore, various sensor materials were studied for the pressure measurement at high temperatures and it was found that, of those studied, YAG:Eu^{3+} was the most suitable for this purpose (ARASHI and ISHIGAME, 1982b). In the present

High-Pressure Research in Mineral Physics, edited by M. H. Manghnani and Y. Syono, pp. 335–339.

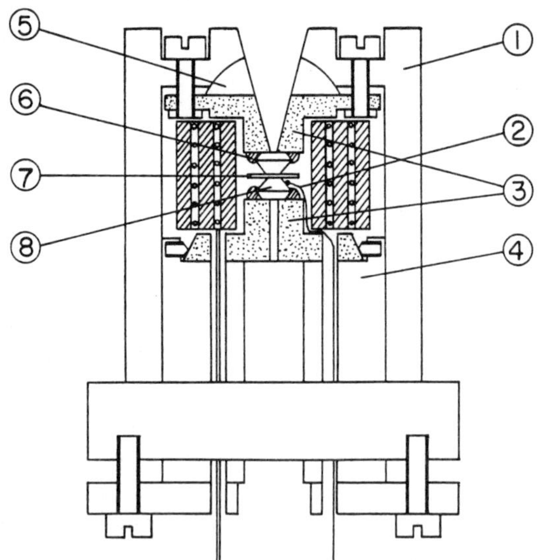

Figure 1. Cross section of the modified DAC for high-temperature use: 1) cylinder, 2) thermocouple, 3) stabilized zirconia ceramic (YTZ) on which diamond anvil is mounted, 4) piston, 5) hemisphere support, 6) ceramic cement, 7) gasket, and 8) diamond anvil.

investigation a small crystal of ruby and YAG:Eu^{3+} as put in the hole of a metal gasket made of 631 stainless steel. Distilled water was used as a pressure-transmitting medium. Using the heating technique described above, temperatures up to 700 °C could be obtained at pressures up to 7 GPa.

Raman spectra of ZrO_2 at high temperatures and high pressures were measured by using this modified diamond-anvil cell. As the phase transition of ZrO_2 is considerably affected by previous stresses to which the sample has been subjected, a small as-grown single crystal of ZrO_2 was used as the specimen in our Raman scattering measurements. Raman spectra were measured by the CT-1000D double monochrometor of 1000 mm in focal length (Nippon Bunko Co., Ltd.). The 514.5 nm line of Ar-ion laser was used as an excitation light. The scattered light from the sample was collected using a back-scattering geometry. Since the thermal radiation from sample did not obscure the Raman spectra in the temperature range below 500 °C, the synchronous single-photon counting method was not necessary. Therefore, Raman spectra were detected by using an ordinary single photon-counting method. The temperature dependence of fluorescence from diamond anvils was measured for three pairs of diamond anvils supplied by D. Drukker & Zn. N. V. Co., Ltd.

Results and Discussion

The pressure dependence of Raman spectra of ZrO_2 at room temperature obtained previously is shown in Figure 2. At atmospheric pressure ZrO_2 belongs to the monoclinic system. In this crystal structure 18 Raman active modes $(9A_g+9B_g)$ are expected from a factor group analysis. Of these 16 Raman active modes are seen in Figure 2. With increasing pressure the monoclinic ZrO_2 makes a phase transition at 3.4 GPa. From the number of Raman bands observed above 3.4 GPa, ARASHI and ISHIGAME (1982a) have indicated that high-pressure phase of ZrO_2 belongs to the orthorhombic system.

Raman spectra measured at high temperatures and high pressures are shown in Figure 3. Pressures was initially increased up to 3.4 GPa at room temperature. The Raman spectra measured under these conditions are affected by the contributions from both the monoclinic and orthorhombic phases. In Figure 3b, Raman bands assigned to the monoclinic phase are indicated by arrows. With increasing temperature the Raman bands assigned to the monoclinic phase disappear while those assigned to the

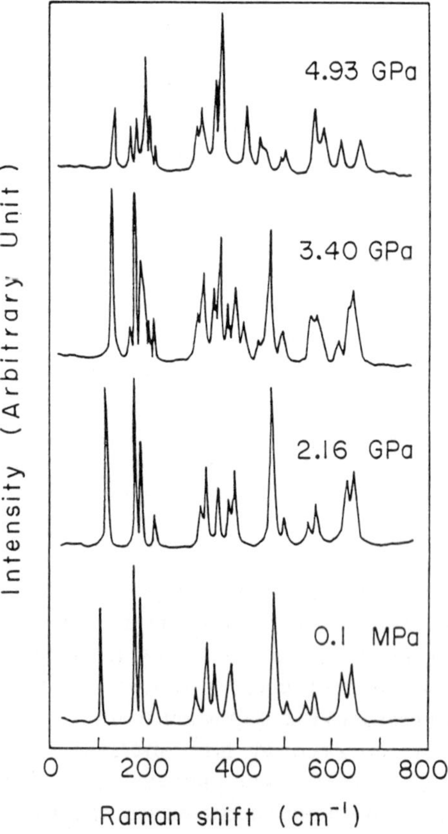

Figure 2. Pressure dependence of Raman spectra of ZrO_2 measured at room temperature.

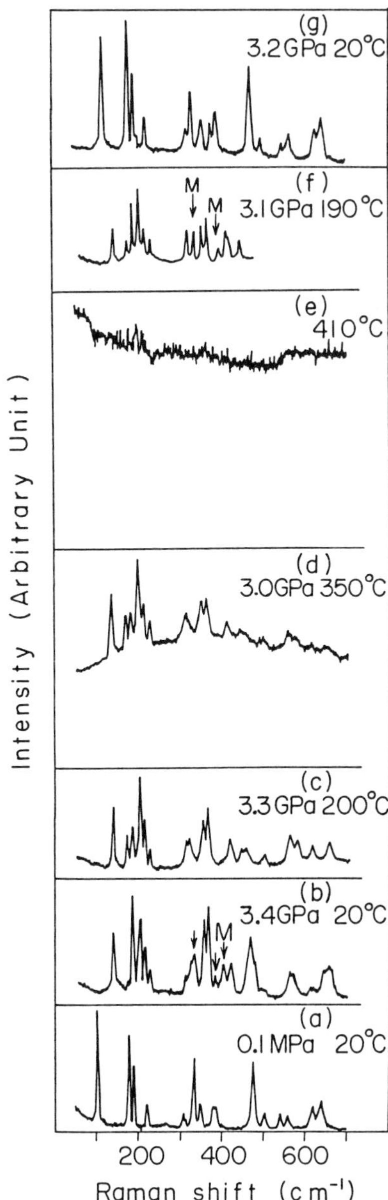

Figure 3. Raman spectra of ZrO_2 measured at high temperatures and high pressures. Zero-lines are shown for each spectra. In b) and f) Raman bands marked by arrow are assigned to monoclinic phase.

orthorhombic phase remain (Figure 3c). With continued increasing temperature, pressure decreases slightly due to the thermal expansion of the DAC. All Raman bands measured at 3.3 GPa and 200 °C are assigned to the orthorhombic phase. As Raman spectra measured at 350 °C and 3 GPa are characteristic of the orthorhombic phase, the orthorhombic phase is still stable in this *P-T* condition. At these temperatures the background of Raman spectra grows considerably (Figure 3d). As the background becomes much stronger with increasing temperature, Raman bands at 410 °C are obscured by the background (Figure 3e). When the temperature decreases, however, the background also diminishes, and Raman bands become observable again in the temperature range below 350 °C (Figure 3f). After temperature decreased to room temperature Raman spectra were measured at 3.2 GPa (Figure 3g). These spectra clearly correspond to those of the monoclinic phase.

To measure Raman spectra at temperatures above 410 °C without background interference the origin of the background must be understood. In Raman spectroscopic study at high temperatures using a DAC five contributions to the signal can be considered besides the Raman scattering of sample. These are as follows: a) thermal radiation from the heated sample and diamond anvils, b) photon scattering of the diamond anvils, c) fluorescence from the sample, d) fluorescence from the pressure-transmitting medium, and e) fluorescence from the diamond anvils.

In the present case, the intensity of light signal becomes zero as a result of interruption of the excitation laser light, which indicates that the strong background can not be attributed to thermal radiation from the heated sample and diamond anvils. It is known that the Raman band originated from one-phonon scattering of diamond has a frequency shift of 1332 cm^{-1}. Since Raman spectra are measured in the spectral region between 0 and 800 cm^{-1} the phonon scattering of diamond anvils may safely be excluded. ZrO_2 has no fluorescence at room temperature and atmospheric pressure. Even at high temperatures fluorescence from ZrO_2 is not observed.

Fluorescence from diamond anvils has been studied by WIJNGAARDEN and SILVERA (1979) and they found the fluorescence strongly sample dependent. However, they did not study the temperature dependence of diamond-anvil fluorescence. To estimate the contribution of this fluorescence on the background of Raman spectra measured at high temperatures, we investigated the temperature dependence of the fluorescence from the diamond anvil.

For this purpose the diamond-anvil cell was set up in the usual way, using stainless steel as a gasket, but without the sample and the pressure-transmitting medium. Laser light was focused on the diamond anvil culet and the collected fluorescence was introduced into the entrance slit of a monochromator using the same optics as in Raman scattering measurements. Since the scattering geometry was not changed at each temperature the intensity of the plasma line of the laser light scattered by diamond anvils was constant during the experiment. Thus, the intensity of the fluorescence from the diamond anvils could be measured by comparison with that of the plasma line.

Results of this experiment in the spectra between 200 and 400 cm^{-1} are shown in Figure 4. Sharp lines are the plasma lines of the laser. At room temperature fluores-

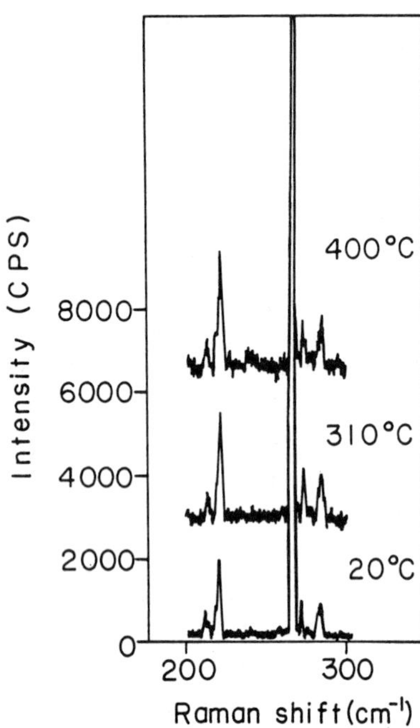

Figure 4. Temperature dependence of the fluorescence spectra of the diamond anvils.

cence is very weak when compared with the plasma line. However, the fluorescence increases with increasing temperature (Figure 4), whereas the intensity of the plasma line held constant at any measured temperature. At 400 °C intensity of the fluorescence reaches 6000 cps, which is four times stronger than that of the most intense Raman band of ZrO_2 measured at high pressure using a DAC. With decreasing temperature fluorescence also decreases and when the experiment returns to room temperature the fluorescence intensity is the same as when measured prior to heating.

Fluorescence levels at high temperature were measured for three pairs of diamond anvils. We found their fluorescence levels to be identical. Fluorescence from pressure-transmitting mediums at high temperatures was not examined in this experiment.

At 400 °C fluorescence levels measured from diamond anvils were at the same level as background levels measured in Raman spectra experiments at the same temperature using distilled water as a pressure-transmitting medium. From this we believe the fluorescence level of distilled water is low at high temperatures and does not affect measurement of Raman spectra at high temperature and high pressure.

We conclude from these experiments that diamond-anvil fluorescence is a main source of the strong background appearing in Raman spectra measurements at high temperatures and high pressures using a DAC. Thus, it is necessary to use diamond anvils with very low fluorescence levels in order to measure Raman spectra at higher temperatures using a DAC.

The optical system for excitation proposed by HEMLEY (this volume) is also a useful method to avoid fluorescence from diamond anvils. In this system laser light for excitation is introduced into the DAC at an angle of 45° to the diamond anvil face and scattered light is collected at an angle of 135°. This configuration significantly reduces spurious interference from the diamond anvils.

A tentative *P-T* diagram of ZrO_2 below 4 GPa and 350 °C can be deduced from identification of Raman spectra measured at high temperatures and high pressures using a modified DAC. As shown in Figure 3b, Raman spectra measured at 3.4 GPa and room temperature are affected by contributions from both the monoclinic and orthorhombic phases. These values of pressure and temperature correspond to a phase boundary between the monoclinic and the orthorhombic phases in *P-T* diagram. We conclude from the Raman spectra that the stable phase of ZrO_2 at 350 °C and 3 GPa is orthorhombic. With decreasing temperature, bands assigned to the monoclinic phase begin to appear in Raman spectra at 190 °C and 3.1 GPa (Figure 3f). This *P-T* condition corresponds to the lower phase boundary between monoclinic and orthorhombic phases. A tentative *P-T* diagram of ZrO_2 (Figure 5) can be given from these experimental results. We conclude that the boundary between the monoclinic and the orthorhombic phases has a negative *P-T* slope. Recently BLOCK et al. (1985) experimentally identified the negative *P-T* slope for ZrO_2.

In summary, a DAC of the lever-arm type was modified for measurement of Raman spectra at high temperatures and high pressures and in situ measurements of the Raman spectra of ZrO_2 were made to investigate phase

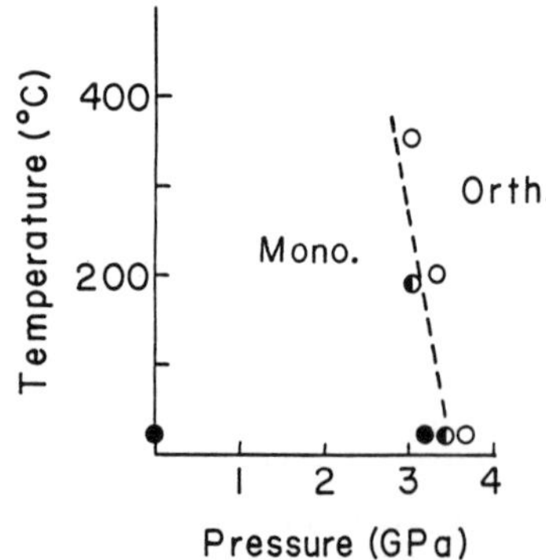

Figure 5. Tentative *P-T* diagram of ZrO_2 obtained from in situ Raman spectroscopic studies at high temperatures and high pressures. The dotted line indicates a phase boundary between monoclinic and orthorhombic phases. ●: monoclinic, ◐: monoclinic + orthorhombic, ○: orthorhombic.

relationships in this compound at high temperatures and high pressures. Raman spectra were observed up to 350 °C. However, above 410 °C Raman spectra could not be observed due to the fluorescence from the diamond anvils. To avoid this fluorescence it will be necessary to use diamond anvils with very low fluorescence levels.

We suggest that Raman spectroscopic measurement using the pico-second technique may also prove useful in removing the influence of diamond-anvil fluorescence because of the slower rise time expected for this fluorescence relative to the Raman scattering process. From Raman spectra measured at high temperatures and high pressures, the *P-T* diagram in the temperature range below 350 °C has been clarified.

Acknowledgment. This research was supported in part by a Grant-in-Aid for Scientific Research from the Japanese Ministry of Education, Science and Culture (Project No. 59540171).

References

ADAMS, D. M., S. K. SHARMA, and R. APPLEBY, Spectroscopy at very high pressures. 14 Laser Raman scattering in ultrasmall samples in a diamond anvil cell, *Appl. Opt., 16*, 2572–2575, 1977.

ARASHI, H., and M. ISHIGAME, Raman spectroscopic studies of the polymorphism in ZrO_2, *Phys. Stat. Sol. (a), 71*, 313–321, 1982a.

ARASHI, H., and M. ISHIGAME, Diamond anvil pressure cell and pressure sensor for high-temperature use, *Jpn. J. Appl. Phys., 21*, 1647–1649, 1982b.

ARASHI, H., A. KAIMAI, M. SASAKI, K. FUNATO, and T. SUGAWARA, Fabrication of lever-arm type diamond-anvil pressure cell and laser micro-optic system for pressure measurements, *Bull. Res. Inst. Sci. Meas. Tohoku Univ., 32*, 27–40, 1984.

BLOCK, S., J. A. H. DA JORNADA, and G. J. PIERMARINI, Pressure-temperature phase diagram of zirconia, *J. Am. Ceram. Soc., 68*, 497–499, 1985.

BOCQUILLON, G., and C. SUSSE, Diagramme de phase de la zircone sous pression, *Rev. Int. Hautes Temper. et Refract., 6*, 263–266, 1969.

GUPTA, T. K., Effect of stress-induced phase transformation on the properties of polycrystalline zirconia containing metastable tetragonal phase, *J. Mater. Sci., 13*, 1464–1470, 1978.

HEMLEY, R. J., Pressure dependence of raman spectra of SiO_2 polymorphs: α-quartz, coesite, and stishovite, this volume.

KULCINSKI, G. L., High pressure induced phase transition in ZrO_2, *J. Am. Ceram. Soc., 51*, 582–584, 1968.

LITYAGINA, L. G., S. S. KABALKINA, T. A. PASHIKINA, and A. I. KHOZYAINOV, Polymorphism of ZrO_2 at high pressures, *Sov. Phys. Solid State, 20*, 2009–2010, 1978.

LIU, L. G., New high pressure phase of ZrO_2 and HfO_2, *J. Phys. Chem. Solids, 41*, 331–334, 1980.

MAMMONE, J. F., S. K. SHARMA, and M. NICOL, Raman study of rutile (TiO_2) at high pressures, *Solid State Commun., 34*, 799–802, 1980.

MAO, H. K., and P. M. BELL, Design and varieties of the megabar cell, *Carnegie Inst. Washington Yearb., 77*, 904–908, 1978.

SHIMOMURA, O., S. YAMAOKA, H. NAKAZAWA, and O. FUKUNAGA, Application of a diamond anvil cell to high-temperature and high-pressure experiments, in *High-Pressure Research in Geophysics*, edited by S. Akimoto and M. H. Manghnani, pp. 49–60, Center for Academic Publication Tokyo, 1982.

WEINSTEIN, B. A., and G. J. PIERMARINI, Raman scattering and phonon dispersion in Si and GaP at very high pressure, *Phys. Rev., B12*, 1172–1186, 1975.

WIJNGAARDEN, R. J., and I. F. SILVERA, Selection of diamonds for Raman scattering in diamond anvil cell, in *High Pressure Science and Technology, Vol. 1*, edited by B. Vodar and Ph. Marteau, pp. 157–159, Pergamon Press, 1979.

RAMAN AND REFLECTION SPECTRA OF PYRITE SYSTEM UNDER HIGH PRESSURE

N. MŌRI

Institute for Solid State Physics, University of Tokyo
Roppongi, Tokyo, Japan

H. TAKAHASHI*

Department of Physics, Faculty of Science, Hokkaido University
Sapporo, Japan

Abstract. Raman and reflection spectra of synthetic MS_2 (M=Fe, Co, and Ni) crystals with the structure type pyrite have been measured under high pressure, up to about 130 kbar, at room temperature. Changes in the Raman frequency of A_g mode with pressure revealed the atomic displacement of sulfur atoms. From the reflectance measurements, changes in the electronic state depending on pressure were obtained. Results indicate that the electronic and magnetic properties of these materials can be interpreted on the basis of band schemes and also throw light on a problem of the stability of ore minerals having the pyrite structure.

Introduction

The most significant approach to understanding geophysical and geochemical aspects of ore minerals is through systematic surveys of their physical properties. Among the most precise modern techniques optical, X-ray, and Mössbauer spectroscopies under high pressure have played an essential role in studying the crystal chemistry and physical properties of minerals.

In recent years various physical properties observed in sulfide minerals of 3d transition metals with the pyrite structure have drawn the attention of many solid-state physicists. One interesting problem concerns electronic states, especially in the pyrite system based on NiS_2. This is because NiS_2 is a particularly convenient material in which to investigate the "Mott-Hubbard" metal-insulator transition caused by the electron-electron correlation effect (WILSON, 1985). NiS_2 shows an insulator-metal transition under high pressure without any change in crystal symmetry (ENDO et al., 1973), a characteristic feature of the Mott-Hubbard transition. Electrical and magnetic properties of NiS_2 under high pressure also lend support to the belief it is a Mott-Hubbard insulator, a belief which cannot be explained by ordinary band theory (WILSON, 1985; MŌRI et al., 1985).

*Now at Institute for Solid State Physics, University of Tokyo, Roppongi, Tokyo, Japan.

Concerning to the stability of the pyrite minerals, on the other hand, CuS_2 and ZnS_2 can only be synthesized under high pressure (BITHER et al., 1966). The covalent interaction between the cation and sulfur ions is thought to be the basis for the stability of these minerals. However, there are no precise data on crystal structural parameters of these minerals under high pressure other than those reported by WILL et al. (1984) measuring the compressibilities of synthetic polycrystals of FeS_2, CoS_2, NiS_2, and CuS_2. These investigators used a belt-type high pressure apparatus, measured the energy dispersive X-ray diffraction, and estimated the pressure dependence of nearest neighbor M-S bond length ($a_{M\text{-}S}$) and S-S bond length ($a_{S\text{-}S}$) of FeS_2 and CoS_2 in pressures ranging up to 40 kbar. From their data they concluded that the $a_{S\text{-}S}$ increased with increasing pressure, whereas the $a_{M\text{-}S}$ decreased, and suggested that covalent mixing increases with pressure causing the pyrites to stabilize.

We have measured the reflectivity and Raman frequency of synthetic single crystals of FeS_2, CoS_2, and NiS_2 under high pressure to investigate these pyrite minerals systematically from both electronic and crystal structural view points.

Experimental

Single crystals of FeS_2, CoS_2, and NiS_2 were grown by a chemical vapour transport method. X-ray examination confirmed a crystal structure of the pyrite type, the space group of which is T_h^6(Pa 3). We used a microscopic spectrometer and a double monochromator assembled with a conventional diamond-anvil cell to measure the reflectivity and Raman spectra under high pressure. Reflectance was measured in the wavelength range of 200 to 2400 nm and Raman scattering measurement was carried out using the 514.5 nm line of an Ar ion laser. Pressure was monitored by measuring the change in wave-length of fluorescence line R_1 of the ruby inside the

High-Pressure Research in Mineral Physics, edited by M. H. Manghnani and Y. Syono, pp. 341–345.

cell. Polymer film was used as the pressure-transmitting medium for reflectance measurement and a methanol-ethanol mixture was used for the Raman scattering measurement. Measurements were made at room temperature. Experimental procedure is described in detail elsewhere (TAKAHASHI, 1987).

Results and Discussions

Reflectance spectra of FeS_2, CoS_2, and NiS_2 under high pressures are shown in Figures 1a, b, and c, respectively.

At ambient pressure almost the same results were obtained as had been reported earlier (SATO, 1984). These spectra reflect the electronic structure of these pyrite minerals. Relatively poor reflectivity observed for FeS_2 in the lower energy range indicates it to be an insulator and the strong reflectivity of CoS_2 shows it to be metallic. Moreover, the peak A seen for FeS_2 and CoS_2 between 1 and 2 eV corresponds to the inter-band transition from t_{2g} to e_g* band. These results are easily understood from an ordinary band model. In the case of NiS_2, however, reflectivity remains weak even below 1 eV and suggests it to be an insulator. Taking into account that the resistivity of NiS_2 shows an insulating behavior (WILSON, 1985), this result anticipates that NiS_2 has another energy gap near the Fermi level in addition to the one between t_{2g} and e_g*, which cannot be explained by simple band theory.

Under high pressure peak A is found to shift to higher energy levels in FeS_2 and CoS_2 as shown in Figures 1a and b. This indicates that the energy gap between t_{2g} and e_g*, which corresponds to 10 Dq, increases with pressure. This result is consistent with what expected from a point charge model where 10 Dq depends on the distance between metal ion and sulfur ion, a_{M-S}, as $(a_{M-S})^{-5}$ (KAMIMURA et al., 1969). Above about 50 kbar, however, for NiS_2 only, reflectivity was found to increase abnormally with pressure in the energy range less than 1 eV (TAKAHASHI et al., 1985). A typical spectrum at 132 kbar is included in Figure 1c. This abnormal behavior of NiS_2 reflectivity corresponds to that observed in the electrical resistivity measurements (MORI et al., 1985).

The real part of dielectric constant, ε_1, is shown in Figures 2a, b, and c for FeS_2, CoS_2, and NiS_2, respectively. These dielectric constants were calculated from the reflectance spectra using the Kramers-Kronig relation (TAKAHASHI, 1986). The positive values of ε_1 in the energy range less than 1 eV seen for FeS_2 and NiS_2 up to 40 kbar confirm their insulating behaviors. The ε_1 of NiS_2 at 132 kbar, however, becomes negative in the same energy range where CoS_2 is observed to behave in a metallic fashion. This is the first confirmation by optical measurement that the insulator-metal transition as shown by electrical resistivity measurements (WILSON and PITT, 1971; MORI et al., 1985) takes place in NiS_2 under high

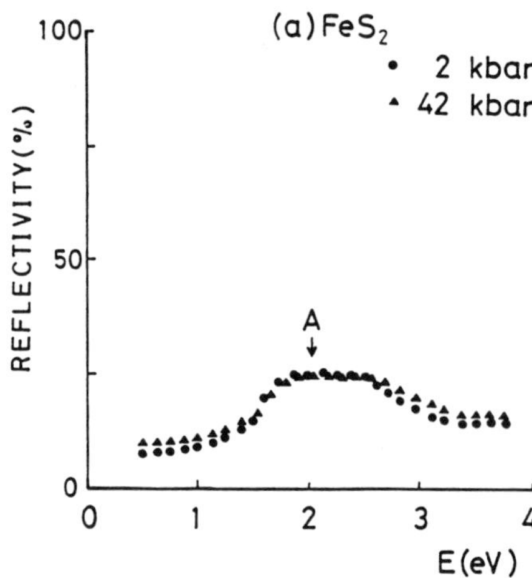

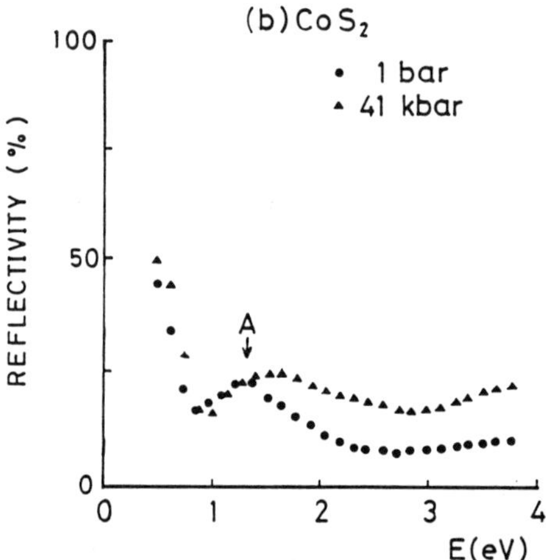

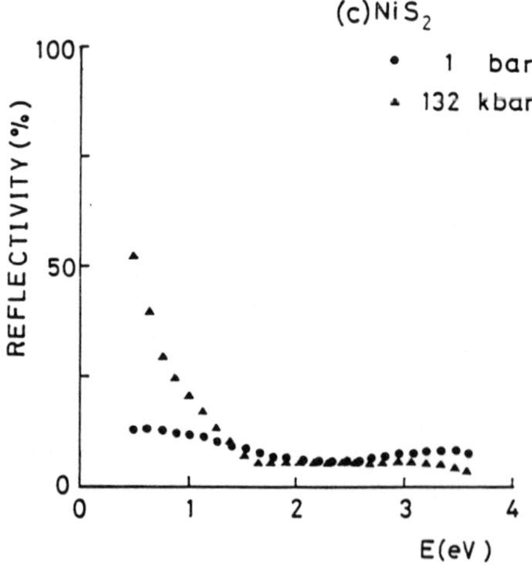

Figure 1. Reflectance spectra of a)FeS_2, b)CoS_2, and c)NiS_2 under high pressure. Arrow A shown in a) and b) indicates the reflectance peak which is assigned to interband transition from t_{2g} to e_g* band for FeS_2 and CoS_2, respectively.

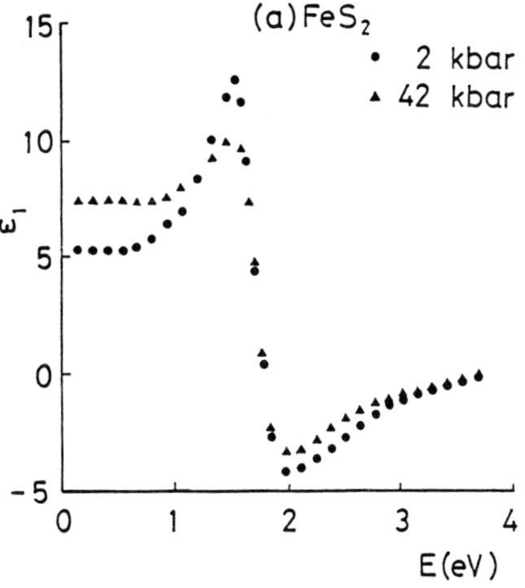

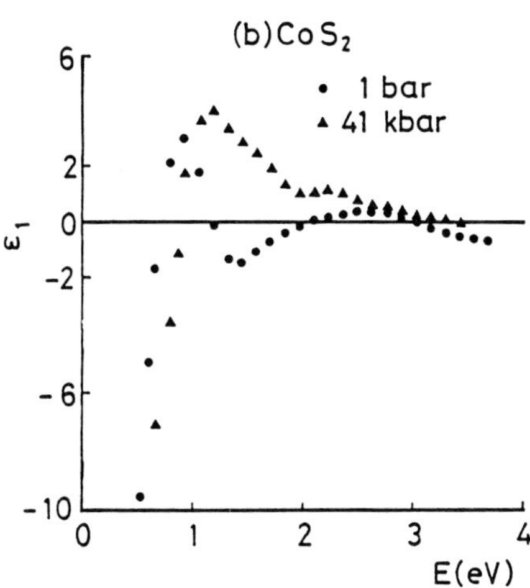

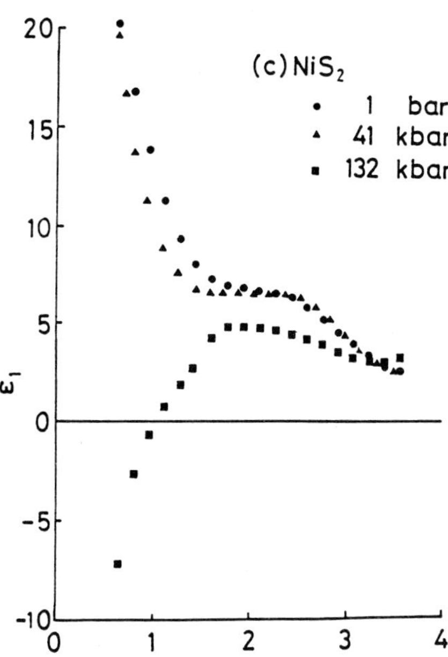

Figure 2. The real part of dielectric constant, ε_1, of a)FeS_2, b)CoS_2, and c)NiS_2 under high pressure, which were calculated from the reflectance spectra with using Kramers-Kronig relation.

pressure. In the energy range more than 1 eV, however, no abnormal change is observed for NiS_2, FeS_2, or CoS_2. We conclude from these optical results that only near the Fermi level is the electronic state of NiS_2 sensitive to pressure. This confirms that the electron-electron interaction near the Fermi surface results in the insulating property of NiS_2, which definitely supports the belief that it is a Mott-Hubbard insulator.

Pressure dependences of the Raman mode A_g of these minerals are shown in Figures 3 and 4. A_g represents the stretching mode between a pair of sulfur atoms (LAUWERS and HERMAN, 1974). We found Raman frequency increases consistently with increasing pressure in FeS_2 but it changes little in the case of CoS_2. For NiS_2, however, a sharp change of about 8 cm^{-1} is observed near 46 kbar where the insulator-metal transition takes place. In the pressure range above and below 46 kbar the A_g mode scarcely depends on pressure, similarly to CoS_2. Assuming that the Raman frequency of A_g mode depends on the distance between paired sulfur atoms, $a_{S\text{-}S}$, as $a_{S\text{-}S}^{-n}$ (ONARI and ARAI, 1979), it is suggested that $a_{S\text{-}S}$ in FeS_2 decreases consistently under pressure, as in normal materials, but remains almost constant in CoS_2 over the pressure range examined here. Moreover, it is noted that $a_{S\text{-}S}$ in the metallic state of NiS_2 appears to become larger than that in insulating state in spite of a decrease in volume of about 0.4% at the tansition (ENDO et al., 1973).

Taking account of the results obtained by the reflectivity measurements described earlier, the pressure dependence of $a_{S\text{-}S}$ in these pyrite minerals as suggested by the Raman shift is interpreted as follows. For these pyrite minerals the electrons in the e_g* band play a main role in their physical properties. FeS_2 has no electron in the e_g* band, generating no effect, and FeS_2 behaves as a normal insulator under pressure, as was shown. In CoS_2 and NiS_2 one and two electrons occupy the e_g* band, respectively. It is important to recognize that the e_g* band is composed of $d\gamma$ state of 3d metal and of antibonding p* state of sulfur molecules which have a higher energy than the $d\gamma$ state. It is easy to assume that the repulsive force due to the electron occupying the antibonding p* state keeps sulfur atoms away from each other. That is, the more the p* state overlaps with the $d\gamma$ state, the larger the distance $a_{S\text{-}S}$ becomes. In fact at ambient pressure the $a_{S\text{-}S}$ of CoS_2, NiS_2, and CuS_2 decreases in order and the overlap between $d\gamma$ and p* state of these minerals is shown also to decrease successively (BULLETT, 1982). From this consideration it is expected that the pressure dependence of $a_{S\text{-}S}$ of CoS_2 and NiS_2 differs from that of FeS_2 since the e_g* electron tends to have more p* character under pressure as a result of the overlapping. Furthermore, the abrupt decrease of the Raman frequency at the insulator-metal transition pressure in NiS_2 is also consistent with this idea: when the insulator to metal transiton takes

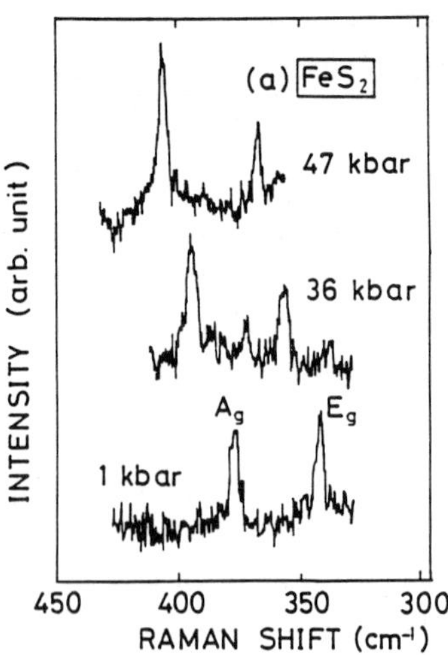

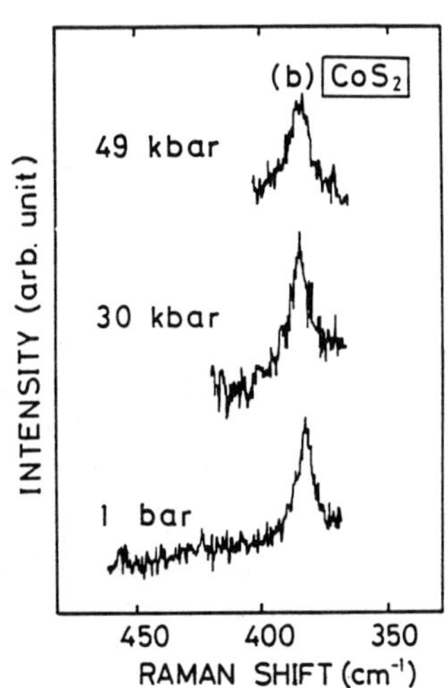

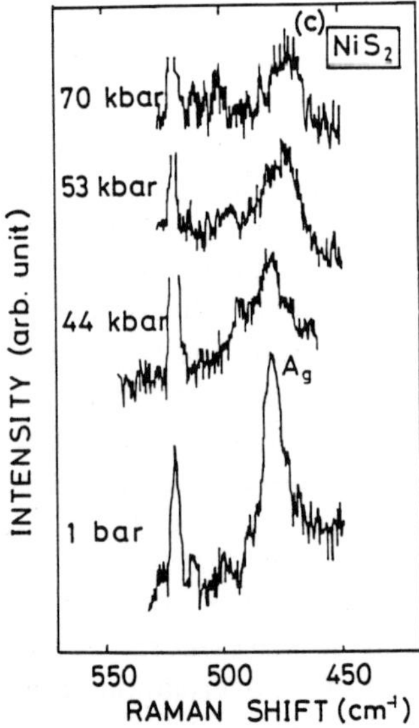

Figure 3. Raman spectra of a)FeS_2, b)CoS_2, and c)NiS_2 under high pressure. E_g mode are also seen in a). Peaks at about 510 cm^{-1} seen in c) come from another laser line.

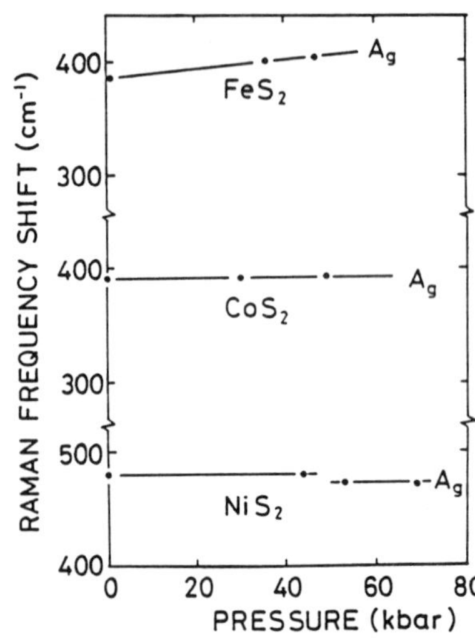

Figure 4. Pressure dependences of the Raman mode A_g of FeS_2, CoS_2, and NiS_2.

place, the electrons localized at the Ni atom become delocalized and migrate toward the S atom such that they have some p^* character in the metallic state, and as a result $a_{S\text{-}S}$ increases abruptly. On the other hand, from the band calculation (BULLETT, 1982), it is shown that the binding energy of these pyrite minerals decreases with increasing atomic number of metal ions. In fact, CuS_2 and ZnS_2 can only be synthesized under high pressure (BITHER et al., 1966) because of the large energy separation between $d\gamma$ and p^* state. From the results discussed above, it is obvious that the metal-sulfur bond is more compressible than the sulfur-sulfur bond in CoS_2 and NiS_2 so that the binding energy can be thought to increase rapidly with increasing pressure when some electrons occupy the e_g^* band. It is then natural to expect that under high pressure $d\gamma$ and p^* states in CuS_2 and ZnS_2 begin to overlap because they have almost the same electronic structure as CoS_2 and NiS_2.

Conclusions

Pressure dependences of the reflectance spectra of FeS_2, CoS_2, and NiS_2 with pyrite type structure can be interpreted on the basis of a band model except for the behavior observed for NiS_2 in the low-energy range. The reflectance peak corresponding to the interband transition between t_{2g} and e_g^* shifts to the higher energy side as pressure increases. This observation indicates that the binding energy increases under high pressure for these minerals and provides an answer to the question why CuS_2 and ZnS_2 are stabilized under high pressure. For NiS_2, the reflectivity is found to change drastically at about 46 kbar only in the low energy side. This is the first confirmation of the insulator-metal transition by optical measurements. Moreover, it shows that the electronic structure changes only near the Fermi level, which supports the belief that the insulator-metal transition is caused by an electron correlation effect.

From the pressure dependence of A_g mode in the Raman spectra of these minerals it is suggested that the S-S distance a_{S-S} decreases with pressure in FeS_2 but changes little in CoS_2 and in NiS_2 except at the insulator-metal transition pressure where the a_{S-S} increases abruptly. As this observation is not entirely consistent with the X-ray results reported earlier (WILL et al., 1984), the more precise single-crystal X-ray data are needed to clarify the discrepancy.

Acknowledgments. We are grateful to S. Minomura for providing facilities for optical measurements during this work. Thanks are also due to S. Ogawa for stimulating discussions and suggestions and for supplying us CoS_2 single crystals. We express our gratitude to K. Sato for providing us FeS_2 single crystals and for helpful information on the Kramers-Kronig analysis.

REFERENCES

BITHER, T. A., C. T. PREWITT, J. L. GILLSON, P. E. BIERSTEDT, R. B. FLIPPEN, and H. S. YOUNG, New transition metal dichalcogenides formed at high pressure, *Solid State Commun., 4*, 533–535, 1966.

BULLETT, D. W., Electric structure of 3d pyrite- and marcasite-type sulfides, *J. Phys. C. Solid State Phys., 15*, 6163–6174, 1982.

ENDO, S., T. MITSUI, and T. MIYADAI, X-ray study of metal-insulator transition in NiS_2, *Phys. Lett., 46A*, 29–30, 1973.

KAMIMURA, H., S. SUGANO, and Y. TANABE, *Ligand Field Theory and Its Applications* (in Japanese), 31 pp., Tokyo, 1969.

LAUWERS, H. A., and M. A. HERMAN, Study on te force field of pyrite, *J. Phys. Chem. Solids, 35*, 1619–1623, 1974.

MŌRI, N., M. KAMADA, H. TAKAHASHI, G. OOMI, and J. SUSAKI, Recent high pressure studies on NiS_2, in *Solid State Physics under Pressure*, edited by S. Minomura, pp. 247–252, D. Reidel Publishing Co., London, 1985.

ONARI, S., and T. ARAI, Infrared lattice vibrations and dielectric dispersion in antiferromagnetic semiconductor $MnSe_2$, *J. Phys. Soc. Jpn., 46*, 184–188, 1979.

SATO, K., Reflectivity spectra and optical constants of pyrites (FeS_2, CoS_2, and NiS_2) between 0.2 and 4.4 eV, *J. Phys. Soc. Jpn., 53*, 1617–1620, 1984.

TAKAHASHI, H., N. MŌRI, S. YOMO, Z. JIN, K. TSUJI, and S. MINOMURA, Reflectance of $M(S_{1-x}Se_x)_2$ (M = Ni, Co) under high pressure, in *Solid State Physics under Pressure,* edited by S. Minomura, pp. 75–80, D. Reidel Publishing Co., London, 1985.

TAKAHASHI, H., Optical study of NiS_2 with pyrite structure under pressure, Ph.D. thesis, Hokkaido University, Sapporo, 1986

WILL, G., J. LAUTERJUNG, H. SCHMITZ, and E. HINZE, The bulk moduli of 3d-transition element pyrites measured with synchrotron radiation in a new belt type apparatus, in *High Pressure in Science and Technology, Vol. 22*, edited by C. Homan et al., pp. 49–52, Mat. Res. Soc. Symp. Proc., 1984.

WILSON, J. A., and G. D. PITT, Metal-insulator transition in NiS_2, *Phil. Mag., 23*, 1297–1310, 1971.

WILSON, J. A., The Mott transition for binary compounds, including a case study of the pyrite system $Ni(S_{1-x}Se_x)_2$, in *The Metallic and Non-Metallic States of Matter*, edited by P. P. Edwards and C. N. R. Rao, pp. 215–260, Taylor and Francis Ltd., London, 1985.

PRESSURE DEPENDENCE OF RAMAN SPECTRA OF SiO_2 POLYMORPHS: α-QUARTZ, COESITE, AND STISHOVITE

R. J. HEMLEY

Geophysical Laboratory, Carnegie Institution of Washington
Washington, D.C., 20008-3898, USA

Abstract. The pressure dependence of the room-temperature Raman spectra of α-quartz, coesite, and stishovite is investigated to probe compression mechanisms, possible soft-mode transitions, and high-pressure thermodynamic properties of these phases. The experiments are performed with a diamond-anvil cell that is optically interfaced with an Ar^+ laser and a triple spectrograph equipped with an optical multichannel analyzer. The pressure dependences are found to be particularly large for the low-frequency modes in quartz and coesite, including the 206 cm^{-1} A_1 phonon in quartz which is associated with the α-β transition. A phase transition is found in coesite at 22–25 GPa, as evidenced by a splitting of the main Raman bands. Above ~30 GPa the Raman bands in quartz and coesite decrease and broaden significantly with increasing pressure, and both spectra closely resemble that of silica glass under these conditions. For stishovite, soft-mode behavior of the B_{1g} optical phonon associated with a pressure-induced shear instability was investigated. The pressure shift of the B_{1g} mode is observed to be negative, whereas those of the other Raman-active bands are positive, in agreement with both crystal-chemical considerations and recent theoretical predictions. The pressure shift and mode-Grüneisen parameter γ_i for the B_{1g} mode in stishovite is considerably smaller than that found for other rutile-type oxides. Implications for the thermodynamic properties of the silica phases at high pressure are examined with the mode-Grüneisen parameters determined from the Raman spectra.

Introduction

The measurement of vibrational spectra at high pressure is useful in several ways for characterizing materials of geological and geophysical interest. The pressure dependence of lattice vibrations reveals detailed information on bonding and crystal chemical properties. Structural changes with compression, including phase transitions driven by soft vibrational modes, can be determined from spectral measurements at high pressure. Such measurements thus complement direct methods of structure determination such as X-ray and neutron diffraction. In addition, vibrational spectroscopy can be used to probe the structure of materials that are amorphous to diffraction techniques. Vibrational Raman spectroscopy is particularly useful for high pressure studies because samples can be probed at extreme pressure and with spatial resolution approaching the diffraction limit of the excitation light that is used (e.g., ~1 μm). In this regard, Raman spectroscopy can serve as a useful technique for phase identification in phase equilibrium studies at high pressures and temperatures. Finally, analysis of vibrational spectra may provide constraints on thermodynamic properties of solids at pressures that are beyond range of direct thermochemical measurements.

Because of the widespread occurrence of silica in mineral assemblages, the behavior of the SiO_2 polymorphs over a broad range of pressures and temperatures has long been of interest to geophysics and geochemistry (see SOSMAN, 1965). The study of the vibrational spectrum of quartz, including its temperature dependence, has a rich history dating back to the early days of vibrational spectroscopy (see SCOTT and PORTO, 1967). KIEFFER (1979) has summarized ambient-condition studies of the vibrational properties of quartz and the other major rock-forming silica phases, including the low-pressure tetrahedrally bonded structures and the early work on stishovite (the high-pressure polymorph with the rutile-type structure and octahedrally coordinated silicon) (STISHOV and POPOVA, 1961; CHAO et al., 1962). Recently, several studies have more fully characterized the vibrational properties of stishovite under ambient conditions (WEIDNER et al., 1982; HEMLEY et al., 1986a).

Comparatively little work has been done to determine the vibrational properties of the silica phases at high pressure. For quartz, the pressure dependences of the Raman spectrum (ASELL and NICOL, 1968; DEAN et al., 1982) and the infrared absorption spectrum (FERRARO et al., 1972) have been studied, but the higher pressure measurements were performed under nonhydrostatic conditions and without in situ pressure determination. Characterization of SiO_2 at mantle pressures and the possibility of post-stishovite phases have also evoked much interest (e.g., MING and MANGHNANI, 1982; LIU, 1982). Because of the possibility that such dense phases of SiO_2 may not be quenchable, an in situ high-pressure probe is required for identification and characterization. A compressional instability in the rutile structure has been predicted in stishovite at high pressure on the basis of studies of analogue compounds (NAGEL and O'KEEFFE, 1971) and theoretical calculations (HEMLEY et al., 1985; STRIEFLER, 1985). This transition has not been observed experimentally. It is ideally suited for study by vibrational Raman spectroscopy because predictions indicate that it is associated with a soft Raman-active phonon.

High-Pressure Research in Mineral Physics, edited by M. H. Manghnani and Y. Syono, pp. 347–359.

In the present study, the pressure dependence of phonon spectra of α-quartz, coesite, and stishovite are investigated by spontaneous Raman spectroscopy of samples pressurized to 30–40 GPa at room temperature in a diamond-anvil, high-pressure cell. For coesite and stishovite, spectra are measured at high pressure for the first time, and the soft mode in stishovite is documented. The experiments on quartz are performed under quasi-hydrostatic conditions at pressures that are approximately an order of magnitude higher than in previous studies. The very high-pressure measurements on quartz and coesite, performed at pressures well outside their respective stability fields, provide an opportunity to examine the intrinsic stability of these structures with respect to compression. The results are compared with recent measurements of the Raman spectrum of SiO_2 glass (vitreous silica) at high pressure (HEMLEY et al., 1986b).

Experimental Methods

The spectra were obtained with a micro-optical spectrometer system designed for optical measurement of samples contained in a diamond-anvil cell at high pressure. The design is shown schematically in Figure 1. The system is built around a Leitz-Ortholux-I microscope equipped with a variable diaphragm for imaging excitation volumes of dimensions as small as 1 μ through the optical system. The microscope also utilizes a pentaprism for simultaneous viewing of the sample, diaphragm, and back-lighted spectrometer slits. This feature is extremely useful for the critical alignment of the micro-optical system.

Excitation is provided by the 514.5, 488.0, and 457.9 nm lines of an Ar^+ laser (Spectra-Physics 165). The laser beam is focused with a microscope objective (Leitz UM or UT series) mounted within either of two beam-steering attachments on the microscope. Typically, the laser radiation is focused to a beam waist of ~3 μm. Figure 1 shows the attachment that provides a ~45° angle of excitation with respect to the diamond face, thus giving a ~135° scattering configuration. A second attachment (not shown) provides vertical illumination of the sample (i.e., 180° scattering). The 135° design is important for optical studies with diamond-anvil cells because of Raman scattering and possible fluorescence from the diamond anvils along the laser path. With this configuration the spurious interference from the anvils is significantly reduced.

The scattered light is imaged through the microscope and focused on the slits of the spectrometer with a Micro-Nikkor camera lens. The light is dispersed with a triple spectrograph (Spex 1877) and detected with an optical scanning multichannel analyzer (OSMA, Princeton Instruments). The spectrograph consists of a 0.22-m double-stage system and 0.6-m single-stage system. In the present experiments 600 gr/mm and 1200 gr/mm gratings in the filter stage were used. In the spectrograph stage 600 gr/mm and 1800 gr/mm gratings were used; these gratings dispersed the signal across the 700 active channels of the OMSA to give Raman windows of 16.1 and 48.3 nm, respectively. Typically, the spectra were measured with the entrance slits of the spectrograph set at 50–100 μm, giving a resolution of 5–10 cm^{-1}.

The experiments are performed with a diamond-anvil, high-pressure cell of the Mao-Bell design (MAO and BELL, 1978). Both type I and low-fluorescence type IIb diamonds with culets of ~0.6 mm were used in these studies. For the initial studies on quartz and coesite, which have relatively strong spectra, measurements could be performed using the 45° scattering configuration with fluorescent type I diamonds. Type IIb diamonds were necessary for the measurements on stishovite and silica glass, because they have significantly weaker spectra. The high-pressure chamber consisted of a 200 μm hole drilled in the gasket (full-hardened T301 stainless steel). The samples were loaded along with a transmitting medium and a few small chips of ruby for pressure determination (MAO et al., 1978).

A variety of pressure-transmitting media were used. In decreasing order of hydrostaticity, these media were argon, ethanol-methanol mixture, water, and silica itself (no medium). Samples of 10–50 μm in length and 10–40 μm in thickness were used. Care was taken to prevent bridging of the sample between the diamond anvils. In lower pressure experiments (i.e., <20 GPa), thicker samples were used to enhance the signal-to-noise ratio of the spectra. The most precise data were obtained with the argon medium, which has been shown to afford quasi-hydrostatic conditions to above 50 GPa (MAO et al., 1986). Argon is particularly useful as a medium in Raman scattering experiments because it lacks a first-order Raman spectrum and therefore does not contribute to the background signal. In contrast, the ethanol-methanol glass has Raman bands in the region of important silica bands (MAMMONE et al., 1980b). Argon was loaded in the high-pressure cell by pre-pressurizing the gas to 0.2 GPa at room temperature in a high-pressure apparatus similar in design to that used by MILLS et al. (1980).

Results

Quartz

Vibrational spectra of crystalline solids can be interpreted by use of factor group analysis (FATELEY et al., 1972). There are two enantiomorphs of α-quartz with space groups $P3_221$ (D_3^5) and $P3_121$ (D_3^6) and $Z=3$ (WYCKOFF, 1963, p. 312; see also, BARRON et al., 1976). The irreducible representation of the optical vibrations for the D_3 factor group may be written

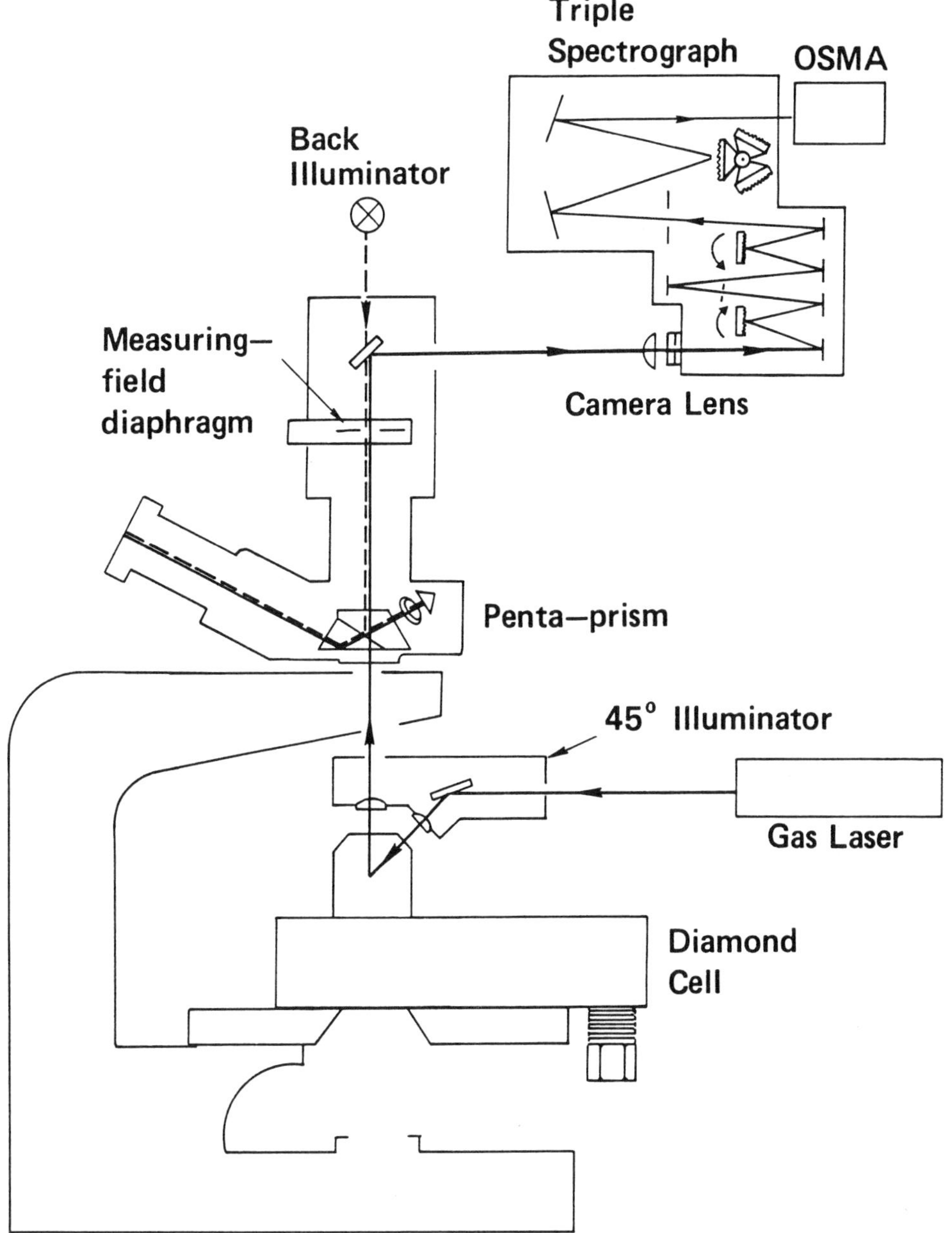

Figure 1. Schematic diagram of micro-optical spectrometer system.

$$\Gamma_{op} = 4\ A_1 + 4\ A_2 + 8\ E$$

where the A_1 modes are Raman-active, A_2 modes are infrared active, and the E modes are both Raman- and infrared-active (FATELEY et al., 1972). The A_2 and degenerate E modes are split by long-range Coulomb forces into longitudinal and transverse optic (LO and TO) components. Numerous studies of the Raman spectrum of quartz have been performed at 0.1 MPa (1 bar). The first complete single-crystal study with laser excitation and polarization measurements was that of SCOTT and PORTO (1967), who also reviewed the early literature. These investigators studied the magnitude of the LO-TO splittings, which were immeasurable for all but one mode (3–4 cm^{-1} for the high-frequency 1230 cm^{-1} vibration), and the dependence of phonon frequencies on direction of propagation.

The pressure dependence of the Raman spectrum was studied up to 4.0 GPa by ASELL and NICOL (1968) using NaCl as a pressure-transmitting medium. In the present experiments the pressure range has been extended and the dependence of the Raman spectrum on various pressure-transmitting media has been examined. Below 10 GPa, no difference in the Raman spectrum of quartz was found with use of either ethanol-methanol or water. With ethanol-methanol, interference from Raman bands due to the medium was somewhat more pronounced. However, most of this spurious scattering could be alleviated with spatial filtering. When no medium was used, a splitting of the LO and TO components of the *E*-symmetry phonons in quartz was significantly enhanced. This effect is illustrated for the 128 cm^{-1} *E*-symmetry phonon at ~12 GPa in Figure 2. The magnitude of the splitting is consistent with the results of uniaxial compression experiments carried out up to 1.0 GPa by TEKIPPE et al. (1973).

Representative spectra of quartz as a function of pressure in an argon medium are shown in Figure 3. Pressure shifts for 13 bands in quartz are obtained; the frequencies are plotted as a function of pressure in Figure 4, and the numerical values for the pressure shifts $(d\nu_i/dP)_T$ at 0.1 MPa are listed in Table 1. Large pressure shifts are observed for the modes that also have large temperature shifts at 0.1 MPa (SHAPIRO et al., 1967; HEMLEY and SHARMA, unpublished). The pressure shift is particularly large at low pressure for the 206 cm^{-1} A_1 phonon; this phonon also displays a large frequency shift as the α-β transition is approached at 0.1 MPa (SHAPIRO et al., 1967). Furthermore, the width of this band decreases markedly with increasing pressure. An analogous broadening of this band observed with increasing temperature at 0.1 MPa suggests that this behavior arises from volume-dependent effects. The pressure dependences of several sets of *E*-symmetry modes are weak. This observation is similar to the observation of ASELL and NICOL (1968). ASELL and NICOL (1968) were unable to detect any pressure dependence up to 4.0 GPa for the 394, 401, 1069, 1162, and 1230 cm^{-1} *E*-symmetry vibrations. Several of the modes have a negative pressure shift, indicating a weakening of the structure along these normal coordinates. The 1162 cm^{-1} *E* mode has the largest effect; it has a negative shift on initial compression (-3.1 ± 0.3 cm^{-1}/GPa) and becomes positive at higher pressure (Figure 4).

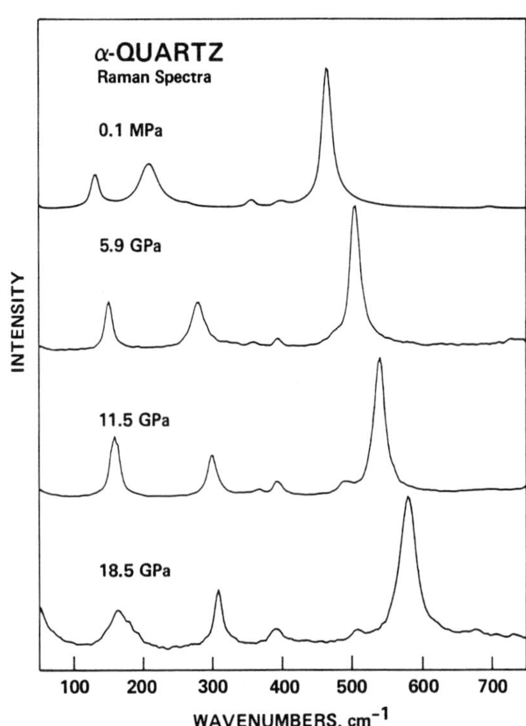

Figure 3. Raman spectrum of α-quartz as a function of pressure (T=298 K). The sample consisted of a randomly oriented single crystal in an argon medium.

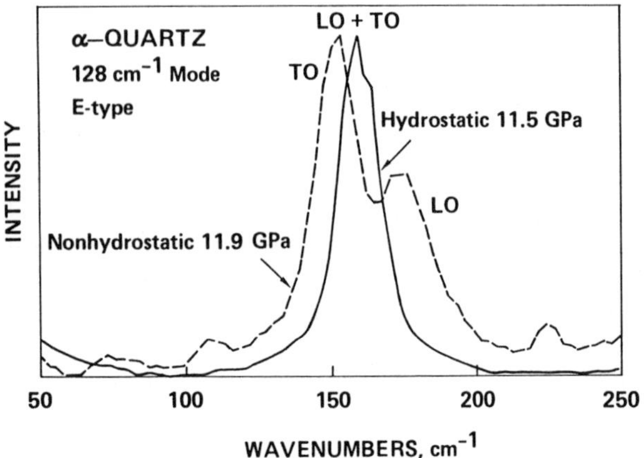

Figure 2. The 128 cm^{-1} *E* band of α-quartz at high pressure with and without a pressure-transmitting medium.

The mode-Grüneisen parameters γ_i, calculated from the pressure shifts and bulk modulus, are listed in Table 2. The numerical values for the γ_i are close to those determined by DEAN et al. (1982) for the three strongest bands (Table 2). In the latter study a simple linear fit to the pressure shifts, which were measured up to 1.0 GPa, was performed. In general, larger γ_i are found for lower frequency modes. Also, the magnitudes of the negative γ_i are small. These observations are germane to the discussion of the thermodynamics implications of the present data (see below). It should be pointed out that the particularly large γ_i obtained for the 206 cm^{-1} A_1 mode (3.58 ± 0.08) decreases by a factor of ~4 at 10 GPa. The γ_i for the intense A_1 band at 464 cm^{-1} is found to increase on compression in this pressure range.

Between 30 and 40 GPa the Raman bands broaden and decrease in intensity with increasing pressure. The precise

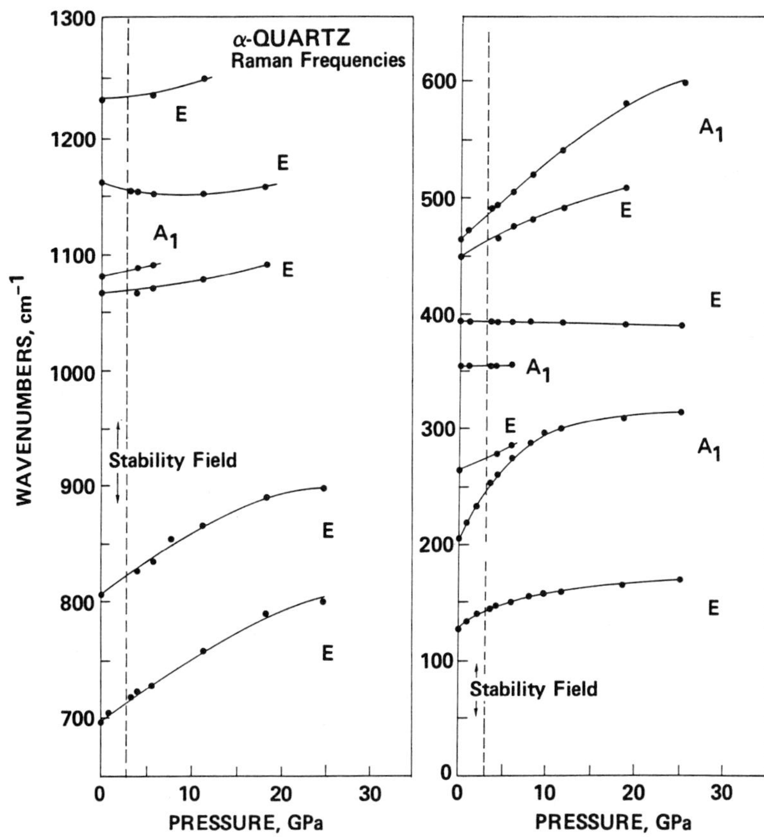

Figure 4. Pressure dependence of the Raman bands of α-quartz (T=298 K), including symmetry assignments (SCOTT and PORTO, 1967). The stability field of α-quartz is indicated (BOHLEN and BOETTCHER, 1982; AKAOGI and NAVROTSKY, 1984).

TABLE 1. Pressure Shifts of the Raman Modes of α-Quartz (0.1 MPa)

Sym.	ν_i, cm^{-1}	$(d\nu_i/dP)_T$, cm^{-1}/GPa			
		This Work (298 K)	ASELL and NICOL (1968) (300 K)	DEAN et al. (1982) (300 K)	BRIGGS and RAMDAS (1977) (4 K)
A_1	206	19.9 ±0.7	18.0	20.4	21.3
	355	−1.2 ±0.4			−0.3
	464	8.0 ±0.2	9.0	8.1	8.0
	1085	1.4 ±0.2			1.5
E (LO+TO)	128	5.5 ±0.2	5.0	5.4	5.8
(LO+TO)	265	3.5 ±0.1			4.8
(LO+TO)	696	7.0 ±0.6	8.0		9.6
(LO+TO)	1162	−3.1 ±0.3			−2.6
E (TO)	394	−0.09±0.03			0.9
(LO)	401				
(TO)	450	4.6 ±0.1			8.2
(LO)	511				
(TO)	796	5.9 ±0.7	8.0		10.9
(LO)	808				
(TO)	1069	1.0 ±0.2			2.0
(LO)	1230				

TABLE 2. Mode-Grüneisen Parameters γ_i of α-Quartz (0.1 MPa)

Sym.	ν_i, cm^{-1}	$\gamma_i=-(V/\nu_i)(d\nu_i/dV)$[a] This Work (298 K)	DEAN et al. (1982) (300 K)
A_1	206	3.58±0.08	3.63
	355	−0.13±0.04	
	464	0.64±0.02	0.64
	1085	0.05±0.01	
E (LO+TO)	128	1.65±0.08	1.55
(LO+TO)	265	0.49±0.01	
(LO+TO)	696	0.37±0.03	
(LO+TO)	1162	−0.10±0.01	
E (TO)	394	−0.01±0.003	
(LO)	401		
(TO)	450	0.38±0.01	
(LO)	511		
(TO)	796	0.27±0.03	
(LO)	808		
(TO)	1069	0.04±0.003	
(LO)	1230		

[a]Calculated from the pressure shifts and the zero-pressure bulk modulus, $\gamma_i=(K_0/\nu_i)(d\nu_i/dP)_T$. For α-quartz K_0=37.1 GPa from ultrasonic measurements is used (MCSKIMIN et al., 1965).

pressure at which these effects occur depends on the nature of the pressure-transmitting medium; in softer media the changes occur at higher pressure. Spectra of samples recovered from high pressure have a strong, diffuse band at ~500 cm^{-1} with a second peak at ~620 cm^{-1}, as discussed below.

Coesite

Coesite is a monoclinic mineral, with space group $B2/b$ (C_{2h}^2) and Z=8 (primitive cell) (WYCKOFF, 1963, p. 319). According to factor group analysis, the irreducible representation of the optical vibrations can be expressed as

$$\Gamma_{op} = 16\ A_g + 17\ B_g + 18\ A_u + 18\ B_u$$

where the A_g and B_g species are Raman-active and the A_u and B_u species are infrared-active (FATELEY et al., 1972).

The room-temperature Raman spectrum of synthetic coesite was first measured at 0.1 MPa by SHARMA et al. (1981). The spectrum has recently been studied by BOYER et al. (1985); that study obtained spectra of both synthetic and natural crystals. The frequencies of the more intense peaks observed by SHARMA et al. (1981) were confirmed in the later study, with the exception of the 220 cm^{-1} and 314 cm^{-1} bands which were not attributed to coesite (or to quartz) by BOYER et al. (1985). In addition, the 466 cm^{-1} band assigned by SHARMA et al. (1981) to coesite was reassigned to quartz. It should be pointed out that, in contrast to the situation for quartz and stishovite, the complete set of Raman-active fundamental vibrations has not yet been determined (at most 20 bands have been observed, in comparison to a total of 33 predicted A_g and B_g vibrations). In addition, specific normal mode assignments of the observed bands on the basis of force field calculations have not yet been made for this polymorph.

For the present study, coesite was synthesized by pressurizing cristobalite at 3.0 GPa and 1100 °C for 3 h in a piston cylinder apparatus (see BOHLEN and BOETTCHER, 1982). Water was added to the starting material to enhance crystal growth, and crystals up to ~50 μm in length were obtained. The intensities of the 0.1 MPa Raman bands measured for these crystals showed a strong dependence on orientation with respect to the incident laser beam (i.e., phonon propagation direction). The peak positions, however, were found to be within ~1 cm^{-1} of those reported by SHARMA et al. (1981), who obtained a powder spectrum. A weak band at 661 cm^{-1}, not reported in previous studies, was also observed. The 0.1 MPa spectrum is compared with those measured at a series of pressures up to 30 GPa in the argon medium in Figure 5. Because of the anisotropy of coesite, spectra of different crystals in random orientations within the diamond-anvil cell were measured to obtain a more complete determination of the pressure shifts. Pressure shifts for 18 bands were determined, as shown in Figure 6. The pressure shifts and mode-Grüneisen parameters are listed in Table 3.

The shifts are continuous to 22 GPa (i.e., through the stability field of the phase), which indicates that the structure is preserved over this pressure range. A large shift is obtained for the 521 cm^{-1} band. This band has been assigned to the ν_s(Si–O–Si) stretching mode which correlates with the 464 cm^{-1} vibration in quartz (SHARMA et al., 1981). The two bands have a similar pressure dependence at high pressure. The pressure dependence of the 466 cm^{-1} band in the coesite spectrum is appreciably smaller than that of the 464 cm^{-1} band in quartz (Tables 1 and 3). Therefore it is concluded that the 466 cm^{-1} band indeed belongs to coesite (in agreement with SHARMA et al. (1981)) and does not result from quartz contamination in the present synthetic sample (see BOYER et al., 1985). Two pairs of bands appear to cross at high pressure (bands with initial frequencies of 116 and 151 cm^{-1}, and 176 and 204 cm^{-1}). This behavior would occur if each member of the pair has a different symmetry (A_g and B_g). It is also possible, however, that an avoided crossing occurs over a very small pressure interval; this would be the case if the modes have identical symmetry (HERZBERG, 1945).

The spectrum of coesite changes discontinuously between 22 and 25 GPa, as shown in Figure 5. The strong band splits into a doublet, along with similar splittings in some of the weaker bands. This change was observed only

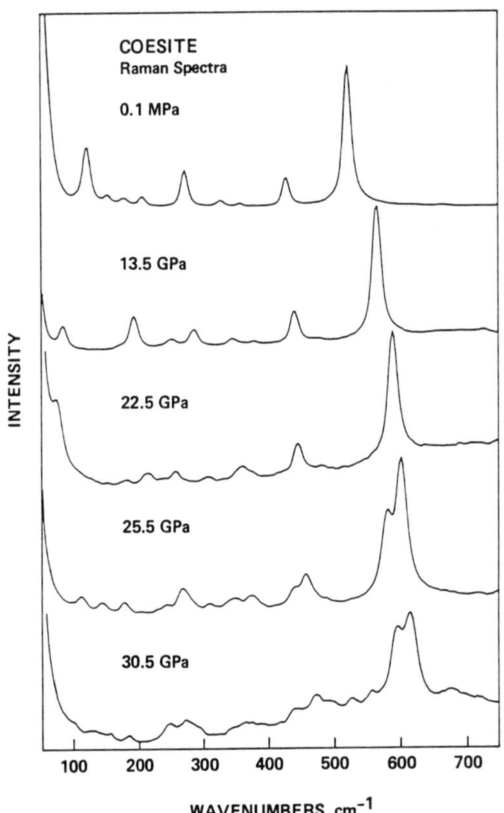

Figure 5. Raman spectrum of coesite as a function of pressure ($T=298$ K), low-frequency portion. The spectra at high pressure were measured on a randomly oriented single-crystal. The 0.1 MPa spectrum is that of a different crystal with a different orientation. Note the splitting of several bands at 22–25 GPa, and the increase in intensity of the diffuse scattering underlying the discrete bands in the highest pressure spectra.

in the case of quasihydrostatic compression with the argon medium, during two separate high-pressure runs. When coesite is compressed in a stronger medium, the observed bandwidths are larger and the splittings tend to be obscured. Note also that the low-frequency mode at 77 cm^{-1} initially has a positive pressure shift, but begins to soften at ~10 GPa (Figure 6).

Stishovite

The irreducible representation of the optical vibrations of stishovite, space group $P4_1/nmm$ (D^7_{4h}) with $Z=2$ (SINCLAIR and RINGWOOD, 1978; HILL et al., 1983), may be written

$$\Gamma_{op} = A_{1g} + A_{2g} + B_{1g} + B_{2g} + 2\ B_{1u} + E_g + 3\ E_u$$

where the A_{1g}, B_{1g}, B_{2g}, and E_g species are Raman-active, and B_{1u} and E_u are infrared-active. Raman spectra of natural and synthetic stishovite were recently measured under room conditions and the bands assigned (HEMLEY et al., 1986a). In the present study the Raman spectrum of synthetic stishovite (grain size of ~20 μm) was measured with the 457.9-nm, Ar^+ laser (see HEMLEY et al., 1986a).

The Raman spectrum of stishovite at 32.8 GPa is compared with that measured at 0.1 MPa in Figure 7. Pressure shifts for three of the four Raman-active fundamentals which were assigned previously are observed at high pressure. Because of the poorer signal-to-noise ratio for the measurement through the diamond cell, the very weak B_{2g} band was not observed. At the highest pressures the B_{1g} mode developed a slight asymmetry. The shift of the A_{1g} and E_g modes is positive, whereas that of the B_{1g} mode is negative. This pressure dependence of the frequencies is shown in Figure 8; the pressure shifts and γ_i (0.1 MPa) are listed in Table 4. The shifts of the Raman bands are close to linear, in contrast to those of quartz and coesite over the same pressure interval. This effect arises from the significantly higher bulk modulus of stishovite (WEIDNER et al., 1982). No irreversible effects on the spectrum with increasing pressure to 32.8 GPa were observed, in marked contrast to the behavior of the lower pressure phases pressurized outside their respective stability fields.

Discussion

Elastic Compression of Quartz and Coesite

Changes in structure that occur upon compression can be inferred from the variation in vibrational spectra with applied pressure. The present study is noteworthy because the pressure range extends well beyond that of current single-crystal diffraction studies of the silica polymorphs. In the case of quartz and coesite, the pressure range is well beyond the respective stability fields of these phases (see Figures 4 and 6). A quantitative analysis of the effects of pressure on individual Raman bands would require the construction of a definitive force field from which atomic displacements could be calculated for each mode. Although unambiguous (a priori) force fields have not been determined for these phases, there is some evidence of the types of atomic motions associated with observed bands in quartz from comparative studies of vibrational spectra (e.g., SHARMA et al., 1981) and lattice dynamics calculations (e.g., ETCHEPARE et al., 1974; STRIEFLER and BARSCH, 1975; BARRON et al., 1976). For example, the high-frequency modes are likely to have large contributions from Si–O stretching motions, whereas the lowest freqeuncy modes (>300 cm^{-1}) involve complex translations and rotations of SiO_4 tetrahedra. Qualitatively, one may expect similar sorts of contributions to the low- and high-frequency vibrations in coesite because of its similar Si coordination. For example, the strongest bands in the spectra of quartz and coesite (frequencies of 464 cm^{-1} and

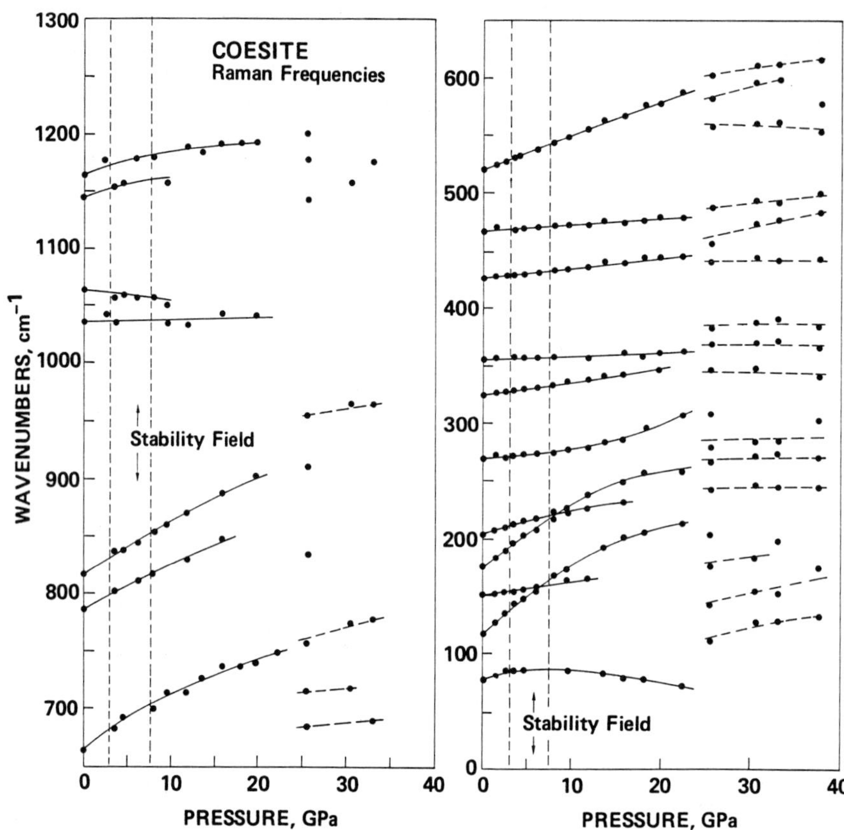

Figure 6. Pressure dependence of the Raman bands of coesite obtained from a series of measurements on different crystals at high pressure (T=298 K). The room-temperature stability field of coesite is indicated (BOHLEN and BOETTCHER, 1982; AKAOGI and NAVROTSKY, 1984). The change in the spectrum at 22–25 GPa is illustrated. Peak positions and pressure shifts of only a few of the bands in the high-pressure form are indicated.

521 cm^{-1}, respectively) have been identified as symmetric Si–O–Si stretching modes, ν_s(Si–O–Si) (SHARMA et al., 1981).

The intense ν_s(Si–O–Si) band has a strong pressure dependence, particularly in quartz (Figures 3 and 4). According to SHARMA et al. (1981), the frequency of the mode depends on the smallest ring size (and therefore smallest θ(Si–O–Si) angle) in the structures for different polymorphs of silica as well as for different aluminosilicate crystals. This study found that this correlation extends to the high-pressure behavior of quartz and coesite. The effect of compression to 6 GPa on the structure of quartz has been studied by single-crystal X-ray diffraction (JORGENSEN, 1978 and LEVIEN et al., 1980). The dominant compression mechanism consists of the reduction of the Si–O–Si angle. This angle has been shown to decrease from 144° under ambient conditions to 134° at 5.9 GPa (LEVIEN et al., 1980). The frequency of the ν_s(Si–O–Si) band at this pressure is 510 cm^{-1}. There is a linear correlation between the frequency of the mode and the angle up to 6 GPa. Assuming this correlation to remain linear to higher pressure, the shift in the band above 20 GPa would give a value for θ(Si–O–Si) in the 120–125° range.

Single-crystal diffraction studies of coesite up to 5.2 GPa indicate that the compression mechanism in this structure is more complex (LEVIEN and PREWITT, 1981). Instead of a single θ(Si–O–Si) angle, as in quartz, the coesite structure has five unique angles (142.7, 144.5, 149.6, 137.4, and 180.0°). Upon compression to 5.2 GPa, three of the intertetrahedral angles decrease (to 136, 141, and 133°). The other two (149 and 180°) are pressure independent, the latter by symmetry. Most of the compression occurs by a decrease in the intertetrahedral distances and not by compression of the Si–O bond lengths. The change in Si–O bond lengths is small (~0.4% reduction in $\langle r$(Si–O)$\rangle$, with a maximum of 0.8%). Moreover, the coesite SiO_4 tetrahedra shows little deformation, even less than quartz, over this pressure range.

The behavior of both quartz and coesite at low pressures (<20 GPa) can be explained in terms of the character of the θ(Si–O–Si) bending potential. This potential surface

TABLE 3. Pressure Shifts and Mode-Grüneisen Parameters (0.1 MPa) for the Observed Raman Bands of Coesite

ν_i (cm^{-1})	$(d\nu_i/dP)_T$, cm^{-1}/GPa	$\gamma_i=-(V/\nu_i)(d\nu_i/dV)_T$[a]
77 s	2.2 ±0.2	2.7 ±0.3
116 s	7.4 ±0.4	6.1 ±0.4
151 m	0.8 ±0.1	0.5 ±0.1
176 s	5.6 ±0.2	3.0 ±0.1
204 m	2.3 ±0.3	1.1 ±0.1
269 s	1.1 ±0.2	0.40±0.08
326 m	1.0 ±0.1	0.30±0.05
355 m	0.44±0.03	0.12±0.01
427 m	0.45±0.04	0.10±0.01
466 m	0.66±0.06	0.14±0.01
521 vs	2.9 ±0.1	0.53±0.02
661 w	6.2 ±0.4	0.90±0.06
795 w	2.2 ±0.2	0.27±0.02
815 w	5.2 ±0.1	0.61±0.03
1036 w	0.2 ±0.1	0.02±0.01
1065 w	−1.4 ±0.1	−0.12±0.01
1144 w	1.6 ±0.3	0.14±0.03
1164 w	1.5 ±0.2	0.13±0.02

[a]Calculated from the pressure shifts and the zero-pressure bulk modulus, $K_0=96\pm3$ GPa measured by static compression (LEVIEN and PREWITT, 1981). This value is 15% lower than that obtained by Brillouin scattering measurements (WEIDNER and CARLETON, 1977); see discussion in LEVIEN and PREWITT (1981).

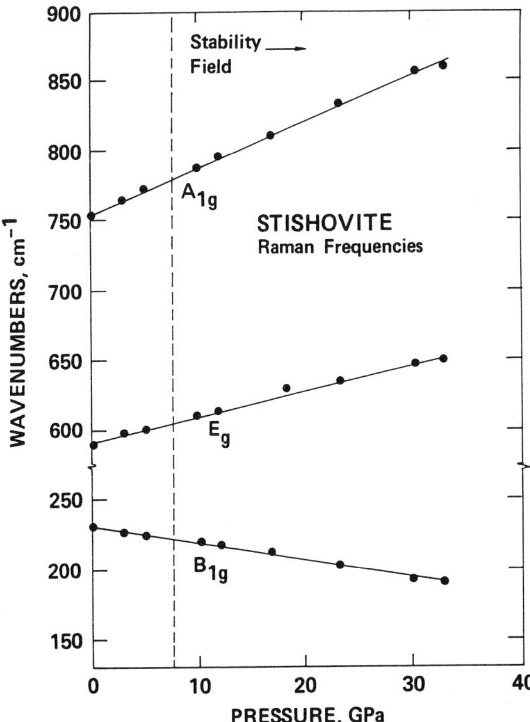

Figure 8. Pressure dependence of the B_{1g}, E_g, and A_{1g} Raman bands of stishovite (T=298 K). The estimated pressure of the coesite-stishovite transition at room temperature is indicated (AKAOGI and NAVROTSKY, 1984).

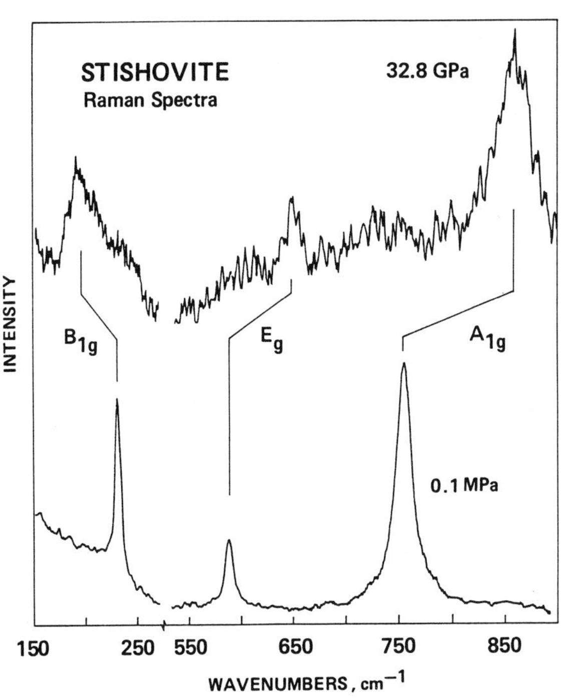

Figure 7. Raman spectrum of stishovite at 0.1 MPa and 32.8 GPa (T=298 K), along with symmetry assignments (HEMLEY et al., 1986a).

TABLE 4. Pressure Shifts and Mode-Grüneisen Parameters (0.1 MPa) for the Raman Bands of Stishovite

Sym.	ν_i, cm^{-1}	$(d\nu_i/dP)_T$, cm^{-1}/GPa	$\gamma_i=-(V/d\nu_i/dV)_T$[a]
A_{1g}	753	3.39±0.08	1.38±0.04
B_{1g}	231	−1.20±0.5	−1.58±0.06
E_g	589	1.93±0.7	1.00±0.03

[a]Calculated from the pressure shifts and the zero-pressure bulk modulus, $K_0=306\pm4$GPa determined from Brillouin scattering measurements (WEIDNER et al., 1982).

has received considerable attention in terms of cluster molecular orbital calculations (GIBBS et al., 1981). According to these calculations, the bending potential in isolated molecules is extremely flat (i.e., soft between 130–150°). This result is also in accord with results from an electron-gas shell model for crystalline quartz (JACKSON, 1986). Molecular orbital calculations reveal that an additional feature, the negative coupling between θ(Si–O–Si) and r(Si–O), is important for understanding the vibrational dynamics of silica structures. In the condensed state, there is an additional increase in the repulsive forces due to the surrounding medium which tends to compress the bond. This effect is generally neglected in the molecular calculations, although there have been some attempts to include this contribution in an approximate way (ROSS and MEAGHER, 1984). It is significant, therefore, that the frequency of the ν_s(Si–O) stretching band at 1162 cm^{-1} initially decreases with pressure. A similar trend is also apparent for the 1065 cm^{-1} band in coesite as well as the broad 1060 cm^{-1} feature in the Raman spectrum of silica glass (HEMLEY et al., 1986b). This behavior suggests an initial expansion of the Si–O bond. In quartz, the repulsive forces appear to counteract the effect by ~10 GPa, where the shift becomes positive. The frequency of a high-frequency A_2 mode in quartz, identified as an ν_{as}(Si–O) stretch, has also been found to decrease with pressure up to 5.0 GPa, according to a recent high-pressure, infrared absorption study (WONG et al., 1986).

Higher Pressure Effects in Quartz and Coesite

A phase transition in coesite is evidenced by the distinct change in the spectrum of coesite at 22–25 GPa (Figure 5). The observed splittings of the bands suggests that the transition is accompanied by a distortion of the coesite structure in the high-pressure phase. A comparison with the spectrum of stishovite shows that there is no evidence for a change to octahedral Si coordination. The apparent increase in the number of Raman-active vibrations further suggests that the new phase has an enlarged unit cell. It is possible that the observed behavior is related to the instability of the 180° intertetrahedral angle at high pressure, as proposed by LEVIEN and PREWITT (1981). The transition may also be associated with the observed softening of the 77 cm^{-1} phonon. This low-frequency vibration is likely to be a complex lattice mode. In the absence of a complete normal coordinate analysis, the possible connection between the mode softening and the structural change cannot be established. In this regard, a structural study of the phase at high pressure, as well as a complete vibrational analysis of coesite including the effects of isotropic substitution, would be useful.

Above ~30 GPa in the argon medium, the spectra of quartz and coesite weaken and broaden, developing into a single diffuse band centered at ~600 cm^{-1} similar to that found in silica glass under these pressures (HEMLEY et al., 1986b). Moreover, spectra of the pressure-quenched quartz, coesite, and silica glass exhibit similar Raman features, displaying bands at ~500 and ~620 cm^{-1}, as shown in Figure 9. These bands correlate with the so-called D_1 and D_2 defect bands identified in the 0.1 MPa spectrum of silica glass (HEMLEY et al., 1986b). Compression enhances the intensity of the 600 cm^{-1} band relative to the sharp peak at 500 cm^{-1}. Similar behavior occurs upon neutron irradiation of both silica glass and quartz (BATES et al., 1974). The Raman profiles of the material produced

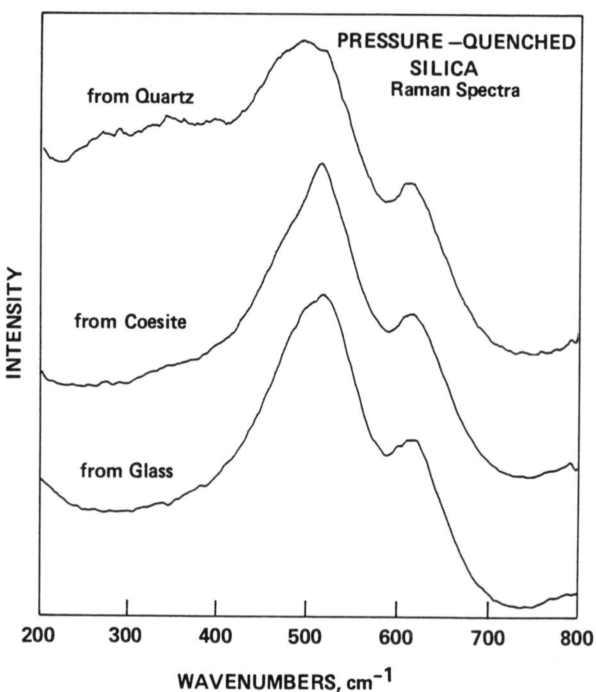

Figure 9. A portion of the Raman spectrum (0.1 MPa) of silica quenched from high pressure (constant temperature, 298 K).

from coesite and from glass are similar throughout the spectrum. In contrast, the spectrum of the material produced from quartz shows more intensity at low frequencies (<400 cm^{-1}).

The deterioration of the spectra occur at a lower pressure when stronger media are used. This observation suggests that the effect is promoted by the presence of nonhydrostatic stresses. In the argon medium the pressure gradients across the sample were estimated to be less than 0.2 GPa at 20–30 GPa. Either the ordered structure is very sensitive to small shear stresses at these pressures, or the irreversible effects will in fact occur under essentially hydrostatic conditions. For glassy phases it has been argued that the latter is the case because, at the atomic level, shear-type stress will be necessarily present upon compression of the glass structure (see PRIMAK, 1975, p. 51). In the crystalline phases the change may also be triggered be an elastic instability; this is a strong possibility for quartz, which has a negative pressure derivative for one of its elastic moduli (C_{66}) (MCSKIMIN et al., 1965). (In fact, a crude linear extrapolation of the pressure shift of C_{66} gives a shear instability at ~15 GPa, roughly in the range of the observed changes in the spectrum.) Whether or not this elastic mode softening is precipitated by softening of an optical vibration (MILLER and AXE, 1967) is not clear at present (e.g., a weak negative pressure shift of the 355 cm^{-1} A_1 mode is observed). The transformations observed here may be similar to those associated with the production of glassy phases from quartz in shock-compression studies (e.g., ASHWORTH and SCHNEIDER, 1985). If so, the transformations observed upon shock-loading are a consequence of the high pressures, and not necessarily the high temperatures, that are achieved in these experiments.

Soft-Mode Behavior in Stishovite

High-pressure transformations in stishovite, including pressure-induced elastic instabilities in the rutile structure, are of particular interest. Powder X-ray diffraction studies at high pressure indicate a strong anisotropy in the compression of this crystal, with the *a* axis being ~50% more compressible than *c* (LIU et al., 1974; OLINGER, 1976; SATO, 1977). According to single-crystal studies of stishovite analogues, this anisotropy is true for virtually all rutile-type phases that have been examined to date (HAZEN and FINGER, 1981). The anisotropic compression of stishovite is reflected in the behavior of the Raman modes with pressure. Modes with motions along the *c* axis have a smaller (but still positive) pressure shift relative to those modes with displacements perpendicular to *c*. This is the case of the A_{1g} and E_g modes, respectively (HEMLEY et al., 1986a, Figure 3). In contrast to the positive shifts observed for these modes, the B_{1g} mode has a negative pressure dependence. The documentation of soft-mode behavior is particularly significant because the drop in frequency of the B_{1g} mode ultimately leads to a shear-type, tetragonal-orthorhombic distortion of rutile to the $CaCl_2$ structure, as pointed out by NAGEL and O'KEEFFE (1971).

Mode-Grüneisen parameters obtained for isostructural oxides are compared to those obtained for stishovite in Table 5. The smaller value of γ_i for the B_{1g} mode in stishovite indicates that SiO_2 in the rutile structure appears to remain stable with respect to the B_{1g} displacements to much higher compression relative to the other oxides. As discussed by STRIEFLER and BARSCH (1976), the B_{1g} displacements contribute to the shear modulus $C_s=(C_{11}-C_{12})/2$. A negative pressure derivative of C_s has been measured for GeO_2 (−0.69) (WANG and SIMMONS, 1973), SnO_2 (−0.74) (CHANG and GRAHAM, 1975), and TiO_2 (−1.32) (MANGHNANI, 1969). MILLER and AXE (1967) have shown that an elastic instability must occur before an optical mode softens completely. Based on the coupling of the B_{1g} mode and C_s and on the negative pressure shift in the optical mode found in the present study, it is expected that dC_s/dP is also negative for stishovite (see also WEIDNER et al., 1982). On the other hand, this trend suggests that the magnitude of dC_s/dP in SiO_2 is less than that for the other oxides.

The negative pressure shift of the B_{1g} mode is in agreement with the predictions of recent lattice-dynamics calculations based on electron-gas potentials (HEMLEY et al., 1985; HEMLEY, unpublished) and semiempirical methods (STRIEFLER, 1985). The stability of the structure at high pressure is supported by the electron-gas calculations, which predict the instability to occur at several hundred GPa (HEMLEY et al., 1985; JACKSON, 1986). In contrast, the semiempirical model of STRIEFLER (1985) found the instability to occur at ~30 GPa. The present data indicate that the latter is not the case; The frequency of the B_{1g} mode dropped by only 40 cm^{-1} (a change of 17%) at 32.8 GPa, and the characteristic rutile-type Raman profile was preserved. The reason that the semiempirical theory appears to underestimate the stability of stishovite is suggested by the fact that the parameters of the model were determined, in part, by fitting to elastic and vibrational data of the analogue compounds GeO_2 and SnO_2; these compounds are appreciably less stable in the rutile structure as a function of pressure.

TABLE 5. Raman Frequencies (cm^{-1}) and Mode-Grüneisen Parameters (0.1 MPa) of Stishovite and Other Rutile-Type Oxides

Sym.	SiO_2[a]		GeO_2[b]		SnO_2[c]		TiO_2[d]	
	ν_i	γ_i	ν_i	γ_i	ν_i	γ_i	ν_i	γ_i
A_{1g}	753	1.38±0.04	700	1.4	613	3.6 1.8[b]	612	1.6 1.5[e]
B_{1g}	231	−1.58±0.06	170	−2.8	121	−10.4	143	−5.0 −5.4[e]
B_{2g}	967	—	873	1.3	776	2.6 1.6[b]	826	—
E_g	589	1.00±0.03	—	—	476	3.2 1.6[b]	447	2.4 2.2[e]

[a]This work.
[b]MAMMONE et al. (1980a).
[c]PEERCY and MOROSIN (1973).
[d]SAMARA and PEERCY (1973).
[e]Calculated from NICOL and FONG (1971).

Thermodynamic Implications

The average value of the mode-Grüneisen parameters $\langle\gamma_i\rangle$ obtained from vibrational spectroscopy provides an approximate means for calculating the thermal Grüneisen parameter γ_{th},

$$\langle\gamma_i\rangle \approx \gamma_{th} = \alpha K_T V / C_V$$

In turn, γ_{th} can be used to calculate thermodynamic properties at high pressure. In the above expression, α is the coefficient of thermal expansion, K_T is the isothermal bulk modulus, V is the volume, and C_V is the specific heat at constant volume. Formally, the average of the mode-Grüneisen parameters $\langle\gamma_i\rangle$ would equal γ_{th} if all of the vibrations of the crystal (i.e., throughout the Brillouin zone) and all anharmonic corrections were included. The averaging procedure for obtaining γ_{th} is most appropriate for structures with a large unit cell, and therefore, a large number of spectroscopically accessible modes (such as coesite), in contrast to those with relatively few optical modes (such as stishovite).

In the absence of complete measurements of elastic moduli and infrared-active modes over the pressure range examined in the present data, an estimate of γ_{th} from vibrational data must be considered preliminary. The complete set of Raman-active modes in quartz was measured as a function of pressure, with the exception of the weaker E(LO) vibrations (Tables 1 and 2). An average $\langle\gamma_i\rangle=0.36$ was obtained for quartz; this value is low relative to experimental determinations of the thermal Grüneisen parameter ($\gamma_{th}=0.57$ (BOEHLER, 1982); $\gamma_{th}=0.68\pm0.07$ (WATANABE, 1982)). A higher value is obtained from the observed modes in coesite ($\langle\gamma_i\rangle=0.94$), which is likely to represent an upper limit of γ_{th} because of the incompleteness of the measured spectra, particularly the absence of high-frequency modes which generally have smaller γ_i. This analysis for coesite is consistent with the low value ($\gamma_{th}=0.352\pm0.001$) estimated by WATANABE (1982). In the calculation of the $\langle\gamma_i\rangle$ for coesite, there is additional uncertainty in K_0 (see Table 3). For stishovite, $\langle\gamma_i\rangle$ is determined to be 0.45, based on Raman data alone. This should represent a lower bound on γ_{th} because of the negative γ_i contribution from the B_{1g} mode. Hofmeister (private communication) calculated $\langle\gamma_i\rangle=1.07\pm0.02$ from high-pressure infrared spectra, estimates of the elastic contribution from the zero-pressure density and bulk modulus, and the present Raman data. This result is still lower than $\gamma_{th}=1.35\pm0.03$ obtained by WATANABE (1982).

Conclusions

Raman spectroscopy is shown to be a useful method of studying the effects of compression on the structure and vibrational dynamics of the crystalline silica phases α-quartz, coesite, and stishovite. Insight is gained into the effect of pressure on the structure of these phases at pressures that are beyond the range of current single-crystal diffraction methods. In quartz and coesite, which have tetrahedrally coordinated silicon, the Raman data indicate that there is a general decrease in the θ(Si–O–Si) angles to ~30 GPa (consistent with observations on silica glass). A possible phase transition in coesite is identified at 22–25 GPa, well into the stability field of stishovite. At higher pressures (>30 GPa) the spectra of quartz and coesite change to resemble that of silica glass at high pressure. This result suggests that a breakdown of the ordered structure in the crystals occurs under pressure. In stishovite, the B_{1g} phonon is found to soften with pressure, in agreement with theoretical and crystal chemical predictions; however, the rutile-type structure is found to be stable to at least 32.8 GPa. The frequency shift of the mode in this pressure range indicates that the compressional instability driven by the B_{1g} phonon occurs at much higher pressure, perhaps in the megabar range. Analysis of the present results for the silica phases suggests that useful thermodynamic constraints on crystals at high pressure may be possible once complete Raman, infrared, and elasticity data are obtained at high pressure.

Acknowledgments. The generous support of H-k. Mao and P. M. Bell during the course of this work is gratefully acknowledged. The author would also like to thank B. O. Mysen for help with the coesite synthesis, Professor S. Akimoto for the gift of the stishovite sample, and A. P. Jephcoat for assistance with measurements and useful discussions. This work has also benefitted from discussions with H. S. Yoder, Jr., R. M. Hazen, R. Jeanloz, R. E. Cohen, and A. Hofmeister.

REFERENCES

AKAOGI, M., and A. NAVROTSKY, The quartz-coesite-stishovite transformations: new calorimetric measurements and calculations of phase diagram, *Phys. Earth Planet. Inter.*, *36*, 124–134, 1984.

ASELL, J. F., and M. NICOL, Raman spectrum of α-quartz at high pressure, *J. Chem. Phys.*, *49*, 5395–5399, 1968.

ASHWORTH, J. R., and H. SCHNEIDER, Deformation and transformation in experimentally shock-loaded quartz, *Phys. Chem. Minerals*, *11*, 241–249, 1985.

BARRON, T. H. K., C. C. HUANG, and A. PASTERNAK, Interatomic forces and lattice dynamics of α-quartz, *J. Phys. C: Solid State Phys.*, *9*, 3925–3940, 1976.

BATES, J. B., R. W. HENDRICKS, and L. B. SHAFFER, Raman spectrum of neutron irradiated SiO_2 glass, *J. Chem. Phys.*, *61*, 4163–4173, 1974.

BOEHLER, R., Adiabats of quartz, coesite, olivine, and magnesium oxide to 50 kbar and 1000 K, and the adiabatic gradient in the earth's mantle, *J. Geophys. Res.*, *87*, 5501–5506, 1982.

BOHLEN, S. R., and A. L. BOETTCHER, The quartz-coesite transformation: a precise determination and the effects of other components, *J. Geophys. Res.*, *87*, 7073–7078, 1982.

BOYER, H., D. C. SMITH, C. CHOPIN, and B. LASNIER, Raman microprobe (RMP) determinations of natural and synthetic coesite, *Phys. Chem. Minerals*, *12*, 45–48, 1985.

BRIGGS, R. J., and A. K. RAMDAS, Piezospectroscopy of the Raman spectrum of α-quartz, *Phys. Rev. B*, *16*, 3815–3826, 1977.

CHANG, E., and E. K. GRAHAM, The elastic constants of cassiterite SnO_2 and their pressure and temperature dependence, *J. Geophys. Res.*, *80*, 2595–2599, 1975.

CHAO, E. C. T., J. J. FAHEY, J. LITTLER, and D. J. MILTON, Stishovite, a very high pressure new mineral from Meteor Crater, Arizona, *J. Geophys. Res.*, *67*, 419–421, 1962.

DEAN, K. J., W. F. SHERMAN, and G. R. WILKINSON, Temperature and pressure dependence of the Raman active modes of vibration of α-quartz, *Spectrochim. Acta*, *38A*, 1105–1108, 1982.

ETCHEPARE, J., M. MERIAN, and L. SMETANKINE, Vibrational normal modes of SiO_2. I. α and β quartz, *J. Chem. Phys.*, *60*, 1873–1876, 1974.

FATELEY, W. G., F. R. DOLLISH, N. T. MCDEVITT, and F. F. BENTLEY, *Infrared and Raman Selection Rules for Molecular and Lattice Vibrations: The Correlation Method*, 222 pp., Wiley-Interscience, New York, 1972.

FERRARO, J. R., M. H. MANGHNANI, and A. QUATTROCHI, Infrared spectra of several glasses at high pressures, *Phys. Chem. Glasses*, *13*, 116–121, 1972.

GIBBS, G. V., E. P. MEAGHER, M. D. NEWTON, and D. K. SWANSON, A comparison of experimental and theoretical bond length and angle variations for minerals, inorganic solids, and molecules, in *Structure and Bonding in Crystals*, edited by M. O'Keeffe and A. Navrotsky, pp. 195–225, Academic Press, New York, 1981.

HAZEN, R. M., and L. W. FINGER, Bulk moduli and high-pressure crystal structures of rutile-type compounds, *J. Phys. Chem. Solids, 42*, 143–151, 1981.

HEMLEY, R. J., M. D. JACKSON, and R. G. GORDON, Lattice dynamics and equations of state of high-pressure mineral phases studied with electron-gas theory, *EOS Trans. AGU, 66*, 357, 1985.

HEMLEY, R. J., H-k. MAO, and E. C. T. CHAO, Raman spectrum of natural and synthetic stishovite, *Phys. Chem. Minerals, 13*, 285–290, 1986a.

HEMLEY, R. J., H-k. MAO, P. M. BELL, and B. O. MYSEN, Raman spectroscopy of SiO_2 glass at high pressure, *Phys. Rev. Lett., 57*, 747–750, 1986b.

HERZBERG, G., *Molecular Spectra and Molecular Structure II. Infrared and Raman Spectra of Polyatomic Molecules*, 632 pp., Van Nostrand Reinhold, New York, 1945.

HILL, R. J., M. D. NEWTON, and G. V. GIBBS, A crystal chemical study of stishovite, *J. Solid State Chem., 47*, 185–200, 1983.

JACKSON, M. D., Theoretical investigations of chemical bonding in minerals, Ph. D. dissertation, 258 pp., Harvard University, 1986.

JORGENSEN, J. D., Compression mechanisms in α-quartz structures: SiO_2 and GeO_2, *J. Appl. Phys., 49*, 5473–5478, 1978.

KIEFFER, S. W., Thermodynamics and lattice vibrations of minerals: 2. Vibrational characteristics of silicates, *Rev. Geophys. Space Phys., 17*, 20–34, 1979.

LEVIEN, L., and C. T. PREWITT, High-pressure crystal structure and compressibility of coesite, *Am. Mineral., 66*, 324–333, 1981.

LEVIEN, L., C. T. PREWITT, and D. J. WEIDNER, Structure and elastic properties of quartz at pressure, *Am. Mineral., 65*, 920–930, 1980.

LIU, L-g., High-pressure transformations of the dioxides: Implications for structures of SiO_2 at high pressure, in *High-Pressure Research in Geophysics*, edited by S. Akimoto and M. H. Manghnani, pp. 349–360, Center for Academic Publications, Tokyo, 1982.

LIU, L-g., W. A. BASSETT, and T. TAKAHASHI, Effect of pressure on the lattice parameters of stishovite, *J. Geophys. Res., 79*, 1160–1164, 1974.

MCSKIMIN, H. J., P. ANDREATCH, and R. N. THURSTON, Elastic moduli of quartz versus hydrostatic pressure at 25 °C and −195.8 °C, *J. Appl. Phys., 36*, 1624–1632, 1965.

MAMMONE, J. F., S. K. SHARMA, and M. NICOL, Raman study of rutile (TiO_2) at high pressures, *Solid State Comm., 34*, 799–802, 1980a.

MAMMONE, J. F., S. K. SHARMA, and M. NICOL, Raman spectra of methanol and ethanol at pressures up to 100 Kbar, *J. Phys. Chem., 84*, 3130–3134, 1980b.

MANGHNANI, M. H., Elastic constants of single-crystal rutile under pressures to 7.5 kilobars, *J. Geophys. Res., 74*, 4317–4328, 1969.

MAO, H-k., and P. M. BELL, Design and varieties of the megabar diamond cell, *Yearbook, Carnegie Inst. Washington, 77*, 904–913, 1978.

MAO, H-k., P. M. BELL, J. SHANER, and D. STEINBERG, Specific volume measurements of Ca, Mo, Pd, and Ag and the calibration of the ruby R_1 fluorescence pressure gauge from 0.06 to 1 Mbar, *J. Appl. Phys., 49*, 3276–3283, 1978.

MAO, H-k., J. XU, and P. M. BELL, Calibration of the ruby pressure gauge to 800 kbar under quasihydrostatic conditions, *J. Geophys. Res., 91*, 4673–4676, 1986.

MILLER, P. B., and J. D. AXE, Internal strain and Raman-active vibrations in solids, *Phys. Rev., 163*, 924–926, 1967.

MILLS, R. L., D. H. LIEBENBERG, J. C. BRONSON, and L. C. SCHMIDT, Procedure for loading diamond cells with high-pressure gas, *Rev. Sci. Instrum., 51*, 891–895, 1980.

MING, L. C., and M. H. MANGHNANI, High-pressure phase transformations in rutile-structured dioxides, in *High-Pressure Research in Geophysics*, edited by S. Akimoto and M. H. Manghnani, pp. 329–347, Center for Academic Publications, Tokyo, 1982.

NAGEL, L., and M. O'KEEFFE, Pressure and stress induced polymorphism of compounds with rutile structure, *Mat. Res. Bull., 6*, 1317–1320, 1971.

NICOL, M., and M. Y. FONG, Raman spectrum and polymorphism of titanium dioxide at high pressure, *J. Chem. Phys., 54*, 3167–3170, 1971.

OLINGER, B., The compression of stishovite, *J. Geophys. Res., 81*, 5341–5343, 1976.

PEERCY, P. S., and B. MOROSIN, Pressure and temperature dependence of the Raman-active phonons in SnO_2, *Phys. Rev. B, 7*, 2779–2786, 1973.

PRIMAK, W., *The Compacted States of Vitreous Silica. Studies in Radiation Effects in Solids*, Vol. 4, edited by G. J. Dienes and L. T. Chadderton, 184 pp., Gordon and Breach, New York, 1975.

ROSS, N. L., and E. P. MEAGHER, A molecular orbital study of $H_6Si_2O_7$ under simulated compression, *Am. Mineral., 69*, 1145–1149, 1984.

SAMARA, G. A., and P. S. PEERCY, Pressure and temperature dependence of the static dielectric constants and Raman spectra of TiO_2 (rutile), *Phys. Rev. B, 7*, 1131–1148, 1973.

SATO, Y., Pressure-volume relationship of stishovite under hydrostatic compression, *Earth Planet. Sci. Lett., 34*, 307–312, 1977.

SCOTT, J. F., and S. P. S. PORTO, Longitudinal and transverse optical lattice vibrations in quartz, *Phys. Rev., 161*, 903–910, 1967.

SHAPIRO, S. M., D. C. O'SHEN, and H. Z. CUMMINS, Raman scattering study of alpha-beta phase transition in quartz, *Phys. Rev. Lett., 19*, 361–364, 1967.

SHARMA, S. K., J. F. MAMMONE, and M. F. NICOL, Raman investigation of ring configurations in vitreous silica, *Nature, 292*, 140–141, 1981.

SINCLAIR, W., and A. E. RINGWOOD, Single crystal analysis of the structure of stishovite, *Nature, 272*, 714–715, 1978.

SOSMAN, R. B., *The Phases of Silica*, 388 pp., Rutgers University Press, Rutgers, 1965.

STISHOV, S. M., and S. V. POPOVA, A new dense modification of silica, *Geochemistry, Engl. Trans., 10*, 923–926, 1961.

STRIEFLER, M. E., On the nature of the structural instability of stishovite, *EOS Trans. AGU, 66*, 358, 1985.

STRIEFLER, M. E., and G. R. BARSCH, Lattice dynamics of α-quartz, *Phys. Rev. B, 12*, 4553–4566, 1975.

STRIEFLER, M. E., and G. R. BARSCH, Elastic and optical properties of stishovite, *J. Geophys. Res., 81*, 2453–2466, 1976.

TEKIPPE, V. J., A. K. RAMDAS, and S. RODRIGUEZ, Piezospectroscopic study of the Raman spectrum of α-quartz, *Phys. Rev. B, 8*, 706–717, 1973.

WANG, H., and G. SIMMONS, Elasticity of some mantle crystal structures 2. Rutile GeO_2, *J. Geophys. Res., 78*, 1262–1273, 1973.

WATANABE, H., Thermochemical properties of synthetic high-pressure compounds relevant to the earth's mantle, in *High-Pressure Research in Geophysics*, edited by S. Akimoto and M. H. Manghnani, pp. 441–464, Center for Academic Publications, Tokyo, 1982.

WEIDNER, D. J., and H. R. CARLETON, Elasticity of coesite, *J. Geophys. Res., 82*, 1334–1346, 1977.

WEIDNER, D. J., J. D. BASS, A. E. RINGWOOD, and W. SINCLAIR, The single-crystal elastic moduli of stishovite, *J. Geophys. Res., 87*, 4740–4746, 1982.

WONG, P. T. T., E. L. BARIDAIS, and D. J. MOFFAT, Hydrostatic pressure effects on TO-LO splitting and softening of infrared active phonons in α-quartz, *J. Chem. Phys., 84*, 671–674, 1986.

WYCKOFF, R. W. G., *Crystal Structures*, Second Edition, Vol. 1, 467 pp., Interscience, New York, 1963.

POLYHEDRAL BULK MODULI FROM HIGH-PRESSURE CRYSTAL FIELD SPECTRA

Roger G. BURNS

Department of Earth, Atmospheric & Planetary Sciences, Massachusetts Institute of Technology Cambridge, Massachusetts 02139, USA

Abstract. A modified Birch-Murnaghan equation of state, $K_{i,0} = (2/3)P\ [(\Delta_p/\Delta_0)^{7/5} - (\Delta_p/\Delta_0)]^{-1}$ is used to evaluate zero-pressure polyhedral bulk moduli, $K_{i,0}$, for transition metal-bearing coordination polyhedra from pressure-induced variations of the crystal field splitting parameter, Δ, obtained from visible-near infrared spectral measurements of oxides and silicates. The $K_{i,0}$ values generally agree with polyhedral bulk moduli derived from high-pressure X-ray data. The spectral method for evaluating polyhedral bulk moduli is best suited to cubic (isometric) phases containing transition metals in regular, or only slightly distorted, octahedral or cubic (8-fold) environments. Such coordination polyhedra occur in garnet, spinel, periclase, and perovskite structure-types in the mantle.

Introduction

Elastic constants of minerals are the key to understanding geophysical properties of the earth's interior. Bulk modulus and rigidity parameters, for example, influence the velocities of seismic waves through the earth. Numerous experimental and semi-empirical approaches have been developed to evaluate bulk moduli or incompressibilities of rock-forming minerals with varying degrees of self consistency between the various methods (ANDERSON and NAFE, 1965; ANDERSON and ANDERSON, 1970; ANDERSON, 1972; HAZEN and FINGER, 1984). Recently, attention has been focused on contributions to the crystal bulk modulus by component polyhedral bulk moduli of individual coordination sites, using X-ray data and crystal structure refinements at elevated pressures (HAZEN and PREWITT, 1977; HAZEN and FINGER, 1979, 1984). An overlooked method for obtaining polyhedral bulk moduli of oxide and silicate minerals (SHANKLAND et al., 1974; ABU-EID, 1976) is based on measurements of pressure-induced variations of crystal field spectra of transition metal-bearing minerals in the visible-near infrared region. This paper reviews such high-pressure optical spectral data, describes how they yield polyhedral bulk moduli of coordination polyhedra surrounding a variety of transition metal cations, and assesses the applicability of the spectrally-derived polyhedral bulk moduli to minerals occurring in the mantle.

Definitions and Derivations

The spectrally-derived polyhedral bulk modulus parameter, K_i, discussed throughout this paper is synonymous with the X-ray determined polyhedral bulk modulus, K_p, described by HAZEN and PREWITT (1977). The bulk modulus of a mineral phase, K, is defined by the pressure-dependence of its molar volume, V, by

$$K = -V(dP/dV) = 1/\beta \tag{1}$$

where β is the volume compressibility of the crystal.

In general K increases as a crystal is compressed at elevated pressures. By convention, crystal bulk moduli of minerals are usually expressed as zero pressure (1 bar or 10^{-4} GPa) values, K_0, and their pressure derivative K_0' (where $K_0' = dK/dP$ at near zero pressure). Several empirical equations of state correlating pressure-molar volume data of solids have been defined, one of which is the Birch-Murnaghan EOS:

$$P = (3/2)K_0(x^7 - x^5)[1 + (3/4)(K_0' - 4)(x^2 - 1)] \tag{2}$$

where

$$x = (V_0/V)^{1/3} \tag{3}$$

in which V and V_0 are the molar volumes of the crystal at elevated P and 1 bar, respectively. The first-order Birch-Murnaghan EOS, corresponding to K_0' approximating 4 for many solids, is

$$P = (3/2)K_0(x^7 - x^5) \tag{4}$$

Equations (2) and (4) indicate that an experimental technique providing pressure variations of a volume-related parameter may be utilized to estimate zero-pressure bulk moduli. One such parameter is derived from crystal field spectra of transition metal-bearing minerals.

Electronic spectra of transition metal compounds in the visible-near infrared region originate from electrons being excited by light between incompletely filled 3d orbital energy levels within the transition metal cation. The crystalline environment of oxygen anions, for example, surrounding the cation splits the 3d orbitals into two or more energy levels, depending on the symmetry and

High-Pressure Research in Mineral Physics, edited by M. H. Manghnani and Y. Syono, pp. 361–369.

distortion of the anion coordination polyhedron. When the transition metal ion is situated in a regular octahedral or cubic environment, produced by nearest-neighbor oxygens in six-fold or eight-fold coordinations, respectively, the five 3d orbitals are split into two groups, t_{2g} and e_g, separated by energies Δ_0 or Δ_c (BURNS, 1985). The magnitude of the 3d orbital splitting parameter, Δ, shows an inverse fifth-law dependence on cation-oxygen interatomic distance, R:

$$\Delta \propto 1/R^5 \tag{5}$$

Since the effect of pressure is to shorten interatomic distances, Δ increases at elevated pressures and

$$\Delta_p/\Delta_0=(R_0/R_p)^5 \tag{6}$$

where Δ_p, Δ_0 and R_p, R_0 are the 3d orbital splittings and average cation-oxygen distances at high pressures and zero pressure (i.e. 1 bar or 0.0001 GPa), respectively. Since $V \propto R^3$, where V is the volume of a coordination site,

$$\Delta \propto V^{-5/3} \text{ or } \ln \Delta \propto -(5/3) \ln V \tag{7}$$

Differentiation of equation 7 and substitution into equation (1) gives

$$K_i = \frac{5\Delta}{3} \frac{dP}{d\Delta} \tag{8}$$

where K_i is the polyhedral bulk modulus. This relationship is the one used in earlier mineral physics applications (SHANKLAND et al., 1974; ABU-EID, 1976).

Substitution into the Birch-Murnaghan EOS (equation (2)) yields

$$P = (3/2)K_{i,0}\ [(\Delta_p/\Delta_0)^{7/5} - (\Delta_p/\Delta_0)] \times [1 + (3/4)K'_{i,0}\ \{(\Delta_p/\Delta_0)^{6/5} - 1\}] \tag{9}$$

where $K_{i,0}$ is the polyhedral bulk modulus at zero pressure, and $K'_{i,0}$ is the pressure derivative of $K_{i,0}$.

Substitution into the Birch-Murnaghan EOS (equation (4)) enables $K_{i,0}$ to be calculated from

$$K_{i,0}=(2/3)P[(\Delta_p/\Delta_0)^{7/5}-(\Delta_p/\Delta_0)]^{-1} \tag{10}$$

Evaluation of Δ

The 3d orbital splitting parameter, Δ, may be readily evaluated when transition metal ions occur in undistorted octahedral, tetrahedral, cubic, or dodecahedral environments. For cations such as Fe^{2+} and Ti^{3+} in regular octahedra, for example, one absorption band corresponding to transitions of 3d electrons between lower-level (ground-state) t_{2g} and upper-level (excited states) e_g orbitals occurs in the visible-near infrared region, and is a direct measure of the octahedral splitting parameter Δ_0. However, when the cations are in distorted coordination sites, uncertainties arise over splittings of lower-level orbital energy levels. The problems are accentuated when the crystal system of the host mineral phase is non-cubic and when cations occur in two or more different coordination sites in the same structure. Three Fe^{2+}-bearing minerals, magnesiowüstite (periclase), almandine (garnet), and fayalite (olivine), illustrate these difficulties.

Magnesiowüstite

The isometric periclase structure consists of a cubic closest packed lattice of O^{2-} anions which provide regular octahedral coordination sites (Figure 1a). The absorption spectrum of a magnesiowüstite, $(Mg_{0.74}Fe_{0.26})O$, illustrated in Figure 1b (GOTO, et al., 1980) shows two absorption maxima, occurring at 10,000 cm^{-1} and 11,600 cm^{-1}, which are attributed to dynamic Jahn-Teller splitting of the upper e_g levels during the electronic transition within Fe^{2+}. Because the $[(Mg, Fe)O_6]$ octahedra in periclase are undistorted, the splitting of lower-level t_{2g} orbitals is negligible. The energy level diagram for Fe^{2+} in periclase derived from the absorption spectrum of magnesiowüstite (Figure 1c) yields the value of 10,800 cm^{-1} for Δ_0 at 1 atm. From the optical spectral measurements of SHANKLAND et al. (1974) Δ_0 at 5 GPa is estimated to be 11,140 cm^{-1}.

Almandine

Although garnet is isometric, the 8-fold coordination polyhedra accommodating Fe^{2+} ions in almandine are distorted cubes (Figure 2a). The 6-fold coordination sites, on the other hand, which are occupied by trivalent cations are less distorted from octahedral symmetry. The absorption spectrum of almandine garnet (Figure 2b) contains three bands, at 7600 cm^{-1}, 5800 cm^{-1}, and 4400 cm^{-1}, corresponding to transitions to the three upper levels of t_{2g} orbital energy levels. The splitting of the lower-level e_g orbitals cannot be obtained from the visible-near infrared spectra, but was estimated to be about 1100 cm^{-1} from calculations based on the temperature variation of almandine Mössbauer spectra (HUGGINS, 1975). The energy level diagram for Fe^{2+} in the distorted cubic site of a garnet with 70% almandine components shown in Figure 2c leads to an estimated value of 5470 cm^{-1} for Δ_c at ambient pressures. SMITH and LANGER (1983), who measured the pressure-induced shifts of the 7600 cm^{-1} and 5800 cm^{-1} of synthetic almandine, estimated that Δ_c had risen from 5380 cm^{-1} to 6352 cm^{-1} at 10.1 GPa.

Fayalite

The olivine structure accommodates Fe^{2+} ions in two

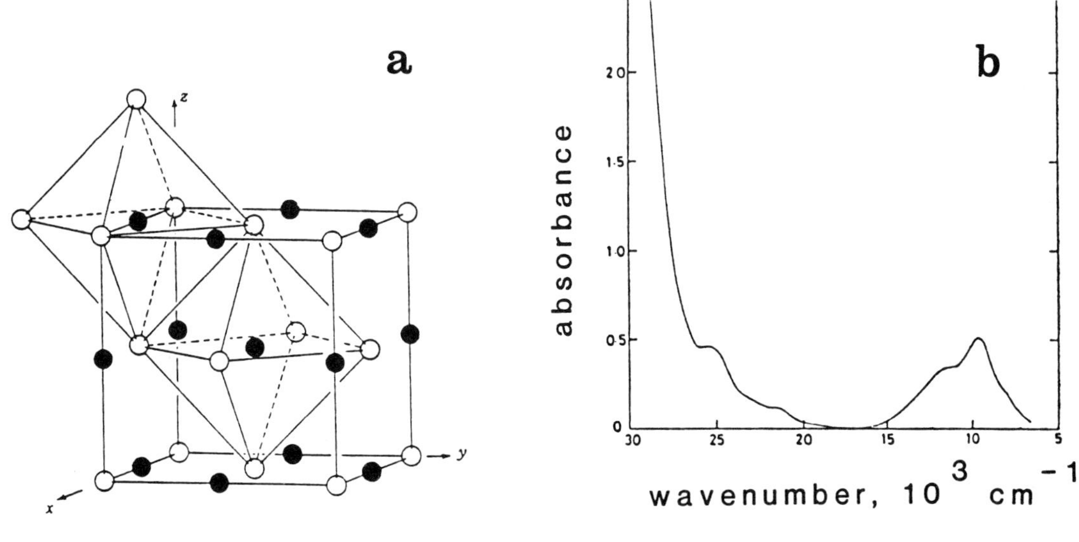

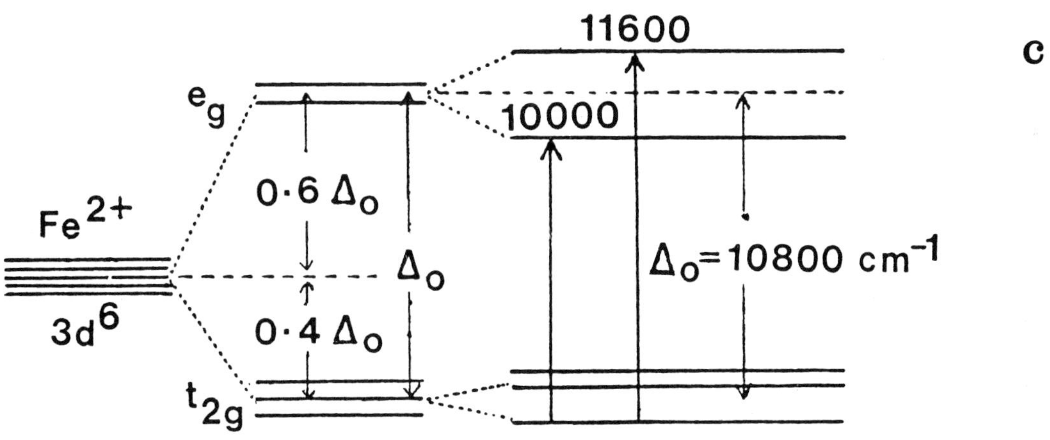

Figure 1. Structural and spectral features of Fe^{2+}-bearing periclase. (a) Crystal structure of MgO. Note that undistorted octahedra containing Fe^{2+} cations share edges with other $[MgO_6]$ octahedra. ●, Mg^{2+}; ○, O^{2-}. (b) Absorption spectrum of magnesiowüstite $(Mg_{0.74}Fe_{0.26})O$ (from GOTO et al., 1980). (c) 3d orbital energy level diagram for Fe^{2+} ions in magnesiowüstite.

distinct six-coordinated sites (designated M1 and M2), both of which are significantly distorted, but in different ways, from octahedral symmetry (BURNS, 1985). This provides a problem of evaluating Δ_0 parameters for Fe^{2+} ions coexisting in two sites in the olivine structure. The problem is further complicated by the needs to identify and to evaluate the splittings of upper-level e_g and lower-level t_{2g} orbitals for Fe^{2+} ions in each site. Further complications arise because the orthorhombic olivine phase exhibits polarization-dependent spectra in which absorption bands in the near-infrared region have different positions and intensities when polarized light is transmitted along the three crystallographic directions (Figure 3a). Nevertheless, despite these complexities, energy level diagrams for Fe^{2+} ions in the distorted M1 and M2 sites may be constructed (Figures 3b and c) leading to approximate Δ_0 parameters for each site at ambient pressures. Although numerous high-pressure studies of olivine crystal field spectra have been made (MAO and BELL, 1971; ABU-EID, 1976; SMITH and LANGER, 1982), insufficient information about changes of ground-state splittings of t_{2g} levels make it very difficult to evaluate Δ_0 at high pressures. This problem is elaborated upon later.

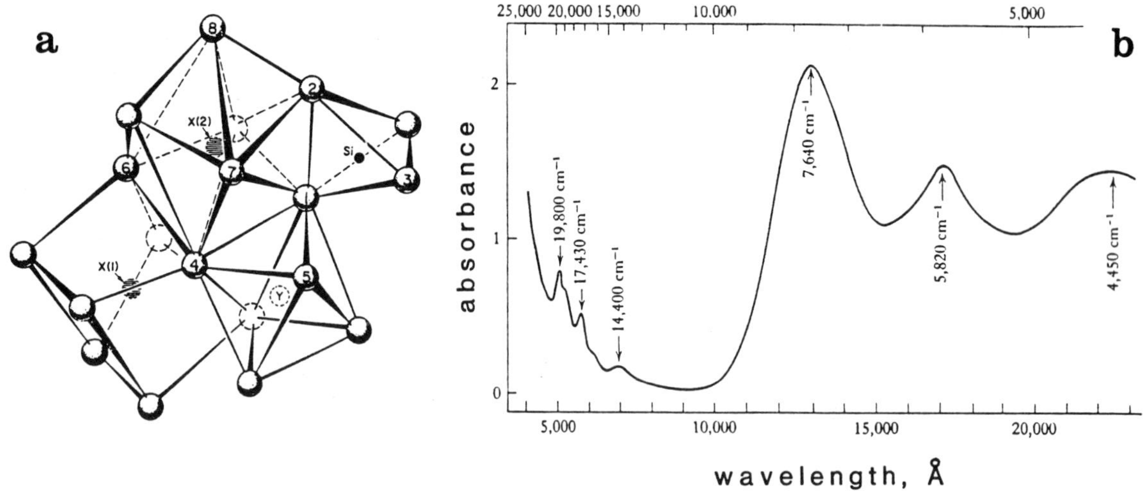

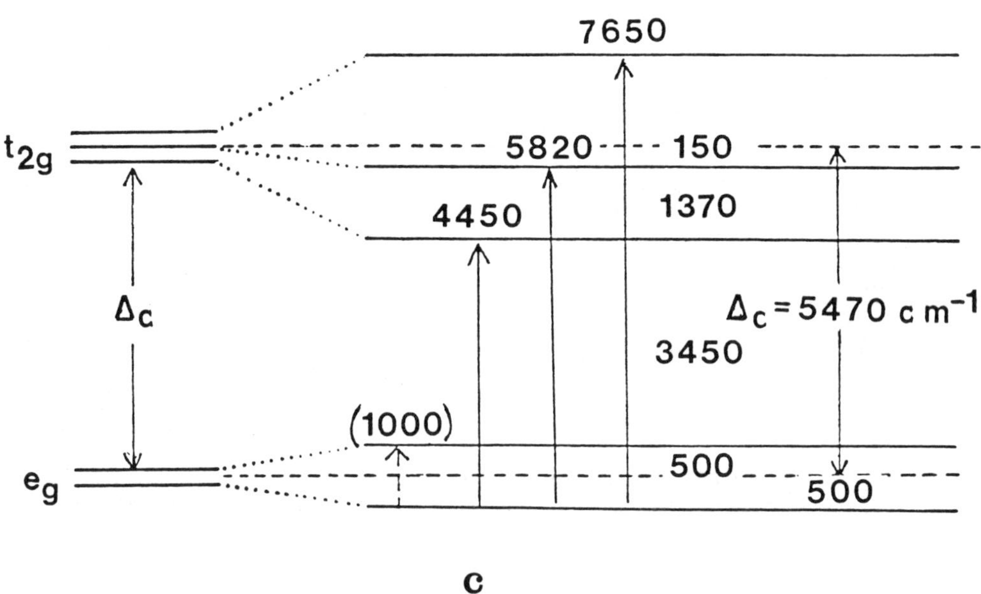

Figure 2. Structural and spectral features of almandine. (a) Portion of the garnet structure showing the configuration of the distorted cubic (8-fold) coordination site, and its relationship to the octahedral and tetrahedral sites. (b) Absorption spectrum of almandine with 70 mol% $Fe_3Al_2Si_3O_{12}$, (c) 3d orbital energy level diagram for Fe^{2+} ions in the almandine garnet 8-fold coordination site.

Pressure-Induced Variations of Δ

Equation (5) suggests that Δ will increase significantly with rising pressure as a result of shortening of cation-oxygen interatomic distances. This is demonstrated by the data for γ-Ni_2SiO_4 (spinel) (YAGI and MAO, 1977) illustrated in Figure 4. YAGI and MAO (1977) estimated from the position of band ν_1, that Δ_o for Ni^{2+} in the octahedral site of the spinel structure increases from 9150 cm^{-1} at 1 bar to 10,100 cm^{-1} at 12.1 GPa in response to a decrease of Ni^{2+}-O^{2-} bondlength from 206.3 pm to 203.2 pm at 12.1 GPa (FINGER et al., 1979). However, BURNS and SUNG(1978) pointed out that trigonal distortion of the spinel octahedral coordination polyhedron leads to splitting of lower-level t_{2g} orbitals by approx. 1100 cm^{-1}. The prominent shoulders at ≈8000 cm^{-1} (1 bar) and 9000

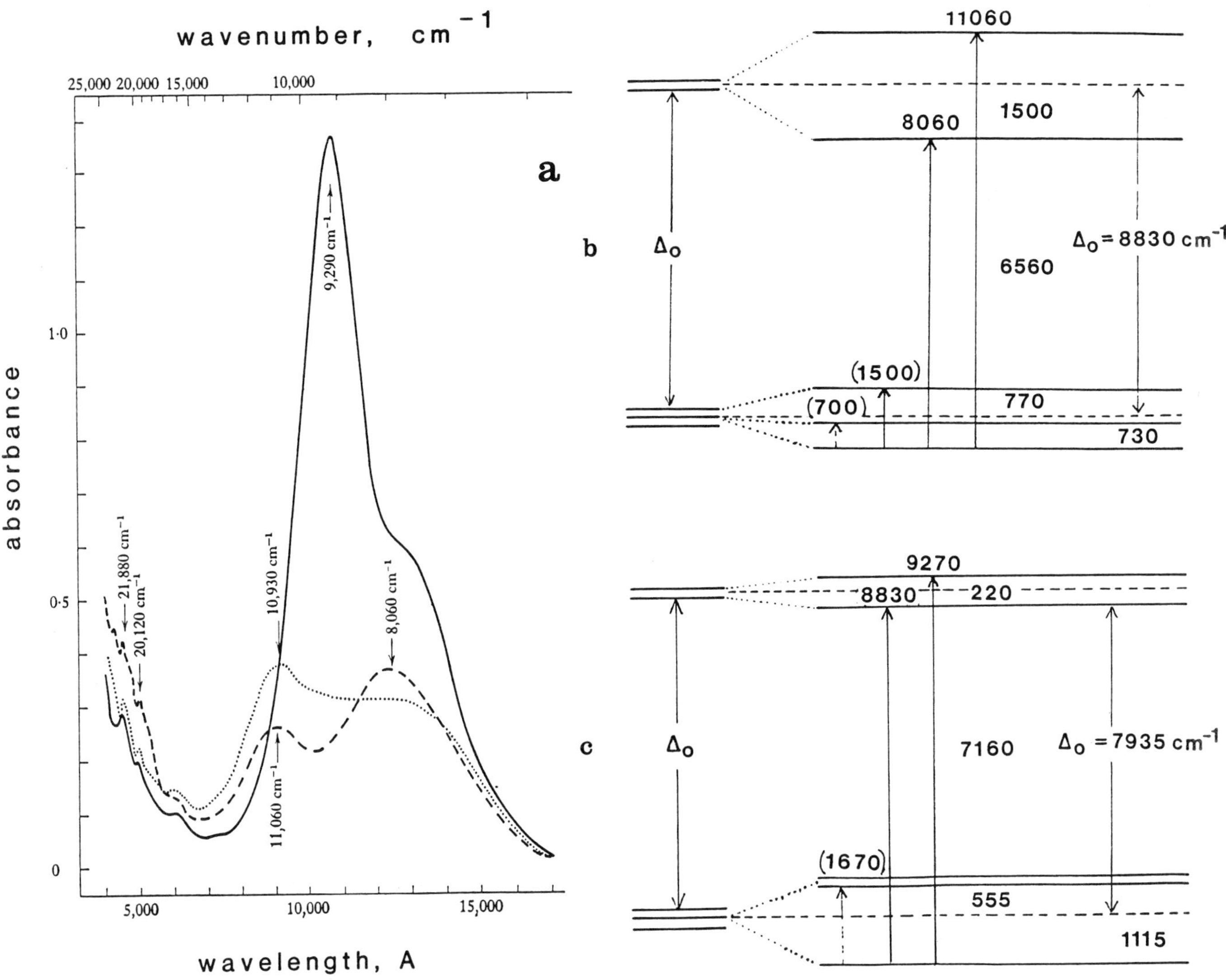

Figure 3. Crystal field spectra and energy levels of fayalite. (a) Polarized absorption spectra of fayalite with 96 mol% Fe_2SiO_4. (b), (c) 3d orbital energy level diagrams for Fe^{2+} ions in the distorted M(1) and M(2) sites of fayalite.

cm^{-1} (120 kbar) in Figure 4 indicate that the ground-state splitting of ≈ 1100 cm^{-1} is retained by the Ni^{2+} t_{2g} orbitals in γ-Ni_2SiO_4 over the entire pressure range.

The data of YAGI and MAO (1977) for γ-Ni_2SiO_4 are tabulated in Table 1. Initial estimates of K_i given by equation (8) indicate a polyhedral bulk modulus of approx. 160 GPa. Calculations based on the Birch-Murnaghan EOS (equation (10)) yield data summarized in Table 1, from which the site incompressibility at zero pressure, $K_{i,0}$, is estimated to be 148 GPa. This value agrees favorably with the polyhedral bulk modulus, K_p, of 150 GPa obtained from crystal structure refinements at elevated pressures (FINGER et al., 1979; HAZEN and FINGER, 1979). The scatter of spectral K_i values in Table 1 may reflect inaccuracies in calibrating pressures and estimating peak maxima in the absorption spectra.

Available high-pressure crystal field spectra (BURNS, 1982) yielding polyhedral bulk modulus data for a number of transition metal-bearing oxides and silicates (BURNS, 1985) are summarized in Table 2. Comparisons are made with crystal and polyhedral bulk moduli for Mg^{2+} and Al^{3+} in host phases (periclase, pyrope, diopside, corundum, grossular), as well as corresponding transition metal phases for which pertinent data are available in the literature. Footnotes to Table 2 indicate the sources of crystal field spectral data and assumptions made to evaluate the polyhedral modulus values.

For Fe^{2+} in the distorted cube environment in almandine

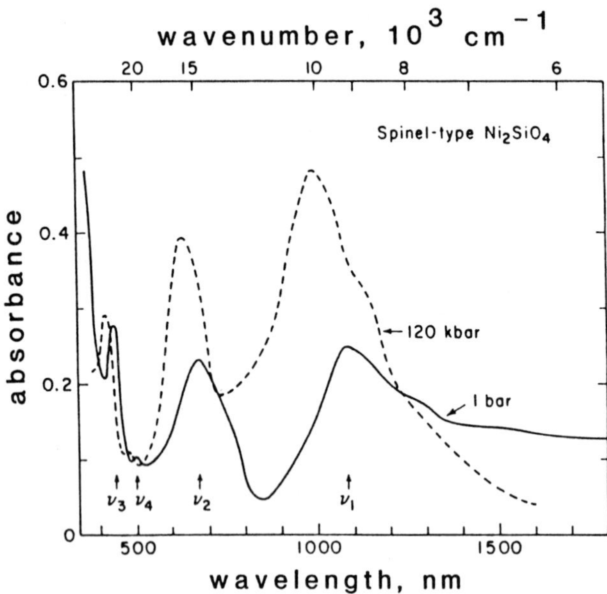

Figure 4. Crystal field spectra of synthetic Ni_2SiO_4 spinel at 1 bar and 120 kbars. Note the prominent shoulders at ≈8000 cm^{-1} (1 bar) and 9000 cm^{-1} (120 kbars), which affect estimates of the crystal field splitting parameters summarized in Table 1.

garnet, SMITH and LANGER (1983) estimated the polyhedral bulk modulus to be 123.5±2.5 Gpa, based on the Δ versus $1/R^5$ proportionality (equation (5)). Using their estimates of Δ_p and Δ_0, and substituting into equation (10) yielded $K_{i,0}$ values of 143 GPa and 83 GPa, respectively, at pressures of 5.4 and 10.1 GPa. The different $K_{i,0}$ values may originate from difficulties of estimating splittings of lower-level e_g orbitals. For Fe^{2+} in the very distorted 6-fold coordination sites of fayalitic olivine and orthopyroxene, drastic assumptions had to be made about the magnitudes of the lower-level t_{2g} orbital splittings at elevated pressures. The polyhedral bulk modulus value of 97 GPa for Fe^{2+} in the fayalite M(2) site was based on the data of SMITH and LANGER (1982) and assuming that the upper-level and lower-level splittings shown in Figure 4 remain constant at elevated pressures. The estimates of Δ_0 by ABU-EID (1976) were used to obtain the $K_{i,0}$ value for Fe^{2+} in the fayalite M(1) site at 4 GPa. The spectral data obtained by MAO and BELL (1971) for orthoferrosilite were used to estimate $K_{i,0}$ values for Fe^{2+} in the pyroxene M(1) and M(2) sites, assuming that orbital splittings calculated at 1 bar (BURNS, 1985) remain constant at 2.5 GPa. The poor agreement with the value estimated for Fe^{2+} in the fassaite (clinopyroxene) M(1) site in Table 2 is a reflection of the crudeness of calculating lower-level orbital splittings for orthoferrosilite at elevated pressures.

TABLE 1. Spectral, Structural, and Derived Thermodynamic Data for Ni^{2+} in Synthetic γ-Ni_2SiO_4 at High Pressures

Parameter	Pressure (GPa)							
	0.0001	1.46	2.10	3.01	4.47	6.86	9.70	12.1
[1]R (pm)	206.3	205.9	205.6	205.4	205.0	204.4	203.7	203.2
[2]ν_1 (cm^{-1})	9150	9240	9260	9350	9540	9710	10000	10100
[3]ν_1' (cm^{-1})	≈8050	(9090)	(9130)	(9200)	(9390)	(9560)	(9850)	≈9000
[4]Δ_p (cm^{-1})	8380	8470	8490	8580	8770	8490	9230	9430
[5]$\delta\Delta$ (cm^{-1})	—	90	110	200	390	560	850	1050
[6]K_i (GPa)	—	226.6	266.6	210.2	160.1	171.1	159.4	160.9
(Δ_p/Δ_0)	1	1.0107	1.0131	1.0239	1.0465	1.0668	1.1014	1.1253
$(\Delta_p/\Delta_0)^{7/5}$	1	1.0151	1.0184	1.0336	1.0657	1.0948	1.1448	1.1797
[7]Diff.	—	0.00433	0.00530	0.00971	0.01922	0.02796	0.04340	0.05441
[8]$K_{i,0}$	—	224.9	264.2	206.8	155.1	163.5	149.0	148.2

[1]Ni^{2+}-O^{2-} interatomic distances in the octahedral site of γ-Ni_2SiO_4 (FINGER et al., 1979).
[2]The position of the absorption peak at highest wavelength (Figure 4) (YAGI and MAO, 1977).
[3]The position of the shoulder on the ν_1 peak, assumed to be offset by 1100 cm^{-1} over the entire pressure range.
[4]Calculated assuming 1100 cm^{-1} for the lower-level splitting of t_{2g} orbitals. For a symmetry C_{3v} octahedron, $\Delta_p=[\nu_1-(2/3)\times1100]$.
[5]$\delta\Delta=\Delta_p-\Delta_0$.
[6]K_i calculated from equation (8).
[7]Diff.$=[(\Delta_p/\Delta_0)^{7/5}-(\Delta_p/\Delta_0)]$.
[8]$K_{i,0}$ calculated from equation (10).

TABLE 2. Polyhedral Bulk Moduli of Transition Metal-Bearing Minerals

Mineral (Structrure)	Cation	Coordination Site	$K_{i,0}$ (GPa)	(Spectral)[1] Ref[4]	K_p (GPa)	(X-ray)[2] Ref[5]	K_0(Bulk Crystal)[3] (GPa)
Periclase	Mg^{2+}	Octahedron	—	—	161	a)	162.2
Magnesiowüstite	Fe^{2+}	Octahedron	148	1)	153(FeO)	a)	
(Periclase)	Ni^{2+}	Octahedron	185	2)	196(NiO)	a)	
(Periclase)	Co^{2+}	Octahedron	139	2)	185(CoO)	a)	
(Periclase)	Cr^{3+}	Octahedron	133	2)	—	—	
Corundum	Al^{3+}	Octahedron	—	—	240	a)	257
Ruby	Cr^{3+}	Octahedron	303	2)	230(Cr_2O_3)	a)	
Ruby	Cr^{3+}	Octahedron	693	3)	—	—	
(Corundum)	V^{3+}	Octahedron	303	2)	180(V_2O_3)	a)	
Spinel	Fe^{2+}	Tetrahedron	263	1)	—	—	196.9($MgAl_2O_4$)
Chromite	Cr^{3+}	Octahedron	145	4)	—	—	
γ-Ni_2SiO_4	Ni^{2+}	Octahedron	148	5)	150	a, b)	223
γ-Fe_2SiO_4	Fe^{2+}	Octahedron	179	6)	—	—	197
Pyrope	Mg^{2+}	8-fold	—	—	130	a)	173
Almandine	Fe^{2+}	8-fold	143–83	7)	—	—	178
Grossular	Al^{3+}	Octahedron	—	—	220	a)	
Uvarovite	Cr^{3+}	Octahedron	453	8)	—	—	
Diopside	Mg^{2+}	M(1)	—	—	105	c)	113
Fassaite	Fe^{2+}	M(1)	145	9)	140	d)	
Fassaite	Ti^{3+}	M(1)	195	10)	—	—	
Ferrosilite	Fe^{2+}	M(1)	≃ 71	11)	—	—	
Ferrosilite	Fe^{2+}	M(2)	≃ 72	11)	—	—	
Fayalite	Fe^{2+}	M(1)	≃140	12)	—	—	137.9
Fayalite	Fe^{2+}	M(2)	≃ 97	13)	—	—	137.9
Piemontite	Mn^{3+}	Al(3)	266	14)	—	—	
Perovskite	Fe^{2+}	A	(70–140)	15)	—	—	260

[1]Calculated from equation (10).
[2]Derived from high-pressure X-ray diffraction data.
[3]Based on data compiled in WATANABE (1982).
[4]References, together with approximations made, are: 1) SHANKLAND et al. (1974); see text and Figure 1; data at 5 GPa. 2) DRICKAMER and FRANK (1973), pp. 74–75; based on Δ_p/Δ_0 values at 5 GPa. 3) GOTO et al. (1979); based on shock loading experiments at 46 GPa. 4) MAO and BELL (1975); for $P=2$ GPa. 5) YAGI and MAO (1977); see text, Figures 4 and 5, and Table 1. 6) YAGI, T. and MAO, H.-K., unpubl. results (1977); for $P=17.1$ GPa. 7) SMITH and LANGER (1983); for $P=5.4$ and 10.1 GPa. 8) ABU-EID (1976); for $P=19.7$ GPa (see BURNS (1985)). 9) HAZEN et al. (1977); MAO et al. (1977); for $P=4.8$ GPa. 10) MAO and BELL (1974); for $P=4$ GPa. (See BURNS (1985)). 11) MAO and BELL (1971); for $P=2.5$ GPa (see text). 12) ABU-EID (1976); for $P=4$ GPa. 13) SMITH and LANGER (1982); assuming 120 cm^{-1}/GPa shift of 9300 cm^{-1} band (see text). 14) ABU-EID (1976); for $P=19.7$ GPa. 15) Estimated (based on values for almandine and pyroxene M(2) sites).
[5]References are: a) HAZEN and FINGER (1979). b) FINGER et al. (1979). c) LEVIEN and PREWITT (1981). d) HAZEN and FINGER (1977).

Discussion

The bulk moduli data summarized in Table 2 indicate that remarkable agreement exists between the spectrally-determined polyhedral bulk moduli and those obtained from high-pressure X-ray data The consistency between the spectral $K_{i,0}$ data is surprising given the diversity of the crystal field spectral measurements and assumptions inherent in the use of equation (10).

The data for Fe^{2+} in structures providing six-fold

coordination sites (periclase, γ-Fe_2SiO_4, olivine, and pyroxenes) and in the spinel (tetrahedral) and garnet (8-fold) structures conform with the pattern of increased incompressibility for sites having lower coordination numbers. The (CrO_6) octahedron is significantly incompressible in the corundum and garnet structures, perhaps reflecting the very large crystal field stabilization energy of Cr^{3+} ions in these minerals. However the discrepancy between the $K_{i,0}$ values obtained from shock-loading (46 GPa) and static (<10 GPa) experiments for ruby is unaccountable.

Approximations embodied in the use of equation (10) include: a) the pressure range over which it is applicable; b) the assumption that $K'_{i,0}=4$ in equation (9); c) the validity of the inverse fifth-law dependence of Δ on R (equation (5)); and d) the accuracy of estimating Δ from spectra of low symmetry environments.

The Birch-Murnaghan EOS (equation (2)) is empirical and is generally regarded to be applicable to P-V data extending to about 15% compression. (HAZEN and FINGER, 1984). The data-base assembled in Table 2 includes results well below this limit.

Although the pressure derivatives of crystal bulk moduli, K'_0, have been determined to lie in the range 3–5 for several oxides and silicates, values for individual sites in mineral structures are unavailable. Therefore the assumption that $K'_{i,0}=4$ used to derive equation (10) from equation (9) remains to be verified.

Although an electrostatic point charge model underlies the proportionality of Δ to R^{-5} (equation (5)), TOSSELL (1976) has pointed out that complete ionicity is not a necessary condition for the existence of such a distance dependence. More significant are the *changes* of ionic-covalent bond character with increasing pressures, which may be determined from spectrally-derived Racah B parameters (BURNS, 1985). For many of the transition metal-bearing minerals assembled in Table 2, the Racah B parameters change by only a few percent for pressures up to 20 GPa (DRICKAMER and FRANK, 1973; ABU-EID, 1976; SMITH and LANGER, 1983). Moreover molecular orbital energy level calculations yielding effective charges on cations in coordination clusters (TOSSELL, 1976; SHERMAN, 1985) indicate relatively small changes of ionicity of Fe-O and Mg-O bonds compressed by 10%. Therefore, equation (5) and its substitution into the Birch-Murnaghan EOS (equations (9) and (10)) is not invalidated by departures from ionicity in the current study. Note that ionic models also underlied semi-empirical expressions for crystal and polyhedral bulk moduli developed previously (HUSH and PRYCE, 1958; ANDERSON, 1972; OHNISHI and MIZUTANI, 1978; HAZEN and FINGER, 1984).

Perhaps the most subjective factor affecting the use of equation (10) to calculate polyhedral bulk moduli is the accuracy of the crystal field splitting parameter Δ derived from optical spectra of transition metal-bearing minerals. The problem centers on evaluations of lower-level orbital splittings for cations in low-symmetry coordination sites. The electronic structures of some cations (e.g., octahedral Cr^{3+}, Mn^{3+}, Ni^{2+}) enable these lower-level splittings to be measured directly, since spin-allowed electronic transitions from split t_{2g} orbitals to upper-level orbitals produce absorption bands in the visible region. Electronic transitions in optical spectra of Fe^{2+} and Ti^{3+} phases, on the other hand, do not originate from all lower-level orbitals, so that ground-state splittings for these cations must be evaluated indirectly.

Additional problems arise for non-cubic phases (e.g., corundum, olivine, pyroxenes) containing transition metal ions in low-symmetry sites because two or more absorption bands may show polarization dependence (dichroism or pleochroism) in visible-region spectra. The difficulties are accentuated for aniosotropic minerals containing cations in two or more distorted sites (e.g., olivine and pyroxene).

As a result the accuracy of the Δ values obtained from crystal field spectra decreases in the order: periclase (Ni^{2+}, Cr^{3+}, Co^{2+}, Fe^{2+}) >spinel tetrahedron (Fe^{2+})>spinel octahedron (chromite, γ-Ni_2SiO_4, γ-Fe_2SiO_4)> garnet octahedron (Cr^{3+})>garnet 8-fold site (Fe^{2+})>corundum (Cr^{3+}, V^{3+})$\gg$pyroxene M1 site (Fe^{2+}, Ti^{3+})>olivine M1 and M2 sites (Fe^{2+}, Ni^{2+}). This order, in turn, underlies the accuracy of spectrally-determined polyhedral bulk moduli cited in Table 2.

Summary

a) The Δ versus $1/R^5$ (equation (5)) and Birch-Murnaghan EOS (equation (10)) relationship are applicable to transition metal-bearing phases with regular, or only slightly distorted, octahedral, cubic or tetrahedral coordination polyhedra. When the coordination polyhedra are very distorted, it becomes increasingly difficult to evaluate splittings of lower-level orbitals and hence the Δ parameter, from crystal field spectra.

b) Zero-pressure polyhedral bulk moduli, $K_{i,0}$, obtained from the Birch-Murnaghan EOS, are more readily evaluated from visible-near infrared spectra of cubic (isotropic) phases which show no polarization dependence. In some cases, dichroism or pleochroism in spectra obtained with polarized light transmitted through anisotropic crystals may facilitate assignment of absorption bands originating from cations in two or more coordination sites.

c) The B-M EOS/Δ method for evaluating polyhedral bulk moduli is in good agreement with values obtained from high-pressure X-ray measurements of transition metal-bearing phases.

d) The spectral method for obtaining $K_{i,0}$ values is best suited to crystal field spectra measured up to ≈20 GPa.

e) Transition metal-bearing periclase, garnet, spinel, and perovskite structure-types occurring in the mantle are most amenable to polyhedral bulk modulus estimates obtained by the B-M EOS/Δ method.

Acknowledgments. Concepts developed here arose from discussions with Frank Huggins, Rateb Abu-Eid, Michael Feves, Chien-Min Sung, Kathleen Parkin, David Sherman, Dan Goldman, Gordon Smith, and George Rossman. I appreciate the editorial assistance of Tina Freudenberger.

Research on the spectra and bonding of transition metal-bearing minerals is supported by grants from the NSF (grant no. EAR83-13585) and NASA (grant no. NSG-7604).

Note added: One reviewer of this paper drew attention to another assumption of the modified Birch-Murnaghan model, namely that one can define a local, microscopic hydrostatic pressure at the site of an ion in a crystal. The viewpoint was expressed that if one chooses to use a macroscopic concept such as stress, one should examine the full stress field and not simply an averaged "hydrostatic" component.

References

Abu-Eid, R. M., Absorption spectra of transition metal-bearing minerals at high pressures, in *The Physics and Chemistry of Minerals and Rocks*, edited by R. G. J. Strens, pp. 641–675, J. Wiley, New York, 1976.

Anderson, D. L., and Anderson, O. L., The bulk modulus-volume relationship for oxides, *J. Geophys. Res., 75*, 3494–3500, 1970.

Anderson, O. L., Patterns in elastic constants of minerals important to geophysics, in *The Nature of the Solid Earth*, edited by E. C. Robertson, pp. 575–613, McGraw-Hill, New York, 1972.

Anderson, O. L., and Nafe, J. E., The bulk modulus-volume relationship for oxide compounds and related geophysical problems, *J. Geophys. Res., 70*, 3951–3963, 1965.

Burns, R. G., Electronic spectra of minerals at high pressures: how the mantle excites electrons, in *High-Pressure Researches in Geoscience*, edited by W. Schreyer, pp. 223–246, E. Schweizerbart'sche Verlagsbuchhandlung, Stuttgart, 1982.

Burns, R. G., Thermodynamic data from crystal field spectra, Chapt. 8 in *Microscopic to Macroscopic: Atomic Environments to Mineral Thermodynamics*, edited by S. W. Kieffer and A. Navrotsky, *Rev. Mineral. 14*, 277–316, 1985.

Burns, R. G., and Sung, C.-M., The effect of crystal field stabilization energy on the olivine→spinel transition in the system Mg_2SiO_4-Fe_2SiO_4, *Phys. Chem. Mineral., 2*, 349–364, 1978.

Drickamer, H. G., and Frank, C. W., *Electronic Transitions and the High Pressure Chemistry and Physics of Solids*, Chapman and Hall, Ltd., London, 1973.

Finger, L. W., Hazen, R. M., and Yagi, T., Crystal structures and electron densities of nickel and iron silicate spinels at elevated temperature or pressure, *Am. Mineral., 64*, 1002–1009, 1979.

Goto, T., Ahrens, T. J., and Rossman, G. R., Absorption spectra of Cr^{3+} in Al_2O_3 under shock compression, *Phys. Chem. Mineral., 4*, 253–264, 1979.

Goto, T., Ahrens, T. J., Rossman, G. R., and Syono, Y., Absorption spectrum of shock-compressed Fe^{2+}-bearing MgO and the radiative conductivity of the lower mantle, *Phys. Earth Planet. Inter., 22*, 277–288, 1980.

Hazen, R. M., and Finger, L. W., Compressibility and crystal structure of Angra dos Reis fassaite to 52 kbar, *Carnegie Inst. Wash. Yearb., 76*, 512–515, 1977.

Hazen, R. M., and Finger, L. W., Bulk modulus-volume relationship for cation-anion polyhedra, *J. Geophys. Res., B84*, 6723–6728, 1979.

Hazen, R. M., and Finger, L. W., *Comparative Crystal Chemistry*, J. Wiley & Sons, New York, 1984.

Hazen, R. M., and Prewitt, C. T., Effects of temperature and pressure on interatomic distances in oxygen-based minerals, *Am. Mineral., 62*, 309–315, 1977.

Hazen, R. M., Bell, P. M., and Mao, H.-K., Polarized absorption spectra of Angra dos Reis fassaite to 52 kbar, *Carnegie Inst. Wash. Yearb., 76*, 515–516, 1977.

Huggins, F. E., The 3d levels of ferrous ions in silicate garnets, *Am. Mineral., 60*, 316–319, 1975.

Hush, N. S., and Pryce, N. S., Influence of the crystal-field potential on interionic separation in salts of divalent iron-group ions, *J. Chem. Phys., 28*, 244–249, 1958.

Levien, L., and Prewitt, C. T., High-pressure structural study of diopside, *Am. Mineral., 66*, 315–323, 1981.

Mao, H.-K., and Bell, P. M., Crystal field spectra, *Carnegie Inst. Yearb., 70*, 207–215, 1971.

Mao, H.-K. and Bell, P. M., Crystal field effects of trivalent titanium in fassaite from the Pueblo de Allende meteorite, *Carnegie Inst. Wash. Yearb., 73*, 488–492, 1974.

Mao, H.-K., and Bell, P. M., Crystal field effects in spinel: oxidation states of iron and chromium, *Geochim. Cosmochim. Acta, 39*, 865–874, 1975.

Mao, H.-K., Bell, P. M., and Virgo, D., Crystal-field spectra of fassaite from the Angra dos Reis meteorite, *Earth Planet. Sci. Lett., 35*, 352–356, 1977.

Ohnishi, S., and Mizutani, H., Crystal field effect on bulk moduli of transition metal oxides, *J. Geophys. Res., 83*, 1852–1856, 1978.

Shankland, T. J., Duba, A. G., and Woronow, A., Pressure shifts of optical absorption bands in iron-bearing garnet, spinel, olivine, and pyroxene, *J. Geophys. Res., 79*, 3273–3282, 1974.

Sherman, D. M., Electronic structures of iron and manganese oxides with applications to their mineralogy, Ph. D. Thesis, MIT, August 1985.

Smith, G., and Langer, K., Single crystal spectra of olivines in the range 40,000–5,000 cm^{-1} at pressures up to 200 kbar, *Am. Mineral., 67*, 343–348, 1982.

Smith, G., and Langer, K., High pressure spectra up to 120 kbars of the synthetic garnet end members spessartine and almandine, *Neues Jahrb. Mineral. Mh.*, 541–555, 1983.

Tossell, J. A., Electronic structures of iron-bearing oxidic minerals at high pressure, *Am. Mineral., 61*, 130–144, 1976.

Watanabe, H., Thermodynamic properties of synthetic high-pressure compounds relevant to the Earth's mantle, in *High-Pressure Research in Geophysics*, edited by M. H. Manghnani and S. Akimoto, pp. 441–464, Center for Academic Publications, Tokyo, 1982.

Yagi, T., and Mao, H.-K., Crystal field spectra of the spinel polymorph of Ni_2SiO_4 at high pressure, *Carnegie Inst. Wash. Yearb., 76*, 505–508, 1977.

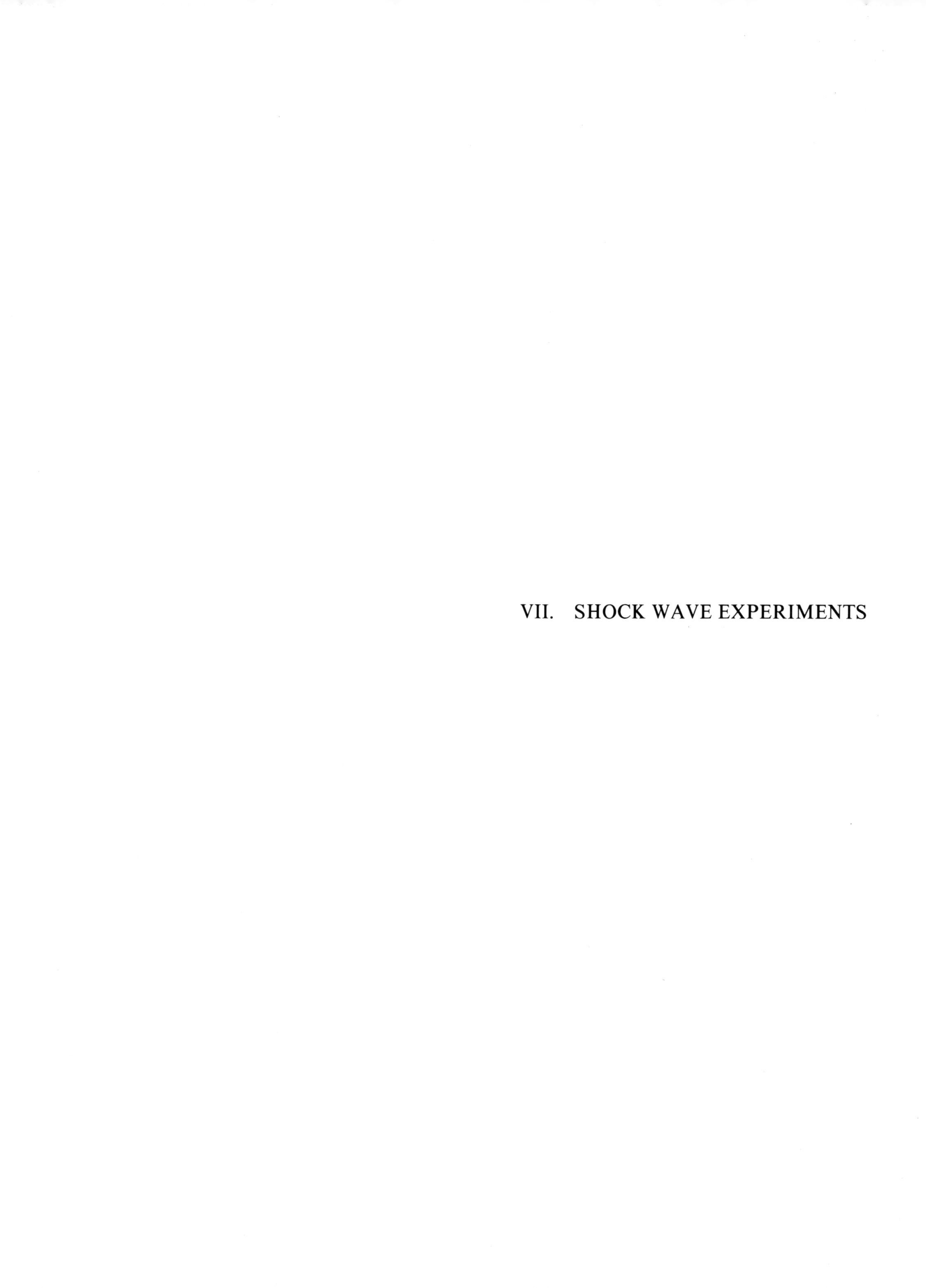

VII. SHOCK WAVE EXPERIMENTS

THERMODYNAMICS FOR $(Mg, Fe)_2SiO_4$ FROM THE HUGONIOT

J. Michael BROWN and Michael D. FURNISH

Graduate Program in Geophysics, University of Washington
Seattle, Washington 98195, USA

Robert G. MCQUEEN

Shock Physics Group, Los Alamos National Laboratory
Los Alamos, New Mexico 87545, USA

Abstract. We review current knowledge of the compression behavior of magnesian olivine ($Mg:(Mg+Fe) \geq 0.90$) and its high-pressure decomposition products. Results from static and dynamic compression experiments are relatively consistent, although equation-of-state parameterizations remain uncertain. The dynamic compression experiments give Hugoniot parameters of $c_0=6.57$ and $s=0.85$ for a low-pressure phase and 3.80 and 1.69, respectively, for the high-pressure phase. The behavior of the isothermal bulk modulus for the high-pressure phase is compatible with the estimated properties of perovskite plus magnesiowustite. Preliminary sound velocities are reported for synthetic forsterite and San Carlos peridot shocked into the high-pressure phase. These data support our previous suggestion that the high-pressure phase of olivine melts near a Hugoniot pressure of 150 GPa. In addition, sound velocities below the inferred melting point match lower mantle longitudinal wave velocity profiles. At 168 GPa the measured velocity is consistent with a bulk sound velocity in a liquid having relatively large values for specific heat and the Grüneisen parameter. The solid-liquid mixed-phase regime probably extends beyond 168 GPa. Since limited data have been obtained for only one silicate phase, conclusions concerning compositional constraints for the lower mantle remain premature.

Introduction

The structural changes which occur in major mantle constituents under lower mantle conditions of pressure and temperature are reasonably well understood. The transformation of upper mantle phases (olivine, pyroxenes, and garnets) to silicate perovskite structures plus oxides at pressures above 24 GPa is now well documented (LIU, 1976, 1979; ITO et al., 1984). To first order, the major seismic discontinuities at 400 and 670 km are usually interpreted in terms of the olivine-spinel (pyroxene-garnet) transition and the spinel (garnet)-perovskite transition, respectively (BASS and ANDERSON, 1984; ANDERSON and BASS, 1984).

Much is still unknown, however, about the physical properties of the major mantle constituents under lower mantle conditions. Bulk and shear moduli, specific heats, and Grüneisen parameters have been measured for only a few materials under ambient conditions. Pressure and temperature derivatives are poorly known or are not known at all (see WEIDNER and ITO, this volume). Therefore, it is not possible using existing laboratory and seismic data to resolve changes in the iron to magnesium ratio (Mg:Fe) or the silicon content (Si:Mg+Fe) at the major discontinuities (JEANLOZ and THOMPSON, 1983). In addition, the cause of the seismically anomalous zone D″ at the base of the mantle may be either chemical or thermal (LAY and HELMBERGER, 1983; STACEY and LOPER, 1984), while sources of the lateral heterogeneities in the lower mantle are tentatively associated with temperature variations of uncertain magnitude (HAGER et al., 1985).

We discuss here an experimental approach to the study of silicates under lower mantle conditions of pressure and temperature. We show how shock-wave data (in some cases augmented with static data) can be used in conjunction with seismic models to provide lower mantle compositional constraints. Although shock-wave data have long been available (e.g., MCQUEEN et al., 1967), it is only recently that the full potential of the experimental techniques has been realized. The combination of results from several shock-wave experimental techniques, including pressure-volume, temperature, and sound velocity determinations, is sufficient to construct a thermodynamically complete description of the shocked state.

Sound velocity determinations along the Hugoniot also provide elasticity data that can be compared directly with seismic velocity profiles. Both longitudinal and shear elastic properties can be determined from shock-wave experiments, although shear moduli are less precise. BROWN and MCQUEEN (1986) noted that previous efforts to constrain composition of the earth's deep interior have frequently been limited by the differing nature of seismic and high-pressure laboratory data. In contrast to density-as-a-function-of-pressure determinations from the laboratory, seismology provides high-resolution elastic-wave velocity profiles. In addition, it is worth noting that experimentally determined Hugoniot states for silicates and oxides extend through the range of pressure-tempera-

High-Pressure Research in Mineral Physics, edited by M. H. Manghnani and Y. Syono, pp. 373–384.

ture conditions for the lower mantle (Figure 1). Thus, data necessary for application to the earth can be interpolated. Long extrapolations in pressure or temperature are not required for comparisons between laboratory and deep earth conditions.

An issue in the use of shock-wave data to address problems of lower mantle composition is the degree of equilibration obtained in these experiments. Silicates are shocked from low-pressure (low coordination) to high-pressure (high coordination) phases in microsecond time scales. Shock-wave equilibrium is discussed in the next section. Although the question is not fully resolved, we believe that the shock-wave data can, in fact, represent "equilibrium behavior." Since decomposition from olivine to silicate perovskite plus oxide remains a difficult process to reconcile with the time scales characteristic of a shock experiment, experiments conducted using high-pressure phases as starting materials should be undertaken.

In this paper we review the current knowledge of the compression behavior of magnesian olivine (Mg:(Mg+Fe)≥0.90) and its high-pressure decomposition products. Results of static and dynamic compression experiments are relatively consistent, although equation-of-state parameterizations remain uncertain. We also report preliminary sound velocities for synthetic forsterite and San Carlos peridot shocked into the high-pressure phase. These data support our previous suggestion that olivine melts near a Hugoniot pressure of 150 GPa. In addition, sound velocities below the inferred melting point match

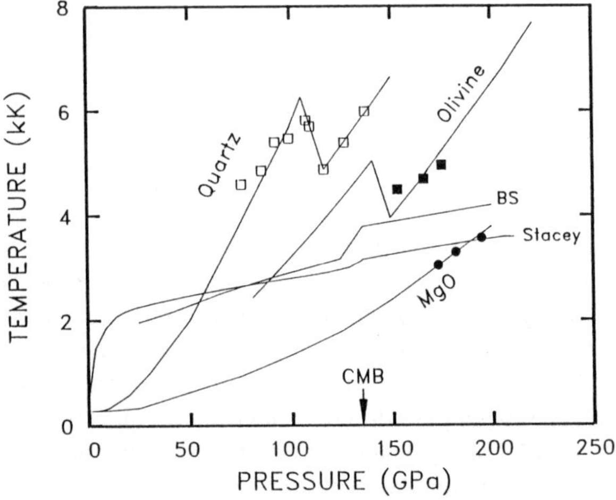

Figure 1. Comparison of model geotherms and pressure-temperature determinations on the Hugoniots of mantle constituents. Geotherms shown are from STACEY (1977) and BROWN and SHANKLAND (1981) (denoted respectively as "Stacey" and "BS"). Experimental temperature determinations for quartz and Mg_2SiO_4 olivine are from LYZENGA and AHRENS (1980), and for MgO, from SVENDSEN and AHRENS (1983). A hypothetical melting kink in the olivine Hugoniot is estimated in analogy with quartz.

mantle compressional wave velocity profiles. Above 150 GPa the measured velocity is consistent with a bulk sound velocity in a liquid having relatively large values for specific heat and the Grüneisen parameter. Since limited data have been obtained for only one silicate phase, conclusions concerning compositional constraints for the lower mantle remain premature. Further experiments for a variety of compositions are clearly necessary.

Thermodynamic Considerations

Although Hugoniot data have long been available for candidate mantle minerals (see e.g., MCQUEEN et al. (1967)), interpretations were limited by an incomplete characterization of the thermodynamic state created in shock experiments. Measurements of shock and particle velocities U_s and U_p constrain density ρ, pressure P, and internal energy E through the Rankine-Hugoniot conservation relationships. However, temperature calculations were previously based on assumed values for specific heat C_V and the Grüneisen parameter γ. With recent experimental innovations, these parameters can now, in fact, be directly observed. The conceptual framework for interpretation of shock-wave data is outlined here. We show that it is possible to combine standard shock data with direct Hugoniot temperature and sound velocity determinations to complete the thermodynamic description of a shocked material. These results, therefore, provide thermophysical data for minerals under lower mantle conditions of pressure and temperature.

The question of whether shock-wave experiments reflect equilibrium behavior of materials provides additional uncertainties in applications to the earth. BROWN and MCQUEEN (1986) reviewed recent efforts to understand the degree of thermodynamic "equilibration" in shocked materials. They identified separate issues of thermalization and metastability. In a thermalized material, temperature is a valid thermodynamic property. However, a thermally equilibrated sample may exist in a metastable state during the duration of a shock experiment. Metastability is a continuing concern in both static and dynamic experiments, and must be considered on a case by case basis. In contrast, theoretical and experimental studies suggest that thermalization can occur rapidly (in the thickness of a shock front) in materials subjected to strong shock waves.

Associated with an assortment of structural phases and large coordination changes at high pressure, the degree of metastability in shocked silicate minerals is of particular concern. However, that four-fold coordinated minerals like quartz and olivine can be shocked to states with densities appropriate for highly coordinated phases like stishovite or silicate perovskite has been established (MCQUEEN et al., 1967; JACKSON and AHRENS, 1979;

SYONO and GOTO, 1982; FURNISH and BROWN, 1986). Although no silicate structures have been determined in situ, pulsed X-ray diffraction studies on shocked boron nitride (JOHNSON and MITCHELL, 1972) showed that a graphite to wurtzite structure transition occurred during a shock-wave experiment. This established that a shock transition to an ordered high-pressure phase is possible. Thus, in the absence of contradictory evidence, we currently assume that Hugoniot data for silicates at very high pressure represent "equilibrium" behavior.

The standard empirical parameterization of Hugoniot data is as a series expansion of shock velocity as a function of particle velocity:

$$U_s = c_0 + sU_p + qU_p^2 \tag{1}$$

Data for most materials can be adequately represented by the linear form of (1). In a few cases a nonzero value for q is required. We show later that piecewise linear fits are sufficient for olivine. Equation (1) allows a one-independent-variable description of the Hugoniot equation of state when combined with the Rankine-Hugoniot relations:

$$\rho = \frac{\rho_0}{(1 - U_p/U_s)} \tag{2}$$

$$P - P_0 = \rho_0 U_s U_p \tag{3}$$

$$E - E_0 = (P + P_0)(V_0 - V)/2 \tag{4}$$

Clearly absent from equations (2)–(4) is temperature.

The two important auxiliary parameters for determination of the thermodynamic state of shocked materials are the constant volume heat capacity

$$C_V = \left(\frac{\partial E}{\partial T}\right)_V \tag{5}$$

and the Grüneisen parameter

$$\gamma = V\left(\frac{\partial P}{\partial E}\right)_V = \frac{\alpha K_T}{\rho C_V} \tag{6}$$

where α is the coefficient of thermal expansion and K_T is the isothermal bulk modulus.

In the absence of phase transitions, temperature increases on the Hugoniot according to the equation given by MCQUEEN (1964):

$$dT = -T\frac{\gamma}{V}\,dV + \frac{1}{2C_V}[(V_0 - V)dP + (P - P_0)dV] \tag{7}$$

where the differentials dP and dV are evaluated along the Hugoniot. The effect of phase transitions on Hugoniot temperatures is discussed by MCQUEEN et al. (1970) and CARTER et al (1975). BROWN and MCQUEEN (1982) note that at constant volume, the difference ΔT between temperatures on the metastable extension of the Hugoniot beyond a phase transition and temperatures for the high-pressure phase is approximately given by

$$\Delta T = -\frac{T\Delta S}{C_V} = -\frac{Q}{C_V} \tag{8}$$

where ΔS is the entropy change at the phase transition and Q is the latent heat of transition. Equation (8) is applicable only to small-volume-change phase transitions, since it ignores the energy associated with the volume change at a phase transition.

The bulk sound velocity on the Hugoniot is $v_B = (K_s/\rho)^{1/2}$, where K_s is the isentropic bulk modulus, which is given as

$$K_s = V\left[\frac{\gamma}{V}\left(\frac{P_H - P_0}{2}\right) - \left(\frac{dP}{dV}\right)_H\left(1 - \frac{\gamma}{V}\frac{V_0 - V}{2}\right)\right] \tag{9}$$

Equations (7) and (9) can be used to calculate expected Hugoniot temperatures and sound velocities for assumed values of C_V and γ. Alternatively, experimental determinations of γ and C_V are based on measurements of Hugoniot sound velocities and temperatures. The equations have been previously derived (MCQUEEN, 1964; SWENSON et al., 1986) and are briefly summarized in the present paper. Parameters γ and C_V can be written as

$$\gamma = 2\left(\frac{K_s - K_H}{P + K_H\left(1 - \frac{V_0}{V}\right)}\right) \tag{10}$$

$$C_V = \frac{K_s - K_H}{\gamma\rho T(\delta + \gamma)} \tag{11}$$

where $K_H = -V(dP/dV)_H$, $\delta = (\partial \ln T/\partial \ln V)_H$, and the subscript H implies evaluation along the Hugoniot. All quantities in these equations can be determined in shock-wave experiments. K_H, the pressure P, and the compression V/V_0 can be calculated using equations (1)–(4); K_s is determined by bulk sound velocity measurements, and the temperature T can be determined from graybody emission spectra from the shock front.

In summary, determination of the Grüneisen parameter requires measurements of both U_s-U_p relations and the isentropic bulk sound velocity. Specific heats require these constraints plus Hugoniot temperature measurements. Thus, a complete thermodynamic description of the shocked state involves measuring the three quantities U_s, T, and v_B as a function of U_p. LYZENGA and AHRENS

(1980) and AHRENS et al. (1982) described measurements of Hugoniot temperatures. The method for measuring Hugoniot sound velocities is described by MCQUEEN et al. (1982) and by BROWN and MCQUEEN (1986). SWENSON et al. (1986) extended the method to transparent samples.

Equation of State of $(Mg,Fe)_2SiO_4$

The polymorphs and decomposition products of $(Mg,Fe)_2SiO_4$ are clearly of geophysical importance because of their assumed abundance in the earth's mantle. However, their compression behavior under extreme pressure-temperature conditions has not been well established. In this section, we clarify what can be inferred of the equation of state from the available static, ultrasonic, and shock data. Furthermore, we establish that static or dynamic compression data for olivine are not easily differentiated to obtain elastic moduli.

A phase diagram featuring two major systems of transitions has been produced from static high-pressure-temperature experiments on $(Mg,Fe)_2SiO_4$ olivine (see ITO et al. (1984) for a summary). The system of transitions from the α (olivine) structure to the γ (spinel) structure, with the β (modified spinel) structure as an intermediate phase for magnesium-rich compositions, probably gives rise to the 400-km discontinuity. At greater pressures, iron-rich compositions transform to magnesiowustite plus stishovite, while magnesium-rich compositions transform to magnesiowustite plus $(Mg,Fe)SiO_3$ perovskite. No further transformations or decompositions have been observed with increasing pressures in magnesium-rich $(Mg,Fe)_2SiO_4$ compositions.

The available Hugoniot data for both pure forsterite and Fo_{90} olivine (Twin Sisters dunite and San Carlos peridot) are summarized in Figures 2 and 3. In both figures, Los Alamos data for dunite and polycrystalline synthetic forsterite have been corrected for their less than 1% initial porosity. The corrections (using equations (39), (4) and (5) of MCQUEEN et al. (1970)) increase shock and particle velocities by less than 1/2% and are insensitive to assumed values for the Grüneisen parameter. The initial

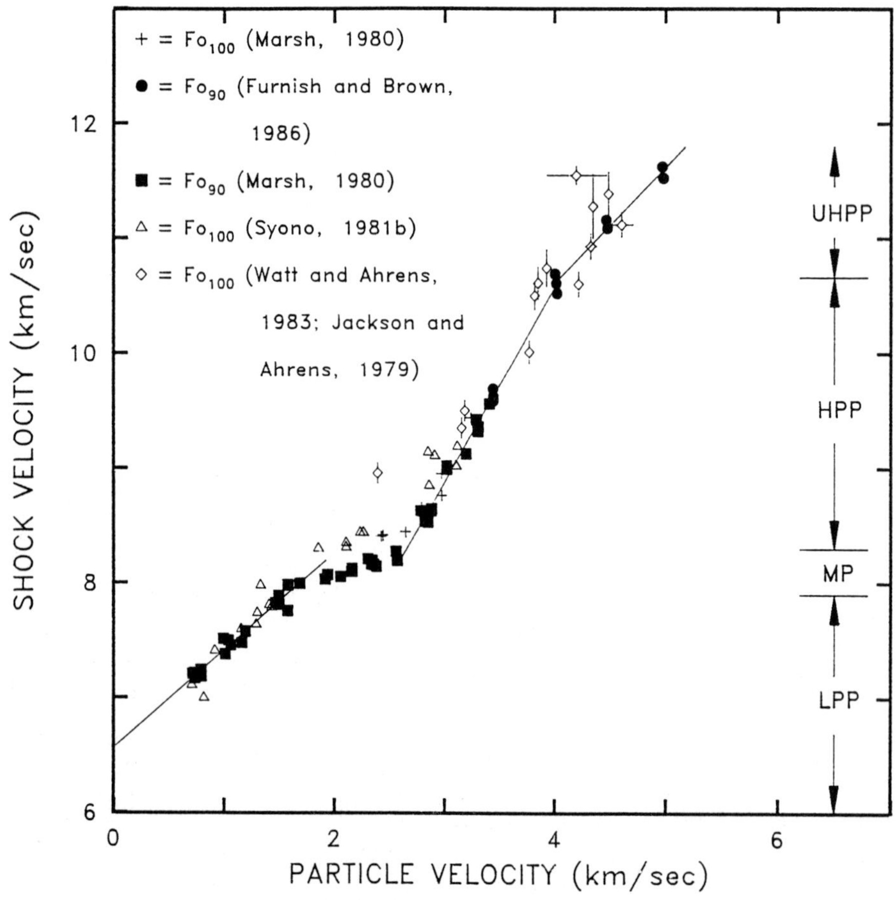

Figure 2. Shock velocities as a function of particle velocity for olivine. LPP: low-pressure phase; MP: mixed phase; HPP: high-pressure phase; UHPP: ultra-high-pressure phase.

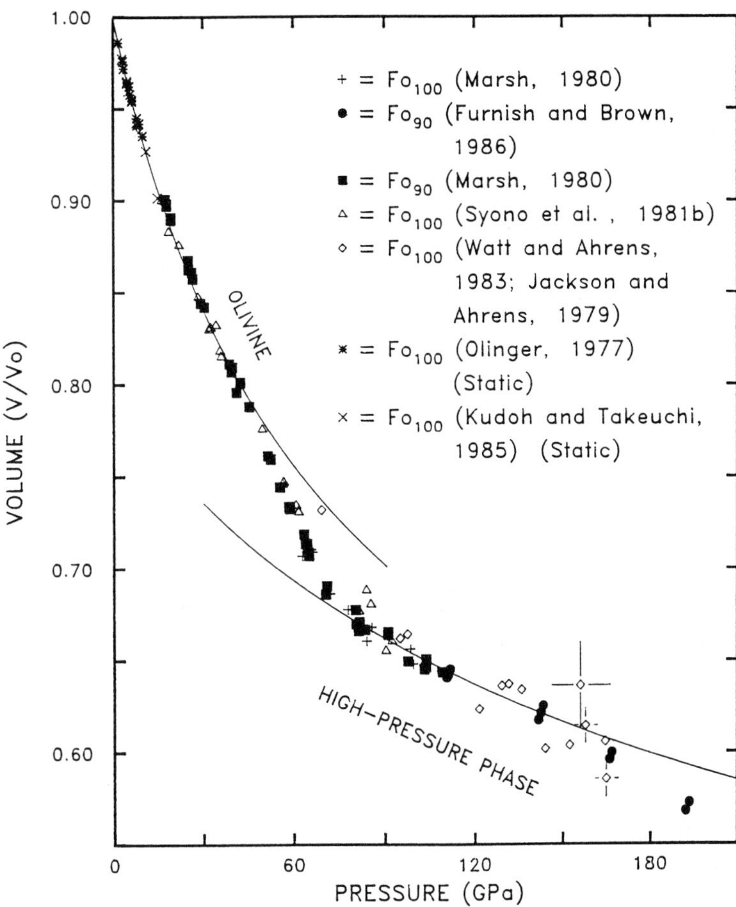

Figure 3. Compression as a function of pressure for olivine. Both static and shock-wave data are included. The solid curves are Hugoniots based on isothermal finite-strain equations of state.

density for the polycrystalline forsterite is 3.20 Mg m^{-3}, while Fo_{90} has a nominal density of 3.348 Mg m^{-3}. The lowest-pressure Los Alamos data for synthetic forsterite have been excluded. Those results were clearly biased by a strong elastic precursor wave.

Several distinct regions appear in the U_s-U_p plot of Figure 2. A low-pressure phase (olivine) persists to approximately U_p=1.75 km/s ($P \approx 50$ GPa), followed by a broad mixed phase region (lasting to U_p=2.9 km/s, or $P \approx 80$ GPa). Above 4.0 km/s ($P \approx 150$ GPa) a softening observed in the peridot Hugoniot (FURNISH and BROWN, 1986) possibly corresponds to the onset of melting. Piecewise linear fits, whose coefficients c_0 and s and ranges are listed in Table 1, are shown in Figure 2. The results of WATT and AHRENS (1983) and JACKSON and AHRENS (1979) indicated that the density of the high-pressure phase of forsterite depends on the orientation of the shocked crystal, although the experiments of FURNISH and BROWN (1986), designed to be sensitive to such an orientation dependence, revealed none for shocked peridot.

The Hugoniot data are shown as compression V/V_0 versus pressure in Figure 3. Static data have been added to this figure; they fall in a region where shock heating is small and the isotherm and Hugoniot are indistinguishable. The phase transition from olivine to the high-pressure phase is clearly delineated by the data. The apparent softening of the Hugoniot above 150 GPa is observed only in the highest-pressure data of FURNISH and BROWN (1986). Within the scatter of the data, increasing iron content does not appear to alter the shock-wave equation of state for magnesium-rich samples. Results from three laboratories using different experimental techniques are in agreement to within several percent. Details of the equation-of-state fits shown in Figure 3 are given later.

That the high-pressure phase of shocked olivine is silicate perovskite and magnesiowustite, as for statically

TABLE 1. Hugoniot Parameters for Magnesium-Rich Olivine

Segment	c_0 (km/s)	s	Range (km/s) U_p (min)	U_p (max)	Comments
LPP	6.57	0.85	0.00	1.75	Olivine
	±0.05	±0.05			(Low-Pressure Phase)
MP	—	—	1.75	2.90	Mixed-Phase Region
HPP	3.80	1.69	2.90	4.00	Perovskite + (Mg,Fe)O?
	±0.14	±0.01			(High-Pressure Phase)
UHPP	6.56	1.01	4.00	?	Solid + Liquid?
	±0.33	±0.08			(Ultra-High-Pressure Phase)

compressed samples, is strongly suggested by the work of SYONO et al. (1981a) and SYONO and GOTO (1982). They recovered samples of forsterite shocked to pressures of greater than 100 GPa and analyzed them by transmission electron microscopy, finding $MgSiO_3$ glass and MgO crystals. This conclusion must be drawn cautiously; a sample jacketed for recovery reaches maximum pressure through a process of shock reverberation, and remains significantly cooler (by 500–800 K in this experiment) than a sample taken to the corresponding pressure by a single-shock (Hugoniot) process. Several attempts have been made to identify the high-pressure phase on the Hugoniot by fitting predicted Hugoniots for the possible decomposed products of olivine to the experimental data (JACKSON and AHRENS, 1979; WATT and AHRENS, 1983; FURNISH and BROWN, 1986). The results are unfortunately nonunique and ambiguous. Hence, a significant uncertainty remains in the nature of the high-pressure phase produced by shocking magnesian olivine to pressures greater than 100 GPa. For our discussions, we assume here that it is a decomposed mixture of silicate perovskite and magnesiowustite, as produced by static experiments in this pressure range.

FURNISH et al. (1986) made high-resolution wave profile measurements in olivine shocked to about 19 GPa, appreciably above the equilibrium pressure for an olivine to β or γ spinel transition. They established that there is no indication of initiation of the olivine-spinel transition. That the transition must be overdriven in shock experiments is supported by the sluggish nature of the transition in static experiments. SUNG and BURNS (1970) and more recently FURNISH and BASSETT (1983) noted that temperatures in excess of 700 °C are necessary to overcome kinetic barriers for transition in magnesian olivine. At 50 GPa we calculate a Hugoniot temperature for olivine of less than 400 °C. Thus, we believe that the initiation of the phase transition for olivine in shock experiments is delayed for kinetic reasons by the very low Hugoniot temperatures in the low-pressure phase regime. As a result, olivine probably transforms directly to the post-spinel phases.

Finite-strain equation-of-state fits for the shock-wave data are shown in Figure 3. The low-pressure data are fit with the Birch-Murnagham relationship, corrected to the Hugoniot (see e.g., JACKSON and AHRENS (1979)), using the ultrasonic determination of K_{0s} of 129 GPa and an initial pressure derivative of the bulk modulus $K'=3.1$. The ultrasonic low-pressure value for K' of 5.2 leads to an unacceptable fit. The high-pressure data are fit using $\rho_{0HP}=4.09$ Mg m^{-3}, $K_0=231$ GPa, and $K_0'=4$ (see, for example, FURNISH and BROWN (1986)). In the following, we discuss the extent to which constraints can be placed on the bulk modulus and its pressure derivatives.

Constraints on the behavior of the bulk modulus are provided with the aid of the finite-strain parameterization proposed by BIRCH (1978), in which an Eulerian strain measure

$$f = \frac{1}{2}\left[\left(\frac{\rho}{\rho_0}\right)^{2/3} - 1\right] \tag{12}$$

is plotted against a normalized pressure

$$F = \frac{P}{3f(1+2f)^{5/2}} \tag{13}$$

The normalized pressure can be expressed as a polynomial in strain, where the bulk modulus and its derivatives at zero pressure are related to the expansion coefficients. Hence, polynomial fits to isothermal data yield values for K_0 and its pressure derivatives. The normalized pressure intercept at $f=0$ is equal to K_0, and the first derivative is $dF/df=-K_0(3/2)(4-K_0')$. Higher-order coefficients in the expansion involve higher derivatives of K_0. Thus, data along a linear trend with zero slope implies $K_0'=4$. Curvature in the trend determines the behavior of K_0''.

The shock-wave data for olivine below 50 GPa were transformed to the 300 K isotherm by calculating Hugoniot temperatures as a function of specific volume (Equation (7)), using the Hugoniot parameters in Table 1 and taking $C_V=3R$ (where R is the gas constant) and $\rho\gamma=4.02$ Mg m^{-3} ($\gamma_0=1.2$ (BOEHLER, 1982)). Thermal pressures were subtracted from the Hugoniot pressure at constant volume. The magnitude of the thermal correction is small (less than 1 GPa at 50 GPa) and is relatively insensitive to assumptions for γ or C_V. Reduced shock-wave data are shown together with static compression data in Figure 4a. Ultrasonic values for K_{0s} and K' at 1 atm and 300 K of 129 GPa and 5.2, respectively (SUMINO and ANDERSON, 1984), are indicated by the short sloped line on the left side of the figure. The lowest-pressure shock-wave data are subject to error associated with the appreciable rise time characteristic of shocks in this pressure range. Time-resolved studies of wave structure in single-crystal peridot show that this rise time to be about 10% of the total transit time in 3-mm-thick crystals shocked to 19 GPa (FURNISH et al., 1986). It is unknown where on the wave front the flash gaps or shorting pins record an arrival. A worst-case overestimate of pressure of about 3 GPa, or of F of about 20 GPa, is possible. The actual systematic error is probably smaller. Furthermore, higher-pressure data will not be significantly affected by this; rise times shorten rapidly with increasing shock strength.

Because of the scatter of the compression data, a detailed evaluation of the behavior of the bulk modulus is not possible. However, general trends can be seen. The shock-wave data lie along a line with a distinctly negative

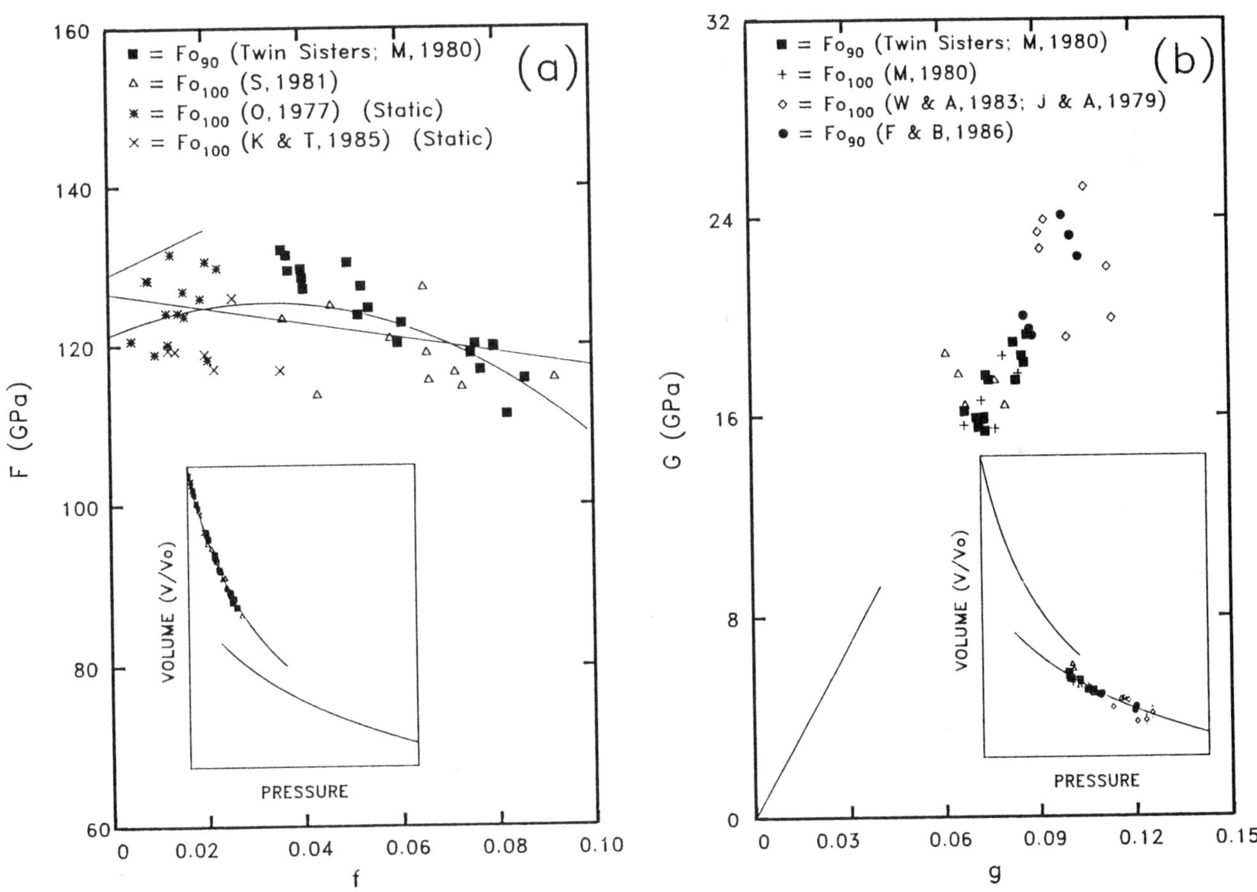

Figure 4. Equation-of-state parameterizations for olivine and its high-pressure phases. a) Isothermal F versus f data for olivine. The inset shows points used, and can be compared with Figure 3. The short line at left indicates ultrasonic measurements of $K_{0S}=129$ GPa and $K_{0S}'=5.2$. Abbreviations: M: MARSH (1980), S: SYONO et al. (1981b), O: OLINGER (1977), K & T: KUDOH and TAKEUCHI (1985). b) Isothermal G versus g for olivine. Density $\rho^*=4.09$ Mg m^{-3}. The slanting line from origin, corresponding to $K_0=231$ GPa and $K_0'=4$, was calculated by combining data for perovskite and magnesiowustite (see FURNISH and BROWN, 1986). Abbreviations: M: MARSH (1980), W & A: WATT and AHRENS (1983), J & A: JACKSON and AHRENS (1979), F & B: FURNISH and BROWN (1986).

slope ($K'<4$). A linear fit through all data in Figure 4a predicts $K_0=127$ GPa and $K'=3.5$. These values nearly duplicate the curve shown for low pressures in Figure 3, following the data to pressures of about 50 GPa. A quadratic fit (also shown in Figure 4a) corresponds to $K_0=121$ GPa, $K'=5.35$, and $K''=-0.23$ GPa^{-1}. This fit gives better agreement with the ultrasonic determinations of K' at the price of a large difference in K_0 values. The pressure-volume behavior predicted by these values closely matches the observed behavior to pressures greater than 60 GPa. However, this match through the mixed-phase region (>50 GPa) is fortuitous.

For the high-pressure phase, the initial density $\rho_{0(HPP)}$ is not strictly known. Thus, the previous parameterization is not appropriate. We use a second parameterization (JEANLOZ, 1981), which avoids the problem by expressing effective strain in terms of a reference density ρ^*:

$$g = \frac{1}{2}\left[\left(\frac{\rho_2}{\rho^*}\right)^{2/3} - 1\right] \tag{14}$$

where subscript 2 refers to the high-pressure phase. Normalized pressure is again given in the form

$$G = \frac{P}{3(1 + 2g)^{5/2}} \tag{15}$$

The direct interpretation of the coefficients of the polynomial expansion of $G(g)$ is straightforward, but complicated. Jeanloz expanded G in a Taylor series about g_0, where $G(g_0)=0$. Since $G=0$ corresponds to $P=0$, the zero-pressure density of the high-pressure phase ρ_{02} is given by setting $g=g_0$ in (14). Writing $\alpha=(\rho^*/\rho_{02})^{1/3}$, the intercept of $G(g)$ at $g=g_0$ is zero, the first derivative is $\alpha^7 K_{02}$, and the second derivative is $-\alpha^9 K_{02}(3/2)(4-K_{02}')$. Thus, for a fit of given order, one less derivative of the bulk modulus is available than with the F-f parameterization of BIRCH (1978) because the zero-pressure density is *a priori* unknown. Deviations of K_0' from 4, for example, are associated with nonlinear trends in the G versus g data.

High-pressure-phase data for shocked olivine, reduced to the 300 K isotherm, are shown in the G-g parameterization in Figure 4b. Data above 150 GPa are excluded since they may lie in a solid-liquid mixed-phase regime (see "Sound Velocities and..."). Hugoniot temperatures for the high-pressure phase were calculated using coefficients from Table 1, $C_V=3R$, and $\rho\gamma=6.06$ Mg m^{-3}. We initiated the temperature calculations in the high-pressure phase at 70 GPa at a temperature of 2000 K. With this choice the calculated Hugoniot temperatures are consistent with those measured by LYZENGA and AHRENS (1980). The calculated temperatures are approximately 1500 K higher than the measured values to allow for a latent heat of melting, where ΔS for melting is assumed to be equal to the gas constant R. It should be noted that the results of the finite-strain analysis depend only weakly on our assumptions for the Hugoniot temperature calculations. A change of 2000 K in Hugoniot temperatures changes G by 12%. Reduction to the isotherm was accomplished using a value of $\rho\gamma=6.06$ Mg m^{-3}. In contrast with Figure 4a, thermal corrections are significant in Figure 4b. Thermal pressures are as large as 44 GPa. Variations in $\rho\gamma$ of 20% lead to uncertainties in G of 5%. The value of $\rho^*=4.09$ Mg m^{-3} is an estimated value for the zero-pressure density of the $MSiO_3$ (pv)+MO (mw) high-pressure assemblage (with $M=(Mg_{0.9}Fe_{0.1})$) (e.g., JEANLOZ and THOMPSON, 1983). This gives $\alpha\approx1$ and $g_0=0$. The line in the lower left corresponds to estimated room pressure-temperature characteristics (see FURNISH and BROWN, 1986) of $K_{02}=231$ GPa and $K_{02}'=4$. Within the uncertainties in our reduction of Hugoniot data to an isotherm, we find the estimate acceptable. No deviation of K' from 4 is resolvable.

These parameterizations make the scatter in the data and the uncertainty in the derived modulus values from compression studies apparent. It is therefore difficult to differentiate pressure-volume data to obtain reliable values of bulk moduli. However, we have shown that ultrasonic data and the shock and static compressional data are in at least general agreement. For purposes of matching seismological data, a more direct measurement of K_s through measuring sound velocities is needed, as discussed in the next section.

Sound Velocities and Temperatures for Shocked Olivine

The method for measuring sound velocities in shocked materials has been described in detail elsewhere (MCQUEEN et al., 1982; BROWN and MCQUEEN, 1986; SWENSON et al., 1986). In a shocked solid, the fastest pressure release wave travels at the longitudinal sound velocity, $v_P=\sqrt{(K_s+(4/3)\mu)/\rho}$, where μ is the shear modulus. A second wave propagates through the shocked sample at the bulk sound velocity, $v_B=\sqrt{K_s/\rho}$. These two elastic wave velocities are related through Poisson's ratio σ:

$$\left(\frac{v_P}{v_B}\right)^2 = 3\left(\frac{1-\sigma}{1+\sigma}\right) \tag{16}$$

In the liquid, only one sound wave, traveling at the bulk sound velocity, can propagate. Hence melting on the Hugoniot is characterized by a discontinuous decrease in elastic wave velocities (ASAY and HAYES, 1975; SHANER et al., 1984; BROWN and SHANER, 1984; BROWN and MCQUEEN, 1986).

Preliminary sound velocities for forsterite and San Carlos peridot shocked into the high-pressure phase regime are shown in Figure 5 as a function of Hugoniot pressure. The synthetic forsterite, with initial porosity of less than 1%, is from the same source as the material reported in MARSH (1980). The two lowest-pressure data may not completely represent behavior of the high-pressure phase since they lie at the upper end of the mixed-phase regime. Current results were obtained with less than optimum quality in the experimental records. Although we have previously obtained data with better than 1% uncertainties, the present data may have uncertainties in excess of 2%. Using better experimental designs, we expect to improve the quality of additional data. In Figure 5 we also plot estimated velocities, corrected to a nominal composition of Fo_{90} using velocity-mean-atomic-weight systematics (SHANKLAND, 1972),

$$v_{Fo_{90}} = v_{Fo_{100}} \left(\frac{\bar{m}_{Fo_{100}}}{\bar{m}_{Fo_{90}}} \right)^{1/2} \tag{17}$$

where $\bar{m}$ is the mean atomic weight.

Below 150 GPa, measured sound velocities are within 8% of the compressional wave velocity profile for the lower mantle. Thus, the high-pressure phase generated in shock-wave experiments appears to be solid. Presently we have data for only one experiment above 150 GPa. This point at 168 GPa, obtained using single-crystal San Carlos peridot, lies at a significantly lower sound velocity than the results for pure forsterite and lies below the trend of bulk sound velocities for the lower mantle. This apparent loss of rigidity qualitatively supports our previous suggestion (FURNISH and BROWN, 1986) that the high-pressure phase of olivine melts near 150 GPa on the Hugoniot. However, interpretation of this apparent Hugoniot melting region is complicated since the equilibrium high-pressure assemblage in our experiments is a multiphase system. Rigidity is probably lost with melting of the lowest melting point phase.

Under the assumption that the high-pressure data represent a bulk sound wave, the isentropic bulk modulus for the liquid high-pressure phase at 168 GPa is 621 GPa.

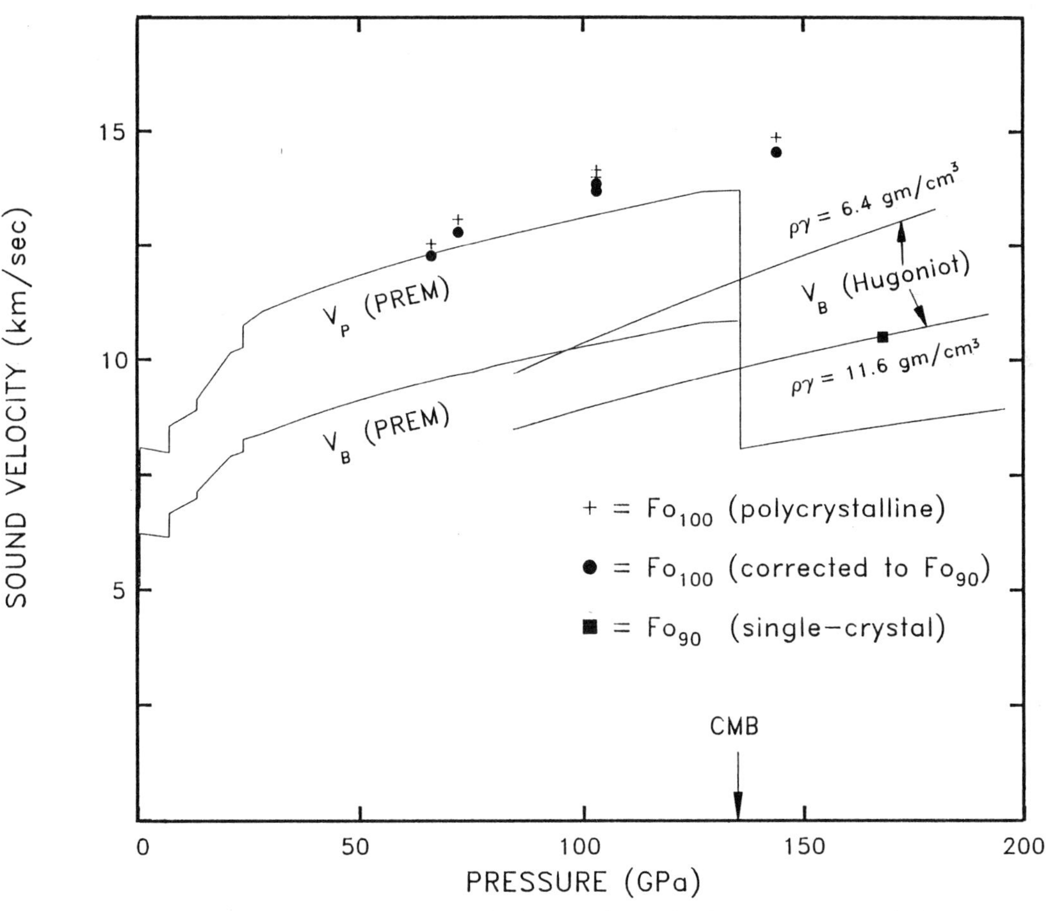

Figure 5. Sound velocity as a function of density for forsterite and for single-crystal Fo_{90}. Seismic model PREM (DZIEWONSKI and ANDERSON, 1981) is shown for comparison.

In contrast, using the Hugoniot coefficients from Table 1 for this pressure regime, the Hugoniot bulk modulus K_H at 168 GPa is 590 GPa. This implies that the Hugoniot is more compressible than the isentrope, whereas K_H should be approximately twice as large as K_s. However, if the highest-pressure data shown in Figures 2 and 3, as well as in Figure 5, are in a mixed-phase (solid-liquid) regime, the Hugoniot bulk modulus reflects both compression and the effects of phase transitions. Therefore, in the absence of better constraints, we use only the (solid) high-pressure phase Hugoniot coefficients from Table 1 to interpret all Hugoniot results above 80 GPa. Still higher-pressure data (above the "solid-liquid" mixed phase regime) must be obtained in order to constrain the "liquid" regime.

Two calculated bulk sound velocity profiles are shown in Figure 5. Both are based on equation (9) with the assumption that $\rho\gamma$ is constant. Using (10), our measured bulk sound velocity at 168 GPa, and Hugoniot parameters from Table 1, we calculate a Grüneisen parameter γ of 2.1 ± 0.1 at 168 GPa for the high-pressure phase of olivine. The uncertainty in this value is based on uncertainties given for the coefficients in Table 1. Since this value is based on the "solid" Hugoniot extrapolated into a "mixed-phase" regime and since the "liquid" equation of state remains uncertain, even larger uncertainties in the Grüneisen parameter exist. For an initial density of 4.09 Mg m^{-3}, our value implies a zero-pressure Grüneisen parameter of 2.8 for the high-pressure (liquid) silicate phase of olivine. In contrast, olivine has $\gamma=1.2$ (BOEHLER, 1982). However, both higher coordination and melting are likely to increase the Grüneisen parameter. For comparison, the second calculated bulk sound velocity curve in Figure 5 is based on the arbitrary choice of $\gamma_0=1.5$. This higher-velocity curve is consistent with bulk sound velocities in the lower mantle. Unfortunately, existing data are insufficient to determine the variation of the Grüneisen parameter with pressure. To preserve a subparallel trend between bulk and compressional velocities, it appears that $\rho\gamma$ increases modestly (approximately 30%) with pressure along the solid portion of the Hugoniot. In analogy with other shocked solids, we expect Poisson's ratio to increase only slowly along the solid Hugoniot (e.g., BROWN and MCQUEEN, 1986). In addition, we believe that the bulk sound velocity profile calculated using the smaller value of γ is more appropriate because the offset between compressional velocities and the lower bulk velocity profile implies an implausibly large shear modulus, comparable to the bulk modulus.

With regard to Hugoniot temperatures, LYZENGA and AHRENS (1980) reported three temperature determinations for shocked forsterite. These data lie near 150 GPa and unfortunately span only a modest range of pressure and volume. This causes large uncertainties in the derivative δ used in equation (11) to determine the specific heat. As a result, our estimate for C_V at 168 GPa of 47 J K^{-1} g $atom^{-1}$ may be uncertain by factors of 2 or more. Converted to units of the gas constant R, the specific heat is $5.7R$, where $3R$ is expected for an insulator above the Debye temperature. One explanation for the anomalously high calculated specific heat is that it includes the latent heat of transition. Furthermore, in analogy to results for CsI (SWENSON et al., 1986), excess specific heat contributions from electrons excited across band gaps may also explain our results at these high temperatures.

Comparisons with the Lower Mantle

Compressional and bulk sound velocities from seismic model PREM (DZIEWONSKI and ANDERSON, 1981) are shown in Figure 5. Sound velocities for forsterite between 70 GPa and 150 GPa on the Hugoniot are within 8% of the seismic compressional wave velocity profile. When corrected to an Fo_{90} composition using equation (17), they match properties of the lower mantle to within 5%. Since the Hugoniot temperature profile in olivine (Figure 1) crosses the geotherm in the lower mantle, relatively small thermal corrections are necessary for direct comparison between seismic observations and laboratory data. Unfortunately, thermal corrections for the compressional velocities are difficult to estimate since we currently have no direct measure of the temperature dependence of shear elastic constants. However, the temperature dependence at constant density of the bulk sound velocity is small. We calculate a maximum thermal correction of less than 3% for v_B between Hugoniot temperatures and mantle conditions. Such differences are nearly negligible.

As previously discussed, bulk sound velocities for the solid high-pressure phase of olivine are not well constrained. However, we noted that the bulk sound velocity curve based on $\gamma_0=1.5$ provided a relatively plausible description of elastic behavior for the high-pressure phase. Therefore, both bulk and shear elastic moduli of the high-pressure phase of olivine are similar to properties of the lower mantle.

Conclusions

An integrated shock-wave experimental approach to the study of candidate mantle constituents has been presented. We reiterated that a combination of measurements in shocked samples (including shock velocities, sound velocities, and temperature) can provide a thermodynamically complete description of the high-pressure state. Furthermore, elastic moduli are directly determined. Such results are clearly needed to interpret the increasingly high-resolution seismic profiles for the mantle. Since Hugoniot temperatures for mantle minerals cross the range of predicted mantle geotherms, only short interpolations from experimental conditions are needed to establish material properties under mantle conditions.

Pressure-volume data for olivine and its high-pressure decomposition products were reviewed. Using finite-strain analysis, an encouraging agreement was found between static and dynamic results for olivine is encouraging. However, current data for the high-pressure phases span a limited compression range and cannot be used to closely constrain elastic behavior.

Sound velocity determinations for forsterite and San Carlos olivine support our previous suggestion that the high-pressure phase of olivine melts near 150 GPa. Our measurements also indicate that the high-pressure phase of olivine has compressional velocities about 5% greater than seismic profiles for the lower mantle. We currently have insufficient data to constrain the bulk velocity of the solid high-pressure phase, and have no data for other compositions. Further experimental measurements of the compressional and bulk velocities for the major mantle constituents under near-geotherm conditions must be undertaken. On the basis of our preliminary sound velocity data, the Grüneisen parameter for "liquid" olivine at 168 GPa is 2.1. Our estimate of $5.7R$ for the specific heat at 168 GPa is based on limited temperature determinations and uncertain values of the Grüneisen parameter. The large value for both thermodynamic quantities may be associated with a mixed-phase regime near the inferred melting point.

Acknowledgments. This work was supported by the National Science Foundation through grant EAR-8509463 and by the Department of Energy.

References

Ahrens, T. J., G. A. Lyzenga, and A. C. Mitchell, Temperatures induced by shock waves in minerals: Applications to geophysics, in *High-Pressure Research in Geophysics*, vol. 12, edited by S. Akimoto and M. H. Manghnani, pp. 579–594, Center for Academic Publications, Tokyo, 1982.

Anderson, D. L., and J. D. Bass, Mineralogy and composition of the upper mantle, *Geophys. Res. Lett., 11*, 637–640, 1984.

Asay, J. R., and D. B. Hayes, Shock-compression and release behavior near melt states in aluminum, *J. Appl. Phys., 46*, 4789–4799, 1975.

Bass, J. D., and D. L. Anderson, Composition of the upper mantle: Geophysical test of two petrological models, *Geophys. Res. Lett., 11*, 229–232, 1984.

Birch, F., Finite strain isotherm and velocities for single-crystal and polycrystalline NaCl at high pressures and 300 degrees K, *J. Geophys. Res., 83*, 1257–1268, 1978.

Boehler, R., Adiabats of quartz, coesite, olivine, and magnesium oxide to 50 kbar and 1000 K, and the adiabatic gradient in the earth's mantle, *J. Geophys. Res., 87*, 5501–5506, 1982.

Brown, J. M., and R. G. McQueen, The equation of state for iron and the earth's core, in *High-Pressure Research in Geophysics*, vol. 12, edited by S. Akimoto and M. H. Manghnani, pp. 611–623, Center for Academic Publications, Tokyo, 1982.

Brown, J. M., and R. G. McQueen, Phase transitions, Grüneisen parameter, and elasticity for shocked iron between 77 GPa and 400 GPa, *J. Geophys. Res., 91*, 7485–7494, 1986.

Brown, J. M., and J. W. Shaner, Rarefaction velocities in shocked tantalum and the high pressure melting point, in *Shock Waves in Condensed Matter*, edited by J. R. Asay, R. A. Graham, and G. K. Straub, pp. 91–94, Elsevier, New York, 1984.

Brown, J. M., and T. J. Shankland, Thermodynamic parameters in the earth as determined from seismic profiles, *Geophys. J. R. Astron. Soc., 66*, 579–596, 1981.

Carter, W. J., J. N. Fritz, S. P. Marsh, and R. G. McQueen, Hugoniot equation of state of the lanthanides, *J. Phys. Chem. Solids, 36*, 741–752, 1975.

Dziewonski, A. M., and D. L. Anderson, Preliminary reference earth model, *Phys. Earth Planet. Inter., 25*, 297–356, 1981.

Furnish, M. D., and W. A. Bassett, Investigation of the mechanism of the olivine spinel transition in fayalite by synchrotron radiation, *J. Geophys. Res., 88*, 10333–10341, 1983.

Furnish, M. D., and J. M. Brown, Shock loading of single-crystal olivine in the 100–200 GPa range, *J. Geophys. Res., 91*, 4723–4729, 1986.

Furnish, M. D., D. Grady, and J. M. Brown, Analysis of shock wave structure in single-crystal olivine using VISAR, in *Shock Waves in Condensed Matter*, edited by Y. Gupta, pp. 595–600, North Holland, New York, 1986.

Hager, B. H., R. W. Clayton, M. A. Richards, R. P. Comer, and A. M. Dziewonski, Lower mantle heterogeneity, dynamic topography and the geoid, *Nature, 313*, 541–545, 1985.

Ito, E., E. Takahashi, and Y. Matsui, The mineralogy and chemistry of the lower mantle: An implication of the ultra high-pressure phase relations in the system MgO–FeO–SiO_2, *Earth Planet. Sci. Lett., 67*, 238–248, 1984.

Jackson, I., and T. J. Ahrens, Shock wave compression of single crystal forsterite, *J. Geophys. Res., 84*, 3039–3048, 1979.

Jeanloz, R., Finite strain equation of state for high pressure phases, *Geophys. Res. Lett., 8*, 1219–1222, 1981.

Jeanloz, R., and A. B. Thompson, Phase transitions and mantle discontinuities, *J. Geophys. Res., 88*, 51–74, 1983.

Johnson, Q., and A. C. Mitchell, First X-ray diffraction evidence for a phase transition during shock compression, *Phys. Rev. Lett., 29*, 1369–1371, 1972.

Kudoh, Y., and Y. Takeuchi, The crystal structure of forsterite Mg_2SiO_4 under high pressure up to 149 kb, *Z. Kristallogr., 171*, 291–302, 1985.

Lay, T., and D. V. Helmberger, The shear-wave velocity gradient at the base of the mantle, *J. Geophys. Res., 88*, 8160–8170, 1983.

Liu, L., Orthorhombic perovskite phases observed in olivine, pyroxene, and garnet at high pressures and temperatures, *Phys. Earth Planet. Inter., 11*, 289–298, 1976.

Liu, L., High pressure phase transitions in the MgO-SiO_2 system: Implications for mantle discontinuities, *Phys. Earth Planet. Inter., 19*, 319–330, 1979.

Lyzenga, G. A., and T. J. Ahrens, Shock temperature measurements in Mg_2SiO_4 and SiO_2 at high pressures, *Geophys. Res. Lett., 7*, 141–144, 1980.

Marsh, S. P., *Shock Hugoniot Data*, 658 pp., University of California Press, Los Angeles, 1980.

McQueen, R. G., Laboratory techniques for very high pressures and the behavior of metals under dynamic loading in *Conference on Metallurgy at High Pressure*, vol. 22, edited by K. A. Gschneider, M. T. Hepworth, and N.A. D. Parlee, pp. 44–132, Gordon and Breach, New York, 1964.

McQueen, R. G., S. P. Marsh, and J. N. Fritz, Hugoniot equation of state of twelve rocks, *J. Geophys. Res., 72*, 4999–5036, 1967.

McQueen, R. G., S. P. Marsh, J. W. Taylor, J. N. Fritz, and W. J. Carter, The equation of state of solids from shock wave studies, in *High-Velocity Impact Phenomena*, edited by R. Kinslow, pp. 244–419,

Academic Press, New York, 1970.

McQueen, R. G., J. W. Hopson, and J. N. Fritz, Optical technique for determining rarefaction wave velocities at very high pressures, *Rev. Sci. Instrum., 53*, 245–250, 1982.

Olinger, B., Compression studies of forsterite (Mg_2SiO_4) and enstatite ($MgSiO_3$), in *High-Pressure Research: Application in Geophysics*, edited by M. H. Manghnani and S. Akimoto, pp. 325–334, Academic Press, New York, 1977.

Shaner, J. W., J. M. Brown, and R. G. McQueen, Melting of metals above 100 GPa, in *High-Pressure in Science and Technology*, vol. 22, edited by C. Homan, R. K. MacCrone, and E. Whalley, pp. 137–141, North Holland, New York, 1984.

Shankland, T. J., Velocity-density systematics: Derivation from Debye theory and the effect of ionic size, *J. Geophys. Res., 77*, 3750–3758, 1972.

Stacey, F. D., A thermal model of the earth, *Phys. Earth Planet. Inter., 15*, 341–348, 1977.

Stacey, F. D., and D. E. Loper, Thermal histories of the core and mantle, *Phys. Earth Planet. Inter., 36*, 99–115, 1984.

Sumino, Y., and O. L. Anderson, Elastic constants of minerals, in *CRC Handbook of Physical Properties of Rocks*, vol. 3, edited by R. S. Carmichael, pp. 39–138, CRC Press, Boca Raton, Florida, 1984.

Sung, C. M., and R. G. Burns, Kinetics of the olivine-spinel transition: Implications to deep-focus earthquake genesis, *Earth Planet. Sci. Lett., 32*, 165-170, 1976.

Svendsen, R., and T. J. Ahrens, Shock-induced radiation and temperatures of MgO, *Eos Trans. AGU, 64*, 848, 1983.

Swenson, C. A., J. M. Brown, and J. W. Shaner, Hugoniot sound velocity measurements on CsI and the insulator-to-metal transition, *Phys. Rev. B, 34*, 7924–7934, 1986.

Syono, Y., and T. Goto, Behavior of single-crystal forsterite under dynamic compression, in *High-Pressure Research in Geophysics*, vol. 12, edited by S. Akimoto and M. H. Manghnani, pp. 563–578, Center for Academic Publications, Tokyo, 1982.

Syono, Y., T. Goto, H. Takei, M. Tokanami, and K. Nobugai, Dissociation reaction in forsterite under shock compression, *Science, 214*, 177–179, 1981a.

Syono, Y., T. Goto, J. Sato, and H. Takei, Shock compression measurements of single-crystal forsterite in the pressure range 15–93 GPa, *J. Geophys. Res., 86*, 6181–6186, 1981b.

Watt, J. P., and T. J. Ahrens, Shock compression of single crystal forsterite, *J. Geophys. Res., 88*, 9500–9512, 1983.

SHOCK-INDUCED PHASE TRANSITIONS IN RUTILE SINGLE CRYSTAL

Yasuhiko SYONO, Keiji KUSABA, Masae KIKUCHI, and Kiyoto FUKUOKA

The Research Institute for Iron, Steel and Other Metals, Tohoku University*
Katahira, Sendai 980, Japan

Tsuneaki GOTO

Institute for Solid State Physics, University of Tokyo
Roppongi, Minato-ku, Tokyo 106, Japan

Abstract. Shock compression experiments on single crystal rutile, TiO_2, were carried to pressure of 123 GPa. Flyer plates, gun-launched to speeds of 4.4 km/s, were used as impactors to induce shock waves. Shock and free surface velocities were measured using streak photography. Pressures and volumes were calculated from the measured shock velocities and particle velocities using conservation relations. Hugoniot elastic limits (HEL) were measured to be 6.2, 6.8, and 7.0 GPa for shock loading along the [100], [110], and [001] directions, respectively. The shock compression data for the (lower pressure) rutile phase was in agreement with static compression, as well as ultrasonic, measurements. Phase transition pressures observed depended strongly on the shock propagation direction, i.e., 13.7±0.3, 16.9±1.8, and 33.8±0.3 GPa for [100], [110], and [001] respectively. A new high-pressure phase (HPP-I), which showed a different pressure-volume trend from that previously reported by MCQUEEN et al. (1967) and AL'TSHULER et al. (1973) (HPP-II), was found between 70 and 100 GPa. The estimated zero-pressure density of HPP-I was 5.00 g/cm^3, 17.6 percent denser than the rutile form. This volume change suggests the phase transition from rutile to fluorite-like phase, consistent with the model, which could explain the observed anisotropy in phase transition pressures. Release adiabat measurements clearly distinguished HPP-II from HPP-I. An extremely high density of HPP-II can be ascribed to the change in bond nature itself at high pressures.

Introduction

High-pressure behavior of rutile, TiO_2, has attracted the attention of geophysicists because it is a crystal-chemical analog of stishovite (SiO_2) and may demonstrate polymorphism which could indicate the occurrence of a post-stishovite phase of SiO_2. Such a post-stishovite phase would have considerable geophysical significance. Rutile is, of course, a representative of MX_2 compounds from the crystal-chemical viewpoint. However, the nature of the high-pressure phase transformations in TiO_2 is still not clear, in spite of previous extensive static and dynamic studies.

Shock-compression Hugoniot measurements were carried out on the single crystals of unspecified orientations with the argon flash gap method by MCQUEEN et al. (1967) in the US and on polycrystalline specimens with the pin-contactor method by AL'TSHULER et al. (1973) in the USSR for pressure range higher than 100 GPa, and a phase transition accompanied by a large volume decrease, about 30 percent, was reported. Shock-recovered products yield a high pressure polymorph with the α-PbO_2 structure (MCQUEEN et al., 1967; LINDE and DECARLI, 1969). These products have a density only 3 percent greater than that of rutile. This suggests that the unusually high density phase observed in Hugoniot measurements was not quenched. More recent measurements by GOTO et al. (unpublished) with the inclined mirror method and by MASHIMO and SAWAOKA (1980) and MASHIMO et al. (1983) with the electromagnetic gage technique as well as the inclined mirror method suggested that apparent discrepancy between the observed phase transition pressures, i.e., about 34 GPa by MCQUEEN et al.(1967) and 20 GPa by AL'TSHULER et al. (1973), may be attributed to anisotropy of the phase transition.

On the other hand, in situ X-ray diffraction (JAMIESON and OLINGER, 1968) and Raman spectra (MAMMONE and SHARMA, 1979) studies under static high pressure initially indicated a phase transition to the α-PbO_2 structure around 7 GPa. Phase transitions corresponding to the dynamically observed ones were noted to occur above 18–25 GPa by in situ X-ray diffraction studies using a diamond-anvil cell with the aid of laser heating technique (LIU, 1975; 1978). However, the structure of the high-pressure phase was not clear. As in the case of the dynamic recovery experiments, the only high-pressure phase recovered was the α-PbO_2 phase (BENDELIANI et al., 1966; DACHILLE et al., 1968).

In the present study precise remeasurements of the shock-compression adiabat (Hugoniot) of single crystals of rutile were planned. Of special interest was the unusual high density phase induced by shock loading. Special attention was also paid to the dependence of the shock-induced phase transition with respect to the shock-propagation direction in the single crystal.

*Now Institute for Materials Research, Tohoku University.

High-Pressure Research in Mineral Physics, edited by M. H. Manghnani and Y. Syono, pp. 385–392.

Experimental Procedure

Large transparent single crystals of rutile grown by Bernouille method were obtained from Nakazumi Crystals Co. and used for shock loading experiments. Tetragonal unit cell dimensions of rutile were determined to be a=4.5929 (3) Å, and c=2.9581 (3) Å, yielding an X-ray density of 4.253 g/cm^3. The bulk density determined by the Archimedean method was 4.248 g/cm^3, in good agreement with the X-ray density. Platelet specimens of 14–20×10 mm in lateral dimension and about 3 mm in thickness were cut out parallel to (100), (110) or (001). Platelets were polished on both parallel surfaces to a tolerance of 2 μm and mounted on 1 mm thick driver plate of copper and tungsten.

Impact experiments were performed using a two-stage light-gas gun (20 mm bore and 10 m length) or a propellant gun (25 mm bore and 4 m length) for projectile velocity ranges up to 4.36 km/s. The condition of symmetrical impacts was fulfilled by using a flyer plate made of the same material as the driver plate. Flyer plates, of thickness 2 and 1 mm for copper and tungsten respectively, were glued on the front surface of the plastic projectile. Velocity of the projectile was determined with a precision of 0.1 percent by measuring the time interval between two coils placed ahead of the gun muzzle with the aid of a small magnet embedded in it.

Hugoniot measurements were carried out by a conventional optical method using a rotating-mirror type streak camera (GOTO and SYONO, 1984). A cross section of the experimental assembly is shown in Figure 1. The front surface of the specimen assembly was illuminated by an intense xenon flash lamp with the aid of an expendable 45° mirror and reflected light was introduced into the slit of the streak camera. Figure 2 shows a typical example of a streak photograph. Arrival of shock waves was detected by observing a sudden decrease in reflectivity of small flat mirrors placed on the surfaces of the specimen and the driver plate. Shock velocity was determined from specimen thickness and travel time of the shock wave across the specimen. To determine the free surface velocity an inclined mirror with a small angle α was placed on the specimen surface. The measured angle, γ, of the slanted image of the inclined mirror yielded the free surface velocity from $u_{fs}= W \tan \alpha / M \tan \gamma$, where W and M are the sweeping speed of the slit image and magnification of the specimen image on the recording film respectively. The inclined mirror method allows continuous recording of the free surface motion with time, resolving the multiwave structure due to phase transitions induced by shock loading. In the case of multiple wave structure due to elastic-plastic transition, velocities of the plastic shock wave were determined neglecting the interaction with the rarefaction wave arising from the preceding elastic wave.

The particle velocity was assumed to be half the free surface velocity (free surface approximation). For the

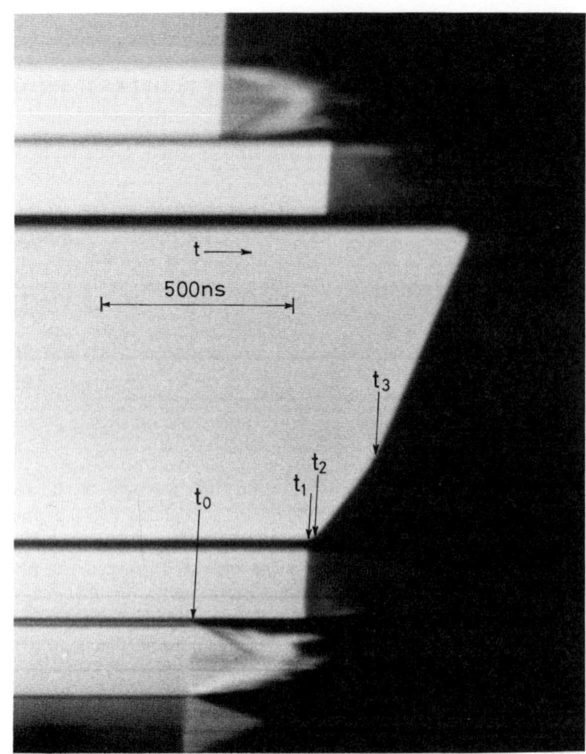

Figure 2. Streak photograph of the inclined mirror run for (100) crystal of rutile (shot no. 82-133), which demonstrates a three wave structure due to elastic-plastic I and plastic I-II transitions. The arrows, t_0 and t_1, indicate the arrival of the elastic precursor wave (HEL) at the rear and front surface of the specimen. Arrows t_2 and t_3 correspond to the arrival of the phase transition (plastic I) and the final deformational (plastic II) wave respectively.

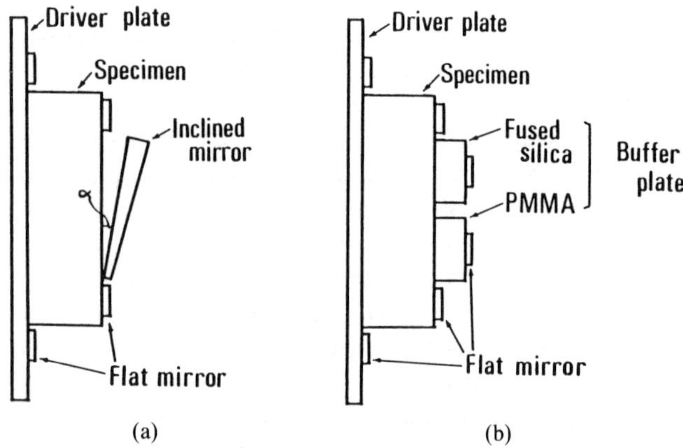

Figure 1. Cross section of the experimental assemblies. a) Inclined mirror assembly for Hugoniot measurements. b) Buffer plates assembly for release adiabat measurements.

final deformational state the particle velocity was also determined from the measured projectile (flyer) velocity on the basis of the impedance-match concept. The shock state was computed from the measured shock and particle velocities by applying the conservation relations of mass, momentum, and energy across the shock front. Accuracy in measurements of shock and free surface velocity was generally 0.5 and 1 percent respectively. Errors in pressure and volume estimation were less than 2 percent.

In several runs the isentropic release path was determined with the buffer technique (AHRENS et al., 1969). Fused silica and PMMA were used as the diagnostic plate. These were placed on the surface of the rutile specimen. The release state in the specimen achieved from the shocked state can be determined by measuring the shock velocity in the buffer plate. The Hugoniot of the buffer plates need to be accurately known. In the present experiments we used the data of MCQUEEN et al. (1970). We assume continuity of pressure and particle velocity across the sample-buffer interface to get the pressure-particle velocity state. The Riemann integral formula was applied to obtain the pressure-volume state along the release isentrope (AHRENS et al., 1969).

Results

Experimental results of Hugoniot and release adiabat measurements are summarized in Table 1 and 2 respectively. Measured shock and particle velocities are plotted in Figure 3. For the final deformational state, particle velocity determined by impedance match solution is adopted. Previous data by MCQUEEN et al. (1967) and AL'TSHULER et al. (1973) are also included in the figure. Velocities of the elastic precursor wave are quite anisotropic with respect to the shock propagation direction. They are found to be very consistent with the corresponding longitudinal sound velocities. The single shock deformational state is not achieved along [110] and [001] in the present experiments owing to quite high sound velocities along these directions.

Seven runs (two for [100], three for [110] and two for [001]) recorded phase transitions, as shown in Figure 2. However, phase transition pressures determined varied drastically with shock propagation direction. Phase transition pressure was highest in the [001] direction, almost twice as high as those in the [100] or [110] directions. A straight line connecting the low-pressure phase points and the bulk-sound velocity at zero pressure is represented by $U_s=7.02+1.24\ u_p$, where the data apparently biased by the loss of shear strength immediately above Hugoniot elastic limit (HEL) are excluded (GOTO et al., 1982).

The U_s-u_p relation for the high pressure phase clearly shows two different trends. The higher trend coincides with previous data by MCQUEEN et al. (1967) and AL'TSHULER et al. (1973) and is referred to HPP-II hereafter. The lower trend, referred to as HPP-I, is definitely different from HPP-II. The newly observed trend HPP-I is particularly well defined in the shock data along [100], in which the phase-transition pressure and the elastic-wave velocity are lowest and is expressed by the relation $U_s=3.63+1.72\ u_p$. The U_s-u_p relation for HPP-II is best represented by $U_s=1.82+2.23\ u_p$.

These U_s-u_p data are converted to pressure-volume relations using the conservation relations of mass and momentum across the shock front and are summarized in Figure 4. Anisotropy in the HEL is also notable. However, the value of HEL itself is almost constant, i.e., 6.2 ± 0.4, 6.8 ± 0.5, and 7.0 ± 0.5 GPa for [100], [110], and [001] directions respectively.

Hugoniot data in the hydrostatic regime are reduced to the isotherm by following the method by WACKERLE (1962). Debye temperature and Grüneisen constant, γ, of rutile are taken to be 758 K and 1.28 respectively (MANGHNANI, 1969). The ratio γ/V is assumed to be constant. The 298 K isotherm obtained for the low pressure rutile phase is also shown in Figure 4 where the static-compression data obtained by high-pressure X-ray diffraction studies are incorporated (SATO, 1977; MING and MANGHNANI, 1979). A least-squares fit for equally weighted dynamic- and static-compression data using the Murnaghan-Birch equation of state gives the zero-pressure bulk modulus, K_0, and its pressure derivatives, K_0' to be 197 GPa and 6.9 respectively. These values, summarized in Table 3, are consistent with the reported values obtained for rutile by ultrasonic measurements and high-pressure X-ray diffraction data.

Striking features of highly anisotropic behavior of the phase transition in rutile are shown most clearly in Figure 4. Pressure values for the phase transition are 13.7 ± 0.3, 17.0 ± 1.7, and 33.8 ± 0.3 GPa for the [100], [110], and [001] shock loading respectively. Phase transition pressure for the [001] direction apparently coincides with that reported by MCQUEEN et al. (1967), whereas values for [100] and [110] directions are close to those determined for polycrystalline specimens by AL'TSHULER et al. (1973).

We observed a wide mixed-phase region extending to about 60 GPa. Present shock-compression data for the new high pressure phase, HPP-I, observed along the [100] direction, were reduced to Murnaghan-Birch isotherm by assuming $K_0'=4$ (Figure 4). Zero-pressure density and bulk modulus for HPP-I are estimated to be 5.00 g/cm^3 and 239 GPa respectively. The volume decrease accompanying phase transition is estimated to be about 15 percent at zero pressure.

The previously reported phase, HPP-II, apparently has a stiffer compression behavior than HPP-I. To distinguish HPP-II from HPP-I in a clearer fashion three

TABLE 1. Summary of Shock Compression Measurements on Single-Crystal Rutile

Shot No.	Projectile Velocity (km/s)	Flyer & Driver Material	Observation Method	Sample Thickness (mm)	Elastic Wave U_1 (km/s)	Elastic Wave u_1 (km/s)	Elastic Wave P_1 (GPa)	V_1/V_0
(100) Crystal								
82-133	1.556	Cu	IM	2.321	8.05	0.18	6.2	0.978
82-154	2.068	W	IM	2.917	8.07	0.17	5.8	0.979
82-015	3.640	Cu	IM	2.435	8.08	0.19	6.5	0.976
82-016	3.976	Cu	IM	2.742	8.09	0.18	6.2	0.978
85-025	3.536	W	FM	2.043				
82-014	4.178	Cu	FM	2.435				
82-018	3.759	W	IM	2.562				
85-026	3.928	W	FM	2.034				
85-027	4.256	W	IM	2.228				
82-019	4.136	W	FM	2.817				
82-020	4.230	W	FM	2.816				
(110) Crystal								
82-134	1.517	Cu	IM	3.020	10.10	0.17	7.3	0.983
82-123	2.109	Cu	IM	3.143	9.99	0.16	6.8	0.984
82-121	2.110	W	IM	3.072	10.06	0.15	6.4	0.985
82-021	3.973	Cu	IM	2.927	10.14	0.16	6.7	0.985
(001) Crystal								
82-117	2.084	W	IM	2.938	10.89	0.15	7.0	0.986
82-022	4.031	Cu	IM	2.966	10.96	0.15	7.0	0.986
82-010	3.996	W	IM	2.873	10.80	0.16	7.4	0.985
82-011	4.362	W	IM	2.263	10.95	0.14	6.5	0.987

IM: Inclined mirror, FM: Flat mirror, fsa: Free surface approximation, ims: Impedance match solution.
*Phase transition wave.

TABLE 2a. Summary of Release Adiabat Measurements for (100) Crystal of Rutile. See the Experimental Conditions in Table 1

Shot No.	Shock State U_s (km/s)	u_p (km/s)	P_s (GPa)	V_s/V_0	Release Adiabat Measurements** Fused Silica Buffer U_{r1} (km/s)	u_{r1} (km/s)	P_{r1} (GPa)	V_{r1}/V_0	PMMA Buffer U_{r2} (km/s)	u_{r2} (km/s)	P_{r2} (GPa)	V_{r2}/V_0
85-025*	7.80	2.65	87.7	0.661	6.37	3.29	46.1	0.703	7.65	3.41	32.0	0.706
85-026	8.30	2.92	102.8	0.648	6.32	3.27	45.4	0.657	8.02	3.52	33.7	0.670
85-027	8.66	3.14	111.5	0.637	7.06	3.70	57.6	0.660	—	—	—	—

*The shock state was not warranted to be achieved by a single shock wave, because the shock velocity was slightly lower than the longitudinal sound velocity along [100].

**Volume of the released state was calculated from the relation $V_r = V_s + (u_{fs}/2 - u_p)^2 / P_s$ where du_p/dP was assumed to be constant.

Plastic I Wave				Plastic II Wave				Reduction Method
U_2 (km/s)	u_2 (km/s)	P_2 (GPa)	V_2/V_0	U_3 (km/s)	u_3 (km/s)	P_3 (GPa)	V_3/V_0	
7.51	0.43	14.0*	0.945	5.41	0.85	23.4	0.865	fsa
				5.41	0.96	26.0	0.842	ims
7.13	0.42	13.4*	0.944	6.41	1.45	41.2	0.782	fsa
				6.41	1.57	44.4	0.763	ims
								fsa
7.56	2.27	73.2	0.701					ims
7.91	2.28	76.7	0.712					fsa
7.91	2.47	83.1	0.688					ims
7.80	2.65	87.7	0.661					ims
8.10	2.59	89.1	0.680					ims
8.41	2.81	100.4	0.666					fsa
8.41	2.78	99.3	0.669					ims
8.30	2.92	102.8	0.648					ims
8.66	2.92	107.4	0.663					fsa
8.66	3.14	111.5	0.637					ims
8.69	3.05	112.6	0.649					ims
8.86	3.11	117.1	0.649					ims
7.25	0.43	15.2*	0.947	5.65	0.74	22.5	0.891	fsa
				5.65	0.91	26.5	0.860	ims
7.48	0.48	16.9*	0.941	6.15	1.08	32.3	0.841	fsa
				6.15	1.29	37.6	0.807	ims
7.63	0.53	18.7*	0.935	6.71	1.52	46.5	0.785	fsa
				6.71	1.58	48.1	0.776	ims
								fsa
7.74	2.46	82.1	0.685					ims
8.14	0.92	33.5*	0.891	6.71	1.35	45.4	0.825	fsa
				6.71	1.53	50.4	0.797	ims
8.21	0.93	34.1*	0.891					fsa
				7.63	2.51	84.5	0.681	ims
								fsa
8.34	2.96	106.4	0.648					ims
								fsa
8.94	3.20	122.7	0.644					ims

TABLE 2b. Volume in the Shock Released State Estimated from Free Surface Velocity Measurements of Rutile Single Crystal

	Shock State*				Released State**	
	U_s (km/s)	u_p (km/s)	P_s (GPa)	V_s/V_0	u_{fs} (km/s)	V_r/V_0
(100)						
82-133	5.41	0.96	26.0	0.842	1.70	0.931
82-154	6.41	1.57	44.4	0.763	2.90	0.932
82-016	7.91	2.47	83.1	0.688	4.56	0.911
82-018	8.41	2.78	99.3	0.669	5.62	1.040
85-027	8.66	3.14	111.5	0.637	5.84	0.915
(110)						
82-134	5.65	0.91	26.5	0.860	1.47	0.910
82-123	6.15	1.29	37.6	0.807	2.16	0.893
82-121	6.71	1.58	48.1	0.776	3.03	0.962
(001)						
82-117	6.71	1.53	50.4	0.797	2.69	0.910

*The shock state determined from impedance match solution.

**Released state calculated from measured free surface velocities by assuming du_p/dP=constant.

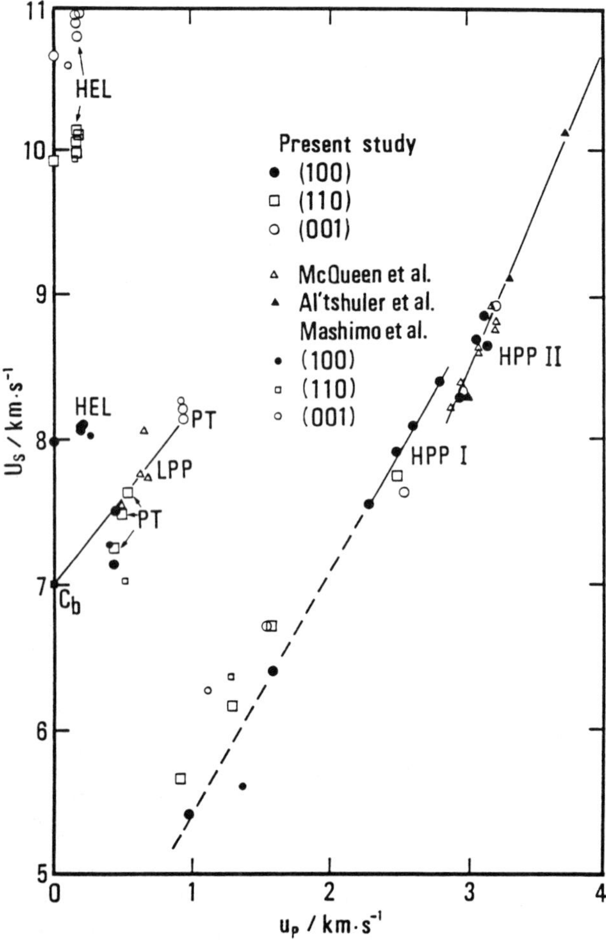

Figure 3. Shock wave velocity (U_s) versus particle velocity (u_p) of rutile single crystals shocked perpendicular to (100), (110), and (001) planes. HEL: Hugoniot elastic limit, C_b: Bulk sound velocity, PT: Phase transition, LPP: Low pressure phase, HPP-I and -II: High pressure phase I and II. Previous data by McQueen et al. (1967), Al'tshuler et al. (1973) and Mashimo et al. (1983) are included. Note clear offset between HPP-I and HPP-II trends.

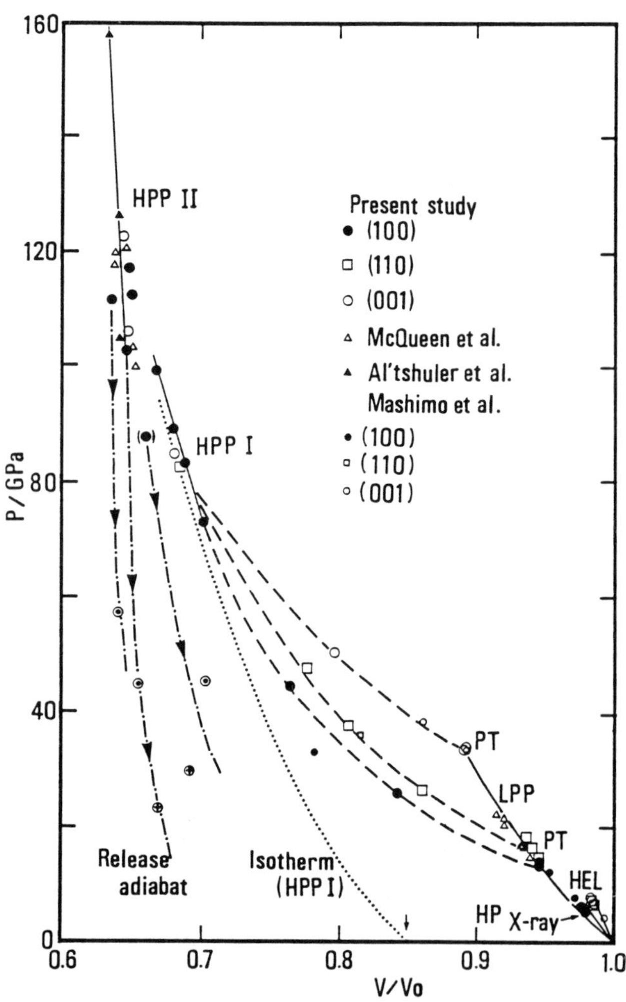

Figure 4. Pressure-volume relation of rutile single crystals determined from present shock compression study. Notation is the same as that shown in Figure 3. Static compression data of rutile (LPP) measured by in situ X-ray diffraction study by Sato (1977) and Ming and Manghnani (1979) are also shown. Note the marked anisotropy in the phase transition pressure (PT). Dotted line is the isotherm for HPP-I reduced from [100] shock data. The arrow shows the estimated zero-pressure volume of HPP-I. Release adiabat path from the HPP-II region is shown by the chained line. Dotted and crossed circles are data points determined from experiments using fused silica glass and PMMA buffers, respectively.

release adiabat measurements were performed above 80 GPa along the [100] direction. The observed release path from the HPP-II region higher than 100 GPa did not follow the HPP-I compression curve but tended toward a denser path, as shown in Figure 4. The release path was found to be close to a simple extrapolation of the HPP-II compression curve and the volume decrease in HPP-II from rutile is estimated to be about 30 percent at zero pressure. The release adiabat from 87.7 GPa was found to be intermediate between HPP-II and HPP-I, indicating that the shock state was still in the mixed phase region of HPP-I and HPP-II.

Discussion

Earlier results from high-pressure phase transformations observed in PbO_2, namely rutile-α-PbO_2-fluorite (Syono and Akimoto, 1968), might lead one to expect similar behavior in TiO_2. However, post α-PbO_2 phases of TiO_2 by in situ X-ray diffraction studies reported by Liu (1975; 1978; 1979) and Ming and Manghnani (1979) have not been structurally identified yet. The newly

TABLE 3. Bulk Modulus and Its Pressure Derivative for Rutile and HPP-I Determined from Shock Compression Experiments. Ultrasonic and High-Pressure X-Ray Determination are also Tabulated for the Sake of Comparison

	K_0 (GPa)	K_0'	Method	Investigator
Rutile	209.5	—	Ultrasonic	VERMA (1960)
	209.0	6.84	Ultrasonic	MANGHNANI (1969)
	207.6	6.97	Ultrasonic	FRITZ (1977)
	197	6.84*	X-ray	SATO (1977)
	188	10.60	X-ray	SATO (1977)
	211	6.84*	X-ray	MING and MANGHNANI (1979)
	197	6.9	Shock†	Present study
HPP-I	239	4.0*	Shock	Present study

*Assumed value.

†Shock data are combined with static X-ray data of SATO (1977) and MING and MANGHNANI (1979), where both data are equally weighted.

discovered phase HPP-I in the present shock study should place a constraint on the interpretation of the X-ray diffraction pattern of the unidentified high-pressure phases of TiO_2. The observed volume decrease of 15 percent at zero pressure accompanied by the phase transition to HPP-I is consistent with the value for the rutile-fluorite transition inferred from the systematic relation between formula volume and bond length for MO_2 compounds (LIU, 1982; MING and MANGHNANI, 1982).

On the other hand, the only high-pressure phase recovered from shock loading is the α-PbO_2 type, only 3 percent denser than the rutile phase. It also shows a very broad X-ray diffraction profile. These facts strongly suggest that the shock-induced phase of the α-PbO_2 type is metastably formed during decompression from the fluorite-like phase produced under shock loading. The instantaneous nature of the retrogressive phase transition from fluorite to α-PbO_2 has been demonstrated in a molecular dynamical computer calculation by MATSUI and KAWAMURA (this volume).

Therefore the remaining problem is to find an instantaneous mechanism for the rutile-fluorite transition which may work under shock loading condition. This possibility has been suggested from the analogy of the shock-induced phase transition observed in zircon to the scheelite structure (KUSABA et al., 1985; 1986). Complete conversion to the scheelite type in the shock recovery experiments above about 50 GPa can be explained by a displacive mechanism which does not involve substantial atomic diffusion; the scheelite-like structure can be easily realized by simple shearing of zircon along the [100] direction. Since the cationic arrangements in rutile and fluorite are quite similar to those in zircon and scheelite respectively (O'KEEFFE and HYDE, 1982; KUSABA et al., 1986), the zircon-scheelite transition can be taken for a model of the rutile-fluorite transition. If this is indeed the case, we can explain why the phase transition pressure to HPP-I is much lower in the shock loading along the [100] or [110] direction than that along the [001] direction.

The present results are not always consistent with those reported previously by LINDE and DECARLI (1969) based on their shock recovery experiments of rutile. Therefore we have made a systematic study of shock recovery experiments on rutile single crystals again, where shock loading directions are varied with respect to the crystallographic orientation, and found favorable evidence for the above mentioned considerations (KUSABA et al., in press). The pressure above which α-PbO_2 phase, presumably formed from the unquenchable high-pressure fluorite-like phase in the pressure releasing process, appears in the shock recovered product coincides with the onset of the phase transition to HPP-I observed in Hugoniot measurements. Furthermore its yield is much larger in the [100], as well as [110], shock loading than in the [001] shock loading. These results lend strong support for the displacive mechanism proposed by KUSABA et al. (1986) for the rutile-fluorite transition under shock loading. Details will be reported shortly.

The fact that the observed release path from above about 80 GPa follows the extrapolated compression curve of HPP-II rather than HPP-I definitely distinguishes HPP-I and HPP-II. The volume decrease of about 30 percent observed in HPP-II from the rutile phase could be explained from conventional extension of systematics in the polymorphic transformations at high pressures. For instance, $PbCl_2$ type structure with the cationic coordination number 9 may be a candidate for HPP-II.

An alternative explanation could be a pressure-induced change of the nature of the bond. This is highly likely for transition metal compounds. A typical example is demonstrated by $LiNbO_3$. This is observed in both static and dynamic compression experiments (DA JORDANA et al., 1985; GOTO and SYONO, 1985). The volume change of about 20 percent observed in the shock-induced phase transition at 32 GPa cannot be explained unless change in the bond nature is assumed to occur because of the closely packed oxygen array in the lithium niobate structure. Evidence for a change in the nature of bonding upon phase transition was obtained via optical absorbance measurements in a diamond-anvil cell. A structural change was also indicated by in situ X-ray measurements (unpublished data by Suzuki and Akimoto quoted in GOTO and SYONO (1985)). Similarity between electronic configurations of Nb^{5+} and Ti^{4+} ions with empty outer d orbitals lends further support for this interpretation. Presumably, partial electron transfer from ligands to cationic ions under high pressure increases the covalency of the bond, resulting in the structural transition.

Acknowledgments. The authors benefited from invaluable discussion with Y. Matsui. They also thank T. J. Ahrens and T. Mashimo for many helpful comments on the manuscript. The work was partially supported by Grant-in-Aid for Scientific Research, Ministry of Education, Science and Culture, Japan.

REFERENCES

AHRENS, T. J., C. F. PETERSEN, and J. T. ROSENBERG, Shock compression of feldspars, *J. Geophys. Res., 74*, 2727–2746, 1969.

AL'TSHULER, L. V., M. A. PODURETS, G. V. SIMAKOV, and R. F. TRUNIN, High-density forms of fluorite and rutile, *Sov. Phys. Solid State, 15*, 969–971, 1973.

BENDELIANI, N. A., S. V. POPOVA, and L. F. VERESHCHAGIN, A new modification of titanium dioxide stable at high pressure, *Geokhimiya, 5*, 499–501, 1966 (Transl. *Geochem. Intern., 3*, 387–390, 1966).

DACHILLE, F., P. Y. SIMONS, and R. ROY, Pressure-temperature studies of anatase, brookite, rutile and TiO_2-II, *Am. Mineral., 53*, 1929–1939, 1968.

DA JORDANA, J. A. H., S. BLOCK, F. A. BAUER, and G. J. PIERMARINI, Phase transition and compression of $LiNbO_3$ under static high pressure, *J. Appl. Phys., 57*, 842–844, 1985.

FRITZ, I. J., Pressure and temperature dependence of the elastic properties of rutile (TiO_2), *J. Phys. Chem. Solids, 35*, 817–826, 1974.

GOTO, T., and Y. SYONO, Technical aspect of shock compression experiments using the gun method, in *Materials Science of the Earth's Interior*, edited by I. Sunagawa, pp. 605–619, Terra/Reidel, Tokyo/Dordrecht, 1984.

GOTO, T., and Y. SYONO, Shock-induced phase transformation in lithium niobate, *J. Appl. Phys., 58*, 2548–2552, 1985.

GOTO, T., T. SATO, and Y. SYONO, Reduction of shear strength and phase transition in shock-loaded silicon, *Jpn. J. Appl. Phys., 21*, L369–L371, 1982.

JAMIESON, J. C., and B. OLINGER, High-pressure polymorphism of titanium dioxide, *Science, 161*, 893–895, 1968.

KUSABA, K., Y. SYONO, M. KIKUCHI, and K. FUKUOKA, Shock behavior of zircon: phase transition to scheelite structure and decomposition, *Earth Planet. Sci. Lett., 72*, 433–439, 1985.

KUSABA, K., T. YAGI, M. KIKUCHI, and Y. SYONO, Structural considerations on the mechanism of shock-induced zircon-scheelite transition in $ZrSiO_4$, *J. Phys. Chem. Solids. 47*, 675–679, 1986.

KUSABA, K., M. KIKUCHI, K. FUKUOKA, and Y. SYONO, Anisotropic phase transition of rutile under shock compression, *Phys. Chem. Minerals*, in press.

LINDE, R. K., and P. S. DECARLI, Polymorphic behavior of titania under dynamic loading, *J. Chem. Phys., 50*, 319–325, 1969.

LIU, L., High pressure phase transformations and compressions of ilmenite and rutile, I. Experimental results, *Phys. Earth Planet. Inter., 10*, 167–176, 1975.

LIU, L., A fluorite isotype of SnO_2 and a new modification of TiO_2: implication for earth's lower mantle, *Science, 199*, 422–424, 1978.

LIU, L., High pressure transformations of dioxides with the rutile structure, in *High-Pressure Science and Technology, Vol. 2*, edited by K. D. Timmerhaus and M. S. Barber, pp. 17–23, Plenum, New York, 1979.

LIU, L., High-pressure phase transformations of the dioxides: implication for structures of SiO_2 at high pressure, in *High-Pressure Research in Geophysics*, edited by S. Akimoto and M. H. Manghnani, pp. 349–360, Center Acad. Publ. Japan/Reidel, Tokyo/Dordrecht, 1982.

MAMMONE, J. F., and S. K. SHARMA, Raman study of TiO_2 under high pressures at room temperature, *Carnegie Inst. Wash. Yearb., 78*, 636–640, 1979.

MANGHNANI, M. H., Elastic constants of single crystal rutile under pressure to 7.5 kilobars, *J. Geophys. Res., 74*, 4317–4328, 1969.

MASHIMO, T., and A. SAWAOKA, Anisotropic behavior of the shock-induced phase transition of rutile phase titanium-dioxide, *Phys. Lett., 78A*, 419–422, 1980.

MASHIMO, T., K. NAGAYAMA, and A. SAWAOKA, Anisotropic elastic limits and phase transitions of rutile phase TiO_2 under shock compression, *J. Appl. Phys., 54*, 5043–5048, 1983.

MATSUI, Y., and K. KAWAMURA, Computer experimental synthesis of silica with the α-PbO_2 structure, this volume.

MCQUEEN, R. G., J. C. JAMIESON, and S. P. MARSH, Shock-wave compression and X-ray studies of titanium dioxide, *Science, 155*, 1401–1404, 1967.

MCQUEEN, R. G., S. P. MARSH, J. W. TAYLOR, J. N. FRITZ, and W. J. CARTER, The equation of state of solids from shock wave studies, in *High Velocity Impact Phenomena*, edited by R. Kinslow, pp. 293–417, Academic, New York, 1970.

MING, L. C., and M. H. MANGHNANI, Isothermal compression of TiO_2 (rutile) under hydrostatic pressure up to 106 kbar, *J. Geophys. Res., 84*, 4777–4779, 1979.

MING, L. C., and M. H. MANGHNANI, High-pressure phase transformation in rutile structured dioxides, in *High-Pressure Research in Geophysics*, edited by S. Akimoto and M. H. Manghnani, pp. 329–347, Center Acad. Publ. Japan/Reidel, Tokyo/Dordrecht, 1982.

O'KEEFFE, M., and B. G. HYDE, Anion coordination and cation packing in oxides, *J. Solid State Chem., 44*, 24–31, 1982.

SATO, Y., Equation of state of mantle minerals determined through high-pressure X-ray study, in *High-Pressure Research: Applications in Geophysics*, pp. 307–323, edited by M. H. Manghnani and S. Akimoto, Academic, New York, 1977.

SYONO, Y., and S. AKIMOTO, High pressure synthesis of fluorite type lead dioxide, *Mater. Res. Bull., 3*, 153–157, 1968.

VERMA, R. K., Elasticity of some high-density crystal, *J. Geophys. Res., 65*, 757–766, 1960.

WACKERLE, J., Shock-wave compression of quartz, *J. Appl. Phys., 33*, 922–937, 1962.

THE TEMPERATURE OF SHOCK COMPRESSED IRON

Jay D. BASS

Department of Geology, University of Illinois, Urbana, Illinois 61801, USA

Bob SVENDSEN and Thomas J. AHRENS

Seismological Laboratory, California Institute of Technology, Pasadena, California 91125, USA

Abstract. Measurements of the temperature of Fe under shock compression were performed to Hugoniot pressures of 300 GPa. The samples consisted of thin Fe films, 0.5 to 9.5 μm in thickness, or Fe foils in contact with a transparent anvil of either single-crystal Al_2O_3 or LiF. Temperatures at the sample/anvil interface were obtained by measuring the spectral radiance of the interface for the duration of the shock transit through the anvil, using a four-color optical radiometer. Results indicate that the Al_2O_3 anvil remains at least partially transparent to pressures of 230 GPa and to temperatures of over 9000 K. The experimental data that yield the lowest temperature at any given pressure define a narrow pressure-temperature trajectory which we infer to be the best estimate of the Hugoniot temperatures of Fe. Although these results must strictly be considered as an upper bound on the Hugoniot temperatures of crystal-density Fe, we have obtained a melting temperature for Fe along the Hugoniot of 6700 ±400 K at 243 GPa. Taken together with recent measurements of the melting temperature to static pressures of 100 GPa (WILLIAMS et al., 1987), our results imply a melting temperature for Fe of 7800 ±500 K at the pressure of the inner core-outer core boundary.

Introduction

The properties of matter at exceedingly high degrees of compression can be investigated using shock wave techniques. In a typical equation-of-state experiment it is usual to determine the shock velocity (U_s), mass or particle velocity (u_p), and the differences in pressure (P), specific volume (V), and internal energy (E) between the initial state and the shock-compressed state. The techniques used to perform such experiments are relatively well developed and have been described in many articles published over the past two decades. However, the parameters mentioned above do not by themselves give a complete thermodynamic description of a material in the shock-compressed, or Hugoniot state. In particular, the temperature along the Hugoniot, the locus of shock-compressed states, is usually not measured. With modern shock wave techniques, pressures on the order of several hundred GPa, corresponding to those of the earth's lower mantle and core, have been attained in solid samples. The concomitant temperatures reached in these experiments are generally many thousands of Kelvins. In order to apply the results of shock wave experiments to states off the Hugoniot (for example, adiabats or isotherms), it is necessary to either measure or calculate Hugoniot temperatures achieved during shock loading. Hugoniot temperatures that are calculated, however, are subject to large uncertainties due to imperfect knowledge of thermal properties such as Grüneisen's parameter and the specific heat. Thus measurement of Hugoniot temperatures is preferred.

Another motivation for performing shock temperature measurements is to identify the existence of phase transitions along the shock compression curve. It has been found that many phase transitions, especially those involving only a small change in density, are not obvious in terms of standard Hugoniot parameters, and are manifest only as subtle changes of slope in the U_s-u_p or P-V Hugoniot relationship of a given material. However, such phase transitions may have a more pronounced signature in the T-P plane. When the Hugoniot intersects a phase boundary there will be, in principle, a substantial offset (or discontinuity) in the Hugoniot T-P curve (KORMER, 1968). As shown in Figure 1, the Hugoniot will coincide with the phase boundary over a pressure interval that is determined by the transition energy of the reaction. Such behavior has been observed from shock temperature

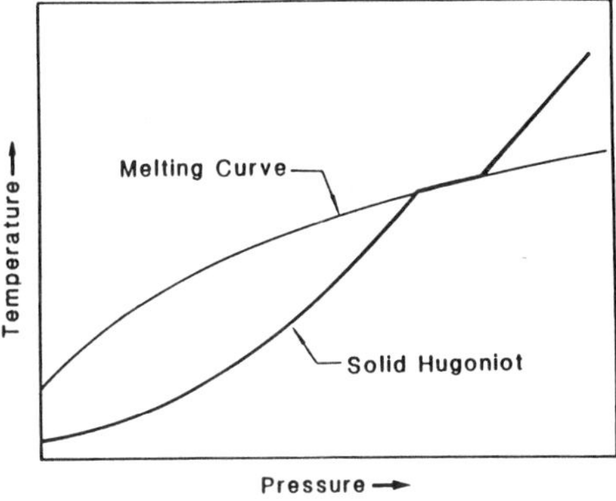

Figure 1. Schematic diagram illustrating the effect of a phase boundary with positive Clapeyron slope, dP/dT, upon the equilibrium Hugoniot temperatures.

High-Pressure Research in Mineral Physics, edited by M. H. Manghnani and Y. Syono, pp. 393–402.

data on NaCl (KORMER et al., 1965; AHRENS et al., 1982).

In this paper we present the results of our initial attempts to measure the temperature of Fe under shock loading, using a four-channel optical radiometric technique. Fe was chosen for our initial experiments because of its geophysical importance as a probable major constituent of the earth's core. Although similar measurements have been successfully made on a variety of transparent materials in recent years, the extension of these methods to the study of opaque materials poses a number of serious experimental difficulties, as discussed below. To the best of our knowledge, the results summarized in this paper represent the most extensive data set obtained to date on the temperature of a shock-compressed opaque material. On the basis of these results, we are convinced that the temperature of shock-compressed opaque samples can indeed be measured. However, we consider our data to be preliminary because of the difficulty that we encountered in accurately characterizing the samples, particularly the initial density of Fe films. The initial density is required to determine the shock pressure, density, and internal energy. Moreover, shock temperature depends on initial density, or porosity, because porous specimens reach higher temperatures than those of ideal crystal density. Because we have not yet measured the densities of our film specimens, we emphasize the preliminary nature of our data.

Experimental Method

All of our experiments were performed using a two-stage, light-gas gun (JONES et al., 1966; JEANLOZ and AHRENS, 1977), in which lexan projectiles bearing Ta flyer plates were accelerated to velocities of up to 6.5 km/s. Impact velocities were measured by taking two flash x-radiographs of the projectile in flight, and are known to better than 0.5% accuracy. Pressures in each part of the sample assembly, described in detail below, were calculated using the impedance matching method (RICE et al., 1958). Necessary equation-of-state parameters are given by MITCHELL and NELLIS (1981, Ta), BROWN and MCQUEEN (1986, Fe), and CARTER (1973, LiF), and the 19 highest-pressure data points that are listed by MARSH (1980) for Al_2O_3.

The basis of the experimental method used in our study is to record the spectral radiance emitted by the sample when it is shock-compressed to high pressure and temperature. Assuming that the sample emits light as a greybody, we fit data obtained at several discrete wavelength bands to the function

$$L(\lambda) = \varepsilon C_1 \lambda^{-5}(\exp(C_2/\lambda T) - 1)^{-1} \quad (1)$$

where C_1 and C_2 are constants, and L, the spectral radiance, is the observed quantity in the experiment. In each experiment data were obtained at the four wavelengths (λ) 450, 600, 750 and 900 nm, and values for the temperature (T) and emissivity (ε) were obtained by a least-squares regression using Equation 1. This technique was initially developed by KORMER et al. (1965), who used a two-color pyrometer to determine the Hugoniot temperature of transparent samples. Later versions of this instrument employing six and four channels in the visible portion of the spectrum were designed by LYZENGA and AHRENS (1979) and by BOSLOUGH (1984), respectively.

In our experiments we used the optical recording apparatus designed by BOSLOUGH (1984). A schematic diagram of the apparatus is shown in Figure 2. Light emitted by the sample was directed to a collimating lens by an expendable front-surface mirror. The lens was positioned at one focal distance (50 cm) from the sample. The collected light was separated into four parts by means of three beamsplitters, and then was demagnified and focused onto four photodiodes. The image of the sample was far smaller than the active area of the photodiodes, so that the photodiodes did not have to be positioned with a high degree of precision. An interference filter was situated in front of each photodiode to pass only a limited band (~40 nm FWHM) about each desired central wavelength. The output voltage of each photodiode was amplified and recorded on an oscilloscope and on a high-speed digital recorder, thus providing redundancy in each

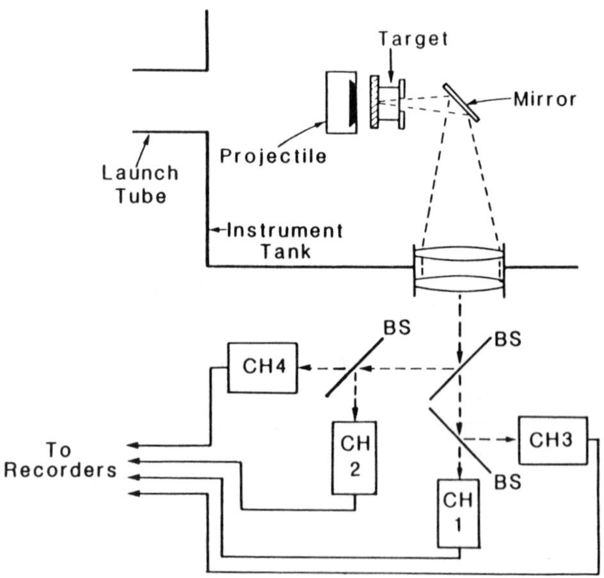

Figure 2. Diagram of the main components of the shock temperature measurement system. The path of light radiated by the sample is indicated by the dashed lines. Each of the four channels (CH#) in the radiometer consist of an interference filter, a lens for demagnifying the image, a photodiode, and an amplifier.

measurement and a backup of each channel. Further details of the system are given by BOSLOUGH (1984).

For shock temperature experiments on opaque materials, the construction of the target assembly is of critical importance. As shown in Figure 3, the main components of the assembly are a 0.5-mm-thick Fe driver plate, the Fe sample in the form of either a film ($<10\ \mu m$ in thickness) or foil (30 μm in thickness), and a disc of single-crystal sapphire or LiF (16 mm in diameter and 3 mm in thickness). The sapphire (or LiF) serves both as an anvil, to maintain the Fe sample at a high pressure after the shock front traverses the Fe-Al_2O_3 interface, and as a window through which thermal radiation can be transmitted during an experiment. Therefore the criteria which are important in choosing an anvil/window material are: 1) that it have a shock impedance as close as possible to that of the metallic sample, to minimize release or reshocking of the sample upon arrival of the shock at the interface, and 2) that the anvil remain transparent when shocked to high pressures. One general conclusion from optical studies on shocked materials in this and other laboratories, is that initially transparent materials seem to radiate as blackbodies, or greybodies with emissivities close to 1, when shocked above phase transition pressures. This implies that phase transitions along the Hugoniot tend to yield an opaque material. Therefore, like LYZENGA and AHRENS (1979) and URTIEW and GROVER (1977), we have chosen Al_2O_3 as our primary anvil material because it does not undergo any known phase transitions along the Hugoniot. Moreover, Al_2O_3 has been observed to remain transparent to static pressures in excess of 500 GPa (XU et al., 1986) and under dynamic loading to at least 100 GPa (URTIEW, 1974). Al_2O_3 also provides the optimal impedance match to Fe, out of all potential window materials.

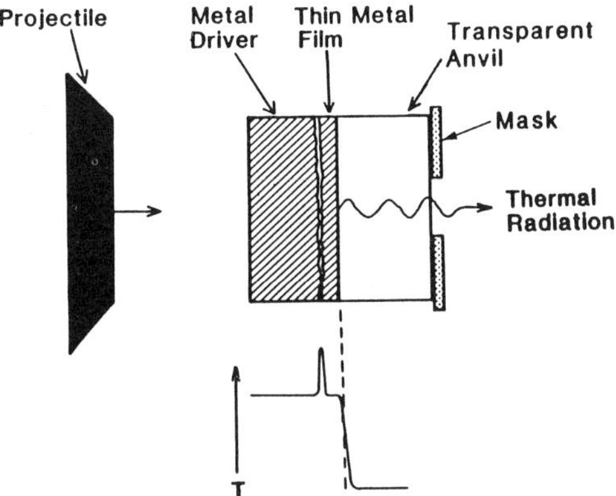

Figure 3. The target assembly used in shock temperature experiments on opaque samples. A foil may be used as a sample in place of the film. A schematic diagram of the temperature profile in various parts of the assembly is shown at the bottom of the figure.

URTIEW and GROVER (1974) have performed a theoretical analysis of the heat generated at the sample/window interface upon passage of a shock wave through the interface. These authors considered the effects of two types of interface imperfections: 1) a small uniform space, or gap, between the two materials, and 2) roughness, or topography, on the surface of the opaque sample. In the first case the metal sample has a free surface where, upon arrival of the shock front, material is released to atmospheric pressure and some elevated temperature in a nearly adiabatic process. The hot, released material at the surface subsequently impacts the anvil surface, thus reshocking the sample to high pressure and a temperature that is greater than would be attained along the initial Hugoniot of the sample. In the second case they modeled the surface roughness as a layer of porous sample material adjacent to the anvil. This also results in temperatures at the interface that are higher than the Hugoniot temperatures of either the anvil or a perfectly dense sample.

From the studies described above, it is clear that the sample must be in nearly perfect contact with the window in order to measure an interface temperature that is directly related to the Hugoniot temperature. This was experimentally verified by LYZENGA and AHRENS (1979), who performed shock temperature measurements on Ag using three different target configurations. They found that the direct impact of a Ag projectile onto a Al_2O_3 window and the use of a Ag foil wrung onto the anvil led to transients in the spectral radiance versus time data. A sample of vapor-deposited Ag on an Al_2O_3 substrate gave a much more stable signal. Based on the experience of LYZENGA and AHRENS (1979) with Ag, we decided to prepare samples by vapor deposition using a Varian electron-beam evaporative coating system. This technique maximizes the chance of obtaining a flawless contact between the sample and window on an atomic scale, thus obviating any thermal signal due to an interfacial gap. Fe was deposited under a total vapor pressure of 3×10^{-7} torr at a rate of approximately 25 Å/s. Films with thicknesses of 0.5 μm (in the first successful run) to 9.5 μm were produced. A calibrated crystal oscillator, with a characteristic frequency that changes as a film is deposited upon it, was positioned near the substrate to monitor the deposition rate and final film thickness. We found that Fe adheres poorly to Al_2O_3 and that a majority of the films peel off of the substrate either during or a short time after coating. This problem became more severe as we tried to increase the film thickness, but was somewhat alleviated by extremely thorough cleaning of the substrate prior to coating.

The thickness of the sample film is an important consideration in this experiment. Because the interface between the driver and film sample cannot be perfect, it is possible to generate significant heat at the interface due to an interfacial gap or surface roughness. If a film is too thin, this heat could diffuse to the sample/anvil interface on the time scale of the experiment, thus yielding an erroneously high temperature that increases with time. For our experiments it was not possible to determine *a priori* what a safe minimum film thickness would be, due to an uncertain knowledge of the appropriate thermal properties at elevated temperatures and pressures. Therefore, we simply tried to obtain as thick a film as possible and found 9 μm to be the approximate upper limit of our technique. Our experimental data indicate that this thickness is satisfactory; the shot records show no consistent evidence for heat diffusion to the sample/anvil interface. In fact, most of our records showed a slight decrease in the intensity of light as the shock front progressed through the window. This can most easily be interpreted as a change in the optical properties of the window material under shock loading (BOSLOUGH, 1985).

As a source of Fe for the films and driver plates, we used a low-carbon steel ("Cor 99", Corey Steel Co., Chicago, IL) with a total impurity content of less than 0.12% (analysis supplied by the manufacturer). The density was determined by the Archimedean method to be 7.84 $\pm$0.02 gm/cm^3, just slightly lower than the X-ray value of 7.874 gm/cm^3 (BERRY, 1967). In our last film experiment, a commercial Fe powder of nominal 99.9% purity was used as a source material. The Fe film from one sample was peeled off the substrate and examined by X-ray powder diffractometry. A well defined peak corresponding to the most intense (110) diffraction line of α-Fe was observed indicating that the films were highly crystalline rather than in an amorphous state.

Vapor deposition is an extremely time-consuming

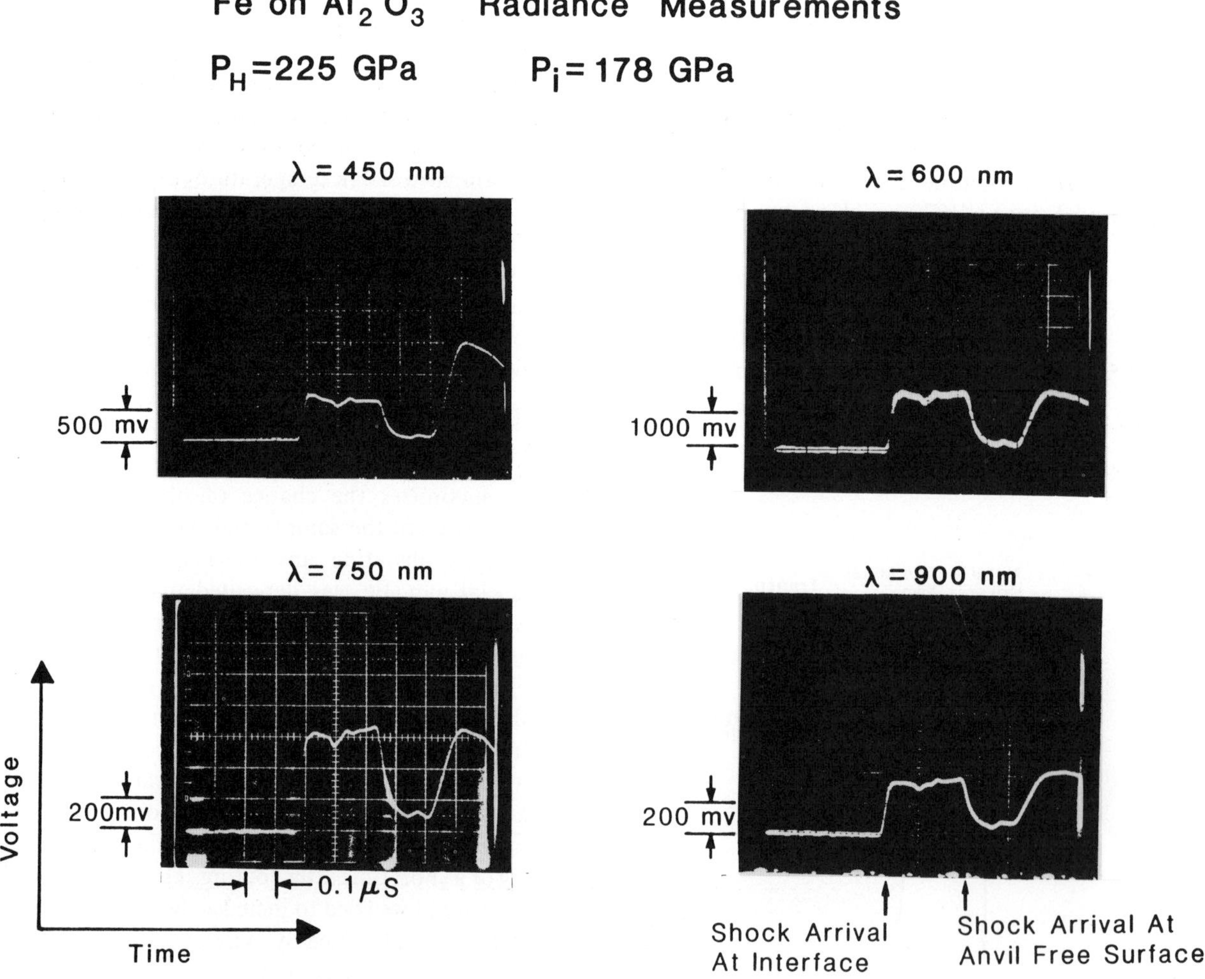

Figure 4. Oscillographic record of voltage as a function of time for one of the shock temperature experiments. The amplitude of the voltage above the baseline seen in the initial ≈400 ns of the record, is proportional to the spectral radiance at the sample/anvil interface.

method of preparing samples. As an alternative, we investigated the possible use of thin Fe foils as samples in five experiments. Fe foils of 0.03 mm in thickness and nominal 99.99% purity were obtained from Alfa Products (Danvers, MA) and used as samples in this series of shots.

Results

Figure 4 shows the raw oscillographic data from one of the shock temperature experiments. A noteworthy feature of these records is that there is no evidence of a "spike", or strong transient in light intensity, when the shock reaches the interface. Such a feature would indicate either the closing of a gap at the interface or the thermal relaxation of a thin Fe layer, which would be extremely hot because of its porosity. It is also important to note that the voltage, or intensity of light, is nearly constant over time, indicating that thermal diffusion from the driver/film interface to the film/anvil interface is probably not significant. All of the voltage records from experiments at higher pressures showed a modest to rapid decrease in light intensity with time, although no spikes were observed. One interpretation of this decrease in light intensity is that the sapphire anvil is appreciably absorbing in the optical wavelength range above pressures of about 225 GPa (BOSLOUGH, 1985). Only experiment No. 167 at 196 GPa exhibited a modest increase of light intensity with time; the reason for this behavior is, as yet, unresolved.

The spectral radiance values obtained from the voltage data of Figure 4 are plotted in Figure 5. Because the spectral radiance is never precisely constant as a function of time, it is important to consider which part of the voltage-time record is appropriate to use for obtaining a Hugoniot temperature. We have chosen to read the initial part of each record, just after the sharp increase in voltage corresponding to the arrival of the shock at the interface. In this way we obtain a measure of the thermal radiance for a sample that is viewed through mostly unshocked, transparent anvil material. This choice should minimize potential problems due to light absorption by the anvil, diffusion of heat from the driver/sample interface to the sample/anvil interface, and signal contributions of the shocked anvil.

To transform the observed voltages to a temperature, it is necessary to calibrate the pyrometer with a standard light source. As described by BOSLOUGH (1984), we used the chopped signal from a tungsten lamp of known spectral irradiance (Optronics Laboratories, Orlando, Florida). The resulting experimental values of spectral radiance could then be fitted to a radiation function, such as Planck's Law, to obtain the temperature and emissivity of the sample/anvil interface. In Figure 5 we show least-squares fits to the experimental data using the emissivity

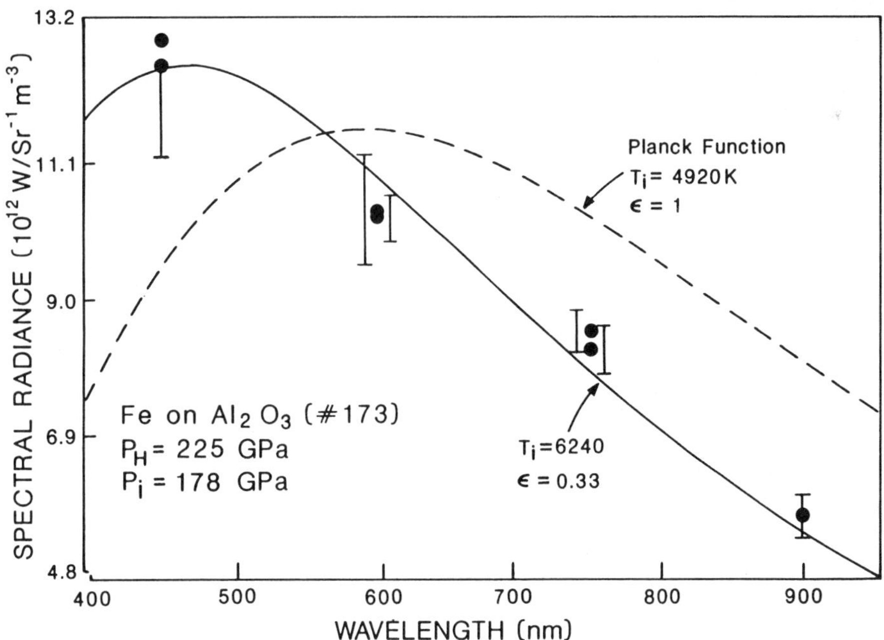

Figure 5. The data from Figure 4, and additional data from digital recorders, plotted in terms of spectral radiance versus wavelength. The solid curve is obtained from a least-squares regression for both temperature and emissivity using Planck's Law; the dashed curve is the least-squares solution for temperature alone with the emissivity fixed at unity.

obtained by regression and a value of unity appropriate for a blackbody. It is clear that the data are described far better by using a greybody with an emissivity <1 than by using a blackbody function; this is quantitatively expressed by emissivity values that are statistically smaller than 1.0 (see Table 1). The errors in interface temperature shown in Figure 4 and listed in Table 1, take into account the estimated uncertainties in reading the baseline and signal voltages on the oscillograms (or transient recorder plots), the calibration voltages, the spectral irradiance of the calibration lamp, and diameter of the mask aperture.

The data from our experiments yield temperature values for that part of the sample material at the sample/anvil interface. In order to obtain the Hugoniot temperature of the bulk sample, it is necessary to correct the interface temperatures for two effects: 1) the influence of the relatively cold anvil, and 2) partial release of the Fe due to the impedance mismatch of the sample and anvil materials. In the ideal situation where the sample has no porosity and is in perfect contact with the anvil, it has been shown (GROVER and URTIEW, 1974) that the interface temperature, T_i, is independent of time and is related to the temperature of the released sample, T_r, by

$$T_i = T_r + (T_a - T_r)/(1 + \alpha) \qquad (2)$$

Here, T_a is the Hugoniot temperature of the anvil, and α is given by

$$\alpha = (\kappa_r/\kappa_a)\cdot(D_a/D_r)^{1/2} = \{(\kappa_r\rho_r C_r)/(\kappa_a\rho_a C_a)\}^{1/2} \qquad (3)$$

where κ is thermal conductivity, D is diffusivity, ρ is the density, C is the specific heat, and the subscripts r and a refer to released Fe and the Hugoniot state of the anvil, respectively.

The thermal properties which are needed to evaluate α in Equation 3 have not been measured under the extreme P and T conditions of our experiments. Therefore α must be estimated from available data and theory. For the anvil materials, we first evaluated the effect of temperature on the lattice contributions to the thermal conductivity, κ_a(P, T). Low pressure conductivity data were fit to expressions

TABLE 1. Results of Shock Temperature Measurements on Fe

Shot No.	Sample Type	Anvil	P_h (GPa) Fe	P_h (GPa) Anvil	T_i (K) Interface	T_a (K) Anvil	T_h (K) Fe	ε
167	Film	Al_2O_3	196	157	4750 ±420	1340	6110	0.29 ±.14
189	Foil	Al_2O_3	202	161	4010 ±420	1380	5200	0.29 ±.14
173	Film	Al_2O_3	226	178	6240 ±170	1550	7910	0.33 ±.05
190	Foil	LiF	227	122	4660 ±420	2790	6180	0.39 ±.19
188	Foil	Al_2O_3	241	188	5390 ±740	1660	6870	1.0 ±.6
191	Film	Al_2O_3	244	190	6990 ±350	1680	8950	0.47 ±.15
183	Foil	Al_2O_3	245	191	6970 ±280	1690	8920	0.34 ±.07
157	Film	Al_2O_3	251	195	6380 ±300	1730	8200	0.70 ±.12
159	Film	LiF	263	140	5270 ±280	3410	7240	0.96 ±.22
192	Film	Al_2O_3	263	203	9220 ±800	1820	11610	0.29 ±.30
174	Film	Al_2O_3	268	207	7580 ±420	1860	9670	0.46 ±.13
181	Foil	Al_2O_3	276	212	9300 ±550	1920	11730	0.32 ±.14
168	Film	Al_2O_3	300	228	6990 ±330	2090	8930	0.86 ±.16

ε=emissivity

All Hugoniot pressures (P_h) are measured with a precision of better than ±1 GPa. Note that P_h for the anvil is also the pressure in the partially released Fe after the shock wave enters the anvil.

in the form $\kappa(0, T)=A_1+B_1/T$, yielding coefficients of: $A_1=-2.599$ (watt/mK), $B_1=1.176\ 10^4$(watt/m), for Al_2O_3 (KINGERY et al., 1954); and $A_1=-0.2023$, $B_1=3.671\ 10^3$, for LiF (MEN' et al., 1974). These equations allowed the calculation of $\kappa_a(0, T)$ at the Hugoniot temperature of the anvil. The effect of pressure on the anvil conductivity was then calculated at the anvil Hugoniot temperature, using the Debye-Grüneisen approximation $\delta\kappa/\kappa \approx 7\delta\rho/\rho$ presented by ROUFOSSE and JEANLOZ (1983). In order to evaluate the Debye-Grüneisen expression, the anvil density at $P=1$ atm and the Hugoniot temperature were obtained from the thermal expansivity in the form $-d\ln\rho/dT=A_2+B_2T$. Values of A_2(at 298 K)$=9.8\ 10^{-5}\ K^{-1}$ and $B_2=1.2\ 10^{-7}\ K^{-2}$ for LiF (PATHAK and VASAVADA, 1972; RAPP and MERCHANT, 1973), and $A_2=1.62\ 10^{-5}\ K^{-1}$ and $B_2=1.1\ 10^{-8}$ were used for Al_2O_3 (TOULOUKIAN et al., 1977). Finally, the heat capacity of the anvil was approximated by the high-temperature DuLong-Petit limit, and the Hugoniot density was determined from the Rankine-Hugoniot relations.

The thermal properties of Fe needed in Equation 3 refer to a partially released state when the shock impedance of the anvil material is lower than that of Fe (such is the case for Al_2O_3 and LiF anvil materials). The released density was calculated using the method of LYZENGA and AHRENS (1978), while the heat capacity was assumed to be the DuLong-Petit value plus an electronic contribution as given by BROWN and MCQUEEN (1986). The incorporation of an electronic contribution to C_v does not affect the corrected Hugoniot temperatures (Figure 6) by more than approximately 50 K. In order to obtain the thermal conductivity of Fe, we chose a different approach than that used for the anvil material. Experimental data have been obtained for the electrical conductivity, σ, of Fe under shock conditions by KEELER (1971). Electrical conductivity can, in turn, be related to the thermal conductivity of metals via the Wiedemann-Franz relation

$$\kappa = L\sigma T \tag{4}$$

where L, the Lorenz number, has a relatively constant value of aobut $2.45\ 10^{-8}$ watt-ohm/K^2 for most metals. A linear least-squares regression of σ versus compression, ρ_0/ρ, yielded an excellent fit, with a correlation coefficient of -0.992. The data used for this curve-fitting was taken from MATASSOV (1977, Figure 7.4). Knowing the compression of Fe in the released state, we obtained κ_r for Fe by assuming that the temperature of the released Fe (needed in the Wiedemann-Franz relation) is given by the observed interface temperature T_i. A value of α (Equation 3) was then calculated, thereby allowing determination of an initial value for the released temperature, T_r, by Equation 2. The entire procedure was repeated itera-

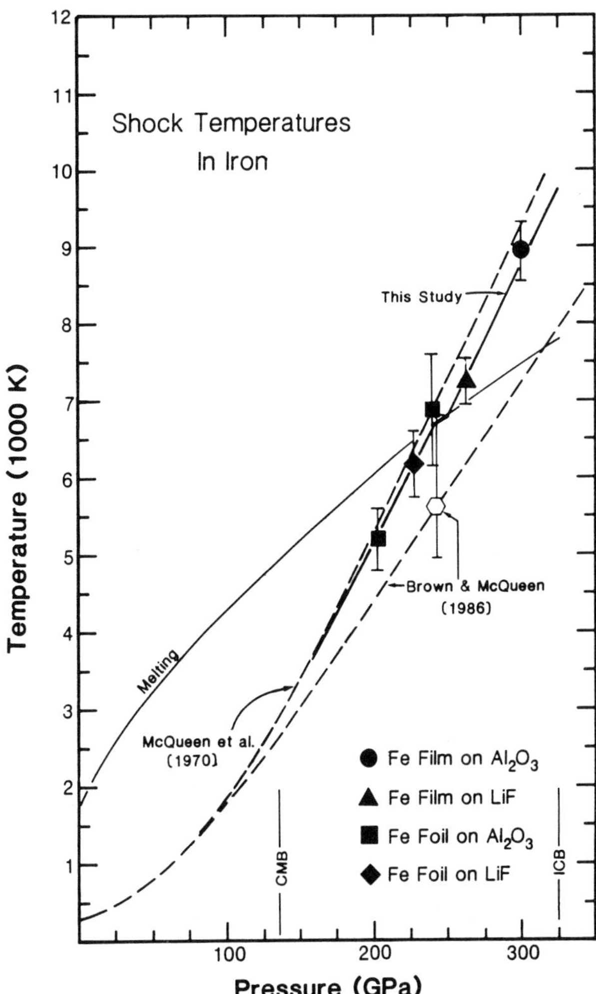

Figure 6. Hugoniot temperatures, deduced from our experimental data, as a function of pressure. The heavy solid line and filled symbols are from the present study. The dashed curves are calculated Hugoniot temperatures that differ mainly in the assumptions made about the specific heat of Fe: MCQUEEN et al. (1970) assume $C_v=3R$, whereas Brown and McQueen incorporate an additional electronic term. The melting curve is consistent with our shock temperature measurements as well as the melting data of WILLIAMS et al. (1987) obtained in a diamond anvil cell at pressures up to 100 GPa. Pressures at the core-mantle (CMB) and inner core-outer core (ICB) boundaries are indicated.

tively, using T_r in Equation 4 to obtain improved values of κ_r, and then recalculating T_r until it converged to a stable value. The values of α obtained by this procedure typically fell in the range of 3.5–4.5.

At this point in the data reduction we have the temperature of Fe in a partially released state of lower pressure than the Hugoniot. To obtain the Hugoniot temperature, it is necessary to correct for the effect of partial release, which we do by using the relation

$$T_r = T_h \exp(-\int_{v_h}^{v_r} (\gamma / v) \mathrm{d}v)$$

where γ is the Grüneisen parameter of Fe. We assume a constant value of $\gamma\rho = 16.7$ (BROWN and MCQUEEN, 1986) and obtain the released volume, v_r, using the method of LYZENGA and AHRENS (1978). Details of the calculations, as well as the theoretical Hugoniot temperature calculations, are given by AHRENS et al. (1982).

Discussion

A complete summary of our results is given in Table 1. Perhaps the most important point to be made is that we have obtained a wide range of inferred Hugoniot temperatures, that is ≈6000 to over 11,000 K. Because the calculated Hugoniot temperatures of Al_2O_3 are lower than our experimental results by several thousand degrees (see Table 1 and SVENDSEN et al., this volume), our data represent compelling evidence that the temperature of the opaque sample, as opposed to that of the anvil material, can be measured using the techniques employed in this study. Although anomalously high temperatures have previously been measured for insulators under shock compression (e.g., SCHMITT and AHRENS, 1984), this appears to primarily be a low pressure phenomenon related to localized "shear band" deformation of the sample which is not operative at high pressures (≈100 GPa). Moreover, the high "shear band" temperatures are usually typified by emissivities of at least one order of magnitude smaller than those measured in the present study. We conclude, therefore, that it is possible to record the temperature of Fe in a shock-induced, high-pressure state. Further, our results indicate that the Al_2O_3 anvil remains at least partially transparent under P and T conditions defined by the Hugoniot pressure in the anvil (≈230 GPa) and the interface temperature (7000–9000 K).

It is apparent from Table 1 that the range of inferred Hugoniot temperatures are larger than would be expected from the precision of the data. Moreover, it is clear that most of the obvious possible sample defects, such as an imperfect sample anvil interface or sample porosity, would yield anomalously high temperatures. Thus the lowest observed temperatures should most closely approximate the actual Hugoniot temperatures of crystal-density Fe. Figure 6 shows our interpretation of the Fe Hugoniot temperatures based on the data obtained thus far. Figure 6 also shows the calculated Hugoniot temperatures of BROWN and MCQUEEN (1986), which take into account possible electronic contributions to the specific heat of Fe, and the calculated Hugoniot temperature of MCQUEEN et al. (1970), which do not include these effects. Electronic contributions increase C_v and thus lower the Hugoniot temperature at any given pressure (see BROWN and MCQUEEN, 1986). The fact that our lowest temperature data define a P-T trend intermediate between these two theoretical bounds strongly suggests that our data closely approximate the true Hugoniot temperatures of Fe.

BROWN and MCQUEEN (1986) have, on the basis of sound velocity measurements, identified two phase transitions along the Hugoniot of Fe at pressures of 200 and 243 GPa; these are inferred to represent ε-γ and γ-melt transitions, respectively. Because the lowest pressure data in Figure 6 exceed 200 GPa, we cannot tell whether or not the $\varepsilon \rightarrow \gamma$ transition has any resolvable effect on the P-T trajectory. However, an offset in the Hugoniot temperatures above 241 GPa is suggested in Figure 6 and is analogous to the effect shown schematically in Figure 1. Thus our data are consistent with the Hugoniot intersecting a melting curve of positive slope at 242 GPa, as suggested by BROWN and MCQUEEN (1986) and shown in Figure 6. This interpretation of our data implies that, in the melt field, the Hugoniot temperatures are decreased by approximately 450 K due to the melting transition; this is in very good agreement with the decrease of 350 K estimated by BROWN and MCQUEEN (1986).

At present, we are unsure why some of the experimental data yield anomalously high temperatures (Table 1). Although great care was taken to produce suitable sample assemblies in a consistent manner, we can only conclude that many of the samples were defective in some way. As discussed previously, the obvious possibilities are an interfacial gap between the foil sample and the anvil, and a porosity of the films. We calculated the temperatures that would be expected in Fe for the case of a uniform interfacial gap and obtained values that far exceed the observed range of inferred Hugoniot temperatures (see SVENDSEN et al., this volume). For example, at 250 GPa, the temperature of Fe which has been released to atmospheric pressure from the Hugoniot state and reshocked upon impact with a Al_2O_3 anvil is calculated as 16,700 K. This value is much larger than the values of 8200–11,700 which were experimentally determined at similar pressures. Therefore we conclude that none of our foils were separated from the anvils by a uniform gap, although imperfect contact over a fraction of the sample area could have produced the high temperatures observed in some of our foil shots.

It is also possible that heat generated at the driver plate/sample interface was able to diffuse through the sample on the time scale of the experiment. We tested this hypothesis by performing two experiments (No. 191 and No. 192) with thin film samples. These samples were sufficiently thin to transmit visible light and thus assured us of detecting a portion of the light generated at the driver/sample interface (see Figure 3); the driver/sample interface should be at a much higher temperature than the Hugoniot state of Fe. These experiments yielded much higher temperatures (Table 1) than those shown in Figure 6. Coupled with the observation that the records from shots utilizing thicker films (Figure 4) did not show an increase of voltage with time, we take these results to be a strong indication that heat diffusion toward the sample/anvil interface did not affect the data shown in Figure 6.

Due to the small mass and delicate nature of the film samples, we have not yet been able to measure the porosity of the films. Therefore we cannot rule out the possiblity of a small amount of film porosity, that varies from one sample to another, to explain the discrepancies between the results in Figure 6 and the higher temperature data listed in Table 1. Nonetheless, we maintain that the interpretation shown in Figure 6, with the set of lowest shock temperatures representing the Hugoniot temperatures of Fe, is reasonable and the most logical conclusion to be drawn based on available data. It is noteworthy that the data shown in Figure 6 are for several types of sample assemblies: both foils and films on Al_2O_3 and LiF substrates. As discussed above, each of these sample configurations has a different experimental problem associated with construction of a suitable target: the foils are most likely to be plagued by interface gaps, whereas the films are probably in perfect contact, but could be slightly porous. However, it is significant that the data in Figure 6 tightly define a Hugoniot P-T trajectory that is within the range of previously calculated theoretical bounds, and is also wholly consistent with the melting transition that has been identified by an independent experimental technique. It is highly unlikely that experiments using different types of sample assemblies would be in error by the same amount. Such a situation would require that the excess temperature produced by interfacial gaps in the foil shots be equal to the excess temperature produced by porosity in the film shots. A simpler explanation is that the data shown in Figure 6 are the Hugoniot temperatures of Fe. This is further supported by the agreement of the shock temperature data with independent measurements of the melting temperature of Fe under static pressures up to 100 GPa (Williams et al., 1987; Strong et al., 1973).

Our shock temperature data limit the melting point of Fe along the Hugoniot to 6700 ±400 K at a pressure of 243 GPa. This value is significantly higher than the recent estimate of 5000–5700 K given by Brown and McQueen (1986), suggesting that electronic contributions to the specific heat of Fe may not be as significant as assumed in their calculations (see also Boness et al., 1986). Based on our results and those of the melting experiments performed in a diamond-anvil cell up to 100 GPa (Williams et al., 1987), we obtain the melting curve for Fe shown in Figure 6. This curve indicates that Fe melts at T=4800 ±200 K when the pressure is 136 GPa, the pressure at the core-mantle boundary, and 7800 ±500 when the pressure is 330 GPa, the inner core-outer core boundary pressure.

Acknowledgments. This research was partially supported by National Science Foundation grants EAR-8419259, EAR-8508969, and EAR-8607784. We thank E. Gelle, M. Long, C. Manning, and W. Miller for assistance with the experiments. Also, comments of J. M. Brown and several anonymous reviewers were extremely useful. Contribution no. 4385, California Institute of Technology.

References

Ahrens, T. J., G. A. Lyzenga, and A. C. Mitchell, Temperatures induced by shock waves in minerals: applications to geophysics, in *High Pressure Research in Geophysics*, edited by S. Akimoto and M. H. Manghnani, pp. 579–594, Academic Press, New York, 1982.

Berry, L. G. (Ed.), *Powder Diffraction File*, Joint Committee On Powder Diffraction Standards, vol. 6, Philadelphia, Pennsylvania, 1967.

Boness, D. A., J. M. Brown, and A. K. McMahan, The electronic thermodynamics of iron under earth core conditions, *Phys. Earth Planet. Inter.*, *42*, 227–240, 1986.

Boslough, Mark B., Shock-wave properties and high-pressure equations of state of geophysically important materials, Ph.D. thesis, 171 pp., California Institute of Technology, Pasadena, California, 1984.

Boslough, M. B., A model for time dependence in shock-induced thermal radiation of light, *J. Appl. Phys.*, *58*, 3394–3399, 1985.

Brown, J. M., and R. G. McQueen, Phase transitions, Grüneisen parameter, and elasticity for shocked iron between 77 GPa and 400 GPa, *J. Geophys. Res.*, *91*, 7485–7494, 1986.

Carter, W. J., Hugoniot equation of state of some alkali halides, *High Temperatures-High Pressures*, *5*, 313–318, 1973.

Grover, R., and P. A. Urtiew, Thermal relaxation at interfaces following shock compression. *J. Appl. Phys.*, *45*, 146–152, 1974.

Jeanloz, R., and T. J. Ahrens, Pyroxenes and olivines: structural implications of shock-wave data for high pressure phases, in *High-Pressure Research: Applications to Geophysics*, edited by M. H. Manghnani and S. Akimoto, pp. 439–461, Academic Press, New York, 1977.

Jones, A. H., W. M. Isbell, and C. J. Maiden, Measurement of the very high-pressure properties of materials using a light-gas gun, *J. Appl. Phys.*, *37*, 3493–3499, 1966.

Keeler, R. N., Electrical conductivity of condensed media at high pressures, in *Physics of High Energy Density, Proc. International School Phys. Enrico Fermi XLVIII*, edited by P. Caldirola and H. Knoepfel, pp. 106–125, Academic Press, New York, N.Y., 1971.

Kingery, W. D., J. Francl, R. L. Coble, and T. Vasilos, Thermal conductivity: x, data for several pure oxide materials corrected to zero porosity, *J. Amer. Cer. Soc.*, *37*, 107–110, 1954.

Kormer, S. B., Optical study of the characteristics of shock-compressed condensed dielectrics, *Soviet Physics USP.*, *11*, 229–254, 1968.

Kormer, S. B., M. V. Sinitsyn, G. A. Kirilov, and V. D. Urlin, Experimental determination of temperature in shock-compressed NaCl and KCl and of their melting curves at pressures up to 700 kbar, *Soviet Phys. JETP*, *21*, 689–700, 1965.

Lyzenga, G. A., and T. J. Ahrens, The relation between the shock-induced free-surface velocity and the postshock specific volume of solids, *J. Appl. Phys.*, *49*, 201–204, 1978.

Lyzenga, G. A., and T. J. Ahrens, Multiwavelength optical pyrometer for shock compression experiments, *Rev. Sci. Instrum.*, *50*, 1421–1424, 1979.

MARSH, S. P. (Ed.), *LASL Shock Hugoniot Data*, pp. 260–263, University of California Press, Berkeley, California, 1980.

MATASSOV, G., *The Electrical Conductivity of Iron-Silicon Alloys at High Pressures and the Earth's Core*, Ph.D. thesis, 180 pp., University of California, Livermore, California. Lawrence Livermore Laboratory Document UCRL-52322, 1977.

MCQUEEN, R. G., S. P. MARSH, J. W. TAYLOR, J. N. FRITZ, and W. J. CARTER, The equation of state of solids from shock wave studies, in *High-Velocity Impact Phenomena*, edited by R. Kinslow, pp. 293–417, Academic Press, New York, 1970.

MEN', A. A., A. Z. CHECHEL'NITSKII, V. A. SOKOLOV, and E. N. SIMUN, Thermal conductivity of lithium fluoride in the 300–1100 K temperature range, *Sov. Phys. Solid State*, *15*, 1844–1845, 1974.

MITCHELL, A. C., and W. J. NELLIS, Shock compression of aluminum, copper, and tantalum, *J. Appl. Phys.*, *52*, 3363–3375, 1981.

PATHAK, P. D., and N. G. VASAVADA, Thermal expansion of LiF by X-ray diffraction and the temperature variation of its frequency spectrum, *Acta Cryst*, *A28*, 30–33, 1972.

RAPP, J. E., and H. D. MERCHANT, Thermal expansion of alkali halides from 70 to 570 K, *J. Appl. Phys.*, *44*, 3919–3923, 1973.

RICE, M. H., R. G. MCQUEEN, and J. M. WALSH, Compression of solids by strong shock waves, in *Solid State Physics*, *6*, edited by F. Seitz D. Turnbull, pp. 1–63, Academic Press, New York, 1958.

ROUFOSSE, M. C., and R. JEANLOZ, Thermal conductivity of minerals at high pressure: the effect of phase transitions, *J. Geophys. Res.*, *88*, 7399–7409, 1983.

TOULOUKIAN, Y. S., R. K. KIRBY, R. E. TAYLOR, and T. Y. R. LEE, *Thermal Expansion, Nonmetallic Solids* (TPRC 13), 1658 pp., IFI/Plenum, New York, 1977.

SCHMITT, D. R., and T. J. AHRENS, Emission spectra of shock compressed solids, in *Shock Waves In Condensed Matter, 1983*, edited by J. R. Asay, R. A. Graham, and G. K. Straub, pp. 313–316, Elsevier, New York 1984.

STRONG, H. M., R. E. TUFT, and R. E. HANNEMAN, The iron fusion curve and the γ-δ-*l* triple point, *Metal. Trans.*, *4*, 2657–2661, 1973.

SVENDSEN, B., T. J. AHRENS, and J. D. BASS, Optical radiation from shock-compressed materials and interfaces, this volume.

URTIEW, P. A., Effect of shock loading on transparency of sapphire crystals, *J. Appl. Phys.*, *45*, 3490–3493, 1974.

URTIEW, P. A., and R. GROVER, Temperature deposition caused by shock interactions with material interfaces, *J. Appl. Phys.*, *45*, 140–145, 1974.

URTIEW, P. A., and R. GROVER, The melting temperature of magnesium under shock loading, *J. Appl. Phys.*, *48*, 1122–1126, 1977.

WILLIAMS, Q., R. JEANLOZ, J. D. BASS, B. SVENDSEN, and T. J. AHRENS, The melting curve of iron to 2.5 Mbar: first experimental constraint on the temperature at the earth's center, *Science*, *236*, 181–182, 1987.

XU, J. A., H. K. MAO, and P. M. BELL, High-pressure ruby and diamond fluorescence: observations at 0.21 to 0.55 terapascal, *Science*, *232*, 1404–1406, 1986.

OPTICAL RADIATION FROM SHOCK-COMPRESSED MATERIALS AND INTERFACES

Bob SVENDSEN and Thomas J. AHRENS

Division of Geological and Planetary Sciences, California Institute of Technology, Pasadena California 91125, USA

Jay D. BASS

Department of Geology, University of Illinois, Urbana, Illinois 61801, USA

Abstract. Recent observations of shock-induced radiation from oxides, silicates, and metals of geophysical interest constrain the shock-compressed temperature of these materials. In these experiments, a projectile impacts a target consisting of a metal driver plate, metal film or foil layer, and transparent window. We investigate the relationships between the temperature inferred from the observed radiation and the temperature of the shock-compressed film or foil and/or window. Changes of the temperature field in each target component away from that of their respective shock-compressed states occur because of: 1) shock-impedance mismatch between target components, 2) thermal mismatch between target components, 3) surface roughness at target interfaces, and 4) conduction within and between target components. In particular, conduction may affect the temperature of the film/foil-window interface on the time scale of the experiments, and so control the intensity and history of the dominant thermal radiation sources in the target. Comparing this model to experiments on Fe-Fe-Al_2O_3 and Fe-Fe-LiF targets, we note that:

1) Fe at Fe-Al_2O_3 interfaces releases from shock-compressed states between 245 and 300 GPa to interface states between 190 and 230 GPa, respectively, with temperatures ≈200–2000 K above model calculations for Fe experiencing no reshock at smooth Fe-Al_2O_3 interfaces. This is so for both Fe foils and films. Below 190 GPa, reshock heating does not apparently affect the temperature of Fe-Al_2O_3 interfaces. In contrast, from the same range of shock states, Fe at Fe-LiF interfaces releases to states between 130 and 160 GPa (because it has a lower shock impedance than Al_2O_3); the data and model imply that Fe experiences little or no reshock at Fe-LiF interfaces up to 140 GPa (where the data end). Both the Fe-Al_2O_3 data and the model suggest that the degree of reshock is strongly pressure dependent above the solid Fe-liquid Fe phase boundary. LiF appears to be a more ideal window than Al_2O_3 also because it is a poorer thermal-inertia match to Fe than is Al_2O_3.

2) In the absence of energy sources and significant energy flux from other parts of the target, the rate of change of the film/foil-window interface temperature, $(\mathrm{d}T_{\mathrm{INT}}/\mathrm{d}t)$, is proportional to $-\mu\exp(-\mu^2)$, where $\mu \equiv \delta_{\mathrm{FW}}/2\sqrt{\kappa_{\mathrm{F}} t}$, δ_{FW} is the thickness of the reshocked zone in the film/foil layer at the film/foil-window interface, κ_{F} is the thermal diffusivity of the film/foil material, and $0 \le t \le t_{\mathrm{exp}}$ (t_{exp} is the time scale of the experiment). On this basis, the temperature of a thin ($\delta_{\mathrm{FW}} \ll 2\sqrt{\kappa_{\mathrm{F}} t_{\mathrm{exp}}}$) reshocked layer relaxes much faster than that of a thick ($\delta_{\mathrm{FW}} \gg 2\sqrt{\kappa_{\mathrm{F}} t_{\mathrm{exp}}}$) layer. We estimate $\sqrt{\kappa_{\mathrm{F}} t_{\mathrm{exp}}} \sim 10$ μm for Fe under the conditions of Fe-Al_2O_3 and Fe-LiF interfaces at high pressure. In this case, a 100-μm-thick reshocked Fe layer would relax very little, remaining near $T_{\mathrm{INT}}(0)$ on the time scale of the experiment, while a 1-μm-thick reshocked Fe layer would relax almost instantaneously (i.e., on a time scale much less than t_{exp}) to a temperature just above $T_{\mathrm{INT}}(\infty)$.

3) Greybody fits to an Fe-Fe film-Al_2O_3 experiment produce a gradually increasing effective greybody emissivity, $\hat{\varepsilon}_{\mathrm{gb}}(t)$, and a gradually decreasing greybody temperature, $T_{\mathrm{gb}}(t)$. This behavior is characteristic of most Fe-Fe-Al_2O_3 experiments. The decrease of $T_{\mathrm{gb}}(t)$ can be explained in terms of the model for the film/foil-window interface temperature, $T_{\mathrm{INT}}(t)$. For this experiment, the model implies that the thickness of the reshocked film layer, δ_{FW}, is approximately equal to the conduction length scale in the film, $\sqrt{\kappa_{\mathrm{F}} t_{\mathrm{exp}}}$ (~10 μm for Fe). Further, assuming $T_{\mathrm{gb}}(t) = T_{\mathrm{INT}}(t)$, the greybody fit constrains the amount of reshock, ΔT_{FW}, to ≤ 2000 K with σ_{WF}, the film/foil-window thermal mismatch, ~0.1, and $\delta_{\mathrm{FW}} \le 2\sqrt{\kappa_{\mathrm{F}} t_{\mathrm{exp}}}$. A slight decrease of the Al_2O_3 absorption coefficient upon shock compression can explain the slight increase of $\hat{\varepsilon}_{\mathrm{gb}}(t)$ with time; this may be consistent with the observation that the refractive index of Al_2O_3 seems to decrease with pressure.

4) In contrast to the Fe-Fe-Al_2O_3 results, greybody fits to data from an Fe-Fe foil-LiF target show a relatively constant greybody temperature and a decreasing greybody emissivity. The constant greybody temperature implies a constant interface temperature, as expected for an interface experiencing minimal reshock, while the decaying $\hat{\varepsilon}_{\mathrm{gb}}(t)$ is consistent with a shock-induced increase in the absorption coefficient of LiF. Setting $T_{\mathrm{INT}}(0) = T_{\mathrm{gb}}(0)$, we fit a simplified version of the full radiation model to these data and obtain an estimate of the absorption coefficient (~100 m^{-1}) of LiF shock-compressed to 122 GPa.

Introduction

Traditional studies of the behavior of shock-compressed materials assess the mechanical response of these materials to shock compression (e.g., the change of density with pressure). Since this approach cannot directly constrain the temperature of the high-pressure state, other means are needed to provide a complete equilibrium thermodynamic description (i.e., pressure-density-temperature) for these materials. To this end, recent studies record shock-induced radiation from initially transparent materials (e.g., alkali halides, summarized by KORMER, 1968; Al_2O_3, URTIEW, 1974; SiO_2 and Mg_2SiO_4, LYZENGA and AHRENS, 1980) and from opaque materials at interfaces viewed through transparent or semitransparent windows (e.g., Mg, URTIEW and GROVER, 1977; Ag, LYZENGA, 1980; Fe, BASS et al., this volume). These recent observations constrain some temperature in the target. In this paper we explore relationships between the experimentally constrained temperature and the temperatures of different high-pressure states achieved in the

High-Pressure Research in Mineral Physics, edited by M. H. Manghnani and Y. Syono, pp. 403–423.

target components and at their interfaces during the experiment. We attempt this in the context of a simple model of energy transfer and transport in the targets. To give the model considerations some weight, we compare model details and results to the recent observations of BASS et al. (this volume) on shock-induced radiation from Fe films and foils.

Model Considerations

Consider the target depicted in Figure 1, representative of that used in the experiments of LYZENGA (1980) and BASS et al. (this volume). This generic target consists of: 1) a 1.5-mm-thick, metallic "driver" plate (DP), 2) a metallic film or foil layer (FL), (1–10 μm and 10–100 μm thick, respectively, and 3) a dielectric, transparent window (TW), 3–4 mm thick. The target is constructed so that the shock impedance (i.e., the product of the initial density and shock wave velocity) of the DP is greater than or equal to that of the FL, which in turn has a shock impedance greater than or equal to that of the TW.

The experiment begins when a projectile impacts the DP (Figure 1), generating a shock wave that propagates through the DP to the DP-FL interface. Since this interface is formed by mechanical juxtaposition of the metallic DP and FL surfaces, it is "rough" on a $\sim$1 μm scale (URTIEW and GROVER, 1974). The shock front thickness is $\leq$0.01 μm in the materials and at the pressures of interest (e.g., KORMER, 1968). With respect to the shock wave, then, the DP and FL surfaces are partially free. Consequently, the shock wave accelerates the DP material at the DP-FL interface across the gap, and simultaneously reflects from the DP surface at the DP-FL interface as a release wave propagating back into the DP and releasing the DP to near-zero pressure. This moving DP surface then impacts the FL surface, generating shock waves of approximately equal magnitude that propagate backward into the just released DP, and forward into the unshocked FL. The former shock wave compresses the just released DP material from its low-pressure, high-temperature release state to one with approximately the same pressure as its previous shock-compressed state. Wave reverberations quickly bring this DP state to a state with a pressure equal to that of the shock-compressed FL and with a temperature well above that of the previous (i.e., first) DP shock-compressed state. If the backward propagating shock wave overtakes the release wave at some distance behind the DP-FL interface, this distance defines the thickness of a reshocked DP material layer at the DP-FL interface. However, if the release wave is faster than the reshock wave, the entire DP may experience low-pressure release and reshock. In either case, subsequent wave reverberations quickly bring the DP to a state with the same normal (to the interface) material velocity and stress fields as the shocked FL.

Since the DP material accelerating across the DP-FL interface impacts a rough FL surface, a thin (on the scale of the surface roughness) layer of film or foil material

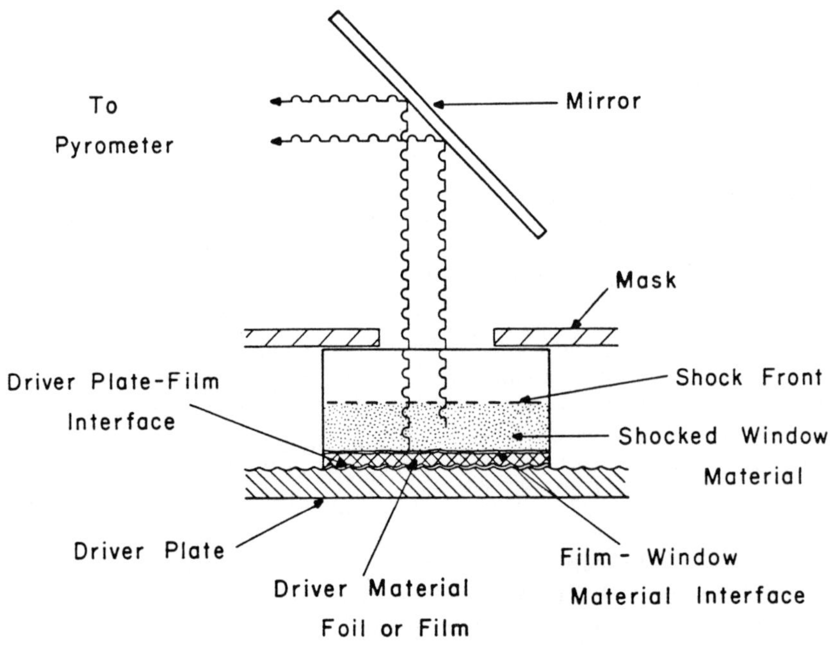

Figure 1. Target assembly.

compresses, much like a porous material (URTIEW and GROVER, 1974), to a much higher temperature than achieved by the shock-compressed FL material beyond this zone. As with the DP material at the DP-FL interface, the shock front traversing the FL reflects from a partially free surface at the FL-TW interface as a low-pressure release wave and accelerates the FL material across the interface to impact with the TW material. Since the TW surface at this interface is smooth relative to the shock front thickness, and is much more incompressible than either the DP or FL, the impacting FL material should not heat a thin layer of TW material, but rather only shock the TW up to high pressure and its Hugoniot temperature. Closure of the FL-TW interface generates backward and forward traveling shock waves; the former wave compresses the low-pressure, high-temperature released FL material to a state with approximately the same pressure as the first FL shock-compressed state. Wave reverberations quickly bring this FL state to a state with a pressure equal to that of the shock-compressed transparent window (shocked window: SW) and with a temperature much higher than the first FL shock-compressed state. If the backward propagating shock wave overtakes the release wave, it cuts off the zone of release/reshock in the FL material. In this case, the combined wave releases the remaining FL material, and then the DP, to a state with approximately the same normal velocity and stress as the SW. Alternatively, if the shock wave does not overtake the low-pressure release wave, the entire FL and/or DP is released and reshocked. In either case, subsequent wave reverberations should quickly bring both the DP and FL to states possessing normal stress and material velocity fields equal to those of the SW.

Since the reshocked layers at each interface are significantly hotter than the surrounding material (see URTIEW and GROVER, 1974, and discussion below), the temperature and radiation histories of targets with smooth versus rough interfaces should be quite different. This is a central issue for the FL-TW interface and so the early observed radiation history of the target (see discussion below). The fact that both foils and films are used as the layer between the DP and TW is directed to this issue. We expect the vacuum-coated film-TW interface to be much smoother than the mechanically formed foil-TW interface. However, this assumption turns out to be somewhat naive, as shown below. Since the TW surface at the FL-TW interface is smooth (defined above), we presume any roughness of this interface is due to roughness of the FL surface there.

As the FL material at the FL-TW interface is compressed, released and possibly reshocked, it heats up and begins to radiate. Consequently, the observed radiation intensity rises sharply (Figure 2a). As the shock wave travels forward into the TW, the thickness of the SW increases (Figure 2b), and consequently, so does its contribution to the total observed radiation (note increase with time in Figure 2c). If the TW is highly absorbing and/or scattering, or shock-compressed to such a state (as is apparent in many experiments: BOSLOUGH, 1985), the radiation intensity from the interface will decay with time (Figure 2c, dash-dot curve labeled "fast decay"); if not, the interface source will dominate the observed radiation history (Figure 2c, continuous curve labeled "slow decay") when the FL at the FL-SW interface is at a higher temperature than the SW. The recorded radiation is the sum of either the interface slow decay or fast decay contribution and the SW contribution. Given these possibilities, we must account for the degree of geometric (interface roughness) and material (shock-impedance) mismatch at each interface, especially at the FL-SW interface, in attempting to constrain the conditions of the FL Hugoniot state from observed radiation.

Even if each interface has little or no roughness, or is passive to energy transport across it, the DP, FL, and TW may shock compress to such different temperatures that the resulting temperature gradients between the layers drive significant relaxation of the FL-SW interface temperature on the time scale of the experiment. Dynamic phase changes or other energy sources and/or sinks present in the FL, FL at the interface, and/or SW on the time scale of the experiment may also introduce time dependence into the temperature and effective emissivity inferred from the radiation observations (GROVER and URTIEW, 1974; SVENDSEN, 1987). Consequently, we must examine whether or not the temperature profile of the compressed/released/reshocked target system relaxes via conductive and/or radiative transport on the time scale of the experiments, leading to time-dependent (thermal) radiation sources. We must also account for the effects of propagation through the SW, shock front, unshocked window (USW), and the TW free surface on the FL-SW interface and SW source radiation (BOSLOUGH, 1984). In this work we focus on the effects of processes at the FL-TW interface on the observed radiation.

Basic Model Assumptions

We assume all sources contributing to the observed radiation intensity are thermal and in local thermodynamic equilibrium. We can then relate the radiation source intensity to the radiation wavelength, λ, and absolute temperature, T, through the Planck function, $I_{Pl}(\lambda, T)$, given by

$$I_{Pl}(\lambda, T) \equiv \frac{2C_1}{\lambda^5(e^{C_2/\lambda T} - 1)}$$

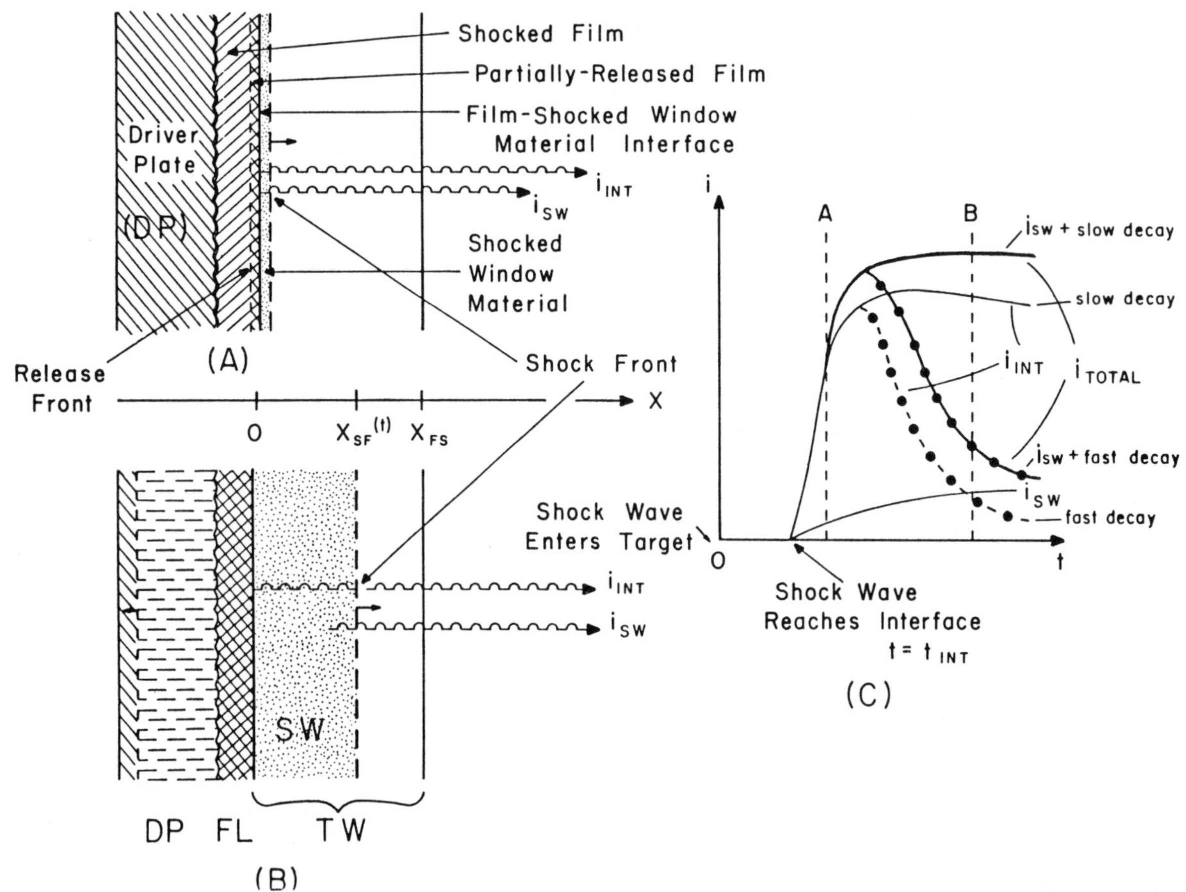

Figure 2. Dynamic model geometry. Shock front reaches film/foil layer (FL)-transparent window (TW) interface ($x=0$) at time t_{INT}, when radiation is first detected. Interface radiation (i_{INT}) dominates the early radiation history (a). If 1) the interface temperature decays slowly, and 2) the FL-TW interface, shocked window, and shock front remain relatively transparent, i_{INT} will dominate the observed radiation history during the experiment ("i_{SW}+slow decay" curve). However, if the FL-TW interface and/or shock front develop significant reflectivity and/or the SW develops significant opacity, i_{INT} will decay quickly (dash-dot curve), and may fall below the radiation intensity of the SW, i_{SW}, on the time scale of the experiment. The total intensity is then represented by the "i_{SW}+fast decay" curve.

where $C_1 \equiv 5.9544 \times 10^{-17}$ W·m^2/sr and $C_2 \equiv 1.4388 \times 10^{-2}$ m·K are constants. Comparison of the observed radiation-wavelength dependence with that of a blackbody source, as represented by $I_{Pl}(\lambda, T)$, implies that materials shock compressed to high pressures are dominantly thermal radiators ($\gtrsim$70 GPa: LYZENGA et al., 1983; BOSLOUGH, 1984). At lower pressures, however, most materials apparently radiate both thermally and non-thermally (SiO_2: KONDO and AHRENS, 1983; BRANNON et al., 1984; SCHMITT et al., 1986). Several of these materials are initially dielectrics (e.g., SiO_2: LYZENGA et al., 1983). The processes responsible for this radiation (defect electronic transitions?) are presently unidentified, but are suggested by spectrometric observations (KONDO and AHRENS, 1983).

In principle, energy transport in the target occurs by both radiation and conduction; our task is much more difficult if both radiation and conduction contribute equally to this process. In simple terms, we can understand the likely relative contribution of radiation and conduction to energy transport within layers and across interfaces via dimensional analysis. On this basis, conduction affects the balance of energy in the target over a length scale of $\lesssim \sqrt{\kappa t_{exp}} \sim 10^{-6}$ m, where κ is a characteristic thermal diffusivity (SVENDSEN, 1987) and $t_{exp} \sim 10^{-7}$ s is the experimental time scale. The radiative component of the energy flux is negligible to the balance of energy in the target if $K \lesssim 10^7$ m^{-1} (SVENDSEN, 1987), where K is a characteristic optical extinction coefficient. We expect the SW to satisfy this condition (LiF, WISE and

CHHABILDAS, 1986; Al_2O_3, BASS et al., this volume, but also see URTIEW, 1974); the FL most likely satisfies it as well (SVENDSEN, 1987). With these considerations in mind, we may reasonably decouple radiative transport from the energy balance in the target components and across their interfaces, and treat radiation separately. We emphasize that this analysis is limited by our ability to estimate the values of many key properties (e.g., thermal conductivity) of the appropriate high-pressure states of the DP and TW.

We assume that a given shock-compressed or -released state of any component of the target contributing to the observed radiation is one of constant and uniform density, stress, and material velocity. Consequently, our model is not directly applicable to the low-pressure regime ($\lesssim$70 GPa) where many shock-compressed materials deform heterogeneously (GRADY, 1980; KONDO and AHRENS, 1983; SCHMITT et al., 1986; SVENDSEN and AHRENS, 1986). Such behavior would require us to consider source distributions, spatially averaged effective emissivities and time-dependent thermomechanical processes, all beyond the scope of the simple equilibrium thermodynamic framework used here.

Initial Conditions

Shock-Compressed State

The model is referenced to the first shock-compressed (Hugoniot) state of each material in the target. Although this state is reached via a nonequilibrium, irreversible process (shock compression), we assume that thermodynamic equilibrium is achieved in the Hugoniot state. This requires the shock-compressed state to be one of constant, uniform density, material velocity, stress, and total energy, as we have already assumed. In this context, we may connect the initial and shock-compressed states via a classical thermodynamic path representing a change in specific internal energy equal to that judged by the general balance of energy across a shock front.

We assume that: 1) the material initially occupies a state with temperature T_i and pressure P_i, 2) shock-compresses adiabatically, and 3) shock-compresses as a fluid. Under these conditions, the classic Rankine-Hugoniot relation (e.g., RICE et al., 1958) for the change in internal energy across the shock front is valid. Equating this relation with an energetically equivalent equilibrium thermodynamic path (MCQUEEN et al., 1967; AHRENS et al., 1969) in density-pressure space, we obtain the following expression for the Hugoniot temperature of a high-pressure phase β, referenced to the initial (T_i, P_i) state of the low-pressure phase α (e.g., SVENDSEN, 1987)

$$\int_{T(s_i,\,\rho_H)}^{T_H} c_v(T, \rho_H)\,dT = \Delta e_v \quad (1)$$

with

$$\Delta e_v \equiv \frac{1}{2\rho_i^\alpha}\,\eta_H\,[P_H + P_i] - [\Delta e_i^{\beta-\alpha} + \Delta e_{s_i}]$$

being the difference in specific internal energy between the Hugoniot and principle isentrope of β at constant volume (density, ρ_H). In (1), $\Delta e_i^{\beta-\alpha} \equiv e(s_i^\beta, \rho_i^\beta) - e(s_i^\alpha, \rho_i^\alpha)$ is the difference in specific internal energy between α and β at T_i and P_i, and $\Delta e_{s_i} \equiv e(s_i, \rho_H) - e(s_i, \rho_i)$ is the change in specific internal energy along an isentrope of β with specific entropy s_i, referenced to a density ρ_i. Also, $T(s_i, \rho_H)$ represents the temperature along this isentrope at a density ρ_H, the Hugoniot-state density of β. In addition, ρ_i^α is the density of α at T_i and P_i, and $\eta_H \equiv 1 - \rho_i^\alpha/\rho_H$ is the relative compression. Finally, P_H is the pressure, and T_H is the temperature of the β-Hugoniot state. The subscripts "i" and "H" stand for the initial and Hugoniot states, respectively, while "s" stands for a state at constant specific entropy. Note that all material quantities in Equation (1) and in the remainder of this section refer to the β-phase, unless otherwise designated. From Equation (1), we see that the specific internal energy remaining after 1) accounting for the energy difference between α and β at T_i and P_i, 2) compressing β isentropically from ρ_i to ρ_H, is manifest as the Hugoniot temperature. We estimate $\Delta e_i^{\beta-\alpha}$ from the difference in density of the two phases at T_i and P_i, or in the case of melting, from the enthalpy of melting. The relative compression, η_H, is calculated from a standard impedance match (RICE et al., 1958) constrained by an experimental shock velocity, U, and material velocity, v, relation. Assuming γ, the equilibrium thermodynamic Grüneisen's parameter, is a function of density only, we calculate Δe_{s_i} and other quantities along the isentrope of β referenced to s_i from the equality of the Rankine-Hugoniot relation and equivalent equilibrium thermodynamic path (SVENDSEN, 1987). This frees us from making any further mechanical constitutive assumptions (e.g., finite strain) beyond that U is related to v. We calculate $T_{s_i} \equiv T(s_i, \rho_H)$ from the relation

$$\left\{\frac{\partial \ln T}{\partial \ln \rho}\right\}_s = \gamma \quad (2)$$

Since both dielectric and metallic solids initially compose our target, we consider equilibrium thermodynamic properties based on a model for the Helmholtz free energy, $F(T, \rho)$, that reflect the influence of both lattice and electronic processes. We use the Debye model (e.g., ZHARKOV and KALININ, 1968; ANDREWS, 1973) for the harmonic lattice free energy. BROWN and MCQUEEN (1986) recently used this model for the lattice contribution to the specific heat at constant volume of Fe. We add the low-temperature (T_H much less than the Fermi temperature) approximation to the electronic free energy, $-(1/2)\Gamma(\rho)T^2$, (e.g., WALLACE, 1972; JAMIESON et al., 1978; BROWN and MCQUEEN, 1986; BONESS et al., 1986) and the high-temperature (T_H greater than the Debye temperature, Θ_D) approximation to the anharmonic free energy, $A_2(\rho)T^2$ (WALLACE, 1972). Note that $\Gamma(\rho)$ describes the electron density-of-states at the Fermi level, while $A_2(\rho)$ is related to the shift in phonon frequency spectrum with temperature at constant pressure (WALLACE, 1972). Neglecting lattice-electron and band-structure contributions to $F(T, \rho)$, we may derive a number of classical thermodynamic properties (WALLACE, 1972). In particular, the form of $c_v(T, \rho)$ resulting from this free energy, is given by

$$c_v(T,\rho) \equiv -T\left\{\frac{\partial^2 F}{\partial T^2}\right\}_\rho = \frac{3\nu R}{M}\left\{4E_D(\xi_D) - \frac{3\xi_D}{[e^{\xi_D}-1]}\right\} + \frac{\Omega(\rho)}{M}T \quad (3)$$

where $\xi_D = \xi_D(T, \rho) \equiv \Theta_D(\rho)/T$ is the Debye similarity parameter,

$$\Omega(\rho) \equiv \Gamma(\rho) - 2A_2(\rho) \quad (4)$$

which combines the density-dependent coefficients of the approximate electronic and anharmonic contributions to F, and where ν is the number of atoms in the chemical formula, M is the molecular weight, and $E_D(\xi)$ is the Debye internal energy function (e.g., GOPAL, 1966). The Debye internal energy function is given by

$$E_D(\xi) \equiv \frac{3}{\xi^3}\int_0^\xi \frac{x^3}{[e^x - 1]}\,dx$$

$\Gamma(\rho)$ is related to the electronic Grüeisen's parameter, γ_e, through

$$\gamma_e \equiv -\left\{\frac{d\ln\Gamma}{d\ln\rho}\right\} \quad (5)$$

(WALLACE, 1972). Assuming γ_e is a constant, we have

$$\Gamma(\rho) \equiv \Gamma(\rho_i)\left\{\frac{\rho_i}{\rho}\right\}^{\gamma_e} \quad (6)$$

Since the approximate anharmonic and electronic contributions have the same temperature dependence, they represent indistinguishable thermal contributions to F and c_v. For the Fe target calculations presented below, we constrain the value of $\Omega(\rho_i)$ and ω empirically by assuming

$$\Omega(\rho) \equiv \Omega(\rho_i)\left\{\frac{\rho_i}{\rho}\right\}^{\omega} \quad (7)$$

which is chosen solely by analogy with Equation (6) for $\Gamma(\rho)$. By substituting Equation (3) into Equation (1), we obtain

$$\frac{3\nu R}{M}[E_D(\xi_{DH})T_H - E_D(\xi_{Ds})T_{s_i}] + \frac{1}{2M}\Omega(\rho_H)[T_H^2 - T_{s_i}^2] = \Delta e_v \quad (8)$$

where $\xi_{Ds} \equiv \Theta_D/T_{s_i}$ and $\xi_{DH} \equiv \Theta_D/T_H$. Equation (8) is an implicit relation for T_H, the temperature of the β state with specific internal energy equal to that of its Hugoniot state; we evaluate it numerically. If $\Omega=0$, (8) reduces to

$$E_D(\xi_{DH})T_H = E_D(\xi_{Ds})T_{s_i} + \frac{M}{3\nu R}\Delta e_v(\rho_H) \quad (9)$$

which is appropriate for a dielectric ($\Gamma=0$) material with $A_2T_{s_i}^2 \lesssim A_2T_H^2 \ll 1$. At high temperature ($\xi_{DH} \to 0$), the electronic and/or anharmonic contribution to c_v is comparable to $3\nu R/M$, and (8) reduces to

$$T_H = \frac{3\nu R}{\Omega(\rho_H)}\left\{\left\{1 + \frac{2}{9}\frac{M\Omega(\rho_H)}{(\nu R)^2}\Lambda_H\right\}^{1/2} - 1\right\} \quad (10)$$

(for $\Omega(\rho_H)>0$), where

$$\Lambda_H \equiv \Delta e_v + \frac{3\nu R}{M}E_D(\xi_{Ds})T_{s_i} + \frac{1}{2M}\Omega(\rho_H)T_{s_i}^2$$

Lastly, if we set both Ω and ξ_{DH} equal to zero in (8), we obtain

$$T_H = E_D(\xi_{Ds})T_{s_i} + \frac{M}{3\nu R}\Delta e_v \quad (11)$$

and this is essentially the relation most commonly used (e.g., JEANLOZ and AHRENS, 1980). For most metals with an electronic contribution to the free energy significantly greater than that from a free electron gas, the electronic and/or anharmonic contribution controls c_v, and so T_H, at high pressure. This is true because, for most materials, $\Theta_D(\rho)$ is a much weaker function of density (pressure) than is T_H. In addition, at low pressure where ξ_{DH} is near or even greater than unity, $T_H \sim T_{s_i}$, ξ_{DH} becomes very small with increasing pressure, and so the harmonic part of c_v approaches the constant Dulong-Petit value of $3\nu R/M$. Only for the most incompressible materials, such as Al_2O_3 and diamond, does ξ_{DH} remain appreciable at high pressure.

We note that, with the parameter set of Table 1, T_H for ε-Fe calculated with (8) is ~200 K higher than T_H

TABLE 1. Standard Temperature-Pressure Parameters

Symbol	ε-Fe	liquid Fe	Al_2O_3	LiF	Units
ρ	8352[a]	7952[b]	3986[c]	2650[d]	kg/m^3
a	4487[e]	4038[e]	8908[f]	5050[d]	m/s
b	1.57[e]	1.58[e]	0.91[f]	1.32[d]	
K_s	168[g]	130[g]	254[h]	68[g]	GPa
K_s'	5.28[i]	5.31[i]	4.32[h]	4.28[i]	
c_p	444[j]		775[c]	1615[d]	J/kg·K
α	4.3[k]		1.6[l]	10.3[l]	$\times 10^{-5}$K^{-1}
γ	1.95[m]		1.32[m]	1.78[m]	
q	1.0[n]		1.0[n]	1.0[n]	
Θ_D	385[j]		1250[h]	580[d]	K
T_M	1809[c]		2345[c]	845[o]	K
k	80[p]		46[p]	3[p]	W/m·K
ρ_e	50[q]		0	0	nΩ·m

[a]JEPHCOAT et al. (1986).
[b]Extrapolated from $\rho(T)=8136.(1-7.608\times10^{-5}T)$: DROTNING (1981).
[c]ROBIE et al. (1978). (assumed the same as δ-Fe)
[d]VAN THIEL (1977).
[e]Estimated from $U=3955+1.58v$ (BROWN and MCQUEEN, 1986).
[f]Fit to data in MARSH (1980).
[g]Calculated assuming $K_s=\rho a^2$.
[h]ANDERSON et al. (1968).
[i]Calculated with $K_s'=4b-1$ (RUOFF, 1967).
[j]ANDREWS (1973).
[k]Assumed the same as α-Fe (TOULOUKIAN et al., 1975).
[l]TOULOUKIAN et al. (1975).
[m]Calculated from $\gamma=\alpha K_s/\rho c_p$.
[n]$\gamma(\rho)=\gamma(\rho_i)[\rho_i/\rho]^q$ assumed in all calculations.
[o]WEAST (1979), p. D-187.
[p]TOULOUKIAN et al. (1970).
[q]Inferred from KEELER (1971).

calculated using (10) at 200 GPa. For Al_2O_3, however, the difference is ~300 K at 200 GPa. We use (8) to calculate T_H (see SVENDSEN, 1987, for further discussion). Having an expression for T_H, the next step is to estimate the effect of release on the shock-compressed state (i.e., T_H, P_H, ρ_H).

Release and Reshock States

In order to construct a classical thermodynamic path between the compressed and released states, we assume that the release process is adiabatic. In addition, we assume any mechanical work done by the system during release is reversible. In the case of a single-phase system, the release path is then both adiabatic and isentropic. The change in temperature with density along this path is related to γ, as given by Equation (2) above. Since the impedance match provides us with the pressure of the release state, P_R, we may calculate the temperature, T_R, and density, ρ_R, along an isentropic path that has not crossed a phase boundary through simultaneous solution of (2) and

$$\rho(T_R, P_R) = \rho(T_{HR}, P_R)\exp\left\{-\int_{T_{HR}}^{T_R}\alpha[T, \rho(T, P_R)]\mathrm{d}T\right\} \tag{12}$$

where $T_R \equiv T(s_H, \rho_R)$ is the temperature, $\rho_R \equiv \rho(T_R, P_R)$ is the density of the release state, and $\rho(T_{HR}, P_R)$ is the density along the Hugoniot of the same phase at a temperature T_{HR} and the pressure of the release state, P_R. The coefficient of thermal expansion, $\alpha(T, \rho)$, in (12) comes from an equilibrium thermodynamic model for the appropriate phase. For example, in the case of solid-state release, α follows from the equilibrium thermodynamic model for $F(T,\rho)$ used to obtain $c_v(T,\rho)$ in the last section, i.e.

$$\alpha \equiv \frac{\rho\left\{\dfrac{\partial^2 F}{\partial T\partial\rho}\right\}_{\rho,T}}{2\rho\left\{\dfrac{\partial F}{\partial\rho}\right\}_T + \rho^2\left\{\dfrac{\partial^2 F}{\partial\rho^2}\right\}_T} \equiv \frac{\rho\gamma c_v}{K_T}$$

where $K_T = K_T(T, \rho)$ is the isothermal bulk modulus, referenced to the isentrope or Hugoniot as discussed in SVENDSEN (1987).

To bound the nature of the release process as initiated at an interface, we focus on the extremes: 1) complete contact (≲shock-front thickness) at the interface, or 2) no contact, in which case each material has a free surface at the interface. We refer to the former interface as the "smooth" interface and to the latter as the "rough" interface. To illustrate the different paths these "end-member" interfaces should take, consider the two examples discussed below and depicted in Figure 3. If we shock-compress the DP(FL) to some point A long its Hugoniot below the Hugoniot-melting curve intersection, it will release to a state having, after one or two wave reverberations, the normal stress and material velocity of

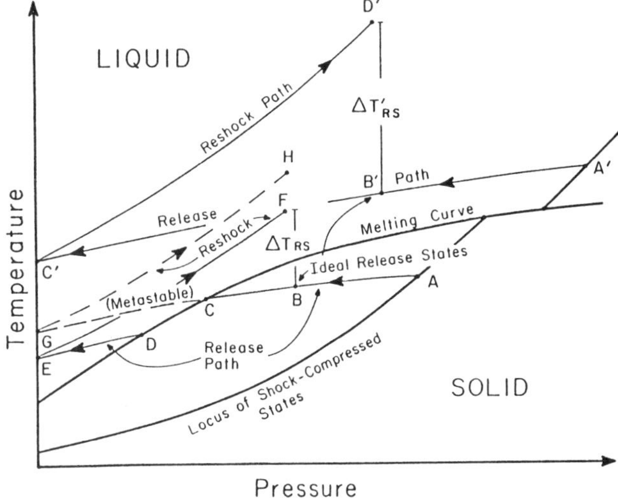

Figure 3. Possible range of T-P paths taken by DP and FL materials near the DP-FL and FL-TW interfaces, respectively, during an experiment.

the shock-compressed FL(TW). If these reverberations are isentropic, the resulting temperature will equal that calculated by direct release to the pressure of the shock-compressed FL(TW). The DP(FL) material at the smooth interface then releases directly to this state, represented by point B in Figure 3. However, the surface of the DP(FL) at the rough interface is partially free; hence the DP(FL) material at this interface releases to near-zero pressure. If we assume the release path is isentropic, its slope will be less steep than that of the melting curve. In this case, the melting curve and release path will intersect (see Figure 3 point C). If the phase transition is slow relative to the rate of decompression, the DP(FL) material will follow the ABCG (metastable) path to low pressure (even though this path is not necessarily isentropic, as discussed below). However, if the transition is uninhibited, the release path will turn along the phase boundary at the intersection point (Figure 3 point C), and the mixed-phase material will decompress along the boundary until the transition is complete or the mixture reaches near-zero pressure. Assuming the transition completes above zero pressure at point D in Figure 3, the now liquid DP(FL) will leave the phase boundary and continue to decompress along DE to zero pressure. As the DP(FL) material closes the interface, it impacts the FL(TW) material and is reshocked and reverberated along a series of paths, collectively symbolized in Figure 3 as the paths lying between EF and GH, up to the smooth-interface, release-state pressure which is that of the shocked FL(TW). Note that the temperature achieved by this set of shock paths is bounded above by the temperature estimated from a single shock compression back up to the Hugoniot pressure of the DP(FL); we use this bound below, along with isentropic release, since it follows directly from the results of the last section. Because the initial state of the reshocked material is at a higher temperature than the unshocked material, the reshocked material attains a higher temperature (ΔT_{RS} higher in Figure 3) than the release state of the DP(FL) material at a smooth interface. If the unshocked DP(FL) material is shocked to a higher-pressure state than A that is still below the melting curve, a smooth interface may release to a state pinned to the melting curve, or be above the melting curve, as for release from A′, in the liquid state. The rough interface released from A′ would follow A′B′C′ and be reshocked along C′D′ to D′. Note that the effect of reshock is much more pronounced as the initial shocked-state pressure increases, regardless of the phase transition.

When the release path encounters a phase boundary, such as the melting boundary shown in Figure 3, Equation (2) is no longer valid. If we believe the release path remains isentropic through this region, then we must require that, in addition to releasing adiabatically and doing or experiencing only reversible work, the material also changes phase in thermodynamic equilibrium. Under these conditions, the isentropic two-phase path for a congruent phase transition from β to π is described in terms of the mass fraction of π, χ, (SVENDSEN, 1987) as

$$\chi(T_{PB}(P), P) = \frac{\rho^{\beta}\rho^{\pi}}{\Delta s \mu \Delta\rho} \int_{P_{PB}^{\beta}}^{P} \frac{\mu(P^*)}{\rho^{\beta}\rho^{\pi}} [T_{PB}(P^*)\alpha^{\beta}\rho^{\pi}\Delta s - c_P^{\beta}\Delta\rho] dP^* \quad (13)$$

with

$$\mu = \mu(P) \equiv \exp\left\{\int^{P} \frac{1}{\Delta s}\left\{\Delta\left(\frac{\alpha}{\rho}\right) - \frac{\Delta\rho}{\rho^{\beta} T_{PB}(P^*)\rho^{\pi}}\left\{1 + \frac{\Delta c_P}{\Delta s}\right\}\right\} dP^*\right\}$$

In Equation (13) and the expression for $\mu(P)$, $T_{PB}=T_{PB}(P)$ is the equilibrium phase boundary, c_P is the specific heat at constant pressure, and $\Delta\psi \equiv \psi^{\pi} - \psi^{\beta}$ is the jump of any quantity ψ across the phase transition. Note that all quantities in (13) and μ are functions only of pressure along the phase boundary. Also, P_{PB}^{β} is the pressure at which the release path intersects the π-β phase boundary, i.e., where $\chi=0$. We evaluate (13) numerically along the phase boundary, $T_{PB}(P)$, with $\chi(P)$ increasing from zero until 1) $\chi=1$, complete transformation, or 2) $T_{PB}=T_{PB}(P_R)$, partial transformation. In the former case, the new phase releases to P_R along a path beginning at the pressure and temperature on the phase boundary where $\chi=1$. At the point where the release path leaves the phase boundary and enters a single-phase region, we may again use Equations (2) and (12) simultaneously.

Equation (13) is valid along any isentropic path through any first-order mixed-phase region; however, we are particularly interested in the solid-liquid phase boundary as discussed above. To utilize (13), we need to estimate solid- and liquid-state properties along the melting boundary, $T_M(P)$. We do this by way of semiempirical models for the solid state (e.g., ANDREWS, 1973; SVENDSEN, 1987) and the liquid state (e.g., STEVENSON, 1980; SVENDSEN, 1987); we require compatibility between the solidus, liquidus, and P-T phase boundary. The details and results of this procedure are discussed elsewhere (SVENDSEN, 1987), and are used in the calculations for Fe-Fe-LiF and Fe-Fe-Al_2O_3 targets as discussed below.

For the DP-FL and FL-TW interfaces with no contact, the DP and FL release to near-zero pressure, and consequently we cannot use the Hugoniot as a reference state. So instead of Equation (12), we solve Equation (2) simultaneously with

$$\rho(T_R, 0) = \rho(T_r, 0)[1 - \alpha(T_r, 0)(T_R - T_r)] \quad (14)$$

where T_r is some reference temperature (e.g., 298 K or T_M), depending on the relevant phase. With the density of the release state, ρ_R, we may estimate the free surface velocity of the DP and FL surfaces at the DP-FL and FL-TW interfaces, respectively, due to isentropic release, via the Riemann integral method (e.g., RICE et al., 1958). We assume, as required by the constraint of thermodynamic equilibrium, that the material velocity is continuous across the phase boundary (i.e., the same for both phases) when calculating the free surface velocity. We then take this free surface velocity as the "projectile" velocity of the DP or FL surface impacting the FL or TW surface, respectively, and use an impedance match to calculate the pressure and density of the reshocked state. To calculate the temperature of the reshock state, we use the appropriate form of (8), but referenced to the temperature and density of the complete release state (SVENDSEN, 1987), rather than to T_i and ρ_i^{α}.

Application to Fe Targets

To exemplify these considerations, we calculate release and reshock states for Fe-Fe-Al_2O_3 and Fe-Fe-LiF targets, as shown in Figures 4a and 4b. The solid and liquid Hugoniot states result from Equations (8) and (10) (with $A_2 \equiv 0$ and $\xi_{Ds}=0$), respectively, as based on the parameter set given in Table 1. The Fe release states result from simultaneous solution of Equations (2) and (12) or (14), together with (13) when appropriate. Solid-state properties along the release path and melting curve and referenced to the ε-Hugoniot via Equations (2), (8), and (12), while the analogous liquid-state properties are referenced to the experimentally constrained Fe melting curve of WILLIAMS and JEANLOZ (1986) via a liquid-state model for Fe (SVENDSEN, 1987). The Fe melting-curve data of WILLIAMS and JEANLOZ (1986), which extend to 100 GPa, are fit to a Lindemann parameterization for the solidus and then extrapolated to 330 GPa; the density along this solidus is referenced to the ε-Hugoniot (metastably above 245 GPa) in the same way as we referenced the density of the release state to this Hugoniot (i.e., through Equation (12) above). In calculating these release paths, we ignore all other solid phases of Fe, save ε-Fe, which is the stable solid phase of Fe along its Hugoniot between 13 GPa (BARKER and HOLLENBACH, 1974) and ~200 GPa, where the sound-speed measurements of BROWN and MCQUEEN (1982, 1986) along the Fe Hugoniot suggest that ε-Fe transforms to γ-Fe (?) or possibly a new solid phase (θ: BOEHLER, 1986). Consequently, γ-Fe and/or another solid phase is in equilibrium with liquid Fe above about 5 GPa to perhaps 280 GPa (e.g., ANDERSON, 1986). In this case, we neglect any effects of an $\varepsilon \rightarrow \gamma$ or $\varepsilon \rightarrow \theta$ transition in referencing compression along the Fe-melting curve to the ε-Fe Hugoniot. As stated above, in calculating the ε-Fe Hugoniot states shown in Figures 4a and 4b, we have constrained $\Omega(\rho_i)$, with $\omega=1.34$, which is the value of γ_e for ε-Fe given by BONESS et al. (1986) and by requiring the parameterized Fe melting curve and ε-Hugoniot to intersect at 245 GPa. On the basis of the parameter set given in Table 1, this fit constrains $\Omega(\rho_i)$ to 0.044 J/kg·K^2. BONESS et al. (1986) calculated a value of 0.090 J/kg·K^2 for $\Gamma(\rho_i)$ (adjusted to STP density for ε-Fe given in Table 1). If we set $\Omega(\rho_i)=0.090$, the ε-Hugoniot based on the parameter set in Table 1 intersects the melting curve at $\approx$280 GPa. We note that BONESS et al. (1986) constrained $\Gamma(\rho_i)=0.09$ J/kg·K^2 and $\gamma_e=1.27$ for γ-Fe, while BUKOWINSKI (1977) constrained $\Gamma(\rho_i)=0.08$ J/kg·K^2 and $\gamma_e=1.5$ for this phase. Both of these values for $\Gamma(\rho_i)$ of γ-Fe are sufficiently close to that of ε-Fe constrained by BONESS et al. (1986) so that, from this point of view, γ-and ε-Fe are indistinguishable. With these values for $\Gamma(\rho_i)$, the value of $\Omega(\rho_i)$ constrained above for ε-Fe implies some competition between anharmonic and electronic contributions to the specific heat of ε-Fe at high pressure.

BROWN and MCQUEEN (1986) conclude that their U-v relation fits the available Fe-Hugoniot data between 13 and 400 GPa. Since their sound-speed measurements also suggest that Fe melts along the Hugoniot above about 245 GPa, their U-v relation should apply to the liquid-solid mixture and pure liquid phase as well as to the solid. On this basis, we use their U-v relation to calculate both the ε-Fe Hugoniot and a metastable liquid-Fe Hugoniot referenced to the extrapolated density of liquid Fe at STP (SVENDSEN, 1987; Table 1). With this U-v relation, Equation (10) above for T_H ($A_2=0$ and $\xi_{Ds}=0$), and Γ as constrained by BONESS et al. (1986), we calculate the metastable Hugoniot of liquid Fe. Using $\Delta e_i^{\beta-\alpha}=0.14$ MJ/kg for Fe (SVENDSEN, 1987, as compared to the enthalpy of melting at standard pressure, 0.25 MJ/kg, from DESAI, 1986), the metastable liquid Fe Hugoniot intersects the melting curve at about 305 GPa. This agrees reasonably well with the results of YOUNG and GROVER (1984), who also ignored all other phases of Fe, save ε and liquid, in their parameterization of the Fe melting curve. We combine this metastable Hugoniot along with the ε-Fe Hugoniot in an ideal mix (e.g., WATT and AHRENS, 1984) to construct the shock-compressed, mixed-phase region shown in Figures 4a and 4b. For the most part, these mixed states return to the phase boundary upon release.

For comparison with the calculations, we have plotted the initial interface temperature results from the Fe film/foil experiments of BASS et al. (this volume) in Figures 4a and 4b. Note that the Fe-Al_2O_3 interface data shown in Figure 4a run almost parallel to the reshock locus, thereby exemplifying the strong pressure dependence of the reshock process (URTIEW and GROVER, 1974). Comparing the data with the smooth-interface release states, shown as squares in Figure 4a, implies that Fe at both film-Al_2O_3 and foil-Al_2O_3 interfaces experiences up to ~2000 K of reshock heating between 190 and

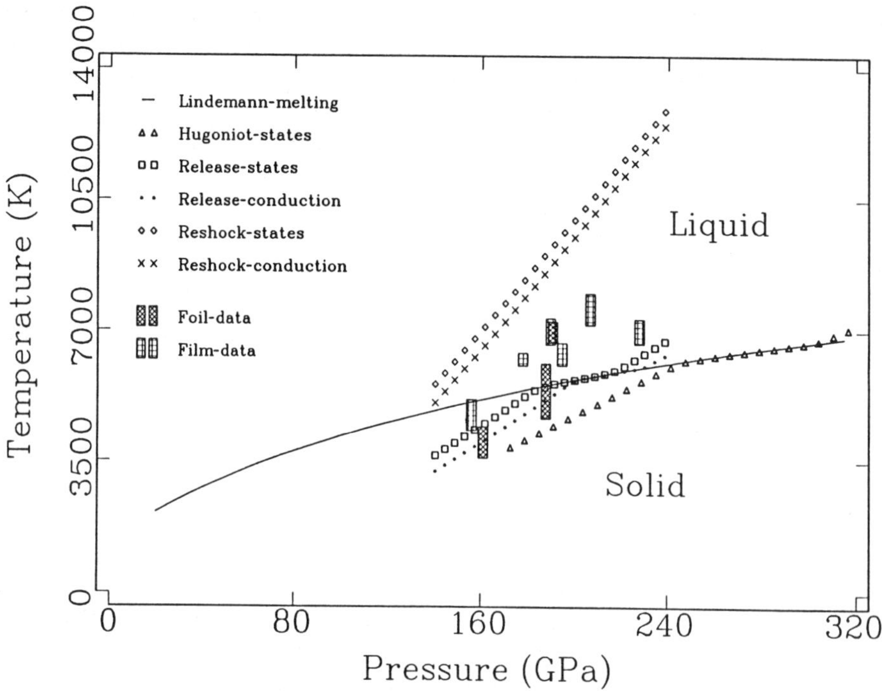

Figure 4a. Release/reshock calculations for Fe film/foil-Al_2O_3 interfaces and initial greybody temperatures inferred from Fe-Fe film/foil-Al_2O_3 radiation data of BASS et al. (this volume). "Release conduction" and "reshock conduction" symbols represent the initial effect of thermal inertia mismatch across the Fe film/foil-Al_2O_3 interface on the indicated states.

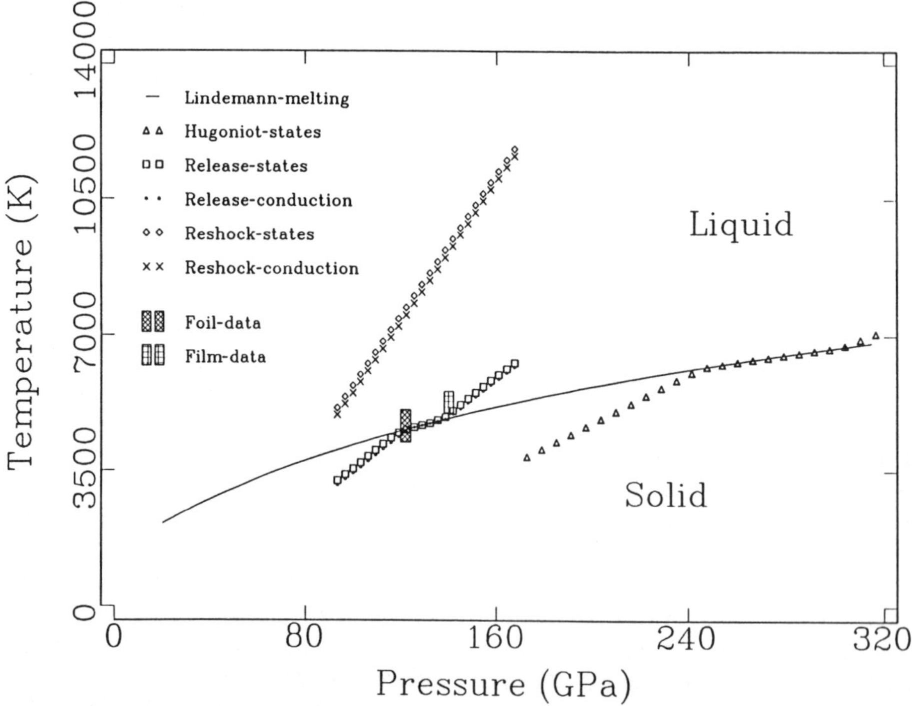

Figure 4b. Release/reshock calculations for Fe film/foil-LiF interfaces and initial greybody temperatures from Fe-Fe film/foil-LiF radiation data. The larger shock-impedance mismatch between Fe and LiF results in a lower release-state pressure at Fe-LiF interfaces than at Fe-Al_2O_3 interfaces, when both release from the same Hugoniot pressure.

230 GPa. As stated above, we naively expected that the film-TW interface would experience consistently less reshocking than the foil-TW interface. The present results contradict this expectation. There appears to be no guarantee that film interfaces will consistently experience any less reshock than the foil interfaces, especially at high pressure. In this case, a well-polished foil surface may actually experience less reshock than a slightly porous film interface. We also note that the films may be slightly porous, in which case they may experience both reshock and extra shock-heating during compaction. In this case, most or all of the deviation of the actual film interface temperatures away from those of the corresponding smooth interface release states may be due to porosity.

Figure 4b displays the results of the calculation for Fe-LiF interfaces. Because of the larger impedance mismatch between Fe and LiF, the Fe-LiF interface reaches a lower release-state pressure than the Fe-Al_2O_3 interface when both release from the same Hugoniot pressure. The data and calculation imply that lower release-state pressure results in less extreme reshocking. Note that the Fe-LiF and low-pressure Fe-Al_2O_3 data fall right on the corresponding smooth-release locus. The points labeled "release conduction" and "reshock conduction" refer to the effect of the contrast or mismatch in "thermal inertia" across the Fe-TW interface on the release and reshock temperatures, as discussed in the following section.

Conduction in the Target

We assume that the temperature profile created by shock compression, release, and/or reshock is established on a time scale short enough to represent the initial conditions for energy transport in the target. URTIEW and GROVER (1974) considered the problem of energy transfer at material interfaces and demonstrated that a rough ($\gtrsim 1$ μm) interface experiences a higher degree of shock heating than a smooth ($\lesssim 1$ μm) interface, much like a porous material experiences relative to its crystalline counterpart. Since the TW surface at the interface is much less rough ($\lesssim 10^{-8}$ m) than the DP surface at the interface, it should experience little, if any, direct reshock heating. However, the DP and FL surfaces at the DP-FL interface, as well as the FL surface at the FL-TW interface, may experience significant reshock heating, as discussed above.

Following GROVER and URTIEW (1974), we assume that: 1) energy transport is parallel to the direction of shock propagation (i.e., one-dimensional), 2) both temperature and heat flux are continuous across each interface in the target, and 3) there are no sources or sinks of energy in any layer or at the interfaces between them. Under these conditions, we may solve the one-dimensional conduction relation (CARSLAW and JAEGER, 1959) for the temperature profile, $T=T(x, t)$, in each target component as a function of position along the direction of shock propagation, x, and of time, t. The time $t=0$ corresponds to coincidence of the shock front and FL-TW interface. Since the temperature profile in the FL, and particularly the temporal variation of temperature at the FL-TW interface, control the source intensity of thermal radiation from the FL-TW interface, we emphasize these in what follows.

We expect a layer of the DP material at the DP-FL interface and a layer of the FL material at the FL-TW interface to experience some degree of reshocking. Also, the rough FL surface at the DP-FL interface should compress into a thin layer with a much higher temperature than the shock-compressed solid FL material. With this structure, the initial ($t=0$) temperature profile of the DP-FL-TW system is of the form shown in Figure 5. Here, d is the FL thickness, and $T_D \equiv T_D(x, 0)$ and $T_F \equiv T_F(x, 0)$ are the temperatures achieved in the DP and FL, respectively, by direct release to the pressure of the shock-compressed TW, which has a temperature $T_W(x, 0)$. Also, $T_D + \Delta T_D$ is the temperature of the reshocked layer with thickness δ_D in the DP at the DP-FL interface, while $T_F + \Delta T_{FW}$ is the temperature of the reshocked layer with thickness δ_{FW} in the FL at the FL-TW interface. If the surface of the FL at the DP-FL interface is also rough on some scale, it will compress like a porous material. Consequently, we assume that a layer with a thickness δ_{FD} and temperature $T_F + \Delta T_{FD}$ forms in the FL at the DP-FL interface. Since the surface of the TW is much smoother than the DP or FL surfaces, we assume that there is no

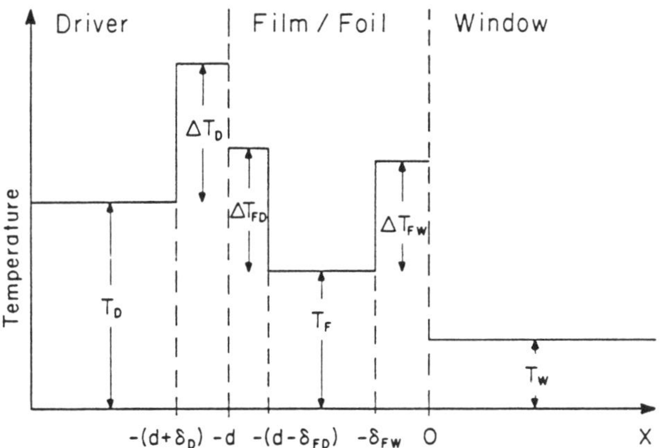

Figure 5. Initial conditions for thermal conduction in target. T_D and T_F represent temperatures achieved in DP and FL, respectively, upon direct release to a state with the pressure of the shock-compressed TW, having a temperature T_W. Variable degrees of reshocking are shown for DP (ΔT_D) and FL (ΔT_{FD}) at the DP-FL interface, and for FL (ΔT_{FW}) at the FL-TW interface; these involve some thickness (δ_D, δ_{FD}, and δ_{FW}) of each target component adjacent to the interface.

reshock heating of the TW material at the FL-TW interface. Note that the DP and TW are idealized as thermal half-spaces, a consequence of our assumption about the rates of shock compression and release relative to conduction. Again, we emphasize that all material properties of each target component are assumed homogeneous and time independent, and are referenced to their respective states at the pressure of the FL-TW interface.

The conduction relation for each layer, combined with the continuity of heat flux and temperature as boundary conditions, and the initial conditions shown in Figure 5 define an initial-boundary value problem for $T(\xi, \tau)$ in the target, where $\xi = x/d$ is the nondimensional position along the direction of shock propagation and $\tau = t/t_{exp}$ is the nondimensional time. We are particularly interested in this profile for the FL and the temporal variation of $T_{INT}(\tau) \equiv T(0, \tau)$, which represents the FL-TW interface temperature. Solving this initial-boundary value problem elsewhere (SVENDSEN, 1987), we obtain expressions for $T_D(\xi, \tau)$, $T_F(\xi, \tau)$, and $T_W(\xi, \tau)$. Here we focus attention on the FL-TW interface temperature, $T_{INT}(\tau) \equiv T_F(0, \tau) = T_W(0, \tau)$, since it is responsible for controlling the interface source radiation. Specializing the relation for $T_{INT}(\tau)$ to the case where the DP and FL are the same material (e.g., Fe), we have $T_D = T_F$; we also assume $\Delta T_D = \Delta T_{FD}$ for simplicity. In this context, the same pressure-temperature state exists on either side of the DP-FL interface, and $T_{INT}(\tau)$ is given by

$$T_{INT}(\tau) = T_F + G(\tau)\Delta T_D + D(\tau)\Delta T_{FW} + \frac{\sigma_{WF}}{(1 + \sigma_{WF})}(T_W - T_F) \quad (15)$$

with

$$G(\tau) = \frac{1}{(1 + \sigma_{WF})}[\text{erfc}\{(1 - \delta_{FW}^*)\omega_F\} - \text{erfc}\{(1 + \delta_D^*)\omega_F\}]$$

and

$$D(\tau) = \frac{1}{(1 + \sigma_{WF})} \text{erf}\{\delta_{FW}^* \omega_F\}$$

In these expressions, $\omega_F \equiv \sqrt{Pe_F/4\tau}$, $Pe_F \equiv d^2/\kappa_F t_{exp}$ is the Péclet number, $\kappa \equiv k/\rho c_P$ is the thermal diffusivity, $\delta^* \equiv \delta/d$, and t_{exp} is the time scale of experiment. Also, we have

$$\sigma_{WF} \equiv \left\{\frac{k_W \rho_W c_{P,W}}{k_F \rho_F c_{P,F}}\right\}^{1/2} \quad (16)$$

as the thermal inertia "mismatch" (CARSLAW and JAEGER, 1959, p. 321) between the TW and FL at the pressure of the FL-TW interface. In Equation (16), k, ρ, and c_P are the thermal conductivity, density, and specific heat at constant pressure, respectively, of the designated material at the pressure of the shock-compressed TW. Noting that the complementary error function, erfc(x), decreases with increasing x, and erf(x) increases with increasing x, we see that $D(\tau)$ will decay with time, while $G(\tau)$ can either grow or decay. Depending on the relative magnitude of those different contributions (investigated elsewhere: SVENDSEN, 1987), $T_{INT}(\tau)$, as given by Equation (15), can either grow or decay with time. In particular, if the entire film/foil layer ($\delta_{FW}^* = 1$) and a significant portion of the DP ($\delta_D > 2\sqrt{\kappa_F t_{exp}}$) are reshocked, $T_{INT}(\tau)$ can actually increase, or decrease and then increase, with time. Most radiation observations constrain a decreasing temperature with time (see discussion below); however, there may be some suggestion of the influence of ΔT_D on $T_{INT}(\tau)$ in the data on Fe-LiF interfaces discussed below.

To estimate the value of σ_{WF} at high pressure, we need the appropriate values of k, ρ, and c_P. Density follows from the impedance match and release calculations, while the specific heat at constant pressure results from the classical thermodynamic models discussed above. Assuming the thermal conductivity of the metallic target components is dominated by the electronic contribution and that the electrical conductivity is constant (Fe: KEELER, 1971), we calculate k_F as a function of temperature with the Wiedemann-Franz-Lorenz relation (e.g., BERMAN, 1976). Assuming the thermal conductivity of the TW material is controlled by lattice processes, we may use the thermal conductivity model of ROUFOSSE and KLEMENS (1974) to estimate k_W. As compared to k_D or k_F, k_W predicted from this model increases much more slowly with pressure, partially accounting for the development of a significant thermal inertia mismatch across the FL-TW target interface. Based on the release/reshock calculations presented in the previous section and on the above considerations, we may calculate σ_{WF} for the Fe-Al_2O_3 and Fe-LiF interfaces using the value of electrical conductivity for ε-Fe given in Table 1, which is estimated from the work of KEELER (1971). We list results of this calculation in Table 2. It is evident that Fe is more closely matched to Al_2O_3 than LiF; since $T_{INT}(\tau)$ is proportional to $(1+\sigma_{WF})^{-1}$ (see Equation (15)), there is a greater adjustment of $T_{INT}(\tau)$ at the Fe-Al_2O_3 interface, as shown below.

If we assume $\Delta T_D = 0 = \Delta T_{FD}$ and/or $\delta_D = 0 = \delta_{FD}$, Equation (15) reduces to

$$T_{INT}(\tau) = T_F + D(\tau)\Delta T_{FW} + \frac{\sigma_{WF}}{(1 + \sigma_{WF})}(T_W - T_F) \quad (17)$$

which is the basic reshock model considered by GROVER and URTIEW (1974). Further, if we let $\delta_{FW} \to 0$, we have, from Equation (17),

TABLE 2. STP and High-Pressure Thermal Mismatches, σ_{WF}, for Fe Targets

	STP	100 GPa	200 GPa
		Ideal interface	
Fe-Al_2O_3	0.56	0.25	0.15
Fe-LiF	0.20	0.11	0.05
		Reshocked interface	
Fe-Al_2O_3	0.56	0.15	0.07
Fe-LiF	0.20	0.08	0.01

$$T_{INT}(\tau) = T_{Id} \equiv T_F + \frac{\sigma_{WF}}{(1+\sigma_{WF})}(T_W - T_F) \quad (18)$$

which relates the temperature of the smooth FL-TW interface, T_{Id}, to the temperature of the direct-release state, T_F. Note that T_{Id} approaches T_F as $\sigma_{WS} \rightarrow 0$, and T_W as $\sigma_{WF} \rightarrow \infty$. Also, note that $T_{INT}(\tau)$, as given by (17), will be time dependent only if the FL-TW interface is reshocked. We use Equation (17) with the reshock-state temperature in the FL at the FL-TW interface ($T_F + \Delta T_{FW}$), the shock-compressed temperature of the TW (T_W), and σ_{WF}, to calculate $T_{INT}(0)$, which is labeled "reshock conduction" in Figures 4a and 4b. Similarly, we use Equation (18) to calculate the "release conduction" temperatures from T_F, T_W, and σ_{WF}. As stated above, the Fe-LiF thermal mismatch is greater (i.e., σ_{WF} is much smaller: see Table 2) than that of the Fe-Al_2O_3 interface, mainly because LiF is more compressible and less conductive (thermally: see Table 1) than Al_2O_3. In this case, the Fe-LiF interface temperature remains closer to the temperature of Fe at the interface than does the Fe-Al_2O_3 interface temperature. Further, the greater compressibility of LiF gives it a much higher shock-compressed temperature than Al_2O_3. For example, T_H for LiF (from Equation (9)) at 160 GPa is $\approx$4200 K (ignoring the possibility of melting), while T_H for Al_2O_3 (also from Equation(9)) at 230 GPa is $\approx$2750 K. The temperature mismatch is much less across the Fe-LiF interface, and the effect of thermal inertia mismatch on T_{INT} is less extreme.

In Figures 6a and 6b, we present a set of calculations for $T_F(\xi, \tau)$ and the associated $T_{INT}(\tau) = T_F(0, \tau)$, with $\Delta T_D = \Delta T_{FD}$, etc., as assumed to write Equation (17), for Fe-Fe-TW targets. To construct these figures, we calculate the compressed/released and reshocked/released states achieved in an Fe-Fe-Al_2O_3 target impacted by a Ta projectile at a velocity of 5.67 km/s; we assume the calculated reshock temperatures at the DP-FL and FL-TW interfaces are the initial values. This impact velocity is that of one of the experiments (Fe-Fe film-Al_2O_3) discussed below. The basic result here is the dependence of the rate of change of $T_{INT}(\tau)$ on Pe_F and δ^*_{FW}. From Equation (15), the change in $T_{INT}(\tau)$ with time is given by

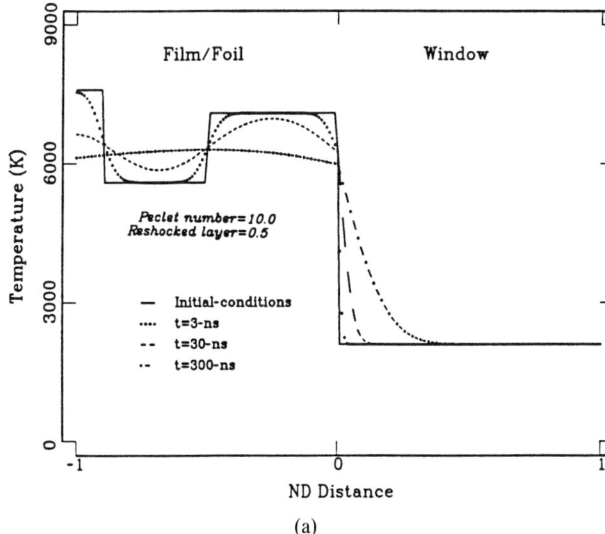

(a)

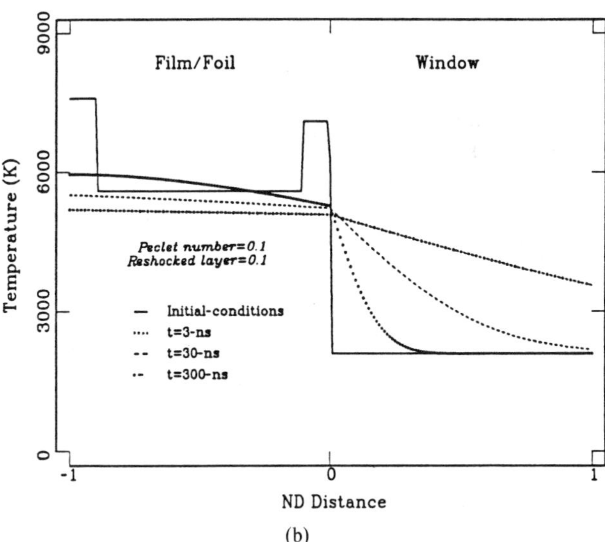

(b)

Figure 6. Variation of the temperature near a reshocked Fe film/foil-Al_2O_3 interface for different values of the FL Péclet number, Pe_F, and nondimensional FL reshocked-layer thickness, δ^*_{FW}. These results are based on the release/reshock calculations discussed in the text, as applied to a Fe-Fe film-Al_2O_3 target impacted by a tantalum projectile at 5.67 km/s. We plot the variation of temperature in the FL layer, $T_F(\xi, \tau)$, as a function of nondimensional (ND) position, ξ, with respect to the FL-TW interface ($\xi = x/d = 0$) at four different times during a 300 ns experiment. The nondimensional range of -1 to 1 corresponds to $-d$ to d, where d is the thickness of the FL layer.

$$\left\{\frac{dT_{INT}(\tau)}{d\tau}\right\} = G'(\tau)\Delta T_D + D'(\tau)\Delta T_{FW} \quad (19)$$

with

$$G'(\tau) = \frac{2}{(1 + \sigma_{\mathrm{WF}})}\left\{\frac{Pe_{\mathrm{F}}}{4\pi\tau^{3}}\right\}^{1/2} \times \{(1 - \delta^{*}_{\mathrm{FD}})e^{-[(1-\delta^{*}_{\mathrm{FD}})\omega_{\mathrm{F}}]^{2}} - (1 + \delta^{*}_{\mathrm{D}})e^{-[(1+\delta^{*}_{\mathrm{D}})\omega_{\mathrm{F}}]^{2}}\}$$

and

$$D'(\tau) = \frac{-2}{(1 + \sigma_{\mathrm{WF}})}\left\{\frac{Pe_{\mathrm{F}}}{4\pi\tau^{3}}\right\}^{1/2} \delta^{*}_{\mathrm{FD}} e^{-[\delta^{*}_{\mathrm{FW}}\omega_{\mathrm{F}}]^{2}}$$

For the particular case we have plotted, and as noted above, unless $\delta^{*}_{\mathrm{FW}} \approx 1$, $\delta^{*}_{\mathrm{D}} \gtrsim 1$ and/or $\Delta T_{\mathrm{D}} \gg \Delta T_{\mathrm{FW}}$, the ΔT_{FW} term dominates $T_{\mathrm{INT}}(\tau)$ (SVENDSEN, 1987). Since $D'(\tau)$, the coefficient of the dominating term, is always negative, the rate of change of $T_{\mathrm{INT}}(\tau)$ will be negative, and $T_{\mathrm{INT}}(\tau)$ will consequently decrease with time. Further, when the $D'(\tau)$ term in (19) is dominant, $\mathrm{d}T_{\mathrm{INT}}/\mathrm{d}\tau$ is proportional to $-\mu\exp(-\mu^{2})$, with $\mu \equiv \delta_{\mathrm{FW}}\sqrt{Pe_{\mathrm{F}}/4\tau} = \delta_{\mathrm{FW}}/2\sqrt{\kappa_{\mathrm{F}}t}$ (where δ_{FW} is the actual dimensional thickness of the reshocked layer). Note that $\mu\exp(-\mu^{2})$ achieves a maximum value near $\mu \sim 1$ and is much smaller (~ 0) for μ much greater or less than unity. On this basis, as shown in Figure 6a, $T_{\mathrm{INT}}(\tau)$ a conductively thick ($Pe_{\mathrm{F}} \sim 10$) FL with half of its thickness reshocked to ΔT_{FW} (i.e., $\delta^{*}_{\mathrm{FW}} = 0.5$) relaxes very little from its initial value on the time scale of the experiment. At the other extreme, as shown in Figure 6b, if $Pe_{\mathrm{F}} \sim 0.1$, the FL is conductively thin. For a thin ($\delta^{*}_{\mathrm{FW}} = 0.1$) reshock layer, relaxation occurs essentially instantaneously to $T_{\mathrm{INT}}(\infty)$; this temperature is also that of the smooth interface, as given by Equation (18). For $Pe_{\mathrm{F}} \sim 1$ and intermediate reshock-layer thicknesses, $T_{\mathrm{INT}}(t)$ is time dependent on an experimentally resolvable time scale, and its variation with time produces a corresponding radiation source time dependence, as we show in the next section.

Radiative Transport in the Target

With a model of the initial temperature profile of the target components and interfaces, we now establish a connection between the radiation intensity of sources at these temperatures and the radiation intensity emerging from the free surface of the TW during the experiment. The target is represented as a series of plane-parallel layers (Figure 2) with Fresnel boundaries (BOSLOUGH, 1985; SVENDSEN, 1987). We assume that: 1) source radiation is collimated by the target geometry; 2) all radiation sources are thermal, and so their intensity is given by the Planck function; 3) sources are located only at the FL-SW and/or uniformly throughout the SW, particularly along the direction of shock propagation; and 4) all optical properties are independent of wavelength. The model spectral intensity of radiation emerging from the free surface of the USW (unshocked window), $I_{\lambda\mathrm{mod}} = I_{\lambda\mathrm{mod}}(\lambda, t)$, as a function of wavelength, λ, and time, t, after the shock front has passed the FL-SW interface, is given by

$$I_{\lambda\mathrm{mod}}(\lambda, t) \equiv \hat{\varepsilon}_{\lambda\mathrm{SW}}(t) I_{\mathrm{Pl}}(\lambda, T_{\mathrm{W}}) + \hat{\varepsilon}_{\lambda\mathrm{INT}}(t) I_{\mathrm{Pl}}[\lambda, T_{\mathrm{INT}}(t)] \quad (20)$$

The Hugoniot temperature of the SW, T_{W}, is homogeneous, uniform, and constant since we assume a uniform distribution of SW sources. The interface temperature, T_{INT}, is a function of time, or constant, in the context of the conduction model discussed above. The "λ" subscript denotes a spectral quantity. In (20), we identify

$$\hat{\varepsilon}_{\lambda\mathrm{SW}}(t) \equiv \Psi_{\lambda}(t)[1 - \tau_{\lambda\mathrm{SW}}(t)][1 + r_{\lambda\mathrm{INT}}\tau_{\lambda\mathrm{SW}}(t)] \quad (21)$$

and

$$\hat{\varepsilon}_{\lambda\mathrm{INT}}(t) \equiv \Psi_{\lambda}(t)\tau_{\lambda\mathrm{SW}}(t)[1 - r_{\lambda\mathrm{INT}}] \quad (22)$$

as the effective normal spectral emissivities of the SW and FL-SW interface, respectively. As evident from (20), $\hat{\varepsilon}_{\lambda\mathrm{SW}}$ and $\hat{\varepsilon}_{\lambda\mathrm{INT}}$ are the properties connecting the intensities of the sources within the target and the intensity emerging from the target. The function $\Psi_{\lambda}(t)$ is defined by

$$\Psi_{\lambda}(t) \equiv [1 - r_{\lambda\mathrm{FS}}]\tau_{\lambda\mathrm{USW}}(t)[1 - r_{\lambda\mathrm{SF}}] \quad (23)$$

and represents the effect on source radiation of propagation through the FS, USW, and SF. In Equations (21) to (23), $r_{\lambda\mathrm{FS}}$, $r_{\lambda\mathrm{SF}}$, and $r_{\lambda\mathrm{INT}}$ are the effective normal spectral reflectivities of the FS, SF, and FL-SW interface, respectively. Further,

$$\tau_{\lambda\mathrm{USW}}(t) \equiv e^{-a^{*}_{\lambda\mathrm{USW}}(1 - t/t_{\mathrm{exp}})} \quad (24)$$

and

$$\tau_{\lambda\mathrm{SW}}(t) \equiv e^{-a^{*}_{\lambda\mathrm{SW}}t/t_{\mathrm{exp}}} \quad (25)$$

are the effective normal spectral transmissivities of SW and USW layers, respectively. The quantities $a^{*}_{\lambda\mathrm{USW}}$ and $a^{*}_{\lambda\mathrm{SW}}$ are nondimensional forms of the effective normal spectral absorption coefficients in the USW and SW, respectively, and they mediate the explicit time dependence of $\hat{\varepsilon}_{\lambda\mathrm{SW}}$ and $\hat{\varepsilon}_{\lambda\mathrm{INT}}$. Noting that, for steady shock propagation, the position of the shock front in the TW (Figure 2) may be written $x_{\mathrm{SF}}(t) \equiv (U - v)t$, and $a^{*}_{\lambda\mathrm{SW}}$ and $a^{*}_{\lambda\mathrm{USW}}$ are related to their dimensional counterparts through

$$a^{*}_{\lambda\mathrm{SW}} \equiv a_{\lambda\mathrm{SW}}(x_{\mathrm{FS}} - v t_{\mathrm{exp}}) \quad (26)$$

and

$$a^{*}_{\lambda\mathrm{USW}} \equiv a_{\lambda\mathrm{USW}} x_{\mathrm{FS}} \quad (27)$$

where x_{FS} is the thickness of the TW, and so the position of the FS with respect to the FL-TW interface (Figure 2).

From Equations (26) and (27), $a^*_{\lambda SW}$ and $a^*_{\lambda USW}$ will be of order unity when

$$a_{\lambda SW} \sim (x_{FS} - vt_{exp})^{-1} \text{ and } a_{\lambda USW} \sim x_{FS}^{-1} \tag{28}$$

Both $a_{\lambda SW}$ and $a_{\lambda USW}$ are $\sim 10^3 m^{-1}$, since the thickness of the TW is generally $\sim 10^{-3}$ m. So for values of $a_{\lambda SW}$ and/or $a_{\lambda USW}$ much larger or smaller than these "geometric" values, source radiation intensity is not resolvable, or is not affected by propagation through the SW and USW, respectively. The USW is usually transparent, so $a_{\lambda USW} \sim 0$; if the SW is transparent as well, then $a_{\lambda SW} \sim 0$ and, from (21) and (22), we have

$$\hat{\varepsilon}_{\lambda INT}(t) \lesssim \hat{\varepsilon}_{\lambda INT}(0) = (1 - r_{\lambda FS})(1 - r_{\lambda SF})(1 - r_{\lambda INT}) \tag{29}$$

and $\hat{\varepsilon}_{\lambda SW}=0$. In this case, $I_{\lambda mod}$ is governed entirely by sources at the FL-SW interface, and any time dependence of the observed radiation history is due solely to T_{INT}. Note that the bound on $\hat{\varepsilon}_{\lambda INT}$ in (29) is also the initial value of $\hat{\varepsilon}_{\lambda INT}$ (i.e., it can only decrease with time). However, if the TW becomes opaque upon shock compression, we have $a_{\lambda SW} \to \infty$. Again, with $a_{\lambda USW} \sim 0$, we have $\hat{\varepsilon}_{\lambda INT} \to 0$ and

$$\hat{\varepsilon}_{\lambda SW}(t) \lesssim (1 - r_{\lambda FS})(1 - r_{\lambda SF}) \tag{30}$$

In this case, observable sources are confined to the shock front (this is the "ideal case" of BOSLOUGH, 1985). The impact of these and other model parameters on $I_{\lambda mod}(\lambda, t)$ is more explicitly depicted by writing the partial derivatives of $I_{\lambda mod}(\lambda, t)$ with respect to λ and t. From the definition of the Planck function, and Equations (20), (21), and (22), these are

$$\lambda\left\{\frac{\partial I_{\lambda mod}}{\partial \lambda}\right\}_t = P(\mu_{SW})\hat{\varepsilon}_{\lambda SW}(t) I_{Pl}(\lambda, T_W) + P(\mu_{INT})\hat{\varepsilon}_{\lambda INT}(t) I_{Pl}[\lambda, T_{INT}(t)] - 5 I_{\lambda mod}(\lambda, t) \tag{31}$$

and

$$t_{exp}\left\{\frac{\partial I_{\lambda mod}}{\partial t}\right\}_\lambda = I_{\lambda mod} a^*_{\lambda USW} + \{[(1 - r_{\lambda INT}) + 2 r_{\lambda INT}\tau_{\lambda SW}(t)] I_{Pl}(\lambda, T_W) - \hat{\varepsilon}_{\lambda INT}(t) I_{Pl}[\lambda, T_{INT}(t)]\} a^*_{\lambda SW} + P(\mu_{INT})\hat{\varepsilon}_{\lambda INT}(t) I_{Pl}[\lambda, T_{INT}(t)]\left\{\frac{d \ln T_{INT}}{dt}\right\} \tag{32}$$

with

$$P(\xi) \equiv \frac{\xi}{1 - e^{-\xi}}$$

and

$$\mu_{SW} \equiv \frac{C_2}{\lambda T_W}, \quad \mu_{INT} = \frac{C_2}{\lambda T_{INT}}$$

Relation (31) exemplifies the fact that the wavelength dependence of $I_{\lambda mod}$ is due solely to that of the Planck function, since we have assumed the optical properties are independent of wavelength. We make this assumption because existing data cannot clearly resolve wavelength-dependent optical properties (SVENDSEN, 1987).

Again, for most TW's, we have $a_{\lambda USW} \sim 0$. If, in addition, $T_{INT}(t)$ is approximately constant with time, which may occur in a conductively thick ($Pe_F \gg 1$) or thin ($Pe_F \ll 1$) FL with a thick ($\delta_{FW} \sim d$) or thin ($\delta_{FW} \ll d$) reshocked layer of FL material, or at the smooth interface, as discussed above, (34) reduces to

$$t_{exp}\left\{\frac{\partial I_{\lambda mod}}{\partial t}\right\}_\lambda = \{[(1 - r_{\lambda INT}) + 2 r_{\lambda INT}\tau_{\lambda SW}] I_{Pl}(\lambda, T_W) - \hat{\varepsilon}_{\lambda INT} I_{Pl}[\lambda, T_{INT}(t)\} a^*_{\lambda SW} \tag{33}$$

This will be positive if

$$[(1 - r_{\lambda INT} + 2 r_{\lambda INT}\tau_{\lambda SW}] I_{Pl}(\lambda, T_W) > \hat{\varepsilon}_{\lambda INT} I_{Pl}[\lambda, T_{INT}(t)] \tag{34}$$

but otherwise will be negative, since $a^*_{\lambda SW}$ is always positive. Consequently, with a finite value of $a^*_{\lambda SW}$ and a time-independent interface temperature, $I_{\lambda mod}$ will grow or decay with time on the basis of the sign of (33). If the TW is initially transparent and remains relatively transparent upon shock compression, we have $a_{\lambda SW} \sim 0$, and (32) simplifies to

$$\left\{\frac{\partial I_{\lambda mod}}{\partial t}\right\}_\lambda = P(\mu_{INT})\hat{\varepsilon}_{\lambda INT} I_{Pl}[\lambda, T_{INT}(t)]\left\{\frac{d \ln T_{INT}}{dt}\right\} \tag{35}$$

and any variation of $I_{\lambda mod}$ with time should reflect that of T_{INT} through the Planck function.

Models and Data

We compare models and data in the context of the standard χ^2 statistic (e.g., BEVINGTON, 1969; PRESS et al., 1986). In our case, it is given by

$$\chi^2(\mathbf{a}) \equiv \sum_{i=1}^{N_\lambda}\sum_{j=1}^{N_t} \frac{1}{\sigma_{ij}^2}\left\{I_{\lambda exp}(\lambda_i, t_j) - I_{\lambda mod}(\lambda_i, t_j; \boldsymbol{a})\right\}^2 \tag{36}$$

In this relation, $I_{\lambda exp}(\lambda_i, t_j)$, $I_{\lambda mod}(\lambda_i, t_j; \boldsymbol{a})$, $\sigma_{ij} \equiv \sigma(\lambda_i, t_j)$ are the experimental and model spectral radiances and the experimental uncertainties, all at a particular wavelength, λ_i, and time, t_j. Also, N_λ and N_t are the number of wavelengths and times sampled, respectively, in the experiment. The five-component "vector" $\boldsymbol{a}$ is the model parameter vector, with components a_k, in our case, given by

$$a_k \equiv \{r_{\lambda SW}, a_{\lambda SW}^*, T_{SW}, r_{\lambda INT}, T_{INT}(t)\} \quad (37)$$

We note that $r_{\lambda FS}$ and $a_{\lambda USW}^*$ are not included in (37), since they may be calculated or determined from index-of-refraction and absorption data for the TW. From the conduction model, we have explicit expressions (e.g., Equation (15)) for the time dependence of T_{INT}, which allow us, in principle, to constrain σ_{WF}, etc., using fit results for T_{INT}. Similarly, the fitted reflectivities allow us to constrain changes in the indices of refraction across boundaries (e.g., the shock front). Note that, in general, the optical properties constrained from Equation (36) cannot be λ-dependent unless we give them, a priori, an explicit λ-dependence, with constants whose values are chosen by the fit (i.e., by the data). Since we have no reasonable expectation for this λ-dependence, we cannot truly constrain it. It is for this reason, plus the limited resolving power of the data itself (BOSLOUGH, 1984; SVENDSEN, 1987), that we assume $a_{\lambda SW}$, $a_{\lambda USW}$, etc., are independent of λ in the previous section. However, we may determine an apparent λ-dependence of the optical properties if we specialize Equation (36) and fit at each wavelength over time, i.e.,

$$\chi^2(\lambda_i; \boldsymbol{a}) \equiv \sum_{j=1}^{N_t} \frac{1}{\sigma_{ij}^2} \left\{ I_{\lambda exp}(\lambda_i, t_j) - I_{\lambda mod}(\lambda_i, t_j; \boldsymbol{a}) \right\}^2 \quad (38)$$

SVENDSEN and AHRENS (1987) constrained $a_{k(min)}$ (i.e., the best fit values of the a_k) in this manner for radiation data from Ta-Ag-MgO targets. We use a very simple version of this approach below with the data of BASS et al. (this volume) to constrain $a_{\lambda SW}$. First, however, it is instructive to consider fits to data using simpler models than that represented by Equations (36) and (38). Most earlier researchers (e.g., KORMER, 1968; URTIEW, 1974; LYZENGA, 1980; LYZENGA et al., 1983) constrained model parameters via the greybody relation

$$L_{gb}(\lambda, \hat{\varepsilon}_{gb}, T_{gb}) \equiv \hat{\varepsilon}_{gb} I_{Pl}(\lambda, T_{gb}) \quad (39)$$

The associated χ^2 statistic is given by

$$\chi_{gb}^2(t_j; \hat{\varepsilon}_{gb}, T_{gb}) = \sum_{i=1}^{N_\lambda} \frac{1}{\sigma_{ij}^2} \left\{ I_{\lambda exp}(\lambda_i, t) - I_{gb}(\lambda_i, t_j; \hat{\varepsilon}_{gb}, T_{gb}) \right\}^2 \quad (40)$$

Since the summation in (40) averages $\hat{\varepsilon}_{gb}$ over all observed wavelengths, it represents a wavelength-averaged (i.e., total) effective emissivity. Given that the only λ-dependence in the greybody model is contained in the Planck function, $I_{Pl}(\lambda, T)$, the more closely the data follow the blackbody wavelength distribution at a given temperature, the better is the fit (i.e., the lower the value of $\chi_{gb}^2(t_j)$). Since both the data and model depend explicitly on λ, the fit proceeds over all observed wavelengths at a given time during the radiation history. As a result, $\hat{\varepsilon}_{gb}$ and T_{gb} are functions of time.

Since I_{gb} depends nonlinearly on T_{gb}, we must find the best fit values of $\hat{\varepsilon}_{gb}$ and I_{gb} iteratively with the minimization constraints on χ_{gb}^2. As starting values of $\hat{\varepsilon}_{gb}$ and T_{gb} for the nonlinear fit, and for comparison, we may use Wien's approximation to $I_{Pl}(\lambda, T)$ in $\chi_{gb}^2(t_j)$, which follows from $I_{Pl}(\lambda, T)$ in the limit $\exp(C_2/\lambda T) \gg 1$, i.e.,

$$I_{wgb}(\lambda, \hat{e}_{wgb}, T_{wgb}) \equiv \hat{\varepsilon}_{wgb} I_{wi}(\lambda, T_{wgb}) = \hat{\varepsilon}_{wgb} \frac{2C_1}{\lambda^5} e^{-C_2/\lambda T_{wgb}} \quad (41)$$

The relative error incurred in approximating I_{Pl} by I_{Wi} is equal to $\exp(-C_2/\lambda T)$; this approximation is accurate to within 1% for $\lambda T < 3 \times 10^{-3}$ m·K (SIEGEL and HOWELL, 1981). Since we can fit Wien's relation to the data in a linear least-squares sense, we can solve for $\hat{\varepsilon}_{wgb}$ and T_{wgb} directly (i.e., without iteration). With these values, we may safely apply an iterative technique to Equation (40) to constrain $\hat{\varepsilon}_{gb}$ and T_{gb} and to be assured of a non-divergent fit (SVENDSEN, 1987). We use the Golden Search (GS) and Levenberg-Marquardt (LM) iterative techniques (PRESS et al., 1986) to obtain three different fits: 1) GS with $\hat{\varepsilon}_{gb}$ variable, 2) GS with $\hat{\varepsilon}_{gb}=1$, and 3) LM with $\hat{\varepsilon}_{gb}$ variable.

We present a greybody fit to the radiation observations from two experiments of BASS et al. (this volume) in Figures 7 and 8. Figure 7 displays a fit to data from an experiment on an Fe-Fe film-Al_2O_3 target impacted by a Ta projectile traveling at 5.67 km/s. The trend in $\chi_{gb}^2(t)$ suggests that the fit gets better with time. Strictly speaking, $\chi^2 \sim \nu \pm 2\sqrt{\nu}$ as $\nu \to \infty$, where ν is the number of degrees of freedom in the fit (i.e., the number of data minus the number of parameters; 2 in this case); we might hope that $\chi_{gb}^2 \sim 2$ represents a reasonable fit for the greybody model. All of the fits show $T_{gb}(t)$ decreasing with time, and for the variable emissivity fits, $\hat{\varepsilon}_{gb}(t)$ increases slightly with time. This behavior is characteristic of most Fe-Fe-Al_2O_3 experiments of BASS et al. (this volume). For all the Fe experiments, we note that $T_{SW} \ll T_{INT}(t)$ (BASS et al., this volume) and that $\hat{\varepsilon}_{gb}(t)$ is inconsistent with $a_{SW} \to \infty$ (BOSLOUGH, 1985; SVENDSEN, 1987). In this case, from Equation (20), we have

$$I_{\lambda mod}(\lambda, t) \approx \hat{\varepsilon}_{\lambda INT}(t) I_{Pl}[\lambda, T_{INT}(t)] \quad (42)$$

and the decrease of $T_{gb}(t)$ with time (Figure 7d) can be explained in terms of $T_{INT}(t)$, as detailed above. Also, the slight increase of $\hat{\varepsilon}_{gb}(t)$ with time (Figure 7c) can be

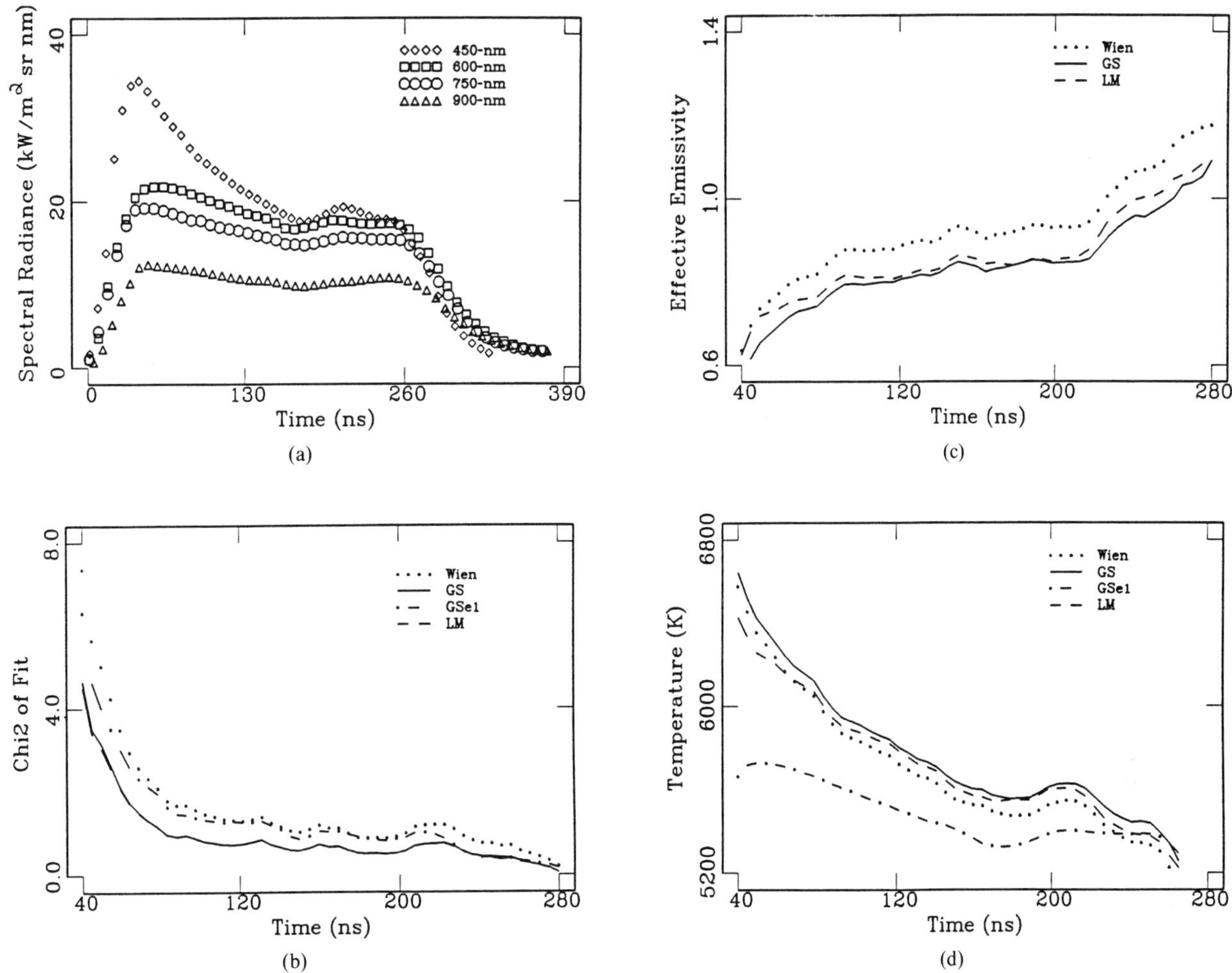

Figure 7. Observed radiation history and greybody model ($\hat{\varepsilon}_{gb}$, T_{gb}) fits for Fe-Fe film-Al_2O_3 target impacted by a tantalum projectile at 5.67 km/s, resulting in an Fe Hugoniot pressure of 244 GPa and an Al_2O_3 Hugoniot pressure of 190 GPa, which is also the Fe-Al_2O_3 interface pressure (BASS et al., this volume). Part (a) of this figure displays the radiation intensity (spectral radiance) data, which is collected at four wavelengths: 450, 600, 750, and 900 nm. Parts (b)–(d) represent the results of fitting the greybody model to this data through the χ^2 statistic, $\chi^2(t)$ given in part (b), to obtain the best-fit effective normal greybody emissivity, $\hat{\varepsilon}_{gb}(t)$, and greybody temperature, $T_{gb}(t)$. Fits using Wien's law, Golden Search (GS), GS with the effective emissivity set to 1 (GSe1), and the Levenburg-Marquardt algorithm (LM) are indicated.

explained most simply by a slight decrease of the Al_2O_3 absorption coefficient upon shock compression (SVENDSEN, 1987). This may be consistent with the observation that the refractive index of Al_2O_3 seems to decrease with pressure (~ -0.001/GPa between 0.1 and 1 GPa: DAVIS and VEDAM, 1967). Since $a_{\lambda USW} \sim 0$ for Al_2O_3, this observation implies that $a_{\lambda SW} \sim 0$ as well. In this case, Equation (42) implies that

$$\begin{aligned} I_{\lambda mod}(\lambda, t) &\approx \hat{\varepsilon}_{\lambda INT}(0) I_{PI}[\lambda, T_{INT}(t)] \\ &= (1 - r_{\lambda FS})(1 - r_{\lambda SF})(1 - r_{\lambda INT}) I_{PI}[\lambda, T_{INT}(t)] \end{aligned} \quad (43)$$

for the Fe-Fe-Al_2O_3 experiments.

In contrast with this last fit, the experimental and greybody fit results displayed in Figures 8a–d, for an Fe-Fe-LiF target impacted by a Ta projectile traveling at 5.41 km/s, exhibit a relatively constant greybody temperature with time (Figure 8d) and a systematically decaying greybody effective emissivity with time (Figure 8c). In this case, $T_{gb}(t)$ implies a relatively constant $T_{INT}(t)$, as we expect for a smooth interface (GROVER and URTIEW, 1974; Equation (18) above) or a reshocked interface with $\delta_{FW} \gg 2\sqrt{\kappa_F t_{exp}}$ (Figure 6b) or $\delta_{FW} \ll 2\sqrt{\kappa_F t_{exp}}$ (Figure 6b). The reshocked interface with $\delta^*_{FW} \gg 2\sqrt{\kappa_F t_{exp}}$ is less likely than the latter possibility, as implied by the data-model comparison in Figure 4b. The behavior of $\hat{\varepsilon}_{gb}(t) \approx \hat{\varepsilon}_{\lambda INT}(t)$

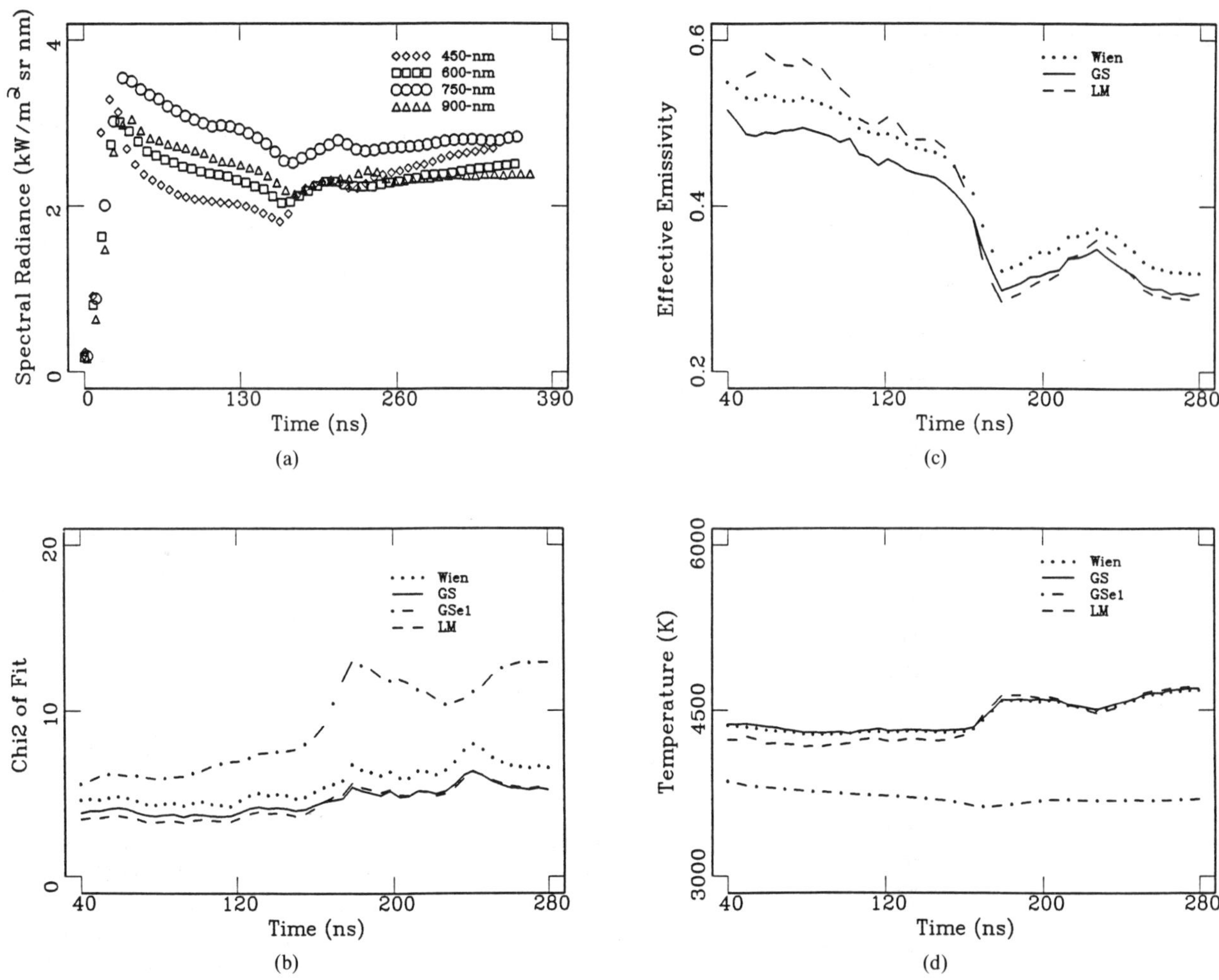

Figure 8. Observed radiation history and greybody model fits for Fe-Fe foil-LiF target impacted by a tantalum projectile at 5.41 km/s, resulting in an Fe foil-Hugoniot pressure of 227 GPa and an LiF-Hugoniot pressure of 122 GPa.

reflects a shock-induced increase in the absorption coefficient (i.e., $a_{\lambda SW} > a_{\lambda USW}$) of LiF via (22), (25), and (33). The fluctuations in the fit after about 160 ns may be due to wave reverberations or other dynamic effects, which are beyond the scope of our model, and/or possibly due to the influence of reshock at the Fe-Fe foil interface, as mentioned above.

As stated above, the interface contribution to the observed intensity dominates the SW contribution (BASS et al., this volume; SVENDSEN, 1987). Based on this observation, we may reasonably fit a simplified version of the full radiation model to the data via Equation (40). We do this for the Fe-LiF data fit to the greybody model in Figure 8. First, we note that, at t=0, $I_{\lambda\mathrm{mod}}$ is, from Equation (20),

$$I_{\lambda\mathrm{mod}}(\lambda, 0) = \hat{\varepsilon}_{\lambda\mathrm{INT}}(0) I_{\mathrm{PI}}(\lambda, T_{\mathrm{INT}}(0)) \qquad (44)$$
$$= [1 - r_{\lambda\mathrm{FS}}] e^{-a_{\lambda\mathrm{USW}}^{*}} [1 - r_{\lambda\mathrm{SF}}][1 - r_{\lambda\mathrm{INT}}] I_{\mathrm{PI}}[\lambda, T_{\mathrm{INT}}(0)]$$

So the magnitude of $I_{\lambda\mathrm{mod}}(\lambda, 0)$ is controlled by the reflectivities, $a_{\lambda\mathrm{USW}}^{*}$, and the initial value of T_{INT}, which is dependent on the values of T_{F}, ΔT_{FW}, T_{W}, and σ_{WF} through Equation (17) in the simplest case. The greybody fits in Figure 8 suggest that, for this experiment at least, $T_{\mathrm{INT}}(t)$ is approximately constant. Assuming this and $a_{\lambda\mathrm{USW}}$=0, we approximate (20) as

$$I_{\lambda\mathrm{mod}}(\lambda, t) = \hat{\varepsilon}_{\lambda\mathrm{INT}}(t) I_{\mathrm{PI}}[\lambda, T_{\mathrm{INT}}(0)] \qquad (45)$$
$$= (1 - r_{\lambda\mathrm{FS}})(1 - r_{\lambda\mathrm{SF}})(1 - r_{\lambda\mathrm{INT}}) e^{-a_{\lambda\mathrm{SW}}^{*} t/t_{\mathrm{exp}}} I_{\mathrm{PI}}[\lambda, T_{\mathrm{INT}}(0)]$$

In this case, the time dependence of $I_{\lambda\mathrm{mod}}$ is due solely to the SW transmissivity. Using Equation (45) in (38), along with $T_{\mathrm{INT}}(0) = T_{\mathrm{gb}}(0)$, we may fit the Fe-LiF data for $a_{\lambda\mathrm{SW}}^{*}$. We present the results of this fit in Figure 9. The data are cut off at 160 ns to reduce the influence of possible dynamics or Fe-Fe interface reshock on the fit. The parameter values resulting from this fit are given in Table

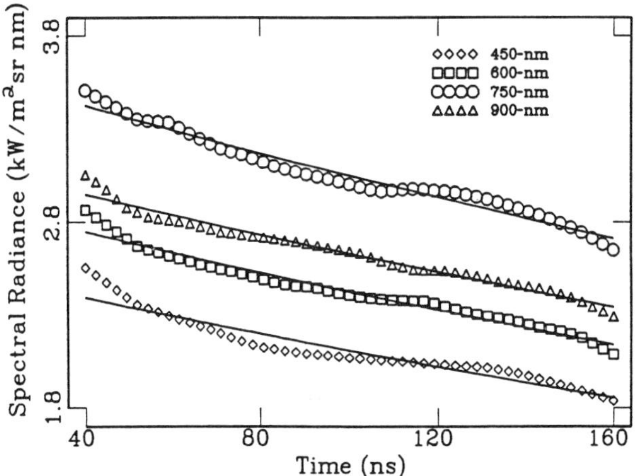

Figure 9. Time-dependent spectral radiance fits using Equation (45) in text to the data displayed in Figure 8 for the Fe-Fe foil-LiF target. We fit (45) to the first half of the Fe-LiF data displayed in Figure 8. The slope of each continuous curve, representing the fit for the corresponding wavelength, constrains the effective normal absorption coefficient of the shocked LiF, while the intercept constrains the Fe-LiF interface and shock-front effective normal reflectivities.

3. We eliminate $r_{\lambda FS}$ from the fit since it is equal to 0.08 for LiF, as estimated from $n=1.39$, the index-of-refraction of LiF at STP (CRC Handbook). Since the index-of-refraction of LiF seems to increase with pressure (~ 0.002/GPa: BURNSTEIN and SMITH, 1948), we expect $r_{\lambda SF} \gtrsim r_{\lambda FS}$ for LiF. Clearly, LiF has lost some transparency upon shock compression. The trend in $a_{\lambda SW}$ toward lower values at longer wavelengths is unresolved but consistent with the Bouguer's law expectation that $a_\lambda = 4\pi\omega_e/\lambda$, if ω_e, the electromagnetic extinction coefficient (SIEGEL and HOWELL, 1981), is constant or varies inversely with λ. As suggested above, similar fits for Fe-Al_2O_3 (SVENDSEN, 1987) imply that $a_{\lambda USW} > a_{\lambda SW}$, an intriguing possibility which we do not yet understand.

Lastly, we take the results of the greybody fit shown in Figure 7 for the Fe-Fe-Al_2O_3 experiment, assume $T_{gb}(t) = T_{INT}(t)$, and use Equation (17) to write

TABLE 3. Simplified Radiation Model Parameters for Fe-Al_2O_3 Experiment[a]

Wavelength (nm)	$(1 - r_{\lambda SF})(1 - r_{\lambda INT})$	$a_{\lambda SW}$ (m^{-1})
450	0.76	137
600	0.56	134
750	0.68	125
900	0.67	122

[a]For this fit, $x_{FS}=4.15$ mm, $t_{exp}=390$ ns, $T_{INT}=4200$ K, and $r_{\lambda SF}=0.08$ at all wavelengths in fit.

$$\Delta T_{FW} = \frac{(1+\sigma_{WF})}{\operatorname{erfc}\{\delta_{FW}/2\sqrt{\kappa_F t_{exp}}\}}\,[T_{gb}(0) - T_{gb}(t_{exp})] \quad (46)$$

With $T_{gb}(0) - T_{gb}(t_{exp}) = 1200$ K from the fit displayed in Figure 7, we may calculate the trade-off between the FL-TW interface temperature due to reshock, ΔT_{FW}, and the ratio of the reshocked-layer thickness, δ_{FW}, to the FL conduction length scale, $\sqrt{\kappa_F t_{exp}}$, for different values of the FL-TW interface mismatch, σ_{WF}. These calculations, displayed in Figure 10, imply that the larger the reshocked-layer thickness relative to the FL conduction length scale, the higher the reshock temperature at a given thermal mismatch. For this particular experiment, we expect $\sigma_{WF} \lesssim 0.1$ from calculations discussed above; we also expect $\Delta T_{FW} \lesssim 1500$–$2000$ K from the calculations presented in Figure 4a. In this case, Figure 10 and model calculations imply that $\sigma_{FW} \lesssim 2\sqrt{\kappa_F t_{exp}} \sim 10^{-5}$ m. Since this is a film experiment, with $d \sim 10^{-6}$ m, we tentatively conclude that the entire film layer experienced reshock in this experiment.

Summary

We consider the effects of release/reshock, phase transitions, and conduction on the shock-compressed temperatures of the target components and their interfaces. Comparison of the model with the results of experiments on Fe-Fe-LiF and Fe-Fe-Al_2O_3 targets suggests the following:

1) Release/reshock calculations for Fe-Fe-Al_2O_3 targets, in comparison with the experimental results of BASS

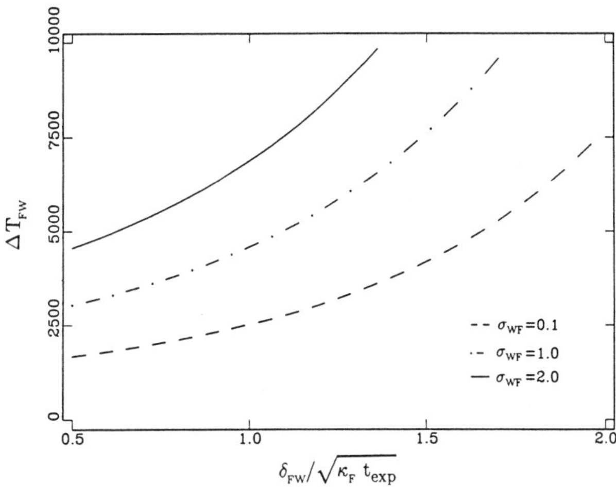

Figure 10. Magnitude of temperature at FL-TW interface due to reshock, ΔT_{FW}, versus the ratio of the reshocked-layer thickness, δ_{FW} to the conduction length scale, $\sqrt{\kappa_F t_{exp}}$, as given by Equation (46) in text. This tradeoff is constrained by the magnitude of $T_{gb}(0) - T_{gb}(t_{exp})$ from the greybody fit for the Fe-Fe-Al_2O_3 data displayed in Figure 7.

et al. (this volume), suggest that Fe experiences approximately 200–2000 K of reshock heating at both Fe foil-Al_2O_3 and Fe film-Al_2O_3 interfaces when released from $\approx$245–300 GPa to interface pressures of 190–230 GPa. Below 190 GPa, reshock for Fe-Al_2O_3 interfaces appears to be minimal. Both the data and calculations suggest that the degree of reshock is strongly pressure dependent, which is consistent with the results of URTIEW and GROVER (1974). In contrast, Fe released from the same range of Hugoniot pressures to Fe-LiF interface pressures between $\approx$130 and 160 GPa experiences little or no reshock. This more ideal nature of Fe-LiF interfaces is enhanced by the fact that, besides being a poorer shock-impedance match to Fe than Al_2O_3, it is also a poorer thermal match, resulting in less change in the interface temperature away from the Fe release-state temperature. Comparison of data and calculations for both of these windows suggest that, while attention to the initial conditions of the interface is essential to minimize reshock, a more important factor may be the choice of window.

2) In the absence of energy sources and significant energy flux from other parts of the target, the rate of change of the interface temperature, (dT_{INT}/dt), is proportional to $-\mu\exp(-\mu^2)$, where $\mu=\delta_{FW}/2\sqrt{\kappa_{F}t}$. For Fe at FL-TW interfaces, $\sqrt{\kappa_{F}t_{exp}}\sim 10\ \mu m$ (SVENDSEN, 1987); consequently, a 100-μm, reshocked Fe layer would relax very little, remaining near $T_{INT}(0)$ on the time scale of the experiment. However, if $\delta_{FW}\sim 1\ \mu m$, $T_{INT}(t)$ relaxes almost instantaneously to its value $T_{INT}(\infty)$. $T_{INT}(t)$ is resolvably time dependent for $\delta_{FW}\sim 2\sqrt{\kappa_{F}t_{exp}}$.

3) Greybody fits to an Fe-Fe film-Al_2O_3 experiment of BASS et al. (this volume) show the greybody effective emissivity, $\hat{\varepsilon}_{gb}(t)$, to increase slightly with time, while the greybody temperature, $t_{gb}(t)$, decreases with time. This behavior is characteristic of most Fe-Fe-Al_2O_3 experiments. The decrease of $T_{gb}(t)$ can be explained in terms of the model for $T_{INT}(t)$, and it implies that $\delta_{FW}\sim 2\sqrt{\kappa_{F}t_{exp}}$ for this experiment. Further, assuming $T_{gb}(t)=T_{INT}(t)$, the greybody fit constrains the amount of reshock, ΔT_{FW}, to be $\lesssim$1500–2000 K with $\sigma_{WF}\sim 0.1$ and $\delta_{FW}\lesssim 2\sqrt{\kappa_{F}t_{exp}}$. A slight decrease of the Al_2O_3 absorption coefficient upon shock compression can explain the slight increase of $\hat{\varepsilon}_{gb}(t)$ with time. This may be consistent with the observation that the refractive index of Al_2O_3 seems to decrease with pressure. In contrast, greybody fits to data from an Fe-Fe foil-LiF target show a relatively constant greybody temperature and decreasing greybody emissivity. The constant greybody temperature implies a constant interface temperature, as we expect for an interface experiencing minimal reshock, while the decaying $\hat{\varepsilon}_{gb}(t)$ is consistent with a shock-induced increase in the absorption coefficient of LiF. Setting $T_{INT}(0)=T_{gb}(0)$, we fit a simplified version of the full radiation model to these data to find $a_{\lambda SW}\sim 100$ m^{-1} (Table 3) for LiF, shocked to 122 GPa in this experiment.

4) Finally, we note that the equilibrium thermodynamic Hugoniot temperature of Fe is strongly influenced by electronic and/or anharmonic contributions to c_v at high pressure, as evidenced by both: 1) the results of BONESS et al. (1986) when used in Equation (8), and 2) by requiring the solid-Fe Hugoniot and an extrapolation of the experimentally constrained Fe melting curve (WILLIAMS and JEANLOZ, 1986) referenced to this Hugoniot, to intersect at 245 GPa (BROWN and MCQUEEN, 1982). This last constraint provides a value of $\Omega(\rho_i)=0.044$ J/kg·K^2, as compared to $\Gamma(\rho_i)=0.090$ J/kg·K^2 from the work of BONESS et al. (1986). These calculations imply a significant anharmonic contribution to the properties of ε-Fe at high T.

Acknowledgments. We thank William W. Anderson, Mark B. Boslough. A. James Friedson, Dion L. Heinz, Douglas R. Schmitt, and James A. Tyburczy for enlightening discussions. We also thank J. Michael Brown and two anonymous reviewers for constructive comments on an earlier version of this work. Support for this work from NSF grants EAR-8608249 and 8608969 is gratefully acknowledged. Contribution #4332, Division of Geological and Planetary Sciences, California Institute of Technology, Pasadena, CA 91125.

REFERENCES

AHRENS, T. J., D. L. ANDERSON, and A. E. RINGWOOD, Equation of state and crystal structure of high-pressure phases of shocked silicates and oxides, *Rev. Geophys.*, *7*, 667–707, 1969.

ANDERSON, O. L., Properties of iron at the earth's core conditions, *Geophys. J. R. Astron. Soc.*, *84*, 561–579, 1986.

ANDERSON, O. L., E. SCHREIBER, R. C. LIEBERMANN, and N. SOGA, Some elastic constant data on minerals relevant to geophysics, *Rev. Geophys.*, *6*, 491–524, 1968.

ANDREWS, D. J., Equation of state of the alpha and epsilon phases of iron, *J. Phys. Chem. Solids*, *34*, 825–840, 1973.

BARKER, L. M., and R. E. HOLLENBACH, Shock wave study of the $\alpha \rightleftharpoons \varepsilon$ transition in iron, *J. Appl. Phys.*, *45*, 4872–4887, 1974.

BASS, J. D., B. SVENDSEN, and T. J. AHRENS, The temperature of shock compressed iron, this volume.

BERMAN, R., *Thermal Conduction in Solids*, 193 pp., Oxford University Press, 1976.

BEVINGTON, P. R., *Data Reduction and Error Analysis for the Physical Sciences*, 336 pp., McGraw-Hill, New York, 1969.

BOEHLER, R., The phase diagram of iron to 430 kbar, *Geophys. Res. Lett.*, *13*, 1153–1156, 1986.

BONESS, D. A., J. M. BROWN, and A. K. MCMAHAN, The electronic thermodynamics of iron under earth core conditions, *Phys. Earth Planet. Inter.*, *42*, 227–240, 1986.

BOSLOUGH, M. B., Shock-wave properties and high-pressure equations-of-state of geophysically-important materials, Ph. D. dissertation, 171 pp., California Institute of Technology, Pasadena, California, 1984.

BOSLOUGH, M. B., A model for time dependence of shock-induced thermal radiation of light, *J. Appl. Phys.*, *56*, 3394–3399, 1985.

BRANNON, P. J., C. H. CONRAD, R. W. MORRIS, E. D. JONES, and J. R. ASAY, Studies of the spectral and spatial characteristics of shock-induced luminescence from x-cut quartz, *J. Appl. Phys.*, *54*, 6374–6381, 1984.

BROWN, J. M., and R. G. MCQUEEN, The equation of state for iron and the earth's core, in *High Pressure Research in Geophysics*, edited by S. Akimoto and M. H. Manghnani, pp. 611–622, Center for Academic Publications, Tokyo, 1982.

BROWN, J. M., and R. G. MCQUEEN, Phase transitions, Gruneisen parameter, and elasticity for shocked iron between 77 GPa and 400 GPa, *J. Geophys. Res.*, *91*, 7485–7494, 1986.

BUKOWINSKI, M. S. T., A theoretical equation of state for the inner core, *Phys. Earth Planet. Inter.*, *14*, 333–344, 1977.

BURNSTEIN, E., and P. L. SMITH, Photoelastic properties of cubic crystals, *Phys. Rev.*, *74*, 229–230, 1948.

CARSLAW, H. S., and J. C. JAEGER, *Conduction of Heat in Solids*, 510 pp., Clarendon Press, Oxford, 1959.

DAVIS, T. A., and K. VEDAM, Photoelastic properties of sapphire, *J. Op. Soc. Am.*, *15*, 4555–4556, 1967.

DESAI, P. D., Thermodynamic properties of iron and silicon, *J. Phys. Chem. Ref. Data*, *15*, 967–983, 1986.

DROTNING, W. D., Thermal expansion of iron, cobalt, nickel and copper at temperatures up to 600 K above melting, *High Temp.-High Pressures*, *13*, 441–458, 1981.

GOPAL, E. S. R., *Specific Heat at Low Temperatures*, 240 pp., Plenum Press, New York, 1966.

GRADY, D. E., Shock deformation of brittle solids, *J. Geophys. Res.*, *85*, 913–924, 1980.

GROVER, R., and P. A. URTIEW, Thermal relaxation at interfaces following shock compression, *J. Appl. Phys.*, *45*, 146–152, 1974.

JAMIESON, J. C., H. H. DEMAREST, Jr., and D. SCHIFERL, A re-evaluation of the Gruneisen parameter for the earth's core, *J. Geophys. Res.*, *83*, 5929–5935, 1978.

JEANLOZ, R., and T. J. AHRENS, Anorthite: Thermal equation of state at high pressure, *Geophys. J. R. Astron. Soc.*, *62*, 529–549, 1980.

JEPHCOAT, A. P., H. K. MAO, and P. M. BELL, Static compression of iron to 78 GPa with rare gas solids as pressure-transmitting media, *J. Geophys. Res.*, *91*, 4677–4684, 1986.

KEELER, R. N., Electrical conductivity of condensed media at high pressures, in *Physics of High Energy Density*, *Proc. Intern. School of Physics "Enrico Fermi", Course XLVII*, pp. 106–125, Academic Press, New York, 1971.

KONDO, K. E., and T. J. AHRENS, Heterogeneous shock-induced thermal radiation in minerals, *Phys. Chem. Minerals*, *9*, 173–181, 1983.

KORMER, S. B., Optical study of the characteristics of shock-compressed condensed dielectrics, *Sov. Phys. Uspekhi*, *11*, 229–254, 1968.

LYZENGA, G. A., Shock temperatures of materials: experiments and applications to the high pressure equation of state, Ph. D. dissertation, 208 pp., California Institute of Technology, Pasadena, California, 1980.

LYZENGA, G. A., and T. J. AHRENS, Shock temperature measurements in Mg_2SiO_4 and SiO_2 at high pressures, *Geophys. Res. Lett.*, *7*, 141–145, 1980.

LYZENGA, G. A., T. J. AHRENS, and A. C. MITCHELL, Shock temperatures of SiO_2 and their geophysical implications, *J. Geophys. Res.*, *88*, 2431–2444, 1983.

MARSH, S. P. (ed.), Al_2O_3, in *Shock Hugoniot Data*, p. 89, UCLA Press, 1980.

MCQUEEN, R. G., S. P. MARSH, and J. N. FRITZ, Hugoniot equation of state of twelve rocks, *J. Geophys. Res.*, *72*, 4999–5036, 1967.

PRESS, W. H., B. P. FLANNERY, S. A. TEUKOLSKY, and W. T. VETTERLING, *Numerical Recipies: The Art of Scientific Computing*, 818 pp., Cambridge University Press, New York, 1986.

RICE, M. H., R. G. MCQUEEN, and J. M. WALSH, Compressibility of solids by strong shock waves, *Solid State Phys.*, *6*, 1–63, 1958.

ROBIE, R. A., B. S. HEMINGWAY, and J. R. FISHER, Thermodynamic properties of minerals and related substances at 298.15 K and 1 bar (10^5 Pascals) pressure and at higher temperatures, USGS Bulletin 1452, 456 pp., 1978.

ROUFOSSE, M. C., and P. G. KLEMENS, Lattice thermal conductivity of minerals at high temperature, *J. Geophys. Res.*, *79*, 703–705, 1974.

RUOFF, A. L., Linear shock-velocity-particle-velocity relationship, *J. Appl. Phys.*, *38*, 4976–4980, 1967.

SCHMITT, D. R., B. SVENDSEN, and T. J. AHRENS, Shock-induced radiation from minerals, in *Shock Waves in Condensed Matter*, edited by Y. M. Gupta, pp. 261–265, Plenum Press, New York, 1986.

SIEGEL, R., and J. HOWELL, *Thermal Radiation Heat Transfer*, 862 pp., McGraw-Hill, New York, 1981.

STEVENSON, D. J., Applications of liquid-state physics to the earth's core, *Phys. Earth. Planet. Inter.*, *22*, 42–52, 1980.

SVENDSEN, B., Optical radiation from shock-compressed materials, Ph. D. dissertation, 250 pp., California Institute of Technology, Pasadena, California, 1987.

SVENDSEN, B., and T. J. AHRENS, Thermal history of shock-compressed solids, in *Shock Waves in Condensed Matter*, edited by Y. M. Gupta, pp. 607–611, Plenum Press, New York, 1986.

SVENDSEN, B., and T. J. AHRENS, Shock-induced temperatures of MgO, *Geophys. J. R. Astron. Soc.*, in press, 1987.

TOULOUKIAN, Y. S., R. W. POWELL, C. Y. HO, and P. G. KLEMENS, *Thermophysical Properties of Matter, Volume 1: Thermal Conductivity*, Thermophysical Properties Research Center of Purdue University, Data Series, 1431 pp., Plenum Publishing Corporation, New York, 1970.

TOULOUKIAN, Y. S., R. H. POWELL, C. Y. HO, and P. G. KLEMENS, *Thermophysical Properties of Matter, Volume 12: Thermal expansion*, Thermophysical Properties Research Center of Purdue University, Data Series, 1321 pp., Plenum Publishing Corporation, New York, 1975.

URTIEW, P. A., Effect of shock loading on transparency of sapphire crystals, *J. Appl. Phys.*, *45*, 3490–3493, 1974.

URTIEW, P. A., and R. GROVER, Temperature deposition caused by shock interactions with material interfaces, *J. Appl. Phys.*, *45*, 140–145, 1974.

URTIEW, P. A., and R. GROVER, The melting temperature of magnesium under shock loading, *J. Appl. Phys.*, *48*, 1122–1126, 1977.

VAN THIEL, M. (ed.), *Compendium of Shock Wave Data, Section A2: Inorganic Compounds, Report UCRL-50108*, 50 pp., Lawrence Livermore Laboratory, Livermore, Calif., 1977.

WALLACE, D. C., *Thermodynamics of Crystals*, 484 pp., John Wiley and Sons, Inc., New York, 1972.

WATT, J. P., and T. J. AHRENS, Shock wave equations of state using mixed-phase regime data, *J. Geophys. Res.*, *89*, 7836–7844, 1984.

WEAST, R. C. (ed.), *CRC Handbook of Chemistry and Physics, 60th edition*, 2449 pp., CRC Press, Boca Raton, Florida, 1979.

WILLIAMS, Q., and R. JEANLOZ, Melting of Fe and FeS to 100 GPa, *EOS*, *67*, 1240, 1986.

WISE, J. L., and L. C. CHHABILDAS, Laser interferometer measurements of refractive index in shock-compressed materials, in *Shock Waves in Condensed Matter-'85*, edited by Y. M. Gupta, pp. 441–454, Plenum Press, New York, 1986.

YOUNG, D. A., and R. GROVER, Theory of the iron equation-of-state and melting curve to very high-pressures, in *Shock Waves in Condensed Matter-1983*, edited by J. A. Asay, R. A. Graham, and G. K. Straub, pp. 65–67, North-Holland, New York, 1984.

ZHARKOV, V. N., and V. A. KALININ, *Equations of State of Solids at High Pressures and Temperatures*, 200 pp., Nauka Press, Moscow, 1968.

VIII. GEOPHYSICAL AND GEOCHEMICAL CONSTRAINTS

MINERALOGY OF MANTLE PERIDOTITE ALONG A MODEL GEOTHERM UP TO 700 KM DEPTH

Eiichi TAKAHASHI and Eiji ITO

Institute for Study of the Earth's Interior, Okayama University Misasa, Tottori-ken 682-02, Japan

Abstract. In order to understand mineralogical constituents of the earth's mantle, a series of synthetic experiments was carried out using a natural garnet peridotite (PHN-1611). The experimental pressure-temperature conditions were along a model geotherm which had a maximum pressure of 26 GPa (equivalent to 720 km in depth) and a maximum temperature of 1600 °C. Quenched run products were analyzed with an X-ray diffractometer for phase identification and with scanning electron microscopes for textural relationships. Chemical compositions of coexisting phases were determined with EPMA. Important phase transformations that have been proposed to occur in the earth's mantle (α to β, β to γ, γ to perovskite (Pv) plus magnesiowüstite (Mw), and pyroxenes to majorite-garnet) are demonstrated to take place in the natural peridotite composition. Two additional phases were found to be present at pressures equivalent to more than 500 km in depth: 1) Ca-P, a Ca-rich phase which is probably an unquenchable, high-pressure polymorph of diopsidic pyroxene, and 2) Al-P, an aluminous phase with unknown crystal structure.

Introduction

Extensive experimental studies have been carried out at very high pressures in order to understand the mineralogical constituents of the earth's deep mantle (see RINGWOOD, 1975; AKIMOTO, this volume). In our laboratory, we have carried out systematic experiments in many simple systems primarily to understand the phase transformations that take place at depths of 500–700 km (for review, see ITO et al., 1984; ITO and TAKAHASHI, this volume). Our experimental studies have established that the seismic discontinuity in the earth's mantle at 400 km is due to the phase transition from olivine (α-phase) to modified spinel (β-phase) (see also RINGWOOD and MAJOR, 1970; AKIMOTO, 1972), and that the discontinuity at 670 km is due to stabilization of $MgSiO_3$-perovskite (see also LIU, 1976; ITO and MATSUI, 1978; YAGI et al., 1979; ITO and YAMADA, 1982). It was also noted that in the transition zone (between the 400 and 670 km discontinuities), majorite-garnet becomes the dominant constituent instead of pyroxenes (RINGWOOD, 1967; LIU, 1977; AKAOGI and AKIMOTO, 1977; KANZAKI, 1987). It is the main purpose of the present study to investigate those high-pressure phase transformations in a natural peridotite composition (PHN-1611).

In the present experiments we tested the use of natural peridotite powder as a starting material. We also attempted to characterize all the experimental charges with extensive SEM studies and X-ray diffractometry. The present experiments used run temperatures 400–600 °C higher than previous work by AKAOGI and AKIMOTO (1979) on the same peridotite composition (Figure 1). We found that all of the above mentioned major phase transformations take place in the natural peridotite. However, it was difficult to achieve chemical equilibrium after 1 h at 1600 °C because of the slow reaction rate between metastable solid phases.

Choice of the Starting Material and Experimental Conditions

Much evidence from research in geology, geochemistry, and geophysics indicates that the earth's upper mantle, at least down to 400 km, consists predominantly of peridotitic composition (RINGWOOD, 1975; ANDERSON, 1984). Among naturally occurring peridotites, garnet lherzolite xenoliths in kimberlite pipes are the rocks derived from the deepest interior of the earth (BOYD, 1973). In the present study, a sheared garnet lherzolite xenolith PHN-1611 (NIXON and BOYD, 1973) was selected as a starting material because: 1) it represents possible asthenospheric components beneath the African lithosphere (BOULLIER and NICOLAS, 1973; NIXON et al., 1973); 2) its bulk chemical composition is considered to be least fractionated (SHIMIZU, 1975); and 3) its high-pressure mineralogy has been investigated experimentally under both subsolidus and melting conditions (AKAOGI and AKIMOTO, 1979; MYSEN and KUSHIRO, 1977; SCARFE and TAKAHASHI, 1986). In a recent study on PHN-1611, however, SMITH and BOYD (1986) discovered that the peridotite xenolith records metasomatism and deformation episodes during the magmatic transportation stage. The fertile nature of PHN-1611, therefore, might not represent the chemistry of the asthenosphere.

Equilibrium pressure and temperature of the garnet lherzolite xenoliths have been much investigated by experimental petrologists, and a number of geothermometers and geobarometers have been proposed (e.g., BOYD, 1973; WOOD and BANNO, 1973; MERCIER and CARTER, 1975; GASPARIK, 1984; BERTRAND et al., 1986). According to these estimates, equilibrium pressure and

High-Pressure Research in Mineral Physics, edited by M. H. Manghnani and Y. Syono, pp. 427–437.

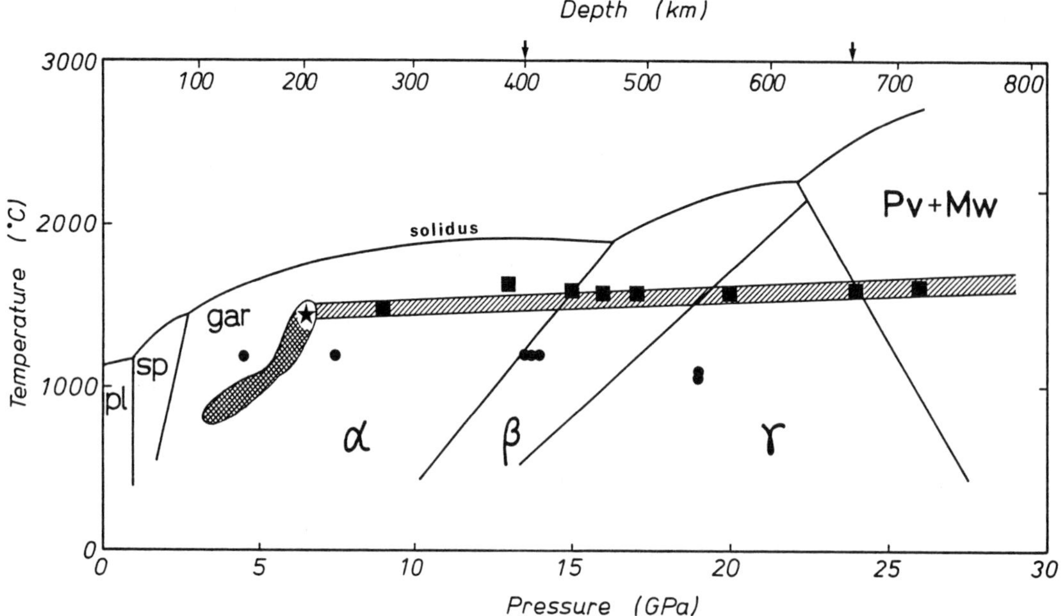

Figure 1. Phase boundaries, melting curve, and geotherm in the earth's mantle relevant to the present study. Cross-hatched area represents the pyroxene geotherm estimated from garnet lherzolite xenoliths in Lesotho kimberlites (BOYD, 1973). Star indicates estimated P-T of equilibration for the starting material PHN-1611. Hatched zone indicates a model geotherm beneath the African continent along which the present experiments were carried out (solid squares). Solidus curves of mantle peridotite is taken from TAKAHASHI (1986). Major phase boundaries for the mantle peridotite (Mg/(Mg+Fe)= 0.88) are drawn from present experiments and the following previous works: SUITO (1977), AKAOGI and AKIMOTO (1979), and ITO and YAMADA (1982). Experimental data points by AKAOGI and AKIMOTO (1979) on the same rock composition are indicated with solid circles using a revised pressure scale.

temperature for the garnet lherzolite PHN-1611 are in the ranges of 6–7 GPa and 1400–1500 °C, respectively. We assume that temperatures within the deep mantle (asthenosphere) are homogenized by convection. Hence the deep mantle has a vertical thermal gradient which is nearly adiabatic (RICHTER and MCKENZIE, 1981). By adopting 0.3 deg/km as the mantle adiabat and the above equilibrium pressure and temperature for PHN-1611 as the highest P-T value for the asthenosphere beneath the African continent 50 Ma b.p., a model mantle geotherm can be constructed (hatched zone in Figure 1). Small humps in the adiabatic geotherm at the high-pressure phase boundaries (NAVROTSKY and AKAOGI, 1984) were disregarded. The present experiments have been carried out at eight different P-T conditions along the model geotherm (solid squares in Figure 1, see also Table 1).

Experimental Procedures

All the experiments were carried out with a 5000-ton uniaxial, split-sphere apparatus (USSA-5000) of the Institute for Study of the Earth's Interior (ITO et al., 1984, Figure 1). In experiments below 14 GPa, an MgO octahedron of 14-mm edge length was used as a pressure medium and compressed in eight cubic tungsten carbide anvils of 32-mm edge length and 8-mm truncation edge length (TAKAHASHI, 1986, Figure 2). In experiments between 15 and 20 GPa, the edge length of the MgO pressure medium was 10 mm and the truncation edge length of the tungsten carbide anvils was 5 mm. In

TABLE 1. Run Details

Run No.	P (GPa)	T (°C)	t (min)	Phase Present
1	9	1500	150	ol, En-px, Di-px, gar
2	13	1650	60	ol, gar, Di-px
3	15	1600	60	ol, gar, metal, (liq?)
4	16	1600	60	ol, gar, metal, (liq?)
5	17	1600	60	β, gar, Di-px, (metal?)
6	20	1600	60	γ, gar, Ca-P*, (metal?)
7	24	1600	30	Pv, Mw, gar, Al-P, Ca-P*
8	26	1600	7	Pv, Mw, gar, Al-P, Ca-P*

*identified by EPMA analysis; ( ), from mass-balance calculation.

experiments above 24 GPa, these values were 7 and 3 mm, respectively. A $LaCrO_3$ furnace (TAKAHASHI, 1986, Figure 2c) was used for the experiments up to 20 GPa, and tantalum metal-foil was used as a heater and sample container for the experiments at 24 and 26 GPa (ITO and YAMADA, 1982, Figure 1). Pressure and temperature were raised gradually and simultaneously until the final experimental conditions were achieved. Temperatures were monitored with a Pt-Rh thermocouple taken out through the gasket. During a typical experimental run at 1600 °C for 60 min, the electric power increased by about 10–15% while maintaining the constant thermocouple emf. Experiments were terminated by shutting off the electric power supply. Applied pressure was released within 2–4 h. Quenched run products were halved; one half was subjected to X-ray diffraction analysis, and the other subjected to scanning electron microscope analysis.

X-Ray Diffraction Analysis

Mineral phases in experimental run products were identified with a powder X-ray diffractometer. In order to avoid ambiguity due to thermal gradients within the small furnace assembly when used at pressures 24 and 26 GPa, phase identification in these run products was made with a microfocus X-ray diffractometer (Rigaku-MDG). For the run at 26 GPa, mineral phases within a 100-μm-diameter area just next to the thermocouple junction were identified. Examples of the X-ray charts are shown in Figures 2a, b. Coexisting mineral phases identified by the X-ray diffraction analysis are summarized in Table 1.

Starting material consisted of olivine (α-phase), Ca-poor orthopyroxene (En-px), diopsidic clinopyroxene (Di-px), and garnet (G). Olivine transformed to modified spinel (β-phase) at a pressure between 16 and 17 GPa in the present experiments, and the modified spinel transformed to spinel (γ-phase) at a pressure below 20 GPa (Figure 2a). The amount of Ca-poor pyroxene decreased considerably at 9 GPa and disappeared from the experimental charges quenched at pressures above 13 GPa. The extensive solubility of the $MgSiO_3$ component in both garnet and diopsidic clinopyroxene solid solutions (see Table 2) was responsible for the absence of the Ca-poor pyroxene at high pressures. The amount of diopsidic clinopyroxene also decreased with increasing pressure and became scarcely identifiable at 17 GPa and absent at 20 GPa (Figure 2a).

In contrast with the decrease in pyroxenes, the amount of garnet increased significantly as a function of increasing pressure. As a consequence, the peridotite quenched at 20 GPa consisted only of γ-spinel and a single garnet phase (Figure 2a). However, as will be described later, a small amount of the Ca-rich phase (Ca-P, an unquenchable high-pressure polymorph of Ca-rich clinopyroxene) was present in the experimental charge quenched at 20 GPa (Figure 3f).

In experimental charges quenched at 24 and 26 GPa, γ-spinel decomposed to $MgSiO_3$-perovskite (Pv) and magnesiowüstite (Mw), and the amount of garnet decreased with increasing pressure. Figure 2b shows an X-ray chart of the run product under lower mantle conditions taken with the micro-focus diffractometer (MFD). Because the X-ray detector of the MFD moves away from the specimen, the diffraction lines at lower angles are relatively weak. Unidentified peaks at about 69° and 71° (d=2.02 and 2.00, respectively) are considered to be due to the Al-rich phase reported by ITO and TAKAHASHI (1986) in the system CaO-MgO-Al_2O_3-SiO_2 at similar P-T conditions.

Scanning Electron Microscope Analysis

Experimental run products were polished and analyzed with scanning electron microscopes of the National Museum of Natural History and the Geological Institute of Tokyo University. Back scattered electron images (BEI) of the run products are shown in Figures 3a, h. The chemical composition of the coexisting phases were determined with an energy-dispersive type of EPMA. Representative mineral analyses are given in Table 2 and are plotted in Figure 4.

It is very peculiar that the low-pressure experimental run products, including α-phase (olivine), are coarse-grained (Figures 3a, d), whereas those consisting of β-phase or γ-phase and those consisting of Pv+Mw assemblage are primarily fine-grained (Figures 3e, h). As shown in Figure 1, the present experimental conditions are relatively close to the dry solidus of the peridotite within the stability range of the α-phase, but become lower than the solidus temperature with increasing pressure above 16 GPa. Because the chemical reaction rate of matter (e.g., chemical diffusion coefficient) is largely dependent on the relative temperature normalized by its melting point, the above-noted contrasting crystallinity may be reasonable.

In the experiments conducted at 15 and 16 GPa, the quenched run products consisted only of olivine and garnet (Figures 3c, d). Mass balance calculation using the composition of the garnet and the bulk chemistry of the peridotite (Table 2) suggest that a Ca-rich component was lost from both of the charges during the experiments. Partial melting and extraction of the partial melt is the most likely reason for the deficiency because: 1) both of the run products were especially coarse-grained (compare Figures 3c, d with Figures 3a, b); and 2) the Mg* value (Mg*=100Mg/(Mg+Fe)) of the coexisting minerals was higher than that of the starting material (Table 1). The following might be responsible for the partial melting

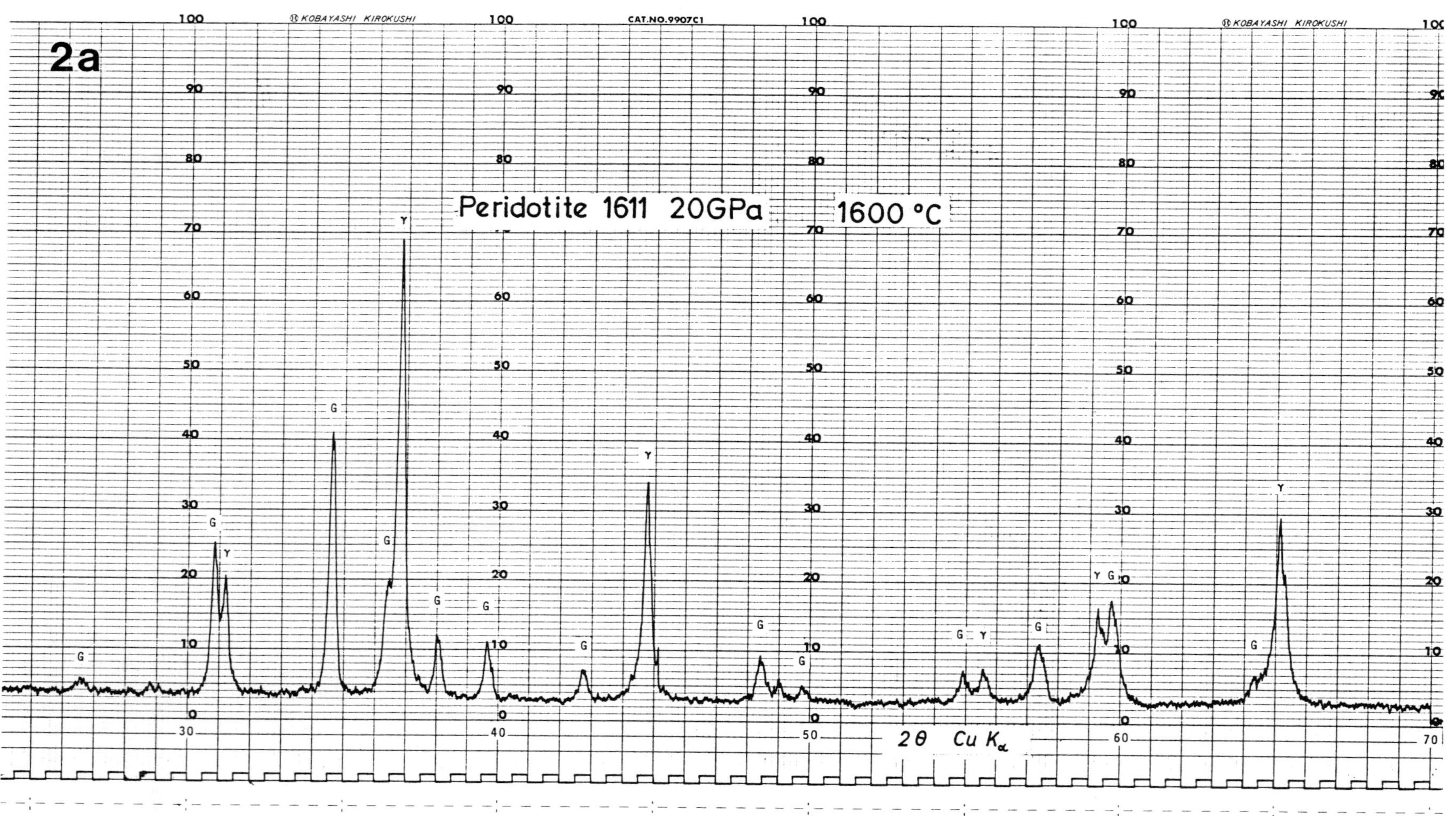
2a
Peridotite 1611 20GPa
1600 °C
2θ Cu Kα
KOBAYASHI KIROKUSHI
CAT.NO.9907C1

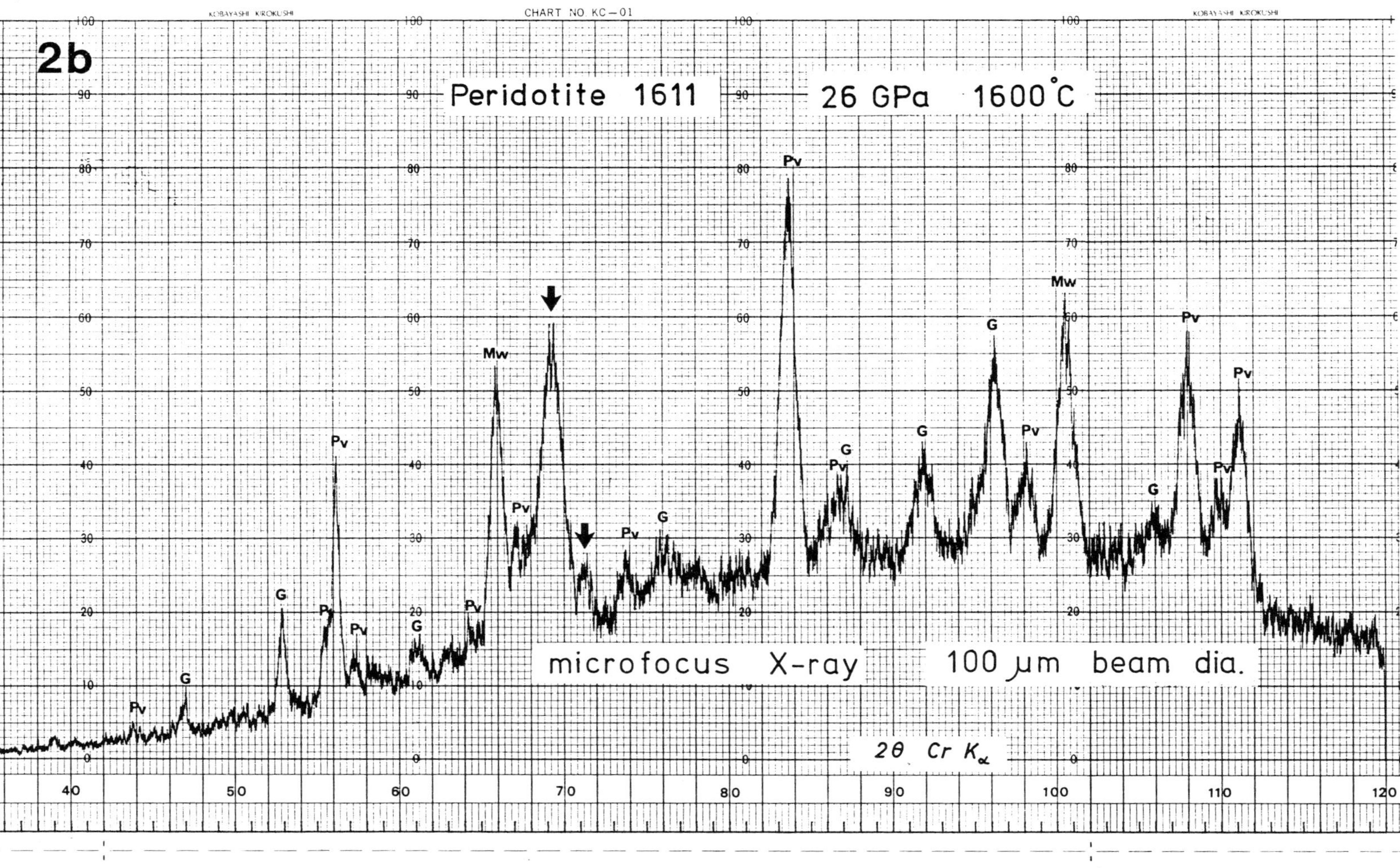

Figure 2. Representative X-ray charts of the experimental run products quenched at 20 GPa/1600 °C (2a, powder X-ray diffractometry) and at 24 GPa/1600 °C (2b, microfocus X-ray diffractometry under reflection mode). See Table 2 for chemical composition of coexisting phases and Figure 3 for their textures. Unidentified peaks in the 24 GPa run products (indicated with arrows) may be due to Al-P, an aluminous phase with unknown crystal structure (ITO and TAKAHASHI, this volume).

TABLE 2. Representative Mineral Compositions of Starting Material and Run Products

	PHN-1611 Starting Material					9 GPa/1500 °C				13 GPa/1650 °C		
	bulk	ol	En-px	Di-px	gar	ol	En-px	Di-px	gar	ol	Di-px	gar
SiO_2	44.55	40.42	56.26	54.79	42.68	40.0	57.3	55.8	43.3	40.6	56.6	46.0
TiO_2	0.25	0.03	0.23	0.30	0.80	n.d.	0.1	0.3	1.0	n.d.	0.2	0.6
Al_2O_3	2.80	0.10	1.34	2.41	21.15	n.d.	0.4	2.5	17.2	n.d.	2.1	15.6
Cr_2O_3	0.29	0.05	0.21	0.49	1.46	n.d.	0.1	0.5	1.6	n.d.	0.5	1.3
FeO*	10.24	11.44	7.04	5.14	8.63	11.5	7.1	5.0	9.0	11.4	5.6	8.8
MnO	0.13	0.14	0.13	0.13	0.26	n.d.	0.1	n.d.	0.2	0.1	0.1	0.2
MgO	37.94	48.06	32.66	20.32	20.66	47.1	34.4	21.3	21.9	48.0	21.6	24.6
CaO	3.32	0.13	1.59	13.03	4.30	0.2	0.8	12.8	4.0	0.2	12.9	3.5
Na_2O	0.34	n.d.	0.34	1.50	0.07	n.d.	0.9	2.2	n.d.	n.d.	1.9	0.5
K_2O	0.14	n.d.	n.d.	n.d.	n.d.	n.d.	n.d.	n.d.	n.d.	n.d.	n.d.	n.d.
Total	100.00	100.77	99.80	98.11	100.01	98.8	101.2	100.4	98.2	100.3	101.5	101.1
Mg*		88.2	89.2	87.6	81.0	87.9	89.7	88.4	81.3	88.3	87.3	83.2
A			1.5	3.2	23.3		0.5	3.3	19.5		2.8	17.1
C			3.0	27.8	8.3		1.5	26.8	7.7		26.5	6.5
M			95.5	69.0	68.4		98.0	69.9	72.8		70.6	76.3

A=50(Al+Cr)/(Mg+Fe+Mn+Ca+0.5(Al+Cr)), C=100Ca/(Mg+Fe+Mn+Ca+0.5(Al+Cr)), M=100(Mg+Fe+Mn)/(Mg+Fe+Mn+Ca+0.5(Al+Cr)), Mg*=100Mg/(Mg+Fe)

TABLE 2. (Continued)

	15 GPa/1600 °C		16 GPa/1600 °C		20 GPa/1600 °C			26 GPa/1600 °C	
	ol	gar	ol	gar	γ	gar	Ca-P	Pv	Ca-P
SiO_2	40.8	47.3	41.1	50.3	43.5	53.6	54.4	55.0	49.1
TiO_2	n.d.	0.2	n.d.	0.2	n.d.	0.2	0.5	0.2	0.8
Al_2O_3	n.d.	14.0	n.d.	12.7	0.1	9.7	2.2	1.2	3.0
Cr_2O_3	n.d.	1.8	0.3	0.8	0.2	0.5	0.4	0.1	0.5
FeO*	6.0	5.6	4.3	3.6	7.7	3.5	3.4	7.7	6.6
MnO	0.1	0.1	0.1	0.1	0.1	n.d.	n.d.	0.1	0.5
MgO	52.0	26.9	52.5	29.3	49.7	26.6	24.7	34.0	18.8
CaO	0.2	3.5	0.2	2.2	0.2	6.2	12.3	1.6	18.4
Na_2O	n.d.	0.6	n.d.	n.d.	n.d.	n.d.	0.7	0.3	0.3
Total	99.1	100.0	98.5	99.2	101.5	100.3	98.6	100.2	98.0
Mg*	93.9	89.5	95.6	93.6	92.0	93.2	92.9	88.7	83.4
A		15.6		13.7		10.7	2.7	1.3	3.5
C		6.5		4.2		12.0	24.3	3.0	35.5
M		77.9		82.2		77.3	73.1	95.8	60.9

which took place about 300 °C lower than the dry solidus (Figure 1): 1) migration of H_2O from the pressure medium during the experiments (KATO and KUMAZAWA, 1985); or 2) reduction of FeO in silicates to Fe-metal by the carbon container which yielded CH_4 and lowered the solidus. Finally, it is also suspected that the actual run temperature became significantly higher than 1600 °C (up to 1800 °C judging from the electric power) due to deterioration of the thermocouple.

In the run product quenched at 20 GPa, a heavily fractured, unknown phase with high reflectivity was found in the BEI photo (Ca-P in Figure 3f). EPMA analysis of this phase shows a composition enriched in Ca (Table 2). Compared with diopsidic clinopyroxene present in the starting material, the Ca-P is slightly depleted in Ca, Al, and Na. We judge that the Ca-P is the unquenchable high-pressure polymorph of diopsidic clinopyroxene, possibly with a perovskite type structure (LIU, 1979),

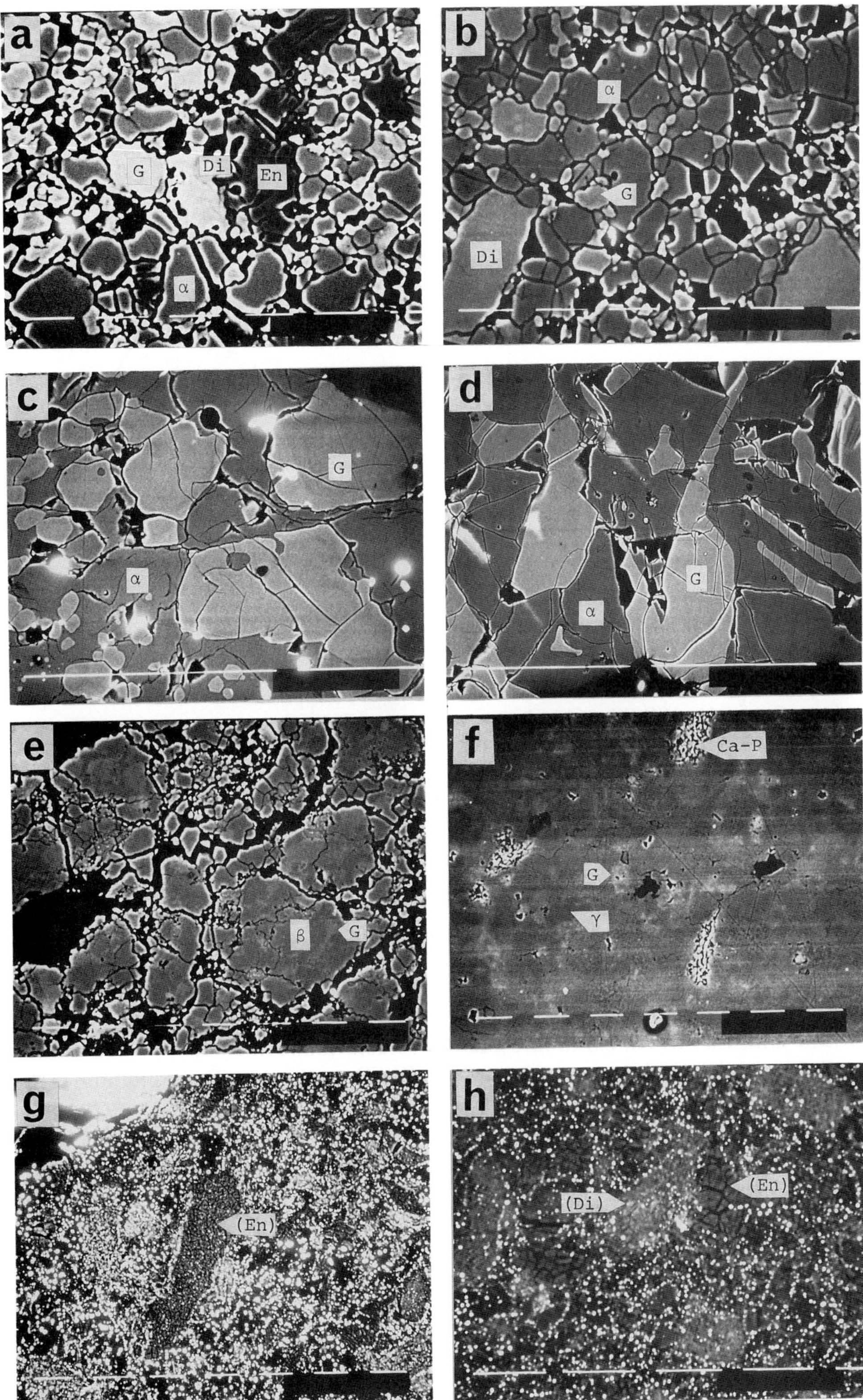

Figure 3. Back scattered electron images (BEI) of the run products quenched at various P-T conditions: a) at 9 GPa/1500 °C; b) at 13 GPa/1650 °C; c) at 15 GPa/1600 °C; d) at 16 GPa/1600 °C; e) at 17 GPa/1600 °C; f) at 20 GPa/1600 °C; g) at 24 GPa/1600 °C; and h) at 26 GPa/1600 °C. Scale bars in photos c and d represent 100 μm; all others represent 10 μm. In the stability field of olivine (α-phase), the run products are coarse-grained and their constituent minerals are homogeneous (a–d). At higher pressures, where β- or γ- or Pv-phase crystallize, the run products consist of very fine-grained mineral mixtures (e–h). A Ca-rich phase (Ca-P) with characteristic fractured texture is present in the experimental charge quenched at 20 GPa (f). In 24 and 26 GPa runs matrix of the peridotite consists of Pv (most dark grains) and Mw (bright spots), and pseudomorphs of Ca-rich pyroxene (Di) and Ca-poor pyroxene (En) are clearly seen (g and h).

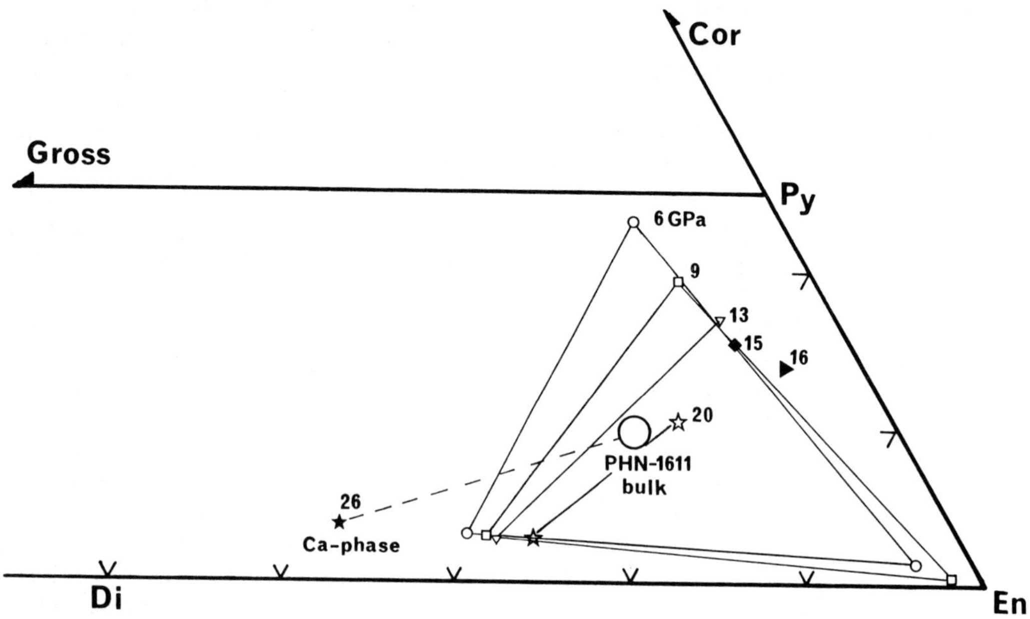

Figure 4. Compositions of coexisting garnet, En-px, Di-px, and Ca-P at various pressures. The EPMA analyses in Table 2 were projected on the plane $CaSiO_3$-(Al, $Cr)_2O_3$-(Mg, $Fe)SiO_3$. The bulk chemical composition of the peridotite PHN-1611 minus M_2SiO_4 component is plotted with a large open circle. From this projection, it is clear that En-Px is absent in the 13 GPa run product and a Ca-rich component (liq?) is missing from the 15 and 16 GPa run products.

because of the following observations: 1) the absence of X-ray diffraction signals due to this phase (Figure 2a), and 2) a chemical composition similar to Ca-rich clinopyroxene. Extensional cracks originating from the Ca-P (Figure 3f) suggest that it transformed to a lower-density amorphous phase (possibly glass) during decompression and after quenching. Such a transformation to the amorphous phase upon decompression has been reported in silicate perovskite with $CaSiO_3$ composition by LIU and RINGWOOD (1975).

IRIFUNE and RINGWOOD (this volume) found the Ca-rich unquenchable phase with $CaSiO_3$ stoichiometry in basalt (MORB) and pyrolite minus olivine compositions at pressures above 20 GPa. Because their run temperatures are 400 °C lower than the present experiments, it is possible that the high-pressure polymorph of diopsidic pyroxene has a stability field only at higher temperatures. Although the Ca-P shows a Ca-rich clinopyroxene-like chemistry, it requires further investigation (such as by transmission electron microscope analysis) in order to check whether it was a single phase under in situ conditions (LIU, 1979) or a mixture of $MgSiO_3$-perovskite and $CaSiO_3$-perovskite (MAO et al., 1977).

In the experimental charges quenched at 24 and 26 GPa (Figures 3g, h), the matrix consists of a very fine aggregate of magnesiowüstite (bright spots) and $MgSiO_3$-perovskite (most dark grains). Pseudomorphs of En-px and Di-px, which are present in the starting material, are clearly seen in the BEI photos (Figures 3g, h). EPMA analysis of such large grains has shown chemical compositions similar to those of the starting material (Table 2).

It is important to note that the perovskite phase obtained at 26 GPa and 1600 °C contains 11 mol% $FeSiO_3$ and 3 mol% $CaSiO_3$ (Table 2). In the MgO-FeO-SiO_2 system, we have determined the maximum Fe-solubility into the $MgSiO_3$-perovskite to be 11 mol% $FeSiO_3$ at the same P-T conditions (ITO et al., 1984). In the CaO-MgO-Al_2O_3-SiO_2 system, we have determined the maximum solubility of $CaSiO_3$ component in $MgSiO_3$-perovskite to be 0.4 mol% at 27 GPa and 1600 °C (ITO and TAKAHASHI, this volume). These observations suggest that the perovskite listed in Table 2 represents a metastable composition directly crystallized from a mineral grain (En-px) which was present in the starting material. If the chemical reaction between the matrix mineral and the $MgSiO_3$-perovskite was completed, it must have had a Mg/(Mg+Fe) ratio of about 0.96 (ITO et al., 1984). In support of this interpretation, precipitation of the magnesiowüstite phase was seen in large $MgSiO_3$-perovskite grains in the run product quenched at 24 GPa with a longer run duration (Figure 3g). These observations indicate that the reaction time was too short in these experiments to achieve chemical equilibrium.

Comparison with Akaogi and Akimoto's Work

AKAOGI and AKIMOTO (1979) have studied the subsolidus phase relations of the peridotite PHN-1611 up to 19 GPa (the pressure scale used in their study has recently been recalibrated M. AKAOGI, personal communication, 1986). Besides the differences in experimental pressures and temperatures (summarized in Figure 1), the present experiments are different from the previous work as follows: 1) a graphite capsule was used in the present study (except for runs 7 and 8), whereas a Mo-metal capsule was employed in the previous work; and 2) the starting material was a synthetic mixture of hydrous silicates in most of the Akaogi and Akimoto's experiments above 10 GPa, whereas we used natural rock powder.

In Akaogi and Akimoto's experiments En-px was found to be a major phase next to orthosilicates (α or β) up to 14 GPa, and it survived in the run products quenched at their highest pressure runs (19 GPa). In the present study, however, En-px was not detected in run products quenched above 13 GPa. Absence of the Ca-poor pyroxene in our run products can be understood from the mass balance consideration based on the EPMA analysis of the coexisting phases (see Figure 4). Our contrasting result on the stability of En-px is primarily due to the lower Ca content in our Di-px because of the higher run temperatures (see TAKAHASHI, 1986, Figure 11). Judging from the large enstatite solubility in the majorite garnet at pressures above 15 GPa and at temperatures of 1000–1200 °C (KANZAKI, 1986; IRIFUNE, 1986; IRIFUNE and RINGWOOD, this volume), En-px in Akaogi and Akimoto's experiments at 19 GPa and 1000–1100 °C may have been relict minerals.

In the present experiments (runs 3–6) in which the run products were mantled with diamond (formed in situ), the Mg* values of the run products were found to be much higher than the Mg* value of the starting material (Table 2). Fe-metal precipitated in these run products due to the very low oxygen fugacity imposed by the diamond container. The precipitated metal phase, however, was very rare in the run products. It may have been absorbed by the carbon container where it acted as a catalyzer to grow diamond. The higher Mg* value in the experimental charges may be part of the reason why olivine did not transform to modified spinel phase in these experiments (particularly at 16 GPa and 1600 °C). However, our run temperature might have been higher than the nominal value; this could also be the reason for the suppression of the above phase transformation. Although the iron loss due to reduction was significant in the present study, the Fe-Mg partitioning behaviors between coexisting garnet and other silicates were very similar to those reported by AKAOGI and AKIMOTO (1979); that is the Mg* value of garnet was much lower than other silicates at low pressures, but increased with pressure and became higher than that of coexisting spinel (γ-phase) at 20 GPa (compare Table 2 with AKAOGI and AKIMOTO, 1979, Figure 5).

Mineral Paragenesis in Mantle Peridotite along a Model Geotherm

On the basis of the present experimental results, mineral paragenesis in the peridotitic mantle (Mg*=88) down to a depth of 700 km is summarized in Figure 5. Major constituent minerals which would be present in more than 30 vol% are indicated with heavy lines. Using a natural peridotite, we observed the following major phase transformations which have been proposed to occur in the earth's mantle: 1) α-phase to β-phase and then β-phase to

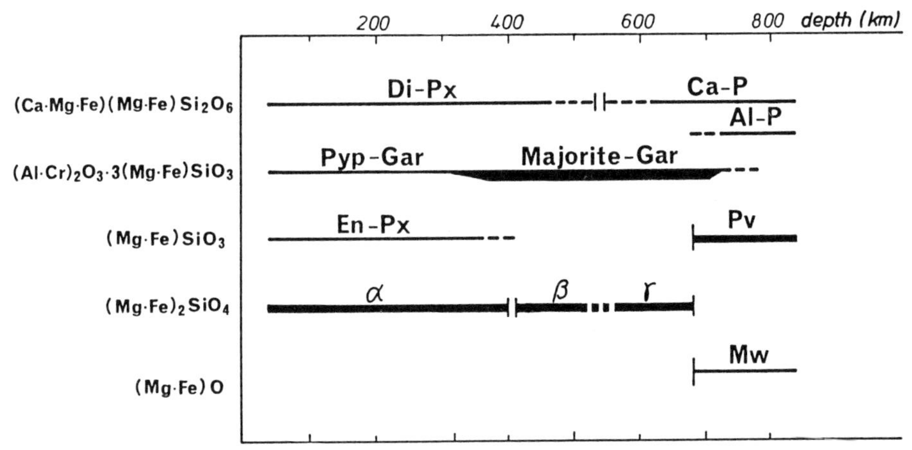

Figure 5. Estimated mineral paragenesis in peridotitic mantle along a model geotherm. Mineral phases which would be present in more than 30 vol% are indicated with heavy lines.

γ-phase in orthosilicate composition (RINGWOOD and MAJOR, 1970; AKIMOTO, 1972); 2) pyroxene to garnet (RINGWOOD, 1967; AKAOGI and AKIMOTO, 1977; KANZAKI, 1986); and 3) γ-spinel to $MgSiO_3$-perovskite plus magnesiowüstite assemblage (LIU, 1976; YAGI et al., 1979; ITO and YAMADA, 1982).

In addition to those major phases, we found two accessory phases, Ca-P and Al-P, in the natural peridotite composition. In a series of experimental studies in the CaO-MgO-Al_2O_3-SiO_2 system which have been carried out in parallel with the present study, we have encountered the same phases (ITO and TAKAHASHI, this volume). However, as described in the preceding sections, our experimental run durations were not long enough to achieve chemical equilibrium and to grow sufficiently large crystals for EPMA analysis. Experimental run products recovered from 24 and 26 GPa exhibit various local phase equilibria reflecting the chemistry of mineral grains in the starting material. Thus it is not clear whether or not the Ca-P phase would be present under experimental conditions above 20 GPa if the bulk system achieved chemical equilibrium. Further experimental investigations are necessary in order to understand the relative proportion of the Ca-P and Al-P phases in the earth's deep mantle as well as their equilibrium chemical compositions.

Acknowledgments. We are grateful to Y. Matsui and S. Akimoto for their encouragement throughout this study. We also appreciate the collaboration of C. M. Scarfe who performed some of the present experiments. Discussions with M. Akaogi and T. Irifune were very helpful, as was the use of their unpublished data. We gratefully acknowledge critical reviews by anonymous reviewers. This research was partially supported by grants 59460048, 61113005, and 61213011 from the Ministry of Education, Science and Culture, Japan. We also thank K. Yokoyama of the National Museum and I. Kushiro of Tokyo University for use of their SEM facilities, and F. R. Boyd of the Carnegie Institute for providing the peridotite specimen.

REFERENCES

AKAOGI, M., and S. AKIMOTO, Pyroxene-garnet solid solution equilibria in the system $Mg_4Si_4O_{12}$-$Mg_3Al_2Si_3O_{12}$ and $Fe_4Si_4O_{12}$-$Fe_3Al_2Si_3O_{12}$ at high pressures and temperatures, *Phys. Earth Planet. Inter., 15*, 90–106, 1977.

AKAOGI, M., and S. AKIMOTO, High-pressure phase equilibria in a garnet lherzolite, with special reference to Mg^{2+}-Fe^{2+} partitioning among constituent minerals, *Phys. Earth Planet. Inter., 19*, 31–51, 1979.

AKIMOTO, S., The system MgO-FeO-SiO_2 at high pressures and temperatures: phase equilibria and elastic properties, *Tectonophys., 13*, 161–187, 1972.

AKIMOTO, S., High-pressure research in geophysics: past, present and future, this volume.

ANDERSON, D. L., The earth as a planet: paradigms and paradoxes, *Science, 223*, 347–355, 1984.

BERTRAND, P., C. SOTIN, J.-C. MERCIER, and E. TAKAHASHI, From the simplest chemical system to the natural one: garnet peridotite barometry, *Contrib. Min. Petrol., 93*, 168–178, 1986.

BOULLIER, A.-M., and A. NICOLAS, Texture and fabric of peridotite nodules from kimberlite at Mothae, Thaba Putsoa and Kimberley, in *Lesotho Kimberlites*, edited by P. H. Nixon, pp. 57–66, Lesotho Nat. Develop. Corp., 1973.

BOYD, F. R., The pyroxene geotherm, *Geochim. Cosmochim. Acta, 37*, 2533–2546, 1973.

GASPARIK, T., Two-pyroxene thermometry with new experimental data in the system CaO-MgO-Al_2O_3-SiO_2, *Contrib. Min. Petrol., 87*, 87–97, 1984.

IRIFUNE, T., An experimental investigation of the pyroxene garnet transformation in a pyrolite composition and its bearing on the constitution of the mantle, *Phys. Earth Planet. Inter.*, in press, 1987.

IRIFUNE, T., and A. E. RINGWOOD, Phase transformations in primitive MORB and pyrolite compositions to 25 GPa and some geophysical implications, this volume.

ITO, E., and Y. MATSUI, Synthesis and crystal-chemical characterization of $MgSiO_3$ perovskite, *Earth Planet. Sci. Lett., 38*, 443–450, 1978.

ITO, E., and E. TAKAHASHI, Ultrahigh-pressure phase transformations and the constitution of the deep mantle, this volume.

ITO, E., and H. YAMADA, Stability relations of silicate spinels, ilmenites, and perovskites, in *High-Pressure Research in Geophysics*, edited by S. Akimoto and M. H. Manghnani, pp. 405–419, D. Reidel, Dordrecht, 1982.

ITO, E., E. TAKAHASHI, and Y. MATSUI, The mineralogy and chemistry of the lower mantle: an implication of the ultrahigh-pressure phase relations in the system MgO-FeO-SiO_2, *Earth Planet. Sci. Lett., 67*, 238–248, 1984.

KANZAKI, M., Ultrahigh-pressure phase relations in the system $MgSiO_3$-$Mg_3Al_2Si_3O_{12}$, *Phys. Earth Planet. Inter.*, in press, 1987.

KATO, T., and M. KUMAZAWA, Stability of phase B, a hydrous magnesium silicate to 2300 °C at 20 GPa, *Geophys. Res. Lett., 12*, 534–535, 1985.

LIU, L., The post-spinel phases of forsterite, *Nature, 262*, 770–772, 1976.

LIU, L., The system enstatite-pyrope at high pressures and temperatures and the mineralogy of the earth's mantle, *Earth Planet. Sci. Lett., 36*, 237–245, 1977.

LIU, L., The system enstatite-wollastonite at high pressures and temperatures, with emphasis on diopside, *Phys. Earth Planet. Inter., 19*, 15–18, 1979.

LIU, L., and A. E. RINGWOOD, Synthesis of a perovskite-type polymorph of $CaSiO_3$, *Earth Planet. Sci. Lett.*, 209–211, 1975.

MAO, H. K., T. YAGI, and P. M. BELL, Mineralogy of the earth's deep mantle: quenching experiments on mineral compositions at high pressure and temperature, *Carnegie Inst. Washington Yearb., 76*, 502–504, 1977.

MERCIER, J.-C., and N. L. CARTER, Pyroxene geotherms, *J. Geophys. Res., 80*, 3349–3362, 1975.

MYSEN, B. O., and I. KUSHIRO, Compositional variations of coexisting phases with degree of melting of peridotite in the upper mantle, *Am. Mineral., 62*, 843–865, 1977.

NAVROTSKY, A., and M. AKAOGI, The α, β, γ phase relations in Fe_2SiO_4-Mg_2SiO_4 and Co_2SiO_4-Mg_2SiO_4: calculation from thermochemical data and geophysical applications, *J. Geophys. Res., 89*, 10,135–10,140, 1984.

NIXON, P. H., and F. R. BOYD, Petrogenesis of the granular and sheared ultrabasic nodule suites in kimberlite, in *Lesotho Kimberlites*, edited by P. H. Nixon, pp. 48–56, Lesotho Nat. Develop. Corp., 1973.

NIXON, P. H., F. R. BOYD, and A. M. BOULLIER, The evidence of kimberlite and its inclusions on the constitution of the outer part of the earth, in *Lesotho Kimberlites*, edited by P. H. Nixon, pp. 312–318, Lesotho Nat. Develop. Corp., 1973.

RICHTER, F. M., and D. P. MCKENZIE, On some consequences and possible causes of layered mantle convection, *J. Geophys. Res., 86*, 6133–6142, 1981.

RINGWOOD, A. E., The pyroxene-garnet transformation in the earth's mantle, *Earth Planet. Sci. Lett., 2*, 255–263, 1967.

RINGWOOD, A. E., *Composition and Petrology of the Earth's Mantle*, 618 pp., McGraw-Hill, New York, 1975.

RINGWOOD, A. E., and A. MAJOR, The system Mg_2SiO_4-Fe_2SiO_4 at high pressures and temperatures, *Phys. Earth Planet. Inter., 3*, 89–108, 1970.

SCARFE, C. M., and E. TAKAHASHI, Melting of garnet peridotite to 13 GPa: implications for the early history of the upper mantle, *Nature, 322*, 354–356, 1986.

SHIMIZU, N., Rare earth elements in garnets and clinopyroxenes from garnet lherzolite nodule in kimberlites, *Earth Planet. Sci. Lett., 25*, 26–32, 1975.

SMITH, D., and F. R. BOYD, Compositional heterogeneities in a high-temperature lherzolite nodule and implications for mantle processes, in *Mantle Xenoliths*, edited by P. H. Nixon, in press, 1987.

SUITO, K., Phase relations of pure Mg_2SiO_4 up to 200 kilobars, in *High Pressure Research-Applications to Geophysics*, edited by M. Manghnani and S. Akimoto, pp. 255–266, Academic Press, New York, 1977.

TAKAHASHI, E., Melting of a dry peridotite KLB-1 up to 14 GPa: implications on the origin of peridotitic upper mantle, *J. Geophys. Res., 91*, 9367–9382, 1986.

WOOD, B. J., and S. BANNO, Garnet-orthopyroxene and orthopyroxene-clinopyroxene relationships in simple and complex systems, *Contrib. Mineral. Petrol., 42*, 109–124, 1973.

YAGI, T., P. M. BELL, and H. K. MAO, Phase relations in the system MgO-FeO-SiO_2 between 150 and 700 kbar at 1000 °C, *Carnegie Inst. Washington Yearb., 78*, 614–618, 1979.

MINERAL PHYSICS CONSTRAINTS ON A UNIFORM MANTLE COMPOSITION

Donald J. WEIDNER

Department of Earth and Space Sciences, State University of New York Stony Brook, New York 11794, USA

Eiji ITO

Institute for Study of the Earth's Interior, Okayama University Misasa, Tottori-ken 682-02, Japan

Abstract. A comparison of seismic data with those from mineral physics demonstrates that the earth's mantle may be of homogeneous chemical composition. In particular, the pyrolite model is compatible with the available data. The test of any model centers on matching the seismic velocity values at the base of the upper mantle, the magnitude of the velocity jump at 400 km, the velocity gradient in the transition zone, and the characteristics of the 670 km discontinuity. When testing a chemical model it is critical to consider the uncertainties of all of the model parameters. Thus the hypothesis includes the chemical variables and the assumptions concerning the physical properties.

Interdependence of chemical and physical variables are examined with reference to various tests afforded by seismic data. The 400 km discontinuity and the transition zone gradient are extremely sensitive to the pressure derivative of the shear moduli of the stable phases and phase transitions in the pyroxene systems. The sensitivity to these parameters is so great as to severely limit the chemical resolution of the seismic data. The expected composition range for the lower mantle results in a mineralogy which ranges from 80% (by volume) of perovskite with 20% magnesio-wustite to pure perovskite. The differences in the resulting velocities is extremely small relative to those introduced from uncertainties in the physical properties.

Further mineral physics data are necessary to resolve the chemical characteristics of the mantle. In particular, the pressure derivatives of the elastic moduli of the high pressure phases are needed. In addition, more data on the phase equilibria of the pyroxene systems, particularly with sodium and calcium, are important. It is also important to evaluate the resolution of the seismic data. The mineralogical models center on the magnitude of velocity discontinuities and velocity gradients. The interdependence of these variables needs to be more clearly elucidated.

Introduction

A complete description of the chemical heterogeneity of the earth's mantle is fundamental to the understanding of the evolution of the earth and the current mode of convection within the mantle. In this paper we address the question of vertical layering. We use the data of seismology and mineral physics to test the hypothesis that the mantle of the earth is chemically homogeneous. As did WEIDNER (1985, 1986), we find that homogeneous composition with the major element abundances of pyrolite is compatible within the uncertainties of the data. However the uncertainties of the data also allow models with significant chemical layering.

The qualitative agreement between the first order seismic discontinuities and the expected phase transformations for a pyrolitic mantle is compelling evidence that these discontinuities are due to phase transformations. While pyrolite has other phase transformations which should also occur they are not expected to create seismic discontinuities. Furthermore the seismic discontinuities do not appear to be greatly disrupted in tectonic settings which probably involve vertical movement of material. Recent body wave studies such as those of GRAND and HELMBERGER (1984), WALCK (1984), BURDICK and HELMBERGER (1978), FUKAO (1977), and GIVEN and HELMBERGER (1980) indicate that the seismic velocities at depths shallower than 400 km differ in ocean ridge areas from ocean basin and continental regions. But the signature of the 400 km discontinuity is strikingly similar for all of these regions. We therefore suggest that the seismic discontinuities are not chemical boundaries but phase transformations. ANDERSON and BASS (1986) and OHTANI (1985) propose that the chemical segregation of the mantle occurred early in its history by melting processes. Thus it is possible to link the chemical boundaries with the phase transformations as the melting curve and buoyancy will reflect the phase transformations. However as the earth further evolves these chemical boundaries will be displaced relative to the phase transformation boundary due to material motion and temperature evolution. Yet the phase transformations will still produce seismic discontinuities whether or not chemical changes contribute. If chemical changes do occur with depth perhaps they are distributed features which are more subtle to identify.

The conclusion that pyrolite satisfies the seismic constraints stands in contrast to the conclusions of ANDERSON and BASS (1986), who prefer to generate all seismic discontinuities by chemical changes. In their rejection of a pyrolite model they made overly restrictive assumptions concerning the unknown physical properties of several of the high-pressure phases. Furthermore, in

High-Pressure Research in Mineral Physics, edited by M. H. Manghnani and Y. Syono, pp. 439–446.

light of the new data on the compressibility of majorite by YAGI et al. (this volume), their proposed transition zone mineralogy, 'piclogite', must be considerably altered if it is to remain a viable alternative for this region of the earth.

The style of hypothesis testing which is afforded by the data allows us only to eliminate incompatible models but never to draw unique conclusions. Furthermore the severity of the tests rests on the accuracy of the model parameters. Relevant mineral physics data have been forthcoming at an increasing rate, as is evidenced by this volume and other recent literature. Still, the voids in the data give considerable latitude to the range of acceptable composition models.

In this paper we evaluate the resolution of the current data base. We use the successful pyrolite model as the standard and evaluate the resolution of the data by perturbing the model in both its chemical and physical variables. We find that uncertainties in physical properties restrict the discrimination between a wide range of chemical models. In this study we focus on the size of the 400 km velocity discontinuity, the velocity gradient in the transition zone, and the velocity at the top of the lower mantle. The result is to emphasize the limitations of the data and suggest the next generation of experimental data which are necessary to more tightly constrain our definition of the mantle composition.

Mantle Model

Calculations presented in this paper follow the methodology and philosophy given in WEIDNER (1985, 1986). In these papers the goal was to test a pyrolite mantle composition in the light of the elastic and phase equilibria data which have been recently obtained. A fundamental element of this approach is the emphasis on the method of hypothesis testing. In such evaluations, one must carefully separate the hypothesis from the test. The hypothesis consists of the unknown physical and chemical parameters along with the more interesting features of the model, such as mantle composition. The test comes by comparing predicted with observed properties, in this case the seismic velocities throughout the mantle. Seismic velocities are the best resolved depth dependent properties of the material of the earth's mantle. They depend on the bulk modulus, the shear modulus, and the density of the constitutive material at the pressure and temperature conditions of the point in question. To effect a test of a mineralogical model we must know these properties for the specific composition of the appropriate phases at the in situ pressure and temperature. Despite the vigorous increase of appropriate data in the past decade, the actual knowledge falls short of achieving this goal and we must resort to estimates or guesses to realize this possibility. Table 1 indicates the mineral phases which are currently expected to constitute the bulk of the mantle along with the physical properties which are necessary to construct an earth seismic model. Indicated in the table by M, E, or G is whether the particular variable has been measured, can be estimated from data, or must be guessed. The variables in the table are the elastic moduli and their first order derivatives with respect to chemical and physical variables. While the second and higher order terms are also very important in defining acoustic velocity with depth in the earth, they must be guessed for all of the phases listed here.

While Table 1 has many entries that are G, we should emphasize the number of M entries. Furthermore, in developing a model, one does not have complete freedom

TABLE 1. Status of Important Elastic Parameter Data for Mantle Phases

Property	Olivine	Ca-poor Pyroxene	Ca Pyroxene	Garnet	Majorite	Spinel	Modified Spinel	Stishovite	Ilmenite	Magnesio-wustite	Perovskite
K	M	M	M	M	E	M	M	M	M	M	E
μ	M	M	M	M	G	M	M	M	M	M	G
$\partial K/\partial P$	M	M	G	M	G	G	G	G	G	M	G
$\partial \mu/\partial P$	M	M	G	M	G	G	G	G	G	M	G
$\partial K/\partial T$	M	M	G	M	G	G	G	G	G	M	G
$\partial \mu/\partial T$	M	M	G	M	G	G	G	G	G	M	G
$\partial^2 K/\partial P \partial T$	G	G	G	G	G	G	G	G	G	M	G
$\partial^2 \mu/\partial P \partial T$	G	G	G	G	G	G	G	G	G	M	G
$\partial K/\partial \mathrm{Fe}$	M	M	M	M	G	E	E	—	G	M	G
$\partial \mu/\partial \mathrm{Fe}$	M	M	M	M	G	E	E	—	G	M	G
$\partial K/\partial \mathrm{Ca}$	—	—	G	M	G	—	—	—	G	—	G
$\partial \mu/\partial \mathrm{Ca}$	—	—	G	M	G	—	—	—	G	—	G
$\partial K/\partial \mathrm{Al}$	—	—	E	—	—	—	—	—	G	—	G
$\partial \mu/\partial \mathrm{Al}$	—	—	E	—	—	—	—	—	G	—	G

M, measured; E, estimated from data; G, guessed.

in the range of guessed values for a particular variable. Ultimately the mantle model that is tested includes a description of the chemical composition of the mantle along with the particular estimates of the physical and chemical parameters. Part of the test comes from the comparison of the deduced seismic model with the observed one, and part of the test is the reasonableness of the estimated parameters. This approach emphasizes the data that are required to make the test more discriminatory.

The values for the variables are mostly unchanged from WEIDNER (1986). In addition the method of calculation and the salient phase diagrams given in WEIDNER (1986) are also used here. There have been a few important changes. The first concerns the pyroxene-garnet equilibrium. The previous study used the results of AKAOGI and AKIMOTO (1977, 1979). The more recent study of KANZAKI (1985) for the enstatite-pyrope portion of the composition regime indicated lower pressures for the phase transformation and a greater solubility of pyroxene in garnet. The suggestion is that the data of AKAOGI and AKIMOTO (1977) may need to be recalibrated in pressure. However, the results of AKAOGI and AKIMOTO (1977, 1979) are internally consistent over a broad range of compositions. We therefore simply change phase diagrams for the pyroxene-garnet system by dividing the pressure by a constant and increasing the solubility of the pyroxene in the garnet. A constant of 1.3 is sufficient to bring the two data sets into agreement.

As ANDERSON and BASS (1986) point out, this would imply that a system with enstatite and pyrope probably completes the phase transformation by the 400 km discontinuity. However, the presence of calcium in the system has the effect of retarding this phase transformation. As WEIDNER (1986) points out, the calcium partitioning into the garnet appears from AKAOGI and AKIMOTO's (1979) data to be insensitive to pressure and hence silicon content of the garnet. On the other hand, the chemical potential of silicon in the garnet is very sensitive to the calcium content. This dependence is reflected in the calculations given here. As a result only about half of the calcium-free pyroxene has transformed to the garnet structure as olivine transforms to the modified spinel phase. While this feature is not critical for the survival of pyrolite, it does contribute to the velocity gradient in the transition zone as will be discussed later.

We have also adopted the bulk modulus reported in this volume by YAGI et al. (this volume) for majorite and retained the guess of WEIDNER (1986) that the shear modulus is the same as for pyrope. Figure 1 illustrates the relative magnitudes of the bulk and shear moduli for several materials in the corundum (or ilmenite) and garnet structures. As shown, the bulk and shear moduli of the Al corundum are greater than for the Cr which in turn are

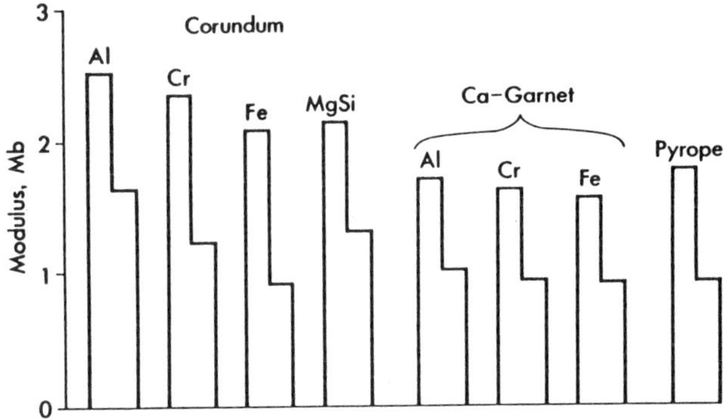

Figure 1. The bulk and shear moduli for several different compounds including the corundum (or ilmenite) structure and the garnet structure.

greater than Fe. The coupled substitution of (MgSi) forming the related ilmenite structure yields a bulk modulus less than that for Cr but greater than for Fe and a shear modulus intermediate between that for Al and Cr. The garnets exhibit the same ordering of elastic moduli as Al, Cr, and Fe occupy the octahedral site, but the magnitude of the effect is reduced, particularly for the shear modulus. It is therefore expected that the elastic moduli of majorite would be slightly less than that of pyrope and the shear modulus approximately equals that of pyrope. As there is a large discrepancy between the data of YAGI et al. (this volume) and those of JEANLOZ (1981), we should view the implications with due reservation until the differences are resolved.

One of the implications of the data from YAGI et al. (this volume) for majorite is that the piclogite model of ANDERSON and BASS (1986) does not fit the seismic data. In the transition zone piclogite is composed of over 50% majorite. A change of 20% in the elastic moduli for majorite from the values that they assume thus completely destroys the quality of the fit. If these newer data are correct then either the mineralogy or some physical property of the piclogite model will need to be greatly changed from the values that they assumed in order for this model to remain viable.

WEIDNER (1986) did not include the phase transformation to perovskite nor the lower mantle in his model, since there is very little data for the perovskite phase. YAGI et al. (1978) have reported a bulk modulus value based on compression studies which we use here. Still, there is not information on the shear modulus, so we simply assume a value consistent with the seismic data. Therefore, the results for the lower mantle should be viewed as qualitative. Recent results of ITO and TAKAHASHI (this volume) indicate the role of calcium and aluminum in the phase transformation of garnet and pyroxene to perovskite. They found that the $(Mg,Fe)SiO_3$ and $(Mg,Fe)_2SiO_4$

components of the system transform to perovskite assemblage in a very narrow pressure interval and that this transformation is followed by the transformation of the calcium- and aluminum-rich phases over a broader pressure interval. We have modeled this by a sharp transition for the olivine and orthopyroxene portion and a transformation linearly distributed over 2 GPa for the remaining portion. We further assume that there is no composition dependence of the perovskite elastic properties. The result is a sharp seismic discontinuity followed by a region with a steep velocity gradient.

Resolution of Model

Results of the calculations are illustrated in Figure 2. Volume percentages of the various phases are illustrated in Figure 3. Seismic models are from GRAND and HELMBERGER (1984) and WALCK (1984) and are given as dotted lines. The calculations correspond to an adiabatic geotherm with a 1400 °C foot. The pyrolite model reflects most of the major features of the seismic model. Here we hope to evaluate the sensitivity of the model to both the chemical and physical variables. We do this by discussion of each region in turn.

Upper 400 km

The upper 400 km exhibit a considerable lateral variation, as evidenced by the tomography studies as well as from the body wave refraction studies of WALCK (1984), BURDICK and HELMBERGER (1978), and GRAND and HELMBERGER (1984). The departure of the observed velocities to lower values than the model velocities may reflect thermally activated relaxation processes including the possibility of partial melting. Thus the significant feature in terms of mantle composition is the deepest section where these processes are probably the least important. Here the pyrolite model matches the observations well within the uncertainty of the elastic modulus data. The calculations of ANDERSON and BASS (1986) agree with ours for this region of the mantle.

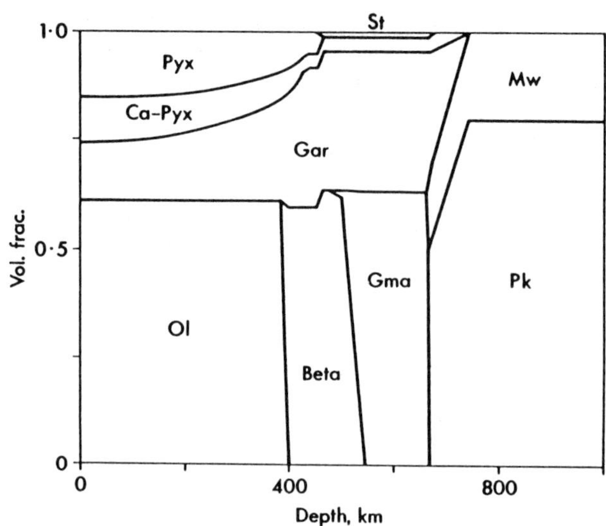

Figure 3. Volume fraction for each mineral phase as a function of depth in the mantle for the pyrolite model.

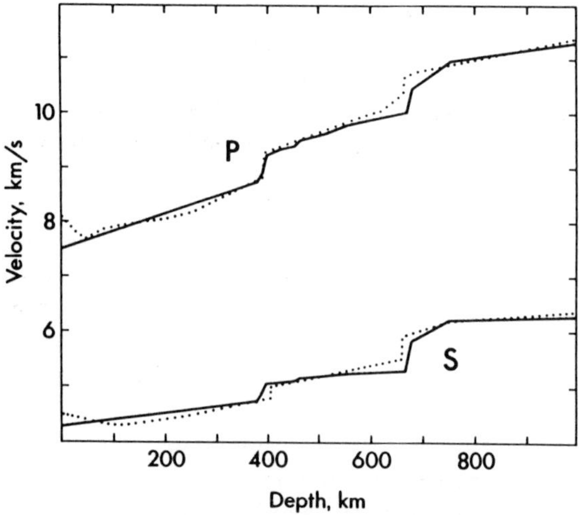

Figure 2. Compressional and shear velocity as a function of depth in the earth's mantle. The dotted lines are from the seismic studies of GRAND and HELMBERGER (1984) and WALCK (1984), the solid lines are for the pyrolite model.

400 km Discontinuity

The 400 km discontinuity in this model results from the transformation of olivine to the modified spinel phase. As the figure indicates, the reported seismic models exhibit a 15 km range in depth for this discontinuity. Presumably this reflects either different ray paths or the uncertainties of the seismic data and is not a physical constraint. The pressure of the phase transformation is subject to uncertainties in pressure calibration, which thereby allows freedom in the actual depth at which the model discontinuity is located. We therefore conclude that the position of the discontinuity is properly matched by the conditions of the olivine to beta transformation.

In this figure the seismic discontinuity is sharper than that for pyrolite. However, the sharpness of the phase transformation induced discontinuity depends critically on the relative values of composition, temperature, and the univariant portion of the phase diagram. WEIDNER (1986) illustrated the effect of changing these relative positions, with the result that slight variations from those assumed here could produce a very sharp discontinuity. The conditions assumed here probably represent the extreme by indicating the broadcast transition region for this transformation. On the other hand, WALCK (1984) indicates that the longitudinal waves, which have the best resolving power of these models, cannot distinguish a discontinuity from an increase distributed over 10 to 20 km. Therefore, pyrolite can reproduce the sharpness of the transition.

The magnitude of the seismic discontinuity is a more restrictive criterion for the phase transformation to match. The reported magnitude is about 40% of the velocity difference between olivine and the modified spinel phase as measured at room conditions. Thus if the relative values of the two velocities do not change as the conditions are changed to those of the mantle, we could only allow about 40% of the mantle to be olivine and the remaining 60% some material to dilute the effect of the phase transformation. However, if the pressure derivative of the shear modulus of the modified spinel phase is lower by an amount of 0.8 to 1.0 from that for the olivine phase, the two velocities will converge sufficiently by the 400 km discontinuity so as to allow a model containing 60% olivine. This value of the pressure derivative of shear modulus is quite reasonable in that aluminate spinels even have lower values. The pressure derivative of the bulk modulus need not be varied to accommodate this difference.

The sensitivity of the 400 km discontinuity to several chemical and physical variables is indicated in Table 2. Here is shown the assumed model value for each variable and the value required to change the magnitude of the velocity jump by 0.1 km/s (note the magnitude of the seismic discontinuity is 0.22 for shear waves and 0.43 for compressional waves). This table is applicable if the 400 km discontinuity is due to the olivine to modified spinel transition. Relatively small changes in the physical properties of the phases can produce an appreciable change in the velocity jump, while rather larger compositional changes are required to produce such a change. In particular, the Si content as reflected in the amount of olivine present is the most sensitive compositional variable. Still, a major change in the olivine to pyroxene ratio is required to substantially change the magnitude of the discontinuity. The iron to magnesium ratio has no effect on the velocity jump at 400 km. On the other hand, the velocity jump is extremely sensitive to the pressure and temperature derivatives of the modified spinel phase. Thus the observed magnitude of the jump cannot accurately define compositional variables until the physical properties are defined. In particular, high quality measurements of the pressure and temperature derivatives of the elastic properties of the modified spinel phase are required.

TABLE 2. Sensitivity of the Magnitude of the 400 km Discontinuity to Model Parameters

Property	Model Value	For ΔV_p (400) to inc. by 0.1	For ΔV_s (400) to inc. by 0.1
$\partial K/\partial P$ modified-spinel	4.5	4.96	*
$\partial \mu/\partial P$ modified-spinel	0.8	1.25	1.14
$\partial K/\partial T$ modified-spinel	−2.0	−1.2	*
$\partial \mu/\partial T$ modified-spinel ($\times 10^{-4}$ Mb/°C)	−1.4	−0.8	−1.0
Fe/Mg	0.11	*	*
Ca/(Fe+Mg+Ca)	0.049	0.17	0.21
Al/Si	0.092	0.45	*
Vol % olivine	61.0	76.0	84.0

Values indicate necessary changes to produce an increase of 0.1 km/s in the velocity discontinuity.

*Variable cannot effect a change in the velocity discontinuity.

Transition Zone: 400–670 km

The transition zone is marked by high velocity gradients which are not reproduced by this pyrolite model as seen in Figure 2. Table 3 summarizes the observed velocity increase with the model values. The increases have been normalized to 10 GPa pressure interval and the increased gradient of the bottom 40 km of WALCK's (1984) compressional velocity model has not been included. It would add another 0.2 km/s to the compressional velocity increase but may be due to the perovskite phase transformation. The table indicates the contribution to the gradient from an homogeneous, adiabatically compressed material with properties of the modified spinel phase. Also indicated are the total velocity increases of the model which include effects of phase transformations of modified spinel to spinel and the loss of pyroxenes to high-pressure phases. The model accounts for about 80% of the compressional velocity gradient and only about 50% of the shear velocity gradient. Many factors can contribute to the velocity increase which are not included in the model. They include greater pressure derivatives for some of the phases, phase transitions which have been incompletely

TABLE 3. Velocity Increase over 10 GPa Pressure Interval of Transition Zone

	Ovserved (km/s)	Compression Mod.-spinel	Total Pyrolite
V_p	0.92*	0.50	0.76
V_s	0.50	0.07	0.24

The observed value is compared with that due to self-compression of a material with the properties of the modified spinel phase and the total increase calculated for the pyrolite model. The difference in these last two terms indicate the portion due to phase transitions.

*Exclude steeper gradient in bottom 40 km of transition zone.

or incorrectly represented, or gradients in chemical composition in this region of the mantle.

If there are no phase transitions and the velocity gradient is due entirely to self compression then the pressure derivative of the bulk modulus must be 5.6 and that for the shear modulus must be 2.5. While this is an extreme case and these values are larger than those measured for olivine, they are not impossibly large. This conclusion must be balanced with the need to restrict the magnitude of the 400 km discontinuity.

In the pyrolite model, a significant portion of the velocity gradient is accommodated by phase transition. The contribution of the phase transition will be larger if the pressure derivatives of the moduli for the high-pressure phases are increased. Figure 4 illustrates the model velocities resulting from changing just the pressure derivative of shear modulus for only the spinel phase from 0.8 to 1.2. Here almost all of the gradient has been matched. If this value is allowed to increase to 1.35, then the entire velocity increase of Table 3 is accommodated. By changing the property of the spinel phase without changing the property of the modified spinel phase, the effective gradient changes without changing the magnitude of the 400 km discontinuity. While a very small change in just this one property can account for the steep gradient, a difference between the properties of the spinel and modified spinel phases are not anticipated on mineral physics grounds. Experimental investigations are required to test this possibility.

Increased amounts of phase transformations can account for a steep gradient in the transition zone. Possible candidates include the pyroxene-garnet transformation and garnet to ilmenite or spinel plus stishovite.

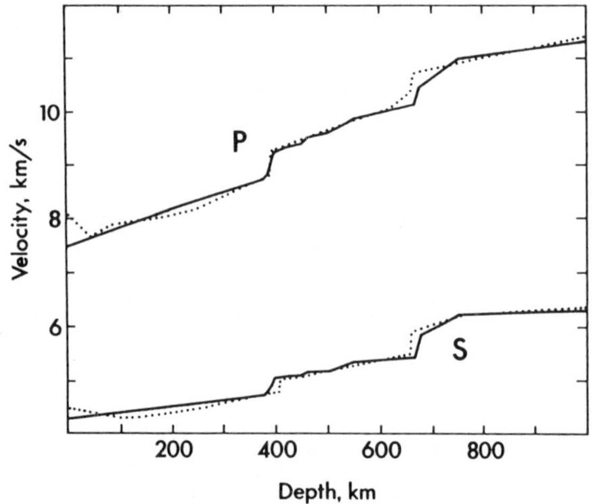

Figure 4. Same as Figure 2, except the pressure derivative of the shear modulus for the spinel phase is 1.2 instead of 0.8.

The importance of the first of these transitions depends on the role of calcium and possibly sodium on the pyroxene to garnet transformation. The phase relations assumed in this study suggest that about half of the calcium free pyroxene transforms to garnet at depths shallower than 400 km with some pyroxene finally transforming to modified spinel plus stishovite well within the transition zone. The calculations are consistent with KANZAKI's (1985) data for magnesium rich systems as well as the data of AKAOGI and AKIMOTO (1977) for end member pyroxenes with a recalibrated pressure scale and AKAOGI and AKIMOTO's (1979) observations for a natural peridotite. In the later paper they report the coexistence of the calcium poor pyroxene and the modified spinel phase. They also report the existence of stishovite at higher pressures indicating silicon saturation of the garnet at the pyroxene stability limit. Nonetheless, this transformation does not contribute significantly to the velocity gradient in the deeper half of the transition zone. The gradient appears equally steep throughout the zone.

This problem is also encountered in the piclogite model of ANDERSON and BASS (1986). The phase transitions for their mineralogy occur either at the same depths or even shallower than for pyrolite, by our calculations, depending on the amount of aluminous garnet in the system. Therefore, we find no advantage of the piclogite model in matching the gradient of the transition zone.

AKAOGI et al. (this volume) report a theoretical study of the pyroxene garnet transformation. They indicate, for pressures higher than the stability of the enstatite, that spinel (or modified spinel) plus stishovite will exsolve from the garnet. SAWAMOTO (this volume) demonstrates a transformation of garnet to ilmenite in the $MgSiO_3$ system. WEIDNER and ITO's (1985) elasticity data show that the elastic properties of spinel plus stishovite and ilmenite are similar. Either of these transitions could account for the excess velocity increase of Table 3 if all of the majorite phase transformed in the transition zone. These possible phase relationships require more experimental data to confirm or exclude.

The final possibility for giving a steep gradient in the transition zone is a broad change in chemical composition. In light of the new data by YAGI et al. (this volume), the system needs to become more silica poor (olivine rich) as depth increases or more magnesium rich. Even if the transition zone changed to a single spinel phase by the base (without changing the iron content), the current model still yields acoustic velocities which are too small at the base to match those observed. On the other hand, replacing all of the iron by magnesium while maintaining the same silicon content is almost capable of producing high enough velocities at the base of the transition zone as well. Thus, if a composition gradient is the reason that the observed and calculated gradients are incompatible, then there must be a substantial decrease in both the iron and silicon content throughout the transition zone.

Of the possibilities considered, a change in pressure derivative of just the shear modulus of the spinel phase from 0.8 to 1.2 produces a larger velocity gradient which agrees better with that observed than replacing all of the iron with magnesium or changing the stoichiometry from pyrolite to olivine in the transition zone. Comparable changes can be accommodated with post garnet phase transformations. Therefore, we emphasize the continued need for high accuracy elasticity and phase equilibria data.

The 670 km Discontinuity

The seismic structure of the 670 km discontinuity includes high velocity gradients. WALCK (1984) suggests that these steep gradients are at both shallower and deeper depths than the discontinuity while GRAND and HELMBERGER (1984) find them only at deeper depths. We feel that these can be modeled by transformation to the perovskite phase. ITO and TAKAHASHI (this volume) determined that the transformation of magnesium silicate phases to perovskite bearing assemblages occurs over an extremely narrow pressure interval, while the transformation of calcium and aluminum bearing phases occurs over a broader pressure range. The former transformation may be responsible for the sharp seismic discontinuity while the latter transformation could cause the steep gradient associated with the 670 km discontinuity. Differing temperatures or compositions may be responsible to change the relative position of the two types of transformations allowing both styles of discontinuities described by WALCK (1984) and GRAND and HELMBERGER (1984).

The Lower Mantle

The lower mantle extends from 700 km to the core mantle boundary. It is believed to consist of perovskite and magnesio-wustite. Very little elasticity data exists for the perovskite phase. Therefore, it is premature to make a detailed analysis of the properties of the lower mantle. We can analyze the sensitivity of the seismic data to the various parameters. An alternative lower mantle composition to pyrolite is a composition based on chondritic meteorites (OHTANI et al., 1986). Pyrolite in the lower mantle will consist of 80% by volume perovskite and 20% magnesio-wustite while chondrite will be divided 90% perovskite and 10% magnesio-wustite. ANDERSON and BASS (1986) propose a lower mantle of 100% perovskite.

Such differences in composition actually make only a small difference in elastic properties. At lower mantle conditions, pyrolite will differ from a chondrite composition by 1.6% in compressional velocity, 1.2% in shear velocity, and 2.3% for the bulk modulus. Pyrolite will differ from the pure perovskite model by roughly twice that. Table 4 indicates the change in physical properties which would have the same effect as changing from pyrolite to chondrite. The point here is that the composition of the lower mantle is much less effective in defining the lower mantle seismic properties than are very small changes in these properties. The conclusion is that these physical properties must be very well determined before we can make detailed inferences about the chemistry of the lower mantle.

Conclusions

During the past decade we have seen a tremendous increase in the data which are necessary to effect a quantitative test of mantle mineralogies. The thesis of this paper is twofold. The first is that a homogeneous chemical composition is consistent with the observed seismic velocities for the earth's mantle. The chemical model represented by RINGWOOD'S (1975) pyrolite is at least one of the successful candidates. Within the uncertainties of the data, all features of the seismic model can be quantitatively described to any degree of accuracy desired.

The second conclusion is that the resolution of chemical composition as a function of depth is still poor. The magnitude of the 400 km discontinuity is as sensitive to an uncertainty in the pressure derivative of the shear modulus of the modified spinel phase of 0.4 as it is to a change in the olivine content of 20%. The gradient of the transition zone can be satisfied by a change of 0.4 in the pressure derivative of the spinel phase as well as by replacing all of the iron by magnesium or all of the garnet and pyroxene by olivine or transforming all of the majorite to a higher pressure phase. A 10% uncertainty in the pressure derivative of the perovskite bulk modulus has an equivalent effect to the difference between a pyrolite and chondrite composition to the lower mantle.

This resolution will substantially improve as more data are gathered on the relevant materials. In particular, pressure and temperature derivatives of the elastic proper-

TABLE 4. Pyrolitic and Chondritic Model Parameters at 800 km Depth (Lower Mantle) and Equivalent Property Changes which Produce the Same Change in the Bulk Modulus

	Pyrolite	Chondrite	Difference (%)
V_P (800 km) km/s	11.1	11.2	1.3
V_s (800 km) km/s	6.2	6.3	1.6
K (800 km) GPa	323.0	330.0	2.3
	Model	For ΔK of 2.3%	
$(\partial K/\partial P)_S$ perovskite	3.7	4.0	
$(\partial K/\partial T)_P$ perovskite	−0.2	−0.15	
Temperature (°C)	—	350	

ties of the spinel and modified spinel phases are very important. All features of the elastic properties of the perovskite phase are necessary. Still, composition gradients in the lower mantle will remain difficult to infer from this data base.

REFERENCES

AKAOGI, M., and S. AKIMOTO, Pyroxene-garnet solid solution equilibria in the systems $Mg_4Si_4O_{12}$ and $Fe_4Si_4O_{12}$ at pressures and temperatures, *Phys. Earth Planet. Inter., 15*, 90–106, 1977.

AKAOGI, M., and S. AKIMOTO, High pressure phase equilibria in a garnet lherzolite, with special reference to Mg^{2+}-Fe^{2+} partitioning among constituent minerals, *Phys. Earth Planet. Inter., 19*, 31–51, 1979.

AKAOGI, M., A. NAVROTSKY, T. YAGI, and S. AKIMOTO, Pyroxene-garnet transformation: Thermochemistry and elasticity of garnet solid solutions, and application to a phrolite mantle, this volume.

ANDERSON, D. L., and J. D. BASS, Transition region of the Earth's upper mantle, *Nature, 320*, 321–328, 1986.

BURDICK, L. J., and D. V. HELMBERGER, The upper mantle P velocity structure of the western United States, *J. Geophys. Res., 83*, 1699–1712, 1978.

FUKAO, Y., Upper mantle P structure on the ocean side of the Japan-Kurile arc, *Geophys. J. R. Astr. Soc., 50*, 575–585, 1977.

GIVEN, J. W., and D. V. HELMBERGER, Upper mantle structure of northwestern Eurasia, *J. Geophys. Res., 85*, 7183–7194, 1980.

GRAND, S., and D. HELMBERGER, Upper-mantle shear structure of North America, *Geophys. J. R. Astr. Soc., 76*, 399–438, 1984.

ITO, E., and E. TAKAHASHI, Ultrahigh-pressure phase transformations and the constitution of the deep mantle, this volume.

JEANLOZ, R., Majorite: Vibrational and compressional properties of a high pressure phase, *J. Geophys. Res., 86*, 6171–6179, 1981.

KANZAKI, M., Ultra high-pressure phase relations in the system $MgSiO_3$-$Mg_3Al_2Si_3O_{12}$, MS thesis, Institute for Study of the Earth's Interior, Okayama Univ., Misasa, Japan, 1985.

OHTANI, E., The primodial terrestrial magma ocean and its implication for stratification in the mantle, *Phys. Earth Planet. Inter., 38*, 70–80, 1985.

OHTANI, E., T. KATO, and H. SAWAMOTO, Melting of a model chondritic mantle to 20 GPa, *Nature, 322*, 352–353, 1986.

RINGWOOD, A. E., *Composition and Petrology of the Earth's Mantle*, 618 pp., McGraw-Hill Book Co., New York, 1975.

SAWAMOTO, H., Phase diagram of $MgSiO_3$ at pressures up to 24 GPa and temperatures up to 2200°C: Phase stability and properties of tetragonal garnet, this volume.

WALCK, M. C., The p-wave upper mantle structure beneath an active spreading center: The Gulf of California, *Geophys. J. R. Astr. Soc., 76*, 697–723, 1984.

WEIDNER, D. J., A mineral physics test of a pyrolite mantle, *Geophys. Res. Lett., 12*, 417–420, 1985.

WEIDNER, D. J., Mantle model based on measured physical properties of minerals, *Advances in Physical Geochemistry*, edited by S. Saxena, Vol. 6, pp. 251–274, Springer-Verlag, New York, 1986.

WEIDNER, D. J., and E. ITO, Elasticity of $MgSiO_3$ in the ilmenite phase, *Phys. Earth Planet. Inter., 40*, 65–70, 1985.

YAGI, T., H. K. MAO, and P. M. BELL, Isothermal compression of perovskite-type $MgSiO_3$, *Carnegie Inst. Wash. Yearbook, 77*, 835–837, 1978.

YAGI, T., M. AKAOGI, O. SHIMOMURA, H. TAMAI, and S. AKIMOTO, High pressure and high temperature equation of state of majorite, this volume.

ERROR ANALYSIS OF PARAMETER-FITTING IN EQUATIONS OF STATE FOR MANTLE MINERALS

P. M. BELL, H. K. MAO, and J. A. XU

Geophysical Laboratory, Washington, DC 20008, USA

Abstract. Experimental data on volume as a function of pressure can now be obtained at maximum pressures of almost twice that of the earth's core-mantle boundary. Thus experimental data can be compared directly with seismic observations. Extrapolated values of experimental data from zero pressure or from any intermediate pressure below the pressure of application are not justified because of significantly high errors. The most accurate derived value of any given data set occurs between 15 and 85% of the maximum experimental pressure range. Tests of these concepts using the Murnaghan and Birch finite strain equations demonstrate the distribution of errors and describe the basis for evaluation of K, K', and V of mantle materials.

Introduction

Seismic observations are used to calculate the bulk modulus defined as $K=V(\partial P/\partial V)$, its first pressure derivative defined as $K'(\partial K/\partial P)$, and the specific volume V (or density $\rho=1/V$), for the earth along the path of a seismic wave. K, K' and V are parameters of the equation of state, which provides the basis of geophysical models of the earth. This paper is concerned with methods for evaluating experimental values of K, K', and V of mantle minerals for comparison with values obtained by seismic earth models. New diamond-anvil techniques are now available for obtaining experimental data on mantle minerals at pressures as high as 5.5 Mbar (XU et al., 1986). Therefore it is important to determine the optimum pressure range for equation of state measurements.

Mantle mineral properties traditionally have been obtained at low pressures because of experimental limitations. X-ray diffraction data for single-crystal and polycrystalline mineral materials obtained at high pressures have been first extrapolated to zero pressure and then extrapolated once more to the much higher pressures required for evaluation of earth models. Although the zero pressure data conditions offer a practical standard state for comparing the results of different types of measurements (e.g., ultrasonic or Brillouin scattering measurements), this approach is misleading with regard to the accuracy of the high-pressure properties that depend heavily on the pressure range in which the original data were obtained.

Direct measurements at the higher pressures can now be obtained, and the data can be interpolated for intermediate values. The interpolation may employ use of a given equation of state for curve fitting to the experimental data instead of for extrapolation. The procedure is also a test of the theoretical equation for the material at experimental conditions.

In the present analysis the errors of V, K, K' as functions of pressure are calculated in detail by fitting synthetic P-V data sets of differing pressure ranges to the Murnaghan and Birch finite strain equations. The resulting principals of error distribution are general, and apply to any functions to which an experimental data set are fitted, whether theoretically based or not. The two equations selected in this analysis have had general acceptance in many applications of high-pressure data, and differ considerably in form and degree of complexity. The formulations of the present paper are based on P-V data. However, the conclusions that a larger range of the experiments leads to better parameters and that the optimum pressure range is twice the pressure of the geophysical application are generally valid for other types of measurements of equations of state.

Equation of State

A general form of the isothermal equation of state can be written as

$$F(V, P, V_x, K_x, K'_x, \ldots) = 0 \qquad (1)$$

where P, the pressure, and V, the volume on the equation of state, are the independent and dependent variables, respectively, and where the V_x, K_x, K'_x, and higher derivatives are the adjustable parameters. Here V_x is the specific volume and K_x and K'_x are the bulk modulus and its pressure derivatives, respectively, at any given fixed pressure X ($X=0$ is the special case of zero pressure).

The individual experimental points are designated P_i and V_{ei}; the total number of points is designated N. A least squares fit of the data points to the equation of state is calculated. This method minimizes the square of the difference between V_{ei}, the experimental volume at P_i, and V_i, the volume calculated from a prescribed form of Equation 1 at P_i. The following analysis is an evaluation of the propagation of errors in fitting the synthetic data sets to the Murnaghan and Birch equations. The errors of V_x, K_x, and K'_x are evaluated at each point (X) along the

High-Pressure Research in Mineral Physics, edited by M. H. Manghnani and Y. Syono, pp. 447–454.

pressure-volume curve.

Error Analysis

The system of error analysis employed is similar to the commonly used method described by CLIFFORD (1973) and to the calculation of BASS et al. (1981). Equation 1 can be written in the more general form as follows

$$F(V_i, P_i, Y_1, Y_2, Y_3, ...) = 0 \quad (2)$$

where $Y_1 = V_x$, $Y_2 = K_x$, $Y_3 = K'_x$. The standard deviations of the parameters can be evaluated by means the transformation matrix

$$(A)ij = (\partial V_i / \partial Y_j)_{Y_{k,(k \neq j)}} \quad (3)$$

and the weighting matrix

$$W = \begin{Bmatrix} 1/\sigma^2(V_1) & 0 & 0 & ..... \\ 0 & 1/\sigma^2(V_2) & 0 & ..... \\ ..... & ..... & ..... & ..... \\ 0 & 0 & ..... & 1/\sigma^2(V_n) \end{Bmatrix} \quad (4)$$

In this notation, the $\sigma(V_i)$ is the standard deviation of the experimental point V_{ei}. The variance-covariance matrix for the parameters thus is defined as

$$M_Y = (A'WA)^{-1} \quad (5)$$

The standard deviation of the parameters is defined as

$$\sigma(Y_i) = [(M_Y)_{ii}]^{1/2} \quad (6)$$

To obtain the $\sigma(Y_i)$, one must derive the particle derivatives of the elements of the transformation matrix, A [i.e., $(\partial V_i/\partial V_x)_{K_x, K'_x}$, $(\partial V_i/\partial K_x)_{K'_x, V_x}$ and $(\partial V_i/\partial K'_x)_{V_x, K_x}$]. The derivations take a form specific to the equation of state selected as follows.

Murnaghan Equation

The Murnaghan equation with zero pressure parameters can be transformed to parameters at pressure X as follows

$$K = K_0 + K'_0 \, P \quad (7)$$

$$K = K_x + K'_0 \, (P - X) \quad (8)$$

In the first-order, $K'_0 = K'_x$=constant, and thus the Murnaghan equation written in the form of Equation 1 is as follows

$$F(V_i, P_i, V_x, K_x, K'_x) = V_i - V_x[1 + (P_i - X)(K'_x/K_x)]^{-(1/K'_x)} = 0 \quad (9)$$

The elements in the transformation matrix, A, of Equation 3 are obtained simply by taking partial derivatives of Equation 9 with respect to the three parameters as follows

$$A_{i1} = (\partial V_i/\partial V_x) = V_i/V_x$$
$$A_{i2} = (\partial V_i/\partial K_x) = V_i(P_i - X)/K_x K_i$$
$$A_{i3} = (\partial V_i/\partial K'_x) = V_i/K'_x[\ln(V_x/V_i) - (P_i - X)/K_i] \quad (10)$$

Birch Equation

In the first order, the Birch equation is defined as

$$P = 1.5\ K_0[(V_0/V)^{7/3} - (V_0/V)^{5/3}]\{1 - \xi[(V_0/V)^{2/3} - 1]\} \quad (11)$$

where the three parameters are V_0, K_0, and ξ. For any given experimental point i, the equation can be restated as

$$F(V_i, P_i, V_0, K_0, \xi) = 0 \quad (12)$$

The Birch equation does not lend itself to simple transformation to an explicit form of Equation 1 other than X=0. Instead, a more general implicit form of Equation 1 is expressed as follows

$$f_1(V_x, X, V_0, K_0, \xi) = 0$$
$$f_2(K_x, V_x, X, V_0, K_0, \xi) = 0$$
$$f_3(K'_x, X, V_0, K_0, \xi) = 0 \quad (13)$$

where f_1 is simple Equation 11 with $P=X$; f_2 and f_3 are the first and second pressure derivatives of Equation 11 at $P=X$, and $\xi = 3/4(4 - K'_0)$. The general form of Equation 13 is rewritten as

$$f_h(Y_1, Y_2, Y_3, X, U_1, U_2, U_3) = 0 \quad (14)$$

where h=1, 2, 3; $Y_1 = V_x$, $Y_2 = K_x$, $Y_3 = K'_x$ as in Equation 1, and $U = V_0$, $U_2 = K_0$, $U_3 = K'_0$.

Taking partial derivative of Equation 14 with respect to Y_j, we obtain the following tensors

$$(\partial f_h)/\partial Y_j = (\partial f_h)/\partial U_m\ ((\partial U_m)/\partial Y_j)_{Y_{k \neq j}} \quad (15)$$

which represents nine equations with h=1, 2, 3 and j=1, 2, 3. Since all the f_h differentials, $(\partial f_h/\partial Y_j)$ and $(\partial f_h/\partial U_m)$ can be obtained directly from Equation 14, the nine variables $(\partial U_m/\partial Y_i)_{Y_{k \neq j}}$ can be solved from Equation 15 and substituted in the following tensor for calculating A_{ij}

$$(\partial V_j)/(\partial Y_j)_{Y_{k \neq j}} = [(\partial F/\partial U_m)\ (\partial U_m/\partial Y_j)_{Y_{k \neq j}}/(\partial F/\partial V_i) \quad (16)$$

The Equation 16 is the differential form of Equation 12.

Again, all the $(\partial F/\partial U_m)$ and $(\partial F/\partial V_i)$ can be obtained directly. The matrix A is solved.

Distribution of Errors

In order to simplify evaluation of the error distribution, the data sets were constructed such that the pressure was scaled to K_0. The maximum pressure of the data sets (P_{max}) was defined as some fraction of K_0, the unit pressure. The zero pressure derivative of the bulk modulus was defined as $K'_0=4$. The data sets had 21 equally distributed pressure-volume points (i.e., $N=21$) along the experimental pressure range from $P=0$ to $P=P_{max}$.

It was assumed in the calculations that the fractional standard deviation of the volume of each data point was constant, so that

$$\sigma(V_i)/V_i = \sigma(V)/V = \text{constant}$$

and therefore the weighting matrix is a scalar quantity. The assumption is valid for X-ray diffraction P-V measurements with inert gas hydrostatic pressure media in a diamond cell up to at least 1 Mbar pressure, which is the practical range for mantle consideration. For other types of measurements in which $\sigma(V)/V$ changes as a function of pressure, the estimated functions can be used in the matrix.

The standard deviation for the set of parameters, normalized by the standard deviation of the volume data, is given by

$$S(Y_i) = [\sigma(Y_i)/Y_i]/[\sigma(V)/V]$$

where the $\sigma(Y_i)$ is calculated from Equation 6.

The number of data points is a scalar quantity, so the standard deviation of the set of parameters is proportional to the number, N, as follows

$$S(Y_i) \propto (N-3)^{-1/2}$$

These relations are utilized in the analysis of the errors $S(V_x)$, $S(K_x)$, and $S(K'_x)$ at each point (X).

The Errors in V

Figures 1a, b show plots of $S(V_x)$ of the Murnaghan and Birch equations as U-shaped functions of X for P_{max} in the range of 0.05–0.8. The broad error minimum lies between a P_{max} of approximately 0.15 and 0.85. Note that the maximum errors within the data sets occur at $X=0$ and P_{max}. The error of V_0, $S(V_0)$, is approximately a factor of two greater than the minimum error of V_x, $S(V_x)_{min}$. The errors along the curves extrapolated beyond the data are shown in Figures 1c, d (expanded scale) to increase exponentially. The magnitude of error of V_0 and V_x is relatively independent of P_{max}, although the range of minimum error increases with increasing values of P_{max}.

The Errors in K

Figures 2a, c show plots of the errors in the parameter set of K_x, $S(K_x)$, for a series of P_{max} values of 0–0.2 for the Murnaghan and Birch equations. Figures 2b, d show the plots in expanded scale for $P_{max}=0$– 0.8. The $S(K_x)$ is a V-shaped function of X, the minimum errors $S(K_{min})$ occuring at pressures slightly less than half the value of P_{max}. The maximum errors of K occur at $X=0$ and P_{max}. The errors on the extrapolated curves beyond the data always increase exponentially.

As P_{max} is increased, the error is reduced over essentially all of the experimental pressure range. Figures 3a, b show plots of $S(K_{min})$, $S(K_0)$, and $S(K_0)/S(K_{min})$ as functions of P_{max} for the two equations of state. Both errors, $S(K_0)$ and $S(K_x)$, decrease rapidly with increasing P_{max}. The error of K_0 is more than a factor of four higher than the minimum error. The ratio of $S(K_0)$ to $S(K_{min})$ has modest increase over the range, $P_{max}=0$–1.0.

It is clear from these results that the error in K decreases as P_{max} increases and that the minimum error is at the middle of the pressure range. Therefore the optimum experimental value of P_{max} of a mantle mineral should be at least a factor of two higher than the pressure required for comparison with a geophysical model of K.

The Errors in K′

The $S(K')$ is independent of X in the Murnaghan equation by definition, and varies only slightly with X in the Birch equation. the $S(K')$, however, has strong dependence on P_{max}, as can be seen in Figures 4a, b. For example, for $N=21$ and $\sigma(V_i)/V_i=0.03$, the percentage standard deviation of K', $\sigma(K')/K'$ is 50% for $P_{max}=0.05$. The standard deviation of K' is reduced by a factor of ten (to 5%) for $P_{max}=0.2$.

Interdependence of K and K′

Among the many errors in reducing equation of state data to zero-pressure initial conditions are those due to the apparent interdependence of the bulk modulus of a material, K, and its first pressure derivative, K'. BASS et al. (1981) and JEANLOZ (1981) observed that fixing one of the two parameters forces dependence of the other at zero pressure. In this regard, BASS et al. (1981) described the strong coupling or covariance of K_0 and K'_0. JEANLOZ (1981) noted that the interdependence is affected to some extent by the choice of equation of state.

The present analysis was made to evaluate the interdependence at the pressure of application of the data $(P=X)$ rather than at zero pressure. Figures 5a, b show separate plots of K_x versus X (within the range $P_{max}=$ 0–0.2) for various values of K'_0, for the Murnaghan and Birch equations. The differences between the two plots

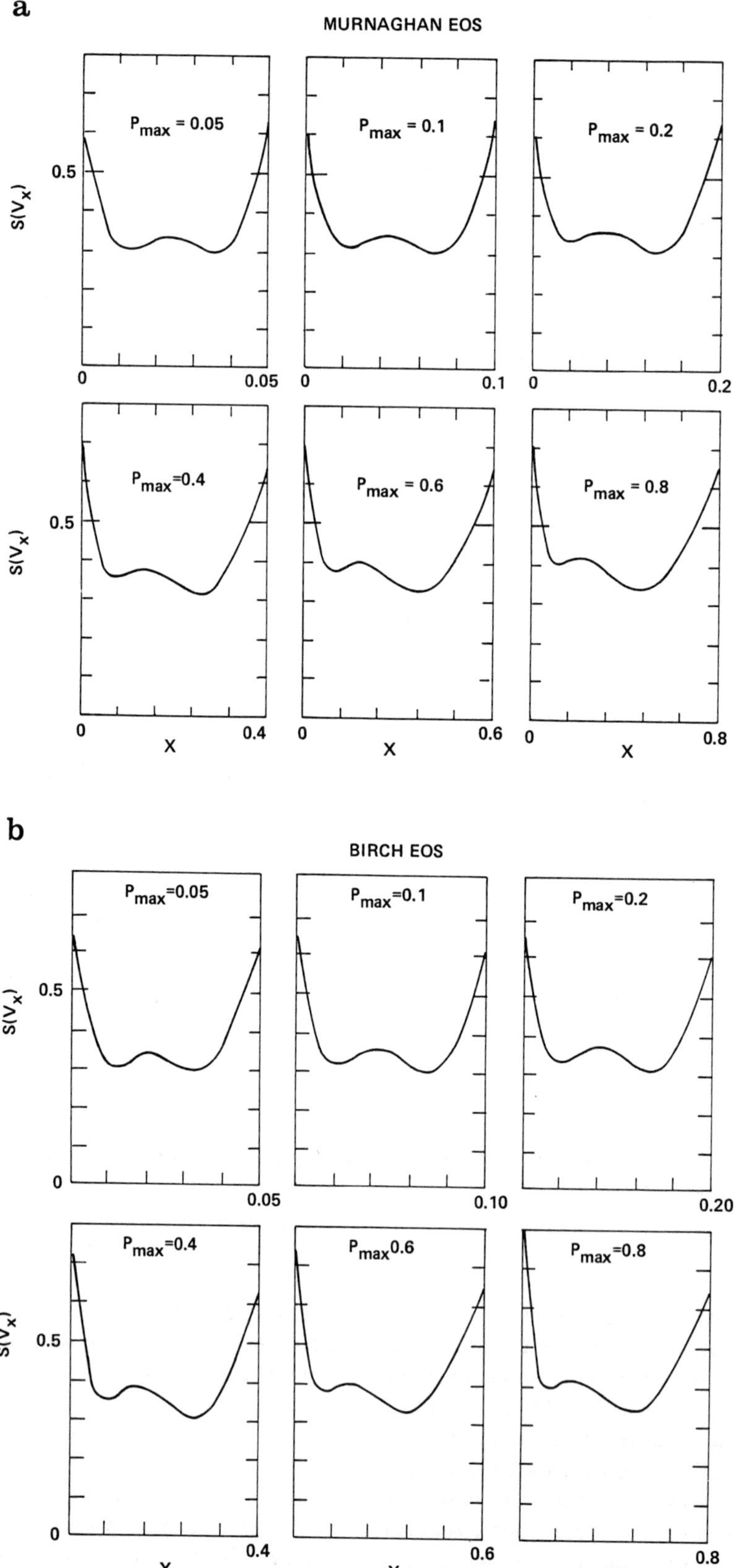

Figure 1a,b. Plots of the errors of volume data $[S(V_x)]$ at pressures (X) along pressure-volume curves of several synthetic data sets, fit to the Murnaghan and Birch equations. P_{max} is the maximum pressure of each set.

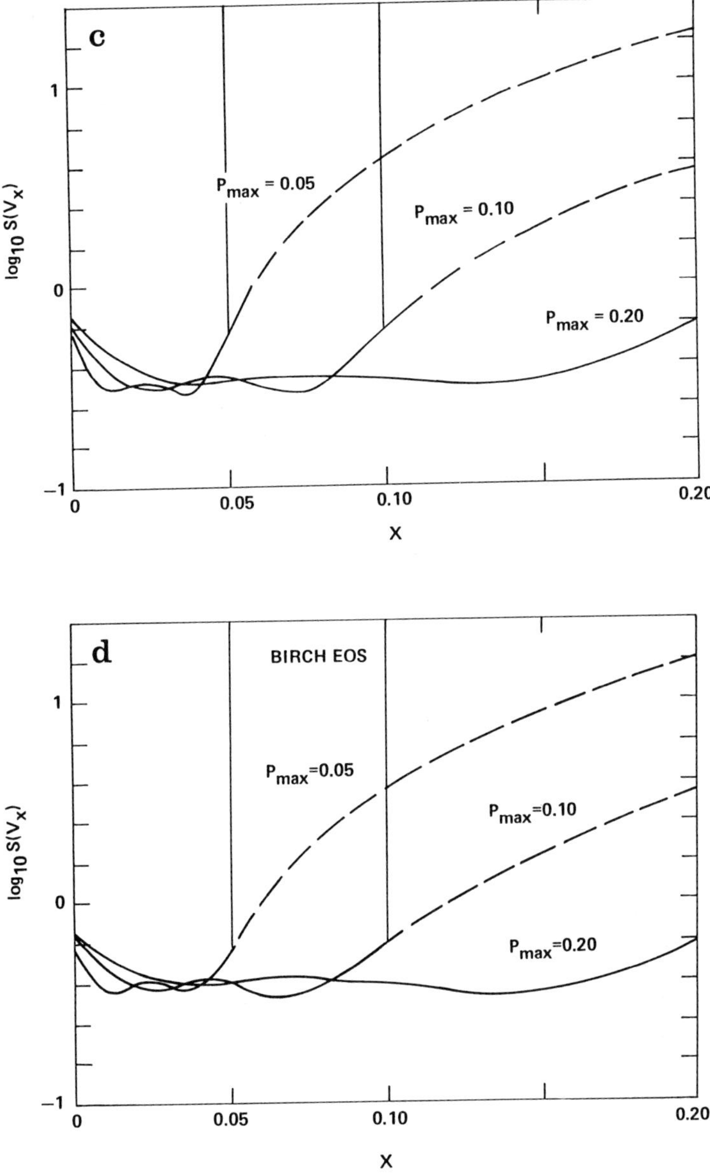

Figure 1c, d. Expanded scale of the plots 1a and b to show the exponential increase of the errors along curves extrapolated beyond the data sets.

that result from the curvature of the Birch equation accentuate the coupling behavior. There is a crossover of the linear plots near the center of the experimental pressure range that are expected to shift location slightly at increasing values of P_{max}. The interdependency vanishes toward the center of the data set, which shows that the results of the present study are, in general, independent of the choice of the fixed value of K'. The interdependency effect is maximized at both ends of the experimental pressure range. Errors arising from interdependency are thus greatly reduced if, as in this case, the experimental pressure range (P_{max}) is selected to be approximately a factor of two higher than the pressure required for application of the data.

Conclusions

The analysis of errors of the parameters utilized in fitting experimental equation of state data has revealed a number of factors concerning selection of the best experimental pressure range and the optimal procedures of data reduction. Procedures previously used to estimate high-a

MURNAGHAN EOS

S(K_x)

P_{max} = 0.05

P_{max} = 0.10

P_{max} = 0.20

X

b

MURNAGHAN EOS

S(K_x)

P_{max}=0.2

P_{max}=0.4

P_{max}=0.6

P_{max}=0.8

X

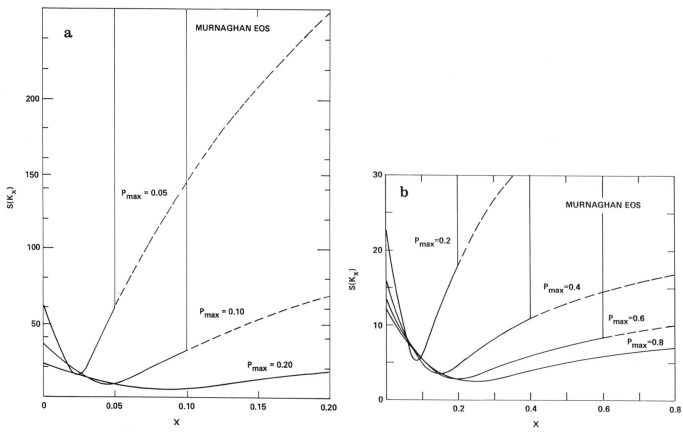

Figure 2a, b. Plots of the errors of bulk modulus data [$S(K_x)$] at pressures (X) along the pressure-volume curves of three synthetic data sets, fit to the Murnaghan and Birch equations. P_{max} is the maximum pressure of each set.

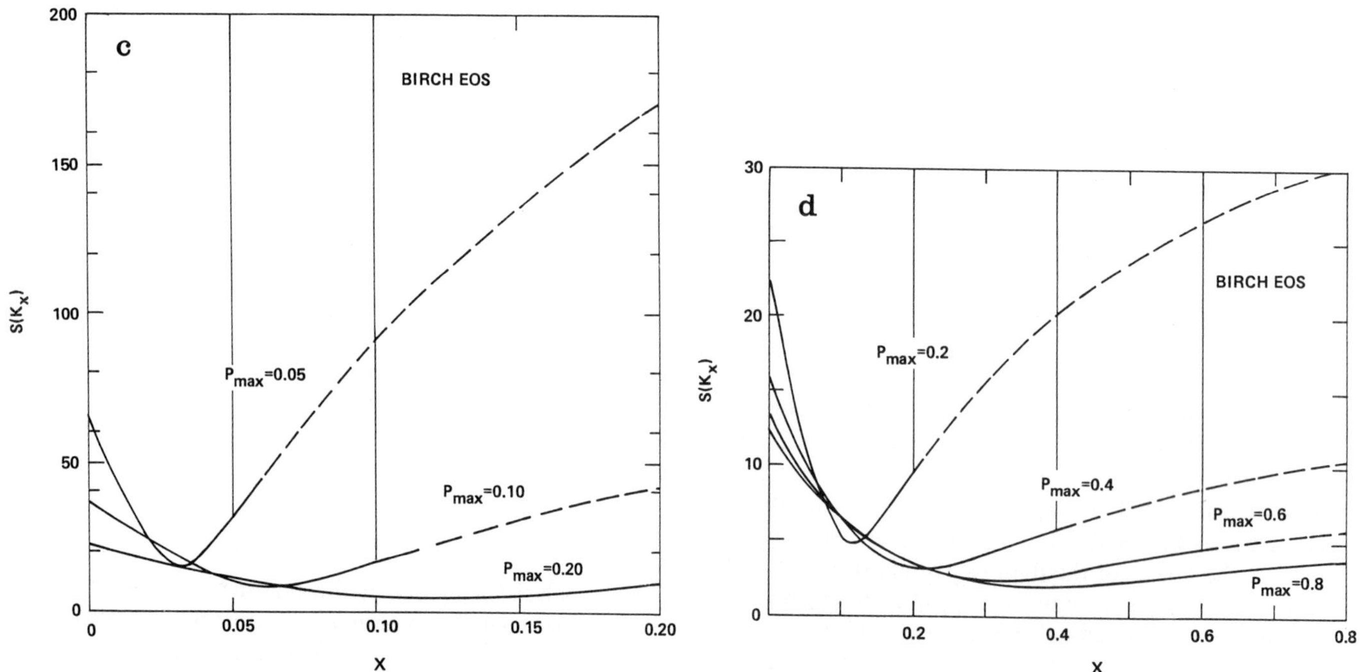

Figure 2c, d. Expanded scale of the plots 2a and b to show the exponential increase of the errors along curves extrapolated beyond the data.

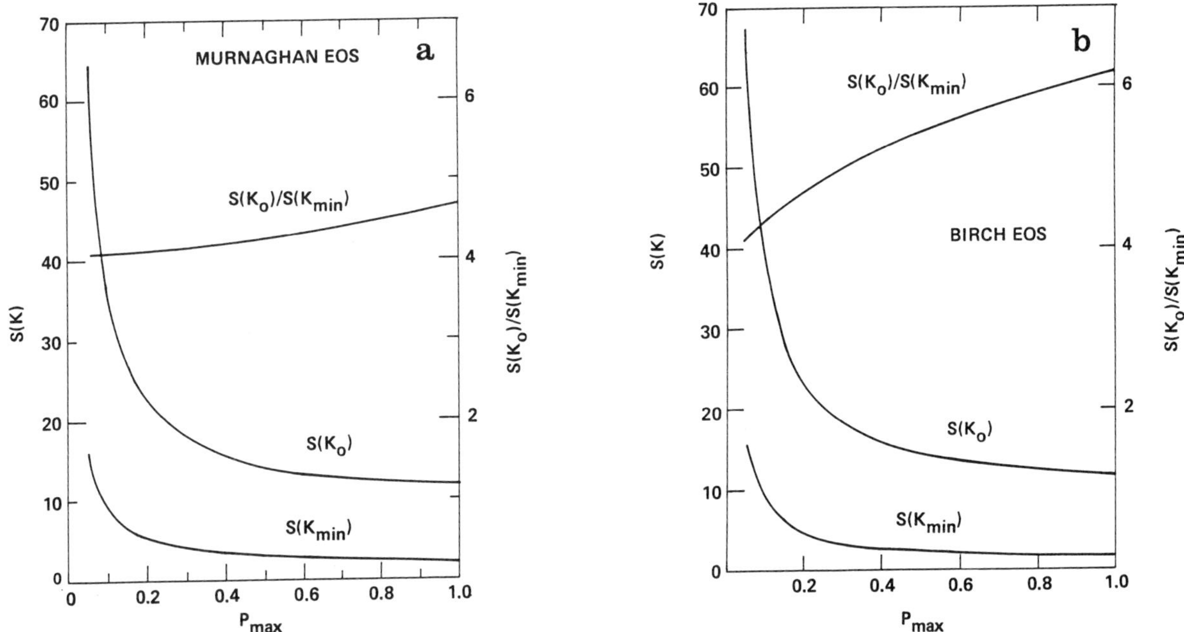

Figure 3. Plots of the errors of the bulk modulus at zero pressure [$S(K_0)$], the minimum error of bulk modulus [$S(K_{min})$], and of their ratio [$S(K_0)/S(K_{min})$] of synthetic data sets fit to the Murnaghan and Birch equations as functions of the maximum pressure of each set (P_{max}).

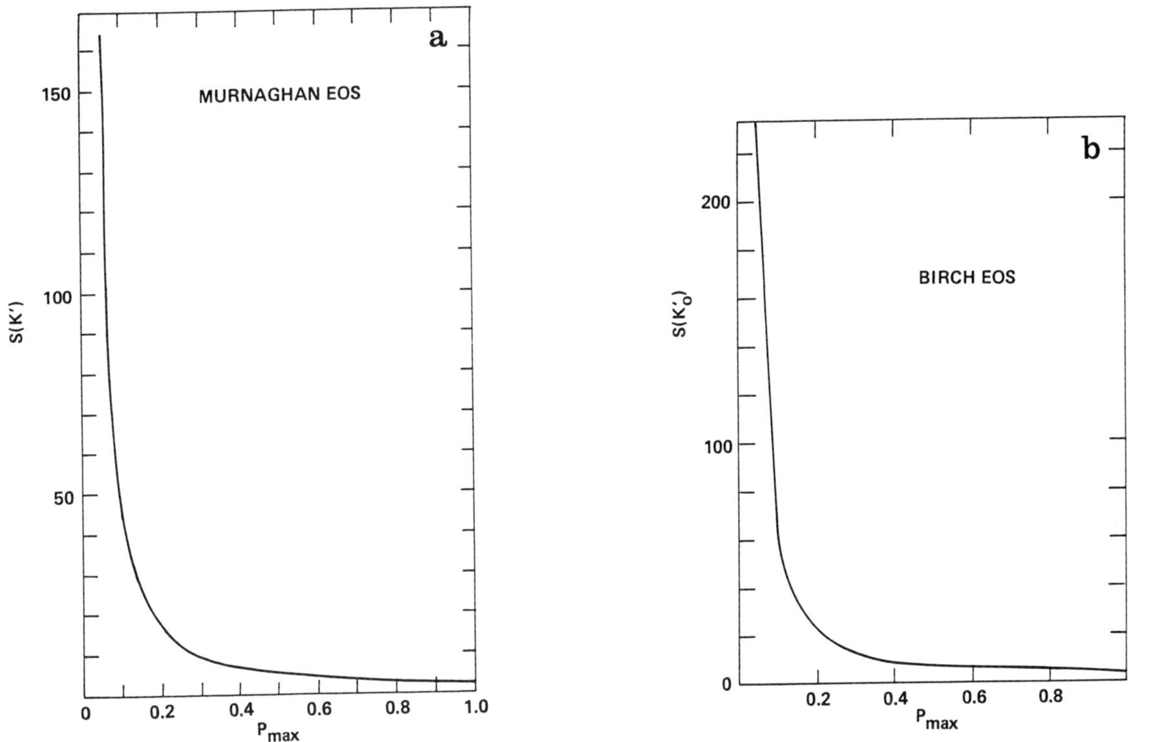

Figure 4. Plots of the errors of the first pressure derivative of the bulk modulus [$S(K')$] of synthetic data sets fit to the a) Murnaghan and b) Birch equations as functions of the maximum pressure of each set (P_{max}).

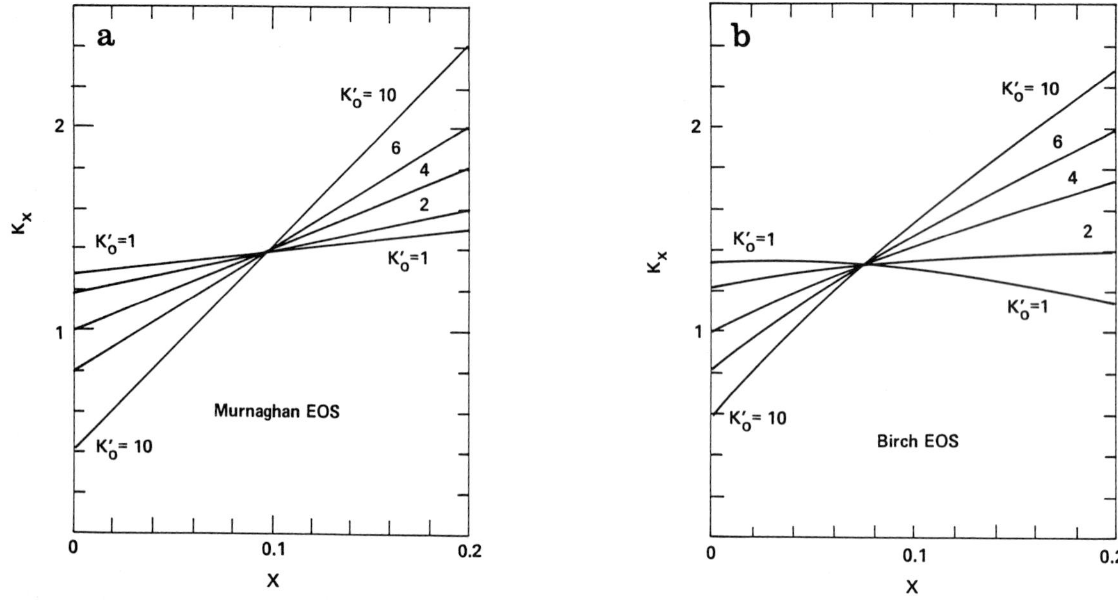

Figure 5. Plots of K_x, the bulk modulus at any pressure X of data for various fixed values of K'_0, over the pressure range 0–0.2 (P_{max}).

pressure properties of mantle minerals from zero pressure carry large errors. Very large errors of a pressure-volume data set occur at zero pressure, whereas the smaller errors occur over a broad range in the middle of the data set. The parameters V_x, K_x, and K'_x (where X=the pressure of a given data point) that were evaluated in the central region of the experimental range, carry errors of at least a factor of two lower than errors of the parameters V_0, K_0, K'_0. In general, the accuracy of the data at any pressure increases with the maximum pressure. Low pressure data could not conceivably be made sufficiently accurate to compare with a large range of higher pressure data. The errors of extrapolation beyond the limits of the data are exponential.

As a separate adjunct to conclusions of the error analysis, it was demonstrated that the K-K′ interdependency phenomenon is greatly reduced or eliminated, if the pressure range of the data extends beyond the desired pressure for application of the data by approximately a factor of two.

Experimental techniques to obtain pressure-volume data of mantle minerals that are at pressures exceeding the pressure of the core-mantle boundary (1.5 Mbar) by a factor of two, are now available. Data on the volume (or density), the bulk modulus, and its first pressure derivative of mantle minerals can be compared to seismic data and geophysical models with the highest accuracy if the experimental pressure range of the data is sufficiently high.

References

Bass, J. D., R. C. Liebermann, D. J. Weidner, and S. J. Finch, Elastic properties from acoustic and volume compression, *Phys. Earth and Planet. Inter., 25*, 140–158, 1981.

Clifford, A. A., *Multivariate Analysis*, 122 pp., John Wiley and Sons, New York, 1973.

Jeanloz, R., Finite-strain equation of state for high-pressure phases, *Geophys. Res. Lett., 8*, 1219–1222, 1981.

Xu, J. A., H. K. Mao, and P. M. Bell, High pressure ruby and diamond fluorescence: observations at 0.21 to 0.55 Terapascal, *Science, 232*, 1404–1406, 1986.

SEISMIC ANISOTROPY DUE TO LATTICE PREFERRED ORIENTATION OF MINERALS: KINEMATIC OR DYNAMIC?

Shun-ichiro KARATO

Ocean Research Institute, University of Tokyo, Nakano, Tokyo 164, Japan

Abstract. The lattice preferred orientation (LPO) of elastically anisotropic minerals is one of the most important anisotropic structures that causes seismic anisotropy. A question is addressed whether seismic anisotropy due to LPO is related to the kinematic framework (shear direction and shear plane) or to the dynamics of flow (orientation of stress). It is shown that the conventional kinematic interpretation of seismic anisotropy based on fabric studies of ultramafic rocks in ophiolites cannot be straightforwardly extrapolated to the deeper mantle, where dynamic recrystallization is important. When dynamic recrystallization occurs, the nature of LPO is determined by the relative importance of plastic deformation (dislocation glide), nucleation, and growth (grain boundary migration (GBM)). Based on experimental studies, a simple microstructural model of dynamically recrystallized materials is proposed. It is shown that the dominant mechanism of LPO is determined by the relative rates of plastic deformation, nucleation, and growth and depends on the conditions of recrystallization. When both nucleation and growth are easy compared to plastic deformation, the LPO will be determined by GBM and related to the stress, if the driving force for GBM reflects the instantaneous stress. When either nucleation or growth (or both) is difficult compared to deformation, the LPO will be determined by plastic deformation and related to the kinematic framework of flow. The transition conditions of the mechanisms of LPO were estimated for olivine, and the LPO mechanism maps were constructed. The results suggest that when dynamic recrystallization occurs, the stress-controlled seismic anisotropy will be formed in relatively cold mantle, while the kinematically controlled seismic anisotropy will be formed in relatively hot mantle.

1. Introduction

Since the pioneering work by HESS (1964), evidence for anisotropic earth structures has been progressively accumulated. It is now well recognized that the earth's structure is anisotropic in both oceanic and continental regions. Also, surface wave studies have revealed anisotropic earth structures on a global scale and to a greater depth than those hitherto studied. For reviews of seismological studies, see FUCHS (1977), ANDO (1984), CRAMPIN et al. (1984), KAWASAKI (1986), and NATAF et al. (1986).

Since the anisotropic structures are no doubt formed by tectonic processes in the earth, these developments in seismology have stimulated the study of tectonic processes in the earth's interior, which is not directly accessible. An important prerequisite for such a study is a thorough understanding of the mechanisms of formation of anisotropic structures.

It is now well appreciated that one of the important anisotropic structures that may cause seismic anisotropy in the mantle is the lattice preferred orientation (LPO) of elastically anisotropic minerals. The relation between anisotropic structures and seismic anisotropy has been well established (for a review, see CRAMPIN (1981)).

One of the most important problems that has not been well understood, but is crucial for any inference on tectonic processes based on seismic anisotropy, is the relation between the anisotropic structures and the tectonic field. This is the main subject of this paper. By the tectonic field, I mean either the kinematic or the dynamic framework (stress system) of flow. The distinction between the two is essential because most of the flow in the earth involves a rotational component (e.g., HOBBS et al., 1976; MCKENZIE, 1979; HOFFMAN and MCKENZIE, 1985). An important example of rotational deformation is simple shear, which is presumably the dominant type of deformation in the asthenosphere. In simple shear, the principal axes of stress are inclined with respect to the kinematic framework of flow (i.e., shear direction and shear plane) by 45°. More general differences between the stress orientation and the kinematic framework of flow will occur in other rotational deformation regimes.

It is well established that the LPO of the uppermost oceanic mantle (i.e., ophiolites) is controlled by the kinematic framework of flow rather than the stress (for a review, see NICOLAS and CHRISTENSEN (1987)). However, the mechanisms of seismic anisotropy and LPO in the deeper upper mantle are not well understood. From the materials science viewpoint, the problem is the controlling factor of LPO formed during dynamic recrystallization, the evidence for which has been found in mantle xenoliths (e.g., MERCIER and NICOLAS, 1975; NICOLAS, 1978; KIRBY and GREEN, 1980; AVÉ LALLEMANT et al., 1980). There has been considerable debate as to which (kinematic or dynamic) interpretation of LPO is appropriate when dynamic recrystallization occurs. AVÉ LALLEMANT and his coworkers have emphasized the importance of stress as a controlling factor (AVÉ LALLEMANT and CARTER, 1970; AVÉ LALLEMANT, 1975; KUNZE and AVÉ LALLEMANT, 1981). On the other hand, Lister and others have argued that the kinematic interpretation of LPO is more appropriate even though

High-Pressure Research in Mineral Physics, edited by M. H. Manghnani and Y. Syono, pp. 455–471.

dynamic recrystallization occurs (LISTER and PRICE, 1978; TORIUMI and KARATO, 1985; URAI et al., 1986). Since the tectonic information one can get from observed seismic anisotropy depends critically on which interpretation is appropriate, it appears very important to understand the physical mechanisms that govern the controlling factors of seismic anisotropy. The purpose of this paper is to clarify the physical principles that determine the relation between LPO and the tectonic field.

2. Lattice Preferred Orientation Due to Plastic Deformation

LPO due to plastic deformation is a result of relative rotation of crystal axes of individual grains (Figure 1a). Therefore the rotational part of deformation plays an essential role. The crystallographic axes rotate in such a way as to compensate the imbalance of vorticity (rotational component of deformation) caused by deformation and vorticity imposed by geometrical constraint on deformation (e.g., LISTER et al., 1978), or they rotate by local bulk rotation of grains due to misfit of deformation at grain boundaries (ETCHECOPAR, 1977; LISTER, 1982). Thus the LPO due to plastic deformation is controlled by the kinematic framework (geometry) of flow rather than the stress (LISTER and HOBBS, 1980). In the case where a single slip system dominates, a simple LPO will be formed (e.g., NICOLAS et al., 1973; ETCHECOPAR, 1977; BOUCHEZ et al., 1983). In pure shear or uniaxial compression, for example, crystallographic axes rotate toward the direction in which the slip plane and slip direction are normal to the shortening direction (an orientation in which deformation is difficult). In simple shear, in contrast, crystallographic axes rotate toward the direction in which the slip plane and slip direction of crystals coincide with the macroscopic shear plane and shear direction (an orientation in which deformation is easy).

In contrast to these kinematic models, IDA (1984) recently proposed a "dynamic" model in which LPO due to dislocation glide is considered to be controlled by stress. The predicted LPO for simple shear by Ida's model is inconsistent with the experimental and the geological observations (e.g., BOUCHEZ et al., 1983; NICOLAS and CHRISTENSEN, 1987). Ida argued that dislocation glide by itself rotates the crystallographic axes (without any specific constraint on displacement) and that the nature of this lattice reorientation is determined by the stress. Neither of these statements is justified. Further and more fundamentally, IDA (1984) did not recognize the important difference between rotational and irrotational deformation.

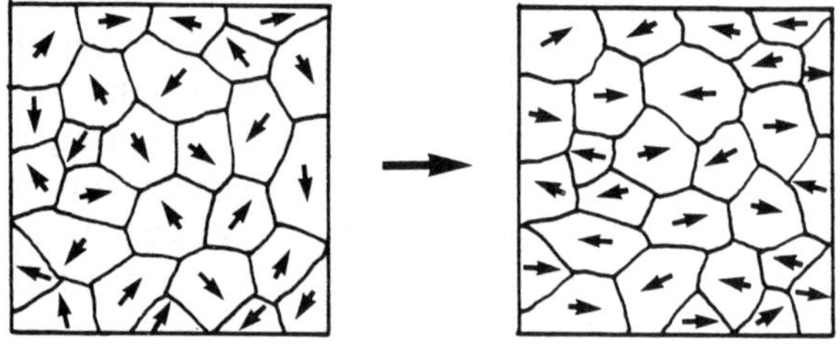

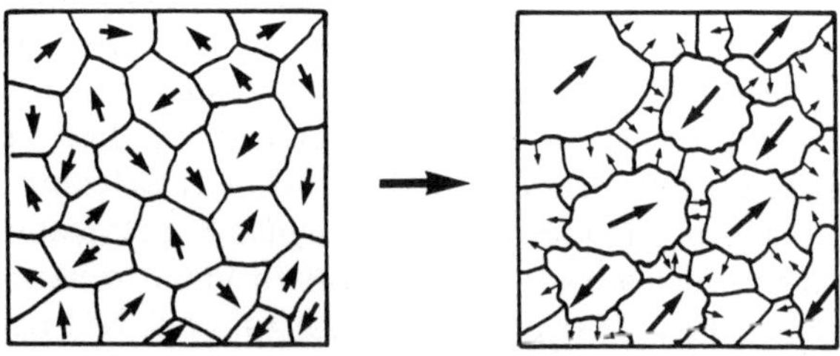

Figure 1. Two mechanisms of lattice preferred orientation. (a) Lattice rotation (reorientation) due to dislocation glide. Lattice orientations of individual grains continuously rotate with respect to the external reference frame. The lattice preferred orientation due to this mechanism is a result of geometrical constraint on deformation and is controlled by the kinematic framework of flow. (b) Grain boundary migration. The driving force for grain boundary migration is the difference in free energies of contacting grains. If the free energy of individual grains depends on lattice orientation with respect to stress axes, this mechanism will produce lattice preferred orientation which is controlled by the stress.

3. Lattice Preferred Orientation Due to Dynamic Recrystallization

3.1. General Remarks

The above kinematic model assumes the absence or unimportance of dynamic recrystallization and applies, in its simplest form, only to "cold-work" situations. At high temperatures (say $T/Tm>0.5$), dynamic recrystallization often occurs at large strain, which significantly alters the microstructure of rocks at the scale of grains (WHITE, 1977; URAI et al., 1986).

Dynamic recrystallization is a process in which new grain boundaries are formed (nucleation) and/or migrate (growth) during deformation. Depending on the details of nucleation-growth processes, a wide range of mechanisms of dynamic recrystallization have been recognized (SELLARS, 1978; GUILLOPÉ and POIRIER, 1979; POIRIER and GUILLOPÉ, 1979; SAKAI and JONAS, 1984; DRURY et al., 1985; URAI et al., 1986).

In this section, I first discuss the role of nucleation and growth (grain boundary migration (GBM)) processes separately, with special reference to their role in the LPO development. Then I propose a simple conceptual model to elucidate the mutual relationships of elementary processes of dynamic recrystallization (i.e., nucleation, growth, and plastic deformation) in developing the LPO.

3.2. Nucleation

Nucleation refers to a process in which new grain boundaries are formed in a crystal or, in a broader sense, to a process in which preexisting boundaries become mobile. Nucleation processes may be classified into two types, depending on the sites of nucleation.

One type is the nucleation at preexisting high-angle boundaries. In this case, new grains are formed due to accumulation of dislocations with the same sign leading to the formation of subgrains whose misfit angles increase with strain (AZUMA, 1986). Alternatively, grain boundaries become mobile, without the appearance of new boundaries, due to the increased driving force for GBM as a result of dislocation accumulation (BAILEY and HIRSCH, 1962; BELLIER and DOHERTY, 1977; MEANS, 1983; AZUMA, 1986; JESSEL, 1986). The nucleation at preexisting boundaries was found to occur at relatively small strain, about 5–20%.

The other type is the progressive misorientation of subgrains due to dislocation accumulation at subboundaries; see, for example, HOBBS (1968), POIRIER and NICOLAS (1975), BELLIER and DOHERTY (1977), GUILLOPÉ and POIRIER (1979), KARATO et al. (1980, 1982), and TORIUMI and KARATO (1985). Subgrain rotation recrystallization occurs at relatively large strain, about 40% or more.

In both cases, nucleation involves the accumulation of dislocations, and plastic deformation plays an important role in LPO. Therefore the LPO due to nucleation is essentially controlled by the kinematic framework of flow, although significant deviation from the simple kinematically controlled LPO (such as predicted by the Taylor model) was found as a result of inherent heterogeneities of deformation in nucleation processes (URAI et al., 1986; KARATO, 1987c).

3.3. Grain Boundary Migration (Growth)

Grain boundaries thus formed or mobilized will migrate when a large enough driving force is exerted. Grain boundary migraion results in LPO, if grains with particular orientations grow at the expense of the others (Figure 1b).

The driving force for GBM is the difference in free energy between grains on both sides of a grain boundary. Among the varieties of free energies that may cause GBM (see, for example, POIRIER and GUILLOPÉ (1979) and URAI et al. (1986)), dislocation energy and elastic strain energy (other than dislocation energy) (see, for example, PATERSON (1973)) are important in LPO because both depend on orientations of crystals with respect to the external reference framework. Both energies have the samc order of magnitude, σ^2/μ (where σ is stress and μ is shear modulus).

KARATO (1987c) made an experimental study to delineate the driving force for GBM in olivine. In the samples deformed by dislocation creep, grain boundaries were found to migrate toward the grains with higher dislocation densities (Figure 2), although the effect of grain boundary energy was locally observed where small grains were present. Thus, the dislocation energy is the main driving force when dislocation creep dominates, and the grains with low dislocation densities will survive and dominate the LPO if grain boundary migration occurs. The observed dominant role of dislocation energy over elastic strain energy may be attributed to the larger heterogeneity of dislocation energy than of elastic strain energy.

The relation between dislocation density and crystal orientation is closely related to the stress distribution in a deforming polycrystal. Note here the distinction between the macroscopic applied stress and the stress at individual grains (the local stress). It is this local stress (or, more precisely, the resolved shear stress on the slip system(s) of individual grains) that determines the dislocation density.

Two idealized end member situations can be distinguished (e.g., KOCKS, 1970; VAN HOUTTE, 1984). If the stress is homogeneously distributed (the Sachs model), grains with low Schmid factor for the soft slip system (hard orientation grains) will have low dislocation densities (and low strain) and dominate the LPO. If the strain is homogeneous (the Taylor model), grains with high Schmid factor (soft orientation grains) will have low dislocation densities and dominate the LPO, and in pure shear or uniaxial compression, for example, the LPO due to GBM will be similar to the LPO due to deformation. In simple shear, however, the LPO due to GBM will be similar to the LPO due to deformation, if the strain is homogeneous. In a realistic case, both stress and strain must be heterogeneous to some extent. The realistic stress and strain distribution will depend on materials and deformation conditions and can not a priori be prescribed.

Recently, some experimental observations have become available regarding the stress/strain distribution in deforming polycrystals. They include the studies by URAI et al. (1980), AZUMA (1986), JESSEL (1986), and Karato (unpublished) on camphor, ice, octachloropropane, and olivine, respectively. As expected, all of these results demonstrated the heterogeneity of strain and suggested the heterogeneity of stress as well. More specifically, URAI et al. (1980) found that the soft orientation grains grew at the expense of the others, suggesting that the stresses (and therefore the dislocation densities) at the soft

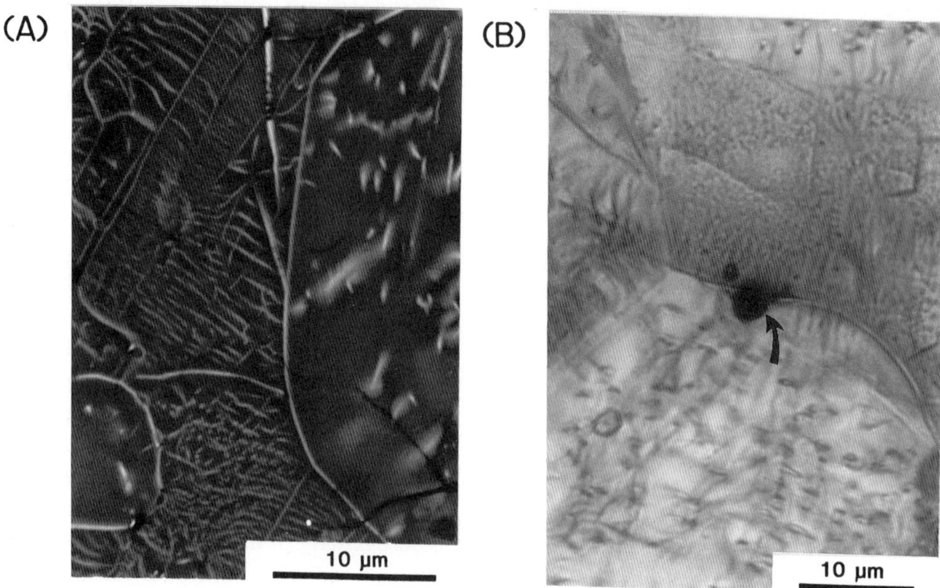

Figure 2. Dislocation distribution across moving grain boundaries in olivine as seen (a) by a scanning electron microscope with backscattered electron image (KARATO, 1987b) and (b) by an optical microscope. The shape of the deformed bubble (in Figure 2b) indicates that the grain boundary is moving toward the grain with higher dislocation density (see also KARATO (1987c)).

orientation grains are relatively small in camphor under the tested conditions. Karato (unpublished), in contrast, found that the soft orientation grains tend to have high dislocation densities (and large strains), although the scatter is significant. Similarly, AZUMA (1986) found that the strains of individual grains increase with increasing Schmid factor. JESSEL's (1986) results are somewhat mixed. He found that grain boundaries migrate toward the grains with higher strains and/or with hard orientations, a result intermediate between the Sachs and the Taylor models. It appears that the stress/strain distribution at the scale of grains depends on the materials. However, except for the study by Karato (unpublished), the dislocation structures were not studied in these works, which makes it difficult to assess the results on a clear physical basis.

Now, the next question to consider is the rate of dislocation multiplication and annihilation relative to the rate of change in stress state. This question arises because, in a rotational deformation history such as simple shear, the principal axes of stress rotate with respect to material points. Thus, the resolved shear stress on slip systems of individual grains will change with progressive deformation. The rates of dislocation multiplication and annihilation have been determined in olivine (DURHAM and GOETZE, 1977; TORIUMI and KARATO, 1978). A simple calculation based on these experimental results indicates that both dislocation multiplication and annihilation are fast in olivine and the dislocation density will reflect the instantaneous stress even in a rotational deformation such as simple shear. However, the characteristic times of multiplication and annihilation of dislocations in other minerals have not been determined. If the characteristic times are not smaller than the characteristic time of stress change, the LPO due to GBM will reflect the deformation history.

Another possible source of LPO due to GBM is the anisotropy of grain boundary mobility. LÜCKE (1975) performed growth selection experiments on aluminum and suggested that selective growth due to anisotropy in mobility is the dominant factor contributing to LPO. So far, little information is available on the anisotropy in grain boundary mobility in minerals. In any case, this effect is not directly related to deformation and cannot determine the LPO by itself, although its effect is possibly important in that it may systematically alter the LPO formed by other mechanisms.

To summarize, GBM (growth) can produce the stress-controlled LPO, if the dislocation density reflects the instantaneous stress. The nature of LPO depends on the stress/strain distribution at the scale of grains, which will depend on materials and on deformation conditions.

The LPO due to GBM will also be important in fabric development due to post deformation annealing (primary recrystallization).

3.4. *The Controlling Factors of Lattice Preferred Orientation During Dynamic Recrystallization*

The arguments in the previous sections have indicated the varieties of mechanisms of LPO due to the elementary processes of dynamic recrystallization. The next question is which of these processes is the most important in the whole process of dynamic recrystallization.

Let us imagine a steady state of dynamic recrystallization in which formation of mobile grain boundaries and their migration occur more or less randomly, with simultaneous plastic deformation. To evaluate the relative importance of the elementary processes in dynamic recrystallization, I first consider the nature of change in lattice orientation across a moving boundary in a deforming material.

Figure 3 schematically illustrates the variation of plastic strain and dislocation density across a moving boundary. When a grain boundary has passed, the plastic strain of the crystal falls to zero, and it will eventually be built up (see also Figure 2). This plastic strain will change the lattice orientation in the manner discussed in Section 2. At the same time, when a grain boundary migrates, some portions of crystals are replaced with crystals of different orientations, thereby changing the lattice orientations (as discussed in Section 3.3). Since the plastic strain of a crystal behind a moving boundary increases with time after the grain boundary has passed, the orientations of crystals close behind a moving boundary are mainly determined by GBM, but the orientations of crystals far from the moving boundary are mainly determined by plastic deformation. Thus the width δ of a zone near a moving grain boundary where the lattice orientation is mainly determined by GBM is given by

$$\delta = \varepsilon_1 v / \dot{\varepsilon} \tag{1}$$

where v is the GBM velocity, $\dot{\varepsilon}$ the strain rate, and ε_1 the strain needed to rotate the lattice orientation via deformation by the amount comparable to the change in lattice orientation due to GBM; ε_1 is estimated to be of the order of unity for random lattice orientation.

When the low-strain parts occupy a significant fraction of the sample, the effect of GBM on LPO will be important. Experimental observations indicate that GBM occurs heterogeneously in a recrystallizing material (MEANS, 1983; URAI, 1983) (see also Figure 7b), and low-strain regions will be formed heterogeneously during a steady state of recrystallization. Figure 4 illustrates schematic models of heterogeneous microstructures during dynamic recrystallization.

When the width of low-strain regions is smaller than average grain size L, the volume fraction of low-strain regions, α, is given by (Figure 4a)

$$\alpha = \beta\delta / L \tag{2}$$

where β is the fraction of grain boundaries that are moving; β depends on nucleation-growth kinetics. Let us consider a grain boundary. When the difference in free energies (dislocation densities) between the two grains on both sides of the boundary exceeds a critical value, the grain boundary migrates (BAILEY and HIRSCH, 1962). GBM continues until impingement with other grain

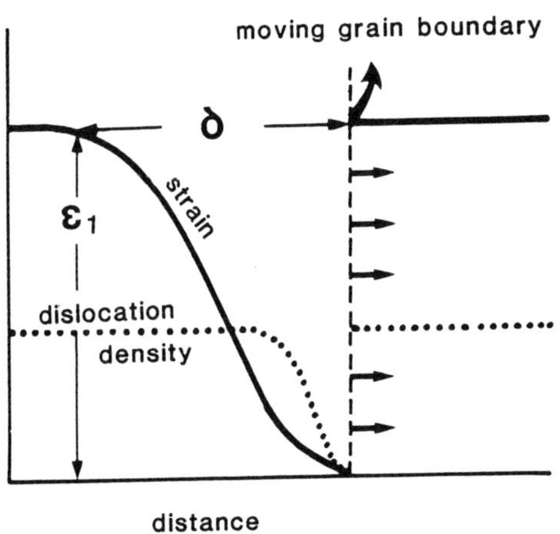

Figure 3. A schematic diagram showing the variation of plastic strain (solid line) and dislocation density (dotted line) across a moving grain boundary (see also Figure 2). Here δ is the thickness of the low-strain zone (strain less than ε_1). The width of the low-dislocation density zone is smaller than that of low-strain zone, because a steady state dislocation density is attained at low strain (in olivine). Therefore the existence of low-dislocation density zones indicates the (more extensive) existence of low-strain zones, although the reverse is not always true. For details, see text.

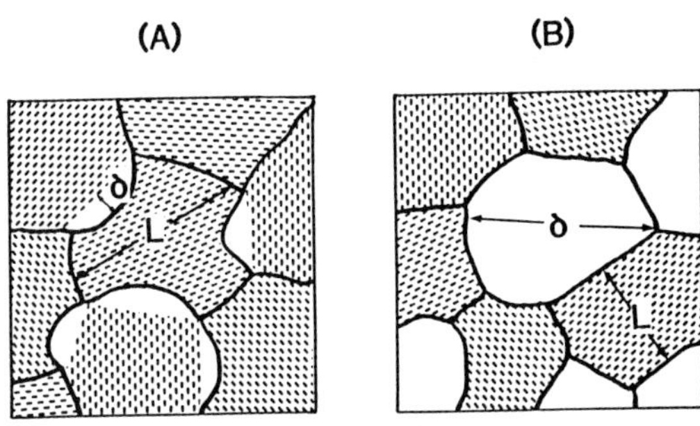

Figure 4. Schematic microstructural models of dynamically recrystallizing polycrystals. Some of the grain boundaries are moving and leave low-strain regions (unshaded regions) behind them. When these regions occupy a significant volume fraction of a sample, the preferred orientation is controlled by grain boundary migration. L is the average grain size, and δ the thickness of the low-strain zone. (a) The case of $\delta < L$. (b) The case of $\delta = L$, where low-strain grains rather than low-strain zones are formed. See text for details.

boundaries occurs. The average duration t_1 of migration for a grain boundary is therefore given by $t_1=L/v$. Immediately after the impingement, the dislocation densities of two contacting grains are generally low, and the difference in dislocation densities is not large enough to start GBM again. Thus a finite time t_2 is necessary to start GBM again. Such an episodic nature of GBM during dynamic recrystallization has been observed by the in situ observations (see, for example, MEANS (1983) and URAI (1983)). Although the details of this remobilization of grain boundaries are not well understood, this process must be similar to the nucleation at preexisting boundaries discussed in Section 3.2. Thus, a finite strain ε_2 (of the order of 10% (see Section 3.2)) is involved in this process; hence, $t_2=\varepsilon_2/\dot{\varepsilon}$. Then we get

$$\beta = t_1/(t_1 + t_2) = 1/(1 + \zeta) \quad (3)$$

with

$$\zeta = (\varepsilon_2/\dot{\varepsilon})(v/L) \quad (4)$$

and from equations (1), (2), and (3),

$$\alpha = 1/(\gamma + \xi) \quad (5)$$

with

$$\gamma = (L/v)(\dot{\varepsilon}/\varepsilon_1) \quad (6)$$

$$\xi = \varepsilon_2/\varepsilon_1 \ (=\gamma\zeta) \quad (7)$$

Thus the microstructures of dynamically recrystallizing materials are determined by two of the three nondimensional parameters (γ, ξ, ζ). Physically, γ is the ratio of rate of change in lattice orientations by deformation to that by GBM, ξ is the ratio of strain rate to nucleation rate (rate of mobilization of grain boundaries), and ζ is the ratio of growth rate (GBM) to nucleation rate (rate of mobilization of grain boundaries). From equations (3) and (5), it is concluded 1) that most of the grain boundaries are moving ($\beta\sim 1$) when nucleation (mobilization of grain boundaries) is easy compared with growth (GBM), and 2) that a significant fraction of the sample is occupied by low-strain regions ($\alpha\sim 1$) and therefore the effect of GBM will dominate the LPO when both nucleation (mobilization of grain boundaries) and growth (GBM) are easy compared with plastic deformation.

Note, however, that equation (2) holds only for $\delta<L$, i.e., $\gamma>1$. Some modification of the model is necessary when the width of the low-strain region becomes equal to average grain size, i.e., $\delta=L$. This occurs when the velocity of grain boundary migration is fast (i.e., $\gamma\leqq 1$). In this case, impingement of grain boundaries occurs before the maximum strain of the region that has been swept by moving boundaries reaches ε_1. Therefore low-strain grains rather than low-strain zones will develop (Figure 4b). The volume fraction of low-strain grains can be estimated by a line of argument similar to that in the case of $\delta<L$, by defining t_1 as a duration for which a grain has low strain ($\varepsilon<\varepsilon_1$). This duration is longer than the duration for that a grain boundary is moving. From these arguments it can be shown that the results for the case $\gamma>1$ can be generalized to the case $\gamma\leqq 1$ as well: the LPO will be determined by GBM if both nucleation and growth are easy relative to deformation (i.e., γ, $\xi\lesssim 1$).

Note that the above model includes the subgrain rotation recrystallization as an extreme case where low-strain zones occupy negligible volume due to low GBM rate compared to the rate of plastic deformation and resultant subgrain rotation (i.e., $\alpha\ll 1$).

To summarize, the microstructures of dynamically recrystallizing materials are controlled by the relative rates of three elementary processes, namely, plastic deformation (due to dislocation glide), nucleation, and growth (GBM) and thus can be specified by two nondimensional parameters. An example is shown in Figure 5, where the microstructures are characterized in terms of the fraction of low-strain regions (α) and the fraction of moving boundaries (β), as a function of the ratios of strain rate to growth rate (γ) and of strain rate to nucleation rate (ξ). Four domains are recognized: in A_1 a small fraction of grain boundaries are moving ($\beta\ll 1$), and LPO is governed by plastic deformation ($\alpha\ll 1$); in A_2 most of the grain boundaries are moving ($\beta\sim 1$), but LPO is governed by plastic deformation because portions of crystals swept by moving boundaries will be deformed immediately ($\gamma\gg 1$); in B_1 a small fraction of the grain boundaries are moving ($\beta\ll 1$), but LPO is controlled by GBM, because most of the grain boundary impingement occurs at low strains ($\gamma\ll 1$); in B_2 most of the grain boundaries are moving ($\beta\sim 1$), and LPO is controlled by GBM. Transitions between these domains are expected to be gradual, and intermediate structures may be found near the transition conditions.

It is important to recognize that the microstructures, notably LPO, are determined by the relative rates of the elementary processes. It is often argued (e.g., LISTER and PRICE, 1978; URAI et al., 1986) that the LPO due to dynamic recrystallization is essentially due to deformation and controlled by the kinematic framework because of the concurrent deformation during GBM and of finite strain involved in nucleation. This conventional view cannot be generally justified. FRIEDMAN and HIGGS (1981) also noted the important role of dynamic recrystallization in forming fabrics different from deformation fabrics.

We now estimate the parameters ξ and γ. From the

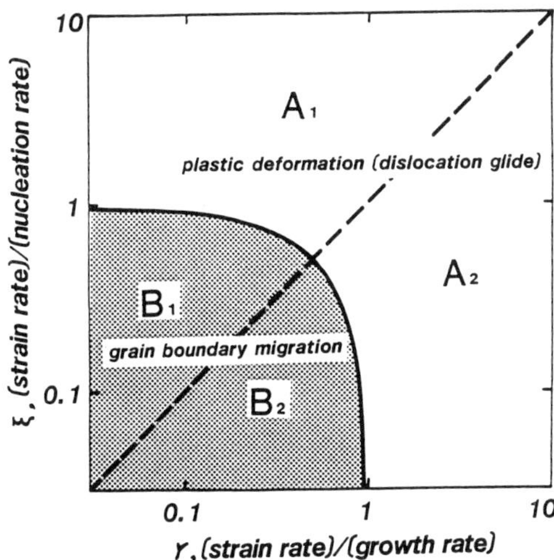

Figure 5. A schematic diagram showing the microstructural characteristics of dynamically recrystallizing materials. Microstructures are mainly controlled by two parameters, γ (the ratio of strain rate to growth rate) and ξ (the ratio of strain rate to nucleation rate). When both nucleation and growth are faster than strain rate (shaded region), most of the sample volume is occupied by low-strain regions, and the lattice preferred orientation is controlled by grain boundary migration. If either nucleation or growth (or both) is slow compared to strain rate (unshaded region), only small portions of the sample volume are occupied by low-strain regions, and the effect of deformation will dominate the lattice preferred orientation. Above the dashed line, where the growth rate is faster than the nucleation rate, only a small fraction of the grain boundaries are moving, while below the line, where nucleation is rapid compared with growth rate, most of the grain boundaries are moving. Thus the microstructural characteristics of dynamically recrystallizing materials may be divided into four domains in this diagram. For details see text.

above arguments, ξ can be estimated to be relatively small, of the order of 0.1, where dynamic recrystallization occurs. Then the most important parameter that will control the dominant factor of LPO is γ.

Since $\varepsilon_1 \sim 1$ in equation (7), the parameter γ can be estimated from the grain size (L), the velocity of GBM (v), and the strain rate ($\dot{\varepsilon}$). All of them depend on the conditions where dynamic recrystallization occurs.

The grain size L at the steady state of dynamic recrystallization depends mainly on applied stress σ as (e.g., TWISS, 1977; EDWARD et al., 1982)

$$L = L_0(\sigma/\mu)^{-m} \tag{8}$$

where L_0 is the grain size at $\sigma=\mu$ (μ is shear modulus) and m is a constant, both of which are nearly independent of temperature but presumably depend on the mechanism of recrystallization (GUILLOPÉ and POIRIER, 1979).

The strain rate $\dot{\varepsilon}$ is given by

$$\dot{\varepsilon} = \dot{\varepsilon}_0 \exp(-Q_1/RT)(\sigma/\mu)^n \tag{9}$$

where $\dot{\varepsilon}_0$ is the preexponential factor, Q_1 the activation energy, and n the stress exponent. The strain rate is often enhanced by dynamic recrystallization (e.g., SELLARS, 1978). In silicates, the presence of water will also enhance the creep rate (e.g., CHOPRA and PATERSON, 1981; KARATO et al., 1986).

The velocity of GBM, v, can be written as (e.g., NICOLAS and POIRIER, 1976).

$$v = MF \tag{10}$$

where M is the mobility and F the driving force for GBM. As discussed in Section 3.3, the dislocation energy is the main driving force; then, $F=F_1+F_2$, where F_1 is the energy of free dislocations and F_2 the energy of dislocations in subgrain boundaries. More precisely, the dislocation energy here is mainly that of the consumed grains, because the portions of crystals just behind the moving boundaries have very low dislocation densities (see Figures 2 and 3). Now, F_1 is proportional to dislocation density ρ, and the energy of a dislocation, E_d; hence, $F_1=\rho E_d \sim \mu(\sigma/\mu)^2$, where use has been made of the relations $\rho \sim b^{-2}(\sigma/\mu)^2$ (b is the magnitude of the Burgers vector) and $E_d \sim \mu b^2$ (e.g., COTTRELL, 1953). F_2 is given by $F_2 \sim \gamma_{SG}/L_s$, where γ_{SG} is the surface energy of sub-boundaries given by $\gamma_{SG} \sim \theta\mu b/4\pi(1-v)$ for a small misfit angle θ(v is Poisson's ratio) (e.g., COTTRELL, 1953), and L_s is the subgrain size given by (e.g., NICOLAS and POIRIER, 1976) $L_s=L_{s0}(\sigma/\mu)^{-1}$. Hence we have

$$F \sim \mu(\sigma/\mu)^2 + (\gamma_{SG}/L_{so})(\sigma/\mu) \tag{11}$$

The grain boundary mobility M is in general a complex function of temperature and of the driving force (e.g., POIRIER and GUILLOPÉ, 1979; URAI et al., 1986). Catastrophic increase in GBM rate is sometimes observed when temperature and/or driving force are increased (GUILLOPÉ and POIRIER, 1979; TUNGATT and HUMPHREYS, 1981). However, within a limited range of temperature and driving force, the following Arrhenius-type dependence is most commonly observed (e.g., YAN et al., 1977):

$$M = M_0 \exp(-Q_2/RT) \tag{12}$$

where M_0 is the preexponential factor and Q_2 the activation energy. Therefore,

$$v = v_0 \exp(-Q_2/RT)[\mu(\sigma/\mu)^2 + (\gamma_{SG}/L_{so})(\sigma/\mu)] \tag{13}$$

where $v_0=M_0$. In silicates, the presence of a small amount of water will also enhance the grain boundary mobility

(TULLIS and YUND, 1982; KARATO, 1987a). Also, the rate of GBM may be sensitive to the presence of secondary phases and/or impurity atoms as found in ceramics (YAN et al., 1977).

Combining equations (8), (9), and (13), one gets

$$\gamma = \gamma_0 \exp[-(Q_1 - Q_2)/RT](\sigma/\mu)^{n-m}/F \qquad (14)$$

with $\gamma_0 = L_0 \dot{\varepsilon}_0 / v_0 \varepsilon_1$ and F given by (11). Note here that in writing equations (11), (13), and (14), the orientation dependence of stress was neglected for simplicity. Some general remarks may be made here. 1) The parameter γ depends on the conditions of dynamic recrystallization, especially temperature, stress, and water content. Therefore the dominant mechanism of LPO will depend on these variables. Observations made under specific conditions should not be generalized without limitation. 2) Under conditions where the temperature dependence of GBM rate and strain rate is of the Arrhenius type, γ will increase with temperature if, as is most often the case, $Q_1 > Q_2$ (e.g., NICOLAS and POIRIER, 1976; YAN et al., 1977). Consequently, the effect of GBM will be important at relatively low temperatures within high-temperature conditions at which dynamic recrystallization occurs. 3) However, when catastrophic increase in GBM rate occurs, γ will decrease, and if it becomes smaller than or close to unity, the effect of GBM will dominate the LPO, although associated increase in recrystallized grain size (GUILLOPÉ and POIRIER, 1979; TUNGATT and HUMPHREYS, 1981) and possibly the strain rate (SELLARS, 1978) will counteract to some extent.

3.5. *Lattice Preferred Orientation Mechanism Maps for Olivine*

Dynamic recrystallization occurs during high-temperature deformation of olivine when strain exceeds a critical value; see, for example, AVÉ LALLEMANT and CARTER (1970), POST (1977), KARATO et al. (1980, 1982), ROSS et al. (1980), CHOPRA and PATERSON (1981), and ZEUCH and GREEN (1984). The strain at the onset of dynamic recrystallization appears to depend on water content. When a small amount of water is present, dynamic recrystallization occurs at small strain, about 10%, via nucleation at grain boundaries (CHOPRA and PATERSON, 1981). When water is absent, the critical strain is somewhat larger (ZEUCH and GREEN, 1984; KARATO et al., 1980, 1982). Apparently contradictory results by NICOLAS et al. (1973) and AVÉ LALLEMANT (1975) (Avé Lallemant found the importance of dynamic recrystallization to be greater than that of dislocation glide when compared at the same strain, but Nicolas et al. found the effect of dislocation glide was more important) can probably be attributed to the effect of water. The water content in the specimens of Avé Lallemant (natural dunites) should have been much larger than in those of Nicolas et al. (synthetic aggregates of forsterite). In any case, at a large strain, say of the order of unity, most of the sample volume will be recrystallized, and dynamic recrystallization should play an important role in LPO. Since the nucleation is not very difficult ($\xi < 1$), the main factor that will determine the mechanism of LPO is the ratio of growth rate to strain rate, γ (see Figure 5).

We now estimate the parameter γ as a function of temperature, stress, and water content using equations (6) and (14) and predict the dominant mechanisms of LPO as a function of these parameters. The relevant parameters needed to estimate γ are listed in Table 1.

The recrystallized grain size versus stress relations in olivine have been determined by POST (1977), KARATO et al. (1980), ROSS et al. (1980), and ZEUCH and GREEN (1984). Although there remain some uncertainties regarding the effects of recrystallization mechanisms and of water, these experimental studies give rather similar results (see also KARATO, 1984). I used, in this calculation, the results of KARATO et al. (1980), for which the recrystallization mechanism has been clearly identified as progressive misorientation of subgrains (KARATO et al., 1980, 1982; TORIUMI and KARATO, 1985).

TORIUMI (1982) and KARATO (1987a) studied the grain boundary migration in olivine. Within the studied conditions (T/Tm=0.70–0.85), no catastrophic behavior in grain boundary mobility was observed, and their results fit the Arrhenius-type relation. KARATO (1987a) also found that the presence of a small amount of water enhances the grain boundary mobility. However, when a large amount of water is present, bubbles of water at grain

TABLE 1. Parameters Used in Estimating the Ratio of Characteristic Time of Grain Boundary Migration to That of Plastic Deformation in Olivine (See Equation (14))

	Dry	Wet
	Strain rate	
$\dot{\varepsilon}_0$ (s^{-1})	3.5×10^{22}	9.7×10^{17}
Q_1 (kJ mol^{-1})	540	420
n	3.5	3
Reference	KARATO et al. (1986)	KARATO et al. (1986)
	Grain Boundary Migration	
v_0 (m s^{-1} Pa^{-1})	1.0×10^{-7}	0.5×10^{-7}
Q_2 (kJ mol^{-1})	210	160
Reference	TORIUMI (1982)	KARATO (1987a)
	Recrystallized Grain size	
L_0 (m)	1.3×10^{-6}	1.3×10^{-6}
m	1.2	1.2
Reference	KARATO et al. (1980)	assumed to be the same as at dry condition

boundaries are found to significantly reduce the grain boundary mobility. Due to this complexity and to the possible effects of other impurities, the parameters for the GBM rate used in this calculation have large uncertainties.

For the strain rate, I used the results of KARATO et al. (1986) for steady state deformation due to dislocation creep. An important point to note here is the possible effect of dynamic recrystallization in enhancing the deformation. This point is important because the strain rate relevant here is for the low-strain regions left behind the moving boundaries (see Figure 3 and equation (1)). The enhancement of deformation due to dynamic recrystallization has been found in many metals and attributed to the formation of low dislocation density grains (e.g., SELLARS, 1978; SAKAI and JONAS, 1984). In olivine, however, such a significant enhancement of creep due to dynamic recrystallization has not been clearly demonstrated, although ZEUCH (1982) suggested that it might be involved in the ductile faulting observed by POST (1977). Absence of significant softening due to recrystallization may be attributed to the fact that steady state deformation in olivine is achieved at very low strains, about 2% (DURHAM and GOETZE, 1977; KARATO et al., 1986).

The transition conditions of mechanisms of LPO were estimated using the experimental results discussed above, assuming that $\gamma=1$ defines the transition conditions (see Figure 5). The results are shown in Figure 6, where the dominant mechanisms of LPO are shown as a function of temperature, stress, and water content (either water-saturated (wet) or water-free (dry) condition). In this calculation, the misfit angle of subboundaries is assumed to be 10^{-3} rad. It is seen 1) that the main factor determining the mechanism of LPO is the temperature (the effect of GBM is important at relatively low temperatures) and 2) that the effect of water is relatively minor. This is because the presence of water enhances both strain rate and GBM rate, and therefore their ratio is not much affected.

It is noted that the predicted transition conditions are close to the conditions where the relevant constants were determined. Therefore the results are not subject to large uncertainties due to a large extrapolation of the experimental results. However, the results shown in Figure 6 must be regarded as only tentative because of the large uncertainties in the parameters used (especially those for GBM velocity) and because of some simplifications in the model.

In Figure 6 the conditions of some experimental studies are also shown. The experiments at high temperatures ($T/Tm=0.92$) by TORIUMI and KARATO (1985) have shown that the LPO of dynamically recrystallized olivines

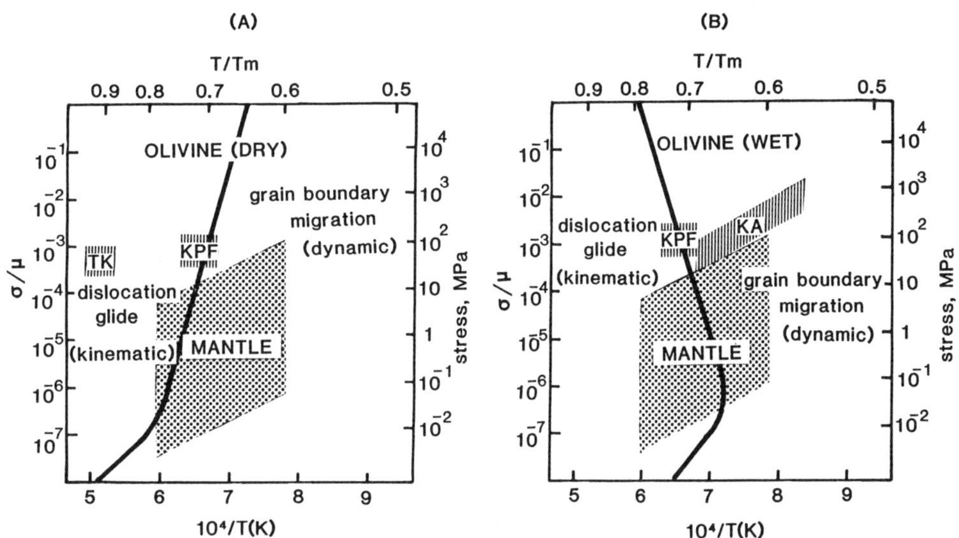

Figure 6. Lattice preferred orientation mechanism maps for olivine. (a) Dry (water free) condition. (b) Wet (water saturated) condition. The heavy lines show the transition conditions between the dislocation glide dominating and the grain boundary migration dominating mechanisms. Lattice preferred orientation is related to the kinematic framework in the former, while it is related to the stress (dynamics) in the latter. TK, KPF, and KA denote the conditions of the experimental studies by TORIUMI and KARATO (1985), KARATO et al. (1986), and KUNZE and AVÉ LALLEMANT (1981), respectively. The temperature and stress conditions in the upper mantle (asthenosphere) are estimated from petrological and microstructural studies of peridotite xenoliths (e.g., AVÉ LALLEMANT et al., 1980).

was controlled by the kinematic framework of flow (or the strain), which is consistent with the present model. KUNZE and AVÉ LALLEMANT (1981) made deformation experiments on olivine with a rotational component (approximately simple shear) at relatively low temperatures (T/Tm=0.6–0.7) and found that the LPO was related to the stress rather than the shear direction/shear plane or the strain. Their fabric diagram (their Figure 4) shows the concentration of [010] axes toward the direction of maximum compressive stress, the [100] axes toward the minimum compression stress direction, and the [001] axes toward the intermediate stress direction. This result is consistent with the present model in that the LPO is related to stress at relatively low temperatures.

The present model also predicts that the experimental condition of KARATO et al. (1986) is close to the boundary between the two mechanisms. Figure 7b shows an optical photomicrograph of a specimen deformed by KARATO et al. (1986). In contrast to the specimens recrystallized at very high temperatures (Figure 7a), dislocation distribution is highly heterogeneous. Moving grain boundaries produce clearly identifiable dislocation-free zones, although their volume fraction is small. To clearly demonstrate the role of GBM, the LPO was measured separately for low dislocation density grains with convex boundaries as well as for all grains (Figure 8). Since the volume fraction of low dislocation density grains is small (a few percent), the LPO of all grains is not appreciably affected by that of the low dislocation density grains.

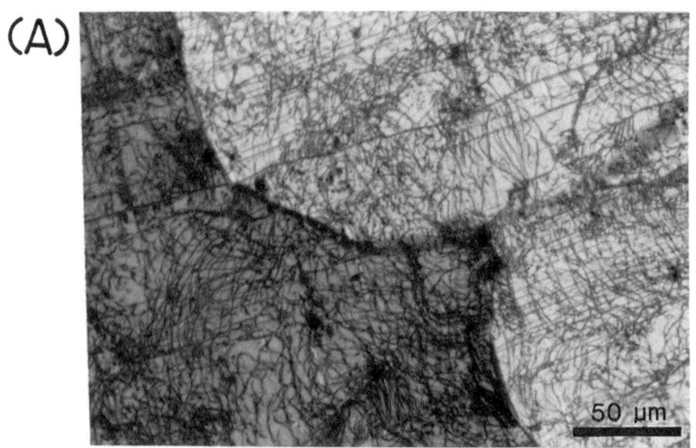

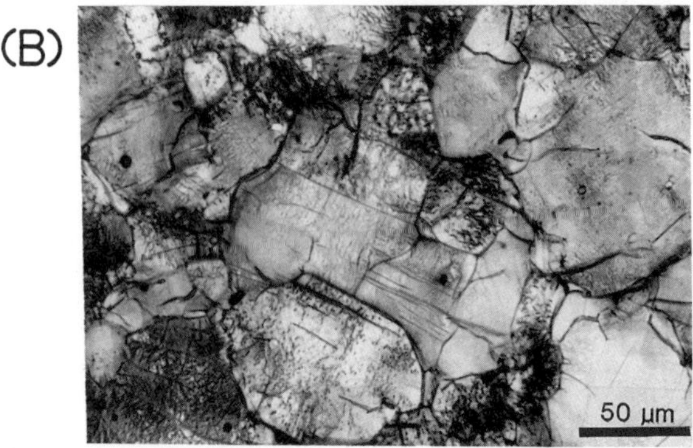

Figure 7. Dislocation distributions in experimentally deformed olivines. (a) Olivines deformed at T/Tm=0.92 (KARATO et al., 1980, 1982). (b) Olivines deformed at T/Tm=0.75 (KARATO et al., 1986). Dislocation distribution is nearly homogeneous at very high temperature but heterogeneous at relatively low temperature.

The LPO of all grains is weak (due presumably to the low strain of 13.2%) but consistent with the slip-induced rotation due mainly to the (010)[100] slip system. In contrast, the LPO of the low dislocation density grains (recrystallized grains) is strong and slightly different from that of all grains: in addition to the strong maxima of [010] axes toward the compression direction, small but significant submaxima of [100] axes toward the compression direction are also seen (such submaxima are also seen in Figure 4 of KUNZE and AVÉ LALLEMANT (1981)). Therefore two mechanisms of LPO, i.e., dislocation glide and grain boundary migration, appear to operate simultaneously under this experimental condition. (Postdeformational original of low dislocation density grains can be discarded from the estimation of grain boundary migration rate based on work by KARATO (1987a)).

These experimental observations are consistent with the present model, which predicts that the effect of GBM, which produces a stress-controlled LPO, will be important at relatively low temperatures. More specifically, the observed LPOs due to recrystallization in both KUNZE and AVÉ LALLEMANT's (1981) and KARATO et al.'s (1986) experiments indicate that the grains with low Schmid factors survive and dominate the LPO. These results are consistent with the orientation dependence of dislocation densities studied by Karato (unpublished). However, the reason for the selection of orientation(s) with [010] (and [100]) parallel to the compression axis out of other low Schmid factor orientations is not clearly understood, although possible explanations include the effect of prerecrystallization LPO, the effect of small but finite deformation needed to create dislocations, and/or effect of elastic strain energy.

To summarize, the present model quite naturally explains the experimental observations so far available. In particular, the model gives a reasonable explanation for often disputed (BOUCHEZ et al.,1983) results by KUNZE and AVÉ LALLEMANT (1981) which apparently conflict with the results by TORIUMI and KARATO (1985). However, well-defined experimental studies on simple shear are necessary to further test the model.

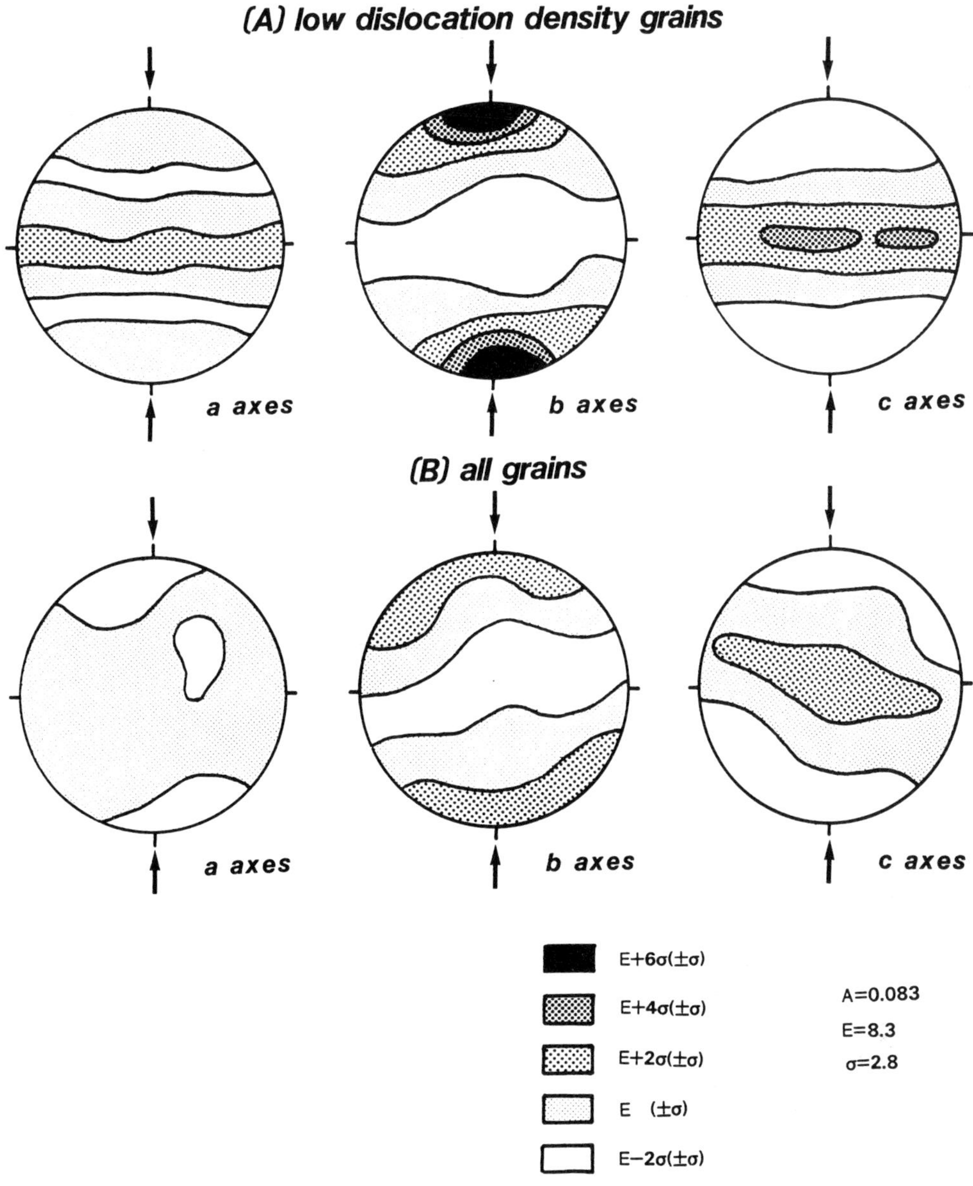

Figure 8. Equal-area projections of the lattice preferred orientation of the two types of olivine shown in Figure 7b. (a) Low dislocation density grains (dislocation density less than half the average density). (b) All grains. KAMB's (1959) method of contouring is used. Arrows indicate the compression direction. Measurements were made on 100 grains for each diagram.

4. Comparison With Geological and Geophysical Observations and Implications for Seismic Anisotropy in the Upper Mantle

Figure 9 shows the simplified olivine LPOs in simple shear (the dominant deformation mode in the oceanic upper mantle) suggested by the present model. It is seen that the stress-controlled seismic anisotropy depends on the sense of shear but the kinematically controlled one does not. If the kinematic model applies, one will be able to infer the convection pattern as has been suggested by D. L. Anderson and coworkers (NATAF et al., 1984, 1986; TANIMOTO and ANDERSON, 1984). On the other hand, if seismic anisotropy is controlled by the stress, one will be able to infer the stress state or the sense of shear in the convecting mantle and therefore the driving mechanism(s) of plate motion. The kinematically controlled LPO will occur where dynamic recrystallization is absent or

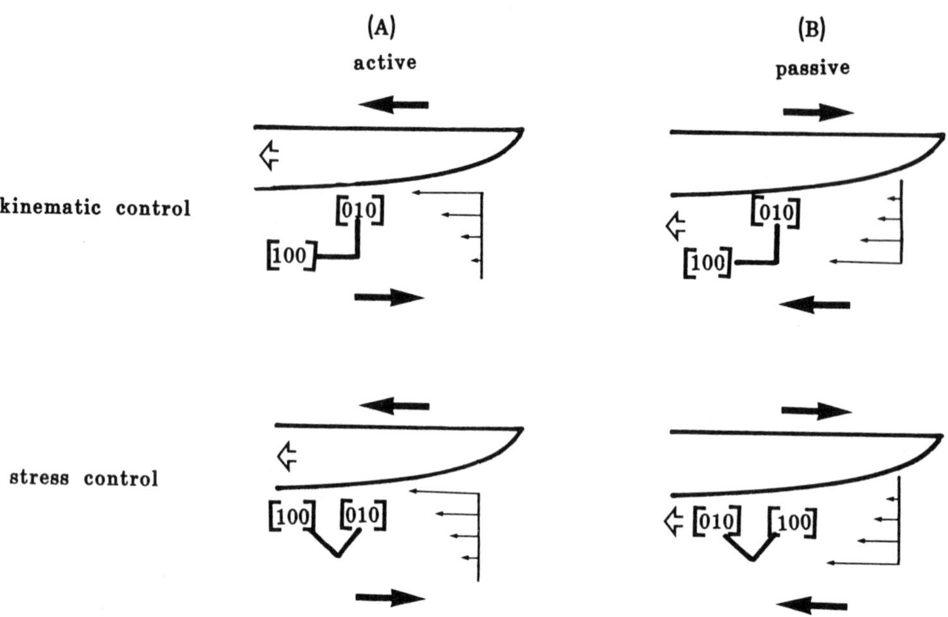

Figure 9. The lattice preferred orientation of olivine in the oceanic upper mantle. Heavy arrows indicate the sense of shear. Light arrows show flow velocities. Only one slip system (010)[100] is considered for simplicity. Note that the lattice preferred orientation depends on the sense of shear if it is controlled by the stress, but not if it is controlled by the kinematic framework. The sense of shear shown in (a) corresponds to the active plate model and that in (b) the passive plate model. The lattice preferred orientation is formed by the flow in the asthenosphere (and is subsequently frozen in when the material is cooled and incorporated into the lithosphere). Lattice preferred orientation (and resultant seismic anisotropy) in the shallow lithosphere is formed near ocean ridges, and will be controlled by the kinematic framework because of the absence of dynamic recrystallization (or of dynamic recrystallization at very high temperatures), while, in the deeper part, stress controlled preferred orientation (and resultant seismic anisotropy) is possible if dynamic recrystallization occurs at relatively low temperatures.

dynamic recrystallization occurs at relatively high temperatures. The stress-controlled LPO will occur if dynamic recrystallization occurs in relatively cold mantle.

In addition to the obvious differences in anisotropy in body waves between the two cases, anisotropy in surface waves will also be different. In a transversely isotropic scheme for horizontal shear flow, $SH>SV$ in the kinematically controlled case, but $SV>SH$ in the stress-controlled case. The azimuthal anisotropy of Rayleigh waves in the stress-controlled case will be smaller than that in the kinematically controlled case, although the orientation will be similar (The orientation of fast Rayleigh waves is parallel to that of fast P waves.)

CHRISTENSEN (1984) reviewed the LPO of naturally deformed ultramafic rocks in ophiolites (rocks of the uppermost oceanic mantle). The LPOs of olivine in most of the ophiolites show orthorhombic to axial symmetry; the axial symmetry axis corresponds to the olivine [100] axis' maxima and is subparallel to paleospreading directions. This result is consistent with the control of LPO by the dislocation glide, olivine slip system being [100](010) or [100]{0*kl*}, which is consistent with the experimental results (CARTER and AVÉ LALLEMANT, 1970).

Seismic body wave studies (e.g., RAITT et al., 1969; SHIMAMURA et al., 1983) clearly demonstrated the existence of seismic anisotropy in the uppermost oceanic mantle (possibly extending to the lower lithosphere), which is consistent with the LPO of olivine (and pyroxenes) of ophiolites (CHRISTENSEN and SALISBURY, 1979; ESTEY and DOUGLAS, 1986; KAWASAKI, 1986; NICOLAS and CHRISTENSEN, 1987) and can be interpreted kinematically. Some surface wave studies (MITCHELL and YU, 1980; NATAF et al., 1984, 1986; REGAN and ANDERSON, 1984) may be interpreted in the same way. Kinematic control of LPO in ophiolites and of seismic anisotropy in the uppermost mantle is presumably the result of the absence of significant dynamic recrystallization due to the relatively low strains as a result of rapid cooling. Alternatively, if the strains are large, as suggested by RABINOWICZ et al. (1984), the kinematically controlled seismic anisotropy will be attributed to the very high temperatures (possibly exceeding the solidus (see

RABINOWICZ et al. 1984)) near ocean ridges at which dynamic recrystallization occurs.

The origin and the nature of LPO of minerals and of seismic anisotropy in the deeper parts of the upper mantle are not well understood. Tectonic processes in the deep upper mantle often involve large strains, and dynamic recrystallization should play an important role in the microstructural development including LPO (e.g., GOETZE, 1975; KIRBY and GREEN, 1980; KARATO, 1984). The results shown in Figure 6 suggest that it is mainly the temperature that controls the mechanism of seismic anisotropy when dynamic recrystallization occurs.

The analysis and interpretation of the LPOs of mantle xenoliths are complicated because of the complex stress and temperature histories and complex deformation geometry. As a result, varieties of LPOs have been observed; see, for example, CARTER et al. (1972), MERCIER and NICOLAS (1975), CHRISTENSEN and LUNDQUIST (1982), and MERCIER (1985). Many xenoliths show the LPOs which can be attributed to the kinematically controlled mechanisms. However, MERCIER (1985) found an important exception in the completely recrystallized peridotite in which the LPO has strong double [100] and [010] axis maxima at 45° to foliation. (In MERCIER's (1985) paper there is a confusion in the LPO of this particular sample. In his Figure 9, the [100] and [001] maxima at 45° are shown, while in the text, [100] and [010] maxima are described. One of them must be incorrect or a misprint. If the figure is correct, it implies the operation of (001)[100] and/or (100)[001] slip systems). Although Mercier gave no explanation for this LPO, this fabric can be quite naturally explained by the stress-controlled recrystallization model, if the mode of flow was simple shear with large strain.

On the seismology side, surface wave studies have revealed anisotropy in the deep upper mantle: azimuthal anisotropy of Rayleigh waves (fast direction parallel to the fast *P* wave direction) (TANIMOTO and ANDERSON, 1984) and polarization anomaly of $SH>SV$ in relatively young (<80 m.y., except near ocean ridges) oceanic upper mantle (FORSYTH, 1975; MITCHELL and YU, 1980; NATAF et al., 1984, 1986; KAWASAKI, 1986). These results can be attributed to kinematically controlled anisotropy similar to that found in ophiolites (e.g., KAWASAKI, 1986). However, in relatively old oceanic mantle (>80 m.y.), anisotropy appears to be different from that in young oceanic mantle; NATAF et al. (1986) found $SV>SH$ anisotropy, and REGAN and ANDERSON (1984) and KAWASAKI (1986) found small polarization anisotropy. These results cannot be explained by LPO similar to that in ophiolites and suggest a change in mechanism of seismic anisotropy with age, possibly from kinematic to dynamic (stress-controlled) mechanism due to decrease in temperatures (see Figure 6) or, as suggested by NATAF et al. (1986), from horizontal to vertical flow (assuming a kinematic control). KIRKWOOD and CRAMPIN (1981) also suggested a stress-controlled mechanism based on the polarization anisotropy of surface waves.

Under the island arcs or the continents, the situation is more complicated. ANDO et al. (1983) found clear evidence of shear wave splitting in the upper mantle beneath Japan. However the interpretation of the results is complicated by the complex geometry of flow under the island arc and by the possible role of partial melting. FUCHS (1983) suggested that the seismic anisotropy under the European continent may be related to the present-day shear stress orientation. BABUŠKA et al. (1984) found seismic anisotropy under the European continent, in which the direction of the fastest *P* wave velocities is significantly inclined with respect to the subhorizontal plane. This anisotropy might be attributed to the stress-controlled LPO near riftzones, although BABUŠKA et al. (1984) suggested fossil subducting slabs, assuming kinematically controlled seismic anisotropy. To summarize, the mechanisms of LPO and the seismic anisotropy in the deep upper mantle are not so well understood as those of the uppermost oceanic mantle. Many studies so far have assumed a kinematic model. However, there are some important observations that suggest a stress-controlled mechanism. Observations so far available cannot give any definitive conclusions on the controlling mechanisms. More detailed studies on the LPO of mantle xenoliths and seismological studies will be required to settle this problem.

5. Discussion

In the present paper, the physical processes that may govern LPO development have been examined with special reference to the role of dynamic recrystallization. The model gives a natural explanation of the apparently conflicting experimental observations on the mechanisms of LPO in olivine (AVÉ LALLEMANT and CARTER, 1970; NICOLAS et al., 1973; AVÉ LALLEMANT, 1975; KUNZE and AVÉ LALLEMANT, 1981; TORIUMI and KARATO, 1985) and the varieties of fabrics of naturally deformed peridotites. However, the model has some limitations and simplified assumptions that need further consideration before applying it to realistic problems. These points will be discussed here.

First, the above argument has been focused on the controlling factors (kinematic or dynamic) of LPO. The details of LPO have not been discussed except in terms of "hard" or "soft" orientations. In particular, the observed choice of particular orientation(s) among equally "hard" or "soft" orientations (orientations with equal Schmid factor) in GBM-controlled LPO will have to be explained or predicted when detailed comparison with observations is to be made.

Second, the model developed here is not satisfactory in that the recrystallized grain size versus stress relation is only empirically incorporated. This relation must be related to the nucleation and growth kinetics and should be predicted by a more satisfactory model of dynamic recrystallization.

Third, only dislocation creep was considered as a deformation mechanism. In practice, as has been suggested by KARATO et al. (1986), diffusion creep may also contribute to plastic deformation in the upper mantle. If diffusion creep dominates, there will be no mechanisms of lattice rotation, and the driving force of grain boundary migration considered here (dislocation energy) will disappear. As a result, little LPO will develop (KARATO, 1987c). The existence of seismic anisotropy in the upper mantle is therefore strong evidence for dislocation creep, although the possibility of diffusion creep at restricted conditions (or in limited regions) cannot be ruled out.

Fourth, I have considered, for the sake of simplicity, a case of pure olivine aggregates. In nature, the earth's mantle is not a pure olivine aggregate but is composed of polyphase rocks. GBM recrystallization does not occur at boundaries between olivine and other phases (e.g., orthopyroxene). Therefore the importance of GBM is overestimated in the present model. Also, the LPO of pyroxenes and other minerals may contribute to seismic anisotropy (e.g., CHRISTENSEN and LUNDQUIST, 1982; ESTEY and DOUGLAS, 1986).

Fifth, the effect of orientation dependence of grain boundary mobility is not taken into consideration in the present model. Experimental studies are needed to assess this point.

It is noted that although the LPO mechanism maps (Figure 6) were constructed only for olivine, the basic concepts developed here should also apply to the other minerals. However, care must be exercised when applying the model to other minerals, particularly regarding the stress/strain distribution and the transient effect, both of which may vary from material to material.

It is perhaps worthwhile to note here that the transition condition of the mechanisms of LPO discussed in this paper is similar to that of the mechanisms of dynamic recrystallization (GUILLOPÉ and POIRIER, 1979). In their discussion, GUILLOPÉ and POIRIER (1979) considered GBM alone and argued that the effect of GBM is dominant where catastrophic increase in GBM velocity occurs. Microstructural change due to catastrophic change in GBM rate can be included in the basic conceptual model (see Figures 4 and 5) developed in this paper as a special case, although LPO mechanism maps for olivine (Figure 6) were constructed assuming no catastrophic behavior. The present model suggests that it is the relative rates of GBM compared to deformation that control the microstructures of recrystallized materials. In particular, it predicts that where GBM occurs following the Arrhenius-type behavior, the effect of GBM will dominate at relatively low temperatures. This is in contrast to MERCIER's (1980) suggestion that GBM recrystallization in olivine occurs at relatively high temperatures (see also KARATO (1984) and AVÉ LALLEMANT (1985)).

Finally, it should be noted that stress-controlled seismic anisotropy will also occur where aligned cracks (NUR and SIMMONS, 1969; ANDERSON et al., 1974), especially those filled with melt (ANDO et al., 1983), are responsible for the anisotropy.

6. Summary

1) Kinematic interpretation of LPO and seismic anisotropy is valid only under restricted conditions, that is, either when dynamic recrystallization is absent or its effect on LPO is unimportant.

2) The controlling factors of LPO formed during dynamic recrystallization are determined by the relative rates of deformation (due to dislocation creep), nucleation, and growth. Therefore the mechanisms of LPO (and the resultant seismic anisotropy) will depend on the conditions at which dynamic recrystallization occurs.

3) When both nucleation and growth are easy compared to deformation, the LPO will be controlled by grain boundary migration. Otherwise, the LPO will be controlled by deformation.

4) The nature of LPO controlled by dislocation glide is related to the kinematic framework of flow (i.e., shear plane and shear direction).

5) The nature of LPO controlled by GBM is related to the stress, if the driving force of GBM (dislocation energy) reflects the instantaneous stress, and depends on the stress/strain distribution at the scale of grains. If stress is homogeneously distributed, grains with hard orientation will dominate the LPO, but if strain is homogeneous, grains with soft orientation will dominate. In a realistic case, both stress and strain must be inhomogeneous, and the resultant LPO will depend on material properties. In olivine, hard orientation grains dominate the LPO due to GBM, resulting in LPO which is different from the LPO due to dislocation glide in simple shear.

6) The LPO of olivine in ophiolites and the seismic anisotropy in the uppermost mantle can be attributed to the kinematically controlled LPO, suggesting the absence of dynamic recrystallization or the dynamic recrystallization at high temperatures. However, the stress-controlled seismic anisotropy is a distinct possibility in the deep lithosphere and/or in the asthenosphere, if dynamic recrystallization occurs at relatively low temperatures.

Acknowledgments. I wish to thank D. L. Anderson, M. Ando, L. H. Estey, Y. Fukao, I. Kawasaki, G. S. Lister, D. H. Mainprice, A. Nicolas, T. Takeshita, and M. Toriumi for reading the manuscript. Thanks are also due to L. H. Estey, N. W. Jessel, I. Kawasaki, W. D. Means, and A. Nicolas for sending preprints of their respective papers. Discussions with M. Ando, H. W. Green II, Y. Ida, I. Kawasaki, D. H. Mainprice, A. Nicolas, T. Sakai, T. Takeshita, and M. Toriumi were helpful in clarifying some of the ideas proposed in this paper. Comments by anonymous reviewers were helpful in improving the presentation of the paper. However, any mistakes in this paper are my own.

References

Anderson, D. L., B. Minister, and D. Cole, The effect of oriented cracks on seismic velocities, *J. Geophys. Res., 79*, 4011–4015, 1974.

Ando, M., *ScS* polarization anisotropy around the Pacific Ocean, *J. Phys. Earth, 32*, 179–195, 1984.

Ando, M., Y. Ishikawa, and F. Yamazaki, Shear wave polarization anisotropy in the upper mantle beneath Honshu, Japan, *J. Geophys. Res., 88*, 5850–5864, 1983.

Avé Lallemant, H. G., Mechanisms of preferred orientations in olivine in tectonite peridotite, *Geology, 3*, 653–656, 1975.

Avé Lallemant, H. G., Subgrain rotation and dynamic recrystallization of olivine, upper mantle diapirism, and extension of the Basin-and-Range province, *Tectonophysics, 119*, 89–117, 1985.

Avé Lallemant, H. G., and N. L. Carter, Syntectonic recrystallization of olivine and modes of flow in the upper mantle, *Geol. Soc. Am. Bull., 81*, 2203–2220, 1970.

Avé Lallemant, H. G., J.-C. C. Mercier, N. L. Carter, and J. V. Ross, Rheology of the upper mantle: Inference from peridotite xenoliths, *Tectonophysics, 70*, 85–113, 1980.

Azuma, N., Experimental studies of fabric development and flow properties of ice from polar ice sheets, Ph. D. thesis, 230 pp., Hokkaido University, 1986.

Babuška, V., J. Plomerová, and J. Šílený, Large-scale oriented structures in the subcrustal lithosphere of central Europe, *Ann. Geophys., 2*, 649–662, 1984.

Bailey, J. E., and P. B. Hirsch, The recrystallization process in some polycrystalline metals, *Proc. R. Soc. London, Ser. A, 267*, 11–30, 1962.

Bellier, S. P., and R. D. Doherty, The structure of deformed aluminum and its recrystallization—Investigations with transmission Kossel diffraction, *Acta Metall., 25*, 521–536, 1977.

Bouchez, J. L., G. S. Lister, and A. Nicolas, Fabric asymmetry and shear sense in movement zones, *Geol. Rundsch., 72*, 401–419, 1983.

Carter, N. L. and H. G. Avé Lallemant, High temperature flow in dunite and peridotite, *Geol. Soc. Am. Bull., 81*, 2181–2202, 1970.

Carter, N. L., D. W. Baker, and R. P. George, Jr., Seismic anisotropy, flow, and constitution of the upper mantle, in *Flow and Fracture of Rocks, Geophys. Monogr. Ser.*, Vol. 16, edited by H. C. Heard, et al., pp. 167–190, AGU, Washington D.C., 1972.

Chopra, P. N., and M. S. Paterson, The experimental deformation of dunite, *Tectonophysics, 78*, 453–473, 1981.

Christensen, N. I., The magnitude, symmetry and origin of upper mantle anisotropy based on fabric analyses of ultramafic tectonites, *Geophys. J. R. Astron. Soc., 76*, 89–111, 1984.

Christensen, N. I., and S. M. Lundquist, Pyroxene orientation within the upper mantle, *Geol. Soc. Am. Bull., 93*, 279–288, 1982.

Christensen, N. I., and M. H. Salisbury, Seismic anisotropy in the oceanic upper mantle: Evidence from the Bay of Islands ophiolite complex, *J. Geophys. Res., 84*, 4601–4610, 1979.

Cottrell, A. H., *Dislocations and Plastic Flow in Crystals*, 224 pp., Clarendon Oxford, England, 1953.

Crampin, S., A review of wave motion in anisotropic and cracked elasticmedia, *Wave Motion, 3*, 343–391, 1981.

Crampin, S., E. M. Chesnokov, and R. G. Hipkin, Seismic anisotropy—The state of art: II, *Geophys. J. R. Astron. Soc., 76*, 1–16, 1984.

Drury, M. R., F. J. Humphreys, and S. H. White, Large strain deformation studies using polycrystalline magnesium as a rock analogue, II, Dynamic recrystallization mechanisms at high temperatures, *Phys. Earth Planet. Inter., 40*, 208–222, 1985.

Durham, W. B., and C. Goetze, Plastic flow of oriented single crystals of olivine, Mechanical data, *J. Geophys. Res., 82*, 5737–5753, 1977.

Edward, G. H., M. A. Etheridge, and B. E. Hobbs, On the stress dependence of subgrain size, *Textures Microstructures, 5*, 127–152, 1982.

Estey, L. H., and B. J. Douglas, Upper mantle anisotropy: A preliminary model, *J. Geophys. Res., 91*, 11,393–11,406, 1986.

Etchecopar, A., A plane kinematic model of progressive deformation in a polycrystalline aggregate, *Tectonophysics, 39*, 121–142, 1977.

Forsyth, D. W., The early structural evolution and anisotropy of the oceanic upper mantle, *Geophys. J. R. Astron. Soc., 43*, 103–162, 1975.

Friedman, M., and N. G. Higgs, Calcite fabrics in experimental shear zones, in *Mechanical Behavior of Crustal Rocks, Geophys. Monogr. Ser.*, Vol. *24*, edited by N. L. Carter et al., pp. 11–27, AGU, Washington, D.C., 1981.

Fuchs, K., Seismic anisotropy of the subcrustal lithosphere as evidence for dynamical processes in the upper mantle, *Geophys. J. R. Astron. Soc., 49*, 167–179, 1977.

Fuchs, K., Recently formed elastic anisotropy and petrological models for the continental subcrustal lithosphere in southern Germany, *Phys. Earth Planet. Inter., 31*, 93–118, 1983.

Goetze, C., Sheared lherzolite: From the point of view of rock deformation, *Geology, 3*, 172–173, 1975.

Guillopé, M., and J. P. Poirier, Dynamic recrystallization during creep of single-crystalline halite: An experimental study, *J. Geophys. Res., 84*, 5557–5567, 1979.

Hess, H., Seismic anisotropy of the uppermost mantle under oceans, *Nature, 203*, 629–631, 1964.

Hobbs, B. E., Recrystallization of single crystals of quartz, *Tectonophysics, 6*, 353–401, 1968.

Hobbs, B. E., W. D. Means, and P. F. Williams, *An Outline of Structural Geology*, 571 pp., John Wiley, New York, 1976.

Hoffman, N. R. A., and D. P. McKenzie, The destruction of geochemical heterogeneities by differential fluid motions during mantle convection, *Geophys. J. R. Astron. Soc., 82*, 163–206, 1985.

Ida, Y., Preferred orientation of olivine and anisotropy of the oceanic lithosphere, *J. Phys. Earth, 32*, 245–257, 1984.

Jessel, N. W., Grain boundary migration and fabric development in experimentally deformed octachloropropane, *J. Struct. Geol., 8*, 543–562, 1986.

Kamb, W. B., Ice petrofabric observation from Blue glacier, Washington, in relation to theory and experiment, *J. Geophys. Res., 64*, 1891–1909, 1959.

Karato, S., Grain-size distribution and rheology of the upper mantle, *Tectonophysics, 104*, 155–176, 1984.

Karato, S., Grain growth kinetics in olivine, *Tectonophysics*, in press, 1987a.

Karato, S., Scanning electron microscope observation of dislocations in olivine, *Phys. Chem. Miner., 14*, 245–248, 1987b.

Karato, S., The role of recrystallization in preferred orientation of olivine, *Phys. Earth Planet. Inter.*, in press, 1987c.

Karato, S., M. Toriumi, and T. Fujii, Dynamic recrystallization of olivine single crystals during high-temperature creep, *Geophys. Res. Lett., 7*, 649–652, 1980.

Karato, S., M. Toriumi, and T. Fujii, Dynamic recrystallization and high-temperature rheology of olivine, in *High Pressure Research in Geophysics*, edited by S. Akimoto and M. H. Manghnani, pp.

171–189, Center for Academic Publications, Tokyo, 1982.

KARATO, S., M. S. PATERSON, and J. D. FITZ GERALD, Rheology of synthetic olivine aggregates: Influence of grain size and water, *J. Geophys. Res.*, *91*, 8151–8176, 1986.

KAWASAKI, I., Azimuthally anisotropic model of the oceanic upper mantle, *Phys. Earth Planet. Inter.*, *43*, 1–21, 1986.

KIRBY, S. H., and H. W. GREEN II, Dunite xenoliths from Hualalai volcano: Evidence for mantle diapiric flow beneath the island of Hawaii, *Am. J. Sci.*, *280A*, 550–575, 1980.

KIRKWOOD, S. C., and S. CRAMPIN, Surface-wave propagation in an ocean basin with an anisotropic upper mantle: Observations of polarization anomalies, *Geophys. J. R. Astron Soc.*, *64*, 487–497, 1981.

KOCKS, U. F., The relation between polycrystal deformation and single-crystal deformation, *Metall. Trans.*, *1*, 1121–1143, 1970.

KUNZE, F. R., and H. G. AVÉ LALLEMANT, Non-coaxial experimental deformation of olivine, *Tectonophysics*, *74*, T1–T13, 1981.

LISTER, G. S., A vorticity equation for lattice reorientation during plastic deformation, *Tectonophysics*, *82*, 351–366, 1982.

LISTER, G. S., and B. E. HOBBS, The simulation of fabric during plastic deformation and its application to quartzite: The influence of deformation history, *J. Struct. Geol.*, *2*, 355–370, 1980.

LISTER, G. S., and G. P. PRICE, Fabric development in a quartz-feldspar mylonite, *Tectonophysics*, *49*, 37–78, 1978.

LISTER, G. S., M. S. PATERSON, and B. E. HOBBS, The simulation of fabric development in plastic deformation and its application to quartzite: The model, *Tectonophysics*, *45*, 107–158, 1978.

LÜCKE, K., The orientation dependence of growth rate of grain boundaries and the formation of recrystallization textures, *J. Phys. Colloq. Orsay Fr.*, *C4*, 339–343, 1975.

MCKENZIE, D., Finite deformation during fluid flow, *Geophys. J. R. Astron. Soc.*, *58*, 689–715, 1979.

MEANS, W. D., Microstructures and micromotion in recrystallization flow of octahloropropane: A first look, *Geol. Rundsch.*, *72*, 511–528, 1983.

MERCIER, J.-C. C., Magnitude of continental lithospheric stress inferred from rheomorphic petrology, *J. Geophys. Res.*, *85*, 6293–6303, 1980.

MERCIER, J.-C. C., Olivine and pyroxenes, in *Preferred Orientation in Deformed Metals and Rocks: An Introduction to Modern Texture Analysis*, edited by H. R. Wenk, pp. 407–430, Academic, Orlando, Fla., 1985.

MERCIER, J.-C. C., and A. NICOLAS, Textures and fabrics of upper-mantle peridotites as illustrated by xenoliths from basalts, *J. Petrol.*, *16*, 454–487, 1975.

MITCHELL, B. J., and G.-K. YU, Surface wave dispersion, regionalized velocity models, and anisotropy of the Pacific crust and upper mantle, *Geophys. J. R. Astron. Soc.*, *63*, 497–514, 1980.

NATAF, H.-C., I. NAKANISHI, and D. L. ANDERSON, Anisotropy and shear wave heterogeneities in the upper mantle, *Geophys. Res. Lett.*, *11*, 109–112, 1984.

NATAF, H.-C., I. NAKANISHI, and D. L. ANDERSON, Measurement of mantle wave velocities and inversion for lateral heterogeneities and anisotropy 3. Inversion, *J. Geophys. Res.*, *91*, 7261–7307, 1986.

NICOLAS, A., Stress estimates from structural studies in some mantle peridotites, *Philos. Trans. R. Soc. London, Ser. A*, *288*, 49–57, 1978.

NICOLAS, A., and N. I. CHRISTENSEN, Formation of anisotropy in upper mantle peridotites—A review, in *The Composition, Structure, and Dynamics of the Lithosphere-Asthenosphere System*, edited by K. Fuchs and C. Froidevaux, pp. 111–123, American Geophysical Union, Washington, D.C., 1987.

NICOLAS, A., and J. P. POIRIER, *Crystalline Plasticity and Solid State Flow in Metamorphic Rocks*, 444 pp., John Wiley, New York, 1976.

NICOLAS, A., F. BOUDIER, and A. M. BOULLIER, Mechanisms of flow in naturally and experimentally deformed peridotites, *Am. J. Sci.*, *273*, 853–876, 1973.

NUR, A., and G. SIMMONS, Stress-induced velocity anisotropy in rock: An experimental study, *J. Geophys. Res.*, *74*, 6667–6674, 1969.

PATERSON, M. S., Nonhydrostatic thermodynamics and its geologic applications, *Rev. Geophys.*, *11*, 355–389, 1973.

POIRIER, J. P., and M. GUILLOPÉ, Deformation induced recrystallization of minerals, *Bull. Mineral.*, *102*, 67–74, 1979.

POIRIER, J. P., and A. NICOLAS, Deformation induced recrystallization due to progressive misorientation of subgrains, with special reference to mantle peridotites, *J. Geol.*, *83*, 707–720, 1975.

POST, R. L., High temperature deformation of Mt. Burnett dunite, *Tectonophysics*, *42*, 75–110, 1977.

RABINOWICZ, M., A. NICOLAS, and J. L. VIGNERESSE, A rolling mill effect in asthenosphere beneath oceanic spreading centers, *Earth Planet. Sci. Lett.*, *67*, 97–108, 1984.

RAITT, R. W., G. G. SHOR, T. J. G. FRANCIS, and G. B. MORRIS, Anisotropy of the Pacific upper mantle, *J. Geophys. Res.*, *74*, 3095–3109, 1969.

REGAN, J., and D. L. ANDERSON, Anisotropic models of the upper mantle, *Phys. Earth Planet. Inter.*, *35*, 227–263, 1984.

ROSS, J. V., H. G. AVÉ LALLEMANT, and N. L. CARTER, Stress dependence of grain and subgrain size in olivine, *Tectonophysics*, *70*, 39–61, 1980.

SAKAI, T., and J. J. JONAS, Dynamic recrystallization: Mechanical and microstructural considerations, *Acta Metall.*, *32*, 189–209, 1984.

SELLARS, S. M., Recrystallization of metals during hot deformation, *Philos. Trans. R. Soc. London, Ser. A*, *288*, 147–158, 1978.

SHIMAMURA, H., T. ASADA, K. SUYEHIRO, T. YAMADA, and H. INATANI, Longshot experiments to study velocity anisotropy in the oceanic lithosphere of the northwestern Pacific, *Phys. Earth Planet. Inter.*, *31*, 348–362, 1983.

TANIMOTO, T., and D. L. ANDERSON, Mapping convection in the mantle, *Geophys. Res. Lett.*, *11*, 287–290, 1984.

TORIUMI, M., Grain boundary migration in olivine at atmospheric pressure, *Phys. Earth Planet. Inter.*, *30*, 26–35, 1982.

TORIUMI, M., and S. KARATO, Experimental studies on the recovery processes of deformed olivine and the mechanical state of the upper mantle, *Tectonophysics*, *49*, 79–95, 1978.

TORIUMI, M., and S. KARATO, Preferred orientation development of dynamically recrystallized olivine during high temperature creep, *J. Geol.*, *93*, 407–417, 1985.

TULLIS, J., and R. A. YUND, Grain growth kinetics of quartz and calcite aggregates, *J. Geol.*, *90*, 301–318, 1982.

TUNGATT, P. D., and F. J. HUMPHREYS, An in-situ optical investigation of the deformation behaviour of sodium nitrate—An analogue for calcite, *Tectonophysics*, *78*, 661–675, 1981.

TWISS, R. J., Theory and applicability of a recrystallized grain size paleopiezometer, *Pure Appl. Geophys.*, *115*, 227–244, 1977.

URAI, J. L., *Deformation of wet salt rocks: An investigation into the interaction between mechanical properties and microstructure processes during deformation of polycrystalline carnallie and bischofite in the presence of a pore fluid*, Ph.D. thesis, 221 pp., Univ. of Utrecht, Utrecht, Netherlands, 1983.

URAI, J. L., F. J. HUMPHREYS, and S. E. BURROWS, In-situ studies of the deformation and dynamic recrystallization of rhombohedral camphor, *J. Mater. Sci.*, *15*, 1231–1240, 1980.

URAI, J. L., W. D. MEANS, and G. S. LISTER, Dynamic recrystallization in minerals, in *Mineral and Rock Deformation: Laboratory Studies, Geophys. Monogr. Ser.*, Vol. 36, edited by B. E. Hobbs and H. C. Heard, pp. 166–199, AGU, Washington, D.C., 1986.

VAN HOUTTE, P., Some recent developments in the theories for deformation texture prediction, in *Textures of Materials*, edited by C. M. Brakman, P. Jongenburger, and E. J. Mittemijer, pp. 7–23, Netherland Society for Materials Science, Zwijndrecht, The Netherlands, 1984.

WHITE, S., Geological significance of recovery and recrystallization processes in quartz, *Tectonophysics, 39*, 143–170, 1977.

YAN, M. F., R. M. CANNON, and H. K. BOWEN, Grain boundary migration in ceramics, in *Ceramic Microstructures '76*, edited by R. M. Fulrath and J. A. Pask, pp. 276–307, Westview Press, Colo., 1977.

ZEUCH, D. H., Ductile faulting, dynamic recrystallization and grain-size-sensitive flow of olivine, *Tectonophysics, 83*, 293–308, 1982.

ZEUCH, D. H., and H. W. GREEN II, Experimental deformation of a synthetic dunite at high temperature and pressure, I, Mechanical behavior, optical microstructure and deformation mechanism, *Tectonophysics, 110*, 233–262, 1984.

STRUCTURE OF THE MANTLE WEDGE AND VOLCANIC ACTIVITIES IN THE ISLAND ARCS

Yoshiaki IDA

Earthquake Research Institute, University of Tokyo, Bunkyo-ku, Tokyo 113, Japan

Abstract. The origin of arc volcanism has not yet been sufficiently investigated to yield a reasonable model of the physical and chemical processes of such volcanism. Seismological observations have revealed that a hot mantle wedge is adjacent to the cool subducting slab in island arcs. Volcanic material and heat flow are largely concentrated near the volcanic front (i.e., the seaward boundary of the volcanic belt) and asymmetrically distributed across the front. There is a systematic variation in the chemical composition of volcanic rocks across the arc, such as the enrichment of incompatible elements in the backarc. All of these observations can not be explained by an interpretation of arc volcanism in which convection is mechanically induced by the slab motion. A mantle diapir containing magma does not meet both the kinematic requirement of a sufficiently high ascending speed and the thermal requirement for sufficient heating of magma. A new model with an ascending continuous flow of mantle material beneath the volcanic belt is proposed to explain the observations mentioned above and the mechanism of magma generation. According to this model, incoming hot material follows the ascending flow and constitutes an upwelling current from the deeper mantle. The subducting slab supplies voltaile components that facilitate partial melting. The convection is maintained by the high temperature state of the incoming flow and by internal partial melting. The flow that has passed the top of the mantle moves out toward the backarc. This outgoing flow is gradually cooled so that the magma partly solidifies. In this way, incompatible elements are concentrated in the backarc.

Introduction

Much geophysical, geochemical, and petrological data on the nature of the subduction zone and arc volcanism have accumulated. Many theories have been proposed to interpret these data and to explain the origin of magmas. Nevertheless, a definitive answer has not yet been given for even the most fundamental questions. For instance, one does not know where the magmas are generated in the mantle wedge nor what the heat source is for them. There is also no satisfactory explanation of how the variety of basaltic to rhyolitic magmas are generated in the island arc. This paper compiles some of the data available in Japan, examines existing theories, and proposes a new model that systematically describes the process of arc volcanism.

Observations Related to Arc Volcanism

A volcanic belt is developed in island and continental arcs along deep sea trenches. This observation indicates that plate subduction results in arc volcanism, even though the physical mechanism linking subduction to volcanism is not well understood. Another fundamental finding is that the continental crust of the arc is generated by arc volcanism. This finding constrains the model of arc volcanism through the constitution and production rate of the entire arc crust; however this issue is not discussed in this paper.

In order to examine the detailed nature of arc volcanism, I have reviewed observations in Japan, especially in northeastern Japan. There are abundant geophysical and petrological data for northeastern Japan, although there may be some doubts that this area is one of the most representative cases of the subduction zone. The distribution of volcanoes in Japan is shown in Figure 1. Some important observations related to arc volcanism are given in Figure 2 for a cross-section of northeastern Japan normal to the Japan trench.

As was first pointed out by SUGIMURA (1960), the distribution of volcanoes in the arc has a quite distinctive character. There is no volcano seaward of the volcanic front (the dashed lines in Figure 1). The volcanic front sharply determines the seaward boundary of the volcanic belt and runs parallel to the trench. A more quantitative study by SUGIMURA et al. (1963) shows that erupted volcanic materials are dominant (in volume) just adjacent to the volcanic front and are less abundant toward the backarc (see Figure 2). Surface heat flow (NAGAO et al., 1985) shows a similar distribution, in magnitude, as a function of the distance normal to the trench. That is, there is a narrow region with extremely high heat flow near the volcanic front. From this narrow region, the heat flow magnitude drops sharply in the seaward direction and decreases more slowly toward the backarc. These observations show that the supply of magma and heat from the mantle is largely concentrated near the volcanic front and is distributed asymmetrically.

Figure 2 also shows the seismic velocity structure. The mantle wedge above the subducting slab is characterized by low velocity and high attenuation (i.e., low Q value), which is in sharp contrast to the high velocity and low attenuation in the interior of the slab (UTSU, 1971). This observation indicates that the mantle wedge is hot and that the slab is cool. According to a more recent study of detailed seismic structure (HASEMI et al., 1984), the

High-Pressure Research in Mineral Physics, edited by M. H. Manghnani and Y. Syono, pp. 473–480.

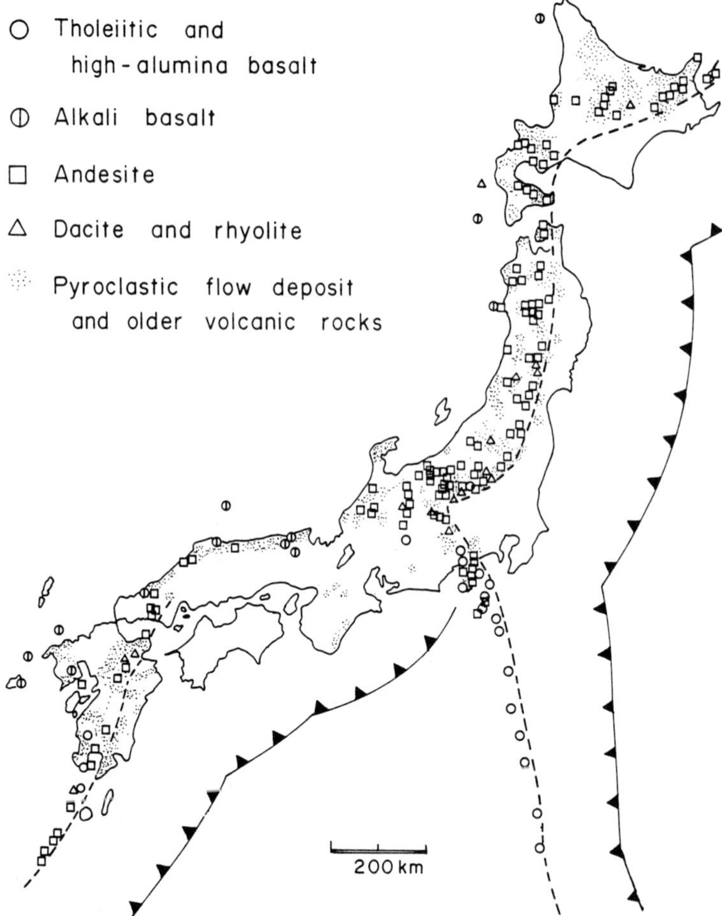

Figure 1. Distribution of volcanoes in Japan. Symbols show Quaternary volcanoes with rock type (ONO et al., 1981). Dashed lines running parallel to the deep sea trenches are the volcanic fronts.

mantle contains a particularly low velocity column just beneath the volcanic belt.

The arc velcanism contains a variety of volcanic rocks, from basalt to rhyolite, with a silica content variable over a wide range (see Figure 1). Nevertheless, a systematic trend in chemical composition, which is probably attributable to the process in the mantle, is found across the arc. Figure 2 shows the average chemical composition of basalts (SUGIMURA, 1961). It has been well established that almost all the incompatible elements, including Na and K, are more abundant with increasing distance from the trench. This pattern in the abundance of incompatible elements is found not only in basalts but also in more felsic volcanic rocks if one compares the composition of rocks with same silica content (NIELSON and STOIBER, 1973).

A Heat Source Paradox

A major problem in investigating plate subduction is explaining the heat source to generate magmas and the high heat flow. Because the cool subducting slab absorbs substantial heat from the surrounding mantle, it is naturally expected that the subduction zone would have the lowest potential for volcanic activity in the world, contrary to reality. The author (IDA, 1983a, b) has discussed this paradox earlier and examines it again here in more detail.

Thermal evolution of the subduction zone can be calculated without a large uncertainty if it is assumed that the process is mainly governed by the conductive heat transfer between the slab and its surrounding environment. The results of such a calculation suggest that the mantle wedge is gradually cooled to below 1000 °C with time (e.g., ANDERSON et al., 1978). Even at such low

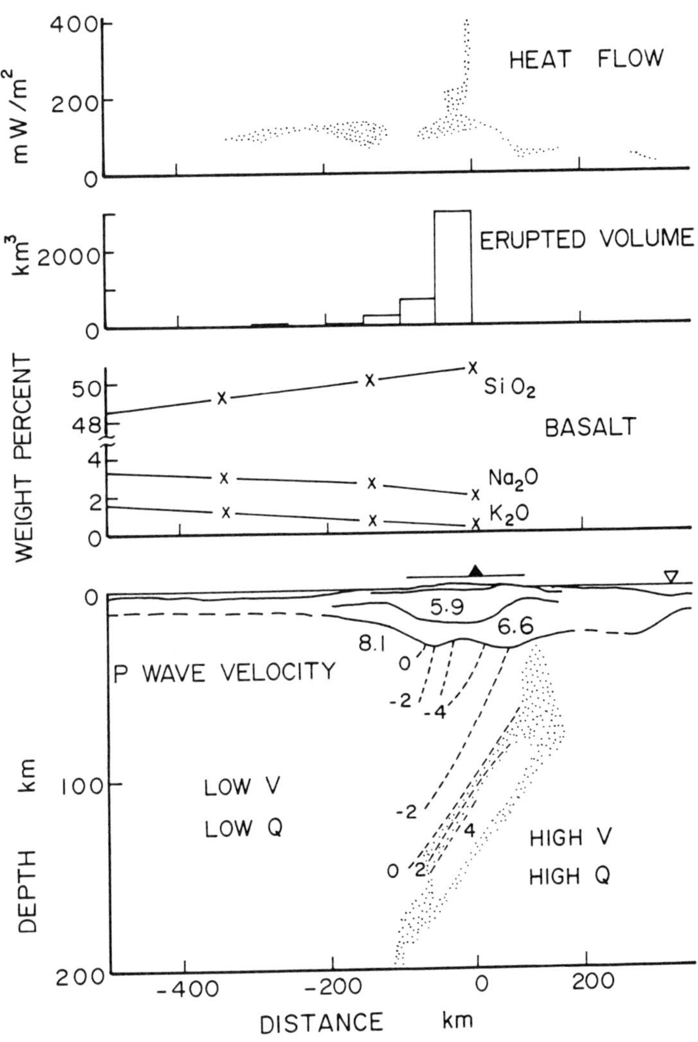

Figure 2. Some observations related to arc volcanism for a cross-section of northeastern Japan normal to the Japan trench. The locations of the volcanic front and the deep sea trench are marked by the solid and open (inverse) triangles, respectively. The common horizontal axis gives the distance from the volcanic front (a positive value is seaward). In the top chart, the range of heat flow data (NAGAO et al., 1985) is shown by the hatched area. The histogram second from the top shows the volume erupted from the quaternary volcanoes (SUGIMURA et al., 1963), and the third represents the chemical composition of basaltic rocks (SUGIMURA, 1961). The bottom chart presents data on the P-wave velocity structure compiled from known results for the crust (YOSHII, 1979) and the upper mantle (HASEMI et al., 1984); dashed lines give the contours of deviatoric velocity in percent (a positive percent indicates higher velocity) in the mantle, and the hatched area indicates high seismic areas in the Benioff-Wadati zone.

temperatures, partial melting may indeed take place in the mantle if some volatile components, such as water, are supplied from the subducting slab. However many magmas are at temperatures higher than 1000 °C, even after the eruption, and must have been substantially hotter in the mantle. Therefore the generation of magmas cannot be attributed to the effect of volatile components alone. It is expected that the mantle wedge is hot, based on thermal and seismological observations described in the last section. Thus a heat source must be available in the mantle wedge.

A candidate heat source that readily comes to mind is the frictional heat that arises from the slip or, more realistically, the shear flow in the boundary zone between the slab and the mantle wedge. The frictional heat is equivalent to the mechanical work done by the slab motion and is given by the shear stress times the slab velocity per unit time and unit area of the boundary surface. Since the slab velocity is known, the frictional heat can be determined if the shear stress is further defined. According to HASEBE et al. (1970), the frictional heat required to explain the observed surface heat flow involves a shear stress as high as several kbars. This shear stress, however, seems to be too high to be realizable.

In the above estimation of the frictional heat, it is assumed that the shear stress and the slab velocity are independent variables. In fact, these two variables are related to each other through the rheological property of the mantle material. Namely, if the slab velocity is given, the corresponding shear stress could be determined from the constitutive equation. In Newtonian flow with a constant viscosity, the heat production is thus proportional to the velocity squared. A more realistic rheology of the mantle material, however, is non-Newtonian and strongly temperature dependent. In this case, a velocity increase is accompanied by a reduced shear stress and results in only a slightly higher heat (Figure 3). This feature reflects a significant variation of viscosity with temperature and supports the prediction that the shear zone is in the asthenosphere (SCHUBERT et al., 1978). In the subduction zone, such an effect makes a large frictional stress even more unrealistic and reduces frictional heat to an insignificant role.

Thus it turns out that heat must come from somewhere outside of the subduction zone. Because thermal conduction would carry heat too slowly, one must consider convective heat transfer.

Convection and Diapirs

Various models of convection in the mantle wedge have been proposed, such as thermal convection driven by the horizontal temperature difference associated with the cold slab (RABINOWICZ et al., 1980). Among these models, the most popular and best-studied one is probably the model of convection mechanically induced by the downgoing motion of the slab (TOKSÖZ and BIRD, 1977; TOKSÖZ and HSUI, 1978). The applicability of this model to the volcanic process in the arc is examined first, even though it was originally proposed to account for the backarc spreading.

According to this model, the slab motion drags the mass adjacent to the slab downward and thus induces a downgoing flow. A return flow is produced by the upwelling of hot material, and the convection more or less raises the temperature in the mantle wedge. In an early model which assumed a constant mantle viscosity, the convection extended over the entire vertical scale specified by the boundary conditions. However, if a more realistic temperature dependence of the viscosity is assumed, the dominant flow tends to occupy a relatively shallow part of the mantle wedge regardless of the prescribed vertical scale (TOKSÖZ and HSUI, 1978). The fact that the major flow is confined to a shallow part of the

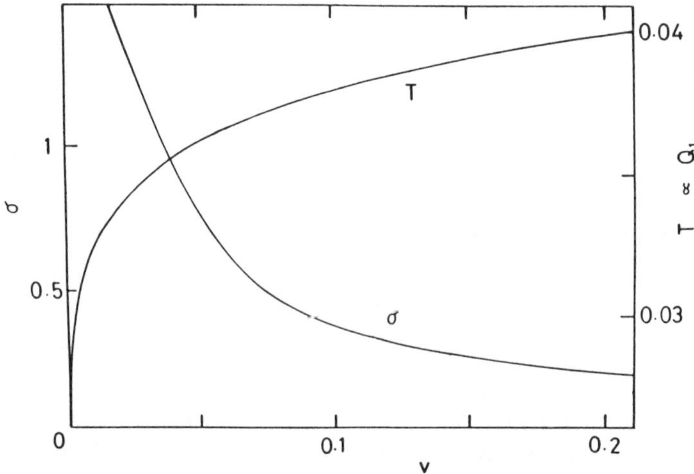

Figure 3. Frictional heat generated in a shear flow for a temperature-dependent, non-Newtonian rheology. When the slab velocity v is given as the boundary condition of the shear flow, the internal distributions of velocity and temperature are determined with a uniform shear stress σ. A significantly large velocity gradient with dominant heat generation is confined to a narrow shear zone. Total frictional heat Q per unit time and unit area is proportional to the maximum temperature T at the center of the shear zone. All variables are scaled by the suitable units involving the rheological parameters.

mantle wedge does not support the heat gain from the deeper mantle. In some calculations for this model, an apparent hot mantle wedge with fairly high temperatures can be obtained, but this result simply reflects the thermal boundary conditions, such as a fixed temperature or heat flow, that are prescribed mathematically for the bottom of the convecting system. Thus this result cannot be considered a solution to the heat source paradox, unless the boundary condition itself is physically justified under the continued subduction of the cool slab.

In this model, the upwelling current is induced indirectly as a return flow and thus spreads over a relatively wide area in the mantle wedge. A high temperature state is therefore also distributed rather broadly. Such broad temperature distribution, however, does not explain the observed sharp peak of thermal and volcanic activity (Figure 2). An additional factor is needed to describe arc volcanism, even though it is accepted that the fundamental thermal structure is governed by the induced return flow.

The idea that the magmas in the arc volcanoes originate from an ascending diapir persists (MARSH, 1979), although the heat source is not always identified. Recently, TATSUMI (1986) proposed a model of magma ascent in a mantle diapir with a thermal constraint. According to TATSUMI (1986), the primary basaltic magma is generated at about 100 km in depth, where the dehydration of hydrous minerals in the subducting slab supplies volatile components to the mantle wedge. The initial melting process takes place under low-temperature conditions. However, the results of an earlier high-pressure melting experiment (TATSUMI et al., 1983) show that the extruded basaltic magmas should have experienced temperatures as high as 1400 °C in equilibrium with the mantle peridotite. TATSUMI (1986) thus inferred that the ascending mantle diapir absorbs heat from the surrounding mantle, which is hot due to convection induced by the slab drag.

TATSUMI's (1986) model provides an attractive explanation of petrological evidence, but presents a problem from the physical viewpoint. Namely, the diapir must be smaller than several kms to absorb enough heat, but simultaneously must be greater than several tens kms to obtain an ascending speed high enough to overcome the downgoing motion of the induced flow. Such a contradiction can be eliminated if one considers an ascending continuous flow of mantle mass, as discussed below, instead of an isolated mantle diapir. If the partial melting that drives the upward motion is associated with the material supply from the continuously subducting slab, a continuous flow might be more likely to occur than an isolated diapir. The ascending continuous flow plays the most fundamental role in the arc volcanism model proposed in the next section of this paper. It is emphasized here, however, that the ascending continuous flow must dominate other convective motions, such as the one induced by the slab drag, if it is to occur.

To examine the possibility of an ascending continuous flow, it is necessary to evaluate the relative importance of the various driving forces of convection. Compare the effects of following factors: 1) the slab drag, 2) horizontal temperature difference, and 3) partial melting. The ascending mantle flow is generated by the internal buoyancy associated with the horizontal temperature difference and partial melting, whereas the external driving force for the induced convection is attributable to slab drag. Although an exact fluid dynamical treatment requires laborious analysis and calculation, the relative importance of these factors can be evaluated more simply in a semiquantitative way (IDA, 1983a). If the path of flow is prescribed, one can derive an analytic expression of the mass flux for each type of convection. These fluxes are compared in Figure 4 for the parameters suitably prescribed. Here the flux J_v is induced by the slab velocity, flux J_T is induced by the horizontal temperature difference ΔT, and flux J_M is induced by the partial melt concentration ζ. As shown in Figure 4, these fluxes depend on the viscosity of the mantle wedge in different ways, and the flow induced by the slab drag could dominate the other types of flows only when the viscosity is relatively high. Because the observed

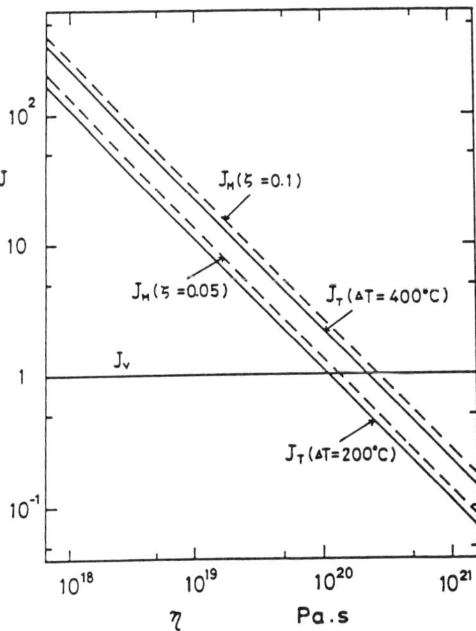

Figure 4. Relative importance of the fluxes J_T, J_M, and J_v which are induced by the horizontal temperature difference ΔT, the melt content ζ, and the downgoing slab velocity, respectively, in the mantle wedge (IDA, 1983a). J_T and J_M are inversely proportional to the mantle viscosity η, whereas J_v is independent of η.

viscosity for the mantle wedge is as low as 10^{19} Pa s (e.g., IDA, 1985), the convection driven by the internal buoyancy of the ascending flow might be more important in the actual mantle wedge.

Model with Buoyant Flow Beneath the Volcanic Belt

To explain the origin and nature of arc volcanism, a new model of the upwelling flow beneath the volcanic belt is proposed and displayed schematically in Figure 5. In this model the upwelling flow is subject to the buoyancy force associated with partial melting and higher temperature conditions. The entire convection is driven by this force. The ascending flow pulls up the mass of deeper mantle and pushes the overlying mass to a shallower part of the mantle wedge. The incoming mass from the deeper mantle becomes hotter than the surrounding medium at the same depth and induces partial melting as it ascends because of a lowered solidus. Volatiles may be provided by the subducting slab, further facilitating partial melting. In this way, the buoyancy to drive the convection is produced in the incoming mass that is consuming internal heat. At the top of the mantle, the flow releases some portion of the melt to the crust and the rest moves out toward the backarc.

The ascending flow has sufficiently high temperature and abundant melt to constitute a low viscosity column in the mantle wedge. As shown in Figure 4, the convection generated by the internal buoyancy can overcome the mechanical drag from the downgoing slab when there is such a low viscosity condition. Because the viscosity is strongly temperature dependent, the column tends to be localized like a narrow shear zone in the same way as shown in Figure 3. In short, a narrow ascending flow is expected to occur in the mantle wedge independent of the slab drag.

One of the most important concepts of this model is that a convection flow maintains itself because of the buoyancy produced in the flow when there is a probable supply of volatiles from the slab. Because such a self-preserving mechanism works only after convection starts, there must be trigger to initiate the process. One of the possible initiation mechanisms is partial melting due to volatile components supplied from the slab. In an early stage of subduction, the mantle wedge is not yet cooled substantially, and thus partial melting that would initiate the convection may take place relatively easily.

There is another initiation mechanism that involves convection induced by slab drag. In an early subduction stage involving only a shallow slab, the upwelling return

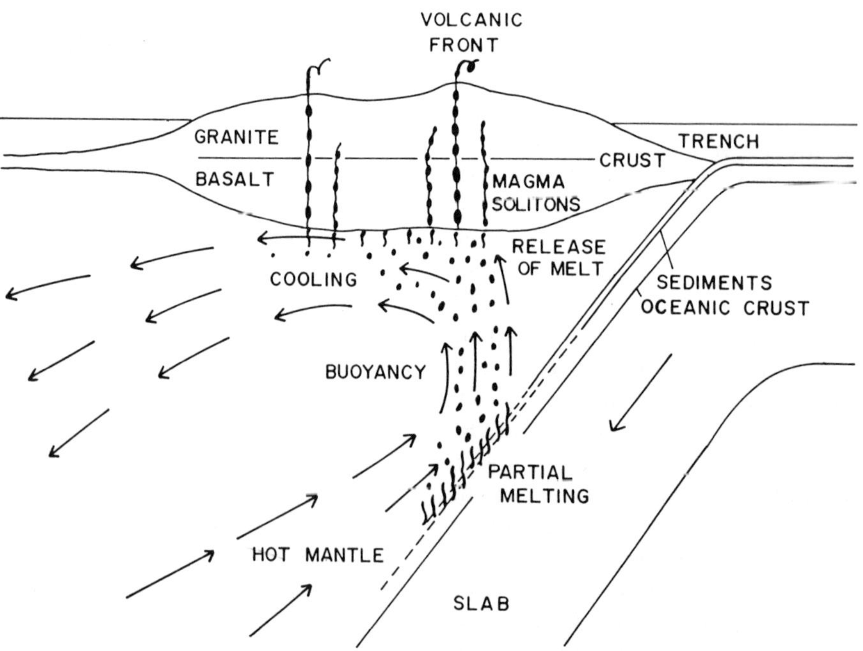

Figure 5. Model of magma generation in an ascending mantle flow. The ascending flow is driven by its internal buoyancy associated with higher temperature conditions and partial melting. The subducting slab supplies volatile components that facilitate partial melting in the flow. The ascending mass requires incoming hot material from deeper mantle, which provides heat to preserve the convection. The flow that has passed the top of the mantle moves toward the backarc. The outgoing flow is gradually cooled so that the magma in it is partly solidified with more incompatible elements.

flow of the induced convection is located relatively close to the trench. It is easier for this upwelling flow to gain buoyancy through partial melting because the mantle adjacent to the slab is still hot. If this internal buoyancy increases to a level comparable with the effect of the slab drag, the flow could form its own convection pattern independent of the slab motion shown in Figure 5. Even after the convection with the upwelling flow is established, a weak form of the induced convection may survive seaward. Such surviving flow could carry hydrous minerals with volatile components downward and thus enhance partial melting, as suggested by TATSUMI (1986).

Observations of arc volcanism in Japan are now used to test the proposed model (see Figure 5). The anomalous structure of the subduction zone, where a hot asthenosphere is adjacent to the cold slab (Figure 2), may be explained by any mechanism that allows sufficient heat to be supplied the mantle wedge. It is noteworthy, however, that in the present model a hot mantle current rises from the slab, whereas in the model of convection induced by the slab drag, hot material arrives in the slab vicinity only after almost completing the convective cycle in the mantle wedge. Therefore the present model is more consistent with seismological evidence that indicates the boundary between the hot mantle and the cold slab is sharp enough to generate reflective and refractive waves (SNOKE et al., 1977; FUKAO et al., 1978). Furthermore, the observed low-velocity column below the volcanic belt (Figure 2) can be directly identified with the area of the upwelling flow in the present model.

The ascending mantle flow has been introduced as part of the present model to explain the observed sharp peak of heat flow and volcanic activity (Figure 2). It is inferred that the hot upwelling flow could discharge substantial amounts of heat and magma beneath the narrow volcanic belt. In the present model, low heat flow and the absence of volcanoes seaward of the volcanic front are attributed to the convection pattern in which the outgoing mantle flows toward the backarc (Figure 5). Such an outgoing flow corresponds to the incoming flow of mass from the deeper mantle on the backarc side. The outgoing flow in the shallow part of the mantle wedge continuously emits heat to the crust so that temperature and heat flow decreases in the backarc direction, consistent with the observed heat flow data (see Figure 2). On the other hand, the rate of magma discharge to the crust is determined by the migration velocity of magma relative to the mantle flow. Some amount of magma is generally left in the outgoing flow, even after a major volume of magma has been subtracted from the ascending part of the flow. The amount of magma supplied to the crust from the outgoing flow is expected to decrease in the backarc direction because the outgoing flow is cooled and contains less magma as it moves toward the backarc.

The chemical composition of igneous rocks across the arc is examined next. The reason why such incompatible elements as K and Na are systematically enriched toward the backarc (see Figure 2) can again be explained by the cooling of the outgoing mantle flow and the resultant solidification of magma. The incompatible elements enter the rock crystals with difficulty and thus are concentrated in the residual melt during the process of solidification.

The enrichment of incompatible elements is usually accompanied by a decrease in silica content (Figure 2), and the chemical composition of magma moves closer to that of alkali basalt. Based on experimental petrology, it is found that the basaltic melt becomes more alkaline in equilibrium with the mantle minerals, as partial melting takes place at a higher pressure. Therefore an alkaline basaltic magma can be produced relatively deep in the mantle. Most of the previous models have attributed the compositional change of basalt to the increasing depth of magma origin. The same idea is also applicable to the present model because the outgoing flow is expected to sink deeper into the mantle due to the loss of buoyancy. The reaction of the magma with the surrounding mantle minerals should make the magma more alkaline in the backarc.

In order for this interpretation of the compositional variation to be valid, however, the magma must be segregated from the mantle rocks at the desired depth and its composition must be maintained without being altered in the process of ascent. There is some doubt that segregation is really possible deep in the mantle with a low degree of partial melting. Therefore this author prefers the following alternative explanation of the magma composition. In the present model (Figure 5), the magmas with a higher silica content than the mantle minerals are subtracted successively from the flow, and the bulk content of silica in the outgoing flow decreases toward the backarc. If this reduction in silica content is combined with the concentration of incompatible elements due to solidification, the observed variation of basalt composition across the arc can be explained.

In connection with this systematic variation of rock chemistry, it is noted that some alkaline volcanoes are distributed in Korea and Northern China, farther from the Japan trench. In the present model (Figure 5), the occurrence of these alkaline rocks is simply explained by assuming that the outgoing flow extends to this distance. However, no Benioff-Wadati zone is observed beneath these volcanoes. Even if the slab does extend this distance, it would have to be extraordinarily deep. Therefore the theories that involve magma origin at different depths in the slab cannot account for these volcanoes.

Conclusion

The generation of magmas in the subduction zone has been examined from geophysical and geochemical viewpoints with special reference to the seismic structure, heat flow, eruption volume, and magma composition observed in the Japan arc. A self-preserving convection that is induced by a buoyant mantle flow beneath the volcanic belt (Figure 5) is proposed to explain the distinctive thermal and volcanic activities there. A proposed model gives a qualitatively consistent picture of arc volcanism and lays the foundation for a more quantitative analysis that will reveal the detailed flow pattern of the convection, the distribution of heat and mechanical forces, and the magma rate of discharge to the crust.

As mentioned above, a wide variability in magma composition is a major characteristic of arc volcanism. The author believes that such variability is mainly attributable to the differentiation process in the crust. Because heat and magma transport in the crust are interrelated, a complete model to explain the observations of thermal and volcanic activity must describe the process in the crust as well as in the mantle. In this context, the author is trying to model the crustal process by using the concept of a moving magma chamber, which is an application of the theoretical analysis of the magma ascent process in a deformable vent (IDA and KUMAZAWA, 1986).

Acknowledgments. The author thanks the reviewers for their helpful comments for the improvement of this manuscript.

REFERENCES

ANDERSON, R. N., S. E. DELONG, and W. M. SCHWARZ, Thermal model for subduction with dehydration in the downgoing slab, *J. Geology*, *86*, 731–739, 1978.

FUKAO, Y., K. KANJO, and I. NAKAMURA, Deep seismic zone as an upper mantle reflector of body waves, *Nature*, *272*, 606–608, 1978.

HASEBE, K., N. FUJII, and S. UYEDA, Thermal processes under island arcs, *Tectonophysics*, *10*, 335–355, 1970.

HASEMI, A. H., H. ISHII, and A. TAKAGI, Fine structure beneath the Tohoku district, northeastern Japan arc, as derived by an inversion of P-wave arrival times from local earthquakes, *Tectonophysics*, *101*, 245–265, 1984.

IDA, Y., Convection in the mantle wedge above the slab and tectonic processes in subduction zones, *J. Geophys. Res.*, *88*, 7449–7456, 1983a.

IDA, Y., Thermal and mechanical processes producing arc volcanism and back-arc spreading, in *Arc Volcanism: Physics and Tectonics*, edited by D. Shimozuru and I. Yokoyama, pp. 165–175, Terra Scientific Publication Company, Tokyo, 1983b.

IDA, Y., Rheology of the mantle, in *Recent Progress of Natural Sciences in Japan*, Vol. 10, pp. 93–99, Science Council of Japan, Tokyo, 1985.

IDA, Y., and M. KUMAZAWA, Ascent of magma in a deformable vent, *J. Geophys. Res.*, *91*, 9297–9301, 1986.

MARSH, B. D., Island arc development: some observations, experiments, and speculations, *J. Geology*, *87*, 687–713, 1979.

NAGAO, T., M. YAMANO, H. FUJISAWA, S. HONDA, and S. UYEDA, Crustal heat flow in Tohoku and Hokkaido, *Programme and Abstracts, Seismological Society of Japan*, *2*, 262, 1985 (in Japanese).

NIELSON, D. R., and R. E. STOIBER, Relationship of potassium content in andesitic lavas and depth to the seismic zone, *J. Geophys. Res.*, *78*, 6887–6892, 1973.

ONO, K., T. SOYA, and K. MIMURA, Volcanoes of Japan (2nd. ed.), *Map Series No. 11*, Geological Survey of Japan, Tsukuba, Japan, 1981.

RABINOWICZ, M., B. LAGO, and C. FROIDEVAUX, Thermal transfer between the continental asthenosphere and the oceanic subducting lithosphere: its effect on subcontinental convection, *J. Geophys. Res.*, *85*, 1839–1853, 1980.

SCHUBERT, G., D. A. YUEN, C. FROIDEVAUX, L. FLEITOUT, and M. SOURIAU, Mantle circulation with partial shallow return flow: effects on stresses in oceanic plates and topography of the sea flow, *J. Geophys. Res.*, *83*, 745–758, 1978.

SNOKE, J. A., I. S. SACKS, and H. OKADA, Determination of the subducting lithosphere boundary by use of converted phases, *Bull. Seismol. Soc. Amer.*, *67*, 1051–1061, 1977.

SUGIMURA, A., Zonal arrangement of some geophysical and petrological features in Japan and its environs, *J. Fac. Sci. Univ. Tokyo, sect. II*, *12*, 133–153, 1960.

SUGIMURA, A., Regional variation of the K_2O/Na_2O ratios of volcanic rocks in Japan and environs (in Japanese with English abstract and captions), *J. Geolog. Soc. Japan*, *67*, 292–300, 1961.

SUGIMURA, A., T. MATSUDA, K. CHINZEI, and K. NAKAMURA, Quantitative distribution of late Cenozoic volcanic materials in Japan, *Bull. Volcanol.*, *26*, 125–140, 1963.

TATSUMI, Y., Role of induced convection and mantle diapir in the generation of subduction zone magmas, *Bull. Volcanol. Soc. Japan*, *31*, 39–44, 1986 (in Japanese with English abstract and captions).

TATSUMI, Y., M. SAKUYAMA, H. FUKUYAMA, and I. KUSHIRO, Generation of arc basalt magmas and thermal structure of the mantle wedge in subduction zones, *J. Geophys. Res.*, *88*, 5815–5825, 1983.

TOKSÖZ, M. N., and P. BIRD, Formation and evolution of marginal basins and continental plateaus, in *Island Arcs, Deep Sea Trenches, and Back-arc Basins, Maurice Ewing Ser., Vol. 1*, edited by M. Talwani and W. C. Pitman III, pp. 379–393, AGU, Washington, D.C., 1977.

TOKSÖZ, M. N., and A. T. HSUI, Numerical studies of back-arc convection and the formation of marginal basins, *Tectonophysics*, *50*, 177–196, 1978.

UTSU, T., Seismological evidence for anomalous structure of island arcs with special reference to the Japanese region, *Rev. Geophys. Space Phys.*, *9*, 839–890, 1971.

YOSHII, T., A detailed cross-section of the deep seismic zone beneath northeastern Honshu, Japan, *Tectonophysics*, *55*, 349–360, 1979.

AUTHOR INDEX

INDEX